The Oxford Handbook of Nonlinear Filtering

The Oxford Handbook of Nonlinear Filtering

Edited by

Dan Crisan
Boris Rozovskiĭ

UNIVERSITY PRESS

OXFORD
UNIVERSITY PRESS

Great Clarendon Street, Oxford OX2 6DP

Oxford University Press is a department of the University of Oxford.
It furthers the University's objective of excellence in research, scholarship,
and education by publishing worldwide in

Oxford New York

Auckland Cape Town Dar es Salaam Hong Kong Karachi
Kuala Lumpur Madrid Melbourne Mexico City Nairobi
New Delhi Shanghai Taipei Toronto

With offices in

Argentina Austria Brazil Chile Czech Republic France Greece
Guatemala Hungary Italy Japan Poland Portugal Singapore
South Korea Switzerland Thailand Turkey Ukraine Vietnam

Oxford is a registered trademark of Oxford University Press
in the UK and in certain other countries

Published in the United States
by Oxford University Press Inc., New York

© Oxford University Press 2011

The moral rights of the authors have been asserted
Database right Oxford University Press (maker)

First published 2011

All rights reserved. No part of this publication may be reproduced,
stored in a retrieval system, or transmitted, in any form or by any means,
without the prior permission in writing of Oxford University Press,
or as expressly permitted by law, or under terms agreed with the appropriate
reprographics rights organization. Enquiries concerning reproduction
outside the scope of the above should be sent to the Rights Department,
Oxford University Press, at the address above

You must not circulate this book in any other binding or cover
and you must impose the same condition on any acquirer

British Library Cataloguing in Publication Data

Data available

Library of Congress Cataloging in Publication Data

Data available

Typeset by SPI Publisher Services, Pondicherry, India
Printed in Great Britain
on acid-free paper by
CPI Antony Rowe, Chippenham, Wiltshire

ISBN 978–0–19–953290–2

1 3 5 7 9 10 8 6 4 2

Preface

In many areas of human endeavour on this planet and beyond, the systems involved are not available for direct measurements. Fortunately, combining mathematical models for their evolution with partial observations of their evolving state allows us to make reasonable inferences about them. Nonlinear/stochastic filtering is the process of using partial observations and a stochastic model to estimate the state of an evolving system.

The scope of applications of stochastic filtering is huge. It includes the study of the global climate, estimating the state of the economy, identifying tumours using noninvasive methods and digital recording. It is an area of research that has been steadily growing both in size and importance since its inception in the late sixties. Despite its importance and the interest in it, there has never been a comprehensive written source for the subject.

The aim of *The Oxford Handbook of Nonlinear Filtering* is to bring together under one cover results both from the classical theory of the subject and also recent ones. This was possible by putting together contributions from authors which rank amongst the leading scientists in the area, many of whom are the pioneers of nonlinear filtering. The comprehensive aspect of the handbook was only possible through the joint efforts and expertise of the contributors. We thank them all for contributing to the handbook and, in some cases, for having to wait such a long time for the project to come to fruition.

Planning the handbook started in August 2006 during the International Congress of Mathematicians in Madrid and with a firm basis put down during the Stochastic PDEs workshop held at Brown University in October 2006. In these early stages of the project the editors benefited greatly from the support of Alison Jones, the Oxford University Press editor in charge of the book whose commitment and enthusiasm we acknowledge wholeheartedly.

During the subsequent three years the editors proceeded to assemble the handbook with contributions from 58 authors. All contributions were refereed and we would like to thank all referees who assisted us, who include Ajay Jasra, Andreas Eberle, Etienne Pardoux, Konstantinos Manolarakis, and Monique Jeanblanc.

In the latter stages of the project, we were assisted by Keith Mansfield, OUP Senior Commissioning Editor for Mathematics. We would like to thank him for his understanding and patience.

Finally we would like to thank our families for their continuous support, without which this project would have never succeeded.

Dan Crisan and Boris Rozovskiĭ
London (UK) and Providence (USA)
December 2009

Contents

List of Contributors xi

1 Introduction 1
  D. Crisan and B. Rozovskiĭ

Part I The Foundations of Nonlinear Filtering

2 Nonlinear Filtering Problems I: Bayes Formulas and Innovations 19
  H. Kunita

3 Nonlinear Filtering Problems II: Associated Equations 55
  H. Kunita

4 Nonlinear Filtering Equations for Stochastic Processes with Jumps 95
  B. Grigelionis and R. Mikulevicius

5 The Filtered Martingale Problem 129
  T. G. Kurtz and G. Nappo

Part II Nonlinear Filtering and Stochastic Partial Differential Equations

6 Filtering Equations for Partially Observable Diffusion Processes
  with Lipschitz Continuous Coefficients 169
  N. V. Krylov

7 Malliavin Calculus Applications to the Study of Nonlinear Filtering .. 195
  M. Chaleyat-Maurel

8 Chaos Approach to Nonlinear Filtering 231
  S. V. Lototsky

Part III Stability and Asymptotic Analysis

9 On Filtering with Unspecified Initial Data for Nonuniformly
  Ergodic Signals 267
  M. L. Kleptsyna and A. Yu. Veretennikov

10 Exponential Decay Rate of the Filter's Dependence on the Initial Distribution 299
    R. Atar

11 Intrinsic Methods in Filter Stability 319
    P. Chigansky, R. Liptser, and R. Van Handel

12 Feller and Stability Properties of the Nonlinear Filter 352
    A. Budhiraja

13 Stability of the Optimal Filter for Nonergodic Signals—A Variational Approach 374
    W. Stannat

Part IV   Special Topics

14 Pathwise Nonlinear Filtering with Correlated Noise 403
    M. H. A. Davis

15 The Innovations Problem 425
    A. J. Heunis

16 Nonlinear Filtering and Fractional Brownian Motion 450
    T. E. Duncan

Part V   Estimation and Control

17 Dual Filters, Path Estimators, and Information 471
    N. J. Newton

18 Filtering for Discrete-Time Markov Processes and Applications to Inventory Control with Incomplete Information 500
    A. Bensoussan, M. Çakanyıldırım, and S. P. Sethi

19 Bayesian Filtering of Stochastic Hybrid Systems in Discrete-Time and Interacting Multiple Model 526
    H. A. P. Blom and Y. Bar-Shalom

Part VI   Approximation Theory

20 Error Bounds for the Nonlinear Filtering of Diffusion Processes 561
    O. Zeitouni

21 Discretizing the Continuous-Time Filtering Problem: Order of Convergence 572
    D. Crisan

| 22 | Large Sample Asymptotics for the Ensemble Kalman Filter | 598 |

F. Le Gland, V. Monbet, and V.-D. Tran

Part VII The Particle Approach

| 23 | Particle Approximations to the Filtering Problem in Continuous Time | 635 |

J. Xiong

| 24 | A Tutorial on Particle Filtering and Smoothing: Fifteen Years Later | 656 |

A. Doucet and A. M. Johansen

| 25 | A Mean Field Theory of Nonlinear Filtering | 705 |

P. Del Moral, F. Patras, and S. Rubenthaler

| 26 | The Particle Filter in Practice | 741 |

T. B. Schön, F. Gustafsson, and R. Karlsson

| 27 | Introducing Cubature to Filtering | 768 |

C. Litterer and T. Lyons

Part VIII Numerical Methods in Nonlinear Filtering

| 28 | Numerical Approximations to Optimal Nonlinear Filters | 799 |

H. J. Kushner

| 29 | Signal Processing Problems on Function Space: Bayesian Formulation, Stochastic PDEs and Effective MCMC Methods | 833 |

M. Hairer, A. Stuart, and J. Voss

| 30 | Robust, Computationally Efficient Algorithms for Tracking Problems with Measurement Process Nonlinearities | 874 |

J. M. C. Clark and R. B. Vinter

| 31 | Nonlinear Filtering Algorithms Based on Averaging Over Characteristics and on the Innovation Approach | 892 |

G. N. Milstein and M. V. Tretyakov

Part IX Nonlinear Filtering in Financial Mathematics

| 32 | Nonlinear Filtering in Models for Interest-Rate and Credit Risk | 923 |

R. Frey and W. Runggaldier

| 33 | An Asset Pricing Model with Mean Reversion and Regime Switching Stochastic Volatility | 960 |

R. J. Elliott, H. Miao, and Z. Wu

34 **Portfolio Optimization Under Partial Observation: Theoretical and Numerical Aspects** 990
H. Pham

35 **Filtering with Counting Process Observations: Application to the Statistical Analysis of the Micromovement of Asset Price** 1019
L. C. Scott and Y. Zeng

Index 1047

List of Contributors

R. Atar Department of Electrical Engineering, Technion, Haifa 32000, Israel

Y. Bar-Shalom University of Connecticut, Electrical and Computer Engineering, 371 Fairfield Road U-2157 Storrs, Connecticut 06269-2157, USA

A. Bensoussan University of Texas at Dallas, School of Management, 800 West Campbell Rd., SM30, Richardson, Texas 75080–3021, USA and the Hong Kong Polytechnic University, Graduate School of Business, Kowloon, Hong Kong

H. A. P. Blom National Aerospace Laboratory NLR, Anthony Fokkerweg 2, PO Box 90502, 1006 BM Amsterdam, The Netherlands

A. Budhiraja Department of Statistics and Operations Research, University of North Carolina, Chapel Hill, NC 27599, USA

M. Çakanyıldırım University of Texas at Dallas, School of Management, 800 West Campbell Rd., SM30, Richardson, Texas 75080–3021, USA

M. Chaleyat-Maurel Université Paris Descartes 45, rue des Saints Peres, 75006, Paris, France

P. Chigansky Department of Statistics, The Hebrew University of Jerusalem, Mount Scopus, Jerusalem 91905, Israel

J. M. C. Clark Department of Electrical and Electronic Engineering, Imperial College London, Exhibition Road, London, SW7 2BT, UK

D. Crisan Department of Mathematics, Imperial College London, 180 Queen's Gate, London, SW7 2AZ, UK

M. H. A. Davis Department of Mathematics, Imperial College London, 180 Queen's Gate, London SW7 2AZ, UK

P. Del Moral Centre INRIA Bordeaux Sud-Quest & Institut de Mathematiques de Bordeaux, Université Bordeaux 1, 351, cours de la Liberation, 33405 Talence Cédex, France

A. Doucet The Institute of Statistical Mathematics, 4-6-7 Minami-Azabu, Minato-ku, Tokyo 106-8569, Japan

T. E. Duncan Department of Mathematics, University of Kansas, 405 Snow Hall, 1460 Jayhawk Blvd, Lawrence, Kansas 66045-7523, USA

R. J. Elliott Haskayne School of Business, University of Calgary, Calgary, AB, Canada T2N 1N4, and School of Mathematics, University of Adelaide SA 5005, Adelaide, Australia

R. Frey Universitat Leipzig, Mathematisches Institut, Fakultat fur Mathematik and Informatik, 04081 Leipzig, Germany

B. Grigelionis Akademijos 4, Vilnius, LT-08663, Lithuania

F. Gustafsson Division of Automatic Control, Linköping University, s-581 83 Linköping, Sweden

M. Hairer Mathematics Institute, University of Warwick, Gibbet Hill Road, Coventry, CV4 7AL, UK

A. J. Heunis Department of Electrical and Computer Engineering, University of Waterloo, 200 University Avenue West Waterloo, Ontario N2L 3G1, Canada

A. M. Johansen Department of Statistics, University of Warwick, Coventry, CV4 7AL, UK

R. Karlsson Division of Automatic Control, Department of Electrical Engineering, Linköping University, s-581 83 Linköping, Sweden

M. L. Kleptsyna Université du Maine, avenue Olivier-Messiaen, 72085 Le Mans cedex 09, France

N. V. Krylov School of Mathematics, 225 VinH, 206 Church St. SE, Minneapolis, University of Minnesota, MN 55455, USA

H. Kunita 3-2-3-606, Kashiihama, Higashiku, Fukuoka, 813-0016, Japan

T. G. Kurtz Department of Mathematics and Statistics, University of Wisconsin-Madison, 480 Lincoln Drive, Madison, WI 53706–1388, USA

H. J. Kushner The Division of Applied Mathematics, Brown University, 182 George Street, Providence, RI 02912, USA

F. Le Gland INRIA Rennes, Campus de Beaulieu, 35042 Rennes Cedex, France

R. Liptser Department of Electrical Engineering Systems, The Iby and Aladar Fleischman Faculty of Engineering, Tel Aviv University, Ramat Aviv 69978, Israel

C. Litterer Mathematical Institute, University of Oxford, 24–29 St Giles, Oxford, OX1 3LB, UK

S. V. Lototsky Department of Mathematics, University of Southern California, 3620 S. Vermont Av., KAP 108, Los Angeles, CA 90089-2532, USA

T. Lyons Mathematical Institute, University of Oxford, 24–29 St Giles, Oxford, OX1 3LB, UK

H. Miao Department of Finance and Real Estate, College of Business, Colorado State University, Fort Collins, CO 80523, USA

R. Mikulevicius USC Department of Mathematics, 3620 Vermont Avenue, KAP108, Los Angeles, CA 90089-2532, USA

G. N. Milstein Ural State University, Lenin Str. 51, 620083 Ekaterinburg, Russia

V. Monbet Université de Rennes 1, Campus de Beaulieu, 35042 Rennes Cédex, France

G. Nappo Department of Mathematics, University "La Sapienza", piazzale Aldo Moro 2, I-00185 Rome, Italy

N. J. Newton School of Computer Science and Electronic Engineering, University of Essex, Wivenhoe Park, Colchester, CO4 3SQ, UK

F. Patras CNRS UMR 6621, Université de Nice, Laboratoire de Mathématiques J. Dieudonné, Parc Valrose, 06108, Nice Cedex 2, France

H. Pham Laboratoire de Probabilités et Modèles Aléatoires CNRS, UMR 7599, Universités Paris 6-Paris 7, 175 Rue du Chevaleret, 75013 Paris, France

B. Rozovskiĭ Division of Applied Mathematics, Brown University, 182 George Street, Providence, RI 02912, USA

S. Rubenthaler CNRS UMR 6621, Université de Nice, Laboratoire de Mathématiques J. Dieudonné, Parc Valrose, 06108, Nice Cedex 2, France

W. Runggaldier Università degli Studi di Padova, Dipartimento di Matematica Pura ed Applicata, 63 Via Trieste, 35121 Padova, Italy

T. B. Schön Division of Automatic Control, Linköping University, s-581 83 Linköping, Sweden

L. C. Scott Department of Mathematics and Statistics, University of Missouri at Kansas City, Kansas City, MO 64110, USA

S. P. Sethi University of Texas at Dallas, School of Management, 800 West Campbell Rd., SM30, Richardson, Texas 75080–3021, USA

W. Stannat Fachbereich Mathematik, TU Darmstadt, Schlossgartenstr. 7, 64289 Darmstadt, Germany

A. Stuart Mathematics Institute, University of Warwick, Gibbet Hill Road, Coventry, CV4 7AL, UK

V.-D. Tran Department of Science and Technology, Hoa Sen University, Ho Chi Minh City, Vietnam

M. V. Tretyakov Department of Mathematics, University of Leicester, University Road, Leicester, LE1 7RH, UK

R. Van Handel Sherrerd Hall, Room 227, Princeton University, Princeton, NJ 08544, USA

A. Yu. Veretennikov School of Mathematics, University of Leeds, Leeds, LS2 9JT, UK and Institute of Information Transmission Problems, B. Karetny 19, 127994, Moscow, Russia

R. B. Vinter Department of Electrical and Electronic Engineering, Imperial College London, Exhibition Road, London, SW7 2BT, UK

J. Voss Mathematics Institute, University of Warwick, Gibbet Hill Road, Coventry, CV4 7AL, UK

Z. Wu Department of Finance and Management Science, N. Murray Edwards School of Business, University of Saskatchewan, Saskatoon, Saskatchewan, Canada S7N 5A7

J. Xiong Mathematics Department, University of Tennessee, 121 Ayres Hall, 1403 Circle Drive, Knoxville, TN 37996-1300, USA

O. Zeitouni Department of Mathematics, University of Minnesota, 206 Church St. SE, Minneapolis, MN 55455, USA and Faculty of Mathematics, Weizmann Institute of Science, Rehovot 76100, Israel

Y. Zeng Department of Mathematics and Statistics, College of Arts and Sciences, University of Missouri at Kansas City, Kansas City, MO 64110-2499, USA

·1·
Introduction

D. Crisan and B. Rozovskiĭ

The increasing complexity of the modern world makes the analysis and synthesis of high-volume data an essential feature in many real problems. Such analysis is addressed by different disciplines and from different perspective and, among them, nonlinear filtering features prominently. Nonlinear filtering is distinguished from other approaches by its probabilistic (in particular, Bayesian) nature. It is a field that combines aspects of stochastic analysis, information theory, and statistical inference. To date, nonlinear filtering is a mature theory that continues to expand by leaps and bounds. The breadth of its applications, firmly established and still emerging, is simply astounding. Early applications of nonlinear filtering such as cryptography, tracking, and guidance were mostly of military nature. Since then, nonlinear filtering has became an extremely potent tool in speech recognition, image and video processing, genetics, financial modeling, Bayesian networks, etc.

The celebrated Kalman–Bucy filter, designed for linear dynamical systems with linearly structured measurements, is probably the most famous Bayesian filter. Its generalizations to nonlinear systems and/or nonlinear observations are collectively referred to as *nonlinear filtering* (NLF). To put it succinctly, nonlinear filtering is an extension of the Bayesian framework to the estimation, prediction, and interpolation of nonlinear stochastic dynamics. Its output is the distribution of the estimated process (the "signal") given the data (the "observations") available. This distribution is commonly known as the posterior distribution of the estimated process. It is a theoretically optimal algorithm in that it provides the best estimate for the quantity of interest, more precisely, it minimises the mean square error of the estimator.

An important special case of the NLF paradigm that addresses Markov type dynamics is often referred to as Hidden Markov Models (HMM). It is an elegant and illuminating example which illustrates the principles of nonlinear filtering so it is worthwhile explaining it here.

Consider the following model: Let $(X_t, \ t = 0, 1, 2, \ldots)$ and $(Y_t, t = 0, 1, 2, \ldots)$ be two random sequences called, respectively, *state* and *observations*. The state is not directly observable. It is modeled as a Markov chain with the transition probability kernel $Q_t(x, y)$ and the initial distribution π_0. The observation sequence is related to the state by the formula

$$Y_t = h(X_t) + \xi_t,$$

where ξ_t is random noise. The a priori information contained in this model consists of the prior distribution π_0, the transition probability kernel $Q_t(x, y)$, and the distribution of ξ. The term "hidden Markov model" alludes to the fact that the Markov chain X_t is hidden from the observer by a possibly nonlinear transformation $h(\cdot)$ and the noise ξ_t. The task of nonlinear filtering is to compute, at each time t, the *posterior* distribution $\pi_{t|t}$ of the state X_t, in other words, the conditional distribution of X_t given the observation $Y_{0|t} = (Y_0, ..., Y_t)$. In this setting, the posterior distribution satisfies a two-step recursion:

$$\text{prediction}: \quad \varphi_{t|t-1}(x) = \int Q_t(x', x) \varphi_{t-1|t-1}(dx')$$
$$\text{and} \quad (1.1)$$
$$\text{correction}: \pi_{t|t}(x) = \Psi_t(x) \pi_{t|t-1}(x) / \int \Psi_t(x') \varphi_{t|t-1}(dx'),$$

where $\Psi_t(x) = g_t(y_t - H(x))$ is the likelihood function. The first step consists in computing the conditional distribution of the state X_t given all but the last observation, i.e., given $(Y_0, ..., Y_{t-1})$. The second step is the well-known Bayes rule.

The early history of HMMs is shrouded in secrecy due to potential applications in cryptography. There exists an anecdotal evidence that the work on this topic was done by engineers in the early sixties but did not really come out in the open until the HMMs were "rediscovered" in late sixties. The crucial part of the developed "technology" is usually referred to as Baum–Welch algorithm (see papers [2] and [21]), particularly for frameworks where the state can take can have only a finite number of values (finite state space).

In the continuous setting, the prediction and correction steps merge and the posterior density $\pi_t(x) = \pi_{t|t}(x)$ is given by the Bayes formula.

$$\pi_t(x) = \frac{\varphi_t(x)}{\int_t \varphi_t(x') dx'}. \quad (1.2)$$

The unnormalized posterior density $\varphi_t(x)$ is a solution of the following equation:

$$\varphi_t(x) = \pi_0(x) + \int_0^t A^*(s, x) \varphi_s(x) dt + \int_0^t h_s(x) \varphi_s(x) dY_s, \quad (1.3)$$

where A^* is the dual of the generator of the signal's transition probability kernel $Q_t(x, y)$ and Y is the observation process. In particular, if the signal is given by the noisy kinematic equation

$$\dot{x}_t = a(t, x_t) + \sigma \dot{w}_t,$$

where $\dot{w}_t$ is white noise, then A^* has the form

$$A\varphi(x) = \frac{\sigma^2}{2}\varphi_t''(x) - (a(t,x)\varphi_t(x))'. \tag{1.4}$$

In the mid-sixties, the continuous time setting nonlinear filtering problem captured the attention of three young mathematicians: Duncan, Mortensen, and Zakai. They studied aspects of nonlinear filtering as a natural generalization of the well known linear filtering results of Kalman and Bucy. In four separate papers (two of them being PhD dissertations), they derived stochastic differential equations for unnormalized posterior distribution of a state process modeled by a continuous time Markov process. In paper [22], the state process was modeled by a continuous time jump Markov process with countable state space. The other three papers ([6], [16], and [23]) dealt with diffusion type state processes. Equation (1.4) and its generalizations are often referred to as Duncan–Mortensen–Zakai equation.

On the other hand, the (normalized) posterior density $\pi_t(x)$ solves a nonlinear stochastic PDE, which is usually referred to as the Kushner equation (see [12] (a corrected version of [11])). A more general version of the Kushner equation was derived by A. Shiryaev ([19]). These papers triggered a surge of activities in NLF during the late sixties and early seventies. The results of this period in the development of NLF were summarized in the influential book by R. Liptser and A. N. Shiryaev [14].

Note that Duncan–Mortensen–Zakai and Kushner equations are *stochastic partial differential equations* (SPDEs). In the sixties and seventies, SPDEs constituted a completely new subject for the stochastic community which quickly became an active area of research. Thus, one unintended but very important effect, triggered by the introduction of Duncan–Mortensen–Zakai and Kushner equations, was the fast development of a general theory of SPDEs. The first comprehensive accounts of these developments were published in [17] and [18]. For more details see the contributions by Krylov and Kunita in this volume.

Another milestone in the evolution of the theoretical side of NLF was related to the introduction of martingale techniques in the paper [7] by Fujisaki, Kallianpur and Kunita (more details can be found in Kunita's contribution in Part I of the Handbook). The martingale approach made it possible to deal with very general and diverse models of state and observations. Later on, Grigelionis and Mikulevicius extended the Fujisaki–Kallianpur–Kunita theory to even larger set of processes, including processes with irregular trajectories (see their contribution in Part I of the handbook).

In the last twenty years, a very impressive progress was made also in the study of asymptotic properties of SPDEs and, in particular, stability of nonlinear filters. This field is covered in great detail in Part III of the handbook.

From the very beginning, Bayesian filtering, both linear and nonlinear, has been an applied field. Numerous practical applications of Kalman filter and Baum–Welch algorithm are well documented.

The simplicity of the nonlinear filtering algorithm, particularly in the discrete setting, is deceptive. While being extremely effective and stable, it is computationally expensive. Both the prediction and the correction step involve computing integrals over the state space and these computations have to be executed every time new observations arrive. Moreover, in many important applications the computations have to be done in real time. Clearly, this is a serious complication since direct quadrature methods are effective in real time only when the dimension D of the state process is comparatively low (in general D should be no larger than 3). In contrast, the Kalman filter can deal in real time with hundreds of states. However, ad hoc extensions of linear Gaussian filters to the nonlinear setting such as the Extended Kalman Filter, are usually unsuccessful.

Fortunately, by the end of eighties two important factors have emerged: (a) a massive increase of computing power; and (b) the proliferation of Bayesian methodology into Monte Carlo simulations and vice versa.

For example, when implementing the Baum–Welch algorithm, one replaces the quadratures in the two steps of (1.1) by two Monte Carlo procedures. Nonlinear filtering algorithms based on Monte Carlo averaging are often called sequential Monte Carlo methods (SMCM) or particle filters. Such algorithms approximate of the posterior distribution using the empirical distribution of a system of n particles which evolve (mutate) according to the law of of the state. After each mutation the system is corrected: each particle is replaced by a random number of particles whose mean is proportional to the likelihood of the position of the particle. The most popular method for correcting the system is to sample with replacement n times from the empirical distribution of the population of particles weighted by their normalized likelihoods. The following is a simple algorithmic description of a garden variety sequential Monte Carlo method:

1. **Initialization** $[t = 0]$.

   For $i = 1, \ldots, n$, sample $x_0^{(i)}$ from π_0.

2. **Iteration** $[t - 1$ to $t]$.

   Let $x_{t-1}^{(i)}$, $i = 1, \ldots, n$ be the positions of the particles at time $t - 1$.

   (a) For $i = 1, \ldots, n$, sample $\bar{x}_t^{(i)}$ from $Q_{t-1}(x_{t-1}^{(i)}, x)dx$. Compute the (normalized) weight $w_t^{(i)} = g_t(\bar{x}_t^{(i)})/(\sum_{j=1}^n g_t(\bar{x}_t^{(j)}))$.

   (b) Pick $x_t^{(i)}$ by sampling with replacement from the set of particle positions $(\bar{x}_t^{(1)}, \bar{x}_t^{(2)}, \ldots, \bar{x}_t^{(n)})$ according to the probability vector of normalized weights $(w_t^{(1)}, w_t^{(2)}, \ldots, w_t^{(n)})$, $i = 1, \ldots, n$.

The approximation π_t^n of the posterior distribution is $\pi_t^n = \dfrac{1}{n}\sum_{i=1}^{n}\delta_{x_t^{(i)}}$, where $x_t^{(i)}$ for $i = 1, \ldots, n$ are the positions of the particles obtained after the second step of the iteration.

Following part (a) of the iteration, each particle changes its position according to the transition kernel of the signal. Step (a) of the algorithm is known as the *importance sampling* step (popular in the statistics literature) or *mutation* step (inherited from the genetic algorithms literature).

Step (b) of the iteration is called the *selection* step. The particles obtained after the first step of the recursion are multiplied or discarded according to the magnitude of the likelihood weights. In turn, the likelihood weights are proportional to the likelihood of the new observation given the corresponding position of the particle. The net effect of part (b) of the iteration is that it discards particles in unlikely positions and multiplies those in more likely ones. The particle filter with this choice of offspring distribution is called the *bootstrap filter* or the *sampling importance resampling* algorithm (SIR algorithm). It was introduced by Gordon, Salmond, and Smith in [8]. Within the context of the bootstrap filter, the second step is called the *resampling* step.

The theory of nonlinear filtering was well prepared for assimilating and expanding the particle approach. In particular, the Lagrangian approach to NLF, developed in the late 1970s and early 1980s (see [13], [10]), turned out to be a helpful framework for particle filters. The Lagrangian representations for forward and backward dynamics of the optimal filter, often called averaging over characteristics formulae, generalize the famous Feynman–Kac representation of solutions for deterministic parabolic equations. The averaging in these formulas is conditioned on the available observations. The Lagrangian characteristics model the stochastic dynamics of the Monte Carlo particles.

Various versions of the optimal nonlinear filters based on Monte Carlo resampling were developed in the 1990s, in particular, interacting particle filter, sampling/importance resampling particles filter, branching particles filter, etc. For a review, see [5], [4], [1] and the contributions in Part VII of the handbook.

The introduction of particle filters has influenced fundamentally the area of nonlinear filtering. It has extended the reach of NLF to higher dimensional applications and, therefore, enlarged the range of practical applicability of nonlinear filtering. It has also posed many new problems and opened new avenues of research. In our opinion, it is the most important development in the last two decades of research in nonlinear filtering.

We complete the introduction with a description of the contributions comprising the handbook.

The first two parts of the handbook contain classical theoretical results related to the filtering equations covered by six contributions. They are as follows:

The contribution of Kunita is a two-part introduction to nonlinear filtering. In the first part, the filtering problem in discrete time is analysed together with some preliminary results on stochastic calculus required for the study of continuous time nonlinear filtering. The Bayes formula for the computation of the nonlinear filter is derived and the existence of the innovation process is discussed. In the second part, the filtering equations are derived for a general class of system processes having the semimartingale property and the Cauchy problem associated with these equations is studied: existence, uniqueness, and smoothness results are included.

The filtering problem corresponding to processes with jumps is analyzed in the chapter of Grigelionis and Mikulevicius. The existence and uniqueness of the solutions of the filtering equations are discussed. Some examples and an application in financial mathematics (volatility tracking) are included.

In the contribution of Kurtz and Nappo, the signal process is defined as the solution of a martingale problem and its conditional distribution with respect to the observation filtration is recast as the solution of a *filtered* martingale problem. It is shown that uniqueness for the signal's martingale problem implies uniqueness for the filtered martingale problem which in turn implies the Markov property for the conditional distribution considered as a probability measure-valued process. Other applications include a Markov mapping theorem and uniqueness for the filtering equations.

In the chapter by Krylov, the smoothness in L_p sense of filtering densities is discussed. The filtering equations are normally considered in terms of formal adjoint of operators in nondivergence form. Here they are rewritten in the divergence form and the smoothness of solutions is established under very general conditions (Lipschitz continuity of the coefficients of the system).

Two different applications of the Malliavin calculus to nonlinear filtering are discussed in the contribution of Chaleyat-Maurel. The first one deals with the existence and smoothness of a density for conditional laws in filtering theory, whereas the second one is concerned with the problem of the existence or nonexistence of finite-dimensional filters. The two applications are different in nature. In the first one, the observation is considered as fixed, and the Malliavin calculus is applied to the signal in a finite-dimensional approach. In the second one, the Malliavin calculus is applied to the observation through the Zakai equation, which is a stochastic partial differential equation, and the setting is thus infinite dimensional.

The chapter by Lototsky discusses various methods of solving the nonlinear filtering problems using expansions of the optimal filter in the chaos space of the observation process. The elements of the expansion can be either multiple integrals or the Cameron–Martin basis. Two particular filtering algorithms are discussed for the time-homogeneous diffusion filtering model with possible correlation between the state process and the observation noise. Both algorithms rely on the Cameron–Martin version of the chaos expansion, and the approximate filter is a finite linear combination of the chaos elements generated by the observation process. The coefficients in the expansion depend only on the deterministic dynamics of the state and observation processes.

Part III of the handbook covers stability properties and asymptotic analysis of the filtering solution. It includes the following:

The chapter by Kleptsyna and Veretennikov presents some of their recent results for the filtering problem for which the signal has an unknown/unspecified initial condition. The authors show that, under suitable conditions, the filtering algorithm forgets its wrong initial data in the long run, that is, the difference between the conditional measures provided by the filtering algorithm with the exact and wrong initial data converges to zero in some suitable topology. Both the discrete and the continuous frameworks are discussed.

The contribution by Atar presents a review of tools from multiplicative ergodic theory and the theory of positive operators and their usage in the analysis of exponential stability of the optimal nonlinear filter. Particularly, in the case of finite state, the chapter studies the filter sensitivity to perturbations in its initial data and its relation to the Lyapunov spectral gap associated with the filtering equation. In a general setting, it is shown how the Hilbert's metric and Birkhoff's contraction coefficient are used to estimate the decay rate of the error.

The chapter by Chigansky, Liptser, and Van Handel presents a survey of some *intrinsic* methods for studying the stability of the nonlinear filter. These intrinsic methods are methods which directly exploit the fundamental representation of the filter as a conditional expectation through classical probabilistic techniques such as change of measure, martingale convergence, coupling, etc. These methods allow one to establish stability of the filter under weaker conditions compared to other methods, e.g., to go beyond strong mixing signals, to reveal connections between filter stability and classical notions of observability, and to discover links to martingale convergence and information theory.

The contribution of Budhiraja describes conditions under which the solution of the filtering problem satisfies the Feller property and the existence of invariant measures. It also studies the ergodicity of the nonlinear filter and gives some sufficient conditions, under which this property holds, phrased in terms of certain stability properties of the nonlinear filter, for example, the finite memory property or asymptotic stability.

The chapter by Stannat gives an overview on results of stability of the optimal filter for nonergodic signal processes with state space $\mathbb{R}^d$ observed with independent additive noise, both in discrete and continuous time. Explicit lower bounds on the rate of stability in terms of the coefficients of the signal and the observation are obtained, using a parabolic ground state transform with respect to log-concave measures of the recursive algorithm for the optimal filter. The lower bounds in the time-continuous case are obtained as limits of lower bounds for appropriate time-discrete approximations. As particular examples, the Kalman and the Kalman–Bucy filters and filters with signals induced by gradient-type stochastic differential equations are discussed.

Part IV of the handbook includes several special topics, which we describe briefly below:

The *pathwise* theory of filtering is discussed in the contribution of Davis. This theory is concerned with casting the filtering equations in a form in which the filtering estimates can be computed separately for each sample path of the observation process. The chapter presents a pathwise theory for the case where the signal is a diffusion on a finite-dimensional manifold and there is correlation with the observation noise. A geometric setting is natural for this problem, which also brings in Kunita's decomposition theorem for solutions of stochastic differential equations and a family of observation-dependent multiplicative functionals of the signal process.

It turns out that the observation process can be decomposed into two components. One is the integral of the expectation of a function of the signal conditioned with respect to the observation data. The second one is a Brownian motion adapted to the observation filtration, called the innovation process. The natural filtration of the innovation process is included in, but not necessarily equal to the observation filtration. Establishing the cases where the two filtrations are equal has become known as the innovation problem. The contribution of Heunis establishes conditions under which the two filtrations are equal.

The paper of Duncan discusses some results for nonlinear filtering problems where the processes satisfy stochastic differential equations driven by fractional Brownian motions. Fractional Brownian motions are a family of Gaussian processes that include the standard Brownian motion and that seem to be appropriate models for many physical phenomena. The paper covers properties of the family of fractional Brownian motions and the explicit expressions for the Radon–Nikodym derivatives appearing in the formulae for the solution of the filtering problem. It also contains: a stochastic integral equation for the evolution of the conditional expectation of a function of the state process, results for the prediction of processes generated by a fractional Brownian motion and two relations between filtering and mutual information.

Part V of the handbook covers the topics of estimation and control in nonlinear filtering.

The contribution of Newton investigates nonlinear filtering from an information theoretic viewpoint. At its heart are two distinct dualities: one is a feature of time reversal, the other is an instance of an abstract Fenchel–Legendre transform for Bayesian estimators. The first duality arises from a time symmetry in the joint dynamics of the signal process and its nonlinear filter. The second duality is that between the *full information* of a log-likelihood function and the *information gain* of the corresponding posterior distribution in the context of Bayesian estimation. By applying this duality to the path estimation problems associated with the forward and reverse time filters, forward and backward stochastic optimal control problems are obtained, in which the two filters appear in the value functions. The second duality, applied in this way, becomes the duality between estimation and control.

The contribution of Bensoussan, Cakanyıldırım, and Sethi develops a general filtering framework for the problem of estimating the state of a system whose dynamics are governed by a discrete-time Markov process. The chapter presents a number of applications to inventory control systems with partial observations. The authors show how one can transform the nonlinear transition equations into linear ones. This transformation facilitates considerably the study of the associated control problem and the corresponding Bellman equation in a convenient functional space.

The contribution of Bar-Shalom and Blom studies stochastic hybrid systems. These are two component Markov processes $\{x_t, \theta_t\}$, where $\{\theta_t\}$ is a Markov chain and $\{x_t\}$ is the solution of a stochastic difference equation (SDE) whose coefficients depend of $\{\theta_t\}$. The chapter covers the exact Bayesian filter recursions and particle filter approximations for these two component Markov processes. During the development of the exact and particle filter recursions, a key role is played by the exact equations that form the basis of the Interacting Multiple Model filter.

Part VI of the handbook includes several topics related to the approximation theory for the filtering problem. The following are covered:

The filtering problem consists, in particular, in finding the best (in the sense of mean-square error) $\mathcal{Y}_t$-measurable estimator $\hat{X}_t$ of the signal X_t, that is the minimizer of the *filtering error*

$$P_t = \mathbb{E}\left[(X_t - \hat{X}_t)(X_t - \hat{X}_t)^T\right],$$

where $(X_t - \hat{X}_t)^T$ is the row vector associated to $(X_t - \hat{X}_t)$. Of course $\hat{X}_t = \mathbb{E}[X_t|\mathcal{Y}_t]$. Explicit expressions for $\hat{X}_t$ and, respectively, P_t are typically hard to obtain. The chapter of Zeitouni investigates a-priori bounds (both upper and lower) of the matrix P_t. Of particular interest are the bounds that are tight when the observation noise is small.

As stated above, the solution of the continuous time filtering problem can be represented as a ratio of two expectations of certain functionals of the signal process. These functionals are parametrized by the observation path $\{Y_s, s \geq 0\}$. However, in practical applications, only the values of the observation corresponding to a discrete time partition are available, i.e., $\{Y_{t_i}, i = 0, 1, ...\}$. This leads to an approximation of the filtering solution in terms of functionals parametrized by these discrete observations. The convergence rate of this approximation as a function of the partition mesh is studied in the contribution of Crisan. It is shown that the two critical factors that influence the order of convergence are the smoothness of the semigroup associated to the signal and the smoothness of the sensor function h.

The chapter by Le Gland, Monbet, and Tran discusses the ensemble Kalman filter (EnKF). Interpreting the ensemble elements as a population of particles with mean-field interactions, the authors prove the convergence of the EnKF, with the classical rate $1/\sqrt{N}$, as the number N of ensemble elements increases to infinity. In the linear case, the limit of the empirical distribution of the ensemble elements is the usual (Gaussian distribution associated with the) Kalman filter, as expected, but in the more general case of a nonlinear state equation with linear observations, this limit differs from the usual Bayesian filter.

Part VII covers the particle approach for solving the filtering problem. It includes the following contributions:

The contribution of Xiong is a survey of recent results on the particle system approximations to filtering problems in continuous time. Firstly, a weighted particle system representation of the optimal filter is given and a numerical scheme based on this representation is presented together with the convergence result to the optimal filter. Secondly, to reduce the estimation error due to the exponential growth of the variance for individual weights, a branching weighted particle system is defined and an approximate filter based on this particle system is included. Its approximate optimality is proved and the rate of convergence is characterized by a central limit type theorem. Thirdly, as an alternative approach in reducing the estimate error, an interacting particle system (with neither branching nor weights) to direct the particles toward more likely regions is proposed and the corresponding convergence result for this system is established. Finally, the weighted branching particle systems is used to approximate the optimal filter for the model with point process observations.

The contribution of Doucet and Johansen is a survey of results on particle system approximations for filtering problems in discrete time. Just as in the continuous time framework, optimal estimation problems for discrete nonlinear non-Gaussian state-space models do not typically admit analytic solutions. Since their introduction in 1993, particle filtering methods have become a very popular class of algorithms to solve these estimation problems numerically in an online manner, i.e. recursively, as observations become available. Particle filtering methods are now routinely used in fields as diverse as computer vision, econometrics, robotics and navigation. The objective of the contribution is to provide a complete, up-to-date survey of this field. Basic and advanced particle methods for filtering, as well as smoothing, are presented.

The contribution by Del Moral, Patras, and Rubenthaler presents a mean field particle theory for the numerical approximation of Feynman–Kac path integrals in the context of nonlinear filtering. The authors show that the conditional distribution of the signal paths given a series of noisy and partial observation data is approximated by the occupation measure of a genealogical tree model associated with mean field interacting particle model. The complete historical model converges to the McKean distribution of the paths of a nonlinear Markov chain dictated by the mean field interpretation model. The chapter also contains a review of the stability properties and the asymptotic analysis of these interacting processes, including fluctuation theorems and large deviation principles and a Laurent type and algebraic tree-based integral representations of particle block distributions.

The chapter by Schön, Gustafsson, and Karlsson contains a number of real-time applications of the particle filter (PF) in both the signal processing and the robotics communities. The authors present several applications to positioning of moving platforms detailing the experiences of using the PF in practice. The applications concern positioning of underwater vessels, surface ships, cars, and aircraft using geographical information systems containing a database with features of the surrounding. In the robotics community, the PF has been developed into one of the main algorithms (FastSLAM) for solving the simultaneous localization and mapping (SLAM) problem. This can be seen as an extension to the aforementioned applications, where the features in the geographical information system are dynamically detected and updated on the fly.

A key problem in filtering, which is only partially addressed by particle filters, is to maintain a good description of the evolving posterior measure using *minimal* computational effort. Recently, it has been shown that a new class of methods developed for approximation distributions of solutions of stochastic differential equations, collectively known as cubatures on Wiener space, can be used to approximate the conditional distribution in the filtering problem. The chapter by Litterrer and Lyons is a survey on cubature on Wiener space and some related algorithms. It also describes how recombination can be added to the basic algorithm as a way to control the number of particles in the approximation when the method is iterated.

Part VIII contains contributions related to numerical methods in nonlinear filtering. The contribution of Kushner considers two types of numerical algorithms for nonlinear filters. The first is based on the Markov chain approximation method, a powerful approach to numerical problems in stochastic control. It yields an approximation to the conditional density and converges in the weak sense as the approximation parameter goes to zero. Various forms are developed and both convergence and robustness results are included. The second type of approximation is called the assumed density approach, where one supposes that the conditional density takes a given parametrized form, and the evolution equations for the parameters are developed. Most typically, this assumed density is Gaussian (more rarely, a Gaussian mixture) and the parameters are the conditional mean and covariance. The method is heuristic, but has been shown to give good results for many problems.

The chapter by Hairer, Stuart, and Voss is an overview of the Bayesian approach to a wide range of signal processing problems in which the goal is to find the signal. In the case of ordinary differential equations (ODEs) this gives rise to a finite dimensional probability measure for the initial condition, which then determines the measure on the signal. In the case of stochastic differential equations (SDEs) the measure is infinite dimensional. The authors derive the posterior measure for these problems, applying the ideas to ODEs and SDEs, with discrete or continuous observations, and with coloured or white noise. The authors highlight the common structure inherent in all of the problems, namely that the posterior measure is absolutely continuous with respect to a Gaussian prior. This structure leads naturally to the study of Langevin equations which are invariant for the posterior measure and they highlight the theory and open questions relating to these S(P)DEs.

The contribution of Clark and Vinter is concerned with an important class of filtering problems referred to as tracking problems, where the objective is to estimate the state of a moving target from noisy sensor measurements. For many tracking problems of interest, the equations for the conditional distribution of the state are computationally intractable and the key challenges therefore relate to their approximation. The chapter identifies an important class of tracking problems, in which the nonlinearities involved in the models of the state and observations processes are confined to the observations process. Special cases involve bearings-only tracking, range-only tracking, and various tracking problems where measurements are suppressed or degraded in some "nonlinear" fashion. A general methodology is presented for constructing filters for such problems which typically provide superior estimates to those obtained by classical linearization techniques. The specific form taken by these filters in four cases of interest is examined in detail. In the case of bearings-only tracking, the filter is known as the shifted Rayleigh filter.

It is well known that numerical methods for nonlinear filtering problems, which directly use the Kallianpur–Striebel formula, can exhibit computational instabilities due to the presence of very large or very small exponents in both the numerator and denominator of the formula. The chapter by Milstein and Tretyakov introduces a class of computationally stable schemes by exploiting the innovation approach. The authors propose Monte Carlo algorithms based on the method of characteristics for linear parabolic stochastic partial differential equations. Convergence and some properties of the considered algorithms are studied and variance reduction techniques are discussed. The chapter also includes results of some numerical experiments.

The last part of the handbook includes a number of applications of nonlinear filtering in financial mathematics:

The chapter by Frey and Runggaldier considers filtering problems that arise in Markovian factor models for the term structure of interest rates and for credit risk. The connections with the filtering problem is based on the fact that investors act on the basis of only incomplete information about the factors. The current state of the factors has to be inferred/filtered from observable financial quantities. The main goal of the chapter is the pricing of derivative instruments in the interest rate and credit risk contexts, but also other applications are discussed.

The contribution of Elliott, Miao, and Wu introduces a generalized stochastic volatility model to help price energy-related assets by capturing two critical features: mean-reverting prices and a volatility which follows different dynamics in different states of the world. Assuming the dynamics of the states are represented by a hidden Markov chain, the authors apply filtering techniques and the EM algorithm to a time-series model for parameter estimation. Several new filters and closed form estimates for all parameters are derived in the paper. Applications of the proposed model in other fields of finance are also discussed.

The contribution of Pham is a survey of the methods involved in portfolio selection with partial observation. The author describes both the theoretical and numerical aspects related to these optimization problems. The presentation is divided in two parts. The first part covers the continuous-time problem: here, the mean rates of return of the asset prices are not directly observable. Investors observe only asset prices. By the method of change of probability and innovation process in filtering theory, the partial observation portfolio selection problem is transformed into a full observation one with the additional filter state variable, for which one may apply the martingale or PDE approach. The following cases for the modeling of the unobservable mean rate of return are investigated: Bayesian, linear-Gaussian, and finite-state Markov chain. The second part covers discrete-time optimization problems: this context includes the case of unobservable volatility. The numerical approximation of the optimization problem under partial observation is studied. Several numerical experiments illustrate the results for hedging problems in the context of partially observed stochastic volatility models.

The chapter by Scott and Zeng surveys the recent developments in a general filtering model with counting process observations for the micromovement of asset price and its related statistical analysis. The normalized and unnormalized filtering equations as well as the system of evolution equations for

Bayes factors are reviewed. A Markov chain approximation method is used to construct recursive algorithms and their consistency is proven. The authors employ a specific micromovement model built upon the model linear stochastic differential equation to show the steps to develop a micromovement model with specific types of trading noises. The model is further utilized to show the steps to construct consistent recursive algorithms for computing the trade-by-trade Bayes estimates and the Bayes factors for model selection.

References

[1] A. Bain and D. Crisan, Fundamentals of Stochastic Filtering. *Stochastic Modeling and Applied Probability* **60** (2009), Springer.
[2] L. E. Baum and T. Petrie, Statistical Inference for Probabilistic functions on finite state Markov chains. *Ann. Math. Stat.* **37** (1966), 1554–63.
[3] V. E. Benesh, Exact finite-dimensional filters for certain diffusions with nonlinear drift, *Stochastics* **5**, (1981), 65–92.
[4] P. Del Moral, *Feyman–Kac Formulae. Genealogical and Interacting Particle Systems with Applications.* Springer-Verlag, New York, (2004).
[5] A. Doucet, N. de Freitas, and N. Gordon (eds), *Sequential Monte Carlo Methods in Practice*, Springer-Verlag, New York, (2001).
[6] T. E. Duncan, Tech. Report 7001-4, Stanford University, Center for Systems Research, May (1967).
[7] M. Fujisaki, G. Kallianpur, and H. Kunita, Stochastic differential equations for the nonlinear filtering problem, *Osaka J. Math.* **9** (1972), no. 1, 19–40.
[8] N. J. Gordon, D. J. Salmon, and A. F. M. Smith. Novel approach to nonlinear/non-Gaussian Bayesian state estimation. *IEEE Proceedings*, Part F, pp. 107–113, (1993).
[9] R. E. Kalman and R. S. Bucy, New results in linear filtering and prediction problems, *J. Basic Engineering, Trans. ASME* **83D** (1961), 95–108.
[10] H. Kunita, *Stochastic Flows and Stochastic Differential Equations*, Cambridge University Press, Cambridge, (1982).
[11] H. J. Kushner, On the differential equations satisfied by conditional probability densities of Markov processes, with applications, *J. Soc. Indust. Appl. Math. Ser. A Control* **2** (1964), 106–19.
[12] H. J. Kushner, Dynamical equations for optimal nonlinear filtering, *J. Differential Equations* **3** (1967), 179–90.
[13] N. V. Krylov and B. L. Rozovskiĭ, Characteristics of second-order degenerate parabolic Itō equations. *Trudy Sem. Petrovsk.* **8** (1982), 153–68.
[14] R. Sh. Liptser and A. N. Shiryayev, *Statistics of Random Processes, I, II*, Springer, New York, (2002).
[15] S. V. Lototsky, R. Mikulevicius, and B. L. Rozovskiĭ, Nonlinear filtering revisited: a spectral approach, *SIAM J. Contr. Optim.* **35** (1997), no. 2, 435–61.
[16] R. E. Mortensen, *Tech. Report ERL–66–1*, UC Berkeley, Electronics Research Laboratory, Aug. (1966).
[17] E. Pardoux, Filtrage non linéaire et équations aux dérivées partielles stochastiques associées. In: *Ecole d'Eté de Probabilités de Saint-Flour XIX*, (1989).
[18] B. L. Rozovskiĭ, *Stochastic Evolution Systems. Linear Theory and Applications to Non-linear Filtering*, Kluwer Academic Publishers, Dordrecht, 1990 (trans. from the 1982 Russian original).

[19] A. N. Shiryaev, On stochastic equations in the theory of conditional Markov process, *Teor. Verojatnost. i Primenen.* **11** (1966), 200–6 in Russian; English trans. *Theor. Probability Appl.* **11** (1966), 179–84.
[20] R. I. Stratonovich, Conditional Markov processes, *Theor. Probab. Appl.* **5** (1960), 156–78.
[21] L. R. Welch, Hidden Markov Models and Baum–Welch Algorithm (The Shannon Lecture). *IEEE Information Theory Society Newsletter* **53**, no. 4, Dec. (2003)
[22] M. Zakai, *The optimal filtering of Markov jump processes in additive white noise*. Research Note No. 563. Applied Research Laboratory, Sylvania Electronic Systems, Waltham, Massachusets. (1965).
[23] M. Zakai, On the optimal filtering of diffusion processes, *Z. Wahrsch. Verw. Gebiete* **11** (1969), no. 3, 230–43.

PART I
The Foundations of Nonlinear Filtering

·2·

Nonlinear Filtering Problems I: Bayes Formulas and Innovations

H. Kunita

2.1 Filtering problems for time series

2.1.1 Introduction

Let $X_t = (X_t^1, ..., X_t^d)$ be a d-dimensional stochastic process called a *system process* or a *signal process*. Let N_t be an e-dimensional process independent of X_t. Suppose that we cannot observe the signal X_t directly and observe the stochastic process disturbed by the *noise* N_t:

$$Y_t = H_t X_t + N_t,$$

where H_t are deterministic matrices. The process Y_t is called an *observation process*. The filtering problem is to estimate the system X_t at each time t based on the observation data $\{Y_s, s \leq t\}$.

Two types of estimations are studied. The one is a linear estimation and the other is a nonlinear estimation. Suppose that the family of random variables Y_t, X_t, N_t are square integrable. Let $\mathcal{H}(t)$ be the Hilbert space spanned by the linear combinations of $\{Y_s^i; s \leq t, i = 1, ..., e\}$, where $Y_t = (Y_t^1, ..., Y_t^e)$. Let $\bar{X}_t^i$ be the unique element of $\mathcal{H}(t)$ which admits the least distance from X_t^i in the sense of L^2-norm. Then $\bar{X}_t^i$ coincides with the orthogonal projection of X_t^i to the subspace $\mathcal{H}(t)$. The stochastic process $\bar{X}_t = (\bar{X}_t^1, ..., \bar{X}_t^d)$ is called the *linear filter* of the process X_t.

Let us denote by $\sigma(Y_s; s \leq t)$ the smallest complete sub-σ-field for which the family of random variables $\{Y_s, s \leq t\}$ is measurable. We set $\mathcal{G}_t = \sigma(Y_s; s \leq t)$. Let $\mathcal{G}(t)$ be the Hilbert space generated by $\mathcal{G}_t$-measurable square integrable random variables. Let $\hat{X}_t^i$ be the unique element of $\mathcal{G}(t)$ which admits the least distance form X_t^i. Then $\hat{X}_t^i$ is the orthogonal projection of X_t^i to the subspace $\mathcal{G}(t)$. Further, it coincides with the conditional expectation of X_t^i with respect to the σ-field $\mathcal{G}_t$, which is denoted by $E[X_t^i | \mathcal{G}_t]$. The stochastic process $\hat{X}_t = (\hat{X}_t^1, ..., \hat{X}_t^d)$ is called the *nonlinear filter* of the process X_t.

These two estimations $\bar{X}_t$ and $\hat{X}_t$ are different. We have the inequality

$$E[|X_t - \hat{X}_t|^2] \leq E[|X_t - \bar{X}_t|^2].$$

Therefore we can say that the nonlinear filter is better than the linear filter. But, generally speaking, the algorithm of the nonlinear filter is more complicated than that of the linear filter.

Assume now that (X_t, N_t) is a Gaussian process. Then Y_t is also a Gaussian process. Then, since $X_t^i - \bar{X}_t^i$ is orthogonal with $\mathcal{H}(t)$, it is independent of any element of $\mathcal{H}(t)$. Therefore it is independent of any element of $\mathcal{G}(t)$. Consequently $E[X_t^i|\mathcal{G}_t] - \bar{X}_t^i = 0$ holds valid, showing that the linear filter and the nonlinear filter coincide. However, we cannot expect the similar fact in the case where the pair (X_t, N_t) is not Gaussian.

The study of the linear filter for stationary process with discrete time parameter (stationary time series) was initiated by Kolmogorov and Wiener. Let $Y_n, n = \cdots, -1, 0, 1, \cdots$ be a stationary time series. Suppose that its mean is 0 and is purely nondeterministic, i.e., $\cap_{k=0}^{\infty} \mathcal{H}(-k) = \{0\}$. Suppose further that the noise $\{N_n\}$ is a stationary orthogonal time series. We set $\mathcal{M}_n = \mathcal{H}(n) \ominus \mathcal{H}(n-1)$. Then elements of $\mathcal{M}_n$ and elements of $\mathcal{M}_m$ are orthogonal if $n \neq m$. Define the time series $I_n = (I_n^1, ..., I_n^e)$ by $I_n = Y_n - H_n\bar{X}_n^-$, where $\bar{X}_n^-$ is the orthogonal projection of X_n to $\mathcal{H}(n-1)$. Then it holds $I_n^i \in \mathcal{M}_n$ and $I_n, n = 0, \pm 1, ...$ is an orthogonal sequence. Further, the closed subspace spanned by linear sums of $\{I_k^i; k \leq n, i = 1, ..., e\}$ coincides with $\mathcal{H}(n)$. By this property, $\{I_n\}$ is called an *innovation* of $\{Y_n\}$. Any element of $\mathcal{H}(n)$, especially $\bar{X}_n$ is then represented by using the innovation $\{I_n\}$ and a matrix function $\{f_n\}$ as

$$\bar{X}_n = \sum_{j=0}^{\infty} f_{n-j} I_{n-j} = \sum_{j=0}^{\infty} f_{n-j}(Y_{n-j} - H_{n-j}\bar{X}_{n-j}^-).$$

Hence if the representing functions $\{f_n\}$ are computed, we can obtain $\{\bar{X}_n\}$. In order to get the algorithms for computing $\{f_n\}$, spectral analysis or theory of Hardy class are used.

Generally it is not an easy task to get the algorithm of computing $\{f_k\}$. Kalman observed the case where the signal process X_n is given by a stochastic difference equation and obtained the algorithm of the linear filter by a stochastic difference equation, so that the filter can be obtained recursively. Kalman's method can be applied easily to nonstationary case.

The program of this chapter is as follows. In Section 2.1 we will discuss the filtering problem for time series, i.e. for stochastic processes with discrete time parameter. In Section 2.1.2, we will consider a linear filter, which is known as the Kalman filter with discrete parameter. In Section 2.1.3, we will consider a problem of nonlinear filtering. We will obtain a kind of Bayes formula for computing the conditional probability distributions $\pi_t(E) = \mathbf{P}(X_t \in E|\mathcal{G}_t)$.

For the study of continuous time parameter case, we need some stochastic calculus including Itô's formula and stochastic differential equations. It will be discussed in Section 2.2. Nonlinear filtering problem for processes with continuous time will be discussed in Sections 2.3 and 2.4 and Chapter 3. In Section 2.3, we derive a Bayes formula for computing the conditional probability distribution $\pi_t(dx)$. In Section 2.4, we study the innovation process. It will be seen that the Bayes formula and innovation property for continuous parameter processes require more complicated arguments than those for discrete parameter processes. In the next chapter, we will discuss stochastic partial differential equations which govern conditional distribution π_t and unnormalized conditional distributions ρ_t.

2.1.2 Kalman filter

Let Z_n, $n = 1, 2, \ldots$ be a Gaussian time series with mean 0 defined on a complete probability space $(\Omega, \mathcal{F}, \mathbf{P})$. If $\{Z_n\}$ are mutually independent, the time series Z_n is called a *white noise*. We introduce

a) The system process X_n, $n = 0, 1, \ldots$ is a d-dimensional Gaussian process. It is determined by the following stochastic difference equation.

$$X_n = A_n X_{n-1} + W_n, \quad n = 1, 2, \ldots \tag{2.1}$$

where A_n, $n = 1, 2, \ldots$ are given deterministic matrices and W_n, $n = 1, \ldots$ is an r-dimensional white noise with the covariance matrices Q_n, $n = 1, 2, \ldots$ respectively, where Q_n are nonnegative definite symmetric matrices. The initial random variable X_0 is Gaussian, independent of W_n, $n = 1, 2, \ldots$.

b) The noise N_n, $n = 1, 2, \ldots$ is an e-dimensional white noise with covariance matrices R_n, $n = 1, 2, \ldots$, which are positive definite invertible symmetric matrices. Further the pair (W_n, N_n) is a white noise such that $E[W_n N_n^T] = S_n$.

c) The observation process is an e-dimensional process given by

$$Y_n = H_n X_n + N_n, \quad n = 1, 2, \ldots \tag{2.2}$$

where H_n, $n = 1, 2, \ldots$ are given deterministic matrices.

We denote by $\bar{X}_n$ the linear filter based on the observation data $\{Y_m; 1 \leq m \leq n\}$. Since the pair (X_n, Y_n) is a Gaussian time series, the linear filter $\bar{X}_n$ coincides with the nonlinear filter $\hat{X}_n$. We denote

$$\mathcal{G}_n = \sigma(Y_k; k \leq n), \quad n = 1, 2, \ldots \tag{2.3}$$

Then $\bar{X}_n = \hat{X}_n = E[X_n|\mathcal{G}_n]$, where $E[X_n|\mathcal{G}_n] := (E[X_n^1|\mathcal{G}_n], \ldots, E[X_n^d|\mathcal{G}_n])$.

The following algorithm of computing the filter $\hat{X}_n$ is applied in many problems. We quote a book by Brockwell and Davis [3] for its theory and applications.

Theorem 2.1 (Kalman) *The filter $\hat{X}_n$ is the solution of the stochastic difference equation described below.*

$$\hat{X}_n = A_n\hat{X}_{n-1} + (P_n^- H_n^T + S_n)(R_n + H_n P_n^- H_n^T)^{-1}(Y_n - H_n A_n \hat{X}_{n-1}), \qquad (2.4)$$

$$\hat{X}_0 = E[X_0], \qquad (2.5)$$

where matrices P_n^- are given by

$$P_n = P_n^- - (P_n^- H_n^T + S_n)(R_n + H_n P_n^- H_n^T)^{-1}(H_n P_n^- + S_n^T), \qquad (2.6)$$

$$P_n^- = A_n P_{n-1} A_n^T + Q_n, \quad n = 1, 2, \ldots \qquad (2.7)$$

$$P_0 = E[(X_0 - E[X_0])(X_0 - E[X_0])^T]. \qquad (2.8)$$

Matrices P_n and P_n^- in the above theorem coincide with the following error covariance matrices.

$$P_n = E[(X_n - \hat{X}_n)(X_n - \hat{X}_n)^T], \qquad (2.9)$$

$$P_n^- = E[(X_n - \hat{X}_n^-)(X_n - \hat{X}_n^-)^T], \qquad (2.10)$$

where $\hat{X}_n^- = E[X_n | \mathcal{G}_{n-1}]$. These matrices can be computed recursively by equations (2.6)–(2.8): If P_{n-1} is computed, then P_n^- is computed by (2.7). Then P_n is computed by (2.6).

The filter is computed in the order of $\hat{X}_1, \hat{X}_2, \ldots$. In order to compute $\hat{X}_n$, we need known coefficients A_n, H_n, R_n, the error covariance matrix P_n^- and the latest past filter $\hat{X}_{n-1}$ and the new observation data Y_n. This means that we can forget the past observation data $Y_k, k = 1, 2, \ldots, n-1$. In this sense, the filter can be computed recursively.

Set

$$I_n = Y_n - H_n \hat{X}_n^- = Y_n - H_n A_{n-1} \hat{X}_{n-1}, \quad n = 1, 2, \ldots \qquad (2.11)$$

Then $I_n, n = 1, 2, \ldots$ is an white noise with covariance $R_n + H_n P_n^- H_n^T$. Let $\mathcal{H}(n)$ be the linear space spanned by $Y_1, \ldots, Y_n$ and let $\mathcal{L}(n)$ be the linear space spanned by $I_1, \ldots, I_n$. Then $\mathcal{H}(n) = \mathcal{L}(n)$ holds for any n. I_n is called the *innovation* of Y_n.

The stochastic difference equation of Theorem 2.1 is written by using innovation I_n as

$$\hat{X}_n = A_n \hat{X}_{n-1} + (P_n^- H_n^T + S_n)(R_n + H_n P_n^- H_n^T)^{-1} I_n, \quad n = 1, 2, \ldots \qquad (2.12)$$

Therefore, the solution $\hat{X}_n$ is a Markov process.

Remark. Kalman's algorithm is valid to linear filter of non-Gaussian time series, too. Suppose that (W_n, N_n) is a (non-Gaussian) orthogonal noises, i.e., these are L^2-time series with means 0 and $E[(W_n, N_n)(W_m, N_m)^T] = 0$ for any $n \neq m$, the linear filter $\bar{X}_n$ satisfies (2.4)–(2.8).

2.1.3 Nonlinear filter

We will consider the nonlinear filtering problem in the case of discrete time parameter. Let X_n, $n = 0, 1, \ldots$ be a system process and let

$$Y_1 = h_1(X_1) + N_1, \qquad (2.13)$$
$$Y_n = h_n(X_n, Y_1, \ldots, Y_{n-1}) + N_n, \quad n = 2, 3, \ldots$$

be an observation process, where $\{N_n\}$ is a white noise independent of the system process. Let $\mathcal{G}_n$, $n = 1, 2, \ldots$ be the filtration of σ-fields defined by (2.3). In non Gaussian case, we cannot expect a recursive formula (like a Kalman filter) for the nonlinear filter $\hat{X}_n$. Instead of $\hat{X}_n$, we shall consider conditional distributions

$$\pi_n(E) = \mathbf{P}(X_n \in E | \mathcal{G}_n). \qquad (2.14)$$

If the above conditional distributions are obtained, the nonlinear filter $\hat{X}_n$ can be computed by the formula $\hat{X}_n = \int x \pi_n(dx)$.

We assume that the triple (N_n, X_n, Y_n) has the following properties.

a) The white noise N_n, $n \geq 1$ is an e-dimensional Gaussian process with covariance R_n, which are invertible matrices.
b) The system X_n, $n \geq 0$ is independent of the white noise N_n, $n \geq 1$.
c) The observation process is an e-dimensional process given by (2.13).

Set $\mathcal{F}_0 = \sigma(X_k; k = 0, 1, \ldots)$ and

$$\mathcal{F}_n = \mathcal{F}_0 \vee \sigma(N_k; k \leq n), \quad n = 1, 2, \ldots$$

and define $\Phi_0 = 1$ and for $n = 1, 2, \ldots$

$$\Phi_n = \exp\left\{-\sum_{k=1}^n (h_k, R_k^{-1} N_k) - \frac{1}{2}\sum_{k=1}^n (h_k, R_k^{-1} h_k)\right\}. \qquad (2.15)$$

Then Φ_n is a positive martingale with respect to $(\mathcal{F}_n)$ with mean 1. Indeed, we have

$$E\left[\exp\left\{-(h_n, R_n^{-1} N_n) - \frac{1}{2}(h_n, R_n^{-1} h_n)\right\} | \mathcal{F}_{n-1}\right]$$
$$= \frac{1}{\sqrt{(2\pi)^d |R_n|}} \int_{\mathbf{R}^e} \exp\left\{-\frac{1}{2}(z + h_n, R_n^{-1}(z + h_n))\right\} dz = 1.$$

Multiply Φ_{n-1} to both sides of the above, then we get $E[\Phi_n | \mathcal{F}_{n-1}] = \Phi_{n-1}$, showing the martingale property of Φ_n.

We can define a probability measure $\mathbf{Q}$ by the formula

$$\mathbf{Q}(B) = E[\Phi_n 1_B], \quad B \in \mathcal{F}_n.$$

We show that the conditional distribution (2.14) is rewritten by

$$\pi_n(E) = \frac{E_Q[\Phi_n^{-1} 1_{X_n \in E} | \mathcal{G}_n]}{E_Q[\Phi_n^{-1} | \mathcal{G}_n]}. \tag{2.16}$$

Let $A \in \mathcal{F}_n$ and $B \in \mathcal{G}_n$. Then we have

$$\mathbf{P}(A \cap B) = E_Q[\Phi_n^{-1} 1_A 1_B] = E_Q[E_Q[\Phi_n^{-1} 1_A | \mathcal{G}_n] 1_B]$$
$$= E[\Phi_n E_Q[\Phi_n^{-1} 1_A | \mathcal{G}_n] 1_B] = E[E[\Phi_n | \mathcal{G}_n] E_Q[\Phi_n^{-1} 1_A | \mathcal{G}_n] 1_B].$$

Therefore,

$$\mathbf{P}(A | \mathcal{G}_n) = E[\Phi_n | \mathcal{G}_n] E_Q[\Phi_n^{-1} 1_A | \mathcal{G}_n].$$

Setting $A = \Omega$, we get $E[\Phi_n | \mathcal{G}_n] E_Q[\Phi_n^{-1} | \mathcal{G}_n.] = 1$. Substitute the last equality to the above and set $A = \{X_n \in E\}$, then we get formula (2.16).

Now we shall compute the right hand side of (2.16). The following is a discrete version of Girsanov's theorem. See Theorem 2.6 and Theorem 2.9.

Lemma 2.1 *With respect to the measure* $\mathbf{Q}$, *the pair* (X_n, Y_n) *has properties stated below.*

1) *The law of* X_n, $n = 0, 1, ...$ *is equal to the law with respect to* $\mathbf{P}$.
2) Y_n, $n = 1, 2, ...$ *is a white noise with covariance* R_n, $n = 1, 2, ...$.
3) X_n *and* Y_n *are independent.*

Proof. Since $\Phi_0 = 1$, $\mathbf{Q} = \mathbf{P}$ holds on $\mathcal{F}_0$. Therefore 1) is clear. For proofs of 2) and 3), we study the law of Y_n by its characteristic function. Let n_0 be a sufficiently large positive integer. For a $n_0 e$-dimensional vector $\xi = (\xi_1, ..., \xi_{n_0})$, we define $\gamma_0^\xi = 1$ and

$$\gamma_n^\xi = \exp\left\{i \sum_{k=1}^n (\xi_k, R_k^{-1}(N_k + h_k)) + \frac{1}{2} \sum_{k=1}^n (\xi_k, R_k^{-1} \xi_k)\right\}, \quad n = 1, 2, ..., n_0.$$

Then for $1 \leq n \leq n_0$,

$$\Phi_n \gamma_n^\xi = \exp\left\{\sum_{k=1}^m (i\xi_k - h_k, R_k^{-1} N_k) - \frac{1}{2} \sum_{k=1}^n (i\xi_k - h_k, R_k^{-1}[i\xi_k - h_k])\right\}.$$

Then the product $\Phi_n \gamma_n^\xi$ is also a (complex) martingale with respect to $\mathbf{P}$, which can be shown similarly as the case of Φ_n. Therefore, for $n > m$ and $A \in \mathcal{F}_m$,

$\int_A \gamma_n^\xi d\mathbf{Q} = \int_A \Phi_n \gamma_n^\xi d\mathbf{P} = \int_A \Phi_m \gamma_m^\xi d\mathbf{P} = \int_A \gamma_m^\xi d\mathbf{Q}$. Therefore $\int_A \gamma_n^\xi (\gamma_m^\xi)^{-1} d\mathbf{Q} = \mathbf{Q}(A)$ for any $A \in \mathcal{F}_m$. Rewriting the above, we get

$$\int_A \exp\left\{i \sum_{k=m+1}^n (\xi_k, R_k^{-1}[N_k + h_k])\right\} d\mathbf{Q} \qquad (2.17)$$

$$= \exp\left\{-\frac{1}{2} \sum_{k=m+1}^n (\xi_k, R_k^{-1} \xi_k)\right\} \mathbf{Q}(A).$$

Consequently $Y_k = h_k + N_k$, $m+1 \leq k \leq n$ and $\mathcal{F}_m$ are independent. In particular, setting $m = 0$, we get 3). The formula (2.17) means that the characteristic function of $Y_n = N_n + h_n$ is equal to that of a Gaussian distribution with mean 0 and covariance R_n, and Y_n, $n = 1, 2, \ldots$ are independent with respect to $\mathbf{Q}$. Therefore 2) follows. □

By the above lemma, we have the formula

$$E_\mathbf{Q}[\Phi_n^{-1} 1_{X_n \in E} | \mathcal{G}_n] = \int_{\{x; x_n \in E\}} \psi_n(\mathbf{x}, Y_0, Y_1, \ldots, Y_n) dF(\mathbf{x}),$$

where $\mathbf{x} = (x_1, \ldots, x_n)$, F is the distribution of nd-dimensional random variable $(X_1, \ldots, X_n)$ and

$$\psi_n(\mathbf{x}, Y_1, \ldots, Y_n) = \exp\left\{\sum_{k=1}^n (h_k, R_k^{-1} Y_k) - \frac{1}{2} \sum_{k=1}^n (h_k, R_k^{-1} h_k)\right\}. \qquad (2.18)$$

Thus we have the following theorem.

Theorem 2.2 *Conditional distribution of the nonlinear filter can be computed by the formula*

$$\pi_n(E) = \frac{\int_{\{\mathbf{x} \in \mathbf{R}^{nd}; x_n \in E\}} \psi_n(\mathbf{x}, Y_0, \ldots, Y_n) dF(\mathbf{x})}{\int_{\mathbf{R}^{nd}} \psi_n(\mathbf{x}, Y_0, \ldots, Y_n) dF(\mathbf{x})}. \qquad (2.19)$$

Finally we consider the case where the system process X_n is Markovian.

Corollary 2.1 *Suppose that the system process X_n is a Markov process with transition probabilities $P_{m,n}(x, E)$. Suppose further that $h_n = h_n(x)$ (not depending on the observation $Y_1, \ldots, Y_{n-1}$). Then we have*

$$\pi_n(E) = \frac{\int_{\mathbf{R}^d} [\int_E g_n(x_n, Y_n) P_{n-1,n}(x_{n-1}, dx_n)] \pi_{n-1}(dx_{n-1})}{\int_{\mathbf{R}^d} [\int_{\mathbf{R}^d} g_n(x_n, Y_n) P_{n-1,n}(x_{n-1}, dx_n)] \pi_{n-1}(dx_{n-1})}, \qquad (2.20)$$

where

$$g_n(x, y) = \exp\left\{(h_n(x), R_n^{-1} y) - \frac{1}{2}(h_n(x), R_n^{-1} h_n(x))\right\}. \qquad (2.21)$$

The above corollary indicates that the filter π_n can be computed by using the transition probability $P_{n-1,n}$, the latest past filter π_{n-1} and the present observation data Y_n. We can forget the past observations $\{Y_k, k \leq n-1\}$. Thus the algorithm of the above filter is recursive.

2.2 Stochastic calculus

2.2.1 Introduction

The Kalman filter with continuous time parameter could be formulated as follows. The system process X_t is governed by a random differential equation

$$\frac{dX_t}{dt} = A(t)X_t + \dot{W}_t, \tag{2.22}$$

where $A(t)$ is a matrix-valued function of t and $\dot{W}_t$ is a white noise with covariance $Q(t)$, i.e., $\dot{W}_t$ are Gaussian random variables such that $E[\dot{W}_t] = 0$ and $E[\dot{W}_t \dot{W}_s] = Q(t)\delta_{t-s}$ where δ_t is the Dirac's delta function supported by the origin 0. Next, the observation process could be

$$\dot{Y}_t = H(t)X_t + \dot{N}_t, \tag{2.23}$$

where $H(t)$ is a matrix function and $\dot{N}_t$ is a white noise with covariance $R(t)\delta_{t-s}$. However, in order to discuss the problem rigorously, we will reformulate equations (2.1) and (2.2) making use of Brownian motions, since white noises with continuous time parameter do not exist.

We first give the definition of a Brownian motion. Loosely speaking, it is a time integral of a white noise like $W_t = \int_0^t \dot{W}_s ds$ or $N_t = \int_0^t \dot{N}_s ds$. Let $W_t = (W_t^1, ..., W_t^d), t \geq 0$ be a d-dimensional stochastic process defined on a probability space $(\Omega, \mathcal{F}, \mathbf{P})$. It is called a *Brownian motion* if it satisfies

a) It has independent increments, i.e., for any $0 \leq t_1 < t_2 < \cdots < t_{n-1} < t_n$ increments $W_{t_2} - W_{t_1}, \cdots, W_{t_n} - W_{t_{n-1}}$ are independent.
b) $W_0 = 0$ and the law of W_t is Gaussian with mean 0.
c) For almost all $\omega \in \Omega$ sample paths $W_t(\omega)$ are continuous in t.

The covariance matrix V_t of W_t is symmetric, nonnegative definite and nondecreasing in t (with respect to the order of nonnegative definiteness). If $V_t = tI$, where I is the identity matrix, W_t is called a *standard Brownian motion*.

In this paper, we will restrict our attention to a Brownian motion such that the covariance matrix V_t has a continuous density $Q(t)$ with respect to t and is written as $V_t = \int_0^t Q(s)ds$, where $Q(s)$ is symmetric and nonnegative definite matrices. If $Q(s)$ is invertible for any s and the inverse $Q(s)^{-1}$ is continuous in s, W_t is called a *nondegenerate Brownian motion*.

Now let (W_t, N_t) be a $d+e$-dimensional Brownian motion such that W_t and N_t have covariances $\int_0^t Q(s)ds$ and $\int_0^t R_s ds$. These two may or may not be independent. Then equation (2.22) can be defined more rigorously as a stochastic differential equation

$$X_t = X_0 + \int_0^t A(s) X_s ds + W_t \qquad (2.24)$$

and the observation process is defined by

$$Y_t = \int_0^t H(s) X_s ds + N_t. \qquad (2.25)$$

Then $\hat{X}_t := E[X_t|\mathcal{G}_t]$ is called the filter of X_t based on the observation data $\mathcal{G}_t = \sigma(Y_s; s \leq t)$. For the study of the filter, we need Itô's stochastic calculus, which will be discussed in this section.

In Sections 3.4 and in the next chapter, we will discuss in details nonlinear filtering problems for stochastic processes with continuous time parameter, where Itô's stochastic calculus plays an important role. In this section we discuss briefly Itô's stochastic calculus including the stochastic integral based on a Brownian motion and related topics such as Itô's formula and Girsanov's theorem. Kalman filter with continuous time will be discussed in Chapter 3.

For literatures on stochastic calculus, we refer to Kallianpur [9], Karatzas–Shreve [11], Kunita [14], Liptzer–Shiryaev [15], Oksendal [16], where stochastic integrals based on *standard* Brownian motions are discussed. In this chapter, we will develop stochastic calculus for a slightly wider class of Brownian motions so that it covers system noise W_t and observation noise N_t mentioned above. In most theorems, proofs are similar to the case of a standard Brownian motion. It will be omitted except Theorem 2.3.

2.2.2 Martingales and quadratic covariations

Let $(\Omega, \mathcal{F}, \mathbf{P})$ be a complete probability space. Let $(\mathcal{F}_t, 0 \leq t < \infty)$ be an increasing family of sub σ-fields of $\mathcal{F}$ such that each $\mathcal{F}_t$ contains all null sets of $\mathcal{F}$. It is called a *filtration* of sub σ-field of $\mathcal{F}$. We assume that the filtration is right continuous, i.e., it holds $\mathcal{F}_t = \mathcal{F}_{t+} \equiv \cap_{\epsilon>0} \mathcal{F}_{t+\epsilon}$ for any $t > 0$. A stochastic process $\{X_t, t \geq 0\}$ is called a *measurable process* if $X_t(\omega)$ is measurable with respect to two variables (t, ω). It is called $(\mathcal{F}_t)$-*adapted* if X_t is an $\mathcal{F}_t$-measurable random variable for any $t \geq 0$. Further it is called $(\mathcal{F}_t)$ *progressively measurable* or simply *progressively measurable*, if $X_s(\omega), 0 \leq s \leq t, \omega \in \Omega$ is $\mathcal{B}([0, t]) \times \mathcal{F}_t$-measurable for any $t > 0$.

Let $\{X_t, t \geq 0\}$ be a one-dimensional stochastic process. It is called a *martingale* (with respect to the filtration $(\mathcal{F}_t)$) if

(1) X_t is $(\mathcal{F}_t)$-adapted.
(2) For any t, X_t is integrable and its conditional expectation with respect to $(\mathcal{F}_s)$ denoted by $E[X_t|\mathcal{F}_s]$ satisfies $E[X_t|\mathcal{F}_s] = X_s$ a.s. for any $s < t$.

It is known that any martingale has a modification such that the sample paths are right continuous with left hand limits.

An important example of martingale is a one dimensional Brownian motion. Let $X_t - X_0$ be a one-dimensional Brownian motion. Let $\mathcal{F}_0$ be a sub σ-field of $\mathcal{F}$ which is independent of the Brownian motion $X_t - X_0$. Let $(\mathcal{F}_t)$ be the filtration generated by the Brownian motion and $\mathcal{F}_0$, i.e., $\mathcal{F}_t = \sigma(X_s - X_0; s \leq t) \vee \mathcal{F}_0$. Then X_t is a martingale with respect to $(\mathcal{F}_t)$. Conversely suppose we are given a filtration $(\mathcal{F}_t)$. Let X_t be a one-dimensional Brownian motion adapted to the filtration $(\mathcal{F}_t)$. It is called an $(\mathcal{F}_t)$-*Brownian motion* if $X_u - X_t$ is independent of $\mathcal{F}_t$ for any $u > t$.

A positive random variable τ taking possibly the value $+\infty$ is called a *stopping time* if $\{\tau \leq t\} \in \mathcal{F}_t$ holds for any $t \geq 0$.

Let M_t be a continuous $(\mathcal{F}_t)$-adapted process. If we can choose an increasing sequence of stopping times τ_n, $n = 1, 2, \ldots$ diverging to infinity such that the stopped sequence of stochastic processes $M_t^n = M_{t \wedge \tau_n}$, $n = 1, 2, \ldots$ are martingales, the process M_t is called a *local martingale*. Further if (M_t^n) are square integrable martingales, it is called a *locally square integrable martingale*. Continuous local martingale is a locally square integrable martingale.

Let A_t be a right continuous $(\mathcal{F}_t)$-adapted process. It is called an *increasing process* if it is increasing with respect to t a.s. and it is called a *process of bounded variation* if it is written as a difference of two increasing processes. A stochastic process X_t is called a *semimartingale* if it is decomposed to the sum of a local martingale and a process of bounded variation.

Let X_t and Y_t be continuous semimartingales. Then the limit

$$\langle X, Y \rangle_t := \lim_{|\Delta| \to 0} \sum_{t_k \leq t} (X_{t_k} - X_{t_{k-1}})(Y_{t_k} - Y_{t_{k-1}}) \qquad (2.26)$$

exists uniformly on compact sets with respect to t in probability, where $\Delta = \{0 = t_0 < t_1 < \cdots t_m = T\}$ and $|\Delta| = \max_k(t_{k+1} - t_k)$. It is a continuous process of bounded variation. It is called the *quadratic covariation* of X_t and Y_t. In the case $X = Y$, we denote $\langle X, X \rangle_t$ by $\langle X \rangle_t$ and call it a *quadratic variation* of X_t. It is a continuous increasing process.

Let M_t, N_t, O_t be continuous local martingales and let A_t, B_t, C_t be continuous processes of bounded variation. Let $X_t = M_t + A_t$, $Y_t = N_t + B_t$ and $Z_t = O_t + C_t$. The quadratic covariation has the following properties.

1) $\langle X, Y \rangle_t$ is a continuous $(\mathcal{F}_t)$-adapted process of bounded variation. It holds

$$\langle X, Y + Z \rangle_t = \langle X, Y \rangle_t + \langle X, Z \rangle_t, \qquad (2.27)$$
$$\langle X, Y \rangle_t = \langle M, N \rangle_t, \quad \langle X, B \rangle_t = \langle A, Y \rangle_t = \langle A, B \rangle_t = 0.$$

2) $M_t N_t - \langle M, N \rangle_t$ is a continuous local martingale.

3) Continuous local martingales M_t, N_t are called *orthogonal* if $\langle M, N \rangle_{t=0}$ *a.s.* holds. In this case, it holds $\langle M + N \rangle_t = \langle M \rangle_t + \langle N \rangle_t$ *a.s.*

4) It satisfies Schwarz's inequality

$$\left| \int_0^t f_s g_s d\langle M, N \rangle_s \right| \leq \left(\int_0^t f_s^2 d\langle M \rangle_s \right)^{1/2} \left(\int_0^t g_s^2 d\langle N \rangle_s \right)^{1/2} \quad (2.28)$$

for any bounded $(\mathcal{F}_t)$-progressively measurable processes f_t, g_t.

5) (Burkholder's inequality) For any $p \geq 2$, there exist positive constants C_1, C_2 such that

$$C_1 E[\langle M \rangle_t^{p/2}] \leq E[|M_t|^p] \leq C_2 E[\langle M \rangle_t^{p/2}]$$

holds for any continuous L^p-martingale M_t.

6) Let $W_t = (W_t^1, ..., W_t^d)$ be a d-dimensional Brownian motion with covariance matrix $(\int_0^t q_s^{ij} ds)$. Then $\langle W^i, W^j \rangle_t = \int_0^t q_s^{ij} ds$.

2.2.3 Stochastic integrals and martingales

Let $(\mathcal{F}_t)$ be a filtration and let W_t be a one dimensional $(\mathcal{F}_t)$-Brownian motion with variance $\int_0^t q_s ds$, where q_s is a nonnegative continuous function. We will define the stochastic integral $\int_0^t f_s dW_s$ for a $(\mathcal{F}_t)$-progressively measurable stochastic process f_t based on the Brownian motion W_t. If the sample path of the Brownian motion $W_t(\omega)$ had the derivative $\dot{W}_t(\omega)$ with respect to t, the Lebesgue integral $\int_0^t f_s(\omega) \dot{W}_s(\omega) ds$ could be well defined for each ω. However, the sample path $W_t(\omega)$ of the Brownian motion has no derivative at any point for almost all ω if q_t is strictly positive. Therefore the above Lebesgue integral is not well defined. In this subsection, we define the stochastic integral (Itô integral) by making use of the probabilistic property of the Brownian motion such as the property of independent increments and the martingale property.

We first consider the case where the integrand $f_t(\omega)$ is a simple process. Here f_t is called a *simple process* if there exists a partition $\Delta = \{0 = t_0 < t_1 < \cdots < t_m < \cdots\}$ of the interval $[0, \infty)$ and a sequence of square integrable $\mathcal{F}_{t_i}$-measurable random variables ϕ_i, $i = 0, 1, ...$ and f_t is written as $f_t = \sum_{i=1}^{\infty} \phi_{i-1} 1_{(t_{i-1}, t_i]}(t)$. For this simple process f_t, we define the *stochastic integral* based on dW_t by

$$\int_0^t f_s dW_s := \sum_{i=1}^{k} \phi_{i-1}(W_{t_i} - W_{t_{i-1}}) + \phi_k(W_t - W_{t_k}),$$

where k is a positive integer such that $t_k \leq t < t_{k+1}$. It admits the following property:

a) The stochastic process $\int_0^t f_s dW_s$, $t \geq 0$ is an L^2-martingale with mean 0.

b) The isometric property is valid for any $t \geq 0$;

$$E\left[\left(\int_0^t f_s dW_s\right)^2\right] = E\left[\int_0^t f_s^2 q_s ds\right]. \tag{2.29}$$

Let f_t be a progressively measurable process satisfying $E[\int_0^T |f_s|^2 q_s ds] < \infty$ for any $T > 0$. We can take a sequence of simple processes (f_t^n) such that $E\left[\int_0^T (f_s - f_s^n)^2 q_s ds\right] \to 0$ for any $T > 0$. Since $\int_0^t f_s^n dW_s - \int_0^t f_s^m dW_s$ is an L^2-martingale, we have by Doob's inequality

$$E\left[\sup_{0 \leq t \leq T}\left(\int_0^t f_s^n dW_s - \int_0^t f_s^m dW_s\right)^2\right] \leq 4E\left[\int_0^T (f_s^n - f_s^m)^2 q_s ds\right]$$

for any $T > 0$. Consequently, letting $n \to \infty$ the sequence of stochastic integrals $\{\int_0^t f_s^n dW_s\}$ converges uniformly on compact sets. We denote the limit process by $\int_0^t f_s dW_s$. It is a continuous martingale and satisfies (2.29).

The stochastic integrals can be defined for a wider class of integrand processes. Let f_t be a progressively measurable process satisfying $\int_0^T f_t^2 q_t dt < \infty$, a.s. for any $T > 0$. Then the stochastic integral $M_t = \int_0^t f_s dW_s$ is well defined as a continuous $(\mathcal{F}_t)$-adapted process. It is not integrable in general. But it is a locally square integrable martingale.

We shall define stochastic integrals based on multidimensional Brownian motion. Let $W_t = (W_t^1, ..., W_t^r)$ be an r-dimensional $(\mathcal{F}_t)$-Brownian motion with mean 0 and covariance $\int_0^t Q(s) ds$. Then each component W_t^i is a one-dimensional Brownian motion and the stochastic integral $\int_0^t f_s^i dW_s^i$ is well defined for progressively measurable process f_s^i satisfying $\int_0^T |f_s^i|^2 q_s^{ii} ds < \infty$ for any $T > 0$.

Theorem 2.3 (Representation of L^2-martingales) *Let W_t be an r-dimensional non-degenerate Brownian motion with covariance $\int_0^t Q(s) ds$. Let $(\mathcal{F}_t)$ be a filtration such that $\mathcal{F}_t = \sigma(W_s - W_0; s \leq t) \vee \mathcal{F}_0$, where $\mathcal{F}_0$ is independent of $W_t - W_0$. Then any square integrable $(\mathcal{F}_t)$-martingale M_t has a modification which is a continuous martingale. Further it is represented by the stochastic integrals based on the Brownian motion W_t, i.e, there exists an r-dimensional $(\mathcal{F}_t)$-progressively measurable process $f_t = (f_t^1, ..., f_t^r)$ satisfying $E[\int_0^T |f_s^i|^2 q_s^{ii} ds] < \infty$ for any i and $T > 0$, and M_t is represented by*

$$M_t = M_0 + \int_0^t (f_s, dW_s) = M_0 + \sum_{i=1}^r \int_0^t f_s^i dW_s^i. \tag{2.30}$$

Proof. Suppose first that $\mathcal{F}_t = \sigma(W_s - W_0; s \leq t)$. The theorem is well known in the case where W_t is a standard Brownian motion. In our case, we transform $W(t)$ to a standard Brownian motion $\tilde{W}_t$ as follows. Let $Q^{1/2}(t)$ be a square root of $Q(t)$, i.e., it is a positive definite symmetric matrix such that $Q^{1/2}(t)Q^{1/2}(t) = Q(t)$. Let $R(t)$ be the inverse matrix of $Q^{1/2}(t)$. Then $\tilde{W}_t = \int_0^t R(r)dW_r$ is a standard Brownian motion. Further, since $\sigma(W_s - W_0; s \leq t) = \sigma(\tilde{W}_s - \tilde{W}_0; s \leq t)$ holds valid, square integrable $\mathcal{F}_t$-martingale M_t is represented by

$$M_t = M_0 + \int_0^t (\tilde{f}_s, d\tilde{W}_s).$$

Since $\tilde{W}_t = \int_0^t Q(r)^{1/2} dW_r$, we get the representation (2.30) by setting $f(r) = Q(r)^{1/2}\tilde{f}(r)$.

Now let X_1 be a bounded $\mathcal{F}_0$-measurable random variable and let X_2 be a bounded $\sigma(W_s - W_0; s \leq T)$-measurable random variable. We consider the martingale $M_t = E[X_1 X_2 | \mathcal{F}_t]$. It is equal to $X_1 E[X_2 | \sigma(W_s - W_0; s \leq t)]$. Then it admits the martingale representation

$$M_t = X_1 E[X_2] + X_1 \int_0^t (f_s, dW_s) = X_1 E[X_2] + \int_0^t (X_1 f_s, dW_s).$$

Since any square integrable $\mathcal{F}_T$-measurable random variable is approximated by a sequence of linear sums of the above $X_1 X_2$, any square integrable $(\mathcal{F}_t)$-martingale is continuous and admits the martingale representation (2.30). □

Remark. Any martingale has a continuous modification. Indeed, let M_t be a martingale. We may choose a sequence of continuous L^2-martingales $M_t^{(n)}$ such that it converges to M_t in L^1 for any $t \leq T$. Then a suitable subsequence of $\{M_t^{(n)}\}$ converges to M_t uniformly a.s. **P** in view of Doob's inequality. Therefore M_t is continuous a.s.

Next, if M_t is a local martingale, there exists a sequence of stopping times τ_n with $\lim \tau_n = \infty$ a.s. such that the stopped process $M_{t \wedge \tau_n}$ is a martingale. Then M_t has a modification of continuous local martingale. Further, we may choose above τ_n such that the stopped process $M_{t \wedge \tau_n}$ is a square integrable martingale and it admits a martingale representation. Then the local martingale M_t admits the martingale representation (2.30), where f_s^i are progressively measurable processes satisfying $\int_0^T |f_s^i|^2 q_s^{ii} ds < \infty$ a.s.

2.2.4 Itô's formula

Let $W_t = (W_t^1, ..., W_t^r)$ be an r-dimensional $(\mathcal{F}_t)$-Brownian motion as in the previous subsection. Let f_t^{ij}, $i = 1, ..., d$, $j = 1, ..., r$ be $(\mathcal{F}_t)$-progressively measurable processes such that $\int_0^T |f_s^{ij}|^2 q_s^{jj} ds < \infty$ holds for any i, j and

$T > 0$. Then the stochastic integrals $\int_0^t f_s^{ij} dW_s^j$ are well defined and are local martingales. Next, let $g_t^i, i = 1, \ldots, r$ be a $(\mathcal{F}_t)$-progressively measurable processes satisfying $\int_0^T |g_s^i| ds < \infty$ for any i and $T > 0$. We define a d-dimensional process $X_t = (X_t^1, \ldots, X_t^d)$ by

$$X_t^i = X_0^i + \sum_{j=1}^r \int_0^t f_s^{ij} dW_s^j + \int_0^t g_s^i ds, \qquad i = 1, \ldots, d \qquad (2.31)$$

and call it an *Itô process based on the $(\mathcal{F}_t)$-Brownian motion W_t*.

Components of Itô process are continuous semimartingales. The quadratic covariations are

$$\langle X^i, X^j \rangle_t = \sum_{k,l=1}^r \int_0^t f_s^{ik} f_s^{jl} q_s^{kl} ds. \qquad (2.32)$$

Now let h_t be a $(\mathcal{F}_t)$-progressively measurable process satisfying $\int_0^T |h_t g_t^i| dt < \infty$ for any T and $\int_0^T |h_t f_t^{ij}|^2 q_t^{jj} dt < \infty$ for any i, j and T. We define the stochastic integral $\int_0^t h_s dX_s^i$ by $\sum_j \int_0^t h_s f_s^{ij} dW_s^j + \int_0^t h_s g_s^i ds$.

Theorem 2.4 *(Itô's formula.)* *Let $F(t, x_1, \ldots, x_d)$ be a continuous function of $(t, x_1, \ldots, x_d)$ which is continuously differentiable with respect to t and twice continuously differentiable with respect to $x_1, \ldots, x_d$. Let $X_t = (X_t^1, \ldots, X_t^d)$ be a d-dimensional Itô process based on an $(\mathcal{F}_t)$-Brownian motion W_t. Then $F(t, X_t)$ is also an Itô process based on the $(\mathcal{F}_t)$-Brownian motion W_t. It is written by*

$$F(t, X_t) = F(0, X_0) + \int_0^t F_t(s, X_s) ds + \sum_{i=1}^d \int_0^t F_{x_i}(s, X_s) dX_s^i$$
$$+ \frac{1}{2} \sum_{i,j=1}^d \int_0^t F_{x_i x_j}(s, X_s) d\langle X^i, X^j \rangle_s$$
$$= F(0, X_0) + \int_0^t F_t(s, X_s) ds + \sum_{i=1}^d \sum_{j=1}^r \int_0^t F_{x_i}(s, X_s) f_s^{ij} dW_s^j$$
$$+ \sum_{i=1}^d \int_0^t F_{x_i}(s, X_s) g_s^i ds + \frac{1}{2} \sum_{i,j=1}^d \int_0^t F_{x_i x_j}(s, X_s) \Big(\sum_{k,l=1}^r f_s^{ik} f_s^{jl} q_s^{kl} \Big) ds.$$

Let W_t be an $(\mathcal{F}_t)$-Brownian motion with covariance $\int_0^t Q(s) ds$. We shall consider the exponential of an Itô process given by

$$\Phi_t = \exp\left\{ \int_0^t (f_s, dW_s) + \int_0^t g_s ds \right\}, \qquad 0 \le t \le T,$$

where f_s, g_s are progressively measurable processes with integrability conditions $\int_0^T (f_s, Q(s)f_s)ds < \infty$, $\int_0^T |g_s|ds < \infty$ a.s. It is a continuous semimartingale.

We shall consider the case where Φ_t becomes a local martingale. Apply Itō's formula to the function $F(x) = e^x$ and $X_t = \int_0^t (f_s, dW_s) + \int_0^t g_s ds$. Then we get

$$\Phi_t = 1 + \int_0^t \Phi_s (f_s, dW_s) + \int_0^t \Phi_s \left[g_s + \frac{1}{2}(f_s, Q(s)f_s) \right] ds.$$

It is a local martingale if and only if the last term of the above is equal to 0, or $\int_0^t g_s ds = -\frac{1}{2}\int_0^t (f_s, Q(s)f_s)ds$. Therefore,

$$\Phi_t = \exp\left\{ \int_0^t (f_s, dW_s) - \frac{1}{2}\int_0^t (f_s, Q(s)f_s)ds \right\} \tag{2.33}$$

is a positive local martingale, called *an exponential (local)martingale*.

We give an exponential representation of a positive local martingale.

Theorem 2.5 *Let W_t be a Brownian motion and let $(\mathcal{F}_t)$ be the filtration given in Theorem 2.5. Let Φ_t be a positive local martingale such that $\Phi_0 = 1$. Then it is represented by (2.33), where f_s is a progressively measurable process such that $\int_0^T (f_s, Q(s)f_s)ds < \infty$ a.s.*

Proof. Set $F(x) = \log x$ and apply Itō's formula to $X_t = \Phi_t$. Since X_t is an Itō process based on W_t (Theorem 2.3), $\log \Phi_t$ is also an Itō process. Therefore Φ_t is written as the exponential of an Itō process. Consequently we get the representation (2.33). □

The positive local martingale (2.33) is a positive supermartingale and the inequality $E[\Phi_t] \leq 1$ holds for any t. The equality holds for any t if and only if Φ_t is a martingale.

Now suppose that Φ_t defined by (2.33) is a martingale. Then it holds $E[\Phi_t] = 1$ for any t. We will fix a terminal time $T > 0$. We can define a new probability $\mathbf{Q}$ by on $(\Omega, \mathcal{F}_T)$ by

$$\mathbf{Q}(B) = E[\Phi_T 1_B], \quad B \in \mathcal{F}_T. \tag{2.34}$$

It holds $\mathbf{Q}(B) = E[\Phi_t 1_B]$ for any $B \in \mathcal{F}_t$ for any $t \leq T$. In particular we have $\mathbf{P} = \mathbf{Q}$ on the σ-field $\mathcal{F}_0$.

Theorem 2.6 *(Girsanov) With respect to the measure $\mathbf{Q}$, the stochastic process $W_t - \int_0^t Q(s)f_s ds, 0 \leq t \leq T$ is an $(\mathcal{F}_t)$-Brownian motion with mean 0 and covariance $\int_0^t Q(s)ds$. Further the Brownian motion $W_t - \int_0^t Q(s)f_s ds$ and $\mathcal{F}_0$ are independent with respect to $\mathbf{Q}$.*

In the next section we will discuss the Bayes formula for continuous stochastic process, where the martingale property of the process Φ_t given by (2.33) plays an important role. It will be studied in detail in Section 2.3.

2.2.5 Stochastic differential equation (SDE)

Let $b(t, x) = (b^1(t, x), ..., b^d(t, x))$, $\sigma(t, x) = (\sigma^{ij}(t, x))_{i=1,...,d, j=1,...,r}$ be continuous functions on $[0, T] \times \mathbf{R}^d$. Let $W_t = (W_t^1, ..., W_t^r)$ be an r-dimensional Brownian motion. Given $t_0 \in [0, T]$ and $\mathcal{F}_{t_0}$-measurable random vector $X_0 = (X_0^1, ..., X_0^d)$, we want to find a continuous $(\mathcal{F}_t)$-adapted stochastic process $X_t = (X_t^1, ..., X_t^d)$, $t_0 \leq t \leq T$ satisfying the following equation.

$$X_t^i = X_0^i + \sum_{j=1}^{r} \int_{t_0}^{t} \sigma^{ij}(s, X_s) dW_s^j + \int_{t_0}^{t} b^i(s, X_s) ds, \quad i = 1, ..., d \quad (2.35)$$

Using vector-matrix notation, we will write it simply as

$$dX_t = \sigma(t, X_t) dW_t + b(t, X_t) dt, \quad X_{t_0} = X_0, \quad (2.36)$$

and call it a *stochastic differential equation (SDE)* and X_t the solution starting from X_0 at time t_0. To show the existence and the uniqueness of the solution, we have to assume some smoothness for coefficients $\sigma(t, x)$, $b(t, x)$.

Here, we shall introduce some classes of smooth functions. A function $\varphi(t, x)$ is said to be *Hölder continuous* if there exists $0 < \gamma \leq 1$ and a positive constant C such that $|\varphi(t, x) - \varphi(t, y)| \leq C|x - y|^\gamma$ holds for all $t \in [0, T]$ and $x, y \in \mathbf{R}^d$. If we can choose $\gamma = 1$ at the above inequality, the function φ is called *Lipschitz continuous*. If the above inequality holds with $\gamma = 1$ locally, the function $\varphi(t, x)$ is called locally Lipschitz continuous. A function $\varphi(t, x)$ is said to be of linear growth, if there exists a positive constant C such that $|\varphi(t, x)| \leq C(1 + |x|)$ holds for any t, x.

Let i, m be nonnegative integers. Let $\varphi(t, x)$ be a continuous function on $[0, T] \times \mathbf{R}^d$. It is said to be of the class $C^{i,m}$ or $C^{i,m}$-function if it is i-times continuously differentiable with respect to t and m-times continuously differentiable with respect to x.

Theorem 2.7 (Blagovescenskii–Freidlin [2], Kunita [14]) *Assume $\sigma(t, x)$ and $b(t, x)$ are locally Lipschitz continuous and of linear growth. Then for any initial value X_0 the solution of equation exists uniquely.*

1) *Let $X_t^{(t_0, x)}$ be the solution starting from x at time t_0. For any real p, there exists a positive constant C_p such that*

$$\sup_{t_0, t \in [0, T]} E[(1 + |X_t^{(t_0, x)}|)^p] \leq C_p (1 + |x|)^p \quad (2.37)$$

holds for all x.

2) Suppose that σ, b are of $C^{0,m}$-class and their derivatives are bounded Hölder continuous. Then the solution $X_t^{(t_0,x)}$ is m-times continuously differentiable with respect to x. For any $p > 1$ and $N > 0$, there exist positive constants C_p and $C_{p,N}$ such that

$$\sup_{t_0,t,x} E[|D^\alpha X_t^{(t_0,x)}|^p] \leq C_p, \quad E[\sup_{t_0,t,|x|\leq N} |D^\alpha X_t^{(t_0,x)}|^p] \leq C_{p,N} \quad (2.38)$$

holds for α with $1 \leq |\alpha| \leq m$, where $D^\alpha = (\frac{\partial}{\partial x^1})^{\alpha_1} \cdots (\frac{\partial}{\partial x^d})^{\alpha_d}$ and $|\alpha| = \sum_{i=1}^d \alpha_i$.

Inequalities (2.37) and (2.38) for $p > 1$ are found in [2],[14] and inequality (2.37) for negative p is found in [14].

In this paper, we will consider several stochastic differential equations and stochastic partial differential equations. In the next section, we will consider system-observation process (X_t, Y_t) which are determined by stochastic differential equations whose coefficients depend on the past $(X_r, Y_r), r \leq s$. In Chapter 3, we derive stochastic partial differential equations satisfied by conditional distributions and unnormalized conditional distributions of the filtering. We will also discuss the Cauchy problem for stochastic partial differential equations connected with the nonlinear filtering.

2.3 Absolute continuity of measures and Bayes formulas

Let $\{X_t, 0 \leq t \leq T\}$ be a stochastic process with state space S, called a *system process*. Suppose that we want to observe the system process but we observe an e-dimensional measurable process $h_t = (h_t^1, ..., h_t^e)$, added by the noise dN_t. Thus we actually observe the following stochastic process, called an *observation process*.

$$Y_t = Y_0 + \int_0^t h_s ds + N_t. \quad (2.39)$$

Here the *noise process* N_t is an e-dimensional nondegenerate Brownian motion with covariance $\int_0^t R(s)ds$ independent of Y_0. Further, h_t is a functional of the system process $\{X_s; s \leq t\}$ or more generally a functional of $\{X_s; s \leq t\}$ and $\{Y_s; s < t\}$. The observation data are given by the filtration of the observation process:

$$\mathcal{G}_t = \sigma(Y_s; s \leq t), \quad t \in [0, T]. \quad (2.40)$$

We consider the conditional distribution of X_t based on the observation data $\mathcal{G}_t$:

$$\pi_t(E) = \mathbf{P}(X_t \in E | \mathcal{G}_t). \quad (2.41)$$

If the above conditional distributions are obtained, the nonlinear filter $\hat{X}_t$ of the system process X_t based on the observation $Y_s; s \leq t$ is computed by the formula $\hat{X}_t = \int x\pi_t(dx)$, provided that the state space S of the system process is an Euclidean space.

For the nonlinear filtering problem, it is important to compute the above conditional distribution. In Section 2.1, we obtained a Bayes formula for computing (2.41) in the case of time series (Theorem 2.2). In this section we will obtain a similar Bayes formula for continuous parameter case. Assuming that matrices $R(t)$ are invertible, a continuous analogue of Φ_n in (2.15) is given by

$$\Phi_t = \exp\left\{-\int_0^t (h_s, R(s)^{-1} dN_s) - \frac{1}{2}\int_0^t (h_s, R(s)^{-1} h_s) ds\right\}. \tag{2.42}$$

It is a positive local martingale by Theorem 2.5. If Φ_t is a martingale, we can define another probability measure $\mathbf{Q}$ by $\mathbf{Q}(A) = E[\Phi_T 1_A]$. Then, similarly as in the case of time series, the conditional distribution π_t is given by the Bayes formula

$$\pi_t(E) = \frac{E_\mathbf{Q}[\Psi_t 1_E(X_t)|\mathcal{G}_t]}{E_\mathbf{Q}[\Psi_t|\mathcal{G}_t]}, \quad a.s. \ \mathbf{P} \tag{2.43}$$

where $\Psi_t = \Phi_t^{-1}$.

However, the martingale property of Φ_t is not so simple as the case of time series. A well known sufficient condition for the martingale property is that the process h_s is uniformly bounded, but it excludes Kalman filter. In this section we will study the martingale property for unbounded h_s.

The martingale property is closely related to the existence, uniqueness of solutions of the system-observation process (X_t, Y_t), and the absolute continuity of the law with respect to the law of system-noise process (X_t, N_t). It will be discussed in Sections 2.3.1–2.3.3.

2.3.1 System-observation processes and Bayes formulas

In this section we shall consider two types of system-observation processes. The first is the case where the system process and the noise process are independent. The second is the case where these two processes are correlated and determined by a stochastic differential equation. The latter case will be discussed in Section 2.3.3.

In the first case, the system process is an arbitrary stochastic process with state space S. Let $\mathbf{T} = [0, T]$ be a finite time interval and let S be a locally compact separable metric space. Let $\{X_t, t \in \mathbf{T}\}$ be a stochastic process with state space S such that the sample paths $X_t(\omega)$ are right continuous with the left-hand limits a.s. It is called a *system process*.

Before we proceed to introduce an observation process, we will prepare a functional space where the law of the pair of system process and observation process should be defined. Let $\mathcal{D}$ be the set of all maps $\mathbf{x} : [0, T] \to S$ which are right continuous with left-hand limits. The value of $\mathbf{x}$ at t is denoted by x_t. For $0 \leq t_1 < \cdots < t_n \leq t$ and Borel sets $E_1, ..., E_n$ of S, the cylinder set is defined by $B = \{\mathbf{x} \in \mathcal{D}; x_{t_1} \in E_1, \ldots, x_{t_n} \in E_n\}$. We denote by $\mathcal{B}_t(\mathcal{D})$ the σ-field generated by the above cylinder set. We set $\mathcal{B}(\mathcal{D}) = \vee_{t \leq T} \mathcal{B}_t(\mathcal{D})$. Let $\mathcal{C}$ be the set of all continuous maps $\mathbf{y} : [0, T] \to \mathbf{R}^e$. The value of $\mathbf{y}$ at t is denoted by y_t. σ-fields $\mathcal{B}_t(\mathcal{C})$ and $\mathcal{B}(\mathcal{C})$ are defined similarly. On the product space $\mathcal{D} \times \mathcal{C}$, we define a σ-field by $\tilde{\mathcal{B}} = \mathcal{B}(\mathcal{D}) \otimes \mathcal{B}(\mathcal{C})$ and a filtration by $\tilde{\mathcal{B}}_t = \wedge_{\epsilon > 0} \mathcal{B}_{t+\epsilon}(\mathcal{D}) \otimes \mathcal{B}_{t+\epsilon}(\mathcal{C})$.

For the system process X_t, we set $\mathbf{X}(\omega) = (X_t(\omega), t \in [0, T])$. It may be regarded as an element of $\mathcal{D}$ for any ω and $\mathbf{X}$ may be regarded as a $\mathcal{D}$-valued random variable. Let Y_0 be a random variable with values in $\mathbf{R}^e$. The joint law $F_{\mathbf{X}, Y_0}$ of the system process X_t and the random variable Y_0 is a probability measure on $\mathcal{B}(\mathcal{D}) \otimes \mathcal{B}(\mathbf{R}^e)$ such that

$$F_{\mathbf{X}, Y_0}(\{(\mathbf{x}, y) \in \mathcal{D} \times \mathbf{R}^e; x_{t_1} \in E_1, \ldots, x_{t_n} \in E_n, y \in E_{n+1}\})$$

$$= \mathbf{P}(X_{t_1} \in E_1, \ldots, X_{t_n} \in E_n, Y_0 \in E_{n+1}).$$

Now, suppose that we are given a progressively $(\tilde{\mathcal{B}}_t)$-measurable vector function $h(t, \mathbf{x}, \mathbf{y}) = (h^1(t, \mathbf{x}, \mathbf{y}), \ldots, h^e(t, \mathbf{x}, \mathbf{y}))$ such that $\int_0^T |h(s, \mathbf{x}, \mathbf{y})| ds < \infty$ holds for any $\mathbf{x}, \mathbf{y}$. We consider a stochastic differential equation on $\mathbf{R}^d$:

$$Y_t = Y_0 + \int_0^t h(s, \mathbf{X}, \mathbf{Y}) ds + N_t, \quad 0 \leq t \leq T, \tag{2.44}$$

where $\mathbf{Y} = (Y_t, t \in [0, T])$ and N_t is an e-dimensional Brownian motion with covariance $\int_0^t R(s) ds$, which is independent of (X_t) and Y_0. Throughout the paper, we assume that the noise process N_t is *nondegenerate*, i.e., the covariance $R(t)$ is invertible and the inverse $R(t)^{-1}$ is continuous with respect to t.

By a *solution of equation* (2.44) (often called a *weak solution*), we mean a suitable probability space $(\Omega, \mathcal{F}, \mathbf{P})$, where the following pair of processes (X_t, Y_t) is defined.

(1) The law of $(\mathbf{X}, Y_0)$ is equal to a given $F_{\mathbf{X}, Y_0}$.
(2) $N_t := Y_t - Y_0 - \int_0^t h(s, \mathbf{X}, \mathbf{Y}) ds$ is a Brownian motion with covariance $\int_0^t R(s) ds$, which is independent of (X_t, Y_0).

If there exists $(X_t, Y_t, \mathbf{P})$ satisfying (1)-(2), Y_t is called an *observation process*. The pair (X_t, Y_t) is called a *system-observation process*. The law of the system-observation process may be defined on $(\mathcal{D} \times \mathcal{C}, \tilde{\mathcal{B}})$.

We say that the *uniqueness of solution for equation (2.44)* holds (in the sense of the law), if whenever (X_t, Y_t) and (X'_t, Y'_t) are two solutions such that $F_{\mathbf{X}, Y_0} = F_{\mathbf{X}', Y'_0}$, the laws of these two system-observation processes coincide.

Remark 2.1 Set $\mathcal{F}_t = \wedge_{\epsilon > 0} \sigma(X_u; 0 \leq u \leq T, Y_s; s \leq t + \epsilon)$. Then the above (1) and (2) are equivalent to the following (1') and (2').

(1') $\mathbf{X}, Y_0$ is $\mathcal{F}_0$-adapted and its law is equal to a given $F_{\mathbf{X}, Y_0}$.
(2') N_t is an e-dimensional $(\mathcal{F}_t)$-Brownian motion with covariance $\int_0^t R(s) ds$.

We are interested in the existence and uniqueness of the solution in the sense of the law, since the conditional distribution π_t of (2.41) is determined by the law of the system-observation process (X_t, Y_t) and the observed sample paths Y_t.

We introduce

Condition (A). *The system process and noise process are independent. Further, there exists a positive constant C and a positive progressively measurable function $f(t, \mathbf{x})$ such that $E[\int_0^T f(t, \mathbf{X})^2 dt] < \infty$ and $h(t, \mathbf{x}, \mathbf{y})$ satisfies*

$$|h(s, \mathbf{x}, \mathbf{y})| \leq C(f(s, \mathbf{x}) + \|\mathbf{y}\|_s), \quad \forall s, \mathbf{x}, \mathbf{y}, \qquad (2.45)$$

where $\|\mathbf{y}\|_s = \sup_{r \leq s} |y_r|$.

Theorem 2.8 *Assume Condition (A). Then the system-observation process exists uniquely. Further Φ_t given by (2.42) is a martingale.*

We first introduce some notations. Let $(\Omega, \mathcal{F}, \mathbf{Q})$ be a probability space and let $(X_t, Y_t, \mathbf{Q})$ is a triple satisfying the following.

(1) The law of $(\mathbf{X}, Y_0)$ with respect to $\mathbf{Q}$ is equal to a given $F_{\mathbf{X}, Y_0}$.
(2) $(Y_t - Y_0)$ is an e-dimensional Brownian motion with covariance $\int_0^t R(s) ds$ independent of (X_t) and Y_0 with respect to $\mathbf{Q}$.

Such a triple exists always. We can define it on the functional space as follows. Let

$$\Omega = \mathcal{D} \times \mathcal{C}, \quad \mathcal{F} = \tilde{\mathcal{B}}, \quad X_t(\omega) = x_t, \quad Y_t(\omega) = y_t, \quad \text{where} \quad \omega = (\mathbf{x}, \mathbf{y}),$$

and $\mathbf{Q}$ is the law of (X_t, Y_t) satisfying (1) and (2). We will fix the triple through this and next subsections. We set $\mathcal{F}_t = \wedge_{\epsilon > 0} \sigma(X_s; 0 \leq s \leq T, Y_s; 0 \leq s \leq t + \epsilon)$.

For a given $(\mathcal{F}_t)$-progressively measurable functional $h(s, \mathbf{x}, \mathbf{y})$, we define

$$\tau_n(\mathbf{x}, \mathbf{y}) = \inf \left\{ t > 0; \int_0^t (h(s, \mathbf{x}, \mathbf{y}), R(s)^{-1} h(s, \mathbf{x}, \mathbf{y})) ds > n \right\}, \qquad (2.46)$$
$$= \infty, \quad \text{if } \{\cdots\} = \phi.$$

It is an increasing sequence of $(\mathcal{F}_t)$-stopping times. We denote the limit by $\tau_\infty(\mathbf{x}, \mathbf{y})$. Note that these stopping times are defined without measures. We set $\tau_n = \tau_n(\mathbf{X}, \mathbf{Y})$, $\tau_\infty = \tau_\infty(\mathbf{X}, \mathbf{Y})$ and $h(s) = h(x, \mathbf{X}, \mathbf{Y})$.

Proof. Consider

$$\Psi_t = \exp\left\{ \int_0^t (h(s), R(s)^{-1} dY_s) - \frac{1}{2} \int_0^t (h(s), R(s)^{-1} h(s)) ds \right\}. \qquad (2.47)$$

It is a positive local martingale with respect to $\mathbf{Q}$ and the stopped processes $\Psi_t^{(n)} = \Psi_{t \wedge \tau_n}$ are martingales. Define $\mathbf{P}_n$ by $\mathbf{P}_n(B) = E_{\mathbf{Q}}[\Psi_T^{(n)} 1_B]$. Then $N_t^{(n)} := Y_t - Y_0 - \int_0^{t \wedge \tau_n} h(s) ds$ is an $(\mathcal{F}_t)$-Brownian motion with respect to $\mathbf{P}_n$ and the joint law of (X_t) and Y_0 with respect to $\mathbf{P}_n$ is equal to $F_{\mathbf{X},Y_0}$ by Girsanov's theorem. Hence $(X_t, Y_t, \mathbf{P}_n)$ is a solution of equation (2.44) associated with $h(s, \mathbf{X}, \mathbf{Y})1_{\{\tau_n > s\}}$. It holds $\mathbf{P}_n = \mathbf{P}_m$ on $\mathcal{F}_{T \wedge \tau_m}$ if $m < n$.

Since
$$Y_{t \wedge \tau_n} = Y_0 + \int_0^{t \wedge \tau_n} h(s, \mathbf{X}, \mathbf{Y})ds + N_{t \wedge \tau_n}^{(n)}$$

holds a.s. $\mathbf{P}_n$, we have

$$E_{\mathbf{P}_n}\left[\sup_{0 \le s \le t} |Y_{s \wedge \tau_n}|^2\right]$$
$$\le 3 E_{\mathbf{P}_n}[|Y_0|^2] + 3 E_{\mathbf{P}_n}\left[\left|\int_0^{t \wedge \tau_n} h(s)ds\right|^2\right] + 3 E_{\mathbf{P}_n}[\sup_{0 \le s \le t}|N_{s \wedge \tau_n}^{(n)}|^2]$$
$$\le 3 E_{\mathbf{P}_n}[|Y_0|^2] + 6C^2 \left(E_{\mathbf{P}_n}\left[\int_0^{t \wedge \tau_n} |f(s, \mathbf{X})|^2 ds\right] + E_{\mathbf{P}_n}\left[\int_0^{t \wedge \tau_n} \|\mathbf{Y}\|_s^2 ds\right]\right)$$
$$+ 12 \sup_{s \le t} \|R(s)\| t$$
$$\le C_1 + C_2 \left(\int_0^t E_{\mathbf{P}_n}[|f(s, \mathbf{X})|^2]ds + \int_0^t E_{\mathbf{P}_n}[\|\mathbf{Y}\|_{s \wedge \tau_n}^2]ds\right)$$
$$\le C_1' + C_2' \int_0^t E_{\mathbf{P}_n}[\|\mathbf{Y}\|_{s \wedge \tau_n}^2]ds,$$

where C_1', C_2' are positive constants not depending on n. Here we used the martingale inequality $E_{\mathbf{P}_n}[\sup_{0 \le s \le t}|N_{s \wedge \tau_n}^{(n)}|^2] \le 4 E_{\mathbf{P}_n}[|N_{t \wedge \tau_n}^{(n)}|^2]$. Therefore we have by Gronwall's inequality $E_{\mathbf{P}_n}[\|\mathbf{Y}\|_{T \wedge \tau_n}^2] \le C_1' e^{C_2' T}$. Then using Condition (A) again, we get

$$E_{\mathbf{P}_n}\left[\int_0^{T \wedge \tau_n} (h(s), R(s)^{-1} h(s))ds\right] \le C_3,$$

where C_3 is a positive constant not depending on n. This implies that $\mathbf{P}_n(\tau_n < T)$ tends to 0 as $n \to \infty$. Then there exists a unique probability $\mathbf{P}$ such that $\mathbf{P} = \mathbf{P}_n$ holds on $\mathcal{F}_{T \wedge \tau_n}$. See Stroock–Varadhan [18]. With respect to $\mathbf{P}$, $N_t := Y_t - \int_0^t h(s, \mathbf{X}, \mathbf{Y})ds$ is an $(\mathcal{F}_t)$-Brownian motion. Therefore, (X_t, Y_t) defined on $\mathbf{P}$ is a solution of equation (2.44).

The uniqueness of the solution and the martingale property of Φ_t are closely related to the absolute continuity of measures $\mathbf{Q}$ and $\mathbf{P}$. Indeed Lemma 2.2 Section 2.3.2 proves the uniqueness of solution, since $\tau_\infty = \infty$ holds a.s. $\mathbf{P}$ and a.s. $\mathbf{Q}$ by Condition (A). The martingale property of Φ_t will follow from Theorem 2.11. □

Theorem 2.9 *Assume Condition (A). Then we have*

$$\pi_t(E) = \frac{\int_{x_t \in E} \psi_t(\mathbf{x}, \mathbf{Y})dF_\mathbf{X}(\mathbf{x})}{\int_D \psi_t(\mathbf{x}, \mathbf{Y})dF_\mathbf{X}(\mathbf{x})}, \quad a.s. \ \mathbf{P}, \tag{2.48}$$

where $\psi_t(\mathbf{x}, \mathbf{Y})$ is given by

$$\psi_t(\mathbf{x}, \mathbf{Y}) = \exp\left\{\int_0^t (h(s), R(s)^{-1}dY_s) - \frac{1}{2}\int_0^t (h(s), R(s)^{-1}h(s))ds\right\}, \quad (2.49)$$

where $h(s) = h(s, \mathbf{x}, \mathbf{Y})$ and $F_\mathbf{X}$ is the law of the system process (X_t).

Proof. Since $Y_t - Y_0$ and X_t are independent with respect to probability $\mathbf{Q}$ and the law of X_t is equal to $F_\mathbf{X}$, we have

$$E_\mathbf{Q}[\psi_t 1_E(x_t)|\mathcal{G}_t] = \int_{x_t \in E} \psi_t(\mathbf{x}, \mathbf{Y}) dF_\mathbf{X}(\mathbf{x}).$$

Then we have the formula (2.48). □

The formula (2.48) is called the *Kallianpur–Striebel formula* [11].

2.3.2 Absolute continuity of measures I: the case of independent noise

In this subsection, assuming the existence of the solution of (2.44), we shall discuss the absolute continuity of the law of the system-observation process with respect to the law of system-noise process and then show the uniqueness of the solution. Our discussion is close to Liptzer–Shiryaev [15], where they considered weak solutions of the SDE of the form

$$d\xi_t = a(t, \xi_t)dt + dN_t, \quad \xi_0 = 0,$$

making use of Girsanov's theorem. Results similar to our Lemma 2.2 and Theorem 2.10 are obtained for the above SDE. See [15] or [9].

Let $(X_t, Y_t, \mathbf{Q})$ be the triple defined in Section 2.3.1. Let $T_n = T_n(X, Y)$ and $T_\infty(X, Y)$. Stochastic integral $\int_0^t 1_{\tau_n > t}(h(s), R(s)^{-1}dY_s)$ is well defined as a continuous martingale with respect to $\mathbf{Q}$. Further, there exists a continuous stochastic process $Z_t, t \in [0, T \wedge \tau_\infty)$ such that $Z_{t \wedge \tau_n} = \int_0^t 1_{\tau_n > t}(h(s), R(s)^{-1}dY_s)$ holds for any $t \leq T$ and n. We denote the process Z_t by $\int_0^t (h(s), R(s)^{-1}dY_s)$.

Lemma 2.2 *Suppose that $(X_t, Y_t, \mathbf{P})$ is a solution of equation (2.44) defined on the same measurable space $(\Omega, \mathcal{F})$ as $\mathbf{Q}$.*

(i) *If $\tau_\infty = \infty$ holds a.s. $\mathbf{P}$, then the measure $(\mathcal{F}, \mathbf{P})$ is absolutely continuous with respect to $(\mathcal{F}, \mathbf{Q})$. The family of Radon–Nikodym densities $\frac{d\mathbf{P}}{d\mathbf{Q}}|\mathcal{F}_t, 0 \leq t \leq T$ is a continuous martingale with respect to $(\mathcal{F}_t, \mathbf{Q})$. It is given a.s. $\mathbf{Q}$ by*

$$\Psi_t = \exp\left\{\int_0^t (h(s), R(s)^{-1}dY_s) - \frac{1}{2}\int_0^t (h(s), R(s)^{-1}h(s))ds\right\}, \quad (2.50)$$
$$\text{if } t < T \wedge \tau_\infty,$$
$$= 0, \quad \text{if } T \wedge \tau_\infty \leq t \leq T.$$

(ii) *Solutions of equation (2.44) such that $\tau_\infty = \infty$ holds a.s. $\mathbf{P}$ are at most unique.*

Proof. Assuming $\tau_\infty = \infty$, a.s. **P**, the stochastic integral $\int_0^t (h(s), R(s)^{-1} dY_s)$ is well defined a.s. **P** and is equal to $\int_0^t (h(s), R(s)^{-1} dN_s) + \int_0^t (h(s), R(s)^{-1} h(s)) ds$, a.s. **P**, where $N_t := Y_t - Y_0 - \int_0^t h(s) ds$ is a Brownian motion with respect to **P**. Consider the exponential process Φ_t defined on $(\Omega, \mathcal{F}, \mathbf{P})$ by

$$\Phi_t = \exp\left\{ -\int_0^t (h(s), R(s)^{-1} dY_s) + \frac{1}{2} \int_0^t (h(s), R(s)^{-1} h(s)) ds \right\}. \qquad (2.51)$$

It is equal to (2.42). Then Φ_t is a positive local martingale with respect to **P** (Theorem 2.5) and the stopped process $\Phi_{t \wedge \tau_n}$ is a positive martingale with mean 1. We define a probability measure $\mathbf{Q}_n$ by $\mathbf{Q}_n = \Phi_{T \wedge \tau_n} \cdot \mathbf{P}$. ($\Phi_{T \wedge \tau_n} \cdot \mathbf{P}$ means a probability $\mathbf{Q}_n$ such that $\mathbf{Q}_n(A) = E_\mathbf{P}[\Phi_{T \wedge \tau_n} 1_A]$ holds for all $A \in \mathcal{F}$.) Then by Girsanov's theorem, $N_t + \int_0^{t \wedge \tau_n} h(s) ds$ is a $(\mathcal{F}_t)$-Brownian motion with respect to $\mathbf{Q}_n$. Hence it is independent of $\mathcal{F}_0$. Since $\Phi_0 = 1$, the equality $\mathbf{Q}_n = \mathbf{Q}$ holds on $\mathcal{F}_{T \wedge \tau_n}$, where $\mathcal{F}_{T \wedge \tau_n}$ is the σ-field defined by

$$\mathcal{F}_{T \wedge \tau_n} = \{ B \in \mathcal{F}_T; B \cap \{\tau_n \leq t\} \in \mathcal{F}_t \text{ holds for any } t \leq T \},$$

where $\mathcal{F}_T = \mathcal{F}$.

We will show that **P** is absolutely continuous with respect to **Q** on $\mathcal{F}_T$. Let B be a set in $\mathcal{F}_T$ such that $\mathbf{Q}(B) = 0$. Since $B \cap \{\tau_n > T\} \in \mathcal{F}_{T \wedge \tau_n}$, we have $\mathbf{P}(B \cap \{\tau_n > T\}) = 0$, because of the equivalence of $\mathbf{Q}_n$ and **P** on $\mathcal{F}_{T \wedge \tau_n}$. Let n tend to ∞. Then we get $\mathbf{P}(B) = 0$, since $\mathbf{P}(\cup_n \{\tau_n > T\}) = 1$. Therefore **P** is absolutely continuous with respect to **Q**.

Let Ψ_t be the Radon–Nikodym density. Then it is a martingale with respect to **Q** and hence it is continuous (see the Remark after Theorem 2.3). For $t \leq T \wedge \tau_n$ it holds $\Psi_t = \Phi_t^{-1}$. Therefore Ψ_t is represented by (2.50) for $t < T \wedge \tau_\infty$.

We will prove that $\Psi_{\tau_\infty} = 0$, a.s. on $\{\tau_\infty < T\}$. Define a process Γ_t by

$$\Gamma_t = \exp\left\{ \frac{1}{2} \int_0^t (h(s), R(s)^{-1} dY_s) - \frac{1}{8} \int_0^t (h(s), R(s)^{-1} h(s)) ds \right\},$$

if $t < T \wedge \tau_\infty$ and by 0 if $T \wedge \tau_\infty \leq t \leq T$. Then we have the formula

$$\Psi_t = \Gamma_t^2 \exp\left\{ -\frac{1}{4} \int_0^t (h(s), R(s)^{-1} h(s)) ds \right\}, \quad t < T \wedge \tau_\infty. \qquad (2.52)$$

By the optional sampling theorem, time series $\{\Psi_{T \wedge \tau_n}\}$ and $\{\Gamma_{T \wedge \tau_n}\}$ are martingales with respect to **Q**. Set $t = T \wedge \tau_n$ at (2.52) and let n tend to infinity. Then L^1-limits $\Psi_{T \wedge \tau_\infty}$ and $\Gamma_{T \wedge \tau_\infty}$ exist and finite a.s. **Q** by the martingale convergence theorem. Further, the last member of (2.52) converges to 0 if $\tau_\infty < T$. Consequently we get $\Psi_{\tau_\infty} = 0$, a.s. **Q** on $\{\tau_\infty < T\}$.

Next, we have by the optional sampling theorem of the supermartingale,

$$E_\mathbf{Q}[\Psi_{T \wedge (t+\tau_\infty)} 1_{\tau_\infty < T}] \leq E_\mathbf{Q}[\Psi_{T \wedge \tau_\infty} 1_{\tau_\infty < T}] = 0.$$

This implies $\Psi_t = 0$ for $\tau_\infty \le t \le T$. Thus we obtain formula (2.50).

Now let $(X_t, Y_t, \mathbf{P}')$ be another solution of equation (2.44) defined on the same measurable space $(\Omega, \mathcal{F})$ such that $\tau_\infty = \infty$ a.s. $\mathbf{P}'$. The first assertion of the lemma shows that the measure $\mathbf{P}'$ is absolutely continuous with respect to $\mathbf{Q}$. Further, the Radon–Nikodym density $\frac{d\mathbf{P}'}{d\mathbf{Q}}|\mathcal{F}_t$ is given by the same Ψ_t. Therefore we have $\mathbf{P} = \mathbf{P}'$, proving the uniqueness of the solution. The proof is complete. $\square$

Now, assume that $\tau_\infty = \infty$ a.s. $\mathbf{Q}$. Then the Radon–Nikodym density Ψ_t given by (2.50) is positive a.s. $\mathbf{Q}$. Then the inverse $\Phi_t = \Psi_t^{-1}$ exists a.s. $\mathbf{Q}$ and a.s. $\mathbf{P}$. Further it holds $\Phi_t \cdot \mathbf{P} = \mathbf{Q}$. Therefore $\mathbf{Q}$ is absolutely continuous with respect to $\mathbf{P}$ and its Radon–Nikodym density is given by Φ_t.

Conversely if two measures $\mathbf{P}$ and $\mathbf{Q}$ are equivalent, Radon–Nikodym densities $\frac{d\mathbf{P}}{d\mathbf{Q}}|\mathcal{F}_t$ and $\frac{d\mathbf{Q}}{d\mathbf{P}}|\mathcal{F}_t$ are continuous martingales and are given by (2.47) and (2.51), respectively. Then we have $\mathbf{P}(\tau_\infty = \infty) = \mathbf{Q}(\tau_\infty = \infty) = 1$. Hence we have the following theorem.

Theorem 2.10 *Let $(X_t, Y_t, \mathbf{P})$ be a solution of equation (2.44). Assume that $\tau_\infty = \infty$ holds a.s. $\mathbf{P}$ and $\mathbf{Q}$. Then two measures are equivalent. Further, the Radon–Nikodym density $\frac{d\mathbf{P}}{d\mathbf{Q}}|\mathcal{F}_t$ is given by (2.47) and the Radon–Nikodym density $\frac{d\mathbf{Q}}{d\mathbf{P}}|\mathcal{F}_t$ is given by (2.51).*

Conversely if $\mathbf{P}$ and $\mathbf{Q}$ are equivalent, $\mathbf{P}(\tau_\infty = \infty) = \mathbf{Q}(\tau_\infty = \infty) = 1$ holds valid.

We will give an example of system-observation process $(X_t, Y_t, \mathbf{P})$ which is absolutely continuous with respect to $\mathbf{Q}$ but these two measures $\mathbf{P}$ and $\mathbf{Q}$ are not equivalent.

Example. Let X_t be a system process and let N_t be a one dimensional standard Brownian motion independent of X_t, both of which are defined on a probability space $(\Omega, \mathcal{F}, \mathbf{P})$. Let $c > 0$ be a positive constant and let $h(s, x, y) = -\frac{g(x)}{|y-c|}$ where $g(x)$ is a bounded continuous function such that $g(x) \ge c_1 > 0$ holds for all x. The function h has a pole at $y = c$. Set $h(s, \mathbf{x}, \mathbf{y}) = h(s, x_t, y_t)$ and consider a one-dimensional SDE (2.44). The solution Y_t (starting from 0 at time 0) exists uniquely for any $0 \le t \le T$. Then $(X_t, Y_t, \mathbf{P})$ is a system-observation process.

We will show that the observation process Y_t never hits the point c. Let Y_t^1 be the solution of equation (2.44) where we replace $h(s, x, y)$ by $h_1(s, x, y) = -\frac{c_1}{|y-c|}$. Since $h \le h_1$ holds for any s, x, y, we have $Y_t \le Y_t^1$ for all t a.s. by the comparison theorem of Yamada [19]. Let ζ be the first hitting time of Y_t^1 to the point c. Then $Y_t^1, t < \zeta$ is a one dimensional diffusion on the half line $(-\infty, c)$ and c is a nonexit boundary point or repelling boundary point, according to the theory of one dimensional diffusion (Dynkin [6]). Then we have $\mathbf{P}(\zeta = \infty) = 1$.

Now, since $Y_t \leq Y_t^1$ holds for any t, Y_t never hits the point c a.s. Therefore we have $\int_0^t \frac{g(X_s)^2}{(Y_s-c)^2} ds < \infty$, a.s. **P**. This proves $\mathbf{P}(\tau_\infty = \infty) = 1$.

On the other hand, with respect to the Brownian motion N_t, we have $\mathbf{P}(\int_0^t \frac{1}{(N_s-c)^2} ds < \infty) < 1$ (or $\mathbf{Q}(\tau_\infty = \infty) < 1$). Indeed, if the above probability were equal to 1, we would have the following formula (Itô's formula).

$$\log |N_t - c| = \log c + \int_0^t \frac{1}{|N_s - c|} dN_s - \frac{1}{2} \int_0^t \frac{1}{|N_s - c|^2} ds,$$

for any $0 \leq t \leq T$. But it does not occur. Indeed, let σ be the first hitting time of N_t to the point c. Then it holds $\mathbf{P}(\sigma < T) > 0$. Let t tend to σ. On the set $\{\sigma < T\}$ the left-hand side of the above tends to $-\infty$, but the right-hand side cannot tend to $-\infty$, since $\int_0^t \frac{1}{|N_s-c|} dN_s$, $0 \leq t \leq T$ is a well-defined local martingale.

Remark. In the above example, Ψ_t defined by (2.50) is a martingale with respect to $\mathbf{Q}$. Since $\mathbf{P}(\tau_\infty = \infty) = 1$, $\Phi_t := \Psi_t^{-1}$ exists for ant $t < T$ a.s. **P**. But it is not a martingale with respect to **P** so that Bayes formula does not hold. Indeed, $\Phi_t \cdot \mathbf{P}$ is a measure whose total mass is less than 1. Further, the measure $\mathbf{Q} - \Phi_t \cdot \mathbf{P}$ is singular with respect to **P**. Therefore $\Phi_t \cdot \mathbf{P}$ is the absolutely continuous part of the measure $(\mathcal{F}_t, \mathbf{Q})$ with respect to $(\mathcal{F}_t, \mathbf{P})$ in the Lebesgue decomposition of the measure $(\mathcal{F}_t, \mathbf{Q})$.

2.3.3 Absolute continuity of measures II: the case of correlated noises

We shall consider a system-observation process determined by a stochastic differential equation. The state space of the system process is $\mathbf{R}^d$. Instead of $\mathcal{D}$ we take $\tilde{\mathcal{C}}$; the space of all continuous maps from $[0, T]$ to $\mathbf{R}^d$. We set $\tilde{\mathcal{B}}_t = \wedge_{\epsilon>0} \mathcal{B}_{t+\epsilon}(\tilde{\mathcal{C}}) \times \mathcal{B}_{t+\epsilon}(\mathcal{C})$ as before. Let $b(t, \mathbf{x}, \mathbf{y}) = (b^1(t, \mathbf{x}, \mathbf{y}), ..., b^d(t, \mathbf{x}, \mathbf{y}))$ and $\sigma(t, \mathbf{x}, \mathbf{y}) = (\sigma^{ij}(t, \mathbf{x}, \mathbf{y}))$, $i = 1, ..., d$, $j = 1, ..., r$ be $(\tilde{\mathcal{B}}_t)$-progressively measurable functions such that $\int_0^T |b(s, \mathbf{x}, \mathbf{y})| ds < \infty$, $\int_0^T |\sigma(s, \mathbf{x}, \mathbf{y})|^2 ds < \infty$ holds for any $\mathbf{x}, \mathbf{y}$.

The system-observation process will be defined as the solution of (2.44) and the following equation

$$X_t = X_0 + \int_0^t b(s, \mathbf{X}, \mathbf{Y}) ds + \int_0^t \sigma(s, \mathbf{X}, \mathbf{Y}) dW_s, \qquad (2.53)$$

where (W_t, N_t) is an $(r+e)$-dimensional Brownian motion with covariance matrix

$$\begin{pmatrix} \int_0^t Q(s) ds & \int_0^t S(s) ds \\ \int_0^t S(s)^T ds & \int_0^t R(s) ds \end{pmatrix}, \qquad (2.54)$$

where $Q(s)$, $S(s)$, $R(s)$ are continuous in s and $Q(s)$ is symmetric and nonnegative definite. W_t is called a *system noise* and N_t is called *observation noise*. If $S(s) = 0$ for any s, the system noise and the observation noise are independent. If $S(s) \neq 0$ for some s, these two noises are correlated.

By a solution of equations (2.44) and (2.53), we mean a $(d+e)$-dimensional stochastic process (X_t, Y_t) defined on a probability space $(\Omega, \mathcal{F}, \mathbf{P})$ with a filtration $(\mathcal{F}_t)$ such that

i) there exists a $(r+e)$-dimensional $(\mathcal{F}_t)$-Brownian motion (W_t, N_t) with covariance (2.52).
ii) (X_t, Y_t) is a continuous $(\mathcal{F}_t)$-adapted process.
iii) 4-ple (X_t, Y_t, W_t, N_t) satisfies (2.44) and (2.53).

We say the uniqueness of equations (2.44) and (2.53) holds if whenever (X_t, Y_t) and (X'_t, Y'_t) are solutions of the equations with the same initial law, the laws of processes (X_t, Y_t) and (X'_t, Y'_t) coincide.

A function $f(t, \mathbf{x}, \mathbf{y})$ is called *locally Lipschitz continuous* with respect to $\mathbf{x}$ if for any $K > 0$ there exists a positive constants C_K such that

$$|f(s, \mathbf{x}, \mathbf{y}) - f(s, \mathbf{x}', \mathbf{y})| \leq C_K \|\mathbf{x} - \mathbf{x}'\|_s,$$

holds for any s and $\mathbf{x}, \mathbf{x}', \mathbf{y}$ with $\|\mathbf{x}\|_s, \|\mathbf{x}'\|_s, \|\mathbf{y}\|_s \leq K$, where $\|\mathbf{x}\|_s$ is defined by $\sup_{0 \leq r \leq s} |x_r|$. It is said of *linear growth* with respect to $\mathbf{x}$ and $\mathbf{y}$ if there exists a positive constant C such that

$$|f(s, \mathbf{x}, \mathbf{y})| \leq C(1 + \|\mathbf{x}\|_s + \|\mathbf{y}\|_s)$$

holds for all s and $\mathbf{x}, \mathbf{y}$.

Condition (B) *Coefficients $b(t, \mathbf{x}, \mathbf{y})$, $\sigma(t, \mathbf{x}, \mathbf{y})$, $h(t, \mathbf{x}, \mathbf{y})$ of the equations are locally Lipschitz continuous with respect to $\mathbf{x}$ and are of linear growth with respect to $\mathbf{x}, \mathbf{y}$. Further, $\sigma(t, \mathbf{x}, \mathbf{y}) S(t)$ is bounded.*

Theorem 2.11 *Assume Condition (B). Then system-observation process exists uniquely. Further Φ_t given by (2.42) is a martingale.*

Proof. Let $(\Omega, \mathcal{F}, \mathbf{Q})$ be another probability space, where a $(d+e)$-dimensional Brownian motion (W_t, Y'_t) is defined. (The measure $\mathbf{Q}$ in this subsection is different from the measure $\mathbf{Q}$ in Section 2.3.2.) Its covariances are $E_{\mathbf{Q}}[W_t W_t^T] = \int_0^t Q(s)ds$, $E_{\mathbf{Q}}[Y'_t(Y'_t)^T] = \int_0^t R(s)ds$ and $E_{\mathbf{Q}}[W_t(Y'_t)^T] = \int_0^t S(s)ds$. Let X_0 be an $\mathbf{R}^d$-valued random variable, independent of (W_t, Y'_t). We define a filtration by $\mathcal{F}_t = \wedge_{\epsilon > 0} \sigma(X_0, Y_0, W_s, Y'_s; s \leq t + \epsilon)$.

Set $Y_t = Y_0 + Y'_t$ and consider an SDE for X_t:

$$X_t = X_0 + \int_0^t \{b(s, \mathbf{X}, \mathbf{Y}) - \sigma(s, \mathbf{X}, \mathbf{Y}) S(s) R(s)^{-1} h(s, \mathbf{X}, \mathbf{Y})\} ds \qquad (2.55)$$
$$+ \int_0^t \sigma(s, \mathbf{X}, \mathbf{Y}) dW_s.$$

Coefficients $b - \sigma S R^{-1} h$, σ are locally Lipschitz continuous with respect to $\mathbf{x}$ and of linear growth with respect to $\mathbf{x}, \mathbf{y}$. Therefore equation (2.55) has a unique solution X_t. The solution X_t is $(\mathcal{F}_t)$-adapted.

Now, we set $Y_t^* = \int_0^t S(s) R(s)^{-1} dY_s'$ and $W_t^* = W_t - Y_t^*$. Then it holds

$$E_{\mathbf{Q}}[W_t^*(Y_t')^T] = E_{\mathbf{Q}}[W_t(Y_t')^T] - E_{\mathbf{Q}}[Y_t^*(Y_t')^T]$$
$$= \int_0^t S(r) dr - \int_0^t S(r) R(r)^{-1} R(r) dr = 0.$$

Therefore, W_t^* and Y_t' are independent Brownian motions with respect to $\mathbf{Q}$. (Y_t^* is the orthogonal projection of W_t to the Brownian motion Y_t'). Then equation (2.55) is written as

$$X_t = X_0 + \int_0^t b(s, \mathbf{X}, \mathbf{Y}) ds + \int_0^t \sigma(s, \mathbf{X}, \mathbf{Y}) dW_s^* \qquad (2.56)$$
$$+ \int_0^t \sigma^*(s, \mathbf{X}, \mathbf{Y})(dY_s - h(s, \mathbf{X}, \mathbf{Y}) ds),$$

where $\sigma^*(s) = \sigma(s) S(s) R(s)^{-1}$.

Let Ψ_t be the functional of (2.50) associated with $h(s) = h(s, \mathbf{X}, \mathbf{Y})$. It is a positive local martingale with respect to $((\mathcal{F}_t), \mathbf{Q})$. Define a sequence of stopping times τ_n^* by

$$\tau_n^*(\mathbf{x}, \mathbf{y}) = \inf\{t < T; \|\mathbf{x}\|_t^2 + \|\mathbf{y}\|_t^2 > n\}, \quad (= \infty \text{ if } \{\cdots\} = \phi).$$

We set $\tau_\infty^* = \lim_{n \to \infty} \tau_n$. These are measure-free stopping times. We denote $\tau_n^*(\omega) = \tau_n(\mathbf{X}(\omega), \mathbf{Y}(\omega))$. Then the stopped processes $\Psi_t^{(n)} = \Psi_{t \wedge \tau_n^*}$ are martingales.

Define a probability $\mathbf{P}_n$ by $\mathbf{P}_n = \Psi_T^{(n)} \cdot \mathbf{Q}$. Then the equality $\mathbf{P}_n = \mathbf{P}_m$ holds on $\mathcal{F}_{T \wedge \tau_m^*}$ if $n \geq m$. Set $N_t^{(n)} := Y_t' - \int_0^{t \wedge \tau_n^*} h(s, \mathbf{X}, \mathbf{Y}) ds$. Then with respect to $\mathbf{P}_n$, W_t^* and $N_t^{(n)}$ are independent Brownian motions and the covariance of $(W_t^*, N_t^{(n)}, \mathbf{P}_n)$ coincides with that of $(W_t^*, Y_t', \mathbf{Q})$, by Girsanov's theorem. Then $W_t - \int_0^{t \wedge \tau_n} S(s) R(s)^{-1} h(s, \mathbf{X}, \mathbf{Y}) ds$ is also a Brownian motion with respect to $\mathbf{P}_n$.

It holds

$$X_{t \wedge \tau_n} = X_0 + \int_0^{t \wedge \tau_n} b(s, \mathbf{X}, \mathbf{Y}) ds + \int_0^{t \wedge \tau_n} \sigma(s, \mathbf{X}, \mathbf{Y}) dW_s^*$$
$$+ \int_0^{t \wedge \tau_n} \sigma^*(s, \mathbf{X}, \mathbf{Y}) dN_s^{(n)},$$
$$Y_{t \wedge \tau_n} = Y_0 + \int_0^{t \wedge \tau_n} h(s, \mathbf{X}, \mathbf{Y}) ds + N_{t \wedge \tau_n}^{(n)},$$

a.s. $\mathbf{P}_n$. Then using the linear growth property of coefficients of the equation, we can show that there exist positive constants C_1, C_2 (not depending on n) such that the inequality

$$E_{\mathbf{P}_n}[\|X\|^2_{t\wedge\tau_n^*} + \|Y\|^2_{t\wedge\tau_n^*}] \le C_1 + C_2 \int_0^t E_{\mathbf{P}_n}[\|X\|^2_{s\wedge\tau_n^*} + \|Y\|^2_{s\wedge\tau_n^*}]ds$$

holds for any n. Then we get $E_{\mathbf{P}_n}[\|X\|^2_{t\wedge\tau_n} + \|Y\|^2_{t\wedge\tau_n}] \le C_1' e^{C_2'T}$. Then it holds $\mathbf{P}_n(\tau_n^* < t) \to 0$ as $n \to \infty$. Hence there exists a probability $\mathbf{P}$ such that $\mathbf{P} = \mathbf{P}_n$ holds on $\mathcal{F}_{T\wedge\tau_n^*}$.

With respect to $\mathbf{P}$, the 4-ple $X_t, Y_t, W_t^*, N_t := Y_t' - \int_0^t h(s, \mathbf{X}^*, \mathbf{Y})ds$ satisfy (2.54). Set

$$\tilde{W}_t := W_t^* + \int_0^t S(s)R(s)^{-1}dN_s = W_t - \int_0^t S(s)R(s)^{-1}h(s, \mathbf{X}, \mathbf{Y})ds.$$

Then the pair $(\tilde{W}_t, N_t, \mathbf{P})$ is a Brownian motion with the same law as that of $(W_t, Y_t', \mathbf{Q})$. Further the 4-ple $(X_t, Y_t, \tilde{W}_t, N_t)$ satisfies

$$X_t = X_0 + \int_0^t b(s, \mathbf{X}, \mathbf{Y})ds + \int_0^t \sigma(s, \mathbf{X}, \mathbf{Y})d\tilde{W}_s,$$
$$Y_t = Y_0 + \int_0^t h(s, \mathbf{X}, \mathbf{Y})ds + N_t.$$

Therefore 4-ple $(X_t, Y_t, \tilde{W}_t, N_t)$ is a solution of equations (2.44) and (2.53) with respect to $\mathbf{P}$.

It holds $\tau_\infty^* = \infty$, a.s. $\mathbf{P}$ and a.s. $\mathbf{Q}$. Then uniqueness of the solution can be verified similarly as in Lemma 2.2. Further, similarly as in Theorem 2.10, the measure $\mathbf{P}$ is equivalent to $\mathbf{Q}$ with respect to $\mathcal{F}_t$ for any t. The Radon–Nikodym density $\frac{d\mathbf{Q}}{d\mathbf{P}}|_{\mathcal{F}_t}$ is given by Φ_t. Hence it is a martingale. We have thus proved the theorem. □

Remark 2.2 In case of correlated noises, if the function h is not of linear growth, Φ_t may or may not be a martingale. Here are examples. Suppose that the system process X_t is a one dimensional Brownian motion N_t and $h = -sgn(x)x^2$. Then the observation process is a one dimensional process $Y_t = Y_0 - \int_0^t sgn(X_s)X_s^2 ds + N_t$. The associated equation (2.53) is given by $X_t = X_0 - \int_0^t sgn(X_s)X_s^2 ds + Y_t$, where Y_t is a Brownian motion with respect to $\mathbf{Q}$. It has a unique solution. Further, the measure $\mathbf{Q}$ is absolutely continuous with respect to $\mathbf{P}$ and the Radon–Nikodym density is given by Φ_t, which is a martingale with respect to $\mathbf{P}$.

On the other hand if $h = x^2$, the observation process is given by $Y_t = Y_0 + \int_0^t X_s^2 ds + N_t$. The associated equation (2.53) is $X_t = X_0 + \int_0^t X_s^2 ds + Y_t'$, where Y_t' is a Brownian motion with respect to $\mathbf{Q}$. The equation has no global solution. The solution should explode at finite random time ζ. See [11]. In this case Φ_t is a supermartingale but is not a martingale. It holds $\mathbf{Q}(\zeta \ge t) = E[\Phi_t]$.

Remark 2.3 In SDE (2.51), assume that coefficients b, σ depend on $(t, \mathbf{x})$ only and that Brownian motion W_t and N_t are independent. Then the system process X_t and the noise process N_t are independent. Further if the initial distribution

ν of X_0 has $2N$-th moment, i.e., $\int |x|^{2N}\nu(dx) < \infty$, then solution X_t has also $2N$-th moment. Hence if the function h is of N-th growth with respect to $\mathbf{x}$, i.e.,

$$|h(t, \mathbf{x}, \mathbf{y})| \leq C(1 + \|\mathbf{x}\|_t^N + \|\mathbf{y}\|_t), \tag{2.57}$$

then the function h satisfies Condition (A). Hence the class of functions h satisfying Condition (A) is larger than the class of functions h satisfying Condition (B).

2.4 Innovation problems

2.4.1 Martingale representation

Let Y_t be an observation process defined on $(\Omega, \mathcal{F}, \mathbf{P})$ given by (2.39), where h_s satisfies $E[|h_s|^2] < \infty$ for any s. Here the system process and the noise process may be or may not be correlated. Let $(\mathcal{G}_t)$ be the filtration generated by the observation process Y_t, i.e., let $\mathcal{G}_t = \sigma(Y_s; s \leq t)$. We define an e-dimensional stochastic process I_t by

$$I_t = Y_t - Y_0 - \int_0^t \hat{h}_s ds, \tag{2.58}$$

where $\hat{h}_s = E[h_s|\mathcal{G}_s] = (E[h_s^1|\mathcal{G}_s], ..., E[h_s^e|\mathcal{G}_s])$.

Lemma 2.3 I_t is a $(\mathcal{G}_t)$-Brownian motion with covariance $\int_0^t R(s)ds$.

Proof. Since $E[I_t^i - I_s^i|\mathcal{G}_s] = \int_s^t E[(h_u^i - \hat{h}_u^i)|\mathcal{G}_s]du = 0$ holds for any $s < t$, I_t^i is a continuous $(\mathcal{G}_t)$-martingale. Consider quadratic covariation of I_t^i and I_t^j. It coincides with the quadratic covariation of N_t^i and N_t^j. Hence it is equal to $\int_0^t r_s^{ij}ds$, which shows that I_t is a $(\mathcal{G}_t)$-Brownian motion with mean 0 and covariance $\int_0^t R(s)ds$, by Lévy's theorem. $\square$

Now let $(\mathcal{G}_t')$ be a subfiltration of the filtration $(\mathcal{G}_t)$ such that $\mathcal{G}_t' = \mathcal{G}_0 \vee \sigma(I_s; s \leq t)$. If M_t is a square integrable $(\mathcal{G}_t')$-martingale, it is represented by stochastic integral based on the Brownian motion I_t (Theorem 2.3). Such a representation can be extended to square integrable $(\mathcal{G}_t)$-martingale. In fact we have the following theorem.

Theorem 2.12 ([7]) *Consider the observation process (2.39), where h_s is a $(\mathcal{F}_t)$-progressively measurable process satisfying $\int_0^T |h(s)|^2 ds < \infty$ a.s. $\mathbf{P}$. Then any $(\mathcal{G}_t)$-martingale M_t has a continuous modification. Further, there exists an $(\mathcal{G}_t)$-progressively measurable process $f_s = (f_s^1, ..., f_s^e)$ with $\int_0^T (f_s^i)^2 r_s^{ii} ds < \infty$ for any i, and M_t is represented by*

$$M_t = M_0 + \int_0^t (f_s, dI_s) := M_0 + \sum_{i=1}^e \int_0^t f_s^i dI_s^i. \qquad (2.59)$$

Proof. We want to show that M_t is an Itô process based on the $(\mathcal{G}_t)$-Brownian motion I_t, i.e., it is written by $M_t = M_0 + \int_0^t (f_s, dI_s) + \int_0^t g_s ds$, where f_t and g_t are $(\mathcal{G}_t)$-progressively measurable processes. Then since M_t is a $(\mathcal{G}_t, \mathbf{P})$-martingale, we have $g_s = 0$. Hence we get a martingale representation of M_t.

Consider a positive continuous local martingale Φ_t given by (2.42). To make the discussion simple, we first assume that Φ_t is a martingale. Define a probability $\mathbf{Q}$ by $\mathbf{Q}(A) = E[\Phi_T 1_A]$. Then $Y_t' = Y_t - Y_0$ is a Brownian motion with respect to $\mathbf{Q}$ by Girsanov's theorem. Therefore any $(\mathcal{G}_t, \mathbf{Q})$-martingale is continuous and is represented by Y_t. Now the Radon–Nikodym density $\Psi_t = \frac{d\mathbf{P}}{d\mathbf{Q}}|_{\mathcal{G}_t}$ is a positive $(\mathcal{G}_t, \mathbf{Q})$-martingale. Since $\mathbf{P} = \mathbf{Q}$ holds on $\mathcal{G}_0 \subset \mathcal{F}_0$, $\Psi_0 = 1$. Then Ψ_t is represented as

$$\Psi_t = 1 + \int_0^t (k_s, dY_s) = 1 + \int_0^t (k_s, dI_s) - \int_0^t (k_s, \hat{h}_s ds), \quad a.s. \ \mathbf{Q},$$

where k_t is a progressively $(\mathcal{G}_t)$-measurable process (Theorem 2.3). The equality is valid with respect to $\mathbf{P}$, since both measures $\mathbf{Q}$ and $\mathbf{P}$ are equivalent. Therefore Ψ_t is an Itô process based on the $(\mathcal{G}_t)$-Brownian motion I_t with respect to $\mathbf{P}$.

Now let M_t be a $(\mathcal{G}_t, \mathbf{P})$-martingale. Then the product $M'_t = M_t \Psi_t^{-1}$ is a $(\mathcal{G}_t, \mathbf{Q})$-martingale. Then it is again an Itô process based on $(\mathcal{G}_t)$-Brownian motion I_t with respect to $\mathbf{P}$, by the same reason as that of the case Ψ_t. Then applying Itô's formula to the product of two Itô processes, we find that $M_t = \Psi_t M'_t$ is also an Itô process based on the $(\mathcal{G}_t)$-Brownian motion I_t with respect to $\mathbf{P}$. Hence we get the representation formula (2.59).

In the case where Φ_t is not a martingale, we can take an increasing sequence of stopping times τ_m, $m = 1, 2, \ldots$ such $\mathbf{P}(\tau_m \geq T) \to 1$ as $m \to \infty$ and each stopped process $\Phi_t^{(m)} = \Phi_{t \wedge \tau_m}$ is a martingale. Define a probability $\mathbf{Q}_m$ by $\mathbf{Q}_m(A) = E[\Phi_T^{(m)} 1_A]$ for $A \in \mathcal{G}_{\tau_m}$. Then $Y_{t \wedge \tau_m}$ is a Brownian motion stopped at τ_m with respect to $\mathbf{Q}_m$. Now let M_t be a $(\mathcal{G}_t, \mathbf{P})$-martingale. Repeating the similar argument to the process $M_{t \wedge \tau_m}$, we can show that $M_{t \wedge \tau_m}$ is represented as $M_0 + \int_0^{t \wedge \tau_m} (f_s^m, dY_s)$. Since this is valid for any m, we can choose a progressively measurable process f_t such that $f_t = f_t^{(m)}$ holds a.e. $dt \times d\mathbf{P}$ for any m. Then M_t is written as (2.59). The proof is complete. $\square$

2.4.2 The innovation property

Let $\mathcal{G}'_t = \mathcal{G}_0 \vee \sigma(I_s; s \leq t)$ as before. If the equality $\mathcal{G}'_t = \mathcal{G}_t$ holds for any t, I_t is called an *innovation process* of Y_t or it is said to have the *innovation property*. We show later (in Chapter 3) that in the case where the system-observation process is Gaussian or conditionally Gaussian, the process I_t defined by (2.58) is an innovation. However, for nonconditionally Gaussian system-observation process, it is a rather difficult problem whether I_t is an innovation or not. We will discuss the problem in the case where the functional h_s is bounded.

Remark. In the linear filtering problem of time series, the observation process Y_n with discrete time has a (linear) innovation I_n, i.e., there exists an orthogonal time series I_n such that for any n the linear span of $I_k; k \leq n$ is equal to the linear span of $Y_k; k \leq n$. However, the existence of nonlinear innovations cannot be expected for time series.

Now we shall restrict our attention to the case where the system process and the noise are independent. The innovation property was shown in Kunita [13][1] in the case where the system process is a Markov process and the function $h(s, x, y) = h(x)$ depends on x only and is bounded continuous. Clark [4] proved that I_t of (2.58) is an innovation in the case where $h(s)$ is bounded and depends on $(s, \mathbf{x})$ only. It is extended as follows.

Theorem 2.13 *Suppose that h_s is given by $h(s, \mathbf{X}, Y_s)$, where $h(t, \mathbf{x}, y), \mathbf{x} \in \mathcal{D}(S), y \in \mathbf{R}^e$ is bounded, differentiable with respect to y and $h_{y^i}(s, \mathbf{x}, y)$ are bounded. Then I_t defined by (2.58) is an innovation.*

The proof of the theorem is complicated. One can skip it at the first reading. It will be given at the next subsection.

The smoothness condition for the function $h(t, \mathbf{x}, y)$ with respect to y can be relaxed by a Lipschitz condition.

$$|h(s, \mathbf{x}, y) - h(s, \mathbf{x}, y')| \leq C|y - y'|.$$

Krylov [12] showed the innovation property for a nondegenerate diffusion process X_t and a possibly unbounded function $h(t) = h(t, x, y)$ such that $|h(t, x, y)| \leq g(x)$ with $g \in L^2$ and satisfies a Lipschitz condition $\int |h(t, x, y) - h(t., x, y')| dx \leq K|y - y'|$.

If the function $h(t, \mathbf{x}, \mathbf{y})$ is not Lipschitz continuous with respect to $\mathbf{y}$, the innovation property may fail to hold. We give such an example.

Example. Let $g(t, \mathbf{y})$ be a bounded progressively measurable function. Consider a one dimensional SDE

$$Y_t = \int_0^t g(s, \mathbf{Y}) ds + B_t, \tag{2.60}$$

where B_t is a one dimensional standard $(\mathcal{F}_t)$-Brownian motion. The equation has a unique solution in the sense of law (a special case of Theorem 2.8). It holds $\sigma(B_s; s \leq t) \subset \sigma(Y_s; s \leq t)$ for any t. If $g(s, \mathbf{y})$ is a smooth function

(Lipschitz continuous with respect to **y** etc), it has a strong solution, i.e., it holds $\sigma(Y_s; s \le t) \subset \sigma(B_s; s \le t)$ for any t. However, if g is a rough function, it might fail to have a strong solution.

We will quote here Chirelson's example [5]. Denote the integer part of $x \in \mathbf{R}$ by $[x]$ and set $\theta(x) = x - [x]$. Let $t_k; k = 0, -1, -2, \ldots$ be a sequence of $[0, T]$ such that $0 < t_{k-1} < t_k < \cdots$ and $\lim_{k \to -\infty} t_k = 0$. Let

$$g(t, \mathbf{y}) = \begin{cases} \theta\left(\frac{y_{t_k} - y_{t_{k-1}}}{t_k - t_{k-1}}\right), & t_k \le t < t_{k+1}, \\ 0, & t \ge t_0. \end{cases} \quad (2.61)$$

Chirelson showed that the associated SDE (2.60) does not have a strong solution, i.e., $\sigma(Y_s; s \le t) \subset \sigma(B_s; s \le t)$ does not hold for some t; there exists $A \in \sigma(Y_s; s \le t)$ with $0 < \mathbf{P}(A) < 1$ such that $A \notin \sigma(B_s; s \le t)$ (See [9] or [15]).

Now we shall introduce an observation process which does not have the innovation property, i.e., $\sigma(Y_s; s \le t) \subset \sigma(I_s; s \le t)$ does not hold for some t. Let $\mathbf{X} = (X_t)$ be a system process with the law $F_\mathbf{X}$ and let $f(s, \mathbf{x})$ be a bounded measurable function. Let $g(t, \mathbf{y})$ be a bounded progressively measurable function such that equation (2.60) has no strong solution. We consider the observation process

$$Y_t = \int_0^t f(s, \mathbf{X}) ds + \int_0^t g(s, \mathbf{Y}) ds + N_t, \quad (2.62)$$

where N_t is a one-dimensional standard Brownian motion, independent of X_t.

We decompose the above equation to two equations.

$$Z_t = \int_0^t f(s, \mathbf{X}) ds + N_t, \quad (2.63)$$

$$Y_t = \int_0^t g(s, \mathbf{Y}) ds + Z_t. \quad (2.64)$$

We may regard (2.63) as an equation of the observation process Z_t. We want to compare two filterings based on the observation process Y_t of (2.62) and the observation process Z_t of (2.63). The filter $\hat{f}(t)$ of $f(t, \mathbf{X})$ based on the observation process Y_t is given by Kallianpur–Striebel formula (Theorem 2.9)

$$\hat{f}(t) = \frac{\int f(t, \mathbf{x}) \psi_t(\mathbf{x}) dF_\mathbf{X}(\mathbf{x})}{\int \psi_t(\mathbf{x}) dF_\mathbf{X}(\mathbf{x})},$$

where $\psi_t(\mathbf{x}) = \exp\{\int_0^t (f(s) + g(s)) dY_s - \frac{1}{2} \int_0^t (f(s) + g(s))^2 ds\}$. The filter $\hat{f}^Z(t)$ of $f(t, \mathbf{X})$ based on the observation data $\mathcal{G}_t^Z = \sigma(Z_s; s \le t)$ is given by

$$\hat{f}^Z(t) = \frac{\int f(t, \mathbf{x}) \psi_t^Z(\mathbf{x}) dF_\mathbf{X}(\mathbf{x})}{\int \psi_t^Z(\mathbf{x}) dF_\mathbf{X}(\mathbf{x})},$$

where $\psi_t^Z(\mathbf{x}) = \exp\{\int_0^t f(s)dZ_s - \frac{1}{2}\int_0^t f(s)^2 ds\}$. A direct computation yields $\psi_t(\mathbf{x}) = \exp\{\int_0^t g(s)dY_s - \frac{1}{2}\int_0^t g(s)^2 ds\}\psi_t^Z(\mathbf{x})$. The exponential functional at the right-hand side does not depend on $\mathbf{x}$. Consequently $\hat{f}(t) = \hat{f}^Z(t)$ a.s. holds from the above two Kallianpur–Striebel formulas.

Let I_t be a $(\mathcal{G}_t)$-Brownian motion constructed from Y_t of (2.62) and let I_t^Z be the $(\mathcal{G}_t^Z)$-Brownian motion constructed from Z_t. Then we have $I_t = I_t^Z$ because $\hat{f}(t) = \hat{f}^Z(t)$ holds a.s. We saw in Theorem 2.13 that I_t^Z is an innovation of the observation process Z_t. Therefore we have $\sigma(Z_s; s \leq t) = \sigma(I_s; s \leq t)$ for any t.

Now since $Z_t = \int_0^t \hat{f}(s)ds + I_t$ holds valid, the process Z_t is a Brownian motion with respect to the measure $\mathbf{Q} = \exp\{-\int_0^t \hat{f}(s)dI_s - \frac{1}{2}\int_0^t \hat{f}(s)^2 ds\} \cdot \mathbf{P}$. Then with respect to $\mathbf{Q}$, equation (2.64) has no strong solution. This means that $\sigma(Y_s; s \leq t) = \sigma(Z_s; s \leq t)$ does not hold for some t. Since $\mathbf{P}$ and $\mathbf{Q}$ are equivalent, the same fact is valid with respect to $\mathbf{P}$. Therefore I_t is not an innovation of Y_t.

2.4.3 Proof of Theorem 2.13

Discussion for the proof of the theorem is rather long. For simplicity, we assume that the noise process N_t is a one-dimensional standard Brownian motion and Y_t is a one-dimensional process defined on $(\Omega, \mathcal{F}, \mathbf{P})$. For a bounded measurable $(\mathcal{G}_t)$-adapted process $\mathbf{Z} = (Z_t)$, we set $h^Z(s, \mathbf{x}) = h(s, \mathbf{x}, Y_0 + I_s + \int_0^s Z_r dr)$. We define a $(\mathcal{G}_t)$-adapted process with parameter $\mathbf{x}$ by

$$K_t^Z(\mathbf{x}) = \int_0^t h^Z(s, \mathbf{x})dI_s + \int_0^t h^Z(s, \mathbf{x})Z_s ds - \frac{1}{2}\int_0^t h^Z(s, \mathbf{x})^2 ds.$$

Set $\psi_t^Z(\mathbf{x}) = \exp K_t^Z(\mathbf{x})$. Then we have $\psi_t^{\hat{h}}(\mathbf{x}) = \psi_t(\mathbf{x}, \mathbf{Y})$ (defined by (2.49)), where $\hat{\mathbf{h}} = (\hat{h}_t)$. Define

$$\pi_t^Z(f) = \frac{\int f(\mathbf{x})\psi_t^Z(\mathbf{x}) F_X(d\mathbf{x})}{\int \psi_t^Z(\mathbf{x}) F_X(d\mathbf{x})}.$$

It may be considered as a perturbation of π_t since $\pi_t^Z(f) = \pi_t(f)$ holds if $Z_t = \hat{h}_t$. So it is expected that π_t^Z tends to π_t as Z_t tends to $\hat{h}_t$. Indeed, we have the following lemma.

Lemma 2.4 *There exists a positive constant C such that the inequality*

$$E[|\pi_t^Z(f) - \pi_t(f)|^2] \leq C\|f\|^2 \int_0^t E[|Z_s - \hat{h}_s|^2]ds \qquad (2.65)$$

holds for any $\mathbf{Z}$ with $\sup_s |Z_s| \leq K$, where K is a positive constant such that $|h(s, \mathbf{x}, y)| \leq K$ holds for any $s, \mathbf{x}, y$

Before we proceed to the proof of the lemma, we shall rewrite $\pi_t^Z(f)$. Set $\tilde{\Psi}_t^Z = \int \psi_t^Z(\mathbf{x}) F_X(d\mathbf{x})$. Apply Itô's formula to the exponential semimartingale ψ_t^Z and then integrate each term by the measure F_X. We get

$$\tilde{\Psi}_t^Z = 1 + \int_0^t \left(\int \psi_s^Z h_s^Z dF_X \right) dI_s + \int_0^t \left(\int \psi_s^Z h_s^Z dF_X \right) Z_s ds,$$

where $h_s^Z = h^Z(s, \mathbf{x})$. Set $\tilde{h}_s^Z = \pi_t^Z(h_t^Z)$. The above is written as $1 + \int_0^t \tilde{h}_s^Z \tilde{\Psi}_s^Z dI_s + \int_0^t \tilde{h}_s^Z \tilde{\Psi}_s^Z Z_s ds$. Consequently we have

$$\tilde{\Psi}_t^Z = \exp\left\{ \int_0^t \tilde{h}_s^Z dI_s + \int_0^t \tilde{h}_s^Z Z_s ds - \frac{1}{2} \int_0^t (\tilde{h}_s^Z)^2 ds \right\}.$$

Then we have $(\tilde{\Psi}_t^Z)^{-1} \psi_t^Z(\mathbf{x}) = \alpha_t^Z(\mathbf{x}) \beta_t^Z(\mathbf{x})$, where

$$\alpha_t^Z(\mathbf{x}) = \exp\left\{ \int_0^t (h_s^Z - \tilde{h}_s^Z) dI_s - \frac{1}{2} \int_0^t (h_s^Z - \tilde{h}_s^Z)^2 ds \right\},$$

$$\beta_t^Z(\mathbf{x}) = \exp\left\{ \int (h_s^Z - \tilde{h}_s^Z)(Z_s - \tilde{h}_s^Z) ds \right\}.$$

Using these notations, we have

$$\pi_t^Z(f) = \int \alpha_t^Z(\mathbf{x}) \beta_t^Z(\mathbf{x}) f(\mathbf{x}) F_X(d\mathbf{x}). \tag{2.66}$$

Since h_s^Z, $\tilde{h}_s^Z$, Z_s are bounded processes, there exist positive constants K_1, K_2 such that the inequality $K_1 \leq \beta_t^Z(\mathbf{x}) \leq K_2$ holds for any $\mathbf{x}$. Then we have the inequality

$$K_1 \int \alpha_t^Z(\mathbf{x}) |f(\mathbf{x})| dF_X \leq |\pi_t^Z(f)| \leq K_2 \int \alpha_t^Z(\mathbf{x}) |f(\mathbf{x})| dF_X, \quad \text{a.s.} \tag{2.67}$$

Proof of Lemma 2.4 Set $h^{Z,\lambda} = h^{\lambda Z + (1-\lambda)\hat{h}}$, $K_t^{Z,\lambda} = K_t^{\lambda Z + (1-\lambda)\hat{h}}$, $\psi_t^{Z,\lambda} = \exp K_t^{Z,\lambda}$ and $\pi_t^{Z,\lambda} = \pi_t^{\lambda Z + (1-\lambda)\hat{h}}$ etc., where $0 \leq \lambda \leq 1$. Instead of (2.65), we want to prove

$$\sup_\lambda E[|D_\lambda \pi_t^{Z,\lambda}(f)|^2] \leq C\|f\|^2 \int_0^t E[|Z_s - \hat{h}_s|^2] ds. \tag{2.68}$$

If it is verified, we get (2.65) because $\pi_t^Z(f) - E_t(f(X_t)) = \int_0^1 D_\lambda \pi^{Z,\lambda}(f) d\lambda$.

A direct calculation yields

$$D_\lambda \pi_t^{Z,\lambda}(f) = \pi_t^{Z,\lambda}(f D_\lambda K_t^{Z,\lambda}) + \pi_t^{Z,\lambda}(f) \pi_t^{Z,\lambda}(D_\lambda K_t^{Z,\lambda}). \tag{2.69}$$

Further $D_\lambda K_t^{Z,\lambda}(\mathbf{x})$ is written as $A_t^{Z,\lambda}(\mathbf{x}) + M_t^{Z,\lambda}(\mathbf{x})$ where,

$$A_t^{Z,\lambda}(\mathbf{x}) = \int_0^t g^{Z,\lambda}(s, \mathbf{x})(Z_s - \hat{h}_s) ds,$$

$$M_t^{Z,\lambda}(\mathbf{x}) = \int_0^t f^{Z,\lambda}(s, \mathbf{x})(Z_s - \hat{h}_s)(dI_s - (h_s^{Z,\lambda} - \tilde{h}_s^{Z,\lambda}) ds),$$

with bounded progressively measurable functions $f^{Z,\lambda}$ and $g^{Z,\lambda}$. Therefore we have

$$E[|D_\lambda \pi_t^{Z,\lambda}(f)|^2] \le 2\|f\|^2 \Big(E[\pi_t^{Z,\lambda}((A_t^{Z,\lambda})^2)] + E[\pi_t^{Z,\lambda}((M_t^{Z,\lambda})^2)] \Big). \qquad (2.70)$$

Since $\pi_t^{Z,\lambda}(1) = 1$ a.s., we have

$$E[\pi_t^{Z,\lambda}((A_t^{Z,\lambda})^2)] \le E[\sup_{s,\mathbf{x},\lambda,Z} |g^{Z,\lambda}(s,\mathbf{x})|^2 \int_0^t |Z_s - \hat{h}_s|^2 ds]$$
$$\le C \int_0^t E[|Z_s - \hat{h}_s|^2] ds.$$

We next consider the second part of (2.70). For any $\mathbf{x}$ and λ, $M_t^{Z,\lambda}(\mathbf{x})$ is a martingale with respect to the conditional measure $\mathbf{Q}_{\mathbf{x},\lambda}^{(\omega)} = a_t^{Z,\lambda}(\mathbf{x}) \cdot \mathbf{P}^{(\omega)}$ where $\mathbf{P}^{(\omega)} = \mathbf{P}(\cdot|\sigma(X_s; s \le T))$, in view of Girsanov's theorem. Then since $(M_t^{Z,\lambda}(\mathbf{x}))^2 - \langle M^{Z,\lambda}(\mathbf{x}) \rangle_t$ is a martingale with respect to $\mathbf{Q}_{\mathbf{x},\lambda}^{(\omega)}$, we have

$$E_{\mathbf{Q}_{\mathbf{x},\lambda}^{(\omega)}}[(M_t^{Z,\lambda}(\mathbf{x}))^2] = E_{\mathbf{Q}_{\mathbf{x},\lambda}^{(\omega)}}[\langle M^{Z,\lambda}(\mathbf{x}) \rangle_t]$$

for any $\mathbf{x}$. Therefore,

$$E[\pi_t^{Z,\lambda}[(M_t^{Z,\lambda})^2]] \le K_2 E[\int a_t^{Z,\lambda}(x) M_t^{Z,\lambda}(\mathbf{x})^2 F_\mathbf{X}(d\mathbf{x})]$$
$$= K_2 E[\int E_{\mathbf{Q}_{\mathbf{x},\lambda}^{(\omega)}}[(M_t^{Z,\lambda})^2]] F_\mathbf{X}(d\mathbf{x})]$$
$$= K_2 E[\int E_{\mathbf{Q}_{\mathbf{x},\lambda}^{(\omega)}}[\langle M_t^{Z,\lambda} \rangle]] F_\mathbf{X}(d\mathbf{x})]$$
$$\le \frac{K_2}{K_1} E[\pi_t^{Z,\lambda}(\langle M^{Z,\lambda} \rangle_t)].$$

Since $\langle M^{Z,\lambda} \rangle_t = \int_0^t (h_s^{Z,\lambda})^2 (Z_s - \hat{h}_s)^2 ds$, the last term of the above is dominated by $C \int_0^t E[|Z_s - \hat{h}_s|^2] ds$. Consequently, we get

$$\sup_\lambda E[|\pi_t^{Z,\lambda}(f D_\lambda K_t^{Z,\lambda})|^2] \le C \|f\|^2 \int_0^t E[|Z_s - \hat{h}_s|^2] ds.$$

By a similar computation we have

$$\sup_\lambda E[|\pi_t^{Z,\lambda}(f)\pi_t^{Z,\lambda}(D_\lambda K_t^{Z,\lambda})|^2] \le 2C \|f\|^2 \int_0^t E[|Z_s - \hat{h}_s|^2] ds$$

These two yield (2.68). $\square$

Proof of Theorem 2.13 We will show that $\hat{h}_t$ is approximated by a sequence of $\sigma(Y_0, I_s; s \le t)$-measurable processes. Set $\hat{h}_t^0 := 0$, $\hat{h}_t^1 := \pi_t^{\hat{h}^0}(h), ..., \hat{h}_t^n := \pi_t^{\hat{h}^{n-1}}(h), ...$ Then $\hat{h}_t^n$ are $\sigma(Y_0, I_s; s \le t)$-measurable. Further,

$$E[|\hat{h}^n_t - \hat{h}_t|^2] \leq C \int_0^t |E[\hat{h}^{n-1}_s - \hat{h}_s|^2] ds \leq \cdots \leq \frac{C^n}{n!} t^n$$

in view of (2.65). The last term converges to 0 as $n \to \infty$. Consequently $\hat{h}^n_t$ converges to $\hat{h}_t$. Then $\hat{h}_t$ is $\sigma(Y_0, I_s; s \leq t)$-measurable. Since $Y_t = Y_0 + \int_0^t \hat{h}_s ds + I_t$, Y_t is also $\sigma(Y_0, I_s; s \leq t)$-measurable. The proof is complete. □

Notes

1. There is a serious gap in Theorem 3.2 in this paper. For details, we refer to [1].

References

[1] Baxendale, P., Chigansky, P., and Liptzer, R. (2004). Asymptotic stability of the Wonham filter: Ergodic and nonergodic signals, *AIAM J. Control Optim.* 43, 643–69.

[2] Blagovescenskii, Ju. N. and Freidlin, M. I. (1961). Some properties of diffusion processes depending on a parameter, *Dokl. Akad. Nauk. SSSR*, 138, 508–11 in Russian, English trans. in *Soviet Math. Doklady*, 2, 3, 633–6.

[3] Brockwell, P. J. and Davis, R. A. (1991). *Time series: Theory and methods*, 2nd edn, Springer.

[4] Clark, J. M. C. (1969). Contribution for the one-to-one correspondence between an observation process and its innovation, Center for computing and automation, Imperial College, London Tech. Rep. I.

[5] Chirelson, D. S. (1975). An example of a stochastic differential equation having no strong solution, *Theory Prob. Appl.* 20, 427–30.

[6] Dynkin, E. B. (1965). *Markov processes II*, Springer.

[7] Fujisaki, M., Kallianpur, G., and Kunita, H. (1972). Stochastic differential equations for the nonlinear filtering problem, *Osaka J. Math.* 9, 19–40.

[8] Ikeda, N. and Watanabe, S. (1981). *Stochastic differential equations and diffusion processes*, North Holland–Kodansha.

[9] Kallianpur, G. (1980). *Stochastic filtering theory*, Springer.

[10] Kallianpur, G. and Striebel, C. (1969). Stochastic differential equations in statistical estimation problems, *Multivariate Analysis*, 2, 367–88.

[11] Karatzas, I. and Shreve, S. E. (1991). *Brownian motion and stochastic calculus*, 2nd edn, Springer.

[12] Krylov, N. V. (1979). On the equivalence of σ-algebras in the filtering problem of diffusion processes, *Teor. Verojatnost. i Primenen* 24, 771–80 in Russian; English trans. in *Theor. Probability Appl.* 24 (1980), 772–81.

[13] Kunita, H. (1971). Asymptotic behavior of the nonlinear filtering errors of Markov processes, *J. Multivariate Anal.* 1, 365–93.

[14] Kunita, H. (1990). *Stochastic flows and stochastic differential equations*, Cambridge Univ. Press.

[15] Liptzer, R. Sh. and Shiryaev, A. V. (2001). *Statistics of random processes I,II*, Springer.

[16] Oksendal, B. (1998). *Stochastic differential equations*, Springer.

[17] Rozovskiĭ, B. L. (1990). *Stochastic Evolution Systems*, Kluwer Academic Publishers.

[18] Stroock, D. W. and Varadhan, S. R. S. (1979). *Multidimensional diffusion processes*, Springer.

[19] Yamada, T. (1973). On a comparison theorem for solutions of stochastic differential equations and its applications, *J. Math. Kyoto Univ.*, 11, 155–67.

·3·
Nonlinear Filtering Problems II: Associated Equations

H. Kunita

3.1 SDE for conditional distributions and unnormalized conditional distributions

3.1.1 Introduction

Let X_t be a system process and let Y_t be an observation process. Let π_t be the conditional distribution of X_t based on the observation data $Y_s; s \le t$. We have seen in Chapter 2, Section 2.3, that conditional laws $\pi_t, 0 \le t \le T$ are given by Bayes formula or Kallianpur–Striebel formula. However, the infinite dimensional law of the system process is hardly computable. Further, the formula is not recursive. For the computation of the π_t at time t, we have to use all observation data $Y_s; 0 \le s \le t$.

To overcome these difficulties, we will derive stochastic partial differential equations for conditional distributions π_t and unnormalized conditional distributions ρ_t for several types of system processes. These system processes include Markov processes, diffusion processes governed by SDE and further conditionally Gaussian processes determined by SDE. All of these processes may be regarded as stochastic processes with semimartingale property. These processes will be discussed in Section 3.1.2.

In Section 3.1.3 we will consider semimartingale system process and we will obtain a stochastic differential equations for conditional expectations $E_t(f(X_t)) = E[f(X_t)|\mathcal{G}_t]$, where $(\mathcal{G}_t)$ is the filtration generated by the observation process. Further in Section 3.1.4, we shall obtain a similar stochastic differential equation for unnormalized conditional expectation.

3.1.2 System processes with semimartingale properties

Let S be a locally compact separable metric space. Let X_t be a system process with state space S adapted to a given filtration $(\mathcal{F}_t, t \in \mathbf{T})$. Let $\mathbf{C}(S)$ be the set of all continuous maps $f : S \to \mathbf{R}$. Suppose that $f \in \mathbf{C}(S)$ satisfies the following property: $f(X_t)$ is an $(\mathcal{F}_t)$-adapted semimartingale decomposed as

$$f(X_t) - f(X_0) = M_t f + \int_0^t A_s f \, ds, \qquad (3.1)$$

where $M_t f$ is a local martingale and $A_t f$ is a progressively measurable process such that $\int_0^t |A_s f| ds < \infty$ for any t. We denote by $\mathbf{D}(S)$ the set of all f with the above property. If the above $\mathbf{D}(S)$ is dense in $\mathbf{C}(S)$ with respect to the compact uniform topology, the system process X_t is called a *semimartingale* (adapted to the filtration $(\mathcal{F}_t)$).

As an example of a semimartingale process, we shall consider a Markov process. Let $P_{s,t}(x, E)$ be transition probabilities on the state space S. A right continuous stochastic process $X_t, t \in [0, T]$ with values in S is called a *Markov process* with transition probabilities $P_{s,t}(x, E)$, if it satisfies for any $s < t$ and bounded continuous function f, $E[f(X_t)|\sigma(X_r; r \leq s)] = P_{s,t}f(X_s)$ a.s., where $P_{s,t}f(x) = \int P_{s,t}(x, dy)f(y)$.

A Markov process is characterized by its generator. Assume that there exists a family of linear operators $A(t) : \mathbf{D}(S) \to \mathbf{C}(S)$ where $\mathbf{D}(S)$ a dense linear subset of $\mathbf{C}(S)$, which satisfy a) $A(t)f(x)$ is continuous in (t, x) for any $f \in \mathbf{D}(S)$, b) If $f \in \mathbf{D}(S)$, $P_{s,t}f \in \mathbf{D}(S)$ and continuously differentiable with respect to s, t and c)

$$\frac{\partial P_{s,t}f}{\partial t} = P_{s,t}A(t)f, \quad \frac{\partial P_{s,t}f}{\partial s} = -A(s)P_{s,t}f \tag{3.2}$$

holds for any $f \in \mathbf{D}(S)$. Then $A(t)$ is called the *generator* of transition probabilities $P_{s,t}$. The former equation is called *Kolmogorov's forward equation* and the latter is called *Kolmogorov's backward equation*. Integrating the forward equation, we get

$$P_{s,t}f(x) = f(x) + \int_s^t P_{s,u}[A(u)f](x)du. \tag{3.3}$$

Let π_0 be the law of X_0. If $f \in \mathbf{D}(S)$ satisfies the integrability conditions $\int P_{0,t}|f|(x)\pi_0(dx) < \infty$ and $\int \int_0^t P_{0,u}|A(u)f|(x)du\pi_0(dx) < \infty$, then

$$M_t f := f(X_t) - f(X_0) - \int_0^t A(s)f(X_s)ds \tag{3.4}$$

is a martingale. Indeed, we have by the Markov property,

$$E[M_t f - M_s f | \sigma(X_r : r \leq s)]$$
$$= P_{s,t}f(X_s) - f(X_s) - \int_s^t P_{s,u}[A(u)f](X_s)du = 0, \quad \text{a.s.}$$

Therefore setting $A_s f = A(s)f(X_s)$, we find that $f(X_t)$ is a semimartingale and hence X_t is a semimartingale system process.

Next, we shall consider the pair of stochastic processes (X_t, Y_t) determined by the following stochastic differential equation

$$X_t = X_0 + \int_0^t b(s, \mathbf{X}, \mathbf{Y})ds + \sigma(s, \mathbf{X}, \mathbf{Y})dW_s, \tag{3.5}$$

$$Y_t = Y_0 + \int_0^t h(s, \mathbf{X}, \mathbf{Y})ds + N_t. \tag{3.6}$$

Here (W_t, N_t) is a Brownian motion with $E[W_t W_t^T] = \int_0^t Q(s)ds$, $E[N_t N_t^T] = \int_0^t R(s)ds$ and $E[W(t) N(t)^T] = \int_0^t S(s)ds$. Coefficients b, σ, h of the equation are functions of $(t, \mathbf{x}, \mathbf{y}) \in [0, T] \times \mathcal{C}^d \times \mathcal{C}^e$. They are assumed to be progressively measurable,[1] locally Lipschitz continuous and are of linear growth. Then for a C^2-function $f(x)$, we have by Itô's formula

$$f(X_t) - f(X_0) = \sum_{i=1}^{d} \sum_{j=1}^{r} \int_0^t f_{x_i}(X_s) \sigma^{ij}(s, \mathbf{X}, \mathbf{Y}) dW_s^j + \int_0^t A_s f ds, \quad (3.7)$$

where

$$A_s f = \sum_{i=1}^{d} b^i(s, \mathbf{X}, \mathbf{Y}) f_{x_i}(X_s) + \frac{1}{2} \sum_{i,j=1}^{d} a^{ij}(s, \mathbf{X}, \mathbf{Y}) f_{x_i x_j}(X_s), \quad (3.8)$$

and

$$(a^{ij}(s, \mathbf{x}, \mathbf{y})) = \sigma(s, \mathbf{x}, \mathbf{y}) Q(s) \sigma(s, \mathbf{x}, \mathbf{y})^T.$$

Therefore X_t is a semimartingale with generator (3.8) with the domain $\mathbf{C}_b^2$ (the set of bounded C^2-functions with bounded derivatives).

Suppose that coefficients of equations (3.5) and (3.6) are Markovian, i.e., there exist functions $b(t, x, y)$, $\sigma(t, x, y)$, $h(t, x, y)$, $x \in \mathbf{R}^d$, $y \in \mathbf{R}^e$, such that coefficients of equations (3.5) and (3.6) are given by $\sigma(t, \mathbf{X}, \mathbf{Y}) = \sigma(t, X_t, Y_t)$ etc. Then the solution (X_t, Y_t) is a Markov process with state space $\mathbf{R}^d \times \mathbf{R}^e$. Let $A(t)$ be its generator. If f is a C^2 function of x only, we have

$$A(s) f(x, y) = \sum_{i=1}^{d} b^i(s, x, y) f_{x_i}(x) + \frac{1}{2} \sum_{i,j=1}^{d} a^{ij}(s, x, y) f_{x_i x_j}(x), \quad (3.9)$$

Further if the functions b, σ depend on (t, x) only, X_t is a Markov process on $\mathbf{R}^d$ and its generator is given by

$$A(s) f(x) = \sum_{i=1}^{d} b^i(s, x) f_{x_i}(x) + \frac{1}{2} \sum_{i,j=1}^{d} a^{ij}(s, x) f_{x_i x_j}(x). \quad (3.10)$$

3.1.3 Stochastic differential equation for conditional distributions

Let X_t be a system process with the semimartingale property. Suppose that the observation process Y_t is given by

$$Y_t = Y_0 + \int_0^t h_s ds + N_t, \quad (3.11)$$

where h_s is a progressively measurable process satisfying

$$\sup_t E[h_t^2] < \infty \quad (3.12)$$

and N_t is a $(\mathcal{F}_t)$-Brownian motion with covariance $\int_0^t R(s)ds$. We assume through Sections 3.1–3.2 that W_t is nondegenerate, i.e, $R(s)$ is invertible and the inverse $R(t)^{-1}$ is continuous with respect to t.

Let $f \in \mathbf{D}(S)$ be a function satisfying integrability conditions.

$$\sup_t E[f(X_t)^2(1+h_t^2)] < \infty, \quad \sup_t E[|A_t f|^2] < \infty. \qquad (3.13)$$

Then $M_t f$ defined by (3.1) is a square integrable $(\mathcal{F}_t)$-martingale. The Brownian motion $N_t = (N_t^1, \ldots, N_t^e)$ and the martingale $M_t f$ are not necessarily orthogonal. Let $\langle N^k, Mf \rangle_t$ be its quadratic covariation.[2] It is absolutely continuous with respect to $\langle N^k \rangle_t$, since $\int 1_A(s) d\langle N^k \rangle_s = 0$ implies $\left| \int_0^t 1_A(s) d\langle N^k, Mf \rangle_s \right| = 0$ by Schwarz's inequality for covariation. We may take its Radon–Nikodym derivative g_s as progressively measurable. Then we have

$$\left| \int_0^t g_s^2 d\langle N^k \rangle_s \right| = \left| \int_0^t g_s d\langle N^k, Mf \rangle_s \right| \le \left(\int_0^t g_s^2 d\langle N^k \rangle_s \right)^{1/2} \langle Mf \rangle_t^{1/2}.$$

This implies $\left(\int_0^t g_s^2 d\langle N^k \rangle_s \right)^{1/2} \le \langle Mf \rangle_t^{1/2}$. Since $E[\langle Mf \rangle_t] < \infty$, we have $E\left[\int_0^t g_s^2 d\langle N^k \rangle_s \right] < \infty$. Now set $D_s^k f := g_s r^{kk}(s)$, where $R(s) = (r^{ij}(s))$. It satisfies $\langle N^k, Mf \rangle_t = \int_0^t D_s^k f ds$. Therefore we have the following Lemma.

Lemma 3.1 *For any $f \in \mathbf{D}(S)$ satisfying (3.13), there exists an $(\mathcal{F}_t)$-progressively measurable e-dimensional process $D_t f = (D_t^1 f, \ldots, D_t^e f)$ such that $E\left[\int_0^t |D_s f|^2 ds \right] < \infty$ and*

$$\langle N^k, Mf \rangle_t = \int_0^t D_s^k f ds, \quad k = 1, \ldots, e. \qquad (3.14)$$

Let $(\mathcal{G}_t)$ be the filtration generated by the observation process Y_t. We denote the conditional expectation $E[Z|\mathcal{G}_t]$ by $E_t(Z)$.

Theorem 3.1 *(Fujisaki–Kallianpur–Kunita [5]). Let X_t be a semimartingale system process with state space S. Then for any $f \in \mathbf{D}(S)$ satisfying the integrability condition (3.13), conditional expectations $E_t(f(X_t))$ satisfy the following stochastic differential equation.*

$$E_t(f(X_t)) = E_0(f(X_0)) + \int_0^t E_s(A_s f) ds \qquad (3.15)$$

$$+ \int_0^t \left(E_s(h_s f(X_s)) - E_s(h_s) E_s(f(X_s)) + E_s(D_s f), R(s)^{-1} dI_s \right),$$

where $I_t = Y_t - Y_0 - \int_0^t E_s(h_s) ds$.

In particular, if the system process X_t and the noise process N_t are independent, the term $E_s(D_s f)$ in the above equation is equal to 0.

Proof. Conditions (3.12) and (3.13) ensure that the stochastic integral

$$M_t^* = \int_0^t (E_s(h_s f(X_s)) - E_s(h_s) E_s(f(X_s)) + E_s(D_s f), R(s)^{-1} d I_s) \qquad (3.16)$$

is well defined as a square integrable $(\mathcal{G}_t)$-martingale. We set

$$M_t = E_t(f(X_t)) - E_0(f(X_0)) - \int_0^t E_s(A_s f) ds. \qquad (3.17)$$

It is also a square integrable $(\mathcal{G}_t)$-martingale. The theorem will be established if we can show $M_t = M_t^*$. Now let $\mathcal{M}^2$ be the space of all continuous square integrable $(\mathcal{G}_t)$-martingales Z_t with $Z_0 = 0$. Then every $Z \in \mathcal{M}^2$ is represented by $Z_t = \int_0^t (g_s, d I_s)$ by Theorem 2.12 in Chapter 2. Therefore if we can show the equality $E[Z_t M_t] = E[Z_t M_t^*]$ for all $Z_t = \int_0^t (g_s, d I_s)$, we will get $M_t = M_t^*$.

Now, we shall compute $E[Z_t M_t]$. M_t is decomposed as $M_t = M_t^1 - M_t^2$, where $M_t^1 = f(X_t) - \int_0^t E_s(A_s f) ds - E_0(f(X_0))$, $M_t^2 = f(X_t) - E_t(f(X_t))$. Then M_t^1 can be regarded as an $(\mathcal{F}_t)$-semimartingale. As to the second term, we have $E[Z_t M_t^2] = 0$. Therefore we have $E[Z_t M_t] = E[Z_t M_t^1]$. Since I_t is written by $N_t + \int_0^t (h_s - E_s(h_s)) ds$, Z_t can be regarded as an $(\mathcal{F}_t)$-semimartingale decomposed as

$$Z_t = \int_0^t (g_s, d N_s) + \int_0^t (g_s, h_s - E_s(h_s)) ds = \int_0^t (g_s, d N_s) + \psi_t.$$

Then applying Itô's formula, we have

$$Z_t M_t^1 = \text{an } (\mathcal{F}_t)\text{-martingale with mean } 0 \qquad (3.18)$$
$$+ \int_0^t M_s^1 d\psi_s + \int_0^t Z_s d\varphi_s^1 + \langle \int (g_s, d N_s), Mf \rangle_t,$$

where φ_t^1 is the bounded variation part of M_t^1, given by $\int_0^t (A_s f - E_s(A_s f)) ds$. Take the expectation of each term of the right-hand side of the above. The expectation of the first term is 0, since it is a martingale with mean 0. The second term is written by

$$\int_0^t (g_s, h_s - E_s(h_s)) f(X_s) ds$$
$$- \int_0^t (g_s, h_s - E_s(h_s)) \left[\int_0^s E_u(A_u f) du + E_0[f(X_0)] \right] ds.$$

Consequently, we have

$$E \left[\int_0^t M_s^1 d\psi_s \right] = \int_0^t E[(g_s, E_s(h_s f(X_s)) - E_s(h_s) E_s(f(X_s)))] ds.$$

The expectation of the third term of (3.18) is

$$E\left[\int_0^t Z_s d\varphi_s^1\right] = \int_0^t E[Z_s(A_s f - E_s(A_s f))]ds = 0.$$

Since $\langle \int_0^t (g_s, dN_s), M_t f \rangle = \sum_i \int_0^t g_s^i D_s^i f ds$, we have

$$E\left[\langle \int_0^t (g_s, dN_s), M_t f \rangle\right] = \int_0^t E[(g_s, D_s f)]ds = \int_0^t E[(g_s, E_s(D_s f))]ds.$$

Therefore we get from (3.18)

$$E[Z_t M_t^1] = \int_0^t E[(g_s, E_s(h_s f(X_s)) - E_s(h_s) E_s(f(X_s)) + E_s(D_s f))]ds$$
$$= E[Z_t M_t^*].$$

Since $E[Z_t M_t^2] = 0$, we get $E[Z_t M_t] = E[Z_t M_t^*]$. This proves $M_t = M_t^*$. The proof is complete. □

We shall consider the case where the system-observation process (X_t, Y_t) is a solution of equations (3.5) and (3.6). Then the functional $D_t^k f$ has more definite meaning.

Theorem 3.2 *Assume that the system-observation process (X_t, Y_t) is determined by (3.5) and (3.6). Then $D_t^k f$ of Lemma 3.1 are first order partial differential operators*

$$D_t^k f = \sum_i d^{ik}(t, \mathbf{X}, \mathbf{Y}) f_{x_i}(X_t), \quad k = 1, \ldots, e, \quad (3.19)$$

where $d(t, \mathbf{x}, \mathbf{y}) = \sigma(t, \mathbf{x}, \mathbf{y}) S(t)$. Further, coefficients d satisfy the inequality

$$(a^{ij}(t, \mathbf{x}, \mathbf{y})) \geq d(t, \mathbf{x}, \mathbf{y}) R(t)^{-1} d(t, \mathbf{x}, \mathbf{y})^T, \quad \forall t, \mathbf{x}, \mathbf{y} \quad (3.20)$$

with respect to the positive definite order.

Proof. By Itô's formula (3.7), we have the formula

$$M_t f = \sum_{i=1}^d \sum_{j=1}^r \int_s^t f_{x_i}(X_u) \sigma^{ij}(u, \mathbf{X}, \mathbf{Y}) dW_u^j.$$

Therefore,

$$\int_0^t D_s^k f ds = \langle N^k, Mf \rangle_t = \sum_{i=1}^d \sum_{j=1}^r \int_0^t f_{x_i}(X_u) \sigma^{ij}(u, \mathbf{X}, \mathbf{Y}) d\langle N^k, W^j \rangle_u.$$

Note $\langle N^k, W^j \rangle_t = \int_0^t s^{jk}(u)du$. Then we get the equality (3.19).

Now let $N_t^* := \int_0^t S(s) R(s)^{-1} d N_s$ and $W_t^* := W_t - N_t^*$. Then

$$\langle W^*, N^T \rangle_t = \langle W, N^T \rangle_t - \langle \int_0^t S(s) R(s)^{-1} d N_s, N^T \rangle_t$$

$$= \int_0^t S(s) ds - \int_0^t S(s) R(s)^{-1} d \langle N, N^T \rangle_s = 0.$$

Therefore W_t^* and N_t are independent Brownian motions. Let $T(t)$ be the covariance of W_t^*, i.e., $E[W_t^*(W_t^*)^T] = \int_0^t T(s) ds$. Then we have $T(s) = Q(s) - S(s) R(s)^{-1} S(s)^T$. Therefore we get

$$(a^{ij}) = \sigma Q \sigma^T = \sigma T \sigma^T + d R^{-1} d^T.$$

This implies the inequality (3.20). □

3.1.4 Stochastic differential equation for unnormalized conditional expectations
Set

$$\Phi_t = \exp\left\{ -\int_0^t (h_s, R(s)^{-1} d N_s) - \frac{1}{2} \int_0^t (h_s, R(s)^{-1} h_s) ds \right\}. \quad (3.21)$$

It is a positive local martingale. If it is a martingale, we may define a new probability $\mathbf{Q}$ by $\mathbf{Q} = \Phi_T \cdot \mathbf{P}$. Then we have a Bayes formula

$$\pi_t(E) = \frac{E_\mathbf{Q}[\Psi_t 1_E(X_t) | \mathcal{G}_t]}{E_\mathbf{Q}[\Psi_t | \mathcal{G}_t]}, \quad \text{a.s. } \mathbf{P}$$

where $\Psi_t = \Phi_t^{-1}$. It is written by

$$\Psi_t = \exp\left\{ \int_0^t (h_s, R(s)^{-1} d Y_s) - \frac{1}{2} \int_0^t (h_s, R(s)^{-1} h_s) ds \right\}. \quad (3.22)$$

Let X be a random variable such that $E_\mathbf{P}[|X|] < \infty$. Then $E_\mathbf{Q}[\Psi_t |X| | \mathcal{G}_t] = E_\mathbf{P}[|X| | \mathcal{G}_t] E_\mathbf{Q}[\Psi_t | \mathcal{G}_t] < \infty$ a.s. Hence the unnormalized conditional expectation $\tilde{E}_t[X] = E_\mathbf{Q}[\Psi_t X | \mathcal{G}_t]$ is well defined a.s.

Theorem 3.3 *Assume that Φ_t of (3.21) is a martingale. Then for any $f \in \mathbf{D}(S)$ satisfying the integrability condition (3.13), $\tilde{E}_t(f(X_t))$ satisfies the linear stochastic differential equation.*

$$\tilde{E}_t(f(X_t)) = \tilde{E}_0(f(X_0)) + \int_0^t \tilde{E}_s(A_s f) ds$$
$$+ \int_0^t \left(\tilde{E}_s(h_s f(X_s)) + \tilde{E}_s(D_s f), R(s)^{-1} d Y_s \right). \quad (3.23)$$

Proof. We shall first show the equality

$$\tilde{E}_t(1) = \exp\left\{ \int_0^t (\hat{h}_s, R(s)^{-1} d Y_s) - \frac{1}{2} \int_0^t (\hat{h}_s, R(s)^{-1} \hat{h}_s) ds \right\}, \text{ where } \hat{h}_s = E_s(h_s)$$
(3.24)

Let $\hat{\Psi}_t$ be the Radon–Nikodym density $\frac{d\mathbf{P}}{d\mathbf{Q}}|\mathcal{G}_t$. Then we have $\hat{\Psi}_t = E_\mathbf{Q}[\Psi_t|\mathcal{G}_t] = \tilde{E}_t(1)$. In view of Girsanov's theorem, Y_t is a Brownian motion with respect to $\mathbf{Q}$. Further, any $(\mathcal{G}_t, \mathbf{Q})$ martingale is continuous and it is represented as the stochastic integral based on the Brownian motion Y_t by Theorem 3.7 in Chapter 2. Since $\hat{\Psi}_0 = 1$ a.s., the positive $(\mathcal{G}_t, \mathbf{Q})$-martingale $\hat{\Psi}_t$ is represented by

$$\hat{\Psi}_t = \exp\left\{\int_0^t (f_s, R(s)^{-1}dY_s) - \frac{1}{2}\int_0^t (f_s, R(s)^{-1}f_s)ds\right\}.$$

We want to show that $f_s = \hat{h}_s$ holds a.e. $dtd\mathbf{P}$. Since $d\mathbf{P} = \hat{\Psi}_T d\mathbf{Q}$ and Y_t is a $(\mathcal{G}_t, \mathbf{Q})$-Brownian motion, the process $Y_t - \int_0^t f_s ds$ is a $(\mathcal{G}_t, \mathbf{P})$-Brownian motion by Girsanov's theorem. On the other hand, since $I_t = Y_t - \int_0^t \hat{h}_s ds$ is a $(\mathcal{G}_t, \mathbf{P})$-Brownian motion, we have $\int_0^t f_s = \int_0^t \hat{h}_s ds$ for any t. Consequently we get formula (3.24).

Next we shall compute $\tilde{E}_t(f(X_t))$. It is equal to $E_t(f(X_t))\hat{\Psi}_t$. By Itô's formula,

$$d(E_t(f(X_t))\hat{\Psi}_t) = \hat{\Psi}_t d E_t(f(X_t)) + E_t(f(X_t))d\hat{\Psi}_t + d\langle \hat{\Psi}_t, E_t(f(X_t))\rangle.$$

Since $E_t(f(X_t))$ satisfies (3.15), we have

$$\hat{\Psi}_t d E_t(f(X_t)) = E_t(A_t f)\hat{\Psi}_t dt$$
$$+(\{E_t(h_t f(X_t)) - \hat{h}_t E_t(f(X_t)) + E_t(D_t f)\}\hat{\Psi}_t, R(t)^{-1}[dY_t - \hat{h}_t dt]),$$

$$E_t(f(X_t))d\hat{\Psi}_t = (\hat{h}_t E_t(f(X_t))\hat{\Psi}_t, R(t)^{-1}dY_t),$$

$$d\langle \hat{\Psi}_t, E_t(f(X_t))\rangle = (R(t)^{-1}\hat{h}_t, \{E_t(h_t f(X_t)) - \hat{h}_t E_t(f(X_t)) + E_t(D_t f)\}\hat{\Psi}_t)dt.$$

Therefore we see that $\tilde{E}_t(f(X_t))$ satisfies

$$d\tilde{E}_t(f(X_t)) = \tilde{E}(A_t f)dt + (\tilde{E}(h_t f(X_t)) + \tilde{E}_t(D_t f), R(t)^{-1}dY_t),$$

proving equation (3.23).

□

3.2 Measure-valued SPDE: Markovian case and conditionally Gaussian case

3.2.1 Introduction

In this section we will apply SDE for (unnormalized) conditional distributions to the case where the system process or system-observation process is a Markov process or a conditionally Gaussian process.

If the system process is a Markov process, stochastic differential equations obtained in the previous section have more definite meaning. It may be regarded as second-order stochastic partial differential equations for the measure-valued process π_t and ρ_t, respectively. We will see that equation for π_t

is a nonlinear equation and equation for ρ_t is a linear equation. If the measure-valued processes have smooth densities a.s., the density process $\pi_t(x)$ of $\pi_t(dx)$ satisfies Kushner–Stratonovitch equation and the density process $\rho_t(x)$ of $\rho_t(dx)$ satisfies Zakai equation (Section 3.2.3). The existence and the smoothness of these density processes will be discussed in Section 3.3.

In Section 3.2.4, we shall consider the case where the pair of system-observation process is Markovian. The filtering equation for π_t is closely related to the one studied by Liptser–Shiryaev [17]. In the last subsection, we consider the case where the pair of system-observation process is conditionally Gaussian, i.e., the conditional distributions π_t are Gaussian. We will derive Liptzer–Shiryaev's algorithm for computing Gaussian distributions π_t. Kalman–Bucy's algorithm for linear filter will be derived from this. Our discussion is based on the uniqueness of the solution of the SPDE for the measure valued process ρ_t. The uniqueness of the solution will be shown in Section 3.3.

We will not discuss the interpolation and extrapolation of the system process. We refer the reader to [17], [22] for these subjects.

3.2.2 Case of Markovian system with independent noise

In this subsection we consider the case where the system process X_t is a Markov process with state space S. Let $h(t, x), t \in [0, T], x \in S$ be an $\mathbf{R}^e$-valued measurable function satisfying

$$\sup_{t \leq T} E[|h(t, X_t)|^2] < \infty. \tag{3.25}$$

We consider the observation process

$$Y_t = Y_0 + \int_0^t h(s, X_s)ds + N_t, \tag{3.26}$$

where X_t and N_t are independent.

Let P_ω^t be the regular version of conditional probability of $\mathbf{P}$ given $\mathcal{G}_t$. It satisfies (i) For each $B \in \mathcal{F}_t$, $P_\omega^t(B)$ is $\mathcal{G}_t$-measurable, and (ii) $\mathbf{P}(A \cap B) = \int_A P_\omega^t(B)d\mathbf{P}(\omega)$ holds for any $A \in \mathcal{G}_t$ and $B \in \mathcal{F}$. (For the existence of the regular conditional distribution, see Stroock–Varadhan [23]). Set $\pi_t(E)(\omega) = P_\omega^t(X_t \in E)$. We may regard that π_t is a stochastic process with values in $\mathcal{M}_1$, where $\mathcal{M}_1$ is the set of all probability measures on S. The conditional expectation $E_t(f(X_t))$ is then equal to $E_t(f(X_t)) := \int f(x)\pi_t(dx)$ a.s.

The system-observation process (X_t, Y_t) satisfies Condition (A) in Chapter 2. Therefore the unnormalized conditional distribution $\rho_t(E) := E_\mathbf{Q}[\Psi_t 1_{X_t \in E}|\mathcal{G}_t]$ is well defined. It may be regarded as a stochastic process with values in $\mathcal{M}$, where $\mathcal{M}$ is the set of all finite measures on S.

The following theorem is a consequence of Theorems 3.1 and 3.3.

Theorem 3.4 *Assume that the function $h(t, x)$ satisfies (3.25).*

(1) π_t satisfies the following nonlinear stochastic partial differential equation

$$\pi_t(f) = \pi_0(f) + \int_0^t \pi_s(A(s)f)ds \qquad (3.27)$$
$$+ \int_0^t (\pi_s(h(s)f) - \pi_s(h(s))\pi_s(f), R(s)^{-1}[dY_s - \pi_s(h(s))ds]),$$

for any $f \in \mathbf{D}(S)$ satisfying (3.13).

(2) ρ_t satisfies the following linear stochastic partial differential equation

$$\rho_t(f) = \pi_0(f) + \int_0^t \rho_s(A(s)f)ds + \int_0^t (\rho_s(h(s)f), R(s)^{-1}dY_s), \qquad (3.28)$$

for any $f \in \mathbf{D}(S)$ satisfying (3.13).

Remark. The formulas (3.27) and (3.28) are recursive. (3.27) implies

$$\pi_{t+h}(f) = \pi_t(f) + \int_t^{t+h} \pi_s(A(s)f)ds$$
$$+ \int_t^{t+h} (\pi_s(h(s)f) - \pi_s(h(s))\pi_s(f), R(s)^{-1}[dY_s - \pi_s(h(s))ds]).$$

Therefore if π_t is computed at time t, we can compute π_{t+h} by using the observation data $\{Y_s; t \leq s \leq t+h\}$, forgetting the past observation data $\{Y_s; 0 \leq s \leq t\}$.

We will consider the case where the system process is a Markov chain. Let S be a finite or countable set with discrete topology and let X_t be a Markov chain with transition probabilities $p_{s,t}(i, j)$, $s < t$, $i, j \in S$. Suppose that limits

$$\lim_{t \to s} \frac{p_{s,t}(i, j) - \delta_{i,j}}{t - s} = q_s(i, j) \qquad (3.29)$$

exists and is bounded for any s, i, j, where $\delta_{i,j}$ is the Kronecker's delta. For $f \in C(S)$, set $P_{s,t}f(i) = \sum_j p_{s,t}(i, j)f(j)$ and $A(s)f(i) = \sum_j q_s(i, j)f(j)$. Let $\mathbf{D}(S)$ be the set of real functions with compact supports. Then $A(s)$ with the domain $\mathbf{D}(S)$ is the generator of the Markov chain. Then the filter $\pi_t(i) = \mathbf{P}(x_t = i | \mathcal{G}_t)$ satisfies the following stochastic differential equation

$$\pi_t(i) = \pi_0(i) + \sum_j \int_0^t \pi_s(j) q_s(i, j)ds + \qquad (3.30)$$

$$\int_0^t (\pi_s(i)h(s, i) - [\sum_j \pi_s(j)h(s, j)]\pi_s(i), R(s)^{-1}[dY_s - (\sum_j \pi_s(j)h(s, j))ds]).$$

3.2.3 Case of Markovian system with correlated noise

We introduce some classes of functions on $\mathbf{R}^d$. For a given integer n, we set $\varphi_n(x) = (1 + |x|)^n$. A continuous function $f(t, x)$ with parameter t is said to be *dominated by* φ_n if there exists a positive constant C_n such that $|f(t, x)| \leq C_n \varphi_n(x)$ holds for any t, x. A function f is dominated by φ_0 if and only if f is a bounded function. If f is dominated by φ_n for some positive n, f is said to be a *slowly increasing function*. If f is dominated by $\varphi_n(x)$ for all negative n, f is said to be a *rapidly decreasing function*. Similarly a function $f(t, x)$ of $C^{0,m}$-class is said to be slowly increasing (or rapidly decreasing) if its derivatives are all slowly increasing (or rapidly decreasing).

We shall consider the case where the system process X_t is a diffusion process on $\mathbf{R}^d$ determined by a stochastic differential equation.

Let $b(t, x)$ and $\sigma(t, x)$ be d vector function and $d \times r$-matrix function respectively. We assume that these are locally Lipschitz continuous and are of linear growth with respect to x. Consider the SDE

$$X_t = X_0 + \int_0^t b(s, X_s)ds + \int_0^t \sigma(s, X_s)dW_s. \tag{3.31}$$

The equation has a unique solution for any initial condition. The solution is a Markov process with the generator (3.10).

We have seen in Theorem 2.7 in Chapter 2 that if f is dominated by $\varphi_n(x)$, the function $P_{s,t}f(x)$ (integral) is well defined and is dominated by the same φ_n. Hence if f is a slowly increasing (or rapidly decreasing) function, $P_{s,t}f$ is also a slowly increasing (or rapidly decreasing) function. Let $A(t)$ be the operator of (3.10). If f is slowly increasing, the function $A(s)f$ is slowly increasing continuous function.

We return to the filtering problem. The system process X_t is determined by (3.31) and the observation process is determined by (3.26), where (W_t, N_t) is a Brownian motion with $E[W_t W_t^T] = \int_0^t Q(s)ds$, $E[N_t N_t^T] = \int_0^t R(s)ds$ and $E[W_t N_t^T] = \int_0^t S(s)ds$.

The integrability conditions for h and f can be replaced by more explicit conditions.

Theorem 3.5 *(1) Assume that the function $h(t, x)$ is dominated by φ_n for some $n \geq 0$. If the initial law π_0 has finite $2n$-moment, the measure-valued process π_t has finite $2n$-moment for any t a.s. It satisfies the stochastic partial differential equation*

$$\pi_t(f) = \pi_0(f) + \int_0^t \pi_s(A(s)f)ds \tag{3.32}$$
$$+ \int_0^t (\pi_s(B(s)f) - \pi_s(h(s))\pi_s(f), R(s)^{-1}[dY_s - \pi_s(h(s))ds]),$$

for any C^2-function f such that f, $B(s)f$, $A(s)f$ are all dominated by φ_n. Here $A(s)$ is the second-order partial differential operator of (3.10) and $B(s) = (B^1(s), \ldots, B^e(s))$ are first-order partial differential operators given by

$$B^k(s)f = \sum_{i=1}^d d^{ik}(s, x) f_{x_i} + h^k(s, x) f, \tag{3.33}$$

where $d(s, x) = (d^{jk}(s, x)) = \sigma(s, x) S(s)$ and it satisfies the inequality

$$(a^{ij}(t, x)) \geq d(t, x) R(t)^{-1} d(t, x)^T, \quad \forall t, x \tag{3.34}$$

with respect to the positive definite order.

(2) Assume that $d(t, x)$ is bounded and $h(t, x)$ is locally Lipschitz continuous and of linear growth with respect to x. Then the measure-valued process ρ_t is well defined. If the initial law π_0 has finite $2n$-moment for some $n \geq 0$, ρ_t has finite $2n$-moment for any t a.s. It satisfies the equation

$$\rho_t(f) = \pi_0(f) + \int_0^t \rho_s(A(s)f) ds + \int_0^t (\rho_s(B(s)f), R(s)^{-1} dY_s), \tag{3.35}$$

for any C^2-function f such that f, $B(s)f$, $A(s)f$ are all dominated by φ_n.

Proof. We first consider the assertion (1). Since the function h is dominated by φ_n, the function $P_{0,t}(h^2)$ is dominated by φ_{2n}. This implies $E[h(t, X_t)^2] \leq C_{2n} \int \pi_0(dx) \varphi_{2n}(x) < \infty$. Therefore integrability condition (3.25) is satisfied. We have also $E[\pi_t(\varphi_{2n})] = E[\varphi_{2n}(X_t)] < \infty$ so that π_t has finite $2n$-moment for any t a.s. Further since f and $A(t)f$ are dominated by φ_n, f satisfies (3.13). Therefore we get equation (3.32) from Theorem 3.1.

Next, if d is bounded and h is of linear growth, the local martingale Φ_t of (3.21) is a martingale by Theorem 2.11 in Chapter 2. Then ρ_t is well defined and satisfies (3.35) by Theorem 3.3. □

Now assume that the initial law $\pi_0(dx)$ has a continuous density $\pi_0(x)$. If there exists a continuous random function $\pi_t(x)$ (or random field) of the class $C^{0,2}$-class such that $\pi_t(dx) = \pi_t(x) dx$ holds for any t a.s., then $\pi_t(x)$ should satisfy the following second-order nonlinear stochastic partial differential equation.

$$\pi_t(x) = \pi_0(x) + \int_0^t A(s)^* \pi_s(x) ds + \int_0^t \Big(B^*(s) \pi_s(x) \tag{3.36}$$
$$- \Big[\int h(s, x') \pi_s(x') dx' \Big] \pi_s(x), R(s)^{-1} \Big[dY_s - \Big(\int h(s, x') \pi_s(x') dx' \Big) ds \Big] \Big).$$

Here $A(s)^*$ is the adjoint of $A(s)$ and $B(s)^*$ is the adjoint of $B(s)$. These are given by

$$A(s)^* f(x) = -\sum_i \frac{\partial}{\partial x_i}(b^i(s,x)f(x)) + \frac{1}{2}\sum_{i,j}\frac{\partial^2}{\partial x_i \partial x_j}(a^{ij}(s,x)f(x)),$$

$$B^k(s)^* f(x) = -\sum_i \frac{\partial}{\partial x^i}(d^{ik}(t,x)f(x)) + h^k(s,x)f(x).$$

Equation (3.36) is called the *Kushner–Stratonovitch equation* ([16]).

If there exists a continuous random function $\rho_t(x)$ of the class $C^{0,2}$ such that $\rho_t(dx) = \rho_t(x)dx$ holds for any t a.s., it satisfies the following linear stochastic partial differential equation.

$$\rho_t(x) = \pi_0(x) + \int_0^t A(s)^* \rho_s(x) ds + \int_0^t (B(s)^* \rho_s(x), R(s)^{-1} dY_s). \tag{3.37}$$

Equation is called *Zakai equation* ([26]). Note that if $\rho_t(x)$ is a solution of Zakai equation, then $\pi_t(x) = \rho_t(x)/\int_{\mathbf{R}^d} \rho_t(x)dx$ is a solution of Kushner–Stratonovitch equation. Therefore the existence and the uniqueness of the solution of Kushner–Stratonovitch equation are reduced to those of Zakai equation.

If the second order partial differential operator $A(t)$ is uniformly elliptic or hypoelliptic, the existence of a smooth density $\rho_t(x)$ is shown by using Malliavin calculus. For details, see Rozovskiĭ's book [22]. We will discuss the existence of the density process in Section 3.3.2, under a certain smoothness condition of $\pi_0(x)$. See Theorem 3.12.

So far we derived SPDE's for conditional distributions π_t and unnormalized conditional distributions ρ_t. In order to compute filtering distributions, we have to show the uniqueness of solutions of these SPDE's. The problem will be discussed in Section 3.3.

3.2.4 Case of Markovian system-observation process

We shall consider the filtering problem in the case where the system-observation process is given by a stochastic differential equation

$$X_t = X_0 + \int_0^t b(s, X_s, Y_s)ds + \int_0^t \sigma(s, X_s, Y_s)dW_s, \tag{3.38}$$

$$Y_t = Y_0 + \int_0^t h(s, X_s, Y_s)ds + \int_0^t \Sigma(s, Y_s)dN_s, \tag{3.39}$$

where coefficients of the equation are continuous in (t, x, y), locally Lipschitz continuous with respect to (x, y) and are of linear growth with respect to (x, y). We assume that $\Sigma(s, y)$ is invertible and the inverse is bounded with respect to (s, y). The pair (W_t, N_t) is a Brownian motion introduced in Section 3.2.3.

We will first show that the filtering problem based on the above observation process could be reduced to the filtering problem based on another observation process defined by

$$Y'_t = Y_0 + \int_0^t \int_0^t h'_s ds + N_t, \tag{3.40}$$

where $h'_s = \Sigma(s, Y_s)^{-1} h(s, X_s, Y_s)$. Set $\mathcal{G}_t = \sigma(Y_s; s \leq t)$, $\mathcal{G}'_t = \sigma(Y'_s; s \leq t)$. Then we have

Lemma 3.2 ([5]) *It holds $\mathcal{G}_t = \mathcal{G}'_t$ for any t.*

Proof. Note that Y_t and Y'_t are related by

$$Y_t = Y_0 + \int_0^t \Sigma(s, Y_s) dY'_s, \tag{3.41}$$

$$Y'_t = Y_0 + \int_0^t \Sigma(s, Y_s)^{-1} dY_s. \tag{3.42}$$

Since $\Sigma(s, Y_s)^{-1}$ is $\mathcal{G}_s$-measurable, it is clear from equation (3.42) that $\mathcal{G}'_t \subset \mathcal{G}_t$. Now let Φ'_t be the exponential local martingale, which is given by (3.21), replacing hs, Y_t by $h's$, Y'_t, respectively. It is a positive martingale since h' satisfies the linear growth condition (Theorem 2.8 in Chapter 2). Set $\mathbf{Q}' = \Phi'_T \cdot \mathbf{P}$. Then Y'_t is a Brownian motion with respect $\mathbf{Q}'$ (Girsanov's theorem). Then equation (3.41) may be considered as a SDE for the process Y_t based on the Brownian motion Y'_t. The coefficient $\Sigma(t, y)$ is locally Lipschitz continuous with respect to y. Then the equation has a path wise unique solution and it is $\mathcal{G}'_t$-measurable. Therefore we get $\mathcal{G}_t \subset \mathcal{G}'_t$, with respect to $\mathbf{Q}'$. Since $\mathbf{P}$ and $\mathbf{Q}'$ are equivalent, the inclusion relation is valid with respect to $\mathbf{P}$. Thus we have proved the assertion of the lemma. □

In view of the above lemma, the filtering problem for the system-observation process (X_t, Y_t) is reduced to that for the system-observation process (X_t, Y'_t).

Let π_t be the conditional measure based on the observation data $\mathcal{G}_t$. For a measurable function $\varphi(t, x, \omega)$, we set $\pi_t(\varphi(t))(\omega) = \int \varphi(t, x, \omega) \pi_t(dx)(\omega)$. Then in view of Theorem 3.1, it satisfies equation

$$\pi_t(f) - \pi_0(f) - \int_0^t \pi_s(A(s)f) ds \tag{3.43}$$

$$= \int_0^t (\pi_s(B'(s)f) - \pi_s(h'(s)) \pi_s(f), R(s)^{-1}[dY'_s - \pi_s(h'(s)) ds]),$$

where $A(s)$ and $B'(s)$ are partial differential operators with random coefficients:

$$A(s)f(x) = \sum_{i=1}^d b^i(s, x, Y_s) f_{x_i}(x) + \frac{1}{2} \sum_{i,j=1}^d a^{ij}(s, x, Y_s) f_{x_i x_j}(x), \tag{3.44}$$

$$B'(s)f(x) = \Sigma(s, Y_s)^{-1} h(s, x, Y_s) f(x) + f_x(x)^T \sigma(s, x, Y_s) S(s),$$

where $f_x^T = (f_{x_1}, \ldots, f_{x_d})$. Set $B(s)f = \Sigma(s, Y_s) B'(s)f$. Then

$$B^k(s)f(x) = h^k(s, x, Y_s)f(x) + \sum_i d^{ik}(s, x, Y_s) f_{x_i}(x), \tag{3.45}$$

$$(d^{ij}(s, x, y)) = \sigma(s, x, y) S(s) \Sigma(s, y)^T.$$

The right-hand side of (3.43) is written as

$$\int_0^t (\Sigma(s)^{-1}(\pi_s(B(s)f) - \pi_s(h(s))\pi_s(f)), R(s)^{-1}\Sigma(s)^{-1}[dY_s - \pi_s(h(s))ds]).$$

Consequently, we obtain

$$\pi_t(f) = \pi_0(f) + \int_0^t \pi_s(A(s)f) ds \tag{3.46}$$

$$+ \int_0^t (\pi_s(B(s)f) - \pi_s(h(s))\pi_s(f), (\Sigma(s) R(s) \Sigma(s)^T)^{-1}[dY_s - \pi_s(h(s))ds]).$$

Further, the unnormalized conditional distributions ρ_t satisfy

$$\rho_t(f) = \pi_0(f) + \int_0^t \rho_s(A(s)f) ds \tag{3.47}$$

$$+ \int_0^t (\rho_s(B(s)f), (\Sigma(s) R(s) \Sigma(s)^T)^{-1} dY_s).$$

Theorem 3.6 (*Liptzer–Shiryaev* [17][3]) *Let (X_t, Y_t) be the pair of system-observation process determined by equations (3.38) and (3.39). Suppose that the initial law π_0 has finite 2n-moment. Then conditional distributions π_t have finite 2n-moments for any t a.s. It satisfy (3.46) for any C^2-function f such that $f(x)$, $B^k f(x)$, $A(s) f(x)$ are dominated by $(1 + |x| + |Y_s|)^n$. Here, $A(s)$ is the random second order operator given by (3.44) and $B^k(s)$ are random first order operators given by (3.45). Their coefficients satisfy the inequality*

$$(a^{ij}(t, x, y)) \geq d(t, x, y) R(t)^{-1} d(t, x, y)^T \quad \forall t, x, y$$

with respect to the positive definite order.

Furthermore, if d is bounded, ρ_t satisfies (3.47) for functions f mentioned above.

Remark. We can extend filtering equations (3.46) and (3.47) to the case of system-observation process

$$X_t = X_0 + \int_0^t b(s, X_s, \mathbf{Y}) ds + \int_0^t \sigma(s, X_s, \mathbf{Y}) dW_s,$$

$$Y_t = Y_0 + \int_0^t h(s, X_s, \mathbf{Y}) ds + \int_0^t \Sigma(s, \mathbf{Y}) dN_s,$$

where coefficients of the equation are functions of $(t, x, y) \in [0, T] \times \mathbf{R}^d \times \mathcal{C}^e$, progressively measurable, locally Lipschitz continuous with respect to (x, y) and are of linear growth with respect to (x, y). Define partial differential operators $A(s)$ and $B^k(s)$ with random coefficients by

$$A(s)f(x) = \sum_{i=1}^{d} b^i(s, x, \mathbf{Y}) f_{x_i}(x) + \frac{1}{2} \sum_{i,j=1}^{d} a^{ij}(s, x, \mathbf{Y}) f_{x_i x_j}(x), \qquad (3.48)$$

$$B^k(s)f(x) = h^k(s, x, \mathbf{Y}) f(x) + \sum_{i} d^{ik}(s, x, \mathbf{Y}) f_{x_i}. \qquad (3.49)$$

Then conditional distributions π_t and unnormalized ones ρ_t satisfies equations (3.46) and (3.47).

3.2.5 Cases of Gaussian and conditionally Gaussian processes

We shall consider the case where the system-observation process (X_t, Y_t) is a stochastic process determined by the equation

$$X_t = X_0 + \int_0^t A_0(s, \mathbf{Y}) ds + \int_0^t A_1(s, \mathbf{Y}) X_s ds + \int_0^t \sigma(s, \mathbf{Y}) dW_s, \qquad (3.50)$$

$$Y_t = Y_0 + \int_0^t H_0(s, \mathbf{Y}) ds + \int_0^t H_1(s, \mathbf{Y}) X_s ds + \int_0^t \Sigma(s, \mathbf{Y}) dN_s,$$

where (W_t, N_t) is a Brownian motion introduced in Section 3.2.3. Coefficients $A_0, A_1, H_0, H_1, \sigma, \Sigma$ are assumed to be progressively measurable and locally Lipschitz continuous with respect to y. Further, A_0, H_0 are of linear growth and A_1, H_1, σ, Σ are bounded with respect to y. Then the equation has a unique global solution.

A feature of the equation is that it is linear with respect to the system process X_t, though coefficients A_0, A_1, σ may depend nonlinearly on the observation process Y_t. If these coefficients are deterministic and the initial random variable (X_0, Y_0) is Gaussian and independent of the Brownian notion (W_t, N_t), the solution (X_t, Y_t) is a Gaussian process. The nonlinear filter of X_t based on the observation $Y_s; s \leq t$ coincides with the linear filter and we can get a Kalman–Bucy filter.

Liptzer–Shiryaev [17] showed that for the system-observation process (3.50), the conditional distributions π_t of the filtering are Gaussian distributions if the initial distribution π_0 is Gaussian, and obtained a stochastic differential equation for its mean process and an ordinary differential equations (with random coefficients) for its covariance. These are closed equations and can be solved completely. Equations are close to those of Kalman–Bucy filter.

We will give an alternative proof for these facts. We study unnormalized conditional laws ρ_t through the characteristic functions. For $\alpha = (\alpha_1, ..., \alpha_d) \in \mathbf{R}^d$, we set $\varphi_\alpha(x) = e^{i(\alpha, x)}$. Then the characteristic function of the unnormalized conditional distribution is given by $\rho_t(\varphi_a) = \int \varphi_a(x)\rho_t(dx)$. It is a complex valued stochastic process. Note that stochastic integrals based on a (real) Brownian motion for complex valued progressively measurable processes may be defined straightforwardly and equation (3.47) is valid for complex function f.

We will see in the next section that ρ_t is uniquely determined by equation (3.47) (Theorem 3.8). Further, the laws $\rho_t(dx)$ are uniquely determined by the characteristic functions. Then any characteristic functions $\rho'_t(\varphi_a)$ satisfying equation (3.47) should be the desired characteristic functions of the unnormalized conditional laws.

Now $A(s)\varphi_a$ of (3.48) and $B(s)\varphi_a$ of (3.49) are computed as

$$A(s)\varphi_a(x) = \{i(a, A_0(s) + A_1(s)x) - \frac{1}{2}(a, \sigma(s)Q(s)\sigma(s)^T a)\}\varphi_a(x),$$

$$B(s)\varphi_a(x) = \{H_0(s) + H_1(s)x + ia^T\sigma(s)S(s)\Sigma(s)^T\}\varphi_a(x),$$

respectively. Therefore, if ρ_t is a measure-valued process with finite $2n$-moment for some $n \geq 1$, satisfying equation (3.47), then $\rho_t(\varphi_a)$ satisfies

$$\rho_t(\varphi_a) = \rho_0(\varphi_a) + i(\int_0^t (\rho_s(\varphi_a)A_0(s) + A_1(s)\rho_s(x\varphi_a))ds, a) \quad (3.51)$$

$$-\frac{1}{2}((\int_0^t \rho_s(\varphi_a)\sigma(s)Q(s)\sigma(s)^T ds)a, a)$$

$$+i(\int_0^t \rho_s(\varphi_a)\sigma(s)S(s)\Sigma(s)^T(\Sigma R\Sigma^T)^{-1}dY_s, a)$$

$$+ \int_0^t (\rho_s(\varphi_a)H_0(s) + H_1(s)\rho_s(x\varphi_a), (\Sigma R\Sigma^T)^{-1}dY_s).$$

The function $v(t, a) := \rho_t(\varphi_a)$ is continuously differentiable with respect to a a.s. Since $\rho_s(x\varphi_a) = \frac{1}{i}\partial_a \rho_s(\varphi_a)$, $v(t, a)$ satisfies the following linear first-order stochastic partial differential equation.

$$v(t, a) = v(0, a) + (\int_0^t (iv(s, a)A_0(s) + A_1(s)\partial_a v(s, a))ds, a) \quad (3.52)$$

$$-\frac{1}{2}(\int_0^t v(s, a)\sigma(s)Q(s)\sigma(s)^T ds\, a, a)$$

$$+i(\int_0^t v(s, a)\sigma(s)S(s)\Sigma(s)^T(\Sigma R\Sigma^T)^{-1}dY_s, a)$$

$$+ \int_0^t (v(s, a)H_0(s) - iH_1(s)\partial_a v(s, a), (\Sigma R\Sigma^T)^{-1}dY_s).$$

Conversely if the characteristic function of a measure-valued process ρ_t with finite moment satisfies (3.52), it satisfies (3.51) and hence ρ_t satisfies (3.48). Consequently it coincides with the unnormalized conditional distributions.

We will show that a solution of the above equation is given by characteristic functions of Gaussian distributions, provided that the initial condition is a characteristic function of a Gaussian distribution.

Lemma 3.3 *Set*

$$v(t, a) = v(t, 0) \exp\left\{i(\hat{X}_t, a) - \frac{1}{2}(P(t)a, a)\right\}, \quad t > 0, \quad (3.53)$$

$$v(0, a) = \exp\left\{i(m, a) - \frac{1}{2}(\gamma a, a)\right\},$$

where m is a d-vector and γ is a symmetric nonnegative definite matrix. Then it is a solution of equation (3.52) if and only if the triple $\hat{X}_t$, $P(t)$, $v(t, 0)$ satisfy the following equations.

$$\hat{X}_t = m + \int_0^t (A_0 + A_1 \hat{X}_s) ds \quad (3.54)$$

$$+ \int_0^t (P(s) H_1^T + \sigma S \Sigma^T)(\Sigma R \Sigma^T)^{-1}[dY_s - (H_0 + H_1 \hat{X}_s) ds],$$

$$\frac{dP(t)}{dt} = \sigma Q \sigma^T + A_1 P(t) + P(t) A_1^T \quad (3.55)$$

$$- (P(t) H_1^T + \sigma S \Sigma^T)(\Sigma R \Sigma^T)^{-1} (P(t) H_1^T + \sigma S \Sigma^T)^T,$$

$$P(0) = \gamma,$$

$$v(t, 0) = \exp\left\{\int_0^t (H_0 + H_1 \hat{X}, (\Sigma R \Sigma^T)^{-1} dY_s)\right\} \times \quad (3.56)$$

$$\exp\left\{-\frac{1}{2} \int_0^t (H_0 + H_1 \hat{X}, (\Sigma R \Sigma^T)^{-1}(H_0 + H_1 \hat{X})) ds\right\}.$$

Here $\sigma = \sigma(t, \mathbf{Y})$, $\Sigma = \Sigma(t, \mathbf{Y})$, $A_0 = A_0(t, \mathbf{Y})$, $A_1 = A_1(t, \mathbf{Y})$, $H_0 = H_0(t, \mathbf{Y})$ and $H_1 = H_1(t, \mathbf{Y})$.

Proof. Assume first that $v(t, a)$ of (3.53) satisfies (3.52). Since $\partial_a v(t, a) = v(t, a)(i \hat{X}_t - P(t)a)$, we get from (3.52)

$$v(t, a) = v(0, a) + i \left(\int_0^t v(s, a)(A_0 + A_1 \hat{X}_s) ds, a \right)$$

$$+ i \left(\int_0^t v(s, a)(P(s) H_1^T + \sigma S \Sigma^T)(\Sigma R \Sigma^T)^{-1} dY_s, a \right)$$

$$- \left(\left(\int_0^t v(s, a)(\frac{1}{2} \sigma Q \sigma^T + A_1 P(s)) ds \right) a, a \right)$$

$$+ \int_0^t (v(s, a)(H_0 + H_1 \hat{X}_s), (\Sigma R \Sigma^T)^{-1} dY_s).$$

It may be regarded as a linear SDE for unknown $v(t, a)$. The solution is given by an exponential process (see the Remark after the proof of this lemma).

$$v(t, a) = v(0, a) \times$$
$$\exp\left\{i\left(\int_0^t (A_0 + A_1 \hat{X}_s)ds + \int_0^t (PH_1^T + \sigma S\Sigma^T)(\Sigma R\Sigma^T)^{-1}dY_s, a\right)\right\} \times$$
$$\exp\left\{-\frac{1}{2}\left(a, \left(\int_0^t (\sigma Q\sigma^T + 2A_1 P)ds\right)a\right)\right\} \times$$
$$\exp\left\{\int_0^t (H_0 + H_1\hat{X}, (\Sigma R\Sigma^T)^{-1}dY_s)\right\} \times$$
$$\exp\left\{-\frac{1}{2}\int_0^t (H_0 + H_1\hat{X} + ia^T(PH_1^T + \sigma S\Sigma^T), (\Sigma R\Sigma^T)^{-1} \times \right.$$
$$\left. \times (H_0 + H_1\hat{X} - ia^T(PH_1^T + \sigma S\Sigma^T)))ds\right\}.$$

This can be rewritten as

$$v(t, a) = v(0, a) \times$$
$$\exp\left\{i\left(\int_0^t (A_0 + A_1 \hat{X}_S)ds + \int_0^t (PH_1^T + \sigma S\Sigma^T)(\Sigma R\Sigma^T)^{-1}dY_S, a\right)\right\} \times$$
$$\exp\left\{-\frac{1}{2}\left(a, \left(\int_0^t (\sigma Q\sigma^T + 2A_1 P)ds\right)a\right)\right\} \times$$
$$\exp\left\{\int_0^t \left(H_0 + H_1\hat{X}, (\Sigma R\Sigma^T)^{-1}dY_S\right)\right\} \times$$
$$\exp\left\{-\frac{1}{2}\int_0^t \left(H_0 + H_1\hat{X} + ia^T(PH_1^T + \sigma S\Sigma^T), (\Sigma R\Sigma^T)^{-1} \times \right.\right.$$
$$\left.\left. \times \left(H_0 + H_1\hat{X} - ia^T\left(PH_1^T + \sigma S\Sigma^T\right)\right)\right)ds\right\}.$$

Consequently $\hat{X}_t$ should satisfy equation (3.54) and $P(t)$ should satisfy (3.55).

Conversely if $(\hat{X}_t, P(t))$ are solutions of equations (3.54) and (3.55) respectively and $v(t, 0)$ is given by (3.56), then $v(t, a)$ of (3.53) satisfies (3.52). Thus we have shown the lemma. □

Remark. If Z_t is a complex valued Itô process satisfying

$$Z_t = Z_0 + \int_0^t Z_s(\varphi_1(s), dX_s) + \int_0^t Z_s \varphi_2(s)ds,$$

where $\varphi_1 = (\varphi_1^1, ..., \varphi_1^d)$, φ_2 are complex valued progressively measurable processes with integrability conditions and X_t is a Brownian motion with covariance $\int_0^t \Gamma(s)ds$. Then it holds

$$Z_t = Z_0 \exp\left\{\int_0^t (\varphi_1(s), dX_s) + \int_0^t \varphi_2(s)ds - \frac{1}{2}\int_0^t (\varphi_1(s), \Gamma(s)\bar{\varphi}_1(s))ds\right\}.$$

Here (x, y) is a complex inner product such that $(x, y) = \sum_{j=1}^d x_j \bar{y}_j$ where $x = (x_1, ..., x_d)$ and $y = (y_1, ..., y_d)$ are complex vectors.

Theorem 3.7 *(Liptzer–Shiryaev [17]) Suppose that the system-observation process is given by (3.50). If the initial conditional law π_0 of X_0 with respect to $\sigma(Y_0)$ is Gaussian with mean 0 and covariance γ, then the conditional laws π_t of the filtering are Gaussian distributions. Its mean $\hat{X}_t = \int x \pi_t(dx)$ and covariance $P(t) = \int (x - \hat{X}_t)(x - \hat{X}_t)^T \pi(dx)$ are solutions of equations (3.54) and (3.55), respectively.*

Proof. Since $\pi_t(dx) = \frac{p_t(dx)}{p_t(1)}$, we have

$$\pi_t(\varphi_a) = \exp\left\{ i\left(\hat{X}_t, a\right) - \frac{1}{2}(P(t)a, a) \right\},$$

for any a by Lemma 3.3. Therefore π_t are Gaussian distributions with mean $\hat{X}_t$ and covariance $P(t)$. These two satisfy equations (3.54) and (3.55), respectively. □

We set

$$I_t := Y_t - Y_0 - \int_0^t (H_0(s) + H_1(s)\hat{X}_s)ds. \tag{3.57}$$

Corollary 3.1 *I_t is an innovation process.*

Proof. Equations (3.54) and (3.57) are written as

$$\hat{X}_t = m + \int_0^t (A_0 + A_1 \hat{X}_s)ds + \int_0^t (P(s)H_1^T + \sigma S \Sigma^T)(\Sigma R \Sigma^T)^{-1} dI_s,$$

$$Y_t = Y_0 + \int_0^t (H_0(s) + H_1(s)\hat{X}_s)ds + I_t.$$

Adjoining equation (3.55) for $P(t)$, these three equations may be regarded as a SDE for the triple $(\hat{X}_t, Y_t, P(t))$ driven by the Brownian motion I_t. Coefficients $A_i = A_i(s, \mathbf{y})$, $H_i = H_i(s, \mathbf{y})$, $\sigma(s, \mathbf{y}) S(s) \Sigma(s, \mathbf{y})$ etc. are locally Lipschitz continuous with respect to $\mathbf{y}$. Then the equation has a path wise unique solution. Thus $(\hat{X}_t, Y_t, P(t))$ is $\sigma(Y_0, I_s; s \leq t)$-measurable. □

Remark. If coefficients $A_0, A_1, \sigma, H_0, H_1, \Sigma$ are deterministic functions of t, equation (3.54) for $\hat{X}_t$ is a linear equation with deterministic coefficients. Hence $\hat{X}_t$ is also a Gaussian process. Further equation (3.55) for $P(t)$ is a differential equation with deterministic coefficients. Hence the solution $P(t)$ is deterministic. These two equations are known as the (extended) Kalman–Bucy filter.

The classical Kalman–Bucy filter is the case where $A_0 = H_0 = 0$, $\sigma = I$, $\Sigma(s) = I$ and Brownian motions W_t and N_t are independent so that matrix $S(t)$ is equal to 0. Then equations for $\hat{X}_t$ and $P(t)$ are written by

$$\hat{X}_t = m + \int_0^t A_1(s)\hat{X}_s ds + \int_0^t P(s)H_1(s)^T R(s)^{-1}[dY_s - H_1(s)\hat{X}_s ds],$$

$$\frac{dP(t)}{dt} = Q(t) + A_1(t)P(t) + P(t)A_1(t)^T - P(t)H_1(t)^T R(t)^{-1} H_1(t) P(t).$$

3.3 Cauchy problems for stochastic partial differential equations: a stochastic approach

3.3.1 Introduction

In this section, we are concerned with the Cauchy problem of a linear stochastic partial differential equation for a measure valued process $\rho_t(dx)$ and a linear stochastic partial differential equation for a smooth function valued process $u(t, x)$. Former equations include the SPDE for unnormalized conditional distributions discussed in Section 3.2. Using test functions $f \in C_0^2(\mathbf{R}^d)$, equation for $\rho_t(f) := \int_{\mathbf{R}^d} f(x) \rho_t(dx)$ is written as

$$\rho_t(f) = \rho_0(f) + \int_0^t \rho_s(A(s)f)ds + \int_0^t (\rho_s(B(s)f), dY_s). \qquad (3.58)$$

Here $A(s)$ is a second-order partial differential operator on $\mathbf{R}^d$ with random (progressively measurable) coefficients:

$$A(s)f(x) = c(s, x)f(x) + \sum_{i=1}^d b^i(s, x) f_{x_i}(x) + \frac{1}{2} \sum_{i,j=1}^d a^{ij}(s, x) f_{x_i x_j}(x), \qquad (3.59)$$

$B(s) = (B^1(s), ..., B^e(s))$, where $B^k(s)$ are first-order partial differential operators with random (progressively measurable) coefficients:

$$B^k(s)f(x) = h^k(s, x)f(x) + \sum_{i=1}^d d^{ik}(s, x) f_{x_i}(x), \quad k = 1, ..., e, \qquad (3.60)$$

and Y_t is an e-dimensional Brownian motion with covariance $\int_0^t \Gamma(s)ds$. The last term of (3.58) is equal to $\sum_k \int_0^t \rho_s(B^k(s)f) dY_s^k$.

In Section 3.2, we saw that unnormalized conditional measures ρ_t of the filtering satisfy equation (3.58) replacing dY_t by $R(t)^{-1}dY_t$. In Kalman's Gaussian system or Liptzer–Shiryaev's conditionally Gaussian system, coefficients $b^i(s, x)$, $h^k(s, x)$ of operators $A(s)$ and $B^k(s)$ are unbounded and of linear growth. Thus we are interested in the case where these coefficients are unbounded.

Existence of solutions of equation (3.58) can be shown by the method of filtering. So we will discuss the uniqueness of the solution in Section 3.3.2. The uniqueness of the solution has been applied to derive filtering equations for conditionally Gaussian system-observation process (Section 3.5).

In Section 3.3.3, we will study the Cauchy problem for a parabolic stochastic partial differential equation:

$$u(t, x) = u_0(x) + \int_0^t A(s)u(s, x)dt + \int_0^t (B(s)u(s, x), dY_s), \qquad (3.61)$$

where $u(t, x)$ is a random function of $C^{0,2}$-class. The above equation may be regarded as an adjoint of equation (3.58). We will see that if $u(t, x)$ is a solution of the above SPDE, $\rho_t(dx) := u(t, x)dx$ is a solution of the SPDE for measure valued process (3.58), replacing $A(s)$ and $B^k(s)$ by adjoint's $A(s)^*$ and $B^k(s)^*$, respectively. Then the uniqueness of the solution of equation (3.61) will be reduced to that of the measure-valued process (3.58). For the existence of solutions, we will assume that coefficients of operators $A(s)$ and $B^k(s)$ are deterministic. We will use inverse stochastic flows.

As an application, we will show that if the initial measure $\rho_0(dx)$ has a C^2-class density, the measure valued process $\rho_t(dx)$ of equation (3.58) has a smooth density $\rho(t, x)$ and the density process is a solution of the adjoint stochastic partial differential equation (3.61) replacing operators $A(s)$, $B^k(s)$ by adjoint operators $A(s)^*$, $B^k(s)^*$, provided that coefficients of these operators are deterministic.

There are a lot of works concerned with the Cauchy problem for equation (3.61). Equations are regarded as stochastic linear evolution equations in a Hilbert space, which fulfil the coercive (nondegenerate case) or dissipative (degenerate case) conditions. The existence and the uniqueness of the solution can be shown by analytic methods and the regularity of the solution by using Sobolev's embedding theorem. See Da Prato–Zabsczyk [4], Krylov–Rozovskiĭ [8], Pardoux [21], Rozovskiĭ [22]. Stochastic approaches using stochastic flows are found in [8] and [22].

The uniqueness problem of the filtering equation is also studied in the framework of filtered martingale problem by Kurtz–Ocone [15] and Bhatt–Kallianpur–Karandikar [1].

In Da Prato–Iannelli–Tubaro [3] the Cauchy problem in a bounded domain is studied. A still other stochastic approach is taken by Walsh [25], where a white noise with multidimensional parameter or a Brownian sheet is utilized.

3.3.2 Measure-valued stochastic partial differential equations: uniqueness of solutions

Let $(\Omega, \mathcal{F}, \mathbf{Q})$ be a probability space equipped with a filtration $(\mathcal{F}_t)$. Let $Y_t, 0 \leq t \leq T$ be an e-dimensional $(\mathcal{F}_t)$-Brownian motion with covariance $\int_0^t \Gamma(s)ds$. For the coefficients of operators $A(s)$ and $B^i(s)$, we will assume that $c(s, x), b(s, x), a^{ij}(s, x)$ and $h(s, x), d(s, x)$ are $(\mathcal{F}_t)$-progressively measurable and that coefficients have *finite moments*, i.e., any coefficient $\varphi(t, x)$ satisfies $\int_0^t E[|\varphi(s, x)|^p]ds < \infty$ for any t, x and $p > 1$. Further, we introduce the following Conditions (a)-(c).

Conditions. (a) $(a^{ij}(s, x))$ is a symmetric nonnegative definite matrix and further there exist $d \times r$-matrices $e(s, x)$ such that

$$e(s, x)e(s, x)^T = (a^{ij}(s, x) - \alpha^{ij}(s, x)),$$

where $(a^{ij}(s, x)) = d(s, x)\Gamma(s)d(s, x)^T$.

(b) $b(s, x), c(s, x), d(s, x), e(s, x), h(s, x), d(s, x)\Gamma(s)h(s, x)$ are of class $C^{0,4}$ and their derivatives up to four are all uniformly bounded[4] and are uniformly Holder continuous.[5]

(c) $d(s, x)$ is uniformly bounded and $c(s, x)$ is uniformly bounded from the above.

Let $\mathcal{M}$ be the set of all finite measures on $\mathbf{R}^d$. Let ρ_0 be an $\mathcal{F}_0$-measurable $\mathcal{M}$-valued random variable. By a solution of equation (3.58), we mean an $\mathcal{M}$-valued continuous $(\mathcal{F}_t)$-adapted process ρ_t such that $\int_0^t |\rho_s(A(s)f)|ds < \infty$, $\int_0^t |\rho_s(h^i(s)f)|^2 ds < \infty$ a.s. for any i and satisfies equation (3.58).

Our goal of this subsection is to prove the following theorem.

Theorem 3.8 *Note that co-efficient of operators $A(s)$ and $B(s)$ are not necessarily uniformly bounded, but they are uniformly of linear growth. If these coefficient are all bounded, we may prove the uniqueness by simple arguments. See [8], [21], [22]. For the proof of the above theorem, we will make use of stochastic flows determined by SDEs. Some basic fact of stochastic flows will be stated in Section 3.4 Appendix. Assume Conditions (a)–(c). Then for any ρ_0, equation (3.58) has a unique solution ρ_t.*

The existence of solutions can be shown by the method of the filtering. We may assume that the initial measure ρ_0 satisfies $\rho_0(1) = 1$. Suppose that coefficients of operators $A(s)$ and $B^i(s)$ satisfy Conditions (a)–(c). Let W_t be a standard d-dimensional Brownian motion independent of $(\mathcal{F}_t)$. Consider a stochastic differential equation

$$X_t = X_0 + \int_0^t b(s, X_s)ds + \int_0^t e(s, X_s)dW_s \qquad (3.62)$$
$$+ \int_0^t d(s, X_s)(dY_s - \Gamma(s)h(s, X_s)ds),$$

where conditional law of X_0 with respect to $\mathcal{F}_0$ coincides with ρ_0 a.s. It has a unique solution X_t. Set

$$\Psi_t = \exp\left\{\int_0^t (h(s, X_s), dY_s) - \frac{1}{2}\int_0^t (h(s, X_s), \Gamma(s)h(s, X_s))ds\right\}. \qquad (3.63)$$

Instead of the unnormalized conditional law, we consider

$$\rho_t'(f) = E_\mathbf{Q}\left[\Psi_t\left(\exp\int_0^t c(s, X_s)ds\right)f(X_t)|\mathcal{F}_t\right]. \qquad (3.64)$$

Then we can show that ρ_t' is a solution of equation (3.58). Since the proof is similar to the case of the filtering equation, we omit it. See [10].

We shall discuss the uniqueness of solutions. Our discussion is close to [13]. We will make use of stochastic flows.

We first prepare a lemma which describes the behavior of a random field $J(x), x \in \mathbf{R}^d$ as $x \to \infty$.

Lemma 3.4 Let $J(x), x \in \mathbf{R}^d$ be a continuous random field such that for any $p > 1$, there exists a positive constant C_p satisfying

$$E[|J(x) - J(y)|^p] \leq C_p |x - y|^p, \quad \forall x, y.$$

If $E[|J(x)|^p] \leq C_p(1 + |x|)^{np}$ is satisfied for $n = 0$ or $n = 1$, then for any $0 < \epsilon < 1$, it holds $\lim_{|x| \to \infty} \frac{|J(x)|}{1+|x|^{n+\epsilon}} = 0$, a.s. Further there exists a positive random variable $K_\epsilon(\omega)$ with finite moment of any order such that

$$|J(x)| \leq K_\epsilon(1 + |x|^{n+\epsilon}), \quad a.s. \tag{3.65}$$

Proof. We give the proof of the lemma in the case $n = 0$ only. For the case $n = 1$, see [12], Exercise 4.5.9. Set $\hat{x} = \frac{x}{|x|^2}$ if $x \neq 0$ and $\hat{x} = 0$ if $x = \infty$. Set

$$\eta(\hat{x}) = \frac{1 + |J(x)|}{1 + |x|^\epsilon}, \quad \text{if } \hat{x} \neq 0 \ (= 0 \text{ if } \hat{X} = 0).$$

Then if $|x| \geq |y|$, we have the inequality

$$|\eta(\hat{x}) - \eta(\hat{y})| \leq \frac{|J(x) - J(y)|}{1 + |x|^\epsilon} + |J(y)| \frac{|x|^\epsilon - |y|^\epsilon}{(1 + |x|^\epsilon)(1 + |y|^\epsilon)}.$$

Therefore,

$$E[|\eta(\hat{x}) - \eta(\hat{y})|^p] \leq C'_p \left\{ \frac{|x - y|^{p\epsilon}}{(1 + |x|^\epsilon)^p} + \left(\frac{|x|^\epsilon - |y|^\epsilon}{(1 + |x|^\epsilon)(1 + |y|^\epsilon)} \right)^p \right\}.$$

Since $\frac{|x-y|}{(1+|x|)(1+|y|)} \leq |\hat{x} - \hat{y}|$ holds, the right-hand side is dominated by

$$C_p^{11} |\hat{x} - \hat{y}|^{\epsilon p}.$$

Take $\epsilon p > d$. Then we find that $\eta(\hat{x})$ is continuous with respect to $\hat{x}$ and $\lim_{\hat{x} \to 0} \eta(\hat{x}) = 0$ a.s. holds by Kolmogorov's theorem. Further, $K'_\epsilon := \sup_{|\hat{x}| \leq 1} |\eta(\hat{x})|$ has a finite moment of any order. This proves the assertion of the lemma. □

Now, we want to transform equation (3.58) to a measure-valued partial differential equation with random coefficients, written by

$$\tilde{p}_t(f(t)) = \tilde{p}_0(f(0)) + \int_0^t \tilde{p}_s((\tilde{A}(s)f(s) + f_t(s))ds, \tag{3.66}$$

with a suitable parabolic partial differential operator $\tilde{A}(s)$ with progressively measurable coefficients. Our aim is to eliminate the term involving the stochastic integral based on Y_t from equation (3.58).

For this purpose, we need some results on stochastic flows. Let $l(s, x)$ be a function such that

$$l(t, x) = \frac{1}{2}(h(t, x), \Gamma(t)h(t, x)) - (1 + |x|^{3/2}). \tag{3.67}$$

Set

$$\beta^i(u, x) = \sum_{j,k,l} d^{ij}_{x_k}(u, x)\gamma^{jl}(u)d^{kl}(u, x), \quad i = 1, ..., d,$$

$$\delta(u, x) = (h(u, x), \Gamma(u)h(u, x)),$$

and consider the $(d + 1)$-dimensional SDE:

$$X_t = x + \int_0^t \beta(u, X_u)du + \int_0^t d(u, X_u)dY_u, \tag{3.68}$$

$$Z_t = z + \int_s^t Z_u(\delta(u, X_u) - l(u, X_u))du + \int_0^t Z_u(h(u, X_u), dY_u).$$

Let $(X_t^x, Z_t^{x,z})$ be the solution. We use the notation of stochastic flows as in Section 3.4. Set $\varphi_t(x) := X_t^x$, $\psi_t(x, z) := Z_t^{x,z}$. They have modifications such that the maps $\varphi_t : \mathbf{R}^d \to \mathbf{R}^d$ and $(\varphi_t, \psi_t) : \mathbf{R}^{d+1} \to \mathbf{R}^{d+1}$ are diffeomorphisms for any t a.s. Let $\hat{\varphi}_t(x) := \varphi_t^{-1}(x)$ and $\hat{\psi}_t(x) := \psi_t^{-1}(x, 1)$ be inverse maps. These are $C^{0,4}$-functions of (t, x). It holds

$$\hat{\psi}_t(x) = \exp\left\{-\int_0^t (h(s, x), dY_s) - (1 + |x|^{3/2})t\right\}.$$

Set $J_t(x) = \int_0^t (h(s, x), dY_s)$. It satisfies inequalities of Lemma 3.4 with $n = 1$. Then for any $\epsilon > 0$, we have $\lim_{|x|\to\infty} \frac{J_t(x)}{1+|x|^{1+\epsilon}} = 0$. Then we have

$$\limsup_{|x|\to\infty} \log \hat{\psi}_t(x) < 0.$$

Therefore $\hat{\psi}_t(x)$ is bounded with respect to x a.s.

Let ρ_t be a measure-valued process satisfying (3.58). Let $f(t, x)$ be a $(\mathcal{G}_t)$-progressively measurable function of the class $C^{1,2}$ such that f and its derivatives are all bounded for any ω a.s. Then $\rho_t(\hat{\psi}_t f(t, \hat{\varphi}_t)) := \int \hat{\psi}_t(x)f(t, \hat{\varphi}_t(x))\rho_t(dx)$ is well defined. We denote $\rho_t(\hat{\psi}_t f(t, \hat{\varphi}_t))$ by $\tilde{\rho}_t(f(t))$.

Lemma 3.5 *If ρ_t is a solution of equation (3.58), then $\tilde{\rho}_t$ satisfies equation (3.66), where $\tilde{A}(s)$ is a second order parabolic partial differential operator with progressively measurable coefficients given by*

$$\tilde{A}(s)f(x) = \hat{\psi}_s(x)^{-1}\bar{A}(s)(\hat{\psi}_s f \circ \hat{\varphi}_s)(\hat{\varphi}_s^{-1}(x)), \tag{3.69}$$

where

$$\bar{A}(s) = A(s) + L(s) - B(s)\Gamma(s)B(s)^T, \tag{3.70}$$

and $L(s) = l(s, x)f + \frac{1}{2}\sum_{i,j} a^{ij}(s, x)f_{x_i x_j}$.

Proof. To avoid the complexity of notations, we assume here and in the proof of the next lemma that Y_t is a standard Brownian motion and $\Gamma(t) = I$ (identity). Let $f(t, x)$ be a function of $C^{1,2}$-class such that f and its derivatives are all bounded. We have by Itô's formula

$$\rho_t(\hat{\psi}_t f(t, \hat{\varphi}_t)) = \int_0^t d\rho_s(\hat{\psi}_s f(s, \hat{\varphi}_s)) + \int_0^t \rho_s(d(\hat{\psi}_s f(s, \hat{\varphi}_s))) + \langle \rho_t, \hat{\psi}_t f(t, \hat{\varphi}_t) \rangle.$$

Since ρ_t satisfies (3.58), we have

$$\int_0^t d\rho_s(\hat{\psi}_s f(s, \hat{\varphi}_s)) = \int_0^t \rho_s(A(s)(\hat{\psi}_s f(s, \hat{\varphi}_s)))ds$$
$$+ \int_0^t (\rho_s(B(s)(\hat{\psi}_s f(s, \hat{\varphi}_s))), dY_s).$$

Apply Itô's formula for the inverse flow $(\hat{\varphi}_t, \hat{\psi}_t)$ (Theorem 3.13). Then we have

$$d(\hat{\psi}_s f(s, \hat{\varphi}_s)) = L(s)(\hat{\psi}_s f(s, \hat{\varphi}_s))ds + \hat{\psi}_s f_t(s, \hat{\varphi}_s)ds - (B(s)(\hat{\psi}_s f(s, \hat{\varphi}_s)), dY_s).$$

Therefore,

$$\int_0^t \rho_s(d\hat{\psi}_s f(s, \hat{\varphi}_s)) = \int_0^t \rho_s(L(s)(\hat{\psi}_s f(s, \hat{\varphi}_s) + \hat{\psi}_s f_t(s, \hat{\varphi}_s))ds$$
$$- \int_0^t (\rho_s(B(s)(\hat{\psi}_s f(s, \hat{\varphi}_s)), dY_s).$$

We have further,

$$\langle \rho_t, \hat{\psi}_t f(t, \hat{\varphi}_t) \rangle = - \int_0^t \rho_s(B(s)B(s)^T(\hat{\psi}_s f(s, \hat{\varphi}_s)))ds.$$

Then we get

$$\rho_t(\hat{\psi}_t f(t, \hat{\varphi}_t)) = \rho_0(f(0))$$
$$+ \int_0^t \rho_s(\{A(s) + L(s) - B(s)B(s)^T\}(\hat{\psi}_s f(s, \hat{\varphi}_s)) + \hat{\psi}_s f_t(s, \hat{\varphi}_s))ds. \quad (3.71)$$

The above equation is written as (3.66). □

We claim:

Lemma 3.6 $\tilde{A}(s)$ *is a second-order linear partial differential operator with progressively measurable $C^{0,2}$-coefficients.*

Let $\tilde{c}(s, x)$ be the coefficient of the 0th-order term of the operator $\tilde{A}(s)$. Then the inequality $\limsup_{|x| \to \infty} \tilde{c}(s, x) < 0$ holds a.s.

Proof. We first consider the operator $\bar{A}(s)$. Note that

$$B(s)B(s)^T f = \left\{\sum_{i,j} d^{ij} h^i_{x_i} + |h|^2\right\} f + \sum_i \left\{\sum_{j,k} d^{jk} d^{ik}_{xj} + \sum_k h^k d^{ik}\right\} f_{x_i} + \sum_{i,j} a^{ij} f_{x_i x_j}.$$

Then a straightforward calculation yields that $\bar{A}(s)$ is also a second-order parabolic partial differential operator written as

$$\bar{A}(s)f(x) = \bar{c}(s,x)f(x) + \sum_i \bar{b}^i(s,x)f_{x_i} + \sum_{i,j} \bar{a}^{ij}(s,x)f_{x_i x_j}(x), \quad (3.72)$$

where $\bar{a}^{ij} = (a^{ij} - a^{ij})$, $\bar{b}^i = b^i - \sum_k d^{ij} h^i$ and $\bar{c} = c + l - \sum_{i,j} d^{ij} h^j_{x_i} - |h|^2$. These are $C^{0,3}$-functions.

We next consider the operator $\tilde{A}(s)$. It is convenient to rewrite $\bar{A}(s)$ as

$$\bar{A}(s) = \bar{c}(s) + X_0(s) + \frac{1}{2}\sum_j X_j(s)^2,$$

where $X_j(s) = \sum_i X^i_j(s,x)\frac{\partial}{\partial x_i}$. Coefficients of operators X_j, $j = 0, ..., d$ and $\tilde{c}$ are $C^{0,3}$-functions and their derivatives are all uniformly bounded. We often identify $X_j(s)$ with the vector field $X_j(s,x) = (X^1_j(s,x), ..., X^d_j(s,x))$. Now define new (random) vector fields $(\hat{\varphi}_s)_* X_j(s,x) = (\partial\hat{\varphi}_s)X_j(\hat{\varphi}_s^{-1}(x))$, where $\partial\hat{\varphi}_s$ is the Jacobian matrix of the map $\hat{\varphi}_s$. These are progressively measurable functions of $C^{0,3}$-class. We denote the first-order partial differential operator with coefficients $(\hat{\varphi}_s)_* X_j(s)$ by the same notation. Then it holds $(\hat{\varphi}_s)_* X_j(s) f(x) = X_j(s)(f \circ \hat{\varphi}_s)(\hat{\varphi}_s^{-1}(x))$. Using these notations, $\tilde{A}(s)$ is written as

$$\tilde{A}(s)f = \left(\bar{c}(s) + \hat{\psi}_s^{-1}\left\{(\hat{\varphi}_s)_* X_0(s)\hat{\psi}_s + \frac{1}{2}\sum_j ((\hat{\varphi}_s)_* X_j(s))^2 \hat{\psi}_s\right\}\right) f \quad (3.73)$$

$$+ (\hat{\varphi}_s)_* X_0(s) f + \sum_j \hat{\psi}_s^{-1}(\hat{\varphi}_s)_* X_j(s)(\hat{\psi}_s)(\hat{\varphi}_s)_* X_j f$$

$$+ \frac{1}{2}\sum_j ((\hat{\varphi}_s)_* X_j(s))^2 f.$$

It is a second-order linear parabolic partial differential operators with $C^{0,2}$-coefficients a.s.

We shall prove the second assertion of the lemma. Set $J_1(x) = \hat{\varphi}_s(x)$ and $J_2(x) = \partial\hat{\varphi}_s(x)$. Then $J_1(x)$ satisfies inequalities of Lemma 3.4 with $n = 1$ and $J_2(x)$ satisfies inequalities of Lemma 3.4 with $n = 0$. Therefore, for any $0 < \epsilon < 1/2$, there exist positive random variables K_1 and K_2 with finite moments of any order such that

$$|\hat{\varphi}_s(x)| \le K_1(1+|x|^{1+\epsilon}), \quad |\partial \hat{\varphi}_s(x)| \le K_2(1+|x|^\epsilon), \quad a.s.$$

in view of Lemma 3.4. Then we have

$$|\hat{\psi}_s^{-1}(x)(\hat{\varphi}_s)_* X_0 \hat{\psi}_s(x)| = |(\hat{\varphi}_s)_* X_0 (\log \hat{\psi}_s)(x)| \le K_3(\omega)(1+|x|^{(1/2+\epsilon)}), \quad a.s.$$

The second order term $((\hat{\varphi}_s)_* X_j)^2 \hat{\psi}_t$ can be estimated similarly as

$$|((\hat{\varphi}_s)_* X_j)^2 \hat{\psi}_t| \le K_4(1+|x|^{1+\epsilon}), \quad a.s.$$

Consequently the coefficient $\tilde{c}$ is estimated as $\tilde{c}(s,x) \le \bar{c}(s,x) + K(1+|x|^{1+\epsilon})$, where $E_Q[K^p] < \infty$ for any $p > 1$. Since

$$\limsup_{|x| \to \infty} \frac{\bar{c}(t,x)}{1+|x|^{3/2}} < 0,$$

$\tilde{c}$ satisfies the same inequality. Then the inequality $\limsup_{|x|\to\infty} \tilde{c}(s,x) < 0$ holds for any s. Therefore $\tilde{c}(s,x)$ is bounded from the above. □

Proof of Theorem 3.8 uniqueness part Consider the backward partial differential equation

$$\left(\tilde{A}(t) + \frac{\partial}{\partial t}\right) f(t,x) = 0, \quad f(T_0, x) = \phi(x), \tag{3.74}$$

where $0 < T_0 \le T$. For almost all ω the equation has a solution $f(t,x)$, for any bounded C^2-functions ϕ with compact support. For almost all ω, it is a bounded function of (t,x), since the coefficient $\tilde{c}$ is bounded from the above. We may take the family of solutions $f(t,x)$ such that it is progressively measurable.

Now let ρ_t and ρ_t' be solutions of equation (3.58) with the same initial condition $\rho_0 = \rho_0'$. Set $\xi_t = \rho_t - \rho_t'$. Let $\tilde{\xi}_t$ be the transformed process of ξ_t by $\hat{\varphi}_t$ and $\hat{\psi}_t$. It satisfies equation (3.66). For a given ϕ, let $f(t,x)$ be the solution of equation (3.74). Then we have $\tilde{\xi}_{T_0}(\phi) = \xi_0(f(0))$ in view of equation (3.66). Since $\xi_0(f(0)) = 0$, we get $\tilde{\xi}_{T_0}(\phi) = 0$. The quality is valid for a countable dense ϕ. Then we get $\tilde{\xi}_{T_0} = 0$. This yields $\xi_{T_0} = 0$ and the uniqueness of the solution of equation (3.58) follows.

Remark. SPDE (3.47) for unnormalized conditional distribution ρ_t obtained in Section 3.2.4 looks different from equation (3.58). However, denote $(\Sigma(s) R(s) \Sigma(s)^T)^{-1} B(s)$ by $B(s)$. Then equation (3.47) can be written as equation (3.58). Coefficients of these operators satisfy Conditions (a)–(c). Therefore the uniqueness of the solution remains valid for equation (3.47).

Finally we consider a nonlinear SPDE for the measure-valued process π_t:

$$\pi_t(f) = \pi_0(f) + \int_0^t \pi_s(A(s)f)ds \qquad (3.75)$$
$$+ \int_0^t (\pi_s(B(s)f) - \pi_s(h(s))\pi_s(f), dY_s - \pi_s(h(s))ds).$$

Theorem 3.9 *Assume that coefficients of operators $A(s)$ and $B(s)$ satisfy Conditions (a)–(c). Assume further that coefficient c of operator $A(s)$ is 0. Then for any initial π_0, equation (3.75) has a unique solution.*

Proof. We prove the uniqueness of the solution only. Let π_t be a solution of equation (3.75). Set

$$\Psi_t^\pi := \exp\left\{ \int_0^t (\pi_s(h(s)), dY_s) - \frac{1}{2}\int_0^t (\pi_s(h(s)), \Gamma(s)\pi_s(h(s)))ds \right\}$$

and $\rho_t^\pi(f) := \pi_t(f)\Psi_t^\pi$. We will prove that ρ_t^π is a solution of equation (3.58). We have

$$d\rho_t^\pi(f) = \pi_t(f)d\Psi_t^\pi + \Psi_t^\pi d\pi_t(f) + d\langle \Psi_t^\pi, \pi_t(f)\rangle.$$

Each term of the right-hand side is computed as

$$\pi_t(f)d\Psi_t^\pi = (\pi_t(f)\pi_t(h(t))\Psi_t^\pi, dY_t),$$
$$\Psi_t^\pi d\pi_t(f) = \pi_t(A(t)f)\Psi_t^\pi dt +$$
$$\qquad (\pi_t(B(t)f)\Psi_t^\pi - \pi_t(f)\pi_t(h(t))\Psi_t^\pi, dY_t - \pi_t(h(t))dt), .$$
$$d\langle a_t, \pi_t(f)\rangle = (\pi_t(h(t)), \pi_t(B(t)f) - \pi_t(h(t))\pi_t(f))\Psi_t^\pi dt.$$

Therefore ρ_t^π satisfies

$$d\rho_t^\pi = \rho_t^\pi(A(t)f)dt + (\rho_t^\pi(B(t)f), dY_t).$$

Therefore ρ_t^π is a solution of equation (3.58).

Now if π'_t is another solution of equation (3.75), we get $\rho_t^\pi = \rho_t^{\pi'}$ by the uniqueness of the solution of equation (3.58). Since $\pi_t(1) = \pi'_t(1) = 1$ holds a.s., we get $\Psi_t^\pi = \Psi_t^{\pi'}$. This implies $\pi_t(f) = \pi'_t(f)$. Therefore we get $\pi_t = \pi'_t$. □

3.3.3 Stochastic partial differential equations: existence and uniqueness of solutions

Let Y_t, $0 \leq t \leq T$ be an e-dimensional $(\mathcal{F}_t)$-Brownian motion with covariance $\int_0^t \Gamma(s)ds$ defined on a probability space $(\Omega, \mathcal{F}, \mathbf{Q})$. In this subsection, we shall consider a stochastic partial differential equation

$$u(t, x) = u_0(x) + \int_0^t A(s)u(s, x)ds + \int_0^t (B(s)u(s, x), dY_s) \qquad (3.76)$$

for a given initial function u_0, where $A(s)$ and $B(s)$ are partial differential operators given by (3.59) and (3.60) satisfying the same conditions as in Section 3.3.2. Let $u_0(x)$ be a $\mathcal{F}_0$-measurable C^2-function. By a solution of equation (3.76) we mean a progressively measurable function $u(t, x)$ of class $C^{0,2}$ such that $\int_0^t |A(s)u(s, x)|ds < \infty$, $\int_0^t |B(s)u(s, x)|^2 ds < \infty$ a.s. and equality (3.76) holds.

We shall first consider the uniqueness of solutions.

Theorem 3.10 *Assume that operators $A(s)$, $B(s)$ satisfy Conditions (a)-(c). Let u_0 be a C^2 function such that $\int |u_0(x)|dx < \infty$. Then solutions of equation (3.76) such that $\int |u(s, x)|dx < \infty$ a.s. for any s are at most unique.*

Proof. The uniqueness of the solution is reduced to the uniqueness of the solution of the measure valued SPDE discussed in Section 3.3.2. Let $u(t, x)$ be a solution of equation (3.76). We define a measure-valued process ρ_t^* by $\rho_t^*(dx) = u(t, x)dx$. Then it satisfies the SPDE

$$\rho_t^*(f) = \rho_0^*(f) + \int_0^t \rho_s^*(A(s)^* f)ds + \int_0^t \left(\rho_s^*(B(s)^* f), dY_s\right),$$

for $f \in C_0^\infty(\mathbf{R}^d)$. Then the solution of the above equation is at most unique by Theorem 3.8. Therefore the solution u of equation (3.76) with the integrability condition is at most unique. $\square$

We shall next discuss the existence of solutions. From now we assume that coefficients of operators $A(s)$ and $B(s)$ are all deterministic. In order to construct a solution, we will introduce a SDE. Let W_t be a standard Brownian motion independent of Brownian motion Y_t. Consider

$$X_s = x - \int_s^t \{b(u, X_u) - \beta(u, X_u)\}du - \int_s^t e(u, X_u)dW_u \quad (3.77)$$
$$- \int_s^t d(u, X_u)(dY_u - \Gamma(u)h(u, X_u)du)$$
$$Z_s = z - \int_s^t Z_u\{c(u, X_u) - \delta(u, X_u)\}du + \int_s^t Z_u(h(u, X_u), dY_u),$$

where

$$\beta^i(u, x) = \sum_{j,k,l} d_{x_k}^{ij}(u, x)\gamma^{jl}(u)d^{kl}(u, x) + \sum_{j,k} e_{x_k}^{ij}(u, x)e^{kj}(u, x), \quad (3.78)$$
$$\delta(u, x) = (h(u, x), \Gamma(u)h(u, x)).$$

Let $(\varphi_{s,t}(x), \psi_{s,t}(x, z))$ be the stochastic flow generated by the above equation. We denote by $(\hat{\varphi}_{t,s}(x), \hat{\psi}_{t,s}(x, z))$ the inverse flow of $(\varphi_{s,t}, \psi_{s,t})$. Then it is a solution of a backward SDE, i.e., $X_s = \hat{\varphi}_{t,s}(x)$, $Y_s = \hat{\psi}_{t,s}(x, z)$ (t being fixed) satisfies the following backward SDE.

$$X_s = x + \int_s^t b(u, X_u)du + \int_s^t e(u, X_u)\partial W_u \qquad (3.79)$$
$$+ \int_s^t d(u, X_u)(\partial Y_u - \Gamma(u)h(u, X_u)du)$$
$$Z_s = z + \int_s^t Z_u c(u, X_u)du + \int_s^t Z_u(h(u, X_u), \partial Y_u).$$

See Theorem 3.14 in the Appendix to this chapter. We set $\hat{\Psi}_{t,s}(x) = \hat{\psi}_{t,s}(x, 1)$. $\hat{\Psi}_{t,s}(x)$ is written by

$$\hat{\Psi}_{t,s}(x) = \exp\left\{ \int_s^t c(u, \hat{\varphi}_{t,u}(x))du \right\} \times \qquad (3.80)$$
$$\exp\left\{ \int_s^t (h(u, \hat{\varphi}_{t,u}(x)), \partial Y_u) - \frac{1}{2}\int_s^t (h(u, \hat{\varphi}_{t,u}(x)), \Gamma(u)h(u, \hat{\varphi}_{t,u}(x)))du \right\}.$$

It is a backward martingale with mean 1. For a slowly increasing C^2-function u_0, the random variable $\hat{\Psi}_{t,s}(x)u_0(\hat{\varphi}_{t,s}(x))$ is integrable for any t, x. Let $(\mathcal{F}_t)$ be the filtration generated by Y_t. We set

$$u(t, x) = E_{\mathbf{Q}}[\hat{\Psi}_{t,0}(x)u_0(\hat{\varphi}_{t,0}(x))|\mathcal{F}_T]. \qquad (3.81)$$

It is well defined for all t, x a.s. Further if $u_0(x)$ is dominated by $\varphi_n(x) = (1 + |x|)^n$, then it satisfies

$$E_{\mathbf{Q}}[|u(t, x)|^2] \leq C(1 + |x|)^{2n}, \quad \forall (t, x) \in [0, T] \times \mathbf{R}^d \qquad (3.82)$$

Remark 3.1 The conditional expectation of (3.81) is determined up to measure 0. The exceptional null set may depend on parameters t, f, x. However, in our discussion, we want to take its modification so that the exceptional null set does not depend on these parameters. It can be done as follows.

Let $(\Omega_1, \mathcal{F}_1, \mathbf{Q}_1)$ be a probability space where a Brownian motion W_t is defined and let $\mathcal{H}$ be the σ-field generated by W_t. Let $(\Omega_2, \mathcal{F}_2, \mathbf{Q}_2)$ be another probability space where the Brownian motion Y_t is defined. Let $\mathbf{Q} = \mathbf{Q}_1 \times \mathbf{Q}_2$ be the product probability measure on $\Omega = \Omega_1 \times \Omega_2$, $\mathcal{F}_1 \otimes \mathcal{F}_2$, so that W_t and Y_t are independent with respect to $\mathbf{Q}$.

Let $\mathcal{F}_T$ be the σ-field generated by Y_t. For an integrable $\mathcal{H} \otimes \mathcal{G}$-measurable random variable $X = X(\omega_1, \omega_2)$, $\omega_1 \in \Omega_1, \omega_2 \in \Omega_2$, we take the expectation by the measure $\mathbf{Q}_1$:

$$E_{\mathbf{Q}_1}[X](\omega_2) := \int_{\Omega_1} X(\omega_1, \omega_2)\mathbf{Q}_1(d\omega_1).$$

Then $E_{\mathbf{Q}_1}[X]$ is a modification of the conditional expectation $E_{\mathbf{Q}}[X|\mathcal{F}_T]$. Taking such a modification, $u(t, x)$ is well defined for all t, x a.s. Another merit of this modification may be that we can handle conditional expectations similarly as usual integrations. We can change the order of the conditional expectation and the derivative under a reasonable condition.

Remark 3.2 We have the equality

$$u(t, x) = E_Q[\hat{\Psi}_{t,0}(x) f(\hat{\varphi}_{t,0}(x))|\mathcal{F}_t].$$

Indeed, we have $\mathcal{F}_T = \mathcal{F}_t \vee \sigma(Y_u - Y_t; t \leq u \leq T)$ and $\hat{\Psi}_{t,0}(x) f(\hat{\varphi}_{t,0}(x))$ is independent of $\sigma(Y_u - Y_t; t \leq u \leq T)$. Hence the conditional expectation with respect to $\mathcal{F}_T$ coincides with that with respect to $\mathcal{F}_t$.

We want to show that the above $u(t, x)$ is a solution of equation (3.76). We begin with a simple lemma. Let $(\mathcal{G}_t)$ be the forward filtration generated by W_t. We set $\mathcal{H}_t = \mathcal{G}_t \otimes \mathcal{F}_t$. Then both Y_t and W_t are $(\mathcal{H}_t)$-Brownian motions.

Lemma 3.7 Let f_t be a $(\mathcal{H}_t)$-progressively measurable process satisfying the inequality $E_Q[|f_s|] < \infty$ for any s and $E_Q[\int_0^T |f_s|^2 ds] < \infty$. Then we have

$$E_Q\left[\int_0^t f_s ds \Big| \mathcal{F}_T\right] = \int_0^t E_Q[f_s|\mathcal{F}_T] ds, \quad a.s. \tag{3.83}$$

$$E_Q\left[\int_0^t (f_s, dW_s) \Big| \mathcal{F}_T\right] = 0, \quad a.s. \tag{3.84}$$

$$E_Q\left[\int_0^t (f_s, dY_s) \Big| \mathcal{F}_T\right] = \int_0^t (E_Q[f_s|\mathcal{F}_T], dY_s), \quad a.s. \tag{3.85}$$

Proof. We consider the one dimensional case only. It is sufficient to consider the case where f_t is a simple process of the form $f_t = \sum \phi_{i-1} 1_{(t_{i-1}, t_i]}(t)$. Here each ϕ_i is equal to $\phi'_i \phi''_i$, where ϕ'_i is bounded $\mathcal{G}_{t_i}$-measurable and ϕ''_i is bounded $\mathcal{F}_{t_i}$-measurable. Then if $t_{i-1} < t < t_i$, it holds $E_Q[f_t|\mathcal{F}_T] = E_Q[\phi'_{i-1}]\phi''_{i-1}$ and hence $E_Q[f_t|\mathcal{F}_T]$ is $(\mathcal{F}_t)$-adapted.

The first equality (3.83) will be obvious. By using properties of conditional expectations, we have

$$E_Q[\phi_{i-1}(W_{t_i} - W_{t_{i-1}})|\mathcal{F}_T] = \phi''_{i-1} E_Q[\phi'_{i-1}(W_{t_i} - W_{t_{i-1}})] = 0.$$

Since this is valid for any i, we get (3.84). Next, we have

$$E_Q[\phi_{i-1}(Y_{t_i} - Y_{t_{i-1}})|\mathcal{F}_T] = E_Q[\phi'_{i-1}]\phi''_{i-1}(Y_{t_i} - Y_{t_{i-1}})$$
$$= E_Q[\phi_{i-1}|\mathcal{F}](Y_{t_i} - Y_{t_{i-1}}).$$

Summing this for $i = 1, 2, \ldots$ we get formula (3.85). $\square$

Now, the function $u(t, x)$ is of $C^{0,2}$-class a.s. Indeed, since

$$E_Q\left[\sup_{|x| \leq N, 0 \leq t \leq T} |D^\alpha(\hat{\Psi}_{t,0}(x) u_0(\hat{\varphi}_{t,0}))| \Big| \mathcal{F}_T\right] < \infty,$$

holds for any α with $|\alpha| \leq 2$ (see Theorem 2.7 in Chapter 2), we can change the order of the conditional expectation and derivative D^α for any α with $|\alpha| \leq 2$. Hence $u(t, x)$ is twice differentiable, derivatives are continuous in (t, x) and satisfies

$$D^\alpha u(t, x) = E_Q[D^\alpha[\hat{\Psi}_{t,0} u_0(\hat{\varphi}_{t,0})](x)|\mathcal{F}_T],$$

for $|\alpha| \leq 2$.

Lemma 3.8 *Assume that h is uniformly bounded. Then for any slowly increasing C^2-function u_0, the stochastic process $u(t, x)$ given by (3.81) is an $L^2(Q)$-solution of SPDE (3.76).*

Proof. Apply Itô's formula for the inverse flow (see Theorem 3.13 in the Appendix to this chapter) to the function $F(t, x, z) = u_0(x)z$ and $(\hat{\varphi}_{t,0}(x), \hat{\Psi}_{t,0}(x))$. To simplify notations, we denote the latter by $(\hat{\varphi}_t(x), \hat{\Psi}_t(x))$. Then we get

$$\hat{\Psi}_t(x)u_0(\hat{\varphi}_t(x)) - u_0(x) = \int_0^t A(s)[\hat{\Psi}_s u_0 \circ \hat{\varphi}_s](x)ds$$

$$+ \int_0^t (B(s)[\hat{\Psi}_s u_0 \circ \hat{\varphi}_s](x), dY_s)$$

$$+ \sum_{i,j} \int_0^t \sigma^{ij}(s, x) \frac{\partial[\hat{\Psi}_s u_0 \circ \hat{\varphi}_s]}{\partial x_i}(x) dW_s^j.$$

Processes $f_s = A(s)[\hat{\Psi}_s u_0 \circ \hat{\varphi}_s](x)$ etc. at the right-hand side of the above satisfy the square integrability conditions of Lemma 3.7 and we may apply the lemma, since $\hat{\Psi}_t$ and $u_0 \circ \hat{\varphi}_s$ have finite moments of any order. Take the conditional expectation of each term of the right-hand side. We have

$$E_Q\left[\int_0^t A(s)[\hat{\Psi}_s u_0 \circ \hat{\varphi}_s](x)ds \Big| \mathcal{F}_T\right] = \int_0^t E_Q[A(s)[\hat{\Psi}_s u_0 \circ \hat{\varphi}_s](x)|\mathcal{F}_T]ds$$

$$= \int_0^t A(s) E_Q[\hat{\Psi}_s(x)u_0(\hat{\varphi}_s(x))|\mathcal{F}_T]ds$$

$$= \int_0^t A(s)u(s, x)ds.$$

Here we first changed the order of the integral $\int_0^t$ and conditional expectation $E_Q[\cdots|\mathcal{F}_T]$ and then next changed the order of the conditional expectation and partial derivatives. The second term is computed as

$$E_Q\left[\int_0^t (B(s)[\hat{\Psi}_s u_0 \circ \hat{\varphi}_s](x), dY_s)\Big|\mathcal{F}_T\right] = \int_0^t (B(s)u(s, x), dY_s).$$

The conditional expectation of the last term is 0, because of the equality (3.84). Therefore we get equation (3.76). □

Next we consider the case where h is of linear growth.

Lemma 3.9 *For any slowly increasing function $u_0(x)$, the function $u(t, x)$ of (3.81) is well defined and is an $L^1(Q)$-solution of equation (3.76).*

Proof. We will consider the case $c = 0$ only. The case for nonzero c will be treated with a slight modification. The process $\hat{\Psi}_{t,s}$ is a positive backward martingale with mean 1. Define a probability $\mathbf{P}$ by $\mathbf{P} = \hat{\Psi}_{T,0} \cdot \mathbf{Q}$. Then with respect to $\mathbf{P}$, the solution $\hat{\varphi}_{t,s}(x)$ has finite moment of any order, since coefficients are of linear growth. ($\hat{\psi}_{t,0}(x, z)$ is integrable but might not be square integrable since coefficients of the equation for $\hat{\psi}_{t,0}$ is not of linear growth.) Then $u_0(\hat{\varphi}_{t,0}(x))$ is integrable with respect to $\mathbf{P}$, which implies that $\hat{\Psi}_{t,0} u_0(\hat{\varphi}_{t,0})$ is integrable with respect to $\mathbf{Q}$. Then the process $u(t, x)$ of (3.81) is well defined.

For a given $h(t, x)$ we define $h_n(t, x) = \max\{-n, \min\{n, h(t, x)\}\}$, where n is a positive integer. The sequence of bounded functions $\{h_n\}$ converges to h point wisely. Let $(\hat{\varphi}_{t,s}^{(n)}, \hat{\psi}_{t,s}^{(n)})$ be the solution of equation (3.79) associated with h_n. We define $u^{(n)}(t, x)$ by (3.81) using $\hat{\psi}_{t,0}^{(n)}, \hat{\varphi}_{t,0}^{(n)}$. Then $u^{(n)}$ satisfies the SPDE (3.76) associated with $B^{(n)}(t)$.

Further, we have the inequality

$$E_\mathbf{Q}[\hat{\psi}_{t,0}^{(n)}(x, 1) \log \hat{\psi}_{t,0}^{(n)}(x, 1) u_0(\hat{\varphi}_{t,0}^{(n)}(x))^2]$$
$$= E_{\mathbf{P}_n}\left[\left(\int_0^t h^{(n)}(u) R(u) h^{(n)}(u) du\right) u_0(\hat{\varphi}_{t,0}^{(n)}(x))^2\right]$$
$$\leq E_{\mathbf{P}_n}\left[\left(\int_0^t h(u) R(u) h(u) du\right)^2\right]^{1/2} E_{\mathbf{P}_n}[u_0(\hat{\varphi}_{t,0}^{(n)})^4]^{1/2}.$$

The right-hand side is bounded with respect to n. Therefore random variables $\{\hat{\psi}_{t,0}^{(n)}(x, 1) u_0(\hat{\varphi}_{t,0}^{(n)})\}$ is uniformly integrable. Then we can change the order of the conditional expectation and the limit and we get

$$\lim_{n \to \infty} E_\mathbf{Q}[\hat{\psi}_{t,0}^{(n)}(x, 1) u_0(\hat{\varphi}_{t,0}^{(n)}(x)) | \mathcal{F}_T] = E_\mathbf{Q}[\hat{\psi}_{t,0}(x, 1) u_0(\hat{\varphi}_{t,0}(x)) | \mathcal{F}_T].$$

Hence the sequence $u^{(n)}(t, x)$ converges to $u(t, x)$ a.s. for any t. Similarly $B(t) u^{(n)}(t)$ converges to $B(t) u(t, x)$ a.s. for any t. Then $\int_0^t (B(s) u^{(n)}(s, x), dY_s)$ converges to $\int_0^t (B(s) u(s, x), dY_s)$. Consequently the limit $u(t, x)$ should satisfy (3.76). □

Theorem 3.11 *Assume that coefficients of operators $A(s)$ and $B(s)$ are deterministic and satisfy Conditions (a)–(c). If u_0 is a slowly increasing C^2-function, equation (3.76) has a solution $u(t, x)$. Further if $u_0(x)$ is integrable, a solution satisfying $\int |u(s, x)| ds < \infty$ a.s. exists uniquely.*

Proof. The first assertion is shown in Lemma 3.9. Assume that u_0 is integrable. We show the existence of the solution with the integrability condition. Assume first that the initial function u_0 is dominated by $\varphi_n(x) = (1 + |x|)^n$ for some negative integer n. Set $\mathbf{P} = \hat{\Psi}_{T,0} \cdot \mathbf{Q}$. Let $u(t, x)$ be a solution given by (3.81). Then we have

$$E_\mathbf{Q}[|u(t, x)|] \leq E_\mathbf{P}[|u_0(\hat{\varphi}_{t,0}(x))|] \leq c_n \varphi_n(x).$$

See Theorem 2.7 in Chapter 2. Therefore, $E_Q[\int_{\mathbf{R}^d} |u(t,x)|dx] \leq c\int_{\mathbf{R}^d} \varphi_n(x)dx$. The last term is finite if $n \leq -d-1$. Then the solution $u(t,x)$ is integrable with respect to dx a.s.

Next we shall consider the case where u_0 is an integrable C^2-function. It is sufficient to consider the case where $u_0(x)$ is a nonnegative function. There exists an increasing sequence $u_0^{(n)}$ of C^2-functions dominated by φ_{-d-1}, which converges to u_0 compact uniformly. Let $u^{(n)}(t,x)$ be the solution with the initial function $u_0^{(n)}$ such that $\int |u^{(n)}(t,x)|dx < \infty$ a.s. Then the sequence of functions $u^{(n)}(t,x)$ increases with n. Let $u(t,x)$ be its limit. Obviously it is a solution of equation (3.76) with the initial function u_0. Now define $\rho_t^{(n)}(f) = \int u^{(n)}(t,x)f(x)dx$. Then $\rho_t^{(n)}$ is a unique solution of equation (3.81) with the initial condition $\rho_0^{(n)}(dx) = u_0^{(n)}(x)dx$. Further, it is represented by (3.63). Then the limit process $\rho_t(f) = \int u(t,x)f(x)dx$ is also represented by (3.63). This implies $\rho_t(1) < \infty$ a.s., proving $\int |u(t,x)|dx < \infty$. □

Finally, let us return to equation (3.58) for measure-valued process $\rho_t(dx)$. Assume that the initial distribution $\rho_0(dx)$ has a C^2-class density function $\rho_0(x)$. Consider the adjoint equation

$$u(t,x) = \rho_0(x) + \int_0^t A(s)^*u(s,x)ds + \int_0^t (B(s)^*u(s,x), dY_s). \tag{3.86}$$

It has a unique solution $u(t,x)$ such that $\int |u(t,x)|dx < \infty$ a.s. by Theorem 3.11. Define $\rho_t(dx) = u(t,x)dx$. Then ρ_t is a measure-valued process and satisfies equation (3.58). In fact it is the unique solution of equation (3.58). Therefore we have the following theorem.

Theorem 3.12 *Assume that coefficients of operators $A(s)$ and $B(s)$ are deterministic and satisfy Conditions (a)–(c). If the initial distribution $\rho_0(dx)$ has a bounded C^2-density $\rho_0(x)$, the measure valued process $\rho_t(dx)$ has a C^2-density $u(t,x)$. Further it satisfies Zakai equation (3.84).*

3.4 Appendix: stochastic flows, inverse flows, and related Itô's formula

3.4.1 Stochastic flows

In some applications, it is convenient to consider the solution of SDE as a stochastic flow. Let $(\Omega, \mathcal{F}, \mathbf{P})$ be a probability space equipped with a filtration $(\mathcal{F}_t)$. Let $\{\varphi_{s,t}(x), 0 \leq s < t \leq T, x \in \mathbf{R}^d\}$ be a family of $\mathbf{R}^d$-valued random variables, adapted to $(\mathcal{F}_t)$, continuous in (s,t,x) a.s., equipped with the following properties. For almost all ω,

1. the map $\varphi_{s,t}: \mathbf{R}^d \to \mathbf{R}^d$ is a homeomorphism for any $s < t$ and
2. it satisfies the cocycle property: $\varphi_{r,t} = \varphi_{s,t} \circ \varphi_{r,s}$ for any $r < s < t$.

The family $\{\varphi_{s,t}\}$ is called a *stochastic flow of homeomorphisms*.

Let W_t is an r-dimensional $(\mathcal{F}_t)$-Brownian motion with covariance $\int_0^t Q(s)ds$. Let $b(s, x)$ and $\sigma(s, x)$ be $(\mathcal{F}_t)$-progressively measurable functionals such that they are Lipschitz continuous and of linear growth with respect to x. Consider the SDE

$$dX_t = b(t, X_t)dt + \sigma(t, X_t)dW_t \tag{3.87}$$

Let $X_t^{(s,x)}$, $t \geq s$ be the solution of the above equation starting from x at time s. Then there exists a stochastic flow of homeomorphisms $\varphi_{s,t}(x)$ such that $\varphi_{s,t}(x) = X_t^{(s,x)}$. Further if coefficients of the equation are m-times continuously differentiable and their derivatives with respect to x are bounded and are Hölder continuous, $\varphi_{s,t}(x)$ is a C^m-function of x (C^m-diffeomorphism) for any $s < t$. For details, see Kunita [12].

3.4.2 Inverse flows and related Itô's formula

Suppose that $\sigma(t, x)$ and $b(t, x)$ are of $C^{0,3}$-class and derivatives are uniformly bounded and uniformly Hölder continuous. Consider the SDE (3.87). Let $\varphi_{s,t}(x)$ be the stochastic flow of C^2-diffeomorphisms associated with the above SDE. For each $s < t$, we set $\hat{\varphi}_{t,s}(x) = \varphi_{s,t}^{-1}(x)$. It has the backward cocycle property $\hat{\varphi}_{t,r} = \hat{\varphi}_{s,r} \circ \hat{\varphi}_{t,s}$ for any $r < s < t$. $\hat{\varphi}_{t,s}$ is called the *inverse flow*.

Theorem 3.13 (*c.f. Theorem 4.4.5 in [12]*) *Let $\varphi_{s,t}(x)$ be the stochastic flow of C^2-diffeomorphisms associated with SDE (3.87). Let $\{\hat{\varphi}_{t,s}\}$ be the inverse flow. Let $F(t, x)$ be a progressively measurable process of $C^{1,2}$-class. Then we have for any $s < t$,*

$$F(t, \hat{\varphi}_{t,s}(x)) - F(s, x) \tag{3.88}$$
$$= \int_s^t \frac{\partial F}{\partial t}(u, \hat{\varphi}_{u,s}(x))du$$
$$- \sum_i \int_s^t \{b^i(u, x) - \beta^i(u, x)\} \frac{\partial}{\partial x^i}[F(u, \hat{\varphi}_{u,s}(x))]du$$
$$- \sum_{i,j} \int_s^t \sigma^{ij}(u, x) \frac{\partial}{\partial x^i}[F(u, \hat{\varphi}_{u,s}(x))]dW_u^j$$
$$+ \frac{1}{2} \sum_{i,j} \int_s^t a^{ij}(u, x) \frac{\partial^2}{\partial x^i \partial x^j}[F(u, \hat{\varphi}_{u,s}(x))]du,$$

where

$$\beta^i(u, x) = \sum_{j,k,l} \sigma_k^{ij}, x) q^{jl}(u) \sigma^{kl}(u, x) \tag{3.89}$$

and $(a^{ij}(u, x)) = \sigma(u, x) Q(u) \sigma(u, x)^T$.

Proof. We will consider the case where $\hat{\varphi}_{t,s}$ is a one-dimensional flow and W_t is a one-dimensional standard Brownian motion. For a partition $\Delta = \{s = u_0 < u_1 < \cdots u_n = t\}$, we have

$$F(t, \hat{\varphi}_{t,s}(x)) - F(s, x) \tag{3.90}$$
$$= \sum_{k=0}^{n-1} \{F(u_{k+1}, \hat{\varphi}_{u_{k+1},s}(x)) - F(u_k, \hat{\varphi}_{u_{k+1},s}(x))\}$$
$$+ \sum_{k=0}^{n-1} \{F(u_k, \hat{\varphi}_{u_{k+1},s}(x)) - F(u_k, \hat{\varphi}_{u_k,s}(x))\}.$$

The first term of the right-hand side converges to $\int_s^t \frac{\partial F}{\partial t}(u, \hat{\varphi}_{u,s}(x)) du$ as $|\Delta| \to 0$. Using the backward cocycle property $\hat{\varphi}_{u_{k+1},s} = \hat{\varphi}_{u_k,s} \circ \hat{\varphi}_{u_{k+1},u_k}$ and Taylor expansion formula, the second term is written as

$$\sum_k \frac{\partial}{\partial x}[F(u_k, \hat{\varphi}_{u_k,s}(x))](\hat{\varphi}_{u_{k+1},u_k}(x) - x) \tag{3.91}$$
$$+ \frac{1}{2} \sum_k \frac{\partial^2}{(\partial x)^2}[F(u_k, \hat{\varphi}_{u_k,s}(x))](\hat{\varphi}_{u_{k+1},u_k}(x) - x)^2 + o(1),$$

where $o(1)$ is a random variable converging to 0 in L^2 as $|\Delta| \to 0$. We have

$$\varphi_{u_k,u_{k+1}}(y) - y = b(u_k, y)(u_{k+1} - u_k)$$
$$+ \sigma(u_k, \varphi_{u_k,u_{k+1}}(y))(W_{u_{k+1}} - W_{u_k})$$
$$+ (\sigma(u_k, y) - \sigma(u_k, \varphi_{u_k,u_{k+1}}(y))(W_{u_{k+1}} - W_{u_k})$$
$$+ o(u_{k+1} - u_k),$$

where $o(u_{k+1} - u_h)$ is a random variable whose L^2-norm is a small order of $u_{k+1} - u_k$. Set $y = \hat{\varphi}_{u_{k+1},u_k}(x)$. Then we have

$$\hat{\varphi}_{u_{k+1},u_k}(x) - x = -b(u_k, \hat{\varphi}_{u_{k+1},u_k}(x))(u_{k+1} - u_k)$$
$$- \sigma(u_k, x)(W_{u_{k+1}} - W_{u_k})$$
$$- (\sigma(u_k, \hat{\varphi}_{u_{k+1},u_k}(x)) - \sigma(u_k, x))(W_{u_{k+1}} - W_{u_k})$$
$$+ o(u_{k+1} - u_k).$$

Now let $|\Delta| \to 0$ at the first term of (3.91). It is decomposed to the following three computations.

$$-\sum_k \frac{\partial}{\partial x}[F(u_k, \hat{\varphi}_{u_k,s}(x))]b(u_k, \hat{\varphi}_{u_{k+1},u_k}(x))(u_{k+1} - u_k)$$
$$\to -\int_s^t \frac{\partial}{\partial x}[F(u, \hat{\varphi}_{u,s}(x))]b(u, x) du,$$

and
$$-\sum_k \frac{\partial}{\partial x}[F(u_k, \hat{\varphi}_{u_k,s}(x))]\sigma(u_k, x)(W_{u_{k+1}} - W_{u_k})$$
$$\to -\int_s^t \frac{\partial}{\partial x}[F(u, \hat{\varphi}_{u,s}(x))]\sigma(u, x) d W_u.$$

Thirdly, since
$$-(\sigma(u_k, \hat{\varphi}_{u_{k+1}, u_k}(x)) - \sigma(u_k, x))(W_{u_{k+1}} - W_{u_k})$$
$$= -\sigma'(u_k, x)(\hat{\varphi}_{u_{k+1}, u_k}(x) - x)(W_{u_{k+1}} - W_k) + o(u_{k+1} - u_k)$$
$$= \sigma'(u_k, x)\sigma(u_k, x)(W_{u_{k+1}} - W_k)^2 + o(u_{k+1} - u_k),$$

we have
$$-\sum_k \frac{\partial}{\partial x}[F(u_k, \hat{\varphi}_{u_k,s}(x))](\sigma(u_k, \hat{\varphi}_{u_{k+1}, u_k}(x)) - \sigma(u_k, x))(W_{u_{k+1}} - W_{u_k})$$
$$\to \int_s^t \frac{\partial}{\partial x}[F(u, \hat{\varphi}_{u,s}(x))]\beta(u, x) du.$$

Finally the second term of (3.91) converges to
$$\frac{1}{2}\int_s^t \frac{\partial^2}{(\partial x)^2}[F(u, \hat{\varphi}_{u,s}(x))]\sigma(u, x)^2 du.$$

Summing up all these computations, we get formula (3.88). □

3.4.3 Backward SDE and backward flows

We will quickly discuss the backward stochastic integral and backward stochastic differential equation. Let W_t be a one dimensional Brownian motion. For any s, t such that $0 \le s < t \le T$, we set $\mathcal{F}_{s,t} = \sigma(W_u - W_v; s \le u, v \le t)$. Then for any fixed t, $(\mathcal{F}_{s,t}; 0 \le s \le t)$ is a backward filtration with respect to s. Namely they are decreasing and left continuous with respect to s. For the moment we will fix $t > 0$. Let $f_s, 0 \le s \le t$ be a continuous stochastic process, measurable with respect to $(\mathcal{F}_{s,t})$ for any s such that $E[\int_0^t |f_s|^2 ds] < \infty$. Then we can define the *backward Itô integral* by
$$\int_r^t f_s \hat{d} W_s = \lim_{|\Delta| \to 0} \sum_{i=0}^{n-1} f_{u_{i+1}}(W_{u_{i+1}} - W_{u_i}),$$
where $\Delta = \{r = u_0 < u_1 < \cdots < u_n = t\}$. For a fixed t, we set $\hat{Y}_r = \int_r^t f_s \hat{d} W_s$. It is a backward martingale adapted to $(\mathcal{F}_{r,t})$. Backward integrals based on a multidimensional Brownian motion are defined similarly.

We consider a backward stochastic differential equation. Let $b(u, x)$ and $\sigma(u, x)$ be $(\mathcal{F}_{u,T})$-progressively measurable functionals of class $C^{0,2}$ with bounded derivatives. We consider a backward SDE:

$$\hat{X}_s = x - \int_s^t (b(u, \hat{X}_u) - \beta(u, \hat{X}_u))du - \int_s^t \sigma(u, \hat{X}_u)\hat{d}W_u, \quad 0 \leq s \leq t, \quad (3.92)$$

where $\beta(u, x)$ is the function given by (3.89).

The equation has a unique solution which we denote by $\hat{X}_s^{(t,x)}$.

Let $\{\hat{\varphi}_{t,s}(x), 0 \leq s < t \leq T, x \in \mathbf{R}^d\}$ be a family of $\mathbf{R}^d$-valued random variables, adapted to $(\mathcal{F}_{s,T})$, continuous in (s, t, x) a.s., equipped with the following properties. For almost all ω,

1) the map $\hat{\varphi}_{t,s} : \mathbf{R}^d \to \mathbf{R}^d$ is a homeomorphism for any $s < t$ and
2) it satisfies the backward cocycle property: $\hat{\varphi}_{t,r} = \hat{\varphi}_{t,s} \circ \hat{\varphi}_{s,r}$ for any $r < s < t$.

The family $\{\hat{\varphi}_{t,s}\}$ is called a *backward stochastic flow of homeomorphisms*.

Associated with the solution $\hat{X}_s^{(t,x)}$ of the backward SDE (3.92), there exists a backward stochastic flow $\hat{\varphi}_{t,s}$ such that $\hat{\varphi}_{t,s}(x) = \hat{X}_s^{(t,x)}$ holds for all s, t, x.

Theorem 3.14 *(c.f. Theorem 4.4.4 in [12])* Assume that coefficients $b(u, x)$ and $\sigma(u, x)$ are deterministic, of $C^{0,2}$-class and their derivatives are bounded and Hölder continuous. Let $\varphi_{s,t}(x)$ be the stochastic flow generated by the SDE (3.87). Then the inverse flow $\hat{\varphi}_{t,s}$ coincides with the backward flow associated with the backward SDE (3.92).

Notes

1. Progressively measurable with respect to $(\tilde{\mathcal{B}}_t)$, where $\tilde{\mathcal{B}}_t = \wedge_{\epsilon>0} \mathcal{B}_{t+\epsilon}(C^d) \otimes \mathcal{B}_{t+\epsilon}(C^e)$ (Chapter 2, Section 2.3).
2. We may define the covariation $\langle N^k, Mf \rangle_t$ for (possibly discontinuous) square integrable martingale $M_t f$ and a continuous martingale N_t^k. It satisfies inequality (2.28) in Chapter 2.
3. The filtering equation obtained in [17] is not equal to ours. They obtain an equation for $\pi_t(f) := E[f(X_t, Y_t, t)|\mathcal{G}_t]$. If f is a function of x only, their equation coincides with ours.
4. There exists a positive constant c such that $|D_x^\alpha \varphi(s, x)| \leq c$ holds for all (s, x), almost all ω and all α with $1 \leq |\alpha| \leq 4$. It holds $|\varphi(s, x)| \leq |\varphi(s, 0)| + c|x|$ and hence $\varphi(s, x)$ is of linear growth.
5. There exist positive constants $0 < \gamma \leq 1$ and c such that inequality $|\varphi(s, x) - \varphi(s, y)| \leq c|x - y|^\gamma$ holds for all s, x, y and almost all ω.

References

[1] Bhatt, A. G., Kallianpur, G., and Karandikar, R. L. (1995). Uniqueness and robustness of solution of measure-valued equations of nonlinear filtering, *Ann. Probab.* 23(4), 1895–1938.

[2] Bucy, R. S. and Joseph, D. D. (1969). Filtering for stochastic processes with applications to guidance, *Interscience*.

[3] Da Prato, G., Iannelli M., and Tubaro, L. (1982). Some results on linear stochastic differential equations in Hilbert spaces, *Stochastics* 6, 105–16.
[4] Da Prato, G. and Zabczyk, J. (1992). *Stochastic equations in infinite dimensions*, Cambridge University Press.
[5] Fujisaki, M., Kallianpur, G., and Kunita, H. (1972). Stochastic differential equations for the nonlinear filtering problem, *Osaka J. Math.* 9, 19–40.
[6] Kalman, R. E. (1960). A new approach to linear filtering and prediction problems, *J. Basic Eng., ASME* 82, 33–45.
[7] Kalman, R. E., Bucy R. S. (1961). New results in linear filtering and prediction theory, *J. Basic Eng., ASME* Ser. D 83, 95–108.
[8] Krylov, N. V. and Rozovskiĭ, B. L. (1977). On the Cauchy problem for linear stochastic partial differential equations, *Math. USSR, Izvestija* 11, 1267–84.
[9] Krylov, N. V. and Rozovskiĭ, B. L. (1981). On the first integrals and Liouville equations for diffusion processes, pp. 117–125 in Stochastic systems, *Proc. 3rd IFIP-WG 7/1 Working Cof., Visegrád, Hungary, Sept. 15–20, 1980*, Lecture Note in Cotr. Inform. Sci., Vol. 36.
[10] Kunita, H. (1981). Cauchy problem for stochastic partial differential equations arizing in nonlinear filtering theory, *Systems and Control Letters* 1, 37–41.
[11] Kunita, H. (1983). Stochastic partial differential equations connected with nonlinear filtering, in *Nonlinear filtering and stochastic control*, eds. S. K. Mitter and A. Moro, Lecture Notes in Math. 972, 100–69.
[12] Kunita, H. (1990). *Stochastic flows and stochastic differential equations*, Cambridge Univ. Press.
[13] Kunita, H. (1994). Generalized solutions of a stochastic partial differential equation, *J. Theoretical Probability* 7, 279–308.
[14] Kunita, H. (2009). Nonlinear filtering problems I: Bayes formulae and innovations (Chapter 2 in this book).
[15] Kurtz, T. G. and Ocone, D. L. (1988). Unique characterization of conditional distributions in nonlinear filtering, *Ann. Probab.* 16(1), 80–107.
[16] Kushner, H. (1967). Dynamical equations for optimal nonlinear filtering, *J. Differential Equations* 30, 209–45.
[17] Liptser, R. S. and Shiryaev, A. V. (2001). *Statistics of random processes I,II*, Springer.
[18] Mikulevicius, R. and Rozovskiĭ, B. L. (1994). Soft solutions of linear parabolic SPDE's and the Wiener Chaos expansion, in *Stochastic analysis on infinite dimensional spaces*, eds. H. Kunita and H. H. Kuo, Pitman Research Notes in Mathematics 310, 211–20, Longman Press.
[19] Mikulevicius, R. and Rozovskiĭ, B. L. (1998). Linear parabolic stochastic PDE's and Wiener chaos, *SIAM J. Anal.* 29, 452–80.
[20] Oksendal, B. (1998). *Stochastic differential equations*, Springer.
[21] Pardoux, E. (1979). Stochastic partial differential equations and filtering of diffusion processes, *Stochastics* 3, 127–67.
[22] Rozovskiĭ, B. L. (1990). *Stochastic evolution systems*, Kluwer Academic Publishers.
[23] Stroock, D. W. and Varadhan, S. R. S. (1979). *Multidimensional diffusion processes*, Springer.
[24] Tubaro, L. (1988). Some results on stochastic partial differential equations by the stochastic characteristic method, *Stoch. Anal. Appl.* 6(2), 217–30.
[25] Walsh, J. B. (1984). An introduction to stochastic partial differential equations, *École d'été de Probabilité de Saint Flour XIV—1984*, ed. P. L. Hennequin, Lect. Notes Math. 1180, 265–439.
[26] Zakai, M. (1969). On the optimal filtering of diffusion processes, *Z.W.* 11, 230–43.

·4·
Nonlinear Filtering Equations for Stochastic Processes with Jumps

B. Grigelionis and R. Mikulevicius

4.1 Introduction

According to the Bayesian approach to the recursive statistical estimation problems, we shall consider a random process $X_t, t \geq 0$, with known a priori distribution defined on some probability space $(\Omega, \mathcal{F}, \mathbb{P})$, adapted to a filtration $\mathbf{F} = (\mathcal{F}_t)_{t \geq 0}$ of σ-subalgebras. The process X_t takes its values in a measurable space $(\mathbb{S}, \mathcal{S})$ and plays a role of the parameters that we want to estimate. As observation data we shall consider a filtration $\mathbf{Y} = (\mathcal{Y}_t)_{t \geq 0}$ such that $\mathcal{Y}_t \subseteq \mathcal{F}_t \subseteq \mathcal{F}, t \geq 0$. The main problem is to construct a posteriori distributions $\mathbb{P}(X_t \in B | \mathcal{Y}_t)$, $B \in \mathcal{S}, t \geq 0$, the evolution of which is described by the so-called stochastic nonlinear filtering equations. Usually it is enough to consider stochastic processes $\mathbb{E}(f(X_t) | \mathcal{Y}_t) = \pi_t(f), t \geq 0$, (which is the optional projection of the process $f(X_t), t \geq 0$, onto $\mathbf{Y} = (\mathcal{Y}_t)_{t \geq 0}$) for a sufficiently wide class of functions $f : \mathbb{S} \to \mathbb{R}$ such that $f(X_t), t \geq 0$, is a $(\mathbb{P}, \mathbf{F})$-special semimartingale, i.e. it can be represented uniquely as a sum

$$f(X_t) = f(X_0) + A_t(f) + L_t(f), t \geq 0,$$

where $A(f)$ is a $(\mathbb{P}, \mathbf{F})$-predictable process of a finite variation in finite time intervals and $L(f)$ is a $(\mathbb{P}, \mathbf{F})$-local martingale (for used terminology see Section 4.2 below or [1]). Under some slight restrictions it is easy to check that $\pi_t(f)$ is a $(\mathbb{P}, \mathbf{Y})$-special semimartingale. If every $(\mathbb{P}, \mathbf{Y})$-local martingale can be represented as a sum of stochastic integrals with respect to some fixed system of local martingales, then the canonical representation of $\pi_t(f)$ leads us directly to the nonlinear filtering equation. This idea was first was used in [2] in the case when $\mathcal{Y}_t = \cap_{\varepsilon > 0} \sigma(Y_s, s \leq t + \varepsilon)$, and the observation process Y satisfies the stochastic differential Itō equation

$$dY_t = a_t(X_t, Y)dt + b_t(Y)dW_t, t \geq 0, \tag{4.1}$$

where W_t is a $(\mathbb{P}, \mathbf{F})$-standard Brownian motion, the coefficients $a_t(x, Y)$ and $b_t(Y)$ satisfy the usual assumptions of measurability smoothness and growth, the matrix $b_t(Y)b_t(Y)^T$ is nondegenerate and for all $t \geq 0$ the σ-algebras

$\sigma(X_s, 0 \leq s \leq t)$ and $\sigma(W_r - W_u, u < r \leq t)$ are independent. This result was generalized in [4] in several aspects, assuming that Y is a $(\mathbb{P}, \mathbf{F})$-semimartingale and the assumptions were formulated in terms of its $(\mathbb{P}, \mathbf{F})$-predictable characteristics.

Sometimes it is important to consider a new measure $\tilde{\mathbb{P}}$ such that $\tilde{\mathbb{P}} \overset{\text{loc}}{\sim} \mathbb{P}$, i.e. the restrictions of $\mathbb{P}$ and $\tilde{\mathbb{P}}$ to $\mathcal{F}_t$ are equivalent for each $t \geq 0$. If

$$Z_t = \frac{d\mathbb{P}}{d\tilde{\mathbb{P}}}|_{\mathcal{F}_t}, t \geq 0,$$

then it is obvious that

$$\mathbb{E}[f(X_t)|\mathcal{Y}_t] = \frac{\tilde{\mathbb{E}}[f(X_t) Z_t|\mathcal{Y}_t]}{\tilde{\mathbb{E}}[Z_t|\mathcal{Y}_t]}, t \geq 0.$$

Applying the same ideas we can obtain the evolution equation for

$$\rho_t(f) = \tilde{\mathbb{E}}[f(X_t) Z_t|\mathcal{Y}_t], t \geq 0,$$

which under some special choice of $\tilde{\mathbb{P}}$ gives us the so called reduced form of stochastic nonlinear filtering equation considered firstly in [5].

Theory of nonlinear filtering was initiated by R. L. Stratonovich in the framework of the conditional Markov processes (see [6], [7]). Partially, it was stimulated by the problems of sequential analysis such as Bayes solutions to the Wald's problem of testing statistical hypothesis or "order–disorder" problem for the continuous time stochastic processes with independent increments. On the other hand, the problem of the Markov property of sufficient statistics in the optimal stopping of stochastic processes led to the theory of locally infinitely divisible processes, the fundamental notion of a random point measure and its compensator, the canonical representation of semimartingales (see, e.g. [1], [8]–[11]). In deriving stochastic recursive nonlinear filtering equations a crucial role is played by the transformation of semimartingales by means of the reduction of filtration and the stochastic integral representation of local martingales. As an illustration let us consider the following model.

Example 4.1 Let a stochastic process $(X_t, Y_t), t \geq 0$, taking values in $\{0, 1, \ldots, N\} \times \mathbb{R}^d$, where X_t is a time homogeneous Markov chain with the transition probability intensity matrix $\Lambda = (\Lambda(j, k))_{0 \leq j, k \leq N}$, $p_j(0) = \mathbb{P}(X_0 = j), 0 < p_j(0) < 1, j = 0, \ldots, N, Y_0 = 0$ for each $0 \leq s < t$ $\mathbb{P}$-a.s.

$$\mathbb{E}\left[\exp\{i(Y_t - Y_s, z\}|\mathcal{Y}_s \vee \mathcal{X}_\infty\right] = \exp\{\Phi_{s,t}^X\},$$

where

$$\Phi_{s,t}^X = i(\int_s^t a_{X_r} dr, z) - \frac{t-s}{2}(z, Cz) + \int_s^t \int_{\mathbb{R}_0^d} (e^{i(z,x)} - 1 \\ -i(z, x)1_{\{|x| \leq 1\}} \nu_{X_r}(dx) dr,$$

$\mathbb{R}_0^d = \mathbb{R}^d \setminus \{0\}$, $\mathcal{X}_\infty = \sigma(X_s, s \geq 0)$, $\mathcal{Y}_s = \cap_{\varepsilon>0} \sigma(Y_r, r \leq s + \varepsilon)$, a_j, C, ν_j are Levy triplets, $\nu_j \sim \nu_0$, $\frac{d\nu_j}{d\nu_0} = \varphi_j(x)$, and

$$\int_{\mathbb{R}_0^d} \left(1 - \sqrt{\varphi_j(x)}\right)^2 \nu_0(dx) < \infty.$$

Assume there exist vectors $h_j \in \mathbb{R}^d$ such that

$$a_j = a_0 + h_j C + \int_{|x| \leq 1} x \left(\varphi_j(x) - 1\right) \nu_0(dx),$$

$j = 1, \ldots, N$.

Let $\mathbf{F} = \left(\mathcal{F}_{t+}^{X,Y}\right)_{t \geq 0}$, $\mathcal{F}_t^{X,Y} = \sigma(X_s, Y_s, s \leq t)$. We have

$$Y_t = Y_0 + \int_0^t a_{X_s} ds + M_t + \int_0^t \int_{|x| \leq 1} x q(ds, dx) + \int_{|x| > 1} x \bar{p}(ds, dx),$$

where $\bar{p}(dt, dx)$ be the jump measure of Y, with $(\mathbb{P}, \mathbf{F})$-dual predictable projection $\Pi(dt, dx) = \nu_{X_t}(dx)dt$,

$$q(dt, dx) = \bar{p}(dt, dx) - \nu_{X_t}(dx)dt = \bar{p}(dt, dx) - \varphi_{X_t}(x)\nu_0(dx)dt,$$

and M_t is a local continuous $(\mathbf{F}, \mathbb{P})$-martingale such that $\left(\langle M^i, M^j \rangle_t\right)_{1 \leq i, j \leq d} = Ct$.

Let $\mathbf{Y} = (\mathcal{Y}_t)_{t \geq 0}$, $\bar{p}_j(t) = \mathbb{P}(X_t = j | \mathcal{Y}_t)$, $j = 0, \ldots, N, t \geq 0$. With respect to to the reduced filtration $\mathbf{Y}$, we have

$$Y_t = Y_0 + \int_0^t \{a_0 + \sum_{j=0}^N [h_j C + \int_{|x| \leq 1} x \left(\varphi_j(x) - 1\right) \nu_0(dx)] \bar{p}_j(s)]\} ds + \bar{M}_t$$
$$+ \int_0^t \int_{|x| \leq 1} x \bar{q}(ds, dx) + \int_{|x| > 1} x \bar{p}(ds, dx),$$

where $\bar{q}(ds, dx) = \bar{p}(dt, dx) - \bar{\Pi}(dt, dx)$, $\bar{\Pi}(dt, dx) = \sum_{j=0}^N \bar{p}_j(t)\varphi_j(x)\nu_0(dx)dt$ is the dual $(\mathbf{Y}, \mathbb{P})$-predictable projection of $\bar{p}(dt, dx)$, and

$$\bar{M}_t = M_t + \int_0^t \left[h_{X_s} - \sum_{j=0}^N h_j \bar{p}_j(s)\right] C ds, t \geq 0,$$

is a local continuous $(\mathbf{Y}, \mathbb{P})$martingale such that $\left(\langle \bar{M}^i, \bar{M}^j \rangle_t\right) = Ct$. Note that

$$\Pi(dt, dx) = \chi(t, x)\bar{\Pi}(dt, dx),$$
$$\chi(t, x) = \frac{\varphi_{X_t}(x)}{\sum_{j=0}^N \bar{p}_j(t)\varphi_j(x)}.$$

The processes X_t and Y_t have no common jumps and for any function $f(x)$ on $\{0, 1, \ldots, N\}$,

$$f(X_t) = f(X_0) + \sum_{s \leq t}[f(X_s) - f(X_{s-})]$$
$$= f(X_0) + \int_0^t \int [f(j) - f(X_{s-})]q^X(ds, dj) \quad (4.2)$$
$$+ \int_0^t \sum_{j=0}^N [f(j) - f(X_{s-})]\Lambda(X_{s-}, j)ds,$$

where $q^X(dt, dj) = p^X(dt, dj) - \sum_{j=0}^N f(j)\Lambda(X_t, j)dt$, $p^X(dt, dj)$ is the jump measure of X_t with $(\mathbf{F}, \mathbb{P})$-dual predictable projection

$$\sum_{j=0}^N f(j)\Lambda(X_s, j)ds.$$

If $f(j) = 1_{\{l\}}(j)$, then

$$1_{X_t = l} = 1_{X_0 = l} + \int_0^t 1_{\{l\}}(j) - 1_{\{l\}}(X_{s-})]q^X(ds, dj)$$
$$+ \int_0^t \Lambda(X_{s-}, l)ds,$$

As a consequence of Theorem 4.1 below, we find that the following nonlinear filtering equations hold

$$\bar{p}_l(t) = \bar{p}_l(0) + \int_0^t \sum_{j=0}^N \Lambda(j, l)\bar{p}_j(s)ds + \int_0^t \bar{p}_l(s-)\left[h_l - \sum_{j=0}^N h_j \bar{p}_j(s)\right] d\bar{M}_s$$
$$+ \int_0^t \int_{\mathbb{R}_0^d} \bar{p}_l(s-)\left(\frac{\varphi_l(x)}{\sum_{j=0}^N \bar{p}_j(t)\varphi_j(x)} - 1\right) \bar{q}(ds, dx),$$

$l = 0, \ldots, N$.

On the other hand, let us define the probability measure $\tilde{\mathbb{P}}$ by means of the equalities

$$\frac{d\tilde{\mathbb{P}}}{d\mathbb{P}}\Big|_{\mathcal{F}_t} = \tilde{Z}_t, t \geq 0,$$

where $\tilde{Z}_t$ solves the linear equation

$$d\tilde{Z}_t = -\tilde{Z}_{t-}\left[h_{X_t} dM_t + \int \left(\frac{1}{\varphi_{X_{t-}}(x)} - 1\right) q(dt, dx)\right], t \geq 0.$$

Then with respect to $\tilde{\mathbb{P}}$, the processes X and Y are independent, (4.2) holds and

$$Y_t = Y_0 + a_0 t + \tilde{M}_t + \int_0^t \int_{|x| \leq 1} x\tilde{q}(ds, dx) + \int_{|x| > 1} x\tilde{p}(ds, dx),$$

where $\tilde{M}$ is continuous $(\mathbf{F},\tilde{\mathbb{P}})$-martingale, $(\langle \tilde{M}^i, \tilde{M}^j\rangle_t)_{i,j} = Ct$, the point measure $\tilde{p}(dt, dx)$ is $(\mathbf{F},\tilde{\mathbb{P}})$-Poisson with compensator $v_0(dx)dt$ and

$$\tilde{q}(dt, dx) = \tilde{p}(dt, dx) - v_0(dx)dt.$$

The inverse density $Z_t = 1/\tilde{Z}_t$ satisfies the linear equation

$$dZ_t = Z_{t-}\left[h_{X_t}d\tilde{M}_t + \int(\varphi_{X_{t-}}(x) - 1)\tilde{q}(dt, dx)\right], t \geq 0.$$

In this case

$$\tilde{p}_l(t) = \frac{\tilde{\mathbb{E}}\left[Z_t 1_{\{X_t=l\}}|\mathcal{Y}_t\right]}{\tilde{\mathbb{E}}[Z_t|\mathcal{Y}_t]} = \frac{\bar{p}_l(t)}{\sum_{j=0}^N \bar{p}_j(t)},$$

and, by Theorem 4.2 below, for $\bar{p}_l(t) = \tilde{\mathbb{E}}\left[Z_t 1_{\{X_t=l\}}|\mathcal{F}_t\right], l = 0, \ldots, N$, the reduced linear filtering equation holds:

$$\bar{p}_l(t) = \bar{p}_l(0) + \int_0^t \sum_{j=0}^N \Lambda(j, l)\bar{p}_j(s)ds + \int_0^t \bar{p}_\ell(s)h_\ell d\hat{M}_s$$
$$+ \int_0^t \int_{\mathbb{R}_0^d} \bar{p}_\ell(s-)(\varphi_\ell(x) - 1)\tilde{q}(ds, dx),$$

$l = 0, \ldots, N$.

In Section 4.2 we shall give a short survey of stochastic calculus connected with semimartingales and point measures (see [1]). Following [2], [4], [11], [12], [13] (see also [14]–[20]), we derive in Section 4.3 a general stochastic nonlinear filtering equation for a semimartingale functional of the signal process. In Section 4.4, a reduced form of stochastic nonlinear filtering equation is obtained. Finally, in Section 4.5 we present some examples and an application in financial mathematics considered in [3], [26].

4.2 Preliminaries

4.2.1 Notation

We introduce some terminology, notation, and results of stochastic calculus which we shall need later (for proofs and details see, e.g., [1]).

Let $(\Omega, \mathcal{F}, \mathbb{P})$ be a probability space with a given right-continuous filtration of σ-algebras $\mathbf{F} = (\mathcal{F}_t)_{t\geq 0}$ of σ-subalgebras of $\mathcal{F}$. Let $(U, \mathcal{U})$ be a Lusin space, i.e. the space homeomorphic to a Borel subset of a metric compact space. We introduce the following notations.

$\mathbb{R}_+ = [0, \infty)$, $\mathbb{R}_0^d = \mathbb{R}^d \setminus \{0\}$;

A process whose paths are right continuous with left limits will be called cadlag;

$\mathcal{P}(\mathbf{F})$ is the σ-algebra of **F**-predictable subsets of $[0, \infty) \times \mathbb{R}_+$;

$\mathcal{O}(\mathbf{F})$ is the σ-algebra of **F**-optional subsets of $[0, \infty) \times \mathbb{R}_+$, generated by **F**-adapted cadlag processes;

$\tilde{\mathcal{P}}(\mathbf{F}) = \mathcal{P}(\mathbf{F}) \times \mathcal{U}$, $\tilde{\mathcal{O}}(\mathbf{F}) = \mathcal{O}(\mathbf{F}) \times \mathcal{U}$;

$\mathcal{T}(\mathbf{F})$ is the class of all **F**-stopping times; $\mathcal{T}_p(\mathbf{F})$ is the class of all predictable **F**-stopping times; $\mathcal{T}_f(\mathbf{F})$ is the set of all sequences of $\mathbb{P}$-a.e. finite **F**-stopping times (T_n) such that $T_n \uparrow \infty$; $\mathcal{T}_{fp}(\mathbf{F})$ is the subset of $\mathcal{T}_f(\mathbf{F})$ consisting of predictable sequences of stopping times.

$\mathcal{M}(\mathbb{P}, \mathbf{F})$ is the class of uniformly integrable $(\mathbb{P}, \mathbf{F})$-martingales; $\mathcal{M}^2(\mathbb{P}, \mathbf{F})$ is the class of square integrable $(\mathbb{P}, \mathbf{F})$-martingales;

$\mathcal{V}^+(\mathbb{P}, \mathbf{F})$ is the class of increasing right-continuous **F**-adapted and $\mathbb{P}$-a.s. finite processes A_t such that $A_0 = 0$;

$\mathcal{A}^+(\mathbb{P}, \mathbf{F}) = \{A \in \mathcal{V}^+(\mathbb{P}, \mathbf{F}) : \mathbb{E}A_\infty < \infty\}$;

$\mathcal{V}(\mathbb{P}, \mathbf{F}) = \mathcal{V}^+(\mathbb{P}, \mathbf{F}) - \mathcal{V}^+(\mathbb{P}, \mathbf{F})$; $\mathcal{A}(\mathbb{P}, \mathbf{F}) = \mathcal{A}^+(\mathbb{P}, \mathbf{F}) - \mathcal{A}^+(\mathbb{P}, \mathbf{F})$;

For $V \in \mathcal{V}(\mathbb{P}, \mathbf{F})$, $t \geq 0$, we denote $|V|_t$ the variation of V in $[0, t]$;

A stochastic process X is of class (D), if the family $\{X_T : T \in \mathcal{T}_f(\mathbf{F})\}$ is uniformly integrable.

For a cadlag process X we denote $\mathcal{F}^X_t = \sigma(X_s, s \leq t)$, $t \geq 0$, $\mathbf{F}^X = (\mathcal{F}^X_{t+})_{t \geq 0}$;

For arbitrary class $\mathcal{C}(\mathbb{P}, \mathbf{F})$ we denote $\mathcal{C}^c(\mathbb{P}, \mathbf{F})$ the subclass of continuous processes and $\mathcal{C}_{loc}(\mathbb{P}, \mathbf{F})$ the class of all processes C_t such that there exist a sequence $T_n \in \mathcal{T}(\mathbf{F})$, $n \geq 1$, $T_n \uparrow \infty$, and $C_{\cdot \wedge T_n} \in \mathcal{C}(\mathbb{P}, \mathbf{F})$ for each $n \geq 1$. Elements of $\mathcal{M}_{loc}(\mathbb{P}, \mathbf{F})$ are called $(\mathbb{P}, \mathbf{F})$-local martingales.

For $M_1, M_2 \in \mathcal{M}_{loc}(\mathbb{P}, \mathbf{F})$ we say that they are orthogonal and denote $M_1 \perp M_2$, if $M_1 M_2 \in \mathcal{M}_{loc}(\mathbb{P}, \mathbf{F})$. For $M_1, M_2 \in \mathcal{M}^2_{loc}(\mathbb{P}, \mathbf{F})$, we denote $\langle M_1, M_2 \rangle$ the unique process in $\mathcal{A}_{loc}(\mathbb{P}, \mathbf{F}) \cap \mathcal{P}(\mathbf{F})$ such that $M_1 M_2 - \langle M_1, M_2 \rangle \in \mathcal{M}_{loc}(\mathbb{P}, \mathbf{F})$; we denote

$$\mathcal{M}^d_{loc}(\mathbb{P}, \mathbf{F}) = \left\{M \in \mathcal{M}_{loc}(\mathbb{P}, \mathbf{F}) : M \perp N \text{ for all } N \in \mathcal{M}^c_{loc}(\mathbb{P}, \mathbf{F})\right\}.$$

The summation convention with respect to the same indices is used in this chapter.

4.2.2 Optional and predictable projections

Given a right-continuous filtration of σ-algebras $\bar{\mathbf{F}} = (\bar{\mathcal{F}}_t)_{t \geq 0}$ in $(\Omega, \mathcal{F})$ and a nonnegative process X, its optional (resp. predictable) projection is the unique optional (resp. predictable) process ${}^1X$ (resp. $\bar{\mathcal{P}}(X) = {}^3X$) such that for all $\tau \in \mathcal{T}(\bar{\mathbf{F}})$ (resp. $\tau \in \mathcal{T}_p(\bar{\mathbf{F}})$)

$${}^1X_\tau = \mathbb{E}\left[X_\tau | \bar{\mathcal{F}}_\tau\right], \text{ (resp. } {}^3X_\tau = \bar{\mathcal{P}}(X)_\tau = \mathbb{E}\left[X_\tau | \bar{\mathcal{F}}_{\tau-}\right])$$

on $\{\tau < \infty\}$ (see, e.g. [1]). For a general stochastic process X_t, we define (see, [1]) for $i = 1, 3$,

$$^i X = \begin{cases} \infty, & \text{if } {}^i(X^+) = {}^i(X^-) = \pm\infty, \\ {}^i(X^+) - {}^i(X^-), & \text{otherwise.} \end{cases}$$

Since for a (D) class process X, $\mathbb{E}[|X_T|1_{T<\infty}] = \lim_{n\to\infty} \mathbb{E}[|X_{T\wedge n}|1_{T<\infty}] < \infty$ for all $T \in \mathcal{T}(\bar{\mathbf{F}})$, its projections $^i X$ ($i = 1, 2$) are finite $\mathbb{P}$-a.s. and

$$^1 X_\tau = \mathbb{E}\left[X_\tau | \bar{\mathcal{F}}_\tau\right]$$

on $\{\tau < \infty\}$ for any $\tau \in \mathcal{T}(\bar{\mathbf{F}})$.

The following characterization of an optional projection holds.

Lemma 4.1 *Assume a stochastic process X and an optional process $\tilde{X}$ are of class (D). Then $\tilde{X} = {}^1 X$ if and only if for any nonnegative bounded martingale M_t, and every bounded $T \in \mathcal{T}(\bar{\mathbf{F}})$*

$$\mathbb{E}[X_T M_T] = \mathbb{E}[\tilde{X}_T M_T]. \tag{4.3}$$

Proof. If $\tilde{X} = {}^1 X$, $T \in \mathcal{T}(\bar{\mathbf{F}})$ is bounded and M is a nonnegative bounded martingale, then

$$\mathbb{E}[\tilde{X}_T M_T] = \mathbb{E}\left\{\mathbb{E}[X_T|\bar{\mathcal{F}}_T]M_T\right\} = \mathbb{E}\left\{X_T M_T\right\}.$$

Assume now that (4.3) holds for all bounded nonnegative martingales and bounded stopping times. Let $T \in \mathcal{T}(\bar{\mathbf{F}})$ is arbitrary, and

$$M_t = \mathbb{E}[1_{\{T<\infty\}}|\bar{\mathcal{F}}_t].$$

Then for each n

$$\mathbb{E}[X_{T\wedge n} M_{T\wedge n}] = \mathbb{E}[\tilde{X}_{T\wedge n} M_{T\wedge n}]$$

Since $\lim_{n\to\infty} M_{T\wedge n} = M_T = 1_{\{T<\infty\}}$ $\mathbb{P}$-a.s.,

$$\lim_n \mathbb{E}[X_{T\wedge n} M_{T\wedge n}] = \mathbb{E}[X_T 1_{\{T<\infty\}}]$$
$$= \lim_n \mathbb{E}[\tilde{X}_{T\wedge n} M_{T\wedge n}] = \mathbb{E}[\tilde{X}_T 1_{\{T<\infty\}}].$$

$\square$

The following projection properties hold.

Lemma 4.2 *(see e.g. [13]) For a stochastic cadlag process X, the following holds.*
  a) If there is $(T_n) \in \mathcal{T}_f(\bar{\mathbf{F}})$ such that $\mathbb{E}\left[\sup_t |X_{t\wedge T_n}|\right] < \infty$ for all n, then $^1 X$ is cadlag;
  b) If there is $(T_n) \in \mathcal{T}_{fp}(\bar{\mathbf{F}})$ such that $\mathbb{E}\left[\sup_t |X_{t\wedge T_n}|\right] < \infty$ for all n, then $(^1 X)_{t-} = {}^3(X_-)_t$, where $(X_-)_t = X_{t-}$, $t \geq 0$.

Assume that X takes values in a Polish space $\mathbb{S}$ and denote $D(\mathbb{S})$ the space of $\mathbb{S}$-valued cadlag functions with Skorohod $\mathcal{J}_1$-topology. Then the following statement is true.

Lemma 4.3 *(see e.g. [13]) Given an $\mathbb{S}$-valued random variable X, there exists a family of $\mathcal{B}([0,\infty)) \otimes \mathcal{F}$-measurable positive measures $\pi_t(dw)$ on $D(\mathbb{S})$ such that $\pi_t(D(\mathbb{S})) = 1$ for $t < \zeta$ and $\pi_t(D(\mathbb{S})) = 0$ for $t \geq \zeta$, $\mathbb{P}(\zeta < \infty) = 0$ and $\pi_t(dw)$ is cadlag in t in the topology of weak convergence and for each continuous bounded function Z on $D(\mathbb{S})$, $\pi_t(Z)$ is the cadlag version of $\mathbb{E}\left(Z(Y)|\tilde{\mathcal{F}}_t\right)$.*

If f is bounded and $\mathcal{O}(\tilde{\mathbf{F}}) \otimes \mathcal{B}(D(\mathbb{S}))$-measurable, then

$$\int f(t,\omega,w)\,\pi_t(dw) = \pi_t\left(f_t(X)\right) =\ {}^1(f(X))_t \, ;$$

If f is bounded and $\mathcal{P}(\tilde{\mathbf{F}}) \otimes \mathcal{B}(D(\mathbb{S}))$-measurable, then

$$\int f(t,\omega,w)\,\pi_{t-}(dw) = \pi_{t-}\left(f_t(X)\right) =\ {}^3(f(X))_t \, ,$$

where $f_t = f_t(X) = f(t,\omega,X)$.

4.2.3 Stochastic Integrals

We remind here the definitions related to stochastic integrals with respect to continuous martingales and compensated integer valued random measures.

Stochastic integral with respect to a continuous local martingale

If we are given $M^1, \ldots, M^m \in \mathcal{M}^c_{loc}(\mathbb{P},\mathbf{F})$ denote $L^2_{loc}(\Gamma,\mathbb{P},\mathbf{F})$ the set of all predictable vector processes $g = (g^1,\ldots,g^m)$, $g^j \in \mathcal{P}(\mathbf{F})$, $j = 1,\ldots,m$, such that for all $t \geq 0$ $\mathbb{P}$-a.s.

$$\int_0^t (g_s, \tilde{\Gamma}_s g_s)\,d\gamma_s < \infty,$$

where $\Gamma_t = (\gamma_t^{jk})$, $\gamma_t^{jk} = \langle M^j, M^k \rangle_t$, $\tilde{\Gamma}_t = \left(\tilde{\gamma}_t^{jk}\right)_{1 \leq j,k \leq m}$, $\tilde{\gamma}_t^{jk} = \frac{d\gamma_t^{jk}}{d\gamma_t}$, $\gamma_t = \sum_{j=1}^m \gamma_t^{jj}$, $1 \leq j,k \leq m$.

For $g \in L^2_{loc}(\Gamma,\mathbb{P},\mathbf{F})$ we can define in the standard way stochastic integrals

$$M_t(g) = \int_0^t g_s^j\,dM_s^j, \, t \geq 0,$$

where $M = (M^1, \ldots, M^m)$.

Stochastic integrals with respect to integer valued random measures

We shall need some notations related to the random measures $\mu : \Omega \times \mathcal{B}(\mathbb{R}_+) \times \mathcal{U} \to [0,\infty]$. A random measure μ is said to be $\mathbf{F}$-optional (resp. $\mathbf{F}$-predictable) if for each $f \in \tilde{\mathcal{O}}_+(\mathbf{F})$ (resp. $f \in \tilde{\mathcal{P}}_+(\mathbf{F})$) we have $f * \mu \in \mathcal{O}_+(\mathbf{F})$ (resp. $f * \mu \in \mathcal{P}_+(\mathbf{F})$), where

$$f * \mu_t = \int_0^t \int_U f_s(x)\mu(ds,dx), \, t \geq 0.$$

If $d\mu d\mathbb{P}|_{\tilde{\mathcal{P}}(F)}$ is σ-finite, then there is a unique **F**-predictable measure ν called $(\mathbb{P}, \mathbf{F})$-dual predictable projection of ν such that for each $Y \in \mathcal{P}_+(\mathbf{F})$

$$\mathbb{E}(f * \mu)_\infty = \mathbb{E}(f * \nu)_\infty.$$

In the case when when $U = \{1\}$ and μ is generated by the increasing process A_t, then ν is also generated by a uniquely defined process $\tilde{A} \in \mathcal{A}^+_{loc}(\mathbb{P}, \mathbf{F}) \cap \mathcal{P}(\mathbf{F})$ and $A - \tilde{A} \in \mathcal{M}_{loc}(\mathbb{P}, \mathbf{F})$.

A random measure p is called **F**-adapted point process if p is **F**-optional, $\tilde{\mathcal{P}}(\mathbf{F})$ σ-finite integer valued and for all $t \geq 0$ we have $p(\{t\} \times U) = 0$ or 1. Denote by Π a $(\mathbb{P}, \mathbf{F})$-dual predictable projection of p (a $(\mathbb{P}, \mathbf{F})$-conditional intensity measure of p) and let $q = p - \Pi$. It is known that there exists a version of Π such that $\mathbb{P}$-a.s. $a_t = \Pi(\{t\} \times U) \leq 1$.

Let $\mathcal{G}_{loc}(\Pi, \mathbb{P}, \mathbf{F})$ be the class of all $\psi \in \tilde{\mathcal{P}}(\mathbf{F})$ such that for all $t \geq 0$ $\mathbb{P}$-a.s.

$$\int_0^t \int_U \frac{(\psi_s(x) - \hat{\psi}_s)^2}{1 + |\psi_s(x) - \hat{\psi}_s|} \Pi(ds, dx) + \sum_{s \leq t}(1 - a_s) \frac{\hat{\psi}_s^2}{1 + |\hat{\psi}_s|}] < \infty,$$

where $\hat{\psi}_s = \int \psi(s, x)\Pi(\{s\} \times dx)$. Let $Q_t(\psi) = \int_0^t \int_U \psi_s(x) q(ds, dx)$ be a stochastic integral defined in the standard way. Denote

$$\mathcal{M}_{loc}(q, \mathbb{P}, \mathbf{F}) = \{Q(\psi), \psi \in \mathcal{G}_{loc}(\Pi, \mathbb{P}, \mathbf{F})\}.$$

Let $\mathcal{G}^2_{loc}(\Pi, \mathbb{P}, \mathbf{F})$ be the class of all $\psi \in \tilde{\mathcal{P}}(\mathbf{F})$ such that for all $t \geq 0$ $\mathbb{P}$-a.s.

$$\int_0^t \int_U (\psi_s(x) - \hat{\psi}_s)^2 \Pi(ds, dx) + \sum_{s \leq t}(1 - a_s)\hat{\psi}_s^2] < \infty.$$

A point process p is $(\mathbb{P}, \mathbf{F})$-Poisson, i.e. for all $\Gamma \in \mathcal{U}, 0 \leq s < t$, the random variable $p((s, t] \times \Gamma)$ is Poisson, if and only if p is quasi-left-continuous and Π is nonrandom.

4.2.4 Semimartingales

A real valued process X_t is called $(\mathbb{P}, \mathbf{F})$-semimartingale (we write $X \in S(\mathbb{P}, \mathbf{F})$), if it can be represented as a sum

$$X_t = X_0 + A_t + L_t, t \geq 0, \tag{4.4}$$

where $A \in \mathcal{V}(\mathbb{P}, \mathbf{F})$ and $L \in \mathcal{M}_{loc}(\mathbb{P}, \mathbf{F})$. The decomposition (4.4) in general is not unique. The process X is said to be a special semimartingale (we write $X \in S_p(\mathbb{P}, \mathbf{F})$) if it has the decomposition (4.4) with $A \in \mathcal{A}_{loc}(\mathbb{P}, \mathbf{F})$. In this case there exists a unique $A \in \mathcal{A}_{loc}(\mathbb{P}, \mathbf{F}) \cap \mathcal{P}(\mathbf{F})$ satisfying (4.4) which will be called the canonical decomposition of the special semimartingale X. It is known that $X \in S_p(\mathbb{P}, \mathbf{F})$ if and only if $X \in S(\mathbb{P}, \mathbf{F})$ and

$$\sup_{s \leq \cdot} |X_s - X_0| \in \mathcal{A}^+_{loc}(\mathbb{P}, \mathbf{F}).$$

Let $X = (X^1, \ldots, X^m) \in S^m(\mathbb{P}, \mathbf{F})$, i.e. $X_j \in S(\mathbb{P}, \mathbf{F})$, $j = 1, \ldots, m$. The point process $p^X = p$ defined by

$$p([0, t] \times B) = \sum_{s \leq t} 1_B(\Delta X_s), t \geq 0, B \in \mathcal{B}(\mathbb{R}_0^m),$$

with $\Delta X_s = X_s - X_{s-}$, is called the jump measure of the m-dimensional semi-martingale X. If Π is the $(\mathbb{P}, \mathbf{F})$-dual predictable projection of p, then for each $t \geq 0$ $\mathbb{P}$-a.s.

$$\int_{\mathbb{R}_0^m} (|x|^2 \wedge 1) \Pi([0, t] \times dx) < \infty.$$

It is easy to check that $X \in S_p^m(\mathbb{P}, \mathbf{F})$ if and only if for each $t \geq 0$ $\mathbb{P}$-a.s.

$$\int_{\mathbb{R}_0^m} (|x|^2 \wedge |x|) \Pi([0, t] \times dx) < \infty.$$

Let $X_t^{(1)} = X_t - \int_0^t \int_{|x|>1} x p(ds, dx)$, $t \geq 0$. Obviously, $X^{(1)} \in S_p^m(\mathbb{P}, \mathbf{F})$. Let $X_t^{(1)} = X_0 + a_t + L_t$, $t \geq 0$, be the canonical decomposition of $X^{(1)}$. It can be shown that

$$L_t = X_t^c + \int_0^t \int_{|x| \leq 1} x q(ds, dx), t \geq 0,$$

where $X^{cj} \in \mathcal{M}_{loc}^c(\mathbb{P}, \mathbf{F})$, $j = 1, \ldots, m$. So we obtain the canonical decomposition of the m-dimensional $(\mathbb{P}, \mathbf{F})$-semimartingale X:

$$X_t = X_0 + a_t + X_t^c + \int_0^t \int_{|x| \leq 1} x q(ds, dx) + \int_0^t \int_{|x|>1} x p(ds, dx), t \geq 0. \quad (4.5)$$

The triplet (a, B, Π), where $B_t = \left(B_t^{jk}\right)_{0 \leq j,k \leq m}$, $B_t^{jk} = \langle X^{cj}, X^{ck} \rangle_t$, $t \geq 0$, is called the $(\mathbb{P}, \mathbf{F})$-predictable characteristics of X. If $X \in S_p^m(\mathbb{P}, \mathbf{F})$ with the triplet (a, B, Π) of $(\mathbb{P}, \mathbf{F})$-predictable characteristics, then in the canonical representation we have

$$A_t = a_t + \int_{|x|>1} x \Pi([0, t] \times dx), t \geq 0.$$

A m-dimensional $(\mathbb{P}, \mathbf{F})$-semimartingale X has independent increments if and only if the triplet (a, B, Π) is nonrandom. Assuming quasi-left-continuity in this case we have that

$$\mathbb{E}\left[\exp\{i(X_t - X_s, z)\} | \mathcal{F}_s\right] = \exp\{\Phi_{s,t}\},$$

where

$$\Phi_{s,t} = i(a_t - a_s, z) - \frac{1}{2}(z, (B_t - B_s)z) + \int_s^t \int_{\mathbb{R}_0^m} [e^{i(z,x)} - 1 - i(z, x) 1_{\{|x| \leq 1\}}] \Pi((s, t] \times dx).$$

If $X \in S^m(\mathbb{P}, \mathbf{F})$ with the triplet (α, B, Π) of $(\mathbb{P}, \mathbf{F})$-predictable characteristics and $\mathbf{F}^X \subseteq \mathbf{G} = (\mathcal{G}_t) \subseteq \mathbf{F}$, then $X \in S^m(\mathbb{P}, \mathbf{G})$ and $(\bar{\alpha}, B, \bar{\Pi})$ is the triplet of the $(\mathbb{P}, \mathbf{G})$-predictable characteristics, where $\bar{\alpha}, \bar{\Pi}$ are the $(\mathbb{P}, \mathbf{G})$-dual predictable projections of α, Π (see [11]).

Let $\mathbb{P} \overset{loc}{\sim} \tilde{\mathbb{P}}$ and $Z_t = \frac{d\tilde{\mathbb{P}}}{d\mathbb{P}}|_{\mathcal{F}_t}$, $t \geq 0$. We can uniquely decompose Z in the following way:

$$Z_t = Z_0 + X_t^c(g Z_-) + Q_t(\psi Z_-) + Z'_t$$

where $(Z_-)_t = Z_{t-}$, $t \geq 0$, $Z' \in \mathcal{M}_{loc}(\mathbb{P}, \mathbf{F})$, Z is orthogonal to each $X^c(h) \in \mathcal{M}_{loc}(X^c, \mathbb{P}, \mathbf{F})$ and each locally bounded $Q(\eta) \in \mathcal{M}_{loc}(q, \mathbb{P}, \mathbf{F})$. It is known that if $X \in S^m(\mathbb{P}, \mathbf{F})$ with the triplet (α, B, Π) of $(\mathbb{P}, \mathbf{F})$-predictable characteristics, $\Pi(\{t\} \times \mathbb{R}_0^m) = 0$, $t \geq 0$, then $X \in S^m(\tilde{\mathbb{P}}, \mathbf{F})$ with the triplet $(\tilde{\alpha}, B, \tilde{\Pi})$ of $(\tilde{\mathbb{P}}, \mathbf{F})$-predictable characteristics, where

$$\tilde{\alpha}_t = \alpha_t + \int_0^t g_s d B_s + \int_0^t \int_{|x| \leq 1} x \psi_s(x) \Pi(ds, dx),$$
$$\tilde{\Pi}(ds, dx) = (\psi_s(x) + 1) \Pi(ds, dx),$$

If, in addition, for each $t \geq 0$ $\mathbb{P}$-a.s.

$$\int_0^t \int_{|x| \leq 1} |x| |\psi_s(x)| \Pi(ds, dx) < \infty,$$

and $X \in S_p^m(\mathbb{P}, \mathbf{F})$ has canonical decomposition (4.4), then $X \in S_p^m(\tilde{\mathbb{P}}, \mathbf{F})$ with the $(\tilde{\mathbb{P}}, \mathbf{F})$-canonical decomposition $X_t = X_0 + \tilde{A}_t + \tilde{L}_t$, where $\tilde{L} \in \mathcal{M}_{loc}(\mathbb{P}, \mathbf{F})$,

$$\tilde{A}_t = A_t + \int_0^t g_s d B_s + \int_0^t \int_{\mathbb{R}_0^m} x \psi_s(x) \Pi(ds, dx), t \geq 0.$$

The following important Itô formula holds. If $X \in S^m(\mathbb{P}, \mathbf{F})$ and $F \in C^2(\mathbb{R}^m)$, then $F(X) \in S(\mathbb{P}, \mathbf{F})$ and

$$F(X_t) = F(X_0) + \int_0^t \partial_j F(X_{s-}) d X_s^j$$
$$+ \frac{1}{2} \int_0^t \partial_{jk}^2 F(X_s) d \langle X^{cj}, X^{ck} \rangle_s$$
$$+ \sum_{s \leq t} [F(X_s) - F(X_{s-}) - \sum_{j=1}^m \partial_j F(X_{s-}) \Delta X_s^j],$$

$t \geq 0$, where the stochastic integrals with respect to $(\mathbb{P}, \mathbf{F})$-semimartingale $X^j = A^j + L^j$ of the form $\int_0^t H_s d X_s^j$ for locally bounded function $H \in \mathcal{P}(\mathbf{F})$ are defined as a sum

$$\int_0^t H_s d A_s^j + \int_0^t H_s d L_s^j, \, j = 1, \ldots, m.$$

For $X \in S(\mathbb{P}, \mathbf{F})$, the Doleans–Dade exponential formula

$$Z_t = \mathcal{E}_t(X) = \exp\left\{X_t - \frac{1}{2}\langle X^c, X^c\rangle_t\right\} \Pi_{s\leq t}(1+\Delta X_s)e^{-\Delta X_s}, \, t \geq 0,$$

represents the solution to the linear stochastic equation

$$Z_t = 1 + \int_0^t Z_{s-} dX_s, \, t \geq 0.$$

Locally infinitely divisible processes

A semimartingale $X \in S^m(\mathbb{P}, \mathbf{F})$ is called a $(\mathbb{P}, \mathbf{F})$-locally infinitely divisible (we write $X \in LID(\mathbb{P}, \mathbf{F})$) if its $(\mathbb{P}, \mathbf{F})$-predictable characteristics (a, B, Π) are absolutely continuous with respect to Lebesgue measure, i.e.

$$a_t = \int_0^t a_s ds, \, B_t = \int_0^t C_s ds, \, \Pi([0,t] \times \Gamma) = \int_0^t \nu(s, \Gamma) ds, \, t \geq 0.$$

The triplet of $\mathcal{O}(\mathbf{F})$-measurable functions (a, C, ν) is called the $(\mathbb{P}, \mathbf{F})$-local characteristics of the process X. If $\mathbb{P} \overset{\text{loc}}{\sim} \tilde{\mathbb{P}}$ with the local density Z satisfying (4.3), then $X \in LID(\mathbb{P}, \mathbf{F})$ with the local characteristics $(\tilde{a}, C, \tilde{\nu})$, where

$$\tilde{a}_t = a_t + g_t C_t + \int_{|x|\leq 1} x\psi_t(x)\nu(t, dx),$$
$$\tilde{\nu}(t, dx) = (\psi_t(x) + 1)\nu(t, dx), \, t \geq 0.$$

If $X \in LID(\mathbb{P}, \mathbf{F})$ and $\mathbf{F}^X \subseteq \mathbf{G} = (\mathcal{G}_t) \subseteq \mathbf{F}$, then $X \in LID(\mathbb{P}, \mathbf{G})$ with local characteristics $(\bar{a}, A, \bar{\nu})$, where $\bar{a}_t = \mathbb{E}(a_t|\mathcal{G}_t)$, $\bar{\nu}(t, \Gamma) = \mathbb{E}(\nu(t, \Gamma)|\mathcal{G}_t), \, t \geq 0, \, \Gamma \in \mathcal{B}(\mathbb{R}_0^m)$, are $(\mathbb{P}, \mathbf{G})$-optional projections.

A process $X \in LID(\mathbb{P}, \mathbf{F})$ if and only if on some extension of the probability space $(\Omega, \mathcal{F}, \mathbb{P})$ a standard Brownian motion $W = (W^j)_{1\leq j \leq m}$ and a Poisson point process $\tilde{p}$ on $[0, \infty) \times \mathbb{R}_0^m$ can be constructed so that the following representation holds

$$X_t = X_0 + \int_0^t b_s ds + \int_0^t \sigma_s^j dW_s^j$$
$$+ \int_0^t \int_{|x|\leq 1} f_s(x)\tilde{q}(ds, dx) + \int_0^t \int_{|x|>1} f_s(x)p(ds, dx),$$

where $\tilde{q}(ds, dx) = \tilde{p}(ds, dx) - \frac{dt dx}{|x|^{m+1}}$ is a martingale measure, the functions b, f and σ_k, $k = 1, \ldots, m$, have explicit formulas expressing them by means of the local characteristics (a, C, ν) (for a more detailed formulation see [1] or [10]).

4.3 Stochastic nonlinear filtering equations

Consider a probability space $(\Omega, \mathcal{F}, \mathbb{P})$ with a right-continuous filtration of σ-algebras $\mathbf{F} = (\mathcal{F}_t)_{t \geq 0}$. Our observation is a filtration $\mathbf{Y} = (\mathcal{Y}_t)_{t \geq 0} \subseteq \mathbf{F}$ of σ-subalgebras of $\mathcal{F}$ such that $\mathcal{Y}_t \subseteq \mathcal{F}_t$ for $t \geq 0$. Let $\bar{M}^j \in \mathcal{M}_{loc}^c(\mathbb{P}, \mathbf{Y})$, $j = 1, \ldots, m$, $\bar{M} = (\bar{M}^j)_{1 \leq j \leq m}$. Denote $\bar{\gamma}_t^{jk} = \langle \bar{M}^j, \bar{M}^k \rangle_t$, $j, k = 1, \ldots, m$, $\Gamma_t = \left(\bar{\gamma}_t^{jk} \right)_{1 \leq j, k \leq m}$, $\gamma_t^{ij} = \frac{d\bar{\gamma}_t^{ij}}{d\gamma_t}$, $\gamma_t = \sum_i \bar{\gamma}_t^{ii}$.

Let $\tilde{p}$ be a $\mathbf{Y}$-adapted $\tilde{\mathcal{P}}(\mathbf{Y})$ σ-finite point process on $\mathbb{R}_+ \times U$ (U is a Lusin space): there is an increasing sequence $\tilde{U}_n \in \tilde{\mathcal{P}}(\mathbf{Y})$ such that

$$\mathbb{E} \int_0^\infty \int_U 1_{\tilde{U}_n}(s, x) \tilde{p}(ds, dx) < \infty.$$

Let $\bar{\Pi}$ be its $(\mathbb{P}, \mathbf{Y})$-conditional intensity measure, $\bar{q} = \tilde{p} - \bar{\Pi}$. Denote $a_s = \bar{\Pi}(\{s\} \times U)$,

$$J = \{(s, \omega) : a_s(\omega) > 0\}, \quad D = \{(s, \omega) : p(\omega, \{s\} \times U) = 1\}.$$

For a measurable function $f(x)$ on U, we denote $\hat{f}_t = \int_U f(x) \bar{\Pi}(\{t\} \times dx)$. Obviously $J = \cup_n [R_n]$, where $R_n \in \mathcal{T}_p(\mathbf{Y})$ are disjoint.

We will need the following assumptions.

A1. Assume that each $M \in \mathcal{M}_{loc}(\mathbb{P}, \mathbf{Y})$ has a form

$$M_t = M_0 + \bar{M}_t(g) + \bar{Q}_t(\psi), \quad t \geq 0,$$

for some $g \in L_{loc}^2(\Gamma, \mathbb{P}, \mathbf{Y})$, $\psi \in \mathcal{G}_{loc}(\bar{\Pi}, \mathbb{P}.\mathbf{Y})$, where $\bar{M}_t(g) = \int_0^t g_s d\bar{M}_s$, $\bar{Q}_t(\psi) = \int_0^t \int \psi_s(x) \bar{q}(ds, dx)$.

A2. Assume that $\bar{M}^j \in \mathcal{S}(\mathbb{P}, \mathbf{F})$ and there is a $\mathcal{P}(\mathbf{F})$-measurable function $H = (H^i)_{1 \leq i \leq m}$ such that

$$\bar{M}_t = \int_0^t H_s d\Gamma_s + M_t,$$

where $M = (M^j)$, $M^j \in \mathcal{M}_{loc}^c(\mathbb{P}, \mathbf{F})$, $j = 1, \ldots, m$.

A3. Assume that there is a $\tilde{\mathcal{P}}(\mathbf{F})$-measurable function $\chi \geq 0$ such that

$$\bar{\Pi}(dt, dx) = \chi_t(x) \Pi(dt, dx),$$

where Π is the $(\mathbb{P}, \mathbf{F})$-conditional intensity measure of $\tilde{p}$.

Assume we are given $\theta \in \mathcal{S}_p(\mathbb{P}, \mathbf{F})$ (a functional of the signal process) such that

$$\theta_t = \theta_0 + A_t + L_t, \quad t \geq 0, \tag{4.6}$$

where $A \in \mathcal{A}_{loc}(\mathbb{P}, \mathbf{F})$ and $L \in \mathcal{M}_{loc}(\mathbb{P}, \mathbf{F})$. According to [1],

$$L_t = M_t(g) + Q_t(F) + K_t + \tilde{L}_t + L_t',$$

where $F \in \mathcal{G}_{loc}(\Pi, \mathbb{P}, \mathbf{F})$, $g \in L^2_{loc}(\Gamma, \mathbb{P}, \mathbf{F})$, $K \in \mathcal{M}^c_{loc}(\mathbb{P}, \mathbf{F})$, $\langle K, M(g) \rangle_t = 0$, $t \geq 0$,

$$\tilde{L}_t = \int_0^t \int V_s(x) p(ds, dx) \in \mathcal{M}^d_{loc}(\mathbb{P}, \mathbf{F}),$$

V is $\tilde{\mathcal{O}}(\mathbf{F})$-measurable, and $L' \in \mathcal{M}^d_{loc}(\mathbb{P}, \mathbf{F})$ is such that

$$\{(s, \omega) : \Delta L'_s(\omega) \neq 0\} \subseteq D^c,$$

where $D = \{(s, \omega) : \bar{p}(\{s\} \times U) \neq 0\}$. We will need the following assumption.

A4. There is a sequence $(T_n) \in \mathcal{T}_f(\mathbf{Y})$ such that for each n $\theta_{\cdot \wedge T_n}$ is of class (D), $\mathbb{E}|A|_{T_n} < \infty$, and

$$\mathbb{E} \int_0^{T_n} |\gamma_s^{ij}(g_s^i + \theta_{s-} H_s^i)| d\gamma_s < \infty;$$

Assume for each k, m there is a sequence $(S_n) \in \mathcal{T}_f(\mathbf{Y})$ such that

$$\mathbb{E} \int_0^{S_n} \int_U |(\theta_{s-} + \Delta A_s 1_J(s) - \widehat{F\chi}_s)[\chi_s(x) - 1]$$
$$+ F_s(x)\chi_s(x) - \widehat{F\chi}_s | 1_{J^c \cup \tilde{R}_m \cup \tilde{U}_k}(s, x) \bar{\Pi}(ds, dx)$$
$$< \infty$$

for all n.

We denote $\bar{\theta}_t = \mathbb{E}[\theta_t | \mathcal{Y}_t]$ the $(\mathbb{P}, \mathbf{Y})$-optional projection of θ_t, $t \geq 0$.

Theorem 4.1 *Under the assumptions A1–A4 the following equality holds for all $t \geq 0$ $\mathbb{P}$-a.s.*

$$\bar{\theta}_t = \bar{\theta}_0 + \bar{A}_t + \int_0^t \bar{g}_s d\bar{M}_s + \int_0^t \int_U \bar{K}_s(x) \bar{q}(ds, dx), \tag{4.7}$$

where

$$\bar{g}_t = \bar{\mathcal{P}}\{\theta_{t-} H_t + g_t\},$$
$$\bar{K}_t(x) = \bar{\mathcal{P}}\left\{[\theta_{t-} + (\Delta A_t - \widehat{F\chi}_t)1_J(t)]\psi_t(x) + F_t(x)\chi_t(x)\right\},$$
$$\psi_t(x) = \chi_t(x) - 1 + \frac{\hat{\chi}_t - a_t}{1 - a_t} 1_{a_t < 1},$$

and $\bar{A}$ is the $(\mathbb{P}, \mathbf{Y})$-dual predictable projection of A and $\bar{\mathcal{P}}$ denotes the $(\mathbb{P}, \mathbf{Y})$-predictable projection.

Proof. Since θ is $\mathbf{Y}$-locally of class (D), its $(\mathbb{P}, \mathbf{Y})$-optional projection $\bar{\theta}_t =^1 \theta_t$ is $\mathbf{Y}$-locally of class (D) as well and

$$\bar{\theta}_t = \bar{\theta}_0 + \bar{A}_t + \bar{L}_t,$$

where $\bar{L} \in \mathcal{M}_{loc}(\mathbb{P}, \mathbf{Y})$. By assumption A1,

$$\bar{L}_t = \bar{M}_t(\tilde{h}) + \bar{Q}_t(\lambda),$$

and $\tilde{h} \in L^2_{loc}(\Gamma, \mathbb{P}, \mathbf{Y})$, $\lambda \in G_{loc}(\bar{\Pi}, \mathbb{P}, \mathbf{Y})$. Localizing we can assume $\lambda \in \mathcal{G}(\bar{\Pi}, \mathbb{P}, \mathbf{Y})$, $\tilde{h} \in L^2(\Gamma, \mathbb{P}, \mathbf{Y})$, θ is of class (D), $\mathbb{E}|A|_\infty < \infty$ and

$$\mathbb{E} \int_0^\infty \left| \gamma_s^{ij} \left(g_s^i + \theta_{s-} H_s^i \right) \right| d\gamma_s < \infty.$$

Therefore $\bar{\theta}_t$ is of class (D), $\mathbb{E}|\bar{A}|_\infty < \infty$. Take a bounded $\bar{N}_t = \bar{N}_0 + \bar{M}_t(h) + \bar{Q}_t(\eta)$ with $h \in L^2(\Gamma, \mathbb{P}, \mathbf{Y})$, $\eta \in \mathcal{G}(\bar{\Pi}, \mathbb{P}, \mathbf{Y})$. Assume $\int_0^\infty \sum_{i,j} \gamma_s^{ij} h_s^i h_s^j d\gamma_s$, h, η are bounded and there are m, k such that

$$\eta = 1_{\tilde{U}_k \cup J^c \cup \tilde{R}_m} \eta,$$

where $\tilde{R}_m = \cup_{n \leq m}[R_n]$. By Itô formula,

$$\bar{\theta}_t \bar{N}_t = \theta_0 \bar{N}_0 + \int_0^t \bar{\theta}_{s-} d\bar{N}_s + \int_0^t \bar{N}_{s-} d\bar{A}_s + \int_0^t \bar{N}_{s-} d\bar{L}_s$$
$$+ \int_0^t \gamma_s^{ij} \tilde{h}_s^i h_s^j d\gamma_s + \sum_{s \leq t} \Delta \bar{A}_s \Delta \bar{N}_s + \int_0^t \int \Delta \bar{L}_s \eta_s(x) \bar{p}(ds, dx)$$
$$- \sum_{s \leq t} \Delta \bar{L}_s \hat{\eta}_s = \bar{\theta}_0 \bar{N}_0 + \bar{m}_t + \int_0^t \bar{N}_{s-} d\bar{A}_s$$
$$+ \int_0^t \gamma_s^{ij} \tilde{h}_s^i h_s^j d\gamma + \int_0^t \int (\lambda_s(x) - \hat{\lambda}_s) \eta_s(x) \bar{p}(ds, dx),$$

where

$$\bar{m}_t = \int_0^t \bar{\theta}_{s-} d\bar{N}_s + \int_0^t \bar{N}_{s-} d\bar{L}_s + \sum_{s \leq t} \Delta \bar{A}_s \Delta \bar{N}_s - \sum_{s \leq t} \Delta \bar{L}_s \hat{\eta}_s, \, t \geq 0,$$

is a $(\mathbb{P}, \mathbf{Y})$-martingale. Let $(\tau_n) \in \mathcal{T}_f(\mathbf{Y})$ be such that $\mathbb{E}\bar{m}_{\tau_n} = 0$ for all n. For any bounded $\tau \in \mathcal{T}(\mathbf{Y})$,

$$\mathbb{E}\bar{\theta}_\tau \bar{N}_\tau = \lim_{n \to \infty} \bar{\theta}_{\tau \wedge \tau_n} \bar{N}_\tau = \mathbb{E}\bar{\theta}_0 \bar{N}_0 + \mathbb{E} \int_0^\tau \bar{N}_{s-} d\bar{A}_s + \mathbb{E} \int_0^\tau \gamma_s^{ij} \tilde{h}_s^i h_s^j d\gamma_s \quad (4.8)$$
$$+ \mathbb{E} \int_0^\tau \int (\lambda_s(x) - \hat{\lambda}_s) \eta_s(x) \bar{\Pi}(ds, dx).$$

Now fix $k \geq 1, m \geq 1$. For a bounded $\mathcal{P}(\mathbf{Y})$-measurable function $\tilde{h}_s = (h_s^i)_{1 \leq i \leq m}$, and a bounded $\tilde{\mathcal{P}}(\mathbf{Y})$-measurable function $\tilde{\eta}_s(x)$ we define

$$\tilde{N}_t = \bar{M}_t(\tilde{h}) + \bar{Q}_t(1_{J^c \cup \tilde{R}_m \cup \tilde{U}_k} \tilde{\eta}),$$

$\tau_p = \inf(t : |\tilde{N}_t| \geq p)$, $h = 1_{[0, \tau_p]} \tilde{h}$, $\eta = 1_{[0, \tau_p]} 1_{J^c \cup \tilde{R}_m \cup \tilde{U}_k} \tilde{\eta}$ and

$$\bar{N}_t = \bar{M}_t(h) + \bar{Q}_t(\eta).$$

Then, $\bar{N}_t = \bar{N}_0 + N_t + D_t$, where

$$N_t = M_t(h) + Q_t(\eta),$$

and

$$D_t = \int_0^t \gamma_s^{ij} H_s^i h_s^j d\gamma_s + \int_0^t \int [\chi_s(x) - 1]\eta_s(x)\tilde{\Pi}(ds, dx).$$

By the Itō formula,

$$\theta_t \bar{N}_t = \theta_0 \bar{N}_0 + \int_0^t \theta_{s-} d\bar{N}_s + \int_0^t \bar{N}_{s-} d\theta_s + \langle \theta^c, \bar{N}^c \rangle_t + \sum_{s \leq t} \Delta\theta_s \Delta \bar{N}_s$$

$$= \theta_0 \bar{N}_0 + \int_0^t \theta_{s-} dN_s + \int_0^t \bar{N}_{s-} dL_s + \int_0^t \theta_{s-} dD_s + \int_0^t \bar{N}_{s-} dA_s$$

$$+ \int_0^t \gamma_s^{ij} g_s^i h_s^j d\gamma_s + \sum_{s \leq t} \Delta A_s \Delta \bar{N}_s + \sum_{s \leq t} \Delta L_s \Delta \bar{N}_s$$

Since

$$\sum_{s \leq \tau} \Delta A_s \Delta \bar{N}_s = \sum_{s \leq \tau} \Delta A_s \Delta N_s + \sum_{s \leq \tau} \Delta A_s \Delta D_s,$$

and

$$\sum_{s \leq \tau} \Delta L_s \Delta \bar{N}_s = \sum_{s \leq \tau} \Delta L_s \left(\int \eta_s(x)(\bar{p}(\{s\} \times dx) - \hat{\eta}_s \right)$$

$$= \int_0^t \int [(F_s(x) - \widehat{F\chi_s}) + V_s(x)]\eta_s(x)\bar{p}(ds, dx) - \sum_{s \leq \tau} \Delta L_s \hat{\eta}_s,$$

we have

$$\theta_t \bar{N}_t = \theta_0 \bar{N}_0 + m_t + \sum_{s \leq t} \Delta A_s \Delta D_s + \int_0^t \sum_{i,j} \gamma_s^{ij} \left(g_s^i + \theta_{s-} H_s^i \right) h_s^j d\gamma_s$$

$$+ \int_0^t \int (F_s(x) - \widehat{F\chi_s})\eta(s, x)\bar{p}(ds, dx)$$

$$+ \int_0^t \int \theta_{s-}[\chi_s(x) - 1]\eta(s, x)\tilde{\Pi}(ds, dx) + \int_0^t \bar{N}_{s-} dA_s,$$

where

$$m_t = \int_0^t \theta_{s-} dN_s + \int_0^t \bar{N}_{s-} dL_s + \sum_{s \leq t} \Delta A_s \Delta N_s$$

$$- \sum_{s \leq t} \Delta L_s \hat{\eta}_s + \int_0^t \int V_s(x)\eta_s(x)\bar{p}(ds, dx),$$

$t \geq 0$, is a local $(\mathbb{P}, \mathbf{F})$-martingale. So,

$$\theta_t \bar{N}_t = \theta_0 \bar{N}_0 + m_t + \int_0^t \int (\theta_{s-} + \Delta A_s 1_J(s))[\chi_s(x) - 1]\eta_s(x)\bar{\Pi}(ds, dx)$$
$$+ \int_0^t \gamma_s^{ij}(g_s^i + \theta_{s-} H_s^i) h_s^j d\gamma_s$$
$$+ \int_0^t \int (F_s(x) - \widehat{F\chi_s})\eta(s, x)\bar{p}(ds, dx) + \int_0^t \bar{N}_{s-} d A_s,$$

$t \geq 0$. By our assumption A4, there is $(S_l) \in \mathcal{T}_f(\mathbf{Y})$ such that for each l

$$\mathbb{E} \int_0^{S_l} \int_U \{|(\theta_{s-} + \Delta A_s 1_J(s) - \widehat{F\chi_s})[\chi_s(x) - 1]$$
$$+ F_s(x) - \widehat{F\chi_s}|\} 1_{J^c \cup \bar{R}_m \cup \tilde{U}_k}(s, x)\bar{\Pi}(ds, dx)$$
$$< \infty.$$

Let $(T_n) \in \mathcal{T}_f(\mathbf{F})$ be such that $\mathbb{E}m_{T_n} = 0$ and

$$\mathbb{E}[\int_0^{T_n} \int_U \frac{(F_s(x) - \widehat{F\chi_s})^2}{1 + |F_s(x) - \widehat{F\chi_s}|} \chi_s(x)\bar{\Pi}(ds, dx) + \sum_{s \leq T_n}(1 - a_s)\frac{\widehat{F\chi_s}^2}{1 + |\widehat{F\chi_s}|}] < \infty$$

for each n. Then for any bounded $\tau \in \mathcal{T}(\mathbf{F})$ and $l \geq 1$,

$$\mathbb{E}\theta_{\tau \wedge S_l} \bar{N}_{\tau \wedge S_l}$$
$$= \lim_{n \to \infty} \mathbb{E}\theta_{\tau \wedge S_l \wedge T_n} \bar{N}_{\tau \wedge S_l \wedge T_n}$$
$$= \mathbb{E}\theta_0 \bar{N}_0 + \mathbb{E} \int_0^{\tau \wedge S_l} \bar{N}_{s-} d A_s + \mathbb{E} \int_0^{\tau \wedge S_l} \gamma_s^{ij}(g_s^i + \theta_{s-} H_s^i) h_s^j d\gamma_s$$
$$+ \mathbb{E} \int_0^{\tau \wedge S_l} \int \{(\theta_{s-} + \Delta A_s 1_J(s))[\chi_s(x) - 1] + (F_s(x) - \widehat{F\chi_s})\chi_s\} \eta_s(x)\bar{\Pi}(ds, dx)$$
$$= \mathbb{E} \int_0^{\tau \wedge S_l} \bar{N}_{s-} d \bar{A}_s + \mathbb{E} \int_0^{\tau \wedge S_l} \gamma_s^{ij} \bar{\mathcal{P}}\{g_s^i + \theta_{s-} H_s^i\} h_s^j d\gamma_s$$
$$+ \mathbb{E} \int_0^{\tau \wedge S_l} \int \bar{\mathcal{P}}\{(\theta_{s-} + \Delta A_s 1_J(s))[\chi_s(x) - 1]$$
$$+ (F_s(x) - \widehat{F\chi_s})\chi_s(x)\}\eta_s(x)\bar{\Pi}(ds, dx).$$

Comparing this formula to (4.8) we see that for all l, p,

$$\mathbb{E}[\int_0^{\tau \wedge S_l \wedge \tau_p} \gamma_s^{ij} \tilde{h}_s^i h_s^j d\gamma_s + \int_0^{\tau \wedge S_l \wedge \tau_p} \int (\lambda_s(x) - \hat{\lambda}_s)\eta_s(x)\bar{\Pi}(ds, dx)]$$
$$= \mathbb{E}[\int_0^{\tau \wedge S_l \wedge \tau_p} \gamma_s^{ij} \bar{\mathcal{P}}\{g_s^i + \theta_{s-} H_s^i\} h_s^j d\gamma_s$$
$$+ \int_0^{\tau \wedge S_l \wedge \tau_p} \int \bar{\mathcal{P}}\{(\theta_{s-} + \Delta A_s 1_J(s))[\chi_s(x) - 1] + (F_s(x) - \widehat{F\chi_s})\chi_s(x)\}$$
$$\eta_s(x)\bar{\Pi}(ds, dx)].$$

Passing to the limit as $p \to \infty$, we have for all l,

$$\mathbb{E}[\int_0^{\tau \wedge S_l} \gamma_s^{ij} \tilde{h}_s^i \tilde{h}_s^j d\gamma_s + \int_0^{\tau \wedge S_l} \int (\lambda_s(x) - \hat{\lambda}_s) 1_{J^c \cup \tilde{R}_m \cup \tilde{U}_k} \tilde{\eta} \tilde{\Pi}(ds, dx)]$$
$$= \mathbb{E}[\int_0^{\tau \wedge S_l} \gamma_s^{ij} \bar{\mathcal{P}} \{g_s^i + \theta_{s-} H_s^i\} \tilde{h}_s^j d\gamma_s$$
$$+ \int_0^{\tau \wedge S_l} \int \bar{\mathcal{P}} \{(\theta_{s-} + \Delta A_s 1_J(s))[\chi_s(x) - 1] + (F_s(x) - \widehat{F\chi_s})\chi_s(x)\}$$
$$1_{J^c \cup \tilde{R}_m \cup \tilde{U}_k} \tilde{\eta} \tilde{\Pi}(ds, dx)].$$

Since $\tilde{h}$ and $\tilde{\eta}$ are arbitrary bounded functions, $d\Gamma d\mathbb{P}$-a.e.

$$\tilde{h}_s = \bar{\mathcal{P}} \{g_s^i + \theta_{s-} H_s^i\},$$

and $d\tilde{\Pi} d\mathbb{P}$-a.e. on $J^c \cup \tilde{R}_m \cup \tilde{U}_k$,

$$\lambda_s(x) - \hat{\lambda}_s = \bar{\mathcal{P}} \{(\theta_{s-} + \Delta A_s 1_J(s))[\chi_s(x) - 1] + (F_s(x) - \widehat{F\chi_s})\chi_s(x)\}.$$

Since k, m are arbitrary, the statement follows. $\square$

Equation (4.7) is simpler if $\mathbb{P}$-a.s. $\Delta A_t = \widehat{F\chi_t}$ for all $t \geq 0$.

Corollary 4.1 *Assume $\mathbb{P}$-a.s. $\Delta A_t = \widehat{F\chi_t}$ for all $t \geq 0$ and A1–A4 hold. Then $\mathbb{P}$-a.s. for all $t \geq 0$,*

$$\bar{\theta}_t = \bar{\theta}_0 + \bar{A}_t + \bar{M}_t(\bar{g}) + \bar{Q}_t(\bar{K}),$$

where

$$\bar{g}_t = \bar{\mathcal{P}}\{\theta_{t-} H_t + g_t\},$$
$$\bar{K}_t(x) = \bar{\mathcal{P}} \left\{ \theta_{t-} \left(\chi_t(x) - 1 + \frac{\hat{\chi}_t - a_t}{1 - a_t} 1_{a_t < 1} \right) + F_t(x) \chi_t(x) \right\},$$

and $\bar{A}$ is the $(\mathbb{P}, \mathbf{Y})$-dual predictable projection of A.

Remark 4.1 Note that **A1** is fulfilled for a given $\bar{M}$ and $\bar{p}$ if for each measure $\mathbb{P}'$ on $\mathcal{Y}_\infty = \vee_{t \geq 0} \mathcal{Y}_t$ such that $\mathbb{P}'|_{\mathcal{Y}_0} = \mathbb{P}|_{\mathcal{Y}_0}$, the $(\mathbb{P}', \mathbf{Y})$-conditional intensity measure of $\bar{p}$ is $\bar{\Pi}$, $\bar{M}^j \in \mathcal{M}_{loc}^c(\mathbb{P}', \mathbf{Y})$ and $\bar{M}^j \bar{M}^k - \tilde{\gamma}^{jk} \in \mathcal{M}^c(\mathbb{P}', \mathbf{Y})$, $j, k, = 1, \ldots, m$, we have that $\mathbb{P}'|_{\mathcal{Y}_\infty} = \mathbb{P}|_{\mathcal{Y}_\infty}$ (see [1]).

Remark 4.2 Instead of $(\bar{M}_j)_{1 \leq j \leq m}$, an infinite dimensional continuous local martingale $\bar{M}_t$ can be considered as well (see [25]).

4.4 Reduced (unnormalized) form of stochastic nonlinear filtering equation

As in the previous section, we consider a probability space $(\Omega, \mathcal{F}, \mathbb{P})$ with a filtration of σ-algebras $\mathbf{F}$. Our observation is a filtration $\mathbf{Y} = (\mathcal{Y}_t)_{t \geq 0} \subseteq \mathbf{F}$ of σ-subalgebras of $\mathcal{F}$ such that $\mathcal{Y}_t \subseteq \mathcal{F}_t$ for $t \geq 0$.

In this section we will make the following assumptions.

C0. *Assume there is a probability measure $\tilde{\mathbb{P}}$ on $(\Omega, \mathcal{F})$ and a sequence $(T_n) \in \mathcal{T}_f(\mathbf{Y})$ such that $d\mathbb{P}|_{\mathcal{F}_{T_n}} \ll d\tilde{\mathbb{P}}|_{\mathcal{F}_{T_n}}$ for all n with the density process Z_t;*

If **C0** holds,

$$\mathbb{E}[\theta_t | \mathcal{Y}_t] = \frac{\tilde{\mathbb{E}}[Z_t \theta_t | \mathcal{Y}_t]}{\tilde{\mathbb{E}}[Z_t | \mathcal{Y}_t]}.$$

We will use Theorem 4.1 to derive an equation for $\tilde{\mathbb{E}}[Z_t \theta_t | \mathcal{Y}_t]$.

Let $\tilde{M}^j \in \mathcal{M}^c_{loc}(\tilde{\mathbb{P}}, \mathbf{Y}) \cap \mathcal{M}^c_{loc}(\tilde{\mathbb{P}}, \mathbf{F})$, $j = 1, \ldots, m$, $\tilde{M} = (\tilde{M}^j)_{1 \leq j \leq m}$. Let $\tilde{p}$ be a $\mathbf{Y}$-adapted $\tilde{\mathcal{P}}(\mathbf{Y})$ σ-finite point process $\tilde{p}$ on $\mathbb{R}_+ \times U$ (U is a Lusin space): there is an increasing sequence $\tilde{U}_n \in \tilde{\mathcal{P}}(\mathbf{Y})$ such that

$$\tilde{\mathbb{E}} \int_0^\infty \int_U 1_{\tilde{U}_n}(s, x) \tilde{p}(ds, dx) < \infty.$$

Let its $(\tilde{\mathbb{P}}, \mathbf{F})$-conditional intensity measure $\tilde{\Pi}$ be $\mathbf{Y}$-adapted. Denote $\tilde{\gamma}^{jk}_t = \langle \tilde{M}^j, \tilde{M}^k \rangle_t$, $j, k = 1, \ldots, m$, $\Gamma_t = (\tilde{\gamma}^{jk}_t)_{1 \leq j,k \leq m}$, $\gamma^{ij}_t = \frac{d\tilde{\gamma}^{ij}_t}{d\gamma_t}$, $\gamma_t = \sum_i \tilde{\gamma}^{ii}_t$.

Let $\tilde{q} = \tilde{p} - \tilde{\Pi}$. Denote $a_s = \tilde{\Pi}(\{s\} \times U)$, $\tilde{q}(ds, du) = \tilde{p}(ds, du) - \tilde{\Pi}(ds, du)$

$$J = \{(s, \omega) : a_s(\omega) > 0\}, \quad D = \{(s, \omega) : \tilde{p}(\omega, \{s\} \times U) = 1\}.$$

For a measurable function $f(x)$ on U we denote $\hat{f}_t = \int_U f(x) \tilde{\Pi}(\{t\} \times dx)$. Obviously $J = \cup_n [R_n]$, where $R_n \in \mathcal{T}_p(\mathbf{Y})$ are disjoint.

C1. *Assume that each $M \in \mathcal{M}_{loc}(\tilde{\mathbb{P}}, \mathbf{Y})$ has a form*

$$M_t = M_0 + \tilde{M}_t(g) + \tilde{Q}_t(\psi), \quad t \geq 0,$$

for some $g \in L^2_{loc}(\Gamma, \tilde{\mathbb{P}}, \mathbf{Y})$, $\psi \in \mathcal{G}_{loc}(\tilde{\Pi}, \tilde{\mathbb{P}}, \mathbf{Y})$, *where* $\tilde{M}_t(g) = \int_0^t g_s d\tilde{M}_s$, $\tilde{Q}_t(\psi) = \int_0^t \int \psi_s(x) \tilde{q}(ds, dx)$.

Assume $\theta \in S_p(\tilde{\mathbb{P}}, \mathbf{F})$ (a functional of the signal process) is such that

$$\theta_t = \theta_0 + A_t + L_t, \quad t \geq 0, \tag{4.9}$$

where $A \in \mathcal{A}_{loc}(\tilde{\mathbb{P}}, \mathbf{F})$ and $L \in \mathcal{M}_{loc}(\tilde{\mathbb{P}}, \mathbf{F})$. According to [1],

$$L_t = \tilde{M}_t(g) + \tilde{Q}_t(F) + K_t + \tilde{L}_t + L'_t,$$

where $F \in \mathcal{G}_{loc}(\tilde{\Pi}, \tilde{\mathbb{P}}, \mathbf{F})$, $g \in L^2_{loc}(\Gamma, \tilde{\mathbb{P}}, \mathbf{F})$, $K \in \mathcal{M}^c_{loc}(\tilde{\mathbb{P}}, \mathbf{F})$, $\langle K, \tilde{M}(g) \rangle_t = 0$, $t \geq 0$,

$$\tilde{L}_t = \int_0^t \int V_s(x) \tilde{p}(ds, dx) \in \mathcal{M}^d_{loc}(\mathbb{P}, \mathbf{F}),$$

V is $\tilde{\mathcal{O}}(\mathbf{F})$-measurable, and $L' \in \mathcal{M}^d_{loc}(\tilde{\mathbb{P}}, \mathbf{F})$ is such that
$$\{(s, \omega) : \Delta L'_s(\omega) \neq 0\} \subseteq D^c,$$
where $D = \{(s, \omega) : \bar{p}(\{s\} \times U) \neq 0\}$.

C2. Let $\chi \geq 0$ is $\tilde{\mathcal{P}}(\mathbf{F})$-measurable, $\hat{\chi}_s \leq 1$, $\{a_s = 1\} \subseteq \{\hat{\chi}_s = 1\}$, where $\hat{\chi}_s = \int \chi_s(x) \tilde{\Pi}(\{s\} \times dx)$, and
$$\psi_t(x) = \chi_t(x) - 1 + \frac{\hat{\chi}_t - a_t}{1 - a_t} 1_{a_t < 1} \in \mathcal{G}_{loc}(\tilde{\Pi}, \tilde{\mathbb{P}}, \mathbf{F}),$$
$$l = (l^i)_{1 \leq i \leq m} \in L^2_{loc}(\Gamma, \tilde{\mathbb{P}}, \mathbf{F}).$$

Let
$$dZ_t = Z_{t-} l_t d\tilde{M}_t + Z_{t-} \int_U \psi_t(x) \tilde{q}(dt, dx),$$
$$Z_0 = 1.$$

C3. Assume there is a sequence $(T_n) \in \mathcal{T}_f(\mathbf{Y})$ such that for each n the product $Z_{\cdot \wedge T_n} \theta_{\cdot \wedge T_n}$ is of class (D),
$$\tilde{\mathbb{E}} \int_0^{T_n} Z_{s-} |\gamma_s^{ij} (g_s^i + \theta_{s-} l_s^i)| d\gamma_s < \infty$$

and $E|\tilde{A}|_{T_n} < \infty$, where
$$\tilde{A}_t = \int_0^t Z_{s-}[dA_s + \int_U (\chi_s(x) - 1) F_s(x) \tilde{\Pi}(ds, dx) + \sum_{i,j} \gamma_s^{ij} l_s^i g_s^j d\gamma_s],$$

$t \geq 0$.

Assume for each k, m there is a sequence $(S_n) \in \mathcal{T}_f(\mathbf{Y})$ such that
$$\tilde{\mathbb{E}} \int_0^{S_n} \int_U Z_{s-} \{[\theta_{s-} + (\Delta A_s - \hat{F}_s) 1_J(s)](\chi_s(x) - 1)$$
$$+ F_s(x) \chi_s(x) - \widehat{F\chi_s}\} 1_{J^c \cup \tilde{R}_m \cup \tilde{U}_k}(s, x) \tilde{\Pi}(ds, dx)$$
$$< \infty$$

for all n, where $\tilde{R}_m = \cup_{j \leq m}[R_j]$.

C4. Assume there is a sequence $(T_n) \in \mathcal{T}_f(\mathbf{F})$ such that for each n and $\tau \in \mathcal{T}(\mathbf{F})$, $\tau \leq T_n$,
$$\tilde{\mathbb{E}} \int_0^{T_n} \int |\chi_s(x) - 1| |F_s(x)| \tilde{\Pi}(ds, dx)$$
$$+ \int_0^{T_n} \int_U |V_s(x)(\psi_s(x) - \hat{\psi}_s)| \bar{p}(ds, dx)$$
$$+ \tilde{\mathbb{E}} |\Delta L'_\tau| |\Delta \hat{\psi}_\tau| < \infty.$$

The following statement holds.

Theorem 4.2 *Under the assumptions C0–C4, we have for all $t \geq 0$ $\mathbb{P}$-a.s.*

$$\tilde{\mathbb{E}}[\theta_t Z_t | \mathcal{Y}_t] = \tilde{\mathbb{E}}[\theta_0 | \mathcal{Y}_0] + \tilde{A}_t + \int_0^t \bar{g}_s \, d\tilde{M}_s + \int_0^t \int_U \bar{K}_s(x) \tilde{q}(ds, dx), \quad (4.10)$$

where

$$\bar{g}_t = \tilde{\mathcal{P}}\{Z_{s-}(\theta_{t-} l_t + g_t)\},$$
$$\bar{K}_t(x) = \tilde{\mathcal{P}}\{Z_{t-}[\theta_{t-} + (\Delta A_t - \hat{F}_t) 1_J(t)] \psi_t(x) + F_t(x) \chi_t(x)\},$$
$$\psi_t(x) = \chi_t(x) - 1 + \frac{\hat{\chi}_t - a_t}{1 - a_t} 1_{a_t < 1},$$

$\tilde{\mathcal{P}}$ *denotes $(\tilde{\mathbb{P}}, \mathbf{Y})$-predictable projection and $\tilde{A}_t$ is the $(\tilde{\mathbb{P}}, \mathbf{Y})$-dual predictable projection of*

$$\int_0^t Z_{s-}[dA_s + \int_U (\chi_s(x) - 1) F_s(x) \tilde{\Pi}(ds, dx) + \sum_{i,j} \gamma_s^{ij} l_s^i g_s^j \, d\gamma_s], \, t \geq 0.$$

Proof. By the Itō formula,

$$Z_t \theta_t = Z_0 \theta_0 + \int_0^t \theta_{s-} \, dZ_s + \int_0^t Z_{s-} \, dA_s + \int_0^t Z_{s-} \, dL_s$$
$$+ \int_0^t Z_s \gamma_s^{ij} l_s^i g_s^j \, d\gamma_s + \int_0^t \int_U Z_{s-} V_s(x) (\psi_s(x) - \widehat{\psi}_s) p(ds, dx)$$
$$+ \sum_{s \leq t} \Delta L_s' Z_{s-} \widehat{\psi}_s + \sum_{s \leq t} (\Delta A_s - \hat{F}_s) \Delta Z_s$$
$$+ \int_0^t \int_U Z_{s-} (\psi_s(x) - \widehat{\psi}_s) F_s(x) \bar{p}(ds, dx)$$

and, collecting terms,

$$Z_t \theta_t = Z_0 \theta_0 + \int_0^t Z_{s-}[dA_s + \int_U (\chi_s(x) - 1) F_s(x) \tilde{\Pi}(ds, dx) + \gamma_s^{ij} l_s^i g_s^j \, d\gamma_s]$$
$$+ \int_0^t Z_{s-} (\theta_{s-} l_s + g_s) d\tilde{M}_s$$
$$+ \int_0^t Z_{s-} \int_U \{F_s(x) \chi_s(x) + [\theta_{s-} + (\Delta A_s - \hat{F}_s) 1_J(s)] \psi_s(x)\} \tilde{q}(ds, dx)$$
$$+ \int_0^t Z_{s-} \, dK_s + \int_0^t \int_U Z_{s-} \chi_s(x) V_s(x) \bar{p}(ds, dx)$$
$$+ \int_0^t Z_{s-} (1 - \hat{\psi}_s) dL_s'.$$

The statement follows by Theorem 4.1. □

4.5 Examples

In this section we consider some applications of Theorem 4.1 and 4.2.

Example 4.2 (cf. [2]) Let $\mathbf{Y}=\mathbf{F}^Y$, where Y is defined by the equation (4.1). Assume θ_t (a functional of the signal X_t) satisfies (4.6). We have obviously, $\bar{p} = 0$,

$$M_t = \int_0^t b_s(Y) d W_s \in \mathcal{M}_{loc}^c(\mathbb{P}, \mathbf{F}),$$

$$\bar{M}_t = M_t + \int_0^t [a_s(X_s, Y) - \bar{\mathcal{P}}\{a_s(X_s, Y)\}]ds$$

$$= M_t + \int_0^t C_s C_s^{-1}[a_s(X_s, Y) - \bar{\mathcal{P}}\{a_s(X_s, Y)\}]ds$$

$$= Y_t - \int_0^t \bar{\mathcal{P}}\{a_s(X_s, Y)\}ds - Y_0 \in \mathcal{M}_{loc}^c(\mathbb{P}, \mathbf{Y}),$$

$$(\langle M^i, M^j \rangle_t) = (\langle \bar{M}^i, \bar{M}^j \rangle_t) = \Gamma_t = \int_0^t C_s ds, \ C_t = b_t(Y)b_t(Y)^T,$$

and

$$H_t = C_t^{-1}[a_t(X_t, Y) - \bar{\mathcal{P}}\{a_t(X_t, Y)\}].$$

Under the assumptions A1–A5, by Theorem 4.1,

$$\mathbb{E}[\theta_t|\mathcal{Y}_t] = \mathbb{E}[\theta_0|\mathcal{Y}_0] + \bar{A}_t + \int_0^t \bar{\mathcal{P}}\left\{\theta_s C_s^{-1}[a_s(X_s, Y) - \right.$$
$$\left. - \bar{\mathcal{P}}\{a_s(X_s, Y)\}] + g_s\right\} d\bar{M}_s.$$

On the other hand, let us define the probability measure $\tilde{\mathbb{P}}$ by means of the equalities

$$\frac{d\tilde{\mathbb{P}}}{d\mathbb{P}}\Big|_{\mathcal{F}_t} = \tilde{Z}_t, \ t \geq 0,$$

where $\tilde{Z}_t$ solves the linear equation

$$d\tilde{Z}_t = -\tilde{Z}_t C_t^{-1} a_t(X_t, Y) d M_t,$$
$$\tilde{Z}_0 = 0, \ t \geq 0.$$

Then with respect to $\tilde{\mathbb{P}}$,

$$Y_t = Y_0 + \tilde{M}_t,$$

where $\tilde{M}$ is continuous $(\mathbb{P}, \mathbf{F})$-martingale, $(\langle \tilde{M}^i, \tilde{M}^j \rangle_t)_{i,j} = \int_0^t C_s^{ij} ds$. The inverse density $Z_t = 1/\tilde{Z}_t$ satisfies the linear equation

$$d Z_t = Z_{t-} C_t^{-1} a(t, X_t, Y) d \tilde{M}_t, \ t \geq 0.$$

In this case

$$\mathbb{E}[\theta_t|\mathcal{Y}_t] = \frac{\tilde{\mathbb{E}}[Z_t\theta_t|\mathcal{Y}_t]}{\tilde{\mathbb{E}}[Z_t|\mathcal{Y}_t]}$$

and, by Theorem 4.2 (under the assumptions C0–C4), for $\tilde{\theta}_t = \tilde{\mathbb{E}}[Z_t\theta_t|\mathcal{Y}_t]$, the reduced linear filtering equation holds:

$$\tilde{\theta}_t = \tilde{\theta}_0 + \bar{A}_t + \int_0^t \tilde{\mathcal{P}}\left\{Z_s\theta_s C_s^{-1} a_s(X_s, Y) + Z_s g_s\right\} d\tilde{M}_s,$$

where $\tilde{\mathcal{P}}$ denotes the $(\tilde{\mathbb{P}}, \mathbf{Y})$-predictable projection, $\bar{A}_t$ is the $(\tilde{\mathbb{P}}, \mathbf{Y})$-dual predictable projection of A_t.

Example 4.3 (cf. [12])a) Assume $\mathbf{Y}=\mathbf{F}^Y$, where Y is m-dimensional quasi-left-continuous $(\mathbb{P}, \mathbf{F})$-semimartingale with the triplet (a, B, Π) of $(\mathbb{P}, \mathbf{F})$-predictable characteristics, i.e. (4.5) holds. Let

$$a_t = \tilde{a}_t + \int_0^t h_s d B_s + \int_0^t \int_{|y|\leq 1} y(\kappa_s(y) - 1)\tilde{\Pi}(ds, dy),$$
$$\Pi(dt, dy) = \kappa_t(y)\tilde{\Pi}(dt, dy), t \geq 0,$$
$$\tilde{a}^j \in \mathcal{A}_{loc}(\mathbb{P}, \mathbf{Y}) \cap \mathcal{P}(\mathbf{Y}), j = 1, \ldots, m,$$

$\tilde{\Pi}$ is $\mathbf{Y}$-predictable and $\mathcal{P}(\mathbf{Y})$ σ-finite, h is $\mathcal{P}(\mathbf{F})$-measurable, κ is $\tilde{\mathcal{P}}(\mathbf{F})$-measurable, $\kappa > 0$ $\tilde{\Pi}$-a.e. and the processes

$$\int_0^t |H_s|d\gamma_s, \int_0^t \int_{|y|\leq 1} |y|(\kappa_s(y) - 1)\tilde{\Pi}(ds, dy)$$

are $(\mathbb{P}, \mathbf{Y})$-locally integrable, $\gamma_t = \sum_{j=1}^m B_t^{jj}$. Obviously, $Y \in S^m(\mathbb{P}, \mathbf{Y})$ with the triplet $(\bar{a}, B, \bar{\Pi})$, where

$$\bar{a}_t = \tilde{a}_t + \int_0^t \tilde{\mathcal{P}}\{h_s\}d B_s + \int_0^t \int_{|y|\leq 1} y(\tilde{\mathcal{P}}\{\kappa_s(y)\} - 1)\tilde{\Pi}(ds, dy),$$
$$\bar{\Pi} = \tilde{\mathcal{P}}\{\kappa_s(y)\}\tilde{\Pi}.$$

We have

$$Y_t = Y_0 + \bar{a}_t + \bar{M}_t + \int_0^t \int_{|y|\leq 1} y\bar{q}(ds, dy) + \int_0^t \int_{|y|>1} y\bar{p}(ds, dy), t \geq 0,$$

where $\bar{p} = p^Y$ is the jump measure of Y, $\bar{q}(ds, dy) = \bar{p}(ds, dy) - \bar{\Pi}(ds, dy)$,

$$\bar{M}_t = Y_t^c + \int_0^t \left(h_s - \tilde{\mathcal{P}}\{h_s\}\right) d B_s.$$

In this case,
$$H_t = h_t - \bar{\mathcal{P}}\{h_t\},$$
$$M_t = Y_t^c,$$
$$\chi_t(y) = \frac{\kappa_t(y)}{\bar{\mathcal{P}}\{\kappa_t(y)\}}.$$

Assume θ_t (a functional of the signal X_t) satisfies (4.6). If A1–A4 holds, we obtain by Theorem 4.1,
$$\mathbb{E}[\theta_t|\mathcal{Y}_t] = \mathbb{E}[\theta_0|\mathcal{Y}_0] + \bar{A}_t + \bar{M}_t(\bar{g}) + \bar{Q}_t(\bar{K}),$$
with
$$\bar{g}_t = \bar{\mathcal{P}}\{\theta_{t-}[h_t - \bar{\mathcal{P}}\{h_t\}] + g_t\},$$
$$\bar{K}_t(x) = \bar{\mathcal{P}}\left\{\theta_{t-}\left(\frac{\kappa_t(y)}{\bar{\mathcal{P}}\{\kappa_t(y)\}} - 1\right) + F_t(y)\frac{\kappa_t(y)}{\bar{\mathcal{P}}\{\kappa_t(y)\}}\right\},$$
where $\bar{\mathcal{P}}$ is a $(\mathbb{P}, \mathbf{Y})$-predictable projection.

b) Assume we have a measure $\tilde{\mathbb{P}}$ on $(\Omega, \mathcal{F})$ such that $Y \in S^m(\tilde{\mathbb{P}}, \mathbf{F})$ is quasi-leftcontinuous with the triplet $(\tilde{a}, B, \tilde{\Pi})$. Let $\mathbf{Y} = (\mathcal{Y}_t)_{t\geq 0} = \mathbf{F}^Y$. Assume $\tilde{a}^j \in \mathcal{A}_{loc}(\mathbb{P}, \mathbf{Y}) \cap \mathcal{P}(\mathbf{Y})$, $j = 1, \ldots, m$, and $\tilde{\Pi}$ is $\mathbf{Y}$-predictable and $\mathcal{P}(\mathbf{Y})$ σ-finite. With respect to $\tilde{\mathbb{P}}$
$$Y_t = Y_0 + \tilde{a}_t + \tilde{M}_t + \int_0^t \int_{|x|\leq 1} x[\tilde{p}(ds, dx) - \tilde{\Pi}(ds, dx)]$$
$$+ \int_0^t \int_{|x|>1} x p^Y(ds, dx).$$

Assume θ_t (a functional of the signal X_t) satisfies (4.9).
Let $\mathbb{P}$ be a probability measure on $(\Omega, \mathcal{F})$ such that $d\mathbb{P}|_{\mathcal{F}_t} = Z_t d\tilde{\mathbb{P}}|_{\mathcal{F}_t}$ and
$$dZ_t = Z_{t-} h_t d\tilde{M}_t + Z_{t-}\int(\chi_t(x) - 1)[\tilde{p}(ds, dx) - \tilde{\Pi}(ds, dx)],$$
$$Z_0 = 1,$$
where $h \in L^2_{loc}(B, \tilde{\mathbb{P}}, \mathbf{F})$, $\chi - 1 \in \mathcal{G}_{loc}(\tilde{\Pi}, \tilde{\mathbb{P}}, \mathbf{F})$.

If C0–C4 hold, then, by Theorem 4.2, for $\tilde{\mathbb{E}}[Z_t\theta_t|\mathcal{Y}_t]$, the reduced linear filtering equation holds:
$$\tilde{\mathbb{E}}[Z_t\theta_t|\mathcal{Y}_t] = \tilde{\mathbb{E}}[\theta_0|\mathcal{Y}_0] + \tilde{A}_t + \tilde{M}_t(\bar{g}) + \tilde{Q}_t(\bar{K})$$
with
$$\bar{g}_s = \tilde{\mathcal{P}}\{Z_{s-}(\theta_s-h_s + g_s)\},$$
$$\bar{K}_s(x) = \tilde{\mathcal{P}}\{Z_{s-}[\theta_{s-}(\chi_s(x) - 1) + F_s(x)\chi_s(x)]\},$$

$\tilde{\mathcal{P}}$ is a $(\tilde{\mathbb{P}}, \mathbf{Y})$-predictable projection and $\tilde{A}_t$ is the $(\tilde{\mathbb{P}}, \mathbf{Y})$-dual predictable projection of

$$\int_0^t Z_{s-}\left[dA_s + \int_U (\chi_s(x)-1)F_s(x)\tilde{\Pi}(ds,dx) + \sum_{i,j}\gamma_s^{ij}l_s^i g_s^j d\gamma_s\right], t \geq 0.$$

Example 4.4 (cf.[4]) a) Let $(\Omega, \mathcal{F}, \mathbb{P})$ be a probability space with the filtration **F**. The signal process $X_t = (X_t^1, \ldots, X_t^d)$ and the observation process $Y_t = (Y_t^1, \ldots, Y_t^m)$ are **F**-adapted. Let $(U, \mathcal{U})$ be a Lusin space, p be **F**-optional point processes on $[0, \infty) \times U$. Assume there is a sequence $U_n \in \mathcal{U}$ such that $U_n \downarrow \emptyset$ and $p([0,t] \times (U \setminus U_n)) < \infty$ for all n and t, $\mathbb{P}$-a.s. Let $\Pi(ds, du) = \nu(s, du)ds$ be $(\mathbb{P}, \mathbf{F})$-conditional intensity measure of p and

$$q(ds, du) = p(ds, du) - \nu(s, du)ds.$$

Let p^Y be the jump measure of Y with $(\mathbb{P}, \mathbf{F})$ conditional intensity measure $\Pi^Y(ds, dy) = \nu^Y(s, dy)ds$ and

$$q^Y(ds, dy) = p^Y(ds, dy) - \nu^Y(s, dy)ds,$$
$$\nu^Y(s, dy) = \kappa_s(X_s, y)\tilde{\nu}(s, dy).$$

Assume p and p^Y do not have any common jumps. Let X_t and Y_t satisfy the equations

$$dX_t = b_t(X_t)\,dt + \sigma_t(X_t)\,dM_t + \int_{U_1} H_t(X_{t-}, u)q(dt, dy) + \int_{U \setminus U_1} H_t(X_{t-}, u)\,p(dt, du)]$$
$$+\lambda_t(X_t)\,dM_t^Y + \int_{|y| \leq 1} K_t(X_{t-}, y)q^Y(dt, dy) + \int_{|y| > 1} K_t(X_{t-}, y)\,p^Y(dt, dy)]$$

and

$$Y_t = Y_0 + \int_0^t \int_{|y|>1} y p^Y(ds, dy) + \int_0^t \int_{|y| \leq 1} y q^Y(ds, dy)$$
$$+ M_t^Y + \int_0^t \int_{|y| \leq 1} y[\kappa_s(X_s, y) - 1]\tilde{\nu}(s, dy) + \int_0^t C_s^Y h_s(X_s)ds + \int_0^t c_s ds,$$

where $M = (M^1, \ldots, M^n)$, $M^Y = (M^{Y,1}, \ldots, M^{Y,m})$, $M^i, M^{Y,j} \in \mathcal{M}_{loc}^c(\mathbb{P}, \mathbf{F})$, $1 \leq i \leq n, 1 \leq j \leq m$, $\langle M^k, M^{Y,j}\rangle_t = 0, t \geq 0, 1 \leq i \leq n, 1 \leq j \leq m$. Let $\mathbf{Y} = (\mathcal{Y}_t) = \mathbf{F}^Y$. Assume

(i)
$$d\langle M^k, M^l\rangle_t = C_t^{kl}dt, k,l = 1, \ldots, n,$$
$$d\langle M^{Y,k}, M^{Y,l}\rangle_t = C_t^{Y,kl}dt, k,l = 1, \ldots, m,$$

and $C^Y = (C^{Y,kl})$ and c are $\mathcal{P}(\mathbf{Y})$-measurable;

(ii) $\nu^Y(s, dy) = \kappa_s(X_s, y)\tilde{\nu}(s, dy)$, $\tilde{\nu}(s, dy) \geq 0$ is $\mathcal{P}(\mathbf{Y})$-measurable, $\kappa_s(X_s, y) > 0$, $\tilde{\Pi}(ds, dy) = \tilde{\nu}(s, dy)ds$ is $\tilde{\mathcal{P}}(\mathbf{Y})$ σ-finite;

In this case the $(\mathbb{P}, \mathbf{Y})$-conditional intensity of $\bar{p}(ds, dy) = p^Y(ds, dy)$ is

$$\bar{\Pi}(ds, dy) = \bar{\mathcal{P}}\{\kappa_s(X_{s-}, y)\}\tilde{\nu}(s, dy)ds$$

and

$$\Pi^Y(ds, dy) = \frac{\kappa_s(X_s, y)}{\bar{\mathcal{P}}\{\kappa_s(X_s, y)\}}\bar{\Pi}(ds, dy),$$

where $\bar{\mathcal{P}}$ is $(\mathbb{P}, \mathbf{Y})$-predictable projection. For $f \in C_0^\infty(\mathbb{R})$, by the Itô formula,

$$f(X_t) = f(X_0) + \int_0^t L_s f(X_s)ds + \int_0^t \int_U [f(X_{s-} + H_s(X_{s-}, u)) - f(X_{s-})] q(ds, du)$$

$$+ \int_0^t \int_{\mathbb{R}^m} [f(X_{s-} + K_s(X_{s-}, y)) - f(X_{s-})] q^Y(ds, dy)$$

$$+ \int_0^t \nabla f(X_s) \cdot \sigma_s(X_s) dM_s + \int_0^t \nabla f(X_s) \cdot \lambda_s(X_s) dM_s^Y]\},$$

where

$$L_s f(x) = \nabla f(x) \cdot b_s(x) + \frac{1}{2}\partial_{kl}^2 f(x)\sigma_s^{kj}(x) \cdot \sigma_s^{lj'}(x) C_s^{jj'}$$

$$+ \frac{1}{2}\partial_{kl}^2 f(x)\lambda_s^{kj}(x) \cdot \lambda_s^{lj'}(x) C_s^{Y,jj'}$$

$$+ \int_U [f(x + H_s(x, u)) - f(x) - \nabla f(x) \cdot H_s(x, u)1_{U_1}(u)]\nu(s, du)$$

$$+ \int_{\mathbb{R}^m} [f(x + K_s(x, y)) - f(x) - \nabla f(x) \cdot K_s(x, y)1_{|y|\leq 1}]\kappa_s(x, y)\tilde{\nu}(s, dy).$$

Let $b_s(x), \lambda_s(x), h_s(x), \sigma_s(x)$ be $\mathcal{P}(\mathbf{Y}) \otimes \mathcal{B}(\mathbb{R}^d)$-measurable, $K_s(x, y), \kappa_s(x, y)$ be $\mathcal{P}(\mathbf{Y}) \otimes \mathcal{B}(\mathbb{R}^d) \otimes \mathcal{B}(\mathbb{R}^m)$-measurable, and $H_s(x, u)$ be $\mathcal{P}(\mathbf{Y}) \otimes \mathcal{B}(\mathbb{R}^d) \otimes \mathcal{U}$-measurable.

Assuming A1–A4 holds with $\theta_t = f(X_t)$, $\bar{q}(ds, dy) = \bar{p}(ds, dy) - \bar{\Pi}(ds, dy)$,

$$\bar{M}_t = M_t^Y + \int_0^t C_s^Y[h_s(X_s) - \bar{\mathcal{P}}\{h_s(X_s)\}]ds,$$

$$\chi_t(y) = \frac{\kappa_t(X_{t-}, y)}{\bar{\mathcal{P}}\{\kappa_t(X_{t-}, y)\}}, \quad H_t = h_t(X_t) - \bar{\mathcal{P}}\{h_t(X_t)\},$$

we obtain for $\pi_t(f) = \mathbb{E}[f(X_t)|\mathcal{Y}_t]$ by Theorem 4.1 and Lemma 4.3,

$$\pi_t(f) = \pi_0(f) + \int_0^t \pi_{s-}(L_s f)ds + \int_0^t \bar{g}_s d\bar{M}_s + \int_0^t \int K_s(y)\bar{q}(ds, dy),$$

with

$$\tilde{g}_t = \pi_{t-}\{f[h_t - \pi_{t-}(h_t)] + \nabla f \cdot \lambda_t\},$$

$$\bar{K}_t(y) = \int_{\mathbb{R}^d} \{f(x)\left(\frac{\kappa_t(x, y)}{\pi_{t-}(\kappa_t(\cdot, y))} - 1\right)$$

$$+ [f(x + K_t(x, y)) - f(x)]\frac{\kappa_t(x, y)}{\pi_{t-}(\kappa_t(\cdot, y))}\}\pi_{t-}(dx).$$

b) (Reduced equation) Let $(\Omega, \mathcal{F}, \tilde{\mathbb{P}})$ be a probability space with the filtration **F**. The signal process $X_t = (X_t^1, \ldots, X_t^d)$ and the observation process $Y_t = (Y_t^1, \ldots, Y_t^m)$ are **F**-adapted. Let $(U, \mathcal{U})$ be a Lusin space, p be **F**-optional point processes on $[0, \infty) \times U$. Assume there is a sequence $U_n \in \mathcal{U}$ such that $U_n \downarrow \emptyset$ and $p([0, t] \times (U \setminus U_n)) < \infty$ for all n and t, $\tilde{\mathbb{P}}$-a.s. Let $\Pi(ds, du) = \nu(s, du)ds$ be $(\tilde{\mathbb{P}}, \mathbf{F})$-conditional intensity measure of p and

$$q(ds, du) = p(ds, du) - \Pi(ds, du)$$
$$= p(ds, du) - \nu(s, du)ds.$$

Let p^Y be the jump measure of Y with $(\tilde{\mathbb{P}}, \mathbf{F})$ conditional intensity measure $\Pi^Y(ds, dy) = \nu^Y(s, dy)ds$ and

$$q^Y(ds, dy) = p^Y(ds, dy) - \nu^Y(s, dy)ds.$$

Assume p and p^Y do not have any common jumps. Let X_t and Y_t satisfy the equations

$$dX_t = b_t(X_t)\,dt + \sigma_t(X_t)\,dM_t + \int_{U_1} H_t(X_{t-}, y)q(dt, dy) + \int_{U \setminus U_1} H_t(X_{t-}, u)\,p(dt, du)]$$

$$+ \lambda_t(X_t)\,dM_t^Y + \int_{|y| \leq 1} K_t(X_{t-}, y)q^Y(dt, dy) + \int_{|y| > 1} K_t(X_{t-}, y)\,p^Y(dt, dy)]$$

and

$$Y_t = Y_0 + \int_0^t \int_{|y|>1} y p^Y(ds, dy) + \int_0^t \int_{|y| \leq 1} y q^Y(ds, dy) + M_t^Y + \int_0^t \tilde{c}_s\,ds,$$

where $M = (M^1, \ldots, M^n)$, $M^Y = (M^{Y,1}, \ldots, M^{Y,m})$, $M^i, M^{Y,j} \in \mathcal{M}_{loc}^c(\tilde{\mathbb{P}}, \mathbf{F})$, $1 \leq i \leq n$, $1 \leq j \leq m$, $\langle M^k, M^{Y,j} \rangle_t = 0$, $t \geq 0$, $1 \leq i \leq n$, $1 \leq j \leq m$.
Let $\mathbf{Y} = (\mathcal{Y}_t) = \mathbf{F}^Y$. Assume

$$d\langle M^k, M^l \rangle_t = C_t^{kl}\,dt,\ k, l = 1, \ldots, n,$$
$$d\langle M^{Y,k}, M^{Y,l} \rangle_t = C_t^{Y,kl}\,dt,\ k, l = 1, \ldots, m,$$

and $C_s^Y = (C_s^{Y,kl}), \tilde{c}_s, \nu^Y(s, dy)$ are $\mathcal{P}(\mathbf{Y})$-measurable.

Let $\mathbb{P}$ be a probability measure on $(\Omega, \mathcal{F})$ such that $d\mathbb{P}|_{\mathcal{F}_t} = Z_t d\tilde{\mathbb{P}}|_{\mathcal{F}_t}$ and

$$dZ_t = Z_{t-}l_t(X_t)dM_t^Y + Z_{t-}\int(\chi_t(X_{t-}, y) - 1)[p^Y(ds, dy) - \Pi^Y(ds, dy)],$$
$$Z_0 = 1,$$

where $l \in L^2_{loc}(B, \tilde{\mathbb{P}}, \mathbf{F})$, $B = (\langle M^{Y,i}, M^{Y,j}\rangle)$, $\chi_t(X_{t-}, y) - 1 \in \mathcal{G}_{loc}(\Pi^Y, \tilde{\mathbb{P}}, \mathbf{F})$, $\chi \geq 0$. For $f \in C_0^\infty(\mathbb{R}^d)$ we define

$$\tilde{L}_s f(x) = \nabla f(x) \cdot \left\{b_s(x) + \lambda_s(x)a_s^Y l_s(x) + \int_{|y|\leq 1} K_s(x, y)(\chi_s(x, y) - 1)\nu^Y(s, dy)\right\}$$

$$+ \frac{1}{2}\sum_{k,l=1}^{d}\sum_{j,j'=1}^{m} \partial_{kl}^2 f(x)\lambda_s^{kj}(x)\cdot\lambda_s^{lj'}(x)C_s^{Y,jj'}$$

$$+ \frac{1}{2}\sum_{k,l=1}^{d}\sum_{j,j'=1}^{n} \partial_{kl}^2 f(x)\sigma_s^{kj}(x)\cdot\sigma_s^{lj'}(x)C_s^{jj'}$$

$$+ \int [f(x + H_s(x, u)) - f(x) - \nabla f(x) \cdot H_s(x, u)1_{U_1}(u)]\nu(s, du)$$

$$+ \int [f(x + K_s(x, y)) - f(x) - \nabla f(x) \cdot K_s(x, y)1_{|y|\leq 1}]\chi_s(x, y)\nu^Y(s, dy).$$

Let $b_s(x), \lambda_s(x), l_s(x), \sigma_s(x)$ be $\mathcal{P}(\mathbf{Y}) \otimes \mathcal{B}(\mathbb{R}^d)$-measurable, $K_s(x, y), \chi_s(x, y)$ be $\mathcal{P}(\mathbf{Y}) \otimes \mathcal{B}(\mathbb{R}^d) \otimes \mathcal{B}(\mathbb{R}_0^m)$-measurable, and $H_s(x, u)$ be $\mathcal{P}(\mathbf{Y}) \otimes \mathcal{B}(\mathbb{R}^d) \otimes \mathcal{U}$-measurable.

Assuming C0–C4 hold with $\theta_t = f(X_t)$, $\tilde{p} = p^Y$, $\tilde{\Pi} = \Pi^Y$, $\tilde{q}(ds, dy) = p^Y(ds, dy) - \Pi^Y(ds, dy)$, $\tilde{M} = M^Y$,

$$l_t = l_t(X_t), \chi = \chi_t(X_{t-}, y),$$

we obtain for $\rho_t(f) = \tilde{\mathbb{E}}[Z_t f(X_t)|\mathcal{Y}_t]$, by Theorem 4.2 and Lemma 4.3, the reduced linear filtering equation

$$\rho_t(f) = \rho_0(f) + \int_0^t \rho_{s-}(\tilde{L}_s f)ds + \int_0^t \tilde{g}_s d\tilde{M}_s + \int_0^t \int \tilde{K}_s(y)\tilde{q}(ds, dy),$$

with

$$\tilde{g}_t = \rho_{t-}\{fl_t + (\nabla f, \lambda_t)\},$$
$$\tilde{K}_t(y) = \int_{\mathbb{R}^d}\{f(x)(\chi_t(x, y) - 1) + [f(x + K_t(x, y)) - f(x)]\chi_t(x, y)\}\rho_{t-}(dx).$$

Note that $Y \in LID(\mathbb{P}, \mathbf{F})$ with local predictable characteristics (c_t, C_t^Y, β), where

$$c_t = \tilde{c}_t + C_t^Y l_t(X_t) + \int_{|y|\leq 1} \gamma(\chi_t(X_t, y) - 1) \nu^Y(t, dy),$$
$$\beta(t, dy) = \chi_t(X_t, y) \nu^Y(t, dy),$$

and the signal X has the generator $\tilde{L} f$ with respect to $\mathbb{P}$.

4.5.1 Tracking volatility: J. Cvitanic, R. Liptser, and B. Rozovskiĭ [3]

A probability space $(\Omega, \mathcal{F}, \mathbb{P})$ is fixed with a filtration of σ-algebras $\tilde{\mathbf{F}} = (\tilde{\mathcal{F}}_t)$. It is assumed that the signal X_t is a cadlag Markov jump diffusion process in $\mathbb{R}$ and satisfies the Itô equation:

$$dX_t = b_t(X_t)\, dt + \sigma_t(X_t)\, d\tilde{W}_t + \int_{\mathbb{R}} u(X_{t-}, x) \left(\mu^X - \nu^X\right)(dt, dx),$$

where $\tilde{W}_t$ is a standard Wiener process and $\mu^X = \mu^X(dt, dx)$ is a Poisson point measure on $\mathbb{R}_+ \times \mathbb{R}$ with the conditional intensity $\nu^X(dt, dx) = K(dx)dt$. The generator of X is denoted $\mathcal{L}$. Let $\mathcal{X}_t = \cap_{\varepsilon>0}\sigma(X_s, s \leq t+\varepsilon)$, $\mathcal{X}_\infty = \sigma(X_s, s \geq 0)$.

Let $r(x)$ and $v(x)$ be measurable bounded functions on $\mathbb{R}$. Consider the log-price process

$$U_t = \int_0^t \left(r(X_s) - \frac{1}{2}v^2(X_s) \right) ds + \int_0^t v(X_s)\, dW_s, \tag{4.11}$$

where W_t is a standard Brownian motion independent of X.

It is assumed that the only observable quantities are the values of U_t at random times $\tau_k, k \geq 0$, so that $\tau_k < \tau_{k+1}$, if $\tau_k < \infty$, and $\tau_k \uparrow \infty$ as $k \to \infty$. The observations are given by the sequence $(\tau_k, U_k) = (\tau_k, U_{\tau_k})$. The σ-algebra generated by $(\tau_k, U_k), k \leq n$, is denoted by $\mathcal{G}(n)$ and the counting measure is introduced

$$\mu(dt, dy) = \sum_{k=1}^\infty \delta_{(\tau_k, U_k)}(t, y)\, dt dy,$$

where $\delta_{(\tau_k, U_k)}$ is the Dirac function on $[0, \infty) \times \mathbb{R}$. Let

$$\mathcal{Y}_t^0 = \sigma\left(\mu\left([0, r] \times \Gamma\right) : r \leq t, \Gamma \in \mathcal{B}(\mathbb{R})\right), \mathcal{Y}_t = \cap_{s>t}\mathcal{Y}_s^0, \mathbf{Y} = (\mathcal{Y}_t)_{t \geq 0}.$$

Equivalently, the observation process in this case is

$$Y_t = \int_0^t \int_{\mathbb{R}} y\, \mu(ds, dy) = \int_0^t U_s\, dL_s,$$

where $L_t = \sum_k \mathbf{1}_{\tau_k \leq t}$.

The following assumptions are made.

B0. For every finite $S \in \mathcal{T}_p(\mathbf{Y})$,

$$\mathbb{P}(L_S - L_{S-} | \mathcal{Y}_{S-}) = 0 \text{ or } 1;$$

B1. W is independent of (X, L);

B2. For each k, there is a $\mathcal{G}(k)$-measurable integrable random measure $\Phi_k(dt)$ supported by $(\tau_k, \infty]$ so that $\mathbb{P}(\tau_{k+1} \in dt | \mathcal{X}_\infty \vee \mathcal{G}(k))$ is absolutely continuous with respect to this measure a.s. with a corresponding density

$$\phi_k(\tau_k, t) = \phi_k(X, Y, t) = \frac{\mathbb{P}(\tau_{k+1} \in dt | \mathcal{X}_\infty \vee \mathcal{G}(k))}{\Phi_k(dt)}.$$

The following two examples illustrate **B2** (see [3]).

Example 4.5 Let (τ_k) be the jump times of a doubly stochastic Poisson process with the intensity $n(X_t)$. Then

$$\mathbb{P}(\tau_{k+1} \leq t | \mathcal{X}_\infty \vee \mathcal{G}(k)) = \begin{cases} 1 - e^{-\int_{\tau_k}^{t} n(X_s)ds}, & t \geq \tau_k, \\ 0, & \text{otherwise.} \end{cases}$$

In this case $\Phi_k(dt) = dt$, $\phi(\tau_k, t) = n(X_t) \exp\left\{-\int_{\tau_k}^{t} n(X_s)ds\right\}$. If n is a constant, we could also choose

$$\Phi_k(dt) = n \exp\{-n(t - \tau_k)\}, \phi(\tau_k, t) = 1.$$

Example 4.6 If the observation moments τ_k are nonrandom, then we can choose $\Phi_k(dt) = \delta_{\tau_{k+1}}(dt)$, and $\phi(\tau_k, t) = 1$.

Set

$$a(s, t) = \int_s^t \left[r(X_u) - \frac{1}{2}v^2(X_u)\right] du,$$

$$\gamma^2(s, t) = \int_s^t v^2(X_u) du,$$

and assume γ^2 is bounded away from zero. Let us denote $\lambda_{s,t}(x) = \lambda_{s,t}(X, x)$ the normal density function with mean $a(s, t)$ and the variance $\gamma^2(s, t)$.

Let $G_k(ds, dx)$ be a regular version of the conditional distribution (τ_{k+1}, U_{k+1}) with respect to $\mathcal{G}(k)$ (it is assumed that $G_k([0.\tau_k], dx) = 0$. Denote $G_k(ds) = G_k(ds, \mathbb{R})$. We have

$$G_k(dt, dy) = 1_{t > \tau_k} \mathbb{E}[\lambda_{\tau_k,t}(X, y - U_k) \phi(\tau_k, t) | \mathcal{G}(k)] dy \Phi_k(dt).$$

Introduce $G_k^X(dt) = \mathbb{P}(\tau_{k+1} \in dt | \mathcal{F}^X \vee \mathcal{G}(k)) = \phi_k(X, Y, t)\Phi_k(dt)$. Notice that

$$\mathbb{P}(\tau_{k+1} \in dt, U_{k+1} \in dx | \mathcal{X}_\infty \vee \mathcal{G}(k)) = 1_{t > \tau_k} \lambda_{\tau_k,t}(x - U_k) G_k^X(dt) dx.$$

Let $\mathbf{F} = (\mathcal{Y}_t \vee \mathcal{X}_t)_{t \geq 0}$. In order to apply our theorem we find the $(\mathbb{P}, \mathbf{Y})$- and $(\mathbb{P}, \mathbf{F})$-conditional intensities of $\bar{p}(dt, dx) = \mu(dt, dx)$. Let $N_t = \sum_k \int_0^t 1_{(\tau_k, \tau_{k+1}]}(r) \Phi_k(dr)$.

Lemma 4.4 *Assume B0–B2 hold. Then*

a) *the $(\mathbb{P}, \mathbf{Y})$-compensator of $\bar{p}$ is*

$$\bar{\Pi}(dt, dx) = \sum_k 1_{(\tau_k, \tau_{k+1}]}(t) \frac{G_k(dt, dx)}{G_k([t, \infty))}$$

$$= \sum_k 1_{(\tau_k, \tau_{k+1}]}(t) \frac{\mathbb{E}[\lambda_{\tau_k, t}(x - U_k) \phi_k(\tau_k, t) | \mathcal{G}(k)]}{\mathbb{E}\left[G_k^X([t, \infty)) | \mathcal{G}(k)\right]} dx d N_t,$$

b) *the $(\mathbb{P}, \mathbf{F})$-compensator of $\bar{p}$ is*

$$\Pi(dt, dx) = \sum_k 1_{(\tau_k, \tau_{k+1}]}(t) \lambda_{\tau_k, t}(x - U_k) \mathcal{P}\left\{1_{(\tau_k, \tau_{k+1}]}(t) \frac{\phi_k(\tau_k, t)}{G_k^X([t, \infty))}\right\} dx d N_t$$

$$= \sum_k 1_{(\tau_k, \tau_{k+1}]}(t) \lambda_{\tau_k, t}(x - U_k) \frac{\mathcal{P}\{\phi_k(\tau_k, t)\}}{\mathcal{P}\{G_k^X([t, \infty))\}} dx d N_t,$$

where $\mathcal{P}\left\{1_{(\tau_k, \tau_{k+1}]}(t) \frac{\phi_k(\tau_k, t)}{G_k^X([t, \infty))}\right\}$, $\mathcal{P}\{\phi_k(\tau_k, t)\}$ are $(\mathbb{P}, \mathbf{F})$-predictable projections, $\mathcal{P}\{G_k^X([t, \infty))\} = \int_{t-}^{\infty} \mathcal{P}\{\phi_k(\tau_k, t)\} d N_t$

Proof. The part a) follows by Proposition 3.41 in [1]. For a $\mathcal{P}(\mathbf{F}) \otimes \mathcal{B}(\mathbb{R})$-measurable $f(t, x) = f(t, x, X, Y)$,

$$\mathbb{E} \int_0^\infty f(t, x,) \bar{p} dt, dx)$$

$$= \sum_{k=1}^\infty f(\tau_k, U_k) = \mathbb{E} \sum_{k=0}^\infty f(\tau_{k+1}, U_{k+1})$$

$$= \mathbb{E} \sum_k \int_{\tau_k}^\infty \int f(t, x) \lambda_{\tau_k, t}(x - U_k) G_k^X(dt) dx$$

$$= \mathbb{E} \sum_k \int_{\tau_k}^\infty \int_t^\infty \int f(t, x) \frac{\lambda_{\tau_k, t}(x - U_k)}{G_k^X([t, \infty))} G_k^X(dt) dx G_k^X(ds)$$

$$\mathbb{E} \sum_k \int_{\tau_k}^\infty \int_{\tau_k}^s \int f(t, x) \frac{\lambda_{\tau_k, t}(x - U_k)}{G_k^X([t, \infty))} G_k^X(dt) dx G_k^X(ds)$$

$$= \mathbb{E} \sum_k \int_{\tau_k}^{\tau_{k+1}} \int f(t, x) \frac{\lambda_{\tau_k, t}(x - U_k)}{G_k^X([t, \infty))} G_k^X(dt) dx$$

$$= \mathbb{E} \sum_k \int_{\tau_k}^{\tau_{k+1}} \int f(t, x) \frac{\lambda_{\tau_k, t}(x - U_k)}{1 - G_k^X([0, t))} \phi_k(\tau_k, t) \Phi_k(dt) dx.$$

Therefore,

$$\Pi(dt, dx) = \sum_k 1_{(\tau_k, \tau_{k+1}]}(t) \lambda_{\tau_k, t}(x - U_k) \mathcal{P}\{1_{(\tau_k, \tau_{k+1}]}(t) \frac{\phi_k(\tau_k, t)}{1 - G_k^X([0, t))}\} dx d N_t.$$

Since

$$\mathbb{E} \sum_k \int_0^\infty \int f(t,x) 1_{(\tau_k,\tau_{k+1}]}(t) \lambda_{\tau_k,t}(x-U_k) \mathcal{P}\{1_{(\tau_k,\tau_{k+1}]}(t) \frac{\phi_k(\tau_k,t)}{1-G_k^X([0,t))}\} dx d N_t$$

$$= \mathbb{E} \sum_k \int_0^\infty \int f(t,x) 1_{(\tau_k,\tau_{k+1}]}(t) \lambda_{\tau_k,t}(x-U_k) \frac{\phi_k(\tau_k,t)}{G_k^X([t,\infty))} \Phi_k(dt) dx$$

$$= \mathbb{E} \sum_k \int_{\tau_k}^\infty \int_0^\infty \int f(t,x) 1_{(\tau_k,s]}(t) \lambda_{\tau_k,t}(x-U_k) \frac{G_k^X(dt)}{G_k^X([t,\infty))} dx G_k^X(ds)$$

$$= \mathbb{E} \sum_k \int_{\tau_k}^\infty \int f(t,x) \lambda_{\tau_k,t}(x-U_k) \phi_k(\tau_k,t) \Phi_k(dt) dx$$

and, similarly,

$$\mathbb{E} \sum_k \int_{\tau_k}^\infty \int f(t,x) \lambda_{\tau_k,t}(x-U_k) \phi_k(\tau_k,t) \Phi_k(dt) dx$$

$$= \mathbb{E} \sum_k \int_{\tau_k}^\infty \int f(t,x) \lambda_{\tau_k,t}(x-U_k) \mathcal{P}\{\phi_k(\tau_k,t)\} \Phi_k(dt) dx$$

$$= \mathbb{E} \sum_k \int_{\tau_k}^\infty \int_{\tau_k}^{\tau_{k+1}} \int f(t,x) \lambda_{\tau_k,t}(x-U_k) \frac{\mathcal{P}\{\phi_k(\tau_k,t)\}}{\mathcal{P}\{G_k^X([t,\infty))\}} \Phi_k(dt) dx.$$

which implies that $dx d N_t d\mathbb{P}$-a.e.

$$\lambda_{\tau_k,t}(x-U_k) \frac{\mathcal{P}\{\phi_k(\tau_k,t)\}}{\mathcal{P}\{G_k^X([t,\infty))\}} = \lambda_{\tau_k,t}(X,x-U_k) \mathcal{P}\left\{1_{(\tau_k,\tau_{k+1}]}(t) \frac{\phi_k(\tau_k,t)}{1-G_k^X([0,t))}\right\}.$$

□

Remark 4.3

1. The assumption B2 implies that X_t and Y_t do not have common jumps.
2. Similarly, as in the last part of the proof, we obtain that $dx d N_t d\mathbb{P}$-a.e.

$$\frac{1_{(\tau_k,\tau_{k+1}]} \bar{\mathcal{P}}\{f(X_{t-}) \lambda_{\tau_k,t}(x-U_k) \phi_k(\tau_k,t)\}}{\bar{\mathcal{P}} G_k^X([t,\infty))}$$

$$= \bar{\mathcal{P}}\left\{f(X_{t-}) 1_{(\tau_k,\tau_{k+1}]}(t) \lambda_{\tau_k,t}(x-U_k) \frac{\phi_k(\tau_k,t)}{G_k^X([t,\infty))}\right\}$$

$$= 1_{(\tau_k,\tau_{k+1}]}(t) \frac{\mathbb{E}[f(X_{t-}) \lambda_{\tau_k,t}(x-U_k) \phi_k(\tau_k,t) | \mathcal{G}(k)]}{\mathbb{E}\left[G_k^X([t,\infty)) | \mathcal{G}(k)\right]},$$

where $\bar{\mathcal{P}}$ denotes **Y**-predictable projection.
3. $\mathbb{P}$-a.s. $\bar{\Pi}(\{t\}, \mathbb{R}) = 0$ or $\bar{\Pi}(\{t\}, \mathbb{R}) = 1$.

Let $\pi_t(f) = \mathbb{E}[f(X_t)|\mathcal{Y}_t]$, $f \in C_k^\infty(\mathbb{R})$ (see Lemma 4.3). Using Theorem 4.1 and Lemma 4.4 we obtain (see J. Cvitanic, R. Liptser, and B. Rozovskiĭ [3])

Theorem 4.3 *Assume B0–B2 hold. Then*

$$\pi_t(f) = \pi_0(f) + \int_0^t \pi_s(\mathcal{L}f)\, ds + \int_0^t \int K f_s\,(x) \bar{q}(ds, dx),$$

where

$$K f_s(x) = \sum_k 1_{(\tau_k, \tau_{k+1}]}(s) \frac{\mathbb{E}[f(X_{s-})\lambda_{\tau_k,s}(x - U_k)\, \phi_k(\tau_k, s))|\mathcal{G}(k)]}{\mathbb{E}[\lambda_{\tau_k,s}(x - U_k)\, \phi_k(\tau_k, s)|\mathcal{G}(k)]} - [\pi_{s-}(f)].$$

References

[1] J. Jacod, *Calcul stochastique et problèmes de martingales*, Lecture Notes in Mathematics, 714, Springer Verlag, Berlin, 1979.

[2] M. Fujisaki, G. Kallianpur, and H. Kunita, Stochastic differential equations for nonlinear filtering problem, *Osaka J. Math.*, 1972, 9(1), pp. 19–40.

[3] J. Cvitanic, R. Liptser, and B. Rozovskiĭ, A filtering approach to tracking volatility from prices observed at random times, *The Annals of Applied Probab.*, 16, 2006, pp. 1633–52.

[4] B. Grigelionis, On stochastic nonlinear filtering of stochastic processes, *Liet. Matematikos Rinkinys*, 1972, 12(4), pp. 37–51 (in Russian).

[5] M. Zakai, On the optimal filtering of diffusion processes, *Z. Wahrscheinlichkeitstheorie verw. Gebiete*, 1969, 11, pp. 230–43.

[6] R. L. Stratonovich, On the nonlinear filtering of random functions, *Theory of Probability and Applications*, 1959, 4(2), pp. 239–43.

[7] R. L. Stratonovich, Conditional Markov processes, *Theory of Probability and Applications*, 1960, 5(2), pp. 172–95.

[8] B. Grigelionis, On Markov property of stochastic processes, *Lietuvos Matem. Rinkinys*, 1968, 8(3), pp. 489–502 (in Russian).

[9] B. Grigelionis, Optimal stopping of stochastic processes, *Proceedings of 6th Math. Summer School: Probability Theory and Math. Statistics, Academy of Sciences of Ukrain. SSR*, Institute of Mathematics, Kiev, 1969 (in Russian).

[10] B. Grigelionis, On the representation of integer valued random measures by means of stochastic integrals with respect to Poisson measure, *Liet. matem. rinkinys*, 1971, 11(3), pp. 93–108 (in Russian).

[11] B. Grigelionis, Stochastic nonlinear filtering equations and semimartingales. In: *Nonlinear filtering and stochastic control* (eds. S.K. Mitter and A. Moro), Lecture Notes in Mathematics, 972, Springer Verlag, Berlin, 1982, pp. 63–99.

[12] B. Grigelionis and R. Mikulevicius, On statistical problems of stochastic processes with penetrable boundaries. In: *Mathematical statistics and probability theory* (eds. W. Klonecki, A. Kozek, and J. Rosinskiĭ), Lecture Notes in Statistics, 2, Springer Verlag, Berlin, 1980, pp. 152–69.

[13] B. Grigelionis and R. Mikulevicius, Stochastic evolution equations and densities of the conditional distributions, *Lecture Notes in Control and Info. Sci.*, 1983, 49, pp. 43–86

[14] R. S. Liptser and A. N. Shiryaev, *Statistics of random processes, I–II*, Springer Verlag, Berlin, 1977.

[15] R. S. Liptser and A. N. Shiryaev, *Theory of martingales*, Kluwer, Dordrecht, 1989.

[16] G. Kallianpur, *Nonlinear filtering*, Springer Verlag, Berlin, 1980.

[17] J. Szpirglas and G. Mazziotto, Modeles générales de filtrage non linéaire et equations diffrerentielles stochastiques associées, *C.R. Acad. Sci, Paris, Ser. A*, 1978, 286, pp. 1067–70.

[18] J. van Schuppen, Stochastic filtering theory: a discussion of concepts, methods and results, *Stochastic Control Theory and Stochastic Differential Systems*, Lecture Notes in Control and Inform Sciences, 16, Springer Verlag, Berlin, 1979, pp. 209–26.
[19] D. I. Hadjiev, On the filtering of semimartingales in case of observations of point processes, *Theory of Probab. and Applications*, 23(1), 1980, pp. 169–78.
[20] L. G. Vetrov, On linearization of stochastic differential equations of optimal nonlinear filtering, *Theory of Probab. and Applications*, 25(2), 1980, pp. 399–407.
[21] P. Bremaud and M. Yor, Changes of filtration and probability measures, *Z. Wahrscheinlichkeitstheorie verw. Geb.*, 1978, 45(4), pp. 269–95.
[22] B. Grigelionis, On nonlinear filtering theory and absolute continuity of measures corresponding to stochastic processes, In: *Proceedings 2nd Japan–USSR Symposium on Probab. Theory* (eds. G. Maruyama and Y. V. Prokhorov), Lecture Notes in Math., 330, Springer Verlag, Berlin, pp. 80–94.
[23] B. Grigelionis, On statistical problems of stochastic processes with boundary conditions, *Liet. matem. rinkinys*, 16(3), pp. 63–87 (in Russian).
[24] B. Grigelionis, On the reduced stochastic equations for nonlinear filtering of random processes, *Liet. matem. rinkinys*, 16(3), 1976, pp. 51–63 (in Russian).
[25] Mikulevicius, R. and B. L. Rozovskiĭ (1999). Martingale problems for stochastic PDEs. In *Stochastic partial differential equations: six perspectives* (eds. R. Carmona and B. Rozovskiĭ). Mathematical Surveys and Monographs 64, 243–325. Amer. Math. Soc., Providence, RI.
[26] H. Shi, Optimal filtering with random observations, PhD Thesis, University of Southern California, Los Angeles, 2006.

·5·
The Filtered Martingale Problem

T. G. Kurtz and G. Nappo

5.1 Introduction

The notion of a *filtered martingale problem* was introduced in Kurtz and Ocone [27] and extended to a more general setting in Kurtz [26]. The basic idea is that the conditional distribution of the state of a Markov process given the information from related observations satisfies a kind of *martingale problem*. The fundamental results give conditions under which every solution of the filtered martingale problem arises from a solution of the original martingale problem, and hence, uniqueness for the original martingale problem implies uniqueness for the filtered martingale problem. These results have a variety of consequences, most notably, the uniqueness for filtering equations and general results on Markov mappings, that is, conditions under which a transformation of a Markov process is still Markov.

The current chapter is concerned with extension of these ideas to further settings. The filtering literature contains a number of results (for example, Bhatt, Budhiraja, and Karandikar [5], Budhiraja [6], Kunita [22]) showing that the conditional distribution for the classical filtering problem is itself a Markov process. In order to address this question for general solutions of filtered martingale problems, we need to generalize the earlier definition to include information available at time zero. We are then able to show the Markov property for the conditional distributions for a large class of partially observed Markov processes.

We also extend the earlier results to local martingale problems which in turn allows us to generalize previous uniqueness results for filtering equations. The basic results can also be extended to constrained martingale problems, that is, martingale problems for processes in which the behavior of the process on the boundary of the state space is determined by a second operator (see Anderson [1], Kurtz [24, 25], Kurtz and Stockbridge [28], Stroock and Varadhan [32]). Reflecting diffusion processes provide one example.

We also relax some technical conditions present in the earlier work.

Throughout this chapter, all filtrations are assumed complete, all processes are assumed to be progressively measurable, $\{\mathcal{F}_t^Y\}$ denotes the completion of the filtration generated by the observed process Y, and assuming Y takes values in $\mathbb{S}_0$, $\widehat{\mathcal{F}}_t^Y$ denotes the completion of $\sigma\left(\int_0^r h(Y(s))ds : r \leq t, h \in B(\mathbb{S}_0)\right) \vee \sigma(Y(0))$.

5.2 Martingale properties of conditional distributions

Let $\mathbb{S}$ be a complete, separable metric space. $C(\mathbb{S})$ will denote the space of $\mathbb{R}$-valued, continuous functions on $\mathbb{S}$, $M(\mathbb{S})$ the Borel measurable functions, $C_b(\mathbb{S})$ the bounded continuous functions, $B(\mathbb{S})$ the bounded measurable functions, and $\mathcal{P}(\mathbb{S})$ the space of probability measures on $\mathbb{S}$. $M_{\mathbb{S}}[0, \infty)$ will denote the space of measurable functions $x : [0, \infty) \to \mathbb{S}$ topologized by convergence in Lebesgue measure, $D_{\mathbb{S}}[0, \infty) \subset M_{\mathbb{S}}[0, \infty)$ the space of $\mathbb{S}$-valued, cadlag functions with the Skorokhod topology, and $C_{\mathbb{S}}[0, \infty) \subset D_{\mathbb{S}}[0, \infty)$ the subspace of continuous functions. We consider martingale problems for operators satisfying the following condition:

Condition 5.1

i) $A : \mathcal{D}(A) \subset C_b(\mathbb{S}) \to M(\mathbb{S})$ with $1 \in \mathcal{D}(A)$ and $A1 = 0$.

ii) Either $\mathcal{R}(A) \subset C(\mathbb{S})$ or there exists a complete separable metric space $\mathbb{U}$, a transition function η from $\mathbb{S}$ to $\mathbb{U}$, and an operator $A_1 : \mathcal{D}(A) \subset C_b(\mathbb{S}) \to C(\mathbb{S} \times \mathbb{U})$ such that

$$Af(x) = \int_{\mathbb{U}} A_1 f(x, z) \eta(x, dz), \quad f \in \mathcal{D}(A). \tag{5.1}$$

iii) There exist $\psi \in C(\mathbb{S})$, $\psi \geq 1$, and constants a_f such that $f \in \mathcal{D}(A)$ implies

$$|Af(x)| \leq a_f \psi(x),$$

or if A is of the form (5.1), there exist $\psi_1 \in C(\mathbb{S} \times \mathbb{U})$, $\psi_1 \geq 1$, and constants a_f such that, for all $(x, z) \in \mathbb{S} \times \mathbb{U}$

$$|A_1 f(x, z)| \leq a_f \psi_1(x, z).$$

(If A is of the form (5.1), then define $\psi(x) \equiv \int_{\mathbb{U}} \psi_1(x, z) \eta(x, dz)$.)

iv) Defining $A_0 = \{(f, \psi^{-1} Af) : f \in \mathcal{D}(A)\}$ (or $\{(f, \psi_1^{-1} A_1 f), f \in \mathcal{D}(A)\}$), A_0 is separable in the sense that there exists a countable collection $\{g_k\} \subset \mathcal{D}(A)$ such that A_0 is contained in the bounded, pointwise closure of the linear span of $\{(g_k, A_0 g_k) = (g_k, \psi^{-1} A g_k)\}$ in $B(\mathbb{S}) \times B(\mathbb{S})$ (or in $B(\mathbb{S}) \times B(\mathbb{S} \times \mathbb{U})$).

v) A_0 is a pregenerator (for each fixed z, if A is of the form (5.1)), that is, A_0 is dissipative and there are sequences of functions $\mu_n : \mathbb{S} \to \mathcal{P}(\mathbb{S})$ and $\lambda_n : \mathbb{S} \to [0, \infty)$ such that for each $(f, g) \in A$

$$g(x) = \lim_{n \to \infty} \lambda_n(x) \int_{\mathbb{S}} (f(y) - f(x)) \mu_n(x, dy) \tag{5.2}$$

for each $x \in \mathbb{S}$.

vi) $\mathcal{D}(A)$ is closed under multiplication and separates points.

Remark 5.1 Suppose that we are interested in a diffusion X in a closed set $\mathbb{S} \subset \mathbb{R}^d$ with absorbing boundary conditions, that is,

$$X(t) = X(0) + \int_0^{t \wedge \tau} \sigma(X(s)) dW(s) + \int_0^{t \wedge \tau} b(X(s)) ds, \tag{5.3}$$

where $\tau = \inf\{t : X(t) \in \partial\mathbb{S}\} = \inf\{t : X(t) \notin \mathbb{S}^\circ\}$, where $\partial\mathbb{S}$ is the topological boundary of $\mathbb{S}$ and $\mathbb{S}^\circ$ is the interior of $\mathbb{S}$. Setting $a(x) = \sigma(x)\sigma(x)^\top$ and $Lf(x) = \sum \frac{1}{2} a_{ij}(x) \partial_i \partial_j f(x) + \sum b_i(x) \partial_i f(x)$ for $f \in C^2(\mathbb{R}^d)$, assuming sufficient smoothness, the natural generator would be Lf with domain being the C^2-functions satisfying $Lf(x) = 0$, $x \in \partial\mathbb{S}$. This domain does not satisfy Condition 5.1(vi). However, if we take $\mathcal{D}(A) = C_b^2(\mathbb{S})$, $\mathbb{U} = \{0, 1\}$, $A_1 f(x, u) = uLf(x)$ and $\eta(x, du) = 1_{\mathbb{S}^\circ}(x)\delta_1(du) + 1_{\partial\mathbb{S}}(x)\delta_0(du)$, where δ_0 and δ_1 are the Dirac measures at 0 and 1 respectively, we have $Af(x) = 1_{\mathbb{S}^\circ}(x) Lf(x)$ with domain satisfying Condition 5.1(vi). Any solution of (5.3) will be a solution of the martingale problem for A, and any solution of the martingale problem for A will be a solution of the martingale problem for the natural generator.

Definition 5.1 Let A satisfy Condition 5.1. A measurable, $\mathbb{S}$-valued process X is a solution of the *martingale problem* for A, if there exists a filtration $\{\mathcal{F}_t\}$ such that X is $\{\mathcal{F}_t\}$-adapted,

$$\mathbb{E}[\int_0^t \psi(X(s)) ds] < \infty, \quad t \geq 0, \tag{5.4}$$

and for each $f \in \mathcal{D}(A)$,

$$f(X(t)) - f(X(0)) - \int_0^t Af(X(s)) ds \tag{5.5}$$

is an $\{\mathcal{F}_t\}$-martingale. For $\nu_0 \in \mathcal{P}(\mathbb{S})$, X is a solution of the *martingale problem* for (A, ν_0), if X is a solution of the martingale problems for A and $X(0)$ has distribution ν_0.

A measurable, $\mathbb{S}$-valued process X and a nonnegative random variable τ are a solution of the *stopped martingale problem* for A, if there exists a filtration $\{\mathcal{F}_t\}$ such that X is $\{\mathcal{F}_t\}$-adapted, τ is a $\{\mathcal{F}_t\}$-stopping time,

$$\mathbb{E}[\int_0^{t \wedge \tau} \psi(X(s)) ds] < \infty, \quad t \geq 0, \tag{5.6}$$

and for each $f \in \mathcal{D}(A)$,

$$f(X(t \wedge \tau)) - f(X(0)) - \int_0^{t \wedge \tau} Af(X(s)) ds \tag{5.7}$$

is an $\{\mathcal{F}_t\}$-martingale.

A measurable, $\mathbb{S}$-valued process X is a solution of the *local martingale problem* for A, if there exists a filtration $\{\mathcal{F}_t\}$ such that X is $\{\mathcal{F}_t\}$-adapted and a sequence $\{\tau_n\}$ of $\{\mathcal{F}_t\}$-stopping times such that $\tau_n \to \infty$ a.s. and for each n, (X, τ_n) is a solution of the stopped martingale problem for A using the filtration $\{\mathcal{F}_t\}$.

Remark 5.2 Note that (5.4) ensures the integrability of (5.5) and similarly for the forward equation (5.8). Furthermore, if $M^f(t)$ denotes the process in (5.5), then (5.4) together with Condition 5.1(iii) imply that M^f is a martingale if and only if it is a local martingale, since

$$\sup_{s\in[0,t]} |M^f(s)| \leq 2\|f\| + \int_0^t \psi(X(s))ds$$

has finite expectation for all $t \geq 0$.

If X is a solution of the local martingale problem for A, then the localizing sequence $\{\tau_n\}$ can be taken to be predictable. In particular, we can take

$$\tau_n = \inf\left\{t : \int_0^t \psi(X(s))ds \geq n\right\}.$$

Definition 5.2 *Uniqueness* holds for the (local) martingale problem for (A, ν_0) if and only if all solutions have the same finite-dimensional distributions. *Stopped uniqueness* holds if for any two solutions, (X_1, τ_1), (X_2, τ_2), of the stopped martingale problem for (A, ν_0), there exists a stochastic process $\widetilde{X}$ and nonnegative random variables $\widetilde{\tau}_1, \widetilde{\tau}_2$ such that $(\widetilde{X}, \widetilde{\tau}_1 \vee \widetilde{\tau}_2)$ is a solution of the stopped martingale problem for (A, ν_0), $(\widetilde{X}(\cdot \wedge \widetilde{\tau}_1), \widetilde{\tau}_1)$ has the same distribution as $(X_1(\cdot \wedge \tau_1), \tau_1)$, and $(\widetilde{X}(\cdot \wedge \widetilde{\tau}_2), \widetilde{\tau}_2)$ has the same distribution as $(X_2(\cdot \wedge \tau_2), \tau_2)$.

Remark 5.3 Note that stopped uniqueness implies uniqueness. Stopped uniqueness holds if uniqueness holds and every solution of the stopped martingale problem can be extended (beyond the stopping time) to a solution of the (local) martingale problem. (See Lemma 4.5.16 of Ethier and Kurtz [18] for conditions under which this extension can be done.)

Definition 5.3 A $\mathcal{P}(\mathbb{S})$-valued function $\{\nu_t, t \geq 0\}$ is a solution of the *forward equation* for A if for each $t > 0$, $\int_0^t \nu_s \psi ds < \infty$ and for each $f \in \mathcal{D}(A)$,

$$\nu_t f = \nu_0 f + \int_0^t \nu_s Af ds. \tag{5.8}$$

A pair of measure-valued functions $\{(\nu_t^0, \nu_t^1), t \geq 0\}$ is a solution of the *stopped forward equation* for A if for each $t \geq 0$, $\nu_t \equiv \nu_t^0 + \nu_t^1 \in \mathcal{P}(\mathbb{S})$ and $\int_0^t \nu_s^1 \psi ds < \infty$, $t \to \nu_t^0(C)$ is nondecreasing for all $C \in \mathcal{B}(\mathbb{S})$, and for each $f \in \mathcal{D}(A)$,

$$\nu_t f = \nu_0 f + \int_0^t \nu_s^1 Af ds. \tag{5.9}$$

A $\mathcal{P}(\mathbb{S})$-valued function $\{v_t, t \geq 0\}$ is a solution of the *local forward equation* for A if there exists a sequence $\{(v^{0,n}, v^{1,n})\}$ of solutions of the stopped forward equation for A such that for each $C \in \mathcal{B}(\mathbb{S})$ and $t \geq 0$, $\{v_t^{1,n}(C)\}$ is nondecreasing and $\lim_{n \to \infty} v_t^{1,n}(C) = v_t(C)$.

Clearly, any solution X of the martingale problem for A gives a solution of the forward equation for A, that is $v_t f = \mathbb{E}[f(X(t))]$, and any solution of the stopped martingale problem for A gives a solution of the stopped forward equation for A, that is, $v_t^0 f = \mathbb{E}[\mathbf{1}_{[\tau,\infty)}(t) f(X(\tau))]$ and $v_t^1 f = \mathbb{E}[\mathbf{1}_{[0,\tau)}(t) f(X(t))]$. The primary consequence of Condition 5.1 is the converse.

Lemma 5.1 *If A satisfies Condition 5.1 and $\{v_t, t \geq 0\}$ is a solution of the forward equation for A, then there exists a solution X of the martingale problem for A satisfying $v_t f = \mathbb{E}[f(X(t))]$.*

If A satisfies Condition 5.1 and $\{(v_t^0, v_t^1), t \geq 0\}$ is a solution of the stopped forward equation for A, then there exists a solution (X, τ) of the stopped martingale problem for A such that $v_t^0 f = \mathbb{E}[\mathbf{1}_{[\tau,\infty)}(t) f(X(\tau))]$ and $v_t^1 f = \mathbb{E}[\mathbf{1}_{[0,\tau)}(t) f(X(t))]$.

If A satisfies Condition 5.1 and $\{v_t, t \geq 0\}$ is a solution of the local forward equation for A, then there exists a solution X of the local martingale problem for A satisfying $v_t f = \mathbb{E}[f(X(t))]$.

Proof. Various forms of the first part of this result exist in the literature beginning with the result of Echeverría [16] for the stationary case, that is, $v_t \equiv v_0$ and $v_0 A f = 0$. Extension of Echeverría's result to the forward equation is given in Theorem 4.9.19 of Ethier and Kurtz [18] for locally compact spaces and in Theorem 3.1 of Bhatt and Karandikar [2] for general complete separable metric spaces. The version given here is a special case of Corollary 1.12 of Kurtz and Stockbridge [28].

The result for stopped forward equations also follows by the same corollary. First enlarge the state space $\widetilde{\mathbb{S}} = \mathbb{S} \times \{0, 1\}$ and define $\widetilde{v}_t h = v_t^0 h(\cdot, 0) + v_t^1 h(\cdot, 1)$. Setting $\mathcal{D}(\widetilde{A}) = \{f(x)g(y) : f \in \mathcal{D}(A), g \in B(\{0,1\})\}$, for $h = fg \in \mathcal{D}(\widetilde{A})$, define $\widetilde{A}h(x, y) = y Ah(x, y) = yg(y) Af(x)$ and $Bh(x, y) = y(h(x, 0) - h(x, y))$. Then

$$0 = v_t^0 h(\cdot, 1) + v_t^1 h(\cdot, 1) - v_0^0 h(\cdot, 1) - v_0^1 h(\cdot, 1) - \int_0^t \widetilde{v}_s \widetilde{A} h \, ds$$

$$= \widetilde{v}_t h - \widetilde{v}_0 h - \int_0^t \widetilde{v}_s \widetilde{A} h \, ds + v_t^0 h(\cdot, 1) - v_t^0 h(\cdot, 0) + v_0^0 h(\cdot, 1) - v_0^0 h(\cdot, 0)$$

$$= \widetilde{v}_t h - \widetilde{v}_0 h - \int_0^t \widetilde{v}_s \widetilde{A} h \, ds - \int_{\mathbb{S} \times \{0,1\} \times [0,t]} Bh(x, y) \mu(dx \times dy \times ds),$$

where, noting that $\nu_t^0(C)$ is an increasing function of t, μ is the measure determined by

$$\mu(C \times \{1\} \times [0, t_2]) = \nu_{t_2}^0(C) - \nu_0^0(C), \quad \mu(C \times \{0\} \times [t_1, t_2]) = 0.$$

Corollary 1.12 of Kurtz and Stockbridge [28] then implies the existence of a process $(\widetilde{X}, Y)$ in $(\mathbb{S} \times \{0, 1\})$ such that $\nu_t^0 f = \mathbb{E}[(1 - Y(t))f(\widetilde{X}(t))]$, $\nu_t^1 f = \mathbb{E}[Y(t)f(\widetilde{X}(t))]$, and

$$f(\widetilde{X}(t)) - f(\widetilde{X}(0)) - \int_0^t Y(s) Af(\widetilde{X}(s)) ds$$

is a martingale for each $f \in \mathcal{D}(A)$. Following the arguments in Section 2 of Kurtz and Stockbridge [28], the process can be constructed in such a way that $Y(s) = 0$ implies $Y(t) = 0$ for $t > s$, and hence $\tau = \inf\{t : Y(t) = 0\}$. Note that $\widetilde{X}(t) = \widetilde{X}(\tau)$ for $t \geq \tau$.

Similarly, suppose $\{(\nu^{0,n}, \nu^{1,n})\}$ is the sequence of solutions of the stopped forward equation associated with a solution of the local forward equation and take $\left(\nu_t^{0,0}, \nu_t^{1,0}\right) \equiv (\nu_0, 0)$. For $f \in B(\mathbb{S} \times \mathbb{Z}_+)$, define

$$\widehat{\nu}_t f = \sum_{n=1}^{\infty} \left(\nu_t^{1,n} f(\cdot, n) - \nu_t^{1,n-1} f(\cdot, n)\right)$$

and

$$\int_{\mathbb{S} \times \mathbb{Z}_+ \times [0,t]} f(x, n) \widehat{\mu}(dx \times dn \times ds) = \sum_{n=1}^{\infty} \nu_t^{0,n} f(\cdot, n).$$

Note that $\widehat{\nu}_t$ is a probability measure with $\mathbb{S}$-marginal $\nu_t = \lim_{n \to \infty} \nu_t^{1,n}$.

Setting

$$\mathcal{D}(\widetilde{A}) = \mathcal{D}(B) = \{gf : g \in C_c(\mathbb{Z}_+), f \in \mathcal{D}(A)\}$$

(where, of course, $C_c(\mathbb{Z}_+)$ is the collection of functions with finite support) and defining $\widetilde{A}gf(x, n) = g(n) Af(x)$ and $Bgf(x, n) = f(x)(g(n+1) - g(n))$,

$$\widehat{\nu}_t gf = \widehat{\nu}_0 gf + \int_0^t \widehat{\nu}_s \widetilde{A}gf \, ds + \int_{\mathbb{S} \times \mathbb{Z}_+ \times [0,t]} Bgf \, d\widehat{\mu}.$$

Let $0 < \psi_0(n) < 1$ satisfy

$$\sum_n \psi_0(n) \int_0^n \nu_s^{1,n} \psi \, ds < \infty.$$

Then $\widetilde{A}$ satisfies Condition (5.1) with ψ replaced by $\widetilde{\psi}(x, n) = \psi(x)\psi_0(n)$, and

$$\int_0^t \widehat{\nu}_s \widetilde{\psi} \, ds < \infty, \quad t > 0.$$

Corollary 1.12 of Kurtz and Stockbridge [28] then implies the existence of a process (X, N) such that $(X(t), N(t))$ has distribution $\widehat{\nu}_t$ and a random measure Γ on $\mathbb{S} \times \mathbb{Z}_+ \times [0, \infty)$ satisfying

$$\mathbb{E}[\int_{\mathbb{S}\times\mathbb{Z}_+\times[0,t]} f(x,n)\Gamma(dx \times dn \times ds)] = \int_{\mathbb{S}\times\mathbb{Z}_+\times[0,t]} f(x,n)\widehat{\mu}(dx \times dn \times ds)$$

such that for each $gf \in \mathcal{D}(\widetilde{A})$,

$$g(N(t))f(X(t)) - \int_0^t g(N(s))Af(X(s))ds - \int_{\mathbb{S}\times\mathbb{Z}_+\times[0,t]} f(x)Bg(n)\Gamma(dx \times dn \times ds) \quad (5.10)$$

is a $\{\mathcal{F}_t^{X,N}\}$-martingale.

Let $\tau_k = \inf\{t : \int_0^t \psi(X(s))ds \geq k\}$. Let $g_m(n) = \mathbf{1}_{[0,m]}(n)$, and consider the limit of the sequence of martingales

$$g_m(N(t \wedge \tau_k))f(X(t \wedge \tau_k)) - \int_0^{t\wedge\tau_k} g_m(N(s))Af(X(s))ds \quad (5.11)$$

$$- \int_{\mathbb{S}\times\mathbb{Z}_+\times[0,t\wedge\tau_k]} f(x)Bg_m(n)\Gamma(dx \times dn \times ds)$$

as $m \to \infty$. The first two terms converge in L^1 by the dominated convergence theorem, and the third term satisfies

$$\mathbb{E}[|\int_{\mathbb{S}\times\mathbb{Z}_+\times[0,t\wedge\tau_k]} f(x)Bg_m(n)\Gamma(dx \times dn \times ds)|] \leq \|f\|\widehat{\mu}(\mathbb{S} \times \{m\} \times [0,t])$$

$$= \|f\|\nu_t^{0,m}(\mathbb{S})$$

and hence converges to zero in L^1. It follows that

$$f(X(t \wedge \tau_k)) - \int_0^{t\wedge\tau_k} Af(X(s))ds$$

is a martingale, and consequently, X is a solution of the local martingale problem for A such that $X(t)$ has distribution ν_t. □

Let X be a solution of the martingale problem for A with respect to a filtration $\{\mathcal{F}_t\}$, and let $\{\mathcal{G}_t\}$ be a filtration with $\mathcal{G}_t \subset \mathcal{F}_t$. Then letting π_t denote the conditional distribution of $X(t)$ given $\mathcal{G}_t$, Lemma A.1 implies that for each $f \in \mathcal{D}(A)$,

$$\pi_t f - \pi_0 f - \int_0^t \pi_s Af \, ds \quad (5.12)$$

is a $\{\mathcal{G}_t\}$-martingale.

Let $\mathbb{S}_0$ and S_0 be complete, separable metric spaces, and let $\gamma : \mathbb{S} \to \mathbb{S}_0$ be Borel measurable. Let X be a solution of the martingale problem for A, and let

Z be a $\mathbb{S}_0$-valued random variable. Assume that $\mathcal{F}_t \supset \sigma(Z)$ for all $t \geq 0$. Define $Y(t) = \gamma(X(t))$,

$$\widehat{\mathcal{F}}_t^Y = \text{completion of } \sigma\left(\int_0^r g(Y(s))ds : r \leq t, g \in B(\mathbb{S}_0)\right) \vee \sigma(Y(0)), \quad (5.13)$$

$\widehat{\mathcal{F}}_t^{Y,Z} = \widehat{\mathcal{F}}_t^Y \vee \sigma(Z)$, $\pi_t(C) = \mathbb{P}\left\{X(t) \in C | \widehat{\mathcal{F}}_t^{Y,Z}\right\}$, where by Theorem A.3 of Kurtz [26], we can assume that π is a progressively measurable, $\mathcal{P}(\mathbb{S})$-valued process and $\pi_{t \wedge \tau}(C) = \mathbb{P}\left\{X(t \wedge \tau) \in C | \widehat{\mathcal{F}}_{t \wedge \tau}^{Y,Z}\right\}$ for every $\left\{\widehat{\mathcal{F}}_t^{Y,Z}\right\}$-stopping time τ.

Remark 5.4 If Y is cadlag with no fixed points of discontinuity, then by Lemma A.5, $\mathcal{F}_t^Y = \widehat{\mathcal{F}}_t^Y$.

Note that

$$\int_0^t \pi_s(g \circ \gamma)ds = \int_0^t g(Y(s))ds, \quad \text{for each } g \in B(\mathbb{S}_0), \quad (5.14)$$

and if (5.4) holds,

$$\pi_t f - \int_0^t \pi_s A f ds$$

is a $\left\{\widehat{\mathcal{F}}_t^{Y,Z}\right\}$-martingale for each $f \in \mathcal{D}(A)$. With these properties in mind, we work with a definition of the filtered martingale problem slightly more general than that of Kurtz [26].

Definition 5.4 Let $\widehat{\mu}_0 \in \mathcal{P}(\mathbb{S} \times \mathbb{S}_0)$. $(\widetilde{Y}, \widetilde{\pi}, \widetilde{Z}, \widetilde{\tau}) \in M_{\mathbb{S}_0}[0, \infty) \times M_{\mathcal{P}(\mathbb{S})}[0, \infty) \times \mathbb{S}_0 \times [0, \infty]$ is a solution of the *stopped, filtered martingale problem* for $(A, \gamma, \widehat{\mu}_0)$, if

$$\mathbb{E}[\widetilde{\pi}_0(C)\mathbf{1}_D(\widetilde{Z})] = \widehat{\mu}_0(C \times D), \quad (5.15)$$

$\widetilde{\pi}$ is $\left\{\widehat{\mathcal{F}}_t^{\widetilde{Y},\widetilde{Z}}\right\}$-adapted, $\widetilde{\tau}$ is a $\left\{\widehat{\mathcal{F}}_t^{\widetilde{Y},\widetilde{Z}}\right\}$-stopping time, for each $g \in B(\mathbb{S}_0)$ and $t \geq 0$,

$$\int_0^t \widetilde{\pi}_s(g \circ \gamma)ds = \int_0^t g(\widetilde{Y}(s))ds, \quad (5.16)$$

$$\mathbb{E}\left[\int_0^{t \wedge \widetilde{\tau}} \widetilde{\pi}_s \psi ds\right] < \infty, \quad t > 0, \quad (5.17)$$

and for each $f \in \mathcal{D}(A)$,

$$\widetilde{M}_f(t \wedge \widetilde{\tau}) \equiv \widetilde{\pi}_{t \wedge \widetilde{\tau}} f - \int_0^{t \wedge \widetilde{\tau}} \widetilde{\pi}_s A f ds \quad (5.18)$$

is a $\left\{\widehat{\mathcal{F}}_t^{\widetilde{Y},\widetilde{Z}}\right\}$-martingale.

If $(\widetilde{Y}, \widetilde{\pi}, \widetilde{Z}, \widetilde{\tau})$ satisfies all the conditions except (5.15), we will refer to it as a solution of the *stopped, filtered martingale problem* for (A, γ).

If $\widetilde{\tau} = \infty$ a.s., then $(\widetilde{Y}, \widetilde{\pi}, \widetilde{Z})$ is a solution of the *filtered martingale problem* for (A, γ).

$(\widetilde{Y}, \widetilde{\pi}, \widetilde{Z}) \in M_{\mathbb{S}_0}[0, \infty) \times M_{\mathcal{P}(\mathbb{S})}[0, \infty) \times \mathbb{S}_0$ is a solution of the *filtered local martingale problem* for (A, γ) if there exists a sequence $\{\widetilde{\tau}_n\}$ of $\{\widehat{\mathcal{F}}_t^{\widetilde{Y}, \widetilde{Z}}\}$-stopping times such that $\widetilde{\tau}_n \to \infty$ a.s. and for each n, $(\widetilde{Y}, \widetilde{\pi}, \widetilde{Z}, \widetilde{\tau}_n)$ is a solution of the stopped, filtered martingale problem for (A, γ).

Remark 5.5 By the optional projection theorem (see Theorem A.3 of Kurtz [26]), there exists a modification of $\widetilde{\pi}$ such that for all $t \geq 0$ and all $\{\widehat{\mathcal{F}}_t^{\widetilde{Y}, \widetilde{Z}}\}$-stopping times τ, $\widetilde{\pi}_{t \wedge \tau}$ is $\widehat{\mathcal{F}}_{t \wedge \tau}^{\widetilde{Y}, \widetilde{Z}}$-measurable. Consequently, we will assume that $\widetilde{\pi}$ has this property.

Remark 5.6 Kurtz and Ocone [27] consider the filtered martingale problem with $\mathbb{S} = \mathbb{S}_1 \times \mathbb{S}_0$ and γ the projection onto $\mathbb{S}_0$.

The following lemma is an immediate consequence of (5.16).

Lemma 5.2 *If $(\widetilde{Y}, \widetilde{\pi}, \widetilde{Z}, \widetilde{\tau})$ is a solution of the stopped, filtered martingale problem for $(A, \gamma, \widehat{\mu}_0)$, then $\widehat{\mathcal{F}}_t^{\widetilde{Y}}$ is contained in the completion of $\sigma(\widetilde{\pi}_s, s \leq t) \vee \sigma(\widetilde{Y}(0))$.*

If X is a solution of the martingale problem for A, then $X^{(r)}$ given by $X^{(r)}(t) = X(r + t)$ is also a solution of the martingale problem for A. The following lemma gives the analogous result for filtered martingale problems. Let $\widetilde{Y}_r$ denote the restriction of $\widetilde{Y}$ to $[0, r]$.

Lemma 5.3 *Suppose $(\widetilde{Y}, \widetilde{\pi}, \widetilde{Z}, \widetilde{\tau}) \in M_{\mathbb{S}_0}[0, \infty) \times M_{\mathcal{P}(\mathbb{S})}[0, \infty) \times \mathbb{S}_0 \times [0, \infty]$ is a solution of the stopped, filtered martingale problem for (A, γ). For $r \geq 0$ such that $\widetilde{Y}(r)$ is $\widehat{\mathcal{F}}_r^{\widetilde{Y}, \widetilde{Z}}$-measurable, let $\widehat{Y}(t) = \widetilde{Y}(r + t)$, $\widehat{Z} = (\widetilde{Z}, \widetilde{Y}_r) \in \mathbb{S}_0 \times M[0, r]$, and $\widehat{\pi}_t = \widetilde{\pi}_{r+t}$. Then*

$$(\widehat{Y}, \widehat{\pi}, \widehat{Z}, (\widetilde{\tau} - r) \vee 0) \in M_{\mathbb{S}_0}[0, \infty) \times M_{\mathcal{P}(\mathbb{S})}[0, \infty) \times \mathbb{S}_0 \times M[0, r] \times [0, \infty]$$

is a solution of the stopped, filtered martingale problem for (A, γ).

Suppose $\widetilde{\tau} = \infty$, a.s. (,that is, $(\widetilde{Y}, \widetilde{\pi}, \widetilde{Z})$ is a solution of the filtered martingale problem for (A, γ)). For $r \geq 0$ such that $\widetilde{Y}(r)$ is $\widehat{\mathcal{F}}_r^{\widetilde{Y}, \widetilde{Z}}$-measurable, let $\widehat{Y}(t) = Y(r + t)$, $\widehat{Z} = \widetilde{\pi}_r = \widetilde{\pi}_0$, and

$$\widehat{\pi}_t = \mathbb{E}\left[\widetilde{\pi}_{r+t} \,\middle|\, \widehat{\mathcal{F}}_t^{\widehat{Y}} \vee \sigma(Y(r)) \vee \sigma(\widetilde{\pi}_r)\right].$$

Then $(\widehat{Y}, \widehat{\pi}, \widehat{Z}) \in M_{\mathbb{S}_0}[0, \infty) \times M_{\mathcal{P}(\mathbb{S})}[0, \infty) \times \mathcal{P}(\mathbb{S})$ is a solution of the filtered martingale problem for (A, γ).

Proof. In the second part of the lemma, the existence of $\hat{\pi}$ as an adapted, $\mathcal{P}(\mathbb{S})$-valued process follows by Theorem A.3 of Kurtz [26] and

$$\mathbb{E}\left[\int_0^t \widehat{\pi}_s \psi ds\right] = \mathbb{E}\left[\int_r^{r+t} \widetilde{\pi}_s \psi ds\right] < \infty.$$

In both parts, the required martingale properties follow by Lemma A.1. □

5.3 Conditional distributions and solutions of martingale problems

Of course, the forward equation is a special case of (5.12) in which the 'martingale' is identically zero. Consequently, the following proposition can be viewed as an extension of Lemma 5.1.

Proposition 5.1 *Let A satisfy Condition 5.1. Suppose that $\widetilde{Y}$ is a cadlag, $\mathbb{S}_0$-valued process with no fixed points of discontinuity, $\{\widetilde{\pi}_t, t \geq 0\}$ is a $\mathcal{P}(\mathbb{S})$-valued process, adapted to $\{\mathcal{F}_t^{\widetilde{Y}}\}$, $\int_0^t \widetilde{\pi}_s \psi ds < \infty$ a.s., $t \geq 0$, and*

$$\widetilde{\pi}_t f - \widetilde{\pi}_0 f - \int_0^t \widetilde{\pi}_s A f\, ds \tag{5.19}$$

is a $\{\mathcal{F}_t^{\widetilde{Y}}\}$-local-martingale for each $f \in \mathcal{D}(A)$. Then there exist a solution X of the local martingale problem for A, a cadlag, $\mathbb{S}_0$-valued process Y, and a $\mathcal{P}(\mathbb{S})$-valued process $\{\pi_t, t \geq 0\}$ such that (Y, π) has the same finite-dimensional distributions as $(\widetilde{Y}, \widetilde{\pi})$ and π_t is the conditional distribution of $X(t)$ given $\mathcal{F}_t^Y$.

For each $t \geq 0$, there exists a Borel measurable mapping $H_t : D_{\mathbb{S}_0}[0, \infty) \to \mathcal{P}(\mathbb{S})$ such that $\pi_t = H_t(Y)$ and $\widetilde{\pi}_t = H_t(\widetilde{Y})$ almost surely.

Proof. As in Kurtz [26], we begin by enlarging the state space so that the current state of the process contains all information about the past of the observation $\widetilde{Y}$. Let $\{b_k\}, \{c_k\} \subset C_b(\mathbb{S}_0)$ satisfy $0 \leq b_k, c_k \leq 1$, and suppose that the spans of $\{b_k\}$ and $\{c_k\}$ are bounded, pointwise dense in $B(\mathbb{S}_0)$. (Existence of $\{b_k\}$ and $\{c_k\}$ follows from the separability of $\mathbb{S}_0$.) Let $a_1, a_2, \ldots$ be an ordering of the rationals with $a_i \geq 1$. For $k, i \geq 1$, let

$$\widetilde{U}_{ki}(t) = c_k(\widetilde{Y}(0)) - a_i \int_0^t \widetilde{U}_{ki}(s)ds + \int_0^t b_k(\widetilde{Y}(s))ds \tag{5.20}$$

$$= c_k(\widetilde{Y}(0))e^{-a_i t} + \int_0^t e^{-a_i(t-s)} b_k(\widetilde{Y}(s))ds.$$

If we assume $c_1 = 1$ and $b_1 = 0$, $\widetilde{U}_{1i}(t) = e^{-a_i t}$ and $\widetilde{U}_{1i}(t)$ determines the value of t. Let $\widetilde{U}(t) = (\widetilde{U}_{ki}(t) : k, i \geq 1) \in [0, 1]^\infty$.

Define $F : (r, y) \in [0, \infty) \times D_{\mathbb{S}_0}[0, \infty) \to u \in [0, 1]^\infty$ by

$$u_{ki}(r, y) = c_k(y(0))e^{-a_i r} + \int_0^r e^{-a_i(r-s)} b_k(y(s))ds,$$

so $\widetilde{U}(t) = F(t, \widetilde{Y})$. Properties of Laplace transforms and the assumption that $\widetilde{Y}$ has no fixed points of discontinuity imply that there are measurable mappings $\Lambda : [0, 1]^\infty \to D_{\mathbb{S}_0}[0, \infty)$ and $\Lambda_0 : [0, 1]^\infty \to \mathbb{S}_0$ such that $\Lambda(\widetilde{U}(t)) = \widetilde{Y}(\cdot \wedge t)$ and $\Lambda_0(\widetilde{U}(t)) = \widetilde{Y}(t)$ almost surely. We can define Λ so that if $u_{1,i} = e^{-a_i t}$, then $y = \Lambda(u)$ satisfies $y(s) = y(t-)$ for $s \geq t$. Note that these observations imply that

the completion of $\sigma(\widetilde{U}(t)) = \widehat{\mathcal{F}}_t^{\widetilde{Y}} = \mathcal{F}_t^{\widetilde{Y}}$,

where the second equality follows by Lemma A.5.

Let $\widehat{\mathbb{S}} = \mathbb{S} \times [0, 1]^\infty$, and let $\mathcal{D}(\widehat{A})$ be the collection of functions on $\widehat{\mathbb{S}}$ given by

$$\{f(x) \prod_{k,i=1}^m g_{ki}(u_{ki}) : f \in \mathcal{D}(A), g_{ki} \in C^1[0, 1], m = 1, 2, \ldots\}.$$

Writing $g(u)$ instead of $\prod_{ki} g_{ki}(u_{ki})$ and denoting the partial derivative with respect to u_{ki} by $\partial_{ki} g$, for $fg \in \mathcal{D}(\widehat{A})$,

$$\widetilde{\pi}_t fg(\widetilde{U}(t)) - \widetilde{\pi}_0 fg(\widetilde{U}(0))$$
$$- \int_0^t \left(g(\widetilde{U}(s))\widetilde{\pi}_s Af + \widetilde{\pi}_s f \sum (-a_i \widetilde{U}_{ki}(s) + b_k(\widetilde{Y}(s))) \partial_{ki} g(\widetilde{U}(s)) \right) ds$$

is a $\{\mathcal{F}_t^{\widetilde{Y}}\}$-local-martingale. Note that without loss of generality, we can take the localizing sequence to be $\widetilde{\tau}_n = \inf\{t : \int_0^t \widetilde{\pi}_s \psi ds \geq n\}$.

Define

$$\widehat{A}_1(fg)(x, u, z) = g(u)Af(x) + f(x)\sum(-a_i u + b_k(z))\partial_{ki} g(u),$$

and

$$\widehat{A}(fg)(x, u) = \int \widehat{A}_1(fg)(x, u, z)\eta(x, u, dz), \quad (5.21)$$

where, with reference to (5.1) and the definition of Λ_0, we define $\eta(x, u, dz) = \delta_{\Lambda_0(u)}(dz)$.

Define $\widetilde{\nu}_t^{0,n}, \widetilde{\nu}_t^{1,n} \in \mathcal{M}(\mathbb{S} \times [0, 1]^\infty)$ by

$$\widetilde{\nu}_t^{0,n} h = \mathbb{E}[\mathbf{1}_{[\widetilde{\tau}_n, \infty)}(t) \int_{\mathbb{S}} h(z, \widetilde{U}(\widetilde{\tau}_n))\widetilde{\pi}_{\widetilde{\tau}_n}(dz)]$$

$$\widetilde{\nu}_t^{1,n} h = \mathbb{E}[\mathbf{1}_{[0, \widetilde{\tau}_n)}(t) \int_{\mathbb{S}} h(z, \widetilde{U}(t))\widetilde{\pi}_t(dz)].$$

Setting $\widetilde{\nu}_t^n = \widetilde{\nu}_t^{0,n} + \widetilde{\nu}_t^{1,n}$, for $fg \in \mathcal{D}(\widehat{A})$,

$$\widetilde{\nu}_t^n(fg) = \mathbb{E}[\widetilde{\pi}_{t\wedge\widetilde{\tau}_n} fg(\widetilde{U}(t \wedge \widetilde{\tau}_n))]$$
$$= \mathbb{E}[\widetilde{\pi}_0 fg(\widetilde{U}(0))]$$
$$+ \mathbb{E}[\int_0^{t\wedge\widetilde{\tau}_n} \left(g(\widetilde{U}(s))\widetilde{\pi}_s Af + \widetilde{\pi}_s f \sum(-a_i \widetilde{U}_{ki}(s) + b_k(\widetilde{Y}(s)))\partial_{ki}g(\widetilde{U}(s))\right) ds]$$
$$= \widetilde{\nu}_0(fg) + \int_0^t \widetilde{\nu}_s^{1,n} \widehat{A}(fg) ds.$$

Consequently, $(\widetilde{\nu}^{0,n}, \widetilde{\nu}^{1,n})$ is a solution of the stopped forward equation for $\widehat{A}$, and $\widetilde{\nu} = \lim_{n\to\infty} \widetilde{\nu}^{1,n}$ is a solution of the local forward equation. By Lemma 5.1, there exists a solution (X, U) of the local martingale problem for $\widehat{A}$, such that

$$\mathbb{E}[f(X(t)) \prod_{k,i=1}^m g_{ki}(U_{ki}(t))] = \widetilde{\nu}_t(f \prod_{k,i=1}^m g_{ki}) \tag{5.22}$$

$$= \mathbb{E}[\widetilde{\pi}_t f \prod_{k,i=1}^m g_{ki}(\widetilde{U}_{ki}(t))].$$

It follows that for each t, $U(t)$ and $\widetilde{U}(t)$ have the same distribution. If we define $Y(\cdot \wedge t) = \Lambda(U(t))$, $Y(\cdot \wedge t)$ and $\widetilde{Y}(\cdot \wedge t)$ have the same distribution on $D_{\mathbb{S}_0}[0, \infty)$.

Define π_t as the conditional distribution of $X(t)$ given $\mathcal{F}_t^Y$. Then, for any bounded measurable function g on $[0, 1]^\infty$

$$\mathbb{E}[f(X(t))g(U(t))] = \mathbb{E}[\pi_t fg(U(t))] \tag{5.23}$$
$$= \mathbb{E}[\widetilde{\pi}_t fg(\widetilde{U}(t))].$$

Since $\mathcal{F}_t^Y$ is the completion of $\sigma(U(t))$ and $\mathcal{F}_t^{\widetilde{Y}}$ is the completion of $\sigma(\widetilde{U}(t))$, for every t, there exist mappings $G_t, \widetilde{G}_t : [0, 1]^\infty \to \mathcal{P}(\mathbb{S})$ such that $\pi_t = G_t(U(t))$ a.s. and $\widetilde{\pi}_t = \widetilde{G}_t(\widetilde{U}(t))$ a.s. By (5.22),

$$\mathbb{E}[G_t(U(t)) fh(U(t))] = \mathbb{E}[\widetilde{G}_t(\widetilde{U}(t)) fh(\widetilde{U}(t))] \tag{5.24}$$
$$= \mathbb{E}[\widetilde{G}_t(U(t)) fh(U(t))]$$

for all $h \in B(\mathbb{S}_0 \times [0, 1]^\infty)$, where the last equality follows from the fact that $U(t)$ and $\widetilde{U}(t)$ have the same distribution. Applying (5.24) with $h = G_t(\cdot)f$ and with $h = \widetilde{G}_t(\cdot)f$, we have

$$\mathbb{E}[G_t(U(t)) f \widetilde{G}_t(U(t)) f] = \mathbb{E}[(\widetilde{G}_t(U(t))f)^2] = \mathbb{E}[(G_t(U(t))f)^2],$$

and it follows that

$$\mathbb{E}[(G_t(U(t))f - \widetilde{G}_t(U(t))f)^2] = 0.$$

Consequently, $\widetilde{\pi}_t f = G_t(\widetilde{U}(t))f$ a.s., and hence $(\pi_t, U(t))$ has the same distribution as $(\widetilde{\pi}_t, \widetilde{U}(t))$.

Since $U(t)$ $(\widetilde{U}(t))$ determines $U(s)$ $(\widetilde{U}(s))$ for $s < t$, U and $\widetilde{U}$ have the same distribution on $C_{[0,1]^\infty}[0, \infty)$. Consequently, (π, Y) and $(\widetilde{\pi}, \widetilde{Y})$ have the same finite-dimensional distributions.

The mapping H_t is given by $H_t(y) \equiv G_t(F(t, y))$. □

Corollary 5.1 *Let A satisfy Condition 5.1. Suppose that $\{\widetilde{\pi}_t, t \geq 0\}$ is a cadlag, $\mathcal{P}(\mathbb{S})$-valued process with no fixed points of discontinuity adapted to a complete filtration $\{\widetilde{\mathcal{G}}_t\}$ such that $\int_0^t \widetilde{\pi}_s \psi ds < \infty$ a.s., $t \geq 0$, and*

$$\widetilde{\pi}_t f - \widetilde{\pi}_0 f - \int_0^t \widetilde{\pi}_s Af ds$$

is a $\{\widetilde{\mathcal{G}}_t\}$-local-martingale for each $f \in \mathcal{D}(A)$. Then there exists a solution X of the local martingale problem for A, a $\mathcal{P}(\mathbb{S})$-valued process $\{\pi_t, t \geq 0\}$ such that $\{\pi_t, t \geq 0\}$ has the same distribution as $\{\widetilde{\pi}_t, t \geq 0\}$, and a filtration $\{\mathcal{G}_t\}$ such that π_t is the conditional distribution of $X(t)$ given $\mathcal{G}_t$.

Proof. The corollary follows by taking $\widetilde{Y}(t) = \widetilde{\pi}_t$ and applying Proposition 5.1. □

The next corollary extends Corollary 3.5 of [26].

Corollary 5.2 *Let A satisfy Condition 5.1. Let $\gamma : \mathbb{S} \to \mathbb{S}_0$ be Borel measurable, and let α be a transition function from $\mathbb{S}_0$ into $\mathbb{S}$ ($y \in \mathbb{S}_0 \to \alpha(y, \cdot) \in \mathcal{P}(\mathbb{S})$ is Borel measurable) satisfying $\alpha(y, \gamma^{-1}(y)) = 1$. Assume that $\widetilde{\psi}(y) \equiv \int_\mathbb{S} \psi(z)\alpha(y, dz) < \infty$ for each $y \in \mathbb{S}_0$ and define*

$$C = \left\{ \left(\int_\mathbb{S} f(z)\alpha(\cdot, dz), \int_\mathbb{S} Af(z)\alpha(\cdot, dz) \right) : f \in \mathcal{D}(A) \right\}.$$

Let $\mu_0 \in \mathcal{P}(\mathbb{S}_0)$, and define $\nu_0 = \int \alpha(y, \cdot)\mu_0(dy)$.

a) *If $\widetilde{Y}$ is a solution of the local martingale problem for (C, μ_0) satisfying $\int_0^t \widetilde{\psi}(\widetilde{Y}(s))ds < \infty$ a.s. for all $t > 0$, then there exists a solution X of the local martingale problem for (A, ν_0) such that $\widetilde{Y}$ has the same distribution on $M_{\mathbb{S}_0}[0, \infty)$ as $Y = \gamma \circ X$. If Y and $\widetilde{Y}$ are cadlag, then Y and $\widetilde{Y}$ have the same distribution on $D_{\mathbb{S}_0}[0, \infty)$.*

b) *If $Y(t)$ is $\widehat{\mathcal{F}}_t^Y$-measurable (which by Lemma A.4 holds for almost every t), then $\alpha(Y(t), \cdot)$ is the conditional distribution of $X(t)$ given $\widehat{\mathcal{F}}_t^Y$.*

c) *If, in addition, uniqueness holds for the martingale problem for (A, ν_0), then uniqueness holds for the $M_{\mathbb{S}_0}[0, \infty)$-martingale problem for (C, μ_0). If $\widetilde{Y}$ has sample paths in $D_{\mathbb{S}_0}[0, \infty)$, then uniqueness holds for the $D_{\mathbb{S}_0}[0, \infty)$-martingale problem for (C, μ_0).*

d) *If uniqueness holds for the martingale problem for (A, ν_0), then Y restricted to $\mathbf{T}^Y = \{t : Y(t) \text{ is } \widehat{\mathcal{F}}_t^Y\text{-measurable}\}$ is a Markov process.*

Proof. We are not assuming that $\widetilde{Y}$ is cadlag, so to apply Proposition 5.1, replace $\widetilde{Y}$ by the continuous process $\widetilde{U}$ given by (5.20). Observing that $\mathcal{F}_t^{\widetilde{U}} = \widehat{\mathcal{F}}_t^{\widetilde{Y}}$, define

$$\widetilde{\pi}_t = \mathbb{E}[\alpha(\widetilde{Y}(t), \cdot)|\mathcal{F}_t^{\widetilde{U}}] = \mathbb{E}[\alpha(\widetilde{Y}(t), \cdot)|\widehat{\mathcal{F}}_t^{\widetilde{Y}}],$$

and note that $\widetilde{\pi}_t = \alpha(Y(t), \cdot)$ for $t \in \mathbf{T}^{\widetilde{Y}}$. Then

$$\widetilde{\pi}_t f - \widetilde{\pi}_0 f - \int_0^t \widetilde{\pi}_s Af \, ds = \widetilde{\pi}_t f - \alpha f(\widetilde{Y}(0)) - \int_0^t \alpha Af(\widetilde{Y}(s)) ds$$

is a $\{\mathcal{F}_t^{\widetilde{U}}\}$-local-martingale for each $f \in \mathcal{D}(A)$ and Proposition 5.1 gives the existence of the processes X and U such that X is a solution of the local martingale problem for A and π_t, the conditional distribution of $X(t)$ given $\mathcal{F}_t^U$, has the same distribution as $\widetilde{\pi}_t$. Consequently, for almost every t, $\pi_t = \alpha(\gamma(X(t)), \cdot)$ and it follows that $Y = \gamma \circ X$ has the same distribution on $M_{\mathbb{S}_0}[0, \infty)$ as $\widetilde{Y}$.

Since the finite-dimensional distributions of X are uniquely determined, the distribution of $\gamma \circ X$ (and hence of $\widetilde{Y}$) on $M_{\mathbb{S}_0}[0, \infty)$ is uniquely determined. If $\widetilde{Y}$ has sample paths in $D_{\mathbb{S}_0}[0, \infty)$, then its distribution on $D_{\mathbb{S}_0}[0, \infty)$ is determined by its distribution on $M_{\mathbb{S}_0}[0, \infty)$.

Since X is the unique solution of a martingale problem, by Lemma A.13, it is Markov. The Markov property for Y for $t \in \mathbf{T}^Y$ follows from the Markov property for X by

$$\mathbb{E}\left[f(Y(t+s))|\widehat{\mathcal{F}}_t^Y\right] = \mathbb{E}\left[\mathbb{E}\left[f(\gamma(X(t+s)))|\mathcal{F}_t^X\right]|\widehat{\mathcal{F}}_t^Y\right]$$
$$= \mathbb{E}\left[h_{f,t,s}(X(t))|\widehat{\mathcal{F}}_t^Y\right]$$
$$= \int_{\mathbb{S}} h_{f,t,s}(x) \alpha(Y(t), dx).$$

$\square$

We will also need a stopped version of Proposition 5.1.

Proposition 5.2 *Let A satisfy Condition 5.1. Suppose that $\widetilde{Y}$ is a cadlag, $\mathbb{S}_0$-valued process with no fixed points of discontinuity, $\widetilde{\tau}$ is a $\{\mathcal{F}_t^{\widetilde{Y}}\}$-stopping time, $\{\widetilde{\pi}_t, t \geq 0\}$ is a $\mathcal{P}(\mathbb{S})$-valued process, adapted to $\{\mathcal{F}_t^{\widetilde{Y}}\}$, $\int_0^{t \wedge \widetilde{\tau}} \widetilde{\pi}_s \psi ds < \infty$ a.s., $t \geq 0$, and*

$$\widetilde{\pi}_{t \wedge \widetilde{\tau}} f - \widetilde{\pi}_0 f - \int_0^{t \wedge \widetilde{\tau}} \widetilde{\pi}_s Af \, ds \tag{5.25}$$

is a $\{\mathcal{F}_t^{\widetilde{Y}}\}$-martingale for each $f \in \mathcal{D}(A)$. Then there exist a solution (X, τ) of the stopped martingale problem for A, a cadlag, $\mathbb{S}_0$-valued process Y, and a $\mathcal{P}(\mathbb{S})$-valued process $\{\pi_t, t \geq 0\}$ such that $\{(Y(t \wedge \tau), \pi_t \mathbf{1}_{\{\tau \geq t\}}), t \geq 0\}$ has the same distribution as $\{(\widetilde{Y}(\cdot \wedge \widetilde{\tau}), \widetilde{\pi}_t \mathbf{1}_{\{\widetilde{\tau} \geq t\}})\}$ and $\pi_{t \wedge \tau}$ is the conditional distribution of $X(t)$ given $\mathcal{F}_{t \wedge \tau}^Y$.

Proof. With $\widehat{A}$ and $\widetilde{U}$ defined as in the proof of Proposition 5.1,

$$\widetilde{\nu}_t^0 h = \mathbb{E}[\mathbf{1}_{[\widetilde{\tau},\infty)}(t) \int_{\mathbb{S}} h(z, \widetilde{U}(\widetilde{\tau}))\widetilde{\pi}_{\widetilde{\tau}}(dz)]$$

$$\widetilde{\nu}_t^1 h = \mathbb{E}[\mathbf{1}_{[0,\widetilde{\tau})}(t) \int_{\mathbb{S}} h(z, \widetilde{U}(t))\widetilde{\pi}_t(dz)]$$

defines a solution of the stopped forward equation martingale problem for $\widehat{A}$. Lemma 5.1 ensures the existence of a solution (X, U, τ) of the stopped martingale problem for $\widehat{A}$ such that

$$\mathbb{E}[f(X(t \wedge \tau))g(U(t \wedge \tau))] = \mathbb{E}[\widetilde{\pi}_{t \wedge \widetilde{\tau}} fg(\widetilde{U}(t \wedge \widetilde{\tau}))].$$

Then for $t \geq 0$, $U(t \wedge \tau)$ has the same distribution as $\widetilde{U}(t \wedge \widetilde{\tau})$ and hence

$$Y(\cdot \wedge \tau) = \lim_{t \to \infty} Y(\cdot \wedge t \wedge \tau) \equiv \lim_{t \to \infty} \Lambda(U(t \wedge \tau))$$

has the same distribution as $\widetilde{Y}(\cdot \wedge \widetilde{\tau})$. With reference to the subsection stopped filtrations and filtrations generated by stopped processes Section A.3,

$$\mathbb{E}[f(X(t \wedge \tau))|\sigma(U(t \wedge \tau))] = \mathbb{E}\left[f(X(t \wedge \tau))|\mathcal{G}_{t \wedge \tau}^Y\right]$$

has the same distribution as

$$\mathbb{E}[\widetilde{\pi}_{t \wedge \widetilde{\tau}} f | \sigma(\widetilde{U}(t \wedge \widetilde{\tau}))] = \mathbb{E}\left[\widetilde{\pi}_{t \wedge \widetilde{\tau}} f | \mathcal{G}_{t \wedge \widetilde{\tau}}^{\widetilde{Y}}\right],$$

and by Lemma A.10, $\widetilde{\pi}_t \mathbf{1}_{\{\widetilde{\tau} \geq t\}}$ has the same distribution as $\pi_t \mathbf{1}_{\{\tau \geq t\}}$, where π_t is the conditional distribution of $X(t \wedge \tau)$ given $\mathcal{F}_{t \wedge \tau}^Y$. □

The only place that Condition 5.1 is used in the proof of Proposition 5.1 is to conclude that every solution of the local forward equation for $\widehat{A}$ defined in (5.21) corresponds to a solution of the local martingale problem. For the filtered martingale problem, $\widehat{A}$ can be given explicitly by

$$\widehat{A}(fg)(x, u) = g(u) Af(x) + f(x) \sum (-a_i u + b_k \circ \gamma(x)) \partial_{k_i} g(u), \qquad (5.26)$$

Consequently, we state the next result under the following hypothesis.

Condition 5.2 *For $\widehat{A}$ defined by (5.26), each solution of the local forward equation for $\widehat{A}$ corresponds to a solution of the local martingale problem for $\widehat{A}$.*

We have the following generalization of Theorem 3.2 of Kurtz [26].

Theorem 5.1 *Let $A \subset B(\mathbb{S}) \times M(\mathbb{S})$, $\widehat{\mu}_0 \in \mathcal{P}(\mathbb{S} \times \mathbb{S}_0)$, and $\gamma : \mathbb{S} \to \mathbb{S}_0$ be Borel measurable, and assume Condition 5.2. Let $(\widetilde{Y}, \widetilde{\pi}, \widetilde{Z})$ be a solution of the local filtered martingale problem for $(A, \gamma, \widehat{\mu}_0)$. Then the following hold:*

  a) *There exists a solution X of the local martingale problem for A and an $\mathbb{S}_0$-valued random variable Z such that $(X(0), Z)$ has distribution $\widehat{\mu}_0$ and $Y = \gamma \circ X$ has the same distribution on $M_{\mathbb{S}_0}[0, \infty)$ as $\widetilde{Y}$.*

b) Let π_t be the conditional distribution of $X(t)$ given $\widehat{\mathcal{F}}_t^{Y,Z}$. For each $t \geq 0$, there exists a Borel measurable mapping $H_t : M_{\mathbb{S}_0}[0, \infty) \times S_0 \to \mathcal{P}(\mathbb{S})$ such that $\pi_t = H_t(Y, Z)$ and $\widetilde{\pi}_t = H_t(\widetilde{Y}, \widetilde{Z})$.

c) If Y and $\widetilde{Y}$ have sample paths in $D_{\mathbb{S}_0}[0, \infty)$, then Y and $\widetilde{Y}$ have the same distribution on $D_{\mathbb{S}_0}[0, \infty)$ and H_t is a Borel measurable mapping from $D_{\mathbb{S}_0}[0, \infty) \times S_0$ to $\mathcal{P}(\mathbb{S})$.

d) If uniqueness holds for the local martingale problem for (A, ν_0), then uniqueness holds for the filtered local martingale problem for $(A, \gamma, \widehat{\mu}_0)$ in the sense that if (Y, π, Z) and $(\widetilde{Y}, \widetilde{\pi}, \widetilde{Z})$ are solutions, then for each $0 \leq t_1 < \cdots < t_m$, $(\pi_{t_1}, \ldots, \pi_{t_m}, Y, Z)$ and $(\widetilde{\pi}_{t_1}, \ldots, \widetilde{\pi}_{t_m}, \widetilde{Y}, \widetilde{Z})$ have the same distribution on $\mathcal{P}(\mathbb{S})^m \times M_{\mathbb{S}_0}[0, \infty) \times S_0$.

Remark 5.7 Note that the theorem does not assume that γ is continuous.

Proof. In the definition of $\widetilde{U}$ in (5.20), replace $c_k(\widetilde{Y}(0))$ by $c_k(\widetilde{Y}(0), \widetilde{Z})$. Note that for a.e. t,

$$f_1(\widetilde{Y}(t \wedge \widetilde{\tau}))\widetilde{\pi}_{t \wedge \widetilde{\tau}} f_2 = \widetilde{\pi}_{t \wedge \widetilde{\tau}}(f_2 f_1 \circ \gamma) \quad a.s. \tag{5.27}$$

(First consider $f_1 = \mathbf{1}_C$, $C \in \mathcal{B}(\mathbb{S}_0)$.)

With $\widetilde{\nu}_t^{n,0}$ and $\widetilde{\nu}_t^{n,1}$ defined as in the proof of Proposition 5.1,

$$\widetilde{\nu}_t^n(fg) = \mathbb{E}[\widetilde{\pi}_{t \wedge \widetilde{\tau}_n} fg(\widetilde{U}(t \wedge \widetilde{\tau}_n))]$$
$$= \mathbb{E}[\widetilde{\pi}_0 fg(\widetilde{U}(0))]$$
$$+ \mathbb{E}\left[\int_0^{t \wedge \widetilde{\tau}_n} \left(g(\widetilde{U}(s))\widetilde{\pi}_s Af + \widetilde{\pi}_s f \sum(-a_i \widetilde{U}_{ki}(s) + b_k(\widetilde{Y}(s)))\partial_{ki}g(\widetilde{U}(s))\right) ds\right]$$
$$= \mathbb{E}[\widetilde{\pi}_0 fg(\widetilde{U}(0))]$$
$$+ \mathbb{E}\left[\int_0^{t \wedge \widetilde{\tau}_n} \left(g(\widetilde{U}(s))\widetilde{\pi}_s Af + \sum(-a_i \widetilde{U}_{ki}(s) + \widetilde{\pi}_s(fb_k \circ \gamma))\partial_{ki}g(\widetilde{U}(s))\right) ds\right]$$
$$= \widetilde{\nu}_0(fg) + \int_0^t \widetilde{\nu}_s^{1,n} \widehat{A}(fg) ds,$$

where the third equality follows from (5.27) and $\widehat{A}$ is defined in (5.26).

We are not assuming that $\widetilde{Y}$ is cadlag, but we still conclude that the completion of $\sigma(\widetilde{U}(t))$ is $\widehat{\mathcal{F}}_t^{\widetilde{Y},\widetilde{Z}}$ and there exist $\Lambda : [0, 1]^\infty \to M_{\mathbb{S}_0}[0, \infty) \times S_0$ and $\Lambda_1 : [0, 1]^\infty \to S_0$ such that $\Lambda(\widetilde{U}(t)) = (\widetilde{Y}(\cdot \wedge t), \widetilde{Z})$ and $\Lambda_1(\widetilde{U}(0)) = \widetilde{Z}$. Condition 5.2 ensures the existence of a solution (X, U) of the local martingale problem for $\widehat{A}$ such that U and $\widetilde{U}$ have the same distribution,

$$U_{ki}(t) = U_{ki}(0) - a_i \int_0^t U_{ki}(s) ds + \int_0^t b_k(Y(s)) ds \tag{5.28}$$
$$= U_{ki}(0)e^{-a_i t} + \int_0^t e^{-a_i(t-s)} b_k(Y(s)) ds,$$

and defining $Z = \Lambda_1(U(0))$, Parts (a) and (b) hold.

Part (c) follows from the fact that the distribution of a cadlag process is determined by its distribution on $M_{\mathbb{S}_0}[0, \infty)$.

Finally, for Part (d), uniqueness for the local martingale problem for (A, ν_0) implies uniqueness for the local martingale problem for $(\widehat{A}, \widehat{\nu}_0)$, where $\widehat{\nu}_0 h = E[\widetilde{\pi}_0 h(\cdot, \widetilde{U}(0))]$. Uniqueness of the distribution of (X, U) in Part (b) implies uniqueness of the distribution of $(\widetilde{Y}, \widetilde{\pi}, \widetilde{Z})$. □

5.3.1 The Markov property

Uniqueness for martingale problems usually implies the Markov property for solutions, and a similar result holds for filtered martingale problems.

Theorem 5.2 *Let $A \subset B(\mathbb{S}) \times M(\mathbb{S})$, $\widehat{\mu}_0 \in \mathcal{P}(\mathbb{S} \times \mathbb{S}_0)$, and $\gamma : \mathbb{S} \to \mathbb{S}_0$ be Borel measurable, and assume Condition 5.2. Let $(\widetilde{Y}, \widetilde{\pi}, \widetilde{Z})$ be a solution of the filtered martingale problem for $(A, \gamma, \widehat{\mu}_0)$. ($\widetilde{\tau} = \infty$.) If uniqueness holds for the martingale problem for (A, ν_0), then $\widetilde{\pi}$ is a $\mathcal{P}(\mathbb{S})$-valued Markov process.*

Proof. Fix $r \geq 0$, and let $(\widehat{Y}, \widehat{\pi})$ be as in the second part of Lemma 5.3. Since $\widehat{\pi}_0 = \widetilde{\pi}_r$, they have the same distribution. By Lemma A.12, a process (Y^*, π^*, Z^*) can be constructed so that $(Y^*(r + \cdot), \pi^*_{r+\cdot}, \pi^*_r)$ has the same distribution on $M_{\mathbb{S}_0 \times \mathcal{P}(\mathbb{S})}[0, \infty) \times \mathcal{P}(\mathbb{S})$ as $(\widehat{Y}, \widehat{\pi}, \widehat{\pi}_0)$, $(Y^*(\cdot \wedge r), \pi^*_{\cdot \wedge r}, Z^*, \pi^*_r)$ has the same distribution on $M_{\mathbb{S}_0 \times \mathcal{P}(\mathbb{S})}[0, r] \times \mathbb{S}_0 \times \mathcal{P}(\mathbb{S})$ as $(\widetilde{Y}(\cdot \wedge r), \widetilde{\pi}_{\cdot \wedge r}, \widetilde{Z}, \widetilde{\pi}_r)$, and

$$\mathbb{E}\left[g\left(Y^*(r+\cdot), \pi^*_{r+\cdot}\right) | Y^*(\cdot \wedge r), \pi^*_{\cdot \wedge r}, Z^*, \pi^*_r\right] = \mathbb{E}\left[g\left(Y^*(r+\cdot), \pi^*_{r+\cdot}\right) | \pi^*_r\right]. \tag{5.29}$$

We claim that (Y^*, π^*, Z^*) is a solution of the filtered martingale problem for $(A, \gamma, \widehat{\mu}_0)$.

(π^*_0, Z^*) has the same distribution as $(\widetilde{\pi}_0, \widetilde{Z})$, so (5.15) holds. Since $(Y^*(\cdot \wedge r), \pi^*_{\cdot \wedge r})$ has the same distribution as $(\widetilde{Y}(\cdot \wedge r), \widetilde{\pi}_{\cdot \wedge r})$, for $g \in B(\mathbb{S}_0)$ and $t \leq r$,

$$\int_0^t \pi^*_s(g \circ \gamma) ds = \int_0^t g(Y^*(s)) ds \quad a.s.$$

For $t > r$, $\left(\int_r^t \pi^*_s(g \circ \gamma) ds, \int_r^t g(Y^*(s)) ds\right)$ has the same distribution as $\left(\int_0^{t-r} \widehat{\pi}_s(g \circ \gamma) ds, \int_0^{t-r} g(\widehat{Y}(s)) ds\right)$, so $\int_r^t \pi^*_s(g \circ \gamma) ds = \int_r^t g(Y^*(s)) ds$ a.s. Consequently, (5.16) follows.

For $f \in \mathcal{D}(A)$, let $M^*_f(t) = \pi^*_t f - \int_0^t \pi^*_s A f ds$. For $r \leq t < t + h$, let H_1 be a bounded random variable measurable with respect to the completion of

$$\sigma\left(\int_r^u h(Y^*(s)) ds : r \leq u \leq t, h \in B(\mathbb{S}_0)\right) \vee \sigma\left(\pi^*_r\right)$$

and H_2 a bounded random variable measurable with respect to $\widehat{\mathcal{F}}_r^{Y^*,Z^*} = \widehat{\mathcal{F}}_r^{Y^*} \vee \sigma(Z^*)$. Then by (5.29),

$$\mathbb{E}\left[(M_f^*(t+h) - M_f^*(t)) H_1 H_2\right] = \mathbb{E}\left[(M_f^*(t+h) - M_f^*(t)) H_1 \mathbb{E}\left[H_2 | \pi_r^*\right]\right],$$

and the right side is zero by the fact that $(Y^*(r + \cdot), \pi_{r+}^*)$ has the same distribution as $(\widehat{Y}, \widehat{\pi})$. It follows that

$$\mathbb{E}\left[M_f^*(t+h) - M_f^*(t) | \widehat{\mathcal{F}}_t^{Y^*,Z^*}\right] = 0. \tag{5.30}$$

If $t < t+h \leq r$, then (5.30) follows from the fact that $(Y^*(\cdot \wedge r), \pi_{\cdot \wedge r}^*, Z^*, \pi_r^*)$ has the same distribution as $(\widetilde{Y}(\cdot \wedge r), \widetilde{\pi}_{\cdot \wedge r}, \widetilde{Z}, \widetilde{\pi}_r)$, and for $t < r < t+h$,

$$\mathbb{E}\left[M_f^*(t+h) - M_f^*(t) | \widehat{\mathcal{F}}_t^{Y^*,Z^*}\right] = \mathbb{E}\left[M_f^*(t+h) - M_f^*(r) + M_f^*(r) - M_f^*(t) | \widehat{\mathcal{F}}_t^{Y^*,Z^*}\right] = 0$$

verifying (5.18).

By uniqueness, (π^*, Y^*, Z^*) and $(\widetilde{\pi}, \widetilde{Y}, \widetilde{Z})$ have the same distribution. Consequently, (5.29) implies

$$\mathbb{E}[g(\widetilde{Y}(r+\cdot), \widetilde{\pi}_{r+\cdot}) | \widetilde{Y}(\cdot \wedge r), \widetilde{\pi}_{\cdot \wedge r}, \widetilde{Z}, \widetilde{\pi}_r] = \mathbb{E}[g(\widetilde{Y}(r+\cdot), \widetilde{\pi}_{r+\cdot}) | \widetilde{\pi}_r],$$

giving the Markov property. □

In the classical setting, $X = (X_1, Y) \in \mathbb{S}_1 \times \mathbb{S}_0$ and $\gamma(X) = Y$, $\pi_t = \pi_t^1 \times \delta_{Y(t)}$, where π_t^1 is the conditional distribution of $X_1(t)$ given $\mathcal{F}_t^Y$. In the observations in additive white noise setting, a number of authors (Kunita [22], Bhatt, Budhiraja, and Karandikar [5], Stettner [31]) have given conditions under which π^1 is Markov. The following example shows that the conclusion does not hold in general, and hence the Markov property for π^1 does not immediately follow from Theorem 5.2.

Example 5.1 Let (X_1, Y) be the Markov process with values in $\{-1, +1\} \times \mathbb{N}$ and generator

$$Af(x, y) = \lambda y[f(-x, y+1) - f(x, y)] + \mu[f(-x, y) - f(x, y)].$$

Given a pure jump Markov counting process (a Yule process) Y with intensity λY and an independent Poisson process Z with intensity μ, the process (X_1, Y) can be represented by

$$X_1(t) = (-1)^{Y(t) - Y(0) + Z(t)} X_1(0).$$

Then the conditional distribution of $X_1(t)$ given $\mathcal{F}_t^Y$ is

$$\pi_t^1(dx) = \mathbf{1}_E(Y(t) - Y(0))\left(\alpha_t \delta_{+1}(dx) + (1 - \alpha_t)\delta_{-1}(dx)\right)$$
$$+ \mathbf{1}_O(Y(t) - Y(0))\left((1 - \alpha_t)\delta_{+1}(dx) + \alpha_t \delta_{-1}(dx)\right),$$

where E is the set of even integers, O the set of odd integers, and

$$a_t = \frac{1 + (2a_0 - 1)e^{-2\mu t}}{2},$$

where $a_0 = \mathbb{P}\{X_1(0) = 1 | Y(0)\}$.

If $a_0 = \frac{1}{2}$, then $a_t = \frac{1}{2}$ and $\pi_t^1(dx) = \frac{1}{2}\delta_{+1}(dx) + \frac{1}{2}\delta_{-1}(dx)$, for all $t \geq 0$, and π^1 is trivially Markov; however, if $\mathbb{P}\{a_0 \neq \frac{1}{2}\} > 0$, in general π^1 is not Markov. Assuming, for example, that $Y(0) = 1$ and $a_0 = \mathbb{P}\{X_1(0) = 1 | Y(0)\} = \mathbb{P}\{X_1(0) = 1\} \neq \frac{1}{2}$, then $\mathcal{F}_t^{\pi^1} \equiv \sigma(\pi_s^1 : s \leq t) = \mathcal{F}_t^Y (= \widehat{\mathcal{F}}_t^Y$ by Lemma A.5). Consequently, a_t is deterministic and $\pi_t^1 f = g(t, Y(t))$, with

$$g(t, y) = \mathbf{1}_E(y - 1)\left(a_t f(1) + (1 - a_t)f(-1)\right) + \mathbf{1}_O(y - 1)\left((1 - a_t)f(1) + a_t f(-1)\right).$$

Taking into account that $a_t' = \mu(1 - 2a_t)$ and that $\mathbf{1}_E(y - 1) = \mathbf{1}_O(y)$, we have

$$\frac{\partial}{\partial t}g(t, y) + \lambda y [g(t, y + 1) - g(t, y)] = (\lambda y + \mu)(1 - 2a_t)[f(-1) - f(1)]\left[\mathbf{1}_E(y) - \mathbf{1}_O(y)\right],$$

and therefore

$$\lim_{h \to 0} h^{-1}\mathbb{E}\left[\pi_{t+h}^1 f - \pi_t^1 f \Big| \mathcal{F}_t^{\pi^1}\right] = (\lambda Y(t) + \mu)(\mathbf{1}_E(Y(t)) - \mathbf{1}_O(Y(t)))(1 - 2a_t)[f(-1) - f(1)].$$

The right side is not just a function of π_t^1, and it follows that π^1 is not a Markov process.

Of course, there is additional structure in the classical example with

$$Y(t) = \sigma W(t) + \int_0^t h(X_1(s))ds, \tag{5.31}$$

W a standard Brownian motion. With this example in mind, we have the following definition.

Definition 5.5 For $\mathbb{S} = \mathbb{S}_1 \times \mathbb{S}_0$ with $\mathbb{S}_0 = \mathbb{R}^d$ and γ the projection of $\mathbb{S}$ onto $\mathbb{S}_0$, the filtered martingale problem for (A, γ) has *additive observations*, if for each solution $(\widetilde{Y}, \widetilde{\pi}, \widetilde{Z})$, each $r \geq 0$, and each $y \in \mathbb{R}^d$,

$$\widehat{Y}(t) = \widetilde{Y}(r + t) - \widetilde{Y}(r) + y \tag{5.32}$$
$$\widehat{\pi}_t^1 = \mathbb{E}\left[\widetilde{\pi}_{r+t}^1 \Big| \widehat{\mathcal{F}}_t^{\widehat{Y}} \vee \sigma(\widetilde{\pi}_r^1)\right]$$

determines a solution $(\widehat{Y}, \widehat{\pi}^1 \times \delta_{\widehat{\gamma}}, \widehat{\pi}_0^1 \times \delta_y)$ of the filtered martingale problem for (A, γ).

Lemma 5.4 *Let $\mathbb{S} = \mathbb{S}_1 \times \mathbb{S}_0$ with $\mathbb{S}_0 = \mathbb{R}^d$ and γ be the projection of $\mathbb{S}$ onto $\mathbb{S}_0$, and suppose that A satisfies Condition 5.1. Assume that every solution $X = (X_1, Y)$ of the martingale problem for A has a version such that Y is cadlag with no fixed points of discontinuity and for each $r \geq 0$ and $y \in \mathbb{R}^d$, $(X_1(\cdot + r), Y(\cdot + r) - Y(r) + y)$ is* a solution of the martingale problem for A. Then the filtered martingale problem for (A, γ) has additive observations.

Remark 5.8 If X_1 is the solution of the martingale problem for a generator L satisfying Condition 5.1 with $\mathbb{S}$ replaced by $\mathbb{S}_1$, $((a_{ij})) = \sigma\sigma^\top$ and

$$A[f_1 f_2] = f_2 L f_1 + f_1(\frac{1}{2}\sum_{i,j} a_{ij}\partial_i\partial_j f_2 + h \cdot \nabla f_2)$$

for $f_1 \in \mathcal{D}(L)$ and $f_2 \in C_c^2(\mathbb{R}^d)$, that is, the $\mathbb{S}_0 = \mathbb{R}^d$ component satisfies (5.31) with W independent of X_1, then the hypotheses of the lemma are satisfied.

Proof. Suppose that $(\widetilde{Y}, \widetilde{\pi}, \widetilde{Z})$ is a solution of the filtered martingale problem for $(A, \gamma, \widehat{\mu}_0)$. Then there exists a solution $X = (X_1, Y)$ of the martingale problem for A and a random variable Z such that $(X_1(0), Y(0), Z)$ has distribution $\widehat{\mu}_0$, Y has the same distribution as $\widetilde{Y}$, and for $t \geq 0$, there exist $H_t : M_{\mathbb{S}_0}[0, \infty) \times S_0 \to \mathcal{P}(\mathbb{S})$ such that $\widetilde{\pi}_t = H_t(\widetilde{Y}, \widetilde{Z})$ and $\pi_t = \pi_t^1 \times \delta_{Y(t)} = H_t(Y, Z)$ is the conditional distribution of $X(t)$ given $\widehat{\mathcal{F}}_t^{Y,Z}$.

By assumption, $(\widehat{X}_1, \widehat{Y}) = (X_1(\cdot + r), Y(\cdot + r) - Y(r) + y)$ is a solution of the martingale problems for A. Consequently, defining $\widehat{\pi}^1$ by

$$\widehat{\pi}_t^1 g = \mathbb{E}\left[g(X_1(r+t)) \Big| \mathcal{F}_t^{\widehat{Y}} \vee \sigma(\pi_r)\right] = \mathbb{E}\left[\pi_{r+t}^1 g \Big| \mathcal{F}_t^{\widehat{Y}} \vee \sigma(\pi_r)\right],$$

$(\widehat{Y}, \widehat{\pi}^1 \times \delta_{\widehat{Y}(\cdot)}, \pi_r)$ is a solution of the filtered martingale problem for (A, γ). Since $(\widetilde{Y}, \pi)$ has the same distribution as $(\widetilde{Y}(\cdot + r) - \widetilde{Y}(r) + y, \widetilde{\pi})$, the filtered martingale problem for (A, γ) has additive observations. $\square$

Theorem 5.3 Let $A \subset B(\mathbb{S}) \times M(\mathbb{S})$ and $\gamma : (x, y) \in \mathbb{S}_1 \times \mathbb{R}^d \to y \in \mathbb{R}^d$, and assume Condition 5.2. Suppose that the filtered martingale problem for (A, γ) has additive observations. Let $(\widetilde{Y}, \widetilde{\pi}^1 \times \delta_{\widetilde{Y}}, \widetilde{Z})$ be a solution of the filtered martingale problem for $(A, \gamma, \widehat{\mu}_0)$. If uniqueness holds for the martingale problem for (A, ν_0), then $\widetilde{\pi}^1$ is a $\mathcal{P}(\mathbb{S}_1)$-valued Markov process.

Proof. As in the proof of Theorem 5.2, fix $r \geq 0$, and let $(\widehat{Y}, \widehat{\pi})$ be as in (5.32) with $y = 0$. Since $\widehat{\pi}_0^1 = \widetilde{\pi}_r^{\,1}$, they have the same distribution. By Lemma A.12, a process (Y^*, π^{1*}, Z^*) can be constructed so that $(Y^*(r + \cdot) - Y^*(r), \pi_{r+\cdot}^{1*}, \pi_r^{1*})$ has the same distribution on $M_{\mathbb{R}^d \times \mathcal{P}(\mathbb{S}_1)}[0, \infty) \times \mathcal{P}(\mathbb{S}_1)$ as $(\widehat{Y}, \widehat{\pi}^1, \widehat{\pi}_0^1)$, $(Y^*(\cdot \wedge r), \pi_{\cdot \wedge r}^{1*}, Z^*, \pi_r^{1*})$ has the same distribution on $M_{\mathbb{R}^d \times \mathcal{P}(\mathbb{S}_1)}[0, r] \times S_0 \times \mathcal{P}(\mathbb{S}_1)$ as $(\widetilde{Y}(\cdot \wedge r), \widetilde{\pi}_{\cdot \wedge r}^1, \widetilde{Z}, \widetilde{\pi}_r^1)$, and

$$\mathbb{E}\left[g\left(Y^*(r+\cdot) - Y^*(r), \pi_{r+\cdot}^{1*}\right) | Y^*(\cdot \wedge r), \pi_{\cdot \wedge r}^{1*}, Z^*, \pi_r^{1*}\right]$$
$$= \mathbb{E}\left[g\left(Y^*(r+\cdot) - Y^*(r), \pi_{r+\cdot}^{1*}\right) | \pi_r^{1*}\right].$$

Employing the assumption of additive observations, the proof that (Y^*, π^*, Z^*) is a solution of the filtered martingale problem for $(A, \gamma, \widehat{\mu}_0)$ and the proof of the Markov property are essentially the same as before. $\square$

Remark 5.9 Theorem 5.3 can be extended to processes in which $\mathbb{S}_0$ is a group and the definition of $\widehat{Y}$ in (5.32) is replaced by $\widehat{Y}(t) = y\widetilde{Y}(r)^{-1}\widetilde{Y}(r+t)$.

5.4 Filtering equations

In the nonlinear filtering literature, a filtering equation is a collection of identities satisfied by $\{\pi_t f, f \in \mathcal{D}\}$ and the observation process Y for a set of test functions $\mathcal{D}$. The set $\mathcal{D}$ should be small enough to handle easily, but large enough to insure that the identities uniquely determine π as a function of Y. Uniqueness means that if $\widetilde{\pi}$ is another $\mathcal{P}(\mathbb{S})$-valued process adapted to $\{\mathcal{F}_t^Y\}$ and $\{\widetilde{\pi} f, f \in \mathcal{D}\}$ and Y satisfy the identities, then $\widetilde{\pi} = \pi$.

The results of Section 5.3 can be exploited to prove uniqueness for a filtering equation provided each solution of the filtering equation has the appropriate martingale properties. Then, Proposition 5.1 ensures that the solution of the filtering equation satisfies

$$\widetilde{\pi}_t = H_t(Y) \quad a.s., \tag{5.33}$$

where H_t is the function that gives the conditional distribution of $X(t)$ given $\mathcal{F}_t^Y$ for *some* solution of the martingale problem for A. H_t (and hence the solution of the filtering equation) is uniquely determined provided the joint distribution of (X, Y) is uniquely determined.

In practice, it frequently turns out that verifying that $\widetilde{\pi}$ has the appropriate martingale properties requires a change of measure, but since the new measure is equivalent to the original measure, (5.33) still holds. In the next section, we illustrate this argument in the classical setting of a signal in additive white noise.

5.4.1 Filtering equations for a signal in additive white noise

We consider the classical *Markov model with additive white noise*, in which

$$Y(t) = Y(0) + W(t) + \int_0^t h(X(s))ds, \tag{5.34}$$

where W is a standard d-dimensional Brownian motion and $h = (h_1, \cdots, h_d)^\top$ is a Borel function. Note that we are assuming $\mathbb{S}_0 = \mathbb{R}^d$. Note that we are not assuming the independence of X and W.

Define

$$M_f(t) := f(X(t)) - \int_0^t Af(X(s))\,ds, \quad f \in \mathcal{D}(A),$$

and assume that, for $i = 1, \cdots, d$,

$$\langle M_f, W_i \rangle_t = \int_0^t C_i f(X(s), Y(s))\, ds,$$

where C_i is an operator mapping $\mathcal{D}(A)$ into $M(\mathbb{S}_0 \times \mathbb{R}^d)$.

Remark 5.10 Note that if $f = 1$, then $M_f(t) = 1$ and therefore

$$\langle M_f, W_i \rangle_t = 0, \qquad t \geq 0,$$

that is, $C_i 1 = 0$.

In the uncorrelated case, that is, when X and W are independent, $C_i f = 0$, for all $f \in \mathcal{D}(A)$.

Example 5.2 Let X be a diffusion process with values in $\mathbb{R}^m$ solving

$$dX(t) = b(X(t))\,dt + \sigma(X(t))\,dW(t) + \overline{\sigma}(X(t))\,d\overline{W}(t), \tag{5.35}$$

with $\overline{W}$ a Wiener process, independent of W. Then

$$C_i f(x, y) = \sum_{j=1}^m \partial_j f(x) \sigma_{j,i}(x), \qquad f \in C_c^2(\mathbb{R}^m),$$

where $\partial_j f$ denotes the partial derivative of f with respect to the j-th component of x.

Setting $Cf = (C_1 f, \cdots, C_p f)$, the pair (X, Y) will be a solution of the martingale problem (or local martingale problem) for $\widehat{A}$ given by

$$Af \otimes \varphi(x, y) = Af(x)\varphi(y) + f(x)\left(\frac{1}{2}\Delta\varphi(y) + h(x) \cdot \nabla\varphi(y)\right) + Cf(x, y)\nabla\varphi(y),$$

with domain

$$\mathcal{D}(\widehat{A}) = \{f \otimes \varphi : f \in \mathcal{D}(A), \varphi \in C_c^\infty(\mathbb{R}^d)\}.$$

Since we need the joint distribution of X and Y to be determined, we need uniqueness for the martingale problem for $\widehat{A}$, and with Condition 5.1(iii) in mind, we assume

Condition 5.3 There exist a function $\psi_0(x, y) \geq 1$ and constants c_f such that for each $i = 1, \cdots, d$ and $f \in \mathcal{D}(A)$,

$$|C_i f(x, y)| \leq c_f \psi_0(x, y). \tag{5.36}$$

Let ν_0 denote the distribution for $X(0)$ and $\widehat{\nu}_0$ the distribution for $(X(0), Y(0))$.

Condition 5.4 The process X is a cadlag process with values in $\mathbb{S}$ and is a solution of the martingale problem for (A, ν_0), where the operator A satisfies Condition 5.1 and uniqueness holds for the stopped martingale problem for $(\widehat{A}, \widehat{\nu}_0)$.

Remark 5.11 Note that under these conditions, $\widehat{A}$ satisfies Condition 5.1 with ψ replaced by

$$\widehat{\psi}(x, y) = \psi(x) + \psi_0(x, y) + |h(x)|.$$

In the uncorrelated case, uniqueness for $\widehat{A}$ is implied by uniqueness for A by a slight modification of Lemma 4.4 of Kurtz and Ocone [27]. In particular, if (X, Y) is a solution of the (local) martingale problem for $\widehat{A}$, then for

$$Z(t) = Y(t) - Y(0) - \int_0^t h(X(s))ds,$$

(X, Z) is a solution of the (local) martingale problem for (X, W), that is, for the operator $\widehat{A}^0$ obtained by setting $h = 0$ in the definition of $\widehat{A}$. Consequently, uniqueness for $\widehat{A}^0$ implies uniqueness for $\widehat{A}$. In the uncorrelated case, uniqueness for $\widehat{A}^0$ follows by Theorem 4.10.1 of Ethier and Kurtz [18].

In the correlated case, a sufficient condition for uniqueness for $\widehat{A}$ can be found in Theorem 4.2 of Bhatt and Karandikar [3] under the assumption that for each i and $f \in \mathcal{D}(A)$, $C_i f(x, y)$ is a function of x alone. In the diffusion case, Example 5.2, Cherny [15] has shown *joint uniqueness in law* for the solution $(X, W, \overline{W})$ of (5.35) which in turn implies uniqueness for the martingale problem for $\widehat{A}^0$ and hence uniqueness for $\widehat{A}$.

Finally, we assume the following integrability conditions.

Condition 5.5 For each $t > 0$,

$$\mathbb{E}\left[\int_0^t (\psi(X(s)) + \psi_0(X(s), Y(s)) + |h(X(s))|) \, ds\right] < \infty, \quad t \geq 0, \quad (5.37)$$

and

$$\int_0^t (\pi_s \psi + (\pi_s \psi_0(\cdot, Y(s)))^2 + (\pi_s |h|)^2) ds < \infty \quad \text{a.s.} \quad (5.38)$$

Under Condition 5.5, the innovation process $I^\pi(t) = Y(t) - Y(0) - \int_0^t \pi_s h \, ds$ is a Brownian motion. (Note that we are not assuming that $\mathbb{E}\left[\int_0^t |h(X(s))|^2 \, ds\right] < \infty$. $\mathbb{E}\left[\int_0^t |h(X(s))| ds\right] < \infty$ is sufficient to ensure that I^π is a martingale.) For all $f \in \mathcal{D}(A)$, π satisfies

$$\pi_t f = \pi_0 f + \int_0^t \pi_s A f \, ds$$
$$+ \int_0^t \left[\pi_s(hf + Cf(\cdot, Y(s))) - \pi_s h \, \pi_s f\right] \left[dY(s) - \pi_s h \, ds\right],$$

where $Cf(x_1, y) = (C_1 f(x_1, y), \cdots, C_p f(x_1, y))$, or equivalently,

$$\pi_t f = \pi_0 f + \int_0^t \pi_s A f \, ds + \int_0^t \left[\pi_s(hf + Cf(\cdot, Y(s))) - \pi_s h \, \pi_s f\right] d I^\pi(s). \quad (5.39)$$

Define the unnormalized filter for X as the $\mathcal{M}(\mathbb{S})$-valued process

$$\rho_t(dx) := \mathcal{Z}_t^\pi \pi_t(dx), \tag{5.40}$$

where

$$\mathcal{Z}_t^\pi := \exp\left\{\int_0^t \pi_s(h)\, dY(s) - \frac{1}{2}\int_0^t |\pi_s(h)|^2\, ds\right\}. \tag{5.41}$$

By Itô's formula, one can show that ρ satisfies the Duncan–Mortensen–Zakai unnormalized filtering equation,

$$\rho_t f = \pi_0 f + \int_0^t \rho_s\, Af\, ds + \int_0^t \rho_s\left(hf + Cf(\cdot, Y(s))\right) dY(s), \quad \forall f \in \mathcal{D}(A).$$

The following result, essentially Theorem 9.1 of Bhatt, Kallianpur, and Karandikar [4], extends Theorems 4.1 and 4.5 of Kurtz and Ocone [27] and gives uniqueness of the Kushner–Stratonovich and Fujisaki–Kallianpur–Kunita equations.

Theorem 5.4 *Assume that Conditions 5.3, 5.4, and 5.5, are satisfied. Let $\{\mu_t\}$ be a $\{\mathcal{F}_t^Y\}$-adapted, cadlag $\mathcal{P}(\mathbb{S})$-valued process satisfying*

$$\int_0^t (\mu_s \psi + (\mu_s \psi_0(\cdot, Y(s)))^2 + (\mu_s|h|)^2)\, ds < \infty, \quad a.s., \tag{5.42}$$

and for $f \in \mathcal{D}(A)$,

$$\mu_t f = \pi_0 f + \int_0^t \mu_s\, Af\, ds$$
$$+ \int_0^t \left[\mu_s\left(hf + Cf(\cdot, Y(s))\right) - \mu_s h \mu_s f\right] dI^\mu(s), \tag{5.43}$$

where $I^\mu(t) = Y(t) - \int_0^t \mu_s h\, ds$. Then $\mu_t = \pi_t$, $t \geq 0$, a.s.

Remark 5.12 There is a large literature on uniqueness for the filtering equations with varying assumptions depending on the techniques used by the authors. Kurtz and Ocone [27], Bhatt, Kallianpur, and Karandikar [4], Lucic and Heunis [29], and Rozovskiĭ [30] provide a reasonable sampling of results and methods. Our introduction and exploitation of the filtered local martingale problem allows us to avoid a number of assumptions that appear in many of the earlier results. In particular, we do not assume that h is continuous. We only require the first moment assumption $\mathbb{E}\left[\int_0^t |h(X(s))|\, ds\right] < \infty$ rather than a second moment assumption. (Note that there is no expectation in (5.38) and (5.42).) There are no a priori moment assumptions on the solution μ (only on the true conditional distribution).

Proof. If μ is a solution of (5.43), then $\widehat{\mu} = \mu \times \delta_Y$ satisfies

$$\widehat{\mu}_t f \otimes \varphi = \varphi(Y(0))\pi_0 f + \int_0^t \widehat{\mu}_s \widehat{A} f \otimes \varphi \, ds$$
$$+ \int_0^t \left[\varphi(Y(s))\mu_s (hf + Cf(\cdot, Y(s))) \right.$$
$$\left. - \varphi(Y(s))\mu_s h \mu_s f C + \mu_s f \nabla \varphi(Y(s))^T \right] dI^\mu(s).$$

If I^μ is a $\{\mathcal{F}_t^Y\}$-local-martingale, then

$$\widehat{\mu}_t f \otimes \varphi - \varphi(Y(0))\pi_0 f - \int_0^t \widehat{\mu}_s \widehat{A} f \otimes \varphi \, ds$$

is a $\{\mathcal{F}_t^Y\}$-local-martingale, and Proposition 5.1 implies that there exist a solution (X^*, Y^*) of the local martingale problem for $(\widehat{A}, \widehat{\nu}_0)$ and a $\mathcal{P}(\mathbb{S})$-valued process π^* such that (π^*, Y^*) and (μ, Y) have the same distribution and π_t^* is the conditional distribution of $X^*(t)$ given $\mathcal{F}_t^{Y^*}$. In addition, there exists H_t such that $\pi_t^* = H_t(Y^*)$ and $\mu_t = H_t(Y)$. Since uniqueness holds for the local martingale problem for $(\widehat{A}, \widehat{\nu}_0)$, (X^*, Y^*) has the same distribution as (X, Y) and $\mu_t = H_t(Y) = \pi_t$.

Unfortunately, it is not immediately clear that I^μ is a local martingale. However, since $I^\pi(t) = Y(t) - Y(0) - \int_0^t \pi_s h \, ds$ is a Brownian motion, if we define

$$\xi(t) = \mu_t h - \pi_t h,$$

$$\tau_n = \inf \left\{ t : \int_0^t (|\xi(s)|^2 + |\mu_s h|^2) ds \geq n \right\},$$

and let Q_n be the probability measure on $\mathcal{F}_{n \wedge \tau_n}^Y$ given by

$$\frac{dQ_n}{dP} = \exp \left\{ \int_0^{n \wedge \tau_n} \xi(s)^T dI^\pi(s) - \frac{1}{2} \int_0^{n \wedge \tau_n} |\xi(s)|^2 ds \right\},$$

then under Q_n,

$$I^\mu(t \wedge \tau_n) = I^\pi(t \wedge \tau_n) - \int_0^{t \wedge \tau_n} \xi(s) ds,$$

is a martingale for $0 \leq t \leq n$. Consequently, under Q_n,

$$\widehat{\mu}_{t \wedge \tau_n} f \otimes \varphi - \varphi(Y(0))\pi_0 f - \int_0^{t \wedge \tau_n} \widehat{\mu}_s \widehat{A} f \otimes \varphi \, ds \qquad (5.44)$$

is a $\{\mathcal{F}_t^Y\}$-martingale, and by Proposition 5.2 and uniqueness of the stopped martingale problem for $\widehat{A}$,

$$\mu_t \mathbf{1}_{\{n \wedge \tau_n \geq t\}} = H_t(Y) \mathbf{1}_{\{n \wedge \tau_n \geq t\}} = \pi_t \mathbf{1}_{\{n \wedge \tau_n \geq t\}}.$$

It follows that $Q_n = P$ on $\mathcal{F}^Y_{n \wedge \tau_n}$, $n = 1, 2, \ldots$ and hence by (5.42), $\tau_n \to \infty$ a.s. and $\mu_t = \pi_t$ a.s. □

Corollary 5.3 *Assume that Conditions 5.3, 5.4, and 5.5 are satisfied. Let $\{\theta_t\}$ be a $\{\mathcal{F}^Y_t\}$-adapted, cadlag $\mathcal{M}(\mathbb{S})$-valued process satisfying*

$$\int_0^t (\theta_s \psi + (\theta_s \psi_0(\cdot, Y(s)))^2 + (\theta_s |h|)^2) ds < \infty, \quad a.s.,$$

such that for every $f \in \mathcal{D}(A)$,

$$\theta_t f = \pi_0 f + \int_0^t \theta_s A f \, ds + \int_0^t \theta_s (hf + Cf(\cdot, Y(s))) \, dY(s). \tag{5.45}$$

Then $\theta_t = \rho_t$ a.s. for all $t \geq 0$.

Proof. For $\epsilon > 0$, define $\beta_\epsilon = \inf\{t > 0 : \theta_t 1 \leq \epsilon\}$ and set

$$\mu_t = \frac{\theta_t}{\theta_t 1} \text{ and } 0 \leq t < \beta_0 \equiv \lim_{\epsilon \to 0} \beta_\epsilon.$$

Then, by Itô's formula, μ satisfies (5.43) on $[0, \beta_0)$. Defining τ_n and the appropriate change of measure as in the proof of the previous theorem, it follows that under the new measure, (5.44), with τ_n replaced by $\tau_n \wedge \beta_\epsilon$ is a martingale, and as before, $\mu_t \mathbf{1}_{\{\tau_n \wedge \beta_\epsilon \geq t\}} = \pi_t \mathbf{1}_{\{\tau_n \wedge \beta_\epsilon \geq t\}}$. Letting $n \to \infty$, $\mu_{t \wedge \beta_\epsilon} = \pi_{t \wedge \beta_\epsilon}$.

Observe that, for $t \leq \beta_\epsilon$,

$$\theta_t 1 = 1 + \int_0^t \theta_s 1 \pi_s h d Y(s),$$

so

$$\theta_t 1 = Z^\pi_t = \exp\left\{\int_0^t \pi_s h d Y(s) - \frac{1}{2} \int_0^t |\pi_s h|^2 ds\right\}, \quad t < \beta_0.$$

Since for $T > 0$, $\inf_{t \leq T} Z^\pi_t > 0$, it follows that $\beta_0 = \infty$ and hence

$$\mu_t = \pi_t \text{ and } \theta_t = \theta_t 1 \pi_t = \rho_t.$$

□

5.4.2 Related results on filtering equations

The filtered martingale problem was first introduced in Kurtz and Ocone [27] in the special case considered here, following a question raised by Giorgio Koch: For all f in the domain of A, the process $\pi_t f - \int_0^t \pi_s A f \, ds$ is a $\{\mathcal{F}^Y_t\}$-martingale; can this observation be used to study nonlinear filtering problems? Under the condition that the state space $\mathbb{S}$ is locally compact, the results on the filtered martingale problem in Kurtz and Ocone [27] give uniqueness of Zakai and Kushner–Stratonovich equations in the natural class of $\{\mathcal{F}^Y_t\}$-adapted measure-valued processes. Stochastic equations relate known random inputs (in our case Y) to unknown random outputs (in our case, the conditional distribution π). Weak uniqueness (or more precisely, joint uniqueness in law) says that the joint distribution of the input and the output is uniquely determined by the distribution of the input. Strong or pathwise uniqueness says that there is a unique, appropriately measurable transformation that maps the input into the output. In our case, since the equations are derived to be satisfied by the conditional distribution, existence of a transformation (H_t of Proposition 5.1 and Theorem 5.1) is immediate. Consequently, it follows by a generalization of a theorem of Engelbert [17] (see Kurtz [23], Theorem 3.14) that weak and strong uniqueness are equivalent.

The uniqueness results derived using the filtered martingale problem in Kurtz and Ocone [27] were extended in Bhatt, Kallianpur, and Karandikar [4], in particular, eliminating the local compactness assumption on the state space of the signal. Still in the framework of signals observed in Gaussian white noise, Bhatt and Karandikar [2] goes beyond the classical Markov model with additive white noise and considers diffusive, non-Markovian signal/observation systems. The systems solve stochastic differential equations with coefficients that may depend on the signal and on the whole past trajectory of the observation process. Therefore, in particular, the signal need not be a Markov process. By enlarging the observation space to a suitable space of continuous functions, such a system can be seen as the solution of a martingale problem, and the filtered martingale problem approach can be used.

In Kallianpur and Mandal [20] the signal is the solution of a stochastic delay-differential equation, and the Wiener processes driving the signal and the observations are independent. In this case the signal state space is enlarged to a suitable space of continuous functions, and the system is seen as the solution of a martingale problem, and again the filtered martingale problem approach can be used.

Filtering models with point process observations can also be analysed using the filtered martingale problem approach. In Kliemann, Koch, and Marchetti [21], the signal/observation system is a Markov process, and the signal is a jump-diffusion process (not necessarily Markovian by itself), while the observation process is a counting process, with unbounded intensity. Strong uniqueness for the filtering equation is obtained in a class of probability-valued processes characterized by a suitable second moment growth condition. In Ceci and Gerardi [11], the observation is still a counting process, while the signal/observation system is a Markov jump process with values in $\mathbb{S} = \mathbb{R}^d \times \mathbb{N}$ (see also Ceci and Gerardi [12], where the signal process itself is a jump Markov process); the weak uniqueness of the Kushner–Stratonovich equation is obtained by the filtered martingale problem approach, while the pathwise uniqueness is obtained by a direct method. (Note that the notion of weak solution considered in this paper places additional restrictions on the solution beyond adaptedness and satisfying the identity. Consequently, the equivalence of weak and strong solutions mentioned above does not hold in this context.) Uniqueness of the filtered martingale problem has also been used for a partially observable control problem of a jump Markov system with counting observations in Ceci, Gerardi, and Tardelli [14] (see also Ceci and Gerardi [13] and [12]). The case of a marked point observation process has been considered in various papers. In Fan [19], the signal/observation system is a continuous-time Markov chain, with $\mathbb{S}$ a finite set. In Ceci and Gerardi [7] and [8], partially observed branching processes are discussed, while Ceci and Gerardi [9] focuses on the financial applications of filtering. In all these examples, the observation state space $\mathbb{S}_0$ is discrete but not necessarily finite.

5.5 Constrained Markov processes

In this section, $\mathcal{M}(\mathbb{S} \times [0, \infty))$ will denote the space of Borel measures μ on $\mathbb{S} \times [0, \infty)$ such that $\mu(\mathbb{S} \times [0, t]) < \infty$ for each $t > 0$.

Let A and B be operators satisfying Condition 5.1 with ψ replaced by ψ_A and ψ_B, respectively, and $\mathcal{D}(A) = \mathcal{D}(B) = \mathcal{D}$, and let $\mathbb{D}$ and $\partial \mathbb{D}$ be closed subsets of $\mathbb{S}$. In many situations, $\partial \mathbb{D}$ will be the topological boundary of $\mathbb{D}$, but that is not necessary.

Definition 5.6 A measurable, $\mathbb{D}$-valued process X and a random measure Γ in $\mathcal{M}(\partial \mathbb{D} \times [0, \infty))$ give a solution of the *constrained martingale problem* for $(A, B, \mathbb{D}, \partial \mathbb{D})$ if there exists a filtration $\{\mathcal{F}_t\}$ such that X and Γ are $\{\mathcal{F}_t\}$-adapted,

$$\mathbb{E}\left[\int_0^t \psi_A(X(s))ds + \int_{\partial \mathbb{D} \times [0,t]} \psi_B(x)\Gamma(dx \times ds)\right] < \infty, \quad t \geq 0,$$

and for each $f \in \mathcal{D}$,

$$f(X(t)) - \int_0^t Af(X(s))ds - \int_{\partial \mathbb{D} \times [0,t]} Bf(x)\Gamma(dx \times ds)$$

is an $\{\mathcal{F}_t\}$-martingale. For $\nu_0 \in \mathcal{P}(\mathbb{D})$, (X, Γ) is a solution of the *constrained martingale problem* for $(A, B, \mathbb{D}, \partial \mathbb{D}, \nu_0)$ if (X, Γ) is a solution of the *constrained martingale problem* for $(A, B, \mathbb{D}, \partial \mathbb{D})$ and $X(0)$ has distribution ν_0.

A $\mathcal{P}(\mathbb{D})$-valued function $\{\nu_t\}$ is a solution of the *forward equation* for $(A, B, \mathbb{D}, \partial \mathbb{D})$ if for each $t > 0$, $\int_0^t \nu_s \psi_A ds < \infty$ and there exists $\mu \in \mathcal{M}(\partial \mathbb{D} \times [0, \infty))$ such that $\int_{\partial \mathbb{D} \times [0,t]} \psi_B d\mu < \infty$, $t > 0$, and for each $f \in \mathcal{D}$,

$$\nu_t f = \nu_0 f + \int_0^t \nu_s Af ds + \int_{\partial \mathbb{D} \times [0,t]} Bf(x)\mu(dx \times ds), \quad t \geq 0.$$

Remark 5.13 By uniqueness for a constrained martingale problem, we mean uniqueness of the finite-dimensional distributions of X. We do not expect Γ to be unique. For example, there may be a measure $\widehat{\mu}$ such that $\int_{\partial \mathbb{D}} Bf d\widehat{\mu} \equiv 0$. If (X, Γ) is a solution of the constrained martingale problem and

$$\widehat{\Gamma}(dx \times ds) = \Gamma(dx \times ds) + \widehat{\mu}(dx)ds,$$

then $(X, \widehat{\Gamma})$ is also a solution.

The most familiar examples of constrained Markov processes are reflecting diffusion processes satisfying equations of the form

$$X(t) = X(0) + \int_0^t \sigma(X(s))d W(s) + \int_0^t b(X(s))ds + \int_0^t m(X(s))d\xi(s),$$

where X is required to remain in the closure of a domain $\mathbb{D} \subset \mathbb{R}^d$ and ξ is a nondecreasing process that increases only when X is on the topological boundary $\partial \mathbb{D}$ of $\mathbb{D}$. Then

$$Af(x) = \frac{1}{2} \sum_{i,j} a_{ij}(x) \frac{\partial^2}{\partial x_i \partial x_j} f(x) + b(x) \cdot \nabla f(x),$$

where $a(x) = ((a_{ij}(x))) = \sigma(x)\sigma(x)^T$,

$$Bf(x) = m(x) \cdot \nabla f(x),$$

and

$$\Gamma(C \times [0, t]) = \int_0^t \mathbf{1}_C(X(s))d\xi(s).$$

As before, let $\gamma : \mathbb{S} \to \mathbb{S}_0$, $Y(t) = \gamma(X(t))$, and

$$\pi_t(\Gamma) = \mathbb{P}\left\{X(t) \in \Gamma | \widehat{\mathcal{F}}_t^{Y,Z}\right\}.$$

Then

$$\pi_t f - \int_0^t \pi_s Af ds - \int_{\partial \mathbb{D} \times [0,t]} Bf(x) \widehat{\Gamma}(dx \times ds)$$

is a $\left\{\widehat{\mathcal{F}}_t^{Y,Z}\right\}$-martingale, where $\widehat{\Gamma}$ is the dual predictable projection of Γ with respect to $\left\{\widehat{\mathcal{F}}_t^{Y,Z}\right\}$. (See Kurtz and Stockbridge [28], Lemma 6.1.)

Definition 5.7 Let $\widehat{\mu}_0 \in \mathcal{P}(\mathbb{D} \times \mathbb{S}_0)$.

$$(\widetilde{Y}, \widetilde{\Gamma}, \widetilde{\pi}, \widetilde{Z}) \in M_{\mathbb{S}_0}[0, \infty) \times \mathcal{M}(\partial \mathbb{D} \times [0, \infty)) \times M_{\mathcal{P}(\mathbb{D})}[0, \infty) \times \mathbb{S}_0$$

is a solution of the *filtered martingale problem* for $(A, B, \mathbb{D}, \partial \mathbb{D}, \gamma, \widehat{\mu}_0)$, if

$$\mathbb{E}[\widetilde{\pi}_0(C) \mathbf{1}_D(\widetilde{Z})] = \widehat{\mu}_0(C \times D), \quad C \in \mathcal{B}(\mathbb{D}), D \in \mathcal{B}(\mathbb{S}_0), \quad (5.46)$$

$\tilde{\pi}_0$ is $\sigma(\tilde{Z})$-measurable, $\tilde{\pi}$ and $\tilde{\Gamma}$ are $\{\widehat{\mathcal{F}}_t^{\tilde{Y},\tilde{Z}}\}$-adapted, for each $g \in B(\mathbb{D})$ and $t \geq 0$,

$$\int_0^t \tilde{\pi}_s(g \circ \gamma)ds = \int_0^t g(\tilde{Y}(s))ds ,$$

$$\mathbb{E}[\int_0^t \tilde{\pi}_s \psi_A ds + \int_{\partial \mathbb{D} \times [0,t]} \psi_B(x)\tilde{\Gamma}(dx \times ds)] < \infty, \quad t \geq 0,$$

and for each $f \in \mathcal{D}$,

$$\tilde{\pi}_t f - \int_0^t \tilde{\pi}_s Af ds - \int_{\partial \mathbb{D} \times [0,t]} Bf(x)\tilde{\Gamma}(dx \times ds)$$

is an $\{\widehat{\mathcal{F}}_t^{\tilde{Y},\tilde{Z}}\}$-martingale.

If $(\tilde{Y}, \tilde{\Gamma}, \tilde{\pi}, \tilde{Z})$ satisfies all the conditions except (5.46), we will refer to it as a solution of the filtered martingale problems for $(A, B, \mathbb{D}, \partial \mathbb{D}, \gamma)$.

The following extension of Lemma 5.1 follows from Corollary 1.12 of Kurtz and Stockbridge [28].

Lemma 5.5 *Let A and B satisfy Condition 5.1 with ψ replaced by ψ_A and ψ_B respectively. If $\{v_t\}$ is a solution of the forward equation for $(A, B, \mathbb{D}, \partial \mathbb{D})$, then there exists a solution (X, Γ) of the constrained martingale problem for $(A, B, \mathbb{D}, \partial \mathbb{D})$ such that $v_t f = \mathbb{E}[f(X(t))]$.*

Define $\widehat{A}$ as in Section 5.3, and define $\widehat{B}$ by

$$\widehat{B}\left[ff_0 \prod_{k,i=1}^m g_{ki}\right](x, z, u) = \left(f_0(z) \prod_{k,i=1}^m g_{ki}(u_{ki})\right) Bf(x) .$$

As before, if A and B satisfy Condition 5.1, then so do $\widehat{A}$ and $\widehat{B}$. Any solution $(\tilde{Y}, \tilde{\Gamma}, \tilde{\pi}, \tilde{Z})$ of the filtered martingale problem for $(A, B, \mathbb{D}, \partial \mathbb{D}, \gamma, \widehat{\mu}_0)$ determines a solution of the forward equation for $(\widehat{A}, \widehat{B}, \mathbb{D} \times S_0 \times [0,1]^\infty, \partial \mathbb{D} \times [0,1]^\infty, \widehat{\nu}_0)$, which, by Lemma 5.5 corresponds to a solution of the constrained martingale problem satisfying (5.22). This observation then gives the following analog of Theorem 5.1.

Theorem 5.5 *Let A and B satisfy Condition 5.1, $\widehat{\mu}_0 \in \mathcal{P}(\mathbb{S} \times S_0)$, and $\gamma : \mathbb{S} \to S_0$ be Borel measurable. Let $(\tilde{Y}, \tilde{\Gamma}, \tilde{\pi}, \tilde{Z})$ be a solution of the filtered martingale problem for $(A, B, \mathbb{D}, \partial \mathbb{D}, \gamma, \widehat{\mu}_0)$. Then the following hold:*

a) *There exists a solution (X, Γ) of the constrained martingale problem for $(A, B, \mathbb{D}, \partial \mathbb{D})$ and an S_0-valued random variable Z such that $(X(0), Z)$ has distribution $\widehat{\mu}_0$ and $\tilde{Y}$ has the same distribution on $M_{S_0}[0, \infty)$ as $Y = \gamma \circ X$.*

b) For each $t \geq 0$, there exists a Borel measurable mapping $H_t : M_{\mathbb{S}_0}[0, \infty) \times S_0 \to \mathcal{P}(\mathbb{S})$ such that $\pi_t = H_t(Y, Z)$ is the conditional distribution of $X(t)$ given $\widehat{\mathcal{F}}_t^{Y,Z}$, and $\widetilde{\pi}_t = H_t(\widetilde{Y}, \widetilde{Z})$ a.s. In particular, $\widetilde{\pi}$ has the same finite-dimensional distributions as π.

c) If Y and $\widetilde{Y}$ have sample paths in $D_{\mathbb{S}_0}[0, \infty)$, then Y and $\widetilde{Y}$ have the same distribution on $D_{\mathbb{S}_0}[0, \infty)$ and H_t is Borel measurable mapping from $D_{\mathbb{S}_0}[0, \infty) \times S_0$ to $\mathcal{P}(\mathbb{S})$.

d) If uniqueness holds for the constrained martingale problem for $(A, B, \mathbb{D}, \partial \mathbb{D}, \nu_0)$, then uniqueness holds for the filtered martingale problem for $(A, B, \mathbb{D}, \partial \mathbb{D}, \gamma, \widehat{\mu}_0)$ in the sense that if (Y, Γ, π, Z) and $(\widetilde{Y}, \widetilde{\Gamma}, \widetilde{\pi}, \widetilde{Z})$ are solutions, then for each $0 \leq t_1 < \cdots < t_m$, $(\pi_{t_1}, \ldots, \pi_{t_m}, Y, Z)$ and $(\widetilde{\pi}_{t_1}, \ldots, \widetilde{\pi}_{t_m}, \widetilde{Y}, \widetilde{Z})$ have the same distribution on $\mathcal{P}(\mathbb{S})^m \times M_{\mathbb{S}_0}[0, \infty) \times S_0$.

The analogs of Theorems 5.2 and 5.3 hold by essentially the same arguments as before.

Appendix

A.1. A martingale lemma

Let $\{\mathcal{F}_t\}$ and $\{\mathcal{G}_t\}$ be filtrations with $\mathcal{G}_t \subset \mathcal{F}_t$.

Lemma A.1 *Suppose U and V are measurable and $\{\mathcal{F}_t\}$-adapted, $\mathbb{E}[|U(t)| + \int_0^t |V(s)|ds] < \infty$, $t \geq 0$, and*

$$U(t) - \int_0^t V(s)ds$$

is an $\{\mathcal{F}_t\}$-martingale. Then

$$\mathbb{E}[U(t)|\mathcal{G}_t] - \int_0^t \mathbb{E}[V(s)|\mathcal{G}_s]ds$$

is a $\{\mathcal{G}_t\}$-martingale, where we take $\mathbb{E}[V(s)|\mathcal{G}_s]$ to be the optional projection of V.

Proof. The lemma follows by the definition and properties of conditional expectations. □

Example A.2 *If X is a solution (wrt $\{\mathcal{F}_t\}$) of the martingale problem for A and*

$$\pi_t(\Gamma) = \mathbb{P}\{X(t) \in \Gamma | \mathcal{G}_t\},$$

then

$$\pi_t f - \int_0^t \pi_s Af \, ds$$

is a $\{\mathcal{G}_t\}$-martingale.

A.2. Filtrations generated by processes

Let Y be a measurable stochastic process. $\mathcal{F}_t^Y$ will denote the completion of $\sigma(Y(s), s \leq t)$ and $\widehat{\mathcal{F}}_t^Y$ will denote the completion of

$$\sigma(\int_0^s h(Y(r))dr, s \leq t, h \in B(\mathbb{S})) \vee \sigma(Y(0)).$$

Lemma A.3 *If Y is $\{\mathcal{F}_t^Y\}$ -progressively measurable, then $\widehat{\mathcal{F}}_t^Y \subset \mathcal{F}_t^Y$.*

Lemma A.4 *For almost every t, $Y(t)$ is $\widehat{\mathcal{F}}_t^Y$-measurable.*

Proof. For $g \in B(\mathbb{S}_0)$,

$$M_g(t) = \int_0^t g(Y(s))ds - \int_0^t \mathbb{E}\left[g(Y(s))|\widehat{\mathcal{F}}_s^Y\right] ds$$

is a continuous $\{\widehat{\mathcal{F}}_t^Y\}$-martingale. (Take $\mathbb{E}\left[g(Y(s))|\widehat{\mathcal{F}}_s^Y\right]$ to be the optional projection of $g \circ Y$.) Since M_g is a finite variation process, it must be zero with probability one, and hence, with probability one

$$g(Y(t)) = \mathbb{E}\left[g(Y(t))|\widehat{\mathcal{F}}_t^Y\right] \text{ for almost every } t,$$

which in turn implies that for almost every t, $g(Y(t)) = \mathbb{E}\left[g(Y(t))|\widehat{\mathcal{F}}_t^Y\right]$ a.s. and $g(Y(t))$ is $\widehat{\mathcal{F}}_t^Y$-measurable. Since $\mathbb{S}_0$ is separable, there exists a countable separating set $\{g_k\}$ such that for almost every t, $g_1(Y(t)), g_2(Y(t)), \ldots$ are $\widehat{\mathcal{F}}_t^Y$-measurable and hence $Y(t)$ is $\widehat{\mathcal{F}}_t^Y$-measurable. □

Lemma A.5 *If Y is cadlag with no fixed points of discontinuity (that is, $\mathbb{P}\{Y(t) = Y(t-)\} = 1$ for all t), then $\mathcal{F}_t^Y = \widehat{\mathcal{F}}_t^Y$, $t \geq 0$.*

Proof. If Y is cadlag, it is $\{\mathcal{F}_t^Y\}$-progressively measurable so $\widehat{\mathcal{F}}_t^Y \subset \mathcal{F}_t^Y$. $Y(0)$ is $\widehat{\mathcal{F}}_0^Y$ measurable by definition. For $t > 0$, since $\mathbb{P}\{Y(t) = Y(t-)\} = 1$, for $g \in C_b(\mathbb{S})$,

$$g(Y(t)) = \lim_{\epsilon \to} \frac{1}{\epsilon} \int_{t-\epsilon}^t g(Y(r))dr \quad a.s.,$$

and hence $Y(t)$ is $\widehat{\mathcal{F}}_t^Y$-measurable. Consequently, $\mathcal{F}_t^Y \subset \widehat{\mathcal{F}}_t^Y$. □

The following lemma implies that most cadlag processes of interest will have no fixed points of discontinuity.

Lemma A.6 *Let U and V be $\mathbb{S}$-valued random variables and let $\mathcal{G}$ be a σ-algebra of events. Suppose that $M \subset C_b(\mathbb{S})$ is separating and*

$$\mathbb{E}[f(U)|\mathcal{G}] = f(V) \tag{A.1}$$

for all $f \in M$. Then $U = V$ a.s.

In particular, if Y is cadlag and adapted to $\{\mathcal{G}_t\}$ and

$$\lim_{s \to t-} \mathbb{E}[f(Y(t))|\mathcal{G}_s] = \mathbb{E}[f(Y(t))| \vee_{s<t} \mathcal{G}_s] = f(Y(t-)), \quad (A.2)$$

for $f \in M$, then $Y(t) = Y(t-)$ a.s.

Remark A.7 Note that if

$$f(Y(t)) - \int_0^t Z_f(s)ds$$

is a martingale for Z_f satisfying $\mathbb{E}\left[\int_0^t |Z_f(s)|ds\right] < \infty$, then (A.2) holds.

Proof. Let Z be a nonnegative $\mathcal{G}$-measurable random variable. Then $\mathbb{E}[f(U)Z] = \mathbb{E}[f(V)Z]$ for $f \in M$ and since M is separating, this identity must hold for all $f \in B(\mathbb{S})$. Consequently, (A.1) holds for all $f \in B(\mathbb{S})$, and replacing f by f^2,

$$\mathbb{E}[f^2(U)] = \mathbb{E}[f^2(V)] = \mathbb{E}[\mathbb{E}[f(U)|\mathcal{G}]f(V)] = \mathbb{E}[f(U)f(V)]$$

and hence that $\mathbb{E}[(f(U) - f(V))^2] = 0$ for all $f \in B(\mathbb{S})$. □

A.3. Stopped filtrations and filtrations generated by stopped processes

Let V be a cadlag process, $\mathcal{F}_t^V$ be the completion of $\sigma(V(s), s \leq t)$, and $\mathcal{S}$ be the collection of finite, $\{\mathcal{F}_t^V\}$-stopping times. For $\tau \in \mathcal{S}$, define $\mathcal{G}_\tau^V$ to be the completion of $\sigma(V(t \wedge \tau) : t \geq 0) \vee \sigma(\tau)$. Of course, $\mathcal{G}_t^V = \mathcal{F}_t^V$, and more generally $\mathcal{G}_\tau^V = \mathcal{F}_\tau^V$ for all discrete stopping times in $\mathcal{S}$ and $\mathcal{G}_\tau^V \subset \mathcal{F}_\tau^V$ for all $\tau \in \mathcal{S}$, but we do not know whether or not $\mathcal{G}_\tau^V = \mathcal{F}_\tau^V$ for all $\tau \in \mathcal{S}$.

Lemma A.8 If $\tau_1, \tau_2 \in \mathcal{S}$, then

$$\mathcal{F}_{\tau_1 \wedge \tau_2}^V = \{(A_1 \cap \{\tau_1 < \tau_2\}) \cup (A_2 \cap \{\tau_1 \geq \tau_2\}) : A_1 \in \mathcal{F}_{\tau_1}^V, A_2 \in \mathcal{F}_{\tau_2}^V\} \quad (A.3)$$

and for $\tau \in \mathcal{S}$ and $t \geq 0$,

$$\mathcal{G}_{\tau \wedge t}^V = \{(A_1 \cap \{\tau < t\}) \cup (A_2 \cap \{\tau \geq t\}) : A_1 \in \mathcal{G}_\tau^V, A_2 \in \mathcal{G}_t^V = \mathcal{F}_t^V\}. \quad (A.4)$$

Remark A.9 The first assertion holds for arbitrary filtrations.

Proof. Since $\mathcal{F}_{\tau_1 \wedge \tau_2}^V \subset \mathcal{F}_{\tau_1}^V \cap \mathcal{F}_{\tau_2}^V$, it follows that $\mathcal{F}_{\tau_1 \wedge \tau_2}^V$ is contained in the right side of (A.3). (Take $A_1 = A_2 \in \mathcal{F}_{\tau_1 \wedge \tau_2}^V$.)

Observe that $\{\tau_1 < \tau_2\} \in \mathcal{F}_{\tau_1 \wedge \tau_2}^V$ since

$$\{\tau_1 < \tau_2\} \cap \{\tau_1 \wedge \tau_2 \leq t\} = \{\tau_1 \leq t < \tau_2\} \cup \bigcup_{r \in \mathbb{Q}, r \leq t}\{\tau_1 \leq r < \tau_2\} \in \mathcal{F}_t^V.$$

Now let $A_1 \in \mathcal{F}_{\tau_1}^V$. Then

$$A_1 \cap \{\tau_1 < \tau_2\} \cap \{\tau_1 \wedge \tau_2 \leq t\} = A_1 \cap \{\tau_1 < \tau_2\} \cap \{\tau_1 \leq t\} \in \mathcal{F}_t^V,$$

so $A_1 \cap \{\tau_1 < \tau_2\} \in \mathcal{F}^V_{\tau_1 \wedge \tau_2}$, and for $A_2 \in \mathcal{F}^V_{\tau_2}$,

$$A_2 \cap \{\tau_1 \geq \tau_2\} \cap \{\tau_1 \wedge \tau_2 \leq t\} = A_2 \cap \{\tau_1 \geq \tau_2\} \cap \{\tau_2 \leq t\} \in \mathcal{F}^V_t,$$

so $A_2 \cap \{\tau_1 \geq \tau_2\} \in \mathcal{F}^V_{\tau_1 \wedge \tau_2}$.

If $A_1 \in \mathcal{G}^V_\tau$, then A_1 differs from a set of the form $\{(V(\cdot \wedge \tau), \tau) \in C\}$, $C \in \mathcal{B}(D_E[0, \infty)) \times \mathcal{B}([0, \infty))$, by an event of probability zero, and, noting that $\{\tau \wedge t < t\} = \{\tau < t\}$,

$$\{(V(\cdot \wedge \tau), \tau) \in C\} \cap \{\tau < t\} = \{(V(\cdot \wedge \tau \wedge t), \tau \wedge t) \in C\} \cap \{\tau < t\} \in \mathcal{G}^V_{\tau \wedge t}.$$

Similarly,

$$\{V(\cdot \wedge t) \in C\} \cap \{\tau \geq t\} = \{V(\cdot \wedge \tau \wedge t) \in C\} \cap \{\tau \geq t\} \in \mathcal{G}^V_{\tau \wedge t},$$

and the right side of (A.4) is contained in the left.

Finally,

$$\{(V(\cdot \wedge \tau \wedge t), \tau \wedge t) \in C\} \cap \{\tau < t\} = \{(V(\cdot \wedge \tau), \tau) \in C\} \cap \{\tau < t\} \in \mathcal{G}^V_\tau$$

and

$$\{(V(\cdot \wedge \tau \wedge t), \tau \wedge t) \in C\} \cap \{\tau \geq t\} = \{(V(\cdot \wedge t), t) \in C\} \cap \{\tau \geq t\} \in \mathcal{G}^V_t,$$

so the left side of (A.4) is contained in the right. □

We have the following consequence of the previous lemma.

Lemma A.10 *Let $\mathbb{E}[|Z|] < \infty$ and $\tau \in \mathcal{S}$. Then*

$$\mathbb{E}\left[Z|\mathcal{F}^V_{\tau \wedge t}\right] = \mathbb{E}\left[Z|\mathcal{F}^V_\tau\right] \mathbf{1}_{\{\tau < t\}} + \mathbb{E}\left[Z|\mathcal{F}^V_t\right] \mathbf{1}_{\{\tau \geq t\}}$$

and

$$\mathbb{E}\left[Z|\mathcal{G}^V_{\tau \wedge t}\right] = \mathbb{E}\left[Z|\mathcal{G}^V_\tau\right] \mathbf{1}_{\{\tau < t\}} + \mathbb{E}\left[Z|\mathcal{G}^V_t\right] \mathbf{1}_{\{\tau \geq t\}},$$

and since $\mathcal{G}^V_t = \mathcal{F}^V_t$,

$$\mathbb{E}\left[Z|\mathcal{F}^V_{\tau \wedge t}\right] \mathbf{1}_{\{\tau \geq t\}} = \mathbb{E}\left[Z|\mathcal{G}^V_{\tau \wedge t}\right] \mathbf{1}_{\{\tau \geq t\}}.$$

A.4. Random probability measures as conditional distributions

Lemma A.11 *Let $\widetilde{\pi}$ be a $\mathcal{P}(\mathbb{S})$-valued random variable and $\widetilde{Z}$ a $\mathbb{S}_0$-valued random variable on a probability space $(\widetilde{\Omega}, \widetilde{\mathcal{F}}, \widetilde{P})$. Then there exists a probability space with random variables (X, Z, π) in $\mathbb{S} \times \mathbb{S}_0 \times \mathcal{P}(\mathbb{S})$, and a sub-$\sigma$-algebra $\mathcal{D}$ such that (Z, π) has the same distribution as $(\widetilde{Z}, \widetilde{\pi})$, Z is $\mathcal{D}$-measurable, and π is the conditional distribution of X given $\mathcal{D}$.*

Proof. For $C \in \mathcal{B}(\mathbb{S})$ and $D \in \mathcal{B}(\mathbb{S}_0 \times \mathcal{P}(\mathbb{S}))$,

$$\nu(C, D) = \mathbb{E}[\widetilde{\pi}(C) \mathbf{1}_D(\widetilde{Z}, \widetilde{\pi})]$$

defines a bimeasure on $\mathbb{S} \times (\mathbb{S}_0 \times \mathcal{P}(\mathbb{S}))$. By Morando's theorem (see, for example, Ethier and Kurtz [18], Appendix 8), ν extends to a probability measure on $\mathbb{S} \times \mathbb{S}_0 \times \mathcal{P}(\mathbb{S})$. Let (X, Z, π) be the coordinate random variables on $(\mathbb{S} \times \mathbb{S}_0 \times \mathcal{P}(\mathbb{S}), \mathcal{B}(\mathbb{S} \times \mathbb{S}_0 \times \mathcal{P}(\mathbb{S})), \nu)$. Then by definition, (Z, π) has the same distribution as $(\widetilde{Z}, \widetilde{\pi})$, and defining $\mathcal{D} = \sigma(Z, \pi)$, the fact that

$$\mathbb{E}[\mathbf{1}_C(X)\mathbf{1}_D(Z, \pi)] = \mathbb{E}[\widetilde{\pi}(C)\mathbf{1}_D(\widetilde{Z}, \widetilde{\pi})] = \mathbb{E}[\pi(C)\mathbf{1}_D(Z, \pi)]$$

implies that π is the conditional distribution of X given $\mathcal{D}$. □

A.5. A coupling lemma

Lemma A.12 *Let $\mathbb{S}_0$, $\mathbb{S}_1$, and $\mathbb{S}_2$ be complete, separable metric spaces, and let $\mu \in \mathcal{P}(\mathbb{S}_0, \mathbb{S}_1)$ and $\nu \in \mathcal{P}(\mathbb{S}_0, \mathbb{S}_2)$ satisfy $\mu(\cdot \times \mathbb{S}_1) = \nu(\cdot \times \mathbb{S}_2)$. Then there exists a probability space and random variables X_0, X_1, X_2 such that (X_0, X_1) has distribution μ, (X_0, X_2) has distribution ν, and*

$$\mathbb{E}[g(X_2)|X_0, X_1] = \mathbb{E}[g(X_2)|X_0], \quad g \in B(\mathbb{S}_2),$$

which is equivalent to

$$\mathbb{E}[g_1(X_1)g_2(X_2)|X_0] = \mathbb{E}[g_1(X_1)|X_0]\mathbb{E}[g_2(X_2)|X_0], \quad g_1 \in B(\mathbb{S}_1), g_2 \in B(\mathbb{S}_2).$$

Proof. The lemma is a consequence of Ethier and Kurtz [18], Lemma 4.5.15. □

A.6. The Markov property

Lemma A.13 *Let $A \subset B(\mathbb{S}) \times M(\mathbb{S})$ and $\nu_0 \in \mathcal{P}(\mathbb{S})$. Suppose that there exists $\psi \geq 1$ such that for each $f \in \mathcal{D}(A)$, there exists a constant a_f such that $|Af(x)| \leq a_f \psi(x)$. Assume that uniqueness holds for the local martingale problem for (A, ν_0). If X is a solution of the local martingale problem for (A, ν_0) with respect to the filtration*

$$\mathcal{F}_t = \sigma(X(s) : s \leq t) \vee \sigma\left(\int_0^s h(X(r))dr : s \leq t, h \in B(\mathbb{S})\right),$$

then X is an $\{\mathcal{F}_t\}$-Markov process.

Proof. Let $Y(t) = \int_0^t \psi(X(s))ds$, and for $f \in \mathcal{D}(A)$ and $g \in C_c^1[0, \infty)$, define

$$\widehat{A}(fg)(x, y) = g(y)Af(x) + \psi(x)f(x)g'(y).$$

Note that since g has compact support, there exists a constant $a_{f,g}$ such that $|\widehat{A}(fg)(x, y)| \leq a_{f,g}(1 + y)^{-2}\psi(x)$, and

$$\int_0^t (1 + Y(s))^{-2}\psi(X(s))ds \leq 1.$$

Consequently, if X is the unique solution of the local-martingale problem for (A, ν_0), then (X, Y) is the unique solution of the martingale problem for $(\widehat{A}, \nu_0 \times$

δ_0). The proof of Theorem A.5 in Kurtz [26] remains valid after replacing the assumption $A \subset B(\mathbb{S}) \times B(\mathbb{S})$ with $A \subset B(\mathbb{S}) \times M(\mathbb{S})$. □

References

[1] Robert F. Anderson, Diffusions with second order boundary conditions. I,II. *Indiana Univ. Math. J.*, 25(4): 367–95,403–41, 1976.

[2] Abhay G. Bhatt and Rajeeva L. Karandikar, Invariant measures and evolution equations for Markov processes characterized via martingale problems. *Ann. Probab.*, 21(4): 2246–68, 1993.

[3] Abhay G. Bhatt and Rajeeva L. Karandikar, Characterization of the optimal filter: the non-Markov case. *Stochastics Stochastics Rep.*, 66(3–4): 177–204, 1999.

[4] Abhay G. Bhatt, G. Kallianpur, and Rajeeva L. Karandikar, Uniqueness and robustness of solution of measure-valued equations of nonlinear filtering. *Ann. Probab.*, 23(4): 1895–1938, 1995.

[5] Abhay G. Bhatt, Amarjit Budhiraja, and Rajeeva L. Karandikar, Markov property and ergodicity of the nonlinear filter. *SIAM J. Control Optim.*, 39(3): 928–49 (electronic), 2000.

[6] Amarjit Budhiraja, Asymptotic stability, ergodicity and other asymptotic properties of the nonlinear filter. *Ann. Inst. H. Poincaré Probab. Statist.*, 39(6): 919–41, 2003.

[7] Claudia Ceci and Anna Gerardi, Multitype branching processes observing particles of a given type. *J. Appl. Probab.*, 42(2): 446–62, 2005.

[8] Claudia Ceci and Anna Gerardi, Modelling a multitype branching Brownian motion: filtering of a measure-valued process. *Acta Appl. Math.*, 91(1): 39–66, 2006a.

[9] Claudia Ceci and Anna Gerardi, A mode for high frequency data under partial information: a filtering approach. *Int. J. Theor. Appl. Finance*, 9(4): 555–76, 2006b.

[10] Claudia Ceci and Anna Gerardi, Filtering of a Markov jump process with counting observations. *Appl. Math. Optim.*, 42(1): 1–18, 2000.

[11] Claudia Ceci and Anna Gerardi, Nonlinear filtering equation of a jump process with counting observations. *Acta Appl. Math.*, 66(2): 139–54, 2001a.

[12] Claudia Ceci and Anna Gerardi, Controlled partially observed jump processes: dynamics dependent on the observed history. In *Proceedings of the Third World Congress of Nonlinear Analysts, Part 4 (Catania, 2000)*, 47, 2449–60, 2001.

[13] Claudia Ceci and Anna Gerardi, Partially observed control of a Markov jump process with counting observations: equivalence with the separated problem. *Stochastic Process. Appl.*, 78(2): 245–60, 1998.

[14] Claudia Ceci, Anna Gerardi, and Paola Tardelli, Existence of optimal controls for partially observed jump processes. *Acta Appl. Math.*, 74(2): 155–75, 2002.

[15] A. S. Cherny, On the uniqueness in law and the pathwise uniqueness for stochastic differential equations. *Theory Probab. Appl.*, 46(3): 406–19, 2003.

[16] Pedro Echeverría, A criterion for invariant measures of Markov processes. *Z. Wahrsch. Verw. Gebiete*, 61(1): 1–16, 1982.

[17] H. J. Engelbert, On the theorem of T. Yamada and S. Watanabe. *Stochastics Stochastics Rep.*, 36(3–4): 205–16, 1991.

[18] Stewart N. Ethier and Thomas G. Kurtz, *Markov processes: Characterization and convergence*. Wiley Series in Probability and Mathematical Statistics: Probability and Mathematical Statistics. John Wiley & Sons, New York, 1986.

[19] Kaisheng Fan, On a new approach to the solution of the nonlinear filtering equation of jump processes. *Probab. Engrg. Inform. Sci.*, 10(1): 153–63, 1996.

[20] G. Kallianpur and P. K. Mandal, Nonlinear filtering with stochastic delay equations. In *Advances on theoretical and methodological aspects of probability and statistics (Hamilton, ON, 1998)*, pp. 3–36. Taylor & Francis, London, 2002.

[21] W. Kliemann, G. Koch, and F. Marchetti, On the unnormalized solution of the filtering problem with counting observations. *IEEE Trans. Inform. Theory*, 36(6): 1415–25, 1990.

[22] Hiroshi Kunita, Asymptotic behavior of the nonlinear filtering errors of Markov processes. *J. Multivariate Anal.*, 1: 365–93, 1971.

[23] Thomas G. Kurtz, The Yamada–Watanabe–Engelbert theorem for general stochastic equations and inequalities. *Electron. J. Probab.*, 12: 951–65 (electronic), 2007.

[24] Thomas G. Kurtz, Martingale problems for constrained Markov problems. In *Recent advances in stochastic calculus (College Park, MD, 1987)*, Progr. Automat. Info. Systems, pp. 151–68. Springer, New York, 1990.

[25] Thomas G. Kurtz, A control formulation for constrained Markov processes. In *Mathematics of random media (Blacksburg, VA, 1989)*, vol. 27 of *Lectures in Appl. Math.*, pp. 139–150. Amer. Math. Soc., Providence, RI, 1991.

[26] Thomas G. Kurtz, Martingale problems for conditional distributions of Markov processes. *Electron. J. Probab.*, 3(9), 29 pp. (electronic), 1998.

[27] Thomas G. Kurtz and Daniel L. Ocone, Unique characterization of conditional distributions in nonlinear filtering. *Ann. Probab.*, 16(1): 80–107, 1988.

[28] Thomas G. Kurtz and Richard H. Stockbridge, Stationary solutions and forward equations for controlled and singular martingale problems. *Electron. J. Probab.*, 6(17), 52 pp. (electronic), 2001.

[29] Vladimir M. Lucic and Andrew J. Heunis, On uniqueness of solutions for the stochastic differential equations of nonlinear filtering. *Ann. Appl. Probab.*, 11(1): 182–209, 2001.

[30] B. L. Rozovskiĭ, A simple proof of uniqueness for Kushner and Zakai equations. In *Stochastic analysis*, pp. 449–58. Academic Press, Boston, MA, 1991.

[31] Łukasz Stettner, On invariant measures of filtering processes. In *Stochastic differential systems (Bad Honnef, 1988)*, vol. 126 of *Lecture Notes in Control and Inform. Sci.*, pp. 279–92. Springer, Berlin, 1989.

[32] Daniel W. Stroock and S. R. S. Varadhan, Diffusion processes with boundary conditions. *Comm. Pure Appl. Math.*, 24: 147–225, 1971.

PART II
Nonlinear Filtering and Stochastic Partial Differential Equations

PART II

·6·

Filtering Equations for Partially Observable Diffusion Processes with Lipschitz Continuous Coefficients

N. V. Krylov

6.1 Introduction

For the author, one of the main motivations for developing the theory of stochastic partial differential equations (SPDEs) is its relation to the filtering problem for partially observable diffusion processes.

This problem's setting is as follows.

Let $(\Omega, \mathcal{F}, P)$ be a complete probability space with an increasing filtration $\{\mathcal{F}_t, t \geq 0\}$ of complete, with respect to $(\mathcal{F}, P)$, σ-fields $\mathcal{F}_t \subset \mathcal{F}$. Denote by $\mathcal{P}$ the predictable σ-field in $\Omega \times (0, \infty)$ associated with $\{\mathcal{F}_t\}$. Let $d \geq 1, d_1 > d$, and $d_2 \geq d_1$ be integers and w_t be a d_2-dimensional Wiener process with respect to $\{\mathcal{F}_t\}$. Let $K, T, \delta > 0$ be fixed finite constants.

Consider a d_1-dimensional two component process $z_t = (x_t, y_t)$ with x_t being d-dimensional and y_t $(d_1 - d)$-dimensional. We assume that z_t is a diffusion process defined as a solution of the system

$$\begin{aligned} dx_t &= b(t, z_t)dt + \theta(t, z_t)dw_t, \\ dy_t &= B(t, z_t)dt + \Theta(t, y_t)dw_t \end{aligned} \quad (6.1)$$

with some initial data.

The coefficients of (6.1) are assumed to be vector- or matrix-valued functions of appropriate dimensions defined on $[0, T] \times \mathbb{R}^{d_1}$. Actually $\Theta(t, y)$ is assumed to be independent of x, so that it is a function on $[0, T] \times \mathbb{R}^{d_1-d}$ rather than $[0, T] \times \mathbb{R}^{d_1}$ but as always we may think of $\Theta(t, y)$ as a function of (t, z) as well.

The component x_t is treated as unobservable and y_t as the only observations available. The problem is to find a way to compute the density $\pi_t(x)$ of the conditional distribution of x_t given $y_s, s \leq t$. Finding an equation satisfied by π_t (filtering equation) is considered to be a solution of the (filtering) problem. Filtering equations turn out to be particular cases of SPDEs.

The history of filtering equations for *diffusion processes* is rather long. Without going through the whole of it we only comment on its beginning and the role of SPDEs. Probably, the first equations were published in [22] even before the famous Kalman–Bucy filter was discovered in [4]. However the arguments in [22] were not completely satisfactory. Therefore, in [14] *different* equations were proposed and later (see [15]) rigorously justified for models without the so-called cross terms.

Meanwhile, in [21] the correct filtering equations were presented in full generality, yet assuming some regularity of the filtering density, and then in [16] they were rigorously proved. In the presence of cross terms these equations are *different* from the ones in [14]. This is the reason we propose to call the filtering equations in the case of partially observable diffusion processes *Shiryaev's equations* and their particular case without cross terms *Kushner's equations*.

The result of [21] is presented in [17] on the basis of the famous Fujisaki–Kallianpur–Kunita theorem (see [1]) about the filtering equations in a very general setting (much more general than in [16]). Some authors even call the filtering equation for diffusion processes the Fujisaki–Kallianpur–Kunita equation. Comments on the subsequent development of filtering theory from probabilistic point of view can be found in [17].

By adding to the Fujisaki–Kallianpur–Kunita theorem some simple facts from *the theory of SPDEs*, the a priori regularity assumption was removed in [9] and under the Lipschitz and uniform nondegeneracy assumption the L_2-version of Theorem 6.1 was proved. The basic result of [9] is that $\pi_t \in H_2^1$. It is also proved that if the coefficients are smoother, $\pi_t(x)$ is smoother too. The nondegeneracy assumption is removed in [20] on the account of assuming that $\theta\theta^*$ is three times continuously differentiable in x. It is again proved that $\pi_t \in H_2^1$ and π_t is even smoother if the coefficients are smoother.

In [6] the results of [9] were improved, $\theta\theta^*$ is assumed to be twice continuously differentiable in x and it is shown that $\pi_t \in H_p^2$ with any $p \geq 2$.

The above-mentioned results of [9], [20], and [6] use the filtering theory in combination with the theory of SPDEs, the latter being stimulated by certain needs of filtering theory. It turns out that the theory of SPDEs alone can be used to obtain the above-mentioned regularity results about π_t without knowing anything from the filtering theory itself. It also can be used to solve other problems from the filtering theory.

The first 'direct' (only using the theory of SPDEs) proof of regularity of π_t is given in [11] in the case that system (6.1) defines a nondegenerate diffusion process and $\theta\theta^*$ is twice continuously differentiable in x. It is proved that $\pi_t \in H_p^2$ with any $p \geq 2$ as in [6]. Advantages of having arbitrary p are seen from results like our Theorem 6.2. Of course, on the way of investigating π_t in [11] filtering equations are derived 'directly' in an absolutely different manner than before (on the basis of an idea from [10]).

In this chapter we relax the smoothness assumption in [11] to the assumption that the coefficients of (6.1) are merely Lipschitz continuous, the assumption which is almost always supposed to hold when one deals with systems like (6.1).

We find that $\pi_t \in H_p^1$. Thus, under the weakest smoothness assumptions we obtain the best (in the author's opinion) regularity result on π_t. In particular, we prove that if the initial data is sufficiently regular, then the filtering density is almost Lipschitz continuous in x and 1/2 Hölder continuous in t. However, we still assume z_t to be nondegenerate. Our approach is heavily based on analytic results. There is also a probabilistic approach developed in [13] and based on explicit formulas for solutions initiated in [18] and later developed in [10] and [12] (also see references therein). This approach cannot give as sharp results as ours in our situation.

It seems to the author that under the same assumptions of Lipschitz continuity, by following an idea from [5] one can solve another problem from filtering theory, the so-called innovation problem, and obtain the equality

$$\sigma\{y_s, s \leq t\} = \sigma\{\check{w}_s, s \leq t\},$$

where $\check{w}_t$ is the innovation Wiener process of the problem (there is a reminder of its definition in Section 6.2). Recall that for degenerate diffusion processes the positive solution of the innovation problem is obtained in [19] again on the basis of the theory of SPDEs under the assumption that the coefficients are more regular.

By the way, in our situation, if the coefficients are more regular, the filtering equation can be rewritten in a nondivergence form and then additional smoothness of the filtering density, existence of which is already established in this article, is obtained on the basis of regularity results from [6].

The chapter is organized as follows. In Section 6.2 we state our main results part of which is proved in the same section. In Sections 6.3 and 6.4 we prove Theorems 6.1 and 6.3, respectively. Section 6.5 contains a collection of results from the theory of SPDEs which we use in the previous sections.

As is traditional in filtering theory we consider finite-dimensional driving Wiener processes. However, our results will be based on the theory of SPDEs, outlined in Section 6.5, with countably many Wiener processes. We leave to the reader to do some trivial modifications in Section 6.5 in order to be able to apply its results in such cases.

The author is sincerely grateful to the referee for useful suggestions.

6.2 Main results

First let us state and discuss our assumptions.

Assumption 6.1 The functions b, θ, B, and Θ are Borel measurable and bounded functions of their arguments. Each of them satisfies the Lipschitz condition in z with constant $K \in (0, \infty)$.

Introduce

$$\tilde{\theta}_t(z) = \begin{pmatrix} \theta(t, z) \\ \Theta(t, y) \end{pmatrix}, \quad \tilde{a}_t(z) = \frac{1}{2}\tilde{\theta}_t\tilde{\theta}_t^*(z), \quad \tilde{b}_t(z) = \begin{pmatrix} b(t, z) \\ B(t, z) \end{pmatrix}, \qquad (6.2)$$

$$\tilde{L}_t(z) = \tilde{a}_t^{ij}(z)\frac{\partial^2}{\partial z^i \partial z^j} + \tilde{b}_t^i(z)\frac{\partial}{\partial z^i}, \qquad (6.3)$$

where $\tilde{\theta}^*$ is the transpose of $\tilde{\theta}$ and the summation convention is imposed.

Remark 6.1 System of equations (6.1) can be now written as

$$dz_t = \tilde{b}(t, z_t)dt + \tilde{\theta}(t, z_t)dw_t. \qquad (6.4)$$

Assumption 6.2 The process z_t is uniformly nondegenerate: for any $\lambda, z \in \mathbb{R}^{d_1}$ and $t \in [0, T]$ we have

$$\tilde{a}_t^{ij}(z)\lambda^i\lambda^j \geq \delta|\lambda|^2.$$

Traditionally, Assumption 6.2 is split into two following assumptions in which some useful objects are introduced. These assumptions were also used in the past to reduce $\tilde{\theta}$ to the so-called triangular form by replacing w_t with a different Brownian motion.

Assumption 6.3 The symmetric matrix $\Theta\Theta^*$ is invertible and

$$\Psi := (\Theta\Theta^*)^{-\frac{1}{2}}$$

is a bounded function of (t, y).

Remark 6.2 Assumption 6.3 follows from Assumption 6.2 and, furthermore, $\Psi \leq \delta^{-1}(\delta^{ij})$.

Assumption 6.4 For any $\xi \in \mathbb{R}^d$, $z = (x, y) \in \mathbb{R}^{d_1}$, and $t > 0$, we have

$$|Q(t, y)\theta^*(t, z)\xi|^2 \geq \delta|\xi|^2,$$

where Q is the orthogonal projector on Ker Θ. In other words,

$$(\theta(I - \Theta^*\Psi^2\Theta)\theta^*\xi, \xi) \geq \delta|\xi|^2. \qquad (6.5)$$

Remark 6.3 From (6.5) we see that $\theta\theta^*$ is uniformly positive definite with constant of positivity δ. Also, it turns out that (6.5) holds under Assumption 6.2.
Indeed, take a $\zeta = (\xi, \Psi\eta) \in \mathbb{R}^d \times \mathbb{R}^{d_1-d}$ with $\eta = -\Psi\Theta\theta^*\xi$ and observe that

$$2\delta|\xi|^2 \leq 2(\tilde{a}\zeta, \zeta) = |\tilde{\theta}^*\zeta|^2 = |\theta\xi|^2 + 2(\tilde{\theta}^*\xi, \Theta^*\Psi\eta) + |\Theta^*\Psi\eta|^2$$

$$= |\theta\xi|^2 + 2(\Psi\Theta\tilde{\theta}^*\xi, \eta) + |\eta|^2 = |\theta\xi|^2 - |\Psi\Theta\tilde{\theta}^*\xi|^2,$$

which is even stronger than (6.5).

Remark 6.4 We have seen that Assumptions 6.4 and 6.3 follow from Assumption 6.2. In turn Assumptions 6.4 and 6.3 in combination with Assumption 6.1 imply Assumption 6.2 perhaps with a different constant in the latter.

To show this, we take $\zeta = (\xi, \eta) \in \mathbb{R}^d \times \mathbb{R}^{d_1-d}$ and observe that

$$2(\tilde{a}\zeta, \zeta) = (\theta\theta^*\xi, \xi) + 2(\Theta\theta^*\xi, \eta) + (\Theta\Theta^*\eta, \eta)$$
$$= |\theta^*\xi|^2 + 2(\Psi\Theta\theta^*\xi, \tilde{\eta}) + \varepsilon(\tilde{\eta}, \tilde{\eta}) + (1-\varepsilon)(\Theta\Theta^*\eta, \eta)$$

where $\tilde{\eta} = \Psi^{-1}\eta$, and $\varepsilon \in (0, 1)$. By using the inequality $2(\mu, \nu) + \varepsilon|\mu|^2 \geq -\varepsilon^{-1}|\nu|^2$ we see that

$$2(\Psi\Theta\theta^*\xi, \tilde{\eta}) + \varepsilon(\tilde{\eta}, \tilde{\eta}) \geq -\varepsilon^{-1}|\Psi\Theta\tilde{\theta}^*\xi|^2,$$

and by taking N such that $\Psi \leq N(\delta^{ij})$, for which $\Theta\Theta^* \geq N^{-2}(\delta^{ij})$, we conclude

$$2(\tilde{a}\zeta, \zeta) \geq |\theta^*\xi|^2 - \varepsilon^{-1}|\Psi\Theta\theta^*\xi|^2 + (1-\varepsilon)N^{-2}|\eta|^2$$
$$\geq \delta|\xi|^2 + (1-\varepsilon^{-1})|\Psi\Theta\theta^*\xi|^2 + (1-\varepsilon)N^{-2}|\eta|^2,$$

where the last inequality follows from (6.5). Finally, $\Psi\Theta\theta^*$ is a bounded function, so that, for a constant N_1,

$$2(\tilde{a}\zeta, \zeta) \geq (\delta + N_1(1-\varepsilon^{-1}))|\xi|^2 + (1-\varepsilon)N^{-2}|\eta|^2.$$

For ε sufficiently close to 1 the last expression is greater than $\delta_1|\zeta|^2$ with a constant $\delta_1 > 0$, which is equivalent to the uniform ellipticity of $\tilde{a}$.

Before stating the next assumption we remind the reader that, for $\gamma \in \mathbb{R}$ and $u \in C_0^\infty(\mathbb{R}^d)$ one introduces $(1-\Delta)^{-\gamma/2}u$ by means of the Fourier transform. Then, for $p \in (1, \infty)$, one defines the spaces of Bessel potential $H_p^\gamma(\mathbb{R}^d)$ as the set of distributions obtained as the closure of $C_0^\infty(\mathbb{R}^d)$ with respect to the norm

$$\|u\|_{H_p^\gamma(\mathbb{R}^d)} := \|(1-\Delta)^{\gamma/2}u\|_{L_p(\mathbb{R}^d)}.$$

One important and highly nontrivial piece of information is that

$$H_p^1(\mathbb{R}^d) = W_p^1(\mathbb{R}^d) := \{u \in L_p(\mathbb{R}^d) : \nabla u \in L_p(\mathbb{R}^d)\}$$

and

$$\|u\|_{H_p^1(\mathbb{R}^d)} \sim \|u\|_{W_p^1(\mathbb{R}^d)} := \|u\|_{L_p(\mathbb{R}^d)} + \|\nabla u\|_{L_p(\mathbb{R}^d)}. \tag{6.6}$$

Assumption 6.5 The random vectors x_0 and y_0 are independent of the process w_t. The conditional distribution of x_0 given y_0 has a density, which we denote by $\pi_0(x) = \pi_0(\omega, x)$. We have $p \geq 2$ and $\pi_0 \in L_p(\Omega, H_p^{1-2/p}(\mathbb{R}^d))$ (actually, we need slightly less, see Remark 6.5).

Next we introduce some more notation. Let

$$\Psi_t = \Psi(t, y_t), \quad \Theta_t = \Theta(t, y_t), \quad a_t(x) = \frac{1}{2}\theta\theta^*(t, x, y_t), \quad b_t(x) = b(t, x, y_t),$$

$$\sigma_t(x) = \theta(t, x, y_t)\Theta_t^*\Psi_t, \quad B_t(x) = \Psi_t B(t, x, y_t).$$

In the remainder of the chapter we use the notation

$$D_i = \frac{\partial}{\partial x^i}$$

only for $i = 1, \ldots, d$ and set

$$L_t(x) = a_t^{ij}(x) D_i D_j + b_t^i(x) D_i, \tag{6.7}$$

$$L_t^*(x)u_t(x) = D_i D_j(a_t^{ij}(x)u_t(x)) - D_i(b_t^i(x)u_t(x))$$

$$= D_j(a_t^{ij}(x) D_i u_t(x) - b_t^j(x)u_t(x) + u_t(x) D_i a_t^{ij}(x)), \tag{6.8}$$

$$\Lambda_t^k(x)u_t(x) = \beta_t^k(x)u_t(x) + \sigma_t^{ik}(x) D_i u_t(x), \tag{6.9}$$

$$\Lambda_t^{k*}(x)u_t(x) = \beta_t^k(x)u_t(x) - D_i(\sigma_t^{ik}(x)u_t(x))$$

$$= -\sigma_t^{ik}(x) D_i u_t(x) + (\beta_t^k(x) - D_i \sigma_t^{ik}(x))u_t(x), \tag{6.10}$$

where $t \in [0, T]$, $x \in \mathbb{R}^d$, $k = 1, \ldots, d_1 - d$, and as above we use the summation convention over all 'reasonable' values of repeated indices, so that the summation in (6.7), (6.8), (6.9), and (6.10) is done for $i, j = 1, \ldots, d$ (whereas in (6.3) for $i, j = 1, \ldots, d_1$). Observe that Lipschitz continuous functions have bounded generalized derivatives and by

$$D_i a_t^{ij}, \quad D_i \sigma_t^{ik}$$

we mean these derivatives. From Remark 6.3 we have that the operator L defined by (6.7) is uniformly elliptic with constant of ellipticity δ.

Finally, by $\mathcal{F}_t^y$ we denote the completion of $\sigma\{y_s : s \leq t\}$ with respect to $P, \mathcal{F}$. Let us consider the following initial value problem

$$d\bar{\pi}_t(x) = L_t^*(x)\bar{\pi}_t(x) dt + \Lambda_t^{k*}(x)\bar{\pi}_t(x)\Psi_t^{kr} dy_t^r, \tag{6.11}$$

$$\bar{\pi}_0(x) = \pi_0(x),$$

where $t \in [0, T]$, $x \in \mathbb{R}^d$, and $\bar{\pi}_t(x) = \bar{\pi}_t(\omega, x)$. Equation (6.11) is called the Duncan–Mortensen–Zakai or just the Zakai equation.

We understand this equation and the initial condition in the following sense. We are looking for a function $\bar{\pi} = \bar{\pi}_t(x) = \bar{\pi}_t(\omega, x)$, $\omega \in \Omega$, $t \in [0, T]$, $x \in \mathbb{R}^d$, such that

(i) For each (ω, t), $\bar{\pi}_t(\omega, x)$ is a generalized function on $\mathbb{R}^d$,
(ii) We have $\bar{\pi} \in L_p(\Omega \times [0, T], \mathcal{P}, H_p^1(\mathbb{R}^d))$,

(iii) For each $\varphi \in C_0^\infty(\mathbb{R}^d)$ with probability one for all $t \in [0, T]$ it holds that

$$(\bar{\pi}_т, \varphi) = (\pi_0, \varphi) - \int_0^t (a_\tau^{ij} D_i \bar{\pi}_\tau - b_\tau^j \bar{\pi}_\tau + \bar{\pi}_\tau D_i a_\tau^{ij}, D_j \varphi)\, d\tau$$
$$- \int_0^t (\sigma_\tau^{ik} D_i \bar{\pi}_\tau + (D_i \sigma_\tau^{ik} - \beta_\tau^k)\bar{\pi}_\tau, \varphi) \Psi_\tau^{kr}(B^r(t, z_\tau)\, dt + \Theta^{rs}\, dw_\tau^s), \quad (6.12)$$

where by (f, φ) we mean the action of a generalized function f on φ, in particular, if f is a locally summable,

$$(f, \varphi) = \int_{\mathbb{R}^d} f(x)\varphi(x)\, dx.$$

Observe that all expressions in (6.12) are well defined due to the fact that the coefficients of $\bar{\pi}$ and of $D_i\bar{\pi}$ are bounded and appropriately measurable and $\bar{\pi}, D_i\bar{\pi} \in L_p(\Omega \times [0, T], \mathcal{P}, L_p(\mathbb{R}^d))$ (see (6.6)).

Hence, equation (6.11) has the same form as (6.40) and the existence and uniqueness part of Lemma 6.1 below follow from Theorem 6.4 and Remark 6.5. The second assertion of the lemma follows from Theorem 6.7.

In all that follows in the main part of the chapter, we suppose that Assumptions 6.1, 6.2, and 6.5 are satisfied.

Lemma 6.1 *There exists a unique solution $\bar{\pi}$ of (6.11) with initial condition π_0 in the sense explained above. In addition, $\bar{\pi}_t \geq 0$ for all $t \in [0, T]$ (a.s.).*

Here is a basic result of filtering theory for partially observable diffusion processes. Its relation to the previously known ones is discussed above.

Theorem 6.1 *Let $\bar{\pi}$ be the function from Lemma 6.1. Then*

$$0 < \int_{\mathbb{R}^d} \bar{\pi}_t(x)\, dx = (\bar{\pi}_t, 1) < \infty \quad (6.13)$$

for all $t \in [0, T]$ (a.s.) and for any $t \in [0, T]$ and real-valued, bounded or nonnegative, (Borel) measurable function f given on $\mathbb{R}^d$

$$E[f(x_t)|\mathcal{F}_t^y] = \frac{(\bar{\pi}_t, f)}{(\bar{\pi}_t, 1)} \quad (a.s.). \quad (6.14)$$

Equation (6.14) shows (by definition) that

$$\pi_t(x) := \frac{\bar{\pi}_t(x)}{(\bar{\pi}_t, 1)}$$

is a conditional density of distribution of x_t given $y_s, s \leq t$. Since, generally, $(\bar{\pi}_t, 1) \neq 1$, one calls $\bar{\pi}_t$ an unnormalized conditional density of distribution of x_t given $y_s, s \leq t$.

The following is a direct corollary of Theorem 6.8.

Theorem 6.2 *Let π_0 be a nonrandom function and $\pi_0 \in H_p^{1-2/p}(\mathbb{R}^d)$ for all $p \geq 2$, which happens for instance, if π_0 is a Lipschitz continuous function with compact support. Then for any $\varepsilon \in (0, 1/2)$ almost surely $\bar{\pi}_t(x)$ is $1/2 - \varepsilon$ Hölder continuous in t with a constant independent of x, $\bar{\pi}_t(x)$ is $1 - \varepsilon$ Hölder continuous in x with a constant independent of t, and the above mentioned (random) constants have all moments.*

In filtering theory usually the following theorem is proved before anything else is done. We do not need it for proving the above results and give the proof just to show that the L_p-theory of SPDEs allows one to get all basic results from filtering theory.

Historically, $P_t[\beta]$ was introduced by (6.17) and shown to have (a modification possessing) appropriate measurability properties. Then $\bar{\pi}_t$ used to be defined as the density of conditional distribution of x_t given $\mathcal{F}_t^y$ divided by an appropriate modification of

$$E(\rho_t \mid \mathcal{F}_t^y), \qquad (6.15)$$

where

$$\rho_t = \exp(-\int_0^t \tilde{\beta}_s \, d\tilde{w}_s - \tfrac{1}{2}\int_0^t |\tilde{\beta}_s|^2 \, ds), \quad \tilde{\beta}_s = \beta_s(x_s), \quad \tilde{w}_t = \int_0^t \Psi_s \Theta_s \, dw_s.$$

In this case $(\bar{\pi}_t, 1)^{-1}$ turns out to be this same appropriate modification of (6.15) (cf. our (6.39)).

The most surprising statements in Theorem 6.3 are assertions (iv) and (v). In (iv) the difference of two Wiener processes $\check{w}_t$ and $\tilde{w}_t$ (that the latter is a Wiener process is checked in the proof of Lemma 6.3) is asserted to be a differentiable nontrivial function.

Assertion (v) shows that (6.15), which is a conditional expectation of a martingale, is again a martingale and, moreover, while evaluating it we can just put conditional expectations of $\tilde{\beta}_s$ given $\mathcal{F}_s^y$ in place of $\tilde{\beta}_s$ in the expression of ρ_t with simultaneous replacement of $\tilde{w}$ with $\check{w}$.

Theorem 6.3

(i) *The process $(\bar{\pi}_t, 1)$ is continuous in t (a.s.) and (a.s.) for all $t \in [0, T]$*

$$(\bar{\pi}_t, 1) = (\pi_0, 1) + \int_0^t (\bar{\pi}_s, \beta_s^k) \Psi_s^{kr} B^r(t, z_s) \, ds + \int_0^t (\bar{\pi}_s, \beta_s^k) \Psi_s^{kr} \Theta_s^{rn} \, dw_s^n. \qquad (6.16)$$

(ii) *The process $\bar{\pi}_t$ is a continuous $L_1(\mathbb{R}^d)$-valued process (a.s.).*

(iii) *Introduce $P_t[\beta] = (P_t[\beta^1], \ldots, P_t[\beta^{d_1-d}])$ by*

$$P_t[\beta] = (\bar{\pi}_t, 1)^{-1} \int_{\mathbb{R}^d} \beta_t(x) \bar{\pi}_t(x) \, dx = (\bar{\pi}_t, 1)^{-1} \Psi(t, y_t) \int_{\mathbb{R}^d} B(t, x, y_t) \bar{\pi}_t(x) \, dx.$$

Then $P_t[\beta]$ is a jointly measurable bounded $\mathcal{F}_t^y$-adapted process on $[0, t]$ (a.s.) and for each $t \in [0, T]$

$$P_t[\beta] = E(\beta_t(x_t) \mid \mathcal{F}_t^y) \quad (a.s.). \tag{6.17}$$

(iv) The process

$$\check{w}_t = \tilde{w}_t + \int_0^t (\beta_s(x_s) - P_s[\beta]) \, ds$$

is a $(d_1 - d)$-dimensional Wiener process with respect to $\mathcal{F}_t^y$ (the so-called innovation process), where

$$\tilde{w}_t = \int_0^t \Psi_s \Theta_s \, dw_s.$$

(v) We have (a.s.) for all $t \in [0, T]$

$$(\bar{\pi}_t, 1) = \exp\left(\int_0^t P_s[\beta] \, d\check{w}_s + \tfrac{1}{2} \int_0^t |P_s[\beta]|^2 \, ds\right), \tag{6.18}$$

so that

$$(\bar{\pi}_t, 1)^{-1} = \exp\left(-\int_0^t P_s[\beta] \, d\check{w}_s - \tfrac{1}{2} \int_0^t |P_s[\beta]|^2 \, ds\right)$$

is an exponential martingale, and for each $m > 0$

$$E \sup_{t \leq T} (\bar{\pi}_t, 1)^m + E \sup_{t \leq T} (\bar{\pi}_t, 1)^{-m} < \infty. \tag{6.19}$$

6.3 Proof of Theorem 6.1

We will use some ideas and results from the theory of SPDEs, which are recalled in Section 6.5. From now on we drop $\mathbb{R}^d$ in notation like $H_p^\gamma(\mathbb{R}^d)$ and $L_p(\mathbb{R}^d)$.

Remark 6.5 The assumption that $\pi_0 \in L_p(\Omega, H_p^{1-2/p})$ is only needed to guarantee (see the proof of Theorem 5.1 of [6]) that there exists a $\psi \in \mathcal{H}_p^1(T)$ such that $\psi_0 = \pi_0$,

$$d\psi_t = \Delta \psi_t \, dt = D_i f_t^i \, dt, \quad (f_t^i = D_i \psi_t),$$

$$\|\psi\|_{\mathbb{H}_p^1(T)}^p \leq N E \|\pi_0\|_{H_p^{1-2/p}}^p$$

with N independent of π_0.

As is mentioned before Lemma 6.1, by Theorem 6.4 and Remark 6.5, there exists a unique solution $\bar{\pi} \in \mathcal{H}_p^1(T)$ of (6.11) with initial condition π_0. By Theorem 6.7, $\bar{\pi}_t \geq 0$ for all $t \in [0, T]$ (a.s.). By Theorem 6.8, $\bar{\pi}_t$ is a continuous L_p-valued process and

$$E \sup_{t \in [0,T]} \|\bar{\pi}_t\|_{L_p}^p \, dt < \infty. \tag{6.20}$$

Finally, for $\tau = T \wedge \inf\{t \geq 0 : \|\bar{\pi}_t\|_{L_p} = 0\}$ we obviously have that $\bar{\pi}_{t \wedge \tau}$ also is a solution of (6.11), which implies that on the set where $\tau < T$, we have (a.s.) $\bar{\pi}_T = 0$.

Now, we prove three auxiliary results.

Lemma 6.2 *Let ξ_t, ξ_t^n, $n = 1, 2, \ldots$, $t \in [0, T]$, be k-dimensional continuous semimartingales such that, for any $t \in [0, T]$, $\xi_t^n \to \xi_t$ in probability as $n \to \infty$. Assume that*

$$\xi_t^n = \xi_0^n + \int_0^t a_s^n \, ds + m_t^n, \quad \xi_t = \xi_0 + \int_0^t a_s \, ds + m_t,$$

where a_t and a_t^n are predictable processes bounded by the same nonrandom constant and m_t and m_t^n are martingales such that

$$\langle m^{ni}, m^{nj} \rangle_t = \int_0^t \gamma_s^{nij} \, ds, \quad \langle m^i, m^j \rangle_t = \int_0^t \gamma_s^{ij} \, ds, \quad i, j = 1, \ldots, k,$$

where $\gamma_t^n := (\gamma_t^{nij})$ and $\gamma_t := (\gamma_t^{ij})$ are predictable matrix-valued processes bounded by the same nonrandom constant and such that $(\gamma_t^n)^{-1}$ and $(\gamma_t)^{-1}$ exist and are also bounded by the same nonrandom constant.

Assume that on $[0, T] \times \mathbb{R}^l \times \mathbb{R}^k$ we are given functions $f_t^n(x, y)$ and $f_t(x, y)$ such that they are uniformly bounded and $f^n \to f$ in measure as $n \to \infty$.

Then $f_t^n(x, \xi_t^n) \to f_t(x, \xi_t)$ in measure on $\Omega \times [0, T] \times \mathbb{R}^l$.

Proof. It suffices to show that any subsequence $\{n'\}$ of integers has a subsequence $\{n''\}$ such that $f_t^{n''}(x, \xi_t^{n''}) \to f_t(x, \xi_t)$ in measure. Since any subsequence $\{n'\}$ has a subsequence $\{n''\}$ such that $f^{n''} \to f$ almost everywhere, by having in mind renumbering if needed, we may assume that for the original sequence we have $f^n \to f$ almost everywhere. In that case for almost any $x \in \mathbb{R}^l$, $f_t^n(x, y) \to f_t(x, y)$ and, if we prove that for each such x we have $f_t^n(x, \xi_t^n) \to f_t(x, \xi_t)$ in measure on $\Omega \times [0, T]$, then

$$E \int_0^T |f_t^n(x, \xi_t^n) - f_t(x, \xi_t)| \, dt \to 0,$$

which after being integrated with respect to x would shows that $f_t^n(x, \xi_t^n) \to f_t(x, \xi_t)$ in measure on $\Omega \times [0, T] \times \mathbb{R}^l$.

It follows that we only need to prove that, if on $[0, T] \times \mathbb{R}^k$ we are given functions $f_t^n(y)$ and $f_t(y)$ such that they are uniformly bounded and $f^n \to f$ (t, y) almost everywhere as $n \to \infty$, then

$$E \int_0^T |f_t^n(\xi_t^n) - f_t(\xi_t)| \, dt \to 0. \tag{6.21}$$

Furthermore, since the coefficients a^n, a, γ^n, and γ are uniformly bounded

$$\sup_n \sup_{t\in[0,T]} P(|\xi_t^n| + |\xi_t| \geq R) \leq R^{-2} \sup_n \sup_{t\in[0,T]} E(|\xi_t^n|^2 + |\xi_t|^2) \to 0$$

as $R \to \infty$. Therefore, if for any $R \in (0, \infty)$ we know that (6.21) is true provided that $f_t^n(y)$ and $f_t(y)$ vanish for $|y| \geq R$, then by applying this result in the general case to $f_t^n(y) I_{|y|<R}$ and $f_t(y) I_{|y|<R}$ we would obtain that

$$\varlimsup_{n\to\infty} E \int_0^T |f_t^n(\xi_t^n) - f_t(\xi_t)|\, dt \leq NR^{-2},$$

where N is independent of R. This would imply (6.21) in the general case. This shows that without restricting generality we may assume that for an $R \in (0, \infty)$ the functions $f_t^n(y)$ and $f_t(y)$ vanish if $|y| \geq R$.

Now observe that the left-hand side of (6.21) is majorated by $I_n + J_n$, where

$$I_n = E \int_0^T |f_t^n(\xi_t^n) - f_t(\xi_t^n)|\, dt, \quad J_n = E \int_0^T |f_t(\xi_t^n) - f_t(\xi_t)|\, dt.$$

We recall a result of [4] implying that for any $g \in L_{k+1}([0, T] \times \mathbb{R}^k)$ we have

$$E \int_0^T (|g_t(\xi_t^n)| + |g_t(\xi_t)|)\, dt \leq N\|g\|_{L_{k+1}([0,T]\times\mathbb{R}^k)},$$

where N is independent of n and g. We apply this result to $g = f^n - f$ and observe that these functions are uniformly bounded, vanish for $|y| \geq R$, and tend to zero in measure. Hence, their $L_{k+1}([0, T] \times \mathbb{R}^k)$-norms tend to zero. This implies that $I_n \to 0$.

Next, notice that for any function g

$$J_n \leq E \int_0^T |g_t(\xi_t^n) - g_t(\xi_t)|\, dt$$
$$+ E \int_0^T |f_t(\xi_t^n) - g_t(\xi_t^n)|\, dt + E \int_0^T |f_t(\xi_t) - g_t(\xi_t)|\, dt$$

implying that

$$\varlimsup_{n\to\infty} J_n \leq \varlimsup_{n\to\infty} E \int_0^T |g_t(\xi_t^n) - g_t(\xi_t)|\, dt + N\|f - g\|_{L_{k+1}([0,T]\times\mathbb{R}^k)}, \qquad (6.22)$$

where N is independent of g. For any $\varepsilon > 0$ we can find a smooth g such that the second term on the right in (6.22) will be less than ε. In addition, the first term vanishes for smooth g since $\xi_t^n \to \xi_t$ in probability for any t. Since ε is arbitrary, it follows that the left-hand side of (6.22) equals zero. The lemma is proved. □

The following result with its proof is an adaptation of Lemma 5.1 of [11] and its proof.

Lemma 6.3 *The function $\bar{\pi}_t$ is $\mathcal{F}_t^y$-adapted.*

Proof. Define

$$\tilde{\beta}_t = \beta_t(x_t) = \Psi_t B(t, z_t), \quad \hat{w}_t = \int_0^t \Psi_s \, dy_s, \quad \tilde{w}_t = \int_0^t \Psi_s \Theta_s \, dw_s.$$

Since Ψ_t is $\mathcal{F}_t^y$-adapted, the process $\hat{w}_t$ is $\mathcal{F}_t^y$-adapted too. Furthermore, $\Psi_s \Theta_s \Theta_s^* \Psi_s$ is a unit matrix so that by Lévy's theorem $\tilde{w}_t$ is a Wiener process. We want to change the probability measure so that $\hat{w}_t$ would become a Wiener process with respect to this new measure. Define

$$\rho_t = \exp(-\int_0^t \tilde{\beta}_s \, d\tilde{w}_s - \tfrac{1}{2} \int_0^t |\tilde{\beta}_s|^2 \, ds), \quad Q(d\omega) = \rho_T(\omega) \, P(d\omega). \quad (6.23)$$

The process ρ_t is an exponential local martingale. Since $\tilde{\beta}$ is bounded, ρ_t is square integrable, so that Q is a probability measure. Since

$$d\hat{w}_t = \tilde{\beta}_t \, dt + d\tilde{w}_t$$

and $\tilde{w}_t$ is a Wiener process on $(\Omega, \mathcal{F}, P)$, by Girsanov's theorem, $\hat{w}_t, t \in [0, T]$, is a Wiener process on $(\Omega, \mathcal{F}, Q)$ with respect to the filtration $\{\mathcal{F}_t\}$. As has been noticed before, it is $\mathcal{F}_t^y$-adapted and, obviously,

$$\mathcal{F}_t^y \subset \mathcal{F}_t,$$

so that $(\hat{w}_t, \mathcal{F}_t^y)$ is a Wiener process. Now rewrite (6.11) as

$$d\bar{\pi}_t(x) = L_t^*(x)\bar{\pi}_t(x) \, dt + \Lambda_t^{k*}(x)\bar{\pi}_t(x) \, d\hat{w}_t^k, \quad (6.24)$$

and consider this equation relative to $(\Omega, \mathcal{F}, \mathcal{F}_t^y, Q)$.

By Theorem 6.4 and Remark 6.5 equation (6.24) with initial data π_0 has a unique $\mathcal{F}_t^y$-adapted solution belonging to $\mathcal{H}_p^1(\mathcal{F}_\cdot^y, Q, T) \subset \mathcal{H}_p^1(\mathcal{F}_\cdot, Q, T)$, where by $\mathcal{H}_p^1(\mathcal{F}_\cdot^y, Q, T)$ we mean the space $\mathcal{H}_p^1(T)$ constructed on the basis of the new probability measure Q and filtration $\mathcal{F}_\cdot^y$. We denote by $\tilde{\pi}_t$ this solution.

We have already mentioned that $\bar{\pi} \in \mathcal{H}_p^1(\mathcal{F}_\cdot, P, T)$. We want to derive that $\bar{\pi}_t$ is $\mathcal{F}_t^y$-adapted from the uniqueness by showing that $\bar{\pi} = \tilde{\pi}$ because both are $\mathcal{F}_t$-adapted solutions of the same equation. The only obstacle is that the norms in $\mathcal{H}_p^1(\mathcal{F}_\cdot, Q, T)$ and $\mathcal{H}_p^1(T)$ are different. To overcome this obstacle, we are going to use stopping times.

For integers n define

$$\tau(n) = T \wedge \inf\{t \geq 0 : \int_0^t \|\tilde{\pi}_s\|_{\mathcal{H}_p^1}^p \, ds \geq n\}.$$

Obviously, $\tau(n)$ are $\mathcal{F}_t^y$-stopping times and $\mathcal{F}_t$-stopping times. Furthermore,

$$\|\tilde{\pi}\|_{\mathbb{H}_p^1(\mathcal{F}., P, \tau(n))}^p = E \int_0^{\tau(n)} \|\tilde{\pi}_s\|_{H_p^1}^p \, ds \leq n < \infty.$$

This and the equation (cf. (6.24))

$$d\tilde{\pi}_t(x) = [L_t^*(x)\tilde{\pi}_t(x) + \tilde{\beta}_t^k \Lambda_t^k(x)\tilde{\pi}_t(x)] \, dt + \Lambda_t^k(x)\tilde{\pi}_t(x) \, d\tilde{w}_t^k$$

show that, $\tilde{\pi} \in \mathcal{H}_p^1(\mathcal{F}., P, \tau(n))$. By the above mentioned uniqueness, $\tilde{\pi}_t = \bar{\pi}_t$ on $(0, \tau(n)]$ (a.e.). Since both functions are continuous in $t \in [0, T]$ (Theorem 6.8 (i)), we have that

$$\tilde{\pi}_t I_{0 < t \leq \tau(n)} \quad \text{and} \quad \bar{\pi}_t I_{0 < t \leq \tau(n)}$$

are indistinguishable, and since one of them is $\mathcal{F}_t^y$-adapted, so is the other. We conclude that $\bar{\pi}_t I_{0 < t \leq \tau(n)}$ is $\mathcal{F}_t^y$-adapted, which after letting $n \to \infty$ yields the result. The lemma is proved.

Assertion of the following lemma is a very particular case of one of the assertions of Theorem 6.3. Before stating the lemma we recall that $\bar{\pi}_t \geq 0$ for all $t \in [0, T]$ (a.s.), so that $(\bar{\pi}_t, 1)$ is well defined (and may be infinite).

Lemma 6.4 *We have*

$$E \sup_{t \in [0, T]} (\bar{\pi}_t, 1)^{1/2} < \infty. \tag{6.25}$$

Proof. For $\varphi \in C_0^\infty(\mathbb{R}^d)$ one can rewrite (6.12) as

$$(\bar{\pi}_t, \varphi) = (\pi_0, \varphi) + \int_0^t (\bar{\pi}_s, L_s\varphi) \, ds$$
$$+ \int_0^t (\bar{\pi}_s, \Lambda_s^k\varphi)\Psi_s^{kr}(B^r(s, z_s) \, ds + \Theta_s^{rn} \, dw_s^n). \tag{6.26}$$

Using (6.20) and an obvious passage to the limit, it is easy to prove that (6.26) holds not only for $\varphi \in C_0^\infty(\mathbb{R}^d)$, but also for $\varphi \in W_q^2$ with $q = p/(p-1)$.

On $\mathbb{R}^d$ for $m = 1, 2, \ldots$ introduce the functions

$$\varphi(x) = (1 + |x|^2)^{-d}, \quad \varphi_m(x) = \varphi(x/m).$$

Observe that for a constant N it holds that

$$|D_i\varphi_m| + m|D_i D_j\varphi_m| \leq Nm^{-1}\varphi_m \tag{6.27}$$

on $\mathbb{R}^d$ for all m. In particular,

$$2L_t\varphi_m \leq N_0\varphi_m, \quad 2|\Psi_t^{kr} B^r(t, z_t)\Lambda_t^k\varphi_m| \leq N_0\varphi_m, \tag{6.28}$$

where N_0 is a constant independent of m and the arguments of the functions involved.

By plugging in (6.26) the function φ_m in place of φ, we obtain

$$(\bar{\pi}_t, \varphi_m) = (\pi_0, \varphi_m) + \int_0^t (\bar{\pi}_s, L_s \varphi_m)\, ds$$
$$+ \int_0^t (\bar{\pi}_s, \Lambda_s^k \varphi_m) \Psi_s^{kr} (B^r(s, z_s)\, ds + \Theta_s^{rn}\, dw_s^n). \tag{6.29}$$

By using Itô's formula for transforming

$$(\bar{\pi}_t, \varphi_m) e^{-N_0 t}, \tag{6.30}$$

and using (6.28) we see that

$$d[(\bar{\pi}_t, \varphi_m) e^{-N_0 t}] = e^{-N_0 t} (\bar{\pi}_t, \Lambda_t^k \varphi_m) \Psi_t^{kr} \Theta_t^{rn}\, dw_t^n$$
$$+ e^{-N_0 t} [(\bar{\pi}_t, L_t \varphi_m) + (\bar{\pi}_s, \Lambda_s^k \varphi_m) \Psi_s^{kr} B^r(s, z_s) - N_0 (\bar{\pi}_s, \varphi_m)] dt$$
$$\leq e^{-N_0 t} (\bar{\pi}_t, \Lambda_t^k \varphi_m) \Psi_t^{kr} \Theta_t^{rn}\, dw_t^n.$$

It follows that process (6.30) is a supermartingale. It is continuous and nonnegative. Therefore,

$$E \sup_{t \in [0,T]} e^{-N_0 t} \Big(\int_{\mathbb{R}^d} \varphi_m \bar{\pi}_t(x)\, dx \Big)^{1/2} \leq 2 \Big(E \int_{\mathbb{R}^d} \varphi_m \bar{\pi}_0(x)\, dx \Big)^{1/2} \leq 2.$$

Upon letting $m \to \infty$ and using the monotone convergence theorem we come to (6.25) and the lemma is proved. □

Proof of Theorem 6.2 Take a nonnegative $\zeta \in C_0^\infty(\mathbb{R}^{d_1})$, which integrates to one and for $n = 1, 2, \ldots$ set

$$\zeta_n(z) = n^{d_1} \zeta(nz).$$

Also introduce mollifications of one of the coefficients of (6.1) by

$$\theta^{(n)}(t, z) = \zeta_n(z) * \theta(t, z),$$

where the convolutions is taken with respect to z.

The function ζ can be considered as the density of a random variable. If needed, we extend our initial probability space in such a way that it would allow us to introduce a new random $\mathbb{R}^{d_1}$-valued vector ξ having density ζ and such that ξ is independent of z_0 and the process $w_t, t \geq 0$.

After that, for $n = 1, 2, \ldots$, we consider the following modification of (6.1):

$$dx_t^{(n)} = b(t, z_t^{(n)}) dt + \theta^{(n)}(t, z_t^{(n)}) dw_t$$
$$dy_t^{(n)} = B(t, z_t^{(n)}) dt + \Theta(t, y_t^{(n)}) dw_t \tag{6.31}$$

with initial data $x_0^{(n)} = x_0 + n^{-1}\xi$, $y_0^{(n)} = y_0$ and $z_t^{(n)} = (x_t^{(n)}, y_t^{(n)})$. Observe that the conditional distribution of $x_0^{(n)}$ given y_0 has a density equal to

$$\pi_0^{(n)} = \zeta_n * \pi_0.$$

Since $\theta(t, x, y)$ is Lipschitz in x (even in (x, y)) we have $|\theta(t, z) - \theta^{(n)}(t, z)| \le Nn^{-1}$, where N is independent of n, t, z. This shows that system (6.31) satisfies Assumption 6.2 for all large n. In addition $\theta^{(n)}$ possesses enough smoothness in order for the results of [11] to be applicable. For all large n, it follows that, for any smooth bounded and nonnegative function $c_t(y)$ on $[0, T] \times \mathbb{R}^{d_1-d}$ and any $\varphi \in C_0^\infty(\mathbb{R}^{d_1})$,

$$E\varphi(z_T^{(n)}) \exp(-\int_0^T c_s(y_s^{(n)}) \, ds)$$
$$= E\rho_T^{(n)} \int_{\mathbb{R}^d} \varphi(x, y_T^{(n)}) \bar{\pi}_T^{(n)}(x) \, dx \, \exp(-\int_0^T c_s(y_s^{(n)}) \, ds), \qquad (6.32)$$

where $\bar{\pi}_t^{(n)}$ is the solution of equation (6.11) corresponding to system (6.31) with initial condition $\bar{\pi}_0^{(n)} = \pi_0^{(n)}$ and $\rho_t^{(n)}$ is introduced as in (6.23) on the basis of (6.31):

$$\rho_t^{(n)} = \exp(-\int_0^t \tilde{\beta}_s^{(n)} \, d\tilde{w}_s^{(n)} - \tfrac{1}{2} \int_0^t |\tilde{\beta}_s^{(n)}|^2 \, ds),$$

$$\tilde{w}_t^{(n)} = \int_0^t \Psi_s^{(n)} \Theta_s^{(n)} \, dw_s, \quad \tilde{\beta}_t^{(n)} = \beta_t^{(n)}(x_t^{(n)}), \quad \beta_t^{(n)}(x) = \Psi_t^{(n)} B(t, x, y_t^{(n)}),$$

$$\Theta_t^{(n)} = \Theta(t, y_t^{(n)}), \quad \Psi_t^{(n)} = \Psi(t, y_t^{(n)}).$$

Later on we will also use the following notation for other coefficients of equation (6.11) corresponding to system (6.31). Introduce

$$a_t^{(n)}(x) = \tfrac{1}{2} \theta^{(n)} \theta^{(n)*}(t, x, y_t^{(n)}), \quad b_t^{(n)}(x) = b(t, x, y_t^{(n)}),$$

$$\sigma_t^{(n)}(x) = \theta^{(n)}(t, x, y_t^{(n)}) \Theta_t^{(n)*} \Psi_t^{(n)}.$$

Since we know that $\bar{\pi}_t^{(n)} \ge 0$, it follows from the validity of (6.32) for all $\varphi \in C_0^\infty(\mathbb{R}^{d_1})$, that it is also valid for all Borel nonnegative or bounded φ. In particular, for any $f \in C_0^\infty(\mathbb{R}^d)$ (independent of y) we have

$$Ef(x_T^{(n)}) \exp(-\int_0^T c_s(y_s^{(n)}) \, ds)$$
$$= E\rho_T^{(n)} \int_{\mathbb{R}^d} f(x) \bar{\pi}_T^{(n)}(x) \, dx \, \exp(-\int_0^T c_s(y_s^{(n)}) \, ds). \qquad (6.33)$$

Our next step is to pass to the limit in (6.33) as $n \to \infty$. It is a standard fact that for any $m > 0$

$$\lim_{n\to\infty} E \sup_{t\leq T} |z_t^{(n)} - z_t|^m = 0, \qquad (6.34)$$

which, in particular, implies that the left-hand sides of (6.33) tend to

$$Ef(x_T)\exp(-\int_0^T c_s(y_s)\,ds).$$

Furthermore, the process $\rho_t^{(n)}$ is the solution of the linear equation

$$d\rho_t^{(n)} = -\rho_t^{(n)} \gamma_t^{(n)} \, dw_t,$$

with initial condition $\rho_0^{(n)} = 1$, where

$$\gamma_t^{(n)} = \Psi(t, y_t^{(n)}) B(t, z_t^{(n)}) \Psi(t, y_t^{(n)}) \Theta(t, y_t^{(n)}).$$

Also introduce

$$\gamma_t = \Psi(t, y_t) B(t, z_t) \Psi(t, y_t) \Theta(t, y_t)$$

and observe that the processes $\gamma_t^{(n)}$ and γ_t are bounded.

Furthermore, it follows from (6.34) that for any $m > 0$

$$\lim_{n\to\infty} E \sup_{t\leq T} |\gamma_t^{(n)} - \gamma_t|^m = 0,$$

which in turn implies that

$$\lim_{n\to\infty} E \sup_{t\leq T} |\rho_t^{(n)} - \rho_t|^m = 0,$$

where ρ_t is the solution of the equation $d\rho_t = -\rho_t \gamma_t \, dw_t$ with initial condition $\rho_0 = 1$ and is given in (6.23).

To investigate the limit of the remaining factor on the right in (6.33) we will use Theorem 6.5. By the well-known properties of convolutions

$$\|\pi_0^{(n)}\|_{H^{1-2/p}}^p \leq \|\pi_0\|_{H^{1-2/p}}^p, \qquad \lim_{n\to\infty} E \|\pi_0^{(n)} - \pi_0\|_{H^{1-2/p}}^p = 0.$$

This and Remark 6.5 show that the assumption of Theorem 6.5 regarding the convergence of the initial data for $\bar\pi_t^{(n)}$ and $\bar\pi_t$ is satisfied. Furthermore, there are no free terms in filtering equations. Therefore, it only remains to check the appropriate convergence of the coefficients. Theorem 6.5 requires the following convergences in measure $P(d\omega)dtdx$ to hold on $\Omega \times [0, T] \times \mathbb{R}^d$:

$$a_t^{(n)}(x) \to a_t(x), \quad b_t^{(n)}(x) \to b_t(x), \quad D_i a_t^{(n)ij}(x) \to D_i a_t^{ij}(x),$$

$$\sigma_t^{(n)}(x) \to \sigma_t(x), \quad \beta_t^{(n)}(x) \to \beta_t(x), \quad D_i \sigma_t^{(n)ik}(x) \to D_i \sigma_t^{ik}(x).$$

Relation (6.34) and the assumption that the coefficients of system (6.1) are Lipschitz continuous show that, actually, apart from cases involving the derivatives of a and σ all the remaining convergences hold uniformly in (t, x) almost surely. It is easy to see that in order to take care of the terms with derivatives it suffices to check that

$$D_i \theta^{(n)}(t, x, y_t^{(n)}) \to D_i \theta(t, x, y_t) \qquad (6.35)$$

in measure for any $i = 1, \ldots, d$. Observe that by the well-known properties of convolutions

$$D_i \theta^{(n)}(t, x, y) \to D_i \theta(t, x, y)$$

for almost all (t, x, y). Therefore, applying Lemma 6.2 shows that (6.35) holds.

Now by Theorem 6.5 and Hölder's inequality we conclude

$$\lim_{n \to \infty} E| \int_{\mathbb{R}^d} f(x) \bar{\pi}_T^{(n)}(x) \, dx - \int_{\mathbb{R}^d} f(x) \bar{\pi}_T(x) \, dx|^p = 0. \qquad (6.36)$$

This along with the above investigation of other terms in (6.33) yields after letting $n \to \infty$ that

$$E f(x_T) \exp(- \int_0^T c_s(y_s) \, ds) = E \rho_T (\bar{\pi}_T, f) \exp(- \int_0^T c_s(y_s) \, ds).$$

The arbitrariness of c leads to

$$E(f(x_T) \mid \mathcal{F}_T^y) = E(\rho_T(\bar{\pi}_T, f) \mid \mathcal{F}_T^y), \quad (\text{a.s.}),$$

which combined with the $\mathcal{F}_T^y$-measurability of $\bar{\pi}_T$ (Lemma 6.3) shows that

$$E(f(x_T) \mid \mathcal{F}_T^y) = (\bar{\pi}_T, f) E(\rho_T \mid \mathcal{F}_T^y) \quad (\text{a.s.}). \qquad (6.37)$$

Observe that on the set of ω where

$$E(\rho_T \mid \mathcal{F}_T^y) = 0 \qquad (6.38)$$

we have (a.s.)

$$E(f(x_T) \mid \mathcal{F}_T^y) = 0.$$

The arbitrariness of f shows that on the said set (a.s.)

$$1 = E(1 \mid \mathcal{F}_T^y) = 0$$

and consequently (6.38) can only happen with probability zero.

Furthermore, by Theorem 6.7 we have $\bar{\pi}_t \geq 0$. A standard measure-theoretic argument then shows that (6.37) holds for all nonnegative Borel f rather than only for $f \in C_0^\infty(\mathbb{R}^d)$. By taking $f \equiv 1$ we see that

$$1 = (\bar{\pi}_T, 1) E(\rho_T \mid \mathcal{F}_T^y) \quad (\text{a.s.})$$

implying that

$$\infty > (\bar{\pi}_T, 1) > 0, \quad E(\rho_T \mid \mathcal{F}_T^y) = (\bar{\pi}_T, 1)^{-1} \quad (\text{a.s.}). \qquad (6.39)$$

Coming back to (6.37) we conclude

$$E[f(x_T)|\mathcal{F}_T^y] = \frac{(\bar{\pi}_T, f)}{(\bar{\pi}_T, 1)} \quad \text{(a.s.)}$$

for any nonnegative and any bounded Borel f as well. Obviously, one can replace here T with any $t \in [0, T]$ and to prove Theorem 6.1 it only remains to show that (a.s.) relation (6.13) holds for all $t \in [0, T]$.

The second inequality in (6.13) holds due to Lemma 6.4. To prove the first one it only remains to observe that by the above for each particular $t \in [0, T]$ with probability one

$$\int_{\mathbb{R}^d} \bar{\pi}_t^p(x)\,dx > 0,$$

by Theorem 6.8 the above integral is continuous in t with probability one, and by Remark 6.5 this integral cannot vanish before T. The theorem is proved.

6.4 Proof of Theorem 6.3

To prove (i) we first show that the right-hand sides of (6.29) converge as $n \to \infty$ uniformly in $t \in [0, T]$ in probability to the right-hand side of (6.16). Owing to (6.27) and (6.25)

$$\int_0^T |(\bar{\pi}_s, L_s\varphi_m)|\,ds \leq NTm^{-1} \sup_{s \in [0, T]} (\bar{\pi}_s, 1) \to 0 \quad \text{(a.s.),}$$

where N is the constant from (6.27). Similarly one takes care of the term with ds containing the derivatives of φ_m in the second integral on the right in (6.29). Observing that by the dominated convergence theorem and again by (6.25)

$$\int_0^T |(\bar{\pi}_s, |\beta_s^k|\,|\varphi_m - 1|)\,ds \to 0 \quad \text{(a.s.),}$$

we conclude that the usual integrals on the right-hand sides of (6.29) converge as $n \to \infty$ uniformly in $t \in [0, T]$ to the usual integral the right-hand side of (6.16) almost surely.

To show the convergence of the stochastic integrals in (6.29) to the stochastic integral in (6.16) uniform in probability it suffices (and is necessary) to show that the quadratic variation of the differences converges to zero in probability. The said quadratic variation is obviously less than a constant times

$$\sum_k \int_0^T (\bar{\pi}_s, \Lambda_s^k(\varphi_m - 1))^2\,ds,$$

which tends to zero (a.s.) by the same reasons as above. Thus, indeed the right-hand sides of (6.29) converge as $n \to \infty$ uniformly in $t \in [0, T]$ in probability to the right-hand side of (6.16). The left-hand sides converge for all $t \in [0, T]$ (a.s.) by the monotone convergence theorem. This proves (i).

Assertion (ii) easily follows from the continuity of $(\bar{\pi}_t, 1)$, the continuity of $\bar{\pi}_t$ as an L_p-valued process, and Scheffé's lemma.

In (iii) that $P_t[\beta]$ is bounded follows from the boundedness of β. The stated measurability properties of $P_t[\beta]$ are obtained by a standard measure-theoretic argument form the fact that if $f(t, x, y) = a(t)\beta(x)\gamma(y)$, where a, β, γ are smooth functions with compact support, then

$$\int_{\mathbb{R}^d} f(t, x, y_t)\bar{\pi}_t(x)\,dx = a(t)\gamma(y_t)\int_{\mathbb{R}^d}\beta(x)\bar{\pi}_t(x)\,dx$$

possesses the measurability properties in (iii) since the last factor is a continuous (a.s.) $\mathcal{F}_t^y$-adapted process.

To prove (6.17) it suffices to use (6.14) which implies that for each $t \in [0, T]$ and $y \in \mathbb{R}^{d_1-d}$

$$E(B(t, x_t, y) \mid \mathcal{F}_t^y) = (\bar{\pi}_t, 1)^{-1}\int_{\mathbb{R}^d} B(t, x, y)\bar{\pi}_t(x)\,dx \quad \text{(a.s.)}$$

and then plug in here y_t in place of y in the argument of B, which is possible because $B(t, x, y)$ is Lipschitz in y (even in (x, y)). This finishes proving assertion (iii).

In (iv) the fact that $\check{w}_t$ is $\mathcal{F}_t^y$-measurable easily follows from an equivalent formula for $\check{w}_t$:

$$\check{w}_t = \int_0^t \Psi(s, y_s)\,dy_s - \int_0^t P_s[\beta]\,ds,$$

where all terms on the right are $\mathcal{F}_t^y$-measurable. Furthermore, $\check{w}_t$ turns out to be an $\mathcal{F}_t^y$-martingale on $[0, T]$. To check this, take any $\mathcal{F}_t^y$-stopping time $\tau \leq T$ and notice that τ is also an $\mathcal{F}_t$-stopping time, so that

$$E\check{w}_\tau = E\int_0^\tau (\beta_t(x_t) - P_t[\beta])\,dt.$$

By using (6.17) and the fact that, by definition, $\{t < \tau\} \in \mathcal{F}_t^y$ we see that the right-hand side equals

$$E\int_0^T I_{t<\tau}(\beta_t(x_t) - P_t[\beta])\,dt = \int_0^T E I_{t<\tau}\beta_t(x_t)\,dt - \int_0^T E I_{t<\tau}P_t[\beta]\,dt$$

$$= \int_0^T E I_{t<\tau}\beta_t(x_t)\,dt - \int_0^T E I_{t<\tau}(E(\beta_t(x_t) \mid \mathcal{F}_t^y))\,dt = 0.$$

Thus, $E\check{w}_\tau = 0$ for any $\mathcal{F}_t^y$-stopping time $\tau \leq T$ which combined with the $\mathcal{F}_t^y$-adaptedness of $\check{w}_t$ and its continuity in t is well known to be equivalent to saying that $\check{w}_t$ is an $\mathcal{F}_t^y$-martingale on $[0, T]$. Its quadratic variation can be evaluated as the limit of sums of products of increments and is, obviously, equal to the quadratic variation of $\tilde{w}_t$, which, as we have seen in the proof of Lemma 6.3, is a Wiener process. Therefore, the quadratic variation of $\check{w}_t$ is that of a Wiener process and by Lévy's theorem $\check{w}_t$ is itself a Wiener process with respect to $\mathcal{F}_t^y$. This proves assertion (iv).

In (v) inequality (6.19) follows from (6.18), the fact that β is bounded, and the well-known properties of exponential martingales. To prove (6.18) observe that (6.16) in terms of $P_t[\beta]$ and $\check{w}_t^k$ is rewritten as

$$d(\bar{\pi}_t, 1) = (\bar{\pi}_t, \beta_t^k)\beta_t^k(x_t)\, dt + (\bar{\pi}_t, \beta_t^k)\, d\tilde{w}_t^k$$
$$= (\bar{\pi}_t, 1)\, P_t[\beta^k]\beta_t^k(x_t)\, dt + (\bar{\pi}_t, 1)\, P_t[\beta^k]\, d\tilde{w}_t^k$$
$$= (\bar{\pi}_t, 1)|P_t[\beta]|^2\, dt + (\bar{\pi}_t, 1)\, P_t[\beta^k]\, d\check{w}_t^k.$$

Hence, $(\bar{\pi}_t, 1)$ satisfies the linear equation

$$d(\bar{\pi}_t, 1) = (\bar{\pi}_t, 1)|P_t[\beta]|^2\, dt + (\bar{\pi}_t, 1)\, P_t[\beta^k]\, d\check{w}_t^k,$$

the unique solution of which with initial data $(\bar{\pi}_0, 1) = (\pi_0, 1) = 1$ is known to be given by (6.18). The theorem is proved.

6.5 Appendix

The setting in this section is somewhat different from that of Section 6.1. Let $(\Omega, \mathcal{F}, P)$ be a complete probability space with an increasing filtration $\{\mathcal{F}_t, t \geq 0\}$ of complete with respect to $(\mathcal{F}, P)$ σ-fields $\mathcal{F}_t \subset \mathcal{F}$. Denote $\mathcal{P}$ the predictable σ-field in $\Omega \times (0, \infty)$ associated with $\{\mathcal{F}_t\}$. Let w_t^k, $k = 1, 2, ...$, be independent one-dimensional Wiener processes with respect to $\{\mathcal{F}_t\}$.

We take a stopping time τ and for $t \leq \tau$ we are considering the following equation in $\mathbb{R}^d$

$$du_t = (L_t u_t - \lambda u_t + D_i f_t^i + f_t^0)\, dt + (\Lambda_t^k u_t + g_t^k)\, dw_t^k, \qquad (6.40)$$

where $u_t = u_t(x) = u_t(\omega, x)$ is an unknown function,

$$L_t \psi(x) = D_j(a_t^{ij}(x) D_i \psi(x) + a_t^j(x)\psi(x)) + b_t^i(x) D_i \psi(x) + c_t(x)\psi(x),$$
$$\Lambda_t^k \psi(x) = \sigma_t^{ik}(x) D_i \psi(x) + \nu_t^k(x)\psi(x),$$

the summation convention with respect to $i, j = 1, ..., d$ and $k = 1, 2, ...$ is enforced and detailed assumptions on the coefficients and the free terms will be given later.

Fix a number

$$p \geq 2$$

and denote $L_p = L_p(\mathbb{R}^d)$. We use the same notation L_p for vector- and matrix-valued or else ℓ_2-valued functions such as $g_t = (g_t^k)$ in (6.40). For instance, if $u(x) = (u^1(x), u^2(x), ...)$ is an ℓ_2-valued measurable function on $\mathbb{R}^d$, then

$$\|u\|_{L_p}^p = \int_{\mathbb{R}^d} |u(x)|_{\ell_2}^p \, dx = \int_{\mathbb{R}^d} \Big(\sum_{k=1}^\infty |u^k(x)|^2\Big)^{p/2} \, dx.$$

As above

$$D_i = \frac{\partial}{\partial x^i}, \quad i = 1, ..., d, \quad \Delta = D_1^2 + ... + D_d^2.$$

By Du and $D^2 u$ we mean the gradient and the matrix of second order derivatives with respect to x of a function u on $\mathbb{R}^d$.

As above, for $\gamma \in \mathbb{R}$ by $H_p^\gamma = (1 - \Delta)^{-\gamma/2} L_p$ we denote the space of Bessel potentials. Observe a slight change of notation. Since we will always be dealing with $\mathbb{R}^d$ we drop this symbol in the notation like $H_p^\gamma(\mathbb{R}^d)$. Most often in this appendix we will use H_p^γ for $\gamma = 0, 1$ and use (6.6).

If τ is a stopping time, then

$$\mathbb{H}_p^\gamma(\tau) := L_p((0, \tau], \mathcal{P}, H_p^\gamma), \quad \mathbb{L}_p(\tau) = \mathbb{H}_p^0(\tau).$$

We also need the space $\mathcal{H}_p^1(\tau)$, which is the space of functions $u_t = u_t(\omega, \cdot)$ on $\{(\omega, t) : 0 \leq t \leq \tau, t < \infty\}$ with values in the space of generalized functions on $\mathbb{R}^d$ having the following properties:

(i) For any $T \in [0, \infty)$, we have $u \in \mathbb{H}_p^1(\tau \wedge T)$ and $u_0 \in L_p(\Omega, \mathcal{F}_0, L_p)$;

(ii) There exist $f^i \in \mathbb{L}_p(\tau)$, $i = 0, ..., d$ and $g = (g^1, g^2, ...) \in \mathbb{L}_p(\tau)$ such that for any $\varphi \in C_0^\infty$ with probability 1 for all finite $t \leq \tau$ we have

$$(u_t, \varphi) = (u_0, \varphi) + \int_0^t (-(f_s^i, D_i \varphi) + (f_s^0, \varphi)) \, ds + \sum_{k=1}^\infty \int_0^t (g_s^k, \varphi) \, dw_s^k. \quad (6.41)$$

The reader can find in [6] a discussion of (i) and (ii), in particular, the fact that the series in (6.41) converges uniformly in probability on every finite subinterval of $[0, \tau)$. On the other hand, it is worth saying that the above introduced space $\mathcal{H}_p^1(\tau)$ are not quite the same as in [6]. There are three differences. One is that there is a restriction on u_0 in [6]. However the most important spaces are $\mathcal{H}_{p,0}^1(\tau)$ which are defined as the subsets of $\mathcal{H}_p^1(\tau)$ consisting of functions with $u_0 = 0$. All other elements of $\mathcal{H}_p^1(\tau)$ are obtained by adding to an element of $\mathcal{H}_{p,0}^1(\tau)$ an appropriate continuation for $t > 0$ of the initial data. Another issue is that in [6] we have $f^i = 0$, $i = 1, ..., d$, and $f^0 \in \mathbb{H}_p^{-1}(\tau)$. Actually, this difference is fictitious because one knows that any $f \in H_p^{-1}$

(a) has the form $D_i f^i + f^0$ with $f^j \in L_p$ and

$$\|f\|_{H_p^{-1}} \leq N \sum_{j=0}^d \|f^j\|_{L_p},$$

where N is independent of f, f^j, and on the other hand,

(b) for any $f \in H_p^{-1}$ there exist $f^j \in L_p$ such that $f = D_i f^i + f^0$ and
$$\sum_{j=0}^{d} \|f^j\|_{L_p} \le N \|f\|_{H_p^{-1}},$$
where N is independent of f.

The third difference is that instead of (i) we require $D^2 u \in \mathbb{H}_p^{-1}(\tau)$ in [6]. However, as it follows from Theorem 3.7 of [6] and the boundedness of the operator $D: L_p \to H_p^{-1}$, this difference disappears if τ is a bounded stopping time.

To summarize, the spaces $\mathcal{H}_{p,0}^1(\tau)$ introduced above and in [6] coincide if τ is bounded and we choose a particular representation of the deterministic part of the stochastic differential just for convenience.

In case that property (ii) holds, we write
$$du_t = (D_i f_t^i + f_t^0)\, dt + g_t^k\, dw_t^k \tag{6.42}$$
for $t \le \tau$ and this explains the sense in which equation (6.40) is understood. Of course, we still need to specify appropriate assumptions on the coefficients and the free terms in (6.40). Before we go to these assumptions we remind the reader that according to [6] and the above discussion, for bounded τ, one introduces a norm in $\mathcal{H}_{p,0}^1(\tau)$ by
$$\|u\|_{\mathcal{H}_{p,0}^1(\tau)} = E \int_0^\tau \Big(\sum_{j=1}^{d} \|D_j u_t\|_{L_p}^p + \sum_{j=0}^{d} \|f_t^j\|_{L_p}^p + \|g_t\|_{L_p}^p \Big)\, dt$$
if u satisfies (6.42). By identifying two elements of $\mathcal{H}_{p,0}^1(\tau)$ if their difference has a zero $\mathcal{H}_{p,0}^1(\tau)$-norm, one obtains a Banach space (see [6]).

We will also identify two elements $u', u'' \in \mathcal{H}_p^1(\tau)$ if and only if the difference $u' - u''$ is in $\mathcal{H}_{p,0}^1(\tau)$ and equals zero.

Assumption 6.6

(i) The coefficients a_t^{ij}, a_t^i, b_t^i, σ_t^{ik}, c_t, and ν_t^k are measurable with respect to $\mathcal{P} \times \mathcal{B}(\mathbb{R}^d)$, where $\mathcal{B}(\mathbb{R}^d)$ is the Borel σ-field on $\mathbb{R}^d$.

(ii) There is a constant K such that for all values of indices and arguments
$$|a_t^i| + |b_t^i| + |c_t| + |\nu|_{\ell_2} \le K, \quad c_t \le 0.$$

(iii) There is a constant $\delta > 0$ such that for all values of the arguments and $\xi \in \mathbb{R}^d$
$$(a_t^{ij} - \alpha_t^{ij})\xi^i \xi^j \ge \delta |\xi|^2, \quad |a_t^{ij}| \le \delta^{-1}, \tag{6.43}$$
where $\alpha_t^{ij} = (1/2)(\sigma^{i\cdot}, \sigma^{j\cdot})_{\ell_2}$. Finally, the constant $\lambda \ge 0$.

Assumption 6.6 (i) guarantees that equation (6.40) makes perfect sense for any constant λ if $u \in \mathcal{H}_p^1(\tau)$. By the way, adding the term $-\lambda u_t$ with constant $\lambda \geq 0$ is one more technically convenient step. One can always introduce this term, if originally it is absent, by considering $v_t := u_t e^{\lambda t}$.

Assumption 6.7 There is a continuous function $\kappa(\varepsilon)$ defined for $\varepsilon \geq 0$ such that $\kappa(0) = 0$ and
$$|\sigma_t^{i\cdot}(x) - \sigma_t^{i\cdot}(x)|_{\ell_2} + |a_t^{ij}(x) - a_t^{ij}(y)| \leq \kappa(|x-y|)$$
for all i, j, t, x, y.

Here are the main results used in the previous sections concerning (6.40). They are taken from [3] and [8]. Generalization of these results to the case of VMO coefficients a_t^{ij} can be found in [8].

Theorem 6.4 *Let $\lambda \geq 0$, let τ be a stopping time, let $f^j, g \in \mathbb{L}_p(\tau)$, and let ψ be a function such that $\psi \in \mathcal{H}_p^1(\tau) \cap \mathbb{H}_p^1(\tau)$. Then equation (6.40) on $[0, \tau)$ has a unique solution $u \in \mathcal{H}_p^1(\tau)$ such that $u_0 = \psi_0$.*
Write
$$d\psi_t = (D_i a_t^i + a_t^0) \, dt + \beta_t^k \, dw_t^k.$$
Then the above solution u satisfies
$$\lambda^{1/2} \|u\|_{\mathbb{L}_p(\tau)} + \|Du\|_{\mathbb{L}_p(\tau)}$$
$$\leq N \Big(\sum_{i=1}^d \|f^i\|_{\mathbb{L}_p(\tau)} + \|g\|_{\mathbb{L}_p(\tau)} + \sum_{i=1}^d \|a^i\|_{\mathbb{L}_p(\tau)} + \|\beta\|_{\mathbb{L}_p(\tau)} + \|\psi\|_{\mathbb{H}_p^1(\tau)} \Big)$$
$$+ N\lambda^{-1/2} \big(\|f^0\|_{\mathbb{L}_p(\tau)} + \|a^0\|_{\mathbb{L}_p(\tau)} + \|\psi\|_{\mathbb{H}_p^1(\tau)} \big) + N\lambda^{1/2} \|\psi\|_{\mathbb{L}_p(\tau)}, \quad (6.44)$$
provided that $\lambda > \lambda_0$, where the constants $N, \lambda_0 \geq 0$ depend only on d, p, K, δ, and the function κ.

Observe that estimate (6.44) shows a good reason for writing the free term in (6.40) in the form $D_i f^i + f^0$, because $f^i, i = 1, ..., d$, and f^0 enter (6.44) differently.

Here is a result about continuous dependence of solutions on the data.

Theorem 6.5 *Assume that for each $n = 1, 2, ...$ we are given functions a_t^{nij}, a_t^{ni}, b_t^{ni}, c_t^n, σ_t^{nik}, v_t^k, f_t^{ni}, g_t^{nk}, and ψ^n having the same meaning and satisfying the same assumptions with the same δ, K, κ as the original ones. Assume that*
$$(a_t^{nij}, a_t^{ni}, b_t^{ni}, c_t^n) \to (a_t^{ij}, a_t^i, b_t^i, c_t),$$
$$|\sigma_t^{ni\cdot} - \sigma_t^{i\cdot}|_{\ell_2} + |v_t^n - v_t|_{\ell_2} \to 0$$

as $n \to \infty$ in measure $P(d\omega)dtdx$. Also let

$$d\psi_t^n = (D_i a_t^{ni} + a_t^{n0})\, dt + \beta_t^{nk}\, dw_t^k$$

and assume that for a stopping time τ

$$\sum_{j=0}^{d}(\|f^{nj} - f^j\|_{\mathbb{L}_p(\tau)} + \|a^{nj} - a^j\|_{\mathbb{L}_p(\tau)})$$

$$+\|g^n - g\|_{\mathbb{L}_p(\tau)} + \|\beta^n - \beta\|_{\mathbb{L}_p(\tau)} + \|\psi^n - \psi\|_{\mathbb{H}_p^1(\tau)} \to 0$$

as $n \to \infty$. Take $\lambda \geq \lambda_0$, take the function u from Theorem 6.4 and let u^n be unique solutions of equations (6.40) constructed from a_t^{nij}, a_t^{ni}, b_t^{ni}, c_t^n, σ_t^{nik}, v_t^k, f_t^{ni}, and g_t^{nk} and having initial values ψ_0^n.

Then for any finite $T \geq 0$ we have

$$\|u^n - u\|_{\mathbb{H}_p^1(\tau \wedge T)} \to 0, \quad E \sup_{t \leq \tau \wedge T} \|u_t^n - u_t\|_{L_p}^p \to 0$$

as $n \to \infty$.

The following result shows that the solution does not depend on p.

Theorem 6.6 *Let $p_1, p_2 \in [2, \infty)$ and let the assumptions of Theorem 6.4 be satisfied with $p = p_1$ and $p = p_2$. Then the solutions corresponding to $p = p_1$ and $p = p_2$ coincide, that is there is a unique solution $u \in \mathcal{H}_{p_1}^1(\tau) \cap \mathcal{H}_{p_2}^1(\tau)$ of equation (6.40) with initial data ψ_0.*

In many situations the following maximum principle is useful.

Theorem 6.7 *Under the assumptions of Theorem 6.4 suppose that $\psi_0 \geq 0$, $f^i = 0$, $i = 1, \ldots, d$, $f^0 \geq 0$, $g = 0$. Then for the solution u almost surely we have $u_t \geq 0$ for all finite $t \leq \tau$.*

Finally, we used the following embedding theorem (see Corollary 4.12 and Remark 4.14 of [7]). For $\kappa \in (0, 1)$, a Banach space X, and a set $A \subset \mathbb{R}^d$ by $C^\kappa(A, X)$ we mean Hölder's space of continuous X-valued functions on A with finite norm $\|\cdot\|_{C^\kappa(A,X)}$ defined by

$$[|u|]_{C^\kappa(A,X)} = \sup_{s,t \in A} |t-s|^{-\kappa}|u(t) - u(s)|_X, \quad \|u\|_{C(A,X)} = \sup_{t \in A} |u(t)|_X,$$

$$\|u\|_{C^\kappa(A,X)} = [|u|]_{C^\kappa(A,X)} + \|u\|_{C(A,X)}.$$

Theorem 6.8 *Let $\tau \leq T$, where the constant $T \in (0, \infty)$ and let $u \in \mathcal{H}_p^1(\tau)$ satisfy (6.42) with $f^j \in \mathbb{L}_p(\tau)$, $g \in \mathbb{L}_p(\tau)$, and $u_0 \in L_p(\Omega, \mathcal{F}_0, H_p^{1-2/p})$, Then:*
(i) Almost surely u_t is a continuous function of t with values in L_p for all $t \in [0, \tau]$.

(ii) (case $p > 2$) Assume that for some numbers α and β we have

$$2/p < \alpha < \beta \leq 1.$$

Then, for any $a > 0$,

$$E[u]^p_{C^{\alpha/2-1/p}([0,\tau], H_p^{1-\beta})} \leq NT^{(\beta-\alpha)/p} a^{\beta-1} I(a), \tag{6.45}$$

$$E\|u\|^p_{C([0,\tau], H_p^{1-\beta})} \leq NE\|u_0\|^p_{H_p^{1-\beta}} + NT^{p\beta/2-1} a^{\beta-1} I(a), \tag{6.46}$$

where the constants N are independent of a, τ, T, and u and

$$I(a) := a\|u\|^p_{\mathbb{H}_p^1(\tau)} + a^{-1}\|D_i f^i + f^0\|^p_{\mathbb{H}_p^{-1}(\tau)} + \|g\|^p_{\mathbb{L}_p(\tau)}.$$

In particular, if $p(1 - \beta) > d$, then

$$E \sup_x [u(\cdot, x)]^p_{C^{\alpha/2-1/p}([0,\tau])} \leq NT^{(\beta-\alpha)/p} a^{\beta-1} I(a), \tag{6.47}$$

$$E \sup_{t \in [0,T]} \|u(t,\cdot)\|^p_{C^{1-\beta-d/p}} \leq NE\|u(0)\|^p_{H_p^{1-\beta}} + NT^{p\beta/2-1} a^{\beta-1} I(a). \tag{6.48}$$

Finally, (6.46) also holds if $p = 2$ and $\beta = 1$.

It is probably worth saying that (6.47) and (6.48) are not stated in [7]. These are just obvious consequences of (6.45) and (6.46) and the embedding theorem: $H_p^\gamma \subset C^{\gamma-d/p}$ if $\gamma - d/p > 0$ and $\gamma - d/p$ is not an integer.

Notes

This chapter was partially supported by NSF Grant DMS-0653121.

References

[1] M. Fujisaki, G. Kallianpur, and H. Kunita, *Stochastic differential equations for the non linear filtering problem*, Osaka J. Math., 9 (1972), 19–40.

[2] R. E. Kalman and R. S. Bucy, *New results in linear filtering and prediction theory*, J. Basic Eng., 83, Ser. D (1961).

[3] Kyeong-Hun Kim, *On L_p-theory of stochastic partial differential equations of divergence form in C^1 domains*, Probab. Theory Related Fields, 130 (2004), No. 4, 473–92.

[4] N. V. Krylov, "Controlled diffusion processes", Nauka, Moscow, 1977, in Russian; English trans.: Springer, 1980.

[5] N. V. Krylov. *On the equivalence of σ-algebras in the filtering problem of diffusion processes*, Teor. Verojatnost. i Primenen, 24 (1979), No. 4, 771–80, in Russian; English trans.: Theor. Probability Appl., 24 (1980), No. 4, 772–81.

[6] N. V. Krylov, *An analytic approach to SPDEs*, pp. 185–242 in Stochastic partial differential equations: Six perspectives, Mathematical Surveys and Monographs, 64, AMS, Providence, RI, 1999.

[7] N. V. Krylov, *Some properties of traces for stochastic and deterministic parabolic weighted Sobolev spaces*, Journal of Functional Analysis, 183, No. 1 (2001), 1–41.

[8] N. V. Krylov, *On divergence form SPDEs with VMO coefficients*, SIAM J. Math. Anal., 40 (2009), No. 6, 2262–85.

[9] N. V. Krylov and B. L. Rozovskiĭ, *On conditional distributions of diffusion processes*, Izvestiya Akademii Nauk SSSR, seriya matematicheskaya, 42, No. 2 (1978), 356–78, in Russian; English trans.: Math. USSR Izvestija, 12 (1978), No. 2, 336–56.

[10] N. V. Krylov and B. L. Rozovskiĭ, *On the first integrals and Liouville equations for diffusion processes*, pp. 117-125 in Stochastic differential systems, Proc. 3rd IFIP-WG 7/1 working conf., Visegrád, Hungary, Sept. 15–20, 1980; Lecture Notes in Contr. Inform. Sci., 36, 1981.

[11] N. V. Krylov and A. Zatezalo, *A direct approach to deriving filtering equations for diffusion processes*, Applied Mathematics and Optimization, 42, No. 3 (2000), 315–32.

[12] H. Kunita, *First order stochastic partial differential equations*, Stochastic analysis (Katata/Kyoto, 1982), 249–69, North-Holland Math. Library, 32, North-Holland, Amsterdam, 1984.

[13] H. Kunita, *Stochastic flows and stochastic differential equations*, Reprint of the 1990 original, Cambridge Studies in Advanced Mathematics, 24, Cambridge University Press, Cambridge, 1997.

[14] H. J. Kushner, *On the differential equations satisfied by conditional probability densities of Markov processes, with applications*, J. Soc. Indust. Appl. Math. Ser. A Control, 2 (1964), 106–19.

[15] H. J. Kushner, *Dynamical equations for optimal nonlinear filtering*, J. Differential Equations, 3 (1967), 179–90.

[16] R. Sh. Liptser and A. N. Shiryayev, *Nonlinear filtration of diffusion Markov processes* (Russian), Trudy Mat. Inst. Steklov, 104 (1968), 135–80; English trans.: Proceedings of the Steklov Institute of Mathematics, 104 (1968), 163–218.

[17] R. Sh. Liptser and A. N. Shiryayev, "Statistics of random processes", "Nauka", Moscow, 1974 in Russian; English trans.: Vols. I, II, Springer-Verlag, New York, 1977–8, 2nd edn. 2001.

[18] E. Pardoux, *Stochastic partial differential equations and filtering of diffusion processes*, Stochastics, 3 (1979), No. 2, 127–67.

[19] O. G. Purtukhia, *The innovation problem for degenerate diffusion processes (growing coefficients)*, Uspekhi Mat. Nauk, 39 (1984), No. 4 (238), 177–8 in Russian; English trans.: Russ. Math. Surv., 39 (1984), No. 4, 137–38.

[20] B. L. Rozovskiĭ, Stochastic evolution systems, Kluwer, Dordrecht, 1990.

[21] A. N. Shiryaev, *On stochastic equations in the theory of conditional Markov process*, Teor. Verojatnost. i Primenen., 11 (1966), 200–6 in Russian; English trans.: Theor. Probability Appl., 11 (1966), 179–84.

[22] R. L. Stratonovich, *Conditional Markov processes*, Teor. Verojatnost. i Primenen, 5 (1960), 172–95 in Russian; English trans.: Theor. Probability Appl., 5 (1960), 156–78.

·7·
Malliavin Calculus Applications to the Study of Nonlinear Filtering

M. Chaleyat-Maurel

The goal of this chapter is to present two different applications of the Malliavin calculus to nonlinear filtering.

The first one deals with the existence and smoothness of a density for conditional laws in filtering theory, whereas the second one is concerned with the problem of the existence or nonexistence of finding finite-dimensional filters.

These two applications are completely different in nature. In the first one, the observation is considered as fixed, and the Malliavin calculus is applied to the signal in a finite-dimensional approach. On the contrary, in the second one, the Malliavin calculus is applied to the observation through the Zakai equation, which is a stochastic partial differential equation, and the setting is thus infinite-dimensional.

Let us comment on the first application mentioned above. The main purpose of the Malliavin calculus is to establish general criteria for a random vector to possess a smooth density. Originally, the Malliavin calculus provided a probabilistic proof of Hörmander's "sum of squares theorem," which implies the existence and smoothness of a density for the probability distribution of a diffusion process. The question of the existence and regularity of a density for conditional laws arises naturally in filtering theory. In this context, the Malliavin calculus has been successfully applied to the filtering problem, in the form of a "partial" or "conditional" Malliavin calculus, which leads to a Hörmander type theorem for the filter. Of course, other approaches to the study of the existence and smoothness of the density for the filter have been developed. They are quoted in the last section.

As for our second application of Malliavin calculus to filtering theory, the main objective is to show that finite-dimensional filters do not exist, except in very special cases. The point is that the existence of a density for the law of the solution of the Zakai equation implies that all projections of this law on finite-dimensional subspaces of $\mathbb{L}^2(\mathbb{R}^n)$ also have a density with respect to Lebesgue measure. This contradicts the existence of finite-dimensional filters.

This chapter is organized as follows. In the first section, we present basic facts of the Malliavin calculus and of the relevant part of the theory of stochastic flows. In the second section, we first prove that, under a local Hörmander condition, both the filter and the unnormalized filter have a density with respect to Lebesgue measure. The remaining part of the section is then devoted to the more technical question of the smoothness of the density: Under the same hypoellipticity type hypothesis, it is proved that the density belongs to the Schwartz space of rapidly decreasing functions. In the third section, we present the application of the Malliavin calculus to the problem of finite-dimensional filters. The last section provides references and a few comments on the history of the subject.

7.1 The Malliavin calculus and the theory of stochastic flows

The Malliavin calculus, also named stochastic calculus of variations, provides a differential calculus on the Wiener space. This differential calculus is well-fitted to functionals such as solutions of stochastic differential equations. It was initiated in the 1970s by Malliavin, and further developed by Bismut, Stroock, Watanabe, and others.

In this section, we give the main results that are necessary for our study. In particular, we restrict ourself to the canonical space of continuous functions on $[0, T]$. One should however note that the Malliavin calculus is available in the general framework of abstract Wiener spaces. We follow mainly the exposition of Nualart in his book on Malliavin calculus.

7.1.1 The operators of the Malliavin calculus

Let $(\Omega, \mathcal{F}, \mathbb{P})$ be the canonical space of a d-dimensional Brownian motion indexed by a finite interval $[0, T]$ ($T > 0$). This means that $\Omega = \mathcal{C}([0, T], \mathbb{R}^d)$, $\mathbb{P}$ is the d-dimensional Wiener measure on Ω and $\mathcal{F}$ is the completion of the Borel σ-field of Ω w.r.t. $\mathbb{P}$. The canonical process on Ω is denoted by $(W_t)_{t \in [0, T]}$. We denote by $(\mathcal{F}_t)_{t \in [0, T]}$ the completion of the canonical filtration of W w.r.t. $\mathbb{P}$.

Let H^1 be the subspace of Ω that consists of all absolutely continuous functions $f : [0, T] \to \mathbb{R}^d$ with square integrable density, i.e., $f(t) = \int_0^t \dot{f}(s)ds$, $\dot{f} \in H := \mathbf{L}^2([0, T]; \mathbb{R}^d)$. We denote by $\langle .,. \rangle$ the scalar product of H.

Let $\mathcal{C}_p^\infty$ be the set of all infinitely continuously differentiable functions $f : \mathbb{R}^n \to \mathbb{R}$ such that f and all its partial derivatives have polynomial growth. Let **S** denote the class of smooth random variables (also called "simple" random variables) of the form

$$F = f(W(h_1), \ldots, W(h_n)) \tag{7.1}$$

where $f \in \mathcal{C}_p^\infty(\mathbb{R}^n)$, $h_1, \ldots, h_n \in H$, $W(h_i) = \sum_{j=1}^d \int_0^T h_i^j(s) dW_s^j$ and $n \geq 1$. Note that **S** is dense in $\mathbf{L}^2(\Omega)$. As usual, $\|F\|_2 = (\mathbb{E}[|F|^2])^{\frac{1}{2}}$ denotes the norm of $\mathbf{L}^2(\Omega)$.

The derivative

Definition 7.1 The derivative of a simple variable F of the form (7.1) is the d-dimensional stochastic process $DF = (D_t F)_{t \in [0,T]}$ with values in the set of H-valued random variables, which is given by

$$D_t F = \sum_{i=1}^n \frac{\partial f}{\partial x_i}(W(h_1), \ldots, W(h_n)) h_i(t). \tag{7.2}$$

The operator D is closable as an operator from $\mathbf{L}^p(\Omega)$ to $\mathbf{L}^p(\Omega; H)$ for any $p \geq 1$. The domain of D in $\mathbf{L}^p(\Omega)$ (denoted by $\mathbb{D}^{1,p}$) is the closure of **S** with respect to the norm

$$\|F\|_{1,p} = \left[\mathbb{E}(|F|^p) + \mathbb{E}\left(\|DF\|_H^p\right)\right]^{\frac{1}{p}}.$$

The higher order derivatives D^k ($k \geq 1$) are defined on **S** in a similar manner. They are also closable and $D^k F$ is an $H^{\otimes k}$-valued random vector. For every integer $p \geq 1$ and $k \geq 1$, we introduce the following norm on **S**

$$\|F\|_{k,p} = [\mathbb{E}(|F|^p) + \sum_{j=1}^k \mathbb{E}\left(\|D^j F\|_{H^{\otimes j}}^p\right)]^{\frac{1}{p}}.$$

The completion of **S** with respect to the norm $\|.\|_{k,p}$ is denoted by $\mathbb{D}^{k,p}$, and one defines $\mathbb{D}^\infty = \cap_{p \geq 1} \cap_{k \geq 1} \mathbb{D}^{k,p}$.

The following "chain rule" will be useful in the sequel.

Proposition 7.1 *Let $\phi \in C^1(\mathbb{R}^n, \mathbb{R})$ with bounded partial derivatives. Let $F = (F^1, \ldots, F^n)$ be a random vector whose components belong to $\mathbb{D}^{1,p}$ for some $p \geq 1$. Then, $\phi(F) \in \mathbb{D}^{1,p}$ and*

$$D\phi(F) = \sum_{i=1}^n \frac{\partial \phi}{\partial x_i}(F) DF^i. \tag{7.3}$$

The Skorohod integral and the integration by parts formula

Definition 7.2 We call "Skorohod integral" or "divergence operator" and denote by δ the adjoint of the operator D. This is an unbounded operator from $\mathbf{L}^2(\Omega; H)$ into $\mathbf{L}^2(\Omega)$ such that

1. Dom(δ) is the set of all $u \in \mathbf{L}^2(\Omega; H)$ such that there exists a positive constant c satisfying

$$\left|\mathbb{E}\left[\int_0^T \langle u_t, D_t F \rangle dt\right]\right| \leq c \|F\|_2, \quad \forall F \in \mathbf{S}.$$

2. If $u \in \text{Dom}(\delta)$, $\delta(u)$ is the element of $\mathbf{L}^2(\Omega)$ characterized by the following "integration by parts formula"

$$\mathbb{E}[F\delta(u)] = \mathbb{E}\left[\int_0^T \langle u_t, D_t F\rangle dt\right] \quad \forall F \in \mathbf{S}. \tag{7.4}$$

If u is progressively measurable, $\delta(u)$ coincides with the Itô stochastic integral. In general, $\delta(u)$ is called the Skorohod integral of u. The integration by parts formula (7.4) expresses the duality between D and δ.

7.1.2 Criteria for absolute continuity and smoothness of probability laws

Malliavin, in his first paper, used the following criteria of real analysis for the existence and smoothness of density. We denote by $C_b^\infty(\mathbb{R}^n)$ the space of all infinitely continuously differentiable functions $f : \mathbb{R}^n \to \mathbb{R}$ such that f and all its partial derivatives are bounded. The supremum norm on this space is $\|\phi\|_\infty = \sup_{x \in \mathbb{R}^n} |\phi(x)|$.

Proposition 7.2 *Let μ be a finite measure on $\mathbb{R}^n$.*

1. *Suppose that there exist positive constants C_i, $i = 1, \ldots, n$ such that for every $\phi \in C_b^\infty(\mathbb{R}^n)$*

$$\left|\int_{\mathbb{R}^n} \frac{\partial \phi}{\partial x_i} d\mu\right| \leq C_i \|\phi\|_\infty$$

    Then μ is absolutely continuous w.r.t. the Lebesgue measure.
2. *Suppose that for any multiindex $a = (a_1, \ldots, a_k) \in \{1, \ldots, n\}^k$, there exist positive constants C_a such that for every $\phi \in C_b^\infty(\mathbb{R}^n)$*

$$\left|\int_{\mathbb{R}^n} \frac{\partial^k \phi}{\partial x_{a_1} \ldots \partial x_{a_k}} d\mu\right| \leq C_a \|\phi\|_\infty$$

    Then μ is absolutely continuous w.r.t. the Lebesgue measure and its density is C^∞.

Definition 7.3 Suppose $F = (F^1, \ldots, F^n)$ is a random vector with $F^i \in \mathbb{D}^{1,2}$ for every $i = 1, \ldots, n$. The Malliavin matrix of F is the following random symmetric nonnegative definite matrix γ_F

$$\gamma_F = (\langle DF^i, DF^j\rangle)_{1 \leq i,j \leq n}.$$

From Proposition 7.2 and the integration by parts formula, we get the following probabilistic criteria.

Proposition 7.3 *Suppose $F = (F^1, \ldots, F^n)$ is a random vector satisfying the following conditions*

1. $F^i \in \mathbb{D}^{1,2}$, for every $i = 1, \ldots, n$
2. The Malliavin matrix γ_F is invertible a.s.

Then the law of F has a density with respect to the Lebesgue measure on $\mathbb{R}^n$.

Proposition 7.4 *Suppose $F = (F^1, \ldots, F^n)$ is a random vector satisfying the following conditions*

1. $F^i \in \mathbb{D}^\infty$, for every $i = 1, \ldots, n$
2. The Malliavin matrix γ_F is invertible a.s. and

$$\det\left(\gamma_F^{-1}\right) \in \cap_{p \geq 1} \mathbf{L}^p(\Omega)$$

Then the law of F has an infinitely differentiable density with respect to the Lebesgue measure on $\mathbb{R}^n$.

7.1.3 Stochastic differential equations and stochastic flows

The proof of the existence and smoothness of the density of the conditional laws of filtering theory given here is due to Bismut and Michel, who used the theory of stochastic flows. The definition and main properties of stochastic flows are summarized below.

Let $A^0, A^1, \ldots, A^d$ be $d + 1$ smooth vector fields on $\mathbb{R}^n$ with coefficients depending on t and x. We suppose that these vector fields are bounded as well as their derivatives in the variable x.

Let $x \in \mathbb{R}^n$. We denote by $X^x = (X_t^x)_{t \in [0,T]}$ the unique solution of the inhomogeneous stochastic differential equation written in Stratonovich form

$$X_t^x = x + \int_0^t A^0(s, X_s^x)\, ds + \sum_{i=1}^d \int_0^t A^i(s, X_s^x) \circ dW_s^i. \tag{7.5}$$

When the vector fields do not depend on the variable t, the process $(X_t^x)_{t \in [0,T]}$ is a homogeneous diffusion which solves the stochastic differential equation

$$X_t^x = x + \int_0^t A^0(X_s^x)\, ds + \sum_{i=1}^d \int_0^t A^i(X_s^x) \circ dW_s^i. \tag{7.6}$$

Let us recall the relation between stochastic integrals in Itô form and in Stratonovich form. If A, B are vector fields on $\mathbb{R}^n$ such that:

$$A = \sum_{j=1}^n A_j \frac{\partial}{\partial x_j}, \qquad B = \sum_{j=1}^n B_j \frac{\partial}{\partial x_j}$$

the covariant derivative of B in the direction of A is defined as the vector field

$$A^\nabla B = \sum_{i=1}^{n} \sum_{k=1}^{n} A_i \frac{\partial B_k}{\partial x_i} \frac{\partial}{\partial x_k}$$

and the Lie bracket between the vector fields A and B is defined by

$$[A, B] = A^\nabla B - B^\nabla A.$$

In the case of a stochastic differential equation, the Stratonovich integral is expressed in terms of the Itô integral by the formula

$$\sum_{i=1}^{d} \int_0^t A^i(X_s^x) \circ dW_s^i = \sum_{i=1}^{d} \int_0^t A^i(X_s^x) dW_s^i + \frac{1}{2} \sum_{i=1}^{d} \int_0^t A^{i\,\nabla} A^i(X_s^x) ds.$$

Definition 7.4 In both inhomogeneous and homogeneous cases, the flow associated with the equations (7.5) and (7.6), denoted by $\Psi = (\Psi_t)_{t \in [0,T]}$, is the unique application from $[0, T] \times \Omega \times \mathbb{R}^n$ into $\mathbb{R}^n$ defined by

$$\Psi_t(\omega, x) = X_t^x(\omega), \quad a.s.$$

moreover, it verifies the following properties

1. For every $x \in \mathbb{R}^n$, $(t, \omega) \to \Psi_t(\omega, x)$ is progressively measurable.
2. $(\Psi_t(\omega, .))_{t \in [0,T]}$ is a family of C^∞-diffeomorphisms from $\mathbb{R}^n$ into $\mathbb{R}^n$, a.s.
3. For every compact subset K of $\mathbb{R}^n$, every $q \geq 1$, and every $N \in \mathbb{N}$

$$\sup_{t \in [0,T], \xi \in K} (\|\Psi_t(X_s^x)\|, \|\Psi_t^{'-1}(X_s^x)\|, \|\Psi_t^{(k)}(X_s^x)\|, k \leq N)$$

belongs to $\mathbf{L}^q(\mathcal{C}([0, T], \mathbb{R}^d))$.

4. Moreover, for any $q \geq 1$, there exist constants (depending on q, T, n) such that

$$\mathbb{E}[\sup_{s \leq t} \|X_s^x - x\|^{2q}] \leq Ct^q \qquad (7.7)$$

for every $t \in [0, T]$.

In the sequel, the dependence of X on x will be omitted. We have the following property.

Proposition 7.5 *Let X be a solution of (7.5). Then $X_t^i \in \mathbb{D}^\infty$ for every $t \in [0, T]$ and for every $i = 1, \ldots n$*

1. *The Malliavin derivative of X_t is calculated as*

$$D_s^i X_t = \Psi_t' \Psi_s^{'-1} A^i(s, X_s) \qquad (7.8)$$

2. The associated Malliavin matrix is

$$Q_t = (\langle DX_t^i, DX_t^j \rangle)_{1 \le i,j \le n} = \sum_{i=0}^{n} \int_0^t \Psi_t' {\Psi_s'}^{-1} A^i(s, X_s) (\Psi_t' {\Psi_s'}^{-1} A^i(s, X_s))^T ds. \tag{7.9}$$

Remark. The invertibility of Q_t reduces to that of the matrix

$$M_t = \sum_{i=0}^{n} \int_0^t {\Psi_s'}^{-1} A^i(s, X_s) ({\Psi_s'}^{-1} A^i(s, X_s))^T ds. \tag{7.10}$$

Itô–Ventzell formula

The so-called Itô–Ventzell formula shows that the image of a continuous semimartingale by the flow of diffeomorphisms associated with a homogeneous stochastic differential equation is also a semimartingale and gives its decomposition.

In this paragraph, we take $d = n$. Let $X = (X_t)_{t \in [0,T]}$ be the unique solution of (7.6) and let Ψ_t be the associated flow.

Theorem 7.1 *Let Z_t be a continuous semimartingale with values in $\mathbb{R}^n$ such that $Z_t = Z_0 + V_t + \sum_{i=1}^{n} \int_0^t H_s^i d W_s^i$ with $Z_0 \in \mathbb{R}^n$, $V = (V_t)_{t \in [0,T]}$ is a continuous adapted bounded variation process such that $V_0 = 0$ and $H = \left(H_s^1, \ldots, H_s^n \right)_{t \in [0,T]}$ is an adapted process such that $\int_0^T |H_s|^2 ds < +\infty$, a.s.*

Then, $\Psi_t(Z_t)$ is a continuous semimartingale whose Itô decomposition is

$$\Psi_t(Z_t) = Z_0 + \int_0^t [(A^0 + \frac{1}{2} \sum_{i=1}^{n} A^{i\nabla} A^i)(\Psi_s(Z_s)).$$

$$+ \sum_{i=1}^{n} (\frac{\partial A^i}{\partial x}(\Psi_s(Z_s))) \frac{\partial \Psi_s}{\partial x}(Z_s) H_s^i + \frac{1}{2} \frac{\partial^2 \Psi_s}{\partial x^2}(Z_s)(H_s^i, H_s^i))]ds$$

$$+ \int_0^t [\frac{\partial \Psi_s}{\partial x}(Z_s) d V_s + \sum_{i=1}^{n} (A^i(\Psi_s(Z_s)) + \frac{\partial \Psi_s}{\partial x}(Z_s) H_s^i) d W_s^i]. \tag{7.11}$$

As usual, this formula becomes simpler with the Stratonovich calculus.

Corollary 7.1 *The formula of Theorem 7.1 can be rewritten as*

$$\Psi_t(Z_t) = Z_0 + \int_0^t A^0(\Psi_s(Z_s)) ds$$

$$+ \int_0^t [\frac{\partial \Psi_s}{\partial x}(Z_s) \circ d Z_s + \sum_{i=1}^{n} A^i(\Psi_s(Z_s)) \circ d W_s^i]. \tag{7.12}$$

Action of the reciprocal flow on stochastic differential equations

One important tool in the proof of the regularity of the filter is the formula giving the action of the reciprocal flow on certain stochastic differential equations. This formula, which follows from the preceding results, provides a factorization of the flow and makes it possible to apply the Malliavin calculus when the observation is fixed. We state this formula in a form suitable for filtering theory.

Suppose that the probability space is the product $\Omega = \mathcal{C}([0, T], \mathbb{R}^n) \times \mathcal{C}([0, T], \mathbb{R}^p)$ endowed with the canonical filtration and the product Wiener measure. We denote by $(W, \tilde{W})$ the corresponding $n + p$ dimensional Wiener process.

Let $A^0, A^1, \ldots, A^n, \tilde{A}^1, \ldots, \tilde{A}^p$ be $n + p + 1$ smooth vector fields on $\mathbb{R}^n \times \mathbb{R}^p$ bounded as well as their derivatives and $\xi_0 \in \mathbb{R}^n$. We suppose that the A^i's have no component on $\mathbb{R}^p$ and that the $\tilde{A}^i$'s have their last p components equal to $\delta_{ij} \frac{\partial}{\partial x_j}$. We denote by $\xi = (\xi_t)_{t \in [0, T]}$ the unique solution of the stochastic differential equation

$$\xi_t = \xi_0 + \int_0^t A^0(\xi_s) ds + \sum_{i=1}^n \int_0^t A^i(\xi_s) \circ dW_s^i + \sum_{i=1}^p \int_0^t \tilde{A}^i(\xi_s) \circ d\tilde{W}_s^i. \quad (7.13)$$

Let us denote by Φ the flow of ξ.

Denote by $\bar{\xi} = (\bar{\xi}_t)_{t \in [0, T]}$ the unique solution of the following stochastic differential equation

$$\bar{\xi}_t = \xi_0 + \sum_{i=1}^p \int_0^t \tilde{A}^i(\bar{\xi}_s) \circ d\tilde{W}_s^i. \quad (7.14)$$

The flow of $\bar{\xi}$ is denoted by $\bar{\Phi}$. Then

Proposition 7.6 *Define $\tilde{\xi}_t = \bar{\Phi}_t^{-1}(\xi_t)$. Then, $\tilde{\xi}_t$ is the unique solution of the stochastic differential equation*

$$\tilde{\xi}_t = \xi_0 + \int_0^t \bar{\Phi}_t'^{-1}(\tilde{\xi}_s) A^0(\bar{\Phi}_s(\tilde{\xi}_s)) ds + \sum_{i=1}^n \int_0^t \bar{\Phi}_t'^{-1}(\tilde{\xi}_s) A^i(\bar{\Phi}_s(\tilde{\xi}_s)) \circ dW_s^i. \quad (7.15)$$

Itô's formula for the vector fields

This formula is fundamental in the proof of the existence of a density for the conditional law, because it provides a kind of Taylor expansion involving recursively all the Lie brackets arising in the hypoellipticity hypothesis.

Proposition 7.7 *Let X be the solution of (7.6) and let Ψ be its flow. Then, for every $i = 1, \ldots n$*

$$\Psi_t^{'-1} A^i(X_t) = A^i(x_0) + \int_0^t \Psi_s^{'-1}[A^0, A^i](X_s)ds$$

$$+ \sum_{j=1}^n \int_0^t \Psi_s^{'-1}[A^j, A^i](X_s) \circ dW_s^j \qquad (7.16)$$

and, in Itô form

$$\Psi_t^{'-1} A^i(X_t) = A^i(x_0) + \int_0^t \Psi_s^{'-1}[A^0, A^i](X_s)ds$$

$$+ \sum_{j=1}^n \int_0^t \Psi_s^{'-1}[A^j, A^i](X_s)dW_s^j$$

$$+ \frac{1}{2} \sum_{j=1}^n \int_0^t \Psi_s^{'-1}[A^j[A^j, A^i]](X_s)ds. \qquad (7.17)$$

7.2 Existence and smoothness of a conditional density for the nonlinear filtering

7.2.1 The filtering model

This section describes the model studied in this chapter.

Suppose that $(\Omega, \mathcal{F}, \mathbb{P})$ is the canonical space associated with a $(n+p)$-dimensional Brownian motion $(W_t, V_t)_{t \in [0,T]}$ on a finite interval $[0, T]$ $(T > 0)$. This means that $\Omega = \mathcal{C}([0, T], \mathbb{R}^n) \times \mathcal{C}([0, T], \mathbb{R}^p)$, $\mathbb{P}$ is the $n+p$-dimensional Wiener measure on Ω and $\mathcal{F}$ is the completion of the Borel σ-field of Ω with respect to $\mathbb{P}$. We denote by $(\mathcal{F}_t)_{t \in [0,T]}$ the completion of the canonical filtration of the $(n+p)$-dimensional Brownian motion (W, V) with respect to $\mathbb{P}$.

The pair signal–observation (X, Y) satisfies:

$$\begin{cases} X_t = x_0 + \int_0^t b(X_s, Y_s)ds + \int_0^t f(X_s, Y_s) \circ dW_s + \int_0^t g(X_s, Y_s) \circ dY_s \\ Y_t = y_0 + \int_0^t h(X_s, Y_s)ds + V_t \end{cases}$$

where

- b, f, g, h are smooth functions from $\mathbb{R}^n \times \mathbb{R}^p$ into $\mathbb{R}^n$, $\mathbb{R}^n \times \mathbb{R}^n$, $\mathbb{R}^n \times \mathbb{R}^p$, $\mathbb{R}^p$ respectively. All these functions are assumed to be bounded as well as their derivatives.

- (x_0, y_0) is a $n+p$-dimensional vector. We may assume that $y_0 = 0$ with no loss of generality (except when we consider the stochastic flows) and we will do so in this section.

- We use the Stratonovich integral which is more fitted to intrinsic geometric calculus. Moreover, we have chosen to write the noise terms (stochastic integrals) appearing in the equation for the signal in terms of W and Y, rather than in terms of W and V, in order to simplify the Girsanov transformation to be introduced in the next subsection.

This model is a general Markovian correlated homogeneous model.

We associate to the previous system the $n + p$-dimensional diffusion process $\xi_t = (X_t, Y_t)$, which solves the equation

$$\xi_t = \xi_0 + \int_0^t A^0(\xi_s)ds + \sum_{i=1}^n \int_0^t A^i(\xi_s) \circ dW_s^i + \sum_{k=1}^p \int_0^t \bar{A}^k(\xi_s) \circ dY_s^k \qquad (7.18)$$

where $\xi_0 = (x_0, y_0)$, $A^0 = (b, 0)$, $A^i = (f_i, 0)$, $\bar{A}^k = (g_k, \mathbf{1}_p)$.

The following notation is used: if M is an $n \times d$ matrix, A^i is the i-th column of the matrix considered as a vector of $\mathbb{R}^n$, for $i = 1, \ldots, d$.

We denote by $(\mathcal{Y}_t)_{t \in [0,T]}$ the completion of the canonical filtration of Y and we set $\mathcal{Y} = \bigvee_{t \in [0,T]} \mathcal{Y}_t$.

The main concern of filtering theory is to study the conditional law of X_t, knowing the past observations until time t (that is $(\mathcal{Y}_t)_{t \in [0,T]}$). We thus introduce the collection of random measures π_t defined by

$$\pi_t \phi = \mathbb{E}[\phi(X_t)|\mathcal{Y}_t]$$

where ϕ is any bounded measurable function on $\mathbb{R}^n$.

7.2.2 The existence of a conditional density

It is well known that the problem of the existence of a density for the conditional law becomes easier under the reference probability measure $\mathring{\mathbb{P}}$ defined by

$$\left.\frac{d\mathring{\mathbb{P}}}{d\mathbb{P}}\right|_{\mathcal{F}_t} = Z_t^{-1}$$

where

$$Z_t = \exp(\sum_{k=1}^p \int_0^t h^k(X_s, Y_s)dY_s^k - \frac{1}{2}\int_0^t |h(X_s, Y_s)|^2 ds)$$

Recall that the unnormalized filter ρ_t is defined by

$$\rho_t \phi = \mathring{\mathbb{E}}[\phi(X_t)Z_t|\mathcal{Y}_t] = \mathring{\mathbb{E}}[\phi(X_t)Z_t|\mathcal{Y}]$$

where ϕ is any bounded measurable function on $\mathbb{R}^n$. The last equality comes from the fact that Y is a $\mathring{\mathbb{P}}$-Wiener process. Moreover since Y and W are independent Wiener processes under $\mathring{\mathbb{P}}$, there is no loss of generality in assuming that (Y, W) is the coordinate process on the canonical Wiener space

$\mathcal{C}([0, T], \mathbb{R}^n) \times \mathcal{C}([0, T], \mathbb{R}^p)$. Denoting the expectation with respect to the law of W under $\mathring{\mathbb{P}}$ by $\mathring{\mathbb{E}}^y$, we write ρ_t as

$$\rho_t \phi = \mathring{\mathbb{E}}^y[\phi(X_t) Z_t]. \tag{7.19}$$

We deduce from the Kallianpur–Striebel formula ($\pi_t \phi = \frac{\rho_t \phi}{\rho_t 1}$) that the problem of the existence of a density is equivalent for π_t and for ρ_t.

For every y in $\mathcal{C}([0, T], \mathbb{R}^p)$, denote by $\bar{Z}_t(y, .)$ the measurable function on $\mathbb{R}^n$ defined by

$$\mathring{\mathbb{E}}^y[Z_t | X_t] = \bar{Z}_t(y, X_t). \tag{7.20}$$

It is straightforward that

$$\rho_t \phi = \mathring{\mathbb{E}}^y[\phi \bar{Z}_t(y, .)(X_t)]. \tag{7.21}$$

for any bounded measurable function ϕ on $\mathbb{R}^n$.

The idea of the proof of the existence of the conditional density is to write X_t as a function (depending on t and y) of a diffusion driven by the Wiener process W and parametrized by y (see Proposition 7.9 below). Then, for fixed y, one can use the Malliavin calculus in $\mathbb{R}^n$ to prove that this inhomogeneous diffusion has a density.

A technical problem arises because the coefficients of the new diffusion and their derivatives are no longer bounded. Everything works well if the support of each map $x \to g(x, y)$, for $y \in \mathbb{R}^p$, is contains in a compact subset of $\mathbb{R}^n$ that does not depend on y. So, we need to approximate $(X_t)_{t \in [0, T]}$ by a sequence of diffusion processes having this property.

Let $(\psi_m)_{m \in \mathbb{N}}$ be a sequence of smooth real functions defined on $\mathbb{R}^n$ such that $\psi_m = 0$ on $\{x \in \mathbb{R}^n, \|x\| > m + 1\}$ and $\psi_m = 1$ on $\{x \in \mathbb{R}^n, \|x\| \leq m\}$.

For every $m \in \mathbb{N}$, define $g_m : \mathbb{R}^n \times \mathbb{R}^p \to \mathbb{R}^n \times \mathbb{R}^p$ by $g_m(x, y) = \psi_m(x) g(x, y)$. Now, it is clear that $g_m(x, y) = 0$ if $(x, y) = \{x \in \mathbb{R}^n, \|x\| > m + 1\} \times \mathbb{R}^p$.

Let X_t^m the solution at time t of the stochastic differential equation

$$X_t^m = x_0 + \int_0^t b(X_s^m, Y_s) \, ds + \int_0^t f(X_s^m, Y_s) \circ dW_s + \int_0^t g_m(X_s^m, Y_s) \circ dY_s$$

Then $(X_s)_{0 \leq s \leq t}$ and $(X_s^m)_{0 \leq s \leq t}$ coincide on the set

$$\{(w, y) \in \mathcal{C}([0, T], \mathbb{R}^n) \times \mathcal{C}([0, T], \mathbb{R}^p) : \sup_{0 \leq s \leq t} \|X_s(w, y)\| \leq m\}$$

whose probability tends to 1 as m goes to infinity. This implies that, for any bounded measurable function ϕ on $\mathbb{R}^n$

$$\rho_t \phi = \lim_{m \to +\infty} \mathring{\mathbb{E}}^y[\phi \bar{Z}_t(y, .)(X_t^m) \mathbf{1}_{\sup_{0 \leq s \leq t} \|X_s\| \leq m}]$$

where the convergence holds in $\mathbf{L}^1(\mathcal{C}([0, T], \mathbb{R}^p))$.

Then, if we prove that for a.e. $y \in \mathcal{C}([0, T], \mathbb{R}^p)$, the law of X_t^m under $\mathring{\mathbb{P}}^y$ has a density for every $m \in \mathbb{N}$, the same conclusion will hold for X_t.

We deduce from the previous discussion that with no loss of generality, we may assume that there exists a compact set of $\mathbb{R}^n$ containing the support of each function $x \to g(x, y)$, for $y \in \mathbb{R}^p$. We make this assumption in the remaining part of this section.

A factorization of X_t via stochastic flows

Consider first the case $g = 0$. Because Y and W are independent under $\mathbb{P}$, the conditional law of X given $Y = u \in \mathcal{C}([0, T], \mathbb{R}^p)$ is the law of the diffusion

$$X_t^u = x_0 + \int_0^t b\left(X_s^u, u_s\right) ds + \int_0^t f\left(X_s^u, u_s\right) \circ dW_s.$$

So, the problem reduces to studying the existence of a density for the law of a inhomogeneous diffusion.

Let us explain how a similar reduction can be done in the case when $g \neq 0$.

Denote by Φ_t, (resp. $\bar{\Phi}_t$, Ξ_t) the stochastic flow on $\mathbb{R}^n \times \mathbb{R}^p$ (resp. $\mathbb{R}^n \times \mathbb{R}^p$ and $\mathbb{R}^n$) associated with the stochastic differential equation (7.18) (resp. with

$$\bar{\xi}_t = \xi_0 + \sum_{k=1}^p \int_0^t \bar{A}^k(\bar{\xi}_s) \circ dY_s^k,$$

and with

$$\bar{X}_t = x_0 + \sum_{k=1}^p \int_0^t g_k(\bar{X}_s, Y_s) \circ dY_s^k).$$

Remark. Φ_t and $\bar{\Phi}_t$ are homogeneous stochastic flows, whereas Ξ_t is not. However we have the relation

$$\bar{\Phi}_t(x, 0) = (\Xi_t(x), Y_t).$$

In the following proposition we recall the main properties of these flows.

Proposition 7.8

The preceding flows satisfy the following properties

1. a.s on $\mathcal{C}([0, T], \mathbb{R}^n) \times \mathcal{C}([0, T], \mathbb{R}^p)$ (resp. $\mathcal{C}([0, T], \mathbb{R}^p), \mathcal{C}([0, T], \mathbb{R}^p)$), for every $t \in [0, T]$, the map Φ_t, (resp. $\bar{\Phi}_t$, Ξ_t) is a $\mathbf{C}^\infty$ diffeomorphism of $\mathbb{R}^n \times \mathbb{R}^p$ (resp. $\mathbb{R}^n \times \mathbb{R}^p$, $\mathbb{R}^n$).
2. For every compact subset K of $\mathbb{R}^n \times \mathbb{R}^p$, every $q \geq 1$, and every $N \in \mathbb{N}$

$$\sup_{t \in [0, T], \xi \in K} (\|\Phi_t(\xi)\|, \|\Phi_t^{'-1}(\xi)\|, \|\Phi_t^{(k)}(\xi)\|, k \leq N)$$

belongs to $\mathbf{L}^q(\mathcal{C}([0, T], \mathbb{R}^n) \times \mathcal{C}([0, T], \mathbb{R}^p))$. Similarly,

$$\sup_{t\in[0,T],\xi\in\mathbb{R}^n\times\mathbb{R}^p} (\|\bar{\Phi}_t(\xi)\|, \|\bar{\Phi}_t^{'-1}(\xi)\|, \|\bar{\Phi}_t^{(k)}(\xi)\|, k \leq N)$$

belongs to $\mathbf{L}^q(\mathcal{C}([0, T], \mathbb{R}^p))$, and

$$\sup_{t\in[0,T],x\in\mathbb{R}^n} (\|\Xi_t(x)\|, \|\Xi_t^{'-1}(x)\|, \|\Xi_t^{(k)}(x)\|, k \leq N)$$

belongs to $\mathbf{L}^q(\mathcal{C}([0, T], \mathbb{R}^p))$.

We now use the flow Ξ and the Itô–Ventzell formula to "factorize" X_t.

Proposition 7.9 *Define $\tilde{X}_t = \Xi_t^{-1}(X_t)$. Then, $\tilde{X}_t$ is the unique solution of the stochastic differential equation*

$$\tilde{X}_t = x_0 + \int_0^t \Xi_t^{'-1}(\tilde{X}_s) b(\Xi_t(\tilde{X}_s), Y_s) ds + \int_0^t \Xi_t^{'-1}(\tilde{X}_s) f(\Xi_t(\tilde{X}_s), Y_s) \circ dW_s.$$
(7.22)

Proof. Formula (7.15) implies that $\tilde{X}_t$ is solution of (7.22). The uniqueness follows from the uniqueness of the equation satisfied by X_t. $\square$

Remark. By Proposition 7.8, for a.e. $y \in \mathcal{C}([0, T], \mathbb{R}^p)$, the coefficients of (7.22) have bounded derivatives, the bounds being uniform with respect to t. This implies that, when y is fixed, equation (7.22) has a unique solution on $\mathcal{C}([0, T], \mathbb{R}^n)$. This solution X coincides with the first component of the solution of equation (7.18) which has also a unique solution on $\mathcal{C}([0, T], \mathbb{R}^n) \times \mathcal{C}([0, T], \mathbb{R}^p)$. Consequently, we may discard a negligible set $\mathcal{N} \subset \mathcal{C}([0, T], \mathbb{R}^p)$ of values of y, and apply the Malliavin calculus to equation (7.22) where Y is replaced by $y \in \mathcal{N}^c$.

It is easier to argue with flows. For this purpose, define $\tilde{\xi}_t = \bar{\Phi}_t^{-1}(X_t)$. This implies that $\tilde{\xi}_t = (\tilde{X}_t, 0)$, and we have the following corollary.

Corollary 7.2 *$\tilde{\xi}_t$ satisfies the equation*

$$\tilde{\xi}_t = (x_0, 0) + \int_0^t \bar{\Phi}_t^{'-1}(\tilde{\xi}_s) A^0(\bar{\Phi}_s(\tilde{\xi}_s)) ds + \sum_{i=0}^n \int_0^t \bar{\Phi}_t^{'-1}(\tilde{\xi}_s) A^i(\bar{\Phi}_s(\tilde{\xi}_s)) \circ dW_s^i$$

From now on, we deal with the homogeneous flow $\bar{\Phi}$ rather than with Ξ, and with the vector fields A^i, rather than with the functions b and f, because the assumptions are more easily written in terms of $\bar{\Phi}$ and the vector fields A^i. So, we will often identify $\tilde{\xi}_t$ with its first component $\tilde{X}_t$ and in the same manner $\bar{\Phi}_t^{'-1}(\tilde{\xi}_s) A^i(\bar{\Phi}_s(\tilde{\xi}_s))$ with its component in $\mathbb{R}^n$ $\left(\Xi_t^{'-1}(\tilde{X}_s) f_i(\Xi_t(\tilde{X}_s), Y_s)\right)$, the component in $\mathbb{R}^p$ being 0.

The existence of a conditional density

We use the relation $\tilde{\xi}_t = (\tilde{X}_t, 0)$: This implies that the unnormalized filter is expressed as

$$\rho_t \phi = \mathring{\mathbb{E}}^y[\phi \bar{Z}_t(y, .) \circ \bar{\Phi}_t(\tilde{X}_t, 0)] \tag{7.23}$$

where ϕ is any bounded measurable function on $\mathbb{R}^n$.

This formula and the preceding remark imply that, in order to prove the existence of a conditional density for $y \notin \mathcal{N}$, it is enough to prove that the law of the inhomogeneous diffusion $\tilde{X}_t$ driven by W has a density.

We calculate the Malliavin matrix associated to this inhomogenous diffusions (see (7.10)).

Proposition 7.10 *Let $y \notin \mathcal{N}$; the law of $\tilde{X}_t(., y)$ has a density w.r.t. the Lebesgue measure on $\mathbb{R}^n$ if the associated Malliavin matrix M_t^y is invertible for a.e. $w \in \mathcal{C}([0, T], \mathbb{R}^n)$ where M_t^y is given by the formula*

$$M_t^y = \sum_{i=0}^{n} \int_0^t \Phi_s'^{-1} A^i(X_s, Y_s) \left(\Phi_s'^{-1} A^i(X_s, Y_s)\right)^T ds.$$

Proof. We check that $\tilde{X}_t$ is in $\mathbb{D}^{1,2}$. We fix $y \notin \mathcal{N}$ and denote by $\tilde{\Phi}_t$ the stochastic flow associated to $\tilde{X}_t$. The Malliavin matrix associated to $\tilde{X}_t$ is given by

$$M_t^y = \sum_{i=0}^{n} \int_0^t \tilde{\Phi}_s'^{-1} \bar{\Phi}_s'^{-1}(\tilde{X}_s, 0) A^i(\bar{\Phi}_s(\tilde{X}_s, 0))$$

$$\left(\tilde{\Phi}_s'^{-1} \bar{\Phi}_s'^{-1}(\tilde{X}_s, 0) A^i(\bar{\Phi}_s(\tilde{X}_s, 0))\right)^T ds \tag{7.24}$$

Clearly, $\tilde{\xi}_s = \bar{\Phi}_s^{-1}(x_0, 0)$. It thus follows from the relation between $\tilde{\xi}_s$ and $\tilde{X}_s$ that $\Phi_s^{-1} = \tilde{\Phi}_s^{-1} \bar{\Phi}_s^{-1}(\tilde{X}_s, 0)$. We differentiate and obtain

$$\tilde{\Phi}_s'^{-1} \bar{\Phi}_s'^{-1}(\tilde{X}_s, 0) = \Phi_s'^{-1}(x_0, 0)$$

We have also $\bar{\Phi}_s(\tilde{X}_s, 0) = (X_s, Y_s)$. The desired formula is proved. □

Now, we give a geometrical condition on the vector fields $A^0, A^1, \ldots, A^n$, $\bar{A}^1, \ldots, \bar{A}^p$ which ensures the invertibility of the Malliavin matrix.

Definition 7.5 Denote by $\mathcal{I}(A^0, \bar{A}^1, \ldots, \bar{A}^p; A^1, \ldots, A^n)$ the ideal generated by the Lie algebra $\mathcal{L}ie(A^1, \ldots, A^n)$ in the Lie algebra $\mathcal{L}ie(A^0, A^1, \ldots, A^n, \bar{A}^1, \ldots, \bar{A}^p)$; if $(x, y) \in \mathbb{R}^n \times \mathbb{R}^p$, we define $\mathcal{I}(A^0, \bar{A}^1, \ldots, \bar{A}^p; A^1, \ldots, A^n)(x, y) = \{A(x, y), A \in \mathcal{I}(A^0, \bar{A}^1, \ldots, \bar{A}^p; A^1, \ldots, A^n)\}$.

Remark. The $\mathbb{R}^p$-component of any element of $\mathcal{L}ie(A^1, \ldots, A^n)$ is 0. To see this, let A be such an element. For $i = 1, \ldots, p$,

$$[A, \bar{A}^i] = [A, \sum_{j=1}^n g_{ji} \frac{\partial}{\partial x_j}] + [A, \sum_{k=1}^p \frac{\partial}{\partial y_k}].$$

The first term has no $\mathbb{R}^p$-component, and the same is true for the second one because there is no derivation in x.

The preceding remark implies that the $\mathbb{R}^p$-component of any element of $\mathcal{I}(A^0, \bar{A}^1, \ldots, \bar{A}^p; A^1, \ldots, A^n)$ is 0. It follows that for $(x, y) \in \mathbb{R}^n \times \mathbb{R}^p$, $\mathcal{I}(A^0, \bar{A}^1, \ldots, \bar{A}^p; A^1, \ldots, A^n)(x, y)$ is a real vector space whose dimension is less than n.

In the following, we identify any element of this ideal with its component on $\mathbb{R}^n$.

Theorem 7.2 *If* $\dim \mathcal{I}(A^0, \bar{A}^1, \ldots, \bar{A}^p; A^1, \ldots, A^n)(x_0, 0) = n$, *then, there exists a set* $\mathcal{N}' \subset \mathcal{C}([0, T], \mathbb{R}^p)$ *of measure 0 such that, for* $y \notin \mathcal{N}'$, $t \in]0, T]$, M_t^y *is invertible for a.e.* $w \in \mathcal{C}([0, T], \mathbb{R}^n)$.

Proof. This is similar to the proof of the analogous result for diffusion processes. Define, for $s \in [0, T]$, $U_s = \text{span}\{\Phi_s'^{-1} A^i(X_s, Y_s), 1 \le i \le n\}$, $V_t = \cup_{0 \le s \le t} U_s$ and $V_t^+ = \cap_{s>t} V_s$. □

By the Blumenthal zero–one law, V_0^+ is not random. Assume that V_0^+ is different from $\mathbb{R}^n$ and take $q \ne 0$, $q \in (V_0^+)^\perp$.

Define $\tau = \inf\{s > 0, V_s \ne V_0^+\}$. Then, $\tau > 0$ a.s. and for $t < \tau$, it holds

$$\langle q, \Phi_t'^{-1} A^i(X_t, Y_t)\rangle = 0 \quad \forall i = 1, \ldots, n.$$

By Proposition (7.7), we get

$$\langle q, \Phi_t'^{-1} A^i(X_t, Y_t)\rangle = \langle q, A^i(x_0, 0)\rangle + \int_0^t \langle q, \Phi_s'^{-1}[A^0, A^i](X_s, Y_s)\rangle ds$$
$$+ \sum_{j=1}^n \int_0^t \langle q, \Phi_s'^{-1}[A^j, A^i](X_s, Y_s)\rangle \circ dW_s^j$$
$$+ \sum_{k=1}^p \int_0^t \langle q, \Phi_s'^{-1}[\bar{A}^k, A^i](X_s, Y_s)\rangle \circ dY_s^k. \quad (7.25)$$

For $t < \tau$, the left-hand side of this equality is 0; the same is true for the martingale part and the bounded variation part of the right-hand side. Because the martingale part is a sum of stochastic integrals w.r.t. independent Wiener processes, we get

$$\forall t < \tau, \quad \langle q, \Phi_t'^{-1}[A^j, A^i](X_t, Y_t)\rangle = 0, \quad \forall i, j = 1, \ldots, n$$
$$\forall t < \tau, \quad \langle q, \Phi_t'^{-1}[\bar{A}^j, A^i](X_t, Y_t)\rangle = 0, \quad \forall i = 1, \ldots, n, \; j = 1, \ldots, p.$$

Now recall that the equation is written in Stratonovich sense and that we need to take into account the correction terms coming from the stochastic integral in W and the one in Y. To this end, writing again Itô's formula for $\langle q, \Phi_t'^{-1}[A^j, A^i](X_t, Y_t)\rangle$ and for $\langle q, \Phi_t'^{-1}[\bar{A}^k, A^i](X_t, Y_t)\rangle$, we have

$$\forall t < \tau, \quad \left\langle q, \Phi_t'^{-1}[A^k, [A^j, A^i]](X_t, Y_t) \right\rangle = 0, \quad \forall i, j, k = 1, \ldots, n$$

and

$$\forall t < \tau, \quad \langle q, \Phi_t'^{-1}[\bar{A}^l, [\bar{A}^k, A^i]](X_t, Y_t)\rangle = 0, \quad \forall i = 1, \ldots, n \; k, l = 1, \ldots, p.$$

From these equalities, we deduce that: $\langle q, \Phi_s'^{-1}[A^0, A^i](X_s, Y_s)\rangle = 0$ for every $i = 1, \ldots, n$.

Iterating this method, we obtain, taking $t = 0$, that q is orthogonal to every element of $\mathcal{I}(A^0, \bar{A}^1, \ldots, \bar{A}^p; A^1, \ldots, A^n)(x_0, 0)$ which is a contradiction with the assumption.

Let $q \in \mathbb{R}^n$, $q \neq 0$, then

$$\langle M_t^y q, q \rangle = \sum_{i=0}^n \int_0^t \left\langle \Phi_s'^{-1} A^i(X_s, Y_s), q \right\rangle^2 ds.$$

The map $s \to \langle \Phi_s'^{-1} A^i(X_s, Y_s), q \rangle$ is a.s. continuous from $]0, T]$ to $\mathbb{R}^+$; and we have just proved that, a.s., it takes its values in $\mathbb{R}^{+*}$. This implies that, a.s., $\langle M_t^y q, q \rangle > 0$, with the exceptional set not depending on q and t. The proof of the theorem is complete. □

Corollary 7.3 *If* $\dim \mathcal{I}(A^0, \bar{A}^1, \ldots, \bar{A}^p; A^1, \ldots, A^n)(x_0, 0) = n$, *then, there exists a set* $\mathcal{N}' \subset \mathcal{C}([0, T], \mathbb{R}^p)$ *of measure* 0 *such that, for* $y \notin \mathcal{N}'$, $t \in]0, T]$, *the measure* $\rho_t(y)$ *has a density w.r.t. the Lebesgue measure on* $\mathbb{R}^n$.

Proof. This is a consequence of Proposition (7.3) and formula (7.23). □

Remark. In order to get a better understanding of the condition involving the ideal $\mathcal{I}(A^0, \bar{A}^1, \ldots, \bar{A}^p; A^1, \ldots, A^n)$, recall the celebrated Hörmander "sum of squares" theorem. This theorem proves the hypoellipticity of the heat operator $\frac{\partial}{\partial t} + V^0 + \sum_{i=1}^d (V^i)^2$ whose coefficients do not depend on t, under the condition that the Lie algebra generated by $V^1, \ldots, V^d$ in the Lie algebra generated by $V^0, V^1, \ldots, V^d$ is of full rank at each point x in $\mathbb{R}^n$.

The same type of condition occurs here. Consider that the drift $\frac{\partial}{\partial t} + V^0$ is formally replaced by $\frac{\partial}{\partial t} + A^0 + \sum_{i=1}^p \bar{A}^i \text{``}\frac{dY_t^i}{dt}\text{''}$; then, the independence of the components of the Wiener process Y implies that the drift splits up into $p + 1$ independent components $A^0, \bar{A}^1, \ldots, \bar{A}^p$.

As a final remark, let us point out that the use of stochastic flows makes it possible to obtain the desired result assuming only that the condition of full rank holds at the starting point.

7.2.3 The smoothness of the conditional density

In this section, we prove a stronger result than Corollary 7.3; namely that under the same assumption, the conditional density exists and is smooth. To this end, we directly use the Malliavin calculus on the Wiener space $\mathcal{C}([0, T], \mathbb{R}^n) \times \mathcal{C}([0, T], \mathbb{R}^p)$; however, because we deal with conditional laws, we only need the Malliavin derivatives along $\mathcal{C}([0, T], \mathbb{R}^n) \times \{0\}$.

Recall that, if ϕ is a bounded measurable fonction on $\mathbb{R}^n$, the unnormalized filter is $\rho_t\phi = \mathring{\mathbb{E}}^y[\phi(X_t)Z_t]$. To get rid of the unboundedness of the exponential map in Z, we introduce a new diffusion process $(\zeta_t)_{t \in [0, T]}$ with values in $\mathbb{R}^{n+p+1}$ defined by

$$\zeta_t = (\xi_t, \ln Z_t + \chi_0) \tag{7.26}$$

where $\chi_0 \in \mathbb{R}$.

Denote by Ψ_t the stochastic flow associated to this diffusion.

With every bounded measurable fonction ϕ on $\mathbb{R}^n$, we associate the real function $\tilde{\phi}$ defined on $\mathbb{R}^{n+p+1}$ given by

$$\tilde{\phi}(x, y, u) = \phi(x)e^u.$$

Remark that $\tilde{\phi}(x, y, u)$ does not depend on y. We introduce this function because we want to deal with the Markov process ζ_t (note that $(X_t, \ln Z_t)$ is not Markov). Then, we have the formula

$$\rho_t\phi = \mathring{\mathbb{E}}^y\left[\tilde{\phi}\left(\zeta_t^0\right)\right] \tag{7.27}$$

where ζ_t^0 is defined by (7.26) with $\chi_0 = 0$.

This subsection is divided into three parts: First, we establish a conditional integration by parts formula; then, we prove that, if the appropriate Malliavin matrix is a.s. invertible, the conditional density exists and is smooth; finally, we prove that the assumption on the ideal implies the a.s. invertibility of the Malliavin matrix.

The conditional integration by parts formula

Proposition 7.11 *For $t \in [0, T]$, $k \in \mathbb{N}^*$, $q \geq 1$, ζ_t is in $\mathbb{D}^{k,q}$. Furthermore,*

1. $D_s^i \zeta_t = \Psi_t'(\zeta_0)\Psi_s'^{-1}(\zeta_s) A^i(X_s, Y_s) \quad i = 1, \ldots, n$
2. $D_s^i X_t = \Phi_t'(\xi_0)\Phi_s'^{-1}(\xi_s) A^i(X_s, Y_s) \quad i = 1, \ldots, n$

    where in the first equality $A^i(X_s, Y_s)$ is identified with $(A^i(X_s, Y_s), 0)$ in $\mathbb{R}^{n+p+1}$ and $D_s^i X_t$ is identified with $\left(D_s^i X_t, 0\right)$ in $\mathbb{R}^{n+p}$ in the second one.

Proof. The first assertion and equality 1 are consequences of Proposition 7.5. For the second assertion, remark that $A^i(X_s, Y_s)$ is in $\mathbb{R}^n \times \{0\} \times \{0\}$ and the component in $\mathbb{R}^p$ of $\Phi_s'^{-1}(\xi_s) A^i(X_s, Y_s)$ is zero because Y_t does not depend on x_0. This implies that $D_s^i \zeta_t = \Phi_t'(\xi_0)\Phi_s'^{-1}(\xi_s)A^i(\xi_s)$ is in $\mathbb{R}^n \times \{0\}$ and can be identified with $D_s^i X_t$. □

Corollary 7.4 *For $t \in [0, T]$ and $\phi \in C_b^\infty(\mathbb{R}^n)$, $\tilde{\phi}(\zeta_t)$ is in $\mathbb{D}^{k,q}(H)$ for all $k \in \mathbb{N}^*$, $q \geq 1$.*

Proof. This is essentially a consequence of the results recalled in section 1. Of course one difficulty comes from the fact that the exponential map is not of polynomial growth. This problem can be handled by approximating the exponential map by a sequence of functions in $C_b^2(\mathbb{R})$ and using the basic fact that Z_t is in all $\mathbf{L}^q$ ($q \geq 1$). □

Proposition 7.12 *Let $\phi \in C_b^\infty(\mathbb{R}^n)$, $H \in \cap_{q \geq 1}\mathbf{L}^q$, G a simple functional on $\mathcal{C}([0, T], \mathbb{R}^p)$ and $u \in \mathbf{L}^2([0, T] \times \mathcal{C}([0, T], \mathbb{R}^n) \times \mathcal{C}([0, T], \mathbb{R}^p)$. It holds for $0 \leq t \leq T$*

$$\mathring{\mathbb{E}}[\langle D(\tilde{\phi}(\zeta_t)GH), u\rangle] = \mathring{\mathbb{E}}[\tilde{\phi}(\zeta_t)GH(\delta u)_t]$$

where $(\delta u)_t = \int_0^t <u_s, dW_s>$.

Proof. This is a particular case of the Malliavin integration by part formula stated in section 1 (see (7.4)). □

Lemma 7.1 *Let ϕ, H, G, u be as in Proposition 7.12. Then,*

$$\mathring{\mathbb{E}}[\langle D(\tilde{\phi}(\zeta_t)GH), u\rangle] = \mathring{\mathbb{E}}[(\sum_{i=0}^n \frac{\partial \phi}{\partial x_i}\langle DX_t^i, u\rangle H + \phi(X_t)\langle D(H\ln Z_t), u\rangle)GZ_t].$$

Proof. We use the computation of the Malliavin derivatives given in the first section and the fact that G is independent of W. □

Corollary 7.5 *Let ϕ, H, u be as before. Then, for a.e. $y \in \mathcal{C}([0, T], \mathbb{R}^p)$, it holds for $t \in [0, T]$*

$$\mathring{\mathbb{E}}^y[\sum_{i=0}^n \frac{\partial \phi}{\partial x_i}\langle DX_t^i, u\rangle HZ_t] = \mathring{\mathbb{E}}^y[\phi(X_t)(H(\delta u)_t - \langle D(H\ln Z_t), u\rangle)Z_t]$$

Proof. This is a consequence of the density of simple functionals in every space $\mathbf{L}^q(\mathcal{C}([0, T], \mathbb{R}^p))$ ($q \geq 1$). □

The conditional Malliavin matrix

We recall the definition of the Malliavin matrix stated in Proposition 7.10

$$M_t^y = \sum_{i=0}^n \int_0^t \Phi_s'^{-1}A^i(X_s, Y_s)\left(\Phi_s'^{-1}A^i(X_s, Y_s)\right)^T ds.$$

Proposition 7.13 *Assume that for $t \in]0, T]$, the Malliavin matrix M_t^y is invertible a.s. on $\mathcal{C}([0, T], \mathbb{R}^n) \times \mathcal{C}([0, T], \mathbb{R}^p)$ and such that for all $\varepsilon > 0$*

$$\sup_{\varepsilon \leq t \leq T} |\det(M_t^y)^{-1})| \in \cap_{q \geq 1} \mathbf{L}^q(\mathcal{C}([0, T], \mathbb{R}^n) \times \mathcal{C}([0, T], \mathbb{R}^p))$$

for any $q \geq 1$.

Then, there exists a continuous process $(C_t)_{t \in]0, T]}$ defined on $\mathcal{C}([0, T], \mathbb{R}^n) \times \mathcal{C}([0, T], \mathbb{R}^p)$ with values in $\mathbb{R}^n$ such that

1. $\forall \varepsilon > 0, \forall q \geq 1, \sup_{\varepsilon \leq s \leq T} \|C_s\| \in \mathbf{L}^q(\mathring{\mathbb{P}})$ and $\mathbf{L}^q(\mathbb{P})$
2. $\forall \phi \in C_b^\infty(\mathbb{R}^n), t \in]0, T]$, for a.e. $y, i = 1, \ldots, n$

$$\rho_t(\frac{\partial \phi}{\partial x_i}) = \mathring{\mathbb{E}}^y[\phi(X_t) C_t^i Z_t].$$

Proof. We apply Corollary (7.5) with

$$u_k^l(s) = (\Phi_s'^{-1}(\xi_s) A^l(\xi_s))^k, \quad k, l = 1, \ldots, n$$
$$H_{k,l} = ((M_t^y)^{-1} \Phi_t'^{-1})_{kl}, \quad k, l = 1, \ldots, n.$$

We verify that such u and H satisfy the assumption of the corollary. For u, this is a direct consequence of Proposition 7.8 and of the uniform boundedness of the vector fields A^i. The assumption on H is obtained from Proposition 7.11, and the hypothesis $\sup_{\varepsilon \leq t \leq T} |\det(M_t^y)^{-1})| \in \cap_{q \geq 1} \mathbf{L}^q$.

We next calculate $\langle DX_t^i, u_k \rangle$ for $i, k = 1, \ldots, n$. We have by Proposition 7.11

$$\langle DX_t^i, u_k \rangle = \sum_{l=1}^n \int_0^t (\Phi_t'(\xi_0)(\Phi_s'^{-1}(\xi_s) A^l(\xi_s))^i (\Phi_s'^{-1}(\xi_s) A^l(\xi_s))^k ds = (\Phi_t'(\xi_0) M_t^y)_{i,k}.$$

Consequently, $\sum_{k=1}^n \langle DX_t^i, u_k \rangle H_{kl} = \delta_{il}$.

We now apply Corollary 7.5 to u_k and H_{kl}, sum over k and take $l = i$. It follows that

$$\mathring{\mathbb{E}}^y[\frac{\partial \phi}{\partial x_i}(X_t) Z_t] = \mathring{\mathbb{E}}^y[\phi(X_t) Z_t C_t^i]$$

with

$$C_t^i = \sum_{k=1}^n \{((M_t^y)^{-1} \Phi_t'^{-1}(\xi_t))_{ki} \delta(\Phi_t^{-1}(\xi_t) A^l(\xi_t))^k$$
$$- \langle D((M_t^y)^{-1} \Phi_t'^{-1}(\xi_t))_{ki} \ln Z_t), \Phi_t^{-1}(\xi_s) A^l(\xi_t))^k \rangle\}.$$

It follows from the assumption on M_t^y and Proposition 7.11 that C_t verifies property 1. □

Theorem 7.3 *Under the assumptions of Proposition 7.13, there exists a subset $\mathcal{N}''$ of $\mathcal{C}([0, T], \mathbb{R}^p)$ of zero measure such that*

$$\forall y \notin \mathcal{N}'', \forall t \in]0, T], \forall \phi \in C_b^\infty(\mathbb{R}^n), \forall i = 1, \ldots, n$$

$$\rho_t\left(\frac{\partial \phi}{\partial x_i}\right) = \mathring{\mathbb{E}}^y\left[\phi(X_t) C_t^i Z_t\right].$$

Proof. The difference with the previous proposition is that the exceptional set in $\mathcal{C}([0, T], \mathbb{R}^p)$ no longer depends on (ϕ, t). The argument is the following.

The space $\mathcal{C}([0, T], \mathbb{R}^p) \times \mathbb{R}^{+*}$ is a separable metric space. Let $(\phi_m, t_m)_{m \in \mathbb{N}}$ be a dense subset of this space. There exists $\mathcal{N}'' \subset \mathcal{C}([0, T], \mathbb{R}^p)$ of zero measure such that

$$\forall y \notin \mathcal{N}'', \forall m \in \mathbb{N}, \; \rho_{t_m}\left(\frac{\partial \phi_m}{\partial x_i}\right) = \mathring{\mathbb{E}}^y[\phi_m(X_{t_m}) C_{t_m}^i Z_{t_m}].$$

Then the desired result is a consequence of the continuity of the two sides of the previous equality with respect to (ϕ, t), together with the dominated convergence theorem. $\square$

Let us introduce some notation. For every multi-index $\alpha = (\alpha_1, \ldots, \alpha_k) \in \{1, \ldots, n\}^k$, we write $x^\alpha = x_{\alpha_1} \ldots x_{\alpha_k}$ and $\frac{\partial^\alpha}{\partial x^\alpha} = \frac{\partial^k}{\partial x_{\alpha_1} \ldots \partial x_{\alpha_k}}$.

Also, we write $\mathcal{S} = \mathcal{S}(\mathbb{R}^n)$ for the Schwartz space of rapidly decreasing functions defined by

$$\mathcal{S} = \left\{ f \in C^\infty(\mathbb{R}^n; \mathbb{R}) : \forall \alpha \in \mathbb{N}^k, \forall q \in \mathbb{N}, \sup_{x \in \mathbb{R}^n}(1 + |x|^2)^q \left|\frac{\partial^\alpha f}{\partial x^\alpha}(x)\right| < +\infty \right\}$$

Corollary 7.6 *Under the assumptions of Proposition 7.13, for any $(t, y) \in]0, T] \times (\mathcal{N}'')^c$ $\rho_t(y, .)$ has a density $q_t(y, .)$ such that*

1. $\forall (t, y) \in]0, T] \times (\mathcal{N}'')^c$, $q_t(y, .) \in \mathcal{S}(\mathbb{R}^n)$
2. *For any $(y, \beta, \alpha) \in (\mathcal{N}'')^c \times \mathbb{N}^n \times \mathbb{N}^n$, $\sup_{\varepsilon \leq t \leq T, x \in \mathbb{R}^n} \left| x^\beta \frac{\partial^\alpha q_t(y,x)}{\partial x^\alpha} \right|$ is finite and in $\mathbf{L}^q(\mathcal{C}([0, T], \mathbb{R}^p)$ for any $q \geq 1$.*
3. *For any $(y, \alpha) \in (\mathcal{N}'')^c \times \mathbb{N}^n$, the map $(t, x) \to \frac{\partial^\alpha q_t(y,x)}{\partial x^\alpha}$ is continuous from $\mathbb{R}^{+*} \times \mathbb{R}^n$ into $\mathbb{R}$.*

Proof. Let us recall the integration by parts formula stated in Theorem 7.3

$$\forall y \notin \mathcal{N}'', \forall t \in]0, T], \forall \phi \in C_b^\infty(\mathbb{R}^n), \forall i = 1, \ldots, n$$

$$\rho_t\left(\frac{\partial \phi}{\partial x_i}\right) = \mathring{\mathbb{E}}^y\left[\phi(X_t) C_t^i Z_t\right].$$

We need to iterate the procedure of integration by parts to prove the existence and smoothness of the conditional densities. The proof of these higher-order integration by parts formulas is similar to the one of the classical case. Consequently, we state the result without proof in the following lemma.

Lemma 7.2 *Under the assumptions of Proposition 7.13, there exists a subset $\mathcal{N}''' \subset \mathcal{C}([0, T], \mathbb{R}^p)$ of zero measure and for each $\alpha \in \mathbb{N}^n$, a continuous process $(C_t^\alpha)_{t \in]0, T]}$ defined on $\mathcal{C}([0, T], \mathbb{R}^n) \times \mathcal{C}([0, T], \mathbb{R}^p)$ with values in $\mathbb{R}$ such that*

1. $\forall \varepsilon > 0, q \geq 1, \sup_{\varepsilon \leq s \leq T} |C_s^\alpha| \in \mathbf{L}^q(\mathring{\mathbb{P}}) \cap \mathbf{L}^q(\mathbb{P})$
2. $\forall \phi \in C_b^\infty(\mathbb{R}^n), t \in]0, T], \forall y \in (\mathcal{N}''')^c, \forall \alpha \in \mathbb{N}^n$

$$\rho_t\left(\frac{\partial^\alpha \phi}{\partial x^\alpha}\right) = \mathring{\mathbb{E}}^y\left[\phi(X_t) C_t^\alpha Z_t\right]. \tag{7.28}$$

We immediately deduce from the previous lemma that all the derivatives of ρ_t (denoted by $D^\alpha \rho_t$) in the distribution sense are bounded measures and verify $\|D^\alpha \rho_t\| \leq \mathring{\mathbb{E}}^y\left[|C_t^\alpha| Z_t\right]$. By Proposition (7.2) this implies that ρ_t has a smooth density.

To prove part 2 of Corollary 7.6 and the fact that the density is in the Schwartz space, we use equation (7.28) for a particular family of functions and apply Fourier analysis techniques. Define, for $\xi \in \mathbb{R}^n$, $\phi^\xi(x) = e^{i\langle x, \xi \rangle}$ and apply equation (7.28) with $\phi = \phi^\xi$. We obtain

$$\rho_t((i\xi)^\alpha e^{i\langle x, \xi \rangle}) = \mathring{\mathbb{E}}^y\left[e^{i\langle X_t, \xi \rangle} C_t^\alpha Z_t\right]. \tag{7.29}$$

As X_t is in $\mathbf{L}^q$ for every $q \geq 1$, both sides of equation (7.29) are C^∞ functions of ξ. Furthermore, we have

$$\forall \beta \in \mathbb{N}^n, \quad \rho_t\left(\frac{\partial^\beta}{\partial x^\beta}[(i\xi)^\alpha e^{i\langle x, \xi \rangle}]\right) = \mathring{\mathbb{E}}^y\left[(iX_t)^\beta e^{i\langle X_t, \xi \rangle} C_t^\alpha Z_t\right]. \tag{7.30}$$

For $\beta \in \mathbb{N}^n$, define $q_t^\beta(y, x) = x^\beta q_t(y, x)$. We now prove by induction on $|\alpha| + |\beta|$ that

$$\forall y \notin \mathcal{N}''', \quad \sup_{\xi \in \mathbb{R}^n} |\xi|^\alpha \left|\widehat{q_t^\beta}(\xi)\right| < +\infty. \tag{7.31}$$

Remark first that, for $\beta = 0$, (7.31) is a consequence of equation (7.30) and Lemma 7.2. For $\alpha = 0$, this is standard.

Assume (7.31) is true for $|\alpha| + |\beta| \leq n_0$ and take $(\alpha, \beta) \in \mathbb{N}^{2n}$ such that $|\alpha| + |\beta| = n_0 + 1$. Denote by $P_{\alpha, \beta}$ the polynomial in the two variables x and ξ, with complex coefficients such that

$$\frac{\partial^{\beta}}{\partial x^{\beta}}[(i\xi)^{a}e^{i\langle x,\xi\rangle}] = [(i\xi)^{a}(ix)^{\beta} + P_{(a,\beta)}(x,\xi)]e^{i\langle x,\xi\rangle}.$$

We deduce from (7.30) that

$$i^{a+\beta}\xi^{a}\widehat{q_{t}^{\beta}}(\xi) = \mathring{\mathbb{E}}^{y}\left[\left\{(iX_{t})^{\beta}C_{t}^{a} - P_{(a,\beta)}(X_{t},\xi)\right\}e^{i\langle X_{t},\xi\rangle}Z_{t}\right].$$

Each term of $P_{(a,\beta)}(X_{t},\xi)$ is of the form $cX_{t}^{\gamma}\xi^{\delta}$ with $|\gamma|+|\delta| \leq n_{0}$. It follows from the induction hypothesis and Lemma 7.2 that the relation (7.31) is true for all multi-indices.

Furthermore, there exists a universal linear function of the quantities $\left\{X_{t}^{\gamma}C_{t}^{\delta}, \ |\gamma|+|\delta| \leq n_{0}\right\}$ (we denote it by $\Phi_{a,\beta}(t)$) such that

$$\xi^{a}\widehat{q_{t}^{\beta}}(\xi) = \mathring{\mathbb{E}}^{y}[\Phi_{a,\beta}(t)e^{i\langle X_{t},\xi\rangle}Z_{t}].$$

By Lemma 7.2, $\Phi_{a,\beta}(t)$ belongs to $\mathbf{L}^{q}(\mathcal{C}([0,T],\mathbb{R}^{n}) \times \mathcal{C}([0,T],\mathbb{R}^{p}))$ for every $q \geq 1$.

By relation (7.31), we have for every $(a,\beta,\gamma) \in \mathbb{N}^{3n}$

$$|\xi|^{\gamma}\frac{\widehat{\partial^{a}q_{t}^{\beta}}}{\partial x^{a}}(\xi) \leq \mathring{\mathbb{E}}^{y}[|\Phi_{a+\gamma,\beta}(t)|Z_{t}].$$

Since $(1+|\xi|^{\gamma})^{-1}$ is in $\mathbf{L}^{1}(\mathbb{R}^{n})$ if $|\gamma| > n$, we get

$$\sup_{x\in\mathbb{R}^{n}}\left|\frac{\partial^{a}q_{t}^{\beta}}{\partial x^{a}}(x)\right| \leq C(\gamma)\mathring{\mathbb{E}}^{y}[|\Phi_{a+\gamma,\beta}(t)|Z_{t}]$$

and this proves part 2 of Corollary 7.6.

A similar calculation shows that

$$\sup_{x\in\mathbb{R}^{n}}\left|\frac{\partial^{\beta}q_{t}}{\partial x^{\beta}}(x) - \frac{\partial^{\beta}q_{t'}}{\partial x^{\beta}}(x)\right| \leq C(\gamma)\mathring{\mathbb{E}}^{y}[|\Phi_{a+\gamma,\beta}(t) - \Phi_{a+\gamma,\beta}(t')|Z_{t}].$$

So, every derivative of q_{t} is continuous with respect to t uniformly in $x \in \mathbb{R}^{n}$. This gives the desired continuity w.r.t. (t,x). □

The invertibility in $\mathbf{L}^{q}$ of the Malliavin matrix

In this subsection, we omit the dependence in y in our notation for the Malliavin matrix, and we write M_{t} for M_{t}^{y}.

We use the Landau notation $o(\ )$ and $O(\ )$ to avoid introducing too many different constants.

Theorem 7.4 *Assume that* $\dim\mathcal{I}(A^{0},\bar{A}^{1},\ldots,\bar{A}^{p};A^{1},\ldots,A^{n})(x_{0},0) = n$. *Then, for every* $\varepsilon \in]0,T]$, $\sup_{\varepsilon \leq t \leq T}|(\det M_{t})^{-1}| \in \mathbf{L}^{q}(\mathcal{C}([0,T],\mathbb{R}^{n}) \times \mathcal{C}([0,T],\mathbb{R}^{p}))$ *for every* $q \geq 1$.

Proof. First step: We recall some elementary facts.

1. Let X be a positive random variable on a probability space $(\Omega, \mathcal{F}, \mathbb{P})$. The formula $\mathbb{E}[X] = \int_0^{+\infty} \mathbb{P}(X > x)dx$ implies that if

$$\mathbb{P}(X > x) = o\left(\frac{1}{x^r}\right), \quad \forall r \geq 1,$$

X has moments of all orders.

2. Fix $t \in]0, T]$, $q \geq 1$, and define $X = \sup_{t \leq s \leq T} |\det M_s|^{-q}$. Then,

$$\mathbb{P}(X > x) = \mathbb{P}(\inf_{t \leq s \leq T} |\det M_s| < x^{-\frac{1}{q}}).$$

Thus, using 1, X is integrable if one has

$$\mathbb{P}(\inf_{t \leq s \leq T} |\det M_s| < \varepsilon) = o(\varepsilon^r), \forall r \geq 1.$$

3. Because M_t is a symmetric positive matrix, it is conjugate to a positive diagonal matrix. This implies that $\det M_t \geq \inf_{\|u\|=1} \langle M_t u, u \rangle^n$. Recall also that the map $t \to \langle M_t u, u \rangle$ is increasing, so $\inf_{t \leq s \leq T} \det M_t \geq \inf_{\|u\|=1} \langle M_t u, u \rangle^n$.

We have thus reduced the initial problem to the proof of the following collection of estimates

$$\mathbb{P}(\inf_{\|u\|=1} \langle M_t u, u \rangle < \varepsilon) = o(\varepsilon^q), \quad \forall q \geq 1. \tag{7.32}$$

Second step: Define the stopping time

$$\tau_a = \inf\{s, \|\xi_s - \xi_0\| \leq a \text{ or } \|\Phi_s^{'-1}(\xi_s) - I\| \leq a\}$$

where a is a positive constant to be fixed later. $\square$

Lemma 7.3 $\tau_a^{-1} \in \mathbf{L}^q(C([0, T], \mathbb{R}^n) \times C([0, T], \mathbb{R}^p))$ *for every* $q \geq 1$.

Proof. We have

$$\mathbb{P}(\tau_a < \varepsilon) = \mathbb{P}(\sup_{s \leq \varepsilon}\{\|\xi_s - \xi_0\|, \|\Phi_s^{'-1}(\xi_s) - I\|\} \geq a)$$

$$\leq \frac{1}{a^{2q}} \mathbb{E}[\sup_{s \leq \varepsilon}\{\|\xi_s - \xi_0\|, \|\Phi_s^{'-1}(\xi_s) - I\|\}^{2q}] = o(\varepsilon^q).$$

The last estimate is a consequence of the properties of the flow and its derivative (see (7.7)). $\square$

We denote by S the unit sphere in $\mathbb{R}^n$. We view S as a subset of $\mathbb{R}^{n+p}$ by identifying it with $S \times \{0\}$. If $v \in S$, we denote by $\mathcal{V}_S(v)$ the collection of all neighbourhoods of v in S.

Proposition 7.14 *Assume that*

$$\forall v \in S, \exists i = 1, \ldots, n, \exists N(v) \in \mathcal{V}_S(v) \text{ such that}$$

$$\sup_{u \in N(v)} \mathbb{P}\left(\int_0^{\tau_a} \langle \Phi_s'^{-1} A^i(\xi_s), u \rangle^2 ds < \varepsilon \right) = o(\varepsilon^q), \forall q \geq 1. \tag{7.33}$$

Then the estimates (7.32) hold.

Proof. This uses the compactness of S and the property that, by the construction of τ_a, the quadratic forms $u \to \int_0^{\tau_a} \langle \Phi_s'^{-1} A^i(\xi_s), u \rangle^2 ds$ are uniformly Lipschitz. □

Third step: We now use our assumption on the vector fields $A^0, A^1, \ldots, A^n$, $\bar{A}^1, \ldots, \bar{A}^p$ in a similar way as in the proof of Theorem 7.2, namely using Itô's formula. However, more precise estimates are needed to get (7.33). These are obtained from a technical lemma on the growth of semimartingales, which is stated at the end of this subsection.

For $l \in \mathbb{N}^*$, let K_l be the set of elements of $\mathcal{I}(A^0, \bar{A}^1, \ldots, \bar{A}^p; A^1, \ldots, A^n)$ appearing as brackets of length at most l. Our assumption is equivalent to : $\exists n_0 \in \mathbb{N}$, $\dim(\text{span } K_{n_0}(x_0, y_0)) = n$, and then we have

$$\delta := \inf_{u \in S} \{ \sup_{K \in K_{n_0}} \langle K(\xi_0), u \rangle^2 \} > 0.$$

Lemma 7.4 $\exists a > 0, \forall v \in S, \exists N(v) \in \mathcal{V}_S(v), \exists K \in K_{n_0}$ *such that*

$$\inf_{s \leq \tau_a, u \in N(v)} \langle \Phi_s'^{-1} K(\xi_s), u \rangle^2 \geq \frac{\delta}{2}.$$

From now on a is fixed as in Lemma 7.4 and we omit the dependence of τ on a in the notation.

Lemma 7.5 $\forall v \in S, \exists N(v), \exists K \in K_{n_0}$ *such that*

$$\sup_{u \in N(v)} \mathbb{P}\left(\int_0^{\tau} \langle \Phi_s'^{-1} K(\xi_s), u \rangle^2 ds < \varepsilon \right) = o(\varepsilon^q), \forall q \geq 1. \tag{7.34}$$

Proposition 7.15 *The estimate (7.34) implies (7.33).*

Proof. First remark that, when $n_0 = 1$, estimates (7.34) and (7.33) are equivalent.

In order to deal with the general case, fix $v \in S$. Then (7.34) yields an element K of K_{n_0}. We write K as $K = [\tilde{A}^{i_{n_0}}, [\tilde{A}^{i_{n_0-1}}, \ldots, [\tilde{A}^{i_2}, A^{i_1}] \ldots]$ where $i_1 \neq 0$ and $\tilde{A}^{i_k} = A^{i_k}$ or $\bar{A}^{i_k}$.

Define by induction

$$H^1 = A^{i_1}, \quad H^j = [\tilde{A}^{i_j}, H^{j-1}], \quad j = 2, \ldots, n_0.$$

Then $K = H^{n_0}$. We prove the following properties by decreasing induction on j

$$(P_j) \quad \sup_{u \in N(v)} \mathbb{P}(\int_0^\tau \langle \Phi_s'^{-1} H^j(\xi_s), u \rangle^2 ds < \varepsilon) = o(\varepsilon^q), \quad \forall q \geq 1.$$

Clearly, (P_{n_0}) is (7.34) and (P_1) is (7.33). Thus our induction will complete the proof.

We assume (P_j) and write Itô's formula for $\Phi_t'^{-1} H^{j-1}(\xi_t)$ in Stratonovich form and in Itô form.

$$\begin{aligned}
\Phi_t'^{-1} H^{j-1}(\xi_t) &= H^{j-1}(\xi_0) + \int_0^t \Phi_s'^{-1}[A^0, H^{j-1}](\xi_s) ds \\
&+ \sum_{i=1}^n \int_0^t \Phi_s'^{-1}[A^i, H^{j-1}](\xi_s) \circ dW_s^i \\
&+ \sum_{k=1}^p \int_0^t \Phi_s'^{-1}[\tilde{A}^k, H^{j-1}](\xi_s) \circ dY_s^k \\
&= H^{j-1}(\xi_0) + \int_0^t \Phi_s'^{-1}[A^0, H^{j-1}](\xi_s) ds \\
&+ \frac{1}{2} \sum_{i=1}^n \int_0^t \Phi_s'^{-1}[A^i[A^i, H^{j-1}]](\xi_s) ds \\
&+ \frac{1}{2} \sum_{k=1}^p \int_0^t \Phi_s'^{-1}[\tilde{A}^k[\tilde{A}^k, H^{j-1}]](\xi_s) ds \\
&+ \sum_{i=1}^n \int_0^t \Phi_s'^{-1}[A^i, H^{j-1}](\xi_s) dW_s^i \\
&+ \sum_{k=1}^p \int_0^t \Phi_s'^{-1}[\tilde{A}^k, H^{j-1}](\xi_s) dY_s^k. \quad (7.35)
\end{aligned}$$

We set $\tilde{H}^{j-1} = [A^0, H^{j-1}] + \frac{1}{2} \sum_{i=1}^n [A^i[A^i, H^{j-1}]] + \frac{1}{2} \sum_{k=1}^p [\tilde{A}^k[\tilde{A}^k, H^{j-1}]]$.

Under assumption (P_j), we get estimates for the probability distribution of the integrands of the (deterministic and stochastic) integrals in the right-hand side of (7.35). In order to get the desired estimates for the left-hand side, we use the following technical lemma, which is stated without proof.

Lemma 7.6 *Let $(a_0, s_0) \in \mathbb{R}^2$, let $(B_t)_{t \in \mathbb{R}^+}$ be a Wiener process with values in $\mathbb{R}^d$, and let $(a_t)_{t \in \mathbb{R}^+}, (b_t)_{t \in \mathbb{R}^+}, (\alpha_t)_{t \in \mathbb{R}^+}, (\beta_t)_{t \in \mathbb{R}^+}$ be processes adapted to the canonical filtration of B, with values respectively in $\mathbb{R}^n \times \mathbb{R}^d$, $\mathbb{R}^n$, $\mathbb{R}^n \times \mathbb{R}^d$, $\mathbb{R}^n$. Assume*

1. $a_t = a_0 + \int_0^t \alpha_s \, dB_s + \int_0^t \beta_s \, ds$
2. There exist a bounded stopping time σ and a positive constant $\bar{C}$ such that

$$\sup_{s \leq \sigma} \{|a_s|, |b_s|, |\alpha_s|, |\beta_s|\} \leq \bar{C} < +\infty$$

Define $S_t = s_0 + \int_0^t a_s \, dB_s + \int_0^t b_s \, ds$. Then, for $r > 8$ and $q \geq 1$, for every $\varepsilon > 0$ small enough,

$$\mathbb{P}\left(\int_0^\sigma S_t^2 \, dt < \varepsilon^r, \int_0^\sigma (\|a_t\|^2 + \|b_t\|^2) \, dt \geq \varepsilon\right) \leq C(q, r, \bar{C}) \varepsilon^q.$$

We apply Lemma 7.6 with $S_t = \langle \Phi_t'^{-1} H^{j-1}(\xi_t), u \rangle$, $B = (W, Y)$, and $\sigma = \tau$. We get

$$\mathbb{P}\left(\int_0^\tau \langle \Phi_s'^{-1} H^{j-1}(\xi_s), u \rangle^2 \, ds < \varepsilon,\right.$$

$$\int_0^\tau \{\sum_{i=1}^n \langle \Phi_s'^{-1}[A^i, H^{j-1}](\xi_s), u \rangle^2 + \sum_{k=1}^p \langle \Phi_s'^{-1}[\bar{A}^k, H^{j-1}](\xi_s), u \rangle^2$$

$$\left. + \langle \Phi_s'^{-1} \tilde{H}^{j-1}(\xi_s), u \rangle^2 \} \, ds > \varepsilon^{\frac{1}{r}}\right) \leq C(q, r) \varepsilon^q. \tag{7.36}$$

This implies that for $q \geq 1$, $r > 8$:

$$\mathbb{P}(\int_0^\tau \langle \Phi_s'^{-1} H^{j-1}(\xi_s), u \rangle^2 \, ds < \varepsilon) \leq C(q, r) \varepsilon^{\frac{q}{r}}$$

$$+ \mathbb{P}(\int_0^\tau \{\sum_{i=1}^n \langle \Phi_s'^{-1}[A^i, H^{j-1}](\xi_s), u \rangle^2 + \sum_{k=1}^p \langle \Phi_s'^{-1}[\bar{A}^k, H^{j-1}](\xi_s), u \rangle^2$$

$$+ \langle \Phi_s'^{-1} \tilde{H}^{j-1}(\xi_s), u \rangle^2 \} \, ds < \varepsilon^{\frac{1}{r}}). \tag{7.37}$$

Denote the last term of this inequality by P. We will use our assumption (P_j) to estimate this term. Recall that $H^j = [\tilde{A}^{ij}, H^{j-1}]$. Thus, two cases have to be considered.

First case: $\tilde{A}^{ij} \neq A^0$. Then, we immediately see, from the fact that $\tilde{A}^{ij}$ is one of the vector fields $A^1, \ldots, A^n, \bar{A}^1, \ldots, \bar{A}^p$, that

$$P \leq \mathbb{P}(\int_0^\tau \langle \Phi_s'^{-1} H^j(\xi_s), u \rangle^2 \, ds < \varepsilon^{\frac{1}{r}}) = o(\varepsilon^q), \forall q \geq 1.$$

Second case: $\tilde{A}^{ij} = A^0$. In that case, we have

$$H^j = \tilde{H}^{j-1} - \frac{1}{2} \sum_{i=1}^n [A^i[A^i, H^{j-1}]] - \frac{1}{2} \sum_{k=1}^p [\bar{A}^k[\bar{A}^k, H^{j-1}]]$$

and we cannot estimate P directly.

For $r, r' > 8$, we introduce the events $\left\{ \int_0^\tau \langle \Phi_s'^{-1}[A^i[A^i, H^{j-1}]](\xi_s), u \rangle^2 \, ds > \varepsilon^{\frac{1}{rr'}} \right\}$ and $\left\{ \int_0^\tau \langle \Phi_s'^{-1}[\bar{A}^k[\bar{A}^k, H^{j-1}]](\xi_s), u \rangle^2 \, ds > \varepsilon^{\frac{1}{rr'}} \right\}$ and we have:

$$P \leq \sum_{i=1}^n \mathbb{P}(\int_0^\tau \langle \Phi_s'^{-1}[A^i, H^{j-1}](\xi_s), u \rangle^2 ds < \varepsilon^{\frac{1}{r}},$$

$$\int_0^\tau \langle \Phi_s'^{-1}[A^i[A^i, H^{j-1}]](\xi_s), u \rangle^2 ds > \varepsilon^{\frac{1}{rr'}})$$

$$+ \sum_{k=1}^p \mathbb{P}(\int_0^\tau \langle \Phi_s'^{-1}[\bar{A}^k, H^{j-1}](\xi_s), u \rangle^2 ds < \varepsilon^{\frac{1}{r}},$$

$$\int_0^\tau \langle \Phi_s'^{-1}[\bar{A}^k[\bar{A}^k, H^{j-1}]](\xi_s), u \rangle^2 ds > \varepsilon^{\frac{1}{rr'}})$$

$$+ \mathbb{P}(\int_0^\tau \langle \Phi_s'^{-1} \tilde{H}^{j-1}(\xi_s), u \rangle^2 < \varepsilon^{\frac{1}{r}}, \int_0^\tau \langle \Phi_s'^{-1}[A^i[A^i, H^{j-1}]](\xi_s), u \rangle^2 ds \leq \varepsilon^{\frac{1}{rr'}},$$

$$\int_0^\tau \langle \Phi_s'^{-1}[\bar{A}^k[\bar{A}^k, H^{j-1}]](\xi_s), u \rangle^2 ds < \varepsilon^{\frac{1}{rr'}}, \forall i = 1, \ldots, n \quad \forall k = 1, \ldots, p).$$

Lemma 7.6 applied to $\Phi_s'^{-1}[A^i, H^{j-1}](\xi_s)$ and $\Phi_s'^{-1}[\bar{A}^k, H^{j-1}](\xi_s)$ implies that the first and second term are $o(\varepsilon^q)$. The last term is bounded by

$$\mathbb{P}(\int_0^\tau < \Phi_s'^{-1} \tilde{H}^j(\xi_s), u >^2 ds \leq C(n, p)\varepsilon^{\frac{1}{rr'}}$$

where $C(n, p)$ is a universal function of n and p. This term is $o(\varepsilon^q)$ for $q \geq 1$ by our assumption (P_j).

We thus have proved that (P_j) implies (P_{j-1}) and this completes the proof of the theorem. □

7.3 Nonexistence of finite dimensionally computable filters

The problem of finding so-called "finite-dimensional filters" has been addressed for a long time because of its interest in applications. It consists in trying to express the filter in terms of a finite number of statistics driven by the observation. In particular, this is the case for the linear Kalman–Bucy filter.

The relation between this question and the structure of the Lie algebra generated by the coefficients has been conjectured in the 1970s by Brockett, Clark, and Mitter. Many authors have contributed to this problem with different methods (see Section 7.4 below). The method using the Malliavin calculus, due to Ocone, is presented here. It says, roughly speaking, that if the Lie algebra associated to the Zakai equation is "large enough", the filter is of infinite type.

Note that the result is true for $t \in\,]0, T]$ fixed. This second application of Malliavin calculus presented here is totally different from the first one because one deals with the Zakai equation, which is a stochastic partial differential equation driven by Y, in contrast with the filtering system where the observation Y is "frozen", which we considered in Section 7.2.

7.3.1 The model and the Zakai equation

We consider a simplified filtering system, which already will lead to nonexistence results for finite-dimensional filters.

The pair signal-observation, $(X, Y) \in \mathbb{R}^n \times \mathbb{R}^p$ satisfies:

$$\begin{cases} X_t = x_0 + \int_0^t A^0(X_s)ds + \sum_{i=0}^n \int_0^t A^i(X_s) \circ dW_s^i \\ Y_t = \int_0^t h(X_s)ds + V_t \end{cases}$$

where $x_0 \in \mathbb{R}^n$, $A^0, A^1, \ldots, A^n$ are $n+1$ vector fields on $\mathbb{R}^n$ and h is a function from $\mathbb{R}^n$ into $\mathbb{R}^p$. Our assumptions on coefficients will be made precise below.

Suppose that the unnormalized filter ρ_t admits a density $q_t(x)$ (we omit the dependence w.r.t. the observation). Then, this density solves the Zakai equation, which we write in Itô form

$$\begin{cases} dq_t(x) = L^* q_t(x)dt + \sum_{i=1}^p h_i q_t(x) dY_t^i \\ q_0(.) = \delta_{x_0} \end{cases}$$

where L^* is the adjoint of $L = \frac{1}{2}\sum_{i=1}^n (A^i)^2 + A^0 - \frac{1}{2}\sum_{i=1}^p h_i^2$.

7.3.2 The Malliavin calculus for the Zakai equation

The method consists in applying the Malliavin calculus to the Zakai equation. The difference with the preceding sections comes from the fact that the solution of this equation evolves in an infinite-dimensional space with no Lebesgue measure. So, the existence of a density for the law of the process $q_t(.)$ will mean that all orthogonal projections of q_t on d-dimensional subspaces of $\mathbb{L}^2(\mathbb{R}^n)$ have a density w.r.t. the Lebesgue measure on $\mathbb{R}^d$.

We introduce the Lie algebra Λ generated by L^* and the zero order differential operators $h_1, \ldots, h_p$. The notation $\Lambda(x_0)$ refers to the vector space where the elements of Λ have their coefficients fixed at x_0.

Let $C_b^\omega(\mathbb{R}^n)$ denote the space of all analytic functions from $\mathbb{R}^n$ into $\mathbb{R}$, that are bounded as well as their derivatives. We denote by (ϕ, ψ) the scalar product $(\phi, \psi) = \int \phi(x)\psi(x)dx$.

For $s \in \mathbb{R}$, let $H^s(\mathbb{R}^n)$ be the Sobolev space of order s with usual norm $\|f\|_s$.

Theorem 7.5 *Suppose that the following assumptions hold*

(i) *all coefficients of L^* and h are in $C_b^\omega(\mathbb{R}^n)$,*
(ii) *$L(x)$ is uniformly elliptic, for every x,*

(iii) $\Lambda(x_0)$ contains all the partial derivative operators of all orders with constant coefficients,

Then, for every $t \in]0, T]$ and $N \in \mathbb{N}$, for every linearly independent sequence $(\phi_1, \ldots, \phi_N)$ in $\mathbf{L}^2(\mathbb{R}^n)$, the probability law of $((q_t(.), \phi_1), \ldots, (q_t(.), \phi_N))$ has a density w.r.t. Lebesgue measure on $\mathbb{R}^N$.

Proof. Under hypotheses (i) and (ii), the theory of stochastic evolution equations in Hilbert spaces implies that the Zakai equation has a unique, adapted solution $q_t(.)$ satisfying

$$\mathring{\mathbb{E}}\left[\int_0^T \|q_t(.)\|_k^2 dt\right] < +\infty, \quad \text{for } k < -\frac{n}{2}.$$

Moreover $q_t(.) \in \mathcal{C}(]0, T]; H^k(\mathbb{R}^n))$ and $q_t(.)$ is smooth in x, this last property beeing also a consequence of the preceding sections.

In Section 2, the Malliavin derivatives were taken along $\mathcal{C}([0, T], \mathbb{R}^n) \times \{0\}$, that is w.r.t. W, with $Y = y$ fixed, to prove the existence and smoothness of a density for the conditional law w.r.t. the starting point. Here, the Malliavin calculus is applied on $(\mathcal{C}([0, T], \mathbb{R}^p), \mathcal{Y}, \mathbb{P}_Y)$ where $\mathbb{P}_Y$ is the law of Y under $\mathring{\mathbb{P}}$, which is the Wiener measure on $\mathcal{C}([0, T], \mathbb{R}^p)$. We denote also by D the derivative operator w.r.t. Y.

We use Proposition 7.3 to prove the existence of a density for

$$((q_t(.), \phi_1), \ldots, (q_t(.), \phi_N)) = (\rho_t \phi_1, \ldots, \rho_t \phi_N).$$

For $i = 1, \ldots, N$, the unnormalized filter is expressed as

$$\rho_t \phi_i = \mathring{\mathbb{E}}^w[\phi_i(X_t) Z_t]$$

where $\mathring{\mathbb{E}}^w$ is the expectation with respect to the law of W under $\mathring{\mathbb{P}}$. Note that, in this case, X does not depend on Y.

The following ingredients are used to prove that $\rho_t \phi_i$ is in $\mathbb{D}^{1,2}$, for every $i = 1, \ldots, N$, and to calculate its Malliavin derivative as

$$D_s \rho_t \phi_i = \mathring{\mathbb{E}}^w[\phi_i(X_t) Z_t h(X_s)].$$

1. In this uncorrelated case, thanks to an integration by parts, Z_t is continuous w.r.t. y and for every fixed y, it is a bounded function of X.
2. $\phi_i(X_t)$ is integrable since the law of X_t has a density which belongs to $\mathbf{L}^2(\mathbb{R}^n)$ (recall that L is uniformly elliptic).
3. $\rho_t \phi_i$ can be approached by a sequence of variables in $\mathbb{D}^{1,2}$.

Next, we prove that the Malliavin derivative associated to the independent sequence $(\rho_t\phi_1, \ldots, \rho_t\phi_N)$, whose generic term is

$$\langle D\rho_t\phi_i, D\rho_t\phi_j \rangle,$$

is invertible. This is a consequence of the following property

$$\|D\rho_t\phi\|_H \neq 0, \quad \text{for every } \phi \in \mathbf{L}^2(\mathbb{R}^n) \text{ such that } \phi \neq 0, \text{ a.s. in } y. \tag{7.38}$$

To check this property, we use another expression for $D\rho_t\phi$, which is obtained by using backward partial stochastic differential equations. It can be proved that

$$D_s\rho_t\phi = \mathring{\mathbb{E}}^w[Z_s h(X_s) v_\phi(s, X_s)] \tag{7.39}$$

where

$$v_\phi(s, x) = \mathring{\mathbb{E}}^w_{s,x}[\phi(X_t)\exp(\sum_{k=1}^p \int_s^t h_k(X_s) dY_s^k - \frac{1}{2}\int_s^t |h(X_s)|^2 ds)]$$

is the unique solution in $\mathbb{L}^2([0, t]; H^1(\mathbb{R}^n))$ of the following equation

$$\begin{cases} dv_\phi(s, x) + L^* v_\phi(s, x) ds + \sum_{i=1}^p h_i v_\phi(s, x) dY_s^i = 0, & 0 < s < t, \quad x \in \mathbb{R}^n \\ v_\phi(t, x) = \phi(x) \end{cases}$$

So, $D_s\rho_t\phi_i = (q_s(.), h(.)v_{\phi_i}(s, .))$. Then we use the following proposition that we state without proof. The arguments are analogous to those of the proof of Theorem 7.2, but now in an infinite-dimensional setting.

Proposition 7.16 *There exists a negligible subset $\mathcal{N}$ of $\mathcal{Y}$ such that if $y \notin \mathcal{N}$, and $\phi \in \mathbb{L}^2(\mathbb{R}^n)$, $\phi \neq 0$, the condition*

$$\int_0^t |(q_s(.), h(.)v_\phi(s, .))|^2(y) ds = 0$$

implies that

$$(v_\phi(s, .), \Gamma q_s(.))(y) = 0, \quad 0 < s \leq t$$

for every differential operator Γ belonging to Λ.

Now, to prove (7.38), let $y \in \mathcal{C}([0, T], \mathbb{R}^p)$ such that $\|D\rho_t\phi\|_H = 0$. This implies by Proposition 7.16

$$(v_\phi(s, .), \Gamma q_s(.))(y) = (\Gamma^* v_\phi(s, .), q_s(.))(y) = 0, \quad 0 < s \leq t, \quad \Gamma \in \Lambda.$$

For all $\Gamma \in \Lambda$, $\exists l \in \mathbb{N}$ such that $\Gamma^* \in \mathcal{L}(H^{r+l}(\mathbb{R}^n); H^r(\mathbb{R}^n))$, $r \in \mathbb{N}$. Moreover, $q_s(.) \to \delta_{x_0}$ in $H^{-r}(\mathbb{R}^n)$ for $r > \frac{n}{2}$. This implies

$$\Gamma^* v_\phi(0, x_0) = 0, \quad \forall \Gamma \in \Lambda(x_0).$$

As hypothesis (iii) of Theorem 7.16 holds also for the adjoints of elements of Λ, one deduces that $v_\phi(0,.)$ and all its partial derivatives vanish at $x = x_0$. In order to get a contradiction with $\phi \neq 0$, it is necessary to introduce an analytic function. To this end, let u_ϕ be the function on $[0, T] \times \mathbb{R}^n$ defined by

$$u_\phi(s, x) = e^{\sum_{i=1}^{p} h^i(x) y^i(s)} v_\phi(s, x).$$

It can be proved that this function is the unique solution of the following parabolic partial differential equation parametrized by y.

$$\begin{cases} \frac{\partial u_\phi}{\partial s}(s, x) = -\left(L_s^y\right)^* u_\phi(s, x), & 0 < s < t, \quad x \in \mathbb{R}^n \\ u_\phi(t, x) = \phi(x) e^{\sum_{i=1}^{p} h^i(x) y^i(t)} \end{cases}$$

where $L_s^y = e^{-\sum_{i=1}^{p} h^i(x) y^i(s)} \left[L - \frac{1}{2} \sum_{i=1}^{p} h_i^2\right] e^{\sum_{i=1}^{p} h^i(x) y^i(s)}$.

The hypotheses (i) and (ii) on the coefficients imply that u_ϕ is analytic, so $u_\phi(0, x) = 0$ for all $x \in \mathbb{R}^n$. The same holds for v_ϕ, which is in contradiction with $v_\phi(t, x) = \phi(x) \neq 0$ by uniqueness. $\square$

7.3.3 Application to the finite-dimensional filters

Definition 7.6 Let $(\phi_1, \ldots, \phi_k, \ldots)$ be a sequence of functions in $L^2(\mathbb{R}^n) \cap C_b(\mathbb{R}^n)$ and fix $t \in]0, T]$. One says that the collection of statistics $(\rho_t \phi_1, \ldots, \rho_t \phi_k, \ldots)$ admits a sufficient regular statistic β in finite dimension r if β is $\mathcal{Y}_t$-measurable and if for every $i \geq 1$, there exists $\theta_i \in C^1(\mathbb{R}^r; \mathbb{R})$ such that

$$\rho_t \phi_i = \theta_i(\beta).$$

Theorem 7.6 *Under the hypothesis of Theorem 7.5, there does not exist an infinite independent sequence $(\phi_1, \ldots, \phi_k, \ldots)$ in $L^2(\mathbb{R}^n) \cap C_b(\mathbb{R}^n)$ such that $(\rho_t \phi_1, \ldots, \rho_t \phi_k, \ldots)$ admits a sufficient regular statistic β in finite dimension for some $t \in]0, T]$.*

Proof. Suppose that there exist β and θ_i satisfying Definition 7.16 such that $\rho_t \phi_i = \theta_i(\beta)$, then its density cannot verify the conclusion of Theorem 7.5. $\square$

Corollary 7.7 *The conclusion of Theorem 7.6 also holds for the filter π_t.*

Proof. The formula

$$\rho_t = \pi_t \exp\left(\sum_{i=1}^{p} \int_0^t \pi_s h^i \, dY_s^i - \frac{1}{2} \int_0^t |\pi_s h|^2 \, ds\right)$$

leads to the same contradiction as the one for ρ_t. $\square$

7.4 Remarks and bibliographical comments

7.4.1 Section 1

The stochastic calculus of variations, usually named "Malliavin calculus" has been introduced in a series of seminal papers by Malliavin ([28], [29], and [30]). See also his book [31]. This theory has inspired a huge literature, with alternative approaches by Bismut [4], Stroock [48], and many others. A good reference is the book [36]. A simplified version can be found in [35]. Note that the refined criterion (Proposition 7.3) is due to [9].

The theory of stochastic flows started with [14] and has been developped also by Kunita ([21], [22], [23]), Bismut ([5], [2]), and many others.

The generalized Itô formula, now called the Itô–Ventzell formula, giving the image of a semimartingale under a flow of diffeomorphism has been introduced in [49] and proved by Rozovskiĭ [44]. Bismut gave another proof in [2]; see also [6] and [5].

7.4.2 Section 2

The question of the existence of a density for the conditional laws of filtering theory has a long story. It goes back to [46] and [47]. Despite the fact that the proofs were not rigorous, this work was a source of inspiration for many authors. For a good summary of the works on this subject before 1972, see [27] (see also [17]).

A new approach came from the method of the reference probability, introduced in [50], which gives a linear equation for the unnormalized filter. Important contributions on the subject are: [18], [19], [40], [41]and [42] (see also the Summer School (Pardoux 1991)). These authors consider a stochastic partial differential equation (typically the Zakai equation) in a weak sense and show that the solution is regular under a partial ellipticity hypothesis.

Let us quote [20], where a regularity result is proved for the Zakai equation when the coefficients do not depend on the observation.

The idea of applying the Malliavin calculus to the existence and smoothness of a density for the filter is due to Michel [32]. This first paper used Stroock's approach to the Malliavin calculus and proved the regularity of the filter under a partial ellipticity hypothesis. Let us point out that in this framework, the coefficients can depend on the whole past of the observation. In their fundamental papers [7] and [8], the authors used Bismut's approach to the Malliavin calculus in order to get the regularity result under a hypoellipticity condition on the starting point. This is the method we have presented in this section.

Lemma 2.6 is due to Norris [35].

Another important contribution is [24], where the authors introduced what they call the "partial Malliavin calculus" and applied it to the filtering theory, obtaining the same result as Bismut–Michel. In the same spirit, Nualart and Zakai developed the formalism of the partial Malliavin calculus in [37]. In [13] regularity results for certain classes of heat equations are established, including the Zakai equation. These regularity results also lead to the smoothness of the density of the filter, under a global hypoellipticity hypothesis. See also [33].

7.4.3 Section 3

The problem of finite-dimensional filters received a lot of interest in the 1980's with the positive result of Beneš [1] and the introduction, in analogy with geometric control theory, of Lie algebraic methods (see [34], [10], [15]). To give a rough idea of the method, the existence of finite-dimensional filters implies a strong restriction on the Lie algebra associated to the Zakai equation, namely to be finite dimensional. Almost all the results are of negative type because of the infinite-dimensional character of this Lie algebra. Let us quote [12], [16], [25].

Starting with the result of the smoothness of the filter in the sense of Malliavin proved in [11], Ocone [38] developed a stochastic calculus of variations for stochastic partial differential equations with the Zakai equation as an example. Then, Ocone and Pardoux [39] (see also [43]) applied this particular Malliavin calculus in the context of finite-dimensional filters and obtained negative results for the filter at a fixed time. These papers contain the proofs of the results of Section 3.

Acknowledgement

We thank Dominique Michel for providing us with an unpublished set of notes, which inspired much of Section 7.2.

References

[1] Beneš, V. E. (1981), 'Exact finite-dimensional filters for certain diffusions with nonlinear drift', *Stochastics* (5)(1–2), 65–92.
[2] Bismut, J.-M. (1981a), 'A generalized formula of Itô and some other properties of stochastic flows', *Z. Wahrsch. Verw. Gebiete* **55**(3), 331–50.
[3] Bismut, J.-M. (1981b), Martingales, the Malliavin calculus and Hörmander's theorem, *in* 'Stochastic integrals (Proc. Sympos., Univ. Durham, Durham, 1980)', Vol. 851 of *Lecture Notes in Math.*, Springer, Berlin, pp. 85–109.
[4] Bismut, J.-M. (1981c), 'Martingales, the Malliavin calculus and hypoellipticity under general Hörmander's conditions', *Z. Wahrsch. Verw. Gebiete* **56**(4), 469–505.
[5] Bismut, J.-M. (1981d), *Mécanique aléatoire*, Vol. 866 of *Lecture Notes in Mathematics*, Springer Berlin. With an English summary.

[6] Bismut, J.-M. (1982), Mécanique aléatoire, *in* 'Tenth Saint Flour Probability Summer School—1980 (Saint Flour, 1980)', Vol. 929 of *Lecture Notes in Math.*, Springer, Berlin, pp. 1–100.

[7] Bismut, J.-M. & Michel, D. (1981), 'Diffusions conditionnelles. I. Hypoellipticité partielle', *J. Funct. Anal.* **44**(2), 174–211.

[8] Bismut, J.-M. & Michel, D. (1982), 'Diffusions conditionnelles. II. Générateur conditionnel. Application au filtrage', *J. Funct. Anal.* **45**(2), 274–92.

[9] Bouleau, N. & Hirsch, F. (1991), *Dirichlet forms and analysis on Wiener space*, Vol. 14 of *de Gruyter Studies in Mathematics*, Walter de Gruyter & Co., Berlin.

[10] Brockett, R. W. & Clark, J. M. C. (1980), The geometry of the conditional density equation, *in* 'Analysis and optimisation of stochastic systems (Proc. Internat. Conf., Univ. Oxford, Oxford, 1978)', Academic Press, London, pp. 299–309.

[11] Chaleyat-Maurel, M. (1986), 'Robustesse du filtre et calcul des variations stochastique', *J. Funct. Anal.* **68**(1), 55–71.

[12] Chaleyat-Maurel, M. & Michel, D. (1984a), 'Des résultats de non existence de filtre de dimension finie', *Stochastics* **13**(1–2), 83–102.

[13] Chaleyat-Maurel, M. & Michel, D. (1984b), 'Hypoellipticity theorems and conditional laws', *Z. Wahrsch. Verw. Gebiete* **65**(4), 573–97.

[14] Elworthy, K. D. (1978), Stochastic dynamical systems and their flows, *in* 'Stochastic analysis (Proc. Internat. Conf., Northwestern Univ., Evanston, Ill., 1978)', Academic Press, New York, pp. 79–95.

[15] Hazewinkel, M. & Marcus, S. I. (1982), 'On Lie algebras and finite-dimensional filtering', *Stochastics* **7**(1–2), 29–62.

[16] Hijab, O. (1983), Finite-dimensional causal functionals of Brownian motion, *in* 'Nonlinear stochastic problems (Algarve, 1982)', Vol. 104 of *NATO Adv. Sci. Inst. Ser. C Math. Phys. Sci.*, Reidel, Dordrecht, pp. 425–35.

[17] Kallianpur, G. (1980), *Stochastic filtering theory*, Vol. 13 of *Applications of Mathematics*, Springer, New York.

[18] Krylov, N. V. & Rozovskiĭ, B. L. (1978), 'Conditional distributions of diffusion processes', *Izv. Akad. Nauk SSSR Ser. Mat.* **42**(2), 356–78, 470.

[19] Krylov, N. V. & Rozovskiĭ, B. L. (1979), Stochastic evolution equations, *in* 'Current problems in mathematics, Vol. 14 (Russian)', Akad. Nauk SSSR, Vsesoyuz. Inst. Nauchn. i Tekhn. Informatsii, Moscow, pp. 71–147, 256.

[20] Kunita, H. (1981/82), 'Densities of a measure-valued process governed by a stochastic partial differential equation', *Systems Control Lett.* **1**(2), 100–4.

[21] Kunita, H. (1984), Stochastic differential equations and stochastic flows of diffeomorphisms, *in* 'École d'été de probabilités de Saint-Flour, XII—1982', Vol. 1097 of *Lecture Notes in Math.*, Springer, Berlin, pp. 143–303.

[22] Kunita, H. (1986), *Lectures on stochastic flows and applications*, Vol. 78 of *Tata Institute of Fundamental Research Lectures on Mathematics and Physics*, Published for the Tata Institute of Fundamental Research, Bombay.

[23] Kunita, H. (1990), *Stochastic flows and stochastic differential equations*, Vol. 24 of *Cambridge Studies in Advanced Mathematics*, Cambridge University Press, Cambridge.

[24] Kusuoka, S. & Stroock, D. (1984), 'The partial Malliavin calculus and its application to nonlinear filtering', *Stochastics* **12**(2), 83–142.

[25] Lévine, J. (1991), 'Finite-dimensional realizations of stochastic PDEs and application to filtering', *Stochastics Stochastics Rep.* **37**(1–2), 75–103.

[26] Liptser, R. S. & Shiryayev, A. N. (1977), *Statistics of random processes. I*, Springer, New York. General theory, trans. from the Russian by A. B. Aries, Applications of Mathematics, Vol. 5.

[27] Liptser, R. S. & Shiryayev, A. N. (1978), *Statistics of random process. II*, Springer-Verlag, New York. Applications, trans. from the Russian by A. B. Aries, Applications of Mathematics, vol. 6.
[28] Malliavin, P. (1978a), C^k-hypoellipticity with degeneracy, *in* 'Stochastic analysis (Proc. Internal. Conf., Northwestern Univ., Evanston, Ill., 1978)', Academic Press, New York, pp. 199–214.
[29] Malliavin, P. (1978b), C^k-hypoellipticity with degeneracy. II, *in* 'Stochastic analysis (Proc. Internal. Conf., Northwestern Univ., Evanston, Ill., 1978)', Academic Press, New York, pp. 327–40.
[30] Malliavin, P. (1978c), Stochastic calculus of variation and hypoelliptic operators, *in* 'Proceedings of the International Symposium on Stochastic Differential Equations (Res. Inst. Math. Sci., Kyoto Univ., Kyoto, 1976)', Wiley, New York, pp. 195–263.
[31] Malliavin, P. (1997), *Stochastic analysis*, Vol. 313 of *Grundlehren der Mathematischen Wissenschaften [Fundamental Principles of Mathematical Sciences]*, Springer, Berlin.
[32] Michel, D. (1981), 'Régularité des lois conditionnelles en théorie du filtrage non-linéaire et calcul des variations stochastique', *J. Funct. Anal.* **41**(1), 8–36.
[33] Michel, D. (1984), Conditional laws and Hörmander's condition, *in* 'Stochastic analysis (Katata/Kyoto, 1982)', Vol. 32 of *North-Holland Math. Library*, North-Holland, Amsterdam, pp. 387–408.
[34] Mitter, S. K. (1978), 'Filtering theory and quantum fields', *Asterisque* **75–6**, 199–206, [?]
[35] Norris, J. (1986), Simplified Malliavin calculus, *in* 'Séminaire de Probabilités, XX, 1984/85', Vol. 1204 of *Lecture Notes in Math.*, Springer, Berlin, pp. 101–30.
[36] Nualart, D. (1995), *The Malliavin calculus and related topics*, Probability and its Applications, Springer, New York.
[37] Nualart, D. & Zakai, M. (1989), The partial Malliavin calculus, *in* 'Séminaire de Probabilités, XXIII', Vol. 1372 of *Lecture Notes in Math.*, Springer, Berlin, pp. 362–81.
[38] Ocone, D. (1988), 'Stochastic calculus of variations for stochastic partial differential equations', *J. Funct. Anal.* **79**(2), 288–331.
[39] Ocone, D. & Pardoux, É. (1989), A Lie algebraic criterion for nonexistence of finite-dimensionally computable filters, *in* 'Stochastic partial differential equations and applications, II (Trento, 1988)', Vol. 1390 of *Lecture Notes in Math.*, Springer, Berlin, pp. 197–204.
[40] Pardoux, E. (1979), 'Stochastic partial differential equations and filtering of diffusion processes', *Stochastics* **3**(2), 127–67.
[41] Pardoux, E. (1981/2), 'Équations du filtrage non linéaire, de la prédiction et du lissage', *Stochastics* **6**(3–4), 193–231.
[42] Pardoux, É. (1985), 'Sur les équations aux dérivées partielles stochastiques, de type parabolique', *Astérisque* (132), 71–87. Colloquium in honor of Laurent Schwartz, Vol. 2 (Palaiseau, 1983).
[43] Pardoux, É. (1991), Filtrage non linéaire et équations aux dérivées partielles stochastiques associées, *in* 'École d'Été de Probabilités de Saint-Flour XIX—1989', Vol. 1464 of *Lecture Notes in Math.*, Springer, Berlin, pp. 37–163.
[44] Rozovskiĭ, B. L. (1973), 'A formula of Itô–Ventzel', *Vestnik Moskou Univ. Se.* **1**, 26–32.
[45] Rozovskiĭ, B. L. & Shimizu, A. (1981), 'Smoothness of solutions of stochastic evolution equations and the existence of a filtering transition density', *Nagoya Math. J.* **84**, 195–208.
[46] Stratonovich, R. L. (1959), 'On the theory of optimal non-linear filtration of random functions', *Theor. Probability Appl.* **4**, 223–5.
[47] Stratonovich, R. L. (1960), 'Conditional markov processes', *Theor. Probability Appl.* **5**, 156–78.

[48] Stroock, D. W. (1981), 'The Malliavin calculus, a functional analytic approach', *J. Funct. Anal.* **44**(2), 212–57.
[49] Ventzell, A. D. (1965), 'On the equations of the theory of conditional processes', *Theory Probab. Appl.* **10**, 357–61.
[50] Zakai, M. (1969), 'On the optimal filtering of diffusion processes', *Z. Wahrscheinlichkeitstheories und Verw. Gebiete* **11**, 230–43.

·8·
Chaos Approach to Nonlinear Filtering

S. V. Lototsky

8.1 Introduction

Let us recall the general mathematical formulation of the filtering problem [32, 35, 51, 71]. Consider two random processes, the *State* $X = X(t)$ and *Observations* $Y = Y(t)$, $t \geq 0$, both defined on a suitable stochastic basis $\mathbb{F} = (\Omega, \mathcal{F}, (\mathcal{F}_t)_{t\geq 0}, \mathbb{P})$. Let $f = f(t, X_{0,t}, Y_{0,t})$ be a square integrable measurable functional, depending at time t on the trajectories $X_{0,t}$ and $Y_{0,t}$ of the processes X and Y up to time t. The filtering problem is to find the *Optimal Filter* estimate of f, that is, the best mean-square estimate $\widehat{f}_t$ of $f(t, X_{0,t}, Y_{0,t})$ on the basis of the observations $Y(s)$, $0 \leq s \leq t$. From the basic probability theory, this estimate is known to be the conditional expectation

$$\widehat{f}_t = \mathbb{E}(f(t, X_{0,t}, Y_{0,t})|\mathcal{Y}_t), \tag{8.1}$$

where $\mathcal{Y}_t$ is the σ-algebra generated by the random variables $Y(s)$, $0 \leq s \leq t$. The subject of the mathematical theory of filtering is finding suitable ways of computing this conditional expectation, either exactly or approximately.

If $f(t, X_{0,t}, Y_{0,t}) = f(X(t))$, then, for a large class of processes X, Y, the optimal filter (8.1) is computable by the formula

$$\widehat{f}_t = \frac{\int_{\mathbb{S}} f(x)\rho(t, x)dx}{\int_{\mathbb{S}} \rho(t, x)dx}, \tag{8.2}$$

where

- $\mathbb{S}$ is the phase space of the unobserved process X;
- $\rho = \rho(t, x)$ is a random field called `unnormalized filtering density`.

In the *diffusion filtering model*, the function ρ is a unique solution of a linear stochastic parabolic equation driven by the observation process Y; the equation is known as the *Zakai equation*.

Below is a general idea of the filtering algorithms discussed in this Chapter.

By definition, the unnormalized filtering density u is a function of the time t, space x, and the elementary outcome ω; by convention, the dependence on ω is usually not shown, but always implied. Assume that the filtering problem is considered on a fixed time interval $[0, T]$. Similar to the usual Fourier series expansion, we can write

$$\rho(t, x, \omega) = \sum_{k,\ell=1}^{\infty} \varphi_{k\ell}(t) e_k(x) \xi_\ell(\omega), \tag{8.3}$$

where $\varphi_{k\ell}$, $k, \ell \geq 1$, are *deterministic* functions of time, $\{e_k, k \geq 1\}$ is an orthonormal basis in $L_2(S)$, and $\{\xi_\ell, \ell \geq 1\}$ is an orthonormal basis in a suitable space of random variables generated by the observation process $Y(s)$, $0 \leq s \leq T$.

Let $0 = t_0 < t_1 < \ldots < t_n = T$, with $t_k = k\Delta$, be a uniform partition of the interval $[0, T]$. We will see that the uniqueness of the unnormalized filtering density and linearity of the Zakai equation imply the following recursive version of (8.3):

$$\rho(t_i, x, \omega) = \sum_{k=1}^{\infty} \varphi_k^i(\omega) e_k(x), \qquad \varphi_k^i(\omega) = \sum_{l,n=1}^{\infty} q_k^{\ell n} \varphi_n^{i-1}(\omega) \xi_l^i(\omega), \tag{8.4}$$

where $q_k^{\ell n}$ are real numbers and $\{\xi_\ell^i, \ell \geq 1\}$ is an orthonormal basis in the suitable space of random variables generated by the observation process $Y(s)$, $t_{i-1} \leq s \leq t_i$. In most applications, the space S is all or part of $\mathbb{R}^d$, so that there is no difficulty in selecting the functions e_k. Construction of the basis in the space of random variables generated by the observation process is less straightforward and relies on a theorem of Cameron and Martin [8]. If the filtering model is time-homogeneous, then the numbers $q_k^{\ell n}$ can be computed in advance, because they depend only on the timestep Δ and the coefficients in the state and observation equations.

A recursive approximation of u is obtained by keeping only finitely many terms in each of the sums in (8.4); We call the corresponding algorithm for computing the approximation of u a *spectral separating scheme*. There are two fundamentally different ways to truncate the expansions in (8.4), leading to the spectral separating schemes of the first and second kind.

This chapter is essentially based on a similar review by the author [54]. The structure of the chapter is as follows. Sections 8.2 and 8.3 contain the background information about nonlinear filtering and Wiener chaos. Section 8.2 introduces the time-homogeneous diffusion filtering model and outlines the derivation of representation (8.2) for the optimal filter. Section 8.3 reviews the construction of the Cameron–Martin basis and presents the corresponding expansion for the solution of a linear stochastic evolution equation. Section 8.4 is the main part of the chapter and is mostly based on papers [53] and [55]. In this section, we derive representation (8.4) for the solution of the Zakai equation in the time-homogeneous diffusion filtering model, study the spectral separating schemes, and preform a comparative analysis of the results from [53] and [55]. Section 8.5 discusses other filtering models and other approaches to constructing the optimal filter, as well as possible connections with the

Wiener chaos method. It appears that the spectral separating schemes have both theoretical and practical interest: (i) these algorithms have a potential for good real-time performance when the dimension of the state process is large, and (ii) the implementation and analysis of the algorithms are not affected by the presence of the observation noise in the state equation.

8.2 The filtering problem

8.2.1 Change of measure

Consider two continuous random processes $X = X(t)$ and $Y = Y(t)$, $0 \leq t \leq T$, both defined on a suitable stochastic basis $\mathbb{F} = (\Omega, \mathcal{F}, (\mathcal{F}_t)_{t \geq 0}, \mathbb{P})$. Let $f = f(t, X_{0,t}, Y_{0,t})$ be a square integrable measurable functional, depending at time t on the trajectories $X_{0,t}$ and $Y_{0,t}$ of the processes X and Y up to time t. The filtering problem is to find the conditional expectation

$$\widehat{f}_t = \mathbb{E}\left(f(t, X_{0,t}, Y_{0,t}) | \mathcal{Y}_t\right), \tag{8.5}$$

where $\mathcal{Y}_t$ is the σ-algebra generated by the random variables $Y(s)$, $0 \leq s \leq t$. A general method of computing this conditional expectation is to simplify the distribution of Y by introducing a new probability measure on $(\Omega, \mathcal{F})$. For example, assume that

$$Y(t) = \int_0^t h(s, X(s)) ds + W(t), \tag{8.6}$$

where h is a bounded measurable function and W is a standard Wiener process. Define

$$Z^o = \exp\left(\int_0^T h(s, X(s)) dY(s) - \frac{1}{2} h^2(s, X(s)) ds\right)$$

and let $\mathbb{P}^o$ be the probability measure on $(\Omega, \mathcal{F})$ such that

$$\frac{d\mathbb{P}}{d\mathbb{P}^o} = Z^o. \tag{8.7}$$

By the Girsanov theorem, under the measure $\mathbb{P}^o$ the process Y is a standard Wiener process.

Accordingly, we need a formula connecting conditional expectations of the type ((8.5)) for different measures

Lemma 8.1 *Let $\mathbb{P}_1$ be a probability measure on $(\Omega, \mathcal{F})$ such that*

$$\frac{d\mathbb{P}}{d\mathbb{P}_1} = Z \tag{8.8}$$

for a $\mathbb{P}_1$-almost everywhere positive and finite random variable Z. Denote by $\mathbb{E}_1$ the expectation with respect to $\mathbb{P}_1$. If $\mathbb{E}_1(Z^2) < \infty$ and η is a random variable such that $\mathbb{E}_1(\eta^2) < \infty$, then $\mathbb{E}|\eta| < \infty$ and, for every sigma-algebra $\mathcal{G} \subset \mathcal{F}$,

$$\mathbb{E}(\eta|\mathcal{G}) = \frac{\mathbb{E}_1(\eta \, Z|\mathcal{G})}{\mathbb{E}_1(Z|\mathcal{G})} \tag{8.9}$$

Proof. Since $\mathbb{E}|\eta| = \mathbb{E}_1(|\eta| \, Z)$, we have $\mathbb{E}|\eta| < \infty$ by the Cauchy–Schwartz inequality. Next, let $\zeta = \mathbb{E}(\eta|\mathcal{G})$. Then

$$\mathbb{E}(\zeta \, \xi) = \mathbb{E}(\eta \, \xi) \tag{8.10}$$

for every bounded, $\mathcal{G}$-measurable random variable ξ. On the other hand, by (8.8),

$$\mathbb{E}(\zeta \, \xi) = \mathbb{E}_1(\zeta \, \xi \, Z), \quad \mathbb{E}(\eta \, \xi) = \mathbb{E}_1(\eta \, \xi \, Z).$$

Since ξ is arbitrary, we conclude that $\mathbb{E}_1(\zeta \, Z|\mathcal{G}) = \mathbb{E}_1(\eta \, Z|\mathcal{G})$, which, because of $\mathcal{G}$-measurability of ζ, is the same as (8.9). □

Remark 8.1 Equality (8.9) is an extension of the Bayes formula; see, for example, Liptser and Shiryaev [51, Lemma 7.4] or Øksendal [68, Lemma 8.6.2].

8.2.2 The diffusion filtering model

The most general model of the processes X, Y for continuous-time filtering problem in Gaussian white noise is

$$\begin{aligned} dX(t) &= A(t, X_{0,t}, Y_{0,t})dt + B(t, X_{0,t}, Y_{0,t})d\mathcal{V}(t), \\ dY(t) &= C(t, X_{0,t}, Y_{0,t})dt + D(t, X_{0,t}, Y_{0,t})d\mathcal{V}(t), \end{aligned} \tag{8.11}$$

where A, B, C, D are measurable functionals of suitable dimensions, and $\mathcal{V}$ is a Wiener process, also of a suitable dimension. Since the process $\mathcal{V}$ is multidimensional and the functions B, D are matrix-valued, the state and observation equations are, in general, driven by *different* noise processes, even though the first look at equation (8.11) might suggest otherwise. Under the linear growth and Lipschitz continuity conditions on A, B, C, D, the system of equations (8.11) has a unique strong solution [51, Theorem 4.6]. In the *diffusion filtering model*, the functionals A, B, C, D at time moment t depend only on $X(t)$ and $Y(t)$.

Solution of the filtering problem for (8.11) requires additional conditions on the coefficients in the equation, and under these conditions the system (8.11) is reduced to an upper triangular form

$$\begin{aligned} dX(t) &= b(t, X_{0,t}, Y_{0,t})dt + \sigma(t, X_{0,t}, Y_{0,t})dV + \varrho(t, X_{0,t}, Y_{0,t})dW(t), \\ dY(t) &= h(t, X_{0,t}, Y_{0,t})dt + H(t, Y_{0,t})dW(t), \end{aligned} \tag{8.12}$$

with modified coefficients b, σ, ϱ, h, H and new Wiener processes V, W; Section 6.0.2 in [71], together with Lemma 10.4 in [51], provide the details of this reduction. The square matrix H does not depend on X and is uniformly positive definite; with no dependence on t and Y, it will be just a constant invertible matrix.

There are specific reasons for the additional assumptions about the function H in (8.12): without these assumptions, the procedure of estimating X from the observations of Y is somewhat different. In particular, if H depends on the state process X, then some information about X can be obtained from the quadratic variation of Y. If H is not uniformly positive definite, then we have filtering problem with degenerate observation noise [5]. Filtering with perfect observations ($H = 0$) is also possible (Joannides, LeGland [34, 33]).

In what follows, we consider a time homogeneous diffusion filtering model with no observation process Y in the coefficients. Let $(\Omega, \mathcal{F}, \mathbb{P})$ be a complete probability space with independent standard Wiener processes $V = V(t)$ and $W = W(t)$ of dimensions d_1 and r respectively. Let X_0 be a random variable independent of W and V. We consider a time-homogeneous diffusion filtering model, in which the unobserved d-dimensional state, or signal, process $X = X(t)$ and the r-dimensional observation process $Y = Y(t)$ satisfy the Itô stochastic ordinary differential equations

$$\begin{aligned} dX(t) &= b(X(t))dt + \sigma(X(t))dV(t) + \varrho(X(t))dW(t), \\ dY(t) &= h(X(t))dt + dW(t), \ 0 < t \leq T; \\ X(0) &= X_0, \ Y(0) = 0, \end{aligned} \quad (8.13)$$

where $b(x) \in \mathbb{R}^d$, $\sigma(x) \in \mathbb{R}^{d \times d_1}$, $\varrho(x) \in \mathbb{R}^{d \times r}$, $h(x) \in \mathbb{R}^r$. We call (8.13) *time-homogeneous* because the functions b, σ, ϱ, and h do not depend on time. Possible extensions of the model are discussed in Section 8.5.

Assumption R1. The components of the matrix functions σ and ϱ are $C_b^3(\mathbb{R}^d)$, that is, bounded and three times continuously differentiable on $\mathbb{R}^d$ so that all the derivatives are also bounded; the components of the functions b and h are $C_b^2(\mathbb{R}^d)$, and the distribution of the random variable X_0 has a density p_0 with respect to the Lebesgue measure on $\mathbb{R}^d$.

Under Assumption **R1** system (8.13) has a unique strong solution [38, Theorems 5.2.5 and 5.2.9].

Let $f = f(x)$ be a scalar measurable function on $\mathbb{R}^d$ so that $\sup_{0 \leq t \leq T} \mathbb{E}|f(X(t))|^2 < \infty$. In what follows, we show that the optimal filter

$$\widehat{f}_t = \mathbb{E}\left(f(X(t))|\mathcal{Y}_t\right)$$

has the representation (8.2) and derive the equation for the function ρ. We start by introducing the unnormalized optimal filter.

Define a new probability measure $\widetilde{\mathbb{P}}$ on $(\Omega, \mathcal{F})$ as follows: for $A \in \mathcal{F}$,

$$\widetilde{\mathbb{P}}(A) = \int_A Z_T^{-1} d\mathbb{P}, \quad (8.14)$$

where

$$Z_T = \exp\left(\int_0^T h^\top(X(s))dY(s) - \frac{1}{2}\int_0^T |h(X(s))|^2 ds\right). \tag{8.15}$$

HERE AND BELOW, IF $\zeta \in \mathbb{R}^k$, THEN ζ IS A *column* VECTOR, $\zeta^\top = (\zeta_1, \ldots, \zeta_k)$, AND $|\zeta|^2 = \zeta^\top \zeta$; SIMILARLY, THE TRANSPOSE OF A MATRIX A IS $A^\top$.

The measures $\mathbb{P}$ and $\widetilde{\mathbb{P}}$ are equivalent because the function h is bounded. The expectation with respect to the measure $\widetilde{\mathbb{P}}$ will be denoted by $\widetilde{\mathbb{E}}$. The boundedness of h implies $\widetilde{\mathbb{E}}(Z_T^2) < \infty$ and $\mathbb{E}(Z_T^{-2}) < \infty$.

Theorem 8.1 *The measure $\widetilde{\mathbb{P}}$ has the following properties:*

P1. *Under the measure $\widetilde{\mathbb{P}}$, the distributions of the Wiener process V and the random variable X_0 are unchanged, the observation process Y is a standard Wiener process, and the state process X satisfies*

$$\begin{aligned} dX(t) &= b(X(t))dt + \sigma(X(t))dV(t) \\ &\quad + \varrho(X(t))\left(dY(t) - h(X(t))dt\right), \ 0 < t \leq T; \\ X(0) &= X_0; \end{aligned} \tag{8.16}$$

P2. *Under the measure $\widetilde{\mathbb{P}}$, the Wiener processes W and Y and the random variable X_0 are independent of one another;*

P3. *The optimal filter $\widehat{f}_t$ satisfies*

$$\widehat{f}_t = \frac{\widetilde{\mathbb{E}}(f(X(t))Z_t|\mathcal{Y}_t)}{\widetilde{\mathbb{E}}(Z_t|\mathcal{Y}_t)}. \tag{8.17}$$

P4. *If $\xi \in L_2(\Omega, \widetilde{\mathbb{P}})$, then $\xi \in L_1(\Omega, \mathbb{P})$ and*

$$\mathbb{E}|\xi| \leq C\sqrt{\widetilde{\mathbb{E}}\xi^2}, \tag{8.18}$$

where C depends only on T and $\sup_{x \in \mathbb{R}^d} |h(x)|$.

Proof. Properties **P1** and **P2** follow from the Girsanov theorem; notice that equation (8.16) is equivalent to the first equation in (8.13) because $dW = dY - hdt$. Properties **P3** and **P4** follow from Lemma 8.1. □

Remark 8.2 Various aspects of the above result are discussed in Kallianpur [35], Liptser and Shiryaev [51], Rozovskiĭ [71], Wong [76], and in many other references. For example, if the observation noise is independent of the state process X, then the distribution of X under $\widetilde{\mathbb{P}}$ does not change, and the processes X, Y become independent. Moreover, the change of measure (8.14) in this *uncorrelated noise model* works without assuming any special structure of the process X and with unbounded h: we still have $\widetilde{\mathbb{E}}Z_T = 1$ by conditioning on X. Thus, if $\varrho = 0$, the statement of the theorem is a particular case of the Kallianpur–Striebel formula (Kallianpur and Striebel [36]).

The random variable $\phi_t[f] = \widetilde{\mathbb{E}}\left[f(X(t))Z_t|\mathcal{Y}_t\right]$ is called the *unnormalized optimal filter*. To describe the time evolution of $\phi_t[f]$, consider the partial differential operators

$$\mathcal{L}g(x) = \frac{1}{2}\sum_{i,j=1}^{d}\left((\sigma(x)\sigma^\top(x))_{ij} + (\varrho(x)\varrho^\top(x))_{ij}\right)\frac{\partial^2 g(x)}{\partial x_i \partial x_j} + \sum_{i=1}^{d} b_i(x)\frac{\partial g(x)}{\partial x_i};$$

$$\mathcal{M}_k g(x) = h_\ell(x)g(x) + \sum_{i=1}^{d}\varrho_{ik}(x)\frac{\partial g(x)}{\partial x_i}, \quad k = 1,\ldots,r.$$

Then the time evolution of $\phi_t[f]$ can be described as follows; for the proof, see, for example, Rozovskiĭ [71, Theorem 5.3.1].

Theorem 8.2 *Under assumptions* **R1**, *for every* $f \in C_b^2(\mathbb{R}^d)$,

$$\phi_t[f] = \int_{\mathbb{R}^d} p_0(x)f(x)dx + \int_0^t \phi_s[\mathcal{L}f]ds + \sum_{k=1}^{r}\int_0^t \phi_s[\mathcal{M}_k f]dY_k. \tag{8.19}$$

Assumptions **R1** are, in fact, strong enough to ensure existence of a density of ϕ_t, that is, representation (8.2) on page 231. To describe the corresponding result, we need some additional constructions. Consider the adjoint operators of $\mathcal{L}$ and $\mathcal{M}_k$:

$$\mathcal{L}^* g(x) = \frac{1}{2}\sum_{i,j=1}^{d}\frac{\partial^2}{\partial x_i \partial x_j}\left((\sigma(x)\sigma^\top(x))_{ij}g(x) + (\varrho(x)\varrho^\top(x))_{ij}g(x)\right) - \sum_{i=1}^{d}\frac{\partial}{\partial x_i}(b_i(x)g(x));$$

$$\mathcal{M}_k^* g(x) = h_\ell(x)g(x) - \sum_{i=1}^{d}\frac{\partial}{\partial x_i}(\varrho_{ik}(x)g(x)), \quad k = 1,\ldots,r.$$

Let $\mathbf{H}^\gamma$, $\gamma \in \mathbb{R}$, be the Sobolev space with the norm

$$\|f\|_\gamma = \left(\int_{\mathbb{R}^d}(1+|y|^2)^{\gamma/2}|\check{f}(y)|^2 dy\right)^{1/2}, \tag{8.20}$$

where $\check{f} = \check{f}(y)$ is the Fourier transform of f; $\mathbf{H}^0 = L_2(\mathbb{R}^d)$. Both the inner product in $L_2(\mathbb{R}^d)$ and the duality between $\mathbf{H}^1$ and $\mathbf{H}^{-1}$ relative to $L_2(\mathbb{R}^d)$ will be denoted by $(\cdot,\cdot)_0$.

Note that the operators $\mathcal{L}, \mathcal{L}^*$ are bounded from $\mathbf{H}^1$ to $\mathbf{H}^{-1}$ and operators $\mathcal{M}, \mathcal{M}^*$ are bounded from $\mathbf{H}^1$ to $L_2(\mathbb{R}^d)$. Moreover, direct calculations show that, under Assumption **R1**, there exists a number $C > 0$ such that, for every $g \in \mathbf{H}^1$,

$$2(\mathcal{L}^* g, g)_0 + \sum_{k=1}^{r}\|\mathcal{M}_k^* g\|_0^2 \leq C\|g\|_0^2. \tag{8.21}$$

If the matrix $\sigma\sigma^\top$ is uniformly positive definite, that is, there exists a $\delta > 0$ so that, for all $x, y \in \mathbb{R}^d$,

$$\sum_{i,j=1}^{d}\sum_{p=1}^{d_1} \sigma_{ip}(x)\sigma_{jp}(x)y_i y_j \geq \delta |y|^2, \tag{8.22}$$

then a stronger version of (8.21) holds:

$$2(\mathcal{L}^* g, g)_0 + \sum_{k=1}^{r} \|\mathcal{M}_k^* g\|_0^2 \leq -\delta \|g\|_1^2 + C \|g\|_0^2. \tag{8.23}$$

The following result follows by combining the above Theorem 8.2 with Theorem 4.2.1 in Rozovskiĭ [71].

Theorem 8.3 *In addition to assumptions* **R1** *suppose that the initial density p_0 belongs to the space* $\mathbf{H}^1$. *Then there exists a unique $\mathcal{Y}_t$-adapted random field $\rho = \rho(t, x)$, $t \in [0, T]$, $x \in \mathbb{R}^d$, with the following properties:*

U1 *The function $\rho(t, x)$ is an element of $L_2(\Omega, \widetilde{\mathbb{P}}; C([0, T], \mathbf{H}^1))$ and, for every $v \in \mathbf{H}^1$, satisfies the equality*

$$(\rho, v)_0(t) = (p_0, v)_0 + \int_0^t (\rho, \mathcal{L}v)_0(s)ds + \int_0^t \sum_{k=1}^{r}(\rho, \mathcal{M}_k v)_0(s) dY_k(s) \tag{8.24}$$

on the same set of $\widetilde{\mathbb{P}}$-measure one for all $0 \leq t \leq T$. In other words, $\rho = \rho(t, x)$ is the unique generalized solution of the stochastic partial differential equation

$$d\rho(t, x) = \mathcal{L}^*\rho(t, x)dt + \sum_{k=1}^{r} \mathcal{M}_k^*\rho(t, x) dY_k(t), \ 0 < t \leq T, \ x \in \mathbb{R}^d; \tag{8.25}$$
$$\rho(0, x) = p_0(x).$$

U2 *The equality*

$$\widetilde{\mathbb{E}}\left(f(X(t))Z_t | \mathcal{Y}_t\right) = \int_{\mathbb{R}^d} f(x)\rho(t, x)dx \tag{8.26}$$

holds for all bounded measurable functions f.

The random field $\rho = \rho(t, x)$ is called the *Unnormalized Filtering Density*, and equation (8.25), the *Zakai (filtering) equation*.[1]

Remark 8.3 If condition (8.22) holds, then equation (8.25) has a unique solution under weaker regularity conditions on the coefficients and initial condition: it is enough to have the components of σ, ϱ in $C_b^2(\mathbb{R}^d)$, the components of b, h in $C_b^2(\mathbb{R}^d)$, and initial density p_0 in $L_2(\mathbb{R}^d)$; see Theorem 4.1.1 in [71].

8.3 Chaos expansions and nonlinear filtering

In the first part of this section, we review the two main versions of the chaos expansion: using the Cameron–Martin basis and multiple integrals. For a linear *stochastic* evolution equation, we show that the coefficients of the Cameron–Martin expansion of the solution satisfy a lower-triangular system of linear *deterministic* evolution equations. In the second part, we derive the chaos expansions of the unnormalized optimal filter and its density.

8.3.1 The Wiener chaos

For a fixed $T > 0$, let $\mathbb{F} = (\Omega, \mathcal{F}, \{\mathcal{F}_t\}_{0 \leq t \leq T}, \mathbb{P})$ be a stochastic basis with the usual assumptions and $W = (W_1(t), \ldots, W_r(t))$, $0 \leq t \leq T$, an r-dimensional Wiener process on $\mathbb{F}$. Denote by $\mathcal{F}_t^W$ the sigma-algebra generated by the random variables $W_k(s)$, $k = 1, \ldots, r$; $s \leq t$, and by $L_2(\mathbb{W})$, the Hilbert space of $\mathcal{F}_T^W$-measurable square integrable random variables. The Cameron–Martin basis is a special orthonormal basis in the space $L_2(\mathbb{W})$.

Let $\mathfrak{m} = \{m_\ell, \ell \geq 1\}$ be an orthonormal basis in $L_2(0, T)$ and define independent standard Gaussian random variables

$$\xi_{k\ell} = \int_0^T m_\ell(s) d W_k(s).$$

Consider the set of multi-indices

$$\mathcal{J} = \left\{\alpha = (a_\ell^k, k = 1, \ldots, r, \ell \geq 1), \ a_\ell^k \in \{0, 1, 2, \ldots\}, \ \sum_{k,\ell} a_\ell^k < \infty \right\}. \quad (8.27)$$

The set $\mathcal{J}$ is countable, and, for every $\alpha \in \mathcal{J}$, only finitely many of a_ℓ^k are not equal to zero. For $\alpha \in \mathcal{J}$, we write

$$|\alpha| = \sum_{k,\ell} a_\ell^k, \ \alpha! = \prod_{k,\ell} a_\ell^k!,$$

and define the collection $\Xi = \{\xi_\alpha, \alpha \in \mathcal{J}\}$ of random variables by

$$\xi_\alpha = \frac{1}{\sqrt{\alpha!}} \prod_{k,\ell} H_{a_\ell^k}(\xi_{k\ell}), \quad (8.28)$$

where

$$H_n(t) = (-1)^n e^{t^2/2} \frac{d^n}{dt^n} e^{-t^2/2} \quad (8.29)$$

is n-th Hermite polynomial. Recall that $H_0(t) = 1$, $H_1(t) = t$, $H_2(t) = t^2 - 1$, $H_3(t) = t^3 - 3t$, etc. The *N-th Wiener chaos* is the linear span of the random variables ξ_α, $|\alpha| = N$. Finally, for every $\alpha \in \mathcal{J}$ we define the set K_α, called the *characteristic set* of α, as the ordered collection of pairs (k_j, ℓ_j), $j = 1, \ldots, |\alpha|$

such that $k_j \leq k_{j+1}$; $\ell_j \leq \ell_{j+1}$ if $k_j = k_{j+1}$, and for every sequence of positive numbers $b_{k\ell}$, $k = 1, \ldots, r$; $\ell \geq 1$,

$$\prod_{k,\ell} b_{k\ell}^{a_\ell^k} = \prod_{(k,\ell) \in K_a} b_{k\ell} \qquad (8.30)$$

As an example, let $r = 2$ and consider the multi-index

$$a = \begin{pmatrix} 0 & 1 & 0 & 2 & 3 & 0 & \cdots \\ 2 & 0 & 0 & 1 & 3 & 0 & \cdots \end{pmatrix}$$

with four nonzero entries $a_2^1 = 1$; $a_4^1 = 2$; $a_5^1 = 3$; $a_1^2 = 2$; $a_2^4 = 1$; $a_5^2 = 3$. Then

$K_a = \{(1, 2), (1, 4), (1, 4), (1, 5), (1, 5), (1, 5), (2, 1), (2, 1), (2, 4), (2, 5), (2, 5), (2, 5)\}$,

and the corresponding basis element is

$$\xi_a = \frac{H_1(\xi_{2,1})}{\sqrt{1!}} \cdot \frac{H_2(\xi_{4,1})}{\sqrt{2!}} \cdot \frac{H_3(\xi_{5,1})}{\sqrt{3!}} \cdot \frac{H_2(\xi_{1,2})}{\sqrt{2!}} \cdot \frac{H_1(\xi_{4,2})}{\sqrt{1!}} \cdot \frac{H_3(\xi_{5,2})}{\sqrt{3!}}$$

$$= \xi_{2,1} \left(\frac{\xi_{4,1}^2 - 1}{\sqrt{2}} \right) \left(\frac{\xi_{5,1}^3 - 3\xi_{5,1}}{\sqrt{6}} \right) \left(\frac{\xi_{1,2}^2 - 1}{\sqrt{2}} \right) \xi_{4,2} \left(\frac{\xi_{5,2}^3 - 3\xi_{5,2}}{\sqrt{6}} \right).$$

Theorem 8.4 *The collection $\Xi = \{\xi_a, a \in \mathcal{J}\}$ is an orthonormal basis in $L_2(\mathbb{W})$.*

Proof. This is a version of the classical result due to Cameron and Martin [8]. Other possible references are [26, Theorem 1.9], [27, Theorem 2.2.3], and [59, Theorem 2.1]. □

If $\eta \in L_2(\mathbb{W})$, then, by Theorem 8.4, we have the following *Cameron–Martin expansion of η*:

$$\eta = \sum_{a \in \mathcal{J}} \mathbb{E}(\eta \xi_a) \xi_a \quad \text{and} \quad \mathbb{E}|\eta|^2 = \sum_{a \in \mathcal{J}} |\mathbb{E}(\eta \xi_a)|^2. \qquad (8.31)$$

Theorem 8.5 *For every $\eta \in L_2(\mathbb{W})$, there exists a collection of nonrandom functions $\eta_{k_1,\ldots,k_n}$, $n \geq 1$, $k_j = 1, \ldots, r$, such that every $\eta_{k_1,\ldots,k_n}$ is an element $L_2((0, T)^n)$ and*

$$\eta = \mathbb{E}\eta + \sum_{k=1}^r \int_0^T \eta_k(s) dW_k(s) + \sum_{k_1,k_2=1}^r \int_0^T \int_0^{s_2} \eta_{k_1,k_2}(s_1, s_2) dW_{k_1}(s_1) dW_{k_2}(s_2)$$

$$+ \sum_{n \geq 3} \sum_{k_1,\ldots,k_n=1}^r \int_0^T \int_0^{s_n} \cdots \int_0^{s_2} \eta_{k_1,\ldots,k_n}(s_1, \ldots, s_n) dW_{k_1}(s_1) \ldots dW_{k_n}(s_n).$$

$$(8.32)$$

The collection of the functions $\eta_{k_1,\ldots,k_n}$ is unique if every $\eta_{k_1,\ldots,k_n}(s_1, \ldots, s_n)$ is required to be a symmetric function of $(s_1, \ldots, s_n)$.

To study the time evolution of the Cameron–Martin expansion, we need the following technical result.

Proposition 8.1 *Define $\xi_a(t)$ by*

$$\xi_a(t) = \mathbb{E}\left(\xi_a | \mathcal{F}_t^W\right). \tag{8.33}$$

Then

$$\xi_a(t) = \begin{cases} 1, & |a| = 0; \\ \int_0^t \sum_{k,\ell} \sqrt{a_\ell^k}\, \xi_{a^-(k,\ell)}(s) m_\ell(s) d\,W_k(s), & |a| > 0, \end{cases} \tag{8.34}$$

where $a^-(k, \ell)$ is the multi-index with components

$$\left(a^-(k,\ell)\right)^i_j = \begin{cases} \max\left(a_\ell^k - 1, 0\right), & \text{if } i = k \text{ and } j = \ell, \\ a^i_j, & \text{otherwise.} \end{cases} \tag{8.35}$$

Proof. If $|a| = 0$, then, by definition, $\xi_a = 1$. Otherwise, let $h = (h_1(t), \ldots, h_r(t))$ be an r-vector such that each h_k is a finite linear combination of the elements of $\mathfrak{m}$. Define

$$\mathcal{E}(t, h) = \exp\left(\sum_{k=1}^r \left(\int_0^t h_k(s) d\,W_k(s) - \frac{1}{2}\int_0^t |h_k(s)|^2 ds\right)\right). \tag{8.36}$$

We also introduce the notations

$$\mathcal{E}(h) = \mathcal{E}(T, h);\ h_{k\ell} = \int_0^T h_k(t) m_\ell(t) dt,\ m_\ell \in \mathfrak{m};\ h^a = \prod_k h_{k\ell}^{a_\ell^k} = \prod_{(k,\ell) \in K_a} h_{k\ell},\ a \in \mathcal{J}.$$

Then $h_k(t) = \sum_{\ell \geq 1} h_{k\ell} m_\ell(t)$, $\int_0^T |h_k^2(t)|^2 dt = \sum_{\ell \geq 1} h_{k\ell}^2$. The generating function formula for the Hermite polynomials $\sum_{k \geq 0} H_k(y) x^k / k! = \exp\left(xy - (x^2/2)\right)$, applied to (8.36), implies

$$\mathcal{E}(h) = \sum_{a \in \mathcal{J}} \frac{h^a}{\sqrt{a!}} \xi_a, \tag{8.37}$$

leading to an alternative representation of ξ_a:

$$\xi_a = \frac{1}{\sqrt{a!}} \frac{\partial^{|a|}}{\partial h^a} \mathcal{E}(h)\bigg|_{h=0}. \tag{8.38}$$

On the other hand, by the Itô formula applied to (8.36),

$$\mathcal{E}(t, h) = 1 + \int_0^t \mathcal{E}(s, h) \sum_{i,k} h_{ik} m_i(s) d\,W_k(s). \tag{8.39}$$

In particular,

$$\mathbb{E}\left(\mathcal{E}(h) | \mathcal{F}_t^W\right) = \mathcal{E}(t, h). \tag{8.40}$$

Taking the conditional expectation with respect to $\mathcal{F}_t^W$ on both sides of (8.38), using the martingale property (8.40) of $\mathcal{E}(t, h)$, and exchanging the expectation and differentiation operations (for example, by Theorem 3.3.3(iii) in Kunita [45]), we conclude that

$$\xi_\alpha(t) = \frac{1}{\sqrt{\alpha!}} \frac{\partial^{|\alpha|}}{\partial h^\alpha} \mathcal{E}(t, h)\bigg|_{h=0}. \tag{8.41}$$

Equality (8.33) now follows from (8.39). □

Corollary 8.1 (cf. Itô [28, Theorem 8.4]) *Let α be a multi-index with $|\alpha| = n \geq 1$ and the characteristic set $\{(k_1, \ell_1), \ldots, (k_n, \ell_n)\}$ and let $\mathcal{P}_n$ be the collection of all permutations of the set $(1, \ldots, n)$. Then*

$$\xi_\alpha = \frac{1}{\sqrt{\alpha!}} \sum_{\sigma \in \mathcal{P}_n} \int_0^T \int_0^{s_n} \cdots \int_0^{s_2} m_{\ell_{\sigma(1)}}(s_1) \ldots m_{\ell_{\sigma(n)}}(s_n) dW_{k_{\sigma(1)}}(s_1) \ldots dW_{k_{\sigma(n)}}(s_n). \tag{8.42}$$

Proof. This follows from (8.34) by induction; the details are left to the reader. □

We are now ready to derive the Cameron–Martin and multiple integral expansions of the solution of a linear stochastic evolution equation.

Theorem 8.6 *Let $U = U(t)$, $0 \leq t \leq T$, be a square integrable, $\mathcal{F}_t^W$-adapted random process with values in a Hilbert space X. Denote by $(\cdot, \cdot)_X$ and $\|\cdot\|_X$ the inner product and the norm in X. Assume that*

$$U(t) = U(0) + \int_0^t \mathcal{A}(s) U(s) ds + \sum_{k=1}^r \int_0^t \mathcal{B}_k(s) U(s) dW_k(s), \tag{8.43}$$

where the initial condition $U(0)$ is independent of $\mathcal{F}_T^W$ and the linear operators $\mathcal{A}(s), \mathcal{B}_k(s)$ are non-random. Then $U(t) = \sum_{\alpha \in \mathcal{J}} U_\alpha(t) \xi_\alpha$, and

$$U_\alpha(t) = U_0 I(|\alpha| = 0) + \int_0^t \mathcal{A}(s) U_\alpha(s) ds + \sum_{k, \ell} \sqrt{\alpha_\ell^k} \mathcal{B}_k(s) U_{\alpha^-(k,\ell)}(s) m_\ell(s) ds, \tag{8.44}$$

with multi-index $\alpha^-(i, k)$ defined in (8.35). If, in addition, the operator $\mathcal{A}$ generates a semigroup $\Psi_{t,s}$, $t \geq s \geq 0$, that is, a family of continuous operator on X such that, for every $u, v \in X$, $(\Psi_{t,s} v, u)_X = (v, u)_X + \int_s^t (\mathcal{A}(\tau) \Psi_{\tau,s} v, u)_X d\tau$, then

$$\sum_{|\alpha|=n} U_\alpha(t) \xi_\alpha = \sum_{k_1, \ldots, k_n = 1}^r \int_0^t \int_0^{s_n} \cdots \int_0^{s_2} \tag{8.45}$$
$$\Psi_{t,s_n} \mathcal{B}_{k_n}(s_n) \cdots \Psi_{s_2, s_1} \mathcal{B}_{k_1}(s_1) \Psi_{s_1, 0} U_0(s_1) dW_{k_1}(s_1) \cdots dW_{k_n}(s_n),$$

and

$$\sum_{|a|=n} \|U_a(t)\|_X^2 = \sum_{k_1,\ldots,k_n=1}^{r} \int_0^t \int_0^{s_n} \cdots \int_0^{s_2} \qquad (8.46)$$
$$\|\Psi_{t,s_n}\mathcal{B}_{k_n}(s_n)\cdots\Psi_{s_2,s_1}\mathcal{B}_{k_1}(s_1)\Psi_{s_1,0}U_0(s_1)\|_X^2 ds_1\cdots ds_n.$$

Proof. By Theorem 8.4 we have $U_a(t) = \mathbb{E}(U(t)\xi_a)$ and then $\mathcal{F}_t^W$-measurability of $U(t)$ implies

$$\mathbb{E}(U(t)\xi_a) = \mathbb{E}\left(U(t)\mathbb{E}\left(\xi_a|\mathcal{F}_t^W\right)\right) = \mathbb{E}(U(t)\xi_a(t)).$$

To derive (8.44), it remains to apply the Itô formula to the product $U(t)\xi_a(t)$ using equation (8.34).

Given the semigroup $\Psi_{t,s}$, the system of equations (8.44) is solvable by induction: using the same notations as in Corollary 8.1,

$$U_a(t) = \frac{1}{\sqrt{a!}} \sum_{\sigma \in \mathcal{P}_n} \int_0^t \int_0^{s_n} \cdots \int_0^{s_2} \Psi_{t,s_n}\mathcal{B}_{k_{\sigma(n)}}(s_n)\cdots\Psi_{s_2,s_1}\mathcal{B}_{k_{\sigma(1)}}(s_1)\Psi_{s_1,0}U_0(s_1)$$
$$m_{\ell_{\sigma(1)}}(s_1)\ldots m_{\ell_{\sigma(n)}} ds_1 \ldots ds_n. \qquad (8.47)$$

After that, (8.45) follows from (8.42), and then (8.46) follows from (8.45). □

Remark 8.4 If $\mathcal{A} = \mathcal{L}$ or $\mathcal{A} = \mathcal{L}^*$, then Assumptions **R1** ensure the existence of the corresponding semigroup $\Psi_{t,s}$. Moreover, in this time-homogeneous situation, $\Psi_{t,s} = \Psi_{t-s,0}$.

Multiple integral expansion is more explicit than the Cameron–Martin expansion, and has been widely used in the study of stochastic equations. For example, Kunita [44] used it to sharpen existence and uniqueness results for the Zakai equation and Krylov and Veretennikov [42], to establish a new criterion for existence and uniqueness of strong solutions for stochastic *ordinary* differential equations. An advantage of the Cameron–Martin version (8.31) is computational: numerically, it is much easier to work with the system of equations (8.44) than with the integrands.

8.3.2 Chaos expansions of the optimal filter

Consider the filtering problem (8.13) on page 235 under the assumptions **R1**, together with the corresponding operators $\mathcal{L}$, $\mathcal{M}_k$, $\mathcal{L}^*$ and $\mathcal{M}_k^*$. Then

- The operator $\mathcal{L}$ generates a semigroup Φ_t in $L_2(\mathbb{R}^d)$;
- The operator $\mathcal{L}^*$ generates a semigroup Φ_t^* in $L_2(\mathbb{R}^d)$.

Let $\{m_\ell,\ \ell \geq 1\}$ be an orthonormal basis in $L_2((0, T))$. Define

$$\xi_{k\ell} = \int_0^T m_\ell(t)\,dY_k(t)$$

and then ξ_α, $\alpha \in \mathcal{J}$ according to (8.28).

Consider the following system of equations:

$$\frac{\partial \rho_\alpha(t, x)}{\partial t} = \mathcal{L}^* \rho_\alpha(t, x) + \sum_{k,\ell} \sqrt{\alpha_\ell^k}\, m_\ell(t) \mathcal{M}_k^* \rho_{\alpha^-(k,\ell)}(t, x),\quad 0 < t \leq T, \qquad (8.48)$$

with initial condition

$$\rho_\alpha(0, x) = \begin{cases} p_0(x), & \text{if } |\alpha| = 0 \\ 0, & \text{otherwise.} \end{cases}$$

If

- $\mathcal{P}_n$ is the collection of the permutations of the set $(1, \ldots, n)$;
- $\{(k_1, \ell_1), \ldots, (k_n, \ell_n)\}$ is the characteristic set of $\alpha \in \mathcal{J}$ with $|\alpha| = n$;
- Φ_t^* is the semigroup generated by the operator $\mathcal{L}^*$,

then, by induction on $|\alpha|$,

$$\rho_\alpha(t, x) = \frac{1}{\sqrt{\alpha!}} \sum_{\theta \in \mathcal{P}_n} \int_0^t \int_0^{s_n} \cdots \int_0^{s_2} \Phi_{t,s_n}^* \mathcal{M}_{k_{\theta(n)}}^* \cdots \Phi_{s_2,s_1}^* \mathcal{M}_{k_{\theta(1)}}^* \Phi_{s_1,0}^* p_0(x) \qquad (8.49)$$

$$m_{\ell_{\theta(1)}}(s_1) \ldots m_{\ell_{\theta(n)}}(s_n)\,ds_1 \ldots ds_n.$$

Theorem 8.7 *Under assumptions* **R1**, *we have the following expansions:*

$$\rho(t, x) = \sum_{\alpha \in \mathcal{J}} \rho_\alpha(t, x)\xi_\alpha, \qquad (8.50)$$

$$\phi_t[f] = \sum_{\alpha \in \mathcal{J}} \left(\int_{\mathbb{R}^d} f(x)\rho_\alpha(t, x)\,dx \right) \xi_\alpha \qquad (8.51)$$

$$\rho(t, x) = \Phi_t^* p_0(x) + \sum_{n=1}^\infty \sum_{k_1,\ldots,k_n=1}^r \int_0^t \int_0^{s_n} \cdots \int_0^{s_2} \qquad (8.52)$$

$$\Phi_{t-s_n}^* \mathcal{M}_{k_n}^* \cdots \Phi_{s_2-s_1}^* \mathcal{M}_{k_1}^* \Phi_{s_1}^* p_0(x)\,dY_{k_1}(s_1)\ldots dY_{k_n}(s_n);$$

$$\widetilde{\mathbb{E}}(f(X(t))|\mathcal{Y}_t; X(0) = x) = \Phi_t f(x) + \sum_{n=1}^\infty \sum_{k_1,\ldots,k_n=1}^r \int_0^t \int_0^{s_n} \cdots \int_0^{s_2} \qquad (8.53)$$

$$\Phi_{s_1}\mathcal{M}_{k_1}\Phi_{s_2-s_1}\mathcal{M}_{k_2}\cdots \mathcal{M}_{k_n}\Phi_{t-s_n} f(x)\,dY_{k_1}(s_1)\ldots dY_{k_n}(s_n).$$

Expansions (8.50), (8.52) *and* (8.53) *converge in* $L_2(\Omega, \mathcal{Y}_T, \widetilde{\mathbb{P}}; L_2(\mathbb{R}^d))$ *for every* $t \in [0, T]$, *expansion* (8.51) *converges in* $L_2(\Omega, \mathcal{Y}_T, \widetilde{\mathbb{P}})$ *for every* $t \in [0, T]$.

Proof. Recall that Y is an r-dimensional standard Brownian motion under the measure $\widetilde{\mathbb{P}}$ (see Theorem 8.1). Equalities (8.50) and (8.52) follow by applying

Theorem 8.6 to equation (8.25); then (8.51) and (8.53) follow from the relation $\phi_t[f] = \int_{\mathbb{R}^d} f(x)\rho(t, x)dx$. □

If the state process X is independent of the observation noise W, then the chaos expansions of $\rho(t, x)$ and $\phi_t[f]$ can be written even if X is not a diffusion process. While a rigorous derivation requires a different approach (see, for example, Ocone [66]), an informal derivation is possible by setting $\varrho \equiv 0$. Then $\mathcal{M}_k = h_k = \mathcal{M}_k^*$ and

$$\Phi_{s_1}\mathcal{M}_{k_1}\Phi_{s_2-s_1}\mathcal{M}_{k_2}\cdots\mathcal{M}_{k_n}\Phi_{t-s_n}f(x)$$
$$= \mathbb{E}\left(f(X(t))h_{k_n}(X(s_n))\cdots h_{k_1}(X(s_1))|X(0)=x\right) \quad (8.54)$$

and

$$\Phi_{t-s_n}^*\mathcal{M}_{k_n}^*\Phi_{s_n-s_{n-1}}^*\mathcal{M}_{k_{n-1}}^*\cdots\mathcal{M}_{k_1}^*\Phi_{s_1}^*p_0(x)$$
$$= \mathbb{E}\left(h_{k_n}(X(s_n))\cdots h_{k_1}(X(s_1))|X(t)=x\right)p_t(x), \quad (8.55)$$

where $p_t(x)$ is the distribution density of the random variable $X(t)$; equality (8.55) is verified by multiplying both sides by a smooth compactly supported function and integrating with respect to x. Notice that the right-hand sides of both (8.54) and (8.55) do not require X to be a diffusion process and can therefore be used in (8.49), (8.52), and (8.53). For example,

$$\rho_\alpha(t, x) = \frac{p_t(x)}{\sqrt{\alpha!}} \sum_{\theta \in P_n} \int_0^t \int_0^{s_n} \cdots \int_0^{s_2} \mathbb{E}\left(h_{k_{\theta(n)}}(X(s_n))\cdots h_{k_{\theta(1)}}(X(s_1))|X(t)=x\right)$$
$$m_{\ell_{\theta(1)}}(s_1)\ldots m_{\ell_{\theta(n)}}(s_n)ds_1\ldots ds_n. \quad (8.56)$$

Here are some other comments concerning the filtering problem in which the state process is independent of the observation noise:

(1) The distribution of X is the same under $\mathbb{P}$ and $\widetilde{\mathbb{P}}$, so either $\mathbb{E}$ or $\widetilde{\mathbb{E}}$ can be used in (8.54), (8.55).
(2) With Z_T defined in (8.15), we have, after conditioning on X, that $\widetilde{\mathbb{E}} Z_T = 1$ for every measurable locally bounded function h. Similarly, $\widetilde{\mathbb{E}}(Z_T)^2 = \mathbb{E}\exp\left((1/2)\int_0^T |h(X(s))|^2 ds\right)$. Thus, convergence of the chaos expansions is ensured by the condition $\mathbb{E}\exp\left((1/2)\int_0^T |h(X(s))|^2 ds\right) < \infty$, which is weaker than boundedness of h (see [66] for details).

8.4 Spectral separating schemes

In this section, we describe two algorithms for solving the Zakai equation (8.25) using the Cameron–Martin expansion of the solution. Consider the filtering model (8.13). By Theorem 8.3, given a suitable function $f = f(x)$, the optimal filter $\widehat{f}_t = \mathbb{E}(f(X(t))|\mathcal{Y}_t)$ has the representation $\widehat{f}_t = \phi_t[f]/\phi_t[1]$, where $\phi[f] = \int_{\mathbb{R}^d} f(x)\rho(t,x)dx$ and u is the solution of the Zakai equation (8.25). Using Theorem 8.6, we will implement the idea described in the Introduction and construct recursive approximations of $\rho(t,x)$ and $\varphi_t[f]$, as well as study the quality of the approximations.

Let $0 = t_0 < t_1 < \ldots < t_M = T$ be a uniform partition of the interval $[0, T]$ with step Δ so that $t_i = i\Delta$, $i = 0, \ldots, M$. Fix an orthonormal basis $\mathfrak{m} = \{m_k(s), k \geq 1\}$ in $L_2([0, \Delta])$ and define random variables

$$\xi^i_{k,\ell} = \int_{t_{i-1}}^{t_i} m_k(s - t_{i-1}) dY_\ell(s) \text{ and } \xi^i_\alpha = \prod_{k,\ell} \left(\frac{H_{a^\ell_k}(\xi^i_{k,\ell})}{\sqrt{a^\ell_k!}} \right), \quad \alpha \in \mathcal{J}, \quad (8.57)$$

where $\mathcal{J}$ is the set of multi-indices (8.27), and H_n is the n-th Hermite polynomial (8.29). We also fix an orthonormal basis $\mathfrak{e} = \{e_k(x), k \geq 1\}$ in $L_2(\mathbb{R}^d)$ and recall that $(\cdot, \cdot)_0$ denotes the inner product in $L_2(\mathbb{R}^d)$.

8.4.1 Spectral separating scheme of the first kind

Consider the following system of equations:

$$\frac{\partial \varphi_\alpha(s, x; g)}{\partial s} = \mathcal{L}^* \varphi_\alpha(s, x; g) + \sum_{k,\ell} \sqrt{a^k_\ell} m_\ell(s) \mathcal{M}^*_k \varphi_{\alpha-(k,\ell)}(s, x; g), \quad 0 < s \leq \Delta,$$
(8.58)

with initial condition

$$\varphi_\alpha(0, x; g) = \begin{cases} g(x), & \text{if } |\alpha| = 0 \\ 0, & \text{otherwise.} \end{cases}$$

Define the numbers

$$q^\ell_{\alpha k} = (\varphi_\alpha(\Delta, \cdot; e_k), e_\ell)_0, \quad (8.59)$$

and then by induction

$$\psi_j(0) = (p_0, e_j)_0; \quad \psi_j(i) = \sum_{\alpha \in \mathcal{J}} \sum_{k=1}^\infty \psi_k(i-1) q^j_{\alpha k} \xi^i_\alpha. \quad (8.60)$$

Theorem 8.8 *The unnormalized filtering density has the representation*

$$\rho(t_i, x) = \sum_{j=1}^\infty \psi_j(i) e_j(x), \quad 0 \leq i \leq M, \quad (8.61)$$

and, for $f \in L_2(\mathbb{R}^2)$, the unnormalized optimal filter $\phi_t[f]$ is

$$\phi_{t_i}[f] = \sum_{j=1}^\infty \psi_j(i) f_j, \text{ where } f_k = \int_{\mathbb{R}^d} f(x) e_k(x) dx. \quad (8.62)$$

Proof. Recall that u is a square integrable $\mathcal{Y}_t$-adapted solution of the Zakai equation (8.25), and Y is an r-dimensional Wiener process under measure $\widetilde{\mathbb{P}}$. By Theorem 8.6 we have $\rho(t, x) = \varphi_a(t, x; \rho(t_{i-1}, x))\xi_a^i$, $t \in [t_{i-1}, t_i]$. To establish (8.61), it remains to write $\rho(t_{i-1}, x) = \sum_{k \geq 1}(\rho(t_{i-1}, \cdot), e_k)_0 e_k(x)$ and use linearity of equations (8.25) and (8.58). Equality (8.62) is a direct consequence of (8.61), because $\phi_t[f] = \int_{\mathbb{R}^d} f(x)\rho(t, x)dx$. $\square$

The integrals $\int_{\mathbb{R}^d} f(x)e_k(x)dx$ may be defined for all $k \geq 1$ even when $f \notin L_2(\mathbb{R}^d)$. In that case, representation (8.62) of the unnormalized optimal filter can still hold; see Theorems 8.10, 8.11, and 8.12 below.

Next, we use Theorem 8.8 to construct recursive approximations of $\rho(t_i, x)$ and $\phi_{t_i}[f]$ for all $j \geq 1$. Let K, n, N be positive integers, and $\mathcal{J}_N^n$, the collection of those multi-indices in $\mathcal{J}$ for which $|\alpha| \leq N$ and $\alpha_\ell^k = 0$ if $\ell > n$; note that $\mathcal{J}_N^n$ is a *finite* set. With the numbers $q_{\alpha k}^\ell$ from (8.59), we define $\psi_\ell^K(i, N, n)$ by truncating the sums in (8.60):

$$\psi_\ell^K(0, N, n) = (p_0, e_\ell)_0, \quad \ell = 1, \ldots, K;$$
$$\psi_\ell^K(i, N, n) = \sum_{\alpha \in \mathcal{J}_N^n} \sum_{k=1}^K \psi_k^K(i-1, N, n) q_{\alpha k}^\ell \xi_\alpha^i, \quad (8.63)$$

and then define the approximations of $\rho(t_i, x)$ and $\phi_{t_i}[f]$ by

$$\rho_{N,K}^n(t_i, x) = \sum_{j=1}^K \psi_j^K(i, N, n)e_j(x), \quad \phi_{N,K;i}^n[f] = \sum_{j=1}^K \psi_j^K(i, N, n)f_j, \quad 0 \leq i \leq M,$$
(8.64)

where we assume that $f_j = \int_{\mathbb{R}^d} f(x)e_j(x)dx$ exists for each $j \geq 1$. The following is an algorithm for computing the approximations of the unnormlized filtering density and filter using (8.64).

1. *Preliminary computations* (before the observations are available)
    (1) Choose suitable basis functions $\{e_k, k = 1, \ldots, K\}$ in $L_2(\mathbb{R}^d)$ and $\{m_i, i = 1, \ldots, n\}$ in $L_2([0, \Delta])$.
    (2) For $\alpha \in \mathcal{J}_N^n$ and $k, \ell = 1, \ldots, K$ compute $q_{\alpha k}^\ell = (\varphi_\alpha(\Delta, \cdot, e_k), e_\ell)_0$ (using (8.58)),

$$f_k = \int_{\mathbb{R}^d} f(x)e_k(x)dx, \quad \psi_k^K(0, N, n) = \int_{\mathbb{R}^d} p_0(x)e_k(x)dx.$$

2. *Real-time computations, i-th step* (as the observations become available): compute ξ_α^i according to (8.57) and update the coefficients ψ:

$$\psi_\ell^K(i, N, n) = \sum_{\alpha \in \mathcal{J}_N^n} \sum_{k=1}^K \psi_k^K(i-1, N, n) q_{\alpha k}^\ell \xi_\alpha^i \quad \ell = 1, \ldots, K;$$

then, if necessary, compute

$$\rho_N^{n,K}(t_i, x) = \sum_{\ell=1}^{K} \psi_\ell^K(i, N, n) e_\ell(x) \tag{8.65}$$

and/or

$$\phi_{N,K;i}^n[f] = \sum_{j=1}^{K} \psi_j^K(i, N, n) f_j, \quad \widehat{f}_{N,K;i}^n = \frac{\phi_{N,K;i}^n[f]}{\phi_{N,K;i}^n[1]}. \tag{8.66}$$

We call this algorithm the *spectral separating scheme of the first kind*. For the filtering model (8.13) with $\varrho \equiv 0$, the algorithm was first suggested in [60] and analyzed in [55]; analysis for general ϱ is in [54]. We discuss the properties of this algorithm in 8.4.3 and analyse its convergence in Section 8.4.4.

8.4.2 Spectral separating scheme of the second kind

We now present an alternative algorithm for solving the Zakai equation (8.25) in the filtering model (8.13). In the spectral separating scheme of the first kind, the truncation of the expansion in $L_2(\mathbb{R}^d)$ is done after the truncation of the Cameron–Martin expansion. Now, we will do the truncation in $L_2(\mathbb{R}^d)$ first.

Let $\mathfrak{e}$ be an orthonormal basis in $L_2(\mathbb{R}^d)$ such that every function $e_k = e_k(x)$ belongs to $\mathbf{H}^1$ (the Sobolev spaces $\mathbf{H}^\gamma$ are defined in (8.20)). As before, $(\cdot, \cdot)_0$ denotes the inner product in $L_2(\mathbb{R}^d)$.

Fix a positive integer number K. Define the matrices $A^K = \left(A_{ij}^K, i, j = 1, \ldots, K\right)$ and $B_\ell^K = \left(B_{\ell,ij}^K, i, j = 1, \ldots, K; \ell = 1, \ldots, r\right)$, by

$$A_{ij}^K = (\mathcal{L}^* e_j, e_i)_0, \quad B_{\ell,ij}^K = (\mathcal{M}_\ell^* e_j, e_i)_0,$$

and consider the Galerkin approximation $u^K(t, x)$ of $\rho(t, x)$:

$$u^K(t, x) = \sum_{i=1}^{K} u_i^K(t) e_i(x), \tag{8.67}$$

where the vector $u^K(t) = \{u_i^K(t), i = 1, \ldots, K\}$ is the solution of the system of stochastic ordinary differential equations

$$du^K(t) = A^K u^K(t) dt + \sum_{\ell=1}^{r} B_\ell^K u^K(t) dY_\ell(t) \tag{8.68}$$

with the initial condition $u_i^K(0) = (p_0, e_i)_0$. Because the matrices B_ℓ^K, $\ell = 1, \ldots, r$, do not, in general, commute with each other even if $\varrho(x) \equiv 0$, system (8.68) has no closed-form solution and must be solved numerically.

Define random variables ξ_α^i according to (8.57). Theorem 8.6 implies the following result.

Theorem 8.9 *For every $i = 1, \ldots, M$, the solution of (8.68) can be written in $L_2(\Omega, \widetilde{\mathbb{P}}; \mathbb{R}^K)$ as*

$$u^K(t_i) = \sum_{a \in \mathcal{J}} \varphi_a^K(\Delta; u^K(t_{i-1})) \xi_a^i, \quad i = 1, \ldots, M, \tag{8.69}$$

where, for $s \in (0, \Delta]$ and $\zeta \in \mathbb{R}^K$, the functions $\varphi_a^K(s; \zeta)$ are the solutions of

$$\frac{\partial \varphi_a^K(s; \zeta)}{\partial s} = A^K \varphi_a^K(s; \zeta) + \sum_{k,l} \sqrt{a_\ell^k}\, m_\ell(s) B_k^K \varphi_{a^-(k,\ell)}^K(s; \zeta), \quad 0 < s \leq \Delta, \tag{8.70}$$

with initial conditions

$$\varphi_a^K(0; \zeta) = \begin{cases} \zeta, & \text{if } |a| = 0, \\ 0, & \text{if } |a| > 0, \end{cases}$$

and $a^-(k, \ell)$ is defined in (8.35).

To construct a recursive approximation of u^K, fix positive integers N and n and define the set $\mathcal{J}_N^n$ as the collection of multi-indices a from $\mathcal{J}$ such that $|a| \leq N$ and $a_k^\ell = 0$ if $k > n$. The approximation $u_N^{K,n}(t_i)$ of $u^K(t_i)$ is defined by

$$u_N^{K,n}(t_0) = u^K(0), \quad u_N^{K,n}(t_i) = \sum_{a \in \mathcal{J}_N^n} \varphi_a^K(\Delta; u_N^{K,n}(t_{i-1})) \xi_a^i, \quad i = 1, \ldots, M. \tag{8.71}$$

To establish a representation of $u_N^{K,n}(t_i)$ similar to (8.65), note that $u_N^{K,n}(t_i)$ is a vector in $\mathbb{R}^K$. Let $\mathfrak{U} = \{\mathbf{u}^j, j = 1, \ldots, K\}$ be a basis in $\mathbb{R}^K$. The vector $\rho_N^{K,n}(t_i)$ can then be written as

$$\rho_N^{K,n}(t_i) = \sum_{j=1}^K \rho_{N,j}^{K,n}(t_i; \mathfrak{U}) \mathbf{u}^j,$$

and by the recursive definition of $\rho_N^{K,n}(t_i)$,

$$\rho_N^{K,n}(t_i) = \sum_{a \in \mathcal{J}_N^n} \varphi_a^K(\Delta; \rho_N^{K,n}(t_{i-1})) \xi_a^i$$

$$= \sum_{a \in \mathcal{J}_N^n} \sum_{j=1}^K \varphi_a^K(\Delta; \mathbf{u}^j) \rho_{N,j}^{K,n}(t_{i-1}; \mathfrak{U}) \xi_a^i.$$

Once again, $\varphi_a^K(\Delta, \mathbf{u}^i)$ is a vector in $\mathbb{R}^K$, so we write

$$\varphi_a^K(\Delta, \mathbf{u}^j) = \sum_{k=1}^K q_{jk}^{K,a}(\mathfrak{U}) \mathbf{u}^k, \tag{8.72}$$

and conclude that

$$\rho_{N,j}^{K,n}(t_i;\mathfrak{U}) = \sum_{\alpha \in \mathcal{J}_N^n} \sum_{k=1}^{K} q_{jk}^{K,\alpha}(\mathfrak{U}) \rho_{N,k}^{K,n}(t_{i-1};\mathfrak{U}) \xi_\alpha^i. \tag{8.73}$$

If $f_k = \int_{\mathbb{R}^d} f(x) e_k(x) dx$ is defined for all $k \geq 1$, then

$$\rho_N^{K,n}(t_i, x) = \sum_{j,k=1}^{K} \rho_{N,j}^{K,n}(t_i;\mathfrak{U}) \mathbf{u}_k^j e_k(x), \quad \phi_{N;i}^{K,n}[f] = \sum_{j,k=1}^{K} \rho_{N,j}^{K,n}(t_i;\mathfrak{U}) \mathbf{u}_k^j f_k \tag{8.74}$$

are the approximations of the unnormalized filtering density and filter.

The following is an algorithm for computing the approximations of the unnormlized filtering density and filter using (8.74).

1. *Preliminary computations (before the observations are available)*:
   (1) *Choose suitable basis functions* $\{e_k, k = 1, \ldots, K\}$ *in* $L_2(\mathbb{R}^d)$, $\{m_k, k = 1, \ldots, n\}$ *in* $L_2([0, \Delta])$, *and a* **standard unit basis** $\{\mathbf{u}^j, j = 1, \ldots, K\}$ *in* $\mathbb{R}^K$, *that is,* $\mathbf{u}_j^j = 1$, $\mathbf{u}_\ell^j = 0$ *otherwise.*
   (2) *For* $\alpha \in \mathcal{J}_N^n$ *and* $j, k = 1, \ldots, K$, *compute* $q_{jk}^{K,\alpha} = \varphi_{\alpha,j}^K(\Delta; \mathbf{u}^k)$ *(using* (8.70)*),* $f_k = \int_{\mathbb{R}^d} f(x) e_k(x) dx$, $\rho_{N,k}^{K,n}(t_0) = \int_{\mathbb{R}^d} p_0(x) e_k(x) dx$.

2. *Real-time computations, i-th step (as the observations become available)*: compute ξ_α^i, $\alpha \in \mathcal{J}_N^n$ *according to* (8.57) *and update the coefficients* $\rho_{N,k}^{K,n}$ *as follows*:

$$Q_{jk}^K(\xi^i) = \sum_{\alpha \in \mathcal{J}_N^n} q_{jk}^{K,\alpha} \xi_\alpha^i, \quad \rho_{N,j}^{K,n}(t_i) = \sum_{k=1}^{K} Q_{jk}^K(\xi^i) \rho_{N,k}^{K,n}(t_{i-1}), \quad j = 1, \ldots, K; \tag{8.75}$$

then, if necessary, compute

$$u_N^{K,n}(t_i, x) = \sum_{j=1}^{K} \rho_{N,j}^{K,n}(t_i) e_j(x) \tag{8.76}$$

and/or

$$\phi_{N;i}^{K,n}[f] = \sum_{j=1}^{K} f_j \rho_{N,j}^{K,n}(t_i), \quad \widehat{f}_{N;i}^{K,n} = \frac{\phi_{N;i}^{K,n}[f]}{\phi_{N;i}^{K,n}[1]}. \tag{8.77}$$

We call this algorithm the *spectral separating scheme of the second kind*. It was suggested and analyzed in [23] when $\varrho = 0$ and in [53] for the general model (8.13). The difference in the notations for the approximations of ρ, ϕ, and $\widehat{f}$ corresponding to the two schemes is the location of the index K: compare (8.74) with (8.64) on 247.

8.4.3 Discussion

The main advantage of the spectral separating schemes, as compared to most other nonlinear filtering algorithms, is that the time-consuming computations, including solving partial differential equations and evaluation of integrals, are performed in advance, while the real-time part is relatively simple even when the dimension d of the state process is large. Here are some other features of the spectral separating schemes:

(1) The overall amount of preliminary computation does not depend on the number of the online timesteps;
(2) Formulas (8.66) and (8.77) can be used to compute an approximation to $\widehat{f}_{t_i}$, for example, conditional moments, without the time-consuming computations of the unnormalized filtering density and the related integrals;
(3) Only the coefficients $\psi_j^K(i, n, N)$ or $u_{N,j}^{K,n}(t_i)$ must be updated at every timestep; the filtering density and/or filter can be computed independently of each other as needed, for example, at the final time moment.
(4) The real-time part of the algorithms can be easily parallelized.
(5) If $n = 1$, then each ξ_a^i depends only on the increments $Y_\ell(t_i) - Y_\ell(t_{i-1})$ of the observation process, and the corresponding algorithms can be used for filtering with discrete-time observations [57]. For $n > 1$ and $k > 1$, the integral $\int_{t_{i-1}}^{t_i} m_k(s - t_{i-1}) dY_\ell(s)$ can be reduced to a usual Riemann integral and then approximated by the trapezoidal rule.
(6) The implementation of both algorithms does not depend on whether the model is noise-correlated ($\varrho \neq 0$) or not.

Successful implementation of the algorithms requires effective numerical methods for solving deterministic parabolic equations [70] and evaluating integrals [14], but no special tools from numerical stochastics. On the other hand, successful testing and tuning of the algorithms will require effective numerical methods for stochastic ODEs to simulate the processes X, Y. The literature [39, 63] describes many such methods.

Theoretical analysis of the algorithms is possible with little or no change if the model is not time homogeneous, that is, the functions b, ϱ, σ, h in (8.13) depend on time. This time dependence certainly decrease the computational advantages, as the number of preliminary computations will grow substantially and will depend on the number of the online timesteps.

The Wiener chaos approach is far less effective if the functions b, ϱ, σ, h in (8.13) depend on the observation process Y, because the corresponding systems (8.58) and (8.70) have a much more complicated structure and are no longer solvable by induction. The corresponding analysis is still an open problem.

8.4.4 Approximation error

The quality of the approximation for the spectral separating schemes is controlled by four numbers: K, n, N, and Δ. On the other hand, the amount of the preliminary computations and the storage space are controlled by the size of the array q; the size of this array is $K^2 |\mathcal{J}_N^n|$, where K is the number of basis functions in $L_2(\mathbb{R}^d)$, and $|\mathcal{J}_N^n|$, the size of the set $\mathcal{J}_N^n$, is the number of the Cameron–Martin basis functions. By construction, it is impossible to improve the quality of approximation without increasing K. While increasing n and N should also lead to better approximation, it is essentially impossible to use large values of n and N because of the prohibitively large size of the set $\mathcal{J}_N^n$. For example, if $r = 1$, the number of the elements in the set $\mathcal{J}_5^{10}$ is 740, and this number *more than doubles* for $r = 2$. A rough asymptotic of $|\mathcal{J}_N^n|$ is $(nN)^r$. Accordingly, the convergence of the approximations must be studied with fixed values of n and N: to improve the quality of approximation, we should decrease the timestep Δ and increase the number K of the spatial basis functions.

The study of convergence of the spectral separating schemes requires a special choice of the bases $\mathfrak{e}$ and $\mathfrak{m}$, as well as extra regularity of the filtering model (8.13).

We begin by specifying the basis $\mathfrak{e}$ in $L_2(\mathbb{R}^d)$. Denote by Γ the set of ordered d-tuples $\gamma = (\gamma_1, \ldots, \gamma_d)$ with $\gamma_j = 0, 1, 2, \ldots$. For $\gamma \in \Gamma$ define

$$\mathcal{H}_\gamma(x) = \prod_{j=1}^{d} \mathcal{H}_{\gamma_j}(x_j),$$

where

$$\mathcal{H}_n(t) = \frac{(-1)^n}{\sqrt{2^n \pi^{1/2} n!}} e^{t^2/2} \frac{d^n}{dt^n} e^{-t^2}, \quad n = 0, 1, 2, \ldots.$$

If Λ is the operator

$$\Lambda = -\nabla^2 + (1 + |x|^2), \tag{8.78}$$

where ∇^2 is the Laplace operator, then, by direct computation,

$$\Lambda \mathcal{H}_\gamma = \lambda_\gamma e_\gamma, \tag{8.79}$$

with $\lambda_\gamma = \left(2 \sum_{j=1}^{d} \gamma_j + d + 1\right)$.

Next, we introduce an ordering of the set Γ as follows: define $|\gamma| = \sum_{j=1}^{d} \gamma_j$ and then say that $\gamma < \tau$ if $|\gamma| < |\tau|$ or if $|\gamma| = |\tau|$ and $\gamma < \tau$ under the lexicographic ordering, that is, $\gamma_{i_0} < \tau_{i_0}$, where i_0 is the first index for which $\gamma_i \neq \tau_i$. Finally, we define the basis $\mathfrak{e}$, known as the *Hermite basis*, as the collection $\{\mathcal{H}_\gamma(x), \ \gamma \in \Gamma\}$ together with the above ordering of Γ. By construction, the elements e_k of $\mathfrak{e}$ satisfy

$$\Lambda e_k = \lambda_k e_k, \tag{8.80}$$

where $c_1 k^{1/d} \leq \lambda_k \leq c_2 k^{1/d}$ and $0 < c_1 < c_2$ do not depend on k. The construction of the Hermite basis implies that each e_k decays at infinity faster than every power of $|x|$, and therefore the number $f_k = \int_{\mathbb{R}^d} f(x) e_k(x) dx$ is defined for every $k \geq 1$ and every measurable function f of polynomial growth.

As far as the basis $\mathfrak{m}$ in $L_2([0, \Delta])$, we use the Fourier cosine basis

$$m_1(s) = \frac{1}{\sqrt{\Delta}}; \quad m_k(s) = \sqrt{\frac{2}{\Delta}} \cos\left(\frac{\pi(k-1)s}{\Delta}\right), \; k > 1; \; 0 \leq s \leq \Delta. \tag{8.81}$$

Definition 8.1 The filtering model (8.13) is called ν-regular for some positive integer ν if the functions σ and ϱ belong to $C_b^{2\nu+3}$, the functions b and h belong to $C_b^{2\nu+2}$, and $\Lambda^\nu p_0$ belongs to $\mathbf{H}^1$, with Λ as in (8.78).

We are now ready to study the convergence of the spectral separating schemes. Recall that the spectral separating scheme of the first kind defines the approximations $u_{N,K}^n(t_i, x)$, $\widetilde{\phi}_{t_i}$ of the unnormalized filtering density and filter according to (8.64). The following theorem presents the quality of these approximations and establishes the convergence in the limit $\lim_{\Delta \to 0} \lim_{K \to \infty}$ for the noise uncorrelated model.

Theorem 8.10 Assume that $N \geq 2$, $\varrho(x) \equiv 0$, and the matrix $\sigma \sigma^\top$ is uniformly positive definite, that is, condition (8.22) holds. If the filtering model (8.13) is ν-regular for some $\nu > d + 1$, then

$$\max_{0 \leq i \leq M} \mathbb{E} \left\| \rho(t_i, \cdot) - u_{N,K}^n(t_i, \cdot) \right\|_0 \leq C_0 \left(\frac{(C_{11}\Delta)^{N/2}}{\sqrt{(N+1)!}} + \frac{C_{12}\Delta}{\sqrt{n}} \right) + \frac{C_2}{K^{(\nu-d-1)/d}\Delta}. \tag{8.82}$$

The number C_0 depends on T and the parameters of the model, that is, the coefficients and the initial condition in the Zakai equation (8.25); the numbers C_{11}, C_{12} depend only on the parameters of the model; the number C_2 depends on ν, T, and the parameters of the model.

If, in addition, $(1+|x|^2)^{-w} f \in L_2(\mathbb{R}^d)$ for some $w \geq 0$ so that $\nu > d + 1 + w$ and $\Lambda^\nu((1+|x|^2)^w p_0) \in \mathbf{H}^1$, then

$$\max_{0 \leq i \leq M} \mathbb{E} \left| \phi_{t_i}[f] - \phi_{N,K;i}^n[f] \right| \leq C_3 \left(\frac{(C_{11}\Delta)^{N/2}}{\sqrt{(N+1)!}} + \frac{C_{12}\Delta}{\sqrt{n}} \right) + \frac{C_4}{K^{(\nu-d-1)/d}\Delta}. \tag{8.83}$$

The numbers C_3, C_4 depend on ν, T, the function f, and the parameters of the model; the numbers C_{11} and C_{12} are the same as in (8.82).

Proof. Consider first the local error $\widetilde{\mathbb{E}} \left\| \rho(\Delta, \cdot) - u_{N,K}^n(\Delta, \cdot) \right\|_0^2$. Define $\rho_N(\Delta, x) = \sum_{\alpha \in \mathcal{J}_N^n} \varphi_\alpha(\Delta, x, p_0)$. By Theorem 2.2 in [55],

$$\widetilde{\mathbb{E}}\left\|\rho(\Delta,\cdot)-\rho_N^n(\Delta,\cdot)\right\|_0^2 \le c_1 e^{c_2\Delta}\left(\frac{(c_3\Delta)^{N+1}}{(N+1)!}\|p_0\|_0^2 + \frac{\Delta^3}{n}\|p_0\|_2^2\right), \qquad (8.84)$$

where the numbers c_1, c_2, c_3 depend only on the coefficients of (8.13); recall that $\|\cdot\|_2$ is the norm in the Sobolev space $\mathbf{H}^2$. Next, by Theorem 8.3 in [55],

$$\widetilde{\mathbb{E}}\left\|\rho_N^n(\Delta,\cdot)-\rho_{N,K}^n(\Delta,\cdot)\right\|_0^2 \le c_4 e^{c_5\Delta} K^{-2(\nu-d-1)/d}\|\Lambda^\nu p_0\|_0^2, \qquad (8.85)$$

where the numbers c_3, c_4 depend on ν and the parameters of the model. We combine (8.84), in which $N \ge 2$, and (8.85) to get the overall local error

$$\widetilde{\mathbb{E}}\left\|\rho(\Delta,\cdot)-\rho_{N,K}^n(\Delta,\cdot)\right\|_0^2 \le \left(c_6\Delta^3 + c_7 K^{-2(\nu-d-1)/d}\right) e^{c_8\Delta};$$

the global error is then

$$\widetilde{\mathbb{E}}\left\|\rho(t_i,\cdot)-\rho_{N,K}^n(t_i,\cdot)\right\|_0^2 \le c_9\Delta^2 + c_{10} K^{-2(\nu-d-1)/d}\Delta^{-2};$$

see [55] for details. Error bound (8.82) now follows from (8.18).

Error bound (8.83) follows from (8.82) by the Cauchy–Schwartz inequality. For more details, see Lototsky et al. [55]. □

The following properties of the functions m_k were essential in the proof of (8.84): if $M_k(t) = \int_0^t m_k(s)\,ds$, then $M_k(\Delta) = 0$, $|M_k(t)| \le \sqrt{\Delta}/n$. Any other basis with these properties can also be used, but for now the Fourier cosine basis (8.81) appears to be the only one for which these properties are easily verified. The Haar basis, while simplifying calculations of ξ_α^i, results in a *local* error bound (8.84) with a slower rate of decay in Δ [6, Corollary 3.8], [7, Corollary 4.5]; the corresponding *global* error bound for the Haar basis has not been derived yet.

The assumption $\varrho \equiv 0$ was also essential for the proof of (8.84); without this assumption, a different error bound holds.

Theorem 8.11 *Assume that the matrix $\sigma\sigma^\top$ is uniformly positive definite. If the filtering model (8.13) is ν-regular for some $\nu > \max(4, d+1)$, then*

$$\max_{0\le i\le M}\mathbb{E}\left\|\rho(t_i,\cdot)-u_{N,K}^n(t_i,\cdot)\right\|_0 \le C_1\left(\frac{1}{(1+\delta)^{N/2}} + \frac{1}{\sqrt{n}}\right)\Delta^{1/2} + \frac{C_2}{K^{(\nu-d-1)/d}\Delta}. \qquad (8.86)$$

The number C_1 depends on T and the parameters of the model, that is, the coefficients and the initial condition in the equation (8.13); the number $\delta > 0$ depends only on the parameters of the model; C_2 depends on ν, T, and the parameters of the model.

If, in addition, $(1+|x|^2)^{-w}f \in L_2(\mathbb{R}^d)$ for some $w \ge 0$ so that $\nu > d+1+w$ and $\Lambda^\nu((1+|x|^2)^w p_0) \in \mathbf{H}^1$, then

$$\max_{0 \le i \le M} \mathbb{E} \left| \phi_{t_i}[f] - \phi_{N,K;i}^n[f] \right| \le C_3 \left(\frac{1}{(1+\delta)^{N/2}} + \frac{1}{\sqrt{n}} \right) \Delta^{1/2} + \frac{C_4}{K^{(\nu-d-1)/d} \Delta}. \tag{8.87}$$

The numbers C_3, C_4 depend on ν, T, the function f, and the parameters of the model.

Proof. Once we establish the local error bound of the type (8.84), which in this case turns out to be

$$\widetilde{\mathbb{E}} \left\| \rho(\Delta, \cdot) - \rho_N^n(\Delta, \cdot) \right\|_0^2 \le c_1 e^{c_2 \Delta} \left(\frac{\Delta^2}{(1+\delta)^N} \|p_0\|_2^2 + \frac{\Delta^2}{n} \|p_0\|_4^2 \right), \tag{8.88}$$

for a suitable $\delta > 0$, the proof is completed by the same arguments as in Theorem 8.10.

To establish (8.88), we use equality (8.46), in which we replace equation (8.43) with the Zakai equation (8.25), and also put $X = \mathbf{H}^\gamma$ for a suitable γ. Then

$$\sum_{|\alpha|=n} \|\varphi_\alpha(t, \cdot, p_0)\|_\gamma^2 = \sum_{k_1,\ldots,k_n=1}^r \int_0^t \int_0^{s_n} \cdots \int_0^{s_2} \tag{8.89}$$
$$\left\| \Phi_{t-s_n}^* \mathcal{M}_{k_n}^* \cdots \Phi_{s_2-s_1}^* \mathcal{M}_{k_1}^* \Phi_{s_1}^* p_0 \right\|_\gamma^2 ds_1 \ldots ds_n,$$

where $\Phi^* = \Phi_t^*$ is the semi-group of the operator $\mathcal{L}^*$. The assumptions of the current theorem imply that the semi-group Φ^* is bounded above by the heat kernel:

$$\left\| \Phi_t^* f \right\|_\gamma \le C_1 \int_{\mathbb{R}^d} e^{-C_2 |y|^2 t} |\check{f}(y)|^2 (1+|y|^2)^\gamma dy \tag{8.90}$$

for some positive numbers C_1, C_2, where $\check{f}$ is the Fourier transform of f; see [19] for details. Notice also that

$$\left\| \mathcal{M}_k^* f \right\|_\gamma \le C_3 (\|f\|_\gamma + \|\nabla f\|_\gamma),$$

where ∇f is the gradient of f. Then direct computations show that

$$\int_0^t \int_0^s \left\| \mathcal{M}_k^* \Phi_{s-s_1}^* f(s_1) \right\|_\gamma^2 ds_1 ds \le C_4 \int_0^t \|f(s)\|_\gamma^2 ds.$$

For $n \ge 2$, we combine the last inequality with Theorem 9.5 in [58] to conclude that

$$\sum_{k_1,\ldots,k_n=1}^r \int_0^\Delta \int_0^{s_n} \cdots \int_0^{s_2} \left\| \Phi_{\Delta-s_n}^* \mathcal{M}_{k_n}^* \cdots \Phi_{s_2-s_1}^* \mathcal{M}_{k_1}^* \Phi_{s_1}^* p_0 \right\|_0^2 ds_1 \ldots ds_n$$
$$\le C_5 \sum_{k_1,\ldots,k_{n-3}}^r \int_0^\Delta \int_0^{s_n} \cdots \int_0^{s_2} \tag{8.91}$$
$$\left\| \mathcal{M}_{k_{n-3}}^* \Phi_{s_{n-2}-s_{n-3}}^* \mathcal{M}_{k_{n-3}}^* \cdots \Phi_{s_2-s_1}^* \mathcal{M}_{k_1}^* \Phi_{s_1}^* p_0 \right\|_2^2 ds_1 \ldots ds_n$$
$$\le C_6 (1+\delta)^{-n} \Delta^2 \|p_0\|_2^2$$

for some $\delta > 0$. Then local error bound (8.88) follows by same arguments as in the proof of Theorem 8.10. The main reason for the factor Δ^2 rather than Δ^3 in (8.91) is that the operators $\mathcal{M}_k^*$ do not commute with one another when $\varrho \neq 0$. □

If the matrix $\sigma\sigma^\top$ is not uniformly positive definite, then the rate of convergence is an open question.

We now establish the rate of convergence for the spectral separating scheme of the second kind. Recall that this algorithm defines the approximations $u_N^{K,n}(t_i, x)$, $\phi_{N;i}^{K,n}$ of the unnormalized filtering density and filter according to (8.74). The following theorem presents the quality of these approximations and establishes the convergence in the limit $\lim_{K\to\infty} \lim_{\Delta\to 0}$.

Theorem 8.12 *If the filtering model (8.13) is ν-regular for some $\nu > d + 1$, then*

$$\max_{0\leq i\leq M} \mathbb{E}\left\|\rho(t_i,\cdot) - u_N^{K,n}(t_i,\cdot)\right\|_0 \leq \frac{C_1}{K^{(\nu-d-1)/d}} + C_2\left(\frac{(C_{21}\Delta)^{N/2}}{\sqrt{(N+1)!}} + \frac{C_{22}\Delta^{1/2}}{\sqrt{n}}\right).$$
(8.92)

The number C_1 depends on ν, T, and the parameters of the model, that is, the coefficients and the initial condition in the Zakai equation (8.25); the number C_2 depends on T, K and the parameters of the model; the numbers C_{21}, C_{22} depend on K and the parameters of the model.

If, in addition, $(1 + |x|^2)^{-w} f \in L_2(\mathbb{R}^d)$ for some $w \geq 0$ so that $\nu > d + 1 + w$ and $\Lambda^\nu((1 + |x|^2)^w p_0) \in \mathbf{H}^1$, then

$$\max_{0\leq i\leq M} \mathbb{E}\left|\phi_{t_i}[f] - \phi_{N;i}^{K,n}[f]\right| \leq \frac{C_3}{K^{(\nu-w-d-1)/d}} + C_4\left(\frac{(C_{21}\Delta)^{N/2}}{\sqrt{(N+1)!}} + \frac{C_{22}\Delta^{1/2}}{\sqrt{n}}\right).$$
(8.93)

The number C_3 depends on ν, T, the function f, and the parameters of the model; the number C_4 depends on K, T, the function f, and the parameters of the model; the numbers C_{21}, C_{22} are the same as in (8.92).

Proof. This theorem is proved in [53]. The reference also contains a more detailed information about the constants C. □

Note that, in the spectral separating scheme of the second kind, the approximation in space is carried out first, and the Wiener chaos expansion is applied to a system of ordinary differential equations (8.68). As a result, unlike Theorems 8.10 and 8.11, the error bound can be established without assuming nondegeneracy of the matrix $\sigma\sigma^\top$. According to [4, 10], the rate of convergence in Δ for an approximation of the optimal filter for (8.13) is, in general, not better than $\Delta^{1/2}$, and both spectral separating schemes achieve this rate. Indeed, for $N \geq 2$, formulas (8.87) and (8.93) can be written as

$$\max_{0\leq i\leq M} \mathbb{E}\left|\phi_{t_i}[f] - \phi^n_{N,K;i}[f]\right| \leq C_3\Delta^{1/2} + \frac{C_4}{K^{(\nu-w-d-1)/d}\Delta} \quad (8.94)$$

and

$$\max_{0\leq i\leq M} \mathbb{E}\left|\phi_{t_i}[f] - \phi^{K,n}_{N;i}[f]\right| \leq \frac{C_3}{K^{(\nu-w-d-1)/d}} + C_4\Delta^{1/2}, \quad (8.95)$$

respectively. Note that the error due to truncation in space is $C_3 K^{-(\nu-w-d-1)/d}$ in both cases, but, since computation of $\phi^n_{N,K;i}[f]$ in (8.94) involves truncation in space on every timestep, this error is multiplied by the number of timesteps, and the number of timesteps is proportional to $1/\Delta$. The rate of convergence in time is still $\Delta^{1/2}$, since we first take the limit $K \to \infty$.

8.5 Other directions

8.5.1 Representations of the optimal filter

In this section, we review the main representations of the optimal filter using the Zakai, Kushner–Stratonovich, and Fujisaki–Kallianpur–Kunita equations, as well as the Kallianpur–Striebel formula. The Wiener chaos expansion can be used to study any of these representations.

Consider the filtering model (8.12). If $f = f(t, X(t), Y_{0,t})$ is a square integrable functional, and the functionals b, σ, ϱ, h at time t depend only on $X(t)$ in a sufficiently smooth way, then the optimal filter $\widehat{f} = \mathbb{E}(f|\mathcal{Y}_t)$ is

$$\widehat{f} = \frac{\int_{\mathbb{R}^d} f(t, x, Y_{0,t})\rho(t, x)dx}{\int_{\mathbb{R}^d} \rho(t, x)dx}, \quad (8.96)$$

where u satisfies the Zakai equation similar to (8.25). The original reference, the paper by Zakai [80], provides the derivation for the diffusion model with the coefficients independent of Y and with $\varrho \equiv 0$. A more general derivation, together with the detailed investigation of the analytical properties of the solution, is in [43, 71]. In the traditional formulation of the problem, the coefficients in the state and observation equations are assumed to be bounded, but this assumption can be relaxed [25].

Some of the results of [80] were obtained independently by Duncan [18] and Mortensen [65]; accordingly, equation (8.25) is also known as the DMZ (Duncan–Mortensen–Zakai) equation. Being a linear equation, (8.25) is well-suited for analysis using various forms of the Wiener chaos expansion, especially if the coefficients do not depend on Y.

If the unnormalized filtering density $\rho = \rho(t, x)$ exists, then the time evolution of the *normalized filtering density* $\pi(t, x) = \rho(t, x)/\int_{\mathbb{R}^d} \rho(t, x)$ is described by a nonlinear integro-differential equation, which for the time-homogeneous diffusion model (8.13) becomes

$$\pi(t, x) = p_0(x) + \int_0^t \mathcal{L}^* \pi(s, x) ds + \sum_{k=1}^r (\mathcal{M}_k^* \pi(s, x) \qquad (8.97)$$
$$- \pi(s, x) \int_{\mathbb{R}^d} \pi(s, x) h_k(x) dx) \left(dY_k - \left(\int_{\mathbb{R}^d} \pi(s, x) h_k(x) dx \right) ds \right).$$

The time evolution of the normalized filtering density was originally derived, in various forms and with various degrees of mathematical rigor, in [47, 49, 73, 74], and is known as the Kushner–Stratonovich equation. Some of the computations in [74] do not agree with the accepted standards of the Itô calculus and were initially dismissed as a mistake. A more careful analysis of the computations later lead to the creation of the now famous Stratonovich stochastic calculus [75]. (This calculus was independently discovered by D.L. Fisk in his PhD dissertation (1963), using motivations other than filtering; see Jarrow and Protter [31] for more details.)

More recent works, such as chapter 6 in Rozovskiĭ [71], or section 8 in Krylov [41], present a modern approach, both to the derivation and to the study of the analytical properties of the filtering density p.

It can happen that representation (8.17) of the optimal filter holds, while representation (8.96) does not; one example is the general model (8.12). In fact, for (8.17) to hold, the process X does not need any particular structure. Papers [52, 61, 66, 76] successfully study the optimal filter using (8.17) and various versions of the Wiener chaos expansion. The results illustrate the power of the Wiener chaos method by providing an insight into the structure of the optimal filter when the Zakai equation is not available. On the other hand, this level of generality prevents a detailed analysis of the potential numerical methods based on the expansions.

Another application of representation (8.17) is the *particle system approximations* of the optimal filter [15, 16, 17], which are deep extensions of the Monte Carlo method.

One disadvantage of (8.17) is that this representation does not provide the time evolution of the optimal filter. It turns out that this time evolution can be written even when the filtering density does not exist. The *Fujisaki–Kallianpur–Kunita equation* [22] describes the time evolution of the conditional expectation $\widehat{F}(t) = \mathbb{E}(F(t)|\mathcal{Y}_t)$, $t \geq 0$, for two semimartingales F, Y and can serve as the starting point in the derivation of all other filtering equations. Assume that $F = F(t)$ is a one-dimensional semimartingale and $Y = Y(t)$ is an r-dimensional semimartingale with representations

$$F(t) = F(0) + \int_0^t B(s) ds + M(t),$$
$$Y_\ell(t) = Y_\ell(0) + \int_0^t h_\ell(s) ds + \sum_{k=1}^r H_{\ell k}(s) dW_k(s), \quad \ell = 1, \ldots, r,$$

and assume that the matrix $H(s)$ is $\mathcal{Y}_t$-measurable and invertible. Define $\overline{W}_k(s) = \int_0^t H_{k\ell}^{-1}(s)(dY_\ell(s) - \widehat{h}_\ell(s)ds)$ and let D_ℓ be the process such that the quadratic covariation $\langle M, W_\ell \rangle$ of the martingale M and the Wiener process W_ℓ is $\langle M, W_\ell \rangle(t) = \int_0^t D_\ell(s)ds$. Then

$$\widehat{F}(t) = \widehat{F}(0) + \int_0^t \widehat{B}(s)ds \\ + \sum_{k=1}^r \int_0^t \left(\widehat{D}(s) + \sum_{\ell=1}^r \left(\widehat{Fh_\ell}(s) - \widehat{F}(s)\widehat{h}_\ell(s) \right) H_{\ell k}^{-1}(s) \right) d\overline{W}_k(s). \quad (8.98)$$

A complicated nonlinear structure of equation (8.98) prevents a direct numerical analysis of this time evolution.

8.5.2 Solving the filtering problem

If the pair of the processes (X, Y) is jointly Gaussian, then, by the Normal Correlation Theorem, the conditional distribution of $X(t)$ given $\mathcal{Y}_t$ is Gaussian and is uniquely determined by the conditional mean $m(t) = \mathbb{E}(X(t)|\mathcal{Y}_t)$ and variance $P(t) = \mathbb{E}((X(t) - m(t))(X(t) - m(t))^\top|\mathcal{Y}_t)$. The system of stochastic ordinary differential equations describing time evolution of m and P is the foundation of the *Kalman–Bucy filter* [37]. Various extensions of this filter to conditionally Gaussian processes have been derived (Liptser and Shiryaev [51, Ch. 12]).

In the Gaussian model, the filtering density is characterized by a finite number of parameters. Even though *exact finite dimensional filters* can exist for non-Guassian processes X, Y (Beneš [2], Mitter [64], Yau [77]), a result of Ocone and Pardoux [67, 69] shows that for most models the optimal filter is infinite-dimensional. Some special infinite-dimensional optimal filters can be computed exactly (Duam [12], Kouritzin [40], Schmidt [72], Yau and Hu [78]), but a typical filtering problem requires an approximation of the optimal filter.

The *extended Kalman filter* [32] is one of the most straightforward approximations and is derived by applying the Kalamn filter to the linearization of the model around the current estimate of X. This approximation preserves the relative analytical and computational simplicity of the Kalman filter and works well in many applications. The main drawback is lack of rigorous justification for this approximation, and, as a consequence, absence of reliable error bounds. More theoretically sound are the geometrically intrinsic filter [11], and the moment approximation of the optimal filter [46, 48]. A large class of numerical methods is based on approximating the original filtering model with a simpler one, for a example, a finite state Markov chain, and using the actual observations in the optimal filter for the approximating model [50].

Still, the most straightforward way to get an approximation together with an error bound is to solve numerically the equation for the filtering density, either normalized or unnormalized. The unnormalized filtering density is a more popular object to study because the corresponding equation is linear. For some numerical methods, the step-by-step solution of the Zakai equation with normalization at every step provides an approximate solution of the Kushner–Stratonovich equation, with approximation error under control [30]. In general, nonlinearity in the Kushner–Stratonovich equation complicates, but does not completely rule out, an investigation of the chaos expansion of the solution. In particular, Mukulevicius and Rozovskiĭ [62] studied the Cameron–Martin expansion for stochastic Navier–Stokes equation, while Ocone [66] investigated the multiple integral expansion of the normalized optimal filter without using the Kushner–Stratonovich equation.

A large class of numerical methods for the Zakai equation employ the corresponding algorithms used for deterministic partial differential equations: Galerkin approximation [3, 29], finite element [24], or operator splitting [4, 20, 21]. All these methods require the solution of certain partial differential equation at every timestep, which is usually impossible to achieve in real time if the dimension of the process X is bigger than three. The spectral separating schemes deal with this 'curse of dimensionality' by doing all the time consuming calculations in advance. Theoretically, many methods based on operator-splitting or on the robust form of the Zakai equation (Davis [13], Yau and Yau [79]) can achieve a similar separation by precomputing the fundamental solution of the corresponding deterministic equation. With a careful choice of the branching mechanism, the particle system approximations [15, 16, 17] also have a potential for the real-time implementation.

For $t < T$, the problems of computing $\mathbb{E}(f(X(t))|\mathcal{Y}_T)$ (interpolation, or smoothing) and $\mathbb{E}(f(X(T))|\mathcal{Y}_t)$ (extrapolation, or prediction) are studied using the same technical tools as in the filtering problem; see [51, Sections 8.4, 8.5] or [71, Section 6.3].

The traditional formulation of the filtering problem assumes that the probability distributions of X and Y are completely known. Multiple model filtering [1] provides a more realistic setting and was studied using Wiener chaos expansions [56]. An even more realistic setting is simultaneous filtering and estimation [9].

The following is a list of some problems in which a chaos approach is possible but has not been studied in detail:

(1) Analysis of the Fujisaki–Kallianpur–Kunita and Kushner–Stratonovich equations (chaos expansion of the normalized optimal filter).
(2) Nonlinear filtering problem with unbounded coefficients in the state and observation equations.
(3) Interpolation and extrapolation.
(4) Simultaneous filtering and estimation.

Notes

The author gratefully acknowledges support from the NSF CAREER award DMS-0237724 and the hospitality and support of the Institut Mittag-Leffler (Djursholm, Sweden).
1. The reader is encouraged to distinguish between the function $\rho = \rho(t, x)$ (the unnormalized filtering density) and the function $\varrho = \varrho(x)$ (one of the coefficients in the equations (8.13)).

References

[1] Y. Bar-Shalom and X.-R. Li, *Estimation and tracking. Principles, techniques, and software*, Artech House, Inc., Boston, MA, 1993.
[2] V. E. Benesh, *Exact finite-dimensional filters for certain diffusions with nonlinear drift*, Stochastics **5** (1981), 65–92.
[3] J. F. Bennaton, *Discrete time Galerkin approximation to the nonlinear filtering solution*, J. Math. Anal. Appl. **110** (1985), no. 2, 364–83.
[4] A. Bensoussan, R. Glowinski, and R. Rascanu, *Approximations of the Zakai equation by splitting-up method*, SIAM Journal on Control and Optimization **28** (1990), no. 6, 1420–31.
[5] A. V. Borisov, *Optimal filtering in systems with degenerate noise in the observations*, Automat. Remote Control **59** (1999), no. 11, 1526–37.
[6] A. Budhiraja and G. Kallianpur, *Approximations to the solution of the Zakai equation using multiple Wiener and Stratonovich integral expansions*, Stochastics Stochastics Rep. **56** (1996), nos. 3–4, 271–315.
[7] A. Budhiraja and G. Kallianpur, *The Feynman–Stratonovich semigroup and Stratonovich integral expansions in nonlinear filtering*, Appl. Math. Optim. **35** (1997), no. 1, 91–116.
[8] R. H. Cameron and W. T. Martin, *The orthogonal development of nonlinear functionals in a series of Fourier–Hermite functions*, Ann. Math. **48** (1947), no. 2, 385–92.
[9] F. Campillo and F. Le Gland, *MLE for partially observed diffusions: direct maximization vs. the EM algorithm*, Stochastic Process. Appl. **33** (1989), no. 2, 245–74.
[10] J. M. C. Clark and R. J. Cameron, *The maximum rate of convergence of discrete approximation for stochastic differential equations*, Springer Lecture Notes in Control and Inform. Sc. **25** (1980), 162–71.
[11] R. W. R. Darling, *Geometrically intrinsic nonlinear recursive filters I: Algorithms*, Tech. Report 494, UC Berkeley, Depertment of Statistics, March 1998, http://www.stat.berkeley.edu/tech-reports/494.pdf.
[12] F. E. Daum, *New exact nonlinear filters*, Bayesian Analysis of Time Series and Dynamic Models (J. C. Spall, ed.), Marsel Dekker, New York, 1988, pp. 199–226.
[13] M. H. A. Davis, *On a multiplicative functional transformation arising in nonlinear filtering theory*, Z. Wahrsch. Verw. Gebiete **54** (1980), no. 2, 125–39.
[14] P. J. Davis and P. Rabinowitz, *Methods of numerical integration*, Academic Press, 1984.
[15] P. Del Moral, *Non-linear filtering: interacting particle resolution*, Markov Proc. Rel. Fields **2** (1996), 555–79.
[16] P. Del Moral, J. Jacod, and P. Protter, *The Monte Carlo method for filtering with discrete-time observations*, Probab. Theory Related Fields **120** (2001), no. 3, 346–68.
[17] P. Del Moral and L. Miclo, *Branching and interacting particle systems approximations of Feynman–Kac formula with applications to non-linear filtering*, Séminaire de Probabilités, XXXIV, Lecture Notes in Mathematics, vol. 1729, Springer, Berlin, 2000, pp. 1–145.
[18] T. E. Duncan, Tech. Report 7001-4, Stanford University, Center for Systems Research, May 1967.

[19] S. D. Eidelman, *Parabolic systems*, Wolters-Noordhoff, Groningen, 1969.
[20] R. J. Elliott and R. Glowinski, *Approximations to solutions of the Zakai filtering equation*, Stoch. Anal. Appl. **7** (1989), no. 2, 145–68.
[21] P. Florchinger and F. LeGland, *Time discretization of the Zakai equation for diffusion processes observed in correlated noise*, Stoch. and Stoch. Rep. **35** (1991), no. 4, 233–56.
[22] M. Fujisaki, G. Kallianpur, and H. Kunita, *Stochastic differential equations for the nonlinear filltering problem*, Osaka J. Math. **9** (1972), no. 1, 19–40.
[23] C. P. Fung and S. V. Lototsky, *Nonlinear filtering: separation of parameters and observations using Galerkin approximation and Wiener chaos decomposition*, Tech. Report 1458, IMA Preprint Series, February 1997.
[24] A. Germani and M. Piccioni, *Semi-discretisation of stochastic partial differential equations on $\mathbf{R}^d$ by a finite element technique*, Stochastics **23** (1988), no. 2, 131–48.
[25] I. Gyöngy and N. V. Krylov, *Stochastic partial differential equations with unbounded coefficients and applications. II*, Stochastics Stochastics Rep. **32** (1990), nos. 3–4, 165–80.
[26] T. Hida, H.-H. Kuo, J. Potthoff, and L. Sreit, *White noise*, Kluwer, 1993.
[27] H. Holden, B. Øksendal, J. Ubøe, and T. Zhang, *Stochastic partial differential equations: A modeling, white noise functional approach*, Birkhäuser, 1996.
[28] K. Itô, *Multiple Wiener integral*, J. Math. Soc. Japan **3** (1951), 157–169.
[29] K. Itô, *Approximation of the Zakai equation for nonlinear filtering*, SIAM J. Control Optim. **34** (1996), no. 2, 620–634.
[30] K. Itô and B. L. Rozovskiĭ, *Approximation of the Kushner equation for nonlinear filtering*, SIAM J. Control Optim. **38** (2000), no. 3, 893–915.
[31] R. Jarrow and P. Protter, *A short history of stochastic integration and mathematical finance: the early years, 1880–1970*, A festschrift for Herman Rubin, IMS Lecture Notes Monogr. Ser., vol. 45, Inst. Math. Statist., Beachwood, OH, 2004, pp. 75–91.
[32] A. H. Jazwinski, *Stochastic processes and filtering theory*, Academic Press, New York, 1970.
[33] M. Joannides and F. LeGland, *Nonlinear filtering with continuous time perfect observations and noninformative quadratic variation*, Proceedings of the 36th IEEE Conference on Decision and Control, San Diego, December 10–12, 1997, pp. 1645–50.
[34] M. Joannides and F. LeGland, *Nonlinear filtering with perfect discrete time observations*, Proceedings of the 34th IEEE Conference on Decision and Control, New Orleans, Dec. 13–15, 1995, pp. 4012–17.
[35] G. Kallianpur, *Stochastic filtering theory*, Applications of Mathematics, vol. 13, Springer, Berlin, 1980.
[36] G. Kallianpur and C. Striebel, *Estimation of stochastic systems: Arbitrary system process with additive white noise observation errors*, Ann. Math. Stat. **39** (1968), 785–801.
[37] R. E. Kalman and R. S. Bucy, *New results in linear filtering and prediction problems*, J. Basic Engineering, Trans. ASME **83D** (1961), 95–108.
[38] I. Karatzas and S. E. Shreve, *Brownian motion and stochastic calculus*, 2nd edn., Springer, New York, 1991.
[39] P. E. Kloeden and E. Platen, *Numerical solution of stochastic differential equations*, Springer, Berlin, 1992.
[40] M. A. Kouritzin, *Exact infinite dimensional filters and explicit solutions*, Stochastic Models, CMC Conference Proceedings, vol. 26, AMS, 2000, pp. 265–82.
[41] N. V. Krylov, *An analytic approach to SPDEs*, Stochastic Partial Differential Equations. Six Perspectives (B. L. Rozovskiĭ and R. Carmona, eds.), Mathematical Surveys and Monographs, AMS, 1999, pp. 185–242.
[42] N. V. Krylov and A. J. Veretennikov, *On explicit formula for solutions of stochastic equations*, USSR Math. Sbornik **29** (1976), no. 2, 239–56.

[43] N. V. Krylov and A. Zatezalo, *A direct approach to deriving filtering equations for diffusion processes*, Appl. Math. Optim. **42** (2000), no. 3, 315–32.

[44] H. Kunita, *Cauchy problem for stochastic partial differential equations arising in nonlinear filtering theory*, Syst. & Cont. Letters **1** (1981), no. 1, 37–41.

[45] H. Kunita, *Stochastic flows and stochastic differential equations*, Cambridge University Press, Cambridge, 1982.

[46] H. Kushner and A. Budhiraja, *A nonlinear filtering algorithm based on an approximation of the conditional distribution*, IEEE Trans. Automat. Control **45** (2000), no. 3, 580–5.

[47] H. J. Kushner, *On the dynamical equations of conditional probability density functions, with applications to optimal stochastic control theory*, J. Math. Anal. Appl. **8** (1964), 332–44.

[48] H. J. Kushner, *Approximations to optimal nonlinear filters*, IEEE Trans. AC **3** (1967), 179–90.

[49] H. J. Kushner, *Dynamical equations for optimal nonlinear filtering*, J. Differential Equations **3** (1967), 179–90.

[50] H. J. Kushner, *Probability methods for approximations in stochastic control and for elliptic equations*, Academic Press, New York, 1977.

[51] R. S. Liptser and A. N. Shiryaev, *Statistics of random processes, I, II*, Springer, New York, 2002.

[52] J. T.-H. Lo and S.-K. Ng, *Optimal orthogonal expansion for estimation I: signal in white Gaussian noise*, Nonlinear Stochastic Problems (R. S. Bucy and J. M. F Moura, eds), D. Reidel Publ. Company, 1983, pp. 291–309.

[53] S. Lototsky, *Nonlinear filtering of diffusion processes in correlated noise: analysis by separation of variables*, Appl. Math. Optim. **47** (2003), no. 2, 167–94.

[54] S. V. Lototsky, *Wiener chaos and nonlinear filtering*, Appl. Math. Optim. **54** (2006), no. 3, 265–91.

[55] S. V. Lototsky, R. Mikulevicius, and B. L. Rozovskiĭ, *Nonlinear filtering revisited: a spectral approach*, SIAM J. Contr. Optim. **35** (1997), no. 2, 435–61.

[56] S. V. Lototsky, C. Rao, and B. L. Rozovskiĭ, *Fast nonlinear filter for continuous-discrete time multiple models*, Proceedings of the 35th IEEE Conference on Decision and Control, Kobe, Japan, 1996, vol. 4, pp. 4071–6.

[57] S. V. Lototsky and B. L. Rozovskiĭ, *Recursive nonliner filter for a continuous-discrete time model: separation of parameters and observations*, IEEE Trans. AC **43** (1998), no. 8, 1154–8.

[58] S. V. Lototsky and B. L. Rozovskiĭ, *Stochastic differential equations: a Wiener chaos approach*, From Stochastic Calculus to Mathematical Finance (Yu. Kabanov, R. Liptser, and J. Stoyanov, eds.), Springer, Berlin, 2006, pp. 433–506.

[59] P. Major, *Multiple Wiener–Itô integrals with applications to limit theorems*, Lecture Notes in Mathematics, vol. 849, Springer, 1981.

[60] R. Mikulevicius and B. L. Rozovskiĭ, *Separation of observations and parameters in nonlinear filtering*, Proceedings of the 32nd IEEE Conference on Decision and Control, San Antonio, TX, 1993, vol. 2, IEEE Control Systems Society, 1993, pp. 1564–9.

[61] R. Mikulevicius and B. L. Rozovskiĭ, *Fourier–Hermite expansion for nonlinear filtering*, Theory Probab. Appl. **44** (2000), no. 3, 606–12.

[62] R. Mikulevicius and B. L. Rozovskiĭ, *Global L_2-solutions of stochastic Navier–Stokes equations*, Ann. Probab. **33** (2005), no. 1, 137–176.

[63] G. Milstein, *Numerical integration of stochastic differential equations*, Kluwer Academic Publishers, Dordrecht, 1995.

[64] S. K. Mitter, *On the analogy between mathematical problems of nonlinear filtering and quantum physics*, Richerche Automat. **10** (1979), no. 2, 163–216.

[65] R. E. Mortensen, Tech. Report ERL-66-1, UC Berkeley, Electronics Research Laboratory, August 1966.

[66] D. Ocone, *Multiple integral expansions for nonlinear filtering*, Stochastics **10** (1983), no. 1, 1–30.

[67] D. Ocone and E. Pardoux, *A Lie algebraic criterion for nonexistence of finite-dimensionally computable filters*, Stochastic partial differential equations and applications, II (Trento, 1988), Lecture Notes in Math., vol. 1390, Springer, Berlin, 1989, pp. 197–204.

[68] B. K. Øksendal, *Stochastic differential equations: an introduction with applications*, 5th edn, Springer, 1998.

[69] E. Pardoux, *Filtrage Non Lineaire et Equations aux Derivees Partielles Stochastiques Associees*, Ecole d'été de Probabilités de Saint-Flour XIX—1989, Lecture Notes in Math., vol. 1464, Springer, Berlin, 1991, pp. 67–163.

[70] A. Quarteroni and A. Vali, *Numerical approximation of partial differential equations*, Springer, 1994.

[71] B. L. Rozovskiĭ, *Stochastic evolution systems*, Kluwer Academic Publishers, Dordrecht, 1990.

[72] G. C. Schmidt, *Designing nonlinear filters based on Daum's theory*, Journal of Guidance, Control, and Dynamics **16** (1993), no. 2, 371–6.

[73] A. N. Shiryaev, *On stochastic equations in theory of conditional Markov processes*, Theor. Probab. Appl. **11** (1966), 179–84.

[74] R. L. Stratonovich, *Conditional Markov processes*, Theor. Probab. Appl. **5** (1960), 156–78.

[75] R. L. Stratonovich, *A new representation for stochastic integrals and equations*, Vestn. Moskov. Univer., Ser. Mat. Mekhan. **1** (1964), 3–12.

[76] E. Wong, *Stochastic processes in information and dynamical systems*, McGraw-Hill Education, 1971.

[77] S. S.-T. Yau, *Finite-dimensional filters with nonlinear drift. I. A class of filters including both Kalman–Bucy filters and Benes filters*, J. Math. Systems Estim. Control **4** (1994), no. 2, 181–203.

[78] S. S.-T. Yau and G.-Q. Hu, *Finite-dimensional filters with nonlinear drift. X. Explicit solution of DMZ equation*, IEEE Trans. Automat. Control **46** (2001), no. 1, 142–8.

[79] S.-T. Yau and S. S.-T. Yau, *Real time solution of nonlinear filtering problem without memory. I*, Math. Res. Lett. **7** (2000), nos 5–6, 671–93.

[80] M. Zakai, *On the optimal filtering of diffusion processes*, Z. Wahrsch. Verw. Gebiete **11** (1969), no. 3, 230–43.

PART III
Stability and Asymptotic Analysis

·9·

On Filtering with Unspecified Initial Data for Nonuniformly Ergodic Signals

M. L. Kleptsyna and A. Yu. Veretennikov

9.1 Introduction

9.1.1 Brief start

The efforts of many researchers in the area have allowed new progress in filtering for unspecified models, in particular, with unspecified initial data. This chapter is a new version of recent results by the authors, with an emphasis on important techniques. The main results are formulated under more general assumptions than earlier, and our main aspiration is that this new presentation may give another insight into the combination of methods and techniques, which, in turn, may help make first steps towards solving new problems in the near future. We cannot pretend that we present all methods in this area, nor that our approach is the best, nor that it may cover all or most earlier results. All we hope to achieve is a better understanding. We emphasize that here we only discuss certain methods, but not the history in all its details; that is to say, from the history of the problem we mention primarily important papers and results related directly to the development of our approach. We may not mention or we mention just briefly many other important and relevant results, because they have not contributed directly to this development. This does not in any way undervalue their importance.

Introducing the problem

A careful formulation of the problem requires several pages of preliminary explanation. In words, we have one process which represents a signal, and another one which we observe; here we consider a Markov signal, and coupled with an observation process, we have a Markov system. Then, the model being known, one can construct a process of Bayesian estimations of a signal, as it were, online, with time running. This is called constructing a filtering measure. Its realization requires, of course, an exact knowledge of the whole model, in particular of the initial distribution of the signal (normally it should not be a fixed point, to allow a good algorithm). If the latter is unknown, the situation is said to be with unspecified initial data. We present below an analysis of how quickly the unknown initial data may be forgotten in a long run in a quite general (although finite-dimensional) noncompact Euclidean state space.

Importance

The problem is of principle importance: its positive solution would allow one to use algorithms with nonexact initial data, while its negative solution would not.

Milestone

Paper [26] was a milestone. Criticism of it can be found in [3] and [7].

The Birkhoff metric in the 'compact' case

Generally, forgetting initial data is one of the main features of ergodic Markov processes. Indeed, even the first ergodic theorem for finite-state irreducible Markov chains tells us straightforwardly that the transition probabilities over a long time period tend to some limiting (stationary) distribution *which does not depend on initial data*: see [11], [14], or any other monograph or advanced textbook on Markov processes. More recent results on nonconditional mixing for *non*uniformly ergodic Markov processes are based usually on 'recurrence' to some 'good' set and 'local mixing' for the process on this set. Forgetting initial data for filtering algorithms is a more subtle phenomenon, as it relates to *conditional* measures and conditional dynamics: first of all, it is not Markov any more. Hence, it was not clear whether the same methods could work equally well in both areas, i.e. for conditional and nonconditional forgetting of errors, and in fact, it appears that they actually do not. First qualitative (exponential) bounds for convergence of conditional filtering measures towards exact ones (i.e. with exact initial data) appeared in [2]. The main success was achieved due to a very important novelty, namely, the use of the *Birkhoff distance*, a powerful method from positive operators (see [25]) and dynamical systems. This distance dominates the total variation one, however it often allows an easier verification of contraction, which means geometric convergence or forgetting. In no doubt it is most suitable for 'linearization' of a nonlinear problem of this type, where nonlinearity is only due to normalization constants. The Birkhoff metric, in fact, may be regarded as the distance not for measures or signed measures themselves, but rather for the 'rays' generated by those measures. Hence, normalization constants being multipliers, they do not affect this distance. It remained to find suitable conditions which would suffice for convergence in this metric, and those conditions turned out to be also most natural. Then, in this metric, local mixing ensured contraction on every step or unit of time in the compact case, which meant a uniform exponential contraction between measures.

Noncompact setting

After this success in the compact case, certain results have been established for noncompact state spaces, too, under conditions of 'small noise' in the observations, or in some way close to that, although those conditions were, of course, different in different papers, so that even the term 'small noise' is not fully precise here: see [1], [5], [37]. Nevertheless, more general noncompact cases remained open until, apparently, [20]–[21]. In particular, one of the most important cases, of Gaussian IID noise in observations with an arbitrary covariance matrix, could not be tackled. In no way do we mean here that the results in [20]–[21] include all earlier cases, nor that our method is the most general, or that it is applicable directly to any model. We only present a further step, a development which allowed us to include a wider class of equations; in particular, for all such equations a general (nondegenerate) Gaussian IID noise in observations is possible, as well as many non-Gaussian ones. As a result, the method shows rates of convergence of the difference between two measures in total variation, which may be exponential or polynomial, depending on the recurrence properties of the signal; a wider range of possible rates could be achieved, including, e.g., subexponential rates.

Harnack inequalities

As a 'local mixing' condition, a specially designed estimate on the transition density has been used in [2]. Probably the most natural here is to use Harnack-type inequalities, similar to the nonconditional mixing of [39], [41], et al. See our comments in Sections 9.4 and 9.5 below.

Robustness

The success achieved in [2] and following papers, made it possible to advance in the parallel important problem of robustness. This was done for discrete models in several papers by P. Del Moral and A. Guionnet [12] and by F. Le Gland with N. Oudjane and other co-authors, [31], [32], et al.

Stopping times do not help much

After all successes, there yet remained the question how to tackle the problem of unspecified initial data in general noncompact cases. Intuitively, a similar idea based on a Birkhoff metric apparently should also have worked, so to speak, on 'good compacts'—i.e., in a domain with a good local mixing, due to recurrence properties. Here, however, there was a surprise waiting the authors and apparently many other researchers: the most natural idea to use stopping times did not work, at least not straightaway. Only quite recently, and yet only in some *very good* cases—i.e. under large deviation type assumptions—was a result close to Theorem 9.2 from [21] realized *without* the Birkoff metric in the forthcoming paper [13], as a deliberate attempt to establish this or similar result *not* using this metric. Nonetheless, the latter already played its role, and probably will remain one of the main tools in this area for a long time, due to the reasons presented above.

Ergodicity of the signal process

The problem of unspecified initial data in filtering of stochastic processes is basic for all Bayesian considerations, and it remains very important in hidden Markov chain models. Indeed, any Bayesian approach is based on the assumption that we know exactly the *prior distribution(s)*: if not, it is not a Bayesian approach and the filter is not the optimal one. In simple cases the problem can be tackled without any difficulty, one may just notice that, at least, a small error in the exact prior does not change the output too much, in suitable probability terms. However, when we deal with processes and when the time horizon in the problem is large enough, small initial errors may accumulate, in some cases exponentially, and destroy any good filter. How to tackle this difficulty? One immediate assumption that we make, nonrigorously, claims that the signal is ergodic; more rigorously, see (A1)–(A2) below; then the signal itself returns frequently and persistently into some (bounded) domain in which, presumably due to local mixing conditions, accumulated errors are forgotten, which makes reasonable a suggestion that possibly the errors may not accumulate at all. In principle, this requirement of ergodicity is not absolutely necessary: see e.g. [33].

Clearly, for nonhomogeneous processes similar forgetting results are possible and even would not require much additional work, despite the fact that there is no ergodicity. So, we could say that for the method which is presented here the most important features are recurrence and local mixing.

Filtering and Bayesian parameter estimation

We should say a few words about some related problems. In particular, the problem of unspecified initial prior in a closely related area of *parameter estimation* has been studied extensively for the IID and some more general cases, cf. [17]. In the Bayesian setting, a parameter is random, but its prior may be unknown. The observations in a standard statistical approach are, of course, a bit different, and the whole problem is not equivalent to ours, but clearly they have some common flavour: for example, we may regard a parameter as a constant process. So it may be interesting to see how the unspecified estimation problem is solved in this situation. A brief, folkloric answer is that a more or less arbitrary Bayesian estimator is asymptotically equivalent to the Maximum Likelihood Estimator, and the latter does not depend on any prior. Hence, all Bayesian estimators—with priors from some appropriate class—are also asymptotically equivalent, under rather mild assumptions. This can be interpreted as asymptotical independence of the estimator from a possibly wrong prior.

A more careful view, however, shows that conditions for consistency of Bayesian estimators are significantly weaker than what is needed for a MLE (see [17]); so, the idea of a reduction to MLE analysis perhaps is not the best possible. Nevertheless the conclusion about asymptotic independence on a wrong prior remains valid even under less restricted assumptions than needed for MLE.

Another more intrinsic link of our setting to parameter estimation is the problem of filtering similar to ours, but with coefficients of the system which depend additionally on some unknown parameter. In a 'compact case' under certain conditions this problem was studied in [35]. In a noncompact case the problem is waiting for its solution.

Local mixing: nonconditional and conditional dynamics

Currently there is an undramatic but significant difference between required local mixing assumptions for nonconditional Markov dynamics and the conditional one which arises in filtering. While for the continuous time case everything may be guaranteed by Harnack-type inequalities—at least for nondegenerate cases—in the discrete time case certain assumptions of nonsingularity are required explicitly. Here we describe briefly the main ideas behind discrete time local mixing.

For the purpose of establishing bounds for the *nonconditional* mixing rate, instead of frequently used mixing conditions of the type

$$\epsilon \lambda(\cdot) \leq Q(x, \cdot) \leq \epsilon^{-1} \lambda(\cdot), \tag{9.1}$$

for some positive measure λ, one may assume the following integral form of the Doeblin–Doob-type condition, due to Dobrushin (cf. [10]),

$$\inf_{\tilde{x}, x} \int \left(\frac{Q(\tilde{x}, dx')}{Q(x, dx')} \wedge 1 \right) Q(x, dx') := \nu > 0,$$

or its localized form, which it would be appropriate to call the *locally uniform Markov–Dobrushin condition*,

$$\inf_{\tilde{x}, x \in B} \int_{B'} \left(\frac{Q(\tilde{x}, dx')}{Q(x, dx')} \wedge 1 \right) Q(x, dx') =: \nu(B, B') > 0, \tag{9.2}$$

for appropriate sets B and B'. If the latter integral and infimum in (9.2) are both taken over the whole space, the value $1 - \nu$ is called Dobrushin's coefficient of ergodicity which was introduced in [10]. For the purpose of further possible induction, it may be appropriate to choose $B' = B$. The latter condition with arbitrary B has been used earlier in most of one of the authors' papers on mixing. This condition is considerably weaker than (9.1), at the same time providing better constants in the bounds of convergence and mixing. The condition (9.1) is frequently used in the literature for establishing mixing or convergence rate to equilibrium distribution, starting from the seminal book [11]. It does provide uniform bounds for such convergence to equilibrium (and mixing) on the class of processes (see again [11]). However, the use of (9.2) instead of (9.1) gives better constants in the bounds, and a wider class of processes is covered; here 'better' may be infinitely better in the sense that the ratio of the two constants under the exponential can be arbitrarily large. It is worth mentioning that the most general Doeblin–Doob condition of uniform ergodicity does provide some exponential convergence rate, however, it does *not* provide any uniform convergence rate on any *class of processes* under this sort of condition. At least, the technique in [11] in the general case which is based on Lebesgue differentiation cannot give any uniform convergence. Hence, for nonconditional dynamics, the condition (9.2) can be considered as a generalization of (9.1) in all meanings. More than that, as was noticed by one of the authors long ago—and this comment still makes sense—it is not clear how the condition like (9.1) may be applied directly to such important class of processes as Markov diffusions—e.g., with bounded coefficients and a nondegenerate diffusion coefficient, but without any smoothness—unlike (9.2), which works perfectly.

Thus, it looks as though condition (9.1) may simply disappear, being replaced in all cases by (9.2). And there is only one point: in its generality, (9.2) so far does *not* work for conditional dynamics. Below we propose, however, some intermediate condition between (9.1) and (9.2); so in any case (9.1) never seems to be the best for establishing convergence or mixing rates.

There is an *open problem*: are there better/weaker conditions that still allow control of the rate of convergence as well as nonconditional or conditional mixing for some appropriate classes of processes? Perhaps some further localization of condition (9.2) could be useful. Another open question is whether assumptions imposed in filtering might approach more closely the assumptions required for nonconditional mixing.

Degeneracy in the signal noise

In the present chapter, we allow distributions with partially vanishing density of the noise in the signal. Technically it is, of course, preferable to work with nondegenerate noises.

The 'mystery' of $1/2$

For the *conditional* setting we will use a *relaxed version* of the following mixing assumption, suggested in [23], intermediate between (9.1) and (9.2): for every $R > 0$ large enough, there exists $A_R \subset B_R := \{|x| \leq R\}$, such that

$$p_R := \inf_{x \in B_R} Q(x, A_R) > 1/2, \quad \& \quad \sup_{x, \tilde{x} \in B_R, \, x' \in A_R} \frac{Q(x, dx')}{Q(\tilde{x}, dx')} < \infty, \quad (9.3)$$

see below Assumption (A3) for more precise details. From the first sight, earlier no condition similar to that, with a probability which should be greater than 1/2, ever showed up in this area. At least, for the authors this became a kind of mystery, even though we in no way insist that it is a mystery for everybody. However, a more careful analysis suggests that in continuous time setting the set $\{\sup_{0 \le s \le 1} |X_s| \le R + 1 \, \& \, |X_0| \le R\}$ actually plays the role of A_R, and, thus, some weaker condition of this type was actually used, but remained hidden because Harnack inequalities provide slightly stronger than 'necessary' properties.

Recurrence conditions

Further concretization of our recurrence conditions (A1) below in terms of Lyapunov functions is possible: see e.g. [41] or Remark 9.4 below. On the other hand side, e.g., for continuous-time models Lyapunov-type conditions are not necessary even for nonconditional mixing, nor they are necessary for conditional ones. Other types of sufficient conditions for (A1) in terms of 'potentials' are available for continuous time, where Lyapunov conditions, generally speaking, may fail; see, for example, [42]. This signifies that conditions (A1) are genuinely more general than Lyapunov's ones.

In turn, further concretization of Lyapunov-type conditions is available in terms of coefficients of stochastic difference or the differential equations in question. In particular examples, it is usually easier to work with coefficients and to construct appropriate Lyapunov functions rather than to start from the latter and ask what is required from coefficients to ensure those functions. From the point of view of some statisticians, conditions in terms of coefficients are often preferable; this is also the point of view of the authors.

Organization of the chapter

The chapter is arranged as follows: Section 9.2 contains the assumptions, the main result, and some auxiliaries for the discrete time case; Section 9.3 is devoted to the proof of the main result; Sections 9.4 and 9.5 present a similar result and its brief proof for the continuous time case.

9.2 Setting, assumptions, main result

9.2.1 Discrete time, the model

We consider a discrete-time filter for a hidden Markov chain (X_n) with values in the Euclidean space R^d, with conditionally Markov observations (Y_n) from

R^ℓ. In principle, more general state spaces are possible; however, we will use some mixing bounds established in finite-dimensional spaces, and we would not like to *assume* such bounds in more general spaces where they are not yet found. The chain (X_n) is assumed to be homogeneous, with a transition kernel $Q(x, dx')$, that is,

$$P(X_{n+1} \in A \mid X_n) = Q(X_n, A). \tag{9.4}$$

Observations (Y_n) satisfy the *memoryless channel* assumption, that is, they are *conditionally independent* given the signal trajectory $(X_n, n \geq 0)$, and the conditional distribution of every Y_n depends only on X_n. In addition, we assume that the distributions $P(Y_n \in \cdot \mid X_n)$ are all dominated by some reference nonnegative measure Λ' (usually *Lebesgue's measure* is considered, although this is not necessary), with a likelihood function $\Psi_n(X_n, Y_n)$, that is,

$$P(Y_n \in B \mid X_n) = \int_B \Psi_n(X_n, y) \, \Lambda'(dy). \tag{9.5}$$

The basic example is given by the equations,

$$X_{n+1} = X_n + b(X_n) + \sigma(X_n)\xi_{n+1}, \quad (n \geq 0), \tag{9.6}$$

$$Y_n = h(X_n) + V_n \quad (n \geq 1), \tag{9.7}$$

where $(\xi_n, V_n, n \geq 1)$ is a family of IID random vectors—in particular, all ξ's are independent from V's—of dimensions d and ℓ correspondingly, with densities $q_\xi(x)$ and $q_V(y)$, $b(\cdot)$ is a d-dimensional vector function, $\sigma(\cdot)$ a $d \times d$ matrix function, $h(\cdot)$ an ℓ-dimensional vector function. Below we will formulate the main result for the model (9.4)–(9.5) under certain assumptions.

It is a fairly standard situation that an exact initial distribution of X_0 denoted by μ_0 is known with some error. Equally standard is that a filtering algorithm is used with this approximate distribution as if it were exact. Explicitly or implicitly—and usually *not* explicitly—this is performed with a hope that a close distribution does not spoil the algorithm too much, or even that in the long run the initial error may be forgotten. Of course, this is not rigorous, because there are many situations, e.g., in dynamical systems, where a small initial error may lead to an exponentially growing error over long periods. A major part of filtering theory belongs to Bayesian statistics with a quadratic error criterion, and the point of all such Bayesian problems is that one *must* use a precise algorithm which is called Bayesian. There are many nice features of Bayesian solutions: e.g., they are automatically optimal *in the Bayesian sense*. But this optimality breaks down straightaway if there are *any* errors in the algorithm, for example, if there is an error in initial distribution. Practitioners usually just hope that a small error may have only a small affect on the solution, but, of course, this is a complicated theoretical issue. The role of the theory here is to justify this belief under appropriate assumptions, but also to indicate possible limitations. We do not construct any counterexamples here; however, it is a desirable issue of a further study in this area—to create counterexamples.

Hence, the main problem addressed in this chapter is whether an initial error is forgotten by the optimal filtering algorithm in the long run under suitable assumptions. More precisely the setting is explained in section 9.2.3 below.

Our main result can be further extended in several directions. For example, one may be interested not in precise bounds, but in some weaker convergence; e.g. the mere fact of forgetting wrong initial data in the sense of convergence to zero of the mean total variation of the difference between the correct and wrong filtering measures may be established under much more relaxed conditions. In particular, we may assume considerably weaker moment assumptions on the measures.

9.2.2 Assumptions

(A1) *Recurrence and moments for the signal.* For every $R \geq R_0$ with R_0 large enough, and every $k > 0$, there exist $m > 0$ and C_1, such that

$$E_x \tau_R^k \leq C_1(1 + |x|^m), \qquad (9.8)$$

where $\tau = \tau_R := \inf(t \geq 0 : |X_t| \leq R)$, and, moreover, for the same m there exists $C_2 > 0$, such that

$$\sup_{t \geq 0} E_x 1(t \leq \tau_R)|X_t|^m \leq C_2(1 + |x|^m), \quad \& \quad \sup_{|x| \leq R_0} E_x |X_1|^m < \infty. \quad (9.9)$$

(A2) *Local mixing.* We assume that for every R large enough there exists $A_R \subset B_R$ and a positive reference measure $\Lambda(dx')$ such that the signal transition measure $Q(x, dx')$ restricted on A_R has a density component $q(x, x')$ with respect to $\Lambda(dx')$ in the sense that for any Borel $A \subset A_R$ and any $x \in B_R$,

$$Q(x, A) \geq \int_A q(x, x') \Lambda(dx'),$$

and, moreover,

$$0 < a_R^- := \inf_{|x| \leq R} \inf_{x' \in A_R} q(x, x')$$

$$\leq \sup_{|x| \leq R} \sup_{x' \in A_R} q(x, x') =: a_R^+ < \infty,$$

and

$$p_R = \inf_{|x| \leq R} \int_{A_R} q(x, x') \Lambda(dx') > 1/2.$$

(A3) *Ψ and h.* The conditional density Ψ is assumed to be positive everywhere. The function h is locally bounded.

(A4) *Equivalence and moments for two initial measures.* The measures μ_0 and ν_0 are equivalent in the sense of absolute continuity, and moreover there exists $C > 0$ such that

$$C^{-1} \leq \frac{d\mu_0}{d\nu_0} \leq C. \tag{9.10}$$

We also assume that both μ_0 and ν_0 possess all polynomial moments.

9.2.3 Setting and main results

The setting is based on *the algorithm* that solves the exact filtering problem, which, like any Bayesian algorithm, depends on the initial data. Hence, we are going to plug into the algorithm a new initial measure instead of the exact one. The sequence of filtering measures is constructed via the observations (Y_n), as a sequence of conditional probabilities, $P_{\mu_0}(X_n \in \cdot \mid \mathcal{F}_n^Y)$, – where $\mathcal{F}_n^Y = \sigma(Y_k : 1 \leq k \leq n)$, – with the initial measure μ_0. Via the Bayes formula, this conditional measure can be represented as a *probability measure for any* Y, via the following nonlinear operator $\bar{S}_n^{Y_1,\ldots,Y_n,\mu_0}$, which we will denote for simplicity by $\bar{S}_n^{Y,\mu_0}$, applied to the measure μ_0,

$$P_{\mu_0}(X_n \in dx_n \mid Y_1, \ldots Y_n) = \int \prod_{i=1}^n Q(x_{i-1}, dx_i) c_i^{\mu_0} \Psi(x_i, Y_i) \mu_0(dx_0)$$

$$= d_n^{\mu_0} \int \prod_{i=1}^n Q(x_{i-1}, dx_i) \Psi(x_i, Y_i) \mu_0(dx_0) =: \mu_0 \bar{S}_n^{Y,\mu_0}(dx_n).$$

Remember that $\Psi(X_i, y_i)$ is a conditional density of Y_i given X_i, computed at y_i, and in the Bayes formula for X given Y we have to replace it by a substitution $\Psi(x_i, Y_i)$. Here the normalization constant $d_i^{\mu_0}$ is defined by

$$d_i^{\mu_0} = \left(E_{\mu_0} \left(\prod_{j=1}^i \Psi(X_j, y_j) \right) \Big|_{y_1=Y_1,\ldots y_i=Y_i} \right)^{-1},$$

and, correspondingly,

$$c_i^{\mu_0} = \frac{d_i^{\mu_0}}{d_{i-1}^{\mu_0}} = \frac{E_{\mu_0}\left(\prod_{j=1}^{i-1} \Psi(X_j, y_j)\right)}{E_{\mu_0}\left(\prod_{j=1}^i \Psi(X_j, y_j)\right)} \Bigg|_{y_1=Y_1,\ldots y_i=Y_i}.$$

The 'wrong initialization' problem can be formulated more precisely as follows. One does not know the measure μ_0 exactly, but only some its approximation ν_0. Hence, one plugs in the observed value Y's and this new measure ν_0 into the procedure. The problem is whether this algorithm forgets its wrong initial data in the long run, that is, whether the difference between the conditional measures provided by the algorithms with the exact and wrong initial data converges to zero in some suitable topology. First of all, is this operation of using ν_0 instead of μ_0 well defined? The answer is that it is not well defined if and only if our actually observed vector Y is impossible under ν_0 for some n, or, equivalently, if the vector value $(X_0, \ldots, X_n)$ starting from the distribution ν_0 is impossible under the observed $(Y_1, \ldots, Y_n)$ for some n. Since clearly any value of $(X_0, \ldots, X_n)$ with $X_0 \in \mathrm{supp}(\mu_0)$ is possible, we have a sufficient condition for our operation to be well defined, $\mathrm{supp}(\nu_0) \subset \mathrm{supp}(\mu_0)$, or, equivalently,

$$\nu_0 << \mu_0. \tag{9.11}$$

This condition is sufficient whatever all other distributions are. In this chapter we actually assume more in (A4), to serve the convergence rate.

Now we shall explain how one can interpret this setting in a probabilistic way. In fact, for the initial distribution ν_0, we have another sequences of measures and observations,

$$d_n^{\nu_0} \int \prod_{i=1}^{n} Q(x_{i-1}, dx_i) \Psi(x_i, \tilde{Y}_i) \nu_0(dx_0) = \nu_0 \bar{S}_n^{\tilde{Y}, \nu_0}(dx_n).$$

This can, indeed, be regarded as another conditional expectation, for the same Markov process starting from another initial distribution ν_0, given some new observations $(\tilde{Y}_1, \ldots, \tilde{Y}_n)$. Without losing a generality, we can and will assume that this pair $(\tilde{X}, \tilde{Y})$ is defined on some *independent probability space*; we will not change our notation for the probability measure, nor for expectation, though both now will apply to the process $(X, Y, \tilde{X}, \tilde{Y})$. However, due to the setting, only original observations Y are available, so that we are obliged to identify $(\tilde{Y}_1, \ldots, \tilde{Y}_n)$ with $(Y_1, \ldots, Y_n)$, that is, we *keep the original observations* that have risen from the original initial data μ_0, as though they were initialized by its substitution ν_0. The result, $\nu_0 \bar{S}_n^{Y,\nu_0}$, is still some conditional probability, namely the conditional distribution of ν_n given $(\tilde{Y}_1, \ldots, \tilde{Y}_n)$, after the values $(\tilde{Y}_1, \ldots, \tilde{Y}_n)$ have been replaced by $(Y_1, \ldots, Y_n)$. This operation is almost surely well defined with respect to the measure P_{μ_0}, due to our assumption $\Psi > 0$ a.s. The main problem addressed in this chapter concerns a difference of the two measures,

$$(\mu_0 \bar{S}_n^{Y,\mu_0} - \nu_0 \bar{S}_n^{Y,\nu_0})(dx_n),$$

whether it is reasonably small for large values of n, in particular, in the mean total variation norm with respect to the original initial measure μ_0. In the theorem, $\rho(\mu_0, \nu_0)$ is the Birkhoff distance between μ_0 and ν_0; see the definition below in (9.23) or (9.24).

Theorem 9.1

1. *Under the assumptions (A1)–(A4) above, the following bounds hold true:*

$$E_{\mu_0} \| \mu_0 \bar{S}_n^{Y,\mu_0} - \nu_0 \bar{S}_n^{Y,\nu_0} \|_{TV} \leq C_m n^{-m} \rho(\mu_0, \nu_0), \quad \forall m > 0. \qquad (9.12)$$

2. *In addition, the following pathwise inequalities hold true: if*

$$E_{\mu_0} \| \mu_0 \bar{S}_n^{Y,\mu_0} - \nu_0 \bar{S}_n^{Y,\nu_0} \|_{TV} \leq C n^{-m} \rho(\mu_0, \nu_0),$$

then, for every $m' < m$, not necessarily integer, there exists a (random) n_0 such that

$$\| \mu_0 \bar{S}_n^{Y,\mu_0} - \nu_0 \bar{S}_n^{Y,\nu_0} \|_{TV} \leq n^{-m'+1} \rho(\mu_0, \nu_0), \quad n \geq n_0. \qquad (9.13)$$

Remark 9.1 The bound (9.12) is uniform for appropriate classes of processes satisfying all the assumptions, say, with the same constants.

Remark 9.2 The bound (9.13) suggests what can be achieved in the compact case—i.e. a uniform exponential bound almost surely for every n with a uniform multiple constant, that is, *without expectation*—at the same time providing non-compact results practically of the same strength. This is interesting primarily because at first sight the basic bound (9.12) looks a bit weaker. The difference between the two cases arises since, firstly, in the compact case contraction in the Birkhoff metric holds true on each step, unlike in the noncompact situation, and secondly, because the Birkhoff metric is not suitable for using any partitions of the state space in the calculus, simply because this metric can be defined only for *equivalent* measures, and more than that, with positive and even bounded densities with respect to each other.

Remark 9.3 Similar *exponential* bounds can be established under *exponential* recurrence, e.g., if for $R \geq R_0$ and some $\alpha > 0$ there exist $a > 0$, $C_1 > 0$ such that

$$E_x \exp(a \tau_R) \leq C_1 \exp(a|x|), \qquad (9.14)$$

and, moreover, there exists $C_2 > 0$, such that

$$\sup_{t \geq 0} E_x 1(t \leq \tau_R) \exp(a|X_t|) \leq C \exp(a|x|), \quad \& \quad \sup_{|x| \leq R} E_x \exp(a|X_1|) < \infty. \qquad (9.15)$$

More general functions ψ may be suitable instead of exponentials $\exp(a|x|)$. *Subexponential* bounds may be also established, with the help of results from [24].

Remark 9.4 An example of possible sufficient conditions for (A1) in terms of coefficients of the model (9.6)–(9.7) with $\sigma \equiv I$ (identity matrix) and b locally bounded, reads,

$$\limsup_{|x|\to\infty} \left(\frac{|x+b(x)|}{|x|} - 1\right)|x|^2 = -\infty.$$

the noise ξ possesses all polynomial moments, and $E\xi_1 = 0$, cf. [21] and [41]. Also, if $E\xi_1 = 0$, then the inequality

$$\limsup_{|x|\to\infty} \left(\frac{|x+b(x)|}{|x|} - 1\right)|x|^1 < 0,$$

is sufficient for (9.14)–(9.15), if ξ_1 and μ_0 and ν_0 possess some exponential moment.

Remark 9.5 Unlike (9.11), Assumption (9.10) is not necessary and actually may be dropped, as shown in [21]. In this case, similar bounds for forgetting are available, with only two small changes. Firstly, in all exponential inequality multiples like $\pi_R^{\epsilon n}$, should be replaced by $\pi_R^{\epsilon n-1}$, just because induction in (9.27) must start from $k = 1$, not from zero. And secondly, there will be no multiple ρ_0 in all bounds like (9.12), (9.27), et al., since in the case where (9.11) fails we actually have $\rho(\mu_0, \nu_0) = +\infty$. Informally, this multiple in all bounds should be replaced by 'ρ_1,' which is not literally '$\rho(\mu_1, \nu_1)$,' but some more complicated coefficient, see the details in [21]. Further extensions are available under continuity of measure conditions considerably weaker than (9.10), however, with stronger recurrence assumptions of large deviation type, see the Theorem 9.2 from [21]. For this presentation we chose a stronger version (9.10) with the multiple $\rho(\mu_0, \nu_0)$.

9.3 Proof, discrete time

1. *(Start; some indicators)*
First of all, let us introduce some indicators. Note that in this proof, x stands for the whole sequence $(x_0, x_1, \ldots, x_n)$, and likewise for $\tilde{x}$, and the same for the random sequences X, $\tilde{X}$, and Y. As suggested above in the setting, we consider independent couples (X, Y) and $(\tilde{X}, \tilde{Y})$, with initial distributions of the first components, $\mathcal{L}(X_0) = \mu_0$ and $\mathcal{L}(\tilde{X}_0) = \nu_0$. We will reduce the original problem to a similar one for product measures,

$$\|P_{\mu_0}(X_n \in \cdot \mid Y) - P_{\nu_0}(X_n \in \cdot \mid Y)\|_{TV} \qquad (9.16)$$

$$\leq \|P_{\mu_0,\nu_0}((X_n, \tilde{X}_n) \in \cdot \mid Y, \tilde{Y})|_{\tilde{Y}=Y} - P_{\nu_0,\mu_0}((X_n\tilde{X}_n) \in \cdot \mid Y)|_{\tilde{Y}=Y}\|_{TV}.$$

At first we are going to tackle a more general, and hence a more difficult, problem; however, the reason why it is useful will be seen below.

For every $i \geq 0$, let $M_i := \max(|X_i|, |\tilde{X}_i|)$. For fixed R and n, we denote by $\delta = (\delta_0, \delta_1, \ldots, \delta_n)$ a (nonrandom) vector of dimension $n+1$ with coordinates 1 or 0 at every place, and by $\delta' = (\delta'_1, \delta'_2, \ldots, \delta'_n)$ any other nonrandom vector of

dimension n with coordinates 1 or 0 such that all zeros in δ remain zeros in δ', or in the other words,
$$(\delta'_i = 1) \implies (\delta_i = 1).$$
Denote by Δ the set of all possible values of the couple (δ, δ').

For every δ, δ' let us define *nonrandom* objects,

$$J(\delta, \delta') := \{i : 1 \leq i \leq n, \delta_{i-1} = 1, \delta'_i = 1\}, \quad \#1(\delta, \delta') := \sum_{j=1}^{n} 1(j \in J(\delta, \delta')), \quad (9.17)$$

$$J(\delta) := \{i : 0 \leq i \leq n, \delta_i = 1\}, \quad \#1(\delta) := \sum_{j=1}^{n} 1(j \in J(\delta)), \quad (9.18)$$

and

$$Q^{i-1:i}_{\delta,\delta'}(x_{i-1}, dx_i) = 1\left(\delta_{i-1} = \delta'_i = 1\right) q(x_{i-1}, x_i) \Lambda(dx_i)$$
$$+ \left(1 - 1\left(\delta_{i-1} = \delta'_i = 1\right)\right) \left(Q(x_{i-1}, dx_i) - q(x_{i-1}, x_i)\Lambda(dx_i)\right),$$

and similarly

$$Q^{i-1:i}_{\delta,\delta'}(\tilde{x}_{i-1}, d\tilde{x}_i) = 1\left(\delta_{i-1} = \delta'_i = 1\right) q(\tilde{x}_{i-1}, \tilde{x}_i) \Lambda(d\tilde{x}_i)$$
$$+ \left(1 - 1\left(\delta_{i-1} = \delta'_i = 1\right)\right) \left(Q(\tilde{x}_{i-1}, d\tilde{x}_i) - q(\tilde{x}_{i-1}, \tilde{x}_i)\Lambda(d\tilde{x}_i)\right).$$

2. (*Symmetrization and doubling space; probabilities as operators*)
Let us define new operators on the spaces of normalized and nonnormalized measures on $R^{2d} = R^d \times R^d$, or rather on the space of pairs of measures, each one on R^d, as follows (we use a double integral notation just to emphasize that we integrate with respect to the variables x and $\tilde{x}$),

$$(\mu, \nu)\, \bar{S}^{Y;\mu_0,\nu_0}_n(A \times B) = \int\!\!\int 1(x_n \in A, \tilde{x}_n \in B)$$
$$\times \left(\prod_{i=1}^{n} c^{\mu_0}_i c^{\nu_0}_i \Psi(x_i, Y_i)\Psi(\tilde{x}_i, Y_i) Q(x_{i-1}, dx_i) Q(\tilde{x}_{i-1}, d\tilde{x}_i)\right) \mu(dx_0)\nu(d\tilde{x}_0),$$

and

$$(\mu, \nu)\, \bar{S}^{Y;R;\delta,\delta';\mu_0,\nu_0}_n(A \times B) = \int\!\!\int 1(x_n \in A, \tilde{x}_n \in B)$$
$$\times \left(\prod_{i=1}^{n} c^{\mu_0}_i c^{\nu_0}_i \Psi(x_i, Y_i)\Psi(\tilde{x}_i, Y_i) Q^{i-1:i}_{\delta,\delta'}(x_{i-1}, dx_i) Q^{i-1:i}_{\delta,\delta'}(\tilde{x}_{i-1}, d\tilde{x}_i)\right) \mu(dx_0)\nu(d\tilde{x}_0),$$

and

$$(\mu, \nu)\, S^{Y;R;\delta,\delta'}_n(A \times B) = \int\!\!\int 1(x_n \in A, \tilde{x}_n \in B)$$
$$\times \left(\prod_{i=1}^{n} \Psi(x_i, Y_i)\Psi(\tilde{x}_i, Y_i) Q^{i-1:i}_{\delta,\delta'}(x_{i-1}, dx_i) Q^{i-1:i}_{\delta,\delta'}(\tilde{x}_{i-1}, d\tilde{x}_i)\right) \mu(dx_0)\nu(d\tilde{x}_0).$$

The nonnormalized operator $S_n^{Y;R;\delta,\delta'}$ can be equivalently presented as

$$(\mu, \nu) S_n^{Y;R;\delta,\delta'}(A \times B) = (\mu, \nu) \prod_{i=0}^{n-1} S_{i:i+1}^{Y;R;\delta,\delta'}(A \times B),$$

with

$$(\mu_i, \nu_i) S_{i:i+1}^{Y;R;\delta,\delta'}(A \times B)$$

$$= \int \int 1(x_{i+1} \in A, \tilde{x}_{i+1} \in B) 1_\delta(x_i, \tilde{x}_i) \chi_{\delta'}(x_{i+1}, \tilde{x}_{i+1})$$

$$\times \Psi(x_{i+1}, Y_{i+1}) \Psi(\tilde{x}_{i+1}, Y_{i+1}) Q_{\delta,\delta'}^{i:i+1}(x_i, dx_{i+1}) Q_{\delta,\delta'}^{i:i+1}(\tilde{x}_i, d\tilde{x}_{i+1}) \mu_i(dx_i) \nu_i(d\tilde{x}_i).$$

We shall now accomplish inequality (9.16), rewritten now in terms of the operators $\bar{S}$, by further bounds,

$$\|\mu_0 \bar{S}_n^{Y;\mu_0,\nu_0} - \nu_0 \bar{S}_n^{Y;\nu_0,\mu_0}\|_{TV} \leq \|(\mu_0, \nu_0) \bar{S}_n^{Y;\mu_0,\nu_0} - (\nu_0, \mu_0) \bar{S}_n^{Y;\nu_0,\mu_0}\|_{TV} \tag{9.19}$$

$$\leq \sum_{\delta,\delta' \in \Delta} \|(\mu_0, \nu_0) \bar{S}_n^{Y;R;\delta,\delta';\mu_0,\nu_0} - (\nu_0, \mu_0) \bar{S}_n^{Y;R;\delta,\delta';\mu_0,\nu_0}\|_{TV}$$

$$= 2 \sum_{\delta,\delta' \in \Delta} \sup_D (e_n^{Y;\delta,\delta';\mu_0,\nu_0}(\mu_0, \nu_0) \hat{S}_n^{Y;R;\delta,\delta';\mu_0,\nu_0}(D) - e_n^{Y;\delta,\delta';\nu_0,\mu_0}(\nu_0, \mu_0) \hat{S}_n^{Y;R;\delta,\delta';\mu_0,\nu_0}(D)).$$

$$= 2 \sum_{\delta,\delta' \in \Delta} e_n^{Y;\delta,\delta';\mu_0,\nu_0} \sup_D ((\mu_0, \nu_0) \hat{S}_n^{Y;R;\delta,\delta';\mu_0,\nu_0}(D) - (\nu_0, \mu_0) \hat{S}_n^{Y;R;\delta,\delta';\mu_0,\nu_0}(D)),$$

where D runs all Borel sets $\mathcal{B}(R^{2d})$. Here the inequality (9.19) is due to the identity,

$$(\mu_0, \nu_0) \bar{S}_n^{Y;\mu_0,\nu_0} - (\nu_0, \mu_0) \bar{S}_n^{Y;\nu_0,\mu_0}$$

$$= \sum_{\delta,\delta' \in \Delta} ((\mu_0, \nu_0) \bar{S}_n^{Y;R;\delta,\delta';\mu_0,\nu_0} - (\nu_0, \mu_0) \bar{S}_n^{Y;R;\delta,\delta';\mu_0,\nu_0}), \tag{9.20}$$

which holds true because in the summation we just count all possibilities of the total split over each variable, due to the triangle inequality. Remember that for any $x_{i-1}, \tilde{x}_{i-1}, x_i, \tilde{x}_i$, we have split the kernels $Q(x_{i-1}, dx_i) Q(\tilde{x}_{i-1}, d\tilde{x}_i)$ into two parts, $q(x_{i-1}, x_i) q(\tilde{x}_{i-1}, \tilde{x}_i) \Lambda(dx_i) \Lambda(d\tilde{x}_i)$ if $\delta_{i-1} = \delta'_i = 1$, and the rest, $Q(x_{i-1}, dx_i) Q(\tilde{x}_{i-1}, d\tilde{x}_i) - q(x_{i-1}, x_i) q(\tilde{x}_{i-1}, \tilde{x}_i) \Lambda(dx_i) \Lambda(d\tilde{x}_i)$, otherwise. By definition of Δ, all possibilities in (9.20) have been counted. The usefulness of such 'coding' by means of δ and δ' will be clarified in the sequel.

3. (Representations and indicators)

For the kernel split procedure above—in short, slightly abusing our notation, $dQ d\tilde{Q} = q\, dx'\, q\, d\tilde{x}' + (dQ\, d\tilde{Q} - q\, dx'\, q\, d\tilde{x}')$—it is convenient to use a suitable representation about how to realize such a procedure via construction of random variables, X'_i and X''_i which would correspond to each part, $q\,dx'\,q\,d\tilde{x}'$ and $(d\,Q\,d\,\tilde{Q} - q\,dx'\,q\,d\tilde{x}')$. Of course, we have to normalize both parts. Hence, for $x \in B_R$, let

$$\alpha(x) := \int_{A_R} q(x, x')\,\Lambda(dx').$$

Now, given $X_{i-1} \in B_R$, let us define an independent random variable

$$\eta_i(X_{i-1}) = \begin{cases} 1, & \text{with probability} \quad \alpha(X_{i-1}), \\ 0, & \text{with probability} \quad 1 - \alpha(X_{i-1}), \end{cases}$$

and new independent random variables X'_i and X''_i: the first one with the following (conditional) density with respect to Λ,

$$\alpha(X_{i-1})^{-1}\,q\,(X_{i-1}, x'_i), \quad x'_i \in A_R,$$

and, if $\alpha(X_{i-1}) \neq 1$, the second one with the (conditional) distribution

$$P\left(X''_i \in dx'\right) = (1 - \alpha(X_{i-1})^{-1})\left(Q(X_{i-1}, dx') - q(X_{i-1}, x'_i)\,\Lambda\,(dx'_i)\right).$$

(If $\alpha(X_{i-1}) = 1$, let, for example, $X''_i = 0$.) If $\tilde{X}_{i-1} \in B_R$, we similarly define random independent variables $\eta_i(\tilde{X}_{i-1})$, $\tilde{X}'_i$ and $\tilde{X}''_i$. Then, given $X_{i-1} \in B_R$, without loss of generality we can assume that

$$X_i = X'_i\,\mathbf{1}(\eta_i(X_{i-1}) = 1) + X''_i\,\mathbf{1}(\eta_i(X_{i-1}) = 0).$$

Indeed, the right-hand side here has a correct conditional distribution $Q(X_{i-1}, \cdot)$, so that using this representation does not change the whole distribution of the process X. In the case if $X_{i-1} \notin B_R$, we do not change the next value X_i. Similarly, if $\tilde{X}_{i-1} \in B_R$, without loss of generality we can use the representation

$$\tilde{X}_i = \tilde{X}'_i\,\mathbf{1}(\eta_i(\tilde{X}_{i-1}) = 1) + \tilde{X}''_i\,\mathbf{1}(\eta_i(\tilde{X}_{i-1}) = 0).$$

Notice that, of course, this representation suggests some extension of our probability space. However, there is no need to change our notation for probability and expectations.

Next, consider the following indicators, with a convention $0^0 = 1$,

$$\mathbf{1}_{\delta,\delta'}(X, \tilde{X}) := \prod_{i=0}^{n}(\mathbf{1}(M_i \le R))^{\delta_i}\,(1 - \mathbf{1}(M_i \le R))^{1-\delta_i}$$

$$\times \prod_{i=1}^{n}\left(\mathbf{1}((X_i, \tilde{X}_i) \in A_R \times A_R,\,\eta_i(X_{i-1}) = \eta_i(\tilde{X}_{i-1}) = 1)\right)^{\delta'_i}$$

$$\times \left(1 - \mathbf{1}((X_i, \tilde{X}_i) \in A_R \times A_R,\,\eta_i(X_{i-1}) = \eta_i(\tilde{X}_{i-1}) = 1)\right)^{1-\delta'_i}$$

$$= \mathbf{1}_\delta(X, \tilde{X})\chi_{\delta'}(X, \tilde{X}),$$

where

$$1_\delta(X, \tilde{X}) := \prod_{i=0}^{n} 1_{\delta_i}(M_i), \qquad \chi_{\delta'}(X, \tilde{X}) := \prod_{i=1}^{n-1} \chi_{\delta'_i}(X_i, \tilde{X}_i),$$

and, in turn,

$$1_{\delta_i}(M_i) = 1(\delta_i = 1)1(M_i \leq R) + 1(\delta_i = 0)1(M_i > R),$$

and

$$\chi_{\delta'_i}(X_i, \tilde{X}_i) = 1\left(\delta'_i = 1\right) 1(X_i, \tilde{X}_i \in A_R, \eta_i(X_{i-1}) = \eta_i(\tilde{X}_{i-1}) = 1)$$
$$+ 1\left(\delta'_i = 0\right) (1 - 1(X_i, \tilde{X}_i \in A_R, \eta_i(X_{i-1}) = \eta_i(\tilde{X}_{i-1}) = 1)).$$

Let

$$\zeta_i = 1(X_{i-1} \in B_R, X_i \in A_R, \eta_i(X_{i-1}) = 1), \qquad (9.21)$$

and

$$\tilde{\zeta}_i = 1(\tilde{X}_{i-1} \in B_R, \tilde{X}_i \in A_R, \eta_i(\tilde{X}_{i-1}) = 1), \qquad (9.22)$$

Let us emphasize that

$$1 = \sum_{\delta, \delta' \in \Delta} 1_{\delta, \delta'}(X, \tilde{X}),$$

where Δ is the set of all possible values of the vector δ, δ'. This is because our 'coding' via δ and δ' covers all possibilities of what might have happened with both components, and no one intersects with any other. Also notice that for every δ, δ',

$$(\mu_0, \nu_0) \bar{S}_n^{Y;R;\delta,\delta';\mu_0,\nu_0}(R^{2d}) = E_{\mu_0,\nu_0}(1_{\delta,\delta'}(Z) \mid Y, \tilde{Y})\Big|_{\tilde{Y}=Y},$$

where $Z = (X, \tilde{X})$, expectation over $(\eta, \tilde{\eta})$ included in the right-hand side.

4. (Restricted doubled measures)
Now we define a notion which will play a crucial role in the sequel. For every δ, δ', let

$$e_n^{Y;\delta,\delta';\mu_0,\nu_0} := (\mu_0, \nu_0) \bar{S}_n^{Y;R;\delta,\delta';\mu_0,\nu_0}(R^{2d}) \equiv E_{\mu_0,\nu_0}(1_{\delta,\delta'}(Z) \mid Y, \tilde{Y})\Big|_{\tilde{Y}=Y},$$

where $Z = (X, \tilde{X})$. Due to the assumption on the density q_V, these random variables are well defined. Notice that the symmetry in the definition of $\bar{S}$ implies an identity,

$$e_n^{Y;\delta,\delta';\mu_0,\nu_0} = e_n^{Y;\delta,\delta';\nu_0,\mu_0}.$$

Next, denote

$$(\mu, \nu) \hat{S}_n^{Y;R;\delta,\delta';\mu_0,\nu_0}(A \times B) := (e_n^{Y;\delta,\delta';\mu_0,\nu_0})^{-1} (\mu, \nu) \bar{S}_n^{Y;R;\delta,\delta';\mu_0,\nu_0}(A \times B),$$

with the natural extension to complete measures, due to the uniqueness of extension theorem. The sense of the last notation is that the result of this action is a *normalized* measure restricted to the event $1_{\delta,\delta'}(X, \tilde{X}) = 1$.

5. *(Estimation via doubled measures)*
The next important step is due to the fact that the distance in total variation for the measures in R^d can be estimated from above as follows (we rewrite below in the first line estimate (9.16) but via the operators $\bar{S}$),

$$\|\mu_0 \bar{S}_n^{Y;\mu_0,\nu_0} - \nu_0 \bar{S}_n^{Y;\nu_0,\mu_0}\|_{TV} \leq \|(\mu_0, \nu_0) \bar{S}_n^{Y;\mu_0,\nu_0} - (\nu_0, \mu_0) \bar{S}_n^{Y;\nu_0,\mu_0}\|_{TV}$$

$$\leq \sum_{\delta,\delta' \in \Delta} \|(\mu_0, \nu_0) \bar{S}_n^{Y;R;\delta,\delta';\mu_0,\nu_0} - (\nu_0, \mu_0) \bar{S}_n^{Y;R;\delta,\delta';\mu_0,\nu_0}\|_{TV}$$

$$= 2 \sum_{\delta,\delta' \in \Delta} \sup_D (e_n^{Y;\delta,\delta';\mu_0,\nu_0} (\mu_0, \nu_0) \hat{S}_n^{Y;R;\delta,\delta';\mu_0,\nu_0}(D) - e_n^{Y;\delta,\delta';\nu_0,\mu_0} (\nu_0, \mu_0) \hat{S}_n^{Y;R;\delta,\delta';\mu_0,\nu_0}(D)).$$

$$= 2 \sum_{\delta,\delta' \in \Delta} e_n^{Y;\delta,\delta';\mu_0,\nu_0} \sup_D ((\mu_0, \nu_0) \hat{S}_n^{Y;R;\delta,\delta';\mu_0,\nu_0}(D) - (\nu_0, \mu_0) \hat{S}_n^{Y;R;\delta,\delta';\mu_0,\nu_0}(D)),$$

where D runs all Borel sets $\mathcal{B}(R^{2d})$.

6. *(Birkhoff metric: definition)*
We will use the Birkhoff metric for positive measures: see [25], and also [2], [32] (where it is called the Hilbert metric; one more synonym is a projective metric),

$$\rho(\mu, \nu) = \begin{cases} \ln \dfrac{(\inf s : \mu \leq s\nu)}{(\sup t : \mu \geq t\nu)}, & \text{if finite,} \\ +\infty, & \text{otherwise.} \end{cases} \qquad (9.23)$$

Another equivalent definition reads,

$$\rho(\mu, \nu) = \begin{cases} \ln \sup(d\mu/d\nu) + \ln \sup(d\nu/d\mu), & \text{if finite,} \\ +\infty, & \text{otherwise.} \end{cases} \qquad (9.24)$$

7. *(Birkhoff and total variation)*
For probability measures we have,

$$\|\mu - \nu\|_{TV} \leq \rho(\mu, \nu). \qquad (9.25)$$

For completeness, let us show the folkloric proof, communicated to the authors by C. Leuridan. Remember that

$$\|\mu - \nu\|_{TV} = \sup_A (\mu - \nu)(A) + \sup_A (\nu - \mu)(A).$$

For equivalent measures, let $S_+ = (x : d\mu/d\nu \geq 1)$, $S_- = (x : d\mu/d\nu < 1)$, then

$$\|\mu - \nu\|_{TV} = (\mu - \nu)(S_+) + (\nu - \mu)(S_-).$$

If $d\mu/d\nu \leq C_1$, we have, $\nu(S_+) \geq C_1^{-1}\mu(S_+)$. So, using $\mu(S_+) \leq 1$,

$$(\mu - \nu)(S_+) \leq \mu(S_+) - C_1^{-1}\mu(S_+) = \left(1 - C_1^{-1}\right)\mu(S_+).$$

Now, since both measures are probability ones, then $C_{1,2} \geq 1$, and the function $f(x) := 1 - x^{-1} - \ln x \leq 0$, $x \geq 1$. Indeed, $f(1) = 0$, and $f'(x) = x^{-2} - x^{-1} \leq 0$, $x \geq 1$. Hence,

$$\left(1 - C_1^{-1}\right)\mu(S_+) \leq \ln C_1.$$

Similarly,

$$(\nu - \mu)(S_-) \leq \left(1 - C_2^{-1}\right)\nu(S_-) \leq \ln C_2.$$

If the two measures are not equivalent, then the distance in total variation does not exceed two, while the Birkhoff distance equals $+\infty$. This completes the proof of (9.25).

8. (Birkhoff metric for the couple)
For the sequel, denote $\rho_0 := \rho(\mu_0, \nu_0)$. For the doubled measures we have,

$$\rho((\mu_0, \nu_0), (\nu_0, \mu_0)) = \ln \sup_{C \in \mathcal{B}^{2d}} \frac{\int_C 1\, \mu_0(dx)\nu_0(d\tilde{x})}{\int_C 1\, \nu_0(dx)\mu_0(d\tilde{x})} + \ln \sup_{C \in \mathcal{B}^{2d}} \frac{\int_C 1\, \nu_0(dx)\mu_0(d\tilde{x})}{\int_C 1\, \mu_0(dx)\nu_0(d\tilde{x})}$$

$$\leq 2 \left(\ln \operatorname{ess\,sup} \frac{d\mu_0}{d\nu_0} + \ln \operatorname{ess\,sup} \frac{d\nu_0}{d\mu_0} \right) = 2\rho(\mu_0, \nu_0) = 2\rho_0,$$

where, of course, only C of positive measures are considered.

9. (Reduction to Birkhoff metric)
Due to the inequality (9.25), and since both measures below—that is, $(\mu_0, \nu_0)\hat{S}_n^{Y;R;\delta,\delta';\mu_0,\nu_0}$ and $(\nu_0, \mu_0)\hat{S}_n^{Y;R;\delta,\delta';\mu_0,\nu_0}$—are normalized, we have,

$$2 \sup_D \left((\mu_0, \nu_0)\hat{S}_n^{Y;R;\delta,\delta';\mu_0,\nu_0}(D) - (\nu_0, \mu_0)\hat{S}_n^{Y;R;\delta,\delta';\mu_0,\nu_0}(D) \right)$$

$$\leq \rho\!\left((\mu_0, \nu_0)\hat{S}_n^{Y;R;\delta,\delta';\mu_0,\nu_0}, (\nu_0, \mu_0)\hat{S}_n^{Y;R;\delta,\delta';\mu_0,\nu_0} \right). \tag{9.26}$$

So,

$$\|\mu_0 \bar{S}_n^{Y;\mu_0,\nu_0} - \nu_0 \bar{S}_n^{Y;\nu_0,\mu_0}\|_{TV}$$

$$\leq 2 \sum_{\delta,\delta' \in \Delta} e_n^{Y;\delta,\delta';\mu_0,\nu_0} \rho\!\left((\mu_0, \nu_0)\hat{S}_n^{Y;R;\delta,\delta';\mu_0,\nu_0}, (\nu_0, \mu_0)\hat{S}_n^{Y;R;\delta,\delta';\mu_0,\nu_0} \right).$$

10. *(Contraction)*

We claim that there exists $\pi_R < 1$ such that for every $k \geq 0$,

$$\rho((\mu_0, \nu_0) \hat{S}_n^{Y;R;\delta,\delta';\mu_0,\nu_0}, (\nu_0, \mu_0) \hat{S}_n^{Y;R;\delta,\delta';\mu_0,\nu_0})$$

$$\equiv \rho((\mu_0, \nu_0) S_n^{Y;R;\delta,\delta'}, (\nu_0, \mu_0) S_n^{Y;R;\delta,\delta'}) \leq 2\rho_0 \pi_R^k, \qquad (9.27)$$

where $k = \#1(\delta, \delta')$, see (9.17).

11. *(Local mixing and induction)*

The inequality (9.27) follows by induction from the following two inequalities: see e.g. [32]; we use here the concise notation $(\mu_i, \nu_i) = (\mu_0 \nu_0) S_i^{Y;R;\delta,\delta'}$.

(1°) For every i,

$$\rho\left((\mu_i, \nu_i) S_{i:i+1}^{Y;R;\delta,\delta'}, (\nu_i, \mu_i) S_{i:i+1}^{Y;R;\delta,\delta'}\right) \leq \rho((\mu_i, \nu_i), (\nu_i, \mu_i)).$$

(2°) there exists $\pi_R < 1$ such that if $i \in J(\delta, \delta')$,

$$\rho\left((\mu_i, \nu_i) S_{i:i+1}^{Y;R;\delta,\delta'}, (\nu_i, \mu_i) S_{i:i+1}^{Y;R;\delta,\delta'}\right) \leq \pi_R \rho((\mu_i, \nu_i), (\nu_i, \mu_i)).$$

The latter can be found in the Proposition 3.9 from [32], with the contraction constant,

$$\pi_R \leq \left(1 - \tilde{C}_R^{-2}\right) / \left(1 + \tilde{C}_R^{-2}\right), \qquad (9.28)$$

due to the 'local mixing condition' for $i \in J(\delta, \delta')$,

$$\operatorname*{ess\,sup}_{\Lambda} \sup_{D_R} \frac{Q_{\delta,\delta'}^{i-1:i}(x_{i-1}, \tilde{x}_{i-1}, dx_i', d\tilde{x}_i')}{Q_{\delta,\delta'}^{i-1:i}(v_{i-1}, \tilde{v}_{i-1}, dx_i', d\tilde{x}_i')} = \sup_{D_R} \frac{q(x_{i-1}, x_i') q(\tilde{x}_{i-1}, \tilde{x}_i')}{q(v_{i-1}, x_i') q(\tilde{v}_0, \tilde{x}')} =: \tilde{C}_R < \infty, \qquad (9.29)$$

where $\tilde{C}_R = \left(a_R^+ / a_R^-\right)^2$, see (A2), with

$$D_R = \{(x_{i-1}, \tilde{x}_{i-1}, v_{i-1}, \tilde{v}_{i-1}, x_i', \tilde{x}_i') :$$

$$|x_{i-1}|, |\tilde{x}_{i-1}|, |v_{i-1}|, |\tilde{v}_{i-1}|, |x_i'|, |\tilde{x}_i'| \leq R, \ x_i', \tilde{x}_i' \in A_R\}.$$

Then, the meaning of inequality (2°) is that the replacement of nonrandom kernels $\hat{Q}$ by random ones $\hat{Q}^\Psi$ does not change the supremum of the derivative of one measure with respect to another. We leave the reader to consult the proof of (9.28) in [25, ch. 2, §10] or [15, Theorem 9.1], or [4, Theorem XVI.3], and only show here the most important inequality—though a very simple one—which links 'local mixing' for kernels Q^Ψ and

Filtering for Nonuniformly Ergodic Signals 287

Q. See the proof of Proposition 3.9 in [32]: it follows from the Assumption (A2) that for any $A \subset A_R$, with $|x|, |\tilde{x}| \leq R$,

$$a_-^2 \int \Psi(x', Y_i) \Psi(\tilde{x}', Y_i) \Lambda(dx') \Lambda(d\tilde{x}')$$

$$\leq \int \Psi(x', Y_i) q(x, x') \Psi(\tilde{x}', Y_i) q(\tilde{x}, \tilde{x}') \Lambda(dx') \Lambda(d\tilde{x}')$$

$$\leq a_+^2 \int \Psi(x', Y_i) \Psi(\tilde{x}', Y_i) \Lambda(dx') \Lambda(d\tilde{x}').$$

This double inequality allows to apply contraction principle with the coefficient π_R to the random operators $\hat{Q}\Psi$.

The induction base $k = 0$ in (9.27) is valid due to the assumption (A4); the induction step follows from (2°) directly.

Now we can estimate as follows,

$$E_{\mu_0} \|(\mu_0, \nu_0) S_n^{Y;R;\delta,\delta'} - (\nu_0, \mu_0) S_n^{Y;R;\delta,\delta'}\|_{TV} \leq \sum_{\delta,\delta' \in \Delta; \#1(\delta,\delta') \geq 1} 2\rho_0 \pi_R^{\#1(\delta,\delta')} E_{\mu_0} e_n^{Y;\delta,\delta';\mu_0,\nu_0}$$

$$+ 2 \sum_{\delta,\delta' \in \Delta; \#1(\delta,\delta')=0} E e_n^{Y;\delta,\delta';\mu_0,\nu_0} \sup_D ((\mu_0, \nu_0) \hat{S}_n^{Y;R;\delta,\delta';\mu_0,\nu_0}(D) - (\nu_0, \mu_0) \hat{S}_n^{Y;R;\delta,\delta';\mu_0,\nu_0}(D))$$

$$\leq 2\rho_0 \sum_{\delta,\delta' \in \Delta} \pi_R^{\#1(\delta,\delta')} E_{\mu_0} e_n^{Y;\delta,\delta';\mu_0,\nu_0}. \quad (9.30)$$

12. *(Split on large and small $\#1(\delta, \delta')$)*
Let us split the sum $\sum_{\delta \in \Delta}$ into three parts: $\{\#1(\delta, \delta') \geq \epsilon'n\}$, $\{\#1(\delta, \delta') < \epsilon'n \,\&\, \#1(\delta) > \epsilon n\}$, and $\{\#1(\delta, \delta') < \epsilon'n \,\&\, \#1(\delta) \leq \epsilon n\}$, where $\epsilon' > 0$ and $\epsilon > 0$ such that $\epsilon' \ll \epsilon$ are to be chosen. This and the following several steps are similar to the corresponding part of the proof from [23]. Nonetheless, notice an important difference in Assumption (A2): in the present chapter, transition density $q(x, x')$ for $x \in B_R$ and $x' \in A_R$ may be only a part of the complete kernel $Q(x, dx')$, that is, it may happen that there is another component of Q with a nonzero mass on A_R. This affects the calculus.

13. *(Estimate the probability that $\#1(\delta, \delta')$ is 'large')*
Whatever $0 < \epsilon' < 1$, we estimate,

$$S^1 = \sum_{\delta,\delta': \#1(\delta,\delta') \geq \epsilon'n} \pi_R^{\#1(\delta,\delta')} E_{\mu_0} e_n^{Y;\delta;\delta'\mu_0,\nu_0}$$

$$= \sum_{\delta: \#1(\delta,\delta') \geq \epsilon'n} \pi_R^{\#1(\delta,\delta')} E_{\mu_0} E_{\mu_0,\nu_0}(1_{\delta,\delta'}(X, \tilde{X}) \mid Y, \tilde{Y})\Big|_{\tilde{Y}=Y}$$

$$\leq \pi_R^{\epsilon'n} E_{\mu_0} P_{\mu_0,\nu_0}(\bigcup_{\delta': \#1(\delta,\delta') \geq \epsilon'n} 1_{\delta,\delta'}(X, \tilde{X}) \mid Y, \tilde{Y})\Big|_{\tilde{Y}=Y} \leq \pi_R^{\epsilon'n}.$$

14. *(Estimate the probability that #1 (δ, δ') is small while #1 (δ) is large)*
This estimation will be based on the following inequalities:

$$\sum_{\delta, \delta': \#1(\delta,\delta') < \epsilon'n \ \& \ \#1(\delta) > \epsilon n} E_{\mu_0}\left(E_{\mu_0,\nu_0}(1_{\delta,\delta'}(X,\tilde{X}) \mid Y, \tilde{Y})\Big|_{\tilde{Y}=Y}\right)$$

$$= E_{\mu_0}\left(E_{\mu_0,\nu_0}\mathbf{1}\left(\sum_{i=1}^n \mathbf{1}(M_{i-1} \leq R) > \epsilon n,\right.\right.$$

$$\left.\left.\sum_{i=1}^n \mathbf{1}(M_{i-1} \leq R \ \& \ \zeta_i = 1) < \epsilon'n\right) \mid Y, \tilde{Y}\right)\Big|_{\tilde{Y}=Y}$$

$$\leq E_{\mu_0}\left(E_{\mu_0,\nu_0}\mathbf{1}\left(\sum_{i=1}^n \mathbf{1}(|X_{i-1}| \leq R) > \epsilon n,\right.\right.$$

$$\left.\left.\sum_{i=1}^n \mathbf{1}(\zeta_i = 1) < \tfrac{1+\epsilon'}{2}n\right) \mid Y, \tilde{Y}\right)\Big|_{\tilde{Y}=Y}$$

$$+ \left(E_{\mu_0,\nu_0}\mathbf{1}\left(\sum_{i=1}^n \mathbf{1}(|\tilde{X}|_{i-1} \leq R) > \epsilon n,\right.\right.$$

$$\left.\left.\sum_{i=1}^n \mathbf{1}(\tilde{\zeta}_i = 1) < \tfrac{1+\epsilon'}{2}n\right) \mid Y, \tilde{Y}\right)\Big|_{\tilde{Y}=Y}$$

$$\leq E_{\mu_0}\mathbf{1}\left(\sum_{i=1}^n \mathbf{1}(|X_{i-1}| \leq R) > \epsilon n, \sum_{i=1}^n \mathbf{1}(\zeta_i = 1) < \tfrac{1+\epsilon'}{2}n\right)$$

$$+ C E_{\nu_0}\left(\mathbf{1}\left(\sum_{i=1}^n \mathbf{1}(|\tilde{X}_{i-1}| \leq R) > \epsilon n, \sum_{i=1}^n \mathbf{1}(\tilde{\zeta}_i = 1) < \tfrac{1+\epsilon'}{2}n\right)\right)$$

$$\leq E_{\mu_0}\mathbf{1}\left(\sum_{i=1}^n \mathbf{1}(|X_{i-1}| \leq R, \zeta_i = 0) \geq \epsilon''n\right)$$

$$+ C E_{\nu_0}\mathbf{1}\left(\sum_{i=1}^n \mathbf{1}(|\tilde{X}_{i-1}| \leq R, \tilde{\zeta}_i = 0) \geq \epsilon''n\right),$$

where $\epsilon'' = \epsilon - \tfrac{1+\epsilon'}{2}$. The last two terms can be estimated similarly, so we deal with only the first one. If we choose ϵ close to 1, and ϵ' close to zero, then ϵ'' will be just slightly less than $1/2$. Then,

$$p_R + \epsilon'' > 1. \tag{9.31}$$

Due to the exponential Bienaimé–Chebyshev, we estimate, with any $\lambda > 0$,

$$E_{\mu_0}\mathbf{1}\left(\sum_{i=1}^n \mathbf{1}(|X_{i-1}| \leq R, \zeta_i = 0) \geq \epsilon''n\right)$$

$$\leq \exp(-\lambda\epsilon''n) \, E_{\mu_0}\exp\left(\lambda \sum_{i=1}^n \mathbf{1}(|X_{i-1}| \leq R, \zeta_i = 0)\right)$$

$$= \exp(-\lambda\epsilon''n) \, E_{\mu_0}\left(\exp\left(\lambda \sum_{i=1}^{n-1} \mathbf{1}(|X_{i-1}| \leq R, \zeta_i = 0)\right)\right.$$

$$\left.\times E_{X_{n-1}}\exp\left(\lambda\mathbf{1}(|X_{n-1}| \leq R, \zeta_n = 0)\right)\right).$$

Here we have,

$$E_{X_{n-1}} \exp\left(\lambda \mathbf{1}(|X_{n-1}| \leq R, \zeta_n = 0)\right)$$

$$= \mathbf{1}(X_{n-1} \in B_R) E_{X_{n-1}} \exp\left(\lambda \mathbf{1}(|X_{n-1}| \leq R, \zeta_n = 0)\right)$$

$$+ \mathbf{1}(X_{n-1} \notin B_R) E_{X_{n-1}} \exp\left(\lambda \mathbf{1}(|X_{n-1}| \leq R, \zeta_n = 0)\right)$$

$$\leq \mathbf{1}(X_{n-1} \in B_R) \sup_{x \in B_R} E_x \exp\left(\lambda \mathbf{1}(\zeta_n = 0)\right) + (1 - \mathbf{1}(X_{n-1} \in B_R))$$

$$\leq 1 + \mathbf{1}(X_{n-1} \in B_R) \left((e^\lambda - 1)(1 - p_R)\right) \leq \left(p_R + e^\lambda(1 - p_R)\right).$$

By induction,

$$\exp(-\lambda \epsilon'' n) E_{\mu_0} \exp\left(\lambda \sum_{i=1}^n \mathbf{1}(|X_{i-1}| \leq R, \zeta_i = 0)\right)$$

$$\leq \exp(-\lambda \epsilon'' n) \left(p_R + e^\lambda(1 - p_R)\right)^n \equiv \left(\exp(-\lambda \epsilon'')(p_R + e^\lambda(1 - p_R))\right)^n.$$

We have,

$$\left.\left(\exp(-\lambda \epsilon'')(p_R + e^\lambda(1 - p_R))\right)\right|_{\lambda=0} = 1,$$

and due to (9.31),

$$\left.\left(\exp(-\lambda \epsilon'')(p_R + e^\lambda(1 - p_R))\right)'_\lambda\right|_{\lambda=0} = -\epsilon'' + 1 - p_R < 0,$$

Hence, λ from some right neighbourhood of zero, we have,

$$q := \left(\exp(-\lambda \epsilon'')(p_R + e^\lambda(1 - p_R))\right) < 1.$$

Let us fix any such $\lambda > 0$ and its corresponding q. Then,

$$E_{\nu_0} \mathbf{1}\left(\sum_{i=1}^n \mathbf{1}(|X_{i-1}| \leq R, \zeta_i = 0) \geq \epsilon'' n\right) \leq q^n.$$

Similarly,

$$E_{\nu_0} \mathbf{1}\left(\sum_{i=1}^n \mathbf{1}(|\tilde{X}_{i-1}| \leq R, \tilde{\zeta}_i = 0) \geq \epsilon'' n\right) \leq q^n.$$

This provides a desired *exponential* upper bound for the probability

$$P(\#\mathbf{1}(\delta, \delta') < \epsilon' n \ \& \ \#\mathbf{1}(\delta) > \epsilon n) \leq C q^n.$$

15. *(Estimating the probability that #1 (δ) is small)*
Finally, let us estimate the last term of the sum,

$$\sum_{\delta,\,\delta':\,\#1(\delta,\delta')<\epsilon'n\,\&\,\#1(\delta)<\epsilon n} \pi_R^{\#1(\delta,\delta')} E_{\mu_0}\left(E_{\mu_0,\nu_0}(1_{\delta,\delta'}(X,\tilde{X})\mid Y,\tilde{Y})\Big|_{\tilde{Y}=Y}\right)$$

$$\leq E_{\mu_0}\left(E_{\mu_0,\nu_0}\left(1\left(\sum_{k=0}^n 1(M_k\leq R)\right)<\epsilon n\right)\mid Y,\tilde{Y}\right)\Big|_{\tilde{Y}=Y}$$

$$\leq E_{\mu_0}\left(E_{\mu_0,\nu_0}\left(1\left(\sum_{k=0}^n 1(|X_k|\leq R)\right)<\tfrac{1+\epsilon}{2}n\right)\mid Y,\tilde{Y}\right)\Big|_{\tilde{Y}=Y}$$

$$+E_{\mu_0}\left(E_{\mu_0,\nu_0}\left(1\left(\sum_{k=0}^n 1(|\tilde{X}_k|\leq R)\right)<\tfrac{1+\epsilon}{2}n\right)\mid Y,\tilde{Y}\right)\Big|_{\tilde{Y}=Y}$$

$$\leq E_{\mu_0}\left(1\left(\sum_{k=0}^n 1(|X_k|\leq R)\right)<\tfrac{1+\epsilon}{2}n\right)$$

$$+C\,E_{\nu_0}\left(1\left(\sum_{k=0}^n 1(|\tilde{X}_k|\leq R)\right)<\tfrac{1+\epsilon}{2}n\right). \tag{9.32}$$

Due to the bounds from [41], for every $0<\epsilon<1$, the latter expectation admits an appropriate bound better than any polynomial, if R is chosen large enough, namely,

$$E_{\mu_0}\left(E_{\mu_0,\nu_0}(\sum_{\delta:\,\#1(\delta)<\epsilon n} 1_\delta(X,\tilde{X})\mid Y,\tilde{Y})\Big|_{\tilde{Y}=Y}\right)\leq C_m n^{-m},$$

for all $m>0$. This is shown below, by using bounds for random sequences with mixing.

16. *(Using recurrence and moment and mixing assumptions)*
Let us introduce some new indicators: let $0<R$ be large enough, and

$$\#1(X)_R := \sum_{k=0}^n 1(|X_k|\leq R),\quad \#0(X)_R = \sum_{k=0}^n 1(|X_k|>R).$$

Let $n\geq n_0$, where n_0 is large enough. Then, we claim the following inequality, for R large enough, and for any $m>0$,

$$E_{\mu_0} 1(\#1(X)_R<\epsilon'n)\leq C_m n^{-m}. \tag{9.33}$$

This follows from the assumption (9.8) on the hitting time τ, due to the inequality $P_{\mu_0}(\#1(X)_R<\epsilon'n)\leq P_{\mu_0}(\tau_{\epsilon'n}>n)$—where $\tau_1=\tau$, and by induction $\tau_{n+1}:=\inf(t\geq \tau_n+1:|X_t|\leq R),\ n\geq 1$.

Indeed, the sequence $(\tau_n-\tau_{n-1},\ n\geq 0)$ possesses *exponential* beta (and alpha) mixing, see [41]; concerning relations between alpha and beta mixing coefficients, see [16, ch. 18]. It should be noticed that the assumptions in [41] involve drift conditions, as in Remark 9.4; however, the latter is only needed to derive inequalities (9.8) and (9.9) from (A1). At the same time, the sequence $\tau_n-\tau_{n-1}$ approaches a stationary regime, exponentially fast, which follows from coupling at each τ_n with a qualitative probability bound from below: see [41]. Hence, due to the moment inequalities for stationary mixing sequences from [19], Lemma 2.1, for every k there exists C such that for all $n \geq 1$,

$$E(\tau_n - n\ell_R)^k \leq C n^{k/2},$$

where ℓ_R is the mean value of $\tau_{n+1} - \tau_n$ in the stationary regime. Next, when R increases to infinity, ℓ_R tends to one, i.e. $\ell_R \approx 1$ for large R, e.g. by virtue of monotone convergence. So, with ϵ' strictly less than one (remember that it is our choice to take $\epsilon' \approx 3/4$),

$$P(\tau_{\epsilon'n} > n) = P(\tau_{\epsilon'n} - \epsilon' n \ell_R > n(1 - \epsilon'\ell_R))$$

$$\leq \frac{E(\tau_{\epsilon'n} - \epsilon' n \ell_R)^k}{(n(1 - \epsilon'\ell_R))^k} \leq C n^{-k/2}.$$

Due to our split procedure, this implies for every m,

$$E_{\mu_0}\left(E_{\mu_0,\nu_0}\left(\sum_{\delta:\,\#1(\delta)<\epsilon n} 1_\delta(X,\tilde{X}) \mid Y, \tilde{Y}\right)\bigg|_{\tilde{Y}=Y}\right) \leq C_m n^{-m}, \quad n > 0.$$

The latter shows (9.33). Combining it with (9.30) and (9.31), we obtain in both cases the final estimate (9.12).

17. *(Borel–Cantelli)*
The nonaveraged bounds (9.13) follow from Chebyshev's inequality and the Borel–Cantelli lemmae. Theorem 9.1 is proved.

9.4 The continuous-time case

For continuous-time Markov diffusions similar results are available. For a compact case see [2], and for a noncompact case see [20], [22]. Here we decided to formulate the setting and a main result, which is slightly changed and generalized in comparison to [22], and to give only a very short sketch of the proof. There are two reasons for this decision, although, clearly, a full proof would be more desirable. Firstly, the assumptions are similar to those of the discrete-time case, of course with some natural changes appropriate for continuous time. Correspondingly, the scheme of the proof is also quite similar, with just a few important changes. Secondly, those important and genuinely continuous time issues have been presented in detail in [22], and they are rather lengthy. The changes in these parts due to a more general equation system are minor, whence, just a reference seems appropriate.

In [22], we have considered a continuous-time filter for a Markov diffusion (X_t) with values in the Euclidean space R^d, with observations (Y_t) from R^ℓ, satisfying the following simplest system of nonlinear Itô equations,

$$dX_t = b(X_t)dt + dW_t, \quad t \geq 0, \tag{9.34}$$

$$dY_t = h(X_t)dt + dB_t \quad t \geq 0, \tag{9.35}$$

with initial data X_0 and $Y_0 = 0$, where (W_t, B_t) is a $(d+\ell)$-dimensional Wiener process. Here $b(\cdot)$ is a d-dimensional vector function, $h(\cdot)$ an ℓ-dimensional vector function, random variable X_0 is independent on W and B. In the present chapter our model is a bit more general,

$$dX_t = b(X_t)dt + \sigma(X_t)dW_t, \quad t \geq 0, \tag{9.36}$$

$$dY_t = h(X_t)dt + dB_t \quad t \geq 0. \tag{9.37}$$

Since we provide only a brief sketch of the proof, further generalizations—such as unbounded coefficients or more involved dependence of coefficients on both components—are not desirable here, and we leave them for further papers. For the system (9.36)–(9.37), we assume the following.

(B1) All functions b, h, σ are bounded. The matrix function σ is uniformly nondegenerate.

(B2) Function h is C_b^2, i.e., twice continuously differentiable with bounded derivatives up to the second order.

(B3) Recurrence drift condition:

$$\lim_{|y| \to \infty} \langle b(x), x \rangle = -\infty. \tag{9.38}$$

(B4) Equivalence condition (9.10) on μ_0 and ν_0 from the discrete-time setting, i.e.

$$C^{-1} \leq d\mu_0/d\nu_0 \leq C.$$

(B5) Both μ_0 and ν_0 possess all polynomial moments.

Now, in order to formulate the main result, we need the Bayes formula,

$$E_{\mu_0}(f(X) \mid Y) = \frac{E_{\mu_0}^\gamma \left(f(X)\gamma_{0,t}^{-1}(X, Y) \mid Y\right)}{E_{\mu_0}^\gamma \left(\gamma_{0,t}^{-1}(X, Y) \mid Y\right)}, \tag{9.39}$$

with Girsanov's exponential

$$\gamma_{0,t}(X, Y) = \exp\left(-\int_0^t h(X_s)\,dB_s - (1/2)\int_0^t \|h(X_s)\|^2\,ds\right),$$

where $h^2 = |h|^2$, $hY \equiv h^*Y$ (here $*$ means transposition), and operators, analogous to the discrete-time case,

$$\mu S_t^Y(A) := \int E_{x_0}^\gamma \left(1(X_t \in A)\gamma_{0,t}^{-1}(X, Y) \mid Y\right)\mu(dx_0).$$

and

$$\mu_0 \bar{S}_t^{Y,\mu_0}(\cdot) := d_t^{\mu_0} \mu_0 S_t^Y(\cdot), \quad \text{where} \quad d_t^{\mu_0}(Y)^{-1} = \mu_0 S_t^Y(R^d).$$

Whenever conditional expectations or probabilities arise, regular versions are considered: see [36], [29]. We have representations,

$$P_{\mu_0}(X_t \in \cdot \mid Y) = \mu_0 \bar{S}_t^{Y,\mu_0}(\cdot), \quad \text{and} \quad E_{\mu_0}^\gamma \left(\gamma_{0,t}^{-1}(X,Y) \mid Y \right) = \langle \mu_0 S_t^Y, 1 \rangle.$$

The weak solution of equation (9.36) with a strong Markov property is considered. It exists due to [27] and [28]. In paricular, weak uniqueness automatically implies that any solution is strong Markov.

Theorem 9.2 *Under Assumptions (B1)–(B5) above, the following bounds hold true:*

$$E_{\mu_0} \| \mu_0 \bar{S}_t^{Y,\mu_0} - \nu_0 \bar{S}_t^{Y,\nu_0} \|_{TV} \le C_m (1+t)^{-m} \rho(\mu_0, \nu_0), \quad \forall m > 0. \tag{9.40}$$

Remark 9.6 For both systems, (9.34)–(9.35) and (9.36)–(9.37), Assumption (B3) may be replaced by a recurrence-moment condition, similar to the discrete-time case.
For each $k > 0$ there exist $C_m > 0$ such that

$$E_x \tau^k \le C_m (1 + |x|^m), \tag{9.41}$$

$$E_x 1(t \le \tau) |X_t|^m \le C_m (1 + |x|^m), \tag{9.42}$$

where $\tau = \inf\{t \ge 0 : |X_t| \le R\}$.

9.5 Sketch of the proof, continuous-time case

The main steps of the proof are similar to the discrete-time case, with just two major differences. One relates to continuous dependence of conditional expectations on the trajectory of Y, and appears more or less indispensable for justification that the operation of substitution of observations Y into the second (wrong) conditional measure is well formulated. This can be justified using methods from [8] or [22]. The second concerns local mixing which now must be derived from some properties of the model—in the discrete-time case it has to be more or less assumed—and, indeed, it follows from the *nonconditional* Harnack inequality: see the proof in [22]. We skip those involved and lengthy parts of the proof. Particularly, the number of important steps in the proof should not be misleading, because we only give a sketch.

1. *(Indicators, doubling space, and split.)*
Let us consider the process $Z_t = (X_t, \tilde{X}_t)$, where the components X and $\tilde{X}$ are independent random processes satisfying the same equation (9.36) with independent versions of Wiener processes, correspondingly, W and $\tilde{W}$, with

different initial distributions, $X_0 \sim \mu_0$ and $\tilde{X}_0 \sim \nu_0$, respectively. Let second components Y and $\tilde{Y}$ also satisfy corresponding second equations with independent B and $\tilde{B}$. Denote $n := [t]$, – the integer part of t. In the sequel, x stands for the whole continuous trajectory on $[0, t]$, and likewise for $\tilde{x}$, and the same for the random trajectories X, $\tilde{X}$, and Y. Let us introduce some indicators. We denote a (nonrandom) vector of dimension n with coordinates 1 or 0 at every place by δ, and

$$D_i := \left\{ \max\left(|X_i|, |\tilde{X}_i|\right) \leq R; \max\left(\sup_{i \leq s \leq i+1} |X_s|, \sup_{i \leq s \leq i+1} |\tilde{X}_s|\right) < R+1 \right\},$$

and

$$1_\delta(X, \tilde{X}) := \prod_{i=0}^{n-1} (1(D_i))^{\delta_i} \times (1 - 1(D_i))^{1-\delta_i},$$

with a convention $0^0 = 1$. We have, similarly to the discrete-time case,

$$\|\mu_0 \bar{S}_t^{Y,\mu_0} - \nu_0 \bar{S}_t^{Y,\nu_0}\|_{TV} \leq 2 \sup_{A \in \mathcal{B}(R^d)} \sum_{\delta \in \Delta} |E_{\mu_0,\nu_0}(1_\delta(X, \tilde{X}) 1(X_t \in A) \mid Y)$$

$$- E_{\mu_0,\nu_0}(1_\delta(X, \tilde{X}) 1(\tilde{X}_t \in A) \mid \tilde{Y})|_{\tilde{Y}=Y}|,$$

(9.43)

where Δ is the whole set of possible values of δ.

2. (Normailized linearization.)
For every δ, let

$$e_t^{Y;\delta;\mu_0,\nu_0} := (\mu_0, \nu_0) \bar{S}_t^{Y;R;\delta}(R^{2d}) = e_t^{Y;\delta;\nu_0,\mu_0}, \quad P_{\mu_0,\nu_0} - \text{a.s.}$$

Next, denote

$$(\mu, \nu) \hat{S}_t^{Y;R;\delta;\mu_0,\nu_0}(A \times B) := \left(e_t^{Y;\delta;\mu_0,\nu_0}\right)^{-1} (\mu, \nu) \bar{S}_t^{Y;R;\delta;\mu_0,\nu_0}(A \times B).$$

We have, with $D \in \mathcal{B}(R^{2d})$,

$$\|\mu_0 \bar{S}_t^{Y;\mu_0} - \nu_0 \bar{S}_t^{\tilde{Y};\nu_0}|_{\tilde{Y}=Y}\|_{TV}$$

$$\leq 2 \sum_{\delta \in \Delta} \sup_{D \in \mathcal{B}(R^{2d})} \left(e_t^{Y;\delta;\mu_0,\nu_0}(\mu_0, \nu_0) \hat{S}_t^{Y;R;\delta;\mu_0,\nu_0}(D) - e_t^{Y;\delta;\nu_0,\mu_0}(\nu_0, \mu_0) \hat{S}_t^{Y;R;\delta;\mu_0,\nu_0}(D)\right)$$

$$= 2 \sum_{\delta \in \Delta} e_t^{Y;\delta;\mu_0,\nu_0} \sup_{D \in \mathcal{B}(R^{2d})} \left((\mu_0, \nu_0) \hat{S}_t^{Y;R;\delta;\mu_0,\nu_0}(D) - (\nu_0, \mu_0) \hat{S}_t^{Y;R;\delta;\mu_0,\nu_0}(D)\right), \quad (9.44)$$

3. (Harnack's inequality and contraction in Birkhoff's metric.)
With the evident natural notations $S_{s_1:s_2}^{Y;R;\delta}$, $s_1 < s_2$, we have a *desintegration property*,

$$(\mu, \nu) S_t^{Y;R;\delta}(A \times B) = (\mu, \nu) \left(\prod_{i=0}^{n-1} S_{i:i+1}^{Y;R;\delta} \right) S_{n:t}^{Y;R;\delta}(A \times B).$$

For normalized measures, we have for any $0 \leq i \leq n-1$,

$$\rho\left((\mu_i, \nu_i) S_{i:i+1}^{Y;R;\delta}, (\nu_i, \mu_i) S_{i:i+1}^{Y;R;\delta} \right) \leq \rho\left((\mu_i, \nu_i), (\nu_i, \mu_i) \right), \qquad (9.45)$$

as well as

$$\rho\left((\mu_n, \nu_n) S_{n:t}^{Y;R;\delta}, (\nu_n, \mu_n) S_{n:t}^{Y;R;\delta} \right) \leq \rho\left((\mu_n, \nu_n), (\nu_n, \mu_n) \right), \qquad (9.46)$$

and for any $L > 0$ there is a constant $\pi_R(L) < 1$, such that for every $i \in J := \{j : \delta_j = \delta_{j+1} = 1\}$,

$$\rho\left((\mu_i, \nu_i) S_{i:i+1}^{Y;R;\delta}, (\nu_i, \mu_i) S_{i:i+1}^{Y;R;\delta} \right) \leq \pi_R(L) \rho\left((\mu_i, \nu_i), (\nu_i, \mu_i) \right), \qquad (9.47)$$

if the trajectory Y on $i \leq t \leq i+1$ satisfies the inequality, $\Delta_i(Y) := \sup_{i \leq s \leq i+1} |Y_s - Y_i| \leq L$. This boundedness condition for Y can be checked with a probability arbitrarily close to one, if L is large enough. Inequality (9.47) follows from the nonconditional Harnack inequality, which can be shown by using several Girsanov's transformations, eventually rewriting the SDE system so that *independent* from the signal trajectory Y becomes a *part of the drift coefficient* of the signal. See the details in [22] for the 'simple case' (9.34)–(9.35).

4. *(Further split.)*
Denote by $\#1(\delta)$ the (nonrandom) number of ones in the vector δ, and by $\#1(\delta, L)$ the (random) number of those of them, say, with an index 'i,' which enjoy the property $\Delta_i(Y) \leq L$ on the corresponding unit intervals. Then, by induction, it follows from (9.47) that

$$\rho\left((\mu_0, \nu_0) \hat{S}_t^{Y;R;\delta;\mu_0,\nu_0}, (\nu_0, \mu_0) \hat{S}_t^{Y;R;\delta;\mu_0,\nu_0} \right) \leq C \rho(\mu_0, \nu_0) \pi_R^{\#1(\delta,L)}. \qquad (9.48)$$

From (9.44) and (9.48), it follows that

$$E_{\mu_0} \left\| \mu_0 \bar{S}_n^{Y;\mu_0} - \nu_0 \bar{S}_n^{Y;\nu_0} \right\|_{TV}$$

$$\leq C \rho(\mu_0, \nu_0) E_{\mu_0,\nu_0} \left(\sum_{\#1(\delta,L) \geq \epsilon n} + \sum_{\#1(\delta,L) < \epsilon n} \right) \pi_R^{\#1(\delta,L)} e_n^{Y;\delta;\mu_0,\nu_0}. \qquad (9.49)$$

5. *(Further split again.)*
Clearly, our main task remains to estimate the second term of the sum,

$$E_{\mu_0,\nu_0} \sum_{\delta: \#1(\delta,L) < \epsilon n} \pi_R^{\#1(\delta,L)} \left(E_{\mu_0,\nu_0}(1_\delta(X, Y) \mid Y, \tilde{Y}) \mid_{\tilde{Y}=Y} \right)$$

$$\leq E_{\mu_0,\nu_0} \sum_{\delta: \#1(\delta,L) < \epsilon n} \left(E_{\mu_0,\nu_0}(1_\delta(X, Y) \mid Y, \tilde{Y}) \mid_{\tilde{Y}=Y} \right).$$

In turn, we will split this sum into two further parts, as follows. Let $\epsilon' > \epsilon$. Then,

$$E_{\mu_0, \nu_0} \sum_{\delta: \#1(\delta, L) < \epsilon n} \left(E_{\mu_0, \nu_0}(1_\delta(X, Y) \mid Y, \tilde{Y}) \mid_{\tilde{Y}=Y} \right)$$

$$\leq E_{\mu_0, \nu_0} \sum_{\delta: \#1(\delta) < \epsilon' n} \left(E_{\mu_0, \nu_0}(1_\delta(X, Y) \mid Y, \tilde{Y}) \mid_{\tilde{Y}=Y} \right)$$

$$+ E_{\mu_0, \nu_0} \sum_{\delta: \#1(\delta, L) < \epsilon n, \#1(\delta) \geq \epsilon' n} 1. \tag{9.50}$$

6. (Weak '1/2 condition'.)

Here for any $\epsilon' < 1$ *small enough*, the first term may be estimated similarly to the discrete-time case—see step 15 of the proof of the Theorem 9.1—it admits a polynomial bound $C_m t^{-m}$ with any m. To use the same method, we notice that for the standard Wiener process,

$$P(\max_{0 \leq s \leq 1} W_s > 1) = 2 P(W_1 > 1) \approx 2 \times 0.15866 < 0.5$$

(from the tables for the normal distribution). Hence, due to comparison theorems, for large R it follows that

$$\sup_{|X_0| \leq R} P_{X_0}(\max_{0 \leq s \leq 1} |X_s| \geq R + 1) < 1/2. \tag{9.51}$$

This looks like a weak analogue of the '1/2-condition' in the discrete-time case, and it suffices to applicability of the same technique in order to show any polynomial bound $C_m t^{-m}$ with any m for the first term in (9.32). Why we call it weak, because actually by choosing $R + k$ with k large enough instead of $R + 1$, we may make this probability arbitrarily close to zero. The latter does not seem achievable for the discrete time case, where an undesirable 'loss of probability' may occur inside the ball B_R as well, which apparently makes the condition (9.3) or likewise more or less indispensable in the discrete-time case.

7. (Binomial estimate.)

A really new term in (9.50) is the second one. To tackle it, denote $p_L = P(\sup_{0 \leq s \leq 1} |Y_s - Y_0| \leq L)$, and $q_L = 1 - p_L$, and notice that $p_L \approx 1$, if L is large enough. We have,

$$E_{\mu_0, \nu_0} \sum_{\delta: \#1(\delta, L) < \epsilon n, \#1(\delta) \geq \epsilon' n} 1 \leq P_{\mu_0, \nu_0} \left(\sum_{i: \Delta_i(Y) > L} 1 \geq (\epsilon' - \epsilon)n \right)$$

$$\leq \sum_{k=(\epsilon'-\epsilon)n}^{n} C_n^k q_L^k p_L^{n-k} \leq \left(2 q_L^{\epsilon'-\epsilon} \right)^n, \tag{9.52}$$

which can be made less than any exponential by choosing L large enough, for any $\epsilon' - \epsilon > 0$.

Acknowledgements

The authors are grateful to the anonymous referee for his comments. The second author thanks the Department of Mathematics of University of Maine at Le Mans, Project MATPYL RFBR grant 08-01-00105a for support.

References

[1] R. Atar, Exponential stability for nonlinear filtering of diffusion processes in a noncompact domain. *Annals of Probab.* 26(4) (1998), 1552–74.

[2] R. Atar and O. Zeitouni, Exponential stability for nonlinear filtering. *Ann. Inst. H. Poincaré, Probab. Statist.* 33(6) (1997), 697–725.

[3] P. Baxendale, P. Chigansky, and R. Liptser, Asymptotic stability of the Wonham filter: ergodic and nonergodic signals. *SIAM J. Control Optim.* 43(2) (2004), 643–69.

[4] G. Birkhoff, *Lattice theory, colloquium publications*, 3rd edn, AMS, Providence, RI, 1967.

[5] A. S. Budhiraja and D. L. Ocone, Exponential stability in discrete-time filtering for bounded observation noise. *Systems Control Lettr.* 30 (1997), 185–93.

[6] A. S. Budhiraja and D. L. Ocone, Exponential stability for discrete-time filtering for nonergodic signals. *Stochastic Process. Appl.* 82 (1999), 245–57.

[7] P. Chigansky and R. Liptser, Stability of nonlinear filters in nonmixing case. *Ann. Appl. Probab.* 14(4) (2004), 2038–56.

[8] J. M. Clark and D. Crisan, On a robust version of the integral representation formula of nonlinear filtering. *Probab. Theory Relat. Fields* 133 (2005), 43–56. DOI 10.1007/s00440-004-0412-5

[9] O. V. Gulinsky and A. Yu. Veretennikov, *Large deviations for discrete-time processes with averaging.* VSP, Utrecht, 1993.

[10] R. L. Dobrushin, Central limit theorem for nonstationary Markov chains, I, II. *Theory Probab. Appl.* 1(1) (1956), 66–80, ibid., 1(4) (1956), 330–85.

[11] J. L. Doob, *Stochastic Processes.* Wiley, NY, 1953.

[12] P. Del Moral and A. Guionnet, On the stability of interacting processes with applications to filtering and genetic algorithms. *Ann. Inst. H. Poincaré* 37(2) (2001), 155–94.

[13] R. Douc, G. Fort, E. Moulines, and P. Priouret, Forgetting of the initial distribution for Hidden Markov Model, *Stoch. Process Appl*, 119(4): 1235–1256, 2009.

[14] B. V. Gnedenko, *Theory of probability*, 6th edn. Gordon & Breach, Amsterdam, 1997.

[15] E. Hopf, An inequality for positive integral linear operators. *J. Math. Mechanics* 12 (1963), 683–92.

[16] I. A. Ibragimov and Y. V. Linnik, *Independent and stationary sequences of random variables.* Wolters-Noordhoff, Groningen, 1971.

[17] I. A. Ibragimov and R. Z. Hasminskii. *Statistical estimation: Asymptotic theory*, Springer, Berlin, 1981.

[18] T. Kaiser, A limit theorems for partly observable Markov chains. *Ann. Probability* 3(4) (1975), 677–96.

[19] R. Z. Khas'minskii, On stochastic processes defined by differential eqautions with small parameter. *Theory Probab. Appl.* 11(3) (1966), 211–28.

[20] M. L. Kleptsyna and A. Yu. Veretennikov, On ergodic filters with wrong initial data. *C.R.Acad. Sci. Paris*, Ser. I, 344 (2007), 727–731.

[21] M. L. Kleptsyna, A. Yu. Veretennikov, On discrete time ergodic filters with wrong initial data. *Probab. Theory Rel. Fields* 141 (2008), 411-444, DOI 10.1007/s00440-007-0089-7.

[22] M. L. Kleptsyna and A. Yu. Veretennikov, On continuous time ergodic filters with wrong initial data. *Theory of Probability and its Applications* 53(2) (2008), 240–76.

[23] M. L. Kleptsyna and A. Yu. Veretennikov, On discrete time ergodic filters with wrong initial data, 2, *Stochastics An International Journal of Probabilitly and Stochastic Processes*, 1744–2516, 2009.

[24] S. A. Klokov and A. Yu. Veretennikov, Subexponential mixing rate for a class of Markov processes (multidimensional case). *Theory Probab. Appl.* 49(1) 2004, 1–13.

[25] M. A. Krasnosel'skii, E. A. Lifshits, and A. V. Sobolev, *Positive linear systems. The method of positive operators*. Heldermann Verlag, Berlin, 1989.

[26] H. Kunita, Asymptotic behavior of the nonlinear filtering errors of Markov processes. *J. Multivariate Anal.* 1 (1971), 365–93.

[27] N. V. Krylov, On Itô's stochastic integral equations. *Theory of Probability and its Applications* 14, (1969), 330.

[28] N. V. Krylov, On the selection of a Markov process from a system of processes and the construction of quasi-diffusion processes. *Math. USSR Izv.* 7(3) (1973), 691–709. DOI 10.1070/IM1973v007n03ABEH001971

[29] N. V. Krylov, The regularity of conditional probabilities for stochastic processes. *Theory Probab. Appl.* 18 (1973), 151–5.

[30] N. V. Krylov and M. V. Safonov, A property of the solutions of parabolic equations with measurable coefficients (Russian). *Izv. Akad. Nauk SSSR Ser. Mat.* 44(1) (1980), 161–75, 239. Engl trans. in *Math USSR Izvestija* 16(1) (1981), 151–64.

[31] F. Le Gland and N. Oudjane, A robustification approach to stability and to uniform particle approximation of nonlinear filters: the example of pseudo-mixing signals. *Stochastic Process. Appl.* 106 (2003), 279–316.

[32] F. Le Gland and N. Oudjane, Stability and uniform approximation of nonlinear filters using the Hilbert metric and application to particle filters. *Ann. Appl. Probab.* 14(1) (2004), 144–87.

[33] D. L. Ocone and E. Pardoux, Asymptotic stability of the optimal filter with respect to its initial condition. *SIAM J. Control Optim.* 34(1) (1996), 226–43.

[34] N. Oudjane and S. Rubenthaler, Stability and uniform particle approximation of nonlinear filters in case of non ergodic signals. *Stochastic Anal. Appl.* 31(3) (2005), 421–48.

[35] A. Papavasiliou, Parameter estimation and asymptotic stability in stochastic filtering. *Stochastic Process. Appl.* 116(7) (2006), 1048–65.

[36] K. R. Parthasarathy, *Probability measures on metric spaces*. Academic Press, New York and London, 1967.

[37] W. Stannat, Stability of the pathwise filter equations for a time dependent signal on R^d. *Appl. Math. Optim.* 52 (2005), 39–71.

[38] A. Yu. Veretennikov, Malliavin calculus and its applications to some limit theorems. *Proc. 1st World Congress Bernoully Soc.*, Tashkent, 1986. Utrecht: VNU Sci. Press, 1987, vol. 1, 557–66.

[39] A. Yu. Veretennikov, Estimates of the mixing rate for stochastic equations (Russian). *Teor. Veroyatnost. Primenen.* 32(2) (1987), 299–308; Engl. trans. in *Theory Probab. Appl.* 32(2) (1987), 273–81.

[40] A. Yu. Veretennikov, The mixing rate and the averaging principle for hypoelliptic stochastic differential equations (Russian). *Izv. Akad. Nauk SSSR Ser. Mat.* 52(5) (1988), 899–908, 1118; Engl. trans. in *Math. USSR-Izv.* 33(2) (1989), 221–31.

[41] A. Yu. Veretennikov, On polynomial mixing and the rate of convergence for stochastic differential and difference equations. *Teor. Veroyatn. Primenen.* 44(2) (1999), 312–27; Engl. trans.: *Theory Probab. Appl.* 44(2) (2000), 361–74.

[42] A. Yu. Veretennikov, On polynomial mixing estimates for stochastic differential equations with a gradient drift. *Teor. Veroyatnost. i Primenen.* 45(1) (2000), 163–6; Engl. trans. in *Theory Probab. Appl.* 45(1) (2001), 160–3.

[43] A. Yu. Veretennikov, On approximations of diffusions with equilibrium. Research Report, Helsinki University of Technology, Institute of Mathematics. Report C017, 2004. http://math.tkk.fi/visitors0405/AVslides.pdf.

·10·
Exponential Decay Rate of the Filter's Dependence on the Initial Distribution

R. Atar

10.1 Introduction

The problem of stability of the nonlinear filter arises in the following practical context. If the transition law of a given Markov process is known, but its initial law is not available, under what conditions does one not lose optimality of the filter when initializing it with an arbitrary (thus wrong) initial data, in the limit when time tends to infinity? This question, first posed by Ocone and Pardoux [35] and Delyon and Zeitouni [20], has attracted much attention. This chapter focuses on the *exponential* rate of decay of the error made by wrong initialization, and reviews results that relate this quantity to multiplicative ergodic theory (MET) on one hand, and to Hilbert's metric and Birkhoff's contraction coefficient on the other hand. MET is instrumental in establishing that, in a finite-state setting, the decay rate is deterministic (roughly speaking). In fact, it identifies the rate with the Lyapunov spectral gap associated with the filtering equation. The set of tools borrowed from the theory of positive operators, are more useful in providing estimates on the decay rate, which in turn enable to establish sufficient conditions for nonvanishing thereof.

Our main goal in this exposition is to present the methods in an elementary and reasonably self-contained way, starting from a simple case and then extending the ideas to a general setting. We make no attempt to present the strongest results to date, or to survey various other techniques to tackle the problem (such as [1, 4–6, 11, 14, 16, 29, 33, 40]).

The chapter is organized as follows. In Section 10.2 we begin by describing the finite state, discrete time setting and define the decay rate; we then review relevant results from MET and make the link to the Lyapunov spectral gap. We also describe analogous results in continuous time. The short Section 10.3 reviews definitions regarding positive operators and explains their relation to Lyapunov exponents, based on a lemma of Peres, which leads to first estimates on the decay rate. Section 10.4 presents bounds on the decay rate in a general state space.

Notation. Given a positive integer d we write $M(d)$ for the set of $d \times d$ matrices with real entries, and $M_+(d)$ for the set of elements of $M(d)$ with nonnegative entries. The set of elements of $\mathbb{R}^d$ with nonnegative entries is denoted by $\mathbb{R}^d_+$. Vectors in $\mathbb{R}^d$ are understood to be column vectors unless otherwise specified. The ith entry of a vector x is denoted by x^i, and the (i, j)th entry of a matrix M is $M^{i,j}$. $\langle \cdot, \cdot \rangle$ and $\|\cdot\|$ are the usual scalar product and norm on $\mathbb{R}^d$. For $M \in M(d)$, $\|M\| = \sup\{\|Mx\| : x \in \mathbb{R}^d, \|x\| = 1\}$ is the operator norm corresponding to $\|\cdot\|$. The transpose of a matrix M is $M^\top$. Expectation with respect to a probability measure denoted by $\mathbb{P}^A_B$ is written as $\mathbb{E}^A_B$, for any set of symbols A and B to be used (particularly $\mathbb{E}$ denotes expectation with respect to $\mathbb{P}$).

10.2 Finite-state filtering and Lyapunov exponents

Let d be a positive integer and set $\bar{d} := \{1, 2, \ldots, d\}$. Denoting $\mathbf{1} = (1, 1, \ldots, 1) \in \mathbb{R}^d$, and

$$\mathcal{P} = \{x \in \mathbb{R}^d : x_i \geq 0, \langle x, \mathbf{1} \rangle = 1\}, \tag{10.1}$$

we will identify members of $\mathcal{P}$ with probability distributions over $\bar{d}$ via $(p^i)_{i \in \bar{d}} = (p(\{i\}))_{i \in \bar{d}}$. Consider a homogeneous Markov process $X = \{X_n, n \geq 0\}$ taking values in $\bar{d}$, on a probability space $(\Omega, \mathcal{F}, \mathbb{P})$. Denote by G the transition matrix

$$G^{i,j} = P(X_{n+1} = j | X_n = i), \quad i, j \in \bar{d}, \; n \geq 0,$$

and by $\pi_0 \in \mathcal{P}$ the initial distribution, regarded as a column vector. Next, fix $\ell \in \mathbb{N}$ and denote $\mathcal{R} = \mathcal{B}(\mathbb{R}^\ell)$. Let a family $\widetilde{G}(i, dy)$, $i \in \bar{d}$ of probability measures on $(\mathbb{R}^\ell, \mathcal{R})$ be given, and assume that for some measurable function $g : \bar{d} \times \mathbb{R}^\ell \to \mathbb{R}$ and a probability measure $\widetilde{G}_0 \in \mathcal{M}(\mathbb{R}^\ell, \mathcal{R})$,

$$\widetilde{G}(i, dy) = g(i, y) \widetilde{G}_0(dy), \quad i \in \bar{d}.$$

We are given a process $\{Y_n, n \geq 1\}$ of noisy observations of X_n, satisfying

$$\mathbb{P}(Y_i \in E_i, i \in \{1, 2, \ldots, n\} | X_i = x_i, i \in \{0, 1, \ldots, n\}) = \prod_{i=1}^{n} \widetilde{G}(x_i, E_i),$$

for all $n \in \mathbb{N}$, $E_1, E_2, \ldots, E_n \in \mathcal{R}$ and $x_1, x_2, \ldots, x_n \in \bar{d}$. For simplicity we assume in this section that g takes positive values. X_n and Y_n are called the *state* and *observation* processes, respectively.

Example 10.1 For some $m \in \mathbb{N}$, $\widetilde{G}_0$ could be the uniform probability measure on the finite set $\{1, 2, \ldots, m\}$, and then $\widetilde{G}(i, \{j\}) = g^{i,j}$, where $\{g^{i,j}\}_{i,j}$ is a positive $d \times m$ matrix. Thus provided $X_n = i$, the conditional probability of the event $\{Y_n = j\}$ is given by $g^{i,j}$. ◊

Example 10.2 Let $\ell = 1$. Let $k : d \to \mathbb{R}$ be a given function. We model Y_n as observations of X_n via the 'sensor' k, perturbed by Gaussian noise, by defining

$$Y_n = k(X_n) + \sigma W_n, \qquad n \geq 1. \tag{10.2}$$

Here $\{W_n, n \geq 1\}$ is an i.i.d. sequence of standard normals, independent of $\{X_n\}$, and the parameter $\sigma > 0$ is the level of noise. This model is seen to fit the above setup if we set $\widetilde{G}_0$ to be the $(0, \sigma^2)$-Gaussian measure on $\mathbb{R}$,

$$\widetilde{G}_0(dy) = \frac{1}{\sqrt{2\pi}\sigma} e^{-\frac{y^2}{2\sigma^2}} dy,$$

and let

$$\widetilde{G}(i, dy) = \frac{1}{\sqrt{2\pi}\sigma} e^{-\frac{(y-k(i))^2}{2\sigma^2}} dy.$$

This example will be referred to as the *one-dimensional additive Gaussian noise*. ◇

We denote by $\mathcal{Y}_n$ the σ-field generated by the observations $\{Y_1, Y_2, \ldots, Y_n\}$, $n \geq 1$, and set $\mathcal{Y}_0$ to be the trivial σ-field. Let

$$\pi_n = (P(X_n = 1|\mathcal{Y}_n), \ldots, P(X_n = d|\mathcal{Y}_n))^\top, \qquad n \geq 0. \tag{10.3}$$

The stochastic process $\{\pi_n, n \geq 0\}$ takes values in $\mathcal{P}$. It represents the conditional law of X_n given $\mathcal{Y}_n$, and is often referred to as the *nonlinear filter* associated with the processes X_n and Y_n. A use of Bayes' rule shows that π_n satisfies the recursion

$$\pi_n = \frac{D_n G^\top \pi_{n-1}}{\langle D_n G^\top \pi_{n-1}, 1 \rangle}, \qquad n \geq 1,$$

where D_n is the diagonal matrix

$$D_n^{i,i} = g(i, Y_n), \qquad i \in d.$$

An equivalent way of writing this recursion is via

$$\rho_n = D_n G^\top \rho_{n-1}, \qquad n \geq 1, \qquad \rho_0 = \pi_0, \tag{10.4}$$

in which case π_n is given by $\rho_n/\langle \rho_n, 1 \rangle$. Thus the filter can be expressed as a normalized version of the solution to a simple linear recursion. The process ρ_n is often referred to as the *unnormalized conditional measure* of X_n given $\mathcal{Y}_n$. Often in practice, the initial law π_0 is not available, and one uses the recursion (10.3) with wrong initial data. Given $p \in \mathcal{P}$, let π_n^p denote the solution to the recursion with initial data p. We will use the term *exact filter* for $\pi_n^{\pi_0}$, and refer to π_n^p as the *filter initialized at* p. As posed by Ocone and Pardoux [35], it is interesting to ask whether the filter 'overcomes' the error made by choosing wrong initial data, as $n \to \infty$ (a question much related to earlier work by Kunita

[24], [25] and Stettner [38], [39] on convergence in law of the exact filter to a unique measure, under suitable ergodic assumptions about the state process). Delyon and Zeitouni [20] suggested that, because of the multiplicative form of (10.4), it is natural to ask when the error resulting from wrong initialization decays exponentially, and to study the rate of decay via multiplicative ergodic theory (MET). Denote by d_{TV} the total variation distance between measures, and note that one has $d_{TV}(p, q) = \|p - q\|_1$ where $\|\cdot\|_1$ denotes ℓ_1 norm and one uses the identification alluded to above. Let

$$\gamma(p, q) = \limsup_{n \to \infty} \frac{1}{n} \log \|\pi_n^p - \pi_n^q\|_1. \tag{10.5}$$

The main point of this section is to recall results based on MET, stating that the quantity γ is, loosely speaking, deterministic and independent of p and q. If $\gamma < 0$ then the filter can be said to be *exponentially stable with respect to perturbations in the initial condition*. Also, it is natural to interpret $-1/\gamma$ as the memory length of the filter, and thus it is useful to quantify γ.

We will need some basic results from MET (the reader is referred to [15, 17, 28] for further reading on the subject). Let $\{T_n, n \geq 1\}$ be a stationary ergodic sequence of $d \times d$ matrices, defined on $(\Omega, \mathcal{F}, \mathbb{P})$, satisfying $\mathbb{E}[\log^+ \|T_1\|] < \infty$. Denote $T_{(n)} = T_n T_{n-1} \cdots T_1$. Oseledec's theorem states that there exist deterministic constants

$$-\infty \leq \lambda_d \leq \lambda_{d-1} \leq \cdots \leq \lambda_1 < \infty,$$

and a full $\mathbb{P}$-measure event, Ω_1, on which the following holds.

(i) The sets

$$V(i) := \{x \in \mathbb{R}^d : \lim \frac{1}{n} \log \|T_{(n)} x\| \leq \lambda_i\}$$

are subspaces, and $\dim V(i) = \#\{j : \lambda_j \leq \lambda_i\}$ (in particular, $V(1) = \mathbb{R}^d$).

(ii) With $V(d + 1) = \{0\}$, one has for every $x \in \mathbb{R}^d \setminus \{0\}$

$$\lim_{n \to \infty} \log \|T_{(n)} x\| = \lambda_i,$$

where i is the unique j for which $x \in V(j) \setminus V(j + 1)$.

(iii) The sequence of matrices $\left(T_{(n)}^\top T_{(n)}\right)^{1/(2n)}$ converges to a matrix T whose eigenvalues are $e^{\lambda_1}, e^{\lambda_2}, \ldots, e^{\lambda_d}$. For $i \in \bar{d}$ such that $V(i) \neq V(i + 1)$, the orthogonal complement of $V(i + 1)$ in $V(i)$ is the eigenspace of T corresponding to e^{λ_i}.

The constants λ_i are called the *Lyapunov exponents* associated with $\{T_n\}$, under $\mathbb{P}$.

One defines the ith exterior power $\wedge^i M$ of a matrix M as the linear operator on the ith exterior power of $\mathbb{R}^d$, for which

$$\wedge^i M(e_{j_1} \wedge e_{j_2} \wedge \cdots \wedge e_{j_i}) = (Me_{j_1}) \wedge (Me_{j_2}) \wedge \cdots \wedge (Me_{j_i})$$

(see e.g. [10]). The only two facts that we need about exterior products here are, first, that given i vectors $x_1, x_2, \ldots, x_i \in \mathbb{R}^d$, the quantity $\bar{v}(x_1, x_2, \ldots, x_i) := \|x_1 \wedge x_2 \wedge \cdots \wedge x_i\| = (\det[\{\langle x_j, x_k \rangle\}_{j,k}])^{1/2}$ equals the i-dimensional volume of the parallelogram generated by these vectors; particularly, $\bar{v}(x_1, x_2)^2 = \|x_1 \wedge x_2\|^2 = \|x_1\|^2 \|x_2\|^2 - \langle x_1, x_2 \rangle^2$. And second, that (as in fact a corollary of Oseledec's theorem) the following holds $\mathbb{P}$-a.s.

$$\lim_{n \to \infty} \frac{1}{n} \log \|\wedge^i T_{(n)}\| = \sum_{j=1}^i \lambda_j, \qquad i \in \bar{d}, \tag{10.6}$$

where $\|\cdot\|$ is used here to denote the corresponding operator norm.

For a nonzero vector x in $\mathbb{R}^d_+$, write $\bar{x}$ for $x/\langle x, \mathbf{1} \rangle$. For x and y such vectors,

$$\bar{x} - \bar{y} = \frac{x}{\langle x, \mathbf{1} \rangle} - \frac{y}{\langle y, \mathbf{1} \rangle} = \frac{\|y\|_1 x - \|x\|_1 y}{\|x\|_1 \|y\|_1},$$

hence

$$\|\bar{x} - \bar{y}\|_1 \leq \frac{\sum_{i,j=1}^d |x^i y^j - y^i x^j|}{\|x\|_1 \|y\|_1}.$$

Also,

$$\bar{v}(x, y)^2 = \|x \wedge y\|^2 = \|x\|^2 \|y\|^2 - \langle x, y \rangle^2 = \frac{1}{2} \sum_{i,j} (x^i y^j - x^j y^i)^2,$$

which gives

$$\|\bar{x} - \bar{y}\|_1 \leq c_d \frac{\bar{v}(x, y)}{\|x\|_1 \|y\|_1}. \tag{10.7}$$

This inequality will help us establish a bound on $\gamma(p, q)$ in term of exponential growth rates of the three objects $\bar{v}(\rho_n^p, \rho_n^q)$, $\|\rho_n^p\|_1$ and $\|\rho_n^q\|_1$. To see that the latter are quite simple to quantify by Lyapunov exponents, consider first an arbitrary sequence $\{T_n\}$ of matrices from $M_+(d)$, satisfying the assumptions of Oseledec's theorem. Let $x, y \in \mathcal{P}$, and denote $x_n = T_{(n)} x$ and $y_n = T_{(n)} y$. By the Perron–Frobenius theorem, the matrix $T_{(n)}^\top T_{(n)} \in M_+(d)$ has an eigenvector $u_n \in \mathbb{R}^d_+$ corresponding to the largest eigenvalue, say μ_n. Consequently $\mu_n^{1/(2n)}$ and u_n are eigenvalue and eigenvector of $(T_{(n)}^\top T_{(n)})^{1/(2n)}$, and thus by item (iii) of Oseledec's theorem, the limit matrix T has an eigenvector $u_0 \in \mathbb{R}^d_+$ corresponding to its largest eigenvalue, e^{λ_1}. If i is such that $V(i) \neq V(1) = \mathbb{R}^d$ then it follows from item (iii) that $V(i)$ is orthogonal to u_0, hence, provided that x is strictly positive, $x \notin V(i)$. Thus by item (ii), for all such x, one has $\mathbb{P}$-a.s.,

$$\lim \frac{1}{n} \log \|x_n\| = \lambda_1.$$

Combining this with (10.6) and (10.7), we have

$$\limsup_{n \to \infty} \frac{1}{n} \log \|\bar{x}_n - \bar{y}_n\|_1 \leq \lambda_1 + \lambda_2 - 2\lambda_1 = \lambda_2 - \lambda_1, \tag{10.8}$$

provided $x, y \in \mathcal{P}_P := \{z \in \mathcal{P} : z^i > 0, i \in \bar{d}\}$. The quantity $\lambda_1 - \lambda_2$ is often referred to as the *spectral gap*.

This analysis is not directly applicable to the filtering situation described above because the matrix process is not necessarily stationary. Of course if $\{X_n\}$ is a stationary ergodic process then so is the matrix process $T_n := D_n G^\top$. We will assume that $\{X_n\}$ is an irreducible aperiodic chain. Thus there exists a unique invariant measure for the chain, π_S, and we denote by $\mathbb{P}_S$ the corresponding measure on $(\Omega, \mathcal{F})$. In the special case where π_0 equals π_S, the sequence $\{T_n\}$ is stationary and ergodic. For general π_0 we let the term *the Lyapunov exponents associated* $\{T_n\}$ mean the Lyapunov exponents associated with $\{T_n\}$ under $\mathbb{P}_S$. The result (10.8) developed above can in fact be improved in the following way [2]:

Theorem 10.1 *Assume that the chain* $\{X_n\}$ *is irreducible and aperiodic. Assume* $\mathbb{E}_S[\log^+ \|D_1 G^\top\|] < \infty$. *Let U denote the uniform measure on $\mathcal{P}$. Then $\mathbb{P}$-a.s., for $U \times U$-a.e. (p, q),*

$$\gamma(p, q) = \lambda_2 - \lambda_1,$$

where $\{\lambda_i\}$ are the Lyapunov exponents associated with $\{D_n G^\top\}$. Moreover, $\mathbb{P}$-a.s., we have for every $(p, q) \in \mathcal{P} \times \mathcal{P}$

$$\gamma(p, q) \leq \lambda_2 - \lambda_1.$$

Remark 10.1 This result is proved in [2, Sec. 10.2] for the case of additive Gaussian noise, but the proof holds in the generality presented here. Also, the second assertion of Theorem 10.1 above is written in [2] in a slightly weaker way, namely that for every p and q, we have, $\mathbb{P}$-a.s., $\gamma(p, q) \leq \lambda_2 - \lambda_1$, but the form presented here is valid according to the same proof (indeed, a review of that proof shows that, in the claim made in (8) and (9) of [2], the full $\mathbb{P}$-measure event on which the inequality holds is what we have denoted by Ω_1 in the above statement of Oseledec's theorem, which in particular does not depend on the initial conditions p and q).

For a continuous-time analog of this result, consider a Markov process X taking values in $\bar{d}$ with intensity matrix $\hat{G}$, and initial distribution π_0. Let the observation process be given by

$$Y_t = \int_0^t k(X_s)ds + \sigma W_t,$$

where W is a standard Brownian motion independent of X and denote by π_t the conditional law of X_t given $\mathcal{Y}_t := \sigma\{Y_s : s \in [0, t]\}$. Denoting by K the diagonal $d \times d$ matrix with $K^{i,i} = k(i)$, and letting $\{\rho_t\}$ be the unique solution to the stochastic differential equation

$$d\rho_t = \hat{G}^\top \rho_t dt + \sigma^{-2} K \rho_t dY_t, \qquad t \geq 0, \qquad \rho_0 = \pi_0,$$

one has $\pi_t = \rho_t/\langle \rho_t, \mathbf{1}\rangle$ (see e.g. [41, equation (10.5)] and use Itō's lemma). Note that this can be written in terms of the $M_+(d)$-valued process $\{T_t\}$ solving

$$dT_t = \hat{G}^\top T_t dt + \sigma^{-2} K T_t dY_t, \qquad t \geq 0, \qquad T_0 = I, \qquad (10.9)$$

where $I \in M(d)$ is the identity matrix. Namely, one has $\rho_t = T_t \pi_0$. For general initial data, $p \in \mathcal{P}$, set $\rho_t^p = T_t p$ and $\pi_t^p = \rho_t^p / \langle \rho_t^p, \mathbf{1}\rangle$.

Oseledec theorem has an analogue in continuous time. We present here the version [15, Sec. IV.2]. Let a probability space $(\Omega, \mathcal{F}, \mathbb{P})$ be endowed with a semigroup $\{\theta_t, t \geq 0\}$ of measure preserving transformations. Let $\{T_t, t \geq 0\}$ be a process on this space, taking values in $GL(d, \mathbb{R})$ (the group of linear automorphisms of $\mathbb{R}^d$). Then $\{T_t\}$ is said to be *multiplicative* if $T_0 = I$, and $T_{t+s} = (T_s \circ \theta_t) T_t$, for all $s, t \geq 0$. Assume that $\{\theta_t\}$ is ergodic, $\{T_t\}$ is a separable multiplicative process, and that $\mathbb{E}[\sup_{t \in [0,1]} \log^+ \|T_t^k\|] < \infty$ for both $k = 1$ and $k = -1$. Then items (i) and (ii) of the discrete version of the theorem, that appears above, hold upon replacing $T_{(n)}$ by T_t, $\frac{1}{n}$ by $\frac{1}{t}$, and $\lim_n$ by $\lim_t$. The way this result is used in the present context is by considering the standard shift transformation. As in the discrete-time case, where the process may not be stationary, the shift transformation may not be measure preserving; however, under an irreducibility assumption, it is so in the case when π_0 equals the invariant measure π_S. For general π_0 we use the same convention and define Lyapunov exponents for the nonstationary process to be the ones for the stationary counterpart.

Analogously to (10.5) let

$$\gamma(p, q) = \limsup_{t \to \infty} \frac{1}{t} \log \|\pi_t^p - \pi_t^q\|_1. \qquad (10.10)$$

The assumptions of the above version of the MET can be verified, and one has the following.

Theorem 10.2 *[2, 20] Assume that the process X is an irreducible continuous-time Markov chain on $\bar{d}$. Then the conclusions of Theorem 10.1 hold for γ of (10.10) and the Lyapunov exponents $\{\lambda_i\}$ associated with the process (10.9).*

The above results assert that the decay rate γ is, roughly stated, deterministic and independent of the initial condition, and identify it with the (negative)

Lyapunov spectral gap. The question of exponential stability can thus be posed as that of determining whether the gap is positive. Unfortunately, the Lyapunov spectrum is in general hard to calculate, if not impossible. However, one can sometimes obtain bounds on the gap from which such information can be extracted. This will be the point of the next section.

10.3 Hilbert's projective metric in finite-state space

We present here Hilbert's projective metric, and its contraction properties under the action of positive matrices, used to obtain a bound on the Lyapunov spectral gap. This will enable us to attain conditions under which the gap, and in view of last section's results, the decay rate, is nonzero. It will also give rise to quantitative information on the gap.

We need some notation regarding positive matrices (we follow Seneta [37]). A matrix that is an element of $M_+(d)$ is said to be *allowable* if it contains no columns or rows whose entries are all zero. *Hilbert's projective metric* is the mapping $h : \mathcal{P}_P^2 \to \mathbb{R}_+$ defined by

$$h(x, y) = \log \max_{1 \le i,j \le d} \frac{x_i y_j}{x_j y_i}. \tag{10.11}$$

An allowable $d \times d$ matrix M can be seen, by normalization of the action of M, as an operator $M : \mathcal{P}_P \to \mathcal{P}_P$. We denote by $M.x$ its action on $x \in \mathcal{P}_P$. This definition turns out to be very useful mainly due to the fact that h makes any allowable matrix a contraction. Namely, $\tau(M) \le 1$ where τ is the *Birkhoff contraction coefficient* of an allowable matrix M, defined by

$$\tau(M) = \sup \left\{ \frac{h(M.x, M.y)}{h(x, y)} : x, y \in \mathcal{P}_P, x \ne y \right\}. \tag{10.12}$$

An explicit formula for τ in terms of M is available [37], namely

$$\tau(M) = \frac{1 - \sqrt{\psi(M)}}{1 + \sqrt{\psi(M)}}, \quad \text{where} \quad \psi(M) = \min_{i,j,k,l} \left\{ \frac{M^{i,k} M^{j,l}}{M^{i,l} M^{j,k}} : M^{i,l} M^{j,k} \ne 0 \right\}. \tag{10.13}$$

Here are two additional elementary properties. As follows directly from (10.11) and (10.12), if D is a diagonal matrix with $D^{i,i} > 0$ for $i \in \bar{d}$ then $\tau(MD) = \tau(DM) = \tau(M)$. If $M \in M_+(d)$ is a matrix whose entries are all positive, then, by (10.13), $\tau(M) < 1$.

We borrow the following from [36].

Lemma 10.1 *Let $\{T_n\}$ be a stationary ergodic sequence of nonnegative, allowable matrices, and assume $\mathbb{E}[\log^+ \|T_1\|] < \infty$. Let λ_1 and λ_2 denote the top two Lyapunov exponents associated with the sequence. Then*

$$\lambda_1 - \lambda_2 \ge -\mathbb{E}[\log \tau(T_1)],$$

where $\lambda_2 = -\infty$ if the right-hand side is infinite.

Combined with Theorems 10.1 and 10.2, this gives a direct relation between the decay rate and the contraction coefficient, not involving the Lyapunov spectrum.

Corollary 10.1 *Under the assumptions of Theorem 10.1, $\mathbb{P}$-a.s., for every $p, q \in \mathcal{P}$, $\gamma(p, q) \leq \mathbb{E}_S[\log \tau(T_1)]$, where $T_1 = D_1 G^\top$. Under the assumptions of Theorem 10.2, $\gamma(p, q) \leq \mathbb{E}_S[\log \tau(T_t)]$, where T_t is given in (10.9).*

By the foregoing discussion on properties of τ, we have in the discrete setting $\tau(T_1) = \tau(D_1 G^\top) = \tau(G^\top) = \tau(G)$. If $G^{i,j} > 0$ for all i, j then $\tau(G) < 1$, and as a consequence we have the following.

Theorem 10.3 *[2] In the discrete-time setting, assume that $G^{i,j} > 0$ for all $i, j \in \bar{d}$. Then $\mathbb{P}$-a.s., for all $p, q \in \mathcal{P}, \gamma(p, q) \leq \log \tau(G) < 0$.*

This provides an estimate that depends only on the transition law of $\{X_n\}$. We refer to [16], this volume, for a bound that holds in greater generality, under which G may be a more general element of $M_+(d)$.

The approach is useful in obtaining estimates on the decay rate in the small noise (large signal-to-noise ratio) asymptotics. For the following result recall the setting of one-dimensional additive Gaussian noise (Example 10.2). For $i \in \bar{d}$ denote

$$\delta(i) = \min_{j \neq i} |k(i) - k(j)|.$$

Denote by γ_σ the negative spectral gap, emphasizing the dependence on σ.

Theorem 10.4 *[2] Consider the setting of Example 10.2. Then under the assumptions of Theorem 10.1,*

$$\limsup_{\sigma \to 0} \sigma^2 \gamma_\sigma \leq -\frac{1}{2} \mathbb{E}_S[\delta(X_1)^2]. \tag{10.14}$$

If, in addition, $\det G \neq 0$, we have

$$\liminf_{\sigma \to 0} \sigma^2 \gamma_\sigma \geq -\frac{1}{2} \mathbb{E}_S\Big[\sum_{i=1}^{d}(k(X_1) - k(i))^2\Big]. \tag{10.15}$$

Note that δ is not identically zero if and only if there exists at least one i for which $k(i)$ is distinct from $k(j)$, all $j \neq i$. Thus the upper bound presented above is meaningful only under this condition; and when the condition holds, the combination of both bounds establish that the order of magnitude of γ_σ is σ^{-2}.

In general it is an open question whether either the upper or lower bounds can be improved.

Let us mention that (10.14) holds with equality when one assumes that X_n is a "nearest neighbor" process, in the following sense: the transition matrix is given by $G = \exp(s\,A)$ for some $s > 0$ and A is an intensity matrix for which $|i - j| > 1$ implies $A(i, j) = 0$ (some additional technical conditions are required; see [3, Theorem 5]).

Analogous bounds hold in continuous time.

Theorem 10.5 *[2] Consider the filtering problem in continuous time, and let the assumptions of Theorem 10.2 hold. Then $\mathbb{P}$-a.s., for all $p, q \in \mathcal{P}$,*

$$\gamma(p, q) \leq -2 \min_{i,j \in \tilde{d}: i \neq j} \sqrt{\hat{G}^{i,j} \hat{G}^{j,i}}.$$

Moreover, the following bounds hold:

$$\limsup_{\sigma \to 0} \sigma^2 \gamma_\sigma \leq -\frac{1}{2} \mathbb{E}_S[\delta(X_0)^2], \qquad \liminf_{\sigma \to 0} \sigma^2 \gamma_s \geq -\frac{1}{2} \mathbb{E}_S\Big[\sum_{i=1}^{d} (k(X_0) - k(i))^2\Big].$$

10.4 Hilbert's projective metric in general-state space

The fact established in Corollary 10.1 deserves a deeper look. We will see that this result has an easy proof not involving Lyapunov exponents or MET, valid in fact in far greater generality. Thus in what follows we shall study the filtering problem in a general setting, and present an extended definition of h, and present an extended version of Corollary 10.1.

Let $\mathbb{S}$ be a Polish space, and let $\mathcal{S}$ denote the corresponding Borel σ-field. Fix a positive integer ℓ. We will introduce a Markov process X_n taking values in $\mathbb{S}$ and an observation process Y_n taking values in $\mathbb{R}^\ell$. We will, in fact, present a Markovian family. To this end, assume we are given a probability kernel $G : \mathbb{S}^2 \to \mathbb{R}$ (that is, for every $a \in \mathbb{S}$, $G(a, \cdot)$ is a probability measure on $(\mathbb{S}, \mathcal{S})$ and for every $E \in \mathcal{S}$, $G(\cdot, E)$ is a measurable map). Also, we are given a probability kernel $\tilde{G} : \mathbb{S} \times \mathbb{R}^\ell \to \mathbb{R}$. We define for every $a \in \mathbb{S}$ a probability measure $\mathbb{P}^{(a)}$ on $(\Omega, \mathcal{F})$ via

$$\mathbb{P}^{(a)}(X_1 \in E_1, X_2 \in E_2, \ldots, X_n \in E_n, Y_1 \in F_1, Y_2 \in F_2, \ldots, Y_n \in F_n)$$

$$= \int_{E_1 \times \cdots \times E_n \times F_1 \times \cdots \times F_n} G(a, dx_1) \prod_{i=2}^{n} G(x_{i-1}, dx_i) \prod_{i=1}^{n} \tilde{G}(x_i, dy_i),$$

for $n \in \mathbb{N}$, $E_i \in \mathcal{S}$, $F_i \in \mathcal{R}$, $i \leq n$, where, throughout, we denote $\mathcal{R} = \mathcal{B}(\mathbb{R}^\ell)$. For $p \in \mathcal{M}(\mathbb{S}, \mathcal{S})$, let

$$\mathbb{P}^p = \int_\mathbb{S} \mathbb{P}^{(a)} p(da).$$

Fix a probability measure $\pi_0 \in \mathcal{M}(\mathbb{S}, \mathcal{S})$, and let $\mathbb{P} := \mathbb{P}^{\pi_0}$. Then under $\mathbb{P}$, X_n is a Markov process starting from initial measure π_0, and Y_n is an observation process. As before, we write $\mathcal{Y}_n$ for the σ-field generated by $(Y_1, Y_2, \ldots, Y_n)$, $n \in \mathbb{N}$. The exact filter is thus

$$\pi_n(\varphi) = \mathbb{E}[\varphi(X_n)|\mathcal{Y}_n], \qquad n \geq 0.$$

We introduce a reference measure on $(\Omega, \mathcal{F})$. To this end we will need the assumption: *There exists a probability measure $\widetilde{G}_0$ on $(\mathbb{R}^\ell, \mathcal{R})$ and a measurable mapping $g : \mathbb{S} \times \mathbb{R}^\ell \to \mathbb{R}$ with respect to $\mathcal{S} \otimes \mathcal{R}$, such that, for every $a \in \mathbb{S}$,*

$$\widetilde{G}(a, F) = \int_F g(a, y) \widetilde{G}_0(dy), \qquad F \in \mathcal{R}.$$

Define for $a \in \mathbb{S}$ the reference probability measure $\mathbb{P}_0^{(a)}$

$$\mathbb{P}_0^{(a)}(X_1 \in E_1, X_2 \in E_2, \ldots, X_n \in E_n, Y_1 \in F_1, Y_2 \in F_2, \ldots, Y_n \in F_n)$$

$$= \int_{E_1 \times \cdots \times E_n \times F_1 \times \cdots \times F_n} G(a, dx_1) \prod_{i=2}^n G(x_{i-1}, dx_i) \prod_{i=1}^n \widetilde{G}_0(dy_i),$$

for $n \in \mathbb{N}$, $E_i \in \mathcal{S}$, $F_i \in \mathcal{R}$, $i \leq n$. Denote $\mathbb{P}_0^p = \int \mathbb{P}^{(a)} p(da)$ and $\mathbb{P}_0 = \mathbb{P}_0^{\pi_0}$. Consider the stochastic process $\{\Lambda_n\}$, $n \geq 0$, where $\Lambda_0 = 1$ and

$$\Lambda_n = \prod_{i=1}^n g(X_i, Y_i), \qquad n \geq 1.$$

Then clearly $\mathbb{P}^p(B) = \int_B \Lambda_n d\mathbb{P}_0^p$, for $B \in \sigma\{X_i, Y_i, i \leq n\}$. A use of Bayes' rule shows

$$\pi_n(\varphi) = \frac{\mathbb{E}_0[\varphi(X_n)\Lambda_n|\mathcal{Y}_n]}{\mathbb{E}_0[\Lambda_n|\mathcal{Y}_n]}.$$

We thus let

$$\rho_n(\varphi) = \mathbb{E}_0[\varphi(X_n)\Lambda_n|\mathcal{Y}_n], \qquad (10.16)$$

and note that $\pi_n = \rho_n/\rho_n(1)$. Toward writing a recursion for ρ_n, we select a regular conditional probability distribution

$$\mathbb{P}_0^{(a)}[\cdot | \mathcal{Y}_n, X_n = \beta],$$

satisfying, for $A \in \mathcal{S}$, $B \in \sigma\{X_i, Y_i, i \leq n\}$,

$$\mathbb{P}_0^{(a)}[B \cap \{X_n \in A\}|\mathcal{Y}_n] = \int_A \mathbb{P}_0^{(a)}(B|\mathcal{Y}_n, X_n = \beta) \mathbb{P}_0^{(a)}(X_n \in d\beta).$$

Letting

$$I_n(a, \beta) = \mathbb{E}_0^{(a)}[\Lambda_n|\mathcal{Y}_n, X_n = \beta],$$

we can write ρ_n (10.16) as $\rho_n^{\pi_0}$, where for $p \in \mathcal{M}(\mathbb{S}, \mathcal{S})$,

$$\rho_n^p(\varphi) = \iint \varphi(\beta) G_n(a, d\beta) I_n(a, \beta) p(da), \qquad (10.17)$$

and we denote $G_n(a, B) = \mathbb{P}_0^{(a)}(X_n \in B)$. Let $\mathcal{V}$ denote the vector space of finite signed measures on $(\mathbb{S}, \mathcal{S})$. Denote by $J_{0,n}$ the (random) mapping from $\mathcal{V}$ to itself, mapping $p \in \mathcal{V}$ to ρ_n^p according to (10.17). Denoting by θ_n the shift transformation, let also

$$J_{m,n} = J_{0,n-m} \circ \theta_m, \qquad 0 \le m \le n.$$

By conditioning, it is clear that for $p \in \mathcal{P}$, $0 \le m \le n$,

$$\rho_n^p = J_{m,n} J_{0,m} p, \tag{10.18}$$

and consequently, $J_{0,n} = J_{m,n} J_{0,m}$. This gives rise to the recursion

$$\rho_n^p = J_{n-1,n} \rho_{n-1}^p, \qquad n \ge 1, \qquad \rho_0^p = p. \tag{10.19}$$

Set $\pi_n^p = \rho_n^p / \rho_n^p(1)$, $p \in \mathcal{P}$ and

$$\gamma(p, q) = \limsup_{n \to \infty} \frac{1}{n} \log d_{TV}\left(\pi_n^p, \pi_n^q\right), \qquad p, q \in \mathcal{P}. \tag{10.20}$$

A closer look at the recursion (10.19) shows that it is possible that the measure ρ_n becomes zero for some n. Here is an example. Consider the degenerate chain $X_n = X_0$, $n \ge 0$, where X_0 is a random variable on $\{1, 2\}$. Assume the observation process is given by $Y_n = X_n$ for $n \ge 1$. With the notation of the previous section, $G = I$ and $D_n^{i,i} = 1_{\{Y_n = i\}}$. If π_0 consists of an atom at 1 then $\mathbb{P}$-a.s., for all n,

$$T_n = D_n G^\top = \begin{pmatrix} 1 & 0 \\ 0 & 0 \end{pmatrix}.$$

Thus if p is an atom at 2 then $\rho_1^p = T_1 p = 0$.

In view of the foregoing discussion, we must define π_n^p and γ more carefully. Thus if for some n_0 $\rho_{n_0}^p = 0$ we let $\pi_n = 0$ for all $n \ge n_0$. We extend d_{TV} to $\mathcal{P} \cup \{0\}$ by defining $d_{TV}(0, \lambda) = d_{TV}(\lambda, 0) = d_{TV}(0, 0) = 1$ for $\lambda \in \mathcal{P}$. Thus $\gamma(p, q) = 0$ on the event that ρ_n^p takes the value zero for some n.

The goal is now to show that γ can be bounded in terms of contraction coefficient for linear operators, in a fashion similar to Section 10.3. To this end we need an extended definition of Hilbert's metric. Define on $\mathcal{V}$ the partial order $\preccurlyeq$ by $\lambda \preccurlyeq \mu$ for $\lambda, \mu \in \mathcal{V}$ if $\lambda(A) \le \mu(A)$ for every $A \in \mathcal{S}$. Denote by $\mathcal{C}_0 \subset \mathcal{V}$ the cone of members λ of $\mathcal{V}$ for which $0 \preccurlyeq \lambda$, where 0 is the zero measure, and by $\mathcal{C}$ the collection of members of $\mathcal{C}$ excluding the zero measure. Two elements $\lambda, \mu \in \mathcal{C}$ are said to be *comparable* if there exist $0 < \alpha, \beta < \infty$ for which $\alpha \lambda \preccurlyeq \mu \preccurlyeq \beta \lambda$. Define $h : \mathcal{C}^2 \to [0, \infty]$ by

$$h(\lambda, \mu) = \log \frac{\sup_{A \in \mathcal{S}_\mu} \lambda(A)/\mu(A)}{\inf_{A \in \mathcal{S}_\mu} \lambda(A)/\mu(A)} \qquad \text{if } \lambda, \mu \in \mathcal{C} \text{ are comparable,} \tag{10.21}$$

and $h(\lambda, \mu) = \infty$ otherwise, where we denoted $\mathcal{S}_\mu = \{A \in \mathcal{S} : \mu(A) > 0\}$. The function h, called Hilbert's metric, is a pseudometric on $\mathcal{C}$, and a metric on the space of members $\lambda \in \mathcal{P}$ that are comparable to a given $\lambda_0 \in \mathcal{P}$ [8, ch. 16]. A linear operator mapping $\mathcal{V}$ into itself is *positive* if it maps $\mathcal{C}$ into itself. As shown by Birkhoff [7] and Hopf [22], any positive linear operator L on $\mathcal{V}$ is a contraction with respect to Hilbert's metric, and

$$\tau(L) := \sup_{0 < h(\lambda, \mu) < \infty} \frac{h(L\lambda, L\mu)}{h(\lambda, \mu)} = \tanh \frac{H(L)}{4}, \tag{10.22}$$

where

$$H(L) = \sup_{\lambda, \mu \in \mathcal{C}} h(L\lambda, L\mu), \tag{10.23}$$

and $\tau = 1$ in case when $H = \infty$ (see also [8] and [30] for these and various additional useful facts on the Hilbert metric). The formula above for τ is an extension of the formula (10.13) of Section 10.3.

We extend h to $\mathcal{C}_0^2$ by letting $h(0, \lambda) = h(\lambda, 0) = h(0, 0) = \infty$ for $\lambda \in \mathcal{C}$. Further, we let $\tau(L) = 1$ for any linear operator L mapping $\mathcal{C}_0$ into itself that is not positive. Such an operator will be called *weakly positive*.

To apply this to the filtering equations, note that $J_{m,n}$ are weakly positive, and by (10.18) that, for $p, q \in \mathcal{P}$, one has for any $n, m \in \mathbb{N}$,

$$h\left(\rho_{nm}^p, \rho_{nm}^q\right) \leq h(p, q) \prod_{i=1}^n \tau(J_{im-m, im}).$$

Now, by definition of h, $h(c_1\lambda, c_2\mu) = h(\lambda, \mu)$ for any $c_1, c_2 \in (0, \infty)$, and thus $h\left(\pi_n^p, \pi_n^q\right) = h\left(\rho_n^p, \rho_n^q\right)$. Note also that, since $\tau \leq 1$, $h\left(\rho_n^p, \rho_n^q\right)$ is monotone in n. By these considerations, as soon as $h(p, q) < \infty$, a bound on a quantity similar to γ (10.20) follows, namely

$$\gamma^h(p, q) := \limsup_n \frac{1}{n} \log h\left(\pi_n^p, \pi_n^q\right) \leq \limsup_n \frac{1}{mn} \sum_{i=1}^n \log \tau(J_{im-m, im}).$$

In fact, it is easy to prove [3, Lemma 1]

$$d_{TV}(\lambda, \mu) \leq \frac{2}{\log 3} h(\lambda, \mu), \qquad \lambda, \mu \in \mathcal{P},$$

whence $\gamma \leq \gamma^h$. We summarize this in the following:

Lemma 10.2 *If $p, q \in \mathcal{P}$ are comparable, or more generally, if $\inf_n h\left(\rho_n^p, \rho_n^q\right) < \infty$ $\mathbb{P}$-a.s., then for any positive integer m, $\mathbb{P}$-a.s.,*

$$\gamma(p, q) \leq \limsup_n \frac{1}{mn} \sum_{i=1}^n \log \tau(J_{im-m, im}) =: \Gamma_m. \tag{10.24}$$

It is natural to apply this lemma to cases where J has some ergodic properties, so that Γ_m can be expressed as expectation of a single term. For example, consider the case where for some $\pi_S \in \mathcal{P}$, the process $\{X_n\}$ is stationary ergodic under $\mathbb{P}_S := \mathbb{P}^{\pi_S}$. Then under the same law, so is the sequence $\{(X_n, Y_n)\}$, and in turn also the process $\{J_n\}$. In this case, Γ_m is $\mathbb{P}_S$-a.s. equal to $\bar{\Gamma}_m := m^{-1}\mathbb{E}_S[\log \tau(J_{0,m})]$. Next, if say $\pi_0 \ll \pi_S$ then also $\mathbb{P} \ll \mathbb{P}_S$ and thus it is true also $\mathbb{P}$-a.s. that $\Gamma_m = \bar{\Gamma}_m$. More generally, the same conclusion will be valid, provided that $\mathbb{P}$ and $\mathbb{P}_S$ agree on the tail σ-field, because Γ_m is measurable on this σ-field. This is recorded in the following.

Theorem 10.6 *[3] Assume there exists $\pi_S \in \mathcal{P}$ for which the corresponding law $\mathbb{P}_S$ makes $\{X_n\}$ stationary and ergodic. Assume also that the restrictions of both $\mathbb{P}$ and $\mathbb{P}_S$, to the tail σ-field, agree. Then $\mathbb{P}$-a.s.,*

$$\gamma(p,q) \leq \frac{1}{m}\mathbb{E}_S[\log \tau(J_{0,m})], \qquad p, q \in \mathcal{P}, \ m \in \mathbb{N}.$$

Let us exhibit a situation where the above gives rise to an exponential stability result. Consider the case where the state process satisfies the strong mixing condition. Namely, for some $\lambda \in \mathcal{P}$ and constants $0 < c_1, c_2 < \infty$,

$$c_1 \lambda \preccurlyeq G(x, \cdot) \preccurlyeq c_2 \lambda, \qquad x \in \mathbb{S}. \tag{10.25}$$

Note that G is a positive operator, and by (10.22), (10.23) that $\tau(G) < 1$. We claim that without any assumptions on the observation process (i.e., on $\widetilde{G}_0$ and g), one has $\tau(J_{0,1}) \leq c < 1$, for some constant c.

Theorem 10.7 *[3] Assume (10.25) holds for some constants $0 < c_1, c_2 < \infty$ and $\lambda \in \mathcal{P}$. Then $\tau(J_{0,1}) \leq c_3 := (c_2 - c_1)/(c_2 + c_1)$, $\mathbb{P}^{(x)}$-a.s., for any $x \in \mathbb{S}$. Consequently, under the hypotheses of Theorem 10.6, $\gamma(p,q) \leq \log c_3 < 0$, for all $p, q \in \mathcal{P}$, $\mathbb{P}$-a.s.*

Proof. Let $\bar{a}(\gamma) = \int_\mathbb{S} g(\beta, \gamma)\lambda(d\beta)$. We first show that for every a, $\mathbb{P}^{(a)}$-a.s. one has $\bar{a}(Y_1) > 0$. To see this, let B denote the set $\{\gamma \in \mathbb{R}^\ell : \bar{a}(\gamma) = 0\}$. Then

$$\mathbb{P}^{(a)}(Y_1 \in B) = \iint 1_B(\gamma) G(\alpha, d\beta) \widetilde{G}(\beta, d\gamma)$$

$$\leq c_2 \iint 1_B(\gamma) \lambda(d\beta) g(\beta, \gamma) \widetilde{G}_0(d\gamma)$$

$$= c_2 \int 1_B(\gamma) \bar{a}(\gamma) \widetilde{G}_0(d\gamma) = 0.$$

Next, given $A \in \mathcal{S}$ let $a_A := \int_A g(\beta, Y_1)\lambda(d\beta)$ and note by (10.17) that $c_1 a_A \leq \rho_1^p(A) \leq c_2 a_A$ for every $p \in \mathcal{P}$. Hence $c_1/c_2 \leq \rho_1^p(A)/\rho_1^q(A) \leq c_2/c_1$, provided $a_A > 0$. Thus, provided there exists A for which $a_A > 0$, by (10.21), $h\left(\rho_1^p, \rho_1^q\right) \leq 2\log(c_2/c_1)$, and by (10.22), (10.23), $\tau(J_{0,1}) \leq \tanh\left[\frac{1}{4}\log(c_2/c_1)\right] = (c_2 - c_1)/(c_2 + c_1) = c_3$. In view of the first paragraph, one has, in fact, $a_\mathbb{S} = \bar{a}(Y_1) > 0$ $\mathbb{P}^{(a)}$-a.s. for arbitrary a, and we conclude that $\tau(J_{0,1}) \leq c_3$, $\mathbb{P}^{(a)}$-a.s.

The second assertion of the theorem is immediate from the first one. □

It is interesting that Lemma 10.2 may also be useful in nonergodic situations. We refer the reader to [13] for a result based on a similar argument in a setup where the state process is transient (this result was improved in a subsequent paper [14] by other techniques; see also [31] and [34] for additional treatments of transient cases).

We now consider a continuous-time Markov process on a Polish space, observed in white noise. The precise setting is as follows.

Equip the space $\Omega^1 = D(\mathbb{R}_+, \mathbb{S})$, of càdlàg mappings from $\mathbb{R}_+$ to $\mathbb{S}$, with the Skorohod J_1 topology, and let $\mathcal{B}^1$ denote the corresponding Borel σ-field. Let $\Omega^2 = C(\mathbb{R}_+, \mathbb{R}^\ell)$ be equipped with the uniform-on-compacts topology, and denote by $\mathcal{B}^2$ the corresponding Borel σ-field. Let $\Omega = \Omega^1 \times \Omega^2$ and $\mathcal{B} = \mathcal{B}^1 \otimes \mathcal{B}^2$. For $\omega = (\omega_1, \omega_2) \in \Omega$, let the processes X and W be defined via $X_t(\omega) = \omega_1(t)$ and $W_t(\omega) = \omega_2(t)$. For $a \in \mathbb{S}$ let $\mathbb{P}^{(a)}$ denote a probability measure on $(\Omega, \mathcal{B})$ under which W and X are independent, W is a standard ℓ-dimensional Brownian motion, and

$$\mathbb{P}^{(a)}(X_{t_1} \in E_1, X_{t_2} \in E_2, \ldots, X_{t_n} \in E_n) = \int_{E_1 \times \cdots \times E_n} G_{t_1}(a, dx_1) \cdots G_{t_n - t_{n-1}}(x_{n-1}, dx_n),$$

for $n \in \mathbb{N}$, $0 < t_1 < t_2 < \cdots < t_n$ and $E_i \in \mathcal{S}$, $i \leq n$. Here, G_t is a given Feller-Markov semigroup. As before, with $p \in \mathcal{P}$, associate $\mathbb{P}^p = \int \mathbb{P}^{(a)} p(da)$ and set $\mathbb{P} = \mathbb{P}^{\pi_0}$ for some fixed π_0.

To describe the observation process let a measurable function $k : \mathbb{S} \to \mathbb{R}^\ell$ be given. We shall assume $\mathbb{E}^{(a)}[\int_0^t \|k(X_s)\|^2 ds] < \infty$, $a \in \mathbb{S}$, $t \geq 0$. The process Y_t is defined via

$$Y_t = \int_0^t k(X_s) ds + W_t, \qquad t \geq 0. \tag{10.26}$$

Let $\mathcal{Y}_t = \sigma\{Y_s : s \in [0, t]\}$, and set

$$\pi_t^p(\varphi) = \mathbb{E}^p[\varphi(X_t) | \mathcal{Y}_t], \qquad t \geq 0.$$

We note on passing that, under various smoothness assumptions on the coefficients [26, theorem 6.3.3], π_t^p solves the *Kushner–Stratonovich equation*

$$\pi_t(\varphi) = p(\varphi) + \int_0^t \pi_s(\mathcal{L}\varphi) ds + \int_0^t \langle \pi_s(\varphi k) - \pi_s(\varphi)\pi_s(k), dY_s - \pi_s(k) ds \rangle,$$

for φ in the domain of $\mathcal{L}$, where $\mathcal{L}$ is the generator of the semigroup G_t; similarly, one has $\pi_t^p = \rho_t^p / \rho_t^p(1)$ where ρ_t solves the *Zakai equation*

$$\rho_t(\varphi) = p(\varphi) + \int_0^t \rho_s(\mathcal{L}\varphi) ds + \int_0^t \langle \rho_s(\varphi k), dY_s \rangle.$$

Consequently, this study can be viewed as one of sensitivity of solutions to these equations with respect to perturbations in their initial conditions.

Let us now describe the flow in a way similar to the discrete-time case. To this end let

$$\Lambda_t = \exp\left\{\int_0^t \langle k(X_s), dY_s\rangle - \frac{1}{2}\int_0^t \|k(X_s)\|^2 ds\right\}, \qquad t \geq 0.$$

Define

$$\rho_t^p(\varphi) = \mathbb{E}_0^p[\varphi(X_t)\Lambda_t | \mathcal{Y}_t],$$

where under $\mathbb{P}_0^p$, X and W are independent, but each of these processes has the same law as under $\mathbb{P}^p$. Then $\pi_t^p = \rho_t^p / \rho_t^p(1)$, $\mathbb{P}^p$-a.s. Let $J_{0,t}$ denote the linear, weakly positive transformation sending p to ρ_t^p. Then a result analogous to Theorem 10.6 holds, by similar considerations.

Theorem 10.8 *[3] Let assumptions analogous to those of Theorem 10.6 hold for the continuous-time setting described above. Then $\mathbb{P}$-a.s.,*

$$\gamma(p, q) \leq \frac{1}{t}\mathbb{E}_s[\log \tau(J_{0,t})], \qquad p, q \in \mathcal{P}, t > 0.$$

The above result is applicable in the case of a diffusion on a compact manifold. Particularly, let X_t be a diffusion process on a compact manifold M of dimension m. To state the assumptions, we embed the manifold in $\mathbb{R}^d$, some $d \in \mathbb{N}$, and assume that the process is given as the solution to the stochastic differential equation

$$dX_t = b(X_t)dt + \bar{\sigma}(X_t)d\bar{W}_t, \qquad X_0 = x,$$

where $\bar{W}$ is independent of the observation noise W. It is assumed that the semigroup associated with X_t is strictly elliptic on M, and thus (as follows from [18, ch. 3]), given $t > 0$ there exist constants $0 < c_1 < c_2 < \infty$ such that

$$c_1 \lambda \preccurlyeq G_t(a, \cdot) \preccurlyeq c_2 \lambda, \qquad a \in \mathbb{S},$$

where λ is the surface measure on M. Arguments similar to the ones described above in the discrete setting, now based on Theorem 10.8, lead to the following [3].

Theorem 10.9 *Under the assumptions above on the process X_t, and assuming also that k of (10.26) is twice continuously differentiable, one has that $\mathbb{P}$-a.s., $\gamma(p, q) \leq -c$, for all $p, q \in \mathcal{P}$, where $c > 0$ is a deterministic constant.*

We now make some further remarks on small noise asymptotics. First, let us mention that the upper bound (10.14) continues to hold in a setting of countable state space; moreover, under suitable assumptions, a lower bound is also valid, that is different from (10.15) but sufficient to deduce that the order of magnitude of γ_σ is σ^{-2}. These facts were proved by other methods in [3, Sec. 5].

Next, in a continuous state space, the following example was studied in [3] (the proof is based on the estimate from Theorem 10.6, with $m = 2$).

Theorem 10.10 *Let $\mathbb{S} = [0, 1]$. Assume that $G(\alpha, d\beta) = \bar{G}(\alpha, \beta)\ell(d\beta)$, where ℓ is the Lebesgue measure and $\bar{G}$ is three times continuously differentiable on $\mathbb{S}^2$. Assume also that the observations are of the form (10.2) (from Example 10.2), and that k is C^4 on $\mathbb{S}$, while the derivative of k is bounded away from zero. Then $\mathbb{P}$-a.s., for every $p, q \in \mathcal{P}$,*

$$\limsup_{\sigma \to 0} \frac{\gamma(p, q)}{\log \frac{1}{\sigma}} \leq -1. \qquad (10.27)$$

A direct computation in the Gaussian case reveals the behavior $1/\sigma$ rather than $\log(1/\sigma)$. In view of this it is plausible that the above results should be possible to improve upon. A similar situation occurs in the case of a diffusion on $\mathbb{R}$, where [1] bounds the rate by $\log(1/\sigma)$, whereas an analogous analysis of the Kalman filter on $\mathbb{R}$ in continuous time [27] shows dependence of the form $1/\sigma$. Under some restricted assumptions, the behavior $1/\sigma$ is established in [4], but the question is open in any reasonable generality.

The techniques involving the Hilbert metric appear to work well in a variety of settings where the state space is compact, but other than in some trivial cases, they usually fail when the space is noncompact. An exception is the contribution [23], where such techniques are used in conjunction with very clever considerations to establish stability properties of the filter under mixing assumptions on the state process, which lies in $\mathbb{R}^d$. See also [21] for a refinement of this result.

Finally we would like to point out the usefulness of Hilbert metric techniques in treating a more general problem, namely the sensitivity of the filter to perturbations in the transition kernel as well as the initial condition. The result is borrowed from [12]. To this end, let us go back to the setting of Theorem 10.7. Namely, we assume that (10.25) holds for some $0 < c_1 < c_2 < \infty$ and $\lambda \in \mathcal{P}$. In addition, assume we are given a sequence G_m of probability kernels that approximate G in the following sense: G and G_m all admit transition probability densities, $g(\cdot, \cdot)$ with respect to λ, on $(\mathbb{S}, \mathcal{S})$; for every m, $g(\cdot, \cdot)$ and $g_m(\cdot, \cdot)$ are zero and positive on the same sets; and $\log g_m$ converge to $\log g$ on the set $\{(x, y) \in \mathbb{S}^2 : g(x, y) > 0\}$. We are also given $\pi_m \in \mathcal{P}$, converging in total variation to π_0. Denote by $\pi_n^{(m)}$ the filter that uses the initial data π_m and the transition kernel G_m, and as before, denote by π_n the exact filter. Furthermore, assume that the observation process is of the form $Y_n = k(X_n) + W_n$, where W_n are $\mathbb{R}^\ell$-valued, i.i.d., with a bounded density with respect to the Lebesgue measure. The proof of the following result is based solely on elementary properties of the Hilbert metric.

Theorem 10.11 *[12] Under the above assumptions one has*

$$\lim_{m \to \infty} \sup_{n \in \mathbb{N}} d_{TV}\left(\pi_n^{(m)}, \pi_n\right) = 0.$$

Notes

Research supported in part by the Israel Science Foundation (Grant 1349/08), and the Technion fund for promotion of research.

References

[1] R. Atar, Exponential stability for non-linear filtering of diffusion processes in a noncompact domain. *Ann. of Probab.*, 26 No. 4, 1552–74 (1998).

[2] R. Atar and O. Zeitouni, Lyapunov exponents for finite state nonlinear filtering. *Siam J. Contr. Opt.* 35 No. 1, 36–55 (1997).

[3] R. Atar and O. Zeitouni, Exponential stability for nonlinear filtering. *Annales de l'Institut H. Poincaré Probab. Statist.*, 33 No. 6, 697–725 (1997).

[4] R. Atar, F. Viens, and O. Zeitouni, Robustness of Zakai's equation via Feynman–Kac representations. *Stochastic analysis, control, optimization and applications: A volume in honor of W. H. Fleming*, eds W. M. McEneaney, G. Yin, and Q. Zhang, 339–52; Birkhauser (1999).

[5] P. Baxendale, P. Chigansky, and R. Liptser, Asymptotic stability of the Wonham filter: ergodic and nonergodic signals. *SIAM J. Control Optim.*, 43(2): 643–69 (electronic) (2004).

[6] A. G. Bhatt, A. Budhiraja, and R. L. Karandikar, Markov property and ergodicity of the nonlinear filter. *SIAM J. Control Optim.*, 39(3): 928–49 (electronic) (2000).

[7] G. Birkhoff, Extensions of Jentzsch's theorem. *Trans. Am. Math. Soc.*, 1957, Vol. 85, pp. 219–27.

[8] G. Birkhoff, *Lattice theory*. Am. Math. Soc. Publ. 25, 3rd edn, 1967.

[9] R. W. Brockett, Nonlinear systems and nonlinear estimation theory. In *Stochastic systems: The mathematics of filtering and identification and applications*, eds M. Hazwinkel and J. C. Willems, D. Reidel, Dordrecht, pp. 441–77, 1981.

[10] P. Bougerol, Théorèmes limites pour les systèmes linéaire à coefficients Markoviens. *Prob. Theory Rel. Fields*. 1988, Vol. 78, pp. 192–221.

[11] A. Budhiraja, Asymptotic stability, ergodicity and other asymptotic properties of the non-linear filter. *Ann. Inst. H. Poincaré Probab. Statist.*, 39(6): 919–41, 2003.

[12] A. Budhiraja and H. J. Kushner, Robustness of nonlinear filters over the infinite time interval. *SIAM J. Control Optim.*, 36, Issue 5, pp. 1618–37 (1998).

[13] A. Budhiraja and D. Ocone, Exponential stability of discrete-time filters for bounded observation noise. *Systems Control Lett.* 30 (1997), no. 4, 185–93.

[14] A. Budhiraja and D. Ocone, Exponential stability in discrete-time filtering for non-ergodic signals. *Stochastic Process. Appl.* 82 (1999), no. 2, 245–57.

[15] R. Carmona and J. Lacroix, *Spectral theory of random Schrödinger operators*, Birkhäuser, Zurich, 1990.

[16] P. Chigansky, R. Liptser, and R. Van Handel, Intrinsic methods in filter stability. Chapter 11 this volume.

[17] J. E. Cohen, H. Kesten, and C. M. Newman, Oseledec's multiplicative ergodic theorem: A proof. In *Random matrices and their applications*, eds J. E. Cohen, H. Kesten, C. M. Newman, Am. Math. Soc., Providence, pp. 23–30, 1986.
[18] Davies, E. B., *Heat kernels and spectral theory*, Cambridge University Press, 1989.
[19] P. Del Moral and A. Guionnet, On the stability of interacting processes with applications to filtering and genetic algorithms. *Ann. Inst. H. Poincaré Probab. Statist.*, 37(2): 155–94, 2001.
[20] B. Delyon and O. Zeitouni, Lyapunov exponents for filtering problems. In *Applied Stochastic Analysis*. eds M. H. A. Davis and R. J. Elliot, Gordon and Breach Science Publishers, London, pp. 511–21, 1991.
[21] R. Douc, G. Fort, E. Moulines, and P. Priouret, Forgetting of the initial distribution for hidden Markov models. To appear in *Stoch. Proc. Appl.*
[22] E. Hopf, An inequality for positive linear integral operators. *Journal of Math. and Mech.*, 1963, Vol. 12, No. 5, pp. 683–92.
[23] M. L. Kleptsyna and A. Yu. Veretennikov, On discrete time ergodic filters with wrong initial data. *Probab. Theory Related Fields* 141 (2008), nos 3–4, 411–44.
[24] H. Kunita, Asymptotic behavior of the nonlinear filtering errors of Markov processes. *J. Multivariate Anal.*, 1 (1971), pp. 365–93.
[25] H. Kunita, Ergodic properties of nonlinear filtering processes. In *Spatial Stochastic Processes*, eds K. S. Alexander and J. C. Watkins, Birkhguser, Boston, 1991.
[26] H. Kunita, *Stochastic flows and stochastic differential equations*. Cambridge Studies in Advanced Mathematics, 24. Cambridge University Press, 1997
[27] H. Kwakernaak and R. Sivan, *Linear optimal control systems*. Wiley-Interscience, 1972.
[28] F. Ledrappier, Quelques propriétés des exposants caractéristiques. *École d'été de probabilités de Saint-Flour*, XII—1982, 305–96, Lecture Notes in Math., 1097, Springer, Berlin, 1984.
[29] F. Le Gland and L. Mevel, Exponential forgetting and geometric ergodicity in hidden Markov models. *Math. Control Signals Systems*, 13(1): 63–93, 2000.
[30] F. Le Gland and N. Oudjane, Stability and uniform approximation of nonlinear filters using the Hilbert metric and application to particle filters. *Ann. Appl. Probab.* 14 (2004), no. 1, 144–87.
[31] F. Le Gland and N. Oudjane, A robustification approach to stability and to uniform particle approximation of nonlinear filters: The example of pseudo-mixing signals. *Stochastic Process. Appl.* 106 (2003), no. 2, 279–316.
[32] R. S. Liptser and A. N. Shiryayev, *Statistics of random processes*, Nauka, Moscow, 1974; English edn, Springer-Verlag, New York, 1977.
[33] D. Ocone, Entropy inequalities and entropy dynamics in nonlinear filtering of diffusion processes. In *Stochastic analysis, control, optimization and applications*, Systems Control Found. Appl., pp. 477–96. Birkhäuser, Boston, 1999.
[34] N. Oudjane and S. Rubenthaler, Stability and uniform particle approximation of nonlinear filters in case of non ergodic signals. *Stoch. Anal. Appl.* 23 (2005), no. 3, 421–48.
[35] D. Ocone and E. Pardoux, Asymptotic stability of the optimal filter with respect to its initial condition. *SIAM J. Control Optim.* 34 (1996), no. 1, 226–43.
[36] Y. Peres, Domains of analytic continuation for the top Lyapunov exponent. *Ann. Inst. Henri Poincaré*, 28 (1992), pp. 131–48.
[37] E. Seneta, *Non-negative matrices and Markov chains*, Springer-Verlag, 1981.
[38] L. Stettner, On Invariant Measures of Filtering Processes. *Stochastic Differential Systems, Proc. 4th Bad Honnef Conf.*, 1988, Leture Notes in Control and Inform. Sci. 126, eds Christopeit, N., Helmes, K. and Kohlmann, M., Springer, pp. 279–92, 1989.

[39] L. Stettner, Invariant Measures of the Pair: State, Approximate Filtering Process. *Colloq. Math.*, 1991, LXII, pp. 347–51.

[40] R. van Handel, The stability of conditional markov processes and markov chains in random environments. Ann. Probab. Volume 37, Number 5 (2009), 1876–1925.

[41] W. M. Wonham, Some applications of stochastic differential equations to optimal nonlinear filtering. *SIAM J. Control Opt.*, 2 (1965), pp. 347–68.

·11·
Intrinsic Methods in Filter Stability

P. Chigansky, R. Liptser, and R. Van Handel

11.1 Introduction

Consider a pair of random sequences $(X, Y) = (X_n, Y_n)_{n \in \mathbb{Z}_+}$, where the signal component X_n takes values in a Polish space[1] $\mathbb{S}$ and the observation component Y_n takes values in $\mathbb{R}^p$ for some $p \geq 1$. The classical filtering problem is to compute the conditional distribution

$$\pi_n(\cdot) = \mathbb{P}\left(X_n \in \cdot | \mathcal{F}^Y_{0,n}\right), \tag{11.1}$$

where $\mathcal{F}^Y_{k,n}$ stands for the σ-algebra of events generated by Y_m, $k \leq m \leq n$ (similarly, we will use below the σ-algebra $\mathcal{F}^X_{k,n}$ generated by X_m, $k \leq m \leq n$). Once π_n is found, the optimal mean square estimate of $f(X_n)$ can be calculated as

$$\mathbb{E}\left(f(X_n) | \mathcal{F}^Y_{0,n}\right) = \int f(x) \, \pi_n(dx)$$

for any function f with $\mathbb{E}|f(X_n)|^2 < \infty$. If both X and (X, Y) are Markov processes, π_n satisfies a recursive *filtering* equation. Specifically, let Λ and ν denote the transition probability and the initial distribution of X, i.e., for $A \in \mathcal{B}(\mathbb{S})$

$$\begin{aligned} \nu(A) &= \mathbb{P}(X_0 \in A), \\ \Lambda(X_{n-1}, A) &= \mathbb{P}\left(X_n \in A | \mathcal{F}^X_{0,n-1}\right) \quad \mathbb{P}\text{-a.s.,} \end{aligned} \tag{11.2}$$

and assume that Y is a sequence of conditionally independent random variables given $\mathcal{F}^X_{0,\infty} := \bigvee_{n \geq 0} \mathcal{F}^X_{0,n}$ with

$$\begin{aligned} \mathbb{P}(Y_0 = 0) &= 1, \\ \mathbb{P}\left(Y_n \in A | \mathcal{F}^X_{0,\infty}\right) &= \int_A g(X_n, y) \, \psi(dy), \quad n \geq 1 \end{aligned} \tag{11.3}$$

where $g(x, y)$ is a probability density with respect to the σ-finite measure ψ (the deterministic choice of Y_0 is only a matter of convenience; it means that all the information about X_0 is contained in its a priori distribution ν). For such a model π_n satisfies the recursive equation (see, e.g., Proposition 3.2.5 in [7])

$$\pi_n(dx) = \frac{g(x, Y_n) \int \Lambda(u, dx) \, \pi_{n-1}(du)}{\int g(x, Y_n) \int \Lambda(u, dx) \, \pi_{n-1}(du)}, \tag{11.4}$$

subject to $\pi_0 = \nu$. Suppose (11.4) can be solved starting from a probability distribution $\tilde{\nu}$ different from ν and denote by $\tilde{\pi}_n$ the resulting sequence of random measures. A typical question of stability is under which conditions (in terms of the model ingredients Λ, g, etc.), the distance between π_n and $\tilde{\pi}_n$ vanishes as $n \to \infty$:

$$\lim_{n\to\infty} \mathbb{E}\|\pi_n - \tilde{\pi}_n\| = 0, \qquad (11.5)$$

where $\|\cdot\|$ denotes the total variation norm.[2] If the filter (11.4) is started from two initial conditions $\bar{\nu}$ and $\tilde{\nu}$, both different from ν, the distance between the corresponding solutions satisfies $\|\bar{\pi}_n - \tilde{\pi}_n\| \le \|\bar{\pi}_n - \pi_n\| + \|\pi_n - \tilde{\pi}_n\|$ and it therefore suffices to consider the case where $\tilde{\nu} = \nu$ is the true initial distribution.

Depending on the way the filtering equation is thought of, different tools can be used to solve this problem. For example, (11.4) can be seen as the iteration of a positive random operator acting on nonnegative measures and consequently filter stability can be treated using the appropriate tools from the theory of positive operators, namely the Birkhoff contraction inequality for the Hilbert projective metric (see, e.g., [2], [5], [22, 23], [20]). Equation (11.4) can also be considered as a random dynamical system with a special projective structure, so that the stability problem can be related to the Lyapunov exponents of the bilinear Zakai equation for the unnormalized conditional law ([1], [6], [11, 10]). In the continuous-time case, when both signal and the observation processes are sufficiently regular diffusions, the filtering equation corresponds to certain stochastic PDE and can be analysed using PDE tools ([30, 31]). These approaches are reviewed elsewhere in this volume.

In contrast to the above techniques, which essentially study the filtering *recursion* (11.4), this Chapter aims to survey results which rely fundamentally on the probabilistic representation (11.1) of the filtering process π_n as a *conditional expectation*. As will shortly become evident, this "intrinsic" approach is particularly transparent when we impose the following condition:

$$\nu \ll \tilde{\nu}. \qquad (A)$$

Though this assumption can be weakened in certain cases, we will generally restrict ourselves to this setting in the following for sake of simplicity (further details on the relevance of this assumption can be found in Section 11.5).

To give the reader an idea about the methods we have in mind, let us begin our investigation of the filter stability problem assuming only (A). Remarkably, significant insight can be gained already at this level of generality. Let $\tilde{\mathbb{P}}$ be the probability measure on $\mathcal{F}$, such that under $\tilde{\mathbb{P}}$ the process (X, Y) has the same transition law as under $\mathbb{P}$, but $X_0 \sim \tilde{\nu}$. Then (A) implies that $\mathbb{P} \ll \tilde{\mathbb{P}}$ with

$$\frac{d\mathbb{P}}{d\tilde{\mathbb{P}}}(X, Y) = \frac{d\nu}{d\tilde{\nu}}(X_0) \qquad \tilde{\mathbb{P}}\text{-a.s.} \qquad (11.6)$$

The random measures $\pi_n(\cdot)$ and $\bar{\pi}_n(\cdot)$ obtained by the recursion (11.4) are regular versions of the conditional probabilities $\mathbb{P}\left(X_n \in \cdot | \mathcal{F}_{0,n}^Y\right)$ and $\bar{\mathbb{P}}\left(X_n \in \cdot | \mathcal{F}_{0,n}^Y\right)$, respectively. Since $\mathbb{P} \ll \bar{\mathbb{P}}$, $\bar{\pi}_n(\cdot)$ is well defined on a set of full $\mathbb{P}$-probability, which means that (11.4) can be solved subject to $\bar{\nu}$ when the actual observations are drawn from $\mathbb{P}$. Using (11.6) and the Bayes formula, we find that there is a set of full $\mathbb{P}$-probability on which for any bounded measurable f

$$\int_{\mathbb{S}} f(x) \, d\pi_n(x) = \mathbb{E}\left(f(X_n) | \mathcal{F}_{0,n}^Y\right) = \frac{\bar{\mathbb{E}}\left(f(X_n) \frac{dv}{d\bar{v}}(X_0) | \mathcal{F}_{0,n}^Y\right)}{\bar{\mathbb{E}}\left(\frac{dv}{d\bar{v}}(X_0) | \mathcal{F}_{0,n}^Y\right)}$$

$$= \bar{\mathbb{E}}\left(f(X_n) \frac{\bar{\mathbb{E}}\left(\frac{dv}{d\bar{v}}(X_0) | \mathcal{F}_{0,n}^Y \vee \sigma\{X_n\}\right)}{\bar{\mathbb{E}}\left(\frac{dv}{d\bar{v}}(X_0) | \mathcal{F}_{0,n}^Y\right)} \bigg| \mathcal{F}_{0,n}^Y\right)$$

$$= \int_{\mathbb{S}} f(x) \frac{\bar{\mathbb{E}}\left(\frac{dv}{d\bar{v}}(X_0) | \mathcal{F}_{0,n}^Y, X_n = x\right)}{\bar{\mathbb{E}}\left(\frac{dv}{d\bar{v}}(X_0) | \mathcal{F}_{0,n}^Y\right)} \, d\bar{\pi}_n(x)$$

(note that the denominator is strictly positive $\mathbb{P}$-a.s.) Thus evidently $\pi_n \ll \bar{\pi}_n$ $\mathbb{P}$-a.s. and the corresponding Radon–Nikodym derivative is given by ([13])

$$\frac{d\pi_n}{d\bar{\pi}_n}(x) = \frac{\bar{\mathbb{E}}\left(\frac{dv}{d\bar{v}}(X_0) | \mathcal{F}_{0,n}^Y, X_n = x\right)}{\bar{\mathbb{E}}\left(\frac{dv}{d\bar{v}}(X_0) | \mathcal{F}_{0,n}^Y\right)} \quad \mathbb{P}\text{-a.s.} \tag{11.7}$$

Therefore, we obtain $\mathbb{P}$-a.s.

$$\|\pi_n - \bar{\pi}_n\| = \int_{\mathbb{S}} \left| \frac{d\pi_n}{d\bar{\pi}_n}(x) - 1 \right| d\bar{\pi}_n(x)$$

$$= \frac{\bar{\mathbb{E}}\left(\left|\bar{\mathbb{E}}\left(\frac{dv}{d\bar{v}}(X_0) | \mathcal{F}_{0,n}^Y \vee \sigma\{X_n\}\right) - \bar{\mathbb{E}}\left(\frac{dv}{d\bar{v}}(X_0) | \mathcal{F}_{0,n}^Y\right)\right| \bigg| \mathcal{F}_{0,n}^Y\right)}{\bar{\mathbb{E}}\left(\frac{dv}{d\bar{v}}(X_0) | \mathcal{F}_{0,n}^Y\right)}. \tag{11.8}$$

By the Markov property of (X, Y), $\mathcal{F}_{0,n-1}^Y \vee \mathcal{F}_{0,n-1}^X$ and $\mathcal{F}_{n+1,\infty}^Y \vee \mathcal{F}_{n+1,\infty}^X$ are conditionally independent given $\sigma\{X_n, Y_n\}$, which implies that

$$\bar{\mathbb{E}}\left(\frac{dv}{d\bar{v}}(X_0) \bigg| \mathcal{F}_{0,n}^Y \vee \sigma\{X_n\}\right) = \bar{\mathbb{E}}\left(\frac{dv}{d\bar{v}}(X_0) \bigg| \mathcal{F}_{0,\infty}^Y \vee \mathcal{F}_{n,\infty}^X\right).$$

Combined with (11.8), this implies that

$$\mathbb{E}\|\pi_n - \bar{\pi}_n\| = \bar{\mathbb{E}}\left(\bar{\mathbb{E}}\left(\frac{dv}{d\bar{v}}(X_0)\bigg|\mathcal{F}_{0,n}^Y\right) \|\pi_n - \bar{\pi}_n\|\right)$$

$$= \bar{\mathbb{E}}\left|\bar{\mathbb{E}}\left(\frac{dv}{d\bar{v}}(X_0)\bigg|\mathcal{F}_{0,\infty}^Y \vee \mathcal{F}_{n,\infty}^X\right) - \bar{\mathbb{E}}\left(\frac{dv}{d\bar{v}}(X_0)\bigg|\mathcal{F}_{0,n}^Y\right)\right|. \tag{11.9}$$

The conditional expectations in the latter expression are nonnegative uniformly integrable martingales with respect to the decreasing and increasing filtrations $\mathcal{F}_{0,\infty}^Y \vee \mathcal{F}_{n,\infty}^X$ and $\mathcal{F}_{0,n}^Y$, respectively. Hence both converge in $L^1(\bar{\mathbb{P}})$, and thus

$$\lim_{n\to\infty} \mathbb{E}\|\pi_n - \tilde{\pi}_n\| = \bar{\mathbb{E}}\left|\mathbb{E}\left(\frac{d\nu}{d\tilde{\nu}}(X_0)\bigg|\bigcap_{n\geq 0}\mathcal{F}^Y_{0,\infty}\vee\mathcal{F}^X_{n,\infty}\right) - \bar{\mathbb{E}}\left(\frac{d\nu}{d\tilde{\nu}}(X_0)\bigg|\mathcal{F}^Y_{0,\infty}\right)\right|.$$

This suggests that the filter is stable if

$$\bar{\mathbb{E}}\left(\frac{d\nu}{d\tilde{\nu}}(X_0)\bigg|\bigcap_{n\geq 0}\mathcal{F}^Y_{0,\infty}\vee\mathcal{F}^X_{n,\infty}\right) = \bar{\mathbb{E}}\left(\frac{d\nu}{d\tilde{\nu}}(X_0)\bigg|\mathcal{F}^Y_{0,\infty}\right) \quad \bar{\mathbb{P}}\text{-a.s.} \tag{11.10}$$

In particular, under assumption (A), (11.5) holds if and only if (11.10) holds.

It is tempting to interchange the supremum and the intersection operations on the σ-algebras in the left-hand side of (11.10)

$$\bigcap_{n\geq 0}\mathcal{F}^Y_{0,\infty}\vee\mathcal{F}^X_{n,\infty} \stackrel{?}{=} \mathcal{F}^Y_{0,\infty}\vee\bigcap_{n\geq 0}\mathcal{F}^X_{n,\infty}, \tag{11.11}$$

as this would imply stability for a large class of signals, namely those with a.s. empty tail field $\bigcap_{n\geq 0}\mathcal{F}^X_{n,\infty}$. Unfortunately, the exchange of intersection and supremum need not be permitted if no further constraints on the model are imposed, as the illuminating example below shows (see [19], [16], [3, 8] for related discussions). This subtle problem was not recognized in the pioneering work of H. Kunita [21], where the relation (11.11) was taken for granted, and was subsequently inherited by a number of contributions that are based on [21]. The insight gained in [3] from the 'intrinsic' perspective on the filter stability problem revealed this as a serious gap in [21], which to date has not yet been completely resolved. The validity of (11.11) was recently verified in [36] under slightly stronger assumptions than imposed in [21]; see a sketch of the ideas in Section 11.3 below.

Example 11.1 Let X be a Markov chain on $\mathbb{S} = \{1, 2, 3, 4\}$ with transition matrix

$$\Lambda = \begin{pmatrix} 1/2 & 1/2 & 0 & 0 \\ 0 & 1/2 & 1/2 & 0 \\ 0 & 0 & 1/2 & 1/2 \\ 1/2 & 0 & 0 & 1/2 \end{pmatrix}.$$

Let $Y_n = 1_{\{X_n\in\{1,3\}\}}$, $n \geq 0$. If one observes Y and X_k for some $k \geq 0$, then the whole trajectory of X is revealed. Indeed, Y reveals exactly when the transitions of X occur, and the knowledge of a single value of X_k then pins down which one of the two possible trajectories of X occurs given the known transition times. Hence $\mathcal{F}^Y_{0,\infty}\vee\sigma\{X_k\} = \mathcal{F}^Y_{0,\infty}\vee\mathcal{F}^X_{0,\infty} = \mathcal{F}^Y_{0,\infty}\vee\mathcal{F}^X_{n,\infty}$ for any $n \geq 0$ and therefore

$$\bigcap_{n\geq 0}\mathcal{F}^Y_{0,\infty}\vee\mathcal{F}^X_{n,\infty} = \mathcal{F}^Y_{0,\infty}\vee\mathcal{F}^X_{0,\infty}. \tag{11.12}$$

Recall that a finite-state Markov chain is ergodic, i.e., irreducible and aperiodic, if and only if its transition matrix is primitive of order m (the entries of Λ^m are positive for some integer $m \geq 1$). An ergodic chain has almost surely trivial tail σ-algebra. The transition matrix defined above is primitive of order 3 and hence $\bigcap_{n\geq 0} \mathcal{F}^X_{n,\infty}$ is empty $\bar{\mathbb{P}}$-a.s. Therefore $\mathcal{F}^Y_{0,\infty} \vee \bigcap_{n\geq 0} \mathcal{F}^X_{n,\infty} = \mathcal{F}^Y_{0,\infty}$ $\bar{\mathbb{P}}$-a.s.

However, observing Y alone does not eliminate the uncertainty about X_0 (and thus about the whole trajectory of X):

$$\bar{\mathbb{P}}\left(X_0 = 0 | \mathcal{F}^Y_{0,\infty}\right) = \frac{\bar{\nu}(1)}{\bar{\nu}(1) + \bar{\nu}(3)} \mathbf{1}_{\{Y_0=1\}} \neq \mathbf{1}_{\{X_0=1\}},$$

which means that $\mathcal{F}^Y_{0,\infty}$ is strictly smaller than $\mathcal{F}^Y_{0,\infty} \vee \mathcal{F}^X_{0,\infty}$ and by (11.12) than $\bigcap_{n\geq 0} \mathcal{F}^Y_{0,\infty} \vee \mathcal{F}^X_{n,\infty}$. Therefore, both (11.10) and (11.11) fail. In this case, equation (11.4) shows that $\|\bar{\pi}_n - \pi_n\| \geq C$ for all $n \geq 0$, where C is a positive constant depending only on ν and $\bar{\nu}$. Thus the filter is not stable.

Evidently, contrary to intuition, ergodicity of the signal (i.e., triviality of the tail σ-field) alone may not be enough to guarantee filter stability. What additional ingredient is needed? We will consider several possibilities below, including:

- In Section 11.2, we will show that the filter is stable regardless of the observation structure if the signal possesses a strong mixing property. This assumption is stronger than ergodicity and holds, for example, if the signal has a uniformly positive transition density.
- In Section 11.3, we will show that ergodicity of the signal is already sufficient for filter stability if the observations are nondegenerate, i.e., $g(x, y) > 0$.
- In Section 11.4, we will show that the filter may be stable even when the signal is not ergodic provided that the observations are "good enough."

These results indicate that stability of the filter emerges as an interaction between the ergodic properties of the signal and the structure of the observations, a complete understanding of which is still lacking. For example, necessary and sufficient conditions for stability are unknown in the general setting, and the existing sufficient conditions are often difficult to verify in terms of the filtering model. Moreover, the difficult quantitative question of how the rate of stability of the filter is affected by the ergodic properties of the signal and the quality of the observations remains largely open. Despite the abundance of open questions, however, the various results reviewed in this paper indicate that significant insight can be obtained by employing an intrinsic analysis of the filter stability problem.

The remainder of the Chapter consists of four sections: Sections 11.2, 11.3, and 11.4 each describes a particular intrinsic argument for (11.5) (and sometimes its stronger/weaker forms) to hold, while Section 11.5 explores when (11.5) cannot hold.

In Section 11.2, inspired by (11.9), we explore the connection between the stability of the filter and *smoothing* conditional expectations of X_0 given $\mathcal{F}^Y_{0,\infty} \vee$

$\mathcal{F}_{n,\infty}^X$ under $\bar{\mathbb{P}}$. The main outcome is that the so called *mixing condition*, which is often imposed on the signal transition law by other methods, can be relaxed regardless of the observation noise density ([3], [8]).

Section 11.3 studies the (filtering) conditional distribution of X_n given $\mathcal{F}_{0,n}^Y$ as the marginal of the law induced on the space of signal trajectories by conditioning on $\mathcal{F}_{0,n}^Y$. The latter is well known to correspond to a time inhomogeneous Markov process on the signal state space, whose transition probability is controlled by the observation path. This fact places at our disposal a number of tools from the theory of Markov processes, including coupling ([18], [35]). If we condition on $\mathcal{F}_{0,\infty}^Y$ instead, the 'conditioned signal' approach can be used ([36]) to verify (11.11).

Section 11.4 deals with the stabilizing role of the observations. It turns out that estimates of particular functions are stable, in the sense that

$$\lim_{n\to\infty} \mathbb{E}|\pi_n(f) - \bar{\pi}_n(f)| = 0 \qquad (11.13)$$

for certain f, even in cases where (11.5) may fail. This is possible when f is *observable* in an appropriate sense [13], [9], [32, 33, 34]. When every function is observable, it follows that the filter is stable in the sense that (11.13) holds for all f. Remarkably, these results do not rely on any ergodic property of the signal as in the earlier sections, but emerge instead when the observations are "sufficiently informative."

Finally, we will show in section 11.5 that there are some inherent limitations to when (11.5) can hold. In particular, we will discuss how far the assumption (A) can be weakened, and we will argue that some form of absolute continuity is in fact necessary for the filter to forget its initial condition in the sense of (11.5).

A notable omission from this article is the pioneering approach to the filter stability problem, due to Ocone and Pardoux [27], who deduce stability of the nonlinear filter from the results of Kunita [21] by weak convergence arguments. Though this is very much an 'intrinsic' approach, it unfortunately appeals directly to the argument in [21] where (11.11) is taken for granted, and this gap is therefore inherited. Nonetheless this approach remains of significant interest, particularly as some of the machinery is of use in applications to Monte Carlo approximations of nonlinear filters (see [14]). The approach of Ocone and Pardoux, and its relation with the work of Kunita and the gap therein, is discussed elsewhere in this volume.

In order to keep the presentation as transparent as possible, we do not formulate the results in the most general form possible and only sketch the proofs, emphasizing the key ideas. We refer the reader to the original articles for the details of the proofs and for the (important!) technicalities.

11.2 Stability via smoothing

Formula (11.8) suggests that the filter is stable only if the conditional expectation of X_0 given $\mathcal{F}_{0,n}^Y \vee \sigma\{X_n\}$ ceases to depend on X_n as $n \to \infty$. The *smoothing* problem of computing the conditional distribution of X_0 given $\mathcal{F}_{0,n}^Y \vee \sigma\{X_n\}$ leads to a linear equation, whose long time behavior can be efficiently studied for strongly mixing signals.

Consider the signal/observation model (11.2) and (11.3), where the signal transition probability $\Lambda(u, dy)$ is assumed to have a density $\lambda(x, y)$ with respect to some σ-finite measure $\varphi(dy)$, i.e.,

$$\Lambda(x, dy) = \lambda(x, y)\varphi(dy) \qquad \forall x \in \mathbb{S}. \tag{11.14}$$

Suppose $\bar{\nu}$ has a density with respect to φ. Then the regular conditional probability $\bar{\mathbb{P}}\left(X_0 \in \cdot | \mathcal{F}_{0,n}^Y \vee \sigma\{X_n\}\right)$ also has a density $q_n(u; x)$:

$$\bar{\mathbb{P}}\left(X_0 \in A | \mathcal{F}_{0,n}^Y \vee \sigma\{X_n\}\right) = \int_A q_n(u; X_n)\varphi(du) \qquad \forall A \in \mathcal{B}(\mathbb{S}) \qquad \bar{\mathbb{P}}\text{-a.s.}$$

A simple calculation shows that q_n satisfies the recursion (see Lemma 3.1 in [8])

$$\begin{aligned} q_1(u; x) &= \frac{\lambda(u, x)\frac{d\bar{\nu}}{d\varphi}(u)}{\int_{\mathbb{S}} \lambda(v, x)\bar{\nu}(dv)}, \\ q_n(u; x) &= \frac{\int_{\mathbb{S}} \lambda(z, x)q_{n-1}(u; z)\bar{\pi}_{n-1}(dz)}{\int_{\mathbb{S}} \lambda(v, x)\bar{\pi}_{n-1}(dv)}, \qquad n \geq 2. \end{aligned} \tag{11.15}$$

Define

$$\hat{q}_n(u) = \sup_{x \in \mathbb{S}} q_n(u; x), \qquad \check{q}_n(u) = \inf_{x \in \mathbb{S}} q_n(u; x).$$

Our goal is to show that $\Delta_n(u) := \hat{q}_n(u) - \check{q}_n(u)$ converges to zero as $n \to \infty$, i.e., that $q_n(u; x)$ ceases to depend on its second argument. Let

$$a_{n-1}(u; z) := \frac{\hat{q}_{n-1}(u) - q_{n-1}(u; z)}{\hat{q}_{n-1}(u) - \check{q}_{n-1}(u)}$$

(with the convention $0/0 = 0$). Then by (11.15), for any $x, x' \in \mathbb{S}$ and $n \geq 2$,

$$q_n(u; x) - q_n(u; x') = \Delta_{n-1}(u)\left(1 - \int_{\mathbb{S}}\left\{\frac{\lambda(z, x)}{\int_{\mathbb{S}} \lambda(v, x)\bar{\pi}_{n-1}(dv)}a_{n-1}(u; z) \right.\right.$$
$$\left.\left. + \frac{\lambda(z, x')}{\int_{\mathbb{S}} \lambda(v, x')\bar{\pi}_{n-1}(dv)}(1 - a_{n-1}(u; z))\right\}\bar{\pi}_{n-1}(dz)\right).$$

Assume that the transition density is uniformly bounded, i.e. $\lambda(x, u) \leq \lambda^* < \infty$ for some constant λ^*. Since $a_n \in [0, 1]$,

$$q_n(u; x) - q_n(u; x') \leq \Delta_{n-1}(u)\left(1 - \frac{1}{\lambda^*}\int_{\mathbb{S}}\{\lambda(z, x) \wedge \lambda(z, x')\}\bar{\pi}_{n-1}(dz)\right),$$

and by the arbitrariness of x and x',

$$\Delta_n(u) \leq \Delta_{n-1}(u)\left(1 - \frac{1}{\lambda^*}\int_{\mathbb{S}} \inf_{x \in \mathbb{S}} \lambda(z, x)\,\bar{\pi}_{n-1}(dz)\right). \tag{11.16}$$

Notice that

$$\bar{\mathbb{E}}\left(\frac{dv}{d\bar{v}}(X_0)\Big|\mathcal{F}_{0,n}^Y \vee \sigma\{X_n\}\right) = \int_{\mathbb{S}} \frac{dv}{d\bar{v}}(u)\,q_n(u;X_n)\,\varphi(du)$$

$$\bar{\mathbb{E}}\left(\frac{dv}{d\bar{v}}(X_0)\Big|\mathcal{F}_{0,n}^Y\right) = \int_{\mathbb{S}}\int_{\mathbb{S}} \frac{dv}{d\bar{v}}(u)\,q_n(u;x)\,\varphi(du)\,\bar{\pi}_n(dx)$$

and assume $\frac{dv}{d\bar{v}}(u) \geq \varepsilon > 0$ for a constant $\varepsilon > 0$. Then by (11.8)

$$\|\pi_n - \bar{\pi}_n\| = \frac{\int_{\mathbb{S}}\left|\int_{\mathbb{S}}\int_{\mathbb{S}} \frac{dv}{d\bar{v}}(u)\,(q_n(u;x') - q_n(u;x))\,\varphi(du)\bar{\pi}_n(dx)\right|\bar{\pi}_n(dx')}{\bar{\mathbb{E}}\left(\frac{dv}{d\bar{v}}(X_0)\big|\mathcal{F}_{0,n}^Y\right)}$$

$$\leq \frac{1}{\varepsilon}\int_{\mathbb{S}}\int_{\mathbb{S}}\int_{\mathbb{S}} \frac{dv}{d\bar{v}}(u)\,|q_n(u;x') - q_n(u;x)|\,\varphi(du)\,\bar{\pi}_n(dx)\,\bar{\pi}_n(dx')$$

$$\leq \frac{1}{\varepsilon}\int_{\mathbb{S}} \frac{dv}{d\bar{v}}(u)\,\Delta_n(u)\,\varphi(du), \tag{11.17}$$

and

$$\mathbb{E}\|\pi_n - \bar{\pi}_n\| \leq \int_{\mathbb{S}} \frac{dv}{d\bar{v}}(u)\,\bar{\mathbb{E}}\Delta_n(u)\,\varphi(du). \tag{11.18}$$

Now, if a constant $\lambda^* > 0$ can be found such that $\lambda(x, u) \geq \lambda_*$ for all $x, u \in \mathbb{S}$, then by the first equation in (11.15)

$$\int_{\mathbb{S}} \frac{dv}{d\bar{v}}(u)\,\Delta_1(u)\,\varphi(du) \leq \int_{\mathbb{S}} \frac{dv}{d\bar{v}}(u) \sup_{x \in \mathbb{S}} q_1(u;x)\,\varphi(du)$$

$$\leq \int_{\mathbb{S}} \frac{dv}{d\bar{v}}(u) \sup_{x \in \mathbb{S}} \frac{\lambda(u, x)\frac{d\bar{v}}{d\varphi}(u)}{\int_{\mathbb{S}} \lambda(v, x)\,\bar{v}(dv)}\,\varphi(du) \leq \lambda^*/\lambda_*.$$

The latter and (11.16)-(11.18) give the the following bounds

Theorem 11.1 *Assume*

$$0 < \lambda_* < \lambda(x, u) \leq \lambda^* < \infty, \tag{11.19}$$

and $v \ll \bar{v}$, then

$$\mathbb{E}\|\pi_n - \bar{\pi}_n\| \leq \frac{\lambda^*}{\lambda_*}\left(1 - \frac{\lambda_*}{\lambda^*}\right)^{n-1}. \tag{11.20}$$

If in addition, $\frac{dv}{d\bar{v}}(x) \geq \varepsilon > 0$ with a constant $\varepsilon > 0$, then

$$\|\pi_n - \bar{\pi}_n\| \leq \frac{1}{\varepsilon}\frac{\lambda^*}{\lambda_*}\left(1 - \frac{\lambda_*}{\lambda^*}\right)^{n-1} \quad \mathbb{P}\text{-a.s.} \tag{11.21}$$

Condition (11.19) forces the transition density of the signal to be bounded away from zero uniformly over $\mathbb{S}$. This can be somewhat relaxed, at the expense of giving up the time uniformity of the bound (11.21). Suppose X is an aperiodic irreducible Markov chain with the unique invariant measure μ and that

$$\lambda_\circ := \int_\mathbb{S} \inf_{x \in \mathbb{S}} \lambda(u, x)\, \mu(du) > 0. \qquad (11.22)$$

In this case $\mu(dx)$ has a density $m(x)$ with respect to φ, satisfying $\lambda_\circ \leq m(u) \leq \lambda^*$. If we assume that $\frac{d\bar{\nu}}{d\varphi}(x) \geq \varepsilon > 0$, then by (11.15)

$$\int_\mathbb{S} \frac{d\nu}{d\bar{\nu}}(u)\, \Delta_1(u)\, \varphi(du) \leq \int_\mathbb{S} \frac{d\nu}{d\bar{\nu}}(u) \sup_{x \in \mathbb{S}} q_1(u; x)\, \varphi(du)$$

$$\leq \frac{(\lambda^*)^2}{\varepsilon} \int_\mathbb{S} \frac{d\nu}{d\bar{\nu}}(u) \left(\int_\mathbb{S} \inf_{x \in \mathbb{S}} \lambda(v, x) m(v) \varphi(dv) \right)^{-1} \bar{\nu}(du) \leq \frac{(\lambda^*)^2}{\varepsilon \lambda_\circ}.$$

This, combined with (11.16) and (11.17), gives

$$\|\pi_n - \bar{\pi}_n\| \leq \frac{(\lambda^*)^2}{\varepsilon^2 \lambda_\circ} \prod_{m=1}^{n-1} \left(1 - \frac{1}{\lambda^*} \int_\mathbb{S} \inf_{x \in \mathbb{S}} \lambda(z, x)\, \bar{\pi}_m(dz) \right). \qquad (11.23)$$

Finally, (11.22) implies (Theorem 11.1 in [8]) that the chain X is geometrically ergodic (i.e., its marginal distribution converges to μ in total variation geometrically fast), and that $\bar{\pi}_n$ satisfies the law of large numbers under $\bar{\mathbb{P}}$ (Theorem 11.2 in [8]):

$$\lim_{n \to \infty} \frac{1}{n} \sum_{m=1}^{n-1} \int_\mathbb{S} \inf_{x \in \mathbb{S}} \lambda(u, x)\, \bar{\pi}_m(du) = \int_\mathbb{S} \inf_{x \in \mathbb{S}} \lambda(u, x)\, \mu(du) \qquad \bar{\mathbb{P}}\text{-a.s.}$$

Since $\mathbb{P} \ll \bar{\mathbb{P}}$ the latter convergence holds $\mathbb{P}$-a.s. as well and (11.23) gives the following asymptotic bound.

Theorem 11.2 (essentially Theorem 11.1 in [8]). *Assume that $\frac{d\nu}{d\varphi}$ and $\frac{d\bar{\nu}}{d\varphi}$ are bounded away from zero and infinity uniformly over $\mathbb{S}$. Suppose that X is irreducible and aperiodic with the unique invariant measure μ, satisfying the following conditions*

$$\lambda(x, u) \leq \lambda^* < \infty$$
$$\lambda_\circ := \int_\mathbb{S} \inf_{x \in \mathbb{S}} \lambda(u, x)\, \mu(du) > 0. \qquad (11.24)$$

Then

$$\overline{\lim_{n \to \infty}} \frac{1}{n} \log \|\pi_n - \bar{\pi}_n\| \leq -\frac{\lambda_\circ}{\lambda^*} \qquad \mathbb{P}\text{-a.s.}$$

Remark 11.1 The statement of Theorem 11.2 remains true under weaker assumptions on $\frac{d\nu}{d\bar{\nu}}$ (see [8]) since only an asymptotic bound is obtained.

Condition (11.24) is significantly weaker than (11.19): for example, in the case of an ergodic chain X which takes a finite number of values, (11.19) requires that all the entries of the transition matrix are positive, while (11.24) holds when there is at least one row with strictly positive entries. For instance, (11.24) holds true and the filter becomes stable, if the transition matrix of the signal chain from Example 11.1 is perturbed with an $\varepsilon \in (0, 1)$ in a single row:

$$\Lambda = \begin{pmatrix} 1/2(1-\varepsilon) & 1/2(1-\varepsilon) & \varepsilon/2 & \varepsilon/2 \\ 0 & 1/2 & 1/2 & 0 \\ 0 & 0 & 1/2 & 1/2 \\ 1/2 & 0 & 0 & 1/2 \end{pmatrix}. \tag{11.25}$$

However, both condition (11.19) and condition (11.24) imply that X has a strong mixing property with geometric rate.

The more serious drawback of both types of mixing, (11.19) and (11.24), is that they are not well suited to noncompact $\mathbb{S}$. For instance, condition (11.19) is equivalent to requiring that for any $x_1, x_2 \in \mathbb{S}$, $\Lambda(x_1, \cdot)$ has a density with respect to $\Lambda(x_2, \cdot)$, which is uniformly bounded from zero and infinity over the pairs (x_1, x_2). Indeed, if the latter is true, one can just take $\varphi(\cdot) = \Lambda(x, \cdot)$ for an $x \in \mathbb{S}$. Conversely, (11.19) means that there exists a σ-finite measure φ such that

$$\lambda_* \varphi(A) \leq \Lambda(x, A) \leq \lambda^* \varphi(A)$$

for all measurable A and all $x \in \mathbb{S}$. If such measure exists then for any $x_1, x_2 \in \mathbb{S}$

$$(\lambda_*/\lambda^*) \Lambda(x_2, A) \leq \Lambda(x_1, A) \leq (\lambda^*/\lambda_*) \Lambda(x_2, A),$$

and hence

$$\lambda_*/\lambda^* \leq \frac{d\Lambda(x_1, \cdot)}{d\Lambda(x_2, \cdot)} \leq \lambda^*/\lambda_* \quad \forall x_1, x_2 \in \mathbb{S}$$

The latter property is relatively easy to check when $\mathbb{S}$ is compact. For noncompact $\mathbb{S}$, such a choice of φ is sometimes impossible.

Example 11.2 (taken from [23]) Let $\mathbb{S} = \mathbb{R}$. Suppose X is generated by the recursion $X_n = h(X_{n-1}) + Z_n$, where h is a bounded function and Z is a sequence of i.i.d. random variables independent of X. Assume Z_1 has a Laplacian distribution, i.e.,

$$\mathbb{P}(Z_1 \leq x) = \int_{-\infty}^{x} \frac{1}{2} e^{-|u|} du, \quad x \in \mathbb{R}.$$

The Lebesgue measure is obviously not the right choice for φ, since the corresponding transition density $\lambda(x, u) = \frac{1}{2} e^{-|u-h(x)|}$ violates the lower bound. However if one chooses $\varphi(du) = \frac{1}{2} e^{-|u|} du$, the density

$$\lambda(x, u) = e^{-|u-h(x)|+|u|}$$

is bounded between $\lambda_* := e^{-\|h\|_\infty}$ and $\lambda^* := e^{\|h\|_\infty}$ and (11.19) becomes applicable. If, however, Z_1 has standard Gaussian distribution, there is no φ which would guarantee (11.19), since

$$\frac{d\Lambda(x_1,\cdot)}{d\Lambda(x_2,\cdot)}(u) = \frac{e^{-(u-h(x_1))^2/2}}{e^{-(u-h(x_2))^2/2}} = e^{(h(x_1)-h(x_2))u-\frac{1}{2}(h^2(x_1)-h^2(x_2))}$$

is not bounded for $h(x_1) \neq h(x_2)$.

Theorem 11.1 can be proved using different methods, including those based on the Birkhoff (as in [2]) or Dobrushin (as in [15]) contraction inequalities. However, we are not aware of an alternative proof of the result stated in Theorem 11.2 and to the best of our knowledge, condition (11.24) is the weakest known ergodic property of the signal which implies filter stability *without* further constraints on the observation model (note, in particular, that no assumptions are needed on the observation density $g(x, y)$ in order for Theorems 11.1 and 11.2 to hold).

11.2.1 Continuous time

Equation (11.8) remains valid when the time parameter is continuous, and thus the approach based on analysis of the smoothing equation is applicable to continuous-time models as well.

Let $X = (X_t)_{t \in \mathbb{R}_+}$ be a Markov chain with values in $\mathbb{S} = \{a_1, \ldots, a_d\}$, transition rates λ_{ij} and initial distribution ν. The real-valued observation process is given by

$$Y_t = \int_0^t h(X_s)ds + \sigma B_t, \quad t \in \mathbb{R}_+, \tag{11.26}$$

with $h : \mathbb{S} \mapsto \mathbb{R}$, $\sigma > 0$ is a constant (noise intensity), and B is a Brownian motion independent of X. The nonlinear filter in this case is finite dimensional and the vector of the conditional probabilities $\pi_t(i) = \mathbb{P}\left(X_t = a_i | \mathcal{F}_t^Y\right)$ solves the Shiryaev–Wonham stochastic differential equation (see Chapter 9 in [24])

$$d\pi_t = \Lambda^\mathsf{T} \pi_t dt + \sigma^{-2} \left(\text{diag}(\pi_t) - \pi_t \pi_t^\mathsf{T}\right) h(dY_t - h^\mathsf{T} \pi_t dt), \quad \pi_0 = \nu, \tag{11.27}$$

where $\text{diag}(x)$, $x \in \mathbb{R}^d$ stands for the diagonal matrix with entries x_i, Λ is the matrix of transition rates, and h is a vector[3] with entries $h(a_i)$, $i = 1, \ldots, d$. Denote by $\bar{\pi}_t$ the strong solution of (11.27) subject to $\bar{\nu} \in \mathcal{P}(\mathbb{S}) = \mathcal{S}^{d-1}$.

Recall that X is ergodic, i.e., irreducible and aperiodic, if and only if the matrix exponential $\exp(\Lambda)$ has strictly positive entries, or, equivalently, if all the entries of Λ communicate. An ergodic chain has a unique invariant measure μ and $\lim_{t\to\infty} \mathbb{P}(X_t = a_i) = \mu_i > 0$, $i = 1, \ldots, d$.

Theorem 11.3 *Assume that X is ergodic and $\sigma > 0$. Then for any $v, \bar{v} \in \mathcal{S}^{d-1}$ the following stability properties hold:*

$$\gamma := \lim_{t \to \infty} \frac{1}{t} \log \|\pi_t - \bar{\pi}_t\| < 0 \qquad \mathbb{P}\text{-a.s.}, \qquad (11.28)$$

$$\lim_{t \to \infty} \frac{1}{t} \log \|\pi_t - \bar{\pi}_t\| \leq -\sum_{i=1}^{d} \mu_i \min_{j \neq i} \lambda_{ij} \qquad \mathbb{P}\text{-a.s.}, \qquad (11.29)$$

and

$$\|\pi_t - \bar{\pi}_t\| \leq C \exp\{-2t \min_{i \neq j} \sqrt{\lambda_{ij}\lambda_{ji}}\}, \qquad (11.30)$$

where $C := 2 \wedge \max_k \left(\frac{1}{v_k} \vee \frac{1}{\bar{v}_k}\right) \|v - \bar{v}\|$.

Inequality (11.28) appeared in Theorem 11.5 [3] and its proof uses the Birkhoff contraction inequality following [1]. It says that the filter is actually exponentially stable if the observation noise is nondegenerate and the signal is ergodic. If $\sigma = 0$, then the filtering equation looks different from (11.27) and can be unstable as in the Example 11.1 (see Section 11.3 in [3]). The existence of the limit in (11.28) and (11.29) follows from Oseledec's Multiplicative Ergodic Theorem (see [1]). Both (11.29) and the time uniform bound (11.30) were derived in [3] (Theorems 4.2 and 4.3) by the same arguments used in the proof of Theorem (11.2) above (the asymptotic version of (11.30) appeared before in [1]). Notice that (11.29) remains nontrivial as long as Λ has at least one row with nonzero entries, unlike (11.30) which requires that none of the transition rates vanish. The particular value of the constant in (11.30), taken from Proposition 3.5 [12] and Corollary 2.3.2 in [35], makes precise the dependence on the initial conditions. Other bounds, which shed light on the dependence of γ on the noise intensity σ, etc., can be found in [16], [1], [10], [11].

11.3 Conditioned signal

11.3.1 Finite horizon conditioning

The ideas outlined in this section are based on the following simple consequence of the Markov property of (X, Y):

$$\mathbb{P}\left(X_m \in A | \mathcal{F}_{0,m-1}^X \vee \mathcal{F}_{0,n}^Y\right) = \mathbb{P}\left(X_m \in A | \sigma\{X_{m-1}\} \vee \mathcal{F}_{0,n}^Y\right) \qquad \mathbb{P}\text{-a.s.} \quad (11.31)$$

For simplicity of notation, we shall consider hereafter the coordinate processes (X, Y) on the canonical space $(\Omega, \mathcal{F})$ (i.e., Ω is the space of semi-infinite sequences of points in $\mathbb{S} \times \mathbb{R}^p$). Denote by $\mathbb{P}_n^Y$ the regular conditional probability measure, induced on the restriction of the signal paths to the time interval $[0, n]$ by conditioning on $\mathcal{F}_{0,n}^Y$:

$$\mathbb{P}_n^Y(\Gamma) = \mathbb{P}\left((X_0, \ldots, X_n) \in \Gamma \big| \mathcal{F}_{0,n}^Y\right), \qquad \Gamma \in \mathcal{B}(\mathbb{S}^{n+1}).$$

Then π_n is nothing but the law of X_n under $\mathbb{P}_n^Y$:

$$\pi_n(f) = \mathbb{E}\left(f(X_n)\big|\mathcal{F}_{0,n}^Y\right) = \mathbb{E}_n^Y f(X_n),$$

where $\mathbb{E}_n^Y$ stands for the expectation with respect to $\mathbb{P}_n^Y$.

The property (11.31) means that the coordinate process X under $\mathbb{P}_n^Y$, referred to hereafter as the *conditioned signal*, is Markov:

$$\mathbb{P}_n^Y\left(X_m \in A \big| \mathcal{F}_{0,m-1}^X\right) = \mathbb{P}_n^Y(X_m \in A | \sigma\{X_{m-1}\}) \qquad \mathbb{P}_n^Y\text{-a.s.}$$

and a simple calculation should convince the reader that for the model (11.2) with (11.14) and (11.3), the transition probability of X under $\mathbb{P}_n^Y$

$$\Lambda_{m|n}^Y(X_{m-1}, A) := \mathbb{P}_n^Y(X_m \in A | \sigma\{X_{m-1}\}), \qquad A \in \mathcal{B}(\mathbb{S}),$$

has a density $\lambda_{m|n}^Y(u, x)$ with respect to $\varphi(dx)$ satisfying the following backward recursion (see, e.g., Proposition 3.3.2 in [7]):

$$\begin{aligned}
\lambda_{m|n}^Y(u, x) &= \frac{\lambda(u, x) g(x, Y_m) Q_m(x)}{\int_{\mathbb{S}} \lambda(u, v) g(v, Y_m) Q_m(v) \varphi(dv)}, \quad m = 1, \ldots, n \\
Q_m(x) &= \int_{\mathbb{S}} \lambda(x, z) g(z, Y_{m+1}) Q_{m+1}(z) \varphi(dz) \\
Q_n(x) &\equiv 1.
\end{aligned} \tag{11.32}$$

The conditioned signal X is a time inhomogeneous Markov process whose transition density at time m depends on $\mathcal{F}_{m,n}^Y$, i.e., on the future of the observed path. Notice that $\lambda_{m|n}^Y(u, x)$ is independent of the initial distribution ν of X_0, while the law of X_0 under $\mathbb{P}_n^Y$ depends on all the ingredients of the model, including ν (and the whole observation path). Let us also stress that $\mathbb{P}_n^Y$ is not the restriction of $\mathbb{P}_{n+1}^Y$ to the first n coordinates, or in other words, increasing the time horizon changes the conditional measure completely. This should not come as a surprise, since when Y_{n+1} is observed the conditional law of the whole $X_0, \ldots, X_n$ changes.

As before we shall rely on the auxiliary probability measure $\bar{\mathbb{P}}$, under which (X, Y) has the same law as under $\mathbb{P}$ but $X_0 \sim \bar{\nu}$. Conditioning on $\mathcal{F}_{0,n}^Y$ under $\bar{\mathbb{P}}$ induces a regular probability measure on the signal paths restricted to $[0, n]$ which will be denoted as $\bar{\mathbb{P}}_n^Y$:

$$\bar{\mathbb{P}}_n^Y(\Gamma) = \bar{\mathbb{P}}\left((X_0, \ldots, X_n) \in \Gamma \big| \mathcal{F}_{0,n}^Y\right), \qquad \Gamma \in \mathcal{B}(\mathbb{S}^{n+1}).$$

Clearly, the conditioned signal has the same transition law under $\mathbb{P}_n^Y$ and $\bar{\mathbb{P}}_n^Y$, but different initial distributions.

As was mentioned above, the filtering conditional distribution is nothing but the restriction of $\mathbb{P}_n^Y$ to the last coordinate and hence

$$\|\pi_n - \bar{\pi}_n\| = \left\|\mathbb{P}_n^Y(X_n \in \cdot) - \bar{\mathbb{P}}_n^Y(X_n \in \cdot)\right\|. \tag{11.33}$$

This interpretation relates the filter stability problem to the mixing properties of the conditioned signal, which in turn places the ergodic theory of Markov processes at our disposal. In particular, the following fact is well known (see e.g. [25]).

Proposition 11.1 *Suppose that $\xi = (\xi_n)_{n\geq 0}$ is an inhomogeneous Markov chain with values in a Polish space $\mathbb{S}$ and transition probabilities $K_n(x, \cdot)$ under the probability measures $\mathbb{P}$ and $\bar{\mathbb{P}}$, such that ξ_0 has distribution ν under $\mathbb{P}$ and $\bar{\nu}$ under $\bar{\mathbb{P}}$. Assume that there exists a sequence of σ-finite measures μ_n such that*

$$\varepsilon \mu_n(A) \leq K_n(x, A) \leq \frac{1}{\varepsilon} \mu_n(A) \qquad \forall A \in \mathcal{B}(\mathbb{S}), \tag{11.34}$$

for some fixed $\varepsilon > 0$. Then

$$\|\mathbb{P}(\xi_n \in \cdot) - \bar{\mathbb{P}}(\xi_n \in \cdot)\| \leq 2(1-\varepsilon)^n, \qquad n \geq 0. \tag{11.35}$$

If the mixing condition (11.19) is satisfied, (11.32) implies

$$\Lambda^Y_{m|n}(u, A) \geq \frac{\lambda_*}{\lambda^*} \frac{\int_A g(x, Y_m) Q_m(x) \varphi(dx)}{\int_\mathbb{S} g(v, Y_m) Q_m(v) \varphi(dv)} =: \frac{\lambda_*}{\lambda^*} \mu_n(A),$$

and, similarly, $\Lambda^Y_{m|n}(u, A) \leq (\lambda^*/\lambda_*) \mu_n(A)$ for any $A \in \mathcal{B}(\mathbb{S})$. Thus (11.34) holds with $\varepsilon := \lambda_*/\lambda^*$ and (11.33) recovers the statement of Theorem 11.1 (even without the assumption $\nu \ll \bar{\nu}$, as long as $\bar{\pi}_n$ is well defined).

Proposition 11.1 can be verified, e.g., by constructing an appropriate coupling using Nummelin's splitting technique ([26]). In the filtering context under consideration, this coupling method can be pushed further to get finer results which go beyond signals with compact state space (see [18]). In particular, the filter can be shown to be stable for linear models which are driven by noises with unimodal probability densities (see Section 11.5.2 [18]). However, the essential limitation of this approach stems from the time inhomogeneity of the conditioned signal. Unfortunately, the ergodic theory of inhomogeneous Markov processes is not as rich as in the homogeneous case (however, this drawback can be mitigated by conditioning on the infinite-time horizon as in the following section).

The property (11.31) remains valid when the time parameter is continuous and, with some caution, the conditioned signal measure can be explicitly constructed. For example, for finite state signals, the conditioned measure corresponds to a finite state Markov chain with time-varying rates which depend on the observation trajectory. Consequently, the bound (11.30) can be derived via coupling of the conditioned chain (see Section 2.3.2 in [35]).

Another variation on the same theme, which combines both the formula (11.8) and the conditioned signal representation, is to look at the conditioned signal backwards in time, i.e., to consider the process $\tilde{X}_m := X_{n-m}$, $m = 0, \ldots, n$. Note that as X_n is a Markov process, $\tilde{X}_m$ is also Markov. Now suppose that for every $x \in \mathbb{S}$, we can construct a stochastic process $\tilde{X}_n(x)$ on the same probability space such that the law of $(\tilde{X}_m(x))_{m \leq n}$ under $\bar{\mathbb{E}}_n^Y$ coincides with the law of $(\tilde{X}_m)_{m \leq n}$ under $\bar{\mathbb{E}}_n^Y(\cdot | \tilde{X}_0 = x)$. This point of view is particularly fruitful when the signal process is obtained from a stochastic differential equation, so that $\tilde{X}_m(x)$ can be obtained from the stochastic flow generated by this equation. In this setting

$$\bar{\mathbb{E}}\left(\frac{d\nu}{d\bar{\nu}}(X_0) \middle| \mathcal{F}_{0,n}^Y, X_n = x\right) = \bar{\mathbb{E}}_n^Y\left(\frac{d\nu}{d\bar{\nu}}(\tilde{X}_n) \middle| \tilde{X}_0 = x\right) = \bar{\mathbb{E}}_n^Y\left\{\frac{d\nu}{d\bar{\nu}}(\tilde{X}_n(x))\right\},$$

In these terms (11.8) reads:

$$\|\pi_n - \bar{\pi}_n\| = \frac{\int_\mathbb{S} \left|\bar{\mathbb{E}}_n^Y\left\{\frac{d\nu}{d\bar{\nu}}(\tilde{X}_n(y))\right\} - \int_\mathbb{S} \bar{\mathbb{E}}_n^Y\left\{\frac{d\nu}{d\bar{\nu}}(\tilde{X}_n(x))\right\} \bar{\pi}_n(dx)\right| \bar{\pi}_n(dy)}{\int_\mathbb{S} \bar{\mathbb{E}}_n^Y\left\{\frac{d\nu}{d\bar{\nu}}(\tilde{X}_n(x))\right\} \bar{\pi}_n(dx)}$$

$$\leq \frac{\int_\mathbb{S} \int_\mathbb{S} \bar{\mathbb{E}}_n^Y\left|\frac{d\nu}{d\bar{\nu}}(\tilde{X}_n(y)) - \frac{d\nu}{d\bar{\nu}}(\tilde{X}_n(x))\right| \bar{\pi}_n(dy) \bar{\pi}_n(dx)}{\int_\mathbb{S} \bar{\mathbb{E}}_n^Y\left\{\frac{d\nu}{d\bar{\nu}}(\tilde{X}_n(x))\right\} \bar{\pi}_n(dx)}.$$

If $\mathbb{S} = \mathbb{R}^q$ and one assumes that $\frac{d\nu}{d\bar{\nu}}(x)$ is a Lipschitz function which is bounded away from zero by $\varepsilon > 0$, then the latter implies

$$\|\pi_n - \bar{\pi}_n\| \leq \frac{1}{\varepsilon} \left\|\frac{d\nu}{d\bar{\nu}}\right\|_{\text{Lip}} \iint_{\mathbb{S} \times \mathbb{S}} \bar{\mathbb{E}}_n^Y |\tilde{X}_n(y) - \tilde{X}_n(x)| \bar{\pi}_n(dy) \bar{\pi}_n(dx).$$

This translates the filter stability problem to the contraction analysis of the stochastic flow generated by the backward conditioned signal.

For example, for the particular type of continuous time models studied by W. Stannat in the papers [30, 31], one can verify the bounds (see Ch. 4, [35]):

$$\bar{\mathbb{E}}_t^Y |\tilde{X}_t(y) - \tilde{X}_t(x)| \leq e^{-\kappa t} |x - y|,$$

with a constant $\kappa > 0$, expressed explicitly in terms of the model ingredients, and

$$\iint_{\mathbb{S} \times \mathbb{S}} |x - y| \bar{\pi}_t(dy) \bar{\pi}_t(dx) \leq \text{const.}$$

This proves the uniform exponential stability of the filter with the rate $\kappa > 0$, thus establishing by probabilistic techniques the stability results obtained by W. Stannat using PDE techniques [30, 31]. It should be stressed that this is one of the few cases where time uniform exponential pathwise filter stability is known for (possibly nonergodic) signals on noncompact domains.

11.3.2 Infinite horizon conditioning

The relation (11.31) still holds if conditioning on the observations is done on the *infinite* horizon, namely:

$$\mathbb{P}\left(X_m \in A \middle| \mathcal{F}_{0,m-1}^X \vee \mathcal{F}_{0,\infty}^Y\right) = \mathbb{P}\left(X_m \in A \middle| \sigma\{X_{m-1}\} \vee \mathcal{F}_{0,\infty}^Y\right) \qquad \mathbb{P}\text{-a.s.} \quad (11.36)$$

This means that the measure $\mathbb{P}_\infty^Y$, induced on the signal path space by conditioning on $\mathcal{F}_{0,\infty}^Y$

$$\mathbb{P}_\infty^Y(\Gamma) := \mathbb{P}\left(X \in \Gamma | \mathcal{F}_{0,\infty}^Y\right), \qquad \Gamma \in \mathcal{B}(\mathbb{S}^\infty)$$

is Markov, i.e.

$$\mathbb{P}_\infty^Y\left(X_m \in A | \mathcal{F}_{0,m-1}^X\right) = \mathbb{P}_\infty^Y\left(X_m \in A | \sigma\{X_{m-1}\}\right) \qquad \mathbb{P}_\infty^Y\text{-a.s.}$$

As before we refer the coordinate process on $\mathbb{S}^\infty$ under $\mathbb{P}_\infty^Y$ as conditioned signal. Though the transition probability in this case can no longer be expressed in a convenient closed form such as (11.32), an advantage of such infinite horizon conditioning is that a *single* conditioned signal process is obtained, rather than a family of processes whose transition law changes when the horizon increases. More importantly, if the signal process is stationary then a form of stationarity (in the sense of Markov chains in random environments) is inherited by the conditioned signal on the infinite time horizon, which is not the case if one conditions on a finite time horizon. This stationarity property brings into relevance the ergodic theory of Markov chains in random environments, which bears much resemblance to the homogeneous case but is not applicable to general time-inhomogeneous Markov chains (see [36]).

Unlike in the previous subsection, the filtering distributions π_n and $\bar{\pi}_n$ cannot be obtained as marginals of the conditional measure $\mathbb{P}_\infty^Y$, and we must therefore make a different connection with the filter stability problem. Somewhat surprisingly, the mysterious relation (11.11) can be restated in terms of the conditioned process, so that we can attempt to establish stability of the filter directly through (11.9). The connection with (11.11) is established using the following general fact (see Lemma 4.II.1 in [37]). Let $\mathcal{G}_1$ and $\mathcal{G}_2$ be sub σ-algebras of $\mathcal{F}$ and let $\mathbb{P}_{\mathcal{G}_1}(\cdot)$ be a regular version of the conditional probability given $\mathcal{G}_1$. If $\mathcal{G}_2$ is countably generated, then

$$\mathbb{P}(\cdot|\mathcal{G}_1 \vee \mathcal{G}_2) = \mathbb{P}_{\mathcal{G}_1}(\cdot|\mathcal{G}_2) \qquad \mathbb{P}\text{-a.s.} \tag{11.37}$$

Since X_n takes values in a Polish space, $\mathcal{F}_{n,\infty}^X$ is countably generated. In the context of the filtering problem (11.37), this means that for any $A \in \mathcal{B}(\mathbb{S}^\infty)$

$$\tilde{\mathbb{P}}\left(A | \mathcal{F}_{0,\infty}^Y \vee \mathcal{F}_{n,\infty}^X\right) = \tilde{\mathbb{P}}_\infty^Y\left(A | \mathcal{F}_{n,\infty}^X\right) \qquad \tilde{\mathbb{P}}\text{-a.s.}$$

Applying the martingale convergence theorem twice, we obtain

$$\tilde{\mathbb{P}}\left(A | \bigcap_{n\geq 0} \mathcal{F}_{0,\infty}^Y \vee \mathcal{F}_{n,\infty}^X\right) = \lim_{n\to\infty} \tilde{\mathbb{P}}\left(A | \mathcal{F}_{0,\infty}^Y \vee \mathcal{F}_{n,\infty}^X\right) =$$
$$= \lim_{n\to\infty} \tilde{\mathbb{P}}_\infty^Y\left(A | \mathcal{F}_{n,\infty}^X\right) = \tilde{\mathbb{P}}_\infty^Y\left(A | \bigcap_{n\geq 0} \mathcal{F}_{n,\infty}^X\right) \qquad \tilde{\mathbb{P}}\text{-a.s.} \tag{11.38}$$

One might be tempted to conclude from (11.37) that

$$\tilde{\mathbb{P}}\left(A | \mathcal{F}_{0,\infty}^Y \vee \bigcap_{n\geq 0} \mathcal{F}_{n,\infty}^X\right) \stackrel{?}{=} \tilde{\mathbb{P}}_\infty^Y\left(A | \bigcap_{n\geq 0} \mathcal{F}_{n,\infty}^X\right),$$

so that (11.11) would follow from (11.38). However, it is well known that the tail σ-algebra $\bigcap_{n\geq 0} \mathcal{F}^X_{n,\infty}$ is *not* countably generated, so this argument is not correct.[4]

Nonetheless this general approach can be rescued due to the following observation: it already suffices to show that the tail σ-algebra $\bigcap_{n\geq 0} \mathcal{F}^X_{n,\infty}$ is $\bar{\mathbb{P}}^Y_\infty$-trivial for $\bar{\mathbb{P}}$-a.e. observation path. Indeed, in this case

$$\bar{\mathbb{P}}^Y_\infty \left(A | \bigcap_{n\geq 0} \mathcal{F}^X_{n,\infty} \right) = \bar{\mathbb{P}}^Y_\infty(A) = \bar{\mathbb{P}} \left(A | \mathcal{F}^Y_{0,\infty} \right) \qquad \bar{\mathbb{P}}\text{-a.s.},$$

so that we can conclude directly from (11.38) and (11.9) that the filter is stable. Of course, the problem remains to establish that $\bigcap_{n\geq 0} \mathcal{F}^X_{n,\infty}$ is indeed $\bar{\mathbb{P}}^Y_\infty$-trivial. This can be done in the framework of ergodic theory of Markov chains in random environments, which leads to the following result.

Theorem 11.4 *(Corollary 5.5, [36])* *Suppose that the observation density $g(x, y)$ in (11.3) is strictly positive and that the signal is positive Harris recurrent and aperiodic. Then (11.5) holds for every $\nu, \bar{\nu}$.*

Note that the nondegeneracy assumption rules out the problem in Example 11.1.

Remark 11.2 It is tempting to assume that the triviality of $\bigcap_{n\geq 0} \mathcal{F}^X_{n,\infty}$ $\bar{\mathbb{P}}$-a.s. already implies its triviality $\bar{\mathbb{P}}^Y_\infty$-a.s. regardless of any other ingredients of the model (e.g., the observation structure). After all, it is elementary that $\bar{\mathbb{P}}(A) = 0$ or $\bar{\mathbb{P}}(A) = 1$ implies $\bar{\mathbb{P}}\left(A | \mathcal{F}^Y_{0,\infty}\right) = \bar{\mathbb{P}}(A)$ $\bar{\mathbb{P}}$-a.s. However, as $\bigcap_{n\geq 0} \mathcal{F}^X_{n,\infty}$ is not countably generated, it may be impossible to choose a regular conditional probability $\bar{\mathbb{P}}^Y_\infty(\cdot)$ such that $\bar{\mathbb{P}}^Y_\infty(A) = \bar{\mathbb{P}}(A)$ for all $A \in \bigcap_{n\geq 0} \mathcal{F}^X_{n,\infty}$ simultaneously on a set of full $\bar{\mathbb{P}}$-probability (i.e., as the number of sets A is uncountable one may not be able to eliminate the dependence of the $\bar{\mathbb{P}}$-null sets $\{\omega : \bar{\mathbb{P}}\left(A | \mathcal{F}^Y_{0,\infty}\right)(\omega) \neq \bar{\mathbb{P}}(A)\}$ on A). Example 11.1 shows that this is a real problem in models which are by no means pathological.

11.4 Observability

In the previous sections, the stability of the filter was essentially inherited from the ergodic properties of the signal. On the other hand, it is evident that the observations may also have a stabilizing effect on the filter: a trivial example is the case where $Y_n = X_n$, so that $\pi_n = \bar{\pi}_n$, $n \geq 1$ for any $\nu, \bar{\nu}$ *regardless* of the properties of the signal. The aim of this section is to outline two approaches which provide a link between the quality of the observations and the stability of the filter.

11.4.1 An information-theoretic bound

The first result of this kind appeared in [13] and is based on the connection with the information-theoretic notion of *relative entropy*. Recall the definition of the relative entropy between two probability measures $\mathbb{P}$ and $\mathbb{Q}$:

$$D(\mathbb{P} \| \mathbb{Q}) = \begin{cases} \int \log \frac{d\mathbb{P}}{d\mathbb{Q}} \, d\mathbb{P}, & \mathbb{P} \ll \mathbb{Q}, \\ \infty, & \mathbb{P} \not\ll \mathbb{Q}. \end{cases}$$

The relative entropy is a pseudodistance in the sense that it is nonnegative and vanishes if and only if the measures are identical. Note that as

$$\frac{d\mathbb{P}_{|\mathcal{G}}}{d\mathbb{Q}_{|\mathcal{G}}} = \mathbb{E}_{\mathbb{Q}} \left(\frac{d\mathbb{P}}{d\mathbb{Q}} \bigg| \mathcal{G} \right),$$

where $\mathbb{P}_{|\mathcal{G}}$ and $\mathbb{Q}_{|\mathcal{G}}$ stand for the restrictions of $\mathbb{P}$ and $\mathbb{Q}$ to the σ-algebra $\mathcal{G}$, it follows easily from Jensen's inequality that

$$D(\mathbb{P}_{|\mathcal{G}} \| \mathbb{Q}_{|\mathcal{G}}) \leq D(\mathbb{P} \| \mathbb{Q}). \tag{11.39}$$

To develop the result of [13], we work in the following continuous-time setting. Suppose that $X = (X_t)_{t \in \mathbb{R}_+}$ is a Markov process and

$$Y_t = \int_0^t h(X_s) \, ds + B_t, \tag{11.40}$$

where B is a p-dimensional Brownian motion independent of X and $h : \mathbb{S} \to \mathbb{R}^p$ is a function such that

$$\mathbb{E}\left(\int_0^T |h(X_s)|^2 ds\right) < \infty, \quad \bar{\mathbb{E}}\left(\int_0^T |h(X_s)|^2 ds\right) < \infty \quad \forall T > 0.$$

Classical filtering theory tells us (see [24]) that $dY_t = \pi_t(h) \, dt + dV_t$, where V_t is the innovation Brownian motion under $\mathbb{P}$. Similarly $dY_t = \bar{\pi}_t(h) \, dt + d\bar{V}_t$, where $\bar{V}_t$ is the innovation Brownian motion under $\bar{\mathbb{P}}$. Hence, the Girsanov theorem shows that the laws of Y under $\mathbb{P}$ and $\bar{\mathbb{P}}$ are equivalent with

$$\frac{d\mathbb{P}_{|\mathcal{F}_{0,t}^Y}}{d\bar{\mathbb{P}}_{|\mathcal{F}_{0,t}^Y}} = \exp\left\{\int_0^t (\pi_s(h) - \bar{\pi}_s(h)) \cdot dY_s - \frac{1}{2}\int_0^t \left(|\pi_s(h)|^2 - |\bar{\pi}_s(h)|^2\right) ds\right\}$$

$$= \exp\left\{\int_0^t (\pi_s(h) - \bar{\pi}_s(h)) \cdot dV_s + \frac{1}{2}\int_0^t |\pi_s(h) - \bar{\pi}_s(h)|^2 ds\right\},$$

and thus

$$D\left(\mathbb{P}_{|\mathcal{F}_{0,t}^Y} \| \bar{\mathbb{P}}_{|\mathcal{F}_{0,t}^Y}\right) = \frac{1}{2} \mathbb{E}\left(\int_0^t |\pi_s(h) - \bar{\pi}_s(h)|^2 ds\right).$$

Since $\frac{d\mathbb{P}}{d\bar{\mathbb{P}}} = \frac{d\nu}{d\bar{\nu}}(X_0)$, the inequality (11.39) gives

Theorem 11.5 (Theorem 3.1, [13]) *Suppose that $D(\nu \| \bar{\nu}) < \infty$. Then*

$$\frac{1}{2} \mathbb{E}\left(\int_0^\infty |\pi_t(h) - \bar{\pi}_t(h)|^2 \, ds\right) \leq D(\nu \| \bar{\nu}) < \infty. \tag{11.41}$$

Remark 11.3 There is in fact a deeper connection between the relative entropy and the filter stability problem in the general setting, which is developed in [13] also. Let us briefly sketch an alternative proof of this result. It is easily established using (11.7) and the Markov property of (X, Y) that

$$D(\pi_n || \bar{\pi}_n) = \mathbb{E}\left(\log \bar{\mathbb{E}}\left(\frac{dv}{d\bar{v}}(X_0) \Big| \mathcal{F}_{0,\infty}^Y \vee \mathcal{F}_{n,\infty}^X \right) \Big| \mathcal{F}_{0,n}^Y \right) - \log \bar{\mathbb{E}}\left(\frac{dv}{d\bar{v}}(X_0) \Big| \mathcal{F}_{0,n}^Y \right).$$

Applying Jensen's inequality and the Bayes formula to this expression, it is not difficult to establish explicitly that

$$\mathbb{E}\left(D(\pi_n || \bar{\pi}_n) | \mathcal{F}_{0,m}^Y \right) \leq D(\pi_m || \bar{\pi}_m) \quad \mathbb{P}\text{-a.s.} \quad \text{for all } m \leq n,$$

i.e., $D(\pi_n || \bar{\pi}_n)$ is an $\mathcal{F}_{0,n}^Y$-supermartingale under $\mathbb{P}$. Thus the relative entropy is a type of 'Lyapunov function' for the filter stability problem, and in particular the quantity $\mathbb{E}[D(\pi_n || \bar{\pi}_n)]$ is nonincreasing. Unfortunately, showing that the relative entropy actually decreases to zero as $n \to \infty$ appears to be not much easier than verifying filter stability in the total variation distance (see, e.g., Theorem 4.2 in [29]).

11.4.2 Observability

The information-theoretic bound in Theorem 11.5 establishes that the filtered estimate of the observation function h is stable in a weak sense virtually without any assumptions on the signal: in particular, neither compactness of the signal state space nor ergodicity of the signal was assumed! Note, however, that (11.41) does not guarantee the convergence of $\|\pi_t - \bar{\pi}_t\|$, and this may in fact very well fail. This raises an interesting possibility: perhaps there are other functions f for which $|\pi_t(f) - \bar{\pi}_t(f)|$ converges to zero as $t \to \infty$ regardless of whether the filter is stable? It turns out that this question has a nice affirmative answer which naturally leads to the notion of *observability* for nonlinear filtering models.

The basic idea is particularly transparent in discrete time for a model whose observations are defined in a slightly different manner from (11.3): we assume that Y_n is a noisy observation of X_{n-1}, rather than of X_n. In other words, the signal is observed with one time step delay. To be precise, Y still forms a sequence of independent random variables when conditioned on X, where (cf. (11.3))

$$\mathbb{P}\left(Y_n \in A | \mathcal{F}_{0,\infty}^X \right) = \int_A g(X_{n-1}, y) \, \psi(dy). \tag{11.42}$$

The filtering equation in this case is the recursion (cf. (11.4))

$$\pi_n(dx) = \frac{\int_S \Lambda(u, dx) g(u, Y_n) \pi_{n-1}(du)}{\int_S g(u, Y_n) \pi_{n-1}(du)}, \quad n \geq 1, \tag{11.43}$$

whose solution is denoted by π_n when the equation is initialized by ν and by $\bar{\pi}_n$ when it is started from $\bar{\nu}$.

Remark 11.4 Though the following results are more naturally formulated in the modified setting (11.42), some additional work allows one to consider the setting of (11.3) as well; see e.g. [34]. Moreover, as will be discussed briefly below, these ideas can also be developed in the continuous time setting where the difference between (11.42) and (11.3) disappears. For simplicity, however, *we will operate in modified setting (11.42) throughout the remainder of this section.*

To develop stability results in this setting, we first make a brief detour. Instead of considering the filters π_n and $\bar{\pi}_n$, let us turn our attention for the moment to the one-step predictors of the observation process:

$$\eta_{n|n-1}(f) := \mathbb{E}\left(f(Y_n)|\mathcal{F}^Y_{0,n-1}\right) = \int_{\mathbb{R}^p}\int_S f(y)\, g(u,y)\, \pi_{n-1}(du)\, \psi(dy),$$

$$\bar{\eta}_{n|n-1}(f) := \bar{\mathbb{E}}\left(f(Y_n)|\mathcal{F}^Y_{0,n-1}\right) = \int_{\mathbb{R}^p}\int_S f(y)\, g(u,y)\, \bar{\pi}_{n-1}(du)\, \psi(dy).$$

It turns out that these predictors are always stable in the following sense.

Proposition 11.2 (Theorem 11.1, [9]) *If $\nu \ll \bar{\nu}$, then for any bounded function f*

$$\lim_{n\to\infty} \mathbb{E}|\eta_{n|n-1}(f) - \bar{\eta}_{n|n-1}(f)| = 0. \tag{11.44}$$

The proof uses the ideas similar to those presented in the introduction to this chapter. By the Bayes formula

$$\eta_{n|n-1}(f) = \mathbb{E}\left(f(Y_n)|\mathcal{F}^Y_{0,n-1}\right) = \frac{\bar{\mathbb{E}}\left(\frac{d\nu}{d\bar{\nu}}(X_0)f(Y_n)|\mathcal{F}^Y_{0,n-1}\right)}{\bar{\mathbb{E}}\left(\frac{d\nu}{d\bar{\nu}}(X_0)|\mathcal{F}^Y_{0,n-1}\right)} \quad \mathbb{P}\text{-a.s.}$$

and hence under $\mathbb{P}$

$$\eta_{n|n-1}(f) - \bar{\eta}_{n|n-1}(f) =$$
$$\frac{\bar{\mathbb{E}}\left(\frac{d\nu}{d\bar{\nu}}(X_0)f(Y_n)|\mathcal{F}^Y_{0,n-1}\right) - \bar{\mathbb{E}}\left(f(Y_n)|\mathcal{F}^Y_{0,n-1}\right)\bar{\mathbb{E}}\left(\frac{d\nu}{d\bar{\nu}}(X_0)|\mathcal{F}^Y_{0,n-1}\right)}{\bar{\mathbb{E}}\left(\frac{d\nu}{d\bar{\nu}}(X_0)|\mathcal{F}^Y_{0,n-1}\right)}.$$

Simple manipulations with conditional expectations give

$$\mathbb{E}|\eta_{n|n-1}(f) - \bar{\eta}_{n|n-1}(f)| = \bar{\mathbb{E}}\left[\bar{\mathbb{E}}\left(\frac{d\nu}{d\bar{\nu}}(X_0)\Big|\mathcal{F}^Y_{0,n-1}\right)|\eta_{n|n-1}(f) - \bar{\eta}_{n|n-1}(f)|\right] =$$
$$\bar{\mathbb{E}}\left[\bar{\mathbb{E}}\left(\frac{d\nu}{d\bar{\nu}}(X_0)f(Y_n)\Big|\mathcal{F}^Y_{0,n-1}\right) - \bar{\mathbb{E}}\left(f(Y_n)|\mathcal{F}^Y_{0,n-1}\right)\bar{\mathbb{E}}\left(\frac{d\nu}{d\bar{\nu}}(X_0)\Big|\mathcal{F}^Y_{0,n-1}\right)\right] =$$
$$\bar{\mathbb{E}}\left(\left\{\bar{\mathbb{E}}\left(\frac{d\nu}{d\bar{\nu}}(X_0)\Big|\mathcal{F}^Y_{0,n}\right) - \bar{\mathbb{E}}\left(\frac{d\nu}{d\bar{\nu}}(X_0)\Big|\mathcal{F}^Y_{0,n-1}\right)\right\}f(Y_n)\right).$$

But as f is bounded (by a constant C, say), we find that

$$\mathbb{E}|\eta_{n|n-1}(f) - \bar{\eta}_{n|n-1}(f)| \leq C\,\bar{\mathbb{E}}\left|\mathbb{E}\left(\frac{d\nu}{d\bar{\nu}}(X_0)\Big|\mathcal{F}^Y_{0,n}\right) - \bar{\mathbb{E}}\left(\frac{d\nu}{d\bar{\nu}}(X_0)\Big|\mathcal{F}^Y_{0,n-1}\right)\right|,$$

which converges to zero as $n \to \infty$ by the martingale convergence theorem.

Proposition 11.2 shows that the one-step predictive estimates of the observation process are stable, but we are ultimately interested in the stability of the filter. To make the connection with the latter problem, let us now consider (in analogy with (11.40)) the *additive* noise observation scenario, i.e., we assume that

$$Y_n = h(X_{n-1}) + \xi_n, \qquad n \geq 1,$$

where $\xi = (\xi_n)_{n \geq 1}$ is an i.i.d. sequence of $\mathbb{R}^p$-valued random variables independent of X and $h: \mathbb{S} \to \mathbb{R}^p$ is a given observation function. In this case Proposition 11.2 can be used to prove the following result:

Theorem 11.6 (Variant of Proposition 3.3, [9]) *Suppose that*

a_1. h is bounded.
a_2. $|\mathbb{E}e^{ik\cdot\xi}| > 0$ for all $k \in \mathbb{R}^p$.

Then for any continuous function f and $\nu \ll \bar{\nu}$

$$\lim_{n\to\infty} \mathbb{E}|\pi_n(f \circ h) - \bar{\pi}_n(f \circ h)| = 0. \tag{11.45}$$

Indeed, in this case

$$\eta_{n|n-1}\left(e^{ik\cdot}\right) = \mathbb{E}\left(e^{ik\cdot Y_n}|\mathcal{F}^Y_{0,n-1}\right) = \pi_{n-1}\left(e^{ik\cdot h(\cdot)}\right)\mathbb{E}e^{ik\cdot\xi_1},$$
$$\bar{\eta}_{n|n-1}\left(e^{ik\cdot}\right) = \bar{\mathbb{E}}\left(e^{ik\cdot Y_n}|\mathcal{F}^Y_{0,n-1}\right) = \bar{\pi}_{n-1}\left(e^{ik\cdot h(\cdot)}\right)\mathbb{E}e^{ik\cdot\xi_1},$$

and as $\nu \ll \bar{\nu}$, we obtain by Proposition 11.2 and assumption (a_2)

$$\mathbb{E}\left|\pi_{n-1}\left(e^{ik\cdot h(\cdot)}\right) - \bar{\pi}_{n-1}\left(e^{ik\cdot h(\cdot)}\right)\right| = \frac{\mathbb{E}\left|\eta_{n|n-1}\left(e^{ik\cdot}\right) - \bar{\eta}_{n|n-1}\left(e^{ik\cdot}\right)\right|}{|\mathbb{E}e^{ik\cdot\xi_1}|} \xrightarrow{n\to\infty} 0.$$

We therefore find that for any finite-order trigonometric polynomial T

$$\lim_{n\to\infty} \mathbb{E}|\pi_n(T \circ h) - \bar{\pi}_n(T \circ h)| = 0. \tag{11.46}$$

Now note that as a consequence of the Weierstrass approximation theorem, any continuous function can be approximated uniformly on compact sets by trigonometric polynomials. As h is bounded it takes values in a compact set. Therefore, given a continuous function f, there is a sequence of trigonometric polynomials T_ℓ such that $\|f \circ h - T_\ell \circ h\|_\infty \leq \ell^{-1}$. But then

$$\mathbb{E}|\pi_n(f \circ h) - \bar{\pi}_n(f \circ h)| \leq \mathbb{E}|\pi_n(T_\ell \circ h) - \bar{\pi}_n(T_\ell \circ h)| + 2\ell^{-1}$$

for all n, so $\limsup_{n\to\infty} \mathbb{E}|\pi_n(f \circ h) - \bar{\pi}_n(f \circ h)| \leq 2\ell^{-1}$. But ℓ was arbitrary, so letting $\ell \to \infty$ completes the proof.

It follows immediately from Theorem 11.6 that the stability of $\pi_n(h)$

$$\lim_{n\to\infty} \mathbb{E}|\pi_n(h) - \bar{\pi}_n(h)| = 0 \qquad (11.47)$$

is recovered by choosing $f(x) = x$. This resembles the result (11.41). Moreover, as in the case of Theorem 11.5, ergodicity of the signal process is not assumed. However, there are essential differences between the two results. On the one hand, (11.41) places minimal assumptions on the observations, whereas (11.47) relies on the restrictive assumption that the observation function h is bounded. On the other hand, (11.41) only provides stability of the observation function h itself, while (11.45) provides stability also for functions of h. The latter class of functions can be quite large: for example, when h is invertible as in the following corollary, we can even conclude stability of the filter in a weak (as opposed to total variation) sense.

Corollary 11.1 *Let* $\mathbb{S} \subset \mathbb{R}^p$ *and let* $h : \mathbb{S} \to \mathbb{R}^p$ *be bounded. Suppose there is a continuous function* $h^{-1} : \mathbb{R}^p \to \mathbb{R}^p$ *such that* $h^{-1}(h(x)) = x$ *for all* $x \in \mathbb{S}$. *Then*

$$\lim_{n\to\infty} \mathbb{E}|\pi_n(g) - \bar{\pi}_n(g)| = 0$$

for every continuous function g and $\nu \ll \bar{\nu}$, *provided* $|\mathbb{E}e^{ik\cdot\xi}| > 0$ *for all* $k \in \mathbb{R}^p$.

The proof is immediate from Theorem 11.6.

Remark 11.5 The assumptions of the previous corollary require that $\mathbb{S}$ is a bounded subset of $\mathbb{R}^p$; indeed, as $h(\mathbb{S})$ is bounded and h^{-1} is continuous, $\mathbb{S} = h^{-1}(h(\mathbb{S}))$ must be bounded. The result is therefore the most natural when $\mathbb{S}$ is compact, in which case the boundedness of h is not restrictive. When $\mathbb{S}$ is not compact, it may still be the case that the signal is outside a compact set with uniformly small probability, i.e., that the sequence $(X_k)_{k\geq 0}$ is tight or uniformly integrable. In this case, it is straightforward to localize the proof by truncating to a compact set, see [9], and one can relax the boundedness of h. Though this is perhaps not surprising, it should be noted that localization is often not so straightfoward in other methods. For example, we showed that the assumption of Theorem 11.1 is most natural when $\mathbb{S}$ is compact; however, the localization of that result is highly nontrivial [20].

When h is not invertible, Theorem 11.6 yields only 'partial' stability, i.e., stability of the estimates of particular functions. However, with a little more work we can establish the stability of a much larger class of functions than we have investigated so far, which opens the possibility of proving stability of the filter (in the spirit of the previous corollary) even when h is not invertible. To this end, let us begin by noting that the arguments for (11.44) apply to predictors of a more general form (for an even more general statement, see the classic paper [4]):

Proposition 11.3 *For an integer $m \geq 1$, let $f_1, \ldots, f_m$ be continuous bounded functions and $k_1, \ldots, k_m$ distinct positive integers. Then if $\nu \ll \bar{\nu}$*

$$\lim_{n \to \infty} \mathbb{E} \left| \mathbb{E} \left(f_1(Y_{n+k_1}) \cdots f_m(Y_{n+k_m}) \big| \mathcal{F}^Y_{0,n} \right) - \bar{\mathbb{E}} \left(f_1(Y_{n+k_1}) \cdots f_m(Y_{n+k_m}) \big| \mathcal{F}^Y_{0,n} \right) \right| = 0. \tag{11.48}$$

Now note that by time homogeneity and the Markov property of (X, Y)

$$\mathbb{E} \left(f_1(Y_{n+k_1}) \cdots f_m(Y_{n+k_m}) \big| \mathcal{F}^Y_{0,n} \right) = \pi_n(f)$$

with

$$f(x) := \mathbb{E} \left(f_1(Y_{k_1}) \cdots f_m(Y_{k_m}) \big| X_0 = x \right), \tag{11.49}$$

and thus (11.48) states that the filtered estimates of such f are always stable. As the number of times m and the functions $f_1, \ldots, f_m$ are arbitrary, this suggests that the class of functions with stable estimates can be quite large. We are interested in characterizing this class of functions in terms of the filtering model.

Such a characterization is indeed possible and can be formulated in terms of *observability* ([32]). Consider the following equivalence relation for probability measures on $\mathbb{S}$: we say that two probability measures ν_1, ν_2 are equivalent $\nu_1 \smile \nu_2$ if they induce the same law of the observation process, i.e.,

$$\nu_1 \smile \nu_2 \quad \text{iff} \quad \mathbb{P}^{\nu_1}|_{\mathcal{F}^Y_{0,\infty}} = \mathbb{P}^{\nu_2}|_{\mathcal{F}^Y_{0,\infty}},$$

where $\mathbb{P}^{\nu_1}$ and $\mathbb{P}^{\nu_2}$ denote the law of (X, Y) when $X_0 \sim \nu_1$ and $X_0 \sim \nu_2$, respectively. As probability measures on $\mathcal{B}((\mathbb{R}^p)^\infty)$ are determined by their finite-dimensional distributions, $\nu_1 \smile \nu_2$ if and only if for any $m \geq 1$, any bounded and continuous functions $f_1, \ldots, f_m$ and time indices $k_1, \ldots, k_m$

$$\int_{\mathbb{S}} \mathbb{E} \left(f_1(Y_{k_1}) \cdots f_m(Y_{k_m}) \big| X_0 = x \right) \nu_1(dx) = \int_{\mathbb{S}} \mathbb{E} \left(f_1(Y_{k_1}) \cdots f_m(Y_{k_m}) \big| X_0 = x \right) \nu_2(dx).$$

In other words, $\nu_1 \smile \nu_2$ whenever the signed measure $\nu_1 - \nu_2$ is orthogonal to the linear subspace $\mathcal{O}^\circ$ spanned by the functions of the form (11.49).

Let us now suppose that the Markov process (X, Y) is Feller, so that all functions of the form (11.49) are continuous. Moreover, let us suppose that the signal state space $\mathbb{S}$ is compact. Then an elementary functional analytic argument (Prop. 3.3, [32]) shows that $\mathcal{O}^\circ \subseteq C_b(\mathbb{S})$ is dense in the subspace

$$\mathcal{O} = \left\{ f \in C_b(\mathbb{S}) : \int f \, d\nu_1 = \int f \, d\nu_2 \text{ whenever } \nu_1 \smile \nu_2 \right\}$$

in the topology of uniform convergence ($C_b(\mathbb{S})$ is the space of bounded continuous functions on $\mathbb{S}$). In other words, for any $f \in \mathcal{O}$ there is a sequence of functions f_n of the form (11.49) such that $f_n \to f$ uniformly (this argument replaces the application of the Weierstrass theorem in the proof of Theorem 11.6). Since the filtered estimates of functions of the form (11.49) are stable, we obtain

Theorem 11.7 (Variant of Theorem 4.4, [32].) *Assume that $\mathbb{S}$ is compact, (X, Y) is Feller, and $\nu \ll \bar{\nu}$. Then for any $f \in \mathcal{O}$*

$$\lim_{n \to \infty} \mathbb{E}|\pi_n(f) - \bar{\pi}_n(f)| = 0. \tag{11.50}$$

By definition, the space $\mathcal{O}$ consists of those functions whose expectation is uniquely determined by the law of the observations. In this sense it is natural to call $\mathcal{O}$ the observable space of the filtering model and $f \in \mathcal{O}$ observable functions. If $\mathcal{O} = C_b(\mathbb{S})$, the model is referred to as *(fully) observable*, and

Corollary 11.2 *Assume that $\mathbb{S}$ is compact and (X, Y) is Feller. Then the model is (fully) observable if and only if $\nu_1 \smile \nu_2$ implies $\nu_1 = \nu_2$, i.e., if the law of the observations uniquely determines the initial law of the signal.* When this is the case

$$\lim_{n \to \infty} \mathbb{E}|\pi_n(f) - \bar{\pi}_n(f)| = 0 \quad \text{for all } f \in C_b, \ \nu \ll \bar{\nu}.$$

This is a generalization of Corollary 11.1 in the present setting. Note that observability does not require h to be invertible; on the other hand, it is not difficult to establish that a sufficient condition for observability is that h is invertible and $|\mathbb{E}e^{ik\cdot\xi}| > 0$ for all $k \in \mathbb{R}^p$, reproducing essentially the result of Corollary 11.1.

Example 11.3 Let X be a Markov chain on $\mathbb{S} = \{1, 2, 3, 4\}$ with transition matrix

$$\Lambda = \begin{pmatrix} 1/2 + \varepsilon & 1/2 - \varepsilon & 0 & 0 \\ 0 & 1/2 & 1/2 & 0 \\ 0 & 0 & 1/2 & 1/2 \\ 1/2 & 0 & 0 & 1/2 \end{pmatrix},$$

where $0 < \varepsilon \le 1/2$, and let $Y_n = 1_{\{X_{n-1} \in \{1,3\}\}}$, $n \ge 1$. This differs from the model of Example 11.1 in that we have perturbed one of the transition probabilities, and that we have introduced one time step delay in the observation model in keeping with the setting of this section. It is easily verified, however, that when $\varepsilon = 0$ the corresponding filter is unstable exactly as in Example 11.1. In contrast, we now show that when $\varepsilon \ne 0$ the model is observable and hence the filter is stable.

To prove observability, note that for $k \ge 1$

$$\begin{pmatrix} \mathbb{E}(Y_k|X_0 = 1) \\ \mathbb{E}(Y_k|X_0 = 2) \\ \mathbb{E}(Y_k|X_0 = 3) \\ \mathbb{E}(Y_k|X_0 = 4) \end{pmatrix} = \Lambda^{k-1} \begin{pmatrix} 1 \\ 0 \\ 1 \\ 0 \end{pmatrix} := f_k.$$

Computing explicitly, we find that

$$f_1 = \begin{pmatrix} 1 \\ 0 \\ 1 \\ 0 \end{pmatrix}, \quad f_2 = \frac{1}{2}\begin{pmatrix} 2\varepsilon + 1 \\ 1 \\ 1 \\ 1 \end{pmatrix}, \quad f_3 = \frac{1}{2}\begin{pmatrix} 2\varepsilon^2 + \varepsilon + 1 \\ 1 \\ 1 \\ \varepsilon + 1 \end{pmatrix}.$$

But the vectors f_1, f_2, f_3 and $f_0 = (1\ 1\ 1\ 1)^\mathsf{T}$ span $\mathbb{R}^4$, provided $\varepsilon \neq 0$. Therefore every function $f : \mathbb{S} \to \mathbb{R}$ can be written as a function of the form $f(x) = \mathbf{E}(a_0 + a_1 Y_1 + a_2 Y_2 + a_3 Y_3 | X_0 = x)$ for some $a_0, \ldots, a_3 \in \mathbb{R}$, so that observability, and consequently stability of the filter, follow.

The same theory works out in continuous time, where the nuance of the observation delay disappears ((11.42) vs. (11.3)). For the additive white noise model of Theorem 11.5, one can show that $\nu_1 \smile \nu_2$ if and only if $(h(X_t))_{t \geq 0}$ has the same law under $\mathbb{P}^{\nu_1}$ and $\mathbb{P}^{\nu_2}$ (see Proposition 5.2 in [32]). Therefore, if $\nu_1 \smile \nu_2$, then in particular $\mathbb{E}^{\nu_1}(f(h(X_0))) = \mathbb{E}^{\nu_2}(f(h(X_0)))$ for every measurable function f so that

$$\nu_1 \smile \nu_2 \quad \Longrightarrow \quad \int_\mathbb{S} f \circ h \, d\nu_1 = \int_\mathbb{S} f \circ h \, d\nu_2.$$

Thus evidently $f \circ h \in \mathcal{O}$ for any f such that $f \circ h \in C_b(\mathbb{S})$ (Lemma 5.6, [32]). The continuous-time counterparts of Theorem 11.6 and of Corollary 11.1 follow directly.

In the general case where h is not invertible, characterizing the observable space $\mathcal{O}$ in terms of the model ingredients remains a nontrivial task. However, in the special case of finite state space signals this can be done explicitly (sec. 6, [32]). Consider the model from subsection 11.2.1 and let $\mathbb{H} = h(\mathbb{S}) := \{h_1, \ldots, h_r\}$, $r \leq d$ be the set of the distinct observations values. Define $d \times d$ diagonal matrices H_{h_k}, $k = 1, \ldots, r$ such that $H_{h_k}(i, j) = 1$ whenever $i = j$ and $h(a_i) = h_k$ and such that the remaining entries are zero. Then again identifying functions on $\mathbb{S}$ with vectors in $\mathbb{R}^d$, writing $\mathbf{1}$ for the $d \times 1$ vector with unit entries and, as before, denoting the transition matrix of the chain by Λ, one can easily prove the following result along the same lines as the computation in Example 11.3.

Proposition 11.4 (Lemma 6.4 in [32].)

$$\mathcal{O} = \mathrm{span}\left\{ H_{n_0} \Lambda\, H_{n_1} \Lambda \cdots \Lambda H_{n_m} \mathbf{1} : m \geq 0,\, n_i \in \mathbb{H} \right\}.$$

In particular, if $\dim(\mathcal{O}) = d$, the filter is stable in the sense of (11.5).

In this finite state setting, one can in fact go one step further and give the complete characterization of filter stability. For this purpose the following notion of *detectability* is introduced: the model is called detectable if $\lim_{t \to \infty} e^{\Lambda^\mathsf{T} t} \mu = 0$ whenever $\mu \perp \mathcal{O}$ (thus every observable model is detectable, but not vice versa). Let us note that detectability, like observability, can be verified algebraically in terms of the model parameters Λ and h.

Theorem 11.8 (Theorem 6.12 in [32]**)** *Assume that the observations are nondegenerate $\sigma > 0$. Then the Shiryaev–Wonham filter (11.27) is stable in the sense of (11.5) whenever $\nu \ll \bar{\nu}$ if and only if the model is detectable.*

The proof of this result is obtained by combining Theorem 11.3 and the continuous time counterpart of Theorem 11.7. This suggests that at least in the finite state setting, the two main structural assumptions that we have imposed in this article—ergodicity of the signal process and observability—are indeed the fundamental mechanisms that conspire to give stability of the filter.

11.4.3 Uniform observability

A drawback of the observability results above is that they rely on compactness of the state space (or uniform integrability of the signal, as in Remark 11.5, which allows truncation to a compact set). This rules out the interesting possibility that the filter may be stable even in models where the signal itself is unstable (i.e., when the signal diverges to infinity), which is known to hold, e.g., in the linear Gaussian case when the Kalman filter is observable [27]. It turns out the the approach of the previous section can be extended to cover also the unstable case, though the analysis is more subtle in this setting.

The reason that compactness was required above is that both Theorem 11.6 and Theorem 11.7 are proved by showing that a class of continuous functions, obtained from the predictor, is dense in the *uniform* topology. Uniform approximation of continuous functions is a natural problem for functions on a compact state space, and can be tackled using elementary functional analytic arguments. However, when the state space is not compact one obtains approximation uniformly on compact sets, which is insufficient for our purposes when the signal is unstable. Nonetheless a more refined argument, using a uniform approximation property of convolution operators, allows one to resolve this problem in the case of additive observations [33, 34]. This gives, for example, the following counterpart of Corollary 11.1. Here

$$\|\nu_1 - \nu_2\|_{\mathrm{BL}} := \sup_{f \in \mathrm{BL}} \left| \int f \, d\nu_1 - \int f \, d\nu_2 \right|,$$

where BL denotes the class of functions f such that $\|f\|_\infty \leq 1$ and $|f(x) - f(z)| \leq d(x,z)$ for all x, z (the unit ball in the space of bounded Lipschitz functions).

Proposition 11.5 (Proposition 3.11 in [33].**)** *Suppose that $Y_n = h(X_{n-1}) + \xi_n$ with*

- (a_1). $h : \mathbb{S} \to \mathbb{R}^p$ *is invertible.*
- (a_2). h^{-1} *is uniformly continuous.*
- (a_3). $|\mathbb{E}e^{ik\cdot\xi}| > 0$ *for all $k \in \mathbb{R}^p$.*

Then
$$\lim_{n\to\infty} \mathbb{E}\|\pi_n - \bar{\pi}_n\|_{BL} = 0$$
whenever $\nu \ll \bar{\nu}$.

A remarkable property of this result is that only assumptions on the observation structure are made, while the signal transition kernel can be completely arbitrary. This is opposite in spirit to the conditions given in section 11.2: there it was shown that the filter is stable regardless of the observation structure if the signal is sufficiently mixing, while we see here that the filter is stable regardless of the signal structure of the observations are sufficiently informative.

A noncompact counterpart to Corollary 11.2 can be obtained if the notion of observability is replaced by the stronger notion of *uniform observability*. Recall that the filtering model is called observable if

$$\mathbb{P}^{\nu_1}|_{\mathcal{F}^Y_{0,\infty}} = \mathbb{P}^{\nu_2}|_{\mathcal{F}^Y_{0,\infty}} \quad \text{implies} \quad \nu_1 = \nu_2.$$

Uniform observability is, in a sense, a quantitative counterpart of observability: the model is said to be uniformly observable if for every $\varepsilon > 0$, there exists a $\delta > 0$ (which depends only on ε) such that

$$\|\mathbb{P}^{\nu_1}|_{\mathcal{F}^Y_{0,\infty}} - \mathbb{P}^{\nu_2}|_{\mathcal{F}^Y_{0,\infty}}\| < \delta \quad \text{implies} \quad \|\nu_1 - \nu_2\|_{BL} < \varepsilon.$$

Using this definition, one obtains the following counterpart of Corollary 11.2.

Theorem 11.9 (Variant of Theorem 3.3, [33]) *Suppose that the filtering model is uniformly observable. Then*

$$\lim_{n\to\infty} \mathbb{E}\|\pi_n - \bar{\pi}_n\|_{BL} = 0 \quad \text{for all } \nu \ll \bar{\nu}.$$

It can be shown that a known result about the stability of the Kalman filter is a special case of this general result, while in the compact case it turns out that observability and uniform observability are equivalent [33]. In general, however, uniform observability remains difficult to verify for specific filtering models.

Finally, we remark that the results in this section do not provide rates of convergence, while many filter stability results give rise to exponential rates. The key to the stability proofs in this section is the martingale convergence theorem, which does not guarantee a rate of convergence. As the following example shows, exponential stability cannot always be expected to hold without further assumptions.

Example 11.4 For real-valued signal and observations, consider the model $Y_n = X_{n-1} + \xi_n$, where ξ_n are i.i.d. $N(0, 1)$ and $X_n = X_0$ for all n. Let $\nu = N(\alpha, \sigma^2)$ and $\bar{\nu} = N(\beta, \sigma^2)$ for some $\alpha, \beta, \sigma \in \mathbb{R}$ (so $\nu \ll \bar{\nu}$). Linear filtering theory shows that π_n is a random Gaussian measure with mean Z_n and variance V_n given by

$$Z_n = \frac{a}{1+\sigma^2 n} + \frac{\sigma^2 n}{1+\sigma^2 n} \cdot \frac{1}{n}\sum_{\ell=1}^{n} Y_\ell, \qquad V_n = \frac{\sigma^2}{1+\sigma^2 n},$$

and similarly for $\bar{\pi}_n$, $\bar{Z}_n$, $\bar{V}_n$ where a is replaced by β. Evidently the conditional mean of the filter is stable with rate $\Omega(n^{-1})$, which is not exponential. (The conditional mean is an unbounded function of the signal; however, it is not difficult to show that $\mathbb{E}\|\pi_n - \bar{\pi}_n\|_{\mathrm{BL}} = \Omega(n^{-1})$ also, see remark 2.8 in [34].)

11.5 Necessary conditions for stability

Almost all the above results (the exception being the approach of Section 11.3.1) appeal directly to the absolute continuity assumption (A), either through (11.9) or through Proposition 11.2. Indeed, in a sense this assumption lies at the heart of the intrinsic approach to filter stability, as it allows to relate the conditional expectations (11.1) for different initial measures ν, $\bar{\nu}$ using the Bayes formula.

Assumption (A) was introduced without fanfare in the Introduction. However, the assumption is not as innocent as it may seem: for example, it implies that $\bar{\nu}$ has an atom at every point ν does, i.e., a suitable choice of $\bar{\nu}$ requires some information about the possibly unknown true distribution ν. Moreover, many filter stability results obtained by other methods hold for arbitrary ν, $\bar{\nu}$. One might therefore wonder whether the assumption (A) is a restriction of the intrinsic method, or whether it has a deeper relevance. In this section, we will outline how the assumption (A) can be weakened in the context of the intrinsic approach, and we will show that the weakened assumption is in fact a necessary condition for stability in the total variation distance. This indicates that some form of absolute continuity is a fundamental ingredient of the filter stability problem. Though this discussion is not of direct practical interest, it sheds some light on the (often hidden) assumptions that are common to all methods of proving filter stability.

11.5.1 Well posedness

The first question one needs to confront in weakening assumption (A) is whether the recursion (11.4) is even well posed. The problem, which was glossed over in the Introduction, is that the denominator in (11.4) may be zero for some observation sequences. This is typically resolved by noting that the denominator of (11.4) is nonzero for $\mathbb{P}$-a.e. observation sequence (this holds by construction as the recursion (11.4) is obtained from the Bayes formula; see e.g. remark 3.1.5 in [7]). However, this need no longer hold if the initial distribution ν is replaced by $\bar{\nu}$, as the following example shows.

Example 11.5 Consider the signal X_n on $\mathbb{S} = \{2, 1, -1\}$ with the transition matrix

$$\Lambda = \begin{pmatrix} 1/2 & 1/2 & 0 \\ 0 & 0 & 1 \\ 1/2 & 0 & 1/2 \end{pmatrix},$$

and initial distribution $\nu(\{1\}) = 1$. Suppose that the observation sequence is

$$Y_n = X_n U_n,$$

where $U = (U_n)_{n \geq 1}$ are i.i.d. random variables with uniform distribution over $[0, 1]$. If the filter is started with $\bar{\nu}(\{2\}) = 1$, the formula (11.4) yields

$$\bar{\pi}_1(\{2\}) = \frac{\frac{1}{2}\mathbf{1}_{\{Y_1 \in [0,2]\}}}{\frac{1}{2}\mathbf{1}_{\{Y_1 \in [0,2]\}} + \mathbf{1}_{\{Y_1 \in [0,1]\}}}$$

$$\bar{\pi}_1(\{1\}) = \frac{\mathbf{1}_{\{Y_1 \in [0,1]\}}}{\frac{1}{2}\mathbf{1}_{\{Y_1 \in [0,2]\}} + \mathbf{1}_{\{Y_1 \in [0,1]\}}}$$

$$\bar{\pi}_1(\{-1\}) = 0.$$

But $Y_1 = X_1 U_1 = -U_1 < 0$ a.s. and hence the right hand side is ill-posed (0/0).

When is the filtering recursion $\mathbb{P}$-a.s. well posed? We can give a general answer to this question. Note that, by definition, the conditional probability $\mathbb{P}\left(X_n \in \cdot | \mathcal{F}^Y_{0,n}\right)$ is defined uniquely up to $\mathbb{P}_{|\mathcal{F}^Y_{0,n}}$-a.s. equivalence. Therefore π_n is well defined for $\mathbb{P}_{|\mathcal{F}^Y_{0,n}}$-a.e. observation path, while $\bar{\pi}_n$ is well defined for $\bar{\mathbb{P}}_{|\mathcal{F}^Y_{0,n}}$-a.e. observation path. In order for $\bar{\pi}_n, n \geq 0$ to be well defined for $\mathbb{P}$-a.e. observation path, we must require

$$\mathbb{P}_{|\mathcal{F}^Y_{0,n}} \ll \bar{\mathbb{P}}_{|\mathcal{F}^Y_{0,n}}, \qquad n \geq 0. \tag{B}$$

This is obviously satisfied under assumption (A), a fact that we have implicitly used throughout the chapter. However, assumption (A) is not necessary for the filtering recursion to be well posed; for example, (B) holds regardless of the choice of $\bar{\nu}$ in the nondegenerate case where $g(x, y) > 0$ for all x, y. Indeed, it is immediately evident from (11.4) that the filtering recursion is always well posed in this case.

11.5.2 Absolute continuity

In the previous sections, we have proved various sufficient conditions for filter stability under assumption (A). However, in most cases, it is enough to impose the weaker assumption

$$\mathbb{P}_{|\mathcal{F}^Y_{0,\infty}} \ll \bar{\mathbb{P}}_{|\mathcal{F}^Y_{0,\infty}}. \tag{C}$$

Let us briefly outline one way to do this. Suppose that we have proved that (11.5) holds under assumption (A). Now suppose that $\nu, \bar{\nu}$ are such that only (C) holds.

Defining $\tilde{\tilde{\nu}} := (\nu + \bar{\nu})/2$, we have $\nu \ll \tilde{\tilde{\nu}}$ and $\bar{\nu} \ll \tilde{\tilde{\nu}}$. Therefore, using (A), we have

$$\lim_{n\to\infty} \mathbb{E}\|\pi_n - \tilde{\pi}_n\| = \lim_{n\to\infty} \tilde{\mathbb{E}}\|\tilde{\pi}_n - \tilde{\tilde{\pi}}_n\| = 0,$$

where $\tilde{\tilde{\pi}}_n$ is defined in the obvious fashion. In particular, $\|\tilde{\pi}_n - \tilde{\tilde{\pi}}_n\| \to 0$ in $\tilde{\mathbb{P}}$-probability. But then (C) implies that $\|\tilde{\pi}_n - \tilde{\tilde{\pi}}_n\| \to 0$ in $\mathbb{P}$-probability, and by dominated convergence $\mathbb{E}\|\tilde{\pi}_n - \tilde{\tilde{\pi}}_n\| \to 0$. Therefore

$$\lim_{n\to\infty} \mathbb{E}\|\pi_n - \tilde{\pi}_n\| \le \lim_{n\to\infty} \mathbb{E}\|\pi_n - \tilde{\tilde{\pi}}_n\| + \lim_{n\to\infty} \mathbb{E}\|\tilde{\pi}_n - \tilde{\tilde{\pi}}_n\| = 0$$

by the triangle inequality, and we find that indeed the result is automatically extended to the weaker setting of assumption (C). Similar considerations apply to the weaker notions of convergence considered in Section 11.4.

Assumption (C), however, is still stronger than the minimal assumption (B) needed for the filtering recursion (and hence the filter stability problem) to be well posed. As we will argue in the next section, assumption (C) cannot be weakened in general any further if the filter is to be stable, at least if we are interested in proving stability in the total variation distance. Indeed, we will prove that *assumption (C) is necessary for filter stability in total variation*. Evidently absolute continuity on the infinite-time horizon is, in a sense, fundamental to the filter stability problem. Though assumption (C) is not commonly stated in the literature on filter stability, it is typically an implicit consequence of the model assumptions. This insight sheds some light on the minimal requirements needed by any method for proving stability. It also reassures us that little is lost by imposing the convenient assumption (A), which was not entirely obvious at the outset.

Let us note that assumption (C) holds in the following special cases:

(c_1). When $\nu \ll \bar{\nu}$ (assumption (A));
(c_2). When $g(x, y) > 0$ for all x, y and $\mathbb{P}(X_n \in \cdot) \ll \bar{\mathbb{P}}(X_n \in \cdot)$ for some $n \ge 0$;
(c_3). When $g(x, y) > 0$ for all x, y and $\|\mathbb{P}(X_n \in \cdot) - \bar{\mathbb{P}}(X_n \in \cdot)\| \to 0$ as $n \to \infty$.

The case (c_1) is immediate from (11.6). The case (c_2) is not difficult to prove, e.g., as in the proof of Proposition 2.5 in [34]. This is the case, for example, when the signal transition kernel has a strictly positive transition density, or in the linear Gaussian filtering model the signal is *controllable*—a typical assumption in stability results for the Kalman filter. The case (c_3) is proved as Lemma 3.7 in [36]. This is typically the case when the signal process is ergodic.

11.5.3 Necessity

We will finally argue that assumption (C) is necessary for filter stability, at least in the sense of total variation.

Proposition 11.6 *Suppose that assumption* (B) *holds, i.e., that the filter stability problem is well posed, but that* (C) *does not hold. Then* $\liminf_{n\to\infty} \mathbb{E}\|\pi_n - \bar{\pi}_n\| > 0$.

To prove the result, assume that (C) does not hold. Then there is set $A \in \mathcal{F}_{0,\infty}^Y$ such that $\bar{\mathbb{P}}(A) = 0$ and $\mathbb{P}(A) > 0$. But note that $\bar{\mathbb{P}}\left(A|\mathcal{F}_{0,n}^Y\right) = 0$ $\bar{\mathbb{P}}$-a.s., and by assumption (B) we also have $\bar{\mathbb{P}}\left(A|\mathcal{F}_{0,n}^Y\right) = 0$ $\mathbb{P}$-a.s. In particular,

$$\left|\mathbb{P}\left(A|\mathcal{F}_{0,n}^Y\right) - \bar{\mathbb{P}}\left(A|\mathcal{F}_{0,n}^Y\right)\right| = \mathbb{P}\left(A|\mathcal{F}_{0,n}^Y\right) \xrightarrow{n\to\infty} I_A \quad \mathbb{P}\text{-a.s.}$$

Using $\|\mu - \mu'\| := 2\sup_B |\mu(B) - \mu'(B)|$, we obtain

$$\left\|\mathbb{P}\left((Y_k)_{k>n} \in \cdot|\mathcal{F}_{0,n}^Y\right) - \bar{\mathbb{P}}\left((Y_k)_{k>n} \in \cdot|\mathcal{F}_{0,n}^Y\right)\right\| \geq 2\left|\mathbb{P}\left(A|\mathcal{F}_{0,n}^Y\right) - \bar{\mathbb{P}}\left(A|\mathcal{F}_{0,n}^Y\right)\right|,$$

so using Fatou's lemma

$$\liminf_{n\to\infty} \mathbb{E}\left\|\mathbb{P}\left((Y_k)_{k>n} \in \cdot|\mathcal{F}_{0,n}^Y\right) - \bar{\mathbb{P}}\left((Y_k)_{k>n} \in \cdot|\mathcal{F}_{0,n}^Y\right)\right\| \geq 2\mathbb{P}(A) > 0.$$

Finally, note that for any $B \in \mathcal{F}_{n+1,\infty}^Y$

$$\mathbb{P}\left(B|\mathcal{F}_{0,n}^Y\right) = \pi_n(f_B), \qquad \bar{\mathbb{P}}\left(B|\mathcal{F}_{0,n}^Y\right) = \bar{\pi}_n(f_B),$$

where $f_B(x) = \mathbb{P}(B|X_n = x)$. Therefore

$$\left\|\mathbb{P}\left((Y_k)_{k>n} \in \cdot|\mathcal{F}_{0,n}^Y\right) - \bar{\mathbb{P}}\left((Y_k)_{k>n} \in \cdot|\mathcal{F}_{0,n}^Y\right)\right\| \leq \|\pi_n - \bar{\pi}_n\|,$$

and the proof is easily completed.

Remark 11.6 The above proof applies only to stability in the total variation norm. In general, it may be the case that (C) can be weakened if one is interested in weaker notions of stability; this is related to the consistency problem in Bayesian statistics [17]. Nonetheless, the necessity of (C) for total variation stability reassures us that our absolute continuity assumptions are not particularly restrictive. In particular, most of the literature to date has been concerned with total variation stability, and we have shown that no approach to the filter stability problem can circumvent the absolute continuity assumption (C) in this setting.

Acknowledgements

The work on this survey was initiated during the visit of P. Ch. in Laboratório Nacional de Computacão Cientifica, Petropolis, Brazil in August 2006 upon the invitation of Prof. Jack Baczynski, whose hospitality is greatly appreciated. The authors also thank a referee for several suggestions that have helped improve the presentation.

Notes

1. Typical choices are a finite or countable set, a subset of $\mathbb{R}^q$ for some $q \geq 1$, or $\mathbb{R}^q$ itself.
2. We will denote the Euclidean norm on $\mathbb{R}^p$ as $|\cdot|$.
3. Functions and measures on finite $\mathbb{S}$ are identified with vectors in $\mathbb{R}^d$ and $S^{d-1} = \{x \in \mathbb{R}^d : \sum_{i=1}^d x_i = 1, x_i \geq 0\}$, respectively.
4. The countable generation requirement can be weakened somewhat, see Lemma 4.II.1 in [37]; in general, however, verifying the weaker requirement appears to be a very hard problem.

References

[1] Atar, Rami; Zeitouni, Ofer. Lyapunov exponents for finite state nonlinear filtering. *SIAM J. Control Optim.* 35 (1997), no. 1, 36–55.

[2] Atar, Rami; Zeitouni, Ofer. Exponential stability for nonlinear filtering. *Ann. Inst. H. Poincaré Probab. Statist.* 33 (1997), no. 6, 697–725.

[3] Baxendale, Peter; Chigansky, Pavel; Liptser, Robert. Asymptotic stability of the Wonham filter: ergodic and nonergodic signals. *SIAM J. Control Optim.* 43 (2004), no. 2, 643–69

[4] Blackwell, D.; Dubins, L. Merging of opinions with increasing information. *Ann. Math. Statist.* 33 (1962), 882–6.

[5] Budhiraja, A.; Ocone, D. Exponential stability of discrete-time filters for bounded observation noise. *Systems Control Lett.* 30 (1997), no. 4, 185–93

[6] Budhiraja, A.; Ocone, D. Exponential stability in discrete-time filtering for non-ergodic signals. *Stochastic Process. Appl.* 82 (1999), no. 2, 245–57.

[7] Cappé, O.; Moulines, E.; Rydén, T. *Inference in hidden Markov models*. Springer Series in Statistics. Springer, New York, 2005.

[8] Chigansky, Pavel; Liptser, Robert. Stability of nonlinear filters in nonmixing case. *Ann. Appl. Probab.* 14 (2004), no. 4, 2038–56.

[9] Chigansky, Pavel; Liptser, Robert. On a role of predictor in the filtering stability. *Electron. Comm. Probab.* 11 (2006), 129–40

[10] Chigansky, Pavel. An ergodic theorem for filtering with applications to stability. *Systems Control Lett.* 55 (2006), no. 11, 908–17

[11] Chigansky, Pavel. Stability of the nonlinear filter for slowly switching Markov chains. *Stochastic Process. Appl.* 116 (2006), no. 8, 1185–94.

[12] Chigansky, Pavel; van Handel, Ramon. Model robustness of finite state nonlinear filtering over the infinite time horizon. *Ann. Appl. Probab.* 17 (2007), no. 2, 688–715.

[13] Clark, J. M. C.; Ocone, D. L.; Coumarbatch, C. Relative entropy and error bounds for filtering of Markov processes. *Math. Control Signals Systems* 12 (1999), no. 4, 346–60.

[14] Del Moral, P. A uniform convergence theorem for the numerical solving of the nonlinear filtering problem. *J. Appl. Probab.* 35 (1998), 873–84.

[15] Del Moral, Pierre; Guionnet, Alice. On the stability of interacting processes with applications to filtering and genetic algorithms. *Ann. Inst. H. Poincaré Probab. Statist.* 37 (2001), no. 2, 155–94.

[16] Delyon, Bernard; Zeitouni, Ofer. *Lyapunov exponents for filtering problems*. Applied stochastic analysis, 511–521, Stochastics Monogr., 5, Gordon and Breach, New York, 1991.

[17] Diaconis, P.; Freedman, D. On the consistency of Bayes estimates. *Ann. Statist.* 14 (1986), 1–67.

[18] Douc, R.; Moulines, E.; Ritov, Y. Forgetting of the initial condition for the filter in general state-space hidden Markov chain: a coupling approach. Eketnow. J. Probab. 14 (2009), no. 2, 27–49.

[19] Kaijser, Thomas. A limit theorem for partially observed Markov chains. *Ann. Probability* 3 (1975), no. 4, 677–96.

[20] Kleptsyna, Marina L.; Veretennikov, Alexander Yu. On ergodic filters with wrong initial data. *C. R. Math. Acad. Sci. Paris* 344 (2007), no. 11, 727–31

[21] Kunita, Hiroshi Asymptotic behavior of the nonlinear filtering errors of Markov processes. *J. Multivariate Anal.* 1 (1971), 365–93.

[22] LeGland, Francois; Oudjane, Nadia. A robustification approach to stability and to uniform particle approximation of nonlinear filters: the example of pseudo-mixing signals. *Stochastic Process. Appl.* 106 (2003), no. 2, 279–316.

[23] Le Gland, Francois; Oudjane, Nadia. Stability and uniform approximation of nonlinear filters using the Hilbert metric and application to particle filters. *Ann. Appl. Probab.* 14 (2004), no. 1, 144–87.

[24] Liptser, Robert S.; Shiryaev, Albert N. *Statistics of random processes. Theory (I) and Applications (II).* Applications of Mathematics (New York), 5,6. Stochastic Modelling and Applied Probability. Springer-Verlag, Berlin, 2001.

[25] Meyn, S. P.; Tweedie, R. L. *Markov chains and stochastic stability.* Communications and Control Engineering Series. Springer-Verlag, London, 1993.

[26] Nummelin, E. A splitting technique for Harris recurrent Markov chains. *Z. Wahrsch. Verw. Gebiete* 43 (1978), no. 4, 309–18.

[27] Ocone, Daniel; Pardoux, Etienne. Asymptotic stability of the optimal filter with respect to its initial condition. *SIAM J. Control Optim.* 34 (1996), no. 1, 226–43.

[28] Ocone, Daniel. Entropy inequalities and entropy dynamics in nonlinear filtering of diffusion processes. (English summary) *Stochastic analysis, control, optimization and applications,* 477–96, Systems Control Found. Appl., Birkhauser Boston, Boston, MA, 1999.

[29] Ocone, D. L. Asymptotic stability of Benes filters. *Stochastic Anal. Appl.* 17 (1999), no. 6, 1053–74.

[30] Stannat, Wilhelm. Stability of the filter equation for a time-dependent signal on $\mathbb{R}^d$. *Appl. Math. Optim.* 52 (2005), no. 1, 39–71.

[31] Stannat, Wilhelm. Stability of the optimal filter via pointwise gradient estimates. Stochastic partial differential equations and applications—VII, 281–93, *Lect. Notes Pure Appl. Math.,* 245, Chapman & Hall/CRC, Boca Raton, FL, 2006.

[32] Van Handel, Ramon. Observability and nonlinear filtering. *Probab. Th. Rel. Fields* 145 (2009), no. 1–2, 35–74.

[33] Van Handel, Ramon. Uniform observability of hidden Markov models and filter stability for unstable signals, *Ann. Appl. Probab.* 19 (2009), no. 3, 1172–1199.

[34] Van Handel, Ramon. Discrete time nonlinear filters with informative observations are stable. *Electr. Commun. Probab.* 13 (2008), 562–75.

[35] Van Handel, Ramon. Filtering, stability, and robustness (Caltech PhD thesis) 2007.

[36] Van Handel, Ramon. The stability of conditional Markov processes and Markov chains in random environments. *Ann. Probab.* 37 (2009), no. 5, 1876–1925.

[37] von Weizsäcker, Heinrich. Exchanging the order of taking suprema and countable intersections of σ-algebras. *Ann. Inst. H. Poincaré Sect. B* (N.S.) 19 (1983), no. 1, 91–100.

·12·

Feller and Stability Properties of the Nonlinear Filter

A. Budhiraja

12.1 Introduction

The classical model of nonlinear filtering consists of a pair of stochastic processes $(X_t, Y_t)_{t\geq 0}$ where (X_t) is called the signal process and (Y_t) the observation process. The signal is taken to be a Markov process with values in some Polish space $\mathbb{S}$ and the observations are given via the relation:

$$Y_t = \int_0^t h(X_s)ds + W_t, \tag{12.1}$$

where (W_t) is a standard d-dimensional Brownian motion independent of (X_t) and h, referred to as the observation function, is a map from $\mathbb{S} \to \mathbb{R}^d$. Nonlinear filtering concerns with the study of the measure-valued process (Π_t) which is the conditional distribution of X_t given $\mathcal{Y}_t = \sigma\{Y_s : 0 \leq s \leq t\}$. This measure-valued process is called the nonlinear filter. In a very influential paper, Kunita[19] has shown, using uniqueness of solutions to Kushner–Stratonovich equations, that in the above filtering model if the signal is Feller–Markov with a compact Hausdorff state space $\mathbb{S}$ then the nonlinear filter is also a Feller–Markov process with state space $\mathcal{P}(\mathbb{S})$, where $\mathcal{P}(\mathbb{S})$ is the space of all probability measures on $\mathbb{S}$. Furthermore, [19] provides a proof of the statement that if the signal has a unique invariant measure μ then, under a minor additional condition, the filter has a unique invariant measure. Recently, Baxendale et al. [4] have found a gap in the proof of this fundamental result. In this chapter we revisit this work of Kunita and several other works that extend or make use of [19], in light of the discovery made in [4].

We first consider the Feller property and existence of invariant measures. It is shown in Sections 12.3 and 12.4 that these properties hold in a much greater generality than the one originally considered in [19] (or subsequently in [25]). We then turn to the study of ergodicity of the nonlinear filter. Section 12.5 begins with a description of the gap pointed out by Baxendale et al. in [4]. The main issue is the equality of certain σ-fields claimed in [19]. To illustrate the difficulty, consider a probability space $(\Omega_0, \mathcal{F}_0, P_0)$ with sub-σ-fields $\mathcal{F}^*$, $(\mathcal{G}_n^*)_{n\geq 0}$, such that all the σ fields are P_0 complete and $\{\mathcal{G}_n^*\}_{n\geq 0}$ is a decreasing family of σ fields such that $\bigcap_{n\geq 0} \mathcal{G}_n^*$ is P_0 trivial. Then the equality claimed in [19] is analogous to the statement that $\mathcal{H}^* \stackrel{.}{=} \bigcap_{n\geq 0} (\mathcal{F}^* \vee \mathcal{G}_n^*) = \mathcal{F}^*$. This equality holds in many circumstances—for instance when $\mathcal{F}^*$ is independent of $\mathcal{G}_0^*$—however, it is false in general. Clearly $\mathcal{H}^* \supseteq \mathcal{F}^*$, but one can construct examples that show that the reverse inclusion can fail. In Section 12.5 of the chapter we give some sufficient conditions under which the equality of the σ-fields stated in [19] can be shown to hold. We also refer the reader to a recent work of van Handel [27] where the equality of σ-fields stated in [19] has been established under a suitable strengthening of the ergodicity assumptions on the signal process. Sufficient conditions presented in the current chapter are phrased in terms of certain stability properties of the nonlinear filter, e.g. finite memory property or asymptotic stability. Previous works by several authors on asymptotic stability of the filter for a variety of filtering models, then identify a rich class of filtering problems for which the uniqueness of invariant measures for the nonlinear filter holds. We in fact find that the equality of σ-fields stated in [19] is equivalent to several asymptotic properties of the nonlinear filter, such as, uniqueness of invariant measure for the nonlinear filter, uniqueness of invariant measure for the signal–filter pair, asymptotic stability and finite memory of the filter. This result is presented in Section 12.6.

None of the results of this chapter are new. They are based on results established in [19, 20, 25, 24, 5, 7, 8]. Much of the material in this chapter is taken from [5] and [8]. I wish to acknowledge several people without whom this work would not have been possible. I will like to thank Dan Ocone, who introduced me to the problem of asymptotic stability of the nonlinear filter; Harold Kushner, who explained to me the use of uniqueness of invariant measures of signal–filter pair in studying long term performance of numerical filtering schemes; Rajeeva Karandikar, who showed me the elegance of working on canonical function spaces in the study of Markov and asymptotic properties of the nonlinear filter; and finally Baxendale, Chigansky, and Liptser, for pointing me to the gap in [19].

12.2 Notation and the filtering model

Let $\mathbb{S}$ be a complete separable metric space and let $(\Omega, \mathcal{F}, \mathbb{P})$ be a probability space. Let (X_t) be a homogeneous Markov process with values in $\mathbb{S}$ and transition probability function $p(x, t, B)$, i.e. for $t, \tau > 0$, $x \in \mathbb{S}$ and $B \in \mathcal{S} = \mathcal{B}(\mathbb{S})$

$$\mathbb{P}(X_{t+\tau} \in B | \sigma(X_u : u \leq \tau)) = p(X_\tau, t, B) \text{ a.s.}, \qquad (12.2)$$

where for a Polish space E, $\mathcal{B}(E)$ denotes the Borel sigma field on E. Denote the distribution of X_0 by γ, i.e.

$$\gamma = \mathbb{P} \circ (X_0)^{-1}. \qquad (12.3)$$

We will refer to X as the signal process. Denote by $\mathcal{D} \doteq D([0, \infty), \mathbb{S})$, the Skorokhod space of $\mathbb{S}$ valued cadlag functions on $[0, \infty)$ and let $\xi_t(\cdot)$ be the coordinate process on $\mathcal{D}$, i.e. $\xi_t(\theta) \doteq \theta(t)$ for $\theta \in \mathcal{D}$.

We will assume that (X_t) admits a cadlag version, i.e for all $(s, x) \in [0, \infty) \times \mathbb{S}$ there exists a probability measure $P_{s,x}$ on $\mathcal{D}$ such that for $0 \leq s < t < \infty$, and $U \in \mathcal{B}(\mathbb{S})$,

$$P_{s,x}(\xi_t \in U | \sigma(\xi_u : u \leq s)) = p(\xi_s, t - s, U) \text{ a.s. } P_{s,x} \tag{12.4}$$

and

$$P_{s,x}(\xi_u = x, 0 \leq u \leq s) = 1. \tag{12.5}$$

For notational simplicity, $P_{0,x}$ will hereafter be denoted as P_x.

We will also assume that the Markov process is Feller, i.e. for all $s \geq 0$ the map $x \mapsto P_{s,x}$ is a continuous map from $\mathbb{S}$ to $\mathcal{P}(\mathcal{D})$, where for a Polish space E, $\mathcal{P}(E)$ denotes the space of probability measures on $(E, \mathcal{B}(E))$ endowed with the usual topology of weak convergence. Let (T_t) denote the semigroup corresponding to the Markov process (X_t), i.e. for $f \in B(\mathbb{S})$ (for a Polish space E, $B(E)$ denotes the space of real bounded measurable functions on E),

$$(T_t f)(x) \doteq \int_{\mathcal{D}} f(\xi_t(\theta)) d P_x(\theta).$$

The Feller property of the Markov process says that for $f \in C_b(\mathbb{S})$, $(T_t f) \in C_b(\mathbb{S})$, where for a Polish space E, $C_b(E)$ denotes the space of continuous and bounded functions on E.

We will consider the classical filtering model where signal is independent of observation noise. More precisely, the observation process is given as follows:

$$Y_t = \int_0^t h(X_u) du + W_t,$$

where $h : \mathbb{S} \to \mathbb{R}^d$ is a continuous mapping and (W_t) is a $\mathbb{R}^d$-valued standard Wiener process, assumed to be independent of (X_t). Denote by Π_t the conditional distribution of X_t given past and current observations, i.e. for $A \in \mathcal{B}(\mathbb{S})$,

$$\Pi_t(A) \doteq P(X_t \in A | \sigma\{Y_u : 0 \leq u \leq t\}). \tag{12.6}$$

Note that Π is a stochastic process with values in $\mathcal{P}(\mathbb{S})$. This measure-valued process is usually referred to as the nonlinear filter.

Kallianpur–Striebel formula

In Section 12.3 we will describe some basic Markov properties of the nonlinear filter. Original approaches to this study, taken in [19, 25, 20], were based on unique solvability results for Kushner–Stratonovich equation. However, Markov property of the nonlinear filter stems from a much more elementary semigroup property (see (12.12)) that can be seen easily using the Kallianpur–Striebel formula. This formula which gives a representation for the nonlinear filter in terms of certain functional integrals on suitable path spaces is one of the cornerstones in the theory of nonlinear filtering. In order to describe the formula we begin with the following canonical setting. Let (β_t) be the coordinate process on $\mathcal{C} \doteq C([0, \infty) : \mathbb{R}^d)$ (the space of continuous functions from $[0, \infty)$ to $\mathbb{R}^d$), i.e. $\beta_t(\eta) \doteq \eta(t)$ for $\eta \in \mathcal{C}$. Let Q be the standard Wiener measure on $(\mathcal{C}, \mathcal{B}(\mathcal{C}))$. Also set

$$(\hat{\Omega}, \hat{\mathcal{F}}) \doteq (\mathcal{D} \times \mathcal{C}, \mathcal{B}(\mathcal{D}) \otimes \mathcal{B}(\mathcal{C})).$$

With an abuse of notation, we will once more denote by ξ and β the extensions of these processes to the above product space. In particular for $(\theta, \eta) \in \hat{\Omega}$, $t \geq 0$,

$$\xi_t(\theta, \eta) = \xi_t(\theta) = \theta(t); \quad \beta_t(\theta, \eta) = \beta_t(\eta) = \eta(t).$$

Define for $\nu \in \mathcal{P}(\mathbb{S})$, $s > 0$, the so-called "reference probability measure;",

$$R_{s,\nu} \doteq P_{s,\nu} \otimes Q,$$

where $P_{s,\nu} \in \mathcal{P}(\mathcal{D})$ is defined as:

$$P_{s,\nu}(B) \doteq \int_{\mathbb{S}} P_{s,x}(B) \nu(dx), \quad B \in \mathcal{B}(\mathcal{D}).$$

We will sometimes write $P_{0,\nu}$, $R_{0,\nu}$ as P_ν and R_ν respectively. From Theorem 3 in [18] it follows that there is a continuous stochastic process (Z_t), $Z_t : \hat{\Omega} \to \mathbb{R}$, such that for all $0 \leq s \leq t$ and $\nu \in \mathcal{P}(\mathbb{S})$:

$$Z_t - Z_s = \int_s^t h(\xi_u) \cdot d\beta_u, \quad a.s. \ R_{s,\nu}.$$

Note that the stochastic integral on the right side of the above display is defined a.e. $R_{s,\nu}$ for each (s, ν). The significance of the above statement is that one can find a *common version* of these stochastic integrals.

Next, for $0 \leq s \leq t$, let

$$q_{st} \doteq \exp\left(Z_t - Z_s - \frac{1}{2}\int_s^t \|h(\xi_u)\|^2 du\right).$$

For a Polish space E let $\mathcal{M}(E)$ denote the space of positive, finite measures on E. For $f \in B(E)$ and $m \in \mathcal{M}(E)$ we will denote $\int_E f(x) dm(x)$ by $\langle m, f \rangle$ or $m(f)$.

For $\nu \in \mathcal{M}(\mathbb{S})$ and $0 \leq s \leq t < \infty$, define a $\mathcal{M}(\mathbb{S})$-valued process $\Gamma_{st}(\nu)$ on $\mathcal{C}$ as

$$\langle \Gamma_{st}(\nu)(\eta), f \rangle \doteq \int_{\mathbb{S}} \int_{\mathcal{D}} f(\xi_t(\theta)) q_{st}(\theta, \eta) d P_{s,x}(\theta) d\nu(x); \quad \eta - a.s. \ [Q]. \tag{12.7}$$

It is easy to check that $\langle \Gamma_{st}(\nu), 1 \rangle > 0$ a.s. Q. Define for $0 \leq s \leq t$ and $\nu \in \mathcal{M}(\mathbb{S})$ a $\mathcal{P}(\mathbb{S})$ valued random variable $\Lambda_{st}(\nu)$ via the normalization of $\Gamma_{st}(\nu)$, i.e.

$$\Lambda_{st}(\nu) \doteq \frac{\Gamma_{st}(\nu)}{\langle \Gamma_{st}(\nu), 1 \rangle}. \tag{12.8}$$

Also with an abuse of notation we will sometimes denote $\Gamma_{0t}(\nu)$ and $\Lambda_{0t}(\nu)$ by $\Gamma_t(\nu)$ and $\Lambda_t(\nu)$ respectively.

The Kallianpur–Striebel formula (see [17]) states that the nonlinear filter Π can be represented as follows: For $f \in B(\mathbb{S})$

$$\langle \Pi_t(\omega), f \rangle = \langle \Lambda_{0t}(\gamma)(Y.(\omega)), f \rangle, \quad \omega - \text{a.s.}\ [\mathbb{P}], \quad t \in [0, \infty). \tag{12.9}$$

The formula in particular says that determining the nonlinear filter amounts to computing functional integrals of the form in (12.7).

12.3 Markov properties of the filter

In this section we will describe Feller–Markov properties of the nonlinear filter. Results of this section for the setting where h is bounded and continuous and $\mathbb{S}$ is locally compact were obtained in [19, 25, 20]. Results presented here are taken from [5] which using Kallianpur–Striebel formula (12.9), along with equations (12.7) and (12.8), extended the results of the above cited papers to the setting of general continuous maps h and Polish state spaces $\mathbb{S}$. We begin with some notation. Let $\tilde{\mathcal{F}}$ be the Q-completion of $\mathcal{B}(\mathcal{C})$ and $\tilde{\mathcal{N}}$ be the class of Q-null sets in $\tilde{\mathcal{F}}$. For $0 \leq s \leq t \leq \infty$, let $\mathcal{A}_s^t$ be the sub σ-fields of $\tilde{\mathcal{F}}$ defined by

$$\mathcal{A}_s^t = \sigma(\sigma(\beta_u - \beta_s : s \leq u \leq t) \cup \tilde{\mathcal{N}}). \tag{12.10}$$

Next we introduce the probability measure on $\mathcal{C}$ under which the canonical process has the same law as the observation process. For an arbitrary $\nu \in \mathcal{P}(\mathbb{S})$ let $Q_\nu \in \mathcal{P}(\mathcal{C})$ be defined by

$$\frac{dQ_\nu}{dQ} = \Gamma_t(\nu)(\mathbb{S}) \text{ on } \mathcal{A}_0^t, \quad t \in [0, \infty). \tag{12.11}$$

It is easy to see that Q_y is the law of the observation process under $\mathbb{P}$, i.e. $\mathbb{P} \circ Y^{-1} = Q_y$. We will denote the expectation operator on $(\mathcal{C}, \mathcal{B}(\mathcal{C}))$ under Q and Q_ν, as $\mathbb{E}_Q$ and $\mathbb{E}_{Q_\nu}$, respectively.

Theorem 12.1 (Π_t) *is a* $\mathcal{P}(\mathbb{S})$ *valued Markov process with associated semigroup* $\{\mathcal{T}_t\}_{0 \leq t < \infty}$ *defined as follows.*

$$(\mathcal{T}_t F)(\nu) \doteq \mathbb{E}_{Q_\nu}(F(\Lambda_t(\nu))); \quad F \in B(\mathcal{P}(\mathbb{S})); \quad \nu \in \mathcal{P}(\mathbb{S}).$$

Furthermore $(\mathcal{T}_t)$ *is a Feller semigroup.*

The above theorem in particular says that for $F \in B(\mathcal{P}(\mathbb{S}))$, $0 \leq s \leq t$,

$$\mathbb{E}(F(\Pi_t) \mid \sigma\{\Pi_u; u \leq s\}) = (\mathcal{T}_t F)(\Pi_s), \quad a.s.$$

The key ingredients in the proof of the Markov property of Π are the following semigroup properties of the operators Γ and Λ: For $0 \leq s < t < \infty$, $\nu \in \mathcal{P}(\mathbb{S})$,

$$\Gamma_{0t}(\nu) = \Gamma_{st}(\Gamma_{0s}(\nu)), \quad \Lambda_{0t}(\nu) = \Lambda_{st}(\Lambda_{0s}(\nu)). \quad (12.12)$$

We refer the reader to Theorem 4.2 of [5] for a proof of (12.12). Feller property is a consequence of the following convergence result: Let $\{\nu_n\} \subset \mathcal{P}(\mathbb{S})$ be a sequence converging to ν weakly. Then for all $t \geq 0$, as $n \to \infty$, $\Lambda_t(\nu_n) \to \Lambda_t(\nu)$ in Q probability and $\Lambda_t(\nu_n, \mathbb{S}) \to \Lambda_t(\nu, \mathbb{S})$ in $L^1(Q)$.

In various approximation problems in nonlinear filtering (see for example [10, 9, 11, 12]) one needs to work with an incorrectly initialized filter. Roughly speaking, an incorrectly initialized filter is the nonlinear filter computed under the assumption that the Markov process X has an initial distribution $\nu \neq \gamma$ while γ is the true initial law of the signal X that produces the physical observations Y. More precisely, for $\nu \in \mathcal{P}(E)$, a filter initialized at ν is defined as

$$\Pi_t^\nu(\omega) \doteq \Lambda_t(\nu)(Y.(\omega)), \quad \omega \in \Omega.$$

Note that, just as (Π_t), (Π_t^ν) is a $\mathcal{P}(\mathbb{S})$ valued stochastic process defined on $(\Omega, \mathcal{F}, \mathbb{P})$, but the latter cannot be in general interpreted as a conditional distribution. Also, unlike (Π_t), (Π_t^ν) is not a Markov process. Nevertheless, the following result (see theorems 4.8 and 5.3 of [5]) shows that when augmented with the signal process, the pair (X_t, Π_t^ν) is indeed a Markov process with a Feller semigroup.

In order to state this Markov property we need the following notation. For $\nu \in \mathcal{P}(\mathbb{S})$ define

$$\mathcal{K}_s^t(\nu) \doteq \sigma(\sigma\{\beta_u - \beta_s : s \leq u \leq t\} \cup \sigma\{\xi_u : s \leq u \leq t\} \cup \mathcal{N}), \quad (12.13)$$

where $\mathcal{N}$ is the class of all $R_{0,\nu}$ null sets. Now for fixed $\nu \in \mathcal{P}(\mathbb{S})$ define $\hat{R}_\nu$ on $(\hat{\Omega}, \hat{\mathcal{F}})$ as follows:

$$\frac{d\hat{R}_\nu}{dR_\nu}(\theta, \eta) \doteq q_{0t}(\theta, \eta) \quad \text{on } \mathcal{K}_0^t(\nu), \ t \geq 0. \quad (12.14)$$

It is easily seen that R_γ and $\hat{R}_\gamma$ are the laws of (X, W) and (X, Y), respectively, under $\mathbb{P}$. I.e.

$$R_\gamma = \mathbb{P} \circ (X., W.)^{-1}, \quad \hat{R}_\gamma = \mathbb{P} \circ (X., Y.)^{-1}.$$

Theorem 12.2 *The pair (X_t, Π_t^ν) is a $\mathbb{S} \times \mathcal{P}(\mathbb{S})$ valued Feller–Markov process on $(\Omega, \mathcal{F}, \mathbb{P})$ with associated semigroup $\{S_t\}_{0 \leq t < \infty}$ defined as follows. For $F \in B(\mathbb{S} \times \mathcal{P}(\mathbb{S}))$,*

$$(\mathcal{S}_t F)(x, \lambda) \doteq \mathbb{E}_{\hat{R}_{0,x}}[F(\xi_t, \Lambda_t(\lambda))]; \quad (x, \lambda) \in \mathbb{S} \times \mathcal{P}(\mathbb{S}).$$

In particular, for all $v \in \mathcal{P}(\mathbb{S})$ and $0 \leq s \leq t$,

$$\mathbb{E}\left(F\left(X_t, \Pi_t^v\right) \mid \sigma\{X_u, \Pi_u^v, u \leq s\}\right) = (\mathcal{S}_t F)\left(X_s, \Pi_s^v\right), \text{ a.s., } \mathbb{P}.$$

12.4 Invariant measures for the filter

In this section we will summarize some results of [19, 25, 20] which give constructions of invariant measures for the filter (and for the signal–filter pair) starting from an invariant measure for the signal, by considering the so-called 'stationary filtering problem.' The following will be the standing assumption for the rest of this chapter.

Assumption 12.1 There is a unique invariant probability measure, μ, for the semigroup (T_t).

The stationary filtering problem is described as follows. Let $\mathcal{D}_\mathbb{R} \equiv D((-\infty, \infty); \mathbb{S})$ denote the space of r.c.l.l. functions from $(-\infty, \infty)$ to $\mathbb{S}$ with Skorokhod topology and $\mathcal{C}_\mathbb{R} \equiv C((-\infty, \infty); \mathbb{R}^d)$ denote the space of continuous functions from $(-\infty, \infty)$ to $\mathbb{R}^d$ with topology of uniform convergence on compact subsets of $(-\infty, \infty)$. Let the coordinate processes on $\mathcal{D}_\mathbb{R}$ and $\mathcal{C}_\mathbb{R}$ be denoted by $(\tilde{\xi}_t(\cdot))$ and $(\tilde{\beta}_t(\cdot))$ respectively. Let $P_\mu^{(1)}$ be the unique measure on $(\mathcal{D}_\mathbb{R}, \mathcal{B}(\mathcal{D}_\mathbb{R}))$ which satisfies for $E_1, \cdots, E_n \in \mathcal{B}(\mathbb{S})$ and $-\infty < t_1 < t_2 \cdots < t_n < \infty$,

$$P_\mu^{(1)}(\tilde{\xi}_{t_1} \in E_1, \cdots, \tilde{\xi}_{t_n} \in E_n)$$
$$= \int_{E_1 \times \cdots \times E_n} \mu(dx_1) p(x_1, t_2 - t_1, dx_2) \cdots p(x_{n-1}, t_n - t_{n-1}, dx_n).$$

Let $Q^{(1)}$ be a probability measure on $(\mathcal{C}_\mathbb{R}, \mathcal{B}(\mathcal{C}_\mathbb{R}))$, such that for $-\infty < t_0 < t_1 \cdots < t_n < \infty$,

$$\left(\frac{1}{\sqrt{t_1 - t_0}}(\tilde{\beta}_{t_1} - \tilde{\beta}_{t_0}), \cdots, \frac{1}{\sqrt{t_n - t_{n-1}}}(\tilde{\beta}_{t_n} - \tilde{\beta}_{t_{n-1}})\right)$$

are independent $N(0, I_{d \times d})$.

Now let $\Omega^1 \doteq \mathcal{D}_\mathbb{R} \times \mathcal{C}_\mathbb{R}$ and define $R_\mu^{(1)} = P_\mu^{(1)} \otimes Q^{(1)}$. Expectation under $R_\mu^{(1)}$ will be denoted as $\mathbb{E}_{R_\mu^{(1)}}$. As before, with an abuse of notation, we will denote once more by $(\tilde{\xi}_t), (\tilde{\beta}_t)$ the extensions of these processes to the product space $\left(\Omega^1, \mathcal{B}(\Omega^1), R_\mu^{(1)}\right)$. We will refer to $\tilde{\xi}$ as the stationary signal process. Let $\mathcal{F}^*$ be the completion of $\mathcal{B}(\Omega^1)$ under $R_\mu^{(1)}$. Define the observation process:

$$a_t - a_s \doteq \int_s^t h(\tilde{\xi}_u) du + \tilde{\beta}_t - \tilde{\beta}_s.$$

Note that a and $\tilde{\beta}$ have stationary increments under $R_\mu^{(1)}$). The processes ($\tilde{\xi}, \tilde{\beta}, a$) form the signal, noise, and observations, respectively, in the stationary filtering problem. Define the sigma fields

$$\mathcal{Z}_s^t \doteq \sigma(a_v - a_u; s \leq u \leq v \leq t) \cup \mathcal{N}^*), \tag{12.15}$$

$$\mathcal{G}_s^t = \sigma(\sigma(\tilde{\xi}_u : s \leq u \leq t) \cup \mathcal{N}^*) \tag{12.16}$$

where $-\infty \leq s < t \leq \infty$ and $\mathcal{N}^*$ is the class of $R_\mu^{(1)}$ null sets in $\mathcal{F}^*$. Finally, let $\mathcal{G}_{-\infty}^{-\infty}$ be defined as

$$\mathcal{G}_{-\infty}^{-\infty} = \bigcap_{-\infty < t < \infty} \mathcal{G}_{-\infty}^t.$$

Now define for $-\infty < s < t < \infty$,

$$\pi_{s,t}^{(0)} \doteq \Lambda_{t-s}(\mu)(a^s),$$

where $a^s : \Omega^1 \to C([0, \infty); \mathbb{R}^d)$ is defined as $a_u^s(\omega) \doteq a_{s+u}(\omega) - a_s(\omega)$. Also define

$$\pi_{s,t}^{(1)} \doteq \Lambda_{t-s}(\delta_{\tilde{\xi}_s})(a^s).$$

Note that $\pi_{s,t}^{(0)}$ and $\pi_{s,t}^{(1)}$ are $\mathcal{P}(\mathbb{S})$ valued random variables defined on $(\Omega^1, \mathcal{B}(\Omega^1), R_\mu^{(1)})$. These measure valued random variables can be interpreted as conditional distributions as follows. For f in $B(\mathbb{S})$, a.s. $R_\mu^{(1)}$,

$$\pi_{s,t}^{(0)}(f) = \mathbb{E}_{R_\mu^{(1)}}\left[f(\tilde{\xi}_t)|\mathcal{Z}_s^t\right] \tag{12.17}$$

and

$$\pi_{s,t}^{(1)}(f) = \mathbb{E}_{R_\mu^{(1)}}\left[f(\tilde{\xi}_t)|\mathcal{Z}_s^t \vee \sigma(\tilde{\xi}_s)\right]. \tag{12.18}$$

(for two sigma fields $\mathcal{L}_1$ and $\mathcal{L}_2$, $\mathcal{L}_1 \vee \mathcal{L}_2 \doteq \sigma(\mathcal{L}_1 \cup \mathcal{L}_2)$.) The following lemma identifies the laws of above filtering processes. Define for $\nu \in \mathcal{P}(\mathbb{S})$ and $t \geq 0$, $m_t^\nu, M_t^\nu \in \mathcal{P}(\mathcal{P}(\mathbb{S}))$ as

$$m_t^\nu(A) \doteq (\mathcal{T}_t \mathcal{I}_A)(\nu) = \mathbb{E}_{Q_\nu}(\mathcal{I}_A(\Lambda_t(\nu, \cdot))), \quad A \in \mathcal{B}(\mathcal{P}(\mathbb{S})) \tag{12.19}$$

and

$$M_t^\nu(A) \doteq \int_{\mathbb{S}} (\mathcal{T}_t \mathcal{I}_A)(\delta_x) \nu(dx), \quad A \in \mathcal{B}(\mathcal{P}(\mathbb{S})) \tag{12.20}$$

where $\mathcal{I}_A$ is the indicator function of the set A.

Lemma 12.1 For $-\infty < s < t < \infty$, $R_\mu^{(1)} \circ \left(\pi_{s,t}^{(1)}\right)^{-1} = M_{t-s}^\mu$ and $R_\mu^{(1)} \circ \left(\pi_{s,t}^{(0)}\right)^{-1} = m_{t-s}^\mu$. In particular, for $i = 0, 1$, the law of $\pi_{s+r,t+r}^{(i)}$ is independent of r.

Proof. We will only prove the first statement. Note that for F in $B(\mathcal{P}(\mathbb{S}))$

$$\mathbb{E}_{R_\mu^{(1)}}\left[F\left(\bar{\pi}_{s,t}^{(1)}\right)\right] = \mathbb{E}_{R_\mu^{(1)}}[F(\Lambda_{t-s}(\delta_{\tilde{\xi}_s})(a^s))]$$

$$= \int_\mathbb{S} \mathbb{E}_{Q_x} F(\Lambda_{t-s}(\delta_x))\mu(dx)$$

$$= \int_\mathbb{S} (\mathcal{T}_{t-s}F)(\delta_x)\mu(dx) \qquad (12.21)$$

$$= M_{t-s}^\mu(F). \qquad (12.22)$$

The result follows. $\square$

Invariant measures for the nonlinear filter are now obtained as follows.

Theorem 12.3 *As $u \to \infty$, M_u^μ and m_u^μ converge weakly to M^μ and m^μ respectively. These probability measures are $(\mathcal{T}_t)$ invariant and correspond to the laws of $\mathcal{P}(\mathbb{S})$ valued random variables $\bar{\pi}_t^{(0)}, \bar{\pi}_t^{(1)}$, respectively, defined as follows. For $f \in B(\mathbb{S})$*

$$\bar{\pi}_t^{(0)}(f) \doteq \mathbb{E}_{R_\mu^{(1)}}\left[f(\tilde{\xi}_t)|\mathcal{Z}_{-\infty}^t\right]. \qquad (12.23)$$

$$\bar{\pi}_t^{(1)}(f) \doteq \mathbb{E}_{R_\mu^{(1)}}\left[f(\tilde{\xi}_t)| \bigcap_{s=-\infty}^\infty (\mathcal{Z}_{-\infty}^t \vee \mathcal{G}_{-\infty}^s)\right]. \qquad (12.24)$$

Proof. An application of martingale convergence theorem in (12.17) shows that as $s \to -\infty$, almost surely, the measure $\bar{\pi}_{s,t}^{(0)}$ converges (weakly) to $\bar{\pi}_t^{(0)}$. Furthermore, using the Markov property of $\tilde{\xi}$, we have that (cf. lemma 3.3 of Kunita [19])

$$\bar{\pi}_{s,t}^{(1)}(f) = \mathbb{E}_{R_\mu^{(1)}}\left[f(\tilde{\xi}_t)|\mathcal{Z}_{-\infty}^t \vee \mathcal{G}_{-\infty}^s\right].$$

Now applying reverse martingale convergence theorem in (12.18) yields that as $s \to -\infty$, a.s., $\bar{\pi}_{s,t}^{(1)}$ converges weakly to $\bar{\pi}_t^{(1)}$. These observations along with Lemma 12.1 yield that M_u^μ and m_u^μ converge weakly as $u \to \infty$ to the law of $\bar{\pi}_t^{(0)}$, $\bar{\pi}_t^{(1)}$ respectively, in particular showing that the laws of $\bar{\pi}_t^{(0)}, \bar{\pi}_t^{(1)}$ are the same for all t. Denote these laws as m^μ and M^μ respectively. Thus we have that

$$m_u^\mu \to m^\mu; \quad M_u^\mu \to M^\mu, \quad \text{as } u \to \infty. \qquad (12.25)$$

Finally, since $(\mathcal{T}_t)$ is a Feller semigroup, it follows from (12.19) and (12.20) that both m^μ and M^μ have to be $(\mathcal{T}_t)$ invariant. $\square$

12.5 Filter ergodicity

Theorem 12.3 shows that under Assumption 12.1 there is at least one invariant probability measure for the signal. In a similar way one can show (cf. [5]) that, under Assumption 12.1, there is at least one invariant probability measure for the signal–filter pair (i.e. for the semigroup (S_t)). A natural question is to ask that under what additional conditions does the uniqueness of invariant measures holds. It is shown in [19] that m^μ and M^μ are minimal and maximal (T_t) invariant probability measures, respectively, in the sense that if Φ is some other (T_t) invariant probability measure, then $m^\mu(F) \leq \Phi(F) \leq M^\mu(F)$ for all convex bounded and continuous functions F on $\mathcal{M}(\mathbb{S})$. Since the collection of all such F is a measure determining class, we have that the uniqueness of (T_t) invariant probability measure is equivalent to the statement $m^\mu = M^\mu$. In order to establish the latter equality, Kunita [19] introduces the following additional condition on the signal process.

Assumption 12.2 For all $f \in C_b(\mathbb{S})$:

$$\limsup_{t \to \infty} \int_\mathbb{S} |T_t f(x) - \langle \mu, f \rangle| \mu(dx) = 0. \tag{12.26}$$

It is well known that this assumption is equivalent to the statement that the signal is purely nondeterministic, namely, $\mathcal{G}_{-\infty}^{-\infty} = \bigcap_{s=-\infty}^{\infty} \mathcal{G}_{-\infty}^{s}$ is $R_\mu^{(1)}$ trivial (cf. [26]).

It is stated in [19] that under Assumptions 12.1 and 12.2, for all $f \in B(\mathbb{S})$, $\overline{\pi}_t^{(0)}(f) = \overline{\pi}_t^{(1)}(f)$. Such a result, in particular will imply that $m^\mu = M^\mu$ yielding the uniqueness of (T_t) invariant probability measure. The paper [19] does not however prove the statement. Implicit in this statement is the assertion that, under Assumptions 12.1 and 12.2:

$$\bigcap_{s=-\infty}^{\infty} \left(\mathcal{Z}_{-\infty}^0 \vee \mathcal{G}_{-\infty}^s \right) = \mathcal{Z}_{-\infty}^0 \vee \mathcal{G}_{-\infty}^{-\infty} = \mathcal{Z}_{-\infty}^0. \tag{12.27}$$

Currently there is no proof available for the above statement. The above gap in Kunita's [19] proof of filter ergodicity was discovered by Baxendale, Chigansky, and Liptser[4] (also [15]). Several papers (e.g. [25, 24, 5, 7]) have appealed to the results of [19] which in light of the discovered gap need to be revisited. Rest of this section will be concerned with the statement in (12.27) and various asymptotic properties of the nonlinear filter that are closely tied to it.

As noted earlier in the Introduction, a sufficient condition for (12.27) is the independence of $\mathcal{Z}_{-\infty}^0$ and $\mathcal{G}_{-\infty}^0$. The latter property, of course does not hold under $R_\mu^{(1)}$. However, this observation leads to the following lemma which notes that (12.27) is true if $\mathcal{Z}_{-\infty}^0$ is replaced with $\mathcal{Z}_{-a}^0$ for any $a \in [0, \infty)$.

Lemma 12.2 Let $a \in (0, \infty)$ be fixed. Then

$$\bigcap_{t \leq 0} \left(\mathcal{Z}_{-a}^0 \vee \mathcal{G}_{-\infty}^t \right) = \mathcal{Z}_{-a}^0 \vee \mathcal{G}_{-\infty}^{-\infty}.$$

Proof. Define $R^* \in \mathcal{P}(\Omega^1)$ by the relation

$$\frac{dR^*}{dR_\mu^{(1)}} \doteq \exp\left\{-\int_{-a}^0 h(\xi_u)d\alpha(u) + \frac{1}{2}\int_{-a}^0 \|h(\xi_u)\|^2 du\right\}.$$

It can be verified that

$$\text{under } R^*, \ \mathcal{G}_{-\infty}^0 \text{ is independent of } \mathcal{Z}_{-a}^0. \tag{12.28}$$

We refer the reader to [8] for a proof of the above statement. One thus has

$$\bigcap_{t \leq 0}\left(\mathcal{Z}_{-a}^0 \vee \mathcal{G}_{-\infty}^t\right) = \mathcal{Z}_{-a}^0 \vee \mathcal{G}_{-\infty}^\infty \ (\text{mod } R^*),$$

where the equality of the two σ– fields (mod R^*) means that their completions under R^* are the same. Since $R_\mu^{(1)}$ and R^* are mutually absolutely continuous, the result follows. □

In order to say when a can be replaced by ∞ in the above result, we now introduce the property of 'finite memory of the filter' that was introduced in the filter stability problem by Ocone and Pardoux [24]. This property says that for large times (t), the filter initialized at any point $x \in \mathbb{S}$ can be well approximated by a suboptimal filter which is constructed using only the observations from the past τ units of time, for sufficiently large τ.

Definition 12.1 We say that the filter has the finite memory property if for all $\phi \in C_b(\mathbb{S})$

$$\limsup_{\tau \to \infty} \limsup_{t \to \infty} \mathbb{E}_{Q_{\delta_x}} |\langle \Lambda_t(\delta_x), \phi \rangle - \langle \Lambda_{t-\tau,t}(\delta_x T_{t-\tau}), \phi \rangle| = 0; \quad x - \text{a.s } [\mu]. \tag{12.29}$$

We introduce one additional condition on the signal process.

Assumption 12.3 For each $\nu_1, \nu_2 \in \mathcal{P}(\mathbb{S})$ there exists $t \in [0, \infty)$ such that $\nu_1 T_t$ is mutually absolutely continuous with respect to $\nu_2 T_t$.

The above condition is satisfied when the signal is a finite-dimensional jump-diffusion process with nondegenerate diffusion coefficient. The following are the main results of this section.

Theorem 12.4 *Suppose that the filter has the finite memory property. Further suppose that Assumptions 12.2 and 12.3 hold. Then (12.27) holds.*

In light of above result it becomes important to identify conditions under which the filter has the "finite memory property." One useful sufficient condition for this property to hold is given in terms of "asymptotic stability" of the filter. Roughly speaking, asymptotic stability of the filter is the property that the distance, at time t, between filter that is incorrectly initialized and the optimal nonlinear filter, converges to 0 in probability, as $t \to \infty$. The precise definition, as introduced by Ocone and Pardoux [24], is as follows.

Definition 12.2 Let $\mu_1, \mu_2 \in \mathcal{P}(\mathbb{S})$. We say that the filter is (μ_1, μ_2)-asymptotically stable if for all $\phi \in C_b(\mathbb{S})$

$$|\langle \Lambda_t(\mu_1), \phi \rangle - \langle \Lambda_t(\mu_2), \phi \rangle|$$

converges to 0 in Q_{μ_1}- probability as $t \to \infty$.

There have been several papers that have established the asymptotic stability of the filter under appropriate conditions on the signal. The first paper in this direction is [24] where asymptotic stability for Kalman filters was proved. In a sequence of papers Atar and Zeitouni [3, 2] identify several important filtering problems, with a compact or countable state space for the signal, for which asymptotic stability holds. Other works on asymptotic stability for compact state space signals are [21, 16]. The papers [13, 14, 22] study some signals in discrete time with noncompact state space for which asymptotic stability holds. Atar [1] considers a continuous-time filtering problem with noncompact state space and establishes asymptotic stability of the filter. Asymptotic stability for Benes filters is proved in [23]. In [10], asymptotic stability of the filter, for a compact state space signal model and point process observations, is proved. The following result shows that the property of asymptotic stability can be used to prove the finite memory property of the nonlinear filter.

Theorem 12.5 *Suppose that Assumptions 12.2 and 12.3 hold. Further suppose that the filter is (δ_x, μ) asymptotically stable for μ- almost every $x \in \mathbb{S}$. Then the filter has the finite memory property in the sense of Definition 12.1; in particular (12.27) holds.*

The rest of this section is devoted to the proofs of Theorems 12.4 and 12.5. We begin with the following lemma.

Lemma 12.3 *Suppose that Assumptions 12.2 and 12.3 hold. Then for all $\nu \in \mathcal{P}(\mathbb{S})$, $\nu T_t \to \mu$ as $t \to \infty$.*

Proof. Fix $\nu \in \mathcal{P}(\mathbb{S})$. From Assumption 12.3 one can find $\epsilon > 0$ such that $\nu T_\epsilon \ll \mu$. Note that for $f \in C_b(\mathbb{S})$ and $t \geq \epsilon$,

$$(\nu T_t)(f) = \int_\mathbb{S} (T_{t-\epsilon} f)(x) \left(\frac{d\nu T_\epsilon}{d\mu}(x) \right) \mu(dx).$$

This implies that for all $K \in (0, \infty)$

$$|(\nu T_t)(f) - \mu(f)| = \left| \int_\mathbb{S} (T_{t-\epsilon} f)(x) - \mu(f)) \left(\frac{d\nu T_\epsilon}{d\mu}(x) \right) \mu(dx) \right|$$

$$\leq K \int_\mathbb{S} |T_{t-\epsilon} f)(x) - \mu(f)| \mu(dx)$$

$$+ 2 \sup_{x \in \mathbb{S}} |f(x)| \int_\mathbb{S} \left(\frac{d\nu T_\epsilon}{d\mu}(x) \right) I_{\frac{d\nu T_\epsilon}{d\mu}(x) > K} \mu(dx).$$

Taking limit as $t \to \infty$, we have from Assumption 12.2 that

$$\limsup_{t \to \infty} |(\nu T_t)(f) - \mu(f)| \le 2 \sup_{x \in S} |f(x)| \int_S \left(\frac{d\nu T_\epsilon}{d\mu}(x)\right) I_{\frac{d\nu T_\epsilon}{d\mu}(x) > K} \mu(dx).$$

The result now follows on taking limit as $K \to \infty$ in the above display. □

The following lemma will be used in the proof of Lemma 12.6.

Lemma 12.4 *Let $\{\nu_n\}$ be a sequence in $\mathcal{P}(\mathbb{S})$ such that $\nu_n \to \nu$ for some $\nu \in \mathcal{P}(\mathbb{S})$. Then for all $a \in (0, \infty)$*

$$\limsup_{K \to \infty} \limsup_{n \to \infty} \int_{\hat{\Omega}} q_{0,a}(\theta, \eta) 1_{q_{0,a}(\theta,\eta) > K} d R_{\nu_n}(\theta, \eta) = 0.$$

Proof. The proof of this result is contained in theorem 5.1 of [5] and theorem 3.2 of [6], however we sketch the argument for the sake of completeness.

Since $\nu_n \to \nu$ weakly as $n \to \infty$, the Feller property of (T_t) implies that $P_{\nu_n} \to P_\nu$ weakly as $n \to \infty$. Now let $(\tilde{X}_t^n)$ and $(\tilde{X}_t)$ be processes with values in $\mathcal{D}$ defined on some probability space $(\overline{\Omega}, \overline{\mathcal{F}}, \overline{P})$ such that $\overline{P} \circ (\tilde{X}^n)^{-1} = P_{\nu_n}$, $\overline{P} \circ (\tilde{X}.)^{-1} = P_\nu$ and $\tilde{X}^n \to \tilde{X}$ a.s. $\overline{P}$. Define :

$$(\Omega_0, \mathcal{F}_0, R) \doteq (\overline{\Omega} \times C, \overline{\mathcal{F}} \otimes \mathcal{B}(C), \overline{P} \otimes Q)$$

and the processes $Z_.^n$, $Z_.$ on this space as

$$Z_a^n(\overline{\omega}, \eta) \doteq q_{0a}(\tilde{X}^n(\overline{\omega}), \eta),$$

$$Z_a(\overline{\omega}, \eta) \doteq q_{0a}(\tilde{X}(\overline{\omega}), \eta).$$

Then from continuity of h it follows that (cf. [6]) $Z_a^n \to Z_a$ in $L^1(R)$. This immediately yields that

$$\limsup_{K \to \infty} \limsup_{n \to \infty} \int_{\hat{\Omega}} q_{0,a}(\theta, \eta) 1_{q_{0,a}(\theta,\eta) > K} d R_{\nu_n}(\theta, \eta)$$
$$= \limsup_{K \to \infty} \limsup_{n \to \infty} \int_{\Omega_0} Z_a^n 1_{Z_a^n > K} d R$$
$$= 0.$$

This proves the lemma. □

Lemma 12.5 *Let ϕ_i; $i = 1, \cdots k$ be in $C_b(\mathbb{S})$. Let $C \in (1, \infty)$ be such that*

$$\prod_{i=1}^{k} |\phi_i(x_i)| \le C, \quad \forall \, x_i \in \mathbb{S}; \; i = 1, 2, \cdots k. \tag{12.30}$$

Let

$$-\infty < t_1 < t_2 \cdots < t_k = 0.$$

Also let $t, \tau_0 \in (-\infty, 0)$ be such that $t < \tau_0 < t_1$. Then

$$\mathbb{E}_{R_\mu^{(1)}} \left| \mathbb{E}_{R_\mu^{(1)}}[\prod_{i=1}^k \phi_i(\tilde{\xi}_{t_i}) \mid \mathcal{Z}_{-\infty}^0 \vee \mathcal{G}_{-\infty}^t] - \mathbb{E}_{R_\mu^{(1)}}[\prod_{i=1}^k \phi_i(\tilde{\xi}_{t_i}) \mid \mathcal{Z}_{\tau_0}^0 \vee \mathcal{G}_{-\infty}^t] \right| \tag{12.31}$$

is bounded above by $C \int_{\mathbb{S}} (U_1(x) + U_2(x)) \mu(dx)$, where $U_i(x)$ for $i = 1, 2$ is defined as:

$$\int_{\mathcal{C}} \mathbb{E}_Q \left(\Gamma_{t^*}(\delta_x)(\mathbb{S}) \left| \Lambda_{t^*}(\delta_x)(\Psi_i(\cdot, \eta')) - \Lambda_{t^*-\tau^*, t^*}(\delta_x T_{t^*-\tau^*})(\Psi_i(\cdot, \eta')) \right| \right) Q(d\eta'), \tag{12.32}$$

$t^* \doteq t_1 - t$, $\tau^* \doteq t_1 - \tau_0$ and $\Psi_i : \mathbb{S} \times \mathcal{C}$ is defined as follows:

$$\Psi_1(x, \eta) = \phi_1(x) \int_{\mathcal{D}} \prod_{i=2}^k \phi_i(\xi_{t_i - t_{i-1}}) q_{0, -t_1}(\theta, \eta) P_x(d\theta), \tag{12.33}$$

$$\Psi_2(x, \eta) \doteq \Gamma_{0,-t_1}(\delta_x)(\mathbb{S})(\eta) = \int_{\mathcal{D}} q_{0,-t_1}(\theta, \eta) P_x(d\theta). \tag{12.34}$$

Proof. We begin by observing that from the Markov property of the signal and the independence between observation noise and the signal, one can replace the two conditioning σ fields in (12.31) with $\mathcal{Z}_t^0 \vee \sigma\{\xi_t\}$ and $\mathcal{Z}_{\tau_0}^0 \vee \sigma\{\xi_t\}$, respectively. Thus, the expression in (12.31) can be rewritten as an expectation on $(\hat{\Omega}, \hat{\mathcal{F}}, \hat{R}_{\delta_x})$ as,

$$\int_{\mathbb{S}} \mathbb{E}_{\hat{R}_{\delta_x}} \left| \mathbb{E}_{\hat{R}_{\delta_x}}[\prod_{i=1}^k \phi_i(\xi_{t_i - t}) \mid \mathcal{A}_0^{-t}] - \mathbb{E}_{\hat{R}_{\delta_x}}[\prod_{i=1}^k \phi_i(\xi_{t_i - t}) \mid \mathcal{A}_{\tau_0 - t}^{-t}] \right| \mu(dx) \tag{12.35}$$

Now an application of Bayes' formula and a further conditioning yields that

$$\mathbb{E}_{\hat{R}_{\delta_x}} \left[\prod_{i=1}^k \phi_i(\xi_{t_i - t}) \mid \mathcal{A}_0^{-t} \right] = \frac{\int_{\mathcal{D}} \prod_{i=1}^k \phi_i(\xi_{t_i - t}) q_{0, -t}(\theta, \eta) P_x(d\theta)}{\int_{\mathcal{D}} q_{0, -t}(\theta, \eta) P_x(d\theta)}$$

$$= \frac{\int_{\mathcal{D}} \Psi_1(\xi_{t_1 - t}, \gamma_{t_1 - t}(\eta)) q_{0, t_1 - t}(\theta, \eta) P_x(d\theta)}{\int_{\mathcal{D}} \Psi_2(\xi_{t_1 - t}, \gamma_{t_1 - t}(\eta)) q_{0, t_1 - t}(\theta, \eta) P_x(d\theta)}$$

$$= \frac{\Gamma_{t_1 - t}(\delta_x)(\Psi_1(\cdot, \gamma_{t_1 - t}(\eta)))}{\Gamma_{t_1 - t}(\delta_x)(\Psi_2(\cdot, \gamma_{t_1 - t}(\eta)))}$$

$$= \frac{\Lambda_{t_1 - t}(\delta_x)(\Psi_1(\cdot, \gamma_{t_1 - t}(\eta)))}{\Lambda_{t_1 - t}(\delta_x)(\Psi_2(\cdot, \gamma_{t_1 - t}(\eta)))} \tag{12.36}$$

where for $s > 0$; $\gamma_s : \mathcal{C} \to \mathcal{C}$ is defined as $\gamma_s(\eta)(u) \doteq \eta(u + s) - \eta(s)$ and $\Psi_i; i = 1, 2$ are as defined in the statement of the proposition. In exactly the same manner it is shown that

$$\mathbb{E}_{\tilde{R}_{\delta_x}}\left[\prod_{i=1}^{k}\phi_i(\xi_{t_i-t}) \mid \mathcal{A}_{\tau_0-t}^{-t}\right] = \frac{\int_{\mathcal{D}}\Psi_1(\xi_{t_1-t},\gamma_{t_1-t}(\eta))q_{\tau_0-t,t_1-t}(\theta,\eta)P_x(d\theta)}{\int_{\mathcal{D}}\Psi_2(\xi_{t_1-t},\gamma_{t_1-t}(\eta))q_{\tau_0-t,t_1-t}(\theta,\eta)P_x(d\theta)}$$

$$= \frac{\Lambda_{\tau_0-t,t_1-t}(\delta_x T_{\tau_0-t})(\Psi_1(\cdot,\gamma_{t_1-t}(\eta)))}{\Lambda_{\tau_0-t,t_1-t}(\delta_x T_{\tau_0-t})(\Psi_2(\cdot,\gamma_{t_1-t}(\eta)))} \qquad (12.37)$$

The above representations show that the term inside the integral in (12.35) is same as

$$\mathbb{E}_Q\left[\Gamma_{0,-t}(\delta_x)(\mathbb{S})\left|\frac{\Lambda_{t_1-t}(\delta_x)(\Psi_1(\cdot,\gamma_{t_1-t}(\eta)))}{\Lambda_{t_1-t}(\delta_x)(\Psi_2(\cdot,\gamma_{t_1-t}(\eta)))} - \frac{\Lambda_{\tau_0-t,t_1-t}(\delta_x T_{\tau_0-t})(\Psi_1(\cdot,\gamma_{t_1-t}(\eta)))}{\Lambda_{\tau_0-t,t_1-t}(\delta_x T_{\tau_0-t})(\Psi_2(\cdot,\gamma_{t_1-t}(\eta)))}\right|\right]. \qquad (12.38)$$

Next observe that if U, V, U', V' are real numbers such that $\left|\frac{U'}{V'}\right| \leq D$ for some $D \geq 1$ then

$$\left|\frac{U}{V}-\frac{U'}{V'}\right| \leq D\left\{\left|\frac{U-U'}{V}\right|+\left|\frac{V-V'}{V}\right|\right\}.$$

Using this inequality, we have that the term in (12.38) is bounded by

$$C\sum_{i=1}^{2}\mathbb{E}_Q\left[\Gamma_{0,-t}(\delta_x)(\mathbb{S})\left|\frac{\Lambda_{t_1-t}(\delta_x)(\Psi_i(\cdot,\gamma_{t_1-t}(\eta)))-\Lambda_{\tau_0-t,t_1-t}(T_{\tau_0-t}\delta_x)(\Psi_i(\cdot,\gamma_{t_1-t}(\eta)))}{\Lambda_{t_1-t}(\delta_x)(\Psi_2(\cdot,\gamma_{t_1-t}(\eta)))}\right|\right]. \qquad (12.39)$$

Next note that

$$\Lambda_{t_1-t}(\delta_x)(\Psi_2(\cdot,\gamma_{t_1-t}(\eta))) = \frac{\Gamma_{t_1-t}(\delta_x)(\Psi_2(\cdot,\gamma_{t_1-t}(\eta)))}{\Gamma_{t_1-t}(\delta_x)(\mathbb{S})}$$
$$= \frac{\Gamma_{0,-t}(\delta_x)(\mathbb{S})}{\Gamma_{t_1-t}(\delta_x)(\mathbb{S})}.$$

Using this equality we have that the term in (12.39) equals

$$C\sum_{i=1}^{2}\mathbb{E}_Q\left[\Gamma_{t_1-t}(\delta_x)(\mathbb{S})\left|\Lambda_{t_1-t}(\delta_x)(\Psi_i(\cdot,\gamma_{t_1-t}(\eta)))\right.\right.$$
$$\left.\left.-\Lambda_{\tau_0-t,t_1-t}(\delta_x T_{\tau_0-t})(\Psi_1(\cdot,\gamma_{t_1-t}(\eta)))\right|\right].$$

Replacing, in the above display $t_1 - t$ by t^* and $t_1 - \tau_0$ by τ^*, and observing that under Q, γ_{t^*} is independent of $\mathcal{A}_0^{t^*}$, we have that the above display equals $C(U_1(x) + U_2(x))$, where $U_i(x); i = 1, 2$ are as defined in (12.32). Combining this observation with (12.35) we have the result. □

Lemma 12.6 *Suppose that the filter has the finite memory property. Further suppose that Assumptions 12.2 and 12.3 hold. Then, for all $k \in \mathbb{N}$, $-\infty < t_1 < t_2 < \cdots <$*

$t_k < \infty$, $\phi_1, \cdots, \phi_k \in C_b(\mathbb{S})$ and $\epsilon > 0$, there exists $\tau_\epsilon \in (-\infty, t_1)$ and $t_\epsilon < \tau_\epsilon$ such that $\forall t \leq t_\epsilon$

$$\mathbb{E}_{R_\mu^{(1)}} \left| \mathbb{E}_{R_\mu^{(1)}} \left[\prod_{i=1}^k \phi_i(\tilde{\xi}_{t_i}) \mid \mathcal{Z}_{-\infty}^0 \vee \mathcal{G}_{-\infty}^t \right] - \mathbb{E}_{R_\mu^{(1)}} \left[\prod_{i=1}^k \phi_i(\tilde{\xi}_{t_i}) \mid \mathcal{Z}_{\tau_\epsilon}^0 \vee \mathcal{G}_{-\infty}^t \right] \right| \leq \epsilon.$$

Proof. Let C be as in (12.30). Since U_i as defined in (12.32) are bounded by C, we have, from Lemma 12.5 and an application of Fatou's lemma, that it suffices to show that for μ- a.e. x and $i = 1, 2$,

$$\limsup_{\tau^* \to \infty} \limsup_{t^* \to \infty} U_i(x) = 0, \tag{12.40}$$

where $U_i(x)$ are defined in (12.32). For $K \in (0, \infty)$, we write

$$q_{0,-t}(\theta, \eta) = q_{0,-t}(\theta, \eta) \wedge K + [q_{0,-t}(\theta, \eta) - K] I_{q_{0,-t}(\theta,\eta) \geq K}$$
$$\equiv q^{(1)}(\theta, \eta) + q^{(2)}(\theta, \eta).$$

Now for $i = 1, 2$, define $\Psi_i^{(1)}$ and $\Psi_i^{(2)}$ by replacing $q_{0,-t}$ in the definition of Ψ_i (See (12.33), (12.34)) by $q^{(1)}$ and $q^{(2)}$ respectively. Clearly, for $i = 1, 2$, $\Psi_i = \Psi_i^{(1)} + \Psi_i^{(2)}$. Observe that for $i = 1, 2$ and Q- a.e. η', $\Psi_i^{(1)}(\cdot, \eta') \in C_b(\mathbb{S})$. This implies that for μ almost every x,

$$\limsup_{\tau^* \to \infty} \limsup_{t^* \to \infty} \int_C \left(\mathbb{E}_Q \Gamma_{t^*}(\delta_x)(\mathbb{S}) \left| \Lambda_{t^*}(\delta_x) \left(\Psi_i^{(1)}(\cdot, \eta') \right) \right. \right.$$
$$\left. \left. - \Lambda_{t^*-\tau^*, t^*}(\delta_x T_{t^*-\tau^*}) \left(\Psi_i^{(1)}(\cdot, \eta') \right) \right| \right) Q(d\eta')$$
$$\leq \int_C \limsup_{\tau^* \to \infty} \limsup_{t^* \to \infty} \left(\mathbb{E}_Q \Gamma_{t^*}(\delta_x)(\mathbb{S}) \left| \Lambda_{t^*}(\delta_x) \left(\Psi_i^{(1)}(\cdot, \eta') \right) \right. \right.$$
$$\left. \left. - \Lambda_{t^*-\tau^*, t^*}(\delta_x T_{t^*-\tau^*}) \left(\Psi_i^{(1)}(\cdot, \eta') \right) \right| \right) Q(d\eta')$$
$$= \int_C \limsup_{\tau^* \to \infty} \limsup_{t^* \to \infty} \left(\mathbb{E}_{Q_{\delta_x}} \left| \Lambda_{t^*}(\delta_x) \left(\Psi_i^{(1)}(\cdot, \eta') \right) \right. \right.$$
$$\left. \left. - \Lambda_{t^*-\tau^*, t^*}(\delta_x T_{t^*-\tau^*}) \left(\Psi_i^{(1)}(\cdot, \eta') \right) \right| \right) Q(d\eta')$$
$$= 0, \tag{12.41}$$

where the first inequality above follows on observing that the integrand in the first line of the display is uniformly bounded in (t^*, τ^*) and applying Fatou's lemma. The last equality is a consequence of the finite memory property of the filter. □

Next note that for $i = 1, 2$,

$$\int_{\mathcal{C}} \left(\mathbb{E}_Q \Gamma_{t^*}(\delta_x)(\mathbb{S}) \left| \Lambda_{t^*}(\delta_x) \left(\Psi_i^{(2)}(\cdot, \eta') \right) \right| \right) Q(d\eta')$$

$$= \int_{\mathcal{C}} \left(\mathbb{E}_Q \left| \Gamma_{t^*}(\delta_x) \left(\Psi_i^{(2)}(\cdot, \eta') \right) \right| \right) Q(d\eta')$$

$$\leq C \int_{\hat{\Omega}} q_{0,t^*}(\theta, \eta) q_{t^*,t^*-t_1}(\theta, \eta) \mathbf{1}_{q_{t^*,t^*-t_1}(\theta,\eta) > K} P_x(d\theta) Q(d\eta)$$

$$= C \int_{\hat{\Omega}} q_{t^*,t^*-t_1}(\theta, \eta) \mathbf{1}_{q_{t^*,t^*-t_1}(\theta,\eta) > K} P_x(d\theta) Q(d\eta)$$

$$= C \int_{\hat{\Omega}} q_{0,-t_1}(\theta, \eta) \mathbf{1}_{q_{0,-t_1}(\theta,\eta) > K} P_{\delta_x T_{t^*}}(d\theta) Q(d\eta). \tag{12.42}$$

Lemma 12.3 gives that $\delta_x T_{t^*} \to \mu$ as $t^* \to \infty$. Using this observation in the above display, along with Lemma 12.4 we have that for $i = 1, 2$ and all $x \in \mathbb{S}$,

$$\limsup_{K \to \infty} \limsup_{t^* \to \infty} \int_{\mathcal{C}} \left(\mathbb{E}_Q \Gamma_{t^*}(\delta_x)(\mathbb{S}) \left| \Lambda_{t^*}(\delta_x) \left(\Psi_i^{(2)}(\cdot, \eta') \right) \right| \right) Q(d\eta') = 0. \tag{12.43}$$

Next consider

$$\int_{\mathcal{C}} \left(\mathbb{E}_Q \Gamma_{t^*}(\delta_x)(\mathbb{S}) \left| \Lambda_{t^*-\tau^*, t^*}(\delta_x T_{t^*-\tau^*}) \left(\Psi_i^{(2)}(\cdot, \eta') \right) \right| \right) Q(d\eta')$$

$$= \int_{\mathcal{C}} \left(\mathbb{E}_Q \Gamma_{t^*-\tau^*, t^*}(\delta_x T_{t^*-\tau^*})(\mathbb{S}) \left| \Lambda_{t^*-\tau^*, t^*}(\delta_x T_{t^*-\tau^*}) \left(\Psi_i^{(2)}(\cdot, \eta') \right) \right| \right) Q(d\eta')$$

$$= \int_{\mathcal{C}} \left(\mathbb{E}_Q \left| \Gamma_{t^*-\tau^*, t^*}(\delta_x T_{t^*-\tau^*}) \left(\Psi_i^{(2)}(\cdot, \eta') \right) \right| \right) Q(d\eta')$$

$$\leq C \int_{\hat{\Omega}} q_{t^*-\tau^*, t^*}(\theta, \eta) q_{t^*, t^*-t_1}(\theta, \eta) \mathbf{1}_{q_{t^*, t^*-t_1}(\theta,\eta) > K} P_x(d\theta) Q(d\eta)$$

$$\leq C \int_{\hat{\Omega}} q_{0,-t_1}(\theta, \eta) \mathbf{1}_{q_{0,-t_1}(\theta,\eta) > K} P_{\delta_x T_{t^*}}(d\theta) Q(d\eta),$$

where the last step follows as in (12.42). Once more, in view of Lemma 12.3 and Lemma 12.4 we have that for $i = 1, 2$

$$\limsup_{K \to \infty} \limsup_{\tau^* \to \infty} \limsup_{t^* \to \infty} \int_{\mathcal{C}} \left(\mathbb{E}_Q \Gamma_{t^*}(\delta_x)(\mathbb{S}) \left| \Lambda_{t^*-\tau^*, t^*}(\delta_x T_{t^*-\tau^*}) \left(\Psi_i^{(2)}(\cdot, \eta') \right) \right| \right) Q(d\eta')$$

$$= 0. \tag{12.44}$$

Finally, combining (12.41), (12.43) and (12.44), we have that for μ almost every x, (12.40) holds. This proves the result. □

We now turn to the proof of Theorem 12.4.

Proof of Theorem 12.1 We begin by observing that

$$\mathcal{Z}_{-\infty}^0 \vee \mathcal{G}_{-\infty}^{-\infty} \subset \bigcap_{t \leq 0} \left(\mathcal{Z}_{-\infty}^0 \vee \mathcal{G}_{-\infty}^t \right).$$

Thus to prove the result it suffices to show that, in presence of finite memory property,

$$\mathcal{Z}^0_{-\infty} \vee \mathcal{G}^{-\infty}_{-\infty} \supseteq \bigcap_{t \leq 0} \left(\mathcal{Z}^0_{-\infty} \vee \mathcal{G}^t_{-\infty} \right). \tag{12.45}$$

In order to prove the above statement it is enough to show that for all $F \in \bigcap_{t \leq 0} \left(\mathcal{Z}^0_{-\infty} \vee \mathcal{G}^t_{-\infty} \right)$,

$$\mathbb{E}_{R^{(1)}_\mu} \left(1_F \mid \mathcal{Z}^0_{-\infty} \vee \mathcal{G}^{-\infty}_{-\infty} \right) = 1_F, \quad \text{a.s. } R^{(1)}_\mu.$$

In fact, we will prove a stronger statement, namely, for all $F \in \mathcal{Z}^0_{-\infty} \vee \mathcal{G}^0_{-\infty} \supseteq \bigcap_{t \leq 0} \left(\mathcal{Z}^0_{-\infty} \vee \mathcal{G}^t_{-\infty} \right)$, a.s. $R^{(1)}_\mu$

$$\mathbb{E}_{R^{(1)}_\mu} \left(1_F \mid \mathcal{Z}^0_{-\infty} \vee \mathcal{G}^{-\infty}_{-\infty} \right) = \mathbb{E}_{R^{(1)}_\mu} \left(1_F \mid \bigcap_{t \leq 0} \left(\mathcal{Z}^0_{-\infty} \vee \mathcal{G}^t_{-\infty} \right) \right).$$

By a monotone class argument, it suffices to establish the above equality with 1_F replaced by $V_1 V_2$, where V_1 and V_2 are bounded random variables such that V_1 is $\mathcal{G}^0_{-\infty}$ measurable and V_2 is $\mathcal{Z}^0_{-\infty}$ measurable. Since V_2 is measurable with respect to both $\mathcal{Z}^0_{-\infty} \vee \mathcal{G}^{-\infty}_{-\infty}$ and $\bigcap_{t \leq 0} \left(\mathcal{Z}^0_{-\infty} \vee \mathcal{G}^t_{-\infty} \right)$, it follows that we can take (without loss of generality) $V_2 = 1$. Furthermore, another use of monotone class argument yields that, it is enough to prove that, given any $\phi_1, \cdots, \phi_k \in C_b(\mathbb{S})$ and $t_1 < t_2 < \cdots < t_k = 0$,

$$\mathbb{E}_{R^{(1)}_\mu} \left(U \mid \mathcal{Z}^0_{-\infty} \vee \mathcal{G}^{-\infty}_{-\infty} \right) = \mathbb{E}_{R^{(1)}_\mu} \left(U \mid \bigcap_{t \leq 0} \left(\mathcal{Z}^0_{-\infty} \vee \mathcal{G}^t_{-\infty} \right) \right), \tag{12.46}$$

where

$$U \doteq \phi_1(\xi_{t_1}) \cdots \phi_k(\xi_{t_k}).$$

Let $\epsilon > 0$ be arbitrary and let t_ϵ and τ_ϵ be as in Lemma 12.6. Then for all $t \leq t_\epsilon \leq \tau_\epsilon < t_1$

$$\mathbb{E}_{R^{(1)}_\mu} \left| \mathbb{E}_{R^{(1)}_\mu} \left[U \mid \mathcal{Z}^0_{-\infty} \vee \mathcal{G}^t_{-\infty} \right] - \mathbb{E}_{R^{(1)}_\mu} \left[U \mid \mathcal{Z}^0_{\tau_\epsilon} \vee \mathcal{G}^t_{-\infty} \right] \right| \leq \epsilon.$$

Let C be as in (12.30), then $|U| \leq C$. This implies, on taking limit $t \to -\infty$ in the above display that

$$\mathbb{E}_{R^{(1)}_\mu} \left| \mathbb{E}_{R^{(1)}_\mu} [U \mid \bigcap_{t \leq 0} (\mathcal{Z}^0_{-\infty} \vee \mathcal{G}^t_{-\infty})] - \mathbb{E}_{R^{(1)}_\mu} [U \mid \bigcap_{t \leq 0} (\mathcal{Z}^0_{\tau_\epsilon} \vee \mathcal{G}^t_{-\infty})] \right| \leq \epsilon.$$

Combining the above observation with Lemma 12.2 we now have that

$$\mathbb{E}_{R^{(1)}_\mu} \left| \mathbb{E}_{R^{(1)}_\mu} [U \mid \bigcap_{t \leq 0} (\mathcal{Z}^0_{-\infty} \vee \mathcal{G}^t_{-\infty})] - \mathbb{E}_{R^{(1)}_\mu} [U \mid \mathcal{Z}^0_{\tau_\epsilon} \vee \mathcal{G}^{-\infty}_{-\infty}] \right| \leq \epsilon.$$

Thus to every $\epsilon > 0$, there exists a $\mathcal{Z}^0_{-\infty} \vee \mathcal{G}^{-\infty}_{-\infty}$ measurable (indeed, $\mathcal{Z}^0_{\tau_\epsilon} \vee \mathcal{G}^{-\infty}_{-\infty}$ measurable) random variable V_ϵ; $|V_\epsilon| \leq C$, such that

$$\mathbb{E}_{R^{(1)}_\mu} \left| \mathbb{E}_{R^{(1)}_\mu} \left[U \mid \bigcap_{t \leq 0} (\mathcal{Z}^0_{-\infty} \vee \mathcal{G}^t_{-\infty}) \right] - V_\epsilon \right| \leq \epsilon.$$

This implies that $\mathbb{E}_{R^{(1)}_\mu} \left[U \mid \bigcap_{t \leq 0} (\mathcal{Z}^0_{-\infty} \vee \mathcal{G}^t_{-\infty}) \right]$ is a L^1 limit of $\mathcal{Z}^0_{-\infty} \vee \mathcal{G}^{-\infty}_{-\infty}$ measurable random variables and therefore itself must be measurable with respect to this σ-field. This proves (12.46) and the result follows. $\square$

Finally we come to the proof of Theorem 12.5.

Proof of Theorem 12.2 Fix $\phi \in C_b(\mathbb{S})$. In view of Definition 12.1 it suffices to show (12.29). From assumed asymptotic stability, for μ almost every $x \in \mathbb{S}$,

$$\lim_{t \to \infty} \mathbb{E}_{Q_{\delta_x}} \left| \langle \Lambda_t(\delta_x), \phi \rangle - \langle \Lambda_t(\mu), \phi \rangle \right| = 0. \tag{12.47}$$

Following [24], we have

$$\mathbb{E}_{Q_\mu} \left| \langle \Lambda_t(\mu), \phi \rangle - \langle \Lambda_{t-\tau,t}(\mu T_{t-\tau}), \phi \rangle \right|^2 = \mathbb{E}_{Q_\mu} (\langle \Lambda_t(\mu), \phi \rangle)^2 + \mathbb{E}_{Q_\mu} (\langle \Lambda_{t-\tau,t}(\mu T_{t-\tau}), \phi \rangle)^2$$
$$- 2 \mathbb{E}_{Q_\mu} (\langle \Lambda_t(\mu), \phi \rangle \langle \Lambda_{t-\tau,t}(\mu T_{t-\tau}), \phi \rangle)$$
$$= \mathbb{E}_{Q_\mu} (\langle \Lambda_t(\mu), \phi \rangle)^2 - \mathbb{E}_{Q_\mu} (\langle \Lambda_{t-\tau,t}(\mu T_{t-\tau}), \phi \rangle)^2$$
$$= \mathbb{E}_{Q_\mu} (\langle \Lambda_t(\mu), \phi \rangle)^2 - \mathbb{E}_{Q_\mu} (\langle \Lambda_\tau(\mu), \phi \rangle)^2$$
$$= m^\mu_t(F_\phi) - m^\mu_\tau(F_\phi),$$

where $F_\phi \in C_b(\mathcal{P}(\mathbb{S}))$ is defined as $F_\phi(\nu) \doteq \langle \nu, \phi \rangle^2$, $\nu \in \mathcal{P}(\mathbb{S})$ and the last step in the display follows from (12.19). Taking limit as $t \to \infty$ and then $\tau \to \infty$ in the above display, we have from (12.25) that

$$\limsup_{\tau \to \infty} \limsup_{t \to \infty} \mathbb{E}_{Q_\mu} \left| \langle \Lambda_t(\mu), \phi \rangle - \langle \Lambda_{t-\tau,t}(\mu T_{t-\tau}), \phi \rangle \right| = 0. \tag{12.48}$$

Next, from Assumption 12.3 it follows that $Q_{\delta_x} \ll Q_\mu$. For a proof of this statement we refer the reader to corollary 3.4 of [7]. Thus observing that

$$\left| \langle \Lambda_t(\mu), \phi \rangle - \langle \Lambda_{t-\tau,t}(\mu T_{t-\tau}), \phi \rangle \right| \leq 2 \sup_{x \in \mathbb{S}} |\phi(x)|,$$

we have from (12.48) that $\forall x \in \mathbb{S}$,

$$\limsup_{\tau \to \infty} \limsup_{t \to \infty} \mathbb{E}_{Q_{\delta_x}} \left| \langle \Lambda_t(\mu), \phi \rangle - \langle \Lambda_{t-\tau,t}(\mu T_{t-\tau}), \phi \rangle \right| = 0. \tag{12.49}$$

Next note that, for μ almost every x and every fixed τ

$$\mathbb{E}_{Q_{\delta_x}} \left| \langle \Lambda_{t-\tau,t}(\mu T_{t-\tau}), \phi \rangle - \langle \Lambda_{t-\tau,t}(\delta_x T_{t-\tau}), \phi \rangle \right| = \mathbb{E}_{Q_{\delta_x T_{t-\tau}}} \left| \langle \Lambda_\tau(\mu T_{t-\tau}), \phi \rangle - \langle \Lambda_\tau(\delta_x T_{t-\tau}), \phi \rangle \right| \tag{12.50}$$

Lemma 12.3 gives that, $\delta_x T_{t-\tau}$ converges to μ as $t \to \infty$. Also, $\mu T_{t-\tau}$, being equal to μ, trivially converges to μ as $t \to \infty$. It now follows from theorem 7.2 of [5] that the expression in the above display converges to 0, as $t \to \infty$. Thus

$$\limsup_{\tau \to \infty} \limsup_{t \to \infty} \mathbb{E}_{Q_{\delta_x}} \left| \langle \Lambda_{t-\tau,t}(\mu T_{t-\tau}), \phi \rangle - \langle \Lambda_{t-\tau,t}(\delta_x T_{t-\tau}), \phi \rangle \right| = 0. \quad (12.51)$$

Finally (12.29) follows on combining (12.47), (12.49), and (12.51) and an application of triangle inequality. The result follows. □

12.6 Equivalence of several stability properties

In Section 12.5 we saw that under Assumptions 12.1, 12.2, and 12.3, asymptotic stability of the filter implies the finite memory property of the filter which in turn implies the ergodicity of the filter. In this section, using results from [19, 24, 5, 7] and those of Section 12.5, we show that, under Assumptions 12.1, 12.2, and 12.3, in fact all these properties are equivalent. Additionally, they are equivalent to the uniqueness of an invariant measure for the signal–filter pair and the stated equality of the σ-fields in (12.27). The following is the main result.

Theorem 12.6 *Suppose that Assumptions 12.1, 12.2, and 12.3 hold. Then the following are equivalent.*

(i) $\bigcap_{s=-\infty}^{\infty} \left(\mathcal{Z}_{-\infty}^{0} \vee \mathcal{G}_{-\infty}^{s} \right) = \mathcal{Z}_{-\infty}^{0} \vee \mathcal{G}_{-\infty}^{-\infty}$.

(ii) *The filter has a unique invariant measure, i.e. there is a unique T_t invariant probability measure.*

(iii) *The signal–filter pair has a unique invariant measure, i.e. there is a unique S_t invariant probability measure.*

(iv) *For all $\nu_1, \nu_2 \in \mathcal{P}(\mathbb{S})$, the filter is (ν_1, ν_2) asymptotically stable.*

(v) *The filter has the finite memory property.*

One can in fact establish a somewhat stronger result. We refer the reader to Theorem 2.7 of [8] for details.

Proof. (i) $\Rightarrow$ (ii) (Kunita [19]). This has been already argued above (12.27).
(ii) $\Rightarrow$ (iv) (Ocone and Pardoux [24]). Lemma 12.3 gives that $\nu_i T_t \to \mu$, for $i = 1, 2$, as $t \to \infty$. Also Assumption 12.3 implies that Q_{ν_1} is mutually absolutely continuous with respect to Q_{ν_2} (cf. corollary 3.4 of [7]). Now Theorem 12.2 of [24] yield the implication. We remark that [24] considers the case when $\mathbb{S}$ is locally compact and h is bounded, however the result holds in the current setting as was shown in Theorem 7.3 of [5].
(iv) $\Rightarrow$ (v) This is a consequence of Theorem 12.5.
(v) $\Rightarrow$ (i) This is shown in Theorem 12.4.
(iv) $\Rightarrow$ (iii) ([9], [7]). This is a direct consequence of Theorem 3.6 of [7].
(iii) $\Rightarrow$ (ii) It is shown in Theorem 6.4 of [5] that if we define

$$\overline{m}^\mu \doteq R_\mu^{(1)} \circ \left(\overline{\pi}_0^{(1)}, \tilde{\xi}_0 \right)^{-1}, \quad \overline{M}^\mu \doteq R_\mu^{(1)} \circ \left(\overline{\pi}_0^{(0)}, \tilde{\xi}_0 \right)^{-1},$$

then both $\overline{m}^\mu$ and $\overline{M}^\mu$ are $(\mathcal{S}_t)$ invariant. Now (iii) implies that $\overline{m}^\mu = \overline{M}^\mu$ and thus $(\overline{m}^\mu)_1 = (\overline{M}^\mu)_1$, where for $\rho \in \mathcal{P}(\mathcal{P}(\mathbb{S}) \times \mathbb{S})$, we denote by $(\rho)_1$ the marginal on $\mathcal{P}(\mathbb{S})$. Since $(\overline{m}^\mu)_1 = m^\mu$ and $(\overline{M}^\mu)_1 = M^\mu$, (ii) follows. □

Notes

Research supported in part by the Army Research Office (Grant W911NF-0-1-0080).

References

[1] R. Atar. Exponential stability for nonlinear filtering of diffusion processes in non-compact domain. *Annals of Probability*, 26: 1552–74, 1998.

[2] R. Atar and O. Zeitouni. Exponential stability for nonlinear filtering. *Annales de l'Institut H. Poincaré Probabilités et Statistique*, 33: 697–725, 1997.

[3] R. Atar and O. Zeitouni. Lyapunov exponents for finite state nonlinear filtering. *SIAM J. of Control and Optimization*, 35: 36–55, 1997.

[4] P. Baxendale, P. Chigansky, and R. Liptser. Asymptotic stability of the Wonham filter: Ergodic and nonergodic signals. *SIAM J. of Control and Optimization*, 43: 643–69, 2004.

[5] A. Bhatt, A. Budhiraja, and R. Karandikar. Markov property and ergodicity of the nonlinear filter. *SIAM J. on Control and Optimization*, 39: 928–49, 2000.

[6] A. Bhatt, G. Kallianpur, and R. Karandikar. Robustness of the nonlinear filter. *Stochastic Process. Appl.*, 81: 247–54, 1999.

[7] A. Budhiraja. Ergodic properties of the nonlinear filter. *Stochastic Process. Appl.*, 95: 1–24, 2001.

[8] A. Budhiraja. Asymptotic stability, ergodicity and other asymptotic properties of the nonlinear filter. *Annales de l'Institut Henri Poincaré Probabilités et Statistiques*, 39: 919–41, 2003.

[9] A. Budhiraja and H. J. Kushner. Approximation and limit results for nonlinear filters over an infinite time interval. *SIAM J. on Control and Optmization*, 37: 1946–79, 1997.

[10] A. Budhiraja and H. J. Kushner. Robustness of nonlinear filters over the infinite time interval. *SIAM J. on Control and Optimization*, 36: 1618–37, 1998.

[11] A. Budhiraja and H. J. Kushner. Approximation and limit results for nonlinear filters over an infinite time interval: Part II, random sampling algorithms. *SIAM J. of Control and Optimization*, 38: 1874–1908, 2000.

[12] A. Budhiraja and H. J. Kushner. *Stochastics in Finite/Infinite Dimensions*, pages 59–87. Trends Math. Birkhäuser Boston, 2001. Monte Carlo algorithms and asymptotic problems in nonlinear filtering.

[13] A. Budhiraja and D. Ocone. Exponential stability of discrete time filters without signal ergodicity. *System and Control Letters*, 30: 185–93, 1997.

[14] A. Budhiraja and D. Ocone. Exponential stability in discrete time filtering for non-ergodic signals. *Stochastic Process. Appl.*, 82: 245–57, 1999.

[15] P. Chigansky and R. Liptser. Private communication, 2002.

[16] P. Del Moral and A. Guionnet. On the stability of measure valued processes with applications to filtering. *C. R. Acad. Sci. Paris Sér. I Math.* 329: 429–34, 1999.

[17] G. Kallianpur. *Stochastic Filtering Theory*. Springer-Verlag, New York, 1980.

[18] R. L. Karandikar. On pathwise stochastic integration. *Stochastic Process. Appl.*, 57: 11–18, 1995.

[19] H. Kunita. Asymptotic behavior of the nonlinear filtering errors of Markov processes. *J. Multivariate Analysis*, 1: 365–93, 1971.
[20] H. Kunita. Ergodic properties of nonlinear filtering processes. In K. C. Alexander and J. C. Watkins, eds, *Spatial Stochastic Processes*, Birkhäuser, Boston, 1991.
[21] F. Le Gland and L. Mevel. Exponential forgetting and geometric ergodicity in hidden Markov models. *Math. Control Signals Systems*, 13: 63–93, 2000.
[22] F. Le Gland and N. Oudjane. A robustification approach to stability and to uniform particle approximation of nonlinear filters: The example of pseudo-mixing signals. Preprint.
[23] D. Ocone. Asymptotic stability of Benes filters. *Stochastic Anal. Appl.*, 17: 1053–74, 1999.
[24] D. Ocone and E. Pardoux. Asymptotic stability of the optimal filter with respect to its initial condition. *SIAM J. of Control and Optimization*, 34: 226–43, 1996.
[25] L. Stettner. On invariant measures of filtering processes. In K. Helmes, N. Christopeit, and M. Kohlmann, eds, *Stochastic Differential Systems, Proc. 4th Bad Honnef Conf., 1988, Lecture Notes in Control and Inform Sci.*, pp. 279–92, 1989.
[26] H. Totoki. A class of special flows. *Z. Wahr. Verw. Geb.*, 15: 157–67, 1970.
[27] R. van Handel. The stability of conditional Markov processes and Markov chains in random environments. *arXiv:0801.4366v1 [math.PR]*.
[28] H. V. Weizsäcker. Exchanging the order of taking suprema and countable intersections of σ algebras. *Annales de l'Institut Henri Poincaré*, B, Vol XIX, n. 1: 91–100, 1983.

·13·
Stability of the Optimal Filter for Nonergodic Signals—A Variational Approach

W. Stannat

13.1 Introduction

Stability results for the optimal filter of a signal process on $\mathbb{R}^d$ observed with independent additive noise are obtained using the method of parabolic ground state transform. Explicit lower bounds on the rate of convergence are given which are uniform in the observation. In the case where the signal is a solution of the d-dimensional stochastic differential equation

$$dX_t = B(X_t)\,dt + C\,dW_t, \; X_0 = \xi,$$

where the drift term is of the type

$$B(x) = B_1 x + Q\nabla \log \varphi(x), \; Q = CC^T,$$

that is observed through $(Y_t)_{t\geq 0}$ given by

$$dY_t = GX_t\,dt + \Gamma\,de_t, \; Y_0 = 0,$$

where $(e_t)_{t\geq 0}$ is a d-dimensional Brownian motion independent of $(X_t)_{t\geq 0}$ and $R := \Gamma\Gamma^T$ is invertible, our stability result concerning the optimal filter, that is the conditional distribution of X_t given the observation $Y_{0:t} := (Y_s)_{s\in[0,t]}$ up to time t, can be formulated as follows: Let

$$V(x) = \langle B_1 x, \nabla \log \varphi(x)\rangle + \frac{1}{2}\frac{\operatorname{tr}(Q\varphi''(x))}{\varphi(x)} + \frac{1}{2}\langle R^{-1}Gx, Gx\rangle,$$

suppose that there exists a positive definite symmetric $d \times d$ matrix K such that

$$V''(x) \geq K, \; x \in \mathbb{R}^d,$$

where '$\geq$' is to be understood in the quadratic form sense, and let K_* be the unique positive definite symmetric solution of the matrix Riccati equation

$$K_* Q K_* + B_1^T K_* + K_* B_1 - K = 0.$$

Then the optimal filter is stable with exponential rate $\frac{\chi_*}{2}$, where

$$\chi_* = \min\left\{\lambda : \lambda \text{ is an eigenvalue of } \sqrt{Q}K_*\sqrt{Q} - \frac{1}{2}(B_1 + B_1^T)\right\}$$

(see Theorem 13.3 for the precise assumptions needed for the distribution μ_0 of the initial condition ξ). Note that given the signal process with

$$\bar{V}(x) = \langle B_1 x, \nabla \log \varphi(x) \rangle + \frac{\text{tr}(Q\varphi''(x))}{\varphi(x)} \geq a\|x\|^2$$

for $a \in \mathbb{R}$, we can always design the observation process, that is, the way we 'measure' the signal, by increasing G or decreasing the covariance matrix R of the measurement error, in such a way that V becomes uniformly stricly convex. In addition, our criterion allows to distinguish between convex directions of $\bar{V}$, which correspond to ergodic directions of the signal, and nonconvex directions. The lower bound tells us that we need to take care of the nonconvex directions of $\bar{V}$, i.e. measure these components of X_t more precisely, than the other components, in order to achieve a certain prescribed exponential rate.

The organization of the chapter is as follows: Section 13.2 contains the abstract stability result in discrete time (Proposition 13.1) with applications to the Kalman filter (Theorem 13.1) and to time-discretized signals induced by stochastic differential equations (Theorem 13.2). Section 13.3 contains the corresponding stability result in continuous time (Theorem 13.3) with applications to the Kalman–Bucy filter and to one-dimensional signals. The results in this section generalize the results obtained in the papers [8, 9, 10] using the same approach. Note that [8, 9] deal with possibly time-dependent coefficients, which is, in principle, also possible here.

13.2 The time-discrete case

Suppose the signal $X = \{X_n, n \geq 0\}$ is given as a Markov chain on the measurable space $(S, \mathcal{S})$ with absolutely continuous transition probabilities

$$P(X_{n+1} \in dx_{n+1} \mid X_n) = p(X_n, x_{n+1}) \, \nu(dx_{n+1})$$

w.r.t. some fixed σ-finite reference measure ν. We assume that the densities are uniformly bounded:

$$\|p\|_\infty := \sup_{x,x' \in S} p(x, x') < \infty. \tag{13.1}$$

Suppose that X is observed through Y, taking values in $\mathbb{R}^p$, with independent noise as follows: Given X_n, the conditional distribution of Y_n can be written as

$$P(Y_n \in dy \mid X_n) = g(X_n, y) \, dy,$$

where dy denotes the p-dimensional Lebesgue measure. We suppose that g is bounded:

$$\|g\|_\infty := \sup_{x \in S, y \in \mathbb{R}^p} g(x, y) < \infty. \tag{13.2}$$

Denote by $\pi_n^{y_{1:n}}$ the regular conditional distribution of X_n given the information $\mathcal{Y}_n := \sigma(Y_1, \ldots, Y_n)$ provided by the observation process up to time n. Here, $y_{1:n} := (y_1, \ldots, y_n) \in \mathbb{R}^{pn}$.

Due to the Markovian structure of X we can write down the following recursive formula for $\pi_n^{y_{1:n}}$

$$\int f(x) \pi_{n+1}^{y_{1:n+1}}(dx) = \frac{\int P(f \cdot g(\cdot, y_{n+1}))(x) \pi_n^{y_{1:n}}(dx)}{\int Pg(\cdot, y_{n+1})(x) \pi_n^{y_{1:n}}(dx)} \\ = \frac{\int f(x) g(x, y_{n+1}) \hat{P} \pi_n^{y_{1:n}}(dx)}{\int g(x, y_{n+1}) \hat{P} \pi_n^{y_{1:n}}(dx)} \quad (13.3)$$

with initial condition $\pi_0 := P \circ X_0^{-1}$. Here, $\hat{P}$ denotes the dual operator of the transition kernel P associated with the signal X.

The recursive formula (13.3) can be written as

$$\pi_{n+1}^{y_{1:n+1}} = T\left(y_{n+1}, \pi_n^{y_{1:n}}\right), \quad (13.4)$$

where T denotes the (nonlinear) transfer operator

$$T : \mathbb{R}^p \times \mathcal{P}(S) \to \mathcal{P}(S), \qquad (y, \mu) \mapsto T(y, \mu),$$

and the probability measure $T(y, \mu)$ is defined by

$$\int f(x) T(y, \mu)(dx) = \frac{\int f(x) g(x, y) \hat{P} \mu(dx)}{\int g(x, y) \hat{P} \mu(dx)}.$$

Iterating (13.4) we can write

$$\pi_{n+1}^{y_{1:n+1}} = T(y_{n+1}, \cdot) \circ T(y_n, \cdot) \circ \ldots \circ T(y_1, \mu_0)$$

and it is our main goal to understand the dependence of $\pi_{n+1}^{y_{1:n+1}}$ w.r.t. the initial condition μ_0.

13.2.1 The transfer operator acting on densities

Taking into account the absolute continuity of the transition probabilities w.r.t. the measure ν it is natural to restrict T to the subset of all probability measures μ having a density $h = \frac{d\mu}{d\nu}$ w.r.t. ν. Identifying μ with its density h we can then write

$$T(y, h)(x) = \frac{g(x, y) \int p(x', x) h(x') \nu(dx')}{\int g(x, y) \int p(x', x) h(x') \nu(dx') \nu(dx)}. \quad (13.5)$$

The next step will be to 'linearize' T w.r.t. the density h. To this end we introduce a parabolic ground state transform as follows:

Fix an initial function $\hat{m}_0 \in L^1(\nu)$, $\hat{m}_0 \geq 0$, and define recursively

$$\hat{m}_n := g(\cdot, y_n) \hat{P} \hat{m}_{n-1}, \quad n = 1, 2, 3, \ldots \quad (13.6)$$

Note that the boundedness assumptions on p and g imply that $\hat{m}_n$ is bounded for $n \geqslant 1$. If the initial condition μ_0 can be written as

$$\mu_0(dx) = h_0(x)\hat{m}_0(x)\,\nu(dx)$$

then

$$\pi_n^{y_{1:n}}(dx) = \frac{h_n(x)\hat{m}_n(x)\,\nu(dx)}{\int h_n(x)\hat{m}_n(x)\,\nu(dx)}$$

with

$$h_n := P_n^* h_{n-1} := \frac{g(\cdot, y_n)\,\hat{P}\,(h_{n-1}\hat{m}_{n-1})}{\hat{m}_n}, \quad n = 1, 2, 3, \ldots \quad (13.7)$$

The operator P_n^* is Markovian and linear. We call it the (parabolic) ground state transform associated with the solution $(\hat{m}_n)$ of the forward equation (13.6).

The stability properties of the conditional densities (h_n) can now be investigated with the help of the contraction properties of the Markovian integral operators P_n^*. The various strategies to obtain estimates on the contraction properties of P_n^* depend on the choice of the function spaces for their domains.

L^p spaces

Fix $N \in \mathbb{N}$ and a terminal condition $m_N \in L^1(\nu)$, $m_N \geqslant 0$, and define recursively:

$$m_n := P(g(\cdot, y_{n+1})m_{n+1}),\, n = N-1, N-2, \ldots, 0. \quad (13.8)$$

We can then define the probability measures

$$\nu_n^*(dx) := Z_n^{-1} m_n(x)\hat{m}_n(x)\nu(dx),\, n = 0, 1, \ldots, N,$$

where

$$Z_n := \int m_n \hat{m}_n \, d\nu.$$

It is then straightforward to prove that

(i) $\hat{P}_n^* \nu_n^* = \nu_{n-1}^*$, $n = 1, 2, \ldots, N$.
(ii) $P_n^*: L^p(\nu_{n-1}^*) \to L^p(\nu_n^*)$ is a contraction for $p \in [1, \infty]$.

Since P_n^* is Markovian, it follows that the constant function 1 is an eigenvector with corresponding eigenvalue 1. The contraction coefficient of P_n^* acting on the densities is then given by

$$\sup_{h \neq 0 \int h\, d\nu_{n-1}^* = 0} \frac{\|P_n^* h\|_{L^p(\nu_n^*)}}{\|h\|_{L^p(\nu_{n-1}^*)}}$$

and this coincides with

$$\sup_{h \neq const} \frac{\mathrm{Var}_{\nu_n^*}(P_n^* h)}{\mathrm{Var}_{\nu_{n-1}^*}(h)}$$

in the particular case $p = 2$. The drawback of the L^p estimates is the requirement to know the solution (m_n) of the backward equation (13.8). Therefore pointwise estimates, although they require stronger assumptions on the signal process and the observation, are easier to obtain since they do not make use of the backward equation. In this chapter, we will study contraction properties w.r.t. function spaces consisting of Lipschitz-continuous functions in the time discrete and, in a second step, also in the time-continuous case.

13.2.2 Contraction w.r.t. Lipschitz norms

From now on we restrict to the case $S = \mathbb{R}^d$ and $\nu =$ Lebesgue measure and fix a solution $(\hat{m}_n)$ of the forward equation (13.6). We will then study the contraction properties of the corresponding parabolic ground state transform P_n^* w.r.t. the Lipschitz norm

$$\|f\|_{\mathrm{Lip}_A} := \sup_{x,y \in \mathbb{R}^d, x \neq y} \frac{|f(x) - f(y)|}{\|A(x-y)\|}$$

where A is such that $A^T A$ is a positive definite symmetric $d \times d$ matrix. Denote by $\mathrm{Lip}_{\mathbb{R}^d}$ the space of all Lipschitz-continuous functions on $\mathbb{R}^d$.

Proposition 13.1 *Suppose that* $\|P_k^* h\|_{\mathrm{Lip}_A} \leq \chi_k^* \|h\|_{\mathrm{Lip}_A}$, $h \in \mathrm{Lip}_{\mathbb{R}^d}$. *Let* $y_{1:n} \in \mathbb{R}^{np}$, *and*

- $\pi_n^{y_{1:n}}$ *be the optimal filter with initial condition*

$$\mu_0 = \left(\int \hat{m}_0 \, dx \right)^{-1} \hat{m}_0 \, dx$$

- $\tilde{\pi}_n^{y_{1:n}}$ *be the optimal filter with initial condition* $\tilde{\mu}_0$, $\tilde{\mu}_0 \ll \mu_0$, *with Lipschitz-continuous density* $h_0 := \frac{d\tilde{\mu}_0}{d\mu_0}$.

Then

$$\left\| \pi_n^{y_{1:n}} - \tilde{\pi}_n^{y_{1:n}} \right\|_{\mathrm{var}} \leq \frac{\hat{\sigma}_n}{2 \int h_n \, d\pi_n^{y_{1:n}}} \cdot \prod_{k=1}^n \chi_k^* \cdot \|h_0\|_{\mathrm{Lip}_A},$$

where

$$\hat{\sigma}_n := \int \int \|Ax - Ax'\| \, \pi_n^{y_{1:n}}(dx) \, \pi_n^{y_{1:n}}(dx').$$

In particular, if $h_0 \geq \delta > 0$, then $h_n \geq \delta > 0$ and thus

$$\left\|\pi_n^{y_{1:n}} - \tilde{\pi}_n^{y_{1:n}}\right\|_{\mathrm{var}} \leq \frac{\hat{\sigma}_n}{2\delta} \cdot \prod_{k=1}^{n} \chi_k^* \cdot \|h_0\|_{\mathrm{Lip}_A}.$$

The constants $\hat{\sigma}_n$ and χ_k^* will be made explicit in the following examples. The proof of Proposition 13.1 is a direct consequence of the following estimate

$$\begin{aligned}
\left\|\pi_n^{y_{1:n}} - \tilde{\pi}_n^{y_{1:n}}\right\|_{\mathrm{var}} &= \frac{1}{2} \int \left| \frac{1}{\int \hat{m}_n \, dx} - \frac{h_n}{\int \hat{m}_n \, dx} \right| \hat{m}_n \, dx \\
&\leq \frac{1}{2 \int \hat{m}_n \, dx \int h_n \hat{m}_n \, dx} \int \int |h_n(x) - h_n(x')| \hat{m}_n(x) \hat{m}_n(x') \, dx \, dx' \\
&\leq \frac{\|h_n\|_{\mathrm{Lip}_A} \hat{\sigma}_n}{\int h_n \, d\pi_n^{y_{1:n}}},
\end{aligned}$$

combined with the obvious estimate

$$\|h_n\|_{\mathrm{Lip}_A} = \|P_n^* h_{n-1}\|_{\mathrm{Lip}_A} \leq \chi_n^* \|h_{n-1}\|_{\mathrm{Lip}_A} \leq \cdots \leq \prod_{k=1}^{n} \chi_k^* \|h_0\|_{\mathrm{Lip}_A}.$$

13.2.3 Kalman filter

For given $m \in \mathbb{R}^d$ and given positive semidefinite symmetric $d \times d$ matrix Q let $N(m, Q)$ denote the Gaussian distribution on $\mathbb{R}^d$ with mean m and covariance Q. Let

$$X_n = BX_{n-1} + CW_{n-1} \tag{13.9}$$

$$Y_n = GX_n + \Gamma e_n \tag{13.10}$$

with

- (W_n), (e_n) independent $N(0, I)$-distributed on $\mathbb{R}^d$, $\mathbb{R}^p$
- B, C are $d \times d$ matrices
- G is $p \times d$ matrix
- Γ is $p \times p$ matrix.

We assume that both, $Q := CC^T$ and $R := \Gamma\Gamma^T$ are positive definite.

To simplify notations in the following, we will write

$$f \propto g$$

for two nonnegative functions f, g, if there exists $\lambda > 0$ with $f(x) = \lambda g(x) \; \forall x$.

Under the above assumptions on (13.9) and (13.10) we have that $P(x, dx') = N(Bx, Q)(dx')$, thus

$$p(x, x') \propto \exp\left(-\frac{1}{2}\langle Q^{-1}(Bx - x'), (Bx - x')\rangle\right)$$

as a function of x', and
$$g(x, y) \propto \exp\left(-\frac{1}{2}\langle R^{-1}(Gx - y), (Gx - y)\rangle\right)$$
as a function of x.

It is a well-known feature of the Kalman filter (see e.g. [3]), that a Gaussian initial condition $\mu_0 = N(m_0, E_0)$ of (13.9) implies that
$$\pi_n^{y_{1:n}} = N\left(m_n^{y_{1:n}}, E_{n,n}\right), \tag{13.11}$$
where $E_{n,n}$ and $m_n^{y_{1:n}}$ are determined by the Kalman filtering equations:
$$\begin{cases} E_{n,n-1} := B E_{n-1,n-1} B^T + Q \\ K_n := E_{n,n-1} G^T \left(G E_{n,n-1} G^T + R\right)^{-1} \\ E_{n,n} := (I - K_n G) E_{n,n-1}, \end{cases} \tag{13.12}$$
for $n = 1, 2, \ldots$. Here, $E_{0,0} := E_0$ and K_n is the Kalman gain matrix. For the mean $m_n^{y_{1:n}}$ we have the recursive formula
$$m_n^{y_{1:n}} := B m_{n-1}^{y_{1:n-1}} + K_n \left(y_n - G B m_{n-1}^{y_{1:n-1}}\right), \tag{13.13}$$
$n = 1, 2, \ldots$, with $m_0^{y_{1:0}} := m_0$.

The explicit representation (13.11) allows one to obtain an explicit representation of the solution $\hat{m}_n$ of the forward equation (13.6) for Gaussian initial conditions. More precisely, if $m_0 \in \mathbb{R}^d$, E_0 is a symmetric positive definite $d \times d$ matrix, then
$$\hat{m}_0(x) \propto \exp\left(-\frac{1}{2}\langle E_0^{-1}(x - m_0), (x - m_0)\rangle\right)$$
implies that
$$\hat{m}_n(x) \propto \exp\left(-\frac{1}{2}\langle E_{n,n}^{-1}(x - m_n^{y_{1:n}}), (x - m_n^{y_{1:n}})\rangle\right),$$
where $E_{n,n}$ is determined by (13.12) and $m_n^{y_{1:n}}$ is determined by (13.13).

It is important here to note that (13.12) implies that $E_{n,n}$ is invertible, if $E_{n-1,n-1}$ is invertible. Indeed, note that $E_{n,n-1} = B E_{n-1,n-1} B^T + Q$ is always invertible by assumption on Q, so that by the following matrix inversion lemma
$$E_{n,n} = (I - K_n G) E_{n,n-1} = \left(E_{n,n-1}^{-1} + G^T R^{-1} G\right)^{-1}$$
is clearly invertible with inverse $E_{n,n}^{-1} = E_{n,n-1}^{-1} + G^T R^{-1} G$.

In the further analysis of the Kalman filter, the following matrix inversion lemma, already mentioned above, will play a crucial role. The proof of the lemma is straightforward.

Lemma 13.1 *(Matrix Inversion Lemma).* Let

- U be an invertible $m \times m$ matrix,
- V be an invertible $n \times n$ matrix,
- C be an $m \times n$ matrix, D be an $n \times m$ matrix.

Then the following statements are equivalent:

(i) $U - CV^{-1}D$ is invertible,
(ii) $V - DU^{-1}C$ is invertible.

In this case,

$$\left(U - CV^{-1}D\right)^{-1} = U^{-1} + U^{-1}C\left(V - DU^{-1}C\right)^{-1}DU^{-1}.$$

It follows that for Gaussian initial conditions $\hat{m}_0$, the densities $p_n^*(x, x')$ of the transition probabilities of the corresponding ground state transforms can be determined explicitly:

$$\begin{aligned} p_n^*(x, x')\,dx' &\propto p(x', x)\hat{m}_{n-1}(x')\,dx' \\ &\propto N\left(Q_n^*\left(B^T Q^{-1}x + E_{n-1,n-1}^{-1} m_{n-1}^{y_{1:n}}\right),\, Q_n^*\right)(dx'), \end{aligned} \qquad (13.14)$$

where

$$Q_n^* := \left(E_{n-1,n-1}^{-1} + B^T Q^{-1} B\right)^{-1}. \qquad (13.15)$$

This explicit representation allows to estimate the contraction coefficient of P_n^* on Lipschitz-continuous functions. To this end fix a positive definite symmetric $d \times d$ matrix A. Then

$$P_n^* h(x) = \int h\left(Q_n^*\left(B^T Q^{-1}x + E_{n-1,n-1}^{-1} m_{n-1}^{y_{1:n}}\right) + x'\right) N\left(0, Q_n^*\right)(dx'),$$

implies that

$$\begin{aligned} |P_n^* h(x_1) - P_n^* h(x_2)| &\leq \|h\|_{\mathrm{Lip}_A} \|AQ_n^* B^T Q^{-1}(x_1 - x_2)\| \\ &\leq \|h\|_{\mathrm{Lip}_A} \|AQ_n^* B^T Q^{-1} A^{-1}\|_{\mathrm{op}} \|A(x_1 - x_2)\|. \end{aligned}$$

Consequently,

$$\|P_n^* h\|_{\mathrm{Lip}_A} \leq \|AQ_n^* B^T Q^{-1} A^{-1}\|_{\mathrm{op}} \|h\|_{\mathrm{Lip}_A}.$$

Here, $\|B\|_{\mathrm{op}}$ denotes the operator norm of a matrix B w.r.t. the euclidean metric. In the following, let $\mathrm{tr}(B)$ denote its trace. Proposition 13.1 now implies the following:

Theorem 13.1 *Let A be a positive definite symmetric $d \times d$ matrix, let*

- $\pi_n^{y_{1:n}}$ *be the optimal filter with initial Gaussian distribution $\mu_0 = N(m_0, E_0)$,*

- $\tilde{\pi}_n^{y_{1:n}}$ be the optimal filter with initial distribution $\tilde{\mu}_0$, $\tilde{\mu}_0 \ll \mu_0$ with Lipschitz-continuous density $h_0 := \frac{d\tilde{\mu}_0}{d\mu_0}$.

Then

$$\|\pi_n^{y_{1:n}} - \tilde{\pi}_n^{y_{1:n}}\|_{\mathrm{var}} \leq \frac{1 + \mathrm{tr}(AQ_n^* A^T)}{2\int h_n\, d\pi_n^{y_{1:n}}} \prod_{k=1}^n \|AQ_k^*(B^T Q^{-1} A^{-1})\|_{\mathrm{op}} \|h_0\|_{\mathrm{Lip}_A}.$$

In particular, $h_0 \geq \delta > 0$ implies that

$$\|\pi_n^{y_{1:n}} - \tilde{\pi}_n^{y_{1:n}}\|_{\mathrm{var}} \leq \frac{1 + \mathrm{tr}(AQ_n^* A^T)}{2\delta} \prod_{k=1}^n \|AQ_k^*(B^T Q^{-1} A^{-1})\|_{\mathrm{op}} \|h_0\|_{\mathrm{Lip}_A}.$$

Limiting Kalman filter

The asymptotic behaviour of the contraction coefficient $\prod_{k=1}^n \|AQ_k^* B^T Q^{-1} A^{-1}\|_{\mathrm{op}}$ can be made more precise for suitable choice of A in the particular case of a limiting Kalman filter (see ch. 6 in [3]). To this end we suppose that the system (13.9), (13.10) is both

- (completely) controllable, i.e., the matrix

$$\begin{bmatrix} C & BC & \ldots & B^{d-1}C \end{bmatrix}$$

has full rank, and
- observable, i.e., the matrix

$$\begin{bmatrix} G & GB & \ldots & GB^{d-1} \end{bmatrix}$$

has full rank.

It follows under these two assumptions that there exists a unique positive definite symmetric $d \times d$ matrix E_∞ solving the matrix Riccati equation

$$E_\infty = B(E_\infty - E_\infty G^T (GE_\infty G^T + R)^{-1} GE_\infty) B^T + Q,$$

such that

$$\lim_{n\to\infty} E_{n,n-1} = E_\infty$$

for any symmetric positive-semidefinite initial condition E_0. In particular, the Kalman gain matrices converge,

$$K_\infty := \lim_{n\to\infty} K_n = E_\infty G^T \left(GE_\infty G^T + R\right)^{-1}.$$

In addition,

$$\chi_* := \max\{|\lambda| : \lambda \text{ eigenvalue of } B(I - K_\infty G)\} < 1,$$

and both, $(E_{n,n-1})$ and (K_n) converge with geometric rate χ_*.

We can now estimate $\|A Q_n^* B^T Q^{-1} A^{-1}\|_{op}$, as follows: the matrix inversion lemma again implies that

$$\begin{aligned}
Q_n^* B^T Q^{-1} &= \left(E_{n-1,n-1}^{-1} + B^T Q^{-1} B \right)^{-1} B^T Q^{-1} \\
&= \left(E_{n-1,n-1} - E_{n-1,n-1} B^T \left(Q + B E_{n-1,n-1} B^T \right)^{-1} B E_{n-1,n-1} \right) B^T Q^{-1} \\
&= E_{n-1,n-1} B^T \left(Q^{-1} - \left(I - \left(Q + B E_{n-1,n-1} B^T \right)^{-1} Q \right) Q^{-1} \right) \\
&= E_{n-1,n-1} B^T \left(Q + B E_{n-1,n-1} B^T \right)^{-1} = E_{n-1,n-1} B^T E_{n,n-1}^{-1}.
\end{aligned}$$

Consequently,

$$\lim_{n \to \infty} Q_n^* B^T Q^{-1} = (I - K_\infty G) E_\infty B^T E_\infty^{-1}$$

with geometric rate. Since

$$\left\| (I - K_\infty G) E_\infty B^T E_\infty^{-1} \right\|_{op} = \left\| E_\infty^{-1} B (I - K_\infty G) E_\infty \right\|_{op}$$

we obtain from Theorem 13.1, applied to $A = E_\infty$, that the Kalman filter is stable with geometric rate χ_*.

13.2.4 Time-discretized signals induced by stochastic differential equations

Suppose the signal is modelled as the solution of a d-dimensional stochastic differential equation

$$dX_t = B(X_t)\, dt + C d W_t, \quad X_0 = \xi. \tag{13.16}$$

Here, $(W_t)_{t \geq 0}$ is a d-dimensional Brownian motion, ξ is a random variable independent from $(W_t)_{t \geq 0}$, C is a $d \times d$ matrix such that $Q = CC^T$ is strictly positive definite and the drift term has the form

$$B(x) = B_1 x + Q \nabla \log \varphi(x) \tag{13.17}$$

for some $d \times d$ matrix B_1 and some positive function $\varphi \in C^2(\mathbb{R}^d)$ that is polynomially bounded.

Suppose that for all $x \in \mathbb{R}^d$ there exists a unique weak solution $(X_t(x))$ of (13.16) with initial condition $X_0(x) = x$, whose transition probabilities $p_t(x, \cdot)$, $t \geq 0$, can be represented as

$$p_t(x, A) = E\left(1_A(U_t(x))\, Z_t(x)\right), \tag{13.18}$$

with

$$U_t(x) = e^{t B_1} \left(x + \int_0^t e^{-s B_1} C\, dW_s \right), \quad t \geq 0,$$

and

$$Z_t(x) = \exp\left(\int_0^t C^T \nabla \log \varphi\left(U_s(x)\right) dW_s - \frac{1}{2}\int_0^t \left\|C^T \nabla \log \varphi\left(U_s(x)\right)\right\|^2 ds\right).$$

Note that we implicitly assume here that $\int_0^t \left\|C^T \nabla \log \varphi(U_s(x))\right\|^2 ds < \infty$ P-as, so that the stochastic integral $\int_0^t C^T \nabla \log \varphi(U_s(x)) dW_s$ is well-defined and that $Z_t(x)$, $t \geq 0$, is a martingale w.r.t. natural filtration induced by the Brownian motion.

The particular representation of the drift term allows to rewrite the transition probabilities as a Feynman–Kac integral. Indeed, Itô's formula implies that

$$\int_0^t C^T \nabla \log \varphi\left(U_s(x)\right) dW_s = \log\left(\frac{\varphi\left(U_t(x)\right)}{\varphi(x)}\right)$$
$$- \int_0^t \left(\langle \nabla \log \varphi, B_1 \cdot \rangle + \frac{1}{2}\mathrm{tr}\left(Q\left(\log \varphi\right)''\right)\right)(U_s(x)) ds,$$

where f'' denotes the Hessian of a function $f \in C^2(\mathbb{R}^d)$, so that we can write

$$E\left(1_D\left(X_{0:t}\right)\right) = E\left(\varphi^{-1}(X_0)(1_D\varphi)(U_{0:t}(X_0)) \exp\left(-\int_0^t \tilde{V}(U_s(X_0)) ds\right)\right) \tag{13.19}$$

for $t \geq 0$ and $D \subset \mathcal{B}(C_{\mathbb{R}^p}([0, t]))$ with the potential

$$\tilde{V}(x) = \langle \nabla \log \varphi(x), B_1 x \rangle + \frac{1}{2}\frac{\mathrm{tr}\left(Q\varphi''(x)\right)}{\varphi(x)}. \tag{13.20}$$

In particular,

$$p_t(x, A) = \varphi^{-1}(x) E\left((1_A\varphi)\left(U_t(x)\right) \exp\left(-\int_0^t \tilde{V}\left(U_s(x)\right) ds\right)\right). \tag{13.21}$$

For given $t > 0$ and $N \in \mathbb{N}$, an appropriate time discretization of (13.20) can be obtained with the help of the kernel

$$p^{(N)}(x, A) := \varphi^{-1}(x) E\left((1_A\varphi)\left(U_{\frac{t}{N}}(x)\right) \exp\left(-\frac{t}{N}\tilde{V}\left(U_{\frac{t}{N}}(x)\right)\right)\right)$$
$$= \varphi^{-1}(x) \int (1_A\varphi)(x') e^{-\frac{t}{N}\tilde{V}(x')} N\left(e^{\frac{t}{N}B_1}x, Q_{\frac{t}{N}}\right)(dx'),$$

where

$$Q_t := \int_0^t e^{sB_1} Q e^{sB_1^T} ds, \quad t \geq 0.$$

It follows that $p^{(N)}(x, \cdot)$ describes approximately the transition probabilities of $X_{\frac{k}{N}t}$, $k = 0, 1, 2, \ldots$. In particular,

$$\lim_{N \to \infty} \underbrace{P^{(N)} \circ \ldots \circ P^{(N)}}_{N\text{-times}} f(x)$$

$$= \lim_{N \to \infty} \varphi^{-1}(x) E\left((f\varphi)(U_t(x)) \exp\left(-\frac{t}{N} \sum_{k=1}^{N} \bar{V}\left(U_{\frac{k}{N}t}(x)\right)\right)\right)$$

$$= \varphi^{-1}(x) E\left((f\varphi)(U_t(x)) \exp\left(-\int_0^t \bar{V}(U_s(x))\, ds\right)\right)$$

for suitable f, if $\bar{V}$ is bounded from below.

Suppose now that the signal is observed with independent additive noise

$$dY_t = GX_t\, dt + \Gamma de_t, \quad Y_0 = 0, \tag{13.22}$$

where $(e_t)_{t \geq 0}$ is a p-dimensional Brownian motion, independent of the signal (13.16), G is a $p \times d$ matrix and Γ is a $p \times p$ matrix such that $R := \Gamma\Gamma^T$ is positive definite.

The increments $\Delta Y_k^{(N)} := Y_k^{(N)} - Y_{k-1}^{(N)}$ of the time-discretized observation $Y_k^{(N)} := Y_{\frac{k}{N}t}$, $k = 0, 1, 2, \ldots$, are given by

$$\Delta Y_k^{(N)} = \int_{\frac{k-1}{N}t}^{\frac{k}{N}t} GX_s\, ds + \Gamma\left(e_{\frac{k}{N}t} - e_{\frac{k-1}{N}t}\right),$$

which can be approximated by

$$\Delta \tilde{Y}_k^{(N)} := \frac{t}{N} GX_{\frac{k}{N}t} + \Gamma \Delta e_k^N,$$

where $\Delta e_k^{(N)} = e_{\frac{k}{N}t} - e_{\frac{k-1}{N}t}$. It follows with this approximation that the distribution of the measurement error is given by

$$g^{(N)}(x, y)\, dy = N\left(\frac{t}{N} Gx, \frac{t}{N} R\right)(dy).$$

For any given sequence $\tilde{y}_{1:n} \in \mathbb{R}^{pn}$, we now define the following time-discretized approximation $\pi_n^{(N), \tilde{y}_{1:n}}$, $n = 0, 1, 2, \ldots$, of the optimal filter similar to (13.3), but with P replaced by $P^{(N)}$ and g replaced by $g^{(N)}$, i.e.,

$$\int f(x) \pi_{n+1}^{(N), y_{1:n+1}}(dx) = \frac{\int P^{(N)}\left(f \cdot g^{(N)}(\cdot, y_{n+1})\right)(x) \pi_n^{(N), y_{1:n}}(dx)}{\int P^{(N)} g^{(N)}(\cdot, y_{n+1})(x) \pi_n^{(N), y_{1:n}}(dx)}$$

$$= \frac{\int f(x) g^{(N)}(x, y_{n+1}) \hat{P}^{(N)} \pi_n^{(N), y_{1:n}}(dx)}{\int g^{(N)}(x, y_{n+1}) \hat{P}^{(N)} \pi_n^{(N), y_{1:n}}(dx)}.$$

In particular, $\pi_n^{(N),\tilde{y}_{1:n}}$ admits the following representation as a normalized Feynman–Kac integral

$$\int f \, d\pi_n^{(N),\tilde{y}_{1:n}} = \frac{E\left(\varphi^{-1}(X_0)(f\varphi)\left(U_{t\frac{n}{N}}(X_0)\right) Z_n^{(N)}(\tilde{y}_{1:n})\right)}{E\left(\varphi^{-1}(X_0)\varphi\left(U_{t\frac{n}{N}}(X_0)\right) Z_n^{(N)}(\tilde{y}_{1:n})\right)}, \quad (13.23)$$

where

$$Z_n^{(N)}(\tilde{y}_{1:n}) = \exp\left(-\frac{t}{N}\sum_{k=1}^n \tilde{V}(U_{t\frac{k}{N}}(X_0)) + \sum_{k=1}^n \langle R^{-1} G U_{t\frac{k}{N}}(X_0), \tilde{y}_k\rangle \right.$$
$$\left. -\frac{t}{2N}\sum_{k=1}^n \langle R^{-1} G U_{t\frac{k}{N}}(X_0), G U_{t\frac{k}{N}}(X_0)\rangle\right).$$

We will show in Section 13.3 below that for Hölder-continuous paths $y : [0, t] \to \mathbb{R}^p$ with discretization $\Delta y^{(N)} := \left(y_{t\frac{1}{N}}, y_{t\frac{2}{N}} - y_{t\frac{1}{N}}, \ldots, y_{t-\frac{N-1}{N}t}\right)$ and under appropriate assumptions on the initial distribution $P_{X_0} = \mu_0$ of the signal, it follows that

$$\lim_{N\to\infty} \int f \, d\pi_N^{(N),\Delta y^{(N)}} = \int f \, d\pi_t^{y_{0:t}}, \quad f \in \mathcal{B}_b(\mathbb{R}^d),$$

and the right-hand side defines a robust version of the Kallianpur–Striebel representation of the optimal filter.

Stability of $\pi_n^{(N),\tilde{y}_{1:n}}$

For the study of the stability properties of the optimal filter, the following potential V, defined by

$$V(x) := \langle B_1 x, \nabla \log \varphi(x)\rangle + \frac{1}{2}\frac{\mathrm{tr}\,(Q\varphi''(x))}{\varphi(x)} + \frac{1}{2}\langle R^{-1} G x, G x\rangle \quad (13.24)$$

will play a crucial role.

In contrast to the Kalman filter, it is not possible to solve the forward equation (13.6) for suitable initial conditions explicitly. However, one can show that the set of functions $\hat{m}$ with

$$T_\varphi(\hat{m}) := \varphi^{-1}\hat{m} \quad (13.25)$$

log-concave, will be invariant under suitable assumptions on V.

Recall that a function $f \in C^2(\mathbb{R}^d)$ is called log-concave, if $f > 0$ and $-(\log f)'' \geq 0$, where "$\geq$" is to be understood in the quadratic form sense. We say that f is log-concave with lower bound K, if K is a symmetric $d \times d$ matrix and $-(\log f)'' \geq K$.

To state our stability result, we assume that $V \in C^2(\mathbb{R}^d)$ and that there exists a positive definite symmetric $d \times d$ matrix K such that

$$V''(x) \geq K \qquad \forall x \in \mathbb{R}^d. \tag{13.26}$$

It follows that for $N \in \mathbb{N}$, $t \geq 0$, there exists a (unique) positive definite $d \times d$ matrix $E_\infty^{(N)}$ solving the matrix Riccati equation

$$E_\infty^{(N)} = B_1^{(N)} \left(E_\infty^{(N)} - E_\infty^{(N)} \left(E_\infty^{(N)} + \frac{N}{t} K^{-1} \right)^{-1} E_\infty^{(N)} \right) B_1^{(N),T} + Q^{(N)},$$

with

$$B_1^{(N)} := e^{\frac{t}{N} B_1} \quad \text{and} \quad Q^{(N)} := \int_0^{\frac{t}{N}} e^{s B_1} Q e^{s B_1^T} ds \tag{13.27}$$

(see Subsection 13.2.3 on the Kalman filter).

It is then easy to see, using the matrix inversion lemma, that

$$K_*^{(N)} := E_\infty^{(N),-1} + \frac{t}{N} K$$

is a symmetric positive definite solution of the matrix equation

$$\begin{aligned} K_*^{(N)} = & \frac{t}{N} K + Q^{(N),-1} \\ & - Q^{(N),-1} B_1^{(N)} \left(K_*^{(N)} + B_1^{(N),T} Q^{(N),-1} B_1^{(N)} \right)^{-1} B_1^{(N),T} Q^{(N),-1}. \end{aligned} \tag{13.28}$$

Theorem 13.2 Let $\hat{m}_0 \in \mathcal{B}_b^+(\mathbb{R}^d)$ be such that $T_\varphi(\hat{m}_0)$ is log-concave with lower bound $K_*^{(N)}$. Let $y_{1:n} \in \mathbb{R}^{pn}$ and

- $\pi_n^{y_{1:n}}$ be the optimal filter with initial condition

$$\mu_0 = \left(\int \hat{m}_0 \, dx \right)^{-1} \hat{m}_0 \, dx$$

- $\tilde{\pi}_n^{y_{1:n}}$ be the optimal filter with initial condition $\tilde{\mu}_0$, $\tilde{\mu}_0 \ll \mu_0$, with Lipschitz continuous density $h_0 := \frac{d\tilde{\mu}_0}{d\mu_0}$.

Let

$$\tilde{K}^{(N)} := 2\sqrt{Q^{(N)}} B_1^{(N),-T} K_*^{(N)} B_1^{(N),-1} \sqrt{Q^{(N)}} + 2I - B_1^{(N),-T} B_1^{(N),-1} \tag{13.29}$$

and

$$\chi_*^{(N)} := \left(\min \{ \lambda : \lambda \text{ is an eigenvalue of } \tilde{K}^{(N)} \} \right)^{\frac{1}{2}}. \tag{13.30}$$

Then

$$\left\|\pi_n^{y_{1:n}} - \tilde{\pi}_n^{y_{1:n}}\right\|_{\mathrm{var}} \leq \frac{1 + \frac{t}{N}\mathrm{tr}\left(B_1^{(N),T}\sqrt{Q^{(N),-1}}K_*^{(N),-1}\sqrt{Q^{(N),-1}}B_1^{(N)}\right)}{2\int h_n\,d\tilde{\pi}_n^{y_{1:n}}}$$
$$\cdot \chi_*^{(N),-n}\|h_0\|_{\mathrm{Lip}_{\sqrt{\frac{t}{N}Q^{(N),-1}B_1^{(N)}}}}.$$

In particular, $h_0 \geq \delta > 0$ implies that

$$\left\|\pi_n^{y_{1:n}} - \tilde{\pi}_n^{y_{1:n}}\right\|_{\mathrm{var}} \leq \frac{1 + \frac{t}{N}\mathrm{tr}\left(B_1^{(N),T}\sqrt{Q^{(N),-1}}K_*^{(N),-1}\sqrt{Q^{(N),-1}}B_1^{(N)}\right)}{2\delta}$$
$$\cdot \chi_*^{(N),-n}\|h_0\|_{\mathrm{Lip}_{\sqrt{\frac{t}{N}Q^{(N),-1}B_1^{(N)}}}}.$$

13.3 Stability in the time-continuous case

The stability result of Subsection 13.2.4 can be carried over to the time-continuous case. To this end let us introduce the robust version of the Kallianpur–Striebel representation formula of the optimal filter that was introduced in [4] as follows: denote by

$$H_{\frac{1}{3}} := \left\{ y_\cdot \in C_{\mathbb{R}^p}([0,t]) \mid \|y_\cdot\|_{\frac{1}{3}} := \sup_{s_1 \neq s_2} \frac{\|y_{s_1} - y_{s_2}\|}{|s_1 - s_2|^{\frac{1}{3}}} < \infty \right\}$$

the space of Hölder paths $y_\cdot : [0,t] \to \mathbb{R}^p$ with Hölder regularity $\frac{1}{3}$, and define for fixed $y_\cdot \in H_{\frac{1}{3}}$ and $N \in \mathbb{N}$

$$I^{(N)}(y_\cdot) := \sum_{k=0}^{N-1} G^T R^{-1} y_{t\frac{k}{N}}\left(X_{t\frac{k+1}{N}} - X_{t\frac{k}{N}}\right). \tag{13.31}$$

It is shown in [4] in the case $R = I$, but the general case follows similarly, that

$$I(y_\cdot) := \limsup_{N\to\infty} I^{(2^N)}(y_\cdot)$$

is a $\tilde{P}$-version of the stochastic integral

$$\int_0^t G^T R^{-1} y_s\, dX_s,$$

where $\tilde{P}$ is the probability measure on $\mathcal{X}_t \vee \mathcal{Y}_t$ defined by

$$\frac{d\tilde{P}}{dP} = \exp\left(-\int_0^t R^{-1} G X_s \Gamma\, de_s - \frac{1}{2}\int_0^t \langle R^{-1} G X_s, G X_s\rangle\, ds\right).$$

We assume here, that

$$E\left(\exp\left(\alpha \int_0^t \|B(X_s)\| \, ds\right)\right) < \infty \qquad (13.32)$$

for all $\alpha > 0$.

Moreover, $I(y.)$ induces a robust version of the optimal filter as follows: if we define

$$\Theta(y.) := \exp\left(\langle R^{-1}GX_t, y_t\rangle - I(y.) - \frac{1}{2}\int_0^t \langle R^{-1}GX_s, GX_s\rangle \, ds\right),$$

then

$$\Phi(f, y) = \frac{\tilde{E}(f(X_t)\Theta(y.))}{\tilde{E}(\Theta(y.))}, \quad f \in \mathcal{B}_+(\mathbb{R}^d)$$

depends continuously on $y.$ (w.r.t. $\|\cdot\|_\infty$-norm) and $\Phi(f, Y.)$ is a version of $E(f(X_t) \mid \mathcal{Y}_t)$. The limit (13.31) exists $\tilde{P}$-a.s. and in $L^2(\tilde{P})$. In the following, let $\pi_t^{y_{0:t}}$, defined by $\int f \, d\pi_t^{y_{0:t}} := \Phi(f, y)$, be the corresponding regular conditional distribution.

We are now going to study the limit of the time discrete approximations introduced in Subsection 13.2.4 under one additional assumption on the initial distribution $P_{X_0} = \mu_0$ of the signal: There exists $\delta > 0$ such that

$$\int (\varphi^{-(1+\delta)}(x) + 1)e^{\alpha|x|} \mu_0(dx) < \infty \qquad \forall \alpha \geq 0. \qquad (13.33)$$

Fix a path $y. \in H_{\frac{1}{3}}$, $y_0 = 0$, and define for given $N \in \mathbb{N}$ its time discretization

$$\Delta y^{(N)} := \left(y_{\frac{1}{N}t}, y_{\frac{2}{N}t} - y_{\frac{1}{N}t}, \ldots, y_t - y_{\frac{N-1}{N}t}\right).$$

To simplify notations in the following let

$$\Theta^{(N)}(y.) := \exp\left(\langle R^{-1}GX_t, y_t\rangle - I^{(N)}(y.) - \frac{t}{2N}\sum_{k=0}^{N-1}\langle R^{-1}GX_{t\frac{k}{N}}, GX_{t\frac{k}{N}}\rangle\right).$$

Taking into account formula (13.19) we can then write

$$\begin{aligned}\tilde{E}\left(f(X_t)\Theta^{(N)}(y.)\right) &- E\left(\varphi^{-1}(X_0)(f\varphi)(U_t(X_0))Z_N^{(N)}(\Delta y^{(N)})\right) \\ &= E\left(\varphi^{-1}(X_0)(f\varphi)(U_t(X_0))\left(\exp\left(-\int_0^t \bar{V}(U_s(X_0))\,ds\right)\right.\right. \\ &\left.\left. - \exp\left(-\frac{t}{2N}\sum_{k=0}^{N-1}\bar{V}(U_{t\frac{k}{N}}(X_0))\right)\right)\tilde{Z}_N^{(N)}(\Delta y^{(N)})\right),\end{aligned} \qquad (13.34)$$

where
$$\tilde{Z}_n^{(N)}(\Delta y^{(N)}) := \exp\left(\sum_{k=1}^n \langle R^{-1} G U_{t\frac{k}{N}}(X_0), \Delta y_k^{(N)} \rangle - \frac{t}{2N} \sum_{k=1}^n \langle R^{-1} G U_{t\frac{k}{N}}(X_0), G U_{t\frac{k}{N}}(X_0) \rangle \right).$$

It follows that
$$\lim_{N\to\infty} \tilde{E}\left(f(X_t)\Theta^{(N)}(y)\right) - E\left(\varphi^{-1}(X_0)(f\varphi)(U_t(X_0)) Z_N^{(N),\Delta y^{(N)}}\right) = 0$$

for all $f \in \mathcal{B}_b(\mathbb{R}^d)$, since the integrand on the right-hand side of (13.34) converges to zero pointwise and can be estimated from above in absolute value by
$$2e^{-t \inf_{x\in\mathbb{R}^d} \tilde{V}(x)} \|f\|_\infty \varphi^{-1}(X_0) \varphi(U_t(X_0)) \tilde{Z}_N^{(N)}(\Delta y^{(N)}),$$

and $\varphi^{-1}(X_0)\varphi(U_t(X_0)) \tilde{Z}_N^{(N)}(\Delta y^{(N)})$, $N \geq 1$, is uniformly integrable, since for $\alpha \geq 0$

$$E\left(\exp\left(\alpha \sum_{k=0}^{N-1} G^T R^{-1} y_{\frac{k}{N}t}\left(U_{t\frac{k+1}{N}}(X_0) - U_{t\frac{k}{N}}(X_0)\right)\right)\right)$$
$$\leq E\left(\exp\left(\alpha \|G^T R^{-1} y\|_\infty e^{t\|B_1\|_{op}} \|X_0\| \right.\right. \tag{13.35}$$
$$\left.\left. + \alpha \int_0^t \Phi_s^{(N)}(y.) e^{(t-s)B_1} C\, dW_s\right)\right)$$

with
$$\Phi_s^{(N)}(y.) = \sum_{k=0}^{N-1} 1_{]\frac{k}{N}t, \frac{k+1}{N}t]}(s) G^T R^{-1} y_{\frac{k}{N}t}.$$

Clearly, the right-hand side of (13.35) is equal to
$$E\left(\exp\left(\alpha \|G^T R^{-1} y\|_\infty e^{t\|B_1\|_{op}} \|X_0\|\right)\right)$$
$$\exp\left(\frac{1}{2}\alpha^2 \int_0^t \Phi_s^{(N)}(y.) e^{(t-s)B_1} Q e^{(t-s)B_1^T} ds\right),$$

which is finite by assumption (13.33). On the other hand, using the upper bound $|\varphi(x)|^{1+\delta} \leq M(1 + \|x\|^p)$ for suitable constants M and p, we can estimate
$$E\left(\varphi^{-(1+\delta)}(X_0)\varphi^{(1+\delta)}(U_t(X_0))\right)$$
$$\leq M \int \varphi^{-(1+\delta)}(x)(1 + E(\|U_t(x)\|^p)) \mu_0(dx)$$
$$\leq M \int \varphi^{-(1+\delta)}(x)(1 + e^{\alpha\|x\|}) \mu_0(dx) < \infty$$

by assumption (13.33) again. It follows that $\left(\varphi^{-1}(X_0)\varphi(U_t(X_0)) Z_N^{(N),\Delta y^{(N)}}\right)_{N\geq 1} \subset L^{1+\delta/2}(P)$ is bounded, which implies the desired uniform integrability.

Using the fact that

$$\int f d\pi_n^{(N),\Delta y^{(N)}} = \frac{E\left(\varphi^{-1}(X_0)(f\varphi)(U_t(X_0))Z_n^{(N)}\left(\Delta y^{(N)}\right)\right)}{E\left(\varphi^{-1}(X_0)\varphi(U_t(X_0))Z_n^{(N)}\left(\Delta y^{(N)}\right)\right)},$$

we can summarize the approximation result as follows: Let $y. \in H_{\frac{1}{3}}$, $y_0 = 0$, and assume that (13.32) and (13.33) hold. Then

$$\lim_{N \to \infty} \int f d\pi_N^{(N),\Delta y^{(N)}} = \int f d\pi_t^{y_{0:t}} \qquad (13.36)$$

for all $f \in \mathcal{B}_b(\mathbb{R}^d)$.

To state our stability result for the time-continuous case, let K_* be the unique positive definite symmetric solution of the matrix Riccati equation

$$K_* Q K_* + B_1^T K_* + K_* B_1 - K = 0 \qquad (13.37)$$

(see Satz 11.2 in [5]). In the following let

$$\chi_* := \min\left\{\lambda : \lambda \text{ is an eigenvalue of } \sqrt{Q}K_*\sqrt{Q} + \frac{1}{2}\left(B_1 + B_1^T\right)\right\}. \qquad (13.38)$$

We then have the following result:

Theorem 13.3 *Let $\hat{m}_0 \in \mathcal{B}_b^+(\mathbb{R}^d)$ be such that $T_\varphi(\hat{m}_0)$ is log-concave with lower bound $> K_*$. Let $y \in H_{\frac{1}{3}}$, $y_0 = 0$, and let*

- *$\pi_t^{y_{0:t}}$ be the optimal filter with initial condition*

$$\mu_0 = \left(\int \hat{m}_0 dx\right)^{-1} \hat{m}_0 \, dx$$

- *$\tilde{\pi}_t^{y_{0:t}}$ be the optimal filter with initial condition $\tilde{\mu}_0$, $\tilde{\mu}_0 \ll \mu_0$ with Lipschitz-continuous density $h_0 := \frac{d\tilde{\mu}_0}{d\mu_0}$, $h_0 \geq \delta$.*

Assume that μ_0 and $\tilde{\mu}_0$ satisfy (13.33) and that (13.32) holds for the signal started with initial distribution μ_0 (or with $\tilde{\mu}_0$). Then

$$\left\|\pi_t^{y_{0:t}} - \tilde{\pi}_t^{y_{0:t}}\right\|_{\text{var}} \leq \frac{1 + \text{tr}\left(\sqrt{Q^{-1}}K_*^{-1}\sqrt{Q^{-1}}\right)}{2\delta} e^{-\chi_* t} \|h_0\|_{\text{Lip}_{\sqrt{Q^{-1}}}}.$$

In the following we consider as particular examples Kalman–Bucy filters and filters for one-dimensional signals.

13.3.1 Kalman–Bucy filter

The time-continuous analogue of the Kalman filter is the well-known Kalman–Bucy filter: Assume that the signal is given as the solution of the linear stochastic differential equation

$$dX_t = B_1 X_t \, dt + C \, dW_t, \quad X_0 = \xi, \tag{13.39}$$

with $d \times d$ matrices B_1 and C. In this case (13.17) is always satisfied with $\varphi \equiv$ *constant*. The observation is given as

$$dY_t = GX_t \, dt + \Gamma \, de_t, \quad Y_0 = 0. \tag{13.40}$$

Similar to the time-discrete case we assume that both, $Q := CC^T$ and $R := \Gamma\Gamma^T$ are positive definite.

The main feature of the Kalman–Bucy filter is that if it is initialized with a Gaussian distribution $\mu_0 = N(m_0, E_0)$ of (13.39), the optimal filter is again Gaussian

$$\pi_t^{Y_{0:t}} = N\left(m_t^{Y_{0:t}}, E_t\right), \tag{13.41}$$

where E_t and $m_t^{Y_{0:t}}$ are now determined by the Kalman–Bucy filtering equations (see Chapter 11 in [5] or [7])

$$\dot{E}_t = B_1 E_t + E_t B_1^T + Q - E_t G^T R^{-1} G E_t \tag{13.42}$$

$$dm_t^{Y_{0:t}} = B_1 m_t^{Y_{0:t}} \, dt + E_t G^T R^{-1} \left(dY_t - Gm_t^{Y_{0:t}} \, dt\right). \tag{13.43}$$

Note that (13.43) is a stochastic differential equation and that $Y_{0:t}$ is of unbounded variation, so that (13.43) does not immediately induce a representation of the regular conditional distribution.

As a side remark we just mention that E_t is continuously differentiable, in particular of bounded variation, so that we could integrate by parts to rewrite the stochastic integral as

$$\int_0^t E_s G^T R^{-1} \, dY_s = E_t G^T R^{-1} Y_t - \int_0^t \dot{E}_t G^T R^{-1} Y_s \, ds,$$

taking into account $Y_0 = 0$. This gives a continuous representation of the solution of (13.43) and thus induces a regular conditional distribution of $\pi_t^{Y_{0:t}}$. From now on we assume that $p = d$ and that the matrix

$$G \quad \text{has full rank}. \tag{13.44}$$

Choosing $\varphi \equiv 1$ in the time discretization, it follows that the potential V of (13.24) has the form $\frac{1}{2}\langle R^{-1}Gx, Gx \rangle$, hence $V''(x) = G^T R^{-1} G > 0$, so that (13.26) holds with $K = G^T R^{-1} G$.

The matrix Riccati equation (13.37) in this example takes the form

$$K_* Q K_* + B_1^T K_* + K_* B_1 - G^T R^{-1} G = 0$$

and this equation has a unique positive definite solution K_*. Theorem 13.3 now implies stability of the optimal filter with exponential rate

$$\chi_* := \min\left\{\lambda : \lambda \text{ is an eigenvalue of } \sqrt{Q}K_*\sqrt{Q} + \frac{1}{2}\left(B_1 + B_1^T\right)\right\}.$$

Example 13.1 Assume that $B_1 = B_1^T$ is symmetric and that B_1, Q, $G^T R^{-1} G$ are simultaneously diagonizable. Then

$$K_* = \sqrt{Q^{-1}}\left(-B_1 + \sqrt{B_1^2 + QG^T R^{-1} G}\right)\sqrt{Q^{-1}}$$

is the unique positive definite solution of (13.37), hence

$$\sqrt{Q}K_*\sqrt{Q} + B_1 = \sqrt{B_1^2 + QG^T R^{-1} G},$$

so that

$$\chi_* = \min\left\{\lambda : \lambda \text{ is an eigenvalue of } \sqrt{B_1^2 + QG^T R^{-1} G}\right\}$$

coincides with the exponential rate identified in [10].

13.3.2 One-dimensional signals

Consider a signal given as the solution of the one-dimensional stochastic differential equation

$$dX_t = B(X_t)\, dt + C\, dW_t, \quad X_0 = \xi, \tag{13.45}$$

with $B \in C^3(\mathbb{R})$. In this case, (13.17) is always satisfied with $B_1 = 0$ and $\varphi(x) = \exp\left(\int_0^x B(s)\, ds\right)$. Assume that the observation is given as

$$dY_t = GX_t\, dt + \Gamma\, de_t, \quad Y_0 = 0. \tag{13.46}$$

Then assumption (13.26) reduces to the following

$$Q(B^2 + B_x)_{xx} + \left(\frac{G}{\Gamma}\right)^2 \geq K^2 \text{ for some } K > 0, \tag{13.47}$$

which is always satisfied for $\left(\frac{G}{\Gamma}\right)^2$ sufficiently large, if $Q(B^2 + B_x)_{xx}$ is bounded from below. Compare this with the Assumption 1 in [1] where it is on the one hand only assumed that $B^2 + B_x$ has a bounded second derivate, but on the other hand, the stability of the filter is shown only for an observation of the type (13.46) with Γ sufficiently small. Our analysis allows to state precise lower bounds on the rate of stability of the optimal filter that are stated explicitly in terms of the coefficients of (13.45) and (13.46).

13.4 Proofs of Theorem 13.2 and Theorem 13.3

13.4.1 Proof of Theorem 13.2

The proof of Theorem 13.2 relies on two Lemmata on the stability properties of Markovian integral operators with log-concave integral kernels.

Lemma 13.2 *Let $\hat{m} \in \mathcal{B}_+(\mathbb{R}^d)$ be such that $T_\varphi(\hat{m})$ is log-concave with lower bound $K_*^{(N)}$. Let $u \in \mathbb{R}^p$ be arbitrary. Then $g^{(N)}(x, u) \hat{P}^{(N)} \hat{m}(x)$ too, is log-concave with lower bound $K_*^{(N)}$.*

Proof. Recall the simplified notations (13.27). Let
$$W(x) := -\log\left(T_\varphi(\hat{m})\right)(x)$$
and
$$L(x, y) := \frac{t}{N} V(x) - \frac{t}{N} \langle R^{-1} G x, u \rangle + W(y) \\ + \frac{1}{2} \langle Q^{(N),-1}\left(B_1^{(N)} y - x\right), \left(B_1^{(N)} y - x\right) \rangle,$$
so that
$$g^{(N)}(x, u) \hat{P}^{(N)} \hat{m}(x) \propto \int \exp(-L(x, y)) \, dy.$$
We can then write
$$L_{xx} = \frac{t}{N} V_{xx} + Q^{(N),-1} \geq \frac{t}{N} K + Q^{(N),-1},$$
$$L_{xy} = L_{yx}^T = -Q^{(N),-1} B_1^{(N)}$$
and
$$L_{yy} = W_{yy} + B_1^{(N),T} Q^{(N),-1} B_1^{(N)} \\ \geq K_*^{(N)} + B_1^{(N),T} Q^{(N),-1} B_1^{(N)}.$$
Consequently, using (13.28)
$$L_{xx} - L_{xy} L_{yy}^{-1} L_{yx} \geq \frac{t}{N} K + Q^{(N),-1} - \\ - Q^{(N),-1} B_1^{(N)} \left(K_*^{(N)} + B_1^{(N),T} Q^{(N),-1} B_1^{(N)}\right)^{-1} B_1^{(N),T} Q^{(N),-1} \\ = K_*^{(N)}.$$
The assertion now follows from Theorem 4.2 in [2]. □

We will fix from now on an initial condition $\hat{m}_0$ as specified in Theorem 13.2. It follows from Lemma 13.2 that $T_\varphi(\hat{m}_1)$ is log-concave too with lower bound

$K_*^{(N)}$. Iterating, we conclude that $T_\varphi(\hat{m}_n)$ is log-concave with lower bound $K_*^{(N)}$ for all n.

Let us now consider the densities $p_n^*(x, x')$ of the corresponding ground state transforms:

$$p_n^*(x, x')\, dx' \propto p^{(N)}(x', x)\hat{m}_{n-1}(x')\, dx'$$
$$\propto \frac{1}{\varphi(x')} \exp\left(-\frac{1}{2}\langle Q^{(N),-1}\left(B_1^{(N)}x' - x\right), \left(B_1^{(N)}x' - x\right)\rangle\right)$$
$$\hat{m}_{n-1}(x')\, dx'.$$

Since we know that $T_\varphi(\hat{m}_{n-1})$ is log-concave with lower bound $K_*^{(N)}$ we can write the ground state transform as

$$p_n^*(x, x') \propto \exp\left(-V_n^*(x') - \frac{1}{2}\langle Q^{(N),-1}\left(B_1^{(N)}x' - x\right), \left(B_1^{(N)}x' - x\right)\rangle\right) \quad (13.48)$$

for some convex function V_n^* satisfying $\left(V_n^*\right)'' \geq K_*^{(N)}$.

To simplify notations in the following let

$$A := \sqrt{\frac{t}{N} Q^{(N),-1}}\, B_1^{(N)}$$

and recall the definition of $\chi_*^{(N)}$ in (13.30).

Lemma 13.3 Let $f \in \mathrm{Lip}(\mathbb{R}^d)$. Then $P_n^* f \in \mathrm{Lip}(\mathbb{R}^d)$ with

$$\|P_n^* f\|_{\mathrm{Lip}_A} \leq \frac{1}{\chi_*^{(N)}} \|f\|_{\mathrm{Lip}_A}.$$

Proof. Denote by (W_t) the d-dimensional Brownian motion, and let $\left(X_t^\theta(x)\right)$, $\theta, x \in \mathbb{R}^d$, be the unique strong solution of the stochastic differential equation

$$dX_t^\theta(x) = \left(A^T A\right)^{-1}\left(-\nabla V_n^*\left(X_t^\theta(x)\right)\right.$$
$$\left. + B_1^{(N),T} Q^{(N),-1}\left(\theta - B_1^{(N)} X_t^\theta(x)\right)\right) dt + A^{-1}\, dW_t \quad (13.49)$$

with initial condition $X_0^\theta(x) = x$. The existence and uniqueness follows from Theorem V.1.1 in [6].

The Markov process associated with (13.49) has an invariant measure $p_n^*(\theta, x')\, dx'$ and

$$\lim_{t \to \infty} E\left(h\left(X_t^\theta(x)\right)\right) = \int h(y)\, p_n^*(\theta, x')\, dx'$$

for all $h \in C_b(\mathbb{R}^d)$ a.e. w.r.t. the Lebesgue measure.

To further simplify notations let

$$\|x\|_A := \|Ax\| \quad \text{and} \quad \Delta_{\theta_1, \theta_2}(t) := X_t^{\theta_1}(x) - X_t^{\theta_2}(x), \; x \in \mathbb{R}^d.$$

For fixed $x \in \mathbb{R}^d$ we then have

$$d\|\Delta_{\theta_1,\theta_2}(t)\|_A^2 = 2\langle B_1^{(N),T} Q^{(N),-1}(\theta_1 - \theta_2), \Delta_{\theta_1,\theta_2}(t)\rangle dt$$
$$- 2\langle \nabla V_n^*(X_t^{\theta_1}(x)) - \nabla V_n^*(X_t^{\theta_2}(x)), \Delta_{\theta_1,\theta_2}(t)\rangle dt$$
$$- 2\langle B_1^{(N),T} Q^{(N),-1} B_1^{(N)} \Delta_{\theta_1,\theta_2}(t), \Delta_{\theta_1,\theta_2}(t)\rangle dt$$
$$\leq -\langle (2K_*^{(N)} + 2B_1^{(N),T} Q^{(N),-1} B_1^{(N)}$$
$$- B_1^{(N),T} \sqrt{Q^{(N),-1}} B_1^{(N),-T} B_1^{(N),-1} \sqrt{Q^{(N),-1}} B_1^{(N)} \Delta_{\theta_1,\theta_2}(t), \Delta_{\theta_1,\theta_2}(t)\rangle dt$$
$$+ \frac{N}{t}\|\theta_1 - \theta_2\|_A^2 dt$$
$$\leq -\frac{N}{t}\left(\chi_*^{(N)}\right)^2 \|X_t^{\theta_1}(x) - X_t^{\theta_2}(x)\|_A^2 dt + \frac{N}{t}\|\theta_1 - \theta_2\|_A^2 dt.$$

Integrating this differential inequality w.r.t. t, we obtain that

$$\|X_t^{\theta_1}(x) - X_t^{\theta_2}(x)\|_A^2 \leq \chi_*^{(N),-2} \|\theta_1 - \theta_2\|_A^2.$$

Consequently, f Lipschitz and bounded implies that

$$\left(P_n^* f(\theta_1) - P_n^* f(\theta_2)\right)^2$$
$$= \lim_{t \to \infty} \int \left(E\left(f(X_t^{\theta_1}(x)) - f(X_t^{\theta_2}(x))\right)\right)^2 N(0, I)(dx)$$
$$\leq \|f\|_{\text{Lip}_A}^2 \chi_*^{(N),-2} \|\theta_1 - \theta_2\|_A^2,$$

which implies the assertion. □

The proof of Theorem 13.2 is completed by observing that

$$\hat{\sigma}_n = \int\int \|A(x - x')\| \pi_n^{y_{1:n}}(dx) \pi_n^{y_{1:n}}(dx') \leq 1 + \text{tr}\left(AK_*^{(N),-1} A^I\right)$$

by Theorem 4.2 in [2].

13.4.2 Proof of Theorem 13.3

Theorem 13.2 implies that for fixed N, the corresponding time-discretized filter is stable with geometric rate $\chi_*^{(N)}$ defined by (13.30), i.e.

$$\chi_*^{(N)} = \left(\min\{\lambda : \lambda \text{ eigenvalue of } \tilde{K}^{(N)}\}\right)^{\frac{1}{2}}$$

where

$$\tilde{K}^{(N)} := 2\sqrt{Q^{(N)}} B_1^{(N),-T} K_*^{(N)} B_1^{(N),-1} \sqrt{Q^{(N)}} + 2I - B_1^{(N),-T} B_1^{(N),-1}.$$

We can write

$$\chi_*^{(N)} = \left(1 + \frac{t}{N} a_N\right)^{\frac{1}{2}}$$

with
$$a_N = \frac{N}{t} \inf_{\|x\|=1} \langle (\tilde{K}^{(N)} - I) x, x \rangle,$$
so that
$$\left(\frac{1}{\chi_*^{(N)}}\right)^N = \left(\frac{1}{1 + \frac{t}{N} a_N}\right)^{\frac{N}{2}} \leq \exp\left(-\frac{t}{2} \frac{a_N}{1 + \frac{t}{N} a_N}\right).$$

For the proof of the Theorem, it suffices to show now that the sequence (a_N) is bounded and $\liminf_{N \to \infty} a_N \geq \chi_*$.

In the following we are going to study the asymptotics of $K_*^{(N)}$ as $N \to \infty$.

Lemma 13.4 Let $K_*^{(N)}$ be as in (13.28) and K_* be as in (13.37). Then
$$\lim_{N \to \infty} \left\| K_*^{(N)} - K_* \right\|_{op} = 0.$$

Proof 1 First note that we can rewrite equation (13.28) for the matrix $K_*^{(N)}$ as
$$K_*^{(N)} = \frac{t}{N} K + \left(B_1^{(N),-T} K_*^{(N)} B_1^{(N),-1} Q^{(N)} + I \right)^{-1} B_1^{(N),-T} K_*^{(N)} B_1^{(N),-1},$$
which is equivalent to
$$B_1^{(N),-T} K_*^{(N)} B_1^{(N),-1} Q^{(N)} K_*^{(N)} + \left(K_*^{(N)} - B_1^{(N),-T} K_*^{(N)} B_1^{(N),-1} \right)$$
$$= \frac{t}{N} \left(B_1^{(N),-T} K_*^{(N)} B_1^{(N),-1} Q^{(N)} + I \right) K. \tag{13.50}$$

It is easy to see that $\| K_*^{(N)} \|_{op}$ is bounded in N. For the proof of the lemma it suffices now to show that any convergent subsequence of $(K_*^{(N)})$ converges to K_*. To this end fix such a convergent subsequence $(K_*^{(N_k)})$ and denote its limit with $\tilde{K}_*$. Using the fact that
$$\lim_{k \to \infty} \frac{N_k}{t} Q^{(N_k)} = \lim_{k \to \infty} \frac{N_k}{t} \int_0^{\frac{t}{N_k}} e^{s B_1^T} Q e^{s B_1} \, ds = Q,$$
$$\lim_{k \to \infty} \frac{N_k}{t} \left(B_1^{(N_k),-T} - I \right) = -B_1^T \text{ and}$$
$$\lim_{k \to \infty} \frac{N_k}{t} \left(B_1^{(N_k),-1} - I \right) = -B_1,$$
we conclude from equation (13.50), taking the limit $k \to \infty$, that $\tilde{K}_*$ is a positive semidefinite symmetric solution of the matrix Riccati equation
$$\tilde{K}_* Q \tilde{K}_* + B_1^T \tilde{K}_* + \tilde{K}_* B_1 - K = 0.$$

$K > 0$ now implies that $\tilde{K}_*$ must be positive definite, because $\tilde{K}_* x = 0$ implies

$$\langle x, Kx \rangle = \langle x, \tilde{K}_* Q \tilde{K}_* x + B_1^T \tilde{K}_* x + \tilde{K}_* B_1 x \rangle = 0,$$

hence $x = 0$. By uniqueness of the positive definite solution of (13.37), the assertion now follows. □

As a consequence of the Lemma it follows that

$$\liminf_{N \to \infty} a_N \geq 2\chi_*,$$

because

$$\lim_{N \to \infty} \frac{N}{t} \left(\tilde{K}^{(N)} - I \right) = 2\sqrt{Q} K_* \sqrt{Q} + B_1^T + B_1$$

w.r.t. the operator norm.

On the other hand, clearly,

$$\frac{t}{N} a_N \leq \left\| \int_0^{\frac{t}{N}} e^{s B_1} Q e^{s B_1^T} ds \right\|_{\mathrm{op}} e^{2 \frac{t}{N} \| B_1 \|_{\mathrm{op}}} \left\| K_*^{(N)} \right\|_{\mathrm{op}}$$

$$+ \left\| \int_0^{\frac{t}{N}} B_1^T e^{-s B_1^T} e^{-s B_1} ds \right\|_{\mathrm{op}} + \left\| \int_0^{\frac{t}{N}} e^{-s B_1^T} e^{-s B_1} B_1 ds \right\|_{\mathrm{op}}$$

converges to 0 as $N \to \infty$, so that indeed

$$\limsup_{N \to \infty} \chi_*^{(N), -N} \leq \exp(-\chi_* t). \tag{13.51}$$

The proof of Theorem 13.3 now follows from the previous remarks, using the time discretization mentioned above, as follows: since $\lim_{N \to \infty} \| K_*^{(N)} - K^* \|_\infty = 0$, we can find N_0 such that $T_\varphi(\hat{m}_0)$ is log-concave with lower bound $\geq K_*^{(N)}$, $N \geq N_0$. For such N we know from Theorem 13.2 that for any initial condition

$$d\tilde{\mu}_0 = h_0 \, d\mu_0$$

with Lipschitz-continuous density $h_0 \geq \delta > 0$ we conclude that

$$\left\| \pi_N^{(N), \Delta y^{(N)}} - \tilde{\pi}_N^{(N), \Delta y^{(N)}} \right\|_{\mathrm{var}} \leq 1 + \frac{t}{N} \frac{\mathrm{tr} \left(B_1^{(N), T} \sqrt{Q^{(N), -1}} K_*^{(N), -1} \sqrt{Q^{(N), -1}} B_1^{(N)} \right)}{2\delta}$$

$$\cdot \chi_*^{(N), -N} \| h_0 \|_{\mathrm{Lip}_{\sqrt{\frac{t}{N} Q^{(N), -1} B_1^{(N)}}}}.$$

We can now take the limit $N \to \infty$ to conclude that

$$\left\| \pi_t^{y_{0:t}} - \tilde{\pi}_t^{y_{0:t}} \right\|_{\mathrm{var}} \leq \frac{1 + \mathrm{tr}\left(\sqrt{Q^{-1}} K_*^{-1} \sqrt{Q^{-1}} \right)}{2\delta} e^{-\chi_* t} \| h_0 \|_{\mathrm{Lip}_{\sqrt{Q^{-1}}}} \quad □$$

References

[1] R. Atar, Exponential stability for nonlinear filtering of diffusion processes in a noncompact domain, *Ann. of Probab.* **26**, 1552–74 (1998).

[2] H. J. Brascamp, E. H. Lieb, On extensions of the Brunn–Minkowski and Prékopa–Leindler theorems, including inequalities for log concave functions, and with an application to the diffusion equations, *J. Funct. Anal.* **22**, 366–389 (1976).

[3] C. K. Chui and G. Chen, *Kalman Filtering with Real-Time Applications*, 3rd edn, Springer (1999).

[4] J. M. C. Clark and D. Crisan, On a robust version of the integral representation formula of nonlinear filtering, *Probab. Theory Relat. Fields* **133**, 43–56 (2005).

[5] H. W. Knobloch and H. Kwakernaak, *Lineare Kontrolltheorie*, Springer (1980).

[6] N. Krylov, *Introduction to the Theory of Diffusion Processes*, American Mathematical Society (1995).

[7] D. Ocone and E. Pardoux, Asymptotic stability of the optimal filter with respect to its initial condition, *SIAM J. Control and Optimzation* **34**, 226–43 (1996).

[8] W. Stannat, Stability of the filter equation for a time-dependent signal on $\mathbb{R}^d$, *Appl. Math. Optim.* **52**, 39–71 (2005).

[9] W. Stannat, Stability of the optimal filter via pointwise gradient estimates. In Stochastic partial differential equations and applications VII, 281–293, *Lect. Notes Pure. Appl. Math.*, 245, Chapman & Hall/CRC (2006).

[10] W. Stannat, On the stability of Feynman–Kac propagators. In *Progress in Probability* **59**, 345–62, Birkhäuser (2007).

PART IV
Special Topics

·14·

Pathwise Nonlinear Filtering with Correlated Noise

M. H. A. Davis

14.1 Introduction

Aside from minor revision and slight improvements, this chapter is a reprint of the author's paper [5], which appeared in a now unavailable conference proceedings volume [13]. The editors of the present volume have been kind enough to include it here because the "pathwise filtering" approach continues to be of both theoretical and practical importance. In 1978, J. M. C. Clark made the key observation that without some continuity of the filter estimate with respect to the observation sample path, the nonlinear filtering equations are of no practical use; the argument is summarized in §14.2 below. He showed how the required continuity could be obtained by a simple integration by parts, re-expressing the filtering equations in a form in which no stochastic integrals with respect to the observation process appear. The filtering equations then make sense for any observation sample path that is merely a continuous function, and continuity can be established. The technical details of this argument were completed in Clark and Crisan [4]. They cover the case of multidimensional observations when the observation noise is independent of the signal process, due to a fortunate commutativity property of certain operators appearing in the problem. The purpose of this chapter is to analyse the case where there is correlation between the signal and observation noise. Then commutativity is lost, and a pathwise theory is only possible for scalar observations. In deriving it we are inevitably led to a geometric approach which is perhaps of some independent interest—apart from the useful byproduct of showing that the filtering equations apply equally well to signals evolving on finite-dimensional manifolds.

The chapter is laid out as follows. The next section gives some background on the filtering equations and the need for a pathwise approach. §14.3 describes the coordinate-free representation of the signal SDE and gives some Lie-algebraic calculations that will be needed later. §14.4 covers the case of independent signal and observation noise and establishes the essential connection between the Kallianpur–Striebel (KS) formula and observation-dependent multiplicative functional transformations of the signal process semigroup. This is the basis for the pathwise approach. The final section, §14.5, addresses the correlated noise case using an SDE decomposition technique due to Kunita [14]. The main results of the chapter are the formulas (14.46), giving the correct formulation of the KS formula in this context, and (14.49),(14.50) expressing the KS formula as an observation-dependent multiplicative functional transformation.

14.2 Background and summary

The setting is the conventional nonlinear filtering problem of calculating recursively estimates

$$\mathbb{E}[f(X_t)|Y_s, 0 \le s \le t]$$

where X_t is a Markov process and Y_t is a real-valued "observation process" given by

$$dY_t = h(X_t)dt + dW_t^0 \tag{14.1}$$

Here h is a bounded function (additional smoothness assumptions will be imposed later) and W_t^0 is a standard Brownian motion. The articles in Part I of this volume can be consulted for general background and most of the standard results in filtering theory used below. Other references are Bain and Crisan [1] and the brief introduction Davis and Marcus [6].

Let us denote $\mathcal{Y}_t = \sigma\{Y_s, s \le t\}$ and

$$\pi_t(f) = \mathbb{E}[f(X_t)|\mathcal{Y}_t] \tag{14.2}$$

The process π_t should be thought of as the conditional distribution of X_t given $\mathcal{Y}_t$, so that[1]

$$\pi_t(f) = \int_{\mathbb{S}} f(x)\pi_t(dx).$$

Here $\mathbb{S}$ is the state space for X_t. It is convenient to calculate an unnormalized form ρ_t of this, π_t then being given by

$$\pi_t(f) = \frac{\rho_t(f)}{\rho_t(1)}.$$

If X_t and W_t^0 are independent, an appropriate unnormalized distribution can be obtained in two alternative forms:

(i) the *Kallianpur–Striebel formula*, giving ρ_t nonrecursively as a function-space integral:

$$\rho_t(f) = \int_{\Xi} f(X_t) \exp\left(\int_0^t h(X_s)dY_s - \frac{1}{2}\int_0^t h^2(X_s)ds\right) \nu(dx) \tag{14.3}$$

(Here (Ξ, ν) is the sample path probability space.)

(ii) the *Zakai equation*, giving ρ_t in recursive form as the solution of a measure-valued stochastic differential equation:

$$d\rho_t(f) = \rho_t(Af)dt + \rho_t(hf)dY_t \qquad (14.4)$$
$$\sigma_0(f) = \pi(f)$$

(A is the differential generator of the X_t process, $hf(x) = h(x)f(x)$ and π is the distribution of X_0).

Clark's [3] argument about the need for a pathwise approach and a continuity result is as follows. It has to do with questions of stochastic modelling. First, recall some facts about the conditional expectation (14.2). For this discussion, fix a time $t > 0$. Since $\mathcal{Y}_t$ is generated by $Y = \{Y_s, 0 \leq s \leq t\}$, $\pi_t(f)$ is a functional of the continuous process Y, i.e. there is a measurable function $\phi : C[0, t] \to \mathbb{R}$ such that

$$\mathbb{E}[f(X_t)|\mathcal{Y}_t] = \phi(Y) \quad \text{a.s.} \qquad (14.5)$$

ϕ is not uniquely defined, in that any other function ϕ' such that $\phi'(Y) = \phi(Y)$ a.s. would be an equally good version of the conditional expectation. Here "a.s." refers to the distribution of Y on $C[0, t]$, and this distribution has the same null sets as Wiener measure. In particular *the set of functions with bounded variation is a null set*. Now in the observation equation (14.1), Y is a mathematical model for $\tilde{Y}_t = \int_0^t Z_s ds$ where Z_t is a physical observation

$$Z_t = h(X_t) + N_t$$

and N_t is (physical) "wide band" noise. As an estimate for $f(X_t)$ we then plan to take $\phi(\tilde{Y})$. But since $\tilde{Y}$ has bounded variation, ϕ is undefined for $\tilde{Y}$, and indeed on the whole set of physical sample paths. Thus nonlinear filtering theory cannot be applied in practice unless we are able to choose a particular version of the conditional expectation which has "nice" properties. Specifically, what is required is a function $\phi : C[0, t] \to \mathbb{R}$ such that

(i) (14.5) holds.
(ii) ϕ is continuous with respect to the supremum norm on $C[0, t]$. $\qquad (14.6)$

Then $\phi(\tilde{Y})$ is a "sensible" estimator, in that the mean square error $\mathbb{E}[(f(X_t) - \phi(\tilde{Y})^2]$ is close to the predicted value $\mathbb{E}[(f(X_t) - \phi(Y))^2]$ as long as the distributions of $\tilde{Y}$ are close to those of Y in the sense of weak convergence, and this certainly includes all the usual bounded-variation approximations to Bownian motion.

In Davis [8, 9] it was shown that such "robust" filtering algorithms could be produced for a very wide class of Markov signal processes X_t, when the signal and observation noise W_t^0 are independent. The main purpose of this chapter is to extend the results to certain cases where there is correlation between signal and observation noise. This cannot be done at the same level of generality as in Davis [8], and the signals we consider are diffusions on finite-dimensional manifolds. Such signals appear in important areas of application of nonlinear filtering theory, for example in alignment problems in inertial navigation where the signal is an orientation, represented by a quaternion vector. Also, the coordinate-free signal description (introduced in §14.3 below) adds insight even for $\mathbb{R}^d$-valued diffusions.

The basis of our approach is that the unnormalized conditional distribution ρ_t can be represented in the form

$$\rho_t(f) = < T^Y_{0,t} U_{Y_t} f, \pi >$$

where $T^Y_{s,t}$ is the Y-dependent semigroup associated with a certain multiplicative functional transformation, U_t is a group of operators and π is the distribution of the initial state X_0. A recursive form of estimator can then be obtained by considering the forward equation corresponding to $T^Y_{s,t}$ (see §14.4.3 below). Our main concern is therefore to calculate the generator of $T^Y_{s,t}$ and this is most readily done by factoring the relevant multiplicative functional (§14.4.2). The relation with "pathwise solutions" of the Zakai equation is explored in §14.4.4.

In the case of independent signal and noise, U_t is the operator of "multiplication by $\exp(t\,h(x))$". Our main result is that, for the type of noise correlation considered in §14.5, the situation is formally analogous to the independent case, but with U_t now being the flow corresponding to a certain differential operator (see (14.48)). Showing this involves decomposing the signal equation in the way described by Kunita [14] in order to elucidate precisely the dependence of X_t on the observations y.

It must, regretfully, be pointed out that the results for correlated noise cannot, unlike those for the independence case, be extended to vector observations. This is because the corresponding operators U^i_t do not in general commute whereas with no noise correlation they are multiplication operators which automatically commute.

14.3 The signal process

The formulation here follows that of Kunita [14]. The signal process X_t evolves on a σ-compact, connected C^∞ manifold $\mathbb{S}$ of dimension d. Suppose $V_0, \ldots, V_r$ are C^∞ vector fields on $\mathbb{S}$ and $W^1, \ldots, W^r$ are independent scalar Brownian motions which are independent of W^0_t of equation (14.1). Then X_t is the solution of the stochastic differential equation

$$dX_t = V_0(X_t)dt + V_j(X_t) \circ dW^j_t. \tag{14.7}$$

This means that X_t is the unique $\mathbb{S}$-valued process satisfying

$$f(X_t) = f(X_0) + \int_0^t V_0 f(X_s) ds + \int_0^t V_j f(X_s) \circ dW_s^j, \quad 0 \leq t \leq \tau$$

for any real valued C^∞ function f. In these equations the $\circ$ denotes the Stratonovich stochastic integral, and the convention of implied summation from $j = 1$ to r is used.[2] The initial point X_0 is supposed to be a random variable with a given distribution π, independent of all Brownian motions. In (14.7), τ is the *lifetime* and we assume conditions are such that $\tau = \infty$ a.s. (automatically true if $\mathbb{S}$ is compact). The relation of (14.7) with the corresponding Ito equation is the following: for any k, $V_k f \in C^\infty(\mathbb{S})$ and hence from (14.7)

$$dV_k f(X_t) = V_0 V_k f(X_t) dt + V_j V_k f(X_t) \circ dW_t^j.$$

Thus the joint quadratic variation of the semimartingales $V_k f(X_t)$ and W_t^k is

$$d<V_k f, W^k>_t = V_k^2 f(X_t) dt$$

and the Itō version of (14.7) is therefore

$$df(X_t) = (V_0 + \frac{1}{2} \sum_j V_j^2) f(X_t) dt + V_j f(X_t) dW_t^j. \tag{14.8}$$

The process X_t is a Markov diffusion process; its *generator* A is an operator acting on C^∞ functions such that the Dynkin formula[3]

$$\mathbb{E}_x f(X_t) - f(x) = \mathbb{E}_x \int_0^t Af(X_s) ds$$

is satisfied for $f \in C_0^\infty(\mathbb{S})$ (the C^∞ functions of compact support). In view of the time-homogeneity this is equivalent to saying that the process

$$C_t^f := f(X_t) - f(x) - \int_0^t Af(X_s) ds$$

is a martingale. Now in (14.8) the stochastic integral is a martingale for $f \in C_0^\infty(\mathbb{S})$ since $V_j f$ is then bounded, and it follows that

$$A = V_0 + \frac{1}{2} \sum_j V_j^2.$$

A is sometimes called the *extended generator* of X_t: it is an extension of the infinitesimal generator of the semigroup of operators on $C(\mathbb{S})$

$$T_t f(x) := \mathbb{E}_x[f(X_t)] \tag{14.9}$$

associated with the process X_t.

For the sequel, we shall need to compute some Lie brackets. We suppose that the observation function h of (14.1) is in $C_b^2(\mathbb{S})$. h will also denote the zeroth-order operator of "multiplication by h", i.e. for $f \in C^\infty(\mathbb{S})$

$$hf(x) = h(x)f(x)$$

and similarly for other functions below. If D is any differential operator, $\mathrm{ad}_h D$ denotes the Lie derivative

$$(\mathrm{ad}_h D)f(x) = [h, D]f(x) = h(x)Df(x) - D(hf)(x)$$

and $\mathrm{ad}_h^2 D = \mathrm{ad}_h(\mathrm{ad}_h D)$, etc. If V is a vector field then we find using the Leibnitz rule that

$$\begin{array}{ll} \mathrm{ad}_h V = -(Vh) & \mathrm{ad}_h V^2 = -2(Vh)V - V^2 h \\ \mathrm{ad}_h^2 V = 0 & \mathrm{ad}_h^2 V^2 = 2(Vh)^2 \\ & \mathrm{ad}_h^3 V^2 = 0. \end{array}$$

Thus in particular

$$\begin{aligned} \mathrm{ad}_h A &= -(V_j h) V_j - Ah \\ \mathrm{ad}_h^2 A &= \sum_j (V_j h)^2 \\ \mathrm{ad}_h^k A &= 0, \quad k > 2. \end{aligned} \tag{14.10}$$

14.4 Independent signal and noise

We consider the filtering problem over a fixed finite interval $[0, T]$. Recall that the observation equation (14.1) is

$$Y_t = \int_0^t h(X_s)\,ds + W_t^0, \quad 0 \le t \le T.$$

We will assume henceforth that

$$h \in C_b^2(\mathbb{R}^r). \tag{14.11}$$

Notation: Y will denote the $C[0, T]$-valued random variable $\{Y_t, t \in [0, T]\}$, so that Y_t is the value of Y at t. We will use the notation y for an arbitrary but fixed element of $C[0, T]$, with $y(t)$ denoting the value at time t (in accordance with the traditions of real analysis). y may or may not be a sample function $Y(\omega)$.

14.4.1 The KS formula as a multiplicative functional transformation

Let us return to the Kallianpur–Striebel formula, (14.3). In view of (14.7), (14.11) the real-valued process $h(X_t)$ is certainly a semimartingale, and we can write

$$\int_0^t h(X_s)dY_s = h(X_t)Y_t - \int_0^t Y_s dh(X_s).$$

Note that the right hand side of this equality involves no stochastic integration with respect to dY and makes sense if Y is replaced by any function $y \in C[0, T]$. Thus (14.3) can be written in the form

$$\rho_t(f) = \mathbb{E}\left[f(X_t)e^{y(t)h(X_t)}a_t^0(y)\right]\Big|_{y=Y} \tag{14.12}$$

where $a_t^s(y)$ is defined for $s \leq t$ by

$$a_t^s(y) = \exp\left(-\int_s^t y(u)dh(X_u) - \frac{1}{2}\int_s^t h^2(X_u)du\right). \tag{14.13}$$

In (14.12) the expectation is taken over the distribution of X (regarded as a random variable taking values in $C([0, T]; \mathbb{R}^r)$). If we now define $\phi(y) = \rho_t(f)/\rho_t(1)$ where $\rho_t(f)$ is given by (14.12), then this is the desired version of the conditional expectation, in that the two conditions of (14.6) hold (Clark and Crisan [4]). It remains to show how $\rho_t(f)$ can be computed recursively. Associated with the process X_t is a semigroup $(T_t)_{t \leq 0}$ of operators on $C(\mathbb{S})$ defined by (14.9) above. Now the process $a_t^s(y)$ of (14.13) is a *multiplicative functional* of X_t, i.e. an adapted process satisfying

$$a_t^r = a_s^r a_t^s \quad \text{for } r \leq s \leq t.$$

It is easily checked that, if we define

$$T_{s,t}^y f(x) = \mathbb{E}_{x,s}\left[f(X_t)a_t^s f(X_t)\right] \tag{14.14}$$

then $T_{s,t}^y$ is another (two-parameter) semigroup of operators on $C(\mathbb{S})$. Note, however, that it is not Markovian, i.e. does not satisfy $T_{s,t}1 = 1$. For $f \in C(\mathbb{S})$ and μ a measure on $\mathbb{S}$, denote $< f, \mu > = \int_\mathbb{S} f(x)\mu(dx)$. Then from (14.7)–(14.9) we see that

$$\rho_t(f) = < T_{0,t}^y(e^{y(t)h} f), \pi > .$$

This provides us, in principle, with a recursive way of computing ρ_t. Let $U_{t,s}^y$ be the adjoint semigroup to $T_{s,t}^y$, defined by

$$< T_{s,t}^y f, \mu > = < f, U_{t,s}^y \mu >$$

and define

$$\pi_t^y = U_{t,0}^y \pi.$$

Then, formally, π_t^y is the solution of the forward (Fokker–Planck) equation

$$\frac{d}{dt}\pi_t^y = \left(A_t^y\right)^* \pi_t^y, \qquad \pi_0^y = \pi. \qquad (14.15)$$

Here A_t^y is the differential generator of $T_{s,t}^y$. Now (14.15) is a recursive equation for π_t^y, and ρ_t is given by

$$\rho_t(f) = <e^{y(t)h}f, \pi_t^y> . \qquad (14.16)$$

Notice that (14.15), (14.16) constitute a recursive filter in a form in which *no stochastic integration is involved*. The forward equation (14.15) has been investigated in detail for the case $\mathbb{S} = \mathbb{R}^d$ in Pardoux [16] and Pardoux [17]. The remainder of this section is devoted to explicit calculation of the generator A_t^y of the semigroup $T_{s,t}^y$.

14.4.2 Factorization of multiplicative functionals

This section follows up some ideas contained in a paper of Mitter [15]. A related discussion will be found in §VIII.3 of Revuz and Yor [18].

We introduce three simple types of multiplicative functional (MF), all relative to the Markov process X_t, called the *Gauge*, *Feynman–Kac*, and *Girsanov* types, and explore the relations between them. Further general information on MFS can be found in the books of Blumenthal and Getoor [2] and Dynkin [11]. The MF $a_t^s(y)$ of (14.13) is of course time-varying in that it depends on the sample path y, but here we shall discuss time-invariant MFS. A MF β_t^s is time invariant if for any r, and $s \leq t$,

$$\beta_{t+r}^{s+r} = \beta_t^s \circ \theta_r$$

where θ_r is the shift operator $(\theta_r X)_s = X_{r+s}$. In particular this implies that $\beta_t^s = \beta_{t-s}^0 \circ \theta_s$, so that β_t^s is really a one-parameter functional; indeed denoting $\beta_t = \beta_t^0$ we can write the multiplicative property as

$$\beta_{t+s} = \beta_t \, \beta_s \circ \theta_t.$$

Let T_t^β be the semigroup corresponding to β, defined by

$$T_t^\beta f(x) = \mathbb{E}_x[f(X_t)\beta_t]. \qquad (14.17)$$

We wish to consider the generator of T_t^β. If β satisfies

$$\mathbb{E}_x[\beta_t] \leq 1 \qquad (14.18)$$

(or, equivalently, $T_t^\beta 1 \leq 1$) then, as shown in Blumenthal and Getoor (1968), one can construct (possibly on an enlarged state space) the *β-subprocess* of X_t, which is a Markov process X_t^β satisfying

$$T_t^\beta f(x) = \mathbb{E}_x[f(X_t^\beta)].$$

As in §14.3 above, the extended generator of X^β is an operator A^β such that

$$f(X_t^\beta) - f(x) - \int_0^t A^\beta f(X_s^\beta) \, ds$$

is a martingale for $f \in C_0^\infty(\mathbb{S})$, and from (14.17) this is equivalent to saying that

$$C_t^{\beta f} := \beta_t f(X_t) - f(x) - \int_0^t \beta_s A^\beta f(X_s) \, ds \qquad (14.19)$$

is a martingale. The latter formulation has however the advantage that it does not involves the β-subprocess or condition (14.18) (which is not satisfied in any of the applications we have in mind). We thus define the extended generator of T_t^β as an operator A^β such that $C_t^{\beta f}$ given by (14.19) is a martingale for all $f \in C_0^\infty(\mathbb{S})$. Here then are the types of multiplicative functional.

(β) *Gauge transformation type.* Suppose $a \in C_b^\infty(\mathbb{S})$ and $a(x) > 0$ for all $x \in \mathbb{S}$. Define

$$\beta_t = \frac{a(X_t)}{a(X_0)}. \qquad (14.20)$$

This is clearly a MF, and from (14.18),

$$T_g^\beta(x) = \frac{1}{a(x)} T_t(ag)(x).$$

Using the signal equation (14.8) with $f = ag$ we see that $C^{\beta g}$ is a martingale if

$$A^\beta g(x) = \frac{1}{a(x)} A(ag)(x)$$

and this is therefore the generator of T_t^β.

(γ) *Feynman–Kac type.* For a given $v \in C_b^\infty(\mathbb{S})$, define

$$\gamma_t = \exp\left(-\int_0^t v(X_s) \, ds\right). \qquad (14.21)$$

Computing the product of the semimartingales $f(X_t)$ and γ_t using (14.8) and the Itō formula shows immediately that

$$A^\gamma f(x) = Af(x) - v(x) f(x).$$

(δ) *Girsanov type.* The above transformations can be applied to any Markov process but this one is specifically tied to the model (14.7). Fix $g \in C_b^\infty(\mathbb{S})$ and define

$$\delta_t = \exp\left(-\int_0^t V_j g(X_u) \, dW_u^j - \frac{1}{2} \sum_j \int_0^t (V_j g(X_u))^2 \, du\right). \qquad (14.22)$$

A standard application of the Girsanov theorem shows that we can define a new measure $\mathbb{P}^\delta$ by taking $d\mathbb{P}^\delta/d\mathbb{P} = \delta_T$, and that under $\mathbb{P}^\delta$

$$d\tilde{W}_u^j := dW_u^j + V_j g(X_u) du, \qquad u \leq t$$

is a standard Brownian motion for $j = 1, 2, \ldots, r$. Thus (14.7) becomes

$$df(X_u) = (V_0 f(X_u) - V_j g(X_u) V_j f(X_u)) du + V_j f(X_u) \circ d\tilde{W}_u^j.$$

Now

$$\mathbb{E}_x[f(X_t)\delta_t] = \mathbb{E}_x^\delta[f(X_t)]$$

and it follows that

$$A^\delta = A - (V_j g) V_j. \tag{14.23}$$

The three transformations are related by the X_t equation written in Itô form (14.8); indeed, from (14.8)

$$\int_s^t V_j g(X_u) dW_u^j = g(X_t) - g(X_s) - \int_s^t Ag(X_u) du,$$

and inserting this in (14.22) we see that δ_t factors in form

$$\delta_t = \beta_t \gamma_t$$

where β, γ are given by (14.20) and (14.21) respectively with

$$a(x) = e^{-g(x)}$$
$$v(x) = -Ag(x) + \frac{1}{2} \sum_j (V_j g(x))^2.$$

Applying β and γ successively (the order is immaterial) we conclude that

$$A^\delta f = e^g A(e^{-g} f) + (Ag - \frac{1}{2} \sum_j (V_j g)^2) f. \tag{14.24}$$

But this is just a disguised form of the Baker–Campbell–Hausdorff formula: using the expression (14.23) for A^δ and the relations (14.10), equation (14.24) becomes

$$e^g A e^{-g} = A - (V_j g) V_j - Ag + \frac{1}{2} \sum_j (V_j g)^2 = A + \text{ad}_g A + \frac{1}{2} \text{ad}_g^2 A. \tag{14.25}$$

(Recall that $\text{ad}_g^k A = 0$ for $k > 2$.)

14.4.3 The generator of $T_{s,t}^y$

Recall from §14.4.1 that the MF appearing in the Kallianpur–Striebel formula is

$$a_t^s(y) = \exp\left(-\int_s^t y(u)dh(X_u) - \frac{1}{2}\int_s^t h^2(X_u)du\right).$$

Using (14.8) we can factor this into the product of a Girsanov MF and a Feyman–Kac MF as follows:

$$a_t^s(y) = \exp\left(-\int_s^t y(u)Ah(X_u)du - \int_s^t y(u)V_j h(X_u)dW_u^j - \frac{1}{2}\int_s^t h^2(X_u)du\right)$$

$$= \exp\left(-\int_s^t y(u)V_j h(X_u)dW_u^j - \frac{1}{2}\int_s^t y^2(u)\sum_j(V_j h(X_u))^2 du\right)$$

$$\times \exp\left(\int_s^t (\frac{1}{2}y^2(u)\sum_j(V_j h(X_u))^2 - y(u)Ah(X_u) - \frac{1}{2}h^2(X_u))du\right).$$

It follows immediately that the corresponding generator is

$$A_s^y f = Af - y(s)(V_j h)V_j f$$
$$+ \left(\frac{1}{2}y^2(s)\sum_j(V_j h)^2 - y(s)Ah - \frac{1}{2}h^2\right)f \quad (14.26)$$
$$= Af + y(s)(\mathrm{ad}_h A)f + \frac{1}{2}y^2(s)\left(\mathrm{ad}_h^2 A\right)f - \frac{1}{2}h^2 f$$
$$= e^{y(s)h}A(e^{-y(s)h}f) - \frac{1}{2}h^2 f, \quad (14.27)$$

the last equality being an application of (14.25) with $g = y(s)h$.

It is clear a priori that (14.27) must be the right formula: in (14.26) the calculation is done for an arbitrary function $y(\cdot)$ but the result depends only on $y(s)$. Therefore $A_s^y = A_s^{\bar y}$ where $\bar y$ is the constant function

$$\bar y(u) = y(s) \quad \text{for } u \geq s.$$

But

$$a_t^s(\bar y) = \exp(-y(s)h(X_t) + y(s)h(X_s))\exp\left(-\frac{1}{2}\int_s^t h^2(X_u)du\right).$$

This factors $a_t^s(\bar y)$ into the product of a gauge MF and a Feyman–Kac MF, and (14.27) is immediate for $\bar y(\cdot)$. But of course some extra work has to be done to show that the same formula works for nonconstant $y(\cdot)$

Note from (14.26) that A_s^y is of the form

$$A_s^y f(x) = \frac{1}{2}\sum_j V_j^2 f(x) + \tilde V_0(y(s))f(x) + \psi(x, y(s))f(x),$$

where $\tilde V_0$ is a y-dependent vector field, i.e. the second-order part of A_s^Y is the same as that of A, and the effect of the MF transformation is only to add

y-dependent "drift" and "potential" terms. Thus essentially the same conditions that ensure smooth solutions of the Fokker-Planck equation of the signal process also ensure smooth solutions of (14.15) (except that these conditions must allow for continuous, but not differentiable, t-dependence in the coefficients). The general conclusion is that computing the *conditional* distribution of X_t given $\mathcal{Y}_t$ is not in any essential way more complicated than computing the *unconditional* distribution. See Pardoux [16] and [17] for the case $\mathbb{S} = \mathbb{R}^d$.

14.4.4 Dossing the Zakai equation

There is another way of looking at the basic formula

$$\rho_t(f) = <T_{0,t}^y(e^{y(t)h}f), \pi> \qquad (14.28)$$

and that is as a Doss–Sussmann "pathwise solution" Doss [10], Sussmann [19] of the Zakai equation (14.4). This was indeed how (14.28) was originally arrived at, and although the MF approach turns out to be more fundamental, the pathwise solution idea is of value in understanding the picture and particularly in unravelling the complexities of the correlated noise case (see §14.5.3 below).

Let us recall the Doss–Sussmann construction for the simplest type of scalar equation

$$dX_t = f(X_t)dt + g(X_t) \circ dM_t, \quad X_0 = x, \qquad (14.29)$$

where M_t is a real-valued continuous semimartingale and f, g are smooth functions. (The same basic idea is used with considerably more elaboration in §14.5.1 below). Let $G(t, x)$ be the flow of g, i.e. the solution of the ordinary differential equation

$$\frac{\partial}{\partial t} G(t, x) = g(G(t, x))$$
$$G(0, x) = x.$$

Then the solution of (14.29) takes the form

$$X_t = G(M_t, \eta_t) \qquad (14.30)$$

where η is the solution of another ODE, parametrized by the sample path (M_t). Indeed, defining X_t by (14.30) we have

$$dX_t = g(X_t) \circ dM_t + G_x(M_t, \eta_t)\dot\eta_t dt \qquad (14.31)$$

But

$$G_x(t, x) = \exp\left(\int_0^t g_x(G(s, x))ds\right), \qquad (14.32)$$

so that (14.29) and (14.31) agree as long as

$$\dot{\eta}_t = \exp\left(-\int_0^{M_t} g_x(G(s, \eta_t))ds\right) f(G(M_t, \eta_t)), \quad \eta_0 = x. \tag{14.33}$$

This is an ordinary differential equation for η, parametrized by the sample path (M_t), and shows that the solution of (14.29) can be calculated separately for each sample path of (M_t): first solve (14.33) and then evaluate (14.30). The same construction works for $X_t \in \mathbb{R}^n$, except for the explicit expression (14.32), but not in general, for vector M_t (see below). Things are particularly simple in the bilinear case: $f(x) = Ax$, $g(x) = Hx$. Then (14.33) and (14.30) become respectively

$$\dot{\eta}_t = e^{-HM_t} A e^{HM_t} \eta_t, \quad \eta_0 = x,$$
$$X_t = e^{HM(t)} \eta_t.$$

Let us now apply the same argument to the bilinear measure-valued Zakai equation (14.4). In Stratonovich form this is

$$d\rho_t(f) = \rho_t((A - \frac{1}{2}h^2)f)dt + \rho_t(hf) \circ dY_t, \quad \rho_0 = \pi. \tag{14.34}$$

If the drift term in (14.34) were absent then the solution would be, as is easily checked,

$$\rho_t(f) = < e^{Y(t)h} f, \pi > .$$

An argument exactly analogous to the above shows that the solution *with* the drift term is given by (14.28), if $T_{0,t}^Y$ is a semigroup with generator

$$A_t^Y f = e^{Y(t)h} A(e^{-Y(t)h} f) - \frac{1}{2} h^2 f. \tag{14.35}$$

But we saw in (14.24) above that this precisely *is* the generator of the semigroup given by the Kallianpur–Striebel formula. Thus the two approaches lead to the same result. If, however, one starts with the Zakai equation, one has somehow to show that there exists a semigroup whose generator is (14.35). The only way to do this that I know of is through a probabilistic argument [7] which leads straight back to the Kallianpur–Striebel formula. This is why I describe the MF approach as 'more fundamental.'

Finally, let us note that all of the above results extend without difficulty to the case of vector observations

$$dY_t^i = h^i(X_t)dt + dW_t^{0i} \quad i = 1, \ldots, m,$$

if X_t and W_t^{0i} are independent for all i. The Zakai equation is

$$d\rho_t(f) = \rho_t(Af)dt + \sum_1^m \rho_t(h^i f)dY^i,$$

and a pathwise solution is constructed from this (or from the Kallianpur-Striebel formula) as before. The reason this "works" is that the operators of 'multiplication by h^i,' which appear in the diffusion term of the Zakai equation, commute: $h^i h^j f(x) = h^j h^i f(x) = h^i(x) h^j(x) f(x)$. Recall that the condition under which the Doss–Sussmann construction for (14.29) can be extended to multiple inputs $\sum g^i(X_t) \circ d M_t^i$ is precisely that the vector fields g^i commute.

14.5 The correlated noise case

We now wish to consider the filtering problem given by (14.1) and (14.7) as before, but allowing for possible correlation between the signal noise $(W^1, \ldots, W^r)$ and the observation noise W^0. We assume the simplest form of correlation; it will be obvious how to extend the results to more general cases. Specifically, we suppose

(i) W_t^i is a standard Brownian motion (i.e. $W_0^i = 0$ and $< W^i >_t = t$, $i = 0, 1, \ldots, r$).
(ii) W^i, W^j are independent for $i \neq j \neq 0$
(iii) $< W^i, W^0 > = a_i t$ for constants a_i such that $\sum_i a_i^2 < 1$.

The first two of these are the same as before, while the condition in (iii) ensures a well-defined and strictly positive definite covariance matrix. We have in particular

$$\mathbb{E}[W_t^i W_s^0] = a_i \, t \wedge s.$$

The Kallianpur–Striebel formula is no longer valid in the form (14.3); we shall derive the correct form in §14.5.2 below. As regards the Zakai equation, it follows directly from the general filtering equation of [17] that (14.4) should be amended to

$$d\rho_t(f) = \rho_t(Af) dt + \rho_t(Df) dY_t \qquad (14.36)$$

where $D = Z + h$, with

$$Z = a_j V_j \qquad (14.37)$$

and h denoting the zeroth-order operator $hf(x) = h(x) f(x)$.

14.5.1 Decomposition of the signal equation

To get the appropriate form of the Kallianpur–Striebel formula, introduce a measure $\mathbb{P}_0$ via the Girsanov transformation

$$\frac{d\mathbb{P}_0}{d\mathbb{P}} = \exp\left(-\int_0^T h(X_s) dW_s^0 - \frac{1}{2}\int_0^T h^2(X_s) ds\right)$$

and for $i = 1, 2, \ldots, r$ define

$$dv^i := dW^i + a_i h(X_t) dt. \tag{14.38}$$

Then, under $\mathbb{P}_0$,

(i) Y_t and v_t^i, $i = 1, \ldots, r$ are standard Brownian motions
(ii) v^i, v^j are independent for $i \neq j$
(iii) $<v^i, Y>_t = a_i t$.

Now project the v^j onto Y, i.e. define

$$b_t^i = v_t^i - a_i Y_t. \tag{14.39}$$

Then each b^i is an unnormalized Brownian motion, which is uncorrelated with, and hence independent of, Y. Denote $b'_t = (b_t^1, \ldots, b_t^r)$, $a' = (a_1, \ldots, a_r)$ and let I denote the $r \times r$ identity matrix. Then

$$<b>_t = (I - aa')t.$$

Now $I - aa'$ is positive definite and can be factored into a product $\Xi \Xi'$ of positive definite matrices. Defining

$$B_t := \Xi^{-1} b_t \tag{14.40}$$

we find that

$$<B>_t = It,$$

i.e. $B^1, \ldots, B^r$ are (under measure $\mathbb{P}_0$) independent standard Brownian motions, independent of Y. Using (14.38)–(14.40), the signal equation (14.7) becomes

$$df(X_t) = L_0 f(X_t) dt + Z f(X_t) \circ dY_t + L_j f(X_t) \circ dB_t^j \tag{14.41}$$

where Z is given by (14.37),

$$L_0 := V_0 - hZ$$

and, in an obvious notation, the L_j are given by

$$L = \Xi' V.$$

Equation (14.41) is the key formula for the filtering problem, as it expresses X_t under measure $\mathbb{P}_0$ in the form of an equation driven by the observation process Y_t and the other 'inputs' $B^1, \ldots, B^r$ which are independent of Y.

To proceed, we follow the approach of Kunita [14]. Essentially, the idea is to use a transformation of the Doss–Sussmann type, as in §14.4.4, to express the solution of (14.41) sample-pathwise in Y. First we need some ideas from differential geometry.

The tangent space $T_p(\mathbb{S})$ at $p \in \mathbb{S}$ consists of the set of derivations, i.e. linear functionals W_p on $C^\infty(U_p)$, where U_p is a neighbourhood of p, such that the Leibnitz rule

$$W_p(fg) = g W_p g + g W_p f$$

is satisfied. Now let $\phi : \mathbb{S} \to \mathbb{S}$ be a diffeomorphism and denote $q = \phi(p)$. Then ϕ defines a map $\phi^* : C^\infty(U_q) \to C^\infty(U_p)$ by composition:

$$\phi^* f := f \circ \phi, \quad f \in C^\infty(U_q),$$

and a map $\phi_* : T_p(\mathbb{S}) \longrightarrow T_q(\mathbb{S})$ as follows:

$$(\phi_* W_p) f = W_p(\phi^* f), \quad f \in C^\infty(U_q).$$

Since ϕ^{-1} is also a diffeomorphism, $\phi_*^{-1} : T_q(\mathbb{S}) \to T_p(\mathbb{S})$ is given likewise by

$$\left(\phi_*^{-1} W_q\right) g = W_q(g \circ \phi^{-1}), \quad g \in C^\infty(U_p).$$

If W is a vector field and W_p denotes its restriction to $p \in \mathbb{S}$ then this relation defines a mapping, also denoted ϕ_*^{-1}, between vector fields, which, since $q = \phi(p)$, we can write

$$\left(\phi_*^{-1} W\right) g(p) = W(g \circ \phi^{-1})(\phi(p)).$$

Let $\zeta_t(x) = \zeta(t, x)$ denote the flow of the vector field Z defined by (14.37), i.e. the unique solution of the equation

$$\frac{d}{dt} f(\zeta_t(x)) = Zf(\zeta_t(x)), \quad f \in C^\infty(\mathbb{S})$$
$$\zeta_0(x) = x.$$

This is a diffeomorphism for each $t \geq 0$. Define

$$\xi_t(x) = \zeta_{Y(t)}(x).$$

As is easily checked, $\xi = \xi_t(x)$ is the solution of

$$d\xi_t = Z(\xi_t) \circ dY_t$$

and, obviously, $\xi_t(\cdot)$ is almost surely a diffeomorphism for each $t > 0$. Now consider the equation

$$df(\eta_t) = \xi_{t*}^{-1} L_0 f(\eta_t) dt + \xi_{t*}^{-1} L_j f(\eta_t) \circ dB_t^j. \tag{14.42}$$

This equation has a unique solution and it follows by applying the Itô formula that

$$X_t(x) = \xi_t \circ \eta_t(x)$$
$$= \zeta(Y(t), \eta_t(x)). \tag{14.43}$$

The representation (14.42), (14.43) describes the behaviour of X_t conditioned on Y under measure $\mathbb{P}_0$. Recall that the map ξ_{t*}^{-1} is parametrized by Y and that Y, B are independent. Thus, conditional on Y, η_t is a diffusion process whose differential generator is

$$A_t^* = \xi_{t*}^{-1} L_0 + \sum_j \left(\xi_{t*}^{-1} L_j\right)^2$$

and, for each $t > 0$, X_t is diffeomorphically related to η_t by equation (14.43).

14.5.2 The Kallianpur–Striebel formula and associated multiplicative functional

It follows from a standard formula of conditional expectations that $\pi_t(f)$ of (14.2) is given in terms of the measure $\mathbb{P}_0$ by

$$\pi_t(f) = \frac{\rho_t(f)}{\rho_t(1)}$$

where

$$\rho_t(f) := \mathbb{E}_0 \left[f(X_t) \exp\left(\int_0^t h(X_s) dY_s - \frac{1}{2} \int_0^t h^2(X_s) ds \right) \bigg| \mathcal{Y}_t \right]. \quad (14.44)$$

It is immediate from (14.41) that

$$d < h(X_\cdot), Y > = Zh(X_t) dt$$

and hence that the Stratonovich version of (14.44) is

$$\rho_t(f) = \mathbb{E}_0 \left[f(X_t) \exp\left(\int_0^t h(X_s) \circ dY_s - \frac{1}{2} \int_0^t Dh(X_s) ds \right) \bigg| \mathcal{Y}_t \right], \quad (14.45)$$

where, as in (14.36) above, $D = Z + h$. Now use (14.43), giving X_t in the form $X_t = \zeta(Y_t, \eta_t)$ where η_t is a functional of the independent processes Y_t and $B'_t = (B^1_t, \dots, B^r_t)$, to express (14.45) in the form

$$\rho_t(f) = \mathbb{E}^b \left[\xi_t^* f(\eta_t) \exp\left(\int_0^t \xi_s^* h(\eta_s) \circ dY_s - \frac{1}{2} \int_0^t \xi_s^* Dh(\eta_s) ds \right) \right] \quad (14.46)$$

where $\mathbb{E}^b$ means integration over the sample space measure for B, i.e. Wiener measure on $C([0, T]; \mathbb{R}^r)$. This is the "Kallianpur–Striebel" formula for the correlated-noise problem. In order to get it in "robust" form we need to calculate the stochastic integral in (14.46) as an explicit functional of y. Introduce the function (we write $\zeta_s(x) = \zeta(s, x)$)

$$H_t(x) = H(t, x) := \int_0^t \zeta_s^* h(x) ds$$

and calculate $H(Y_t, \eta_t)$ using the Itō formula and (14.43). This gives

$$H(Y_t, \eta_t) = \int_0^t h(X_s) \circ dY_s + \int_0^t \left(\xi_{s*}^{-1} L_0\right) H_{Y_s}(\eta_s) ds$$
$$+ \int_0^t \left(\xi_{s*}^{-1} L_j\right) H_{Y_s}(\eta_s) \circ dB^j_s. \quad (14.47)$$

The stochastic integral with respect to B^j in (14.47) can be re-expressed in Itô form in the standard way using (14.42). Do this and introduce the notation

$$g_s(x) = H_{Y_s}(x) \qquad (14.48)$$
$$L_j^* = \xi_{s*}^{-1} L_j$$
$$U_s f(x) = \exp\left(\int_0^s \zeta_u^* h(x) du\right) \zeta_s^* f(x).$$

Then using (14.47) in (14.45) gives

$$\rho_t(f) = \mathbb{E}\left[U_{Y_t} f(\eta_t) a_t^0(Y)\right] \qquad (14.49)$$

where, for $y \in C([0, T])$,

$$a_t^s(y) = \exp\left(-\int_s^t L_j^* g_u(\eta_u) d B_u^j - \frac{1}{2} \sum_j \int_s^t (L_j^*)^2 g_u(\eta_u) du \right.$$
$$\left. - \int_s^t L_s^* g_u(\eta_u) du - \frac{1}{2} \int_s^t \xi_u^* Dh(\eta_u) du \right). \qquad (14.50)$$

Equation (14.49) is the desired multiplicative functional formula. For each sample path of y, η_t is a diffusion process governed by vector fields L_j^* as in (14.42), and a_t^s given by (14.50) is a MF of η. The expectation in (14.49) is taken over the distribution of η for a fixed path y.

It is possible to show that $\rho_t(f)$ given by (14.49) is continuous in $y \in C[0, T]$, i.e. that this is a "robust" version in the sense of (14.6), but we do not give the details here. In outline, one starts with functions $f \in C_0^\infty(\mathbb{S})$ whose support is contained within a single chart of $\mathbb{S}$. Then, working in local coordinates, equation (14.42) satisfies the standard Itô conditions and continuous dependence (in the mean square sense) of $f(\eta_t)$ on $y \in C[0, t]$ follows from known results on parametric dependence of solutions of stochastic differential equations; see Theorem 2, §2.7, of Gihman and Skorohod [12]. One completes the argument for $f \in C_b^\infty(\mathbb{S})$ by considering a decomposition of the form

$$\rho_t(f) = \sum_i \rho_t(\lambda_i f)$$

where (λ_i) is a partition of unity: $\lambda_i \in C_0^\infty(N)$ for all i and $\sum_i \lambda_i(x) = 1$.

As in §14.4.3 above, we can compute the generator A_t^y corresponding to the MF $a_t^s(y)$ by factorization. Indeed

$$a_t^s(y) = \exp\left(-\int_s^t L_j^* g_u(\eta_u) db_u^j - \frac{1}{2} \sum_j \int_s^t (L_j^* g_u(\eta_u))^2 du\right)$$
$$\times \exp\left(\int_s^t \left(\frac{1}{2} \sum_j (L_j^* g_u(\eta_u))^2 - \frac{1}{2} \sum_j (L_j^*)^2 g_u(\eta_u) - L_0^* g_u(\eta_u) - \frac{1}{2} \xi_u^* Dh(\eta_u)\right) du\right).$$

It now follows as before that

$$A_t^y f = A_t^* f - L_j^* g_t L_j^* f + \left(\frac{1}{2}\sum_j (L_j^* g_t)^2 - A_t^* g_t - \frac{1}{2}\xi_t^* Dh\right) f \quad (14.51)$$

where A_t^* is the generator for η_t, i.e.

$$A_t^* = L_0^* + \frac{1}{2}\sum_j (L_j^*)^2. \quad (14.52)$$

A_t^y can be expressed in somewhat more explicit form by noting that[4]

$$L_j^* g_t(x) = \int_0^{y(t)} \left(\xi_{t*}^{-1} L_j\right) \zeta_u^* h(x) du$$

$$= \int_0^{y(t)} L_j \zeta_{-y(t)}^* h(\zeta_u(x)) du$$

$$= \int_0^{y(t)} L_j h(\zeta_u \circ \zeta_{y(t)}^{-1}(x)) du$$

$$= \int_0^{y(t)} L_j h(\zeta_{u-y(t)}(x)) du$$

$$= \int_0^{y(t)} L_j h(\zeta_{-v}(x)) dv.$$

The similarity of (14.51) to (14.26) is obvious (of course, (14.51) reduces to (14.26) if $\alpha = 0$) and similar remarks are pertinent: A_t^y differs from A^* only in the "drift" and "potential" terms and therefore the complexity of computing the conditional distribution is essentially that of computing the distribution of the decomposition η_t of the signal process X_t.

14.5.3 Solution of the Zakai equation

The Zakai equation for the correlated noise problem was given in (14.36); in Stratonovich form it is

$$d\rho_t(f) = \rho_t((A - \frac{1}{2}D^2)f)dt + \rho_t(Df) \circ dY_t. \quad (14.53)$$

Now

$$A - \frac{1}{2}D^2 = \frac{1}{2}\sum_j V_j^2 - \frac{1}{2}(a_j V_j + h)^2 + V_0$$

and a completion-of-squares calculation shows that

$$A - \frac{1}{2}D^2 = L_0 + \sum_j L_j^2 - \frac{1}{2}Dh \quad (14.54)$$

where L_0, L_j are as in (14.41). The operator $D = Z + h$ is the generator of the group U_t on $C^\infty(\mathbb{S})$ given by (14.48). An argument analogous to that of §14.4.4 shows that the Doss–Sussmann solution of (14.53) is

$$\rho_t(f) = \langle T_{0,t}^Y(U_{Y_t} f), \pi \rangle$$

where $T_{0,t}^Y$ is the semigroup whose generator is

$$\tilde{A}_t^Y = U_{Y_t}\left(A - \frac{1}{2}D^2\right)U_{-Y_t}. \qquad (14.55)$$

We propose to show, using the Baker–Campbell–Hausdorff formula, that this coincides with A_t^Y given by (14.51) above. Denote by C_t the multiplication operator

$$C_t f(x) = f(x) \exp\left(\int_0^{Y_t} \zeta_s^* h(x) ds\right)$$

so that

$$U_{Y_t} = C_t \xi_t^* \qquad (14.56)$$

and

$$U_{-Y_t} = U_{Y_t}^{-1} = \left(\xi_t^{-1}\right)^* C_t^{-1}. \qquad (14.57)$$

(Note that $C_t^{-1} \neq C_{-t}$.) Using (14.54)–(14.57) we can see that

$$\tilde{A}_t^Y = C_t \xi_t^*\left(L_0 + \frac{1}{2}\Sigma Y_j^2 - \frac{1}{2}Dh\right)\left(\xi_t^{-1}\right)^* C_t^{-1}.$$

Now

$$\xi_t^* Dh \left(\xi_t^{-1}\right)^* f(x) = \xi_t^* Dh(x) f(x)$$

and

$$\xi_t^*\left(L_0 + \frac{1}{2}\sum_j L_j^2\right)\left(\xi_t^{-1}\right)^* = A_t^*$$

where A_t^* is given by (14.52). Thus

$$\tilde{A}_t^Y = C_t A_t^* C_t^{-1} - \frac{1}{2}\xi_t^* Dh. \qquad (14.58)$$

Since C_t is a multiplication operator, we can expand the right-hand side using the Baker–Campbell–Hausdorff formula (14.25). We obtain

$$C_t A_t^* C_t^{-1} = A_t^* - \left(L_j^* g_t\right) L_t^* - A_t^* g_t + \frac{1}{2}\sum_j \left(L_j^* g_t\right)^2.$$

Using this expression in (14.58) we see that $\tilde{A}_t^Y$ coincides with A_t^Y given by (14.51). Thus, as claimed in §14.1, the results are formally analogous to the independent case with the operator U_t replacing the operator of multiplication by $\exp(t\,h(x))$. However, while it is (with a bit of hindsight) in the independent case fairly obvious from the Kallianpur–Striebel formula that the appropriate generator is (14.27), the interpretation of (14.51) in the correlated case is by no means obvious and it seems essential to look at the Zakai equation to get the full picture.

Finally, if there are vector observations then the last term in (14.53) will be of the form

$$\sum_i \rho_t(D^i f) \circ dy_t^i$$

where

$$D^i = Z^i + h^i$$

for some vector fields Z^i. There cannot be a pathwise solution of (14.53) unless the D^i commute, but this only happens under extremely artificial conditions. If the D^i do not commute a decomposition of the type (14.43) is still possible, where ξ_t is almost surely a diffeomorphism (Kunita [14]), but no continuous dependence of ξ_t on Y can be expected. Thus the present results are essentially limited to the scalar case.

Acknowledgements

J. M. C. Clark initiated the whole concept of pathwise filtering and I have had many conversations on these topics with him. Another primary influence was the set of lectures given by H. Kunita at the 1980 London Mathematical Society Durham Symposium on Stochastic Integration. On the algebraic and geometric side, I had helpful discussions with S. K. Mitter and a preview of his paper Mitter [15], and some patient instruction in the basics of differential geometry from A. J. Krener.

It is extraordinary, but true, that both the original 1981 version of this paper and the present 2009 version were word-processed by Doris Abeysekera (using an IBM golf-ball typewriter and LaTeX respectively). I take this opportunity to express my gratitude to her for many years of friendly cooperation.

Notes

1. It is of course not obvious a priori that a regular conditional distribution exists.
2. and similarly sums over other repeated indices k etc. All sums are over 1 to r unless otherwise specified.
3. $\mathbb{E}_x$ is the expectation starting at $X_0 = x$.
4. I thank the referee of this chapter for this observation.

References

[1] Bain, A. and Crisan, D., 2008. *Fundamentals of Stochastic Filtering*. Springer.

[2] Blumenthal, R. M. and Getoor, R. K., 1968. *Markov Processes and Potential Theory*. Academic Press, New York.

[3] Clark, J. M. C., 1978. The design of robust approximations to the stochastic differential equations of nonlinear filtering. In *Communication Systems and Random Process Theory*, ed. J. K. Skwirzynski. Sijthof and Noordhoff, the Netherlands.

[4] Clark, J. M. C. and Crisan, D., 2005. On a robust version of the integral representation formula of nonlinear filtering. *Probability Theory and Related Fields* 133, 43–56.

[5] Davis, M., 1981. Pathwise nonlinear filtering. In Hazewinkel and Willems, pp. 505–28.

[6] Davis, M. and Marcus, S., 1981. An introduction to nonlinear filtering. In Hazewinkel and Willems, pp. 53–75.

[7] Davis, M. H. A., 1979. Pathwise solutions and multiplicative functionals in nonlinear filtering. *Proc. 18th IEEE Conference on Decision and Control*, Ft. Lauderdale.

[8] Davis, M. H. A., 1980. On a multiplicative functional transformation arising in nonlinear filtering theory. *Z. Wahrscheinlichkeitstheorie verw. Geb.* 54, 125–39.

[9] Davis, M. H. A., 1982. A pathwise solution of the equations of nonlinear filtering. *Teoria Veroyatnostei i ee Prim.* 27, 160–7.

[10] Doss, H., 1977. Liens entre équations differéntielles stochastiques et ordinaires. *Annales de l'Institut Henri Poincaré* 13, 99–125.

[11] Dynkin, E. B., 1965. *Markov Processes*. Springer-Verlag, Berlin.

[12] Gihman, I. I. and Skorohod, A., 1972. *Stochastic Differential Equations*. Springer-Verlag, Berlin.

[13] Hazewinkel, M. and Willems, J. (eds), 1981. *Stochastic Systems: The Mathematics of Filtering and Identification and Applications*, Les Arcs 1980. Vol. C78 of NATO ASI Series, Sub-Series C: Mathematical and Physical Sciences. D. Reidel, Dordrecht.

[14] Kunita, H., 1981. On the decomposition of solutions of stochastic differential equations. In Williams, D. (ed.), *Stochastic Integrals*. Vol. 851 of Lecture Notes in Mathematics. Springer-Verlag.

[15] Mitter, S. K., 1979. On the analogy between mathematical problems of nonlinear filtering and quantum physics. *Ricerche di Automatica* 10, 163–216.

[16] Pardoux, E., 1979. Backward and forward stochastic partial differential equations associated with a nonlinear filtering problem. *Proc. 18th IEEE Conference on Decision and Control*, Ft. Lauderdale, Florida.

[17] Pardoux, E., 1982. Equations du filtrage non linéaire, de la prédiction et du lissage. *Stochastics* 6, 193–231.

[18] Revuz, D. and Yor, M., 1999. *Continuous Martingales and Brownian Motion*, 3rd edn. Springer.

[19] Sussmann, H. J., 1978. On the gap between deterministic and stochastic ordinary differential equations. *Annals of Probability* 6, 19–41.

·15·
The Innovations Problem
A. J. Heunis

15.1 Introduction

The innovations approach to filtering and estimation has its origins in a fundamental result of Wold [30], which basically asserts that a discrete-time 'nondeterministic' stationary process can always be represented as the output of a time-invariant linear system, for which the input is a stationary *orthogonal* process. It was soon noticed by Kolmogorov [17] that the orthogonality of the 'driving input' in the Wold representation was the key to a powerful and comprehensive method for solving general problems of linear least-squares estimation for discrete-time stationary stochastic processes. During the 1950s the challenge of launching and tracking spacecraft led to an interest in recursively computed linear least-squares estimators for *nonstationary* processes. Kalman [16] used the approach of Kolmogorov to obtain the celebrated Kalman filter for recursively computing linear estimators of the (intrinsically nonstationary) state process of a *discrete-time* stochastic linear system. Adaptation of the Kolmogorov–Wold approach to the analogous problem of state estimation in *continuous-time* was the next obvious step, but this proved to be a definite challenge, mainly because it was not at all clear how to carry over the discrete-parameter computations of [16] to a continuous-parameter setting. These difficulties were elegantly resolved by Kailath [11], who showed that it was indeed possible to adapt the ideas of Kolmogorov and Wold to continuous time problems and in this way obtain a continuous-time Kalman filter (the Kalman–Bucy filter).

The essence of the approach of Kailath [11] may be summarized as follows. One is given an $\mathbb{R}^D$-valued standard Wiener process $\{W_t, t \in [0, T]\}$, together with a jointly-measurable $\mathbb{R}^D$-valued 'signal process' $\{\beta_t, t \in [0, T]\}$, defined on the common probability space $(\Omega, \mathcal{F}, P)$, over the interval $t \in [0, T]$ for some fixed finite 'horizon' $T \in (0, \infty)$. The signal $\{\beta_t\}$ satisfies the 'finite energy' condition

$$E_P \int_0^T |\beta_t|^2 \, dt < \infty, \tag{15.1}$$

and $\{\beta_t\}$ and $\{W_t\}$ are subject to a partial independence condition which effectively requires that future increments in $\{W_t\}$ be independent of the joint past of $\{\beta_t\}$ and $\{W_t\}$, namely

$\sigma\{\beta_\tau, W_\tau : 0 \leq \tau \leq t\}$ and $\sigma\{W_{\tau_2} - W_{\tau_1} : t \leq \tau_1 < \tau_2 \leq T\}$ are P-independent, \hfill (15.2)

for each $t \in [0, T)$. Define the observation process $\{Y_t\}$ and its associated filtration $\{\mathcal{F}_t^Y\}$ in the usual way, namely

$$Y_t := W_t + \int_0^t \beta_s \, ds, \qquad \mathcal{F}_t^Y := \sigma\{Y_s, s \in [0, t]\}, \qquad t \in [0, T]. \qquad (15.3)$$

Put $\hat{\beta}_t := E_P[\beta_t \mid \mathcal{F}_t^Y]$, and define the 'innovations process' $\{I_t\}$ and filtration $\{\mathcal{F}_t^I\}$ as

$$I_t := Y_t - \int_0^t \hat{\beta}_s \, ds, \qquad \mathcal{F}_t^I := \sigma\{I_s, s \in [0, t]\}, \qquad t \in [0, T]. \qquad (15.4)$$

Also, put $\mathbb{H} := L^2(\Omega, \mathcal{F}, P)$, and, for each $t \in [0, T]$, let $\mathbb{H}_t^Y$ and $\mathbb{H}_t^I$ denote the closed linear subspaces of $\mathbb{H}$ spanned by the random variables $\{Y_\tau^n, \tau \in [0, t], n = 1, 2, \ldots, D\}$ and $\{I_\tau^n, \tau \in [0, t], n = 1, 2, \ldots, D\}$ respectively. The central result of Kailath (see appendix II of [11]) is that, subject to the preceding conditions, one has

$$\mathbb{H}_t^Y = \mathbb{H}_t^I, \qquad t \in [0, T]. \qquad (15.5)$$

This result is the essential step in Kailath's derivation [11] of the Kalman–Bucy filter, and may be viewed as the 'correct' adaptation of the basic approach of Kolmogorov and Wold to a continuous-time setting (in fact, the derivation of (15.5) in [11] is based on a rather informal 'white-noise' calculus which certainly makes 'engineering sense' but otherwise leaves something to be desired from a mathematical viewpoint; an elegant and rigorous proof is given by Davis ([7], theorem 1)). In connection with the innovations process at (15.4), one should also note a further result (established by Kailath [12]), to the effect that $\{I_t\}$ is *a standard Brownian motion*. This result is a definite surprise, since, although $\{W_t\}$ is a Brownian motion, nothing at all has been postulated about the law of the signal $\{\beta_t\}$, which may well be non-Gaussian. The fact that the innovations process $\{I_t\}$ is a standard Brownian motion will be essential later in the chapter.

Equality of the subspaces $\mathbb{H}_t^Y$ and $\mathbb{H}_t^I$ given by (15.5) is completely satisfactory when the process $\{(\beta_t, W_t)\}$ is jointly Gaussian, as is the case when establishing the Kalman–Bucy filter. However, when the signal $\{\beta_t\}$ is non-Gaussian, then linear estimators, based on projection onto the subspace $\mathbb{H}_t^Y$, are usually of little practical interest, and one is much more concerned with conditional expectations, that is projections onto the (larger) subspace of $\mathbb{H}$ given by $\mathbb{G}_t^Y := L^2(\Omega, \mathcal{F}_t^Y, P)$. It is now more relevant to compare, not $\mathbb{H}_t^Y$ and $\mathbb{H}_t^I$, but rather the subspaces $\mathbb{G}_t^Y$ and $\mathbb{G}_t^I := L^2(\Omega, \mathcal{F}_t^I, P)$, and, if possible, establish the natural analogue of (15.5) for these subspaces, namely equality of $\mathbb{G}_t^Y$ and $\mathbb{G}_t^I$, or equivalently, equality of the σ-algebras $\mathcal{F}_t^Y$ and $\mathcal{F}_t^I$ for each

$t \in [0, T]$. The assertion of the equality of these σ-algebras is commonly known as the *innovations conjecture*.

Establishing the innovations conjecture is a much more challenging problem than proving a result such as (15.5). Indeed, under the previously postulated conditions (namely (15.1) and (15.2)), the innovations conjecture is not even true, as is demonstrated by an example of Benes ([2], pp. 257–8), which in turn is based upon a subtle extension ([2], theorem 5, p. 247) of Tsirel'son's example [29] of a functional stochastic differential equation which fails to have a strong solution. It is therefore necessary to formulate supplementary conditions under which the innovations conjecture can be shown to hold. One of the earliest results of this kind is due to Clark [5], who established the conjecture for the case in which the signal $\{\beta_t\}$ is uniformly bounded and independent of $\{W_t\}$, using an argument based on the Kallianpur–Striebel representation for the conditional expectation $\hat{\beta}_t$ (a statement and proof of the main result of Clark [5] appears as theorem 11.4.1 of Kallianpur ([13], p. 284), and may also be found in Meyer ([24], pp. 244–6)). A similar result was obtained by Kunita ([20], theorem 2.2, p. 372) in the case where the signal is of the form $\beta_t := h(X_t)$ for some time-homogeneous Markov process $\{X_t\}$ which is independent of $\{W_t\}$ and takes values in a compact Hausdorff space S, and $h : S \to \mathbb{R}^D$ is a continuous mapping. Although this result is a special case of that of Clark [5], it is established by a very different method which makes essential use of the Markov structure of $\{X_t\}$ and the normalized stochastic differential equation for nonlinear filtering [9].

During the late 1970s a new and very elegant approach to the innovations conjecture was worked out by Allinger, Benes, Clark, Erzhov, and Mitter. The basic idea is to use the Kallianpur–Striebel formula to write the conditional expectation $\hat{\beta}_s := E_P\left[\beta_s \mid \mathcal{F}_s^Y\right]$ as a nonanticipative functional of the paths of $\{Y_t\}$; then the relation at (15.4) can be regarded as a stochastic differential equation for which the 'driving process' is the innovation $\{I_t\}$ (which, as we have seen, is a standard Brownian motion), and the 'solution process' is the observation $\{Y_t\}$. Under appropriate conditions a fundamental result of Yamada and Watanabe [32] on stochastic differential equations can be used to show that $\{Y_t\}$ is actually a *strong* solution, and this is enough to establish the innovations conjecture. This approach has been used in [1] to show that the innovations conjecture holds under rather general conditions, namely the finite-energy condition (15.1) and P-independence of $\{\beta_t\}$ and $\{W_t\}$ (this strengthens the partial independence (15.2)). This result is a genuine *pièce de résistance* among contributions to the innovations problem, but, despite this, does not appear to have received the attention that it fully merits, possibly because of the rather elevated concision of the presentation in [1]. One of the main goals of the present chapter is to make this central contribution accessible to readers having only a fairly elementary knowledge of stochastic calculus. We establish the result in full in Section 15.4, having introduced some necessary preliminaries and mathematical tools in Section 15.2 and Section 15.3.

Our comments on the innovations conjecture have so far mainly concerned the case in which $\{\beta_t\}$ and $\{W_t\}$ are independent. We turn next to the the weaker independence condition (15.2), and first indicate a simple but important example for which this partial independence holds, but full independence of $\{\beta_t\}$ and $\{W_t\}$ fails. Suppose that, in addition to the $\mathbb{R}^D$-valued standard Brownian motion $\{W_t\}$ on $(\Omega, \mathcal{F}, P)$, we also have an $\mathbb{R}^d$-valued random vector ξ_0 and another standard $\mathbb{R}^q$-valued Brownian motion $\{\tilde{W}_t\}$ on $(\Omega, \mathcal{F}, P)$, such that ξ_0, $\{W_t\}$, and $\{\tilde{W}_t\}$ are P-independent, and suppose that the processes $\{X_t\}$ and $\{Y_t\}$ are determined by the stochastic differential system

$$dX_t = AX_t\, dt + B\, dW_t + C\, d\tilde{W}_t, \quad X_0 = \xi_0, \qquad (15.6)$$

$$dY_t = HX_t\, dt + dW_t, \quad Y_0 = 0, \qquad (15.7)$$

in which A, B, C and H are matrices of appropriate dimension. Now (15.7) is of the form (15.3) when we put $\beta_t := HX_t$, and it is clear that $\{\beta_t\}$ and $\{W_t\}$ are P-independent when $B = 0$, while this independence fails, but the partial independence (15.2) still holds, when $B \neq 0$. We can regard the process $\{X_t\}$ as the 'state' of a system; the second term on the right of (15.6) results from 'feedback' of the observations to the state dynamics (15.6). This situation is typical of stochastic control problems with 'feedback' of the observations; such problems arise quite naturally in aerospace engineering (e.g. the inertial guidance of satellites and aircraft) as well as other applications. Returning to discussion of the general relations (15.3) and (15.4), let us first suppose that the process $\{(\beta_t, W_t)\}$ is *jointly* Gaussian (for example, this is the case for the pair (15.6), (15.7), with $\beta_t := HX_t$, when the initial state ξ_0 is Gaussian). Now the innovations conjecture follows from the equality of subspaces at (15.5) (see theorem 2 of Davis [7]). Establishing conditions for the innovations conjecture to hold in the non-Gaussian case, with only the partial independence (15.2) in force, is a very challenging problem indeed, and typically relies on formulating explicit models which specify the structure of the correlation between $\{\beta_t\}$ and $\{W_t\}$. In this regard Krylov [18] establishes the innovations conjecture for a model which is a far-reaching 'nonlinearization' of the simple linear model (15.6)–(15.7). Much like the resolution of the innovations conjecture established in [1], the work of Krylov [18] constitutes another central contribution to the innovations problem which really deserves a wider audience. We conclude the chapter by summarizing the main elements of Krylov's approach in Section 15.5.

15.2 A Bayes formula and other preliminaries

Sections 15.2, 15.3, and 15.4 are devoted to establishing the innovations conjecture in the case where $\{\beta_t\}$ and $\{W_t\}$ are independent, following the approach of [1]. In the present section we set forth the basic hypotheses, state the main result on the innovations conjecture (see Theorem 15.2), and deal with some of the technical preliminaries needed for this result. To facilitate easy reference we summarize all of our notation in one place as follows:

Notation 15.1 (I) For a positive integer q write $\mathbb{R}^q$ for the space of all real q-dimensional column vectors, write x^i or $[x]^i$ for the i-th scalar entry and $|x| := \left[\sum_{i=1}^q (x^i)^2\right]^{1/2}$ for the Euclidean norm of $x \in \mathbb{R}^q$. Likewise, for positive integers q and r, write $\mathbb{R}^{q \times r}$ for the space of all q by r matrices with real entries, write A^{ij} or $[A]^{ij}$ for the (i, j)-th scalar entry, A' for the transpose, and $\|A\| := \max_{x \in \mathbb{R}^r, |x|=1} |Ax|$ for the operator norm of $A \in \mathbb{R}^{q \times r}$.

(II) For a fixed $T \in (0, \infty)$ let $C[0, T; \mathbb{R}^q]$ denote the space of all continuous mappings from $[0, T]$ into $\mathbb{R}^q$, with the usual supremum norm. For notational brevity we also denote this space by C_q when there is no possibility of confusion, and put $\mathcal{B}_t(C_q) = \sigma\{\psi(s), s \in [0, t]\}$, $t \in [0, T]$, for the canonical filtration on C_q (with ψ a generic member of C_q).

(III) $\mathcal{B}(S)$ denotes the Borel σ-algebra of a separable metric space (S, ρ), and, for a $\mathcal{F}/\mathcal{B}(S)$-measurable mapping ξ from a probability space $(\Omega, \mathcal{F}, P)$ into S, let $P\xi^{-1}$ denote the probability measure on $\mathcal{B}(S)$ given by $P\xi^{-1}(A) := P[\xi \in A]$ for each $A \in \mathcal{B}(S)$.

(IV) For an $\mathbb{R}^q$-valued process $\{\eta_t\}$ put $\mathcal{F}_t^\eta := \sigma\{\eta_s, s \in [0, t]\}$ for the raw filtration.

(V) For a measure space $(E, \mathcal{S}, \mu)$ (not necessarily complete) write $\mathbf{Z}^\mu[\mathcal{S}]$ for the collection $\{N \subset E : N \subset H \text{ for some } H \in \mathcal{S} \text{ with } \mu(H) = 0\}$. For collections of sets $\mathcal{A}_1 \subset \mathcal{S}$ and $\mathcal{A}_2 \subset \mathcal{S}$ write $\sigma\{\mathcal{A}_1, \mathcal{A}_2\}$ or $\mathcal{A}_1 \vee \mathcal{A}_2$ for the minimal σ-algebra on E which includes all members of $\mathcal{A}_1$ and $\mathcal{A}_2$.

From now on fix a constant $T \in (0, \infty)$, which determines a finite 'time horizon' $t \in [0, T]$ of interest. The innovations problem will be formulated in terms of the basic ingredients outlined in the following condition:

Condition 15.1 *We are given a filtered probability space* $(\Omega, \mathcal{F}, P; \{\mathcal{F}_t\})$ *satisfying the 'usual conditions' (see e.g. definition II (67.1) of Rogers and Williams [26], p. 172), together with an* $\mathbb{R}^D$-valued $\mathcal{F}_t$-*Wiener process* $\{W_t, t \in [0, T]\}$ *and an* $\mathbb{R}^D$-valued $\mathcal{F}_t$-*progressively measurable process* $\{\beta_t, t \in [0, T]\}$ *for which the 'finite energy' condition (15.1) holds.*

Remark 15.1 From Condition 15.1 we see that $\sigma\{\beta_\tau, W_\tau : 0 \leq \tau \leq t\} \subset \mathcal{F}_t$, and the σ-algebras $\mathcal{F}_t$ and $\sigma\{W_{\tau_2} - W_{\tau_1} : t \leq \tau_1 < \tau_2 \leq T\}$ are P-independent for each $t \in [0, T)$. Thus the partial independence (15.2) holds.

We shall regard $\{\beta_t\}$ as the 'signal' and $\{W_t\}$ as the 'observation noise'. Condition 15.1 ensures that the 'observation process' $\{Y_t, t \in [0, T]\}$ at (15.3) is well defined, and is an $\mathbb{R}^D$-valued, continuous and $\mathcal{F}_t$-adapted process. We must now define the innovations process $\{I_t\}$, and this requires us to deal with some measure-theoretic technicalities which are best settled once and for all at this point. In order to avail ourselves of the tools of stochastic calculus we need to 'minimally enlarge' the raw filtration $\{\mathcal{F}_t^Y\}$ (see (15.3)) by including all P-null members of $\mathcal{F}$, and to this end we define the observation filtration $\{\mathcal{Y}_t\}$ as

$$\mathcal{Y}_t := \mathcal{F}_t^Y \vee \mathbf{Z}^\mathbf{P}[\mathcal{F}] \subset \mathcal{F}_t, \qquad t \in [0, T], \tag{15.8}$$

the set-inclusion following since $\{Y_t\}$ is $\mathcal{F}_t$-adapted. From now on we compute conditional expectations with respect to $\mathcal{Y}_t$ rather than $\mathcal{F}_t^Y$. Even with $\{\hat{\beta}_t\}$ computed in terms of the filtration $\{\mathcal{Y}_t\}$, namely $\hat{\beta}_t := \mathrm{E}_P[\beta_t \mid \mathcal{Y}_t]$, the definition of the innovations process at (15.4) is not entirely satisfactory, since each $\mathrm{E}_P[\beta_t \mid \mathcal{Y}_t]$ is unique only to within a P-null event which depends on $t \in [0, T]$, and hence the process of conditional expectations $\{\mathrm{E}_P[\beta_t \mid \mathcal{Y}_t]\}$ is not completely determined. Condition 15.1 is nevertheless ample to ensure that there does exist an $\mathbb{R}^D$-valued process $\{\hat{\beta}_t\}$ such that

$$\{\hat{\beta}_t\} \text{ is } \mathcal{Y}_t\text{-progressively. meas. and } \hat{\beta}_t = \mathrm{E}_P[\beta_t \mid \mathcal{Y}_t], \ P - \text{a.s. for each } t \in [0, T], \tag{15.9}$$

(this follows e.g. from proposition 7.1.1 of Wong and Hajek [31], p. 251). With the process $\{\hat{\beta}_t\}$ at hand we can define the innovations process $\{I_t\}$ by

$$I_t := Y_t - \int_0^t \hat{\beta}_\tau \, d\tau, \qquad t \in [0, T]. \tag{15.10}$$

Then it follows that $\{I_t\}$ is $\mathbb{R}^D$-valued, continuous and $\mathcal{Y}_t$-adapted (from the Fubini theorem and the progressive measurability at (15.9)), and therefore

$$\mathcal{I}_t \subset \mathcal{Y}_t \text{ for } \mathcal{I}_t := \mathcal{F}_t^I \vee \mathbf{Z}^\mathbf{P}[\mathcal{F}]. \tag{15.11}$$

Remark 15.2 Notice that, if $\{\tilde{\beta}_t\}$ is another $\mathbb{R}^D$-valued and $\mathcal{Y}_t$-progressively measurable process such that $\tilde{\beta}_t = \mathrm{E}_P[\beta_t \mid \mathcal{Y}_t]$, $P - $ a.s. for each $t \in [0, T]$, then $\{\tilde{\beta}_t\}$ is necessarily a modification of $\{\hat{\beta}_t\}$. Consequently, by the Fubini theorem, $\int_0^t \hat{\beta}_\tau \, d\tau = \int_0^t \tilde{\beta}_\tau \, d\tau$, $P - $ a.s. for each $t \in [0, T]$, that is, these $d\tau$-integrals give indistinguishable processes (since each integral defines a continuous process). It follows that the innovations process at (15.10) is uniquely defined to within P-indistinguishability, regardless of which $\mathcal{Y}_t$-progressively measurable integrand we use.

In Section 15.1 we noted that the innovations process $\{I_t\}$ is a standard Brownian motion. This result is so essential for later developments that we establish it next, following Rogers and Williams ([27], p. 323):

Theorem 15.1 *Suppose Condition 15.1. Then $\{I_t\}$ defined at (15.10) is an $\mathbb{R}^D$-valued standard $\mathcal{Y}_t$-Wiener process on $(\Omega, \mathcal{F}, P)$.*

Proof. From (15.10) and (15.3) we have

$$I_t = W_t + \int_0^t [\beta_s - \hat{\beta}_s]\, ds, \qquad t \in [0, T], \qquad (15.12)$$

so that

$$\max_{0 \le t \le T} |I_t| \le \max_{0 \le t \le T} |W_t| + \int_0^T |\beta_s|\, ds + \int_0^T |\hat{\beta}_s|\, ds. \qquad (15.13)$$

From Jensen's inequality and (15.9) we get $|\hat{\beta}_s|^2 \le E_P[|\beta_s|^2 \mid \mathcal{Y}_s]$. Then, from (15.1), the Cauchy–Schwarz inequality, Doob's L_2-inequality, and the Fubini theorem, it is easily seen that the random variable on the right of (15.13) is square-integrable (with respect to P). Consequently, fixing an arbitrary $\mathcal{Y}_t$-stopping time such that $U \le T$, one has $E_P|I_U| < \infty$, and (from (15.12))

$$E_P[I_U] = E_P[W_U] + E_P \int_0^T I\{s < U\}[\beta_s - \hat{\beta}_s]\, ds. \qquad (15.14)$$

Now $\{\mathcal{Y}_t\}$ is a subfiltration of $\{\mathcal{F}_t\}$ (see (15.8)), thus U is a bounded $\mathcal{F}_t$-stopping time, hence, by optional sampling and Condition 15.1, we have $E_P[W_U] = 0$. Moreover, $I\{s < U\} \in \mathcal{Y}_s$ and therefore, from (15.9), one sees that $E_P[\beta_s I\{s < U\}] = E_P[\hat{\beta}_s I\{s < U\}]$. From these facts and (15.14) we obtain $E_P[I_U] = 0$. Since the $\mathcal{Y}_t$-stopping time U is arbitrary, it follows from ([26], II (77.6), p. 190) that $\{I_t\}$ is a $\mathcal{Y}_t$-martingale on $(\Omega, \mathcal{F}, P)$. Moreover, from (15.12), the quadratic variation process of $\{I_t\}$ is identical to that of the Wiener process $\{W_t\}$, and now the result follows from ([27], (IV) (33.1), p. 63). □

The next result asserts the validity of the innovations conjecture: the set-inclusion at (15.11) becomes a *set-equality* when $\{\beta_t\}$ and $\{W_t\}$ are independent:

Theorem 15.2 *Suppose (i) Condition 15.1, and (ii) $\{\beta_t\}$ and $\{W_t\}$ are P-independent. Then $\mathcal{I}_t = \mathcal{Y}_t, t \in [0, T]$ (for $\mathcal{I}_t$ and $\mathcal{Y}_t$ given by (15.11) and (15.8)).*

Our goal is to establish Theorem 15.2 by following the approach of [1]. To accomplish this we shall rely on two essential tools, namely the Kallianpur–Striebel (or Bayes) formula which represents the process $\{\hat{\beta}_t\}$ at (15.9) as an explicit functional of the paths of the observation $\{Y_t\}$, and a modification due to Clark [6] of a basic theorem on stochastic differential equations of Yamada and Watanabe [32]. We recall the Kallianpur–Striebel formula next, deferring discussion of Clark's modification of the Yamada–Watanabe theorem to Section 15.3. Our purpose in summarizing the Kallianpur–Striebel formula is to keep the presentation reasonably self-contained and to emphasize some ideas which will be essential for establishing Theorem 15.2. First we define some notation and make some preliminary observations which will be useful throughout the chapter.

Recalling the Notation 15.1 (IV) and (V), define

$$\mathfrak{X}_T := \mathcal{F}_T^\beta \vee \mathbf{Z}^\mathbf{P}[\mathcal{F}] \qquad \text{and} \qquad \mathcal{G}_t := \mathfrak{X}_T \vee \mathcal{F}_t^W, \qquad t \in [0, T], \tag{15.15}$$

$$\Gamma_t := \exp\left[-\int_0^t \beta_\tau' \, dW_\tau - \frac{1}{2}\int_0^t |\beta_\tau|^2 \, d\tau\right], \qquad t \in [0, T]. \tag{15.16}$$

From the P-independence of $\{\beta_t\}$ and $\{W_t\}$ postulated in Theorem 15.2 we know that

$$\{W_t\} \text{ is a } \mathcal{G}_t\text{-Wiener process on } (\Omega, \mathcal{F}, P), \tag{15.17}$$

and of course $\{\beta_t\}$ is $\mathcal{G}_t$-adapted and $\mathcal{B}[0, T] \otimes \mathcal{F}$-measurable. It then follows that $\{\Gamma_t\}$ is a $\mathcal{G}_t$-local martingale, and hence (by Fatou's lemma and nonnegativity of Γ_t) is also a $\mathcal{G}_t$-supermartingale on $(\Omega, \mathcal{F}, P)$. Moreover, since $\mathcal{G}_0 = \mathfrak{X}_T$, we have that $\mathcal{G}_0$ and $\{W_t\}$ are P-independent, and from this it readily follows that $\mathrm{E}_P[\Gamma_t \mid \mathcal{G}_0] = 1$, from which we see that $\mathrm{E}_P\Gamma_t = 1$ for all $t \in [0, T]$, and thus

$$\{\Gamma_t\} \text{ is a } \mathcal{G}_t\text{-martingale on } (\Omega, \mathcal{F}, P). \tag{15.18}$$

In view of (15.18) we can define the *probability* measure

$$P_0(A) := \mathrm{E}_P[\Gamma_T; A], \qquad A \in \mathcal{F}, \tag{15.19}$$

and then it follows from (15.19), (15.17), (15.16), (15.3), and the Girsanov theorem, that

$$\{Y_t, \, t \in [0, T]\} \text{ is a } \mathcal{G}_t\text{-Wiener process on } (\Omega, \mathcal{F}, P_0). \tag{15.20}$$

Remark 15.3 From (15.16) we have $P[\Gamma_T > 0] = 1$, thus from (15.19) one sees that P and P_0 are *equivalent* measures on $\mathcal{F}$, i.e. $P(A) = 0$ if and only if $P_0(A) = 0$ for each $A \in \mathcal{F}$. Therefore we can, and will, use the notation a.s. without specifying which of the measures P or P_0 are involved.

For each $(t, y) \in [0, T] \times C_D$ (recall Notation 15.1(II)) and $\omega \in \Omega$, define

$$\rho_t(\omega, y) := \exp\left[\int_0^t \beta_\tau'(\omega) \, dy_\tau - \frac{1}{2}\int_0^t |\beta_\tau(\omega)|^2 \, d\tau\right]. \tag{15.21}$$

Remark 15.4 In (15.21) it is understood that $\rho_t(\omega, y) := 0$ for each $y \in C_D$ for which the (Paley–Wiener) dy_τ-integral is undefined. The collection of all such y comprises a set of zero Wiener measure, and will not have any effect on subsequent developments.

Again, for each $(t, y) \in [0, T] \times C_D$ define

$$f_t(y) := \int_\Omega \beta_t(\omega) \rho_t(\omega, y) \, dP(\omega), \tag{15.22}$$

$$g_t(y) := \int_\Omega \rho_t(\omega, y) \, dP(\omega), \tag{15.23}$$

$$\gamma_t(y) := \frac{f_t(y)}{g_t(y)}. \tag{15.24}$$

With the preceding in place we are now able to establish a representation of the optimal filter as an explicit function of the paths of the observation process:

Theorem 15.3 *Suppose that the conditions of Theorem 15.2 hold, and recall (15.9). Then*

$$\hat{\beta}_t = \gamma_t(Y), \qquad \text{a.s. for each } t \in [0, T].$$

Proof. In view of (15.20) we know that the σ-algebras $\mathcal{F}_T^Y$ and $\mathcal{G}_0$ are P_0-independent, and since $\mathcal{F}_T^\beta \subset \mathcal{G}_0$ (see (15.15)), we then find that

$$\{Y_t, t \in [0, T]\} \text{ and } \{\beta_t, t \in [0, T]\} \text{ are } P_0\text{-independent.} \tag{15.25}$$

Moreover, for each $A \in \mathcal{G}_0$ we have that $P_0(A) = E_P[\Gamma_T; A] = E_P[E_P[\Gamma_T \mid \mathcal{G}_0]; A] = P(A)$ (where we have used (15.19) and (15.18)), hence we conclude

$$P(A) = P_0(A), \qquad A \in \mathcal{F}_T^\beta. \tag{15.26}$$

Now put

$$\Lambda_t := \frac{1}{\Gamma_t} = \exp\left[\int_0^t \beta_\tau' \, dY_\tau - \frac{1}{2} \int_0^t |\beta_\tau|^2 \, d\tau\right], \qquad t \in [0, T]. \tag{15.27}$$

For each $t \in [0, T]$ one has that $E_{P_0}[\Lambda_t : A] = P(A)$ for all $A \in \mathcal{G}_t$ (as follows from (15.18)), and $\mathcal{Y}_t \subset \mathcal{G}_t$ (as follows from (15.15) and (15.3)). Consequently, from a standard formula for manipulating Radon–Nikodym derivatives (see e.g. lemma 7.7.1 on p. 243 of Wong and Hajek [31], or VIII.27.4(1') on p. 11 of Loève [22]), we have

$$E_P[\beta_t \mid \mathcal{Y}_t] = \frac{E_{P_0}[\beta_t \Lambda_t \mid \mathcal{Y}_t]}{E_{P_0}[\Lambda_t \mid \mathcal{Y}_t]}, \quad \text{a.s.} \quad \text{for each } t \in [0, T]. \tag{15.28}$$

Now define

$$f_t^1(y) := \int_\Omega \beta_t(\omega) \rho_t(\omega, y) \, dP_0(\omega), \qquad (t, y) \in [0, T] \otimes C_D. \tag{15.29}$$

In view of (15.29), (15.27), and (15.25), for each $t \in [0, T]$ we have

$$E_{P_0}[\beta_t \Lambda_t \mid \mathcal{Y}_t] = f_t^1(Y), \qquad \text{a.s.} \tag{15.30}$$

and, from (15.26), we can replace the measure P_0 in (15.29) with the measure P, so that from (15.22) we have

$$f_t(y) = f_t^1(y), \qquad (t, y) \in [0, T] \otimes C_D. \tag{15.31}$$

Exactly as for (15.30), and recalling (15.23), for each $t \in [0, T]$ we have

$$E_{P_0}[\Lambda_t \mid \mathcal{Y}_t] = g_t(Y), \qquad \text{a.s.} \tag{15.32}$$

The result follows upon combining (15.32), (15.31), (15.30), (15.28), (15.24), and (15.9). □

Remark 15.5 The representation of the optimal filter established by Theorem 15.3 is due to Kallianpur and Striebel [15], and is usually known as the *Kallianpur–Striebel formula*. The approach that we have followed in the proof of Theorem 15.3 is somewhat less classical than that of [15], and is guided by Bhatt, Kallianpur, and Karandikar [3] and Kallianpur and Karandikar ([14], pp. 572–5). Other derivations and extensive discussion of the Kallianpur-Striebel formula can be found in Liptser and Shiryayev ([23], §7.9), Meyer ([24], §3), and Kallianpur ([13], §11.3).

Remark 15.6 From now on denote Wiener measure on $\mathcal{B}(C_D)$ by Q and put

$$\overline{\mathcal{B}(C_D)}^Q := \mathcal{B}(C_D) \vee \mathbf{Z}^Q[\mathcal{B}(C_D)], \qquad \overline{\mathcal{B}_t(C_D)}^Q := \mathcal{B}_t(C_D) \vee \mathbf{Z}^Q[\mathcal{B}(C_D)], \tag{15.33}$$

(recall Notation 15.1 (II) and (V)). In order to use the representation (15.9) we shall need the the filtered Wiener space $(C_D, \overline{\mathcal{B}(C_D)}^Q, Q; \{\overline{\mathcal{B}_t(C_D)}^Q\})$, as well as the filtration $\{\mathcal{H}_t\}$ in the probability space $(\Omega \times C_D, \mathcal{F} \otimes \overline{\mathcal{B}(C_D)}^Q, P \otimes Q)$ defined by

$$\mathcal{H}_t := \mathfrak{X}_T \otimes \overline{\mathcal{B}_t(C_D)}^Q, \tag{15.34}$$

(recall (15.15)). We shall use $\{y_t\}$ to denote the usual $\mathbb{R}^D$-valued canonical 'coordinate' process on the probability space $(C_D, \overline{\mathcal{B}(C_D)}^Q, Q)$. It will also be convenient to regard $\{\beta_t\}$ and $\{y_t\}$ as $\mathbb{R}^D$-valued processes on the product space $(\Omega \times C_D, \mathcal{F} \otimes \overline{\mathcal{B}(C_D)}^Q, P \otimes Q)$ (this will always be clear from the context).

From Remark 15.6 it follows that $\{\beta_t\}$ is an $\mathcal{H}_t$-adapted and $\mathcal{B}[0, T] \otimes (\mathcal{F} \otimes \overline{\mathcal{B}(C_D)}^Q)$-measurable (i.e. jointly measurable) process, while $\{y_t\}$ is an $\mathcal{H}_t$-Wiener process, on the product space $(\Omega \times C_D, \mathcal{F} \otimes \overline{\mathcal{B}(C_D)}^Q, P \otimes Q)$. Then, exactly as for (15.18), we see that

$$\{\rho_t\} \text{ is a continuous } \mathcal{H}_t\text{-martingale on } (\Omega \times C_D, \mathcal{F} \otimes \overline{\mathcal{B}(C_D)}^Q, P \otimes Q), \tag{15.35}$$

(recall (15.21)). Define

$$\chi_t(\omega, y) := \left[\int_0^T |\beta_\tau(\omega)|^2 \, d\tau\right] \rho_t(\omega, y), \qquad (t, y) \in [0, T] \times \Omega. \qquad (15.36)$$

Since the term in [...] on the right of (15.36) is $\mathcal{H}_0$-measurable (see Remark 15.6), it follows from (15.35) and Condition 15.1 that

$$E_{P \otimes Q}[\chi_t] = E_P \int_0^T |\beta_\tau|^2 \, d\tau < \infty, \qquad t \in [0, T], \qquad (15.37)$$

$\{\chi_t\}$ is a continuous $\mathcal{H}_t$-martingale on $(\Omega \times C_D, \mathcal{F} \otimes \overline{\mathcal{B}(C_D)}^Q, P \otimes Q)$.
$\qquad(15.38)$

Now put

$$a_t(y) := \int_\Omega \chi_t(\omega, y) \, dP(\omega), \qquad (t, y) \in [0, T] \otimes \Omega. \qquad (15.39)$$

The next result will be essential for establishing Theorem 15.2:

Proposition 15.1 *Suppose the conditions of Theorem 15.2. Then the processes $\{g_t\}$ and $\{a_t\}$ defined by (15.23) and (15.39) are $\overline{\mathcal{B}_t(C_D)}^Q$-martingales on $(C_D, \overline{\mathcal{B}(C_D)}^Q, Q)$, and are Q-a.s. continuous (i.e. for some $N \in \overline{\mathcal{B}(C_D)}^Q$, with $Q(N) = 0$, the paths $t \to a_t(y)$ and $t \to g_t(y)$ are continuous on $[0, T]$ for each $y \notin N$).*

Proof. The assertion will be shown for $\{a_t\}$ only, since the proof for $\{g_t\}$ is an obvious simplification of that for $\{a_t\}$. Since $\{\chi_t\}$ is continuous $\mathcal{H}_t$-adapted (thus progressively measurable) the Fubini theorem ensures that $\{a_t\}$ is $\overline{\mathcal{B}_t(C_D)}^Q$-adapted. Now fix $0 \le t_1 < t_2 \le T$ and $A \in \overline{\mathcal{B}_{t_1}(C_D)}^Q$. From (15.39) and the Fubini theorem we have

$$E_Q[a_{t_i}; A] = E_{P \otimes Q}[\chi_{t_i}; \Omega \times A], \qquad i = 1, 2. \qquad (15.40)$$

Since $\Omega \times A \in \mathcal{H}_{t_1}$ (see (15.34)), it follows from (15.40) and (15.38) that $\{a_t\}$ is a $\overline{\mathcal{B}_t(C_D)}^Q$-martingale on $(C_D, \overline{\mathcal{B}(C_D)}^Q, Q)$. It remains to establish Q-a.s. continuity of $\{a_t\}$. To this end we first observe that $\{a_t\}$ is in fact $\overline{\mathcal{B}_t(C_D)}^Q$-previsible. Indeed, put

$$a_t^n := \int_\Omega n \wedge \chi_t(\omega, y) \, dP(\omega), \qquad (t, y) \in [0, T] \times \Omega, \qquad n = 1, 2, \ldots \qquad (15.41)$$

Then the Fubini theorem and dominated convergence theorem ensure that $\{a_t^n\}$ is a continuous and $\overline{\mathcal{B}_t(C_D)}^Q$-adapted process, hence is $\overline{\mathcal{B}_t(C_D)}^Q$-previsible. But $\lim_{n \to \infty} a_t^n(y) = a_t(y)$ for all $(t, y) \in [0, T] \times \Omega$ (by monotonic convergence), and thus $\{a_t\}$ is $\overline{\mathcal{B}_t(C_D)}^Q$-previsible. Since Q is Wiener measure on $(C_D, \mathcal{B}(C_D))$, it follows from (15.33) that there is a *continuous* $\overline{\mathcal{B}_t(C_D)}^Q$-martingale $\{a_t^*\}$ on $(C_D, \overline{\mathcal{B}(C_D)}^Q, Q)$ which is a modification of $\{a_t\}$ (see Theorem V(3.4) of Revuz and Yor [25], p.192). Denote by $\mathcal{R}$ the countable set comprising all rationals in $[0, T]$ as well as the end-point T, and put

$$N_1 := \bigcup_{r \in \mathcal{R}} \{y \in C_D : a_r(y) \neq a_r^*(y)\}. \tag{15.42}$$

Then $N_1 \in \overline{\mathcal{B}(C_D)}^Q$ and $Q(N_1) = 0$. Fix some $\overline{\mathcal{B}_t(C_D)}^Q$-stopping time $S \leq T$, and put $S_n(y) := 2^{-n}[2^n S(y) + 1] \wedge T$, $n = 1, 2, \ldots$, $y \in C_D$ (in which $[x]$ denotes the integer part of $x \in [0, \infty)$). Then S_n is a $\overline{\mathcal{B}_t(C_D)}^Q$-stopping time with values in $\mathcal{R}$ for each n, thus $a(S_n(y), y) = a^*(S_n(y), y)$ for each $y \notin N_1$ and $n = 1, 2, \ldots$ (we write $a(S_n(y), y)$ for $a_{S_n(y)}(y)$ and similarly for a^*). Now fix some $y \notin N_1$: from the preceding observation and the fact that $\{a_t^*\}$ is continuous, we have

$$a^*(S(y), y) = \lim_{n \to \infty} a^*(S_n(y), y) = \lim_{n \to \infty} a(S_n(y), y) = \liminf_{n \to \infty} \int_\Omega \chi(S_n(y), \omega, y) \, dP(\omega)$$
$$\geq \int_\Omega \liminf_{n \to \infty} \chi(S_n(y), \omega, y) \, dP(\omega) = \int_\Omega \chi(S(y), \omega, y) \, dP(\omega) = a(S(y), y),$$

(here we have used (15.39), Fatou's lemma, $\lim_{n \to \infty} S_n(y) = S(y)$, and the fact that $\{\chi_t\}$ is continuous). Since this holds for each $y \notin N_1$, and $Q(N_1) = 0$, we have

$$a_S^* \geq a_S, \qquad Q-\text{a.s.} \tag{15.43}$$

From (15.39) and the Fubini theorem we have

$$E_Q[a_S] = E_{P \otimes Q}[\chi_S]. \tag{15.44}$$

In view of Remark 15.6 we can regard S as a $\mathcal{H}_t$-stopping time, and then, from (15.38), (15.36), and the optional sampling theorem,

$$E_{P \otimes Q}[\chi_S] = E_{P \otimes Q}[\chi_0] = E_P \int_0^T |\beta_\tau|^2 \, d\tau. \tag{15.45}$$

Similarly, since $\{a_t^*\}$ is a continuous $\overline{\mathcal{B}_t(C_D)}^Q$-martingale on $(C_D, \overline{\mathcal{B}(C_D)}^Q, Q)$, it follows from the optional sampling that

$$E_Q[a_S^*] = E_Q[a_0^*] = E_Q[a_0] = E_P \int_0^T |\beta_\tau|^2 \, d\tau, \tag{15.46}$$

(the final equality following from (15.36) and (15.39)). Upon combining (15.44), (15.45) and (15.46), we obtain $E_Q[a_S^*] = E_Q[a_S]$, and this, together with (15.43), establishes that

$$a_S^* = a_S, \qquad Q-\text{a.s.} \tag{15.47}$$

Now $\{a_t\}$ has been shown to be $\overline{\mathcal{B}_t(C_D)}^Q$-previsible, and $\{a^*\}$ is of course $\overline{\mathcal{B}_t(C_D)}^Q$-previsible (being continuous adapted); since (15.47) holds for all $\overline{\mathcal{B}_t(C_D)}^Q$-stopping times S, it follows that $\{a_t\}$ and $\{a_t^*\}$ are Q-indistinguishable (see e.g. lemma VI(5.2) of Rogers and Williams [27], p. 317). This shows that $\{a_t\}$ is Q-a.s. continuous. □

Remark 15.7 The elegant proof of Q-a.s. continuity of $\{a_t\}$ given above is due to Bhatt and Karandikar ([4], p. 46).

Remark 15.8 In this remark we highlight some bounds which will be needed for the proof of Theorem 15.2. From Proposition 15.1 we immediately have

$$Q\left\{\gamma \in C_D : \sup_{0 \leq t \leq T} a_t(\gamma) < \infty\right\} = 1, \quad Q\left\{\gamma \in C_D : \sup_{0 \leq t \leq T} g_t(\gamma) < \infty\right\} = 1. \tag{15.48}$$

Moreover, from (15.23) and (15.21), we also have $Q\{g_T > 0\} = 1$, and then it follows from Proposition 15.1, together with an application of the optional-sampling theorem (see e.g. proposition II.2.15 on p. 62 of Ethier and Kurtz [8]), that

$$Q\left\{\gamma \in C_D : \inf_{0 \leq t \leq T} g_t(\gamma) > 0\right\} = 1. \tag{15.49}$$

Next define

$$m_t(\gamma) := \int_\Omega |\beta_t(\omega)|^2 \, p_t(\omega, \gamma) \, dP(\omega), \quad (t, \gamma) \in [0, T] \times \Omega. \tag{15.50}$$

Then, since β_t is $\mathcal{H}_0$-measurable (Remark 15.6), from (15.35) and the Fubini theorem we have

$$E_Q[m_t] = E_{P \otimes Q}[|\beta_t|^2 \, p_t] = E_{P \otimes Q}[|\beta_t|^2 \, E_{P \otimes Q}[p_t \mid \mathcal{H}_0]] = E_{P \otimes Q}[|\beta_t|^2] = E_P[|\beta_t|^2],$$

and hence, from Condition 15.1 and the Fubini theorem,

$$E_Q\left[\int_0^T m_t \, dt\right] = E_P\left[\int_0^T |\beta_t|^2 \, dt\right] < \infty,$$

from which we get

$$Q\left\{\gamma \in C_D : \int_0^T m_t(\gamma) \, dt < \infty\right\} = 1. \tag{15.51}$$

The Q-a.s.-boundedness properties of the processes $\{a_t\}$, $\{g_t\}$ and $\{m_t\}$, given by (15.48), (15.49), and (15.51), will be essential for establishing Theorem 15.2 in Section 15.4.

15.3 Pathwise uniqueness and strong solutions

The goals of this section are to write the defining relation (15.10) for the innovations process in the form of a stochastic differential equation for which the 'solution' is the observation process $\{Y_t\}$, and to introduce another of the essential tools we shall need to establish Theorem 15.2, namely a variant due to Clark [6] of a fundamental theorem on stochastic differential equations of Yamada and Watanabe [32].

Our first task is to make sure that we can use $\{\gamma_t\}$ at (15.24) to get a nonanticipating path functional on C_D. To this end, we require the following result of Meyer ([24], pp. 241–3):

$$\{\gamma_t\} \text{ is an } \mathbb{R}^D\text{-valued } \overline{\mathcal{B}_t(C_D)}^{\,Q}\text{-previsible process on } (C_D, \overline{\mathcal{B}(C_D)}^{\,Q}, Q; \{\overline{\mathcal{B}_t(C_D)}^{\,Q}\}),$$
(15.52)

(recall Remark 15.6). We would like to write the relation (15.10) in the form of a stochastic differential equation with unit covariance and drift function given by $\{\gamma_t\}$. As the discussion in Rogers and Williams ([27], pp. 122–3) makes clear, the drift and covariance functions of a stochastic differential equation must be $\mathcal{B}_t(C_D)$-previsible (note the *unaugmented* filtration!), and we see from (15.52) that $\{\gamma_t\}$ does not quite have this property, being only $\overline{\mathcal{B}_t(C_D)}^{\,Q}$-previsible. Fortunately it is easy to 'slightly adjust' $\{\gamma_t\}$ to get a drift function $\{\gamma_t^0\}$ with the desired previsibility. In fact, from (15.52), Remark 15.6, and Rogers and Williams ([27], V(10.12), p.128), we have the following: there exists an $\mathbb{R}^D$-valued and $\mathcal{B}_t(C_D)$-previsible process $\{\gamma_t^0\}$ on $(C_D, \overline{\mathcal{B}(C_D)}^{\,Q}, Q)$ which is Q-indistinguishable from $\{\gamma_t\}$, that is

$$Q\{y \in C_D : \gamma_t^0(y) = \gamma_t(y), \ t \in [0, T]\} = 1.$$
(15.53)

Now P and P_0 are equivalent probabilities on $\mathcal{F}$ (by Remark 15.3), thus PY^{-1} and $P_0 Y^{-1}$ are equivalent probability measures on $\mathcal{B}(C_D)$; but $P_0 Y^{-1} = Q$ on $\mathcal{B}(C_D)$ (by (15.20)), thus

$$PY^{-1} \text{ and } Q \text{ are equivalent probability measures on } \mathcal{B}(C_D).$$
(15.54)

From (15.54), (15.53), and (15.9), we obtain

$$\hat{\beta}_t = \gamma_t^0(Y), \quad \text{a.s.} \quad \text{for each } t \in [0, T],$$
(15.55)

and then, from (15.55), (15.9), and Jensen's inequality, we have $E_P\left[|\gamma_t^0(Y)|^2\right] \le E_P[|\hat{\beta}_t|^2]$ for each $t \in [0, T]$. From this, together with (15.1) and the Fubini theorem, we find

$$\int_0^T |\gamma_t^0(Y)|^2 \, dt < \infty, \quad \text{a.s.},$$
(15.56)

$$Y_t = I_t + \int_0^t \gamma_\tau^0(Y) \, d\tau, \quad \text{a.s.} \quad \text{for each} \quad t \in [0, T],$$
(15.57)

(where we have used (15.55) and (15.10) at (15.57)). The relation (15.57) certainly has the form of a stochastic differential equation with unit covariance and drift function $\{\gamma_t^0\}$, in which the innovation $\{I_t\}$ is the 'input' or 'driving' process, and the observation $\{Y_t\}$ is the 'solution' process. To make this precise we next formulate exactly what is meant by a 'solution' of this stochastic differential equation, and also introduce the essential idea of a *strong* solution:

Definition 15.1 *A pair* $\{(\bar{\Omega}, \bar{\mathcal{F}}, \bar{P}; \{\bar{\mathcal{F}}_t\}), (\bar{Y}_t, \bar{I}_t)\}$ *is a solution of the stochastic differential equation with drift function* $\{\gamma_t^0\}$ *and unit covariance when* $(\bar{\Omega}, \bar{\mathcal{F}}, \bar{P}; \{\bar{\mathcal{F}}_t\})$ *is a filtered probability space satisfying the usual conditions,* $\{\bar{Y}_t\}$ *is an* $\mathbb{R}^D$-*valued, continuous and* $\bar{\mathcal{F}}_t$-*adapted process, and* $\{\bar{I}_t\}$ *is an* $\mathbb{R}^D$-*valued standard* $\bar{\mathcal{F}}_t$-*Wiener process on* $(\bar{\Omega}, \bar{\mathcal{F}}, \bar{P})$, *such that*

$$\int_0^T |\gamma_t^0(\bar{Y})| \, dt < \infty, \quad \bar{P}-a.s., \tag{15.58}$$

$$\bar{Y}_t = \bar{I}_t + \int_0^t \gamma_\tau^0(\bar{Y}) \, d\tau, \quad \bar{P}-a.s. \text{ for each } t \in [0, T]. \tag{15.59}$$

Furthermore, a solution $\{(\bar{\Omega}, \bar{\mathcal{F}}, \bar{P}; \{\bar{\mathcal{F}}_t\}), (\bar{Y}_t, \bar{I}_t)\}$ *is said to be a* strong *solution when* $\{\bar{Y}_t\}$ *is adapted to the filtration* $\{\bar{\mathcal{I}}_t\}$ *defined by* $\bar{\mathcal{I}}_t := \mathcal{F}_t^{\bar{I}} \vee \mathbf{Z}^{\bar{P}}[\bar{\mathcal{F}}]$, *or equivalently, when* $\bar{\mathcal{Y}}_t \subset \bar{\mathcal{I}}_t$, $t \in [0, T]$, *for* $\bar{\mathcal{Y}}_t := \mathcal{F}_t^{\bar{Y}} \vee \mathbf{Z}^{\bar{P}}[\bar{\mathcal{F}}]$ *(as usual we use* $\{\mathcal{F}_t^{\bar{I}}\}$ *and* $\{\mathcal{F}_t^{\bar{Y}}\}$ *to denote the raw filtrations of the processes* $\{\bar{I}_t\}$ *and* $\{\bar{Y}_t\}$ *respectively—see Notation 15.1(IV)).*

Remark 15.9 Since $\{\gamma_t^0\}$ is $\mathcal{B}_t(C_D)$-previsible, the process $\{\gamma_t^0(\bar{Y})\}$ is $\bar{\mathcal{F}}_t$-previsible, hence, in particular, the integral on the right of (15.59) is $\bar{\mathcal{F}}_t$-measurable, as required for measurability of the two sides of (15.59) to 'match up' (see Rogers and Williams [27], V(8.5)–(8.6), p. 123),

Remark 15.10 Observe that $\{(\Omega, \mathcal{F}, P; \{\mathcal{Y}_t\}), (Y_t, I_t)\}$ (for $\{Y_t\}$ at (15.3), $\mathcal{Y}_t$ defined at (15.8), and $\{I_t\}$ at (15.10)) is a solution of the stochastic differential equation with drift function $\{\gamma_t^0\}$ and unit covariance. Indeed, from Remark 15.3 we have $\mathbf{Z}^P[\mathcal{F}] = \mathbf{Z}^{P_0}[\mathcal{F}]$ (recall Notation 15.1(V)), thus, from (15.8), one has $\mathcal{Y}_t = \mathcal{F}_t^Y \vee \mathbf{Z}^{P_0}[\mathcal{F}]$. Then, since $\{Y_t\}$ is a standard Brownian motion on $(\Omega, \mathcal{F}, P_0)$ (recall (15.20)), it follows that $\mathcal{Y}_t = \mathcal{Y}_{t+}$ (see e.g. Revuz and Yor [25], proposition III(2.10), p. 87), so the filtered probability space $\{(\Omega, \mathcal{F}, P; \{\mathcal{Y}_t\})\}$ satisfies the usual conditions. Moreover $\{I_t\}$ is a $\mathcal{Y}_t$-Wiener process on $(\Omega, \mathcal{F}, P)$ (by Theorem 15.1). Now it follows from (15.56) and (15.57) that $\{(\Omega, \mathcal{F}, P; \{\mathcal{Y}_t\}), (Y_t, I_t)\}$ is a solution as required.

It remains to demonstrate that $\{(\Omega, \mathcal{F}, P; \{\mathcal{Y}_t\}), (Y_t, I_t)\}$ is a *strong solution*, for then, from Definition 15.1, we have that $\mathcal{Y}_t \subset \mathcal{I}_t$, as required to establish Theorem 15.2. To this end, we shall use a variant due to Clark [6] of a basic theorem of Yamada and Watanabe [32], which provides a very useful criterion for determining when a given solution of a stochastic differential equation is actually a strong solution. Although this criterion pertains to any designated solution of a completely general stochastic differential equation with nonanticipating functional coefficients (see Clark [6], corollary, p. 157), we intend to state only a rather special case of this result, which is particular to the solution $\{(\Omega, \mathcal{F}, P; \{\mathcal{Y}_t\}), (Y_t, I_t)\}$ of the stochastic differential equation with drift function $\{\gamma_t^0\}$ and unit covariance (see Remark 15.10). To this end, it is useful to have the following restricted notion of the property of pathwise-uniqueness:

Definition 15.2 *The stochastic differential equation with drift $\{\gamma_t^0\}$ and unit covariance has the property of pathwise-uniqueness within the class of solutions having the same law as the solution $\{(\Omega, \mathcal{F}, P; \{\mathcal{Y}_t\}), (Y_t, I_t)\}$ when the following holds: given any pair of solutions $\{(\bar{\Omega}, \bar{\mathcal{F}}, \bar{P}; \{\bar{\mathcal{F}}_t\}), (\bar{Y}_t^i, \bar{I}_t)\}$, $i = 1, 2$ (recall Definition 15.1), each having joint law identical to that of the solution $\{(\Omega, \mathcal{F}, P; \{\mathcal{Y}_t\}), (Y_t, I_t)\}$ (that is $P(I, Y)^{-1} = \bar{P}(\bar{I}, \bar{Y}^1)^{-1} = \bar{P}(\bar{I}, \bar{Y}^2)^{-1}$), it necessarily follows that $\{\bar{Y}_t^1\}$ and $\{\bar{Y}_t^2\}$ are $\bar{P}$-indistinguishable.*

The special case of the criterion of Clark [6] that we shall need can now be stated as follows:

Theorem 15.4 *Suppose that the stochastic differential equation with drift $\{\gamma_t^0\}$ and unit covariance has the property of pathwise uniqueness within the class of solutions having the same law as the solution $\{(\Omega, \mathcal{F}, P; \{\mathcal{Y}_t\}), (Y_t, I_t)\}$. Then the latter is a strong solution, that is $\mathcal{Y}_t \subset \mathcal{I}_t$ for each $t \in [0, T]$.*

It is therefore necessary only to verify that the stochastic differential equation with drift function $\{\gamma_t^0\}$ and unit covariance has the property of pathwise uniqueness formulated at Definition 15.2 in order to establish Theorem 15.2. This will be shown in Section 15.4.

Remark 15.11 Some remarks on the relationship between the classical theorem of Yamada and Watanabe [32] and Clark's modification of this theorem are in order. The theorem of Yamada and Watanabe asserts (among other things) that if a stochastic differential equation has the property of pathwise-uniqueness (in the sense that any pair of putative solutions on a common filtered probability space with common initial value and common 'driving' Wiener process are necessarily indistinguishable) then *each and every* solution of the stochastic differential equation is a strong solution (see e.g. definition IX(15.3)(1) and theorem IX(15.7)(ii) of [25] for a very clear rendition of this result). In contrast, Clark's modification [6] of this theorem postulates a weaker and less universal notion of pathwise uniqueness, in which one compares only putative solutions *having the same joint law as some designated solution*, and in return gives the weaker conclusion that just the designated solution is strong. For many applications this conclusion is all that is wanted (the present one being a case in point), and the weaker hypothesis is often much easier to verify, since pathwise-uniqueness need be established only among solutions with the same law as the solution whose strength must be demonstrated, rather than among all postulated solutions. It is also worth noting that the more universal notion of pathwise-uniqueness that is postulated for the Yamada–Watanabe theorem is in fact strong enough to ensure that the stochastic differential equation actually has the property of *uniqueness-in-law* (see definition IX(15.3)(2) and theorem IX(15.7)(i) of [25]), whereas the weaker notion of pathwise-uniqueness that is appropriate to Clark's version of this result does not of course guarantee that uniqueness-in-law holds. Again, as the innovations problem illustrates, there are genuine applications where one does not need to establish uniqueness-in-law, and it is only the strength of a designated solution which is of interest.

15.4 Proof of Theorem 15.2

In accordance with Theorem 15.4 we fix solutions $\{(\bar{\Omega}, \bar{\mathcal{F}}, \bar{P}; \{\bar{\mathcal{F}}_t\}), (\bar{Y}^i_t, \bar{I}_t)\}$, $i = 1, 2$, of the stochastic differential equation with drift function $\{\gamma^0_t\}$ and unit covariance, such that

$$P(I, Y)^{-1} = \bar{P}(\bar{I}, \bar{Y}^1)^{-1} = \bar{P}(\bar{I}, \bar{Y}^2)^{-1}, \tag{15.60}$$

in which $\{(\Omega, \mathcal{F}, P; \{\mathcal{Y}_t\}), (Y_t, I_t)\}$ is the solution of Remark 15.10. It remains to establish

$$\bar{P}\{\bar{Y}^1_t = \bar{Y}^2_t, \quad t \in [0, T]\} = 1, \tag{15.61}$$

for then we have pathwise uniqueness in the sense of Definition 15.2, and Theorem 15.2 follows from Theorem 15.4. From (15.60) and (15.54) we have

$\bar{P}(\bar{Y}^i)^{-1}$ and Q are equivalent probability measures on $\mathcal{B}(C_D)$ for $i = 1, 2$.

$$\tag{15.62}$$

Then, from (15.62) together with (15.48), (15.49), (15.51), and (15.53), we get

$$\bar{P}\{\bar{\omega} \in \bar{\Omega} : \sup_{0 \le t \le T} a_t(\bar{Y}^i(\bar{\omega})) < \infty\} = 1, \quad i = 1, 2, \tag{15.63}$$

$$\bar{P}\{\bar{\omega} \in \bar{\Omega} : \sup_{0 \le t \le T} g_t(\bar{Y}^i(\bar{\omega})) < \infty\} = 1, \quad i = 1, 2, \tag{15.64}$$

$$\bar{P}\{\bar{\omega} \in \bar{\Omega} : \inf_{0 \le t \le T} g_t(\bar{Y}^i(\bar{\omega})) > 0\} = 1, \quad i = 1, 2, \tag{15.65}$$

$$\bar{P}\{\bar{\omega} \in \bar{\Omega} : \int_0^T m_t(\bar{Y}^i(\bar{\omega})) \, dt < \infty\} = 1, \quad i = 1, 2, \tag{15.66}$$

$$\bar{P}\{\bar{\omega} \in \bar{\Omega} : \gamma_t(\bar{Y}^i) = \gamma_t^0(\bar{Y}^i), \ t \in [0, T]\} = 1, \qquad i = 1, 2. \tag{15.67}$$

Moreover, from (15.60) and (15.56),

$$\bar{P}\{\bar{\omega} \in \bar{\Omega} : \int_0^T |\gamma_t^0(\bar{Y}^i(\bar{\omega}))|^2 \, dt < \infty\} = 1, \qquad i = 1, 2. \tag{15.68}$$

Put

$$L_t := \gamma_t^0(\bar{Y}^1) - \gamma_t^0(\bar{Y}^2), \qquad t \in [0, T]. \tag{15.69}$$

Then, since (15.59) holds when $\bar{Y}$ is replaced with $\bar{Y}^i$, we have

$$\bar{Y}_t^1 - \bar{Y}_t^2 = \int_0^t L_\tau \, d\tau, \qquad t \in [0, T]. \tag{15.70}$$

Remark 15.12 From now on fix some $\bar{\omega}$ in the intersection of the events of unit $\bar{P}$-measure at (15.63)–(15.68). A nonprobabilistic computation will establish $\bar{Y}_t^1(\bar{\omega}) = \bar{Y}_t^2(\bar{\omega})$, $t \in [0, T]$, for this choice of $\bar{\omega}$, which then gives (15.61) (to lighten the notation $\bar{\omega}$ will not be specifically indicated when this causes no confusion).

In view of (15.69), (15.67), and (15.24), we have

$$L_t = \frac{f_t(\bar{Y}^1) - f_t(\bar{Y}^2)}{g_t(\bar{Y}^1)} + f_t(\bar{Y}^2)\left(\frac{1}{g_t(\bar{Y}^1)} - \frac{1}{g_t(\bar{Y}^2)}\right), \qquad t \in [0, T],$$

thus

$$|L_t| \leq \frac{|f_t(\bar{Y}^1) - f_t(\bar{Y}^2)|}{g_t(\bar{Y}^1)} + \frac{|f_t(\bar{Y}^2)| |g_t(\bar{Y}^2) - g_t(\bar{Y}^1)|}{g_t(\bar{Y}^1)g_t(\bar{Y}^2)}, \qquad t \in [0, T]. \tag{15.71}$$

From (15.71) and (15.65) there is some $K_1 \in [0, \infty)$ (depending on $\bar{\omega}$) such that

$$|L_t| \leq K_1 \left(|f_t(\bar{Y}^1) - f_t(\bar{Y}^2)| + |f_t(\bar{Y}^2)| |g_t(\bar{Y}^2) - g_t(\bar{Y}^1)|\right), \qquad t \in [0, T]. \tag{15.72}$$

We next establish upper bounds for the quantities on the right of (15.72). From (15.22)

$$|f_t(\bar{Y}^1(\bar{\omega})) - f_t(\bar{Y}^2(\bar{\omega}))| \leq \int_\Omega |\beta_t(\omega)| |\rho_t(\omega, \bar{Y}^1(\bar{\omega})) - \rho_t(\omega, \bar{Y}^2(\bar{\omega}))| \, dP(\omega). \tag{15.73}$$

From the inequality $|e^x - e^y| \leq (1/2) |x - y| (e^x + e^y)$, $x, y \in \mathbb{R}$, with (15.70) and (15.21),

$$\left|\rho_t(\omega, \bar{Y}^1(\bar{\omega})) - \rho_t(\omega, \bar{Y}^2(\bar{\omega}))\right| \qquad (15.74)$$

$$\leq \frac{1}{2}[\rho_t(\omega, \bar{Y}^1(\bar{\omega})) + \rho_t(\omega, \bar{Y}^2(\bar{\omega}))]\left|\int_0^t \beta_\tau(\omega) L_\tau(\bar{\omega})\,d\tau\right|$$

$$\leq \frac{1}{2}[\rho_t(\omega, \bar{Y}^1(\bar{\omega})) + \rho_t(\omega, \bar{Y}^2(\bar{\omega}))]\left(\int_0^t |\beta_\tau(\omega)|^2\,d\tau\right)^{1/2}\left(\int_0^t |L_\tau(\bar{\omega})|^2\,d\tau\right)^{1/2}.$$

Now put

$$\psi_t^1(\bar{\omega}) := \frac{1}{4}\left[\int_\Omega |\beta_t(\omega)|\left(\int_0^t |\beta_\tau(\omega)|^2\,d\tau\right)^{1/2}[\rho_t(\omega, \bar{Y}^1(\bar{\omega})) + \rho_t(\omega, \bar{Y}^2(\bar{\omega}))]\,dP(\omega)\right]^2.$$
(15.75)

Upon combining (15.73), (15.74) and (15.75), we get

$$\left|f_t(\bar{Y}^1(\bar{\omega})) - f_t(\bar{Y}^2(\bar{\omega}))\right|^2 \leq \psi_t^1(\bar{\omega})\left(\int_0^t |L_\tau(\bar{\omega})|^2\,d\tau\right), \qquad t \in [0, T]. \quad (15.76)$$

Similarly, from (15.74) and (15.23), we obtain

$$\left|g_t(\bar{Y}^1(\bar{\omega})) - g_t(\bar{Y}^2(\bar{\omega}))\right|^2 \leq \psi_t^2(\bar{\omega})\left(\int_0^t |L_\tau(\bar{\omega})|^2\,d\tau\right), \qquad t \in [0, T], \quad (15.77)$$

in which

$$\psi_t^2(\bar{\omega}) := \frac{1}{4}\left[\int_\Omega \left(\int_0^t |\beta_\tau(\omega)|^2\,d\tau\right)^{1/2}[\rho_t(\omega, \bar{Y}^1(\bar{\omega})) + \rho_t(\omega, \bar{Y}^2(\bar{\omega}))]\,dP(\omega)\right]^2.$$
(15.78)

Now put

$$\psi_t(\bar{\omega}) := \psi_t^1(\bar{\omega}) + \left|f_t(\bar{Y}^2(\bar{\omega}))\right|^2 \psi_t^2(\bar{\omega}), \qquad t \in [0, T], \quad (15.79)$$

and combine (15.79), (15.77), (15.76) and (15.72) to get

$$|L_t(\bar{\omega})|^2 \leq 2\,|K_1(\bar{\omega})|^2\,\psi_t(\bar{\omega})\left(\int_0^t |L_\tau(\bar{\omega})|^2\,d\tau\right), \qquad t \in [0, T]. \quad (15.80)$$

Now define

$$\eta_t(\bar{\omega}) := \int_0^t |L_\tau(\bar{\omega})|^2\,d\tau, \qquad t \in [0, T], \quad (15.81)$$

and combine (15.81) and (15.80) to obtain

$$\eta_t(\bar{\omega}) \leq 2\,|K_1(\bar{\omega})|^2 \int_0^t \psi_\tau(\bar{\omega})\eta_\tau(\bar{\omega})\,d\tau, \qquad t \in [0, T]. \quad (15.82)$$

Since $\bar{\omega}$ is a member of the sets of $\bar{P}$-unit measure at (15.68), we see from (15.69) that $\int_0^T |L_\tau(\bar{\omega})|^2\,d\tau < \infty$, and thus the function $t \to \eta_t(\bar{\omega})$ is uniformly bounded (in fact absolutely continuous) over the interval $t \in [0, T]$. We will shortly establish that

$$\int_0^T \psi_t(\bar{\omega}) \, dt < \infty. \tag{15.83}$$

It then follows from (15.83), (15.82), the uniform-boundedness of $\eta(\bar{\omega})$ on $[0, T]$, and the Gronwall inequality (see e.g. [8], theorem 5.1, p. 498), that $\eta_t(\bar{\omega}) = 0$, $t \in [0, T]$, which, in view of (15.81) and (15.70), gives $\bar{Y}_t^1(\omega) = \bar{Y}_t^2(\bar{\omega})$, $t \in [0, T]$. As noted at Remark 15.12, this in turn gives (15.61), as required to establish Theorem 15.2.

It remains to show (15.83). From (15.75), the Cauchy–Schwarz inequality, (15.50), (15.39), and (15.36),

$$\psi_t^1(\bar{\omega}) \le \frac{1}{4} \left[\int_\Omega |\beta_t(\omega)|^2 [\rho_t(\omega, \bar{Y}^1(\bar{\omega})) + \rho_t(\omega, \bar{Y}^2(\bar{\omega}))] \, dP(\omega) \right] \tag{15.84}$$
$$\left[\int_\Omega \left(\int_0^t |\beta_\tau(\omega)|^2 \, d\tau \right) [\rho_t(\omega, \bar{Y}^1(\bar{\omega})) + \rho_t(\omega, \bar{Y}^2(\bar{\omega}))] \, dP(\omega) \right]$$
$$\le \frac{1}{4} \left\{ m_t(\bar{Y}^1(\bar{\omega})) + m_t(\bar{Y}^2(\bar{\omega})) \right\} \left\{ a_t(\bar{Y}^1(\bar{\omega})) + a_t(\bar{Y}^2(\bar{\omega})) \right\}, \quad t \in [0, T].$$

Since $\bar{\omega}$ is a member of the events of unit $\bar{P}$-measure at (15.66) and (15.63), it follows from (15.84) that

$$\int_0^T \psi_t^1(\bar{\omega}) \, dt < \infty. \tag{15.85}$$

As for the second term on the right side of (15.79) put

$$\psi_t^3(\bar{\omega}) := \left| f_t(\bar{Y}^2(\bar{\omega})) \right|^2 \psi_t^2(\bar{\omega}), \quad t \in [0, T]. \tag{15.86}$$

Then, from (15.86), (15.78), (15.22), and the Cauchy–Schwarz inequality,

$$\psi_t^3(\bar{\omega}) \le \frac{1}{4} \left[\int_\Omega |\beta_t(\omega)|^2 \rho_t(\omega, \bar{Y}^2(\bar{\omega})) \, dP(\omega) \right] \left[\int_\Omega \rho_t(\omega, \bar{Y}^2(\bar{\omega})) \, dP(\omega) \right] \tag{15.87}$$
$$\left[\int_\Omega \left(\int_0^t |\beta_\tau(\omega)|^2 \, d\tau \right) [\rho_t(\omega, \bar{Y}^1(\bar{\omega})) + \rho_t(\omega, \bar{Y}^2(\bar{\omega}))] \, dP(\omega) \right]$$
$$\left[\int_\Omega [\rho_t(\omega, \bar{Y}^1(\bar{\omega})) + \rho_t(\omega, \bar{Y}^2(\bar{\omega}))] \, dP(\omega) \right]$$
$$\le \frac{1}{4} m_t(\bar{Y}^2(\bar{\omega})) g_t(\bar{Y}^2(\bar{\omega})) \left\{ a_t(\bar{Y}^1(\bar{\omega})) + a_t(\bar{Y}^2(\bar{\omega})) \right\}$$
$$\left\{ g_t(\bar{Y}^1(\bar{\omega})) + g_t(\bar{Y}^2(\bar{\omega})) \right\}, \quad t \in [0, T],$$

where we have used (15.50), (15.39), and (15.23) at the final inequality. Now, since $\bar{\omega}$ is in the events of unit $\bar{P}$-measure at (15.66), (15.64), (15.63), we get

$$\int_0^T \psi_t^3(\bar{\omega}) \, dt < \infty, \tag{15.88}$$

and (15.83) follows from (15.88), (15.86), (15.85), and (15.79).

15.5 The correlated case: results of N. V. Krylov

In this section we look at the innovations problem when full independence of $\{\beta_t\}$ and $\{W_t\}$ in (15.3) fails (that is, we discard hypthesis (ii) of Theorem 15.2), but the partial independence (15.2) holds. The lack of full independence makes the innovations problem substantially more challenging, and to make any headway it seems necessary to postulate a concrete model for the correlation between $\{\beta_t\}$ and $\{W_t\}$. The most interesting results of this kind are due to Krylov [18], who establishes the innovations conjecture for a model of considerable generality; our goal in the remainder of this chapter is to convey some idea of the basic approach of [18], which is simple and elegant, while sidestepping most of the attendant technicalities, which are quite extensive. To this end we focus on the following simple model

$$dX_t = a(X_t)\,dt + b(X_t)\,dW_t + c(X_t)\,d\tilde{W}_t, \quad X_0 = \xi_0, \tag{15.89}$$

$$dY_t = h(X_t)\,dt + dW_t, \quad Y_0 = 0, \tag{15.90}$$

which is an obvious 'nonlinearization' of the model (15.6)–(15.7). This is much less general than the model of Krylov (see (1) of [18], p. 773), but is nevertheless good enough for our purposes. Notice that (15.90) is just the observation equation (15.3) with $\beta_t := h(X_t)$. The conditions which pertain to the model (15.89)–(15.90) are as follows:

Condition 15.2 (i) $\{(W_t, \tilde{W}_t)\}$ is an $\mathbb{R}^{D+q}$-valued standard Brownian motion on the complete probability space $(\Omega, \mathcal{F}, P)$; (ii) The mappings $a : \mathbb{R}^d \to \mathbb{R}^d$, $b : \mathbb{R}^d \to \mathbb{R}^{d \times D}$, $c : \mathbb{R}^d \to \mathbb{R}^{d \times q}$ and $h : \mathbb{R}^d \to \mathbb{R}^D$, are uniformly bounded on $\mathbb{R}^d$, a and h are twice continuously differentiable with all derivatives uniformly bounded, while b and c are third-order continuously differentiable with all derivatives uniformly bounded; (iii) the least eigenvalue of $cc'(x)$ is lower-bounded by some $\mu > 0$ uniformly in $x \in \mathbb{R}^d$; (iv) the scalar entries of the $\mathbb{R}^D$-valued 'sensor function' h are members of $L^2(\mathbb{R}^d)$; (v) the law of the initial state ξ_0 has a continuous density π_0 on $\mathbb{R}^d$, and $\pi_0 \in W_2^2(\mathbb{R}^d) \cap W_p^3(\mathbb{R}^d)$ for some integer $p > d$.

Remark 15.13 As usual, for $p \in [1, \infty)$ write $L^p(\mathbb{R}^d)$ for the space of all Borel-measurable mappings $u : \mathbb{R}^d \to \mathbb{R}$ with $\|u\|_{L^p}^p := \int_{\mathbb{R}^d} |u(x)|^p\,dx < \infty$. For $v \in L^2(\mathbb{R}^d)$ define $(v, h) \in \mathbb{R}^D$ by $(v, h) := \int_{\mathbb{R}^d} h(x)v(x)\,dx$. For $p \in [1, \infty)$ and positive integers k, write $W_p^k(\mathbb{R}^d)$ for the Sobolev space of all $u \in L^p(\mathbb{R}^d)$ with weak derivatives $D^\alpha u \in L^p(\mathbb{R}^d)$ for each multi-index $|\alpha| \le k$, and with the norm $\|u\|_{W_p^k} := \Sigma_{|\alpha| \le k} \|D^\alpha u\|_{L^p}$.

Remark 15.14 With Condition 15.2 in force, the $\mathcal{Y}_t$-conditional distribution of X_t has a density $\pi_t(x, \omega)$, $(t, x, \omega) \in [0, T] \times \mathbb{R}^d \times \Omega$, with the following properties: (i) $\pi_t(\cdot, \omega) \in W_2^2(\mathbb{R}^d)$ for each (t, ω), (ii) $\omega \to \pi_t(x, \omega)$ is $\mathcal{Y}_t$-measurable for each (t, x); (iii) $(t, x) \to \pi_t(x, \omega)$ is continuous for each ω; (iv) $x \to \pi_t(x, \omega)$ is second-order differentiable for each (t, ω), and the first and second x-derivatives are continuous in (t, x) for each ω; (v) $t \to \pi_t(\cdot, \omega) : [0, T] \to W_2^2(\mathbb{R}^d)$ is $\|\cdot\|_{W_2^2}$-continuous for each ω; (vi) $E_P \left[\sup_{t \in [0,T]} \|\pi_t\|_{W_2^2}^2 \right] < \infty$; (vii) $\{\pi_t\}$ satisfies the stochastic partial differential equation

$$\pi_t(x) = \pi_0(x) + \int_0^t (L\pi_\tau)(x) \, d\tau \qquad (15.91)$$
$$+ \int_0^t [(M\pi_\tau)(x) - (\pi_\tau, h)\pi_\tau(x)] \, dI_\tau, \quad (t, x) \in [0, T] \times \mathbb{R}^d.$$

In (15.91) the operator M is defined by $(M^j u)(x) := h^j(x)u(x) - \sum_{i=1}^d D_{x_i}(b^{i,j}u)(x)$, $x \in \mathbb{R}^d$, $j = 1 - D$, and L is the usual formal adjoint of the second-order linear differential operator for the diffusion (15.89). The relation (15.91) is effectively the normalized filter equation (see Theorem 4.1 of [9]), but written as a forward equation in "density form", and originates with Stratonovich [28] and Kushner [21]. The existence of the density process $\{\pi_t\}$ and the properties (i)–(vii) indicated previously, have been established by Krylov and Rozovskiĭ ([19], theorem 1.2, pp. 341–2).

Remark 15.15 Denote by $\mathcal{L}$ the set of all mappings $\psi : [0, T] \times \mathbb{R}^d \times \Omega \to \mathbb{R}$ with the following properties: (i) $\omega \to \psi_t(x, \omega)$ is $\mathcal{I}_t$-measurable for each (t, x); (ii) $(t, x) \to \psi_t(x, \omega)$ is continuous for each ω; (iii) $\psi_t(\cdot, \omega) \in L^2(\mathbb{R}^d)$ for each (t, ω); (iv) $E_P \left\{ \sup_{t \leq T} |\psi_t|_{L^2}^2 \right\} < \infty$. Furthermore, let $\chi(t)$, $t \in [0, \infty)$, be a smooth nonincreasing function with $\chi(t) := 1$ for all $0 \leq t \leq 1$ and $\chi(t) = 0$ for all $t \in [2, \infty)$, and for each $m = 1, 2, \ldots$, put $\chi_m(t) := \chi(t/m^2)$. Next, fix some $\mathbb{R}^d$-valued Gaussian random vector $\tilde{w}$ on some $(\tilde{\Omega}, \tilde{\mathcal{F}}, \tilde{P})$, with zero mean and unit covariance, and for each $n = 1, 2, \ldots$ and $f \in L^2(\mathbb{R}^d)$, define the mollifier $\Gamma_n f(x) := \tilde{E} f(x + n^{-1}\tilde{w})$, $x \in \mathbb{R}^d$. Finally, motivated by (15.91), for each $m, n = 1, 2, \ldots$ and $\psi \in \mathcal{L}$ define

$$(\mathbb{A}_{m,n}\psi)_t := \pi_0 + \int_0^t \chi_m \left(\|\psi_\tau\|_{L^2}^2 \right) \Gamma_n(L(\Gamma_n\psi_\tau)) \, d\tau \qquad (15.92)$$
$$+ \int_0^t \chi_m \left(\|\psi_\tau\|_{L^2}^2 \right) [\Gamma_n(M(\Gamma_n\psi_\tau)) - (\psi_\tau, h)\Gamma_n\psi_\tau] \, dI_\tau.$$

The essential technical result needed to obtain the innovations conjecture is the following:

Lemma 15.1 *Suppose Condition 15.2. With reference to Remark 15.15 we have the following: for each $m, n = 1, 2, \ldots$ the operator $\mathbb{A}_{m,n}$ maps $\mathcal{L}$ into itself, and* there exists a unique $\pi^{m,n} \in \mathcal{L}$ such that $\pi^{m,n} = \mathbb{A}_{m,n}\pi^{m,n}$. Morever, $t \to \pi_t^{m,n}(\cdot, \omega) : [0, T] \to L^2(\mathbb{R}^d)$ is $\|\cdot\|_{L^2}$-continuous for each ω.

We make no attempt to prove Lemma 15.1, and just note that it is basically established by the method of Picard iterations: one defines a sequence $\{\psi_t^k\}$ according to $\psi^{k+1} := \mathbb{A}_{m,n}\psi^k$, with $\psi_t^0 := \pi_0$, $t \in [0, T]$, and then shows that the sequence converges (in mean square) to a limit $\pi^{m,n} \in \mathcal{L}$ with the stated properties (see [18], section 2, pp. 779–80). For each $m, n = 1, 2, \ldots$, put $T_m := T \wedge \inf\{t \in [0, T] : \|\pi_t\|_{W_2^2} \geq m\}$ and

$$Y_t^{m,n} := I_t + \int_0^t \chi_m\left(\|\pi_\tau^{m,n}\|_{L^2}^2\right) (\pi_\tau^{m,n}, h) \, d\tau, \qquad \rho_t^{m,n} := \pi_{t \wedge T_m} - \pi_{t \wedge T_m}^{m,n}. \tag{15.93}$$

Then, from Remark 15.14, it follows that T_m is a $\mathcal{Y}_t$-stopping time, and, from the definitions of $\chi_m(\cdot)$ (see Remark 15.15) and T_m, we have

$$I[0, T_m](t)\chi_m\left(\|\pi_{t \wedge T_m}\|_{L^2}^2\right) = I[0, T_m](t), \qquad t \in [0, T]. \tag{15.94}$$

From Remark 15.14 and the fact that $\beta_t := h(X_t)$ (recall (15.90)), we have $\hat{\beta}_t = (\pi_t, h)$ (for $\hat{\beta}_t$ given by (15.9)); using this in (15.10), together with (15.94), we obtain

$$Y_{t \wedge T_m} = I_{t \wedge T_m} + \int_0^t I[0, T_m](\tau)\chi_m\left(\|\pi_\tau\|_{L^2}^2\right) (\pi_\tau, h) \, d\tau. \tag{15.95}$$

From Condition 15.2 one sees that the mapping $f \to \chi_m\left(\|f\|_{L^2}^2\right)(f, h) : L^2(\mathbb{R}^d) \to \mathbb{R}^D$ is globally Lipschitz continuous, and then it follows easily from (15.95) and (15.93) that, for each $m = 1, 2, \ldots$, there is a constant $c_m \in [0, \infty)$ (depending on m) such that

$$\mathrm{E}_P\left[|Y_{t \wedge T_m} - Y_{t \wedge T_m}^{m,n}|^2\right] \leq c_m \mathrm{E}_P \int_0^t \|\rho_\tau^{m,n}\|_{L^2}^2 \, d\tau, \qquad t \in [0, T]. \tag{15.96}$$

Since $\Gamma_n f \to f$ in $L^2(\mathbb{R}^d)$ for each $f \in L^2(\mathbb{R}^d)$, upon comparing $\pi_{t \wedge T_m}$ given by (15.91) with $\pi_{t \wedge T_m}^{m,n}$ given by Lemma 15.1 and (15.92), it seems plausible that, as $n \to \infty$ (with m fixed), so $\pi_{t \wedge T_m}^{m,n}$ should tend towards $\pi_{t \wedge T_m}$, which further suggests that the quantity on the right of (15.96) should tend to zero. Indeed, this plausible reasoning can be made precise (although with some technical effort): using Itô's formula, together with (15.91) and the expression for $\pi^{m,n}$ given by (15.92) and Lemma 15.1, one can expand $|\rho_t^{m,n}(x, \omega)|^2$, then integrate over $x \in \mathbb{R}^d$ to get a bound on $\|\rho_t^{m,n}(\cdot, \omega)\|_{L^2}^2$, and then take expectations to obtain an integral inequality over $t \in [0, T]$. This integral inequality can then be used to show that the right side of (15.96) converges to zero as $n \to \infty$, for each $m = 1, 2, \ldots$ and $t \in [0, T]$ (see [18], pp. 776–9). Now fix some $t \in [0, T)$. From

Remark 15.14(v) and the definition of T_m, we have $\lim_{m\to\infty} I[0, T_m](t) = 1$, and then, from the L^2-convergence of $Y^{m,n}_{t\wedge T_m}$ to $Y_{t\wedge T_m}$ given by (15.96) (as $n \to \infty$ with m held fixed), it is easily seen that

$$Y_t = \lim_{m\to\infty} \left\{ P - \lim_{n\to\infty} Y^{m,n}_t \right\}. \qquad (15.97)$$

Now $\omega \to Y^{m,n}_t(\omega)$ is $\mathcal{I}_t$-measurable (from (15.93), Lemma 15.1, and Remark 15.15), thus the limit-in-probability in braces is $\mathcal{I}_t$-measurable for each m, and (15.97) then gives that $\omega \to Y_t(\omega)$ is $\mathcal{I}_t$-measurable for each $t \in [0, T)$. The innovations conjecture for (15.89)–(15.90) follows.

Remark 15.16 In a recent study of the robustness of nonlinear filters Bhatt and Karandikar [4] have obtained a very useful analogue of the Bayes formula (15.24)–(15.3) for a model which is a non-Markov generalization of (15.89)–(15.90), with path-dependent and nonanticipative coefficients. This Bayes formula, together with Clark's variant [6] of the theorem of Yamada and Watanabe [32], has been used in [10] to establish the innovations conjecture for this generalized model following an approach which is motivated by that used for Theorem 15.2.

References

[1] D. F. Allinger and S. K. Mitter, New results on the innovations problem of non-linear filtering, *Stochastics*, pp. 339–48, v.4 (1981).

[2] V. E. Benes, Nonexistence of strong nonanticipating solutions to stochastic DEs: implications for functional DEs, filtering and control, *Stoch. Proc. Appl.* pp. 243–63, v.5 (1977).

[3] A. G. Bhatt, G. Kallianpur, and R. L. Karandikar, Robustness of the nonlinear filter, *Stoch. Proc. Appl.* pp. 247–254, v.81 (1999).

[4] A. G. Bhatt and R. L. Karandikar, Robustness of the nonlinear filter: the correlated case, *Stoch. Proc. Appl.* pp. 41–58, v.97 (2002).

[5] J. M. C. Clark, *Conditions for the one-to-one correspondence between an observation process and its innovation*, Technical Report, Centre for Computing and Automation, Imperial College, London (1969).

[6] J. M. C. Clark, A remark on a strong solution corollary of Watanabe and Yamada. In *Recent Advances in Communication and Control Theory* (volume in honor of A. V. Balakrishnan; R. E. Kalman, G. I. Marchuk, A. Ruperti, and A. J. Viterbi, eds), Optimization Software Inc., New York, pp. 155–8 (1987).

[7] M. H. A. Davis, A direct proof of innovations–observations equivalence for Gaussian processes, *IEEE Trans. Inform. Theory*, pp. 252–4, v.24 (1978).

[8] S. N. Ethier and T. G. Kurtz, *Markov Processes: Characterization and Convergence*. J. Wiley & Sons, New York (1986).

[9] M. Fujisaki, G. Kallianpur, and H. Kunita, Stochastic differential equations for the non-linear filtering problem. *Osaka J. Math.*, pp. 19–40, v.9 (1972).

[10] A. J. Heunis and V. M. Lucic, On the innovations conjecture of nonlinear filtering with dependent data, *Electronic J. Probab.* pp. 2190–2216, v. 13 (2008).

[11] T. Kailath, An innovations approach to least-squares estimation Part I: Linear filtering in additive white noise, *IEEE Trans. Autom. Control*, pp. 646–55, v.13 (1968).
[12] T. Kailath, A general likelihood-ratio formula for random signals in Gaussian noise, *IEEE Trans. Inform. Theory*, pp. 350–61, v.5 (1969).
[13] G. Kallianpur, *Stochastic Filtering Theory*, Springer-Verlag, New York (1980).
[14] G. Kallianpur and R. L. Karandikar, *White Noise Theory of Prediction, Filtering and Smoothing*, Stochastics Monographs No. 3, Gordon and Breach, New York (1988).
[15] G. Kallianpur and C. Striebel, Estimation of stochastic systems: arbitrary system process with additive white noise observation errors, *Annals Math. Stats.*, pp. 785–801, v.39 (1968).
[16] R. E. Kalman, A new approach to linear filtering and prediction problems, *Trans. ASME. J. Basic Eng.*, pp. 34–45, v.82 (1960).
[17] A. N. Kolmogorov, Interpolation and extrapolation of stationary random sequences, *Bull. Acad. Sci. URSS, Ser. Math.*, pp. 3–14, v.5 (1941).
[18] N. V. Krylov, On the equivalence of σ-algebras in the filtering problem of diffusion processes, *Theor. Probab. Appl.* pp. 772–81, v.24 (1979).
[19] N. V. Krylov and B. L. Rozovskiĭ, On conditional distributions of diffusion processes, *Math. USSR Izvestia*, pp. 336–56, v. 12 (1978).
[20] H. Kunita, Asymptotic behaviour of the nonlinear filtering errors of Markov processes, *J. Multivar. Analysis*, pp. 365–93, v.1 (1971).
[21] H. J. Kushner, Dynamical equations for optimal nonlinear filtering, *J. Diff. Equations*, pp. 179–90, v.3 (1967).
[22] M. Loève, *Probability Theory II*, 4th edn. Springer-Verlag, New York (1978).
[23] R. S. Liptser and A. N. Shiryayev, *Statistics of Random Processes I*. Springer-Verlag, New York (1977).
[24] P. A. Meyer, Sur un problème de filtration, *Séminaire de Probabilités VII*, Lecture Notes in Mathematics no. 321, pp. 223–47. Springer-Verlag, Berlin (1973).
[25] D. Revuz and M. Yor, *Continuous Martingales and Brownian Motion*, 2nd edn. Springer-Verlag, Berlin (1994).
[26] L. C. G Rogers and D. Williams, *Diffusions, Markov Processes and Martingales: Volume 1 Foundations*, 2nd edn. Cambridge University Press, Cambridge (2000).
[27] L. C. G Rogers and D. Williams, *Diffusions, Markov Processes and Martingales: Volume 2, Itō Calculus*, 2nd edn. Cambridge University Press, Cambridge (2000).
[28] R. L. Stratonovich, Conditional Markov processes, *Theor. Probab. Appl.*, pp. 156–78, v.5 (1960).
[29] B. S. Tsirel'son, An example of a stochastic differential equation having no strong solution, *Theor. Probab. Appl.* pp. 416–18, v.20 (1975).
[30] H. Wold, *A Study in the Analysis of Stationary Time Series*. Almquist and Wicksell, Uppsala, Stockholm (1938).
[31] E. Wong and B. Hajek, *Stochastic Processes in Engineering Systems*. Springer-Verlag, New York (1985).
[32] T. Yamada and S. Watanabe, On the uniqueness of solutions of stochastic differential equations, *J. Math. Kyoto Univ.* pp. 155–67, v.11 (1971).

·16·
Nonlinear Filtering and Fractional Brownian Motion

T. E. Duncan

16.1 Introduction

The estimation or filtering of a stochastic process, often called the signal or state process, from an observation of this process with noise has a relatively long history. For discrete-time processes where the observation is a sum of a Gaussian signal and an independent sequence of Gaussian random variables, this estimation or filtering problem was solved by Wold [36] and Kolmogorov [23]. During World War II, Wiener [35] solved a filtering problem for continuous time processes where the observation is the sum of a stationary Gaussian signal and a stationary Gaussian noise by using a factorization of their spectral densities (the Fourier transforms of their correlation functions) using his previous methods from potential theory, especially the Wiener–Hopf equation. In the late 1950s, with the emergence of time domain methods to describe linear control systems via multidimensional linear differential equations, Kalman–Bucy [20] formulated and solved a filtering problem where the signal satisfied a linear differential equation driven by a white Gaussian noise and the observation is the signal plus another white Gaussian noise. The conditional mean of the state given the observations satisfies a linear stochastic differential equation so the conditional mean can be computed recursively. This recursive property arises because the signal as well as the pair of signal and observations are Gauss–Markov processes. In the latter part of the 1960s, filtering problems were investigated where the signal process satisfied a nonlinear (stochastic) differential equation driven by a white Gaussian noise and the observation satisfied a nonlinear stochastic differential equation whose drift depends on a function of the signal and is driven by a white Gaussian noise. Such problem formulations were called nonlinear filtering problems.

Even with the early studies of filtering problems, it was recognized that white Gaussian noise (the formal derivative of a Brownian motion) was only an approximation for the noise that occurred empirically. Other noise models were considered but they usually lacked a martingale property so that the solutions seemed to lack any natural dynamical structure.

More than a half-century ago, Hurst [19] sought to model the rainfall in the Nile river valley based on approximately 850 years of rainfall data. With this extensive history of rainfall data, he computed the difference of the cumulative rainfall in k years with the linear approximation from the cumulative rainfall in $n \geq k$ years. The range of these differences normalized by the sample standard deviation behaved as cn^H where $H = 0.7$. Furthermore, the cumulative rainfall in n years had approximately the same distribution as $n^H X$, where X is the rainfall in one year. These empirical results showed that it was inappropriate to assume that the annual rainfalls are independent, identically distributed random variables.

Mandelbrot noted in the 1960s that Hurst's computation was an estimation for the exponent in the covariance of a fractional Brownian motion. These processes had been defined by Kolmogorov [22] and some important properties of these processes were given by Mandelbrot and van Ness [27]. The use of fractional Brownian motion as a stochastic model continued in hydrology after the initial work of Hurst, and Mandelbrot [26] noted the appropriateness of these processes for some economic time series. More recently, it was noted by Leland, Taqqu, Willinger, and Wilson [24] that the conventional models for internet traffic in telecommunications were inappropriate because the covariances of these models decayed to zero too rapidly and failed to describe the observed burstiness of the traffic as well as the self-similarity with respect to different time scales. These features are obtained by modelling with a fractional Brownian motion. The range of applications of a fractional Brownian motion has continued to broaden to diverse fields such as device noise [21] and epilepsy [31]. The various applications of fractional Brownian motion (FBM) was one motivation for the study of a stochastic calculus for a fractional Brownian motion which commenced in the late 1990s.

Given the wide variety of possible applications to physical phenomena, it was natural to consider a nonlinear filtering problem where the 'classical' formulation is modified by replacing the Brownian motions (white Gaussian noises) in the model by an arbitrary fractional Brownian motion (fractional Gaussian noise). For $H \neq 1/2$, a fractional Brownian motion is neither a Markov process nor a semimartingale. The loss of these two properties makes the analysis of a nonlinear filtering problem significantly more difficult and does not allow for the possibility of some desirable features that are possessed by the solution of a nonlinear filtering problem for processes satisfying stochastic differential equations with a Brownian motion. Nonetheless, some results have been obtained and some aspects of the results are naturally analogous to the filtering problem with Brownian motions. In this chapter, the solutions of some nonlinear filtering problems are given which provide generalizations of the known results with Brownian motions.

Some previous work on the nonlinear filtering problem with fractional Brownian motions has been done by Coutin and Decreusefond [4], Amirdjanova [2], and Xiong and Zhao [37]. These authors have typically restricted consideration to $H \in [1/2, 1)$. In this paper, $H \in (0, 1)$ is allowed and the conditions for some of the results are often given under weaker assumptions than have been available.

An outline of the chapter is given now. In Section 16.2, some properties of a fractional Brownian motion are described. In Section 16.3, a nonlinear filtering problem is formulated with fractional Brownian motions and some specific results are given. In Section 16.4, some prediction problems for a fractional Brownian motion are solved. In Section 16.5, the mutual information of a signal and an observation of the signal plus a fractional Brownian motion and the rate of change of such a mutual information are expressed as an integral of a filtering error and a smoothing error respectively.

16.2 Preliminaries on fractional Brownian motion

Fractional Brownian motion has a close association with some aspects of the fractional calculus of Riemann and Liouville—especially resulting from the fact that a factorization of the covariance of a fractional Brownian motion is naturally described in terms of fractional integrals and fractional derivatives. So, initially, a few facts from the theory of fractional calculus (e.g. [34]) are described.

If $\phi \in L^1([0, T], \mathbb{R}^n)$, then the left-sided and the right-sided fractional (Riemann–Liouville) integrals of φ are defined for $\alpha > 0$ (for almost all $t \in [0, T]$) by

$$\left(I_{0+}^\alpha \varphi\right)(t) = \frac{1}{\Gamma(\alpha)} \int_0^t (t-s)^{\alpha-1} \varphi(s)\, ds$$

and

$$\left(I_{T-}^\alpha \varphi\right)(t) = \frac{1}{\Gamma(\alpha)} \int_t^T (s-t)^{\alpha-1} \varphi(s)\, ds$$

respectively, where $\Gamma(\cdot)$ is the gamma function. For $\alpha \in (0, 1)$ the inverse operators of these fractional integrals are called fractional derivatives and can be given by their respective Weyl representations

$$\left(D_{0+}^\alpha \psi\right)(t) = \frac{1}{\Gamma(1-\alpha)} \left(\frac{\psi(t)}{t^\alpha} + \alpha \int_0^t \frac{\psi(t) - \psi(s)}{(t-s)^{\alpha+1}}\, ds \right)$$

and

$$\left(D_{T-}^\alpha \psi\right)(t) = \frac{1}{\Gamma(1-\alpha)} \left(\frac{\psi(t)}{(T-t)^\alpha} + \alpha \int_t^T \frac{\psi(s) - \psi(t)}{(s-t)^{\alpha+1}}\, ds \right)$$

where $\psi \in I_{0+}^\alpha \left(L^1 \left([0, T], \mathbb{R}^n \right) \right)$ and $\psi \in I_{T-}^\alpha \left(L^1 \left([0, T], \mathbb{R}^n \right) \right)$ respectively.

Let $K_H(t, s)$ for $0 \leq s \leq t \leq T$ be the real-valued kernel function

$$K_H(t, s) = \frac{\tilde{c}_H (t - s)^{H-1/2}}{\Gamma(H + 1/2)} + \frac{\tilde{c}_H (1/2 - H)}{\Gamma(H + 1/2)} \int_s^t (u - s)^{H-3/2} \left(1 - \left(\frac{s}{u} \right)^{1/2 - H} \right) du \qquad (16.1)$$

for $H \in (0, 1/2)$. If $H \in (1/2, 1)$, then K_H has a simpler form as

$$K_H(t, s) = \frac{\hat{c}_H}{\Gamma(H - 1/2)} s^{1/2 - H} \int_s^t (u - s)^{H-3/2} u^{H-1/2} du. \qquad (16.2)$$

The terms $\tilde{c}_H$ and $\hat{c}_H$ are constants that only depend on H.

Define the integral operator $\mathbb{K}_H$ induced from the kernel K_H by

$$\mathbb{K}_H h(t) = \int_0^t K_H(t, s) h(s) \, ds \qquad (16.3)$$

for $h \in L^2 \left([0, T], \mathbb{R}^n \right)$. It is well-known ([34]) that

$$\mathbb{K}_H : L^2 \left([0, T], \mathbb{R}^n \right) \to I_{0+}^{H+1/2} \left(L^2 \left([0, T], \mathbb{R}^n \right) \right)$$

is a bijection and $\mathbb{K}_H$ can be described as

$$\mathbb{K}_H h(s) = \bar{c}_H I_{0+}^{2H} \left(u_{1/2 - H} I_{0+}^{1/2 - H} \left(u_{H - 1/2} h \right) \right) (s) \qquad (16.4)$$

for $H \in (0, 1/2]$ and

$$\mathbb{K}_H h(s) = c_H I_{0+}^1 \left(u_{H - 1/2} I_{0+}^{H - 1/2} \left(u_{1/2 - H} h \right) \right) (s) \qquad (16.5)$$

for $H \in [1/2, 1)$ where

$$c_H = \left[\frac{2 H \Gamma(H + 1/2) \Gamma(3/2 - H)}{\Gamma(2 - 2H)} \right]^{1/2}, \qquad (16.6)$$

$$\bar{c}_H = c_H \Gamma(2H),$$

and

$$u_a(s) = s^a$$

for $s \geq 0$ and $a \in \mathbb{R}$. The inverse operator

$$\mathbb{K}_H^{-1} : I_{0+}^{H+1/2} \left(L^2 [0, T], \mathbb{R}^n \right) \to L^2 \left([0, T], \mathbb{R}^n \right)$$

is given by

$$\mathbb{K}_H^{-1} \varphi(s) = \bar{c}_H^{-1} s^{1/2 - H} D_{0+}^{1/2 - H} \left(u_{H - 1/2} D_{0+}^{2H} \varphi \right) (s) \qquad (16.7)$$

for $H \in (0, 1/2]$ and

$$\mathbb{K}_H^{-1} \varphi(s) = c_H^{-1} s^{H - 1/2} D_{0+}^{H - 1/2} \left(u_{1/2 - H} D \varphi \right) (s) \qquad (16.8)$$

for $H \in [1/2, 1)$ and $\varphi \in I_{0+}^{H+1/2}\left(L^2\left([0, T], \mathbb{R}^n\right)\right)$. Note that if $\varphi \in H^1\left([0, T], \mathbb{R}^n\right)$, the Sobolev space, then

$$\mathbb{K}_H^{-1}\varphi(s) = c_H^{-1} s^{H-1/2} I_{0+}^{1/2-H}\left(u_{1/2-H}\varphi'\right)(s) \tag{16.9}$$

for $H \in (0, 1/2]$.

Since the linear operator $\mathbb{K}_H^{-1}$ occurs naturally in an expression for the conditional expectation of a function of the signal given the observation process, as well as for some subsequent Radon–Nikodym derivatives, it is useful to have some information about its domain $I_{0+}^{H+1/2}\left(L^2\left([0, T], \mathbb{R}^n\right)\right)$. Some immediate facts about this domain are for $H \in (0, 1/2)$,

$$I_{0+}^{H+1/2}\left(L^2\left([0, T], \mathbb{R}^n\right)\right) \supset C^\beta\left([0, T], \mathbb{R}^n\right)$$

for $\beta > 1/2 - H$ and for $H \in (1/2, 1)$,

$$I_{0+}^{H+1/2}\left(L^2\left([0, T], \mathbb{R}^n\right)\right) \supset L^2\left([0, T], \mathbb{R}^n\right).$$

To introduce precisely a fractional Brownian motion a definition is given.

Definition 16.1 A real-valued process $(B(t), t \geq 0)$ on the complete probability space $(\Omega, \mathcal{F}, \mathbb{P})$ is called a (real-valued) standard fractional Brownian motion (FBM) with Hurst parameter $H \in (0, 1)$ if B is a Gaussian process with continuous sample paths that satisfies

$$\mathbb{E}\left[B(t)\right] = 0$$

and

$$\mathbb{E}\left[B(s) B(t)\right] = \frac{1}{2}\left[t^{2H} + s^{2H} - |t - s|^{2H}\right]$$

for all $s, t \in \mathbb{R}_+$.

Definition 16.2 The process $(B(t), t \geq 0)$ is an $\mathbb{R}^k$-valued standard fractional Brownian motion with Hurst parameter $H_0 \in (0, 1)$ if B is a k-vector of independent real-valued standard fractional Brownian motions each with the Hurst parameter H_0.

Initially, some useful properties of a fractional Brownian motion are noted. If $(B(t), t \geq 0)$ is a real-valued (standard) fractional Brownian motion with Hurst parameter H, then the sample paths of B are Hölder continuous of order $H - \epsilon$ for each $\epsilon > 0$. Let $r_H(n)$ be the correlation between the first and the $(n+1)$st increment of B, that is,

$$r_H(n) = \mathbb{E}\left[(B(1) - B(0))(B(n+1) - B(n))\right]. \tag{16.10}$$

If $H \in (1/2, 1)$, then $r_H(n) > 0$ and

$$\sum_{n=1}^\infty r_H(n) = +\infty. \tag{16.11}$$

If $H \in (0, 1/2)$, then $r_H(n) < 0$ and

$$\sum_{n=1}^{\infty} |r_H(n)| < \infty. \tag{16.12}$$

The equality (16.11) implies a long range dependence of the FBM. It follows from the covariance of B that if $a > 0$ then $(B(at), t \geq 0)$ and $(a^H B(t), t \geq 0)$ have the same probability law where B has Hurst parameter H. This property is called self-similarity. A FBM $(B(t), t \geq 0)$ has sample paths of unbounded variation so these paths are nondifferentiable (a.s.) but a more refined statement can be made considering the pth variation of the process. Let $(B(t), t \geq 0)$ be a FBM with Hurst parameter H. It follows by an application [33] of the pointwise ergodic theorem that

$$\sum_i \left| B\left(t_{i+1}^{(n)}\right) - B\left(t_i^{(n)}\right) \right|^p = \begin{cases} 0 & pH > 1 \\ c(p) & pH = 1 \\ +\infty & pH < 1 \end{cases} \tag{16.13}$$

where $c(p) = \mathbb{E}\left|B(1)\right|^p$ and $(t_i^{(n)}, i = 1, \ldots, n)$ is a sequence of nested partitions of $[0, 1]$. Thus for $H \in (1/2, 1)$, the quadratic variation ($p = 2$) is zero and for $H \in (0, 1/2)$ the quadratic variation is infinite. These quadratic variation results imply that B is not a semimartingale for $H \neq 1/2$. Thus, the stochastic calculus for semimartingales is not available for B if $H \neq 1/2$. Another asymptotic property of a FBM, $B(\cdot)$, is given by the Law of the Iterated Logarithm [18]:

$$\limsup_{t \to \infty} \frac{B(t)}{\left(2t^{2H} \log \log t\right)^{1/2}} = C_H \quad \text{a.s}$$

where C_H is a constant that only depends on the Hurst parameter H of $B(\cdot)$.

Since an FBM, $B(\cdot)$ for $H \neq 1/2$ is not a semimartingale, it is necessary to define stochastic integration for a fractional Brownian motion. For applications, it is desirable that an integral with respect to a FBM has mean zero because such an integral often arises from a process that is the noise or perturbations of a (deterministic) model for a physical phenomenon. Two natural and equivalent ways to construct such an integral use either the Wick product or the divergence integral from Malliavin calculus [1, 5, 13, 8]. Many properties of a fractional Brownian motion have also been given in two recent monographs [3, 29].

The covariance function for a fractional Brownian motion $(B(t), t \in [0, T])$ with $H \in (0, 1)$ can be expressed using fractional integrals or fractional derivatives as

$$\mathbb{E}\left[B(s) B(t)\right] = \rho(H) \int_0^T u_{1/2-H}^2(r) \left(I_{T-}^{H-1/2} u_{H-1/2} \mathbb{1}_{[0,s]}\right)(r) \left(I_{T-}^{H-1/2} u_{H-1/2} \mathbb{1}_{[0,t]}\right)(r) \, dr \tag{16.14}$$

where $u_a(s) = s^a$ for $s \geq 0$ and $a \in \mathbb{R}$ and I_{T-}^a denotes a fractional integral for $a > 0$ and a fractional derivative for $a < 0$,

$$\rho(H) = \frac{2H\Gamma(H+1/2)\Gamma(3/2-H)}{\Gamma(2-2H)} \qquad (16.15)$$

and Γ is the gamma function.

The equality (16.14) defines a square root of the covariance of B in $L^2([0, T])$. Thus, the representation for a fractional Brownian motion in terms of a Brownian motion is verified. Let $(B(t), t \in [0, T])$ be a fractional Brownian motion with $H \in (0, 1)$. The process B can be represented as an integral of a Brownian motion as

$$B(t) = \sqrt{\rho(H)} \int_0^t u_{1/2-H}(r) \left(I_{T-}^{H-1/2} u_{H-1/2} 1_{[0,t]} \right)(r) \, dW(r) \qquad (16.16)$$

where $(W(s), s \in [0, t])$ is a Brownian motion (Wiener process). Using fractional calculus, the representation (16.16) can be "inverted" as

$$W(t) = \frac{1}{\sqrt{\rho(H)}} \int_0^t u_{H-1/2} I_{T-}^{1/2-H} \left(u_{1/2-H} 1_{[0,t]} \right) dB. \qquad (16.17)$$

The Hilbert space associated with a FBM can be given explicitly from the equality (16.14).

Definition 16.3 Let $H \in (0, 1)$ be fixed. The space $L^2_H([0, T])$ is the Hilbert space of real-valued distributions such that $F \in L^2_H$ if $u_{1/2-H} I_{T-}^{H-1/2} \left(u_{H-1/2} F \right)$ is (Lebesgue) square-integrable where

$$\|F\|_{L^2_H} = \rho(H) \int_0^T \left(u_{1/2-H}(s) I_{T-}^{H-1/2} \left(u_{H-1/2} F \right)(s) \right)^2 ds, \qquad (16.18)$$

$\rho(H)$ is given in (16.15), and $u_a(s) = s^a$ for $s \geq 0$ and $a \in \mathbb{R}$.

It is elementary to verify using (16.14) that for $F \in L^2_H$, the Wiener integral $\int_0^T F \, dB$ is a well-defined Gaussian random variable with zero mean and second moment $\|F\|^2_{L^2_H}$.

For the filtering models that are used in this chapter, it is not necessary to use the more refined stochastic calculus for a fractional Brownian motion to define integration for stochastic integrands. The relation between a fractional Brownian motion and a Brownian motion given by (16.16) and (16.17) plays a basic role in describing Radon–Nikodym derivatives associated with the solution of a nonlinear filtering problem.

Before considering stochastic differential equations to describe a nonlinear filtering problem with fractional Brownian motions, it is important to consider the question of solutions of stochastic differential equations with fractional Brownian motions.

Let $(B(t), t \geq 0)$ be a FBM with $H \in (0, 1)$. Consider the following stochastic differential equation:

$$dX(t) = a(t, X(t))\, dt + b(t, X(t))\, dB(t) \qquad (16.19)$$
$$X(0) = X_0.$$

If $H \neq 1/2$, then the results for solutions are more problematic than for $H = 1/2$. The stochastic differential equation (16.19) is understood as the corresponding integral equation

$$X(t) = X_0 + \int_0^t a(s, X(s))\, ds + \int_s^t b(t, X(s))\, dB(s). \qquad (16.20)$$

Unfortunately, no general results exist for (strong) probabilistic solutions of (16.20) when $b(t, x)$ is a nontrivial function of x. A probabilistic solution means that the stochastic integral term has expectation zero and a second moment that can be given explicitly. Another type of solution is called pathwise (or nonprobabilistic) (e.g. [25, 38]). This approach defines the solution for each Hölder continuous function of a specified order that includes (almost surely) the sample paths of B. For this type of solution, the first two moments of the stochastic integral term are not known in general.

For a nonlinear filtering problem, B is considered as the noise or perturbation so that it is important to have the expectation of the stochastic integral term be zero. Thus for a general model considered for the filtering problem, it is assumed that the diffusion coefficient b satisfies $b \equiv 1$. With $b \equiv 1$, the results for strong solutions follow from the solution of the associated deterministic equation, that is,

$$\frac{dx}{dt} = a(t, x). \qquad (16.21)$$

However, for the computation of conditional moments, it suffices to consider the notion of a weak solution of a stochastic differential equation for the observations. Both weak and strong solutions of a stochastic differential equation are defined.

Definition 16.4 The process $(Z(t), t \geq 0)$ is a strong solution of the stochastic differential equation

$$dZ(t) = a(t, Z(t))\, dt + \beta(t, Z(t))\, dB(t) \qquad (16.22)$$
$$Z(0) = Z_0$$

where $(B(t), t \geq 0)$ is a fractional Brownian motion with Hurst parameter $H \in (0, 1)$ on the probability space $(\tilde{\Omega}, \tilde{\mathcal{F}}, \tilde{\mathbb{P}})$ if Z satisfies the integral equation

$$Z(t) - Z(0) = \int_0^t a(s, Z(s))\, ds + \int_0^t \beta(s, Z(s))\, dB(s) \qquad (16.23)$$

for each $t > 0$ and the stochastic integral in (16.22) is of Itô-type [1, 13].

Definition 16.5 A weak solution of the equation (16.22) with $H_0 \in (0, 1)$ fixed is a triple

$$\left(Z(t), B(t), (\tilde{\Omega}, \tilde{\mathcal{F}}, \tilde{\mathbb{P}})\, , t \geq 0\right)$$

where $(\tilde{\Omega}, \tilde{\mathcal{F}}, \tilde{\mathbb{P}})$ is a probability space with processes B and Z, $(B(t), t \geq 0)$ is a fractional Brownian motion with Hurst parameter H_0 and $(Z(t), t \geq 0)$ is a process satisfying

$$Z(t) = Z_0 + \int_0^t a(s, Z(s))\, ds + \int_0^t \beta(s, Z(s))\, dB(s) \tag{16.24}$$

and the stochastic integral in (16.24) is of Itô-type.

The equation (16.22) has a unique strong solution if for any two strong solutions Z_1 and Z_2 on $(\tilde{\Omega}, \tilde{\mathcal{F}}, \tilde{\mathbb{P}})$, $\tilde{\mathbb{P}}(Z_1(t) = Z_2(t), t \geq 0) = 1$. The equation (16.22) has a unique weak solution if for any two weak solutions

$$\left(Z_1(t), B(t), (\Omega, \mathcal{F}, \mathbb{P})\, , t \geq 0\right)$$

and

$$\left(Z_2(t), \tilde{B}(t), (\tilde{\Omega}, \tilde{\mathcal{F}}, \tilde{\mathbb{P}})\, , t \geq 0\right)$$

the processes $(Z_1(t), t \geq 0)$ and $(Z_2(t), t \geq 0)$ have the same probability law on $(C(\mathbb{R}_+), \mathcal{B})$, where $\mathcal{B}$ is the Borel σ-algebra on $C(\mathbb{R}_+)$ with the uniform norm.

16.3 Nonlinear filtering

Consider a nonlinear filtering problem whose processes are described by the following stochastic differential equations

$$\begin{aligned} dX(t) &= a(t, X(t))\, dt + dB_1(t) \\ X(0) &= X_0 \end{aligned} \tag{16.25}$$

$$\begin{aligned} dY(t) &= c(t, X(t), Y(t))\, dt + dB_2(t) \\ Y(0) &\equiv 0 \end{aligned} \tag{16.26}$$

where $X(t) \in \mathbb{R}^n$, $Y(t) \in \mathbb{R}^p$, $(B_1(t), t \geq 0)$ is an $\mathbb{R}^m$-valued standard fractional Brownian motion with Hurst parameter $H_1 \in (0, 1)$, $(B_2(t), t \geq 0)$ is an $\mathbb{R}^p$-valued standard fractional Brownian motion with Hurst parameter $H_2 \in (0, 1)$, B_1 and B_2 are independent processes, X_0 is an $\mathbb{R}^n$-valued random variable independent of B_1 and B_2, and $a(\cdot, \cdot)$ and $c(\cdot, \cdot, \cdot)$ are suitable functions satisfying conditions that are specified subsequently.

To obtain conditional expectations for filtering problems for the model (16.25) and (16.26), it suffices to verify that (16.26) has one and only one weak solution. Weak solutions are often sufficient because only the probability law for Y is required. The sufficient conditions for the existence and the uniqueness of a weak solution are different for the two cases of $H \in (0, 1/2)$ and $H \in (1/2, 1)$. The "boundary" case of $H = 1/2$ is well known.

Analogous to the nonlinear filtering for processes generated by Brownian motions ($H = 1/2$), an absolute continuity property for the measures of the observation process and the observation noise is important because it allows for the use of the associated Radon–Nikodym derivative and a Bayes formula to compute explicitly conditional expectations of functions of the state process X given the observations Y. The two cases of $H \in (0, 1/2)$ and $H \in (1/2, 1)$ are treated separately because the sufficient conditions for absolute continuity are different.

The following conditions are used subsequently:

C1. The function $c \colon [0, T] \times \mathbb{R}^n \times \mathbb{R}^p \to \mathbb{R}^p$ in (16.26) is Borel measurable and satisfies

$$|c(t, x, y)| \leq k_1(x)(1 + |y|)$$

for all $t \in [0, T]$, $x \in \mathbb{R}^n$, $y \in \mathbb{R}^p$, and $k_1 \colon \mathbb{R}^n \to \mathbb{R}_+$ is Borel measurable and bounded on bounded sets and $|\cdot|$ is the Euclidean norm in $\mathbb{R}^p$.

C2. For $H_2 \in (1/2, 1)$, the function $c \colon [0, T] \times \mathbb{R}^n \times \mathbb{R}^p \to \mathbb{R}^p$ in (16.26) satisfies

$$|c(s, x_1, y_1) - c(t, x_2, y_2)| \leq k_2(x_1, x_2)|y_1 - y_2|^\gamma$$

for all $x_1, x_2 \in \mathbb{R}^n$, $y_1, y_2 \in \mathbb{R}^p$, and $s, t \in [0, T]$, some $\gamma > 1 - 1/2H$ and some $k_2 \colon \mathbb{R}^n \times \mathbb{R}^n \to \mathbb{R}$ that is Borel measurable and bounded on bounded sets.

C3. The function $a \colon [0, T] \times \mathbb{R}^n \to \mathbb{R}^n$ in (16.25) is Borel measurable and satisfies

$$|a(t, x_1) - a(s, x_2)| \leq K|x_1 - x_2|$$

where K is a constant and $|\cdot|$ is the Euclidean norm in $\mathbb{R}^n$.

C4. The function $c \colon [0, T] \times \mathbb{R}^n \times \mathbb{R}^p \to \mathbb{R}^p$ in (16.26) satisfies

$$|c(s, x_1, y_1) - c(t, x_2, y_2)| \leq g(x_1, x_2)|y_1 - y_2|$$

for all $s, t \in [0, T]$, $x_1, x_2 \in \mathbb{R}^n$ and $y_1, y_2 \in \mathbb{R}^p$ and g is bounded on bounded sets of $\mathbb{R}^n \times \mathbb{R}^n$.

Let μ_{B_2}, μ_X, and μ_Y be the probability measures induced on the measurable spaces $(C([0, T], \mathbb{R}^p), \mathcal{B}^p)$, $(C([0, T], \mathbb{R}^n), \mathcal{B}^n)$, and $(C([0, T], \mathbb{R}^p), \mathcal{B}^p)$ by the processes B_2, X, and Y respectively that are given in (16.25) and (16.26) and let μ_{XY} be the probability measure on $(C([0, T], \mathbb{R}^{n+p}), \mathcal{B}^{n+p})$ for the pair (X, Y) where, for $k \in \mathbb{N}$, $\mathcal{B}^k$ is the Borel σ-algebra on $C([0, T], \mathbb{R}^k)$ and the topologies on these spaces of continuous functions are determined by the uniform norm.

For the probability measures μ_1 and μ_2 on the same measurable space, let $\mu_1 \ll \mu_2$ denote that the measure μ_1 is absolutely continuous with respect to the measure μ_2 and let $d\mu_1/d\mu_2$ be the associated Radon–Nikodym derivative. If $\mu_Y \ll \mu_{B_2}$, then there is a weak solution of (16.26) and the mutual absolute continuity of μ_Y and μ_{B_2} can be used to verify the uniqueness of a weak solution [10]. Furthermore, $\mu_Y \ll \mu_{B_2}$ provides a method to obtain explicit expressions for conditional expectations for the filtering problem given by (16.25) and (16.26).

The following theorem provides sufficient conditions for the aforementioned absolute continuity. It has been given in various cases [6, 10, 30], and is a complement to the result for Brownian motion [15]. Note that the conditions differ for $H_2 \in (0, 1/2)$ and $H_2 \in (1/2, 1)$.

Theorem 16.1 *Let H_1, $H_2 \in (0, 1)$ be fixed and let*

$$((B_1(t), B_2(t), X(t), Y(t)), t \in [0, T])$$

be the processes in (16.25) and (16.26). Let C3 be satisfied. If $H_2 \in (0, 1/2)$, then assume that C1 is satisfied and if $H_2 \in (1/2, 1)$, then assume that C1 and C2 are satisfied. Then $\mu_{XY} \ll \mu_X \otimes \mu_{B_2}$ and the Radon–Nikodym derivative is

$$\frac{d\mu_{XY}}{d(\mu_X \otimes \mu_{B_2})} = \tilde{M}(T) = \exp\left[\int_0^T \langle \mathbb{K}_{H_2}^{-1}(Z), dW \rangle - \frac{1}{2}\int_0^T |\mathbb{K}_{H_2}^{-1}(Z)(t)|^2 dt\right]$$
(16.27)

where $\mathbb{K}_{H_2}^{-1}$ is defined in (16.7) or (16.8), W is the standard Wiener process (Brownian motion) associated with B_2 given by (16.17) and $Z(\cdot)$ is given by

$$Z(t) = \int_0^t c(s, X(s), Y(s)) \, ds.$$
(16.28)

While the proof of this theorem follows by a natural extension of the results in [10], an important ingredient of the proof is given here. A basic idea of the proof is to obtain an appropriate bound on

$$|\mathbb{K}_{H_2}^{-1}(Z)|^2_{L^2([0,T])}$$

to verify uniform integrability of a family of Radon–Nikodym derivatives. For $H_2 \in (0, 1/2)$, there is the inequality

$$|\mathbb{K}_{H_2}^{-1}(Z)|^2_{L^2([0,T])} \le c_T \sup_{0 \le t \le T} k_1^2(X(t)) \left(1 + |B|^2_{C([0,T])}\right)$$

and for $H \in (1/2, 1)$, there is the inequality

$$|\mathbb{K}_{H_2}^{-1}(Z)|^2_{L^2([0,T])} \le c_T \left(1 + \sup_{0 \le t \le T} k_1^2(X(t))|Y|^2_{C([0,T])}\right)$$
$$+ c_T \sup_{(s,t) \in [0,T]^2} k_2^2(X(s), X(t))|Y|^2_{C^\beta([0,T])} \int_0^T \left(\int_0^s \frac{(s-r)^\beta}{(s-r)^{H-1/2}} dr\right)^2 ds.$$

By the integration with respect to μ_X in (16.27), it follows immediately that $\mu_Y \ll \mu_{B_2}$ and provides one expression for $d\mu_Y/d\mu_{B_2}$. However, it is useful to compute the integration with respect to μ_X more explicitly to obtain a more useful expression for $d\mu_Y/d\mu_{B_2}$.

Theorem 16.2 *Let C3 be satisfied. Let μ_{B_2}, μ_X, and μ_{XY} be the measures on the spaces of continuous functions for the processes $(B_2(t), t \in [0, T])$, $(X(t), t \in [0, T])$, and $((X(t), Y(t)), t \in [0, T])$ respectively. If $H_2 \in (0, 1/2)$, then let C1 be satisfied and if $H_2 \in (1/2, 1)$, then let C1 and C2 be satisfied. Then $\mu_Y \ll \mu_{B_2}$ and the associated Radon–Nikodym derivative is*

$$M(T) = \mathbb{E}_X\left[\tilde{M}(T)\right] = \frac{d\mu_Y}{d\mu_{B_2}}$$
$$= \exp\left[\int_0^T \left\langle \widehat{\mathbb{K}_{H_2}^{-1}(Z)}(t, t), dW(t)\right\rangle - \frac{1}{2}\int_0^T \left|\widehat{\mathbb{K}_{H_2}^{-1}(Z)}(t, t)\right|^2 dt\right] \quad (16.29)$$

where

$$\widehat{\mathbb{K}_{H_2}^{-1}(Z)}(t, t) = \mathbb{E}\left[\mathbb{K}_{H_2}^{-1}(Z)(t) \mid Y(u), 0 \le u \le t\right].$$

The proof of this result follows by an application of the change of variables formula of K. Itô as for $H = 1/2$ [11].

The following result uses the Radon–Nikodym derivatives $\tilde{M}$ and M in (16.27) and (16.29) to provide useful expressions for conditional expectations of functions of X given the observations Y. The verification follows directly from the definition of conditional expectation and the relation of integration with respect to μ_Y and μ_{B_2} using the mutual absolute continuity of μ_{B_2} and μ_Y.

Theorem 16.3 *Let $f: \mathbb{R}^n \to \mathbb{R}$ be a Borel measurable function and let $t \in [0, T]$ be fixed. Let (X, Y) be the processes given by (16.25) and (16.26). Let C3 be satisfied. If $H_2 \in (0, 1/2)$ then let C1 be satisfied and if $H_2 \in (1/2, 1)$ then let C1 and C2 be satisfied. Assume that*

$$\mathbb{E}\left|f(X(t))\right| < \infty. \quad (16.30)$$

Then

$$\mathbb{E}\left[f(X(t)) \mid Y(u), 0 \le u \le t\right] = \frac{\mathbb{E}_X\left[\tilde{M}(t) f(X(t))\right]}{M(t)} \quad (16.31)$$

where $\mathbb{E}_X$ denotes integration with respect to μ_X and $\tilde{M}$ and M are given by (16.27) and (16.29) respectively.

Corollary 16.1 *If the conditions of Theorem 16.3 are satisfied and $0 \le t \le T$, then*

$$\mathbb{E}\left[f(X(t)) \mid Y(u), 0 \le u \le T\right] = \frac{\mathbb{E}_X\left[\tilde{M}(T) f(X(t))\right]}{M(T)} \quad (16.32)$$

The proofs of Theorem 16.3 and Corollary 16.1 follow directly from the proofs for Brownian motion ($H = 1/2$).

Given the explicit expression (16.31) for the conditional expectation of a function of the state given the observations, a stochastic integral equation can be given for this conditional expectation. One form of this result is given in the following theorem. This result has $H_1 = 1/2$.

Theorem 16.4 *Let $H_1 = 1/2$ and $H_2 \in (0, 1)$ and let (X, Y) be processes given by (16.25) and (16.26). Assume that C1, C3, and C4 are satisfied. Let $a(\cdot, \cdot)$ in (16.25) be continuously differentiable in the first variable and twice continuously differentiable in the second variable, and let these derivatives be uniformly bounded. The conditional moment*

$$\widehat{f}(t) = \mathbb{E}\left[f(X(t)) \mid Y(u), 0 \le u \le t\right] \tag{16.33}$$

satisfies the following equality:

$$\widehat{f}(t) = \widehat{f}(0) + \int_0^t \widehat{Dfa}(s)\,ds + \frac{1}{2}\int_0^t \widehat{D^2 f}(s)\,ds$$
$$+ \int_0^t \left(\widehat{f\mathbb{K}_{H_2}^{-1}(Z)}(s) - \widehat{f}(s)\widehat{\mathbb{K}_{H_2}^{-1}(Z)}(s)\right)\left(dZ(s) - \widehat{\mathbb{K}_{H_2}^{-1}(Z)}(s)\,ds\right) \tag{16.34}$$

where Z is given in (16.28) and $\widehat{}$ denotes conditional expectation as in (16.28).

Since both the state equation and the Radon–Nikodym derivatives M and $\tilde{M}$ are expressed in terms of only Brownian motions, the methods for Brownian motion to obtain an integral equation for the conditional moment (16.33) can be used. Furthermore, if $H_1 \ne 1/2$, then applying the operator $\mathbb{K}_{H_1}^{-1}$ to the state equation (16.25) converts B_1 to a Brownian motion. Since the process $\mathbb{K}_{H_1}^{-1}(X)$ is equivalent to X, that is $\mathbb{K}_{H_1}$ and $\mathbb{K}_{H_1}^{-1}$ are bijections, this filtering problem can be changed to one with only a Brownian motion in the state equation.

16.4 Prediction and fractional Brownian motion

For the model (16.25) and (16.26) with $H_1 = H_2 = 1/2$, that is, B_1 and B_2 are Brownian motions, the prediction of X given the observations has a relatively simple form. Specifically, if $t > s > 0$, then

$$\mathbb{E}\left[X(t) \mid Y(u), 0 \le u \le s\right]$$
$$= \mathbb{E}\left[X(s) + \int_s^t a\,dr + B_1(t) - B_1(s) \mid Y(u), 0 \le u \le s\right]$$
$$= \widehat{X}(s) + \mathbb{E}\left[\int_s^t a\,dr \mid Y(u), 0 \le u \le s\right]$$
$$= \widehat{X}(s) + \int_s^t \mathbb{E}\left[a(r, X(r)) \mid Y(u), 0 \le u \le s\right] dr,$$

or for a more elementary prediction problem,

$$\mathbb{E}\left[X(t) \mid X(u), u \in [0, s]\right]$$
$$= X(s) + \int_s^t \mathbb{E}\left[a(r, X(r)) \mid X(u), u \in [0, s]\right] dr.$$

However, if $H_1 \neq 1/2$, then it is necessary to predict $(B_1(t) - B_1(s))$ given $Y(u), 0 \leq u \leq s$ in the former case, or $X(u), 0 \leq u \leq t$ in the latter case. This conditional expectation is not zero, because for $H_1 \neq 1/2$, the increments of B_1 are not independent.

A few results are given for some special cases of this prediction problem. Each of these prediction problems has observations that are equivalent to the fractional Brownian motion that generates the process to be predicted.

The first result is an elementary modification of the prediction of a fractional Brownian motion given in [16, 32].

Lemma 16.1 *If $0 < s < t$ and $c : [s, t] \to \mathbb{R}$ is an element of L_H^2 given in (16.18), then*

$$\mathbb{E}\left[\int_s^t c\,dB \mid B(r), r \in [0, s]\right] = \int_0^s u_{1/2-H}\left(I_{s-}^{1/2-H}\left(I_{t-}^{H-1/2} u_{H-1/2} c\right)\right) dB. \tag{16.35}$$

Let $(X(t), t \geq 0)$ be the real-valued Gaussian process that is the solution of the stochastic differential equation

$$dX(t) = a(t)X(t)\,dt + dB(t) \tag{16.36}$$
$$X(0) = x_0$$

where $x_0 \in \mathbb{R}$, $a : \mathbb{R}_+ \to \mathbb{R}$ is bounded and Borel measurable and $(B(t), t \geq 0)$ is a standard fractional Brownian motion with $H \in (0, 1)$. It follows directly from ordinary differential equations that the solution of (16.36) is

$$X(t) = e^{\int_0^t a} x_0 + \int_0^t e^{\int_s^t a}\,dB(s) \tag{16.37}$$

and

$$\overline{\sigma(X(u), u \in [0, t])} = \overline{\sigma(B(u), u \in [0, t])}.$$

Using Lemma 16.1, it easily follows that the least mean-square error prediction of X in (16.36) can be given explicitly.

Proposition 16.1 *Let $(X(t), t \geq 0)$ be the process given by (16.37) and let $0 < s < t$. Then the following equality is satisfied:*

$$\mathbb{E}\left[X(t) \mid X(r), r \in [0, s]\right] = e^{\int_s^t a} X(s)$$
$$+ \int_0^s u_{1/2-H}\left(I_{s-}^{1/2-H}\left(I_{t-}^{H-1/2} u_{H-1/2}\, v\mathbb{1}_{[s,t]}\right)\right)(dX - aX\,dr) \tag{16.38}$$

where $u_\alpha(r) = r^\alpha$ for $\alpha \in \mathbb{R}$ and $r \geq 0$ and $v(r) = e^{\int_r^t a}$.

Now, consider the stochastic differential equation for a geometric fractional Brownian motion with $H \in (0, 1)$

$$dX(t) = X(t)(a(t)\,dt + b(t)\,dB(t)) \qquad (16.39)$$
$$X(0) = x_0$$

where $a: \mathbb{R}_+ \to \mathbb{R}$ and $b: \mathbb{R}_+ \to \mathbb{R} \setminus \{0\}$ are bounded, Borel measurable functions and $x_0 > 0$. A solution of (16.39) is

$$X(t) = x_0 \exp\left[\int_0^t a\,ds + \int_0^t b\,dB - \frac{1}{2}|b\mathbf{1}_{[0,t]}|^2_{L^2_H}\right] = x_0 \exp\left[\int_0^t a\,ds\right] \exp^{\diamond}\left(\int_0^t b\,dB\right) \qquad (16.40)$$

where

$$\exp^{\diamond}\left(\int_0^t b\,dB\right) := \sum_{n=0}^{\infty} \frac{1}{n!}\left(\int_0^t b\,dB\right)^{\diamond n} \qquad (16.41)$$

and $(\cdot)^{\diamond n}$ is the nth Wick product and $\exp^{\diamond}$ is called the Wick exponential [13].

An important feature of the Wick product is related to conditional expectation, that is, if $\mathcal{G} \subset \mathcal{F}$ is a σ-algebra and Y and Z are bounded (or integrable) random variables, then

$$\mathbb{E}[Y \diamond Z \mid \mathcal{G}] = \mathbb{E}[Y \mid \mathcal{G}] \diamond \mathbb{E}[Z \mid \mathcal{G}]. \qquad (16.42)$$

Using Equation (16.41) and Equation (16.42), the following result [7] is obtained.

Proposition 16.2 *Let $(X(t), t \geq 0)$ be the process given by (16.40) and let $0 < s < t$. Then the following equality is satisfied:*

$$\mathbb{E}[X(t) \mid X(r), r \in [0, s]] = X(s) \exp\left[\int_s^t a\,dr - \langle b\mathbf{1}_{[0,s]}, b\mathbf{1}_{[s,t]}\rangle_{L^2_H}\right]$$
$$\exp^{\diamond}\left(\mathbb{E}\left[\int_s^t b\,dB \mid B(r), r \in [0, s]\right]\right). \qquad (16.43)$$

Lemma 16.1 can be used to evaluate explicitly the conditional expectation in the Wick exponential of (16.43). The previous results provide an indication of some useful methods to solve prediction problems for processes associated with a fractional Brownian motion.

16.5 Estimation and mutual information

The solutions of nonlinear filtering problems have found applications to other areas that are not clearly directly related to the estimation of stochastic processes. The mutual information between two processes is an application that is given here.

To describe this application to mutual information, initially the model for the processes is explicitly given. Let $(X(t), t \in [0, T])$ and $(B(t), t \in [0, T])$ be two independent real-valued processes on the complete probability space $(\Omega, \mathcal{F}, \mathbb{P})$ where B

is a standard fractional Brownian motion with the Hurst parameter $H \in (0, 1)$. Let $(Y(t), t \in [0, T])$ be the process that satisfies

$$dY(t) = X(t)\,dt + dB(t) \qquad (16.44)$$
$$Y(0) \equiv 0.$$

The mutual information between $(X(t), t \in [0, T])$ and $(Y(t), t \in [0, T])$ is denoted $I(X, Y)$ and is defined [14] as

$$I(X, Y) = \int N \log N\, d(\mu_X \otimes \mu_Y) \qquad (16.45)$$

where

$$N = \frac{d\mu_{XY}}{d(\mu_X \otimes \mu_Y)} \qquad (16.46)$$

and μ_X, μ_Y, and μ_{XY} are the probability measures for $(X(t), t \in [0, T])$, $(Y(t), t \in [0, T])$, and $((X(t), Y(t)), t \in [0, T])$ respectively. The mutual information $I(X, Y) = +\infty$ if the Radon–Nikodym derivative N does not exist.

The following assumptions on X are selectively used subsequently:

A1. For $H \in (1/2, 1)$, the process $(X(t), t \in [0, T])$ in (16.44) has $C^\beta([0, T])$ sample paths for some $\beta > H - 1/2$ where $C^\beta([0, T])$ is the family of β-Hölder-continuous functions on $[0, T]$. The process $(X(t), t \in [0, T])$ satisfies

$$\mathbb{E}\left[|X|^2_{C^\beta([0,T])}\right] < \infty$$

where $|\cdot|_{C^\beta([0,T])}$ is the uniform norm on $C^\beta([0, T])$ for some $\beta > H - 1/2$.

A2. For $H \in (0, 1/2)$, the process $(X(t), t \in [0, T])$ in (16.44) has continuous sample paths. The process $(X(t), t \in [0, T])$ satisfies

$$\mathbb{E}\left[\|X\|^2\right] < \infty$$

where $\|\cdot\|$ is the uniform norm on $C([0, T])$.

The mutual information $I(X, Y)$ can be expressed as an estimation error where X is the signal and Y is the observations. This result for $H = 1/2$ is given in [12] and for $H \neq 1/2$ in [9].

Theorem 16.5 *If $H \in (1/2, 1)$ then let A1 be satisfied, and if $H \in (0, 1/2)$ then let A2 be satisfied. The mutual information $I(X, Y)$ is given by*

$$I(X, Y) = \frac{1}{2}\mathbb{E}\left[\int_0^T \left|\mathbb{K}_H^{-1}(Z)(t) - \widehat{\mathbb{K}_H^{-1}(Z)}(t, t)\right|^2 dt\right] \qquad (16.47)$$

where

$$Z(t) = \int_0^t X(s)\,ds$$

and

$$\widehat{\mathbb{K}_H^{-1}(Z)}(t,t) = \mathbb{E}\left[\mathbb{K}_H^{-1}(Z)(t) \mid Y(u), 0 \le u \le t\right]. \tag{16.48}$$

The conditional expectation $\widehat{\mathbb{K}_H^{-1}(Z)}(t,t)$ can be computed using (16.31).

The next result for mutual information has been given for some types of white Gaussian noise [17], for a Gaussian noise described in an abstract Wiener space [28, 39], and for a fractional Brownian motion [9]. The equation for Y, (16.44), is slightly modified.

Let $a > 0$, let $T > 0$ be fixed and let $(Y_a(t), t \in [0, T])$ be the process that satisfies

$$\begin{aligned} dY_a(t) &= aX(t)\,dt + d\,B(t) \\ Y_a(0) &\equiv 0 \end{aligned} \tag{16.49}$$

where X and B are the processes in (16.44).

The rate of change of the mutual information with respect to a is expressed as an estimation error that differs from (16.47).

Theorem 16.6 *Let $(Y_a(t), t \in [0, T])$ be the process given by (16.49) where $a > 0$ is a parameter. If $H \in (1/2, 1)$, then let A1 be satisfied and if $H \in (0, 1/2)$ then let A2 be satisfied. Then the mutual information $I(X, Y_a)$ satisfies the following equality:*

$$\frac{d\,I(X, Y_a)}{d\,a} = a\mathbb{E}\left[\int_0^T \left|\mathbb{K}_H^{-1}(Z)(t) - \widehat{\mathbb{K}_H^{-1}(Z)}(t, T)\right|^2 dt\right] \tag{16.50}$$

where

$$\widehat{\mathbb{K}_H^{-1}(Z)}(t, T) = \mathbb{E}\left[\mathbb{K}_H^{-1}(Z)(t) \mid Y_a(s), 0 \le s \le T\right] \tag{16.51}$$

and

$$Z(t) = \int_0^t X(s)\,ds.$$

The results for $H = 1/2$ in Theorem 16.5 and Theorem 16.6 follow easily by a formal limit letting $H \to 1/2$.

Notes

Research supported in part by NSF grant DMS 0505706.

References

[1] Elisa Alòs, Olivier Mazet, and David Nualart. Stochastic calculus with respect to Gaussian processes. *Ann. Probab.*, 29(2):766–801, 2001.
[2] Anna Amirdjanova. Nonlinear filtering with fractional Brownian motion. *Appl. Math. Optim.*, 46(2–3):81–8, 2002. Special issue dedicated to the memory of Jacques-Louis Lions.

[3] Francesca Biagini, Yaozhong Hu, Bernt Øksendal, and Tusheng Zhang. *Stochastic Calculus for Fractional Brownian Motion and Applications. Probability and its Applications* (New York). Springer-Verlag, London, 2008.

[4] L. Coutin and L. Decreusefond. Abstract nonlinear filtering theory in the presence of fractional Brownian motion. *Ann. Appl. Probab.*, 9(4):1058–90, 1999.

[5] L. Decreusefond and A. S. Üstünel. Stochastic analysis of the fractional Brownian motion. *Potential Anal.*, 10(2):177–214, 1999.

[6] T. E. Duncan. Some processes associated with a fractional Brownian motion. In *Mathematics of Finance*, vol. 351 of *Contemp. Math.*, pages 93–101. Amer. Math. Soc., Providence, RI, 2004.

[7] T. E. Duncan. Prediction for some processes related to a fractional Brownian motion. *Statist. Probab. Lett.*, 76(2):128–34, 2006.

[8] T. E. Duncan, J. Jakubowski, and B. Pasik-Duncan. Stochastic integration for fractional Brownian motion in a Hilbert space. *Stoch. Dyn.*, 6(1):53–75, 2006.

[9] T. E. Duncan. Mutual information for stochastic signals and fractional Brownian motion. *IEEE Trans. Inf. Theory*, 54(10):4432–8, 2008.

[10] T. E. Duncan, B. Maslowski, and B. Pasik-Duncan. Semilinear stochastic equations in a Hilbert space with a fractional Brownian motion. SIAM J. Math. Analysis 40 (2009), 2286–2315.

[11] Tyrone E. Duncan. Evaluation of likelihood functions. *Information and Control*, 13:62–74, 1968.

[12] Tyrone E. Duncan. On the calculation of mutual information. *SIAM J. Appl. Math.*, 19:215–20, 1970.

[13] Tyrone E. Duncan, Yaozhong Hu, and Bozenna Pasik-Duncan. Stochastic calculus for fractional Brownian motion. I. Theory. *SIAM J. Control Optim.*, 38(2):582–612 (electronic), 2000.

[14] I. M. Gel'fand and A. M. Yaglom. Calculation of the amount of information about a random function contained in another such function. *Amer. Math. Soc. Transl. (2)*, 12:199–246, 1959.

[15] I. V. Girsanov. On transforming a class of stochastic processes by absolutely continuous substitution of measures. *Teor. Verojatnost. i Primenen.*, 5:314–30, 1960.

[16] Gustaf Gripenberg and Ilkka Norros. On the prediction of fractional Brownian motion. *J. Appl. Probab.*, 33(2):400–10, 1996.

[17] Dongning Guo, Shlomo Shamai, and Sergio Verdú. Mutual information and minimum mean-square error in Gaussian channels. *IEEE Trans. Inform. Theory*, 51(4):1261–82, 2005.

[18] G. A. Hunt. Random Fourier transforms. *Trans. Amer. Math. Soc.*, 71:38–69, 1951.

[19] H. E. Hurst. Long-term storage capacity in reservoirs. *Trans. Amer. Soc. Civil Eng.*, 116:400–10, 1951.

[20] R. E. Kalman and R. S. Bucy. New results in linear filtering and prediction theory. *Trans. ASME Ser. D. J. Basic Engrg.*, 83:95–108, 1961.

[21] M. S. Keshner. 1/f noise. *Proc. IEEE*, 70:212–18, 1982.

[22] A. N. Kolmogoroff. Wienersche Spiralen und einige andere interessante Kurven im Hilbertschen Raum. *C. R. (Doklady) Acad. Sci. URSS (N.S.)*, 26:115–18, 1940.

[23] A. N. Kolmogoroff. Sur l'interpolation et extrpolation des suites stationnaires. *C.R. Acad. Sci.*, 208:2043–5, 1939.

[24] W. Leland, M. Taqqu, W. Willinger, and D. Wilson. On the self-similar nature of ethernet traffic. *IEEE/ACM Trans. Networking*, 2:1–15, 1994.

[25] Terry J. Lyons. Differential equations driven by rough signals. *Rev. Mat. Iberoamericana*, 14(2):215–310, 1998.

[26] B. B. Mandelbrot. The variation of certain speculative prices. *Journal of Business*, 36:394–419, 1963.
[27] Benoit B. Mandelbrot and John W. van Ness. Fractional Brownian motions, fractional noises and applications. *SIAM Rev.*, 10:422–37, 1968.
[28] Eddy Mayer-Wolf and Moshe Zakai. Some relations between mutual information and estimation error in Wiener space. *Ann. Appl. Probab.*, 17(3):1102–16, 2007.
[29] David Nualart. *The Malliavin Calculus and Related Topics*. Probability and its Applications (New York). Springer-Verlag, Berlin, 2nd edn, 2006.
[30] David Nualart and Youssef Ouknine. Regularization of differential equations by fractional noise. *Stochastic Process. Appl.*, 102(1):103–16, 2002.
[31] Ivan Osorio and Mark G. Frei. Hurst parameter estimation for epileptic seizure detection. *Commun. Inf. Syst.*, 7(2):167–76, 2007.
[32] Vladas Pipiras and Murad S. Taqqu. Are classes of deterministic integrands for fractional Brownian motion on an interval complete? *Bernoulli*, 7(6):873–97, 2001.
[33] L. C. G. Rodgers. Arbitrage with fractional Brownian motion. *Math. Finance*, 7:95–105, 1997.
[34] Stefan G. Samko, Anatoly A. Kilbas, and Oleg I. Marichev. *Fractional Integrals and Derivatives*. Gordon and Breach Science Publishers, Yverdon, 1993. Theory and Applications, ed. and with a foreword by S. M. Nikol'skiĭ, trans. from the 1987 Russian original, rev. by the authors.
[35] Norbert Wiener. *Extrapolation, Interpolation, and Smoothing of Stationary Time Series. With Engineering Applications*. The Technology Press of the Massachusetts Institute of Technology, Cambridge, MA, 1949.
[36] Herman Wold. *A Study in the Analysis of Stationary Time Series*. 2nd edn, With an appendix by Peter Whittle. Almqvist and Wiksell, Stockholm, 1954.
[37] Jie Xiong and Xingqiu Zhao. Nonlinear filtering with fractional Brownian motion noise. *Stoch. Anal. Appl.*, 23(1):55–67, 2005.
[38] M. Zähle. Integration with respect to fractal functions and stochastic calculus. I. *Probab. Theory Related Fields*, 111(3):333–74, 1998.
[39] Moshe Zakai. On mutual information, likelihood ratios, and estimation error for the additive Gaussian channel. *IEEE Trans. Inform. Theory*, 51(9):3017–24, 2005.

PART V
Estimation and Control

·17·
Dual Filters, Path Estimators, and Information

N. J. Newton

17.1 Introduction

This chapter investigates nonlinear filtering from an information theoretic viewpoint. At its heart are two distinct dualities: one a feature of time reversal, the other an instance of an abstract Fenchel–Legendre transform for Bayesian estimators. The first duality arises from a time symmetry in the joint dynamics of the signal process and its nonlinear filter. A dual system, with the same structure as the original, is obtained by time reversal if the signal and filter processes exchange roles. *Flows* of Shannon information can be identified for primal and dual systems; these comprise two components dubbed *supply* and *dissipation*. The former is the mutual information between particular components of the signal and observation processes, and is thus a property of the underlying partially observed system; the latter, on the other hand, depends on the filter dynamics, and characterizes filter optimality. Despite this conceptual difference, supply and dissipation have similar local characteristics; in fact the rate of supply of one system coincides with the rate of dissipation of its dual.

The concept of dual filters first appeared in the context of discrete-time Kalman filtering, where it was obtained by change of basis techniques in the finite-dimensional Hilbert spaces supporting the signal and observation processes. (See Ch. 15 in [14].) It was developed for continuous-time linear systems in [20], by the somewhat different technique of time-reversal; this has the advantage that it applies equally to *nonlinear* systems. The case of nonlinear filters for discrete-state processes in continuous time was developed in detail in [21], and a more general class of nonlinear systems (including filters for diffusions) was treated in a more formal sense in [22]; the latter also makes connections with nonequilibrium statistical mechanics. (See also [19] which develops statistical mechanical interpretations of the Kalman–Bucy filter.)

The second duality of this chapter is that between the *full information* of a log-likelihood function, and the *information gain* of the corresponding posterior distribution in the context of Bayesian estimation, [18]. It concerns variational characterizations of the log-likelihood function and posterior distribution that have natural information-theoretic interpretations. By applying this duality to the path estimation problems associated with the forward and reverse time filters, we obtain forward and backward stochastic optimal control problems, in which the two filters appear in the value functions. The second duality, applied in this way, becomes the "duality between estimation and control" first studied in the linear case in [13]. (See also [14].) It places this duality in a more fundamental framework. The duality between estimation and control was used in [11], in the context of nonlinear filters for diffusions, to obtain bounds on the filter density process. (See [18] for its Bayesian interpretation in this context.) The "forward" characterization of [18] (log-likelihood function to posterior distribution) was applied to discrete-state, continuous-time filters in [25], in both forward and reverse time. (The results on time reversal in [25], however, concern the *primal* system of this chapter with time-reversed observations, rather than the dual system investigated here.)

We restrict attention to (primal) systems comprising discrete-state, continuous-time signal processes, and "signal-plus-white-noise" observations. The filter for this type of system is reviewed in Section 17.2, where some properties of the joint signal–filter process are also summarized. The corresponding dual system is derived in Section 17.3; it comprises a multidimensional diffusion signal with nonlinear dynamics, and observations of the point process variety; it gives rise to a finite-dimensional nonlinear filter. Section 17.4 defines and evaluates the primal and dual information supply and dissipation processes, and connects them with the log-likelihood functions of the filters. It is based on dual transformations of the Cameron–Martin–Girsanov type. Section 17.5 develops the Fenchel–Legendre duality of Bayesian estimators in this context; the abstract case of [18] is reviewed in Subsection 17.5.1, and then applied in Subsections 17.5.2 and 17.5.3 to the primal and dual systems, respectively. In both applications the "forward" transform yields a problem in stochastic optimal control, and the "inverse" transform becomes a deterministic optimization problem. The Markov nature of the signal-filter pair allows the Bayesian duality to be localized. This gives rise to dynamic programming equations (and their adjoints) that characterize local versions of the information supply and dissipation rates of both systems. Section 17.6 discusses the results and makes some concluding remarks.

17.2 Continuous-time, discrete-state filtering

One of the earliest continuous-time nonlinear filters to be derived is that for a partially observed Markov jump signal process, with observations contaminated by white noise [26], [24]. We review this filter here, and establish a number of properties of the joint signal–filter process that will be needed later. In the model we consider, both signal and observation processes evolve over the symmetric time interval $[-T, T]$. This simplifies the material in later sections where time reversal plays a major role. All random quantities are defined on the complete probability space $(\Omega, \mathcal{F}, \mathbb{P})$.

The signal, X, is a Markov process taking values in the finite set $\mathbb{S} = \{1, 2, \ldots, n\}$, and having finite rate matrix A, so that

$$\mathbb{P}(X_{t+\epsilon} = i \mid X_t = j) = (\mathbb{I}_n + A\epsilon)_{i,j} + o(\epsilon), \tag{17.1}$$

where $\mathbb{I}_n$ is the $n \times n$ matrix identity. We assume that the sample paths of X belong to the Skorokhod space $D([-T, T]; \mathbb{S})$; ie. the space of functions $x : [-T, T] \to \mathbb{S}$ that have left and right limits at all points $t \in (-T, T)$, and that are left or right continuous at all $t \in [-T, T]$. (We do not adopt the common convention of right-continuity, since we shall be interested in jump semimartingales derived from X in both time directions.) $D([-T, T]; \mathbb{S})$ is topologized by the Skorokhod metric. (See, for example, [10].)

We shall be particularly interested in probability measures μ on $\mathbb{S}$ that are mutally absolutely continuous with respect to the counting measure. By identifying $\mu(\{i\})$ with q_i, we can place these in one-to-one correspondence with members of the simplex:

$$\mathbb{F} := \{q \in \mathbb{R}^n : q_i \in (0, 1) \text{ for all } i;\ \Sigma_i q_i = 1\}. \tag{17.2}$$

This association leads to a filter variable that takes values in a subset of $\mathbb{R}^n$, thereby bringing to bear the well developed theory of stochastic processes in $\mathbb{R}^n$. We shall frequently abuse terminology by referring to elements of $\mathbb{F}$ as *distributions*. A standing assumption is that the initial value of the signal, X_{-T}, has a distribution belonging to $\mathbb{F}$; the prior distribution of X_t, p_t, will then also be in $\mathbb{F}$ for all t.

We consider observations comprising *initial* and *running* parts. The details of the initial observation are not important; we require only that it should be X_{-T}-conditionally independent of X, that it should give rise to a posterior distribution (for X_{-T}), $\zeta : \Omega \to \mathbb{F}$, and that

$$\sum_{i \in \mathbb{S}} \mathbb{E} \log\left(\zeta_i^{-1}\right) < \infty. \tag{17.3}$$

(The last condition ensures the finiteness of two of the information flow rates derived in Section 17.4.) The initial observation can be thought of as summarizing observations prior to time $-T$ in a version of the system extending beyond the interval $[-T, T]$. The *running* observation takes the form

$$Y_t^r = \int_{-T}^{t} g(X_s)\, ds + W_t \quad \text{for } t \in [-T, T], \tag{17.4}$$

where $g : \mathbb{S} \to \mathbb{R}$, and W is a *shifted* Brownian motion, independent of (X, ζ); i.e. ($W_t = B_{T+t}$, $t \in [-T, T]$), where B is a standard Brownian motion independent of (X, ζ). The *full* observation is the $\mathbb{F} \times \mathbb{R}$-valued process $Y = ((\zeta, Y_t^r),\ t \in [-T, T])$.

Remark 17.1 The results of this chapter apply equally to systems with multidimensional running observations. As with the initial observation, the details of the running observation process are unimportant in what follows. What matters is the filter process to which they give rise. (See [21] for further discussion of multidimensional observations.)

The notation $\mathcal{F}^{\xi}_{s,t}$ will be used to signify the sub-σ-field of $\mathcal{F}$ generated by a process ξ over the time interval $[s, t]$:

$$\mathcal{F}^{\xi}_{s,t} := \sigma(\xi_r, \, r \in [s, t]). \tag{17.5}$$

Thus the observation filtration is denoted $(\mathcal{F}^{Y}_{-T,t}, \, t \in [-T, T])$.

Let $(Z_t, \, t \in [-T, T])$ be a continuous $\mathbb{F}$-valued process solving the following Itô equation:

$$Z_t = \zeta + \int_{-T}^{t} (AZ_s - \sigma(Z_s)\bar{g}(Z_s))\,ds + \int_{-T}^{t} \sigma(Z_s)\,dY^r_s, \tag{17.6}$$

where $\bar{g} : \mathbb{F} \to \mathbb{R}$ and $\sigma : \mathbb{F} \to \mathbb{T}$ are defined as follows:

$$\bar{g}(q) := \sum_i g(i) q_i \quad \text{and} \quad \sigma(q)_i := q_i(g(i) - \bar{g}(q)), \tag{17.7}$$

and $\mathbb{T}$ ($:= \{\theta \in \mathbb{R}^n : \Sigma_i \theta_i = 0\}$) is the tangent space to $\mathbb{F}$ at all points. That the process Z is well defined is proven in proposition 2.1 of [21]; it is the nonlinear filter for X in the sense that $\mathbb{P}\left(X_t = i \mid \mathcal{F}^{Y}_{-T,t}\right) = Z_{t,i}$ for all t, i. (See theorem 9.1 in [16] or section 3 of [26].)

Remark 17.2 The filtering problems in [26] and [24] (and most papers on nonlinear filtering) do not involve an initial observation. However, their extension to the problem addressed here is straightforward. See the discussion in [21].

Various properties of the joint process (X, Z), are summarized in the following Proposition, which is proved (as proposition 2.1) in [21]. (See also [2] on the Markov property in a more general setting.)

Proposition 17.1

(i) The processes X, Z and (X, Z) are Markov, and have generators defined (for appropriate f) as follows:

$$(\mathcal{L}^X f)(i) = \sum_j A_{j,i}(f(j) - f(i))$$

$$(\mathcal{L}^Z f)(q) = \sum_j (Aq)_j \frac{\partial f}{\partial q_j}(q) + \frac{1}{2} \sum_{j,k} a(q)_{j,k} \frac{\partial^2 f}{\partial q_j \partial q_k}(q)$$

$$(\mathcal{L}^{X,Z} f)(i, q) = \sum_j b(i, q)_j \frac{\partial f}{\partial q_j}(i, q) + \frac{1}{2} \sum_{j,k} a(q)_{j,k} \frac{\partial^2 f}{\partial q_j \partial q_k}(i, q) \tag{17.8}$$

$$+ \sum_j A_{j,i}(f(j, q) - f(i, q)),$$

where
$$a(q) := \sigma(q)\sigma(q)^T \quad \text{and} \quad b(i, q) := Aq + \sigma(q)(g(i) - \bar{g}(q)). \tag{17.9}$$

(ii) For each $t \in [-T, T]$, $\mathcal{F}^X_{-T,T}$ and $\mathcal{F}^Y_{-T,t}$ are $\mathcal{F}^Z_{-T,t}$-conditionally independent.
(iii) For each $t \in [-T, T]$, $\mathcal{F}^{X,Z}_{-T,t}$ and $\mathcal{F}^X_{t,T}$ are X_t-conditionally independent.
(iv) For each $t \in [-T, T]$, $\mathcal{F}^Z_{-T,t}$ and $\mathcal{F}^{X,Z}_{t,T}$ are Z_t-conditionally independent.
(v) $\sup_{i \in \mathbb{S}, t \in [-T,T]} \mathbb{E} \log(Z_{t,i}^{-1}) < \infty$.

Remark 17.3 Proposition 17.1(ii) shows that the filter process Z can be thought of as a *surrogate* observation; for any t, $(Z_s, s \in [-T, t])$ has the same worth as $(Y_s, s \in [-T, t])$ as far as estimating any part of X is concerned. The main focus of this chapter is the dependency between the two processes X and Z, and so we shall not be too concerned with Y in what follows.

17.3 The dual filter

The Markov property of (X, Z) is invariant under time reversal, and the two properties in parts (iii) and (iv) of Proposition 17.1 transform into each other if time is reversed and X and Z are interchanged. Because of this, the time-reversed dynamics of the joint process (X, Z) can be considered to be the forward-time dynamics of a dual system, in which Z is a signal process and X a parametrization of its filter. In particular, the time-reversed filter process can be thought of as a dual signal, which is Markov because of Proposition 17.1(iv). This is true without additional assumptions; however, in order to characterize the dynamics of the dual signal, it is convenient to make the following technical assumption:

(H1) ζ (= Z_{-T}) has a square-integrable density with respect to $(n-1)$-dimensional Lebesgue measure on $\mathbb{F}$, $\lambda_\mathbb{F}$.

It then follows that the distribution of Z_t has a density, $p_Z(\cdot, t)$, for almost all t, and that this density is differentiable (at least in the sense of distributions). (See lemma 3.1 in [21], theorem 3.1 in [12], and Remark 17.4 below.)

For each $t \in [-T, T]$ and $q \in \mathbb{F}$, let
$$\begin{aligned} X^*_t &:= Z_{-t}, \\ Z^*_t &:= X_{-t}, \\ \bar{b}(q, t) &:= -Aq + \text{vec}_i \left\{ \sum_j \frac{\partial a_{i,j}}{\partial q_j} \right\}(q) + \sigma(q) D_q \log p_Z(q, -t)(\sigma(q)), \end{aligned} \tag{17.10}$$

where $\bar{b}$ is taken to be zero at any (q, t) for which $p_Z(q, -t)$ is zero, and $D_q \log p_Z(q, -t) : \mathbb{T} \to \mathbb{R}$ is the directional derivative of $\log p_Z$ with respect to its first argument: for $V \in \mathbb{T}$,

$$D_q \log p_Z(q, -t)(V) := \frac{d}{dr} \log p_Z(q + r V, -t)\Big|_{r=0}. \qquad (17.11)$$

For each $t \in [-T, T]$ and $q \in \mathbb{F}$, let $\bar{A}(t)$ and $\check{A}(q)$ be $n \times n$ Markov rate matrices defined as follows:

$$\bar{A}(t)_{i,j} := A_{j,i} \frac{p_{-t,i}}{p_{-t,j}} \quad \text{and} \quad \check{A}(q)_{i,j} := A_{j,i} \frac{q_i}{q_j} \quad \text{if } j \neq i$$

$$\bar{A}(t)_{i,i} := -\sum_{j \neq i} \bar{A}(t)_{j,i} \quad \text{and} \quad \check{A}(q)_{i,i} := -\sum_{j \neq i} \check{A}(q)_{j,i}, \qquad (17.12)$$

where p_{-t} is the distribution of X_{-t}. The following theorem is proved (as theorem 3.2) in [21].

Theorem 17.1 *The processes X^*, Z^* and (X^*, Z^*) are Markov, and have generators defined (for appropriate f) as follows*

$$\left(\mathcal{L}_t^{X^*} f\right)(q) = \sum_j \bar{b}(q, t)_j \frac{\partial f}{\partial q_j}(q) + \frac{1}{2} \sum_{j,k} a(q)_{j,k} \frac{\partial^2 f}{\partial q_j \partial q_k}(q)$$

$$\left(\mathcal{L}_t^{Z^*} f\right)(i) = \sum_j \bar{A}(t)_{j,i}(f(j) - f(i)) \qquad (17.13)$$

$$\left(\mathcal{L}_t^{X^*, Z^*} f\right)(q, i) = \sum_j \bar{b}(q, t)_j \frac{\partial f}{\partial q_j}(q, i) + \frac{1}{2} \sum_{j,k} a(q)_{j,k} \frac{\partial^2 f}{\partial q_j \partial q_k}(q, i)$$

$$+ \sum_j \check{A}(q)_{j,i}(f(q, j) - f(q, i)).$$

In what follows $p_t^* \, (:= p_Z(\cdot, -t))$ will denote the prior density of X_t^*. (cf. the prior distribution of X_t, p_t.)

Remark 17.4 In [21], condition (H1) is expressed in terms of the $\mathbb{R}^{n-1}$-valued random variable $\Gamma \zeta$, where $\Gamma = [\mathbb{I}_{n-1} \, 0]$. It is then shown that the distribution of $\tilde{Z}_t \, (:= \Gamma Z_t)$ has a density $p_{\tilde{Z}}(\cdot, t)$ for almost all t. The directional derivative of (17.11) then takes the form:

$$D_q \log p_Z(q, -t)(V) = V^\mathsf{T} \Gamma^\mathsf{T} (\nabla_{\tilde{q}} \log p_{\tilde{Z}})(\Gamma q, -t). \qquad (17.14)$$

We regard X^* and Z^* as being the signal and filter processes of a dual system. The dual signal X^* is a Markov diffusion process in its own right, evolving on the state space $\mathbb{F}$, and having initial density p_{-T}^* and generator $(\mathcal{L}_t^{X^*}, t \in [-T, T])$. The $\mathbb{S}$-valued process Z^* is a filter for this dual signal in the sense that Z_t^* is a sufficient statistic for estimating the future of X^* from the past of Z^*, as is shown by Proposition 17.1(iii). Of course Z_t^* does not take values in the set of probability measures on $\mathbb{F}$, but (like Z) it is a *parametrization* of the conditional distribution; in fact an application of Proposition 17.1(iii), and Bayes' formula between X_t^* and Z_t^* shows that, for any Borel subset $B \subseteq \mathbb{F}$,

$$\mathbb{P}\left(X_t^* \in B \mid \mathcal{F}_{-T,t}^{Z^*}\right) = \mathbb{P}\left(X_t^* \in B \mid Z_t^*\right) = \int_B \frac{q_i}{p_{-t,i}} p_t^*(q,t) \lambda_\mathbb{F}(dq)\Big|_{i=Z_t^*}. \quad (17.15)$$

Clearly we could also regard Z^* as being the dual observation process; however, there are other possibilities. Any dual observation process $\left(Y_t^*, t \in [-T, T]\right)$ must satisfy the following two conditions:

(O1) $\mathcal{F}_{-T,t}^{Y^*} \supseteq \mathcal{F}_{-T,t}^{Z^*}$ for all $t \in [-T, T]$;
(O2) X^* and $\mathcal{F}_{-T,t}^{Y^*}$ are $\mathcal{F}_{-T,t}^{Z^*}$-conditionally independent for all $t \in [-T, T]$.

The first of these requires that the dual filter should be derivable from the dual observation, the second that any randomness in Y^* but not in Z^*, should bear no additional information about X^*. The observation process of the primal system of Section 17.2 obviously satisfies the equivalent of (O1); that it also satisfies the equivalent of (O2) is demonstrated in Proposition 17.1(ii). (See also Remark 17.3.)

The following is an example of a dual observation process satisfying (O1) and (O2). Let $\left(Y_t^*, t \in [-T, T]\right)$ be the n-vector process whose i'th component is defined as follows:

$$Y_{t,i}^* = \mathbf{1}_{\{i\}}\left(Z_{-T}^*\right) + N(Z^*, i)_t, \quad (17.16)$$

where $\mathbf{1}$ is the indicator function and, for $x \in D([-T, T]; \mathbb{S})$ and $i \in \mathbb{S}$,

$$N(x, i)_t := \sum_{s \in (-T, t]} (1 - \mathbf{1}_{\{i\}}(x_{s-}))\mathbf{1}_{\{i\}}(x_{s+}). \quad (17.17)$$

Y^*, thus defined, is a vector of right-continuous counting processes. Like the observation process of the primal system it comprises initial and running parts. The i'th component of its running part $\left(Y_{t,i}^* - Y_{-T,i}^*, t \in [-T, T]\right)$ increments by one whenever Z^* jumps into state i. Conditions (O1) and (O2) are trivially satisfied for this process since $\mathcal{F}_{-T,t}^{Y^*} = \mathcal{F}_{-T,t+}^{Z^*}$ and $\mathbb{P}\left(Z_t^* = Z_{t+}^*\right) = 1$ for all t. An example of a dual observation process generating a larger filtration than $\left(\mathcal{F}_{-T,t}^{Z^*}\right)$ is given in [21].

The dual filter is a very particular example of a *finite-dimensional* nonlinear filter. The dual signal is a multidimensional diffusion process with nonlinear dynamics, and the components of the dual observation process are typically Markov-modulated point processes. In general, problems of this nature lead to infinite-dimensional nonlinear filters. Here, because of the special connection between the prior distribution of X^*, and the observation mechanism, the nonlinear filter is not only of finite dimension, but even evolves on a finite space! Of course, any implementation of the dual filter would need to compute the prior density of the dual signal, p_t^*, as this appears in (17.15). However, this would not need to be done in "real time" as it does not depend on the observations.

17.4 Information quantities

This section evaluates the mutual information between various components of the processes X and Z, and shows that there exist well-defined *flows* of Shannon information in both primal and dual systems. The mutual information between the signal and observation processes can be found by means of transformations of the Cameron–Martin–Girsanov type, under which signal and observation processes become independent. This approach was first investigated in the context of nonlinear filters for diffusions in [9]. We start by defining forward- and reverse-time exponential semimartingales.

Let $(L_t, t \in [-T, T])$ and $(L_t^*, t \in [-T, T])$ be $\mathbb{R}^+$-valued processes defined as follows:

$$L_t := l_T(Z_{-T}, X_{-T}) \exp\left(\int_{-T}^{t} (g(X_s) - \bar{g}(Z_s))\, dW_s \right.$$
$$\left. + \frac{1}{2} \int_{-T}^{t} (g(X_s) - \bar{g}(Z_s))^2\, ds\right),$$
$$L_t^* := l_{-T}(X_{-T}^*, Z_{-T}^*) \prod_i \exp\left(\int_{(-T, t]} \log\left(\frac{l_s(X_s^*, i)}{l_s(X_s^*, Z_{s-}^*)}\right) dY_{s,i}^* \right. \tag{17.18}$$
$$\left. - \int_{-T}^{t} \left(\frac{l_s(X_s^*, i)}{l_s(X_s^*, Z_s^*)} - 1\right) \bar{A}(s)_{i, Z_s^*}\, ds\right),$$

where Y^* is the dual observation process of (17.16), and $l_t : \mathbb{F} \times \mathbb{S} \to \mathbb{R}+$ is defined as follows

$$l_t(q, i) := q_i / p_{-t, i}. \tag{17.19}$$

Proposition 17.2 Let $\mathbb{P}^M$ be the measure on $\mathcal{F}$ defined by $d\mathbb{P}^M = L_T^{-1} d\mathbb{P}$.

(i) $\mathbb{P}^M$ is a probability measure, and is mutually absolutely continuous with respect to $\mathbb{P}$.
(ii) $\mathbb{P}(L_T = L_T^*) = 1$.
(iii) Under $\mathbb{P}^M$, $\left(L_t, \mathcal{F}_{-T,t}^{X,Z}\right)$ is a martingale.
(iv) Under $\mathbb{P}^M$, $\left(L_t^*, \mathcal{F}_{-T,t+}^{X^*,Z^*}\right)$ is a martingale.
(v) Under $\mathbb{P}^M$, X and Z are independent processes, but retain the marginal distributions they have under $\mathbb{P}$.

Proof. Parts (i), (iii) and (v) rely on straightforward applications of the Girsanov theorem. The details appear in the proof of parts (vi), (vii) and (viii) of proposition 2.1 in [21]. (See also theorem 7.23 in [16].) Parts (ii) and (iv) are proved by a primal-time development of l_t; according to Itô's rule:

$$\log\left(\frac{l_{-T}(Z_T, X_T)}{l_T(Z_{-T}, X_{-T})}\right) = \int_{-T}^{T}\left(\sum_i(\check{A}(Z_t) - \bar{A}(-t))_{i,X_t} + \frac{1}{2}(g(X_t) - \bar{g}(Z_t))^2\right)dt$$

$$+ \int_{-T}^{T}(g(X_t) - \bar{g}(Z_t))\,dW_t + \sum_{t\in(-T,T)}\log\left(\frac{l_{-t}(Z_t, X_{t+})}{l_{-t}(Z_t, X_{t-})}\right),$$

and so

$$L_T = l_{-T}\left(X^*_{-T}, Z^*_{-T}\right)\exp\left(-\int_{-T}^{T}\sum_i\left(\check{A}\left(X^*_t\right) - \bar{A}(t)\right)_{i,Z^*_t}dt\right.$$

$$\left. - \sum_{t\in(-T,T)}\log\left(\frac{l_t\left(X^*_t, Z^*_{t-}\right)}{l_t\left(X^*_t, Z^*_{t+}\right)}\right)\right) = L^*_T,$$

which proves part (ii). The same argument shows that $L^*_t = l_t(Z_{-t}, X_{-t})L_T/L_{-t}$. Now

$$\mathbb{E}^M\left(L^*_T \mid \mathcal{F}^{X^*,Z^*}_{-T,t}\right) = \mathbb{E}^M(L_T \mid \mathcal{F}^{X,Z}_{-t,T}) = \mathbb{E}^M(L_{-t} \mid \mathcal{F}^{X,Z}_{-t,T})L_T/L_{-t}$$

$$= \frac{\mathbb{E}^M(L_{-t} \mid X_{-t}, Z_{-t})}{l_t(Z_{-t}, X_{-t})}L^*_t,$$

where we have used the Markov property in the final step. However, for any $B \in \mathcal{F}^{X,Z}_{-t,-t}$

$$\int_B \mathbb{E}^M(L_{-t} \mid X_{-t}, Z_{-t})\,d\mathbb{P}^M = \int_B L_{-t}\,d\mathbb{P}^M = \mathbb{P}(B) = \int_B l_t(Z_{-t}, X_{-t})\,d\mathbb{P}^M,$$

where we have used the facts that $L_{-t} = \mathbb{E}^M(L_T|\mathcal{F}^{X,Z}_{-T,-t})$ and $l_t(Z_{-t}, X_{-t}) = \mathbb{E}^M(L_T|\mathcal{F}^{X,Z}_{-t,-t})$. This proves part (iv). $\square$

Remark 17.5 The two representations for $L_T\,(= L^*_T)$ in (17.18) express the dependency between X and Z *progressively* in different time directions. In both cases, the term involving l expresses an initial dependency, and the exponential term expresses a dependency having its origins in the joint generator, $\mathcal{L}^{X,Z}$ or $\mathcal{L}^{X^*,Z^*}_t$. The use of Itô's rule in the proof of Proposition 17.2 moves the initial dependency from one end of the time interval to the other. This technique will be used again in Section 17.5.

We define (primal) information *supply* (S), *storage* (C) and *dissipation* (D) processes as follows: for each $t \in [-T, T]$,

$$S(t) := I(X; (Y_s, s \in [-T, t]))$$
$$C(t) := I((X_s, s \in [t, T]); (Y_s, s \in [-T, t])) \qquad (17.20)$$
$$D(t) := S(t) - C(t),$$

where, for random variables Θ and Φ taking values in measurable spaces and having joint and marginal distributions $\mathbb{P}_{\Theta,\Phi}$, $\mathbb{P}_\Theta$ and $\mathbb{P}_\Phi$, $I(\Theta\,;\Phi)$ is the *mutual information*:

$$I(\Theta\,;\Phi) := \begin{cases} \int \log\left(\frac{d\mathbb{P}_{\Theta,\Phi}}{d(\mathbb{P}_\Theta \otimes \mathbb{P}_\Phi)}\right) d\,\mathbb{P}_{\Theta,\Phi} & \text{if the integral exists} \\ +\infty & \text{otherwise.} \end{cases} \qquad (17.21)$$

We define *dual* information supply (S^*), storage (C^*) and dissipation (D^*) processes in a similar way: for each $t \in [-T, T]$,

$$\begin{aligned} S^*(t) &:= I\left(X^*\,;\,(Y_s^*, s \in [-T, t])\right) \\ C^*(t) &:= I\left((X_s^*, s \in [t, T])\,;\,(Y_s^*, s \in [-T, t])\right) \\ D^*(t) &:= S^*(t) - C^*(t). \end{aligned} \qquad (17.22)$$

Remark 17.6 Where one or both of the random quantities in $I(\cdot\,;\,\cdot)$ is a *stochastic process*, as is the case in (17.20) and (17.22), it is regarded as being a random variable taking values in the appropriate path space. For example, $(X_s, s \in [-T, t])$ takes values in the Skorokhod space $D([-T, t]; \mathbb{S})$, and $(Y_s, s \in [-T, t])$ takes values in the space $\mathbb{F} \times C([-T, t]; \mathbb{R})$ (which is topologized by the supremum metric). The mutual information quantities depend only on the finite-dimensional distributions of these processes.

The reader should think of $S(t)$ (respectively $C(t)$) as being the information about X (respectively $(X_s, s \in [t, T])$) made available to the filter by the observation $(Y_s, s \in [-T, t])$. It follows from Proposition 17.1(ii), (O1) and (O2) that Y and Y^* can be replaced by Z and Z^* in (17.20) and (17.22). Furthermore, it follows from Proposition 17.1(iii,iv), that

$$C(t) = C^*(-t) = I(X_t\,;\,Z_t) = \mathbb{E}\log \mathbb{E}^M\left(L_T | \mathcal{F}_{-t,-t}^{X,Z}\right) = \mathbb{E}\log l_{-t}(Z_t, X_t), \quad (17.23)$$

and from Proposition 17.1(iii) and Proposition 17.2(iii,v) that

$$\begin{aligned} S(t) &= I((X_s, s \in [-T, t])\,;\,(Z_s, s \in [-T, t])) \\ &= \mathbb{E}\log \mathbb{E}^M\left(L_T | \mathcal{F}_{-T,t}^{X,Z}\right) \\ &= \mathbb{E}\log L_t \\ &= C(-T) + \frac{1}{2}\mathbb{E}\int_{-T}^{t} (g(X_s) - \tilde{g}(Z_s))^2\,ds, \end{aligned} \qquad (17.24)$$

and from Proposition 17.1(iv,v) and Proposition 17.2(iv,v) that

$$\begin{aligned} S^*(t) &= I\left((X_s^*, s \in [-T, t])\,;\,(Z_s^*, s \in [-T, t])\right) \\ &= \mathbb{E}\log L_t^* \\ &= C^*(-T) + \mathbb{E}\int_{-T}^{t}\sum_i \alpha\left(\frac{l_s(X_s^*, i)}{l_s(X_s^*, Z_s^*)}\right) \bar{A}(s)_{i, Z_s^*}\,ds, \end{aligned} \qquad (17.25)$$

where $\alpha : (0, \infty) \to [0, \infty)$ is defined as follows,

$$a(r) := r \log r - r + 1. \tag{17.26}$$

Similar calculations express $S^*(-t)$ in terms of S and D:

$$\begin{aligned} S^*(-t) &= I((X_s, s \in [t, T]); (Z_s, s \in [t, T])) \\ &= \mathbb{E} \log \circ \mathbb{E}^M \left(L_T \mid \mathcal{F}_{t,T}^{X,Z} \right) \\ &= \mathbb{E} \log \left(l_{-T}(Z_t, X_t) L_T / L_t \right) \\ &= C(t) + \frac{1}{2} \mathbb{E} \int_t^T (g(X_s) - \tilde{g}(Z_s))^2 \, ds \\ &= S(T) - D(t), \end{aligned} \tag{17.27}$$

and so

$$D(t) = S^*(T) - S^*(-t) = \mathbb{E} \int_{-t}^T \sum_i a \left(\frac{l_s(X_s^*, i)}{l_s(X_s^*, Z_s^*)} \right) \tilde{A}(s)_{i, Z_s^*} \, ds. \tag{17.28}$$

Similarly,

$$D^*(t) = S(T) - S(-t) = \frac{1}{2} \mathbb{E} \int_{-t}^T (g(X_s) - \tilde{g}(Z_s))^2 \, ds. \tag{17.29}$$

It follows from the elementary properties of mutual information that all these quantities are nonnegative. Furthermore, (17.24), (17.25), (17.28), and (17.29) show that S, D, S^*, and D^* have nonnegative time derivatives.

Since $D([-t, T]; \mathbb{S})$ is a Polish space, the distribution of $(Z_s^*, s \in [-t, T])$ can be factorized into the marginal of Z_{-t}^* and a regular Z_{-t}^*-conditional distribution of $(Z_s^*, s \in [-t, T])$. (See, for example, [8].) By factorizing the expectation operator in (17.28) in this way, we see that $D(t)$ is the average over the distribution of Z_{-t}^* (p_t) of the mutual information between X^* and a process $(Z_s^{*i,-t}, s \in [-t, T])$ with initial value $Z_{-t}^{*i,-t} = i$, and rate matrix $(\check{A}(X_s^*), s \in [-t, T])$:

$$D(t) = \sum_i I\left(X^*; \left(Z_s^{*i,-t}, s \in [-t, T]\right)\right) p_{t,i}. \tag{17.30}$$

Expressing this in terms of the primal processes, we see that $D(t)$ is the average mutual information between Z and a version of $(X_s, s \in [-T, t])$ *pinned* at the final value $X_t = i$. A similar interpretation can be given to the dual dissipation process D^*. The supply processes S and S^* are (by definition) *additive*; this means, for example, that the increment $S^*(T) - S^*(t)$ represents *new* information connecting X^* and Z^*, over and above that in $S^*(t)$. Since $S^*(T) - S^*(t) = D(-t)$, this new information is associated with the pinned process $Z^{*i,t}$.

The function l of (17.19) can be used to compute the *likelihood filter* in both time directions. In fact $l_{-t}(Z_t, \cdot)$ is the Radon–Nikodym derivative between the posterior and prior distributions of X_t, and $l_t(\cdot, Z_t^*)$ is the Radon–Nikodym derivative between the posterior and prior distributions of X_t^*. These likelihood filters govern the flow of information; it easily follows from (17.23), and the fact that $\mathbb{E}\partial \log l_t/\partial t\, (X_t^*, Z_t^*) = 0$, that

$$\dot{C}(t) = \mathbb{E}(\mathcal{L}^{X,Z} \log l_{-t})(Z_t, X_t) \quad \text{and} \quad \dot{C}^*(t) = \mathbb{E}\left(\mathcal{L}_t^{X^*,Z^*} \log l_t\right)(X_t^*, Z_t^*).$$

It follows from Proposition 17.1 and Theorem 17.1 that Z is X-conditionally Markov with generator $\mathcal{L}_t^{Z|X}$, and that Z^* is X^*-conditionally Markov with generator $\mathcal{L}_t^{Z^*|X^*}$, where

$$(\mathcal{L}_t^{Z|X} f)(q) = \sum_j b(X_t, q)_j \frac{\partial f}{\partial q_j} + \frac{1}{2} \sum_{j,k} a(q)_{j,k} \frac{\partial^2 f}{\partial q_j \partial q_k}$$

$$(\mathcal{L}_t^{Z^*|X^*} f)(i) = \sum_j \breve{A}(X_t^*)_{j,i} (f(j) - f(i)).$$

Straightforward calculations show how information supply is associated with the conditional generators of the filters, and how its dissipation is associated with the marginal generators of the signals:

$$\dot{S}(t) = \mathbb{E}\left(\mathcal{L}_t^{Z|X} \log l_{-t}(\cdot, X_t)\right)(Z_t),\ \dot{S}^*(t) = \mathbb{E}\left(\mathcal{L}_t^{Z^*|X^*} \log l_t(X_t^*, \cdot)\right)(Z_t^*),$$
$$\dot{D}(t) = -\mathbb{E}\left(\mathcal{L}^X \log l_{-t}(Z_t, \cdot)\right)(X_t),\ \dot{D}^*(t) = -\mathbb{E}\left(\mathcal{L}_t^{X^*} \log l_t(\cdot, Z_t^*)\right)(X_t^*). \tag{17.31}$$

The connection between the Shannon and Fisher information quantities for diffusion filters was investigated in [17]. It was shown there that the rate of change of the equivalent of $D(t)$ for a diffusion filter is a Fisher information quantity for the likelihood function of the filter. This is also true of the dual filter here; in fact

$$\dot{D}^*(t) = \frac{1}{2}\mathbb{E} \left\| \sigma^\mathsf{T} \nabla_q \log l_t\left(X_t^*, Z_t^*\right) \right\|.$$

Of course, it immediately follows from (17.28) and (17.29) that

$$\dot{D}(t) = \dot{S}^*(-t) \quad \text{and} \quad \dot{D}^*(t) = \dot{S}(-t), \tag{17.32}$$

which suggests that the processes of supply and dissipation are *locally* indistinguishable bar the direction of time. This will be investigated further in Section 17.5.

17.5 Path estimation

This section develops variational characterizations of the observation-conditional *path* distributions of the two signal processes X and X^*. Since the primal and dual systems are Markov, the variational characterizations become problems in stochastic optimal control, whose dynamic programming equations characterize the information flows of Section 17.4. In this context, the information flows can be attributed to action of controls. The variational characterizations are based on a more abstract result appearing in [18]. This is reviewed next.

17.5.1 A variational formulation of Bayesian estimation

In its most general form, this concerns an estimand X and an observation Y taking values in measurable spaces $(\mathbf{X}, \mathcal{X})$ and $(\mathbf{Y}, \mathcal{Y})$, and defined on a common probability space $(\Omega, \mathcal{F}, \mathbb{P})$. The marginal and joint distributions of X and Y are denoted P_X, P_Y, and $P_{X,Y}$. It is assumed that:

(H2) $I(X; Y) < \infty$.

Let λ_Y be a σ-finite (reference) measure on $\mathcal{Y}$ such that $P_{X,Y} \ll P_X \otimes \lambda_Y$. P_Y is an example of such a measure because of (H2) (and is, in fact, the measure used in the application of what follows to the dual systems of Sections 17.2 and 17.3). Let $\tilde{Q}$ be a version of the Radon–Nikodym derivative $d P_{X,Y}/d(P_X \otimes \lambda_Y)$, and let

$$\bar{\mathbf{Y}} := \left\{ y \in \mathbf{Y} : 0 < \int_X \tilde{Q}(x, y) P_X(dx) \text{ and } \int_X \tilde{Q}(x, y) \log \tilde{Q}(x, y) P_X(dx) < \infty \right\}.$$

Then $\bar{\mathbf{Y}} \in \mathcal{Y}$ and $P_Y(\bar{\mathbf{Y}}) = 1$. Let $Q(x, y) := \tilde{Q}(x, y)\mathbf{1}_{\bar{\mathbf{Y}}}(y) + \mathbf{1}_{\mathbf{Y}\setminus\bar{\mathbf{Y}}}(y)$ and

$$P_{X|Y}(B, y) := \frac{\int_B Q(x, y) P_X(dx)}{\int_X Q(x, y) P_X(dx)} \quad \text{for any } B \in \mathcal{X}. \tag{17.33}$$

$P_{X|Y}$ is a regular conditional distribution for X: $P_{X|Y}(\cdot, y)$ is a probability measure on $\mathcal{X}$ for each y, $P_{X|Y}(B, \cdot)$ is $\mathcal{Y}$-measurable for each B, and $P_{X|Y}(B, Y) = \mathbb{P}(X \in B \mid Y)$.

Let $\mathcal{P}(\mathcal{X})$ be the set of probability measures on $\mathcal{X}$, and $\mathcal{H}(\mathbf{X})$ the set of measurable $(-\infty, +\infty]$-valued functions on $(\mathbf{X}, \mathcal{X})$. For $\tilde{P}_X \in \mathcal{P}(\mathcal{X})$ and $\tilde{H} \in \mathcal{H}(\mathbf{X})$, let

$$
\begin{aligned}
h(\tilde{P}_X \mid P_X) &:= \begin{cases} \int_X \log\left(\dfrac{d\tilde{P}_X}{dP_X}(x)\right) \tilde{P}_X(dx) & \text{if } \tilde{P}_X \ll P_X \\ +\infty & \text{otherwise,} \end{cases} \\
\langle \tilde{H}, \tilde{P}_X \rangle &:= \begin{cases} \int_X \tilde{H}(x) \tilde{P}_X(dx) & \text{if the integral exists} \\ +\infty & \text{otherwise,} \end{cases} \\
i(\tilde{H}) &:= \begin{cases} -\log \mathbb{E} \exp(-\tilde{H}(X)) & \text{if the expectation is nonzero} \\ -\infty & \text{otherwise.} \end{cases}
\end{aligned} \tag{17.34}
$$

h is the relative entropy (Kullback–Leibler divergence) of two measures. In the context of probability measures, it can be thought of as the *information gain* of the first measure over the second. If we interpret $\exp(-\tilde{H})$ as a likelihood function for X, associated with some (unspecified) observation, then $\tilde{H}(x)$ is the *residual information* in that observation if we already know that $X = x$, and $i(\tilde{H})$

is the *full information* in that observation, i.e. the information in the unspecified observation if all we know about X is its prior P_X. The following proposition characterizes $h(P_{X|Y}(\cdot, y) \mid P_X)$ in terms of $i(H(\cdot, y))$ and vice versa, where

$$H(x, y) := -\log Q(x, y); \qquad (17.35)$$

a simple proof appears (as proposition 2.1) in [18].

Proposition 17.3 Suppose that (H2) is satisfied, and H and $P_{X|Y}$ are as defined in (17.35) and (17.33). Then, for any y:

(i)
$$i(H(\cdot, y)) = \min_{\tilde{P}_X \in \mathcal{P}(\mathcal{X})} \left\{ h(\tilde{P}_X \mid P_X) + \langle H(\cdot, y), \tilde{P}_X \rangle \right\}; \qquad (17.36)$$

(ii)
$$h(P_{X|Y}(\cdot, y) \mid P_X) = \max_{\tilde{H} \in \mathcal{H}(X)} \left\{ i(\tilde{H}) - \langle \tilde{H}, P_{X|Y}(\cdot, y) \rangle \right\}; \qquad (17.37)$$

(iii) $P_{X|Y}(\cdot, y)$ is the unique minimizer in (17.36);
(iv) if $\hat{H}$ is a maximizer in (17.37), then there exists a real constant K such that $\hat{H}(X) = H(X, y) + K$ a.s.

The term to be minimized in (17.36) is called in [18] the *apparent information* of the distribution $\tilde{P}_X$. It is the sum of the information gain of $\tilde{P}_X$ over the prior and the average residual information in the true likelihood function, and is thus the information *apparently* possessed by an estimator that proposes $\tilde{P}_X$ as a posterior distribution for X. This is greater than or equal to the full information in the true likelihood function, with equality if and only if $\tilde{P}_X$ is the true posterior distribution. The two components of the apparent information show the tension in Bayesian estimation between accommodating the prior and posterior information.

The term to be maximized in (17.37) is called in [18] the *compatible information* of the log-likelihood function $\tilde{H}$. It is the difference between the full information and the average (over the true posterior) residual information in the likelihood function $\exp(-\tilde{H})$; it is thus the amount of information in $\exp(-\tilde{H})$ compatible with the true posterior. This is less than or equal to the information gain of the true posterior, with equality if and only if $\exp(-\tilde{H})$ is equivalent to the true likelihood function in the sense of part (iv).

Remark 17.7 The compatible information, like the relative entropy, is an *absolute* information quantity in the sense that it does not depend on the reference measure λ_Y. All the other information quantities are *differential* information quantities. It is because of this that there is uniqueness in (17.36) but not in (17.37).

In the next two subsections we apply Proposition 17.3 to the problem of estimating the path X from the path Z, and that of estimating the path X^* from the path Z^*. The abstract spaces $\mathbf{X}$ and $\mathbf{Y}$ of this subsection become, there, the path spaces $D([-T, T]; \mathbb{S})$ and $C([-T, T]; \mathbb{F})$. Let $(x_t^d, t \in [-T, T])$ and $(x_t^c, t \in [-T, T])$ be the coordinate functions on $D([-T, T]; \mathbb{S})$ and $C([-T, T]; \mathbb{F})$, respectively, and, for any $s \le t$, let

$$\mathcal{D}_{s,t} := \sigma\left(x_r^d, r \in [s, t]\right) \quad \text{and} \quad \mathcal{C}_{s,t} := \sigma\left(x_r^c, r \in [s, t]\right). \tag{17.38}$$

The roles of $\mathcal{X}$ and $\mathcal{Y}$ will be played by $\mathcal{D}_{-T,T}$ and $\mathcal{C}_{-T,T}$.

17.5.2 Estimating X from Z

We apply Proposition 17.3 first to the problem of estimating the primal signal X from the surrogate observation Z. We do this recursively in *reverse* time, effectively estimating Z^* from X^*. The abstract estimand space $(\mathbf{X}, \mathcal{X})$ of Subsection 17.5.1 becomes, here, the space $(D([-T, T]; \mathbb{S}), \mathcal{D}_{-T,T})$, and the observation space $(\mathbf{Y}, \mathcal{Y})$ becomes $(C([-T, T]; \mathbb{F}), \mathcal{C}_{-T,T})$. Taking the marginal distribution of X^* as the reference measure λ_Y, we set

$$Q^*(x, z) := l_T(z_{-T}, x_{-T}) \exp\left(\int_{-T}^T \frac{(\dot{z}_t - Az_t)_{x_t}}{z_{t, x_t}} dt\right), \tag{17.39}$$

for $x \in D([-T, T]; \mathbb{S})$ and $z \in C^1([-T, T]; \mathbb{F})$. This is a *pathwise* version of the Radon–Nikodym derivative L_T of (17.18). It is defined, here, for *differentiable* paths z only. Its value for other paths in $C([-T, T]; \mathbb{F})$ can be defined by continuous extension. It then follows that $L_T = Q^*(X, Z)$ a.s. (See, for example, [4] and [6].)

In order to obtain a problem in stochastic optimal control, we need to express the associated log-likelihood function as an integral in reverse time, and so we define the reverse-time paths $z^* \in D([-T, T]; \mathbb{S})$ and $x^* \in C^1([-T, T]; \mathbb{F})$ by the relations $z_t^* = x_{-t}$ and $x_t^* = z_{-t}$. Let $h^* : \mathbb{T} \times \mathbb{F} \times \mathbb{S} \to \mathbb{R}$ and (for any $s \le t$) $H_{s,t}^* : D([-T, T]; \mathbb{S}) \times C^1([-T, T]; \mathbb{F}) \to \mathbb{R}$ be defined as follows:

$$H_{s,t}^*(z^*, x^*) := \int_s^t \frac{(\dot{x}_r^* + Ax_r^*)_{z_r^*}}{x_{r, z_r^*}^*} dr =: \int_s^t h^*\left(\dot{x}_r^*, x_r^*, z_r^*\right) dr. \tag{17.40}$$

Then $-\log Q^*(x, z) = H_{-T,T}^*(z^*, x^*) - \log l_T(x_T^*, z_T^*)$.

Proposition 17.3 characterizes the regular $(X^* = x^*)$-conditional distribution of Z^* constructed from Q^* as the unique minimizer of the following apparent information: for $\tilde{P} \in \mathcal{P}(\mathcal{D}_{-T,T})$,

$$F_T^*(\tilde{P}, x^*) := h(\tilde{P} \mid P_{Z^*}) + \langle H_{-T,T}^*(\cdot, x^*) - \log l_T\left(x_T^*, x_T^d\right), \tilde{P}\rangle, \tag{17.41}$$

where P_{Z^*} is the path distribution of Z^*. In what follows, we restrict the class of probability measures here to the distributions of *controlled* versions of Z^*.

Let $\mathbf{U}^*$ be the set of functions $v : \mathbb{S} \times \mathbb{S} \to (0, \infty)$ for which $v(i, j) = 1$ whenever $A_{i,j} \leq 0$. (In particular $v(i, i) = 1$ for all i.) For any "control process" $u \in C([-T, T]; \mathbf{U}^*)$, let $\bar{A}^u(t)$ be the following rate matrix:

$$\bar{A}^u(t)_{i,j} := \bar{A}(t)_{i,j} u_t(i, j) \quad \text{if } j \neq i, \quad \bar{A}^u(t)_{i,i} := -\sum_{j \neq i} \bar{A}^u(t)_{j,i}. \tag{17.42}$$

Let $(Z_t^{*u,q}, t \in [-T, T])$ be a Markov process with sample paths in $D([-T, T]; \mathbb{S})$, initial distribution $q \in \mathbb{F}$, and rate matrix $(\bar{A}^u(t), t \in [-T, T])$. (We regard $Z^{*u,q}$ as being a controlled version of Z^*.) Let $P^{*u,q}$ be the distribution of $Z^{*u,q}$, and $P_t^{*u,q}$ its restriction to $\mathcal{D}_{-T,t}$. For any $u \in C([-T, T]; \mathbf{U}^*)$, $z^* \in D([-T, T]; \mathbb{S})$, and $s \leq t$, let

$$M_{s,t}^{*u}(z^*) := \prod_i \exp\left(\int_{(s,t]} \log u_r(i, z_{r-}^*) \, dN(z^*, i)_r - \int_s^t (u_r(i, z_r^*) - 1) \bar{A}(r)_{i, z_r^*} \, dr\right), \tag{17.43}$$

where N is as defined in (17.17). The following lemma shows that the path distribution of the controlled process is absolutely continuous with respect to that of the uncontrolled process, and characterizes the associated Radon–Nikodym derivative.

Lemma 17.1 For any $q \in \mathbb{F}$, $u \in C([-T, T]; \mathbf{U}^*)$ and $t \in [-T, T]$, $P_t^{*u,q} \ll P_{Z^*,t}$ and

$$\frac{dP_t^{*u,q}}{dP_{Z^*,t}}(Z^{*u,q}) = l_{-T}(q, Z_{-T}^{*u,q}) M_{-T,t}^{*u}(Z^{*u,q}), \tag{17.44}$$

where $P_{Z^*,t}$ is the restriction of P_{Z^*} to $\mathcal{D}_{-T,t}$.

Proof. Since u is continuous $\Lambda := \max_{i,t}\{-\bar{A}^u(t)_{i,i}\} < \infty$, $\max_{i,j,t} |\log \circ u_t(i, j)| < \infty$, and, for any $K < \infty$,

$$\mathbb{E} \exp\left(K \sum_i N(Z^{*u,q}, i)_T\right) \leq \exp(2T\Lambda(\exp(K) - 1)) < \infty.$$

Equation (17.44) now follows from theorems T10 and T11 of chapter VIII in [3]. $\square$

The apparent information of (17.41) evaluated at $P^{*u,q}$ can be expressed in the form:

$$F_T^*(P^{*u,q}, x^*) = h(q \mid p_T) + \sum_i J_{-T}^*(u, i, x^*) q_i, \tag{17.45}$$

where, for $t \in [-T, T]$,

$$J_t^*(u, i, x^*) := E_D^{i,t,u} \left(\log M_{t,T}^{*u} + H_{t,T}^*(\cdot, x^*) - \log l_T \left(x_T^*, x_T^d\right) \right)$$
$$= E_D^{i,t,u} \int_t^T \left(R_{h,s}^* \left(u_s, \chi_s^d\right) + h^* \left(\dot{x}_s^*, x_s^*, \chi_s^d\right) \right) ds \qquad (17.46)$$
$$- E_D^{i,t,u} \log l_T \left(x_T^*, \chi_T^d\right).$$

Here $E_D^{i,t,u}$ is expectation with respect to the distribution on $\mathcal{D}_{t,T}$ of a Markov process with initial value i at time t, and rate matrix $(\bar{A}^u(s), s \in [t, T])$,

$$R_{h,s}^*(v, i) := \sum_j a \circ v(j, i) \bar{A}(s)_{j,i} \qquad (17.47)$$

and a is as defined in (17.26).

The problem of minimizing J_{-T}^* over u (for fixed i) is one of stochastic optimal control, involving *running* and *final* costs. The running cost (the first term on the right-hand side of (17.46)) includes the cost of the control (as measured by the information gain of the distribution of the controlled process) and a component corresponding to the closeness of fit of $Z^{*u,q}$ with the sample path of the running observation Y^r of (17.4). The final cost (the last term on the right-hand side of (17.46)) corresponds to the closeness of fit of $Z^{*u,q}$ with the initial observation ζ. According to Proposition 17.3(iii), the optimal control is as follows:

$$u_t^{*\circ}(j, k; x^*) := \frac{l_t\left(x_t^*, j\right)}{l_t\left(x_t^*, k\right)} \quad \text{for all } j, k \text{ such that } A_{j,k} > 0. \qquad (17.48)$$

This is because the resulting rate matrix is that of the true $X^* = x^*$-conditional distribution of Z^*, $(\check{A}(x_t^*), t \in [-T, T])$. (See (17.12).) Furthermore, according to Proposition 17.3(i), the value function for this control problem is as follows:

$$V_t^*(i, x^*) := \min_{u \in C([t, T]; U^*)} J_t^*(u, i, x^*)$$
$$= -\log E_D^{i,t} l_T\left(x_T^*, \chi_T^d\right) \exp\left(-H_{t,T}^*(\cdot, x^*)\right) \qquad (17.49)$$
$$= -\log l_t\left(x_t^*, i\right),$$

where $E_D^{i,t}$ is the equivalent of $E_D^{i,t,u}$ with no control applied ($u \equiv 1$).

If the optimal control is applied over the interval $[-T, T]$, then the apparent information in (17.45) becomes

$$F_T^*(P^{*\circ,q}, x^*) = h(q \mid p_T) + \langle V_{-T}^*(\cdot, x^*), q \rangle. \qquad (17.50)$$

This is the apparent information for the problem of estimating X_T given $X^* = x^*$, for which the posterior distribution is x_{-T}^*. It is thus minimized by the choice $q = x_{-T}^*$. The variational characterization thus splits into two steps: the first involves a stochastic optimal control problem to characterize a regular conditional *generator* for Z^*; the second uses the value function of this control problem and the initial prior to characterize a regular conditional *initial distribution*.

Remark 17.8 In practice, one is often interested in the single-time marginals of the posterior path distribution. These can be obtained by solving the forward equation for the optimally controlled process. Of course, this gives rise to the usual "backward" equations of nonlinear interpolation. (See, for example, Ch. 9 in [16].)

The optimal control at any time t is characterized by the dynamic programming equation of this control problem. In what follows, we show that this equation is a localized version of (17.36) and characterizes the rate matrix $\check{A}(x_t^*)$ as the minimizer of a *rate* of apparent information.

We assume now that the initial distribution q has been chosen optimally, and drop the superscript q. Thus Z^{*u} is a Markov process with paths in $D([-T, T]; \mathbb{S})$, initial distribution x^*_{-T}, and rate matrix $(\bar{A}^u(t), t \in [-T, T])$, and P_t^{*u} is its path distribution restricted to $\mathcal{D}_{-T,t}$. For any $u \in C([-T, T]; U^*)$, $z^* \in D([-T, T]; \mathbb{S})$, $x^* \in C^1([-T, T]; \mathbb{F})$, and $t \in [-T, T)$, let

$$I_t^{*u}(z^*, x^*) := \log\left(\frac{dP_t^{*u}}{dP_{Z^*,t}}(z^*)\right) + H^*_{-T,t}(z^*, x^*) - \log l_t(x_t^*, z_{t+}^*) \tag{17.51}$$

$$= \log M^{*u}_{-T,t}(z^*) + H^*_{-T,t}(z^*, x^*) \tag{17.52}$$

$$+ \int_{-T}^{t} \frac{\partial V_s^*}{\partial s}(z_s^*, x^*) \, ds + \sum_{s \in (-T,t]} \left(V_s^*(z_{s+}^*, x^*) - V_s^*(z_{s-}^*, x^*) \right)$$

$$= \sum_i \int_{(-T,t]} \log\left(\frac{u_s(i, z_{s-}^*)}{u_s^{*o}(i, z_{s-}^*; x^*)}\right) (dN(z^*, i)_s - \bar{A}^u(s)_{i,z_s^*} ds) \tag{17.53}$$

$$+ \int_{-T}^{t} \sum_i \alpha\left(\frac{u_s(i, z_s^*)}{u_s^{*o}(i, z_s^*; x^*)}\right) \check{A}(x_s^*)_{i,z_s^*} \, ds,$$

where we have used (17.48) and (17.49) to develop the value function, and α is as defined in (17.26). I_t^{*u} is the (unaveraged) cost of applying the control u over the time interval $[-T, t]$, and the optimal control thereafter. Now $\alpha(r) \geq 0$ with equality if and only if $r = 1$, and so, according to (17.53), $\left(I_t^{*u}(Z^{*u}, x^*), \mathcal{F}^{Z^{*u}}_{-T,t+}, t \in [-T, T)\right)$ is a submartingale; it is a martingale (in fact $\equiv 0$) if $u = u^{*o}$. This is the Davis–Varaiya characterization of the optimal control [7].

Now $F_t^*(P_t^{*u}, x^*)\left(:= \mathbb{E}I_t^{*u}(Z^{*u}, x^*)\right)$ is the apparent information for the path estimation problem over the partial time interval $[-T, t]$. (This is because the minimal apparent information corresponding to controls applied after time t is zero.) According to (17.51) this comprises three components: the first is the information gain of the putative posterior P_t^{*u} over the prior $P_{Z^*,t}$; the second is the *running* residual information, i.e. that component of the residual information coming from the running observation $(Y_s^r - Y_{-t}^r, s \in [-t, T])$; the third is the component of residual information coming from $(Y_s,\ s \in [-T, -t])$. It follows from (17.49) that the third component is also the average (over the distribution of Z_t^{*u}) of the *full* information in the observation $(Y_s,\ s \in [-T, -t])$ in the context of the Bayesian problem of estimating the pinned process $(Z_s^{*i,t},\ s \in [t, T])$ of (17.30). In this sense it is the information *remaining* in the observation at time t. At time t, *all* the information in $(Y_s^r - Y_{-t}^r,\ s \in [-t, T])$ relevant to X has been extracted, but that part of the information in $(Y_s,\ s \in [-T, -t])$ useful for estimating the pinned process remains in the observation. As t increases this remaining information decreases, and the *running* apparent information (i.e. the sum of the first two terms on the right-hand side of (17.51)) increases. If the control is optimal, then these changes are balanced, in which case we can think of information *flowing* from the observation into the running apparent information; otherwise there is an apparent information excess, and information is not conserved.

A *local* characterization of these information exchanges can be found from (17.52), as follows. Let $R_{F,t}^{*u}$ be the following *local rate of apparent information*:

$$R_{F,t}^{*u}(i, x) := \lim_{\epsilon \downarrow 0} \epsilon^{-1} E_D^{i,t,u}\left(I_{t+\epsilon}^{*u}(\cdot, x^*) - I_t^{*u}(\cdot, x^*)\right)$$
$$= R_{h,t}^*(u_t, i) + h^*\left(\dot{x}_t^*, x_t^*, i\right) + \frac{\partial V_t^*}{\partial t}(i, x^*) + \mathcal{L}_{u_t,t} V_t^*(i, x^*), \qquad (17.54)$$

where $R_{h,t}^*$ is the *local rate of information gain* as defined in (17.47) and, for any $v \in \mathbf{U}^*$ and $f : \mathbb{S} \to \mathbb{R}$,

$$(\mathcal{L}_{v,t} f)(i) = \sum_j (f(j) - f(i)) \bar{A}(t)_{j,i} v(j, i). \qquad (17.55)$$

(The four terms on the right-hand side of (17.54) correspond to the four terms on the right-hand side of (17.52). The first of the terms in (17.52) has a differentiable mean since $\sup_{t,i,j} |\log u_t(i,j)| < \infty$. See (17.43). The remaining terms have differentiable means since x^* is differentiable and $\inf_{t,i} x_{t,i}^* > 0$.) $R_{F,t}^{*u}(i, x)$, thus defined, is the rate of accrument of cost at time t resulting from the instantaneous control u_t. The dynamic programming equation involves its minimization over instantaneous controls in $\mathbf{U}^*$:

$$\frac{\partial V_t^*}{\partial t}(i, x^*) + h^*\left(\dot{x}_t^*, x_t^*, i\right) + \min_{v \in \mathbf{U}^*} \left\{R_{h,t}^*(v, i) + \mathcal{L}_{v,t} V_t^*(i, x^*)\right\} = 0. \qquad (17.56)$$

The two terms in the curly brackets, here, can be interpreted as the two components of the rate of apparent information of a *local* Bayesian problem, whose estimand takes values in $\mathbb{S}$. The estimand is the final value of a time-homogeneous Markov process $(\xi_s^*,\ s \in [0, \epsilon])$, which has initial value i, and (time-invariant) generator $\mathcal{L}_{v,t}$ of (17.55) with $v \equiv 1$ (no control). (ϵ is small.) For any $v \in \mathbf{U}^*$, let T_ϵ^{*v} be the transition probability, over the time interval $[0, \epsilon]$, of a time-homogeneous Markov process with initial value i and (time-invariant) generator $\mathcal{L}_{v,t}$ of (17.55). In particular, T_ϵ^* ($:= T_\epsilon^{*v}|_{v\equiv 1}$) is the transition probability of ξ^*. Let

$$H^{*\mathrm{loc}}\left(j, x_t^*\right) := V_t^*(j, x^*) - V_t^*(i, x^*) = \log l_t\left(x_t^*, i\right) - \log l_t\left(x_t^*, j\right).$$

If we regard $H^{*\mathrm{loc}}\left(\,\cdot\,, x_t^*\right)$ as being a log-likelihood function corresponding to some observation of ξ_ϵ^*, then the apparent information of the putative posterior T_ϵ^{*v} for ξ_ϵ^* is:

$$h\left(T_\epsilon^{*v} \mid T_\epsilon^*\right) + \left\langle H^{*\mathrm{loc}}\left(\,\cdot\,, x_t^*\right), T_\epsilon^{*v}\right\rangle = \left(R_{h,t}^*(v, i) + \mathcal{L}_{v,t} V_t^*(i, x^*)\right)\epsilon + o(\epsilon). \tag{17.57}$$

Comparing this with (17.56), we see that the first term in curly brackets in (17.56) is the rate of information gain of T_ϵ^{*v} over T_ϵ^* in this local problem, and the second term is the *rate of residual information* of T_ϵ^{*v} in this local problem. The local log-likelihood function $H^{*\mathrm{loc}}$ has its origins in the primal likelihood filter $l_t(Z_{-t}, i)$; the log-likelihood differences $(\log l_t\left(x_t^*, i\right) - \log l_t\left(x_t^*, j\right), j \in \mathbb{S})$ are the adjoint variables in (17.56).

$R_{h,t}^*(v, i)$ can be thought of as a *generalized rate of information supply* for the dual filtering problem. Its optimal value is a local version of $\dot{S}^*(t)$, in the sense that

$$\dot{S}^*(t) = \mathbb{E}\, R_{h,t}^*\left(u_t^{*\circ}(\,\cdot\,, \cdot\,; X^*), Z_t^*\right). \tag{17.58}$$

Viewed in this way, the associated flow of information is the result of a *control action* rather than an observation mechanism. Information is not conserved if a suboptimal control is used.

We turn now to the inverse problem of characterizing the log-likelihood function $H^*_{-T,T}(\,\cdot\,, x^*) - \log l_T\left(x_T^*, x_T^d\right)$ of (17.40) as a maximizer of compatible information. Let $\mathcal{H}(D([-T, T]; \mathbb{S}))$ be the space of measurable $(-\infty, \infty]$-valued functions on $D([-T, T]; \mathbb{S})$. We restrict the $\tilde{H} \in \mathcal{H}(D([-T, T]; \mathbb{S}))$ over which we maximize to the following class. Let Φ^* be the set of functions $\psi : \mathbb{S} \times \mathbb{S} \to \mathbb{R}$ for which $\exp(\psi) \in \mathbf{U}^*$. For any $q \in \mathbb{F}$, $\phi \in C([-T, T]; \Phi^*)$, $z \in D([-T, T]; \mathbb{S})$, and $t \in [-T, T]$, let

$$H_{t,T}^{*\phi,q}(z^*) := -\mathbf{1}_{\{-T\}}(t) \log l_{-T}\left(q, z_{-T}^*\right) + \sum_i \int_{(t,T]} \phi_s\left(i, z_{s-}^*\right) dN(z^*, i)_s \tag{17.59}$$
$$+ \int_t^T \sum_i \left(\exp\left(-\phi_s\left(i, z_s^*\right)\right) - 1\right) \tilde{A}_{i,z_s^*}\, ds.$$

A development of the value function of the type used in (17.51)–(17.53) transforms $H^*_{-T,T}(z^*, x^*) - \log l_T\left(x_T^*, z_T^*\right)$ into this form with $t = -T$, $q = x^*_{-T}$, and

$$\phi_t(j, i) = \phi_t^{*\circ}(j, i; x^*) := V_t^*(j, x^*) - V_t^*(i, x^*). \tag{17.60}$$

This transformation moves all the observation information in the likelihood function that is relevant to Z^*_{-T} to the initial time. (cf. Remark 17.5.) Proposition 17.3(ii) now characterizes $H^*_{-T,T}(\,\cdot\,, x^*) - \log l_T(x^*_T, \chi^d_T)$ as a maximizer of the compatible information functional $G^*(\,\cdot\,, x^*) : \mathcal{H}(D([-T, T]; \mathbb{S})) \to [-\infty, \infty)$ of the path estimation problem:

$$G^*\left(H^{*\phi,q}_{t,T}, x^*\right) = 0 - \left\langle H^{*\phi,q}_{t,T}, P^{*o}\right\rangle$$
$$= \left(h\left(x^*_{-T} \mid p_T\right) - h\left(x^*_{-T} \mid q\right)\right) \mathbf{1}_{\{-T\}}(t) \qquad (17.61)$$
$$+ \mathbb{E} \int_t^T \left(R^*_{i,s}(\phi_s, Z^{*o}_s) - (\mathcal{L}_{u^{*o}_s,s}\phi_s(\,\cdot\,, Z^{*o}_s))(Z^{*o}_s)\right) ds,$$

where Z^{*o} is Z^{*u} with the optimal control, P^{*o} is its distribution,

$$R^*_{i,s}(\phi_s, i) := \sum_j \beta \circ \phi_s(j, i) \bar{A}(s)_{j,i} \qquad (17.62)$$

is the *local rate of full information* of $H^{*\phi,q}_{t,T}$, and $\beta : \mathbb{R} \to \mathbb{R}$ is defined as follows:

$$\beta(s) := 1 - \exp(-s). \qquad (17.63)$$

(The first term in the integral in (17.61) is the negative of the mean of the ordinary integral in (17.59); the second term in the integral in (17.61) is the negative of the mean of the jump integral in (17.59).)

Remark 17.9 Log-likelihood functions having the form (17.59) are *normalized* in the sense that they have full information quantities equal to zero. (This does not mean that they bear no information. The full information is a *differential* information quantity, as discussed in Remark 17.7.) By restricting the class over which we optimize in this way, we obtain uniqueness in (17.37).

Log-likelihood functions of the form in (17.59) can be associated with many different types of observation. We could, for example, choose $\phi_t(j, i) = \tilde{V}^*_t(j) - \tilde{V}^*_t(i)$, where

$$\tilde{V}^*_t(i) := -\log E^{i,t}_D \tilde{l}_T \exp\left(-\int_t^T \tilde{h}^*_s \, ds\right),$$

and $\tilde{h}^*$ and $\tilde{l}_T$ are approximations to h^* of (17.40) and l_T of (17.19). Such a choice of ϕ represents a transformed version of a log-likelihood function with the same structure as the true log-likelihood $H^*_{-T,T}(\,\cdot\,, x^*) - \log l_T(x^*_T, \chi^d_T)$; the amount by which the compatible information of this putative log-likelihood falls short of that of $H^*_{-T,T}(\,\cdot\,, x^*) - \log l_T(x^*_T, \chi^d_T)$ is then an information theoretic measure of the effects of observation measurement or modelling errors. (See [18] for further discussion of this issue.)

Equation (17.61) can be interpreted progressively in *reverse* time as follows. When $t = T$, $H^{*\phi,q}_{t,T}$ is zero, and so there is no compatible information; all the information is still in the source (which in the inverse problem of Proposition 17.3(ii) is the regular conditional distribution $P^{*\circ}$). As t decreases, the information remaining in this source decreases, and the compatible information (which is stored in the likelihood function) increases. If $\phi = \phi^{*\circ}$, as defined in (17.60), then these changes balance, in which case we can think of information *flowing* from the conditional distribution into the likelihood function. However, if ϕ is suboptimal then information is lost. At time $t = -T$, there is an impulsive drop in the remaining source information of size $h\left(x^*_{-T} \mid p_T\right)$, and an impulsive increase in compatible information of size $h\left(x^*_{-T} \mid p_T\right) - h\left(x^*_{-T} \mid q\right)$. The latter is strictly smaller than the former unless q takes the optimal value, x^*_{-T}.

We assume now that q has been chosen optimally, and drop the superscript q. As with the minimization of apparent information, the inverse problem of Proposition 17.3(ii) can be *localized*. For any $t > -T$, let $R^{*\phi}_{G,t}$ be the following *local rate of compatible information*:

$$R^{*\phi}_{G,t}(i, x^*) := \lim_{\epsilon \downarrow 0} \epsilon^{-1} \left(G^* \left(H^{*\phi}_{t-\epsilon, T} \right) - G^* \left(H^{*\phi}_{t, T} \right) \right)$$
$$= R^*_{i,t}(\phi_t, i) - \left(\mathcal{L}_{u^{*\circ}_t, t} \phi_t(\,\cdot\,, i) \right)(i). \quad (17.64)$$

The local version of compatible information maximization is then

$$-R^*_{h,t}\left(u^{*\circ}_t, i\right) + \max_{\psi \in \Phi^*} \left\{ R^*_{i,t}(\psi, i) - \left(\mathcal{L}_{u^{*\circ}_t, t} \psi(\,\cdot\,, i) \right)(i) \right\} = 0, \quad (17.65)$$

which is the adjoint optimization problem to that of (17.56). The term to be maximized here is the local rate of compatible information of the putative likelihood function $\exp(-\psi(\,\cdot\,, i))$ in the inverse of the local Bayesian problem of (17.57).

The rate of change of compatible information can be thought of as a *generalized rate of information dissipation* for the *primal* problem. Its maximal value is a local version of $\dot{D}(-t)$, in the sense that

$$\dot{D}(-t) = \mathbb{E} \max_{\psi \in \Phi^*} \left\{ R^*_{i,t}\left(\psi, Z^*_t\right) - \left(\mathcal{L}_{u^{*\circ}_t(\,\cdot\,, \,\cdot\,; X^*), t} \psi\left(\,\cdot\,, Z^*_t\right) \right)\left(Z^*_t\right) \right\}. \quad (17.66)$$

Remark 17.10 *$\dot{D}(-t)$ is the maximal rate of compatible information here, whereas $\dot{S}^*(t)$ is merely one of the components of the minimal rate of apparent information in (17.58). This distinction is associated with the fact that compatible information, unlike apparent information, has an absolute meaning. (See Remark 17.7.)*

Since $R^*_{h,t}$, $R^*_{i,t}$ and $\mathcal{L}_{v,t}$ all comprise sums over $j \in \mathbb{S}$ weighted by the terms $\bar{A}(t)_{j,i}$, the dual local optimization problems of (17.56) and (17.65) can be carried out separately for each $j \neq i$. This reveals the Fenchel–Legendre pair at the heart of the local equations:

$$\beta(s) = \min_{r \in (0, \infty)} \{a(r) + sr\}, \qquad a(r) = \max_{s \in \mathbb{R}} \{\beta(s) - rs\}. \quad (17.67)$$

17.5.3 Estimating X^* from Z^*

We now apply Proposition 17.3 to the problem of estimating the dual signal process X^* from the surrogate observation Z^*. Once again, we do this in reverse (dual) time, effectively estimating Z from X. The development is identical in structure to that of the last subsection, and so will only be sketched. The abstract estimand space $(\mathbf{X}, \mathcal{X})$ of Subsection 17.5.1 becomes here the space $(C([-T, T]; \mathbb{F}), \mathcal{C}_{-T,T})$, and the abstract observation space $(\mathbf{Y}, \mathcal{Y})$ becomes the space $(D([-T, T]; \mathbb{S}), \mathcal{D}_{-T,T})$. Taking the marginal distribution of Z^* as the reference measure λ_Y, we set

$$H_{s,t}(z, x) := \int_s^t h_r(x_r, z_r)\, dr + \sum_i \int_{(s,t]} \log\left(\frac{l_{-r}(z_r, i)}{l_{-r}(z_r, x_{r-})}\right) dN(x, i)_r \quad (17.68)$$

where

$$h_t(i, q) := \sum_j A_{i,j}\left(\frac{q_j}{q_i} - \frac{p_{t,j}}{p_{t,i}}\right).$$

Then $Q(z, x)$ $(:= \exp(-H_{-T,T}(z, x) + l_{-T}(z_T, x_T)))$ is the version of L_T from which we construct the regular $X = x$-conditional distibution of Z.

Apparent information is minimized over the distributions of controlled versions of Z. Let $\mathbf{U}$ be the subset of those $v \in C(\mathbb{F}; \mathbb{R}^n)$ for which $\sup_{q \in \mathbb{F}} |\sigma^T v(q)| < \infty$, and let $(Z_t^u,\ t \in [-T, T])$ be a diffusion process with sample paths in $C([-T, T]; \mathbb{F})$, initial density

$$\eta(q) := l_T(q, x_{-T}) p^*_{-T}(q), \quad (17.69)$$

and generator $(\mathcal{L}_{u_t},\ t \in [-T, T])$, where $u \in D([-T, T]; \mathbf{U})$, and, for $v \in \mathbf{U}$,

$$(\mathcal{L}_v f)(q) := \sum_j (Aq + av(q))_j \frac{\partial f}{\partial q_j}(q) + \frac{1}{2} \sum_{j,k} a(q)_{j,k} \frac{\partial^2 f}{\partial q_j \partial q_k}(q). \quad (17.70)$$

(The control enters the drift coefficient of Z^u through the map $u_t \mapsto au_t$.) Let P^u be the distribution of Z^u, and P_t^u its restriction to $\mathcal{C}_{-T,t}$.

Remark 17.11 By initializing Z^u with the correct regular conditional distribution in (17.69) we are skipping the step in the minimization of apparent information associated with the initial condition. See the discussion following (17.50).

For any $u \in D([-T, T]; \mathbf{U})$ and $s \leq t$, let

$$M^u_{s,t}(Z) := \exp\left(\int_s^t u_r(Z_r)^T (dZ_r - AZ_r\, dr) - \int_s^t R_h(u_r, Z_r)\, dr\right), \quad (17.71)$$

where $R_h : \mathbf{U} \times \mathbb{F} \to [0, \infty)$ is defined as follows

$$R_h(v, q) := \frac{1}{2}(\sigma^\mathsf{T} v(q))^2. \tag{17.72}$$

Since $\sup_{t,q} R_h(u_t, q) < \infty$ the Novikov criterion is satisfied, and so it follows from the Girsanov Theorem that $P^u_t \ll P_{Z,t}$ and

$$\frac{d P^u_t}{d P_{Z,t}}(Z) = l_T(Z_{-T}, x_{-T}) M^u_{-T,t}(Z),$$

where $P_{Z,t}$ is the distribution of Z restricted to $\mathcal{C}_{-T,t}$. (See, for example, [16].)

Remark 17.12 Unlike $M^{*u}_{s,t}$ of (17.43), $M^u_{s,t}$ is not defined in a pathwise sense, but only almost surely with respect to the distribution of Z. This is unimportant, as it is evaluated here, only for processes whose laws are mutually absolutely continuous with respect to this distribution.

The problem is to find the control $u \in D([-T, T]; \mathbf{U})$ that minimizes the following apparent information:

$$F_T(P^u, x) = h\left(\eta \mid p^*_{-T}\right) + \int_\mathbb{F} J_{-T}(u, q, x)\eta(q)\lambda_\mathbb{F}(dq). \tag{17.73}$$

where, for $t \in [-T, T]$,

$$\begin{aligned} J_t(u, q, x) &:= E^{q,t,u}_C \left(\log M^u_{t,T} + H_{t,T}(\cdot, x) - \log l_{-T}\left(\chi^c_T, x_T\right)\right) \\ &= E^{q,t,u}_C \int_t^T \left(R_h\left(u_s, \chi^c_s\right) + h_s\left(x_s, \chi^c_s\right)\right) ds \\ &\quad + E^{q,t,u}_C \sum_i \int_{(t,T]} \log\left(\frac{l_{-s}\left(\chi^c_s, i\right)}{l_{-s}\left(\chi^c_s, x_{s-}\right)}\right) dN(x, i)_s \\ &\quad - E^{q,t,u}_C \log l_{-T}\left(\chi^c_T, x_T\right). \end{aligned} \tag{17.74}$$

Here $E^{q,t,u}_C$ is expectation with respect to the distribution on $\mathcal{C}_{t,T}$ of a diffusion process with initial value q at time t, and generator $(\mathcal{L}_{u_s}, s \in [t, T])$. According to Proposition 17.3(iii), the optimal control is as follows:

$$u^o_t(q; x)_i := \delta_{i, x_t} q^{-1}_{x_t}, \tag{17.75}$$

where $\delta_{i,j}$ is the Krönecker delta. This is because the resulting drift coefficient at time t with this control is $b(x_t, \cdot)$ of (17.9), and this endows Z with the correct $X = x$-conditional distribution. Furthermore, according to Proposition 17.3(i), the value function for this control problem is as follows:

$$\begin{aligned} V_t(q, x) &:= \min_{u \in D([t,T];\mathbf{U})} J_t(u, q, x) \\ &= -\log \mathbb{E}^{q,t}_C l_{-T}\left(\chi^c_T, x_T\right) \exp\left(-H_{t,T}(\cdot, x)\right) \\ &= -\log l_{-t}(q, x_t), \end{aligned} \tag{17.76}$$

where $E_C^{q,t}$ is the equivalent of $E_C^{q,t,u}$ with no control applied ($u \equiv 0$). The equivalent of I_t^{*u} of (17.51) is

$$I_t^u(Z, x) := \log\left(\frac{dP_t^u}{dP_{Z,t}}(Z)\right) + H_{-T,t}(Z, x) - \log l_{-t}(Z_t, x_t) \tag{17.77}$$

$$= \int_{-T}^{t} (u_s(Z_s) + \nabla_q V_s(Z_s; x))^\mathsf{T} (dZ_s - (AZ_s + au_s(Z_s))ds)$$

$$+ \int_{-T}^{t} \left(\frac{1}{2}(g(x_s) - \tilde{g}(Z_s))^2 + R_h(u_s, Z_s)\right. \tag{17.78}$$

$$\left. + u_t^\mathsf{T} a(Z_s) \nabla_q V_s(Z_s, x)\right) ds$$

$$= \int_{-T}^{t} (u_s(Z_s) - u_s^o(Z_s; x))^\mathsf{T} (dZ_s - (AZ_s + au_s(Z_s))ds)$$

$$+ \int_{-T}^{t} R_h\left(u_s(Z_s) - u_s^o(Z_s; x), Z_s\right) ds, \tag{17.79}$$

where R_h is as defined in (17.72). Like I_t^{*u}, $\left(I_t^u(Z^u, x), \mathcal{F}_{-T,t}^{Z^u}, t \in [-T, T]\right)$ is a submartingale; moreover, it is a martingale (in fact $\equiv 0$) if u is optimal.

The dynamic programming equation can be found from (17.78):

$$\frac{1}{2}(g(x_t) - \tilde{g}(q))^2 + \min_{v \in U} \left\{ R_h(v, q) + v^\mathsf{T} a(q) \nabla_q V_t(q, x) \right\} = 0, \tag{17.80}$$

Like (17.56), this is associated with a local estimation problem. In this, the estimand is the final value of a time-homogeneous Markov diffusion process $(\xi_s, s \in [0, \epsilon])$ with initial value q, and generator $\mathcal{L}_v$ of (17.70), with $v = 0$. For any $v \in U$ let T_ϵ^v be the tansition probability of a time-homogeneous diffusion process with initial value q and generator $\mathcal{L}_v$ of (17.70). In particular, T_ϵ $(:= T_\epsilon^v|_{v=0})$ is the transition probability of ξ. Let

$$H^{\text{loc}}(y, x_t) := y^\mathsf{T} \nabla_q V_t(q, x) = -y^\mathsf{T} (\nabla_q \log l_{-t})(q, x_t).$$

If we regard $H^{\text{loc}}(\cdot, x_t)$ as being a log-likelihood function corresponding to some observation of ξ_ϵ, then the apparent information of the putative posterior T_ϵ^v for ξ_ϵ is:

$$h\left(T_\epsilon^v \mid T_\epsilon\right) + \langle H^{\text{loc}}(\cdot, x_t), T_\epsilon^v \rangle = \left(R_h(v, q) + v^\mathsf{T} a(q) \nabla_q V_t(q, x)\right) \epsilon + o(\epsilon). \tag{17.81}$$

Comparing this with (17.80), we see that the first term in curly brackets in (17.80) is the rate of information gain of T_ϵ^v over T_ϵ in this local problem, and the second term is the rate of residual information of the local problem. As with $H^{*\text{loc}}$ of Subsection 17.5.2, the local log-likelihood function H^{loc} has its origins in the dual likelihood filter $l_{-t}(q, X_t)$; the gradient of the log-likelihood function is the vector of adjoint variables in (17.80).

$R_h(v, q)$ can be thought of as a generalized rate of information supply for the primal problem. in the sense that

$$\dot{S}(t) = \mathbb{E} R_h\left(u_t^o(\cdot; X), Z_t\right). \tag{17.82}$$

We turn now to the inverse problem of characterizing $H_{-T,T}(\cdot, x) - \log l_{-T}(\chi_T^c, x_T)$ as a maximizer of compatible information. We restrict the $\tilde{H} \in \mathcal{H}(C([-T, T]; \mathbb{F}))$ over which we maximize to the following class.

$$H_{t,T}^{\phi}(Z) := -\mathbf{1}_{\{-T\}}(t) \log l_T(Z_{-T}, x_{-T}) + \int_t^T \phi_s^\mathsf{T}(dZ_s - AZ_s \, ds) + \int_t^T R_h(\phi_s, Z_s) \, ds, \tag{17.83}$$

where $\phi \in D([-T, T]; \mathbf{U})$. By initializing $H_{-T,T}$ with the true log-likelihood function, we are skipping the step in compatible information maximization associated with the initial condition. (See Remark 17.11.) The true overall log-likelihood function is obtained by choosing $\phi = -u^\circ$ in (17.83). The inverse problem is that of maximizing the compatible information functional $G(\cdot, x) : \mathcal{H}(C([-T, T]; \mathbb{F})) \to [-\infty, \infty)$ of the path estimation problem:

$$G\left(H_{t,T}^{\phi}, x\right) = 0 - \left\langle H_{t,T}^{\phi}, P^\circ \right\rangle \tag{17.84}$$

$$= \mathbb{E} \int_t^T \left(R_i\left(\phi_s, Z_s^\circ\right) - u_s^\circ\left(Z_s^\circ; x\right)^\mathsf{T} a\phi_s\left(Z_s^\circ\right) \right) ds,$$

where $R_i := -R_h$, Z° is Z^u with the optimal control, and P° is its distribution. This can be localized in the same way as was (17.61) in (17.65).

$$-R_h\left(u_t^\circ, q\right) + \max_{\psi \in \mathbf{U}} \left\{ R_i(\psi, q) - u_t^\circ(q; x)^\mathsf{T} a\psi(q) \right\} = 0. \tag{17.85}$$

This is the adjoint optimization problem to that of (17.80). The term to be maximized here is the *compatible information rate* of the putative likelihood function $\exp(-H^{\text{loc}}(\cdot, x_t))$ in the inverse of the local Bayesian estimation problem of (17.81). The local rate of change of compatible information is a generalized rate of information dissipation for the dual problem; in fact

$$\dot{D}^*(-t) = \mathbb{E} \max_{\psi \in \mathbf{U}} \left\{ R_i(\psi, Z_t) - u_t^\circ(Z_t; X)^\mathsf{T} a\psi(Z_t) \right\}. \tag{17.86}$$

Unlike those of Subsection 17.5.2, the two local optimization problems in (17.80) and (17.85) are *self-adjoint* in the sense that the local rate of full information of H^{loc} in (17.85) is the negative of the local rate of information gain in (17.80).

17.6 Discussion and concluding remarks

This chapter has shown how a system comprising a partially-observed Markov signal process and its nonlinear filter can be interpreted, in reverse time, as a dual system with a finite-dimensional filter. It has defined information *supply*, *storage* and *dissipation* quantities for both primal and dual systems, and shown that information is *conserved* by the filters, thereby introducing a notion of information *flow*.

Kalman's "duality between estimation and control" has been shown to be a special instance of an abstract variational transform, that characterizes Bayesian posterior distributions. Applied to the primal and dual *path* estimation problems, this yields problems in stochastic optimal control, whose dynamic programming equations are local instances of the same variational transform. The dynamic programming equations express the information conserving properties of the primal and dual filters in terms of optimally chosen controls acting between the signals and filters.

Although these ideas have been presented in the context of systems with a specific structure (finite-state signals, and observations contaminated by white noise), they can be applied in far greater generality [22].

Since the definitions of information supply and dissipation are based on *mutual* information, they are, like the Shannon information of a discrete random variable, *absolute* information quantities. The main achievement of classical Shannon theory is in the connections it makes between such quantities and the practical issues of source coding, and reliable communication. These connections are mostly based in one way or another on "limits of large systems", and reveal the dichotomy between reliability and timeliness. For example, reliable communication can be achieved through unreliable channels, but only if complex encoding is used, with the inherent loss of timeliness. The nonlinear filter is firmly on the "unreliable but timely" side of this dichotomy, and the path estimator leans in this direction. Of course, this is appropriate if an estimator forms part of a feedback control system that has to address time sensitive issues such as stability. Nevertheless, there is growing interest, stimulated by problems of networked control and communication, in systems that are both reasonably reliable and reasonably timely, [1]. See also [23], where a classification of information types for control problems is made, according to their transmission requirements. A unifying concept in such studies, which connects them with the contents of this chapter, is that of *dynamic information*.

Similarly, recent studies in nonequilibrium statistical mechanics have pointed to the need for a concept of dynamic *entropy*. (See, for example, [15].) Many of the classical notions of equilibrium statistical mechanics require new interpretations in the non-equilibrium setting. The problems posed are very similar in nature to those of dynamic information theory. (See [19] and [22] for specific connections with nonlinear filtering.)

The apparent and compatible information quantities of Section 17.5 are natural measures of suboptimality in nonlinear filtering problems when there is no specific cost metric, such as mean-square error of a particular statistic. For example, the results of [5], on the supermartingale property of the relative entropy between correctly and incorrectly initialized filters, is a statement about the compatible information deficit of the incorrectly initialized filter. The local variational characterizations of (17.81) and (17.65) go beyond this, by providing a means of studying suboptimal filter *dynamics*.

17.7 Glossary of frequently used notation

Numbers in brackets indicate the defining equation where such an equation exists.

A the rate matrix of the primal signal (17.1)
$\bar{A}$ the marginal rate matrix of the dual filter (17.12)
$\check{A}$ the conditional rate matrix of the dual filter (17.12)
a the diffusion matrix of the primal filter/dual signal (17.9)
b the drift coefficient of the primal filter (17.9)
$\bar{b}$ the drift coefficient of the dual signal (17.10)
$\mathbb{F}$ the state space of the primal filter/dual signal (17.2)
g the observation function (17.4)
$\mathcal{H}(\mathbf{X})$ the space of measurable $(-\infty, \infty]$-valued functions on $\mathbf{X}$ (Subsection 17.5.1)
$I(\Theta, \Phi)$ the mutual information between Θ and Φ (17.21)
i the generic element of $\mathbb{S}$
L_t, L_t^* exponential martingales (17.18)
l_t the likelihood function between X_{-t} and Z_{-t} (17.19)
$N(x, i)_t$ a counting processes derived from $x \in D([-T, T]; \mathbb{S})$ (17.17)
$\mathcal{P}(\mathcal{X})$ the set of probability measures on $\mathcal{X}$ (Subsection 17.5.1)
p_t the prior distribution of X_t (Section 17.2)
p_t^* the prior density of X_t^* (Section 17.3)
q the generic element of $\mathbb{F}$
$\mathbb{S}$ the state space of the primal signal/dual filter (Section 17.2)
$\mathbb{T}$ the tangent space to $\mathbb{F}$ (Section 17.2)
x, z^* generic elements of $D([-T, T]; \mathbb{S})$
x^*, z generic elements of $C^1([-T, T]; \mathbb{F})$
α the primal convex function of Subsection 17.5.2 (17.26)
β the adjoint convex function of Subsection 17.5.2 (17.63)
$\lambda_\mathbb{F}$ Lebesgue measure on $\mathbb{F}$ (Section 17.2)
σ the diffusion coefficient of the primal filter/dual signal (17.7)

References

[1] P. Antsaklis and J. Baillieul, eds, Networked control systems, *IEEE Trans. AC*, 49 (no. 9) (2004).
[2] A. G. Bhatt, A. Budhiraja, and R. L. Karandikar, Markov property and ergodicity of the nonlinear filter, *SIAM J. Control Optim.*, 39 (2000), pp. 928–49.

[3] P. Brémaud, *Point Processes and Queues, Martingale Dynamics*, Springer, New York, 1981.
[4] J. M. C. Clark and D. Crisan, On a robust version of the integral representation formula of nonlinear filtering, *Probability Theory and Related Fields*, 133 (2005), pp. 43–56.
[5] J. M. C. Clark, D. Ocone, and C. Coumarbatch, Relative entropy and error bounds for filtering of Markov processes, *Mathematics of Control, Signals and Systems*, 12 (1999), pp. 346–60.
[6] M. H. A. Davis, A pathwise solution of the equations of nonlinear filtering, *Th. Pob. Appl.*, 27 (1983), pp. 167–75.
[7] M. H. A. Davis and P. P. Varaiya, Dynamic programming conditions for partially observable stochastic systems, *SIAM J. Control Optim.*, 11 (1973), pp. 226–61.
[8] P. Dupuis and R. S. Ellis, *A Weak Convergence Approach to the Theory of Large Deviations*, Wiley, New York, 1997.
[9] T. E. Duncan, On the calculation of mutual information, *SIAM J. Appl. Math.*, 19 (1970), pp. 215–20.
[10] S. N. Ethier and T. G. Kurtz, *Markov Processes: Characterisation and Convergence*, Wiley, New York, 1986.
[11] W. H. Fleming and S. K. Mitter, Optimal control and nonlinear filtering for nondegenerate diffusion processes, *Stochastics*, 8 (1982), pp. 63–77.
[12] U. G. Haussmann and E. Pardoux, Time reversal of diffusions, *Annals of Probability*, 14 (1986), pp. 1188–1205.
[13] R. E. Kalman, A new approach to linear filtering and prediction problems, *ASME J. Basic Eng.*, 82 (1960), pp. 34–45.
[14] T. Kailath, A. H. Sayed, and B. Hassibi, *Linear Estimation*, Prentice-Hall, New York, 2000.
[15] J. L. Lebowitz and H. Spohn, A Gallavotti–Cohen type symmetry in the large deviation functional for stochastic dynamics, *J. Statistical Physics*, 95 (1999), pp. 333–66.
[16] R. S. Liptser and A. N. Shiryayev, *Statistics of Random Processes 1—General Theory*, Springer-Verlag, 1977.
[17] E. Mayer-Wolf and M. Zakai, On a formula relating the Shannon information to the Fisher information for the filtering problem, in *Filtering and Control of Random Processes*, H. Korezlioglu, G. Mazziotto, S. Szpirglas (eds), Lecture Notes in Control and Information Sciences 61, Springer, 1984, pp. 164–71.
[18] S. K. Mitter and N. J. Newton, A Variational approach to nonlinear estimation, *SIAM J. Control Optim.*, 42 (2003), pp. 1813–33.
[19] S. K. Mitter and N. J. Newton, Information and entropy flow in the Kalman–Bucy filter, *J. Stat. Phys.*, 118 (2005), pp. 145–76.
[20] N. J. Newton, Dual Kalman–Bucy filters and interactive entropy production, *SIAM J. Control Optim.*, 45 (2006), pp. 998–1016.
[21] N. J. Newton, Dual nonlinear filters and entropy production, *SIAM J. Control Optim.*, 46 (2007), pp. 1637–63.
[22] N. J. Newton, Interactive statistical mechanics and nonlinear filtering, *J. Stat. Phys.*, 133 (2008), pp. 711–37.
[23] A. Sahai and S. K. Mitter, The necessity and sufficiency of anytime capacity for stabilization over a noisy communication link: Part I: scalar systems, *IEEE Trans. IT*, 52 (2006), pp. 3369–95.
[24] A. N. Shiryayev, Stochastic equations of nonlinear filtering of jump Markov processes, *Problemy peredachi informatsii*, II, 3 (1966), pp. 3–22.
[25] R. van Handel, *Filtering, Stability and Robustness*, PhD Thesis, California Institute of Technology, Pasadena, CA, 2007.
[26] W. M. Wonham, Some applications of stochastic differential equations to optimal nonlinear filtering, *SIAM J. Control Optim.*, 2 (1965), pp. 347–69.

·18·
Filtering for Discrete-Time Markov Processes and Applications to Inventory Control with Incomplete Information

A. Bensoussan, M. Çakanyıldırım, and S. P. Sethi

18.1 Introduction

Many real-life decision-making situations in engineering and management are modeled by using controlled Markov processes. For a given discrete-time Markov process defined by its transition probability function, a set of controls, and an initial state, we can use Kolmogorov's (forward) equations to obtain the probability measure of the process to be at a certain state in a future period. These state probabilities can be used to compute the expected cost of specified control actions, and thereby, to obtain optimal controls so as to minimize the expected cost.

These tasks become quite challenging when the state of the system is partially observed. We develop a nonlinear filtering framework to perform these tasks. For this, let us denote by X, Z, U, respectively, the space of states, of observations, and of controls. They are metric spaces equipped with their Borel σ-algebras. The basic probability space is $\Omega = (X \times Z)^N$, where N is the set of integers. It is equipped with the Borel σ-algebra denoted by $\mathcal{A}$. The canonical process is

$$\begin{pmatrix} x_n(\omega) \\ z_n(\omega) \end{pmatrix} = \omega_n$$

where $\omega \in \Omega$ and ω_n is the n-th component of ω. We define the σ-algebras

$$\mathcal{F}^n = \sigma(x_k, z_k, 1 \leq k \leq n) \text{ and } \mathcal{Z}^n = \sigma(z_k, 1 \leq k \leq n),$$

and set $\mathcal{F}^0 = \mathcal{Z}^0 = \{\Omega, \emptyset\}$. We interpret x_n as the state of the system at time n and z_n as the observation at time n. The primary quantity of interest is the conditional probability of the partially observed state given the observation σ-algebra $\mathcal{Z}^n$.

The evolution of the conditional probability is presented in a fairly general context in the next section. It can be used in many applications. We focus on partially observed inventory problems which present interesting specific aspects. A comprehensive discussion of why inventories may be partially observed and the references to the associated empirical work can be found in Bensoussan et al. [2].

In Sections 18.3 and 18.4, special cases of one reference and two reference measures are presented. These measures are unnormalized in Section 18.5 to linearize the evolution equations. This facilitates considerably the study of the associated control problem and the corresponding Bellman equation in a convenient functional space. Further examples of partially observed inventories are presented in Section 18.6. Sections 18.7 and 18.8 are devoted to studying the optimal control problem for the zero-balance walk model described in Section 18.6.1. Section 18.9 concludes the chapter.

18.2 Conditional probability

18.2.1 Probability measure

We define a control sequence as $V = (v_0, v_1, \cdots, v_n, \cdots)$, where $v_0 \in U$, $v_n : Z^n \to U$, a Borel function, $n \geq 1$. Let $x_0 \in X$ be fixed, representing the initial state. We consider a family of transition probability functions $\pi(x, v; d\eta \otimes d\zeta)$, which, for each pair (x, v) is a probability measure on $X \times Z$, and

$$(x, v) \mapsto \pi(x, v; \Delta) \text{ is a Borel map} \tag{18.1}$$

for any fixed (x, v) and any Borel subset Δ of $X \times Z$. To x_0 and V, we associate a probability P^{V, x_0} on $(\Omega, \mathcal{A})$ such that if $\phi(x_1, z_1; \cdots; x_n, z_n)$ is a bounded Borel function, then

$$E^{V, x_0} \phi(x_1, z_1; \cdots; x_n, z_n) = \int \pi(x_0, v_0; d\eta_1 \otimes d\zeta_1) \int \pi(\eta_1, v_1(\zeta_1); d\eta_2 \otimes d\zeta_2) \int \cdots$$
$$\cdots \int \pi(\eta_{n-1}, v_{n-1}(\zeta_1, \cdots, \zeta_{n-1}); d\eta_n \otimes d\zeta_n) \phi(\eta_1, \zeta_1; \cdots; \eta_n, \zeta_n). \tag{18.2}$$

18.2.2 Joint probabilities

We denote by P_n^{V, x_0}, the joint probability of the variables $(z_1, \cdots, z_n; x_n)$. By Kolmogorov Consistency Theorem (see, e.g., Chow and Teicher [8]), it is defined by the relation

$$E^{V, x_0} \psi(z_1, \cdots, z_n; x_n) = E_n^{V, x_0} \psi(z_1, \cdots, z_n; x_n) \tag{18.3}$$

for any Borel function $\psi(z_1, \cdots, z_n; x_n)$. From the definition of P^{V, x_0}, it is easy to check the iteration formula

$$P_{n+1}^{V, x_0}(d\zeta_1 \otimes \cdots d\zeta_{n+1} \otimes d\eta_{n+1}) = \int_X P_n^{V, x_0}(d\zeta_1 \otimes \cdots d\zeta_n \otimes d\eta) \pi(\eta, v_n; d\eta_{n+1} \otimes d\zeta_{n+1}), \tag{18.4}$$

and we have the initial condition

$$P_1^{V,x_0}(d\zeta_1 \otimes d\eta_1) = \pi(x_0, v_0; d\eta_1 \otimes d\zeta_1). \tag{18.5}$$

18.2.3 Derivation

The marginal probability density of the variables $(\zeta_1, \cdots, \zeta_n)$ is given by

$$\varpi_n^{V,x_0}(d\zeta_1 \otimes \cdots \otimes d\zeta_n) = \int_X P_n^{V,x_0}(d\zeta_1 \otimes \cdots d\zeta_n \otimes d\eta) \tag{18.6}$$

Let $\mathbb{1}_\Gamma(\eta)$ denote an indicator function which equals 1 when $\eta \in \Gamma$ and equals 0 otherwise. If Γ is a Borel subset of X, the conditional probability of the random variable $\mathbb{1}_\Gamma(x_n)$ given $\mathcal{Z}^n$ is given by

$$\pi_n^{V,x_0}(z_1, \cdots, z_n; \Gamma) := E^{V,x_0}[\mathbb{1}_\Gamma(x_n)|\mathcal{Z}^n].$$

It is easy to check that this conditional probability is given by a Radon–Nikodym derivative

$$\pi_n^{V,x_0}(\zeta_1, \cdots, \zeta_n; \Gamma) = \frac{\int_X \mathbb{1}_\Gamma(\eta) P_n^{V,x_0}(d\zeta_1 \otimes \cdots d\zeta_n \otimes d\eta)}{\varpi_n^{V,x_0}(d\zeta_1 \otimes \cdots \otimes d\zeta_n)}. \tag{18.7}$$

We can obtain the following evolution of the conditional probability directly from formulas (18.4) and (18.7).

Theorem 18.1 *The conditional probability evolves according to*

$$\pi_{n+1}^{V,x_0}(\zeta_1, \cdots, \zeta_n, \zeta_{n+1}; \Gamma) = \frac{\int_X \pi_n^{V,x_0}(\zeta_1, \cdots \zeta_n; d\eta) \int_X \mathbb{1}_\Gamma(\mu) \pi(\eta, v_n; d\mu \otimes d\zeta_{n+1})}{\int_X \pi_n^{V,x_0}(\zeta_1, \cdots, \zeta_n; d\eta) \int_X \pi(\eta, v_n; d\mu \otimes d\zeta_{n+1})} \tag{18.8}$$

with the initial condition

$$\pi_1^{V,x_0}(\zeta_1; \Gamma) = \frac{\int_X \mathbb{1}_\Gamma(\mu) \pi(x_0, v_0; d\mu \otimes d\zeta_1)}{\int_X \pi(x_0, v_0; d\mu \otimes d\zeta_1)}. \tag{18.9}$$

It is important to point out that evolution (18.8) gives us the conditional probability π_{n+1}^{V,x_0} in terms of π_n^{V,x_0} and the latest observation ζ_{n+1}. This then provides a recursive equation for the conditional probability π_{n+1}^{V,x_0}, where the dependence on the prior observations $(\zeta_1, \ldots, \zeta_n)$ enters into the equation via π_n^{V,x_0}. In optimization problems that we shall consider, the objective function consists of conditional expectations given the prior and the current observation. These expectations can be computed using the conditional probabilities in (18.8). Thus, we can use the conditional probability as the state of the dynamic program for these problems.

18.3 One reference measure

We assume that the transition probability $\pi(x, v; d\eta \otimes d\zeta)$ has a density with respect to a fixed positive measure on Z, $m(d\zeta)$, not necessarily finite. More precisely, we assume that

$$\pi(x, v; d\eta \otimes d\zeta) = \theta(x, v, \zeta; d\eta) \otimes m(d\zeta), \tag{18.10}$$

where for fixed (x, v) and ζ, $\theta(x, v, \zeta; d\eta)$ is a finite measure on X. In this case, we derive from (18.8) and (18.9), the formulas

$$\pi_{n+1}^{V, x_0}(\zeta_1, \cdots, \zeta_n, \zeta_{n+1}; \Gamma) = \frac{\int_X \pi_n^{V, x_0}(\zeta_1, \cdots \zeta_n; d\eta)\theta(\eta, v_n, \zeta_{n+1}; \Gamma)}{\int_X \pi_n^{V, x_0}(\zeta_1, \cdots, \zeta_n; d\eta)\theta(\eta, v_n, \zeta_{n+1}; X)}, \tag{18.11}$$

$$\pi_1^{V, x_0}(\zeta_1; \Gamma) = \frac{\theta(x_0, v_0, \zeta_1; \Gamma)}{\theta(x_0, v_0, \zeta_1; X)}, \tag{18.12}$$

where naturally $\theta(x, v, \zeta; \Gamma) = \int_X \mathbb{1}_\Gamma(\eta)\theta(x, v, \zeta; d\eta)$.

Example. We sketch the classical Kalman filter in discrete time (see Kalman [10]). Suppose we consider the dynamic system

$$\begin{aligned} x_{n+1} &= Fx_n + Bv_n + Ga_{n+1}, \\ z_{n+1} &= Hx_{n+1} + \beta_{n+1}, \end{aligned} \tag{18.13}$$

with the initial state x_0 given. The variables $a_1, \cdots, a_n, \cdots$ and $\beta_1, \cdots, \beta_n, \cdots$ are independent standard Gaussian random variables. We assume that $x_n \in R^{d_1}$ and $z_n \in R^{d_2}$. Setting $Q = GG^T$, where G^T denotes the transpose of G, we can see that this model enters into the general framework above with

$$\pi(x, v; d\eta \otimes d\zeta)$$
$$= \frac{\exp\left(-\frac{1}{2}\left[|\zeta - H\eta|^2 + (\eta - (Fx + Bv))^T Q^{-1}(\eta - (Fx + Bv))\right]\right) d\eta \otimes d\zeta}{|Q|^{\frac{1}{2}}}. \tag{18.14}$$

It is a tedious but not a difficult exercise to check from formulas (18.11) and (18.12) that $\pi_n^{V, x_0}(\zeta_1, \cdots \zeta_n; d\eta)$ is Gaussian with covariance Γ_n and mean μ_n with the recursion

$$\begin{aligned} \Gamma_{n+1}^{-1} &= HH^T + (Q + F\Gamma_n F^T)^{-1}, \\ \mu_{n+1} &= F\mu_n + Bv_n + \Gamma_{n+1} H^T(\zeta_{n+1} - H(F\mu_n + Bv_n)), \\ \Gamma_0 &= 0, \quad \mu_0 = x_0. \end{aligned} \tag{18.15}$$

18.4 Two reference measures

18.4.1 The model

We present here a more complex example of an inventory problem with random spoilage. The state x_n is the inventory level at time n. The control v_n is the order quantity at time n. Both the state and the control are one-dimensional quantities. The inventory dynamics is described by

$$x_{n+1} = (x_n + v_n - D_n - S_n)^+ := \max(x_n + v_n - D_n - S_n, 0), \tag{18.16}$$

where D_n is the random demand and S_n is the random spoilage. We assume that they are independent random variables, between themselves and also for different n. Let D, S represent generic demand and spoilage random variables. They are positive and have probability densities f and g, and cumulative densities F and G, respectively. Also set $\bar{F} = 1 - F$ and $\bar{G} = 1 - G$.

Equation (18.16) represents the natural evolution of the inventory stock, when no shortage is allowed. The observation is given by

$$z_{n+1} = \min(x_n + v_n, D_n), \tag{18.17}$$

which represents the recorded sales at the end of period n. Note that this model assumes that spoilage takes places after the the demand has been satisfied. The transition probability $\pi(x, v; d\eta \otimes d\zeta)$ is the probability of the pair

$$((x + v - D - S)^+, \min(x + v, D)).$$

It is easy to check that the transition probability is

$$\begin{aligned}\pi(x, v; d\eta \otimes d\zeta) = {} & \bar{F}(x + v)\delta(\eta) \otimes \delta(\zeta - x - v) \\ & + \mathbb{1}_{\{0 \leq \zeta < x+v\}} f(\zeta) \left[\bar{G}(x + v - \zeta)\delta(\eta) \right. \\ & \left. + \mathbb{1}_{\{0 < \eta \leq x+v-\zeta\}} g(x + v - \zeta - \eta) d\eta \right] \otimes d\zeta,\end{aligned} \tag{18.18}$$

where $\delta(\eta)$ denotes the delta function. While writing (18.18), we first note that the sum of the sales and the ending inventory in a period cannot be more than the sum of the beginning inventory and the order quantity in that period. Thus, while writing the transition probability, it suffices to consider transitions only into $\{(\eta, \zeta) \in (R^+)^2 : \eta + \zeta \leq x + v\}$. This set can be split into three sets $\{(\eta, \zeta) \in (R^+)^2 : (\eta, \zeta) = (0, x + v)\}$, $\{(\eta, \zeta) \in (R^+)^2 : \eta = 0, \zeta < x + v\}$ and $\{(\eta, \zeta) \in (R^+)^2 : 0 < \eta \leq x + v - \zeta, \zeta < x + v\}$, which respectively lead to the first, the second and the third term in (18.18).

Inserting the transition probability of (18.18) into (18.8), we obtain the recurrence

$$\pi_{n+1}^{V,x_0}(\zeta_1, \cdots, \zeta_n, \zeta_{n+1}; \Gamma) \times \left\{ \mathbb{1}_{\zeta_{n+1} \geq v_n} \bar{F}(\zeta_{n+1}) \pi_n^{V,x_0}(\zeta_1, \cdots, \zeta_n; d(\zeta_{n+1} - v_n)) \right.$$
$$\left. + d\zeta_{n+1} f(\zeta_{n+1}) \int_{(\zeta_{n+1}-v_n)^+}^{\infty} \pi_n^{V,x_0}(\zeta_1, \cdots, \zeta_n; d\lambda) \right\}$$

$$= \mathbb{1}_\Gamma(0) \mathbb{1}_{\zeta_{n+1} \geq v_n} \bar{F}(\zeta_{n+1}) \pi_n^{V,x_0}(\zeta_1, \cdots, \zeta_n; d(\zeta_{n+1} - v_n))$$

$$+ d\zeta_{n+1} f(\zeta_{n+1}) \left[\mathbb{1}_\Gamma(0) \int_{(\zeta_{n+1}-v_n)^+}^{\infty} \bar{G}(\lambda + v_n - \zeta_{n+1}) \pi_n^{V,x_0}(\zeta_1, \cdots, \zeta_n; d\lambda) \right.$$

$$\left. + \int_{(\zeta_{n+1}-v_n)^+}^{\infty} \left(\int_0^{\lambda + v_n - \zeta_{n+1}} \mathbb{1}_\Gamma(\mu) g(\lambda + v_n - \zeta_{n+1} - \mu) d\mu \right) \pi_n^{V,x_0}(\zeta_1, \cdots, \zeta_n; d\lambda) \right]. \tag{18.19}$$

18.4.2 Solution of the recurrence

To solve equation (18.19), we need two reference measures: the Dirac measure at 0 and the Lebesgue measure on $(0, \infty)$; so we write

$$\pi_n^{V,x_0}(\zeta_1, \cdots, \zeta_n; \Gamma) = A_n(\zeta_1, \cdots, \zeta_n) \mathbb{1}_\Gamma(0) + \int_0^\infty B_n(\zeta_1, \cdots, \zeta_n; \eta) \mathbb{1}_\Gamma(\eta) d\eta. \tag{18.20}$$

Since this is a probability measure, we must have the normalization condition

$$A_n(\zeta_1, \cdots, \zeta_n) + \int_0^\infty B_n(\zeta_1, \cdots, \zeta_n; \lambda) d\lambda = 1, \forall \zeta_1, \cdots, \zeta_n. \tag{18.21}$$

After lengthy calculations (see Bensoussan et al. [7]), we arrive at the following result.

Theorem 18.2 *Functions A_n and B_n are obtained recursively as follows:*

$$A_{n+1}(\zeta_1, \cdots, \zeta_n, \zeta_{n+1}) = \mathbb{1}_{\zeta_{n+1} = v_n}$$
$$+ \mathbb{1}_{\zeta_{n+1} < v_n} \left[A_n(\zeta_1, \cdots, \zeta_n) \bar{G}(v_n - \zeta_{n+1}) + \int_0^\infty B_n(\zeta_1, \cdots, \zeta_n; \lambda) \bar{G}(\lambda + v_n - \zeta_{n+1}) d\lambda \right]$$
$$+ \mathbb{1}_{\zeta_{n+1} > v_n} \frac{\bar{F}(\zeta_{n+1}) B_n(\zeta_1, \cdots, \zeta_n; \zeta_{n+1} - v_n) + f(\zeta_{n+1}) \int_{\zeta_{n+1}-v_n}^\infty B_n(\zeta_1, \cdots, \zeta_n; \lambda) \bar{G}(\lambda + v_n - \zeta_{n+1}) d\lambda}{\bar{F}(\zeta_{n+1}) B_n(\zeta_1, \cdots, \zeta_n; \zeta_{n+1} - v_n) + f(\zeta_{n+1}) \int_{\zeta_{n+1}-v_n}^\infty B_n(\zeta_1, \cdots, \zeta_n; \lambda) d\lambda} \tag{18.22}$$

and

$$B_{n+1}(\zeta_1, \cdots, \zeta_n, \zeta_{n+1}; \eta) = \mathbb{1}_{\zeta_{n+1} < v_n} \left[\mathbb{1}_{0 \leq \eta < v_n - \zeta_{n+1}} A_n(\zeta_1, \cdots, \zeta_n) g(v_n - \zeta_{n+1} - \eta) \right.$$
$$\left. + \int_{(\eta + \zeta_{n+1} - v_n)^+}^\infty B_n(\zeta_1, \cdots, \zeta_n; \lambda) g(\lambda + v_n - \zeta_{n+1} - \eta) d\lambda \right]$$
$$+ \mathbb{1}_{\zeta_{n+1} > v_n} \frac{f(\zeta_{n+1}) \int_{\eta + \zeta_{n+1} - v_n}^\infty B_n(\zeta_1, \cdots, \zeta_n; \lambda) g(\lambda + v_n - \zeta_{n+1} - \eta) d\lambda}{\bar{F}(\zeta_{n+1}) B_n(\zeta_1, \cdots, \zeta_n; \zeta_{n+1} - v_n) + f(\zeta_{n+1}) \int_{\zeta_{n+1}-v_n}^\infty B_n(\zeta_1, \cdots, \zeta_n; \lambda) d\lambda} \tag{18.23}$$

with the initial conditions

$$A_1(\zeta_1) = 1\!\!1_{\zeta_1 = x_0 + v_0} + 1\!\!1_{0 \leq \zeta_1 < x_0 + v_0} \bar{G}(x_0 + v_0 - \zeta_1)$$
$$B_1(\zeta_1; \eta) = 1\!\!1_{0 \leq \zeta_1 < x_0 + v_0} 1\!\!1_{0 \leq \eta < x_0 + v_0 - \zeta_1} g(x_0 + v_0 - \zeta_1 - \eta). \quad (18.24)$$

18.5 Unnormalized conditional probabilities

18.5.1 Case: one reference measure

Consider formulas (18.11) and (18.12). We associate the linear recurrence

$$\rho_{n+1}^{V,x_0}(\zeta_1, \cdots, \zeta_n, \zeta_{n+1}; \Gamma) = \int_X \rho_n^{V,x_0}(\zeta_1, \cdots \zeta_n; d\eta) \theta(\eta, v_n, \zeta_{n+1}; \Gamma), \quad (18.25)$$

with the initial condition

$$\rho_1^{V,x_0}(\zeta_1; \Gamma) = \theta(x_0, v_0, \zeta_1; \Gamma), \quad (18.26)$$

and also define

$$\lambda_n^{V,x_0}(\zeta_1, \cdots, \zeta_n) = \rho_n^{V,x_0}(\zeta_1, \cdots, \zeta_n; X). \quad (18.27)$$

Then the sequence $\lambda_n^{V,x_0}(\zeta_1, \cdots, \zeta_n)$ evolves as follows:

$$\lambda_{n+1}^{V,x_0}(\zeta_1, \cdots, \zeta_{n+1}) = \lambda_n^{V,x_0}(\zeta_1, \cdots, \zeta_n) \int_X \pi_n^{V,x_0}(\zeta_1, \cdots \zeta_n; d\eta) \theta(\eta, v_n, \zeta_{n+1}; X). \quad (18.28)$$

Also, it is easy to check that

$$\rho_n^{V,x_0}(\zeta_1, \cdots, \zeta_n; \Gamma) = \lambda_n^{V,x_0}(\zeta_1, \cdots, \zeta_n) \pi_n^{V,x_0}(\zeta_1, \cdots, \zeta_n; \Gamma). \quad (18.29)$$

The measure $\rho_n^{V,x_0}(\zeta_1, \cdots, \zeta_n; \Gamma)$ on X is called the unnormalized conditional probability. The evolution equation (18.25) is linear, and that considerably simplifies the functional analysis framework.

18.5.2 Case: two reference measures

We can also derive the evolution equations for unnormalized conditional probabilities in the case of the model of Section 18.4. We want again to write the property (18.29) in this case together with the fact that $\rho_n^{V,x_0}(\zeta_1, \cdots, \zeta_n; \Gamma)$ evolves according to a linear evolution equation. To simplify notation, we no longer mention the symbols $V, x_0, \zeta_1, \cdots, \zeta_n$, and just refer to $\pi_n(\Gamma)$. We have

$$\pi_n(\Gamma) = A_n 1\!\!1_\Gamma(0) + \int_0^\infty B_n(\eta) 1\!\!1_\Gamma(\eta) d\eta.$$

We look for a similar formula for $\rho_n(\Gamma)$, namely,

$$\rho_n(\Gamma) = a_n \mathbb{1}_\Gamma(0) + \int_0^\infty \beta_n(\eta) \mathbb{1}_\Gamma(\eta) d\eta,$$

so we have

$$a_n = \lambda_n A_n, \quad \beta_n = \lambda_n B_n.$$

Note that we must have $\lambda_n = a_n + \int_0^\infty \beta_n(\eta) d\eta$.

From the recurrence for A_n and B_n and the preceding definitions, one can easily obtain the following linear evolution equations for a_n and β_n:

$$a_{n+1} = \mathbb{1}_{z_{n+1}=v_n} + \mathbb{1}_{z_{n+1}<v_n}\left[a_n \bar{G}(v_n - z_{n+1}) + \int_0^\infty \bar{G}(\lambda + v_n - z_{n+1})\beta_n(\lambda) d\lambda\right] \tag{18.30}$$

$$+ \mathbb{1}_{z_{n+1}>v_n}\left[\bar{F}(z_{n+1})\beta_n(z_{n+1} - v_n) + f(z_{n+1})\int_{z_{n+1}-v_n}^\infty \bar{G}(\lambda + v_n - z_{n+1})\beta_n(\lambda) d\lambda\right],$$

$$\beta_{n+1}(\eta) = \mathbb{1}_{z_{n+1}<v_n}\left[\mathbb{1}_{0\leq\eta<v_n-z_{n+1}} a_n g(v_n - z_{n+1} - \eta)\right.$$

$$\left.+ \int_{(z_{n+1}+\eta-v_n)^+}^\infty g(\lambda + v_n - z_{n+1} - \eta)\beta_n(\lambda) d\lambda\right] \tag{18.31}$$

$$+ \mathbb{1}_{z_{n+1}>v_n} f(z_{n+1})\int_{\eta+z_{n+1}-v_n}^\infty g(\lambda + v_n - z_{n+1} - \eta)\beta_n(\lambda) d\lambda,$$

with initial the conditions

$$a_1 = A_1, \quad \beta_1(\eta) = B_1(\eta). \tag{18.32}$$

Then, λ_n evolves according to the formula

$$\lambda_{n+1} = \mathbb{1}_{z_{n+1}=v_n} + \lambda_n \mathbb{1}_{z_{n+1}<v_n}$$

$$+ \mathbb{1}_{z_{n+1}>v_n}\left[\bar{F}(z_{n+1})\beta_n(z_{n+1} - v_n) + f(z_{n+1})\int_{z_{n+1}-v_n}^\infty \beta_n(\lambda) d\lambda\right] \tag{18.33}$$

with $\lambda_1 = 1$.

18.6 Other models in inventory theory

The inventory models of this section do not accept the two-reference-measure representation of the conditional probability as in (18.20). Thus, Theorem 18.2 does not apply. However, these models still fall into the general framework of the joint conditional probability evolution in Theorem 18.1.

Theorem 18.1 can also be used to derive the conditional probability of demands in Bensoussan et al. [4]. There the inventory is fully observed while the demands, modelled by a Markov process, are partially observed via sales.

For brevity, the application of Theorem 18.1 to the model in Bensoussan et al. [4] is not illustrated here.

18.6.1 The zero-balance walk model

In this model there is no spoilage, so the evolution of the inventory stock is given by

$$x_{n+1} = (x_n + v_n - D_n)^+. \tag{18.34}$$

However, the stock is not observable. We can simply observe a binary variable

$$z_n = \mathbb{1}_{x_n=0}. \tag{18.35}$$

We know that the stock is 0 or that it is strictly positive, and when it is strictly positive we do not know the amount. Thanks to the very particular character of the observation, one can proceed without using the general theory developed above (see Bensoussan et al. [2]). However, we will see how this problem is a particular case of the general theory. We just notice that the observation z_{n+1} can be written as

$$z_{n+1} = \mathbb{1}_{x_n+v_n-D_n \leq 0}, \tag{18.36}$$

and we can easily define the Markov process which governs the process (x_n, z_n). We have $X = R^+$, $Z = \{1, 0\}$. To be consistent with the general notation used in Sections 18.2 and 18.3, we will still denote by $d\zeta$, a probability measure on Z, although it is simply a number representing the probability of 1. We will denote by $\delta_{\{\zeta=1\}}$, the probability for which this number is 1 (analogous to the Dirac probability in Section 18.4.2). The transition probability $\pi(x, v; d\eta \otimes d\zeta)$ is simply the probability of the pair $((x + v - D)^+, \mathbb{1}_{x+v-D\leq 0})$. It is a simple exercise to show that

$$\pi(x, v; d\eta \otimes d\zeta) = \overline{F}(x+v)\delta(\eta) \otimes \delta_{\{\zeta=1\}} + f(x+v-\eta)\mathbb{1}_{0 \leq \eta < x+v} d\eta \otimes \delta_{\{\zeta=0\}}. \tag{18.37}$$

We can then apply Theorem 18.1 to obtain the relation

$$\pi_{n+1}(\Gamma) = \frac{\int_0^\infty \pi_n(d\eta) \int_0^\infty \mathbb{1}_\Gamma(\mu) \pi(\eta, v_n; d\mu \otimes d\zeta_{n+1})}{\int_0^\infty \pi_n(d\eta) \int_0^\infty \pi(\eta, v_n; d\mu \otimes d\zeta_{n+1})} \tag{18.38}$$

with the initial condition

$$\pi_1(\Gamma) = \frac{\int_0^\infty \mathbb{1}_\Gamma(\mu) \pi(x_0, v_0; d\mu \otimes d\zeta_1)}{\int_0^\infty \pi(x_0, v_0; d\mu \otimes d\zeta_1)}. \tag{18.39}$$

If we begin with the initial condition, we easily see that

$$\pi_1(\Gamma) = \mathbb{1}_{\{\zeta_1=1\}} \mathbb{1}_\Gamma(0) + \mathbb{1}_{\{\zeta_1=0\}} \frac{\int_0^{x_0+v_0} \mathbb{1}_\Gamma(\mu) f(x_0+v_0-\mu) d\mu}{F(x_0+v_0)}.$$

More generally, we look for a conditional probability of the form

$$\pi_n(\Gamma) = \mathbb{1}_{\{\zeta_n=1\}}\mathbb{1}_\Gamma(0) + \mathbb{1}_{\{\zeta_n=0\}}\int_0^\infty \mathbb{1}_\Gamma(\mu)B_n(\mu)d\mu. \tag{18.40}$$

The motivation of this formula is clear. If $\zeta_n = 1$, we know that the inventory is 0 and B_n is the conditional probability (given the observations up to n) of the value of the stock, knowing that the stock is positive. We can write (18.40) in an equivalent way as follows:

$$\pi_n(d\eta) = \mathbb{1}_{\{\zeta_n=1\}}\delta(\eta) + \mathbb{1}_{\{\zeta_n=0\}}B_n(\eta)d\eta.$$

So equation (18.38) implies

$$\delta_{\{\zeta_{n+1}=1\}}\mathbb{1}_\Gamma(0)\left[\mathbb{1}_{\zeta_n=1}\bar{F}(v_n) + \mathbb{1}_{\zeta_n=0}\int_0^\infty B_n(\eta)\bar{F}(\eta+v_n)d\eta\right] + \delta_{\{\zeta_{n+1}=0\}}$$
$$\times \left[\mathbb{1}_{\zeta_n=1}\int_0^{v_n}\mathbb{1}_\Gamma(\mu)f(v_n-\mu)d\mu + \mathbb{1}_{\zeta_n=0}\int_0^\infty d\eta\, B_n(\eta)\int_0^{\eta+v_n}\mathbb{1}_\Gamma(\mu)f(\eta+v_n-\mu)d\mu\right]$$
$$= \left(\mathbb{1}_{\{\zeta_{n+1}=1\}}\mathbb{1}_\Gamma(0) + \mathbb{1}_{\{\zeta_{n+1}=0\}}\int_0^\infty \mathbb{1}_\Gamma(\mu)B_{n+1}(\mu)d\mu\right) \times$$
$$\left[\delta_{\{\zeta_{n+1}=1\}}\left(\mathbb{1}_{\zeta_n=1}\bar{F}(v_n) + \mathbb{1}_{\zeta_n=0}\int_0^\infty B_n(\eta)\bar{F}(\eta+v_n)d\eta\right)\right.$$
$$\left.+\delta_{\{\zeta_{n+1}=0\}}\left(\mathbb{1}_{\zeta_n=1}F(v_n) + \mathbb{1}_{\zeta_n=0}\int_0^\infty B_n(\eta)F(\eta+v_n)d\eta\right)\right]. \tag{18.41}$$

Identifying similar terms, we easily get the recurrence in Bensoussan et al. [2]

$$B_{n+1}(\mu) = \mathbb{1}_{\zeta_n=1}\frac{f(v_n-\mu)\mathbb{1}_{\mu<v_n}}{F(v_n)} + \mathbb{1}_{\zeta_n=0}\frac{\int_{(\mu-v_n)^+}^\infty B_n(\eta)f(\eta+v_n-\mu)d\eta}{\int_0^\infty B_n(\eta)F(\eta+v_n)d\eta}. \tag{18.42}$$

18.6.2 The rain check model

In this model of Bensoussan et al. [5], the stock can become negative, and we observe the value of negative stock because it is recorded exactly representing a commitment by issuing of a rain check. Analytically, we have

$$\begin{aligned}x_{n+1} &= x_n + v_n - D_n \\ z_{n+1} &= (x_n + v_n - D_n)^- := \max(-x_n - v_n + D_n, 0).\end{aligned} \tag{18.43}$$

We have $X = R$ and $Z = R^+$. The associated Markov process is described by its transition probability

$$\pi(x, v; d\eta \otimes d\zeta) = \mathbb{1}_{0\leq\eta\leq x+v}f(x+v-\eta)d\eta \otimes \delta(\zeta)$$
$$+ \mathbb{1}_{\zeta>0}\mathbb{1}_{x+v+\zeta\geq 0}f(x+v+\zeta)\delta(\eta+\zeta) \otimes d\zeta. \tag{18.44}$$

The first conditional probability $\pi_1(\Gamma)$ is expressed by

$$\pi_1(\Gamma) = \mathbb{1}_{\zeta_1=0}\mathbb{1}_{x_0+v_0\geq 0}\frac{\int_0^{x_0+v_0}\mathbb{1}_\Gamma(\eta)f(x_0+v_0-\eta)d\eta}{F(x_0+v_0)} + \mathbb{1}_\Gamma(-\zeta_1)\mathbb{1}_{\zeta_1>0}\mathbb{1}_{x_0+v_0+\zeta_1\geq 0}. \tag{18.45}$$

We look for a general formula of the form

$$\pi_n(\Gamma) = \mathbb{1}_{\zeta_n=0} \int_0^\infty \mathbb{1}_\Gamma(\eta) B_n(\eta) d\eta + \mathbb{1}_\Gamma(-\zeta_n)\mathbb{1}_{\zeta_n>0}, \qquad (18.46)$$

where the density $B_n(\eta)$ represents the probability density of the stock at level η knowing that $\eta > 0$. It is a $\mathcal{Z}^{n-1}$ measurable function. The second element on the right-hand side of (18.46) comes from the fact that we know the value of the stock when it is negative. To derive recursive formulas for $B_n(\eta)$, we use the general formula (18.8). So we write the Radon–Nikodym derivative

$$\pi_{n+1}(\Gamma) = \frac{N_n(d\zeta_{n+1})}{D_n(d\zeta_{n+1})},$$

where

$$N_n(d\zeta_{n+1}) = \int_{-\infty}^\infty \pi_n(d\eta) \int_{-\infty}^\infty \mathbb{1}_\Gamma(\mu)\pi(\eta, v_n; d\mu \otimes d\zeta_{n+1})$$

and

$$D_n(d\zeta_{n+1}) = \int_{-\infty}^\infty \pi_n(d\eta) \int_{-\infty}^\infty \pi(\eta, v_n; d\mu \otimes d\zeta_{n+1}).$$

Using formula (18.44) and the representation (18.46), we can compute, after some transformations,

$$N_n(d\zeta_{n+1}) = \delta(\zeta_{n+1}) \int_0^\infty d\mu \mathbb{1}_\Gamma(\mu) \Big[\mathbb{1}_{\zeta_n>0} \mathbb{1}_{v_n \geq \zeta_n} f(v_n - \zeta_n - \mu) \mathbb{1}_{\mu \leq v_n - \zeta_n}$$
$$+ \mathbb{1}_{\zeta_n=0} \int_{(\mu-v_n)^+}^\infty f(\eta + v_n - \mu) B_n(\eta) d\eta \Big]$$
$$+ d\zeta_{n+1} \mathbb{1}_\Gamma(-\zeta_{n+1}) \mathbb{1}_{\zeta_{n+1}>0} \Big[\mathbb{1}_{\zeta_n>0} \mathbb{1}_{v_n-\zeta_n+\zeta_{n+1}\geq 0} f(v_n - \zeta_n + \zeta_{n+1})$$
$$+ \int_0^\infty f(\eta + v_n + \zeta_{n+1}) \mathbb{1}_{v_n+\eta+\zeta_{n+1}\geq 0} B_n(\eta) d\eta \Big]$$

and

$$D_n(d\zeta_{n+1}) = \delta(\zeta_{n+1}) \Big[\mathbb{1}_{\zeta_n>0} \mathbb{1}_{v_n \geq \zeta_n} F(v_n - \zeta_n) + \mathbb{1}_{\zeta_n=0} \int_0^\infty F(\eta + v_n) B_n(\eta) d\eta \Big]$$
$$+ d\zeta_{n+1} \mathbb{1}_{\zeta_{n+1}>0} \Big[\mathbb{1}_{\zeta_n>0} \mathbb{1}_{v_n-\zeta_n+\zeta_{n+1}\geq 0} f(v_n - \zeta_n + \zeta_{n+1})$$
$$+ \int_0^\infty f(\eta + v_n + \zeta_{n+1}) \mathbb{1}_{v_n+\eta+\zeta_{n+1}\geq 0} B_n(\eta) d\eta \Big].$$

Computing the ratio N_n/D_n and updating (18.46) to $n+1$, it is easy to derive the recurrence in Bensoussan et al. [5]

$$B_{n+1}(\mu) = \mathbb{1}_{\zeta_n>0}\mathbb{1}_{v_n-\zeta_n-\mu\geq 0}\frac{f(v_n-\zeta_n-\mu)}{F(v_n-\zeta_n)} + \mathbb{1}_{\zeta_n=0}\frac{\int_{(\mu-v_n)^+}^{\infty} f(\eta+v_n-\mu)B_n(\eta)d\eta}{\int_0^{\infty} F(\eta+v_n)B_n(\eta)d\eta} \quad (18.47)$$

with the initial condition, derived from (18.45),

$$B_1(\mu) = \frac{f(x_0+v_0-\mu)\mathbb{1}_{0\leq\mu\leq x_0+v_0}}{F(x_0+v_0)}. \quad (18.48)$$

18.6.3 The case of two inventory distributions

We consider the evolution of the rain check model, but instead of observing the value of the stock when it is negative, we only are aware of the fact that it is negative, without knowing its exact value. This model, developed in Bensoussan et al. [6], is described as follows

$$x_{n+1} = x_n + v_n - D_n, \quad (18.49)$$

$$z_{n+1} = \mathbb{1}_{x_n+v_n-D_n\leq 0}, \quad (18.50)$$

The corresponding transition probability is given by

$$\pi(x,v;d\eta\otimes d\zeta) = f(x+v-\eta)\mathbb{1}_{\eta\leq -(x+v)^-}d\eta\otimes\delta_{\{\zeta=1\}}$$
$$+ f(x+v-\eta)\mathbb{1}_{0\leq\eta<(x+v)^+}d\eta\otimes\delta_{\{\zeta=0\}}. \quad (18.51)$$

We look for a conditional probability distribution of the form

$$\pi_n(\Gamma) = \mathbb{1}_{\{\zeta_n=1\}}\int_{-\infty}^{0}\mathbb{1}_\Gamma(\eta)B_n^-(\eta)d\eta + \mathbb{1}_{\{\zeta_n=0\}}\int_0^{\infty}\mathbb{1}_\Gamma(\eta)B_n^+(\eta)d\eta, \quad (18.52)$$

and it is easy to find the recurrence equations

$$B_{n+1}^+(\mu) = \mathbb{1}_{\zeta_n=0}\frac{\int_{(\mu-v_n)^+}^{\infty} B_n^+(\eta)f(\eta+v_n-\mu)d\eta}{\int_0^{\infty} B_n^+(\eta)F(\eta+v_n)d\eta} + \mathbb{1}_{\zeta_n=1}\frac{\int_{-(\mu-v_n)^-}^{0} B_n^-(\eta)f(\eta+v_n-\mu)d\eta}{\int_{-v_n}^{0} B_n^-(\eta)F(\eta+v_n)d\eta}, \mu>0,$$

$$B_{n+1}^-(\mu) = \mathbb{1}_{\zeta_n=0}\frac{\int_0^{\infty} B_n^+(\eta)f(\eta+v_n-\mu)d\eta}{\int_0^{\infty} B_n^+(\eta)\bar{F}(\eta+v_n)d\eta} + \mathbb{1}_{\zeta_n=1}\frac{\int_{\mu-v_n}^{0} B_n^-(\eta)f(\eta+v_n-\mu)d\eta}{\int_{-v_n}^{0} B_n^-(\eta)\bar{F}(\eta+v_n)d\eta}, \mu<0$$
$$(18.53)$$

with the the initial conditions,

$$B_1^+(\mu) = \frac{\mathbb{1}_{x_0+v_0\geq 0}f(x_0+v_0-\mu)\mathbb{1}_{0\leq\mu\leq x_0+v_0}}{F(x_0+v_0)},$$

$$B_1^-(\mu) = \frac{f(x_0+v_0-\mu)\mathbb{1}_{\mu\leq -(x_0+v_0)^-}}{\bar{F}((x_0+v_0)^+)}.$$
$$(18.54)$$

In each of the models of Sections 18.6.1–18.6.3, it is possible to derive the evolution equation of the unnormalized probability.

18.7 Stochastic control with incomplete information

18.7.1 Presentation of the general problem

Consider the model of Section 18.3, where the transition probability is given by (18.10). We will need to have an initial state which is random, instead of a given element of X. So we consider a probability distribution on X, denoted by ϖ. The probability P^{V,x_0} on $(\Omega, \mathcal{A})$ will be changed to $P^{V,\varpi}$ such that for any integrable random variable H, one has

$$E^{V,\varpi} H = \int_X \varpi(dx) E^{V,x} H.$$

In the same way we have derived the conditional probability $\pi_n^{V,x_0}(\zeta_1, \cdots, \zeta_n; \Gamma)$ of the state x_n given $\mathcal{Z}^n$ for an initial state x_0, we may obtain the conditional probability of x_n given $\mathcal{Z}^n$ for an initial state which is random and distributed according to the probability distribution ϖ. We will denote it by $\pi_n^{V,\varpi}(\zeta_1, \cdots, \zeta_n; \Gamma)$. The evolution of this conditional probability is described by

$$\pi_{n+1}(\Gamma) = \frac{\int_X \pi_n(d\eta) \theta(\eta, v_n, \zeta_{n+1}; \Gamma)}{\int_X \pi_n(d\eta) \theta(\eta, v_n, \zeta_{n+1}; X)}, \tag{18.55}$$

$$\pi_0(\Gamma) = \varpi(\Gamma). \tag{18.56}$$

To simplify notation, we have not written explicitly the indices V, ϖ in $\pi_n(\Gamma)$.

18.7.2 Stochastic control

Let us now denote the running cost by a Borel function

$$f(x, v) \geq 0, \tag{18.57}$$

and assume that the controls v_n take values in a subset U_{ad} of U. We define for a pair (V, ϖ), a payoff

$$J^{V,\varpi} = E^{V,\varpi} \sum_{n=0}^{\infty} a^n f(x_n, v_n); \tag{18.58}$$

note that, thanks to the nonnegativity of f, this quantity is well defined, although it may take the value ∞. We set

$$u(\varpi) = \inf_V J^{V,\varpi}. \tag{18.59}$$

We have defined a functional on the space of probabilities on X. The stochastic control we have formulated is a partially observed stochastic control problem.

Filtering for Discrete-Time Markov Processes

We can reformulate it as a stochastic control with full information, in which the state of the system is π_n. Indeed,

$$J^{V,\varpi} = E^{V,\varpi} \sum_{n=0}^{\infty} a^n \int_X f(\eta, v_n) \pi_n(d\eta). \tag{18.60}$$

We now write the Bellman equation of dynamic programming for this functional. We proceed formally and obtain

$$u(\varpi) = \inf_{v \in U_{ad}} \left[\int_X f(\eta, v) \varpi(d\eta) + a \int_X \int_Z m(d\zeta) \varpi(d\eta) \theta(\eta, v, \zeta; X) \right.$$
$$\left. \times u\left(\frac{\int_X \varpi(d\eta)\theta(\eta, v, \zeta; \cdot)}{\int_X \varpi(d\eta)\theta(\eta, v, \zeta; X)} \right) \right]. \tag{18.61}$$

Direct analysis of this Bellman equation is not easy, since it is a highly nonlinear functional equation in an infinite dimensional space. We will use the unnormalized conditional probability to simplify it conveniently.

18.7.3 Unnormalized conditional probability

Let us consider a measure on X (not a probability) defined by the recurrence

$$\rho_{n+1}(\Gamma) = \int_X \rho_n(d\eta) \theta(\eta, v, \zeta_{n+1}; \Gamma) \tag{18.62}$$

with initial condition

$$\rho_0(\Gamma) = \varpi(\Gamma) \tag{18.63}$$

and the sequence $\lambda_n = \rho_n(X)$. It is easy to check that

$$\rho_n(\Gamma) = \lambda_n \pi_n(\Gamma) \tag{18.64}$$

and that λ_n is the solution of

$$\lambda_{n+1} = \lambda_n \int_X \pi_n(d\eta) \theta(\eta, v_n, \zeta_{n+1}; X). \tag{18.65}$$

By analogy with the Kushner and Zakai equations for the nonlinear filtering problem of diffusions, we will call equation (18.55) a Kushner equation and (18.62) a Zakai equation.

18.7.4 Bellman equation for the unnormalized conditional probability

We consider the space $\mathcal{M}(X)$ of finite measures on X, not necessarily continuous, equipped with the norm

$$||\rho|| = \sup_{\Gamma} |\rho(\Gamma)|, \tag{18.66}$$

where Γ is a Borel subset of X. If ρ is a positive measure, then simply

$$||\rho|| = \rho(X).$$

The space $\mathcal{M}(X)$ is a Banach space. Naturally a probability ϖ on X is an element of $\mathcal{M}(X)$ for which $||\varpi|| = 1$. We extend the set of functions of the form u defined in (18.61) to the larger set of functions of the form Ψ, whose arguments belong to $\mathcal{M}(X)$ but they are not necessarily probability distributions, by setting

$$\Psi(\rho) = \rho(X) u\left(\frac{\rho}{\rho(X)}\right). \tag{18.67}$$

From equation (18.61), one checks easily that $\Psi(\varpi)$ satisfies the much simpler equation

$$\Psi(\rho) = \inf_{v \in U_{ad}} \left[\int_X f(\eta, v) \rho(d\eta) + \alpha \int_Z m(d\zeta) \Psi\left(\int_X \rho(d\eta) \theta(\eta, v, \zeta; .) \right) \right]. \tag{18.68}$$

The space of continuous functionals $\Upsilon(\rho)$ on $\mathcal{M}(X)$ can be structured as a Banach space with the norm

$$||\Upsilon|| = \sup_{\rho \in \mathcal{M}(X)} \frac{|\Upsilon(\rho)|}{||\rho||}. \tag{18.69}$$

We call B this Banach space. We can now state the following results.

Theorem 18.3 *Assume that $f(\eta, v)$ is continuous and bounded, then the functional equation (18.68) has a unique solution in B.*

Proof. Consider the map $\Psi \mapsto T(\Psi)$ defined by the right-hand side of (18.68). Since

$$|T(\Psi)(\rho) - T(\Psi)(\tilde{\rho})| \leq (\alpha ||\Psi|| + ||f||) ||\rho - \tilde{\rho}||,$$

the functional $T(\Psi)$ is continuous. So T maps B into itself. It is easy to check that T is a contraction. Hence it has a unique fixed point, which proves the result. $\square$

Theorem 18.4 *Assume that $f(\eta, v)$ is continuous and bounded, $\theta(\eta, v, \zeta, \Gamma)$ is bounded, continuous uniformly with respect to Γ, and U_{ad} is compact. Then there exists an optimal feedback control, which is a Borel function on $\mathcal{M}(X)$.*

Proof. From the assumptions, we see for fixed η and ζ that $v \mapsto \theta(\eta, v, \zeta; .)$ is a continuous map from U to $\mathcal{M}(X)$. We have all that is needed to check that

$$\int_X f(\eta, v) \rho(d\eta) + \alpha \int_Z m(d\zeta) \Psi\left(\int_X \rho(d\eta) \theta(\eta, v, \zeta; .) \right)$$

is continuous in v. Since U_{ad} is compact, the result follows. $\square$

From the optimal Borel feedback control and traditional arguments of dynamic programming, it is standard to check the existence of an optimal control $\hat{V}$ for the problem (18.59).

18.8 Control of inventories with incomplete information

18.8.1 Discussion of the zero-balance walk model

Although the inventory models are particular cases of the general theory, we cannot assume for these models that the corresponding one-stage cost $f(\eta, v)$ is bounded. So we cannot apply directly Theorems 18.3 and 18.4. We have to proceed differently and use other techniques, not pertaining to contraction mappings. We first derive the Zakai form of (18.42). We set

$$\rho_n(\eta) = \lambda_n B_n(\eta)$$

with

$$\lambda_n = \int_0^\infty \rho_n(\eta) d\eta.$$

We obtain

$$\rho_{n+1}(\mu) = \mathbb{1}_{\zeta_n=1} \mathbb{1}_{\mu \leq v_n} f(v_n - \mu) + \mathbb{1}_{\zeta_n=0} \int_{(\mu-v_n)^+}^\infty f(\eta + v_n - \mu) \rho_n(\eta) d\eta \quad (18.70)$$

and

$$\lambda_{n+1} = \mathbb{1}_{\zeta_n=1} F(v_n) + \mathbb{1}_{\zeta_n=0} \lambda_n \int_0^\infty F(\eta + v_n) B_n(\eta) d\eta \quad (18.71)$$

with the initial conditions

$$\rho_0(\mu) = \varpi(\mu), \lambda_0 = 1. \quad (18.72)$$

Note that we have introduced an initial probability distribution, which is not just $\delta(\eta - x_0)$, but it represents the distribution of the values of the initial state, provided it is positive. It is completed by a binary variable $\zeta_0 = \zeta$, which is 1 if the initial stock is 0, and it is 0 if the initial stock is strictly positive. We define the payoff by

$$J^{V,\zeta,\varpi} = E^{V,\zeta,\varpi} \sum_{n=0}^\infty a^n l(x_n, v_n), \quad (18.73)$$

where $l(x, v)$ is continuous but not bounded. We assume the growth condition

$$c_2 + cv < l(x, v) < c_0 + c_1 v + hx. \quad (18.74)$$

18.8.2 Dynamic programming for the Kushner equation

We define the value function

$$u(\zeta, \varpi) = \inf_V J^{V,\zeta,\varpi}. \qquad (18.75)$$

Using dynamic programming, we can write

$$u(\zeta, \varpi) = \inf_v \left[\mathbb{1}_{\zeta=1} l(0, v) + \mathbb{1}_{\zeta=0} \int_0^\infty l(\eta, v) \varpi(\eta) d\eta + a E u(\zeta_1, B_1) \right].$$

Moreover,

$$E u(\zeta_1, B_1) = u(1, B_1) \left(\mathbb{1}_{\zeta=1} \bar{F}(v) + \mathbb{1}_{\zeta=0} \int_0^\infty \varpi(\eta) \bar{F}(\eta + v) d\eta \right)$$

$$+ u(0, B_1) \left(\mathbb{1}_{\zeta=1} F(v) + \mathbb{1}_{\zeta=0} \int_0^\infty \varpi(\eta) F(\eta + v) d\eta \right).$$

From (18.42), we can write

$$B_1(\mu) = \mathbb{1}_{\zeta=1} \frac{f(v-\mu) \mathbb{1}_{\mu<v}}{F(v)} + \mathbb{1}_{\zeta=0} \frac{\int_{(\mu-v)^+}^\infty \varpi(\eta) f(\eta + v - \mu) d\eta}{\int_0^\infty \varpi(\eta) F(\eta + v) d\eta}.$$

Collecting results, we can find a system of functional equations for the pair $(u(1, \varpi), u(0, \varpi))$. In fact, it turns out that $u(1, \varpi)$ does not depend on ϖ. To save notation, we set

$$w = u(1, \varpi), \quad u(\varpi) = u(0, \varpi).$$

We derive the following system of functional equations:

$$u(\varpi) = \inf_v \left[\int_0^\infty l(\eta, v) \varpi(\eta) d\eta + a w \int_0^\infty \varpi(\eta) \bar{F}(\eta + v) d\eta + \right. \qquad (18.76)$$

$$\left. + a \int_0^\infty \varpi(\eta) F(\eta + v) d\eta \, u \left(\frac{\int_{(.-v)^+}^\infty \varpi(\eta) f(\eta + v - .) d\eta}{\int_0^\infty \varpi(\eta) F(\eta + v) d\eta} \right) \right],$$

$$w = \inf_v \left[l(0, v) + a \bar{F}(v) w + a F(v) u \left(\frac{f(v - .) \mathbb{1}_{.<v}}{F(v)} \right) \right]. \qquad (18.77)$$

18.8.3 Dynamic programming for the Zakai equation

We then derive simpler functional equations by using the same extension as that discussed in Section 18.7.4; see (18.67). By defining

$$W(\rho) = \rho(X) u \left(\frac{\rho}{\rho(X)} \right),$$

we get the system

$$W(\rho) = \inf_v \left[\int_0^\infty l(\eta, v)\rho(\eta)d\eta + aw \int_0^\infty \rho(\eta)\bar{F}(\eta+v)d\eta \right.$$
$$\left. + aW\left(\int_{(.-v)^+}^\infty \rho(\eta) f(\eta+v-.)d\eta \right) \right], \quad (18.78)$$

$$w = \inf_v [l(0, v) + a\bar{F}(v)w + aW(f(v-.)\mathbb{1}_{.<v})]. \quad (18.79)$$

To study these functional equations, we need a convenient functional space for the argument, which is not the space of finite measures on $[0, \infty)$ because the function $l(\eta, v)$ is not bounded. We introduce

$$\mathcal{H} = \left\{ \rho(.) : [0, \infty) \to R \,\middle|\, ||\rho|| = \int_0^\infty |\rho(x)|dx + \int_0^\infty x|\rho(x)|dx < \infty \right\},$$

which is a Banach space. Let $\mathcal{H}^+$ be the subspace of positive functions. We look for Borel functionals

$$\Psi : \mathcal{H}^+ \to R \text{ with } |\Psi(\rho)| \leq ||\Psi||.||\rho||.$$

The space of these functionals, called $\mathcal{B}$, is structured as a Banach space with the norm

$$||\Psi|| = \sup_{\rho \neq 0} \frac{|\Psi(\rho)|}{||\rho||}.$$

Although nonlinear, these functionals vanish at 0.

We introduce the functional $W_0(\rho)$ corresponding to the control 0, i.e., never ordering, we see that it is the solution of

$$W_0(\rho) = \int_0^\infty l(\eta, 0)\rho(\eta)d\eta + aw_0 \int_0^\infty \rho(\eta)\bar{F}(\eta)d\eta + aW\left(\int_.^\infty \rho(\eta) f(\eta-.)d\eta \right), \quad (18.80)$$

in which

$$w_0 = \frac{l(0, 0)}{1 - a}. \quad (18.81)$$

Note that w_0 is what we obtain if we take $v = 0$ in the right-hand side of (18.79). If fact $W_0(\rho)$ is linear. Indeed, we can write

$$W_0(\rho) = \int_0^\infty \rho(x) V_0(x) dx, \quad (18.82)$$

where $V_0(x)$ is the solution of

$$V_0(x) = l(x, 0) + aw_0 \bar{F}(x) + a \int_0^x f(x-y) V_0(y) dy. \quad (18.83)$$

If we call $\tilde{l}(x) = l(x, 0) + aw_0 \bar{F}(x)$, then the solution of (18.83) is defined uniquely as a series. Indeed if Φ is the linear operator

$$\Phi g(x) = \int_0^x f(x - y)g(y)dy,$$

then

$$V_0 = \sum_{n=0}^{\infty} a^n \Phi^n \tilde{l}.$$

This series is well defined. Indeed,

$$\tilde{l}(x) \leq \frac{c_0}{1-a} + hx \text{ and } \Phi\tilde{l}(x) \leq \frac{c_0}{1-a} + hx,$$

and thus the series converges with

$$0 \leq V_0(x) \leq \frac{c_0}{(1-a)^2} + \frac{hx}{1-a}. \tag{18.84}$$

18.8.4 Main result

In this section, we state the main result and provide its proof.

Theorem 18.5 *Assume (18.74). Then there exists a unique pair $W(\rho)$ in $\mathcal{B}$ and w a real number, which together solve the system (18.78) and (18.79) in the interval defined by*

$$0 \leq W(\rho) \leq W_0(\rho), \quad 0 \leq w \leq w_0.$$

This solution is continuous in ρ, and it coincides with the value function (18.75). More precisely,

$$W(\varpi) = u(0, \varpi), \quad w = u(1, \varpi).$$

There exists an optimal feedback.

Proof. The proof is long and it is divided into four segments. □

Preliminaries: In the first part we set $w = 0$ for simplicity, and consider the functional equation

$$W(\rho) = \inf_v \left[\int_0^{\infty} l(\eta, v)\rho(\eta)d\eta + aW\left(\int_{(.-v)^+}^{\infty} \rho(\eta) f(\eta + v - .)d\eta\right)\right]. \tag{18.85}$$

The same reasoning that will be developed for (18.85) can be used for the system (18.78) and (18.79). We next define some operators. Let Φ be an element of $\mathcal{B}$ which is continuous. Let us define the operators

$$T_v(\Phi)(\rho) = \int_0^{\infty} l(\eta, v)\rho(\eta)d\eta + a\Phi\left(\int_{(.-v)^+}^{\infty} \rho(\eta) f(\eta + v - .)d\eta\right)$$

and
$$T(\Phi)(\rho) = \inf_v T_v(\Phi)(\rho).$$

Then the function $T_v(\Phi)(\rho)$ is continuous in the pair (v, ρ). This property is easily proven. Next we consider $T_0(\Phi)(\rho)$ and the functional $W_0(\rho)$, which is the solution of
$$W_0 = T_0(\rho).$$

As we have seen for (18.80), W_0 is linear and
$$W_0(\rho) = \int_0^\infty \rho(\eta) V_0(\eta) d\eta,$$

in which
$$V_0(\eta) = l(\eta, 0) + a \int_0^\eta f(\eta - \mu) V_0(\mu) d\mu.$$

We have the estimate
$$0 \le V_0(\eta) \le \frac{c_0 + h\eta}{1 - a}.$$

Hence,
$$0 \le W_0(\rho) \le \frac{c_0}{1-a} \int_0^\infty \rho(\eta) d\eta + \frac{h}{1-a} \int_0^\infty \eta \rho(\eta) d\eta. \tag{18.86}$$

We consider now functionals Φ such that $\Phi(\rho) \le W_0(\rho)$. Since
$$T(\Phi)(\rho) \le T_0(\Phi)(\rho) \le T_0(W_0)(\rho) = W_0(\rho),$$
we can restrict the value of v to satisfy $T_v(\Phi)(\rho) \le W_0(\rho)$, which implies on v, thanks to the assumption (18.74), the constraint
$$0 \le v \le \frac{c_0}{c(1-a)} + \frac{h}{c(1-a)} \frac{\int_0^\infty \eta \rho(\eta) d\eta}{\int_0^\infty \rho(\eta) d\eta}. \tag{18.87}$$

Since for a fixed ρ the range of v is bounded, there exists a Borel functional $\hat{v}(\rho)$ such that
$$T(\Phi)(\rho) = T_{\hat{v}(\rho)}(\Phi)(\rho). \tag{18.88}$$

By standard arguments, it then follows that
$$\rho \mapsto T(\Phi)(\rho) \text{ is continuous} \tag{18.89}$$

for any $\Phi \le W_0$ continuous. Now, if Φ is only l.s.c., the same property extends for the functions $\rho, v \to T_v(\Phi)(\rho)$ and $\rho \to T(\Phi)(\rho)$, provided that in the second case $\Phi \leq W_0$. Moreover, there exists a Borel functional $\hat{v}(\rho)$ such that the minimum in the definition of $T(\Phi)(\rho)$ is attained.

Monotone sequences: We define two sequences

$$T(W^n) = W^{n+1}, \quad W^0 = W_0, \tag{18.90}$$

$$T(W_n) = W_{n+1}, \quad W_0 = 0. \tag{18.91}$$

Then we have the inequalities $0 \leq W_n \leq W^n \leq W_0$. We define in this way two monotone sequences. The sequence W_n is increasing and the sequence W^n is decreasing. They are continuous functions (thanks to the properties presented in the preliminaries). So we have $W_n \uparrow \underline{W}$ l.s.c. and $W^n \downarrow \overline{W}$ u.s.c., and naturally $\underline{W} \leq \overline{W}$. Let us check that $\underline{W}$ and $\overline{W}$ are fixed points of T and thus solutions of (18.85).

Concerning the decreasing sequence, we can first write $T(\overline{W}) \leq W^{n+1}$, and thus $T(\overline{W}) \leq \overline{W}$. On the other hand, for any v we have $W^{n+1} \leq T_v(W^n)$, from which we deduce $\overline{W} \leq T_v(\overline{W})$. Hence, $\overline{W} \leq T(\overline{W})$. This proves that $\overline{W}$ is a fixed point.

Consider next the increasing sequence. We first notice that $W_{n+1} \leq T(\underline{W})$, from which we get $\underline{W} \leq T(\underline{W})$. Now the minimum is attained for $T(W_n)$, so we have $T(W_n) = T_{\hat{v}_n}(W_n)$. Fix N and let $n > N$. From the monotone increasing property we can assert that

$$W_{n+1} \geq T_{\hat{v}_n}(W_N), \forall n > N.$$

We can let n tend to ∞ with N fixed. Remember that $\hat{v}_n$ is bounded. So we can assert that there exists a Borel map $\hat{v}(\rho)$ towards which a subsequence of $\hat{v}_n(\rho)$ converges. To simplify notation, we denote the subsequence like the sequence. From the continuity of the function $T_v(W_N)(\rho)$, we can write $\underline{W} \geq T_{\hat{v}}(\underline{W}) \geq T(\underline{W})$, and thus $\underline{W}$ is also a fixed point of T.

Now $\underline{W}$ is l.s.c., and thus an optimal feedback control can be constructed. It is also the function obtained as the limit of $\hat{v}_n(\rho)$ above. Furthermore, it can be seen easily that $\underline{W}$ and $\overline{W}$ are respectively the minimum and the maximum solution of (18.85) in the interval $[0, W_0]$.

Comparison between the value function and the minimum solution: We need to go back to the original system (18.78) and (18.79). Repeating the arguments in the last two segments, we can show that there exist a maximum and a minimum pair of solutions denoted, respectively, by $\overline{W}, \overline{w}$ and $\underline{W}, \underline{w}$ in the intervals $[0, W_0]$ and $[0, w_0]$ for $W_0, w_0 > 0$ defined by (18.81) and (18.82). We know that the minimum solution is l.s.c. and the maximum solution is u.s.c. Moreover there exists an optimal feedback for the minimum solution. This optimal feedback control is made up of two components denoted by $\hat{v}(\rho)$ and $\hat{v}$. They attain the infimum in (18.78) and (18.79), respectively, when $W(\rho)$ and w is the minimum solution pair. Using the optimal feedback control in the

Zakai equation, we define

$$\hat{v}_n = \hat{v}(\rho_n),$$

and we have the evolution equations (18.70) and (18.71) with v_n in place for $\hat{v}_n$. We next use (18.78) and (18.79) with $\rho = \rho_n$, and get

$$\underline{W}(\rho_n) = \int_0^\infty l(\eta, \hat{v}_n)\rho_n(\eta)d\eta + a\underline{w}\int_0^\infty \rho_n(\eta)\bar{F}(\eta + \hat{v}_n)d\eta$$
$$+ a\underline{W}(\int_{(.-\hat{v}_n)^+}^\infty \rho_n(\eta)f(\eta + \hat{v}_n - .)d\eta),$$

from which we deduce immediately, recalling that $\underline{W}(\rho)$ is homogeneous of order 1, that

$$\underline{W}(B_n) = \int_0^\infty l(\eta, \hat{v}_n)B_n(\eta)d\eta + a\underline{w}\int_0^\infty B_n(\eta)\bar{F}(\eta + \hat{v}_n)d\eta$$
$$+ a\int_0^\infty F(\eta + v_n)B_n(\eta)d\eta\,\underline{W}(\frac{\int_{(.-\hat{v}_n)^+}^\infty B_n(\eta)f(\eta + \hat{v}_n - .)d\eta}{\int_0^\infty F(\eta + \hat{v}_n)d\eta}),$$
$$\underline{w} = l(0, \hat{v}) + a\bar{F}(\hat{v})\underline{w} + a\underline{W}(f(\hat{v} - .)\mathbb{1}_{.<\hat{v}}).$$

We then combine these two relations to compute $\underline{W}(B_n)\mathbb{1}_{z_n=0} + \underline{w}\mathbb{1}_{z_n=1}$. One recognizes the relation

$$\underline{W}(B_n)\mathbb{1}_{z_n=0} + \underline{w}\mathbb{1}_{z_n=1} = \hat{E}[l(x_n, \hat{v}_n) + a(\underline{W}(B_{n+1})\mathbb{1}_{z_{n+1}=0} + \underline{w}\mathbb{1}_{z_{n+1}=1})|\mathcal{Z}^n],$$

in which $\hat{E}$ is the expectation with respect to the probability $\hat{P} = P^{\hat{V},\zeta,\varpi}$. In this notation, $\hat{V}$ corresponds to the control policy resulting from the optimal feedback, and ζ and ϖ are the initial conditions. We multiply by a^n and sum up to obtain easily the inequality

$$\underline{W}(\varpi)\mathbb{1}_{\zeta=0} + \underline{w}\mathbb{1}_{\zeta=1} \geq J^{\hat{V},\zeta,\varpi} \geq u(\zeta, \varpi), \qquad (18.92)$$

where $u(\zeta, \varpi)$ is the value function; see (18.75). We want now to establish a reverse inequality that will lead to uniqueness. We cannot use the same reasoning for the maximum solution. This is because it is u.s.c. and, therefore, it does not guarantee the existence of an optimal feedback.

Comparison between the value function and the maximum solution: We need to consider bounded controls. We replace (18.78) and (18.79) by

$$W^m(\rho) = \inf_{0 \le v \le m} [\int_0^\infty l(\eta, v)\rho(\eta)d\eta + aw^m \int_0^\infty \rho(\eta)\bar{F}(\eta + v)d\eta$$

$$+ aW^m(\int_{(.-v)^+}^\infty \rho(\eta)f(\eta + v - .)d\eta)], \qquad (18.93)$$

$$w^m = \inf_{0 \le v \le m} [l(0, v) + a\bar{F}(v)w^m + aW^m(f(v - .)\mathbb{1}_{.<v})]. \qquad (18.94)$$

Of course, we can assert that there are solutions with a maximum and a minimum solution in the intervals $[0, W_0]$ and $[0, w_0]$, respectively. But this time we will find solutions W^m that are Lipschitz (not uniform in m) and prove that they are unique. So we look for solutions that satisfy

$$|W^m(\rho) - W^m(\rho')| \le A^m \int_0^\infty |\rho(x) - \rho'(x)|dx + B^m \int_0^\infty x|\rho(x) - \rho'(x)|dx. \qquad (18.95)$$

We will not give all of the details to prove this kind of estimate, but we will provide the values of A_m and B_m by the technique of a priori estimates. We assume that such an estimate holds, and thus we deduce the consequences implying the possible values of the constants. The minimum values for which such an estimate can hold are easily seen to be

$$A^m = \frac{c_0}{(1-a)^2} + \frac{m}{1-a}(c_1 + \frac{ah}{1-a}) \text{ and } B^m = \frac{h}{1-a}.$$

To get these formulas, we have used the fact that $v \le m$. We see that A_m is not uniform in m (unlike B_m). The rigorous proof requires going through the iterative sequence, and showing that these estimates hold uniformly throughout the iteration process. The maximum and the minimum solutions will consequently satisfy these estimates. Now, if W^m is any solution satisfying these estimates, we can proceed as we did in the previous part of the proof. Thanks to the continuity of W^m, we can define the optimal feedbacks $\hat{v}^m(\rho)$ and v^m and the sequences ρ_n^m and λ_n^m with the controls

$$\hat{v}_n^m = \hat{v}^m\left(\rho_n^m\right).$$

Proceeding as in the previous part of the proof, we can establish the relation

$$W^m(\varpi)\mathbb{1}_{\zeta=0} + w^m\mathbb{1}_{\zeta=1} = \sum_{n=0}^{N-1} a^n E^m l\left(x_n, \hat{v}_n^m\right) + a^N E^m \left[W^m\left(B_N^m\right)\mathbb{1}_{z_N=0} + w^m\mathbb{1}_{z_N=1}\right],$$

where the expectation refers to the probability $P^m = P^{\hat{V}^m, \zeta, \varpi}$ in which $\hat{V}^m$ refers to the optimal feedback defined above. Since

$$W^m\left(B_N^m\right) \le W_0\left(B_N^m\right) \le \frac{c_0}{(1-a)^2} + \frac{h}{1-a}\int_0^\infty \eta B_N^m(\eta)d\eta,$$

we deduce

$$E^m\left[W^m\left(B_N^m\right)\mathbb{1}_{z_N=0} + w^m\mathbb{1}_{z_N=1}\right] \leq \frac{c_0}{(1-a)^2} + \frac{h}{1-a}E^m x_N^m,$$

in which x_N^m is obtained through the sequence $x_{n+1}^m = \left(x_n^m + \hat{v}_n^m - D_n\right)^+$ and the initial value $x_0^m = x_0$. From the bound on $\hat{v}_n^m$, we can claim that $x_{n+1}^m \leq x_n^m + m$, and so $x_N^m \leq x_0 + mN$, and

$$a^N E^m x_N^m \leq a^N(E^m x_0 + mN) \to 0, \text{ as } N \to \infty.$$

Therefore, we obtain

$$W^m(\varpi)\mathbb{1}_{\zeta=0} + w^m\mathbb{1}_{\zeta=1} = \sum_{n=0}^{\infty} a^n E^m l\left(x_n, \hat{v}_n^m\right) = J^{\hat{V}^m,\zeta,\varpi}.$$

But a similar reasoning applied to any decision policy V^m, which satisfies the constraint $v_n^m \leq m$, will lead to the inequality

$$W^m(\varpi)\mathbb{1}_{\zeta=0} + w^m\mathbb{1}_{\zeta=1} \leq J^{V^m,\zeta,\varpi}.$$

This proves that the maximum and the minimum solutions coincide with the value function of the problem with control bounds above by m. Hence (18.93) and (18.94) have a unique solution in the interval $[(0,0),(W_0,w_0)]$ and

$$W^m(\varpi)\mathbb{1}_{\zeta=0} + w^m\mathbb{1}_{\zeta=1} = \inf_{V|v_n\leq m} J^{V,\zeta,\varpi}. \tag{18.96}$$

Now $W^m(\rho), w^m(\rho)$ are decreasing as m increases. We can then assert that

$$W^m(\rho) \downarrow W^*(\rho) \quad w^m \downarrow w^*,$$

and therefore,

$$W^*(\rho) = \tilde{W}(\rho), \quad w^* = \tilde{w},$$

which is the maximum solution pair of (18.78) and (18.79). The proof of this fact consists in comparing the decreasing sequences $(W^m)^n, (w^m)^n$ and W^n, w^n approximating, respectively, W^m, w^m and W, w. It is easy to see that

$$(W^m)^n \geq W^n, \quad (w^m)^n \geq w^n.$$

Therefore,

$$W^* \geq \overline{W}, \quad w^* \geq \overline{w}.$$

Independently, one can verify that W^*, w^* are solutions of (18.78) and (18.79). So they coincide with the maximum solution pair.

Now from (18.96), we can then assert that

$$\overline{W}(\varpi)\mathbb{1}_{\zeta=0} + \overline{w}\mathbb{1}_{\zeta=1} = \inf_V J^{V,\zeta,\varpi} = u(\zeta,\varpi). \tag{18.97}$$

Comparing (18.92) and (18.97) we see that the minimum and the maximum solution pairs coincide and are equal to the value function. Furthermore, the unique solution is continuous since it is both l.s.c. and u.s.c. and, therefore, an optimal feedback exists. This completes the proof. □

18.9 Concluding remarks

This chapter studies the joint conditional distribution of the partially observed state of a system given the history of observations. In particular, it provides evolution equations for the joint conditional distribution, which can be used to obtain the next period's distribution with the knowledge of the current distribution and the observation made in the current period.

This chapter also illustrates how unnormalized probabilities can be used to simplify Bellman equations. These simplifications facilitate proofs of the existence of feedback policies. This is illustrated by sketching an existence proof in the special case of the zero-balance walk model.

The evolution of the conditional distribution is given in a general context, so it can be used to study a number of partially observed inventory/demand problems [2, 3, 4, 5, 6, 7] as well as the Kalman filter [10]. It is expected that this evolution, by simplifying and streamlining the treatment of partial observability, will encourage other applications modelled as partially observed discrete-time Markov processes.

Acknowledgements

This research is supported in part by NSF grant DMS-0509278 and ARP grant 2-23259.

References

[1] Bensoussan, A. (1992). *Stochastic Control of Partially Observable Systems*. Cambridge University Press, Cambridge, UK.

[2] Bensoussan, A., Çakanyıldırım, M., and Sethi, S.P. (2007). Partially Observed Inventory Systems: The Case of Zero Balance Walk. *SIAM Journal of Control and Optimization*. Vol. 46: 176–209.

[3] Bensoussan, A., Çakanyıldırım, M., and Sethi, S.P. (2007). A Multi-period Newsvendor Problem with Partially Observed Demands. *Mathematics of Operations Research*. Vol. 32: 322–344.

[4] Bensoussan, A., Çakanyıldırım, M., Minjarez-Sosa, J. A., Royal, A., and Sethi, S. P. (2007). Inventory Problems with Partially Observed Demands and Lost Sales. School of Management Working Paper, University of Texas at Dallas, TX. *Journal of Optimization Theory and Applications*. Vol. 136, No. 3, 2008: 321–340.

[5] Bensoussan, A., Çakanyıldırım, M., Minjarez-Sosa, J. A., Sethi, S. P., and Shi, R. (2007). Partially Observed Inventory Systems: The Case of Rain Checks. *SIAM Journal of Control and Optimization* Vol. 47, Iss. 5, 2008: 2490–2519.

[6] Bensoussan, A., Çakanyıldırım, M., Minjarez-Sosa, J. A., Sethi, S. P., and Shi, R. (2007). An Incomplete Information Inventory Model with Presence of Inventories or Backorders as only Observations. School of Management Working Paper, University of Texas at Dallas, TX. *Journal of Optimization Theory and Application.* Vol. 146, No. 3, 2010.

[7] Bensoussan, A., Çakanyıldırım, M., Sethi, S. P. (2007). A Cash Register Inventory Model: Unobserved Shrinkages. School of Management Working Paper, University of Texas at Dallas, TX.

[8] Chow, Y. S. and Teicher, H. (1978). *Probability Theory: Independence Interchangeability, Martingales.* Springer, New York.

[9] Elliott, R. J., Aggoun, L., and Moore, J. B. (1995). *Hidden Markov Models: Estimation and Control.* Springer-Verlag, New York.

[10] Kalman, R. E. (1960). A New Approach to Linear Filtering and Prediction Problems. *ASME Journal of Basic Engineering*, Vol. 1: 35–45.

[11] Kushner, H. J. (1967). Dynamic Equations for Nonlinear Filtering. *Journal of Differential Equations*, Vol. 3: 179–90.

[12] Zakai, M. (1969). *On the Optimal Filtering of Diffusion Processes.* Z. Wahrsch. verw. Gebiete, Vol. 11: 230–43.

·19·
Bayesian Filtering of Stochastic Hybrid Systems in Discrete-Time and Interacting Multiple Model

H. A. P. Blom and Y. Bar-Shalom

19.1 Introduction

In the area of filtering of Markov jump linear systems, the Interacting Multiple Model (IMM) filter ([6, 12]) has become one of the most popular nonlinear filters in maneuvering target tracking, e.g. [3], [28], [4], [26]. As a kind of side-effect of this success, in hybrid state estimation literature, the interest in IMM is typically focused on IMM's value for specific applications. However, a key property of the IMM filter equations is that they capture the exact interaction between the Markov switching mode and the state of the jump linear system. The objective of this chapter is to exploit this exact relation for the development of exact recursive equations for a range of stochastic hybrid systems. It will also be shown how particle filtering of stochastic hybrid systems can take advantage of the exact characterization.

The stochastic hybrid systems considered are defined as Euclidean valued solutions of a stochastic difference equation (SDE) whose coefficients jump according to a finite state Markov process (chain). These equations give rise to a problem of nonlinear filtering when the finite state Markov chain is not, or not fully, observed, which implies the following cases:

1. Markov chain is not observed, SDE solution is fully observed;
2. Markov chain is not observed, SDE solution is partially observed, i.e. imperfectly and/or incompletely;
3. Both the Markov chain and the SDE solution are partially observed, i.e. imperfectly and/or incompletely.

The first case is typically referred to as a Hidden Markov Model (HMM), which has been developed and become popular within the signal processing research community. The second, more demanding case, has been developed and become popular within the target tracking research community, and has more recently been picked up by the signal processing community. The third case has been developed within the target tracking community, but is not as widely known. For each of these three cases, an exact Bayesian characterization and main approximate recursions of the nonlinear filtering equations will be given, and their main applications will be addressed.

In Section 19.2, the SDE considered is, for $t = 0, 1, 2, \ldots$:

$$x_{t+1} = a(\theta_{t+1}, x_t, w_t) \tag{19.1}$$

with $\{x_t\}$ a Euclidean valued stochastic process (sequence) and $\{\theta_t\}$ a discrete valued stochastic process (sequence). The process $\{\theta_t\}$ is an $\mathbb{M}$-valued Markov chain, with transition probability matrix Π, the components of which are

$$\Pi_{\eta\theta} = P\{\theta_{t+1} = \theta \mid \theta_t = \eta\} \tag{19.2}$$

The process $\{x_t\}$ is observed, and the process $\{\theta_t\}$ is not. We study the exact conditional probability function. System (19.1)–(19.2) forms a Hidden Markov Model (HMM). The mapping a and the probability matrix Π are assumed to be t-invariant for brevity only.

In Sections 19.3 and 19.4, the process $\{x_t\}$ is no longer observed, though another Euclidean valued process $\{y_t\}$ is. The relation between $\{y_t\}$ and $\{x_t\}$ is given by the following equation:

$$y_t = h(\theta_t, x_t, v_t) \tag{19.3}$$

This nonlinear filtering problem is further evaluated in Section 19.5, by introducing a stepwise approach towards better understanding the nature of the evolution of the exact conditional density. As before, the mapping h is assumed to be t-invariant for reasons of brevity only.

Section 19.6 studies the Markov jump linear special case, with exact characterizations in terms of Gaussian mixture conditional density. Subsequently, Sections 19.7 and 19.8 exploit this better understanding of the evolution of the exact equations for the Markov jump linear special case, with exact characterizations in terms of an IMM particle filter and the IMM filter.

In Section 19.9, the following extra observation equation is included

$$\kappa_t = d(\theta_t, v_t) \tag{19.4}$$

Now both the Euclidean valued process $\{y_t\}$ and the discrete valued process $\{\kappa_t\}$ are observed. The stepwise approach of Section 19.5 is extended to cover this situation as well, including an explanation how this impacts the key results of Sections 19.6, 19.7, and 19.8. Again, the mapping d is assumed to be t-invariant for reasons of brevity only.

In Section 19.10, we show through a simple suddenly maneuvering target tracking problem how the IMM and particle filters perform with focus on their deviation from the exact conditional density solution.

In Section 19.11, we summarize the main results and refer to follow-up developments of the presented material.

Throughout the chapter we assume a probability space $(\Omega, \mathcal{F}, P)$ with $(\Omega, \mathcal{F})$ a measurable space and P a probability measure defined on the σ-algebra $\mathcal{F}$. In line with this, all increasing sequences of σ-algebras that we consider in the sequel are assumed to fall within the σ-algebra $\mathcal{F}$. We use the following notations for conditional density of a random process $\{x_t\}$ with measurable state space X:

$p_{x_t}(\cdot)$ denotes the conditional density of x_t, such that $\int_U p_{x_t}(x)dx = P\{x_t \in U\}$ for any measurable $U \subset X$. $p_{x_t|\theta_t}(\cdot|\theta)$ denotes the conditional density of x_t given the event $\theta_t = \theta$. $p_{x_t|Y_s}(\cdot)$ represents the conditional density of x_t given the σ-algebra Y_s. $p_{x_t|\theta_s,Y_s}(\cdot|\theta)$ represents the conditional density of x_t given $\theta_s = \theta$ and the σ-algebra Y_s.

19.2 Bayesian filtering of a Hidden Markov Model

The process to be estimated is an $\mathbb{M}$-valued Markov chain $\{\theta_t\}$, where $\mathbb{M}$ is a set of M discrete modes, with transition probability matrix Π, with components

$$\Pi_{\eta\theta} = P\{\theta_{t+1} = \theta \mid \theta_t = \eta\} \tag{19.5}$$

The observed process $\{x_t\}$ is the solution of the following stochastic difference equation, on $[0, T]$, for an integer $T < \infty$,

$$x_{t+1} = a(\theta_{t+1}, x_t, w_t) \tag{19.6}$$

where $\{w_t\}$ is a sequence of i.i.d. standard Gaussian vectors of dimension[1] n, and independent of $\{\theta_t\}$ and the $\mathbb{R}^n$ valued initial condition x_0. Furthermore, a is a measurable mapping of $\mathbb{M} \times \mathbb{R}^n \times \mathbb{R}^n$ into $\mathbb{R}^n$, and is assumed to be time-invariant for brevity only.

By an iterative way of working from $t = 0$ to $t = 1$, then to $t = 2$, and so on to $t = T$, it can be verified that for every initial condition (θ_0, x_0) equation (19.6) has a unique measurable solution $\{\theta_t, x_t\}$ assuming values in $\mathbb{M} \times \mathbb{R}^n \times [0, T]$, $T < \infty$.

The filtering problem is to estimate the conditional probability function $p_{\theta_t|X_t}(\theta)$, $\theta \in \mathbb{M}$, of θ_t given the sequence of observations $X_t = \{x_s; s \leq t\}$. We develop the exact recursive equations for the joint conditional probability function $p_{\theta_t|X_t}(\theta)$, where X_t denotes the σ-algebra generated by the process $\{x_t\}$ up to and including moment t.

Because $\{\theta_t\}$ is a Markov process, the characterization of this conditional density consists of two steps:

- Chapman–Kolmogorov (CK) evolution from t to $t+1$, i.e. characterize $p_{\theta_{t+1}|X_t}(\theta)$ as a function of $p_{\theta_t|X_t}(\theta)$.

- Bayes measurement update, i.e. characterize $p_{\theta_{t+1}|X_{t+1}}(\theta)$ as a function of $p_{\theta_{t+1}|X_t}(\theta)$.

For the characterization of the CK step, we start from the total probability theorem, and subsequently evaluate:

$$p_{\theta_{t+1}|X_t}(\theta) = \sum_{\eta \in \mathbb{M}} p_{\theta_{t+1},\theta_t|X_t}(\theta, \eta)$$

$$= \sum_{\eta \in \mathbb{M}} p_{\theta_{t+1}|\theta_t, X_t}(\theta|\eta)\, p_{\theta_t|X_t}(\eta)$$

$$= \sum_{\eta \in \mathbb{M}} \Pi_{\eta\theta}\, p_{\theta_t|X_t}(\eta) \qquad (19.7)$$

The Bayesian measurement updating satisfies:

$$p_{\theta_{t+1}|X_{t+1}}(\theta) = p_{x_{t+1}|\theta_{t+1}, x_t}(x_{t+1}|\theta, x_t)\, p_{\theta_{t+1}|X_t}(\theta)/c_t \qquad (19.8)$$

Substitution of (19.7) into (19.8) yields the exact Bayesian filter recursion in Theorem 19.1.

Theorem 19.1 *For each initial $(\theta_0, x_0) \in \mathbb{M} \times \mathbb{R}^n$, the conditional probability function $p_{\theta_t|X_t}(\theta)$ satisfies the recursion:*

$$p_{\theta_{t+1}|X_{t+1}}(\theta) = p_{x_{t+1}|\theta_{t+1}, x_t}(x_{t+1}|\theta, x_t) \sum_{\eta \in \mathbb{M}} \left[\Pi_{\eta\theta}\, p_{\theta_t|X_t}(\eta)\right]/c_t \qquad (19.9)$$

with c_t a normalization constant.

Remark 19.1 For the characterization of $p_{x_{t+1}|\theta_{t+1}, x_t}(x_{t+1}|\theta, x_t)$ in eq. (19.9) one typically makes use of a change of variables as follows. If the mapping $x = a(\theta, x', w)$ has for each (θ, x', x) an inverse $w = a^{-1}(\theta, x', x)$, which is differentiable in x, then w_t satisfies $w_t = a^{-1}(\theta_{t+1}, x_t, x_{t+1})$ and the following characterization of $p_{x_{t+1}|\theta_{t+1}, x_t}(x_{t+1}|\theta, x_t)$ applies:

$$p_{x_{t+1}|\theta_{t+1}, x_t}(x|\theta, x') = p_{w_t}(a^{-1}(\theta, x', x)) \left|\frac{\partial a^{-1}(\theta, x', x)}{\partial x}\right| \qquad (19.10)$$

In the signal processing literature the Hidden Markov Model has been studied along many directions. The recursive Bayesian filter characterization above just forms one of the results. For an overview of HMM estimation and illustrative applications we refer to [31].

19.3 Bayesian filtering of Markov jump nonlinear systems

The process $\{\theta_t\}$ is an $\mathbb{M}$-valued Markov chain, where $\mathbb{M}$ is a set of M discrete modes, with transition probability matrix Π, with components

$$\Pi_{\eta\theta} = P\{\theta_{t+1} = \theta \mid \theta_t = \eta\} \tag{19.11}$$

We consider the following system of stochastic difference equations, on $[0, T]$, $T < \infty$,

$$x_{t+1} = a(\theta_{t+1}, x_t, w_t) \tag{19.12}$$

where $\{w_t\}$ is a sequence of i.i.d. standard Gaussian vectors of dimension n, and independent of $\{\theta_t\}$ and the $\mathbb{R}^n$ valued initial condition x_0.

The pair (θ_t, x_t) represents the hybrid system state, which is observed through the process $\{y_t\}$ which satisfies the following equation:

$$y_t = h(\theta_t, x_t, v_t) \tag{19.13}$$

where $\{v_t\}$ is a sequences of i.i.d. standard Gaussian variables of dimension m, and independent of $\{w_t\}$, $\{\theta_t\}$ and the initial condition x_0.

Furthermore, a and h are measurable mappings of $\mathbb{M} \times \mathbb{R}^n \times \mathbb{R}^n$ into $\mathbb{R}^n$ and $\mathbb{M} \times \mathbb{R}^n \times \mathbb{R}^m$ into $\mathbb{R}^m$ respectively. Mappings a and h are time-invariant for brevity only.

The filtering problem is to characterize a recursive equation for the evolution of the joint conditional density-probability function $p_{x_t,\theta_t|Y_t}(x, \theta)$, $x \in \mathbb{R}^n$, $\theta \in \mathbb{M}$ of the pair (x_t, θ_t) given the σ-algebra of observations $Y_t = \sigma\{y_s; s \leq t\}$. For short we refer to $p_{x_t,\theta_t|Y_t}(x, \theta)$ as the conditional density. The characterization of this conditional density is accomplished through the following two steps:

- Chapman–Kolmogorov (CK) evolution from t to $t + 1$ which characterizes $p_{x_{t+1},\theta_{t+1}|Y_t}(x, \theta)$ as a function of $p_{x_t,\theta_t|Y_t}(x, \theta)$.
- Bayes measurement update, i.e. characterize $p_{x_{t+1},\theta_{t+1}|Y_{t+1}}(x, \theta)$ as a function of $p_{x_{t+1},\theta_{t+1}|Y_t}(x, \theta)$ and measurement y_{t+1}.

Subsequent combination of these two steps yields the following characterization of the exact Bayesian recursion for the joint conditional density, which expresses $p_{x_{t+1},\theta_{t+1}|Y_{t+1}}(x, \theta)$ as a function of $p_{x_t,\theta_t|Y_t}(x, \theta)$ and measurement y_{t+1}.

Theorem 19.2 *Let the mappings h and a be such that the densities $p_{y_t|x_t,\theta_t}(\cdot|x, \theta)$ and $p_{x_{t+1}|x_t,\theta_{t+1}}(\cdot|x, \theta)$ are measurable for all $(\theta, x) \in \mathbb{M} \times \mathbb{R}^n$, and let $\{\theta_0, x_0\}$ admit a measurable density $P_{\theta_0,x_0}(\cdot, \cdot)$. Then the Markov process $\{\theta_t, x_t\}$ admits a measurable transition density*

$$p_{x_{t+1},\theta_{t+1}|x_t,\theta_t}(x, \theta|x', \eta) = p_{x_{t+1}|\theta_{t+1},x_t}(x|\theta, x')\Pi_{\eta\theta},$$

for all $(\eta, x'), (\theta, x) \in \mathbb{M} \times \mathbb{R}^n$, and the conditional density satisfies the recursion

$$p_{x_{t+1},\theta_{t+1}|Y_{t+1}}(x,\theta)$$
$$= p_{y_{t+1}|x_{t+1},\theta_{t+1}}(y_{t+1}|x,\theta) \sum_{\eta \in M} \left[\Pi_{\eta\theta} \int_{\mathbb{R}^n} p_{x_{t+1}|\theta_{t+1},x_t}(x|\theta,x') \, p_{x_t,\theta_t|Y_t}(x',\eta) dx' \right] / c_t \quad (19.14)$$

with c_t a normalization constant.

Proof. Application of (generalized) Bayes theorem (e.g. [5], p. 129) yields for $(x, \theta) \in \mathbb{R}^n \times M$:

$$p_{x_{t+1},\theta_{t+1}|Y_{t+1}}(x,\theta) = p_{y_{t+1}|x_{t+1},\theta_{t+1}}(y_t|x,\theta) p_{x_{t+1},\theta_{t+1}|Y_t}(x,\theta)/c_t \quad (19.15)$$

Because $\{x_t, \theta_t\}$ is a Markov process, the evolution from t to $t+1$ satisfies the Chapman–Kolmogorov equation

$$p_{x_{t+1},\theta_{t+1}|Y_t}(x,\theta) = \int_{\mathbb{R}^n} \sum_{\eta \in M} p_{x_{t+1},\theta_{t+1}|x_t,\theta_t}(x,\theta|x',\eta) p_{x_t,\theta_t|Y_t}(x',\eta) dx' \quad (19.16)$$

Factoring the Markov transition density in (19.16) yields

$$p_{x_{t+1},\theta_{t+1}|x_t,\theta_t}(x,\theta|x',\eta) = p_{x_{t+1}|\theta_{t+1},x_t,\theta_t}(x|\theta,x',\eta) p_{\theta_{t+1}|x_t,\theta_t}(\theta|x',\eta)$$
$$= p_{x_{t+1}|\theta_{t+1},x_t,\theta_t}(x|\theta,x',\eta) \Pi_{\eta\theta} \quad (19.17)$$

Because w_t is independent of θ_t, from (19.12) follows that x_{t+1} is conditionally independent of θ_t given x_t and θ_{t+1}. Using this in (19.17) yields

$$p_{x_{t+1},\theta_{t+1}|x_t,\theta_t}(x,\theta|x',\eta) = p_{x_{t+1}|\theta_{t+1},x_t}(x|\theta,x') \Pi_{\eta\theta} \quad (19.18)$$

Substituting (19.18) into (19.16) and reorganizing summation yields

$$p_{x_{t+1},\theta_{t+1}|Y_t}(x,\theta) = \sum_{\eta \in M} \left[\Pi_{\eta\theta} \int_{\mathbb{R}^n} p_{x_{t+1}|\theta_{t+1},x_t}(x|\theta,x') p_{x_t,\theta_t|Y_t}(x',\eta) dx' \right] \quad (19.19)$$

Substituting this into (19.15) yields (19.14). □

Remark 19.2 The derivation path used above follows the approach that was taken in the original IMM paper [[6], Appendix], but now extended to Markov jump nonlinear systems. Eq. (19.19) has also been developed in [2] under the implicit assumption that $p_{\theta_{t+1}|Y_t}(\theta) > 0$ for all $\theta \in M$.

19.4 Particle filtering of Markov jump nonlinear system

In general it is challenging to evaluate the recursion (19.14) in a numerical way. One of the most general approaches is to use a SIR (Sampling Importance Resampling) particle filter. At moment t, a SIR particle filter starts with the set of N_p weighted particles $\left\{\theta_t^j, x_t^j, \mu_t^j; j \in \{1, \ldots, N_p\}\right\}$. This set of weighted particles forms (spans) the empirical density

$$\tilde{p}_{x_t, \theta_t | Y_t}(x, \theta) = \sum_{j=1}^{N_p} \mu_t^j \chi\left(\theta, \theta_t^j\right) \delta\left(x - x_t^j\right) \tag{19.20}$$

as an approximation of the exact density $p_{x_t, \theta_t | Y_t}(x, \theta)$, were χ and δ are indicator and dirac functions respectively. The SIR particle filter cycle for $\tilde{p}_{x_t, \theta_t | Y_t} \longrightarrow \tilde{p}_{x_{t+1}, \theta_{t+1} | Y_{t+1}}$ follows from [24] and is specified in Table 19.1. This SIR filter can approximate the conditional density arbitrarily well by increasing the number of particles. Illustrative applications of this SIR particle filter for stochastic hybrid systems are given by [29], [30] and [25] for the case that $\{\theta_t\}$ is a Markov chain.

Table 19.1 SIR particle filter (SIR PF)

Initialization at $t = 0$: For $j = 1, \ldots, N_p$:

$\mu_0^j = 1/N_p$

Simulate (x_0^j, θ_0^j) from $p_{x_0, \theta_0 | Y_0}(\cdot, \cdot)$

Particles and conditional density at moment t:

$\{\mu_t^j \in [0, 1], \theta_t^j \in \mathbb{M}, x_t^j \in \mathbb{R}^n; j = 1, \ldots, N_p\}$

$\tilde{p}_{x_t, \theta_t | Y_t}(x, \theta) = \sum_{j=1}^{N_p} \mu_t^j \chi(\theta, \theta_t^j) \delta(x - x_t^j)$

SIR particle filter cycle:

- Evolution for $j = 1, \ldots, N_p$:

    Simulate $\bar{\theta}_{t+1}^j \sim p_{\theta_{t+1} | \theta_t}(\theta | \theta_t^j) = \Pi_{\theta_t^j \theta}$

    Simulate $w_t^j \sim p_{w_t}(w)$

    $\bar{x}_{t+1}^j = a(\bar{\theta}_{t+1}^j, x_t^j, w_t^j)$

- Bayes update correction for $j = 1, \ldots, N_p$:

    $\bar{\mu}_{t+1}^j = \mu_t^j\, p_{y_{t+1} | x_{t+1}, \theta_{t+1}}(y_{t+1} | \bar{x}_{t+1}^j, \bar{\theta}_{t+1}^j) / c_t$

    with c_t such that $\sum_{j=1}^{N_p} \bar{\mu}_{t+1}^j = 1$

Resampling for $j = 1, \ldots, N_p$:

$\mu_{t+1}^j = 1/N_p$

$(x_{t+1}^j, \theta_{t+1}^j) \sim \sum_{j'=1}^{N_p} \bar{\mu}_{t+1}^{j'} \chi(\theta, \bar{\theta}_{t+1}^{j'}) \delta(x - \bar{x}_{t+1}^{j'})$

Set $t := t + 1$ and repeat the filter cycle above when $t \leq T$

For the marginal empirical densities $\tilde{p}_{\theta_t|Y_t}(\theta)$ and $\tilde{p}_{x_t|\theta_t,Y_t}(x|\theta)$, eq. (19.20) implies

$$\tilde{p}_{\theta_t|Y_t}(\theta) = \sum_{j=1}^{N_p} \mu_t^j \chi(\theta, \theta_t^j) \tag{19.21}$$

$$\tilde{p}_{x_t|\theta_t,Y_t}(x|\theta) = \sum_{j=1}^{N_p} \mu_t^j \chi(\theta, \theta_t^j)\delta(x - x_t^j)/\tilde{p}_{\theta_t|Y_t}(\theta) \text{ if } \tilde{p}_{\theta_t|Y_t}(\theta) > 0 \tag{19.22}$$

In order to simplify this empirical representation, [13] suggested to span an empirical density $\tilde{p}_{x_t,\theta_t|Y_t}(x, \theta)$ with the help of a set of $N_p = MS$ weighted particles $\{x_t^{\theta,j}, \mu_t^{\theta,j}; \theta \in \mathbb{M}, j \in \{1, \ldots, S\}\}$ as follows:

$$\tilde{p}_{x_t,\theta_t|Y_t}(x, \theta) = \sum_{j=1}^{S} \mu_t^{\theta,j}\delta(x - x_t^{\theta,j}) \tag{19.23}$$

For the marginal empirical densities $\tilde{p}_{\theta_t|Y_t}(\theta)$ and $\tilde{p}_{x_t|\theta_t,Y_t}(x|\theta)$ this means

$$\tilde{p}_{\theta_t|Y_t}(\theta) = \sum_{j=1}^{S} \mu_t^{\theta,j} \tag{19.24}$$

$$\tilde{p}_{x_t,\theta_t,Y_t}(x|\theta) = \sum_{j=1}^{S} \mu_t^{\theta,j}\delta(x - x_t^{\theta,j})/\tilde{p}_{\theta_t|Y_t}(\theta) \text{ if } \tilde{p}_{\theta_t|Y_t}(\theta) > 0 \tag{19.25}$$

Straightforward extension of the SIR particle filter with this modified set of particles yields the SIR Hybrid PF in Table 19.2.

19.5 Decomposition of the exact filter recursion

In order to better understand the structure of recursive equation (19.14), we study the conditional densities $p_{\theta_t|Y_t}(\theta)$ and $p_{x_t|\theta_t,Y_t}(x|\theta)$. The product of these two equals the joint conditional density, i.e.

$$p_{x_t,\theta_t|Y_t}(x, \theta) = p_{x_t|\theta_t,Y_t}(x|\theta) \, p_{\theta_t|Y_t}(\theta) \tag{19.26}$$

The other way around, $p_{\theta_t|Y_t}(\theta)$ and $p_{x_t|\theta_t,Y_t}(x|\theta)$ can be characterized in terms of the joint conditional density $p_{x_t,\theta_t|Y_t}(x, \theta)$ as follows:

$$p_{\theta_t|Y_t}(\theta) = \int_{\mathbb{R}^n} p_{x_t,\theta_t|Y_t}(x, \theta) dx \tag{19.27}$$

$$p_{x_t|\theta_t,Y_t}(x|\theta) = p_{x_t,\theta_t|Y_t}(x, \theta)/p_{\theta_t|Y_t}(\theta) \text{ if } p_{\theta_t|Y_t}(\theta) > 0 \tag{19.28}$$

If the condition $p_{\theta_t|Y_t}(\theta) > 0$ is not satisfied for one or more θ-values, then for those θ-values, $p_{x_t|\theta_t,Y_t}(x|\theta)$ is not uniquely defined. The condition to be satisfied

Table 19.2 SIR Hybrid particle filter (HPF)

Initialisation at $t = 0$, for $j = 1, \ldots, S, \theta \in \mathbb{M}$:

$\mu_0^{\theta,j} = p_{\theta_0|Y_0}(\theta)/S$, and simulate $x_0^{\theta,j} \sim p_{x_0|\theta_0,Y_0}(\cdot|\theta)$

Particles and empirical density at moment t:

$\left\{ \mu_t^{\theta,j} \in [0,1], x_t^{\theta,j} \in \mathbb{R}^n; \theta \in \mathbb{M}; j = 1, \ldots, S \right\}$

$\tilde{p}_{x_t,\theta_t|Y_t}(x,\theta) = \sum_{j=1}^{S} \mu_t^{\theta,j} \delta\left(x - x_t^{\theta,j}\right)$

SIR Hybrid particle filter (HPF) cycle:

- Mode switching. For $j = 1, \ldots, S, \theta \in \mathbb{M}$:
  Simulate $\bar{\theta}_{t+1}^{j} \sim p_{\theta_{t+1}|\theta_t}\left(\theta|\theta_t^j\right) = \Pi_{\theta_t^j \theta}$

- Prediction. For $j = 1, \ldots, S, \theta \in \mathbb{M}$:
  Simulate $w_t^{\theta,j} \sim p_{w_t}(w)$

  $\bar{x}_{t+1}^{\theta,j} = a\left(\bar{\theta}_{t+1}^{\theta,j}, x_t^{\theta,j}, w_t^{\theta,j}\right)$

- Bayesian update correction: For $j = 1, \ldots, S, \theta \in \mathbb{M}$:

  $\bar{\mu}_{t+1}^{\theta,j} = \mu_t^{\theta,j} \cdot p_{y_{t+1}|x_{t+1},\theta_{t+1}}(y_{t+1}|\bar{x}_{t+1}^{\theta,j}, \bar{\theta}_{t+1}^{\theta,j})/c_t$

  with c_t such that $\sum_{j=1}^{S} \sum_{\theta \in \mathbb{M}} \bar{\mu}_{t+1}^{\theta,j} = 1$

  $\tilde{p}_{\theta_{t+1}|Y_{t+1}}(\theta) = \sum_{j=1}^{S} \sum_{\eta \in \mathbb{M}} \bar{\mu}_{t+1}^{\eta,j} \chi\left(\bar{\theta}_{t+1}^{\eta,j}, \theta\right)$

- Resampling: For $j = 1, \ldots, S, \theta \in \mathbb{M}$:

  $\mu_{t+1}^{\theta,j} = \tilde{p}_{\theta_{t+1}|Y_{t+1}}(\theta)/S$

  $x_{t+1}^{\theta,j} \sim \sum_{j'=1}^{S} \sum_{\eta \in \mathbb{M}} \bar{\mu}_{t+1}^{\eta,j'} \chi\left(\bar{\theta}_{t+1}^{\eta,j'}, \theta\right) \delta\left(x - \bar{x}_{t+1}^{\eta,j'}\right)/\tilde{p}_{\theta_{t+1}|Y_{t+1}}(\theta)$ if $\tilde{p}_{\theta_{t+1}|Y_{t+1}}(\theta) > 0$

  If $\tilde{p}_{\theta_{t+1}|Y_{t+1}}(\theta) = 0$ then $x_{t+1}^{\theta,j} = \bar{x}_{t+1}^{\theta,j}$

Set $t := t+1$ and repeat the filter cycle above when $t \leq T$

---

is that $p_{x_t,\theta_t|Y_t}(x,\theta) = 0$ for those θ-values, which means that it is allowed to assume an arbitrarily bounded density for $p_{x_t|\theta_t,Y_t}(x|\theta)$ when $p_{\theta_t|Y_t}(\theta) = 0$.

The evolution in time of the conditional density from $p_{\theta_t|Y_t}$ to $p_{\theta_{t+1}|Y_{t+1}}$ can be decomposed in the following two steps:

$$p_{\theta_t|Y_t} \xrightarrow{\text{Mode Switching}} p_{\theta_{t+1}|Y_t} \xrightarrow{\text{Bayes Mode Update}} p_{\theta_{t+1}|Y_{t+1}}$$

The evolution in time of the mode conditional density $p_{x_t|\theta_t,Y_t}(x|\theta)$ to $p_{x_{t+1}|\theta_{t+1},Y_{t+1}}(x|\theta)$ can be decomposed in the following three steps:

$$p_{x_t|\theta_t,Y_t} \xrightarrow{\text{Mixing}} p_{x_t|\theta_{t+1},Y_t} \xrightarrow{\text{State Prediction}} p_{x_{t+1}|\theta_{t+1},Y_t} \xrightarrow{\text{Bayes State Update}} p_{x_{t+1}|\theta_{t+1},Y_{t+1}}$$

In each of these five steps, only one of the stochastic entities moves from t to $t+1$. Next, for each of these five steps a characterization is developed, in the following sequence:

$$p_{\theta_t|Y_t} \xrightarrow{\text{Mode Switching}} p_{\theta_{t+1}|Y_t}$$

$$p_{x_t|\theta_t,Y_t} \xrightarrow{\text{Mixing}} p_{x_t|\theta_{t+1},Y_t}$$

$$p_{x_t|\theta_{t+1},Y_t} \xrightarrow{\text{State Prediction}} p_{x_{t+1}|\theta_{t+1},Y_t}$$

$$p_{x_{t+1}|\theta_{t+1},Y_t} \xrightarrow{\text{Bayes State Update}} p_{x_{t+1}|\theta_{t+1},Y_{t+1}}$$

$$p_{\theta_{t+1}|Y_t} \xrightarrow{\text{Bayes State Update}} p_{\theta_{t+1}|Y_{t+1}}$$

The two mode-related steps are at the top and the end of this sequence, whereas the three state-related steps are in the middle.

Mode switching

From the law of total probability, followed by evaluation, we get:

$$\begin{aligned} p_{\theta_{t+1}|Y_t}(\theta) &= \sum_{\eta \in \mathbb{M}} p_{\theta_{t+1},\theta_t|Y_t}(\theta,\eta) \\ &= \sum_{\eta \in \mathbb{M}} p_{\theta_{t+1}|\theta_t,Y_t}(\theta|\eta)\, p_{\theta_t|Y_t}(\eta) \\ &= \sum_{\eta \in \mathbb{M}} \Pi_{\eta\theta}\, p_{\theta_t|Y_t}(\eta) \end{aligned} \quad (19.29)$$

State mixing

By law of total probability we get

$$p_{x_t|\theta_{t+1},Y_t}(x|\theta) = \sum_{\eta \in \mathbb{M}} p_{x_t,\theta_t|\theta_{t+1},Y_t}(x,\eta|\theta)$$

Next application of Bayes' theorem and subsequent evaluation yields for $p_{\theta_{t+1}|Y_t}(\theta) > 0$:

$$\begin{aligned} p_{x_t|\theta_{t+1},Y_t}(x|\theta) &= \sum_{\eta \in \mathbb{M}} \left[p_{\theta_{t+1}|x_t,\theta_t}(\theta|x,\eta)\, p_{x_t,\theta_t|Y_t}(x,\eta)/p_{\theta_{t+1}|Y_t}(\theta) \right] \\ &= \sum_{\eta \in \mathbb{M}} \left[p_{\theta_{t+1}|\theta_t}(\theta|\eta)\, p_{x_t|\theta_t,Y_t}(x|\eta)\, p_{\theta_t|Y_t}(\eta)/p_{\theta_{t+1}|Y_t}(\theta) \right] \\ &= \sum_{\eta \in \mathbb{M}} \left[\Pi_{\eta\theta} \frac{p_{\theta_t|Y_t}(\eta)}{p_{\theta_{t+1}|Y_t}(\theta)} p_{x_t|\theta_t,Y_t}(x|\eta) \right] \end{aligned} \quad (19.30)$$

State prediction

From the law of total probability, and $\{x_t,\theta_t\}$ being a Markov process, we get:

$$p_{x_{t+1}|\theta_{t+1},Y_t}(x|\theta) = \int_{\mathbb{R}^n} p_{x_{t+1}|x_t,\theta_{t+1}}(x|x',\theta)\, p_{x_t|\theta_{t+1},Y_t}(x'|\theta)dx' \quad (19.31)$$

Remark 19.3 Mode switching, state mixing, and state prediction equations (19.29)–(19.31) characterize the interaction between mode and state evolution. This exact form has initially been developed for Markov jump linear systems in [6]. The decomposition of the exact filter recursion in five steps is from [12], which makes use of the fact that for Markov jump nonlinear systems the five-step decomposition is exactly the same as it is for Markov jump linear systems.

Bayes state update

From Bayes' (generalized) theorem we get:

$$p_{x_{t+1}|\theta_{t+1},Y_{t+1}}(x|\theta) = p_{y_{t+1}|x_{t+1},\theta_{t+1}}(y_{t+1}|x,\theta)\, p_{x_{t+1}|\theta_{t+1},Y_t}(x|\theta)/c_t(\theta) \qquad (19.32)$$

where $c_t(\theta)$ satisfies

$$c_t(\theta) = \int_{\mathbb{R}^n} p_{y_{t+1}|x_{t+1},\theta_{t+1}}(y_{t+1}|x,\theta)\, p_{x_{t+1}|\theta_{t+1},Y_t}(x|\theta)\,dx \qquad (19.33)$$

Bayes mode update

From the law of total probability we get

$$p_{\theta_{t+1}|Y_{t+1}}(\theta) = \int_{\mathbb{R}^n} p_{x_{t+1},\theta_{t+1}|Y_{t+1}}(x,\theta)\,dx \qquad (19.34)$$

Substitution of (19.15) and subsequent evaluation yield:

$$p_{\theta_{t+1}|Y_{t+1}}(\theta) = \int_{\mathbb{R}^n} p_{y_{t+1}|x_{t+1},\theta_{t+1}}(y_{t+1}|x,\theta)\, p_{x_{t+1},\theta_{t+1}|Y_t}(x,\theta)\,dx/c_t$$

$$= \int_{\mathbb{R}^n} p_{y_{t+1}|x_{t+1},\theta_{t+1}}(y_{t+1}|x,\theta)\, p_{x_{t+1}|\theta_{t+1},Y_t}(x|\theta)\, p_{\theta_{t+1}|Y_t}(\theta)\,dx/c_t \qquad (19.35)$$

Together with (19.32), this implies

$$p_{\theta_{t+1}|Y_{t+1}}(\theta) = c_t(\theta)\, p_{\theta_{t+1}|Y_t}(\theta)/c_t \qquad (19.36)$$

Combining the above characterizations of the five steps yields the following theorem.

Theorem 19.3 Let the assumptions of Theorem 19.2 hold true, and let $p_{\theta_t|Y_t}(\theta) > 0$ for all $\theta \in \mathbb{M}$. Then the conditional densities $p_{x_t|\theta_t,Y_t}(x|\theta)$ and $p_{\theta_t|Y_t}(\theta)$ satisfy the following recursions:

$$p_{x_{t+1}|\theta_{t+1},Y_{t+1}}(x|\theta) = c_t(\theta)^{-1} p_{y_{t+1}|x_{t+1},\theta_{t+1}}(y_{t+1}|x,\theta)$$

$$\cdot \int_{\mathbb{R}^n} p_{x_{t+1}|x_t,\theta_{t+1}}(x|x',\theta) \frac{\sum_{\eta \in \mathbb{M}}\left[\Pi_{\eta\theta}\, p_{\theta_t|Y_t}(\eta)\, p_{x_t|\theta_t,Y_t}(x'|\eta)\right]}{\sum_{\eta' \in \mathbb{M}}\left[\Pi_{\eta'\theta}\, p_{\theta_t|Y_t}(\eta')\right]}dx' \qquad (19.37)$$

$$p_{\theta_{t+1}|Y_{t+1}}(\theta) = \frac{c_t(\theta)}{c_t} \sum_{\eta \in \mathbb{M}}\left[\Pi_{\eta\theta}\, p_{\theta_t|Y_t}(\eta)\right] \qquad (19.38)$$

where $c_t(\theta)$ and c_t normalize $p_{x_{t+1}|\theta_{t+1},Y_{t+1}}(x|\theta)$ and $p_{\theta_{t+1}|Y_{t+1}}(\theta)$ respectively.

In comparison with Theorem 19.2, the system of equations (19.37)–(19.38) look more complicated than eq. (19.14). The advantage of the decomposed representation in Theorem 19.2 is that it shows the difference of the nature of the conditional evolution of $\{x_t\}$ and $\{\theta_t\}$. In the sequel this decomposed representation of the exact conditional density will be further exploited.

19.6 Filtering of Markov jump linear systems

In this section we consider the Markov jump linear system case, i.e. the system of difference equations, on $[0, T]$, $T < \infty$,

$$x_{t+1} = A(\theta_{t+1})x_t + B(\theta_{t+1})w_t \tag{19.39}$$

$$y_t = H(\theta_t)x_t + G(\theta_t)v_t \tag{19.40}$$

where $\{\theta_t\}$ is an $\mathbb{M}$-valued Markov chain, with transition probability matrix Π, the components of which are $\Pi_{\eta\theta} = P\{\theta_{t+1} = \theta | \theta_t = \eta\}$, y_t represents the observation at moment t, $\{w_t\}$ and $\{v_t\}$ are independent sequences of i.i.d. standard Gaussian variables of dimension n and m respectively, $\{w_t, v_t\}$ is independent of $\{\theta_t\}$ and of the $\mathbb{R}^n$ valued initial condition x_0. Furthermore, A, B, H and G are matrices of dimensions $n \times n$, $n \times n$, $m \times n$ and $m \times m$ respectively, the components of which are mappings of $\mathbb{M}$ into $\mathbb{R}$. Moreover the initial mode conditional density satisfies

$$p_{x_0|\theta_0}(x|\theta) = N\{x; \hat{x}_0(\theta), \hat{R}_0(\theta)\} \tag{19.41}$$

with $\hat{R}_0(\theta)$, $\theta \in \mathbb{M}$, positive definite $n \times n$ matrices.

In general $p_{x_t|\theta_t,Y_t}(x|\theta)$, $t \geq 0$, is at moment t a mixture of N^t Gaussians, which are characterized as follows.

Theorem 19.4 *Let the intial mode conditional density satisfy eq. (19.41). Then for the Markov jump linear system (19.39)–(19.40) the conditional densities $p_{x_{t+1}|\theta_{t+1},Y_{t+1}}(x|\theta)$ and $p_{\theta_{t+1}|Y_{t+1}}(\theta)$ satisfy:*

$$p_{x_{t+1}|\theta_{t+1},Y_{t+1}}(x|\theta) = \sum_{\eta \in \mathbb{M}} \sum_{\Gamma \in \mathbb{H}_t} \hat{a}_{t+1}(\Gamma, \eta, \theta) N\{x; \hat{x}_{t+1}(\Gamma, \eta, \theta), \hat{R}_{t+1}(\Gamma, \eta, \theta)\} \tag{19.42}$$

with $\mathbb{H}_0 = \{\}$, $\mathbb{H}_t = \{(\bar{\theta}_0, \ldots, \bar{\theta}_{t-1}); \bar{\theta}_s \in \mathbb{M}, s = 0, \ldots, t-1\}$ for $t \geq 1$,

$$\hat{a}_{t+1}(\Gamma, \eta, \theta) = \frac{f_{t+1}(\Gamma, \eta, \theta)\hat{a}_t(\Gamma, \eta)p_{\theta_t|Y_t}(\eta)\Pi_{\eta\theta}}{c_t(\theta)\sum_{\eta' \in \mathbb{M}}\left[\Pi_{\eta'\theta}p_{\theta_t|Y_t}(\eta')\right]} \tag{19.43}$$

$$p_{\theta_{t+1}|Y_{t+1}}(\theta) = \frac{c_t(\theta)}{c_t}\sum_{\eta \in \mathbb{M}}\left[\Pi_{\eta\theta}p_{\theta_t|Y_t}(\eta)\right] \tag{19.44}$$

where c_t is normalizing $p_{\theta_{t+1}|Y_{t+1}}(\theta)$, and $c_t(\theta)$ such that

$$\sum_{\eta \in \mathbb{M}} \sum_{\Gamma \in \mathbb{H}_t} \hat{a}_{t+1}(\Gamma, \eta, \theta) = 1 \tag{19.45}$$

Furthermore:

$$\hat{x}_{t+1}(\Gamma, \eta, \theta) = \bar{x}_{t+1}(\Gamma, \eta, \theta) + K_{t+1}(\Gamma, \eta, \theta) v_{t+1}(\Gamma, \eta, \theta) \tag{19.46}$$

$$\hat{R}_{t+1}(\Gamma, \eta, \theta) = \bar{R}_{t+1}(\Gamma, \eta, \theta) - K_{t+1}(\Gamma, \eta, \theta) H(\theta) \bar{R}_{t+1,j}(\Gamma, \eta, \theta) \tag{19.47}$$

$$f_{t+1}(\Gamma, \eta, \theta) = \left[(2\pi)^m Det\{ Q_{t+1}(\Gamma, \eta, \theta) \} \right]^{-1/2}$$
$$\cdot \exp\{ -1/2 v_{t+1}(\Gamma, \eta, \theta)^T Q_{t+1}(\Gamma, \eta, \theta)^{-1} v_{t+1}(\Gamma, \eta, \theta) \} \tag{19.48}$$

with:

$$v_{t+1}(\Gamma, \eta, \theta) = y_{t+1} - H(\theta) \bar{x}_{t+1}(\Gamma, \eta, \theta) \tag{19.49}$$

$$K_{t+1}(\Gamma, \eta, \theta) = \bar{R}_{t+1}(\Gamma, \eta, \theta) H(\theta)^T Q_{t+1}(\Gamma, \eta, \theta)^{-1} \tag{19.50}$$

$$Q_{t+1}(\Gamma, \eta, \theta) = H(\theta) \bar{R}_{t+1}(\Gamma, \eta, \theta) H(\theta)^T + G(\theta) G(\theta)^T \tag{19.51}$$

$$\bar{x}_{t+1}(\Gamma, \eta, \theta) = A(\theta) \hat{x}_t(\Gamma, \eta) \tag{19.52}$$

$$\bar{R}_{t+1}(\Gamma, \eta, \theta) = A(\theta) \hat{R}_t(\Gamma, \eta) A^T(\theta) + B(\theta) B^T(\theta) \tag{19.53}$$

Proof. Equation (19.44) equals (19.38). Because the initial condition satisfies the general form of eq. (19.42), we only have to show that the equations of Theorem 19.4 hold true when:

$$p_{x_t|\theta_t, Y_t}(x|\theta) = \sum_{\Gamma \in \mathbb{H}_t} \hat{a}_t(\Gamma, \theta) N\{x; \hat{x}_t(\Gamma, \theta), \hat{R}_t(\Gamma, \theta)\} \tag{19.54}$$

with $\mathbb{H}_0 = \{\}$, $\mathbb{H}_t = \{(\bar{\theta}_0, \ldots, \bar{\theta}_{t-1}); \bar{\theta}_s \in \mathbb{M}, s = 0, \ldots, t-1\}$ for $t \geq 1$, and $\sum_{\Gamma \in \mathbb{H}_t} \hat{a}_t(\Gamma, \theta) = 1$.

From eqs. (19.39)–(19.40) we get:

$$p_{x_{t+1}|x_t, \theta_{t+1}}(x|x', \theta) = N\{x; A(\theta) x', B(\theta) B^T(\theta)\} \tag{19.55}$$

$$p_{y_{t+1}|x_{t+1}, \theta_{t+1}}(y|x, \theta) = N\{y; H(\theta) x, G(\theta) G^T(\theta)\} \tag{19.56}$$

Substituting (19.54) and these equations into (19.37) yields:

$$p_{x_{t+1}|\theta_{t+1}, Y_{t+1}}(x|\theta) = c_t(\theta)^{-1} N\{y_{t+1}; H(\theta) x, G(\theta) G^T(\theta)\}$$
$$\cdot \int_{\mathbb{R}^n} N\{x; A(\theta) x', B(\theta) B^T(\theta)\} \frac{\sum_{\eta \in \mathbb{M}} [\Pi_{\eta\theta} p_{\theta_t|Y_t}(\eta) \sum_{\Gamma \in \mathbb{H}_t} \hat{a}_t(\Gamma, \eta) N\{x'; \hat{x}_t(\Gamma, \eta), \hat{R}_t(\Gamma, \eta)\}]}{\sum_{\eta' \in \mathbb{M}} [\Pi_{\eta'\theta} p_{\theta_t|Y_t}(\eta')]} dx'$$

Evaluation of the integral yields

$$p_{x_{t+1}|\theta_{t+1}, Y_{t+1}}(x|\theta) = c_t(\theta)^{-1} N\{y_{t+1}; H(\theta) x, G(\theta) G^T(\theta)\}$$
$$\cdot \frac{\sum_{\eta \in \mathbb{M}} [\Pi_{\eta\theta} p_{\theta_t|Y_t}(\eta) \sum_{\Gamma \in \mathbb{H}_t} \hat{a}_t(\Gamma, \eta) N\{x; \hat{x}_t(\Gamma, \eta), \hat{R}_t(\Gamma, \eta)\}]}{\sum_{\eta' \in \mathbb{M}} [\Pi_{\eta'\theta} p_{\theta_t|Y_t}(\eta')]} \tag{19.57}$$

Evaluating the product of the two Gaussian densities yields:

$$N\{y_{t+1}; H(\theta)x, G(\theta)G^T(\theta)\} N\{x; \bar{x}_{t+1}(\Gamma, \eta, \theta), \bar{R}_{t+1}(\Gamma, \eta, \theta)\}$$
$$= f_{t+1}(\Gamma, \eta, \theta) N\{x; \hat{x}_{t+1}(\Gamma, \eta, \theta), \hat{R}_{t+1}(\Gamma, \eta, \theta)\} \quad (19.58)$$

Substituting (19.58) into (19.57) yields:

$$p_{x_{t+1}|\theta_{t+1}, Y_{t+1}}(x|\theta)$$
$$= c_t(\theta)^{-1} \frac{\sum_{\eta \in \mathbb{M}} [\Pi_{\eta\theta} p_{\theta_t|Y_t}(\eta) \sum_{\Gamma \in \mathbb{H}_t} \hat{a}_t(\Gamma, \eta) f_{t+1}(\Gamma, \eta, \theta) N\{x; \hat{x}_{t+1}(\Gamma, \eta, \theta), \hat{R}_{t+1}(\Gamma, \eta, \theta)\}]}{\sum_{\eta' \in \mathbb{M}} [\Pi_{\eta'\theta} p_{\theta_t|Y_t}(\eta')]}$$

This yields (19.45)–(19.47).
Integrating left and right-hand sides of (19.58) over x yields:

$$f_{t+1}(\Gamma, \eta, \theta) = \int_{\mathbb{R}^n} N\{y_{t+1}; H(\theta)x, G(\theta)G^T(\theta)\} N\{x; \bar{x}_{t+1}(\Gamma, \eta, \theta), \bar{R}_{t+1}(\Gamma, \eta, \theta)\} dx$$
$$= N\{y_{t+1}; H(\theta)\bar{x}_{t+1}(\Gamma, \eta, \theta), H(\theta)\bar{R}_{t+1}(\Gamma, \eta, \theta) H(\theta)^T + G(\theta)G^T(\theta)\} \quad (19.59)$$

This implies (19.48). □

Remark 19.4 The characterization of $p_{x_t|\theta_t, Y_t}(\cdot|\theta)$ as a mixture of N^t Gaussian densities has been done by [33] and also by [23].

From Theorem 19.4 follows the so-called family of Generalized Pseudo Bayesian (GPB) filters [33], with family members GPBd, $d \geq 1$. GPBd filter equations satisfy the set of equations in Theorem 19.4 in combination with the approximation that at the end moment $t+1$ of the filter cycle from t to $t+1$, Gaussian densities in the sum, having equal hypothesis $(\theta_{t+2-d}, \ldots, \theta_{t+1})$, are being merged in a single Gaussian density, with matching first- and second-order central moments.

19.7 IMM filtering of Markov jump linear systems

In this section we characterize the five steps of Section 19.5 in terms of conditional mode probability function $p_{\theta_t|Y_t}(\theta)$ and first and second order central moments of the mode conditional density $p_{x_t|\theta_t, Y_t}(x|\theta)$.

For $p_{x_t|\theta_t, Y_t}(x|\theta)$ the first- and second-order central moments are:

$$\hat{x}_t(\theta) \triangleq E\{x_t|\theta_t = \theta, Y_t\} \quad (19.60)$$

$$\hat{R}_t(\theta) \triangleq E\{[x_t - \hat{x}_t(\theta)][x_t - \hat{x}_t(\theta)]^T|\theta_t = \theta, Y_t\} \quad (19.61)$$

For $p_{x_t|\theta_{t+1}, Y_t}(x|\theta)$ the first- and second-order central moments are:

$$\hat{x}_{t|t+1}(\theta) \triangleq E\{x_t|\theta_{t+1} = \theta, Y_t\} \quad (19.62)$$

$$\hat{R}_{t|t+1}(\theta) \triangleq E\{[x_t - \hat{x}_{t|t+1}(\theta)][x_t - \hat{x}_{t|t+1}(\theta)]^T|\theta_{t+1} = \theta, Y_t\} \quad (19.63)$$

For $p_{x_{t+1}|\theta_{t+1}, Y_t}(x|\theta)$ the first- and second-order central moments are:

$$\bar{x}_{t+1}(\theta) \triangleq E\{x_{t+1}|\theta_{t+1} = \theta, Y_t\} \quad (19.64)$$

$$\bar{R}_{t+1}(\theta) \triangleq E\{[x_{t+1} - \bar{x}_{t+1}(\theta)][x_{t+1} - \bar{x}_{t+1}(\theta)]^T|\theta_{t+1} = \theta, Y_t\} \quad (19.65)$$

In terms of these first- and second-order central moments, the five steps of Section 19.5 are written as follows:

$$p_{\theta_t|Y_t} \xrightarrow{\text{Mode Switching}} p_{\theta_{t+1}|Y_t} \xrightarrow{\text{Bayes Mode Update}} p_{\theta_{t+1}|Y_{t+1}}$$

$$\hat{x}_t(\theta), \hat{R}_t(\theta) \xrightarrow{\text{Mixing}} \hat{x}_{t|\theta_{t+1}}(\theta), \hat{R}_{t|\theta_{t+1}}(\theta) \xrightarrow{\text{State Prediction}} \bar{x}_{t+1}(\theta), \bar{R}_{t+1}(\theta)$$

$$\bar{x}_{t+1}(\theta), \bar{R}_{t+1}(\theta) \xrightarrow{\text{Bayes State Update}} \hat{x}_{t+1}(\theta), \hat{R}_{t+1}(\theta)$$

By characterizing each of these five steps in terms of first- and second-order central moments, we arrive at the Interacting Multiple Model (IMM) filter, the equations of which are specified in Table 19.3.

Table 19.3 IMM filter

- Initialization. Set $t := 0$ and start with initial values for the mode probabilities $p_{\theta_0|Y_0}(\theta)$, and the mode conditional first and second moments $\hat{x}_0(\theta)$ and $\hat{R}_0(\theta)$

IMM filter cycle

- Mode switching:
$$p_{\theta_{t+1}|Y_t}(\theta) = \sum_{\eta \in \mathbb{M}} \Pi_{\eta\theta} p_{\theta_t|Y_t}(\eta) \quad (19.66)$$
- State mixing:
$$\hat{x}_{t|\theta_{t+1}}(\theta) = \sum_{\eta \in \mathbb{M}} \left[\Pi_{\eta\theta} \frac{p_{\theta_t|Y_t}(\eta)}{p_{\theta_{t+1}|Y_t}(\theta)} \hat{x}_t(\eta) \right] \quad (19.67)$$
$$\hat{R}_{t|\theta_{t+1}}(\theta) = \sum_{\eta \in \mathbb{M}} \left[\Pi_{\eta\theta} \frac{p_{\theta_t|Y_t}(\eta)}{p_{\theta_{t+1}|Y_t}(\theta)} \left[\hat{R}_t(\eta) + [\hat{x}_t(\eta) - \hat{x}_{t|\theta_{t+1}}(\theta)][\hat{x}_t(\eta) - \hat{x}_{t|\theta_{t+1}}(\theta)]^T \right] \right] \quad (19.68)$$
- Prediction:
$$\bar{x}_{t+1}(\theta) = A(\theta)\hat{x}_{t|\theta_{t+1}}(\theta) \quad (19.69)$$
$$\bar{R}_{t+1}(\theta) = A(\theta)\hat{R}_{t|\theta_{t+1}}(\theta)A^T(\theta) + B(\theta)B^T(\theta) \quad (19.70)$$
- Bayes state update:
$$\hat{x}_{t+1}(\theta) = \bar{x}_{t+1}(\theta) + K_{t+1}(\theta)\nu_{t+1}(\theta) \quad (19.71)$$
$$\hat{R}_{t+1}(\theta) = \bar{R}_{t+1}(\theta) - K_{t+1}(\theta)H(\theta)\bar{R}_{t+1}(\theta) \quad (19.72)$$
$$\nu_{t+1}(\theta) = y_{t+1} - H(\theta)\bar{x}_{t+1}(\theta) \quad (19.73)$$
$$K_{t+1}(\theta) = \bar{R}_{t+1}(\theta)H(\theta)^T Q_{j+1}(\theta)^{-1} \quad (19.74)$$
$$Q_{j+1}(\theta) = H(\theta)\bar{R}_{t+1}(\theta)H(\theta)^T + G(\theta)G(\theta)^T \quad (19.75)$$
- Bayes mode update:
$$p_{\theta_{t+1}|Y_{t+1}}(\theta) = \left[(2\pi)^m Det\{Q_{j+1}(\theta)\}\right]^{-\frac{1}{2}} \cdot \exp\left\{-\frac{1}{2}\nu_{t+1}(\theta)^T Q_{j+1}(\theta)^{-1}\nu_{t+1}(\theta)\right\} p_{\theta_{t+1}|Y_t}(\theta)/c_t \quad (19.76)$$
- MMSE output:
$$\hat{x}_{t+1} = \sum_{\theta \in \mathbb{M}} \left[p_{\theta_{t+1}|Y_{t+1}}(\theta)\hat{x}_{t+1}(\theta) \right] \quad (19.77)$$
$$\hat{R}_{t+1} = \sum_{\theta \in \mathbb{M}} \left[p_{\theta_{t+1}|Y_{t+1}}(\theta) \left[\hat{R}_{t+1}(\theta) + [\hat{x}_{t+1}(\theta) - \hat{x}_{t+1}][\hat{x}_{t+1}(\theta) - \hat{x}_{t+1}]^T \right] \right] \quad (19.78)$$

Set $t := t + 1$ and repeat the filter cycle above when $t \leq T$.

The derivation of these IMM filter equations works as follows. Eq. (19.66) is a Chapman–Kolmogorov equation for the Markov chain $\{\theta_t\}$. Straightforward evaluation of eq. (19.30) yields eqs. (19.67)–(19.68) for the mode conditional first and second order central moments. The state prediction equations (19.69)–(19.70) follow immediately from eq. (19.39). So far this derivation shows that eqs. (19.66)–(19.70) are obtained as exact equations without adopting any specific assumption.

In order to characterize the Bayes state and mode update equations (19.71)–(19.76) in terms of conditional mode probabilities and mode conditional first- and second-order central moments, we introduce some restrictive assumptions and show that all IMM filter equations are exact under these assumptions.

Assumption A1:

For each $\theta \in \mathbb{M}$, $p_{x_0|\theta_0, Y_0}(x|\theta) = N\{x; \hat{x}_0, \hat{R}_0\}$

Assumption A2:

For each $\theta \in \mathbb{M}$,

$$A(\theta) = \begin{bmatrix} A^1 & 0 \\ A^3(\theta) & A^2(\theta) \end{bmatrix}, B(\theta) = \begin{bmatrix} B^1 & 0 \\ B^3(\theta) & B^2(\theta) \end{bmatrix}, H(\theta) = \begin{bmatrix} H^1 & 0 \end{bmatrix}, G(\theta) = G^1$$

where A^1, B^1, H^1, and G^1 are matrices of dimensions $n' \times n'$, $n' \times n'$, $m \times n'$, and $m \times m$ respectively, with $n' < n$.

Theorem 19.5 *Let Assumptions A1 and A2 hold true, and let the mode probabilities $p_{\theta_0|Y_0}(\theta)$ and the mode conditional first and second central moments $\hat{x}_0(\theta)$ and $\hat{R}_0(\theta)$ be given for $\theta \in \mathbb{M}$. Then for the Markov jump linear system (19.39)–(19.40) the conditional probabilities $p_{\theta_t|Y_t}(\theta)$ and the mode conditional first- and second-order statistics satisfy the system of IMM filter equations in Table 19.3.*

Proof. Eqs. (19.66)–(19.70) have already been shown to hold true. Let x_t^1 denote the first $n' < n$ components of x_t. Under assumption A2, $p_{x_t^1|\theta_t, Y_t}(x^1|\theta) = N\{x^1; \hat{x}_t^1, \hat{R}_t^1\}$ implies $p_{x_{t+1}^1|\theta_{t+1}, Y_{t+1}}(x^1|\theta) = N\{x^1; \hat{x}_{t+1}^1, \hat{R}_{t+1}^1\}$ for each $\theta \in \mathbb{M}$. Due to assumption A1, $p_{x_0^1|\theta_0, Y_0}(x^1|\theta) = N\{x^1; \hat{x}_0^1, \hat{R}_0^1\}$, thus $p_{x_t^1|\theta_t, Y_t}(x^1|\theta) = N\{x^1; \hat{x}_t^1, \hat{R}_t^1\}$ for each $\theta \in \mathbb{M}$ and all $t \in [0, T]$. From this, equations (19.71) through (19.78) follow straightforwardly. □

Corollary 19.1 *Let the mode probabilities $p_{\theta_0|Y_0}(\theta)$ and the mode conditional first and second central moments $\hat{x}_0(\theta)$ and $\hat{R}_0(\theta)$ be given for all $\theta \in \mathbb{M}$. Then for the Markov jump linear system (19.39)–(19.40) the conditional probabilities $p_{\theta_t|Y_t}(\theta)$ and the mode conditional first and second order statistics satisfy the system of equations (19.66)–(19.70) and (19.77)–(19.78). Equations (19.71)–(19.76), however, are now approximations.*

Remark 19.5 Although the computational load of the IMM filter is similar to the computational load of GPB1 filter, IMM in general performs significantly better than GPB1 [12]. For target tracking applications it has been shown that IMM may perform amostly as well as GPB2 ([10], pp. 38–39, Figs. 5 & 7). However, there also are examples of Markov jump linear systems where GPB2 performs significantly better than IMM [12].

19.8 Rao–Blackwellization and IMM particle filter

In this section, we continue the study of the Markov jump nonlinear system eqs. (19.11)–(19.13) of Section 19.3, and use the five-step decomposition of Section 19.5 for the development of an improved particle filter for Markov jump nonlinear systems using the idea of Rao–Blackwellization.

The idea of Rao–Blackwellization [17] is to use exact filter equations for the conditional density of a part of the state components, and to use a particle spanned empirical density approximation for the other state components. [20] applies Rao–Blackwellization to Markov jump linear system by using exact equations for the $\{x_t\}$ part and particles for the $\{\theta_t\}$ part. [14]–[15] apply Rao–Blackwellization the other way around, i.e using exact equations for the $\{\theta_t\}$ part and particles for the $\{x_t\}$ part. The resulting particle filter is referred to as IMM particle filter.

Similar as the Hybrid PF in section 19.4, at moment t the IMM particle filter has the following set of weighted particles:
$\left\{x_t^{\theta,j}, \mu_t^{\theta,j}; \theta \in \mathbb{M}, j \in \{1, \ldots, S\}\right\}$, thus with a total number of $N_P = MS$ particles. This set of weighted particles spans the empirical density

$$\tilde{p}_{x_t,\theta_t|Y_t}(x,\theta) = \sum_{j=1}^{S} \mu_t^{\theta,j} \delta\left(x - x_t^{\theta,j}\right) \tag{19.79}$$

as an approximation of the exact density $p_{x_t,\theta_t|Y_t}(x,\theta)$. This is equal to eq. (19.23) behind the Hybrid particle filter. For the development of the IMM particle filter, however, we now use the characterizations of the previous section to develop recursive equations for the evolution of the set of particles.

Mode switching step of IMMPF:

By law of total probability and subsequent substitution of (19.79) we get:

$$\begin{aligned}\tilde{p}_{\theta_t|Y_t}(\theta) &= \int_{\mathbb{R}^n} \tilde{p}_{x_t,\theta_t|Y_t}(x,\theta)dx \\ &= \int_{\mathbb{R}^n} \sum_{j=1}^{S} \mu_t^{\theta,j} \delta\left(x - x_t^{\theta,j}\right)dx \\ &= \sum_{j=1}^{S} \mu_t^{\theta,j}\end{aligned}$$

Substituting this into (19.29) yields:

$$\tilde{p}_{\theta_{t+1}|Y_t}(\theta) = \sum_{\eta \in \mathbb{M}} \sum_{j=1}^{S} \Pi_{\eta\theta} \mu_t^{\eta,j} \qquad (19.80)$$

This forms the mode switching step of the IMM particle filter in Table 19.4.
Interaction resampling step of IMMPF:

Next, substituting approximation (19.79) into (19.30) and subsequent evaluation yields:

$$\tilde{p}_{x_t|\theta_{t+1},Y_t}(x|\theta) = \sum_{\eta \in \mathbb{M}} \left[\Pi_{\eta\theta} \sum_{j=1}^{S} \mu_t^{\eta,j} \delta\left(x - x_t^{\eta,j}\right) \right] / \tilde{p}_{\theta_{t+1}|Y_t}(\theta)$$

$$= \sum_{\eta \in \mathbb{M}} \sum_{j=1}^{S} \Pi_{\eta\theta} \mu_t^{\eta,j} \delta\left(x - x_t^{\eta,j}\right) / \tilde{p}_{\theta_{t+1}|Y_t}(\theta) \qquad (19.81)$$

Eq. (19.81) makes clear that $\tilde{p}_{x_t|\theta_{t+1},Y_t}(x|\theta)$ is an empirical density spanned by $N_P = MS$ particles, whereas (19.79) shows that $\tilde{p}_{x_t|\theta_t,Y_t}(x|\theta)$ was an empirical density spanned by S particles. A logical way to reduce the number of particles which span $\tilde{p}_{x_t|\theta_{t+1},Y_t}(x|\theta)$, is to perform a resampling of $\tilde{p}_{x_t|\theta_{t+1},Y_t}(x|\theta)$. This forms the interaction resampling step of the IMM particle filter. Choosing the moment of interaction as the right moment of resampling is similar to IMM's merging of hypothesis in combination with the interaction step [Blom & Bar-Shalom, 1988] rather than after measurement update. By resampling of $\tilde{p}_{x_t|\theta_{t+1},Y_t}(x|\theta)$, i.e. $\bar{x}_t^{\theta,j} \sim \tilde{p}_{x_t|\theta_{t+1},Y_t}(x|\theta)$ for $\theta \in \mathbb{M}$, $j \in \{1, \ldots, S\}$, the resampled version of $\tilde{p}_{x_t|\theta_{t+1},Y_t}(x|\theta)$ becomes

$$\tilde{p}_{x_t|\theta_{t+1},Y_t}(x|\theta) = \sum_{j=1}^{S} \tfrac{1}{S} \delta\left(x - \bar{x}_t^{\theta,j}\right) \qquad (19.82)$$

Prediction step of IMMPF

Substituting eq. (19.82) into (19.31) yields:

$$\tilde{p}_{x_{t+1}|\theta_{t+1},Y_t}(x|\theta) = \int_{\mathbb{R}^n} p_{x_{t+1}|\theta_{t+1},x_t}(x|\theta, x') \sum_{j=1}^{S} \tfrac{1}{S} \delta\left(x' - \bar{x}_t^{\theta,j}\right) dx'$$

Subsequent evaluation of $p_{x_{t+1}|\theta_{t+1},x_t}(\cdot|\theta, x)$, using $\tilde{p}_{w_t}(w) = \delta\left(w - w_t^{\theta,j}\right)$, yields:

$$\tilde{p}_{x_{t+1}|\theta_{t+1},Y_t}(x|\theta) = \int_{\mathbb{R}^n} \int_{\mathbb{R}^{n'}} \delta\left(x - a\left(\theta, x', w\right)\right) \delta\left(w - w_t^{\theta,j}\right) dw \cdot \sum_{j=1}^{S} \tfrac{1}{S} \delta\left(x' - \bar{x}_t^{\theta,j}\right) dx'$$

$$= \sum_{j=1}^{S} \tfrac{1}{S} \delta\left(x - a\left(\theta, \tilde{x}_t^{\theta,j}, w_t^{\theta,j}\right)\right)$$

$$= \sum_{j=1}^{S} \tfrac{1}{S} \delta\left(x - x_{t+1}^{\theta,j}\right) \qquad (19.83)$$

with

$$x_{t+1}^{\theta,j} = a\left(\theta, \tilde{x}_t^{\theta,j}, w_t^{\theta,j}\right) \qquad (19.84)$$

This forms the prediction step of the IMM particle filter in Table 19.4.

<u>Bayesian update correction step of IMMPF:</u>

The predicted joint conditional empirical density satisfies:

$$\tilde{p}_{x_{t+1},\theta_{t+1}|Y_t}(x, \theta) = \tilde{p}_{x_{t+1}|\theta_{t+1}, Y_t}(x|\theta) \tilde{p}_{\theta_{t+1}|Y_t}(\theta)$$

Together with (19.15) this yields

$$\tilde{p}_{x_{t+1},\theta_{t+1}|Y_{t+1}}(x, \theta) = p_{y_{t+1}|x_{t+1},\theta_{t+1}}(y_t|x, \theta) \tilde{p}_{x_{t+1}|\theta_{t+1}, Y_t}(x|\theta) \tilde{p}_{\theta_{t+1}|Y_t}(\theta)/c_t$$

Substitution of (19.83), and subsequent evaluation yields

$$\tilde{p}_{x_{t+1},\theta_{t+1}|Y_{t+1}}(x, \theta) = p_{y_{t+1}|x_{t+1},\theta_{t+1}}(y_t|x, \theta) \sum_{j=1}^{S} \tfrac{1}{S} \delta\left(x - x_{t+1}^{\theta,j}\right) \tilde{p}_{\theta_{t+1}|Y_t}(\theta)/c_t$$

$$= \sum_{j=1}^{S} \tfrac{1}{S} p_{y_{t+1}|x_{t+1},\theta_{t+1}}(y_{t+1}|x_{t+1}^{\theta,j}, \theta) \delta\left(x - x_{t+1}^{\theta,j}\right) \tilde{p}_{\theta_{t+1}|Y_t}(\theta)/c_t$$

$$= \sum_{j=1}^{S} \mu_{t+1}^{\theta,j} \delta\left(x - x_{t+1}^{\theta,j}\right) \qquad (19.85)$$

with

$$\mu_{t+1}^{\theta,j} = \tfrac{1}{S} p_{y_{t+1}|x_{t+1},\theta_{t+1}}\left(y_{t+1}|x_{t+1}^{\theta,j}, \theta\right) \tilde{p}_{\theta_{t+1}|Y_t}(\theta)/c_t \qquad (19.86)$$

Equations (19.85) and (19.86) characterize the Bayesian mode and state update correction step of the IMM particle filter.

For output purposes we also characterize $\tilde{p}_{\theta_{t+1}|Y_{t+1}}(\theta)$ and $\tilde{p}_{x_{t+1}|\theta_{t+1}, Y_{t+1}}(x/\theta)$ in terms of the IMMPF particles. From the law of total probability we have:

$$\tilde{p}_{\theta_{t+1}|Y_{t+1}}(\theta) = \int_{\mathbb{R}^n} \tilde{p}_{x_{t+1},\theta_{t+1}|Y_{t+1}}(x, \theta) dx$$

Substitution of (19.85) and subsequent evaluation yields

$$\tilde{p}_{\theta_{t+1}|Y_{t+1}}(\theta) = \sum_{j=1}^{S} \mu_{t+1}^{\theta,j} \qquad (19.87)$$

Table 19.4 IMM particle filter (IMMPF)

Initialization at $t = 0$, for $j = 1, \ldots, S$, $\theta \in \mathbb{M}$:
$\mu_0^{\theta,j} = p_{\theta_0|Y_0}(\theta)/S$, and simulate $x_0^{\theta,j}$ from $p_{x_0|\theta_0,Y_0}(\cdot|\theta)$.

Particles and empirical density at moment t:
$$\left\{ \mu_t^{\theta,j} \in [0,1], x_t^{\theta,j} \in \mathbb{R}^n; \theta \in \mathbb{M}, j = 1, \ldots, S \right\}$$
$$\tilde{p}_{x_t,\theta_t|Y_t}(x,\theta) = \sum_{j=1}^{S} \mu_t^{\theta,j} \delta\left(x - x_t^{\theta,j}\right) \qquad (19.79)$$

IMM particle filter (IMMPF) cycle:
- Mode switching: For $\theta \in \mathbb{M}$:
$$\tilde{p}_{\theta_{t+1}|Y_t}(\theta) = \sum_{\eta \in \mathbb{M}} \sum_{j=1}^{S} \Pi_{\eta\theta} \mu_t^{\eta,j} \qquad (19.80)$$

- Interaction resampling: $j \in \{1, \ldots, S\}$, $\theta \in \mathbb{M}$
$$\bar{\mu}_t^{\theta,j} = \tilde{p}_{\theta_{t+1}|Y_t}(\theta)/S$$
If $\tilde{p}_{\theta_{t+1}|Y_t}(\theta) = 0$ then $\bar{x}_t^{\theta,j} = x_t^{\theta,j}$, else:
$$\bar{x}_t^{\theta,j} \sim \tilde{p}_{x_t|\theta_{t+1},Y_t}(x|\theta) = \sum_{\eta \in \mathbb{M}} \sum_{j'=1}^{S} \Pi_{\eta\theta} \mu_t^{\eta,j'} \delta\left(x - x_t^{\eta,j'}\right) / \tilde{p}_{\theta_{t+1}|Y_t}(\theta) \qquad (19.81)$$

- Prediction; $j \in \{1, \ldots, S\}$, $\theta \in \mathbb{M}$
$$w_t^{\theta,j} \sim p_{w_t}(w)$$
$$x_{t+1}^{\theta,j} = a\left(\theta, \bar{x}_t^{\theta,j}, w_t^{\theta,j}\right) \qquad (19.84)$$

- Bayesian update correction; $j \in \{1, \ldots, S\}$, $\theta \in \mathbb{M}$
$$\mu_{t+1}^{\theta,j} = \frac{1}{S} p_{Y_{t+1}|x_{t+1},\theta_{t+1}}\left(y_{t+1}|x_{t+1}^{\theta,j},\theta\right) \tilde{p}_{\theta_{t+1}|Y_t}(\theta)/c_t \qquad (19.86)$$
with c_t such that $\sum_{j=1}^{S} \sum_{\theta \in \mathbb{M}} \mu_{t+1}^{\theta,j} = 1$
$$\tilde{p}_{x_{t+1},\theta_{t+1}|Y_{t+1}}(x,\theta) = \sum_{j=1}^{S} \mu_{t+1}^{\theta,j} \delta\left(x - x_{t+1}^{\theta,j}\right) \qquad (19.85)$$

Set $t := t + 1$ and repeat the filter cycle above when $t \leq T$

Through dividing the left- and right-hand terms of (19.85) by $\tilde{p}_{\theta_{t+1}|Y_{t+1}}(\theta)$ we get, if $\tilde{p}_{\theta_{t+1}|Y_{t+1}}(\theta) > 0$:

$$\tilde{p}_{x_{t+1}|\theta_{t+1},Y_{t+1}}(x|\theta) = \sum_{j=1}^{S} \mu_{t+1}^{\theta,j} \delta\left(x - x_{t+1}^{\theta,j}\right) / \tilde{p}_{\theta_{t+1}|Y_{t+1}}(\theta) \qquad (19.88)$$

The main changes of the IMMPF over the SIR PF are:

- IMMPF resamples a fixed number of particles per mode;
- IMMPF uses probabilities for $\{\theta_t\}$ instead of particles for $\{\theta_t\}$;
- IMMPF resamples after interaction/mixing rather than after measurement updating.

Table 19.5 provides an overview of similarities and differences between SIR PF, SIR Hybrid PF, IMM PF, and IMM filters.

Table 19.5 Comparison of particle filter aspects

Aspect	SIR PF (Table 1)	SIR HPF (Table 2)	IMMPF (Table 4)	IMM (Table 3)
Particles with mode samples	Yes	No	No	No
Mode switching	Simulation	Simulation	Analytical	Analytical
State mixing	Combined with state prediction	Combined with state prediction	Analytical	Analytical
State prediction	Simulation	Simulation	Simulation	Analytical
Bayes update correction	Standard	Standard	Standard	Mode conditional Gaussian
Timing of resampling	After correction	After correction	Combined with state mixing	n.a.
Resampling type	Equal weight particles	Equal weight particles per mode value	Equal weight particles per mode value	n.a.

Remark 19.6 [15] developed the IMM particle filter for the more general problem that $\{\theta_t\}$ has a state dependent transition matrix $\Pi_{\eta\theta}(x_t)$ where:

$$\Pi_{\eta\theta}(x) = P\{\theta_{t+1} = \theta | \theta_t = \eta, x_t = x\} \tag{19.89}$$

The impact upon the IMM particle filter in Table 19.4 is that in eqs. (19.80) and (19.81), $\Pi_{\eta\theta}$ should be changed into $\Pi_{\eta\theta}\left(x_t^{\eta,j}\right)$ and $\Pi_{\eta\theta}\left(x_t^{\eta,j'}\right)$ respectively. The impact upon the standard particle filter in Table 19.1 and the Hybrid particle filter in Table 19.2, is that $\bar{\theta}_{t+1}^j$ should be simulated from $\Pi_{\theta_t^j\theta}\left(x_t^j\right)$ rather than from $\Pi_{\theta_t^j\theta}$.

19.9 Bayesian filtering when Markov chain is partially observed

We re-address the filtering problem of Section 19.3 when in addition to the Euclidean valued observations there are discrete-valued observations of the Markov chain. As before, the process $\{\theta_t\}$ is an $\mathbb{M}$-valued Markov chain, where $\mathbb{M}$ is a set of M discrete modes, with transition probability matrix Π, with components

$$\Pi_{\eta\theta} = P\{\theta_{t+1} = \theta | \theta_t = \eta\} \tag{19.90}$$

The process $\{\theta_t\}$ influences the system of stochastic difference equations, on $[0, T]$, $T < \infty$,

$$x_{t+1} = a(\theta_{t+1}, x_t, w_t) \tag{19.91}$$

where $\{w_t\}$ is a sequence of i.i.d. standard Gaussian vectors of dimension n, and independent of $\{\theta_t\}$ and the $\mathbb{R}^n$ valued initial condition x_0. The pair (θ_t, x_t) represents the hybrid system state, which is observed through the process $\{y_t\}$ which satisfies the following equation:

$$y_t = h(\theta_t, x_t, v_t) \tag{19.92}$$

where $\{v_t\}$ is a sequences of i.i.d. standard Gaussian variables of dimension m, and independent of $\{\theta_t\}$, $\{w_t\}$ and the initial condition x_0.

Furthermore, a and h are measurable mappings of $\mathbb{M} \times \mathbb{R}^n \times \mathbb{R}^n$ into $\mathbb{R}^n$ and $\mathbb{M} \times \mathbb{R}^n \times \mathbb{R}^m$ into $\mathbb{R}^m$ respectively The mappings a and h are time-invariant for notational simplicity only.

In addition to the $\{y_t\}$ observation process, there is a discrete-valued observation process $\{\kappa_t\}$ which satisfies the following equation

$$\kappa_t = d(\theta_t, u_t) \tag{19.93}$$

where $\{u_t\}$ is a sequence of i.i.d. standard uniform random variables, which are independent of $\{\theta_t\}$, $\{x_t\}$, and $\{v_t\}$. We assume that both the Euclidean valued process $\{y_t\}$ and the discrete valued process $\{\kappa_t\}$ are observed. The stepwise approach of Section 19.5 is extended to cover this situation as well.

The filtering problem is to characterize the joint conditional density-probability $p_{x_t, \theta_t | Y_t, K_t}$, $x \in \mathbb{R}^n$, $\theta \in \mathbb{M}$, of the pair (x_t, θ_t) given the σ-algebras of observations $Y_t = \sigma\{y_s; s \leq t\}$ and $K_t = \sigma\{\kappa_s; s \leq t\}$. For this, [16] developed the exact recursive equations for the joint conditional density $p_{x_t, \theta_t | Y_t, K_t}(x, \theta)$.

Theorem 19.6 *Let the conditions of Theorem 19.2 hold true and let $d(\theta, v)$ be such that $p_{\kappa_t | \theta_t}(\cdot | \theta)$ is measurable for all $\theta \in \mathbb{M}$. Then the conditional density $p_{x_t, \theta_t | Y_t, K_t}(x, \theta)$ satisfies the following recursion*

$$p_{x_{t+1}, \theta_{t+1} | Y_{t+1}, K_{t+1}}(x, \theta) = p_{y_{t+1} | x_{t+1}, \theta_{t+1}}(y_{t+1} | x, \theta) p_{\kappa_{t+1} | \theta_{t+1}}(\kappa_{t+1} | \theta)$$
$$\cdot \sum_{\eta \in \mathbb{M}} \left[\Pi_{\eta\theta} \int_{\mathbb{R}^n} p_{x_{t+1} | \theta_{t+1}, x_t}(x | \theta, x') p_{x_t, \theta_t | Y_t, K_t}(x', \eta) dx' \right] / c_t \tag{19.94}$$

with c_t a normalization constant.

Proof. Application of (generalized) Bayes theorem (e.g. [5] p. 129) yields:

$$p_{x_{t+1}, \theta_{t+1} | Y_{t+1}, K_{t+1}}(x, \theta) = p_{y_{t+1}, \kappa_{t+1} | x_{t+1}, \theta_{t+1}}(y_{t+1}, \kappa_{t+1} | x, \theta) p_{x_{t+1}, \theta_{t+1} | Y_t, K_t}(x, \theta) / c_t$$
$$= p_{y_{t+1} | \kappa_{t+1}, x_{t+1}, \theta_{t+1}}(y_{t+1} | \kappa_{t+1}, x, \theta) p_{\kappa_{t+1} | x_{t+1}, \theta_{t+1}}(\kappa_{t+1} | x, \theta) p_{x_{t+1}, \theta_{t+1} | Y_t, K_t}(x, \theta) / c_t$$
$$= p_{y_{t+1} | x_{t+1}, \theta_{t+1}}(y_{t+1} | x, \theta) p_{\kappa_{t+1} | \theta_{t+1}}(\kappa_{t+1} | \theta) p_{x_{t+1}, \theta_{t+1} | Y_t, K_t}(x, \theta) / c_t$$

From this point on, the derivation of (19.94) works similar to the proof of Theorem 19.2. □

In order to gain further in sight, the evolution in time of the conditional density $p_{\theta_t|Y_t,K_t}$ to $p_{\theta_{t+1}|Y_{t+1},K_{t+1}}$ can be decomposed in the following three steps:

$$p_{\theta_t|Y_t,K_t} \xrightarrow{\text{Mode Switching}} p_{\theta_{t+1}|Y_t,K_t} \xrightarrow{\kappa\text{-based mode update}} p_{\theta_{t+1}|Y_t,K_{t+1}} \xrightarrow{\text{Bayes Mode Update}} p_{\theta_{t+1}|Y_{t+1},K_{t+1}}$$

The evolution in time of the conditional density $p_{x_t|\theta_t,Y_t}(x|\theta)$ can be decomposed in the following three steps:

$$p_{x_t|\theta_t,Y_t,K_t} \xrightarrow{\kappa\text{-based mixing}} p_{x_t|\theta_{t+1},Y_t,K_{t+1}} \xrightarrow{\text{State Prediction}} p_{x_{t+1}|\theta_{t+1},Y_t,K_{t+1}}$$

$$p_{x_{t+1}|\theta_{t+1},Y_t,K_{t+1}} \xrightarrow{\text{Bayes State Update}} p_{x_{t+1}|\theta_{t+1},Y_{t+1},K_{t+1}}$$

With exception of the κ-based mixing step, in each of these steps only one of the stochastic entities moves from t to $t + 1$. Next, for each of these basic steps a characterization is developed in the following sequence:

$$p_{\theta_t|Y_t,K_t} \xrightarrow{\text{Mode Switching}} p_{\theta_{t+1}|Y_t,K_t}$$
$$p_{x_t|\theta_t,Y_t,K_t} \xrightarrow{\kappa\text{-based mixing}} p_{x_t|\theta_{t+1},Y_t,K_{t+1}}$$
$$p_{\theta_{t+1}|Y_t,K_t} \xrightarrow{\kappa\text{-based mode update}} p_{\theta_{t+1}|Y_t,K_{t+1}}$$
$$p_{x_t|\theta_{t+1},Y_t,K_{t+1}} \xrightarrow{\text{State Prediction}} p_{x_{t+1}|\theta_{t+1},Y_t,K_{t+1}}$$
$$p_{x_{t+1}|\theta_{t+1},Y_t,K_{t+1}} \xrightarrow{\text{Bayes State Update}} p_{x_{t+1}|\theta_{t+1},Y_{t+1},K_{t+1}}$$
$$p_{\theta_{t+1}|Y_t,K_{t+1}} \xrightarrow{\text{Bayes Mode Update}} p_{\theta_{t+1}|Y_{t+1},K_{t+1}}$$

Two of the three mode-related steps are at the top of this sequence, and the third one is at the end of this sequence, whereas the three state-related steps are in the middle. In comparison with section 19.5, the novel transitions are κ-based mixing and κ-based mode updating.

κ-based mixing

By law of total probability we get

$$p_{x_t|\theta_{t+1},Y_t,K_{t+1}}(x|\theta) = \sum_{\eta \in \mathbb{M}} p_{x_t,\theta_t|\theta_{t+1},Y_t,K_{t+1}}(x,\eta|\theta)$$

Application of Bayes theorem and subsequent evaluation yields, if $p_{\theta_{t+1}|Y_t,K_t}(\theta) > 0$:

$$p_{x_t|\theta_{t+1},Y_t,K_{t+1}}(x|\theta) = \sum_{\eta \in \mathbb{M}} p_{\theta_{t+1},\kappa_{t+1}|x_t,\theta_t,Y_t,K_t}(\theta,\kappa_{t+1}|x,\eta) p_{x_t,\theta_t|Y_t,K_t}(x,\eta|\theta) / p_{\theta_{t+1},\kappa_{t+1}|Y_t,K_t}(\theta,\kappa_{t+1})$$

Substitution of the following two equations

$$p_{\theta_{t+1},\kappa_{t+1}|x_t,\theta_t,Y_t,K_t}(\theta,\kappa_{t+1}|x,\eta) = p_{\kappa_{t+1}|\theta_{t+1}}(\kappa_{t+1}|\theta)p_{\theta_{t+1}|\theta_t}(\theta|\eta) \quad (19.95)$$

$$p_{\theta_{t+1},\kappa_{t+1}|Y_t,K_t}(\theta,\kappa_{t+1}) = p_{\kappa_{t+1}|\theta_{t+1}}(\kappa_{t+1}|\theta)p_{\theta_{t+1}|Y_t,K_t}(\theta) \quad (19.96)$$

yields

$$p_{x_t|\theta_{t+1},Y_t,K_{t+1}}(x|\theta) = \sum_{\eta \in \mathbb{M}} \left[p_{\theta_{t+1}|\theta_t}(\theta|\eta) p_{x_t,\theta_t|Y_t,K_t}(x,\eta) \right] / p_{\theta_{t+1}|Y_t,K_t}(\theta)$$

$$= \sum_{\eta \in \mathbb{M}} \left[\Pi_{\eta\theta} p_{x_t,\theta_t|Y_t,K_t}(x,\eta) \right] / p_{\theta_{t+1}|Y_t,K_t}(\theta)$$

$$= \sum_{\eta \in \mathbb{M}} \left[\Pi_{\eta\theta} \frac{p_{\theta_t|Y_t,K_t}(\eta)}{p_{\theta_{t+1}|Y_t,K_t}(\theta)} p_{x_t|\theta_t,Y_t,K_t}(x|\eta) \right] \quad (19.97)$$

κ-based mode updating

Application of (generalized) Bayes theorem yields:

$$p_{\theta_{t+1}|Y_t,K_{t+1}}(\theta) = p_{\kappa_{t+1}|\theta_{t+1},Y_t,K_t}(\kappa_{t+1}|\theta) p_{\theta_{t+1}|Y_t,K_t}(\theta)/c'_t$$

$$= p_{\kappa_{t+1}|\theta_{t+1}}(\kappa_{t+1}|\theta) p_{\theta_{t+1}|Y_t,K_t}(\theta)/c'_t \quad (19.98)$$

with $p_{\kappa_{t+1}|\theta_{t+1}}(\kappa|\theta) = \int_{[0,1]} \chi(\kappa, d(\theta, u)) p_{u_{t+1}}(u) du = \int_{[0,1]} \chi(\kappa, d(\theta, u)) du$

Combining the above characterizations with the characterizations of the other five basic steps yields the following Theorem.

Theorem 19.7 *Let the assumptions of Theorem 19.6 hold true, and let $p_{\theta_t|Y_t,K_t}(\theta) > 0$ for all $\theta \in \mathbb{M}$. Then the conditional densities $p_{x_t|\theta_t,Y_t,K_t}(\cdot|\theta)$ and $p_{\theta_t|Y_t,K_t}(\cdot)$ satisfy the following recursions:*

$$p_{x_{t+1}|\theta_{t+1},Y_{t+1},K_{t+1}}(x|\theta) = c_t(\theta)^{-1} p_{y_{t+1}|x_{t+1},\theta_{t+1}}(y_{t+1}|x,\theta)$$

$$\cdot \int_{\mathbb{R}^n} p_{x_{t+1}|x_t,\theta_{t+1}}(x|x',\theta) \frac{\sum_{\eta \in \mathbb{M}} \left[\Pi_{\eta\theta} p_{\theta_t|Y_t,K_t}(\eta) p_{x_t|\theta_t,Y_t,K_t}(x'|\eta) \right]}{\sum_{\eta' \in \mathbb{M}} \left[\Pi_{\eta'\theta} p_{\theta_t|Y_t,K_t}(\eta') \right]} dx' \quad (19.99)$$

$$p_{\theta_{t+1}|Y_{t+1},K_{t+1}}(\theta) = \frac{c_t(\theta)}{c_t} p_{\kappa_{t+1}|\theta_{t+1}}(\kappa_{t+1}|\theta) \sum_{\eta \in \mathbb{M}} \left[\Pi_{\eta\theta} p_{\theta_t|Y_t,K_t}(\eta) \right] \quad (19.100)$$

where $c_t(\theta)$ and c_t normalize $p_{x_{t+1}|\theta_{t+1},Y_{t+1},K_{t+1}}(x|\theta)$ and $p_{\theta_{t+1}|Y_{t+1},K_{t+1}}(\theta)$ respectively.

From the characterization in Theorem 19.7 it follows straightforwardly how the extra $\{\kappa_t\}$ observations extend the Gaussian mixture equations in Theorem

19.4, the IMM filter equations in Table 19.3, and the IMM particle filter in Table 19.4.

Extension of Theorem 19.4 (exact Gaussian mixture)

Given the additional observation eq. (19.93), in Theorem 19.4, the conditioning upon Y_t should be replaced by conditioning upon (Y_t, K_t), and eq. (19.44) should be replaced by

$$p_{\theta_{t+1}|Y_{t+1},K_{t+1}}(\theta) = \frac{c_t(\theta)}{c_t} p_{\kappa_{t+1}|\theta_{t+1}}(\kappa_{t+1}|\theta) \sum_{\eta \in \mathbb{M}} \left[\Pi_{\eta\theta} p_{\theta_t|Y_t,K_t}(\eta) \right] \quad (19.101)$$

This inclusion of extra mode observation eq. (19.93) has originally been developed by [22] using a reference probability measure approach.

Extension of IMM (Table 19.3):

Given the additional observation eq. (19.93), in the IMM filter equations of Table 19.3, the conditioning upon Y_t should be replaced by conditioning upon (Y_t, K_t), and eq. (19.76) should be replaced by

$$\begin{aligned} p_{\theta_{t+1}|Y_{t+1},K_{t+1}}(\theta) &= \left[(2\pi)^m Det\{Q_{t+1}(\theta)\}\right]^{-1/2} \\ &\cdot \exp\{-1/2 v_{t+1}(\theta)^T Q_{t+1}(\theta)^{-1} v_{t+1}(\theta)\} p_{\theta_{t+1}|Y_t,K_{t+1}}(\theta) p_{\kappa_{t+1}|\theta_{t+1}}(\kappa_{t+1}|\theta)/c_t \end{aligned} \quad (19.102)$$

Extension of IMM particle filter (Table 19.4):

Given the additional observation eq. (19.93), in the IMM particle filter equations of Table 19.4, the conditioning upon Y_t should be replaced by conditioning upon (Y_t, K_t), and eq. (19.86) should be replaced by

$$\mu_{t+1}^{\theta,j} = \frac{1}{S} p_{y_{t+1}|x_{t+1},\theta_{t+1}}\left(y_{t+1}|x_{t+1}^{\theta,j}, \theta\right) p_{\kappa_{t+1}|\theta_{t+1}}(\kappa_{t+1}|\theta) \tilde{p}_{\theta_{t+1}|Y_t,K_t}(\theta)/c_t \quad (19.103)$$

with c_t such that $\sum_{j=1}^{S} \sum_{\theta \in \mathbb{M}} \mu_{t+1}^{\theta,j} = 1$

19.10 Manoeuvring target tracking example

In this section some Monte Carlo simulation results by Blom & Bloem[15] are given for the SIR particle filter (PF) of Table 19.1, the SIR Hybrid particle filter (HPF) of Table 19.2, the IMM particle filter (IMMPF) of Table 19.4, and the IMM filter of Table 19.3. For each of the particle filters we used a total of $N_p = 10000$ and $N_p = 1000$ particles respectively. In the example scenarios there is an object moving with two possible modes, i.e. $M = 2$. One mode is constant velocity and the other mode is constant acceleration. The object starts with zero velocity and continues this for 40 scans. After scan 40 the object starts to accelerate at a value equal to the standard deviation σ_a of acceleration values. The object continues with constant velocity after scan 60. The model considered is a Markov jump linear system:

$$x_{t+1} = A(\theta_{t+1})x_t + B(\theta_{t+1})w_t \quad (19.104)$$

$$y_t = Hx_t + \sigma_m v_t \quad (19.105)$$

with $\mathbb{M} = \{1, 2\}$ and

$$A(1) = \begin{bmatrix} 1 & T_s & 0 \\ 0 & 1 & 0 \\ 0 & 0 & 0 \end{bmatrix}, \; A(2) = \begin{bmatrix} 1 & T_s & \frac{1}{2}T_s^2 \\ 0 & 1 & T_s \\ 0 & 0 & a \end{bmatrix}, \; B(1) = \sigma_a \begin{bmatrix} 0 \\ 0 \\ 1 \end{bmatrix},$$

$$B(2) = \sigma_a \begin{bmatrix} 0 \\ 0 \\ \sqrt{1-a^2} \end{bmatrix}, \; \Pi = \begin{bmatrix} 1 - \frac{T_s}{\tau_1} & \frac{T_s}{\tau_1} \\ \frac{T_s}{\tau_2} & 1 - \frac{T_s}{\tau_2} \end{bmatrix}, \; H = [1\; 0\; 0]$$

where σ_a represents the standard deviation of acceleration noise, σ_m represents the standard deviation of the measurement error, T_s is the time duration between two successive observation moments t and $t + 1$, τ_1 and τ_2 are the mean durations of modes 1 and 2 respectively, the parameter $a \in (0, 1]$ allows the acceleration in mode 2 to vary randomly in time. Table 19.6 gives the scenario parameter values that are being used for the Monte Carlo simulations.

Table 19.6 Scenario parameter values

Scenario	a	σ_a [m/s^2]	σ_m [m]	τ [m]	τ_2 [s]	T_s [s]
1	0.9	50	30	50	5	1
2	0.9	50	30	5000	5	1

For each of the scenarios Monte Carlo simulations containing 100 runs have been performed for each of the filters. In each simulation, the filters start with perfect estimates. To make the comparison more meaningful, for all filters the same random number streams were used. The position RMS error results of the Monte Carlo simulations of the two scenarios are shown in figures 19.1 and 19.2. The computational load is shown in Table 19.7. For speed and acceleration RMS error results we refer to [15].

Scenario 1:

Table 19.7 Computational load per scan (10^{-3} s)

N_p	IMM	PF	HPF	IMMPF
10^4	4	138	115	96
10^3	4	19	13	11

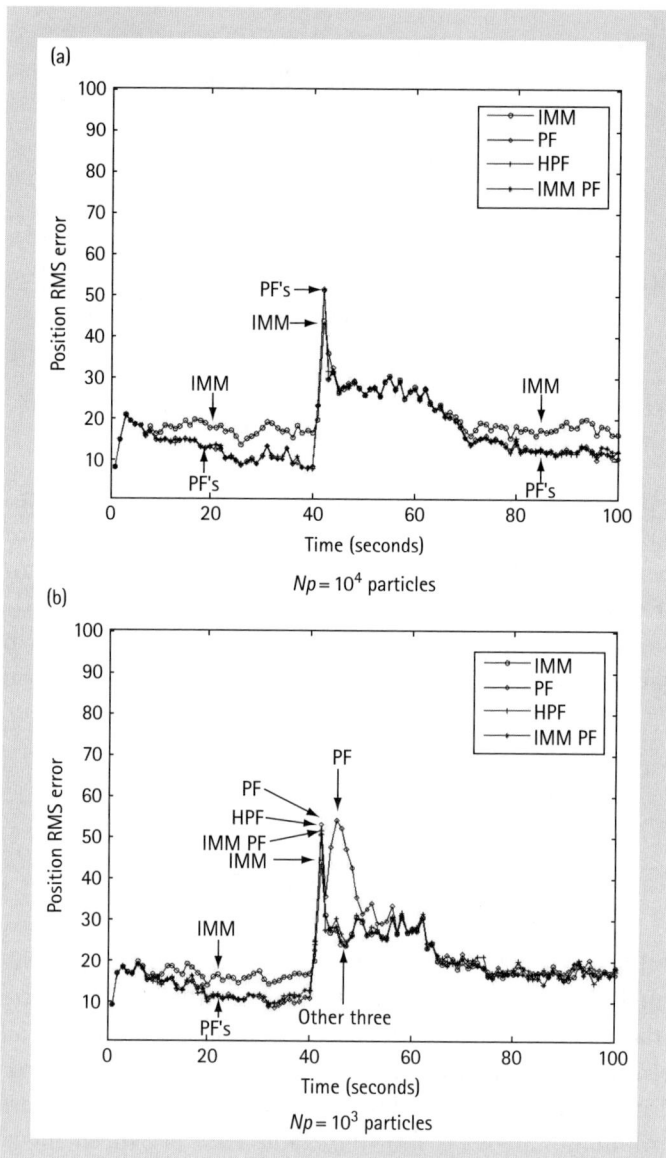

Fig. 19.1 Scenario 1. The target accelerates with 5g between 40s and 60s. The particle filter parameters are $\sigma_a = 5g$, $\tau_1 = 50s$, and $\tau_2 = 5s$. From [15], Fig. 1. (© 2007 IEEE.)

In this scenario, the target accelerates with 5g between 40s and 60s. The τ_1 and τ_2 values used by the particle filters mean that accelerations are expected to happen about once per minute. With $N_p = 10^4$ particles, all three particle filters perform similarly well; they converge to a lower value during uniform motion than IMM does. As a side effect, the peak RMS error at the start of acceleration is for the particle filters slightly higher than it is for IMM. These results agree

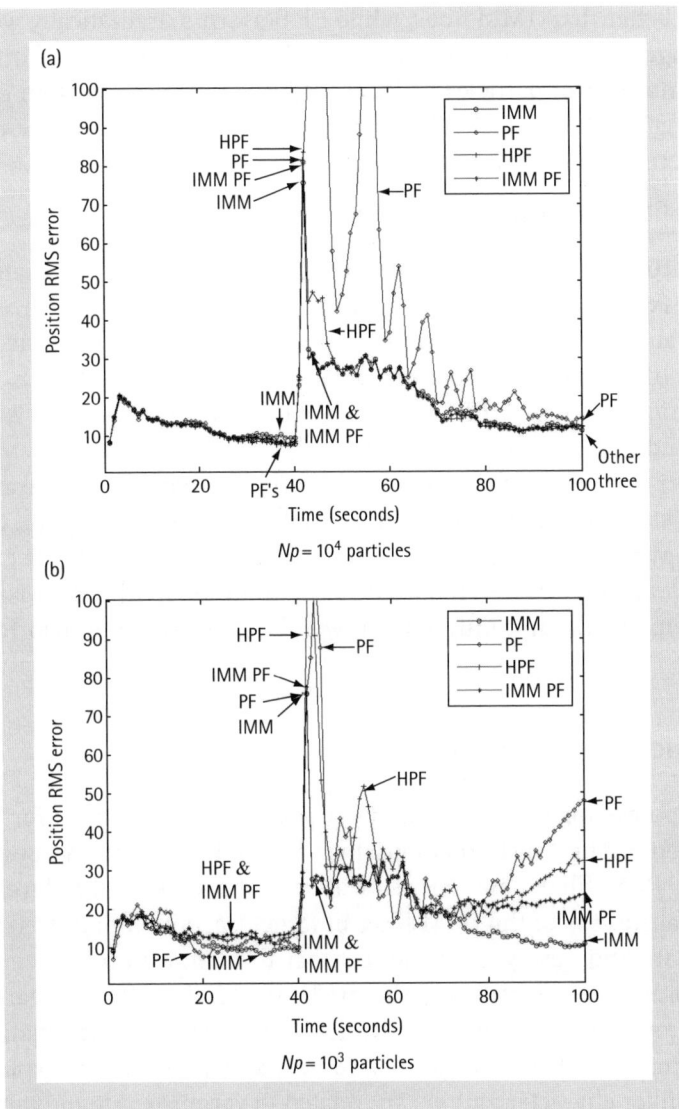

Fig. 19.2 Scenario 2. The target accelerates with 5g between 40s and 60s. The particle filter parameters are $\sigma_a = 5g$, $\tau_1 = 5000s$, and $\tau_2 = 5s$. From [15] Fig. 2. (© 2007 IEEE.)

well with those of [29, 30]. Reduction of the number of particles to $N_p = 10^3$ affects PF dramatically, but has negligible impact on HPF and IMMPF.

Scenario 2:

In this scenario, the target also accelerates with 5g between 40s and 60s. The τ_1 and τ_2 values used by the particle filters mean that accelerations are expected to happen less than once per hour. With $N_p = 10^4$ particles, IMMPF performs marginally better than IMM does, while PF performs dramatically worse. HPF performs significantly worse during the initial acceleration period only. Reduction of the number of particles to $N_p = 10^3$ has a negative effect on the convergence during constant velocity for all three particle filters. Moreover, during the period of acceleration, PF and HPF worsen dramatically in performance.

Summary of Monte Carlo simulation results:

With $N_p = 10^4$ particles, all three particle filters perform better than IMM for scenario 1. For scenario 2, however, IMM and IMMPF perform similarly well, while the standard PF and Hybrid PF perform less good on sudden acceleration. With $N_p = 10^3$ particles the performance of PF degrades for both scenarios, and the performance of HPF and IMMPF degrades for scenario 2 only. (See Figs. 19.1 and 19.2.)

Altogether this means that even for the simple target tracking example considered the standard PF and the HPF are outperformed by both the IMM filter and the IMM particle filter. This illustrates that exploiting the power of using the exact interaction between mode and state evolution in filtering Markov jump linear systems works not only for the well known IMM but also for particle filtering.

19.11 Concluding remarks

This chapter developed exact recursive Bayesian filter equations for stochastic hybrid systems. The exact equations that are underlying to the power of the well-known IMM filter have formed the central theme of the developments. The practical success of the IMM filter in target tracking literature has actually moved the attention away from the fact that the interaction between the conditional evolution of the state and the mode of a Markov jump linear system is captured through exact equations in the IMM filter. These exact equations easily extend to situations of Markov jump nonlinear systems. It also is explained how these exact filter characterizations are related to various exact and approximate nonlinear filtering approaches towards Markov jump linear (or nonlinear) systems, such as Gaussian mixture densities, IMM filtering, GPB filtering, and particle filtering. The main results obtained are as follows:

- Section 19.2 developed an exact recursion for the conditional density of the Markov switching mode given exact observations of the Markov jump nonlinear system.
- Section 19.3 developed an exact recursion for the conditional density of the joint mode-state, given partial observations of the Markov jump nonlinear system.
- Section 19.4 developed standard kind of particle filters for this problem.

- Section 19.5 developed a decomposition of the exact filter recursion into five separate steps.
- Sections 19.6 and 19.7 used the characterizations of these five steps for the elaboration of filtering a Markov jump linear system. This leads to a characterization of the exact conditional density in terms of a Gaussian mixture, and a characterization of the IMM filter equations being exact under some specific conditions.
- Section 19.8 used the decomposition of Section 19.5 for the development of a Rao–Blackwellized particle filter, which explicitly makes use of the exact equations for the interactions between mode and state evolution. Because this Rao–Blackwellized particle filter has several similarities with the IMM filter, it is referred to as IMM particle filter.
- Section 19.9 extended the results of Sections 19.3 through 19.8 to the situation of an extra observation of the mode switching process. The extensions of the exact Gaussian mixture, the IMM filter, and of the IMM particle filter was shown to impact the Bayesian update equation of the mode conditional density only.
- Section 19.10 illustrated the working of the IMM filter of Section 19.7, and the particle filters of Sections 19.4 and 19.8 for a simple example of tracking a suddenly manoeuvring target. The filter results illustrate the value of IMM and the IMM particle filter relative to the SIR particle filter.

Throughout the chapter, the focus was on providing a sound introduction to the baseline theory background of filtering for stochastic hybrid systems. This implied that there are many valuable complementary developments left untouched, such as:

- State-dependent Markov switching ([11], [14], [15]);
- IMM particle filter convergence ([19], [18]);
- Complementary particle filtering improvements, e.g. ([1]);
- Hybrid jumps [8, 9];
- Continuous time hybrid system setting ([7], [10], [27], [21], [32]).

Notes

1. For notational simplicity only, the dimension of $\{w_t\}$ is equal to those of $\{x_t\}$.

References

[1] Arulampam, M., Maskell, S., Gordon, N., and Clapp, T. (2002). A tutorial on particle filters for online nonlinear/non-Gaussian Bayesian tracking, *IEEE Tr. Signal Processing* 50: 174–88.

[2] Bar-Shalom, Y., Challa, S., and Blom, H. (2005). IMM estimator versus optimal estimator for hybrid systems, *IEEE Tr. Aerospace and Electronic Systems* **41**: 986–91.
[3] Bar-Shalom, Y. and Li, X. R. (1993). *Estimation and tracking: principles, techniques and software*, Artech House.
[4] Bar-Shalom, Y., Li, X. R., and Kirubarajan, T. (2001). *Estimation, tracking and navigation*, Wiley-Interscience, New York.
[5] Bernardo, J. and Smith, A. (1994). *Bayesian theory*, Wiley, New York.
[6] Blom, H. (1984). An efficient filter for abruptly changing systems, *Proc. of the 23rd IEEE CDC, Las Vegas*, pp. 656–8.
[7] Blom, H. (1984). Markov jump-diffusion models and decision-making-free filtering, in A. Bensoussan and J. Lions (eds), *Proc. Analysis and Optimization of Systems, Nice, France*, Part 1, Springer, pp. 568–80.
[8] Blom, H. (1985). An efficient decision-making-free filter for processes with abrupt changes, *Proc. 7th IFAC Symp. on Identification and System Pararmeter Estimation, York*, pp. 631–6.
[9] Blom, H. (1986). Overlooked potential of systems with Markovian switching coefficients, *Proc. 25th IEEE CDC1986, Athens, Greece*, pp. 1758–64.
[10] Blom, H. (1990). *Bayesian estimation for decision-directed stochastic control*, PhD thesis, Delft University of Technology.
[11] Blom, H. (1991). Hybrid state estimation for systems with semi-Markov switching coefficients, *Proc. 1st European Control Conf., Grenoble*, pp. 1132–7.
[12] Blom, H. and Bar-Shalom, Y. (1988). The Interacting Multiple Model algorithm for systems with Markovian switching coefficients, *IEEE Tr. on Automatic Control* **33**: 780–3.
[13] Blom, H. and Bloem, E. (2003). Joint IMMPDA particle filter, *Proc. 6th Int. Conf. on Information Fusion*, pp. 785–92.
[14] Blom, H. and Bloem, E. (2004). Particle filtering for stochastic hybrid systems, *Proc. IEEE CDC2004, Bahamas*.
[15] Blom, H. and Bloem, E. (2007). Exact Bayesian and particle filtering of stochastic hybrid systems, *IEEE Tr. AES* **43**: 55–70.
[16] Boers, Y. and Driessen, H. (2002). Hybrid state estimation: a target tracking application, *Automatica* **38**: 2153–8.
[17] Casella, G. and Robert, C. (1996). Rao–Blackwellisation of sampling schemes, *Biometrika* **83**: 81–94.
[18] Daum, F. and Huang, J. (2003). The curse of dimensionality for particle filters, *Proc. IEEE Conf. on Aerospace, Big Sky, MT*.
[19] Del Moral, P. and Guionnet, A. (1998). Large deviations for interacting particle systems—application to nonlinear filtering problems, *Stochastic Processes and their Applications*, Vol. 78, Elsevier, pp. 69–95.
[20] Doucet, A., Godsill, S., and Andrieu, C. (2000). On sequential Monte Carlo sampling methods for Bayesian filtering, *Statistics and Computing*, Vol. 10, Springer, pp. 197–208.
[21] Elliott, R., Aggoun, L., and Moore, J. (1995). *Hidden Markov models: estimation and control*, Springer.
[22] Elliott, R., Dufour, F., and Sworder, D. (1996). Exact hybrid filters in discrete time, *IEEE Tr. Automatic Control* **41**: 1807–10.
[23] Elliott, R. and Van der Hoek, J. (1998). A finite-dimensional filter for hybrid observations, *IEEE Tr. Automatic Control* **43**: 736–9.
[24] Gordon, N., Salmond, D., and Smith, A. (1993). Novel approach to nonlinear/non-Gaussian Bayesian state estimation, *IEE Proceedings-F*, Vol. 140, pp. 107–13.

[25] Koutsoukis, X., Kurien, J., and Zhao, F. (2002). Monitoring and diagnosis of hybrid systems using particle filtering methods, *Proc. Mathematical Theory of Networks and Systems (MTNS)*.

[26] Li, X. and Jilkov, V. (2005). A survey of maneuvering target tracking—Part V: Multiple model methods, *IEEE Tr. Aerospace and Electronic Systems* **41**: 1255–1321.

[27] Mariton, M. (1990). *Jump linear systems in automatic control*, Marcel Dekker, New York.

[28] Mazor, E., Averbuch, A., Bar-Shalom, Y., and Dayan, J. (1998). Interacting multiple model methods in target tracking: a survey, *IEEE Transactions Aerospace and Electronic Systems* **43**: 103–23.

[29] McGinnity, S. and Irwin, G. (2000). Multiple model bootstrap filter for maneuvering target tracking, *IEEE Tr. on Aerospace and Electronic Systems* **36**: 1006–1–12.

[30] McGinnity, S. and Irwin, G. (2001). Maneuvering target tracking using a multiple-model bootstrap filter, *in* A. Doucet, N. de Freitas, and N. Gordon (eds), *Sequential Monte Carlo methods in practice*, Springer, New York, pp. 479–97.

[31] Rabiner, L. (1998). A tutorial on hidden Markov models and selected applications in speech recognition, *Proc. of the IEEE*, Vol. 77, pp. 257–86.

[32] Sworder, S. and Boyd, J. (1999). *Estimation problems in hybrid systems*, Cambridge University Press, Cambridge.

[33] Tugnait, J. and Haddad, A. (1979). A detection-estimation scheme for state estimation in switching environments, *Automatica* **15**: 477–81.

PART VI
Approximation Theory

·20·

Error Bounds for the Nonlinear Filtering of Diffusion Processes

O. Zeitouni

20.1 Introduction

In this short chapter we discuss the following. Consider the standard filtering problem of diffusion processes

$$dx_t = m(x_t)dt + \sigma(x_t)dw_t, \quad x_t \in \mathbb{R}^d$$
$$dy_t = h(x_t)dt + \sqrt{\epsilon}dv_t, \quad y_t \in \mathbb{R}^p \quad (20.1)$$

where w_t, v_t are independent Brownian motions of dimensions n, p, respectively, h, m, σ are (vector/matrix valued) functions that are twice differentiable with bounded derivatives (with σ also bounded), and $\epsilon > 0$. We assume that x_0 is distributed according to a (known) law p_0 (and often even consider $x_0 = 0$). Let $\mathcal{Y}_t = \sigma(y_s, s \leq t)$ be the sigma-algebra generated by the observations. The nonlinear filtering problem consists of evaluating the conditional (on $\mathcal{Y}_t$) law of x_t, and in particular, finding the best (in the sense of mean-square error) $\mathcal{Y}_t$-measurable estimator $\hat{x}_t$ of x_t, that is the minimizer of the *filtering error*

$$P_t = E\left((x_t - \hat{x}_t)(x_t - \hat{x}_t)^T\right).$$

Of course, $\hat{x}_t = E(x_t|\mathcal{Y}_t)$. Since explicit expressions for $\hat{x}_t$ are typically hard to obtain, our goal in this chapter is to investigate a priori bounds (both upper and lower) on the matrix P_t. We will be particularly interested in bounds that are tight when ϵ is small.

Before discussing bounds on the nonlinear filtering error, it is appropriate to recall the linear case, where P_t can be evaluated as the solution of a Riccati equation. An asymptotic evaluation of the error (in small observation noise) then follows. Thus, consider the linear filtering problem

$$dx_t = Ax_t dt + Cdw_t, \quad (20.2)$$
$$dy_t = Bx_t dt + Ddv_t,$$

where we assume that DD^T is a nondegenerate matrix, and $w.$, $v.$ are independent Brownian motions of dimension n, k, respectively. Suppose x_0 is Gaussian, of covariance P_0 (possibly degenerate). The Kalman–Bucy filter [15] can be used to evaluate $\hat{x}_t$, and the error covariance matrix P_t solves the Riccati equation

$$\dot{P}_t = AP_t + P_t A + CC^T - P_t B^T (DD^T)^{-1} B P_t. \tag{20.3}$$

It is particularly instructive to consider what happens when $DD^T = \epsilon I_p$, with I_p the identity matrix. In particular, we say the filter is asymptotically exact if $P_t \to 0$ as $\epsilon \to 0$. It is then not hard to derive the following.

Lemma 20.1 (Kwakernaak–Sivan [17]) *Assume B, C are full rank. If $n > k$ then the filter is not asymptotically exact. If $n = k$, then the filter is asymptotically exact if and only if the function $h(s) := \det B(sI - A)^{-1}C$ does not possess zeroes in the right half of the complex plane.*

No such general result characterizing the asymptotic exactness seems to exist in the nonlinear filtering problem, motivating the search for upper and lower bounds on the filtering error. In terms of sufficient conditions, the following result is derived by analyzing the extended Kalman filter, obtained via a formal linearization of the equations of the optimal filter.

Lemma 20.2 (Picard [24]) *Consider the nonlinear filtering problem (20.1). Suppose further that $h : \mathbb{R}^d \to \mathbb{R}^p$ is linear and that h is strongly injective in the sense that $\inf_{|x|=1} |h(x)| > 0$. Then, $P_t = O(\sqrt{\epsilon})$ as $\epsilon \to 0$.*

It is worthwhile to note that if $h(x)$ is not linear, but is strongly injective in the sense that $\inf_{x \neq y} |h(x) - h(y)|/|x - y| > 0$, the transformation $z_t = h(x_t)$ leads to a system of the form (20.1), to which one may attempt to apply Lemma 20.2. If in addition the inverse of h is Lipschitz, and the mean square filtering error of z_t given $\mathcal{Y}_t$ is of order $\sqrt{\epsilon}$, the same conclusion extends to the filtering error of x_t.

Some weakening of the strong injectivity condition is possible. To state a rather general result, we introduce the following.

Definition 20.1 A (square) matrix valued adapted process A_t is called exponentially stable if the solution of the matrix differential equation

$$\frac{d}{dt} z_t = A_t z_t, \quad z_0 = I$$

satisfies

$$\|z_t z_s^{-1}\| \leq C e^{-c(t-s)},$$

for some deterministic constants c, C and all $0 \leq s \leq t$.

Of course, when A_t does not depend on t, the property of exponential stability can be checked from the transfer matrix $(zI - A)^{-1}$.

The proof of the following theorem combines stochastic analysis with large deviation (upper) bounds.

Theorem 20.1 (Picard [24]) *Consider the nonlinear filtering problem (20.1). Further assume the existence of a bounded adapted matrix $G_t = G_t(\epsilon)$ (of dimension $p \times d$) so that the $p \times p$ dimensional process process $\epsilon \nabla m(z_t) - G_t \nabla h(z_t)$ is exponentially stable (uniformly in ϵ) for all bounded adapted processes z_t. Then, for some deterministic constants C_1, C_2,*

$$P_t \leq C_1 \sqrt{\epsilon} + C_2 P_0 e^{-t/C_2 \sqrt{\epsilon}}.$$

Here, ∇m and ∇h are matrix valued, with e.g. $(\nabla m(x))_{ij} = \partial m_i(x)/\partial x_j$. Of course, in order to apply Theorem 20.1, one has to come up with an appropriate matrix G_t.

Comparing Lemma 20.2 with Lemma 20.1, the condition of strong injectivity present in the former requires that $d \leq p$, and thus does not yield Lemma 20.1 when reduced to the linear case. While Theorem 20.1 dispenses with the condition of strong injectivity, similar dimensional requirements remain. This is not surprising: even in the linear case, being asymptotically exact is *not* equivalent to P_t being of order $\sqrt{\epsilon}$; for the latter, one needs $d \leq p$.

Sufficient conditions for the nonlinear filter to be asymptotically exact are derived in [34], for a class of filtering problems with linear observations and "cone bounded nonlinearities" in the drift (of the form (20.19), discussed in Section 20.4 below), that however do not necessarily satisfy strong injectivity. In particular, these sufficient conditions apply to certain nonlinear filtering problems with $d > p = n$. The derivation is based on the analysis of a suboptimal filter, and of a (high dimension) Riccati-like equation. An partial counterpart to Lemma 20.1 in the fully non-linear case was recently derived in [14].

20.2 Cramér–Rao type lower bounds

We begin by recalling the derivation of the classical (Bayesian) nonparametric Cramér–Rao bound for the estimation of random variables [6, 27, 30, 3]. Let X, Y be random variables possessing the joint density $p(x, y)$, which is everywhere positive and smooth. Then, for any bounded measurable f, and $\delta > 0$,

$$0 = \frac{1}{\delta} \int_{\mathbb{R}^2} f(y) \left(1 - \frac{p(x+\delta, y)}{p(x, y)}\right) p(x, y) dx dy, \qquad (20.4)$$

$$1 = \frac{1}{\delta} \int_{\mathbb{R}^2} x \left(1 - \frac{p(x+\delta, y)}{p(x, y)}\right) p(x, y) dx dy. \qquad (20.5)$$

In particular, for any such function f,

$$1 = E\left((X - f(Y)) \cdot \frac{1}{\delta}\left(1 - \frac{p(X+\delta, Y)}{p(X, Y)}\right)\right).$$

Applying the Cauchy–Schwarz inequality, one concludes that

$$E(X - f(Y))^2 \geq \frac{1}{\frac{1}{\delta^2} E \left(1 - \frac{p(X+\delta, Y)}{p(X, Y)}\right)^2} . \qquad (20.6)$$

Formally taking limits as $\delta \to 0$, one obtains

$$E(X - f(Y))^2 \geq \frac{1}{\int_{\mathbb{R}^2} p(x, y) \left(\frac{\partial}{\partial x} \log p(x, y)\right)^2 dx dy} =: 1/J_p . \qquad (20.7)$$

(Of course, at this stage one may relax the condition f bounded to $f(Y)$ square integrable.) The scalar J_p is the *Fisher information* of the density $p(x, y)$ with respect to x. Note that one may rewrite the Fisher information as

$$J_p = E \left(\frac{\partial \log p(x|Y)}{\partial x} \bigg|_{x=X}\right)^2 .$$

This form makes it clear that in the above derivation, Y could take values in an arbitrary measure space, as long as the conditional density $p(x|Y)$ is well defined (as a regular conditional probability density) and smooth in x.

Since (20.7) holds for *any* function f, we have obtained a lower bound on the estimation error of X based on observing Y, that is

$$E((X - E(X|Y))^2 \geq 1/J_p . \qquad (20.8)$$

It is straightforward to check that equality holds in (20.8) if and only if the joint density $p(x, y)$ is Gaussian (and then, the conditional expectation $E(X|Y)$ is an affine function of Y).

The above considerations readily extend to a situation where the scalar random variable X is replaced by a column vector $\mathbf{X}$. In that case, write

$$\mathcal{E}_Y := E((\mathbf{X} - E(\mathbf{X}|Y))(\mathbf{X} - E(\mathbf{X}|Y))^T)$$

for the covariance matrix of the estimation error, and define the Fisher information matrix

$$\mathbf{J}_p = E \left(\nabla \log p(x|Y)^T \nabla \log p(x|Y) \big|_{x=\mathbf{X}}\right).$$

(Here, ∇ denotes differentiation w.r.t. the components of x, and for a scalar function $f(x)$, ∇f is a row vector.) Then, (20.8) takes the form

$$\mathcal{E}_Y \geq \mathbf{J}_p^{-1}, \qquad (20.9)$$

where for symmetric matrices A, B, we say that $A \geq B$ if $A - B$ is non-negative definite. Again, equality holds in case the conditional density of $\mathbf{X}$ is Gaussian. Thus, if the Fisher information matrix $\mathbf{J}_p$ of a density $p(\cdot)$ coincides with the Fisher information matrix of a Gaussian density $p'(\cdot)$, the covariance matrix of the estimation error in the Gaussian case is a lower bound on the covariance matrix of the estimation error under p.

By the argument described above and an appropriate limit procedure, one gets the following.

Theorem 20.2 (Bobrovsky–Zakai [2]) *Consider the standard nonlinear filtering problem*

$$dx_t = m(x_t)dt + Cdw_t, \quad x_0 = 0, \, x_t \in \mathbb{R}^d$$
$$dy_t = g(x_t)dt + Ddv_t, \quad y_t \in \mathbb{R}^p \quad (20.10)$$

where w_t, v_t are independent Brownian motions of dimensions n, k, C, D are deterministic matrices, and m, g twice differentiable with bounded derivatives. Let ∇m and ∇g denote the matrices of partial derivatives of m, g. Suppose $x_0 = 0$, set $A_t = E[\nabla m(x_t)]$, and let B_t satisfy

$$B_t^T (DD^T)^{-1} B_t =$$
$$E\left((\nabla m(x_t) - A_t)^T (CC^T)^{-1} (\nabla m(x_t) - A_t) + \nabla g(x_t)^T (DD^T)^{-1} \nabla g(x_t)\right).$$

Let $\bar{x}_t$, $\bar{y}_t$ correspond to the linear filtering problem

$$d\bar{x}_t = A_t \bar{x}_t dt + Cdw_t, \quad \bar{x}_0 = 0, \, \bar{x}_t \in \mathbb{R}^d$$
$$d\bar{y}_t = B_t \bar{x}_t dt + Ddv_t, \quad \bar{y}_t \in \mathbb{R}^p. \quad (20.11)$$

Let $\hat{x}_t = E\left(x_t | y_0^t\right)$, $\hat{\bar{x}}_t = E\left(\bar{x}_t | \bar{y}_0^t\right)$, where $y_0^t = \sigma(y_s, s \in [0, t])$ and $\bar{y}_0^t = \sigma(\bar{y}_s, s \in [0, t])$. Then

$$E\left((x_t - \hat{x}_t)(x_t - \hat{x}_t)^T\right) \geq E\left((\bar{x}_t - \hat{\bar{x}}_t)(\bar{x}_t - \hat{\bar{x}}_t)^T\right). \quad (20.12)$$

The same bound holds if x_0 is a random variable (independent of the Brownian motions $w.$, $v.$) with density p_0 possessing a Fisher information matrix J_0, if one chooses $\bar{x}_0$ to be Gaussian of zero mean and covariance $R_0 = J_0^{-1}$.

(A similar statement holds for time-dependent deterministic matrices $C = C_t$, $D = D_t$). Note that the covariance matrix R_t of the filtering error of the system (20.11) satisfies the Riccati differential equation, compare with (20.3),

$$\dot{R}_t = A_t R_t + R_t A_t + CC^T - R_t B_t^T (DD^T)^{-1} B_t R_t.$$

The proof of Theorem 20.2 follows the approach described above for the finite dimensional case: fix a family of functions $\phi_i(t)$ satisfying $\phi_i(0) = 0$ and $\int_0^t \|\dot{\phi}_i(s)\|^2 ds < \infty$. Considers the map $x_t \mapsto x_t + \delta\phi_i(t) := x_t^{i,\delta}$. Because the diffusion matrix C is fixed, the measures on path-space induced by $(x_\cdot, y_\cdot)$ and $(x_\cdot^{i,\delta}, y_\cdot)$ are mutually absolutely continuous, and the evaluation of the Fisher information involves differentiating the Radon–Nikodym derivative with respect to the parameter δ. One then checks that the Fisher information matrix coincides with that of the linear system (20.11), leading to the conclusion.

Theorem 20.2 is not applicable in either one of the following cases:

- The diffusion matrix C is state dependent, that is $C = C(x_t)$.
- The matrix CC^T is singular (and, if nonsingular but small, the resulting bound is not tight).

By considering perturbations of the Brownian motion w_t (by functions ϕ_i) and of the initial condition x_0, lower bounds of Cramér–Rao type can be derived, that address these deficiencies, see [4]. However, in these cases, the expression for the lower bound is rather cumbersome and typically does not lead to closed-form expressions.

We conclude this section by noting that comparing Theorems 20.1 and 20.2 in the case $d = n = p = 1$ and $\sigma = 1$, one concludes that the lower bound (20.12) is asymptotically tight (as $\epsilon \to 0$) when h is strongly injective.

20.3 Information-based lower bounds

We describe in this section an approach toward lower bounds on the filtering error based on information-theoretic ideas. We begin by introducing the notion of *mutual information*; see [25] for an accessible introduction to this topic.

Definition 20.2 Let $\mathbf{X}, \mathbf{Y}$ denote two (Polish) spaces, with their Borel σ-algebra $\mathcal{X}, \mathcal{Y}$. Let $P_{X,Y}$ denote a probability distribution on the product space $\mathbf{X} \times \mathbf{Y}$, with marginals P_X, P_Y, and let $P_{X|Y}$ denote the (regular) conditional probability distribution of X given Y. Let (X, Y) denote a pair of random variables distributed according to $P_{X,Y}$. Suppose $P_{X|Y}$ is almost surely absolutely continuous with respect to P_X.
a) The mutual information between X and Y is defined as

$$I(X, Y) = E \log \frac{d P_{X|Y}}{d P_X}. \tag{20.13}$$

b) Assume $\mathbf{X} = \mathbf{Y} = \mathbb{R}$. The rate distortion function associated with P_X is the solution of variational principle

$$R(D) = \inf I(X, Y),$$

where the infimum is over all laws $P_{X,Y}$ of marginal P_X satisfying $\int (x - y)^2 d P_{X,Y}(x, y) \leq D$.

By Shannon's bound, see [11], if $\mathbf{X} = \mathbb{R}$ and P_X possesses a density p_X with respect to Lebesgue measure such that the differential entropy $h_X = -\int p_X(z) \log p_X(z) dz$ is finite, then

$$R(D) \geq h_X - \frac{1}{2} \log(2\pi e\, D). \tag{20.14}$$

For our needs, the most important property of mutual information is the following.

Lemma 20.3 (Pinsker [25]) *Let* **X**, **Y** *and* (X, Y) *be as in Definition 20.2, let* **Z** *be another Polish space (equipped with its Borel σ-algebra), and let* $g : \mathbf{X} \mapsto \mathbf{Z}$ *be a measurable map. Set* $Z = g(X)$. *Then*

$$I(Z, Y) \leq I(X, Y). \tag{20.15}$$

The link between mutual information and nonlinear filtering in Gaussian additive noise is described in the following theorem. Let

$$dy_t = f(h_t, y_\cdot, t)dt + dv_t, \, t \in [0, T], \tag{20.16}$$

where $h_\cdot$ is a square integrable, adapted process independent of the p-dimensional Brownian motion $v_\cdot$, and f is measurable and such that $E \int_0^T |f(h_s, y_\cdot, s)|^2 ds < \infty$. Recall that $\mathcal{Y}_t = \sigma(y_s : s \leq t)$.

Theorem 20.3 (Kadota, Ziv, and Zakai [20]) *The mutual information between the paths* $y_\cdot$ *and* $h_\cdot$ *equals*

$$I(y_\cdot, h_\cdot) = \frac{1}{2} E \left(\int_0^T |h_s - E(h_s|\mathcal{Y}_s)|_2^2 ds \right). \tag{20.17}$$

(The statement in [20] is concerned with the case $p = 1$, but the proof carries over to the formulation of Theorem 20.3.) We mention in passing that a relation between mutual information, and causal and noncausal filtering error, is derived in [13], see also [33]. In particular, it is shown in these references that if in (20.16), $f(h_t, y_\cdot, t) = \sqrt{\gamma} f(h_t)$ is scalar then the derivative of the mutual information $I(y_\cdot, h_\cdot)$ with respect to γ equals to the right side of (20.17), with $E(h_s|\mathcal{Y}_s)$ replaced by $E(h_s|\mathcal{Y}_T)$.

We return to the filtering problem, and obtain lower bound on the filtering error from information concepts.

Theorem 20.4 (Ziv and Zakai [32]) *Consider the system (20.10) with* $D = \mathbf{I}_p$. *Suppose* $x_j(t)$ *possesses a density* $p_{j,t}(\cdot)$ *with respect to Lebesgue measure, with (differential) entropy* $H_j(t) = - \int p_{j,t}(z) \log p_{j,t}(z) dz$. *Let* $\hat{x}_j(t) = E(x_j(t)|\mathcal{Y}_t)$. *Then,*

$$E(x_j(t) - \hat{x}_j(t))^2 \geq \frac{1}{2\pi e} \exp \left(2 H_j(t) - \int_0^t \sigma_g^2(s) ds \right), \tag{20.18}$$

where $\sigma_g^2(t) = E|g(x_t) - Eg(x_t)|_2^2$.

(Using Jensen's inequality, one obtains a similar result if x_0 is random and $p_{j,t}$ is replaced by the density $p_{j,t}(\cdot|x_0)$ of $x_j(t)$ conditioned on x_0, $H_j(t)$ is replaced by the expectation of the differential entropy of $p_{j,t}(\cdot|x_0)$, and $\sigma_g^2(t)$ is replaced by $E|g(x_t) - E(g(x_t)|x_0)|_2^2$.)

We provide a sketch of the proof of Theorem 20.4. By Lemma 20.3, the mutual information between $x_j(t)$ and $\hat{x}_j(t)$ is bounded above by the mutual information between x and y, and by Theorem 20.3, the latter is bounded above by $\int_0^t \sigma_g^2(s)ds$. On the other hand, the mutual information between $x_j(t)$ and $\hat{x}_j(t)$ is bounded below by the rate distortion function of $x_j(t)$ (with respect to the mean square fidelity criterion) evaluated at $E(x_j(t) - \hat{x}_j(t))^2$. By Shannon's bound (20.14), the latter is bounded below by

$$H_{j,t} - \frac{1}{2}\log\left(2\pi e\, E\left(x_j(t) - \hat{x}_j(t)\right)^2\right).$$

Combining these facts yields the bound of Theorem 20.4.

20.4 Upper bounds on filtering error

Upper bounds on the filtering error can be obtained by evaluating the filtering error of a suboptimal filter, often linear (in the asymptotics of small ϵ, this approach led to Lemma 20.2 and Theorem 20.1). A typical example of this approach, not restricted to small noise, is the following bound, derived under a condition of 'cone bounded nonlinearity'.

Theorem 20.5 (Gilman–Rhodes [12]) *Consider the system (20.10), and assume that m, g satisfy, for some matrices A, B, the following:*

$$|m(x + x') - m(x) - Ax'|_2 \le a|x'|_2,$$
$$|g(x + x') - g(x) - Bx'|_2 \le b|x'|_2. \quad (20.19)$$

Let P_t satisfy the following nonlinear (matrix) differential equation

$$\dot{P}_t = V_t + \bar{A}P_t + P_t \bar{A}^T - P_t B^T W_t B P_t, \qquad P_0 = \mathrm{Cov} x_0, \quad (20.20)$$

where

$$\bar{A} = A + \frac{1}{2}a\mathbf{I},$$
$$V_t = CC^T + a\mathrm{trace}(P_t) \cdot \mathbf{I},$$
$$W_t = (DD^T)^{-1} + b\mathrm{trace}(P_t) \cdot \mathbf{I}.$$

Then,

$$E((x_t - E(x_t|\mathcal{Y}_t))(x_t - E(x_t|\mathcal{Y}_t))^T) \le P_t. \quad (20.21)$$

The bound (20.21) is achieved by upper-bounding the covariance of $x_t - \bar{x}_t$, where $\bar{x}_t$ is the suboptimal filter that solves the equation

$$d\bar{x}_t = m(\bar{x}_t)dt + K_t[dy_t - g(\bar{x}_t)dt], \quad \bar{x}_0 = E x_0,$$

with the *gain* K_t optimized to yield the smallest possible bound (resulting in $K_t = P_t B^T W_t^{-1}$).

It is worthwhile to note that the above technique applies to other systems that do not necessarily satisfy (20.19). We do not attempt to cover here all possible such situations, but rather chose to bring a representative (and early!) such result.

Bibliographical notes: As mentioned above, Lemma 20.2 and Theorem 20.1 are due to Picard [24]. In that paper, he considers the case where $\sigma = \epsilon^\beta \tilde{\sigma}$ in (20.1), with $\tilde{\sigma}$ independent of ϵ, and with possibly ϵ-dependent drift. An earlier treatment of the performance of suboptimal filters, and comparisons with the optimal filter, in the one-dimensional case, can be found in [23], see also [35] for a-priori large deviations estimates for the one-dimensional filtering problem with linear observation. Another approach to the latter, via (formal) asymptotic expansions, can be found in [16]. Finally, Gaussian limits for the filtering error are derived in [35] and [21], in the case of linear h and constant σ.

Theorem 20.2 is due to [2] (see also [1] and [10] for a discrete time, formal version). An earlier treatment (valid for the case of Gaussian state process x_t) is contained in [26], and the idea of using Bayesian Cramér–Rao bounds for the estimation of continuous time processes (and comparisons with Gaussian systems) can be traced back to [29]. Related bounds using "nuisance parameters" are also contained in [5]. More recent variants pertaining to discrete time can be found in [7, 28]. Of course, we have not reviewed the extensive literature on Cramér–Rao bounds for parameter estimation, see [31] for a guided tour.

The link between mutual information and filtering error was first explored in [8], and then developed in [20] and [32], see also [18] for further development. A general discussion of the link between filtering errors and the mutual information can be found in [22].

Several extensions of the bound of Theorem 20.4 have appeared. Unfortunately, as pointed out in [19], an often quoted extension [9], contains a mistake in the derivation (in (A10) there).

References

[1] B. Z. Bobrovsky and M. Zakai, A lower bound on the estimation error for Markov processes, *IEEE Trans. Aut. Cont.* **20** (1975), pp. 785–8.

[2] B. Z. Bobrovsky and M. Zakai, A lower bound on the estimation error for certain diffusion processes, *IEEE Trans. Inf. Theory* **22** (1976), pp. 45–52.

[3] B. Z. Bobrovsky, E. Mayer-Wolf, and M. Zakai, Some classes of global Cramer–Rao bounds, *Annals Stat.* **15** (1987), pp. 1421–38.

[4] B. Z. Bobrovsky, M. Zakai, and O. Zeitouni, Error bounds for the nonlinear filtering of signals with small diffusion coefficients, *IEEE Trans. Inf. Theory* **34** (1988), pp. 710–21.

[5] C. B. Chang, Two lower bounds on the covariance for nonlinear estimation, *IEEE Trans. Aut. Cont.* **26** (1981), pp. 1294–7.
[6] H. Cramér, *Mathematical methods of statistics*, Princeton Univ. Press (1946).
[7] P. C. Doerschuk, Cramér–Rao bounds for discrete time nonlinear filtering problems, *IEEE Trans. Aut. Control* **40** (1995), pp. 1465–9.
[8] T. Duncan, On the calculation of mutual information, *SIAM J. Appl. Math.* **19** (1970), pp. 215–20.
[9] J. I. Galdos, A lower bound on filtering error with application to phase demodulation, *IEEE Trans. Inf. Theory* **25** (1979), pp. 452–62.
[10] J. I. Galdos, A Cramér–Rao bound for multidimensional discrete time dynamical systems, *IEEE Trans. Aut. Control* **25** (1980), pp. 117–19.
[11] R. G. Gallager, *Information theory and reliable communication*, Wiley (1968).
[12] A. S. Gilman and I. B. Rhodes, Cone bounded nonlinearities and mean-square bounds—estimation upper bound, *IEEE Trans. Aut. Cont.* **18** (1973), pp. 260–5.
[13] D. Guo, S. Verdu, and S. Shamai, Mutual information and minimum mean-square error in Gaussian channels, *IEEE Trans. Inf. Theory* **51** (2005), pp. 1261–82.
[14] R. van Handel, When do nonlinear filters achieve maximal accuracy? *SIAM J. Control Optim.* **48**, 3151–3168 (2009).
[15] R. E. Kalman and R. S. Bucy, New results in linear filtering and prediction theory, *Trans. of the ASME—Journal of Basic Engineering* **83** (1961), pp. 95–107.
[16] R. Katzur, B. Z. Bobrovsky, and Z. Schuss, Asymptotic analysis of the optimal filtering problem for one dimensional diffusions measured in a low noise channel, *SIAM J. Applied Math* **44** (1984), pp. 591–601, 1176–91.
[17] H. Kwakernaak and R. Sivan, The maximally achievable accuracy of linear optimal regulators and linear optimal filters, *IEEE Trans. Aut. Control* **17** (1972), pp. 79–86.
[18] R. Liptser and A. Shiryayev, *Statistic of random processes: Theory and applications II*, Springer (1978).
[19] S. Lu and P. C. Doerschuk, Performance bounds for nonlinear filtering, *IEEE Trans. Aerospace Elect. Sys.* **33** (1997), pp. 316–18.
[20] T. Kadota, M. Zakai, and J. Ziv, Mutual information of the white Gaussian channel with and without feedback, *IEEE Trans. Inf. Theory* **17** (1971), pp. 368–71.
[21] E. Mayer-Wolf, A central limit theorem in nonlinear filtering, *Stochastics Stochastics Reports* **35** (1991), pp. 191–211.
[22] E. Mayer-Wolf and M. Zakai, Some relations between mutual information and the estimation error in Wiener space, *Ann. Appl. Probab.* **17** (2007), pp. 1102–16.
[23] J. Picard, Nonlinear filtering of one-dimensional diffusions in the case of high signal-to-noise ratio, *SIAM J. Appl. Math.* **46** (1986), pp. 1098–1125.
[24] J. Picard, Efficiency of the extended Kalman filter for nonlinear systems with small noise, *SIAM J. Appl. Math.* **51** (1991), pp. 843–85.
[25] M. S. Pinsker, *Information and information stability of random variables and processes*, Holden Day (1964).
[26] D. L. Snyder and I. B. Rhodes, Filtering and control performance bounds with implications on asymptotic separation, *Automatica* **8** (1972), pp. 747–53.
[27] C. Rao, Information and the accuracy attainable in the estimation of statistical parameters, *Bull. Calcutta Math. Soc.* **37** (1945), pp. 81–9.
[28] P. Tichavsky, C. H. Muravchik, and A. Nehorai, Posterior Cramer-Rao bounds for discrete time nonlinear filtering, *IEEE Trans. Signal Proc.* **46** (1998), pp. 1386–96.
[29] H. L. Van Trees, Bounds on the accuracy attainable in the estimation of continuous time random processes, *IEEE Trans. Inf. Theory* **12** (1966), pp. 298–305.

[30] H. L. Van Trees, *Detection, estimation and modulation theory, I*, Wiley (1968).
[31] H. L. Van Trees and K. L. Bell, *Bayesian bounds for parameter estimation and nonlinear filtering/tracking*, IEEE Press (2007).
[32] M. Zakai and J. Ziv, Lower and upper bounds on the optimal filtering error of certain diffusion processes, *IEEE Trans. Inf. Theory* **18** (1972), pp. 325–31.
[33] M. Zakai, On mutual information, likelihood ratios, and estimation error for the additive Gaussian channel, *IEEE Trans. Inf. Theory* **51** (2005), pp. 3017–24.
[34] O. Zeitouni and A. Dembo, On the maximum achievable accuracy in the nonlinear filtering problem, *IEEE Trans. Aut. Control* **33** (1988), pp. 965–7.
[35] O. Zeitouni, Approximate and limit results for nonlinear filters with small observation noise: the linear sensor and constant diffusion coefficient case, *IEEE Trans. Aut. Control* **33** (1988), pp. 595–9.

·21·
Discretizing the Continuous-Time Filtering Problem: Order of Convergence

D. Crisan

21.1 The filtering framework

Let $(\Omega, \mathcal{F}, \mathbb{P})$ be a probability space together with a filtration $(\mathcal{F}_t)_{t \geq 0}$ which satisfies the usual conditions. On $(\Omega, \mathcal{F}, \mathbb{P})$ we consider an $\mathcal{F}_t$-adapted process $X = \{X_t, t \geq 0\}$ (the *signal* process) which is the solution of a d-dimensional stochastic differential equation driven by the d-Brownian motion V:

$$X_t = X_0 + \int_0^t f(X_s) \, ds + \int_0^t \sigma(X_s) \, dV_s. \qquad (21.1)$$

We denote by π_0 the distribution of X_0. We assume that both $f : \mathbb{R}^d \to \mathbb{R}^d$ and $\sigma : \mathbb{R}^d \to \mathbb{R}^{d \times d}$ are bounded and globally Lipschitz. Next let W be a standard $\mathcal{F}_t$-adapted m-dimensional Brownian motion defined on $(\Omega, \mathcal{F}, \mathbb{P})$ and independent of X. Let Y be the process (called the *observation* process) satisfying the following evolution equation

$$Y_t = \int_0^t h(X_s) \, ds + W_t, \qquad (21.2)$$

where $h : \mathbb{R}^d \to \mathbb{R}^m$ is a measurable function with linear growth. Let $\{\mathcal{Y}_t, t \geq 0\}$ be the usual augmentation of the filtration associated with the process Y, viz

$$\mathcal{Y}_t = \sigma(Y_s, s \in [0, t]) \vee \mathcal{N}, \qquad (21.3)$$

where $\mathcal{N}$ comprises all $\mathbb{P}$-null sets of $(\Omega, \mathcal{F}, \mathbb{P})$. Note that since Y is $\mathcal{F}_t$-adapted (by the measurability of h) it follows that $\mathcal{Y}_t \subset \mathcal{F}_t$.

The filtering problem consists in determining π_t, the conditional distribution of the signal X at time t given the information accumulated from observing Y in the interval $[0, t]$, that is, for φ Borel bounded function, computing

$$\pi_t(\varphi) = \mathbb{E}[\varphi(X_t) \mid \mathcal{Y}_t]. \qquad (21.4)$$

Under conditions described below, the process $\tilde{Z} = \{\tilde{Z}_t, t > 0\}$ defined by

$$\tilde{Z}_t = \exp\left(-\sum_{i=1}^m \int_0^t h^i(X_s) \, dW_s^i - \frac{1}{2} \sum_{i=1}^m \int_0^t h^i(X_s)^2 \, ds\right), \quad t \geq 0. \qquad (21.5)$$

is an $\mathcal{F}_t$-adapted martingale. For fixed $t \geq 0$, we may define a probability measure $\tilde{\mathbb{P}}^t$ on $\mathcal{F}_t$ by defining its Radon–Nikodym derivative with respect to $\mathbb{P}$ to be given by $\tilde{Z}_t$ viz

$$\left.\frac{d\tilde{\mathbb{P}}^t}{d\mathbb{P}}\right|_{\mathcal{F}_t} = \tilde{Z}_t.$$

From the martingale property of $\tilde{Z}$, the probability measures $\tilde{\mathbb{P}}^t$ form a consistent family. Therefore we can define a probability measure $\tilde{\mathbb{P}}$ which is equivalent to $\mathbb{P}$ on $\bigcup_{0 \leq t < \infty} \mathcal{F}_t$ and we will suppress the superscript t in subsequent calculations. Under $\tilde{\mathbb{P}}$, the observation process Y is a Brownian motion independent of X; additionally the law of the signal process X under $\tilde{\mathbb{P}}$ is the same as its law under $\mathbb{P}$.

Let $Z = \{Z_t, t > 0\}$ be the process defined as $Z_t = \tilde{Z}_t^{-1}$ for $t \geq 0$. Then

$$Z_t = \exp\left(\sum_{i=1}^{m} \int_0^t h^i(X_s)\, dY_s^i - \frac{1}{2}\sum_{i=1}^{m}\int_0^t h^i(X_s)^2\, ds\right). \tag{21.6}$$

The process $Z = \{Z_t,\ t \geq 0\}$ is an $\mathcal{F}_t$-adapted martingale.[1] Next define $\rho = \{\rho_t, t \geq 0\}$ to be the *unnormalized* conditional distribution the signal defined as

$$\rho_t(\varphi) = \tilde{\mathbb{E}}\left[\varphi(X_t) Z_t \mid \mathcal{Y}_t\right] \tag{21.7}$$

for every bounded Borel measurable function φ. Then, by the Kallianpur–Striebel's formula, we have

$$\pi_t(\varphi) = \frac{\rho_t(\varphi)}{\rho_t(\mathbf{1})} \quad \tilde{\mathbb{P}}(\mathbb{P}) - \text{a.s.}, \tag{21.8}$$

where $\mathbf{1}$ is the constant function $\mathbf{1}(x) = 1$ for any $x \in \mathbb{R}^d$. The Kallianpur–Striebel formula explains the usage of the term *unnormalized* for ρ_t as the denominator $\rho_t(\mathbf{1})$ can be viewed as the normalizing factor for ρ_t. For further details of the filtering framework see for example [1] and the references therein.

We observe now that $\rho_t(\varphi)$ and $\rho_t(\mathbf{1})$ are conditional expectations of functionals that depend on the stochastic integrals

$$\int_0^t h^i(X_s)\, dY_s^i, \quad i = 1, \ldots, m.$$

For practical reasons (the values of the observation corresponding to certain discrete times are available) and to permit the numerical approximation of $\rho_t(\varphi)$ and $\rho_t(\mathbf{1})$ and implicitly of $\pi_t(\varphi)$ the stochastic integrals

$$\int_0^t h^i(X_s)\, dY_s^i, \quad i = 1, \ldots, m$$

are replaced in many applications by certain discrete approximations, the most popular being the Itô approximation of the integral, i.e.,

$$\sum_{k=0}^{\infty} h^i(X_{t_k})\left(Y^i_{t_{k+1}\wedge t} - Y^i_{t_k\wedge t}\right),$$

where $\tau = \{0 = t_0 < \cdots < t_k < \cdots\}$ is a given partition of the time interval $[0, \infty)$ with partition mesh $\delta = \max_k (t_{k+1} - t_k)$.

Let now $Z^\tau = \left\{Z^\tau_{t_j}, j \geq 0\right\}$ be the (discrete) process defined as $Z^\tau_{t_0} = 1$ and

$$Z^\tau_{t_j} = \exp \sum_{k=0}^{j} \left(\langle h(X_{t_k}), \delta Y_{t_k}\rangle - \frac{1}{2}\|h(X_{t_k})\|^2 \delta_k\right), \quad j > 0, \tag{21.9}$$

where $\delta Y_{t_k} = Y_{t_{k+1}} - Y_{t_k}$, $\delta_k = t_{k+1} - t_k$, $\langle a, b\rangle$ is the inner product between two vectors $a = (a_i)_{i=1}^m$, $b = (b_i)_{i=1}^m \in \mathbb{R}^m$, $<a, b> = \sum_{i=1}^m a_i b_i$ and $\|a\|^2 = <a, a>$. Next let $\rho^\delta = \left\{\rho^\delta_{t_k}, k \geq 0\right\}$ be the corresponding measure valued process

$$\rho^\tau_{t_k}\varphi = \tilde{\mathbb{E}}\left[\varphi(X_{t_k}) Z^\tau_{t_k} \mid \mathcal{Y}_{t_k}\right]. \tag{21.10}$$

In the following, we analyse the rate of convergence of $\rho^\tau_t(\varphi)$ to $\rho_t(\varphi)$ for $t = t_i$, $i \geq 0$ by obtaining upper bounds for the corresponding mean-square error. For example, (see Theorems 21.3, 21.4, 21.5 below), we show that there exists a constant[2] $c_1 = c_1(\varphi)$ independent of δ such that

$$\tilde{\mathbb{E}}[|\rho^\tau_{t_k}\varphi - \rho_{t_k}\varphi|^2] \leq c_1 \delta^\beta,$$

where $\beta = 1, 2 - \varepsilon$ and 2 depending on the choice of the conditions imposed on the filtering framework. This, in turns, leads to the bound

$$\mathbb{E}[|\pi^\tau_{t_k}\varphi - \pi_{t_k}\varphi|] \leq c_2 \delta^{\frac{\beta}{2}},$$

where, again, $c_2 = c_2(\varphi)$ is a constant independent of δ and $\pi^\tau = \{\pi^\tau_{t_k}, k \geq 0\}$ is the normalized version of ρ^τ, that is defined by $\pi^\tau_{t_k} = \rho^\tau_{t_k}/\rho^\tau_{t_k}(1)$. We show that the two critical factors that influence the rate of convergence are the smoothness of the semigroup associated to the signal and the smoothness of the sensor function h. In particular, we recover a classical results by Picard (see theorem 1 and corollary 2 in [28]) with the case $\beta = 2$ in Theorem 21.5 below. See Remark 21.7 for brief discussion on the connection with Picard's result.

The filtering problem can be looked at in two ways, both just as valid and useful. One is that we have a stream of observation out of which we want to "filter out" the signal. The second is that we want to compute a certain integral of a functional (that depends on the given observation stream) with respect to the law of the signal. When attempting to numerically estimate the integral, two procedures are required. The first procedure is the (time) discretization of the functional and this chapter describes one which is most commonly used.

The second one is the approximation of the law of the signal which can then be done in various ways. Among the possible ways to do perform this procedure, we enumerate:

- the Euler method (as in [28] or the more recent works [26], [27]) or any other classical weak approximation (see [19] and the references therein), possibly with the added variance reduction step of a particle filter (see [6], [7], [9], [10], [11], [29]).
- "exact sampling" (see [2], [3], [4]).
- cubature methods (see [8], [25]).
- partial differential equations methods (see [5], [13], [14], [20], [21], [24]).

Obviously, the error for the second procedure depends on the choice of the method, while the error of the first procedure is common to all of them. Of course in practice both procedures are applied simultaneously. However, when it comes to estimating the error it helps to separate them. This chapter deals with the analysis of the error of the first procedure. It turns out that the accuracy of the approximation depends essentially on two factors: the degree of differentiability of the functional and the smoothness of the signal's distribution.

We start the analysis by proving a number of general approximation results in the following section, which are then applied in the context of filtering to the deduction of the main results.

21.2 General approximation results

In the following we will be assuming the following condition:
M. All moments of X_0 are finite. The functions $f : \mathbb{R}^d \to \mathbb{R}^d$, $\sigma : \mathbb{R}^d \to \mathbb{R}^{d \times d}$ are Lipschitz continuous.

Under condition **M**, the signal process, i.e., the solution of the equation (21.1), has moments of all orders. Moreover, for any $p > 0$, we have

$$\tilde{\mathbb{E}}\left[\sup_{s \in [0,t]} |X_t|^p\right] < \infty, \tag{21.11}$$

$$\sup_{u,v \in [0,t]} \tilde{\mathbb{E}}[|X_u - X_v|^p] \leq c_3 |u - v|^{\frac{p}{2}}, \tag{21.12}$$

where $c_3 = c_3(p, t)$ is a constant independent of $u, v \in [0, t]$. For a proof of (21.11) and (21.12) see, for example, [16] page 156.

To simplify the notation, we will add an additional component to the Brownian motion Y. Let Y^0 be the process $Y^0_s = s$, for all $s \geq 0$ and consider the

$(m + 1)$-dimensional process $Y = (Y^i)_{i=0}^m$. Let $\xi = (\xi_i)_{i=0}^m$ be the random vector with entries

$$\xi_i = \int_0^t a_i(X_s)\,dY_s^i, \quad i = 0, 1, \ldots, m,$$

where $a = (a_i)_{i=0}^m : \mathbb{R}^d \to \mathbb{R}^{m+1}$ is a Lipschitz-continuous function. Next let τ be a given partition of the time interval $[0, t]$,

$$\tau = \{0 = t_0 < \cdots < t_k < \cdots < t_n = t\}$$

and denote by $\delta = \max_{k=1,\ldots,n}(t_k - t_{k-1})$ the size of the partition. Next let $\xi^\tau = (\xi_i^\tau)_{i=0}^m$ be the random vector with entries

$$\xi_i^\tau = \sum_{k=1}^n a_i(X_{t_{k-1}})\left(Y_{t_k}^i - Y_{t_{k-1}}^i\right) = \int_0^t a_i(X_{\tau(s)})\,dY_s^i, \quad i = 0, 1, \ldots, m,$$

where $\tau(s)$ is the largest element of the partition smaller than or equal to s, that is,

$$\tau(s) = t_{k-1}, \quad s \in [t_{k-1}, t_k), \quad k = 1, \ldots, n.$$

The following remark is proved in the Appendix (see Lemma 21.1):

Remark 21.1 For any $p > 0$, there exists a constant $c_4 = c_4(p, t)$ independent of $i = 0, 1, \ldots, m$ and $\delta > 0$ such that

$$\tilde{\mathbb{E}}\left[|\xi_i - \xi_i^\tau|^p\right] \leq c_4 \delta^{\frac{p}{2}}. \tag{21.13}$$

Moreover, if ν is a random variable such that $\tilde{\mathbb{E}}[|\nu|^q] < \infty$, then for any $p \in (0, q)$, there exists a constant $c_5 = c_5(p, t, \nu)$ independent of $i = 0, 1, \ldots, m$ and $\delta > 0$ such that

$$\tilde{\mathbb{E}}\left[|\nu(\xi_i - \xi_i^\tau)|^p\right] \leq c_5 \delta^{\frac{p}{2}}. \tag{21.14}$$

Let $\varphi : \mathbb{R}^d \to \mathbb{R}$ be a measurable function and $\psi : \mathbb{R}^{m+1} \to \mathbb{R}$ be a continuously differentiable function. Proposition 21.1 and Theorems 21.1 and 21.2 from below give us increasingly finer bounds for the second moment of the following quantity, called henceforth the approximation error,

$$q(X_t, \xi, \xi^\tau) \triangleq \tilde{\mathbb{E}}[\varphi(X_t)\psi(\xi) \mid \mathcal{Y}_t] - \tilde{\mathbb{E}}[\varphi(X_t)\psi(\xi^\tau) \mid \mathcal{Y}_t]$$

under increasingly stronger conditions. Firstly we assume that the following condition holds true:

Lp. There exists a constant $p > 2$ such that

$$\sup_\tau \sup_{s \in [0,1]} \tilde{\mathbb{E}}\left[|\varphi(X_t)\partial_i\psi(s\xi + (1-s)\xi^\tau)|^p\right] < \infty, \quad i = 0, 1, \ldots, m. \tag{21.15}$$

Remark 21.2 Since both ξ and ξ^τ have moments of all orders, condition (21.15) holds true if the function φ and the partial derivatives of ψ have polynomial growth. If ξ and ξ^τ have exponential moments such that

$$\tilde{\mathbb{E}}\left[\exp\left(r\xi\right)\right] < \infty, \quad \sup_\tau \tilde{\mathbb{E}}\left[\exp\left(r\xi^\tau\right)\right] < \infty$$

for some $r > 0$, condition (21.15) holds true also if the partial derivatives of ψ have exponential growth. This is the case if the functions a_i $i = 0, 1, \ldots, m$ are bounded or if X_0 has exponential moments, σ is bounded and t is sufficiently small (see, for example, [17]).

The following result gives an upper bound for the approximation error which holds true under conditions **M** and **Lp**:

Proposition 21.1 *Under conditions **M** and **Lp**, there exists a constant $c_6 = c_6(t)$ independent of the partition τ such that*

$$\tilde{\mathbb{E}}\left[q\left(X_t, \xi, \xi^\tau\right)^2\right] \leq c_6 \delta. \tag{21.16}$$

Proof. Using the mean-value theorem

$$\varphi\left(X_t\right)\left(\psi\left(\xi^\tau\right) - \psi\left(\xi\right)\right) = \sum_{i=0}^m \eta_i \left(\xi_i^\tau - \xi_i\right), \tag{21.17}$$

where $\eta = (\eta_i)_{i=0}^m$ is the random vector defined as

$$\eta_i = \int_0^1 \varphi\left(X_t\right) \partial_i \psi\left(s\xi + (1-s)\xi^\tau\right) ds. \tag{21.18}$$

From condition (21.15) it follows that $\eta_i \in L_p(\Omega, \mathcal{F}_t, \tilde{P})$. Therefore, from (21.14) with $v = \eta_i$ we get that

$$\tilde{\mathbb{E}}\left[\left(\varphi\left(X_t\right)\left(\psi\left(\xi^\tau\right) - \psi\left(\xi\right)\right)\right)^2\right] \leq c_6 \delta,$$

where $c_6 = c_5(2, t, \eta_i)$ and the result follows by the (conditional) Jensen's inequality. $\square$

The proof of following remark can be found in the appendix (see Lemma 21.2).

Remark 21.3 The random variables $\eta_i \in L_p(\Omega, \mathcal{F}_t, \tilde{P})$ admit the following integral representation

$$\eta_i = \eta_i' + \sum_{j=1}^d \int_0^t \psi_s^{ij} dW_s^j, \quad i = 0, 1, \ldots, m \tag{21.19}$$

In (21.19), the random variables $\eta_i' \in L_p(\Omega, \mathcal{F}_0 \vee \mathcal{Y}_t, \tilde{P})$ are given by

$$\eta_i' = \tilde{\mathbb{E}}\left[\eta_i \mid \mathcal{F}_0 \vee \mathcal{Y}_t\right], \quad i = 0, 1, \ldots, m$$

and $\psi^{ij} = \{\psi^{ij}_s, s \in [0, t]\}$ are progresively measurable $\mathcal{F}_s \vee \mathcal{Y}_t$-adapted processes such that

$$\tilde{\mathbb{E}}\left[\left(\int_0^t (\psi^{ij}_s)^2 \, ds\right)^{\frac{p}{2}}\right] < \infty \quad (21.20)$$

for $i = 0, 1, \ldots, m$ and $j = 1, \ldots, d$.

The bound in Proposition 21.1 says that the second moment of the approximation error has an upper bound of an order equal to size of the partition mesh. A finer bound can be obtained provided we impose an additional assumption:

AP. Let $(P_s)_{s \geq 0}$ be the semigroup associated to the Markov process X. We will assume that, for any Lipschitz continuous function $\psi : \mathbb{R}^d \to \mathbb{R}$, $P_s\psi$ is twice differentiable for any $s \in [0, t]$. Moreover, if

$$P_{a,b}\psi \triangleq P_a\psi - P_b\psi, \ a, b \in [0, t],$$

we will assume that there exists a constant $c_7 = c_7(t)$ independent of a and b such that

$$\sup_{x \in \mathbb{R}^d} |P_{a,b}\psi(x)| \leq c_7 k_\psi \left(\sqrt{a} - \sqrt{b}\right) \quad (21.21)$$

$$\sup_{x \in \mathbb{R}^d} |\partial_i P_{a,b}\psi| \leq \frac{c_7}{b} k_\psi (a - b), \ i = 1, \ldots, d, \quad (21.22)$$

where k_ψ is the Lipschitz constant of ψ. We will also assume that the processes ψ^{ij} $i = 0, 1, \ldots, m$ and $j = 1, \ldots, d$ satisfy

$$\tilde{\mathbb{E}}\left[\sup_{s \in [0,t]} (\psi^{ij}_s)^2 \bigg| \mathcal{Y}_t\right] < \infty. \quad (21.23)$$

The proof of following remark can be found in the Appendix (see Lemma 21.4).

Remark 21.4 Inequalities (21.21) and (21.22) are satisfied if, for example, the functions $f : \mathbb{R}^d \to \mathbb{R}^d$, $\sigma = (\sigma^i)_{i=1}^d : \mathbb{R}^d \to \mathbb{R}^{d \times d}$ are bounded continuous with bounded partial derivatives of all orders and the vector fields $(\sigma^i)_{i=1}^d$ satisfy the Hörmander condition.

We show in the Appendix (see Lemma 21.5) that the bound (21.23) holds true if $\psi : \mathbb{R}^{m+1} \to \mathbb{R}$ is the exponential function, i.e.,

$$\psi(x_0, \ldots, x_m) = \exp(x_0 + \cdots + x_m), \ (x_0, \ldots, x_m) \in \mathbb{R}^{m+1}$$

as is the case encountered within the filtering context. See also Remark 21.8 for a general class of functions ψ for which (21.23) holds true.

Theorem 21.1 *Under conditions **M**, **Lp**, and **AP**, there exists $\delta_0 > 0$ such that for all $\delta \in (0, \delta_0)$ and $\varepsilon \in (0, 1)$, there exists a constant $c_8 = c_8(t, \delta_0, \varepsilon)$ independent of the partition τ such that*

$$\tilde{\mathbb{E}}\left[q\left(X_t, \xi, \xi^\tau\right)^2\right] \leq c_8 \delta^{2-\varepsilon}. \tag{21.24}$$

Proof. Using the decomposition (21.19), we get that

$$q\left(X_t, \xi, \xi^\tau\right) = \sum_{i=0}^m \int_0^t \tilde{\mathbb{E}}\left[\eta_i' F_i(s) \mid \mathcal{Y}_s\right] dY_s^i + \sum_{i=0}^m \sum_{j=1}^d \int_0^t \left(\Lambda_{ij}(s) + \Psi_{ij}(s)\right) dY_s^i, \tag{21.25}$$

where

$$F_i(s) = \tilde{\mathbb{E}}\left[\left(a_i(X_s) - a_i(X_{\tau(s)})\right) \mid \mathcal{F}_0 \vee \mathcal{Y}_t\right] = P_{s,\tau(s)} a_i(X_0) \tag{21.26}$$

$$\Lambda_{ij}(s) = \tilde{\mathbb{E}}\left[\left(a_i(X_s) - a_i(X_{\tau(s)})\right) \int_{\max(0,\tau(s)-\delta)}^s \psi_r^{ij} dW_r^j \mid \mathcal{Y}_t\right] \tag{21.27}$$

$$\Psi_{ij}(s) = \tilde{\mathbb{E}}\left[\left(a_i(X_s) - a_i(X_{\tau(s)})\right) \int_0^{\max(0,\tau(s)-\delta)} \psi_r^{ij} dW_r^j \mid \mathcal{Y}_t\right] \tag{21.28}$$

for $i = 0, 1, \ldots, m$ and $j = 1, \ldots, d$. The result will follow by finding upper bounds of the order $\delta^{2-\varepsilon}$ (or better) for the second moments of each of the terms in (21.25).

First, from (21.21) we get that

$$|F_i(s)| \leq \sup_{x \in \mathbb{R}^d} \left|P_{s,\tau(s)} a_i(x)\right| \leq c_7 k_{a_i} \left(\sqrt{s} - \sqrt{\tau(s)}\right),$$

hence

$$\tilde{\mathbb{E}}\left[\left(\int_0^t \tilde{\mathbb{E}}\left[\eta_i' F_i(s) \mid \mathcal{Y}_s\right] dY_s^i\right)^2\right] = \int_0^t \tilde{\mathbb{E}}\left[\left(\eta_i' F_i(s)\right)^2\right] ds$$

$$\leq c_7^2 k_{a_i}^2 \tilde{\mathbb{E}}\left[\left(\eta_i'\right)^2\right] \int_0^t \left(\sqrt{s} - \sqrt{\tau(s)}\right)^2 ds$$

and we get the required bound by using the estimate (21.47) proved in the Appendix.

Secondly observe that, using the the Lipschitz property of a_i and (21.12), we get

$$\Lambda_{ij}(s)^2 = \tilde{\mathbb{E}}\left[\left(a_i(X_s)\right) - \left(a_i(X_{\tau(s)})\right) \int_{\max(0,\tau(s)-\delta)}^s \psi_r^{ij} dW_r^j \mid \mathcal{Y}_t\right]^2$$

$$\leq \tilde{\mathbb{E}}\left[\left(a_i(X_s)\right) - \left(a_i(X_{\tau(s)})\right)^2 \mid \mathcal{Y}_t\right] \tilde{\mathbb{E}}\left[\left(\int_{\max(0,\tau(s)-\delta)}^s \psi_r^{ij} dW_r^j\right)^2 \mid \mathcal{Y}_t\right]$$

$$\leq k_{a_i}^2 \tilde{\mathbb{E}}\left[|X_s - X_{\tau(s)}|^2\right] \int_{\max(0,\tau(s)-\delta)}^s \tilde{\mathbb{E}}\left[\left(\psi_r^{ij}\right)^2 \mid \mathcal{Y}_t\right] dr$$

$$\leq k_{a_i}^2 c_3 \delta \left(s - \max(0,\tau(s)-\delta)\right) \sup_{r \in [0,t]} \tilde{\mathbb{E}}\left[\left(\psi_r^{ij}\right)^2 \mid \mathcal{Y}_t\right]$$

Next, from (21.47)

$$\tilde{\mathbb{E}}\left[\left(\int_0^t \Lambda_{ij}(s)\, dW_s^j\right)^2\right] = \int_0^t \tilde{\mathbb{E}}\left[(\Lambda_{ij}(s))^2\right] ds$$

$$\leq k_{a_i}^2 c_3 \delta^2 t \tilde{\mathbb{E}}\left[\sup_{r\in[0,t]} \tilde{\mathbb{E}}\left[(\psi_r^{ij})^2 \middle| \mathcal{Y}_t\right]\right]$$

and the required bound follows by using (21.23).

Lastly, it remains to find a bound for the term $\tilde{\mathbb{E}}\left[\left(\int_0^t \Psi_{ij}(s)\, dW_s^j\right)^2\right]$. If s is such that $\tau(s) \leq \delta$ then $\Psi_{ij}(s)$ is zero. If $\tau(s) > \delta$ then

$$\Psi_{ij}(s) = \tilde{\mathbb{E}}\left[\left(a_i(X_s) - a_i(X_{\tau(s)})\right)\int_0^{\tau(s)-\delta} \psi_r^{ij}\, dW_r^j \middle| \mathcal{Y}_t\right]$$

$$= \tilde{\mathbb{E}}\left[P_{s-\tau(s)+\delta,\delta} a_i\left(X_{\tau(s)-\delta}\right)\int_0^{\tau(s)-\delta} \psi_r^{ij}\, dW_r^j \middle| \mathcal{Y}_t\right]. \quad (21.29)$$

By Itô's rule we have that

$$P_{s-\tau(s)+\delta,\delta} a_i\left(X_{\tau(s)-\delta}\right) = P_{s,\tau(s)} a_i(X_0)$$
$$+ \sum_{u,v=1}^d \int_0^{\tau(s)-\delta} \partial_u P_{s-r,\tau(s)-r} a_i(X_r)\, \sigma_{uv}(X_r)\, dW_r^v,$$

hence,

$$\Psi_{ij}(s) = \sum_{u=1}^d \int_0^{\tau(s)-\delta} \tilde{\mathbb{E}}\left[\partial_u P_{s-r,\tau(s)-r} a_i(X_r)\, \sigma_{uj}(X_r)\, \psi_r^{ij} \middle| \mathcal{Y}_t\right] dr.$$

Next, by (21.22) and Cauchy–Schwarz's inequality

$$\left|\tilde{\mathbb{E}}\left[\partial_u P_{s-r,\tau(s)-r} a_i(X_r) \sigma_{uj}(X_r) \psi_r^{ij} \middle| \mathcal{Y}_t\right]\right| \leq \frac{c_7}{\tau(s)-r} k_\psi(s-\tau(s)) \sqrt{\tilde{\mathbb{E}}\left[(\sigma_{uj}(X_r))^2\right]}$$

$$\times \sqrt{\tilde{\mathbb{E}}\left[(\psi_r^{ij})^2 \middle| \mathcal{Y}_t\right]}.$$

Hence

$$\Psi_{ij}(s)^2 \leq c_7^2 k_\psi \delta^2 \left(\ln\frac{s}{\delta}\right)^2 \sup_{r\in[0,t]} \tilde{\mathbb{E}}\left[(\sigma_{uj}(X_r))^2\right] \sup_{r\in[0,t]} \tilde{\mathbb{E}}\left[(\psi_r^{ij})^2 \middle| \mathcal{Y}_t\right].$$

Finally observe that the Lipschitz property of σ and (21.11) implies that

$$\sup_{r\in[0,t]} \tilde{\mathbb{E}}\left[(\sigma_{uj}(X_r))^2\right] < \infty,$$

which in turn together with (21.23) implies that

$$\tilde{\mathbb{E}}\left[\left(\int_0^t \Psi_{ij}(s)\, dW_s^j\right)^2\right] = \int_0^t \tilde{\mathbb{E}}\left[\left(\Psi_{ij}(s)\right)^2\right] ds$$
$$\leq c_8 \delta^2 \left(\ln \frac{t}{\delta}\right)^2,$$

where

$$c_8 = c_7^2 k_\psi t \sup_{r\in[0,t]} \tilde{\mathbb{E}}\left[\left(\sigma_{uj}(X_r)\right)^2\right] \tilde{\mathbb{E}}\left[\sup_{r\in[0,t]} \tilde{\mathbb{E}}\left[\left(\psi_r^{ij}\right)^2 \Big| \mathcal{Y}_t\right]\right].$$

Since $\lim_{x\to\infty} \frac{\ln x}{x^{\varepsilon/2}} = 0$, there exists a constant $c_9 = c_9(\varepsilon)$ such that

$$\ln\left(\frac{t}{\delta}\right) \leq c_9(\varepsilon)\, t^{\frac{\varepsilon}{2}} \delta^{-\frac{\varepsilon}{2}},$$

which give a bound of order of order $\delta^{2-\varepsilon}$ for the second moment of the last term in (21.25). □

The following example shows that the bound (21.24) cannot be improved to one of order δ^2:

Example 21.1 Assume that $d=1$ and that $f = a_0 = 0$, $\sigma = \varphi = 1$, $\psi: \mathbb{R}^2 \to \mathbb{R}$, $\psi(x,y) = x+y$ and $a_1: \mathbb{R} \to \mathbb{R}$, $a_1(x) = |x|$ for any $x \in \mathbb{R}$. In other words, we wish to approximate

$$\iota \triangleq \tilde{\mathbb{E}}\left[\int_0^t |W_s|\, dY_s \,\Big|\, \mathcal{Y}_t\right]$$

with

$$\iota^n \triangleq \tilde{\mathbb{E}}\left[\sum_{k=1}^n |W_{t_{k-1}}|\left(Y_{t_k} - Y_{t_{k-1}}\right) \Big| \mathcal{Y}_t\right].$$

Let also τ be an equidistant partition with partition mesh $\delta = \frac{T}{n}$. Then there exists constants c_{10}, c_{11} and $\delta_0 > 0$ such that for all $\delta \in (0, \delta_0)$ we have

$$c_{10}\delta^2 \ln\frac{1}{\delta} \leq \tilde{\mathbb{E}}\left[(\iota - \iota^n)^2\right] \leq c_{11}\delta^2 \ln\frac{1}{\delta}. \tag{21.30}$$

where c_{10} and c_{11} are independent of δ.

Proof. We have that

$$\iota - \iota^n = \int_0^t \tilde{\mathbb{E}}[|W_s| | \mathcal{Y}_t] - \tilde{\mathbb{E}}[|W_{[\frac{s}{\delta}]\delta}| | \mathcal{Y}_t]\, dY_s$$
$$= \sqrt{\frac{2}{\pi}} \int_0^t \left(\sqrt{s} - \sqrt{\left[\frac{s}{\delta}\right]\delta}\right) dY_s,$$

and (21.30) follows from the estimate (21.48) proved in the appendix. □

We introduce next the following condition which will replace the cumbersome condition **Ap**:

D2. The function $a = (a_i)_{i=0}^m : \mathbb{R}^d \to \mathbb{R}^{m+1}$ is twice continuously differentiable with bounded first and second order derivatives.

Under this condition, we can improve on the bound obtained in Theorem 21.1:

Theorem 21.2 *Under conditions* **M**, **Lp**, *and* **D2** *there exists a constant* $c_{12} = c_{12}(t)$ *independent of the partition* τ *such that*

$$\tilde{\mathbb{E}}\left[q\left(X_t, \xi, \xi^\tau\right)^2\right] \leq c_{12}\delta^2. \tag{21.31}$$

Proof. By Itô's rule

$$a_i(X_s) - a_i(X_{\tau(s)}) = \int_{\tau(s)}^s Aa_i(X_r)dr + \sum_{u,v=1}^d \int_{\tau(s)}^s \sigma_{uv}(X_r)\partial_u a_i(X_r)dW_r^v. \tag{21.32}$$

From **D2** we get that the functions Aa_i and $\sigma_{uv}\partial_u a_i$ have linear growth (using the Lipschitz property of f and σ and the boundedness of the partial derivatives of a_i). Hence, by (21.11), the random variables $Aa_i(X_r)$ and $\sigma_{uv}(X_r)\partial_u a_i(X_r)$ have moments of all orders and, implicitly, so have both terms on the right-have side of (21.32). Moreover the stochastic integral on the right-hand side of (21.32) is a genuine martingale.

We use next a decomposition similar to (21.25). That is,

$$q\left(X_t, \xi, \xi^\tau\right) = \sum_{i=0}^m \int_0^t \tilde{\mathbb{E}}\left[\eta_i' F_i(s) \middle| \mathcal{Y}_s\right] dY_s^i + \sum_{i=0}^m \sum_{j=1}^d \int_0^t \left(\Lambda_{ij}(s) + \Psi_{ij}(s)\right) dY_s^i, \tag{21.33}$$

where

$$F_i(s) = \tilde{\mathbb{E}}\left[\left(a_i(X_s) - a_i\left(X_{\tau(s)}\right)\right) \middle| \mathcal{F}_0 \vee \mathcal{Y}_t\right]$$

$$= \int_{\tau(s)}^s \tilde{\mathbb{E}}[Aa_i(X_r)|\mathcal{F}_0 \vee \mathcal{Y}_t]dr,$$

$$\Lambda_{ij}(s) = \tilde{\mathbb{E}}\left[\left(a_i(X_s) - a_i\left(X_{\tau(s)}\right)\right) \int_{\tau(s)}^s \psi_r^{ij} dW_r^j \middle| \mathcal{Y}_t\right],$$

$$\Psi_{ij}(s) = \tilde{\mathbb{E}}\left[\left(a_i(X_s) - a_i\left(X_{\tau(s)}\right)\right) \int_0^{\tau(s)} \psi_r^{ij} dW_r^j \middle| \mathcal{Y}_t\right]$$

$$= \tilde{\mathbb{E}}\left[\int_{\tau(s)}^s Aa_i(X_r)dr \int_0^{\tau(s)} \psi_r^{ij} dW_r^j \middle| \mathcal{Y}_t\right]$$

for $i = 0, 1, \ldots, m$ and $j = 1, \ldots, d$. We find bounds of order δ^2 for all the terms in (21.33).

First, by Cauchy–Schwarz inequality

$$\tilde{\mathbb{E}}\left[F_i(s)^2\right] \le (s-\tau(s)) \int_{\tau(s)}^{s} \tilde{\mathbb{E}}[Aa_i(X_r)^2]dr$$

$$\le \delta^2 \sup_{r\in[0,t]} \tilde{\mathbb{E}}\left[(Aa_i(X_r))^2\right],$$

which gives the bound for the first term since $\sup_{r\in[0,t]} \tilde{\mathbb{E}}\left[(Aa_i(X_r))^2\right] < \infty$ by the linear growth of Aa_i and (21.11).

Secondly observe that, using the the Lipschitz property of a_i and (21.12), we get

$$\Lambda_{ij}(s)^2 = \tilde{\mathbb{E}}\left[(a_i(X_s)) - (a_i(X_{\tau(s)})) \int_{\tau(s)}^{s} \psi_r^{ij} dW_r^j \middle| \mathcal{Y}_t\right]^2$$

$$\le \tilde{\mathbb{E}}\left[(a_i(X_s)) - (a_i(X_{\tau(s)}))^2 \middle| \mathcal{Y}_t\right] \tilde{\mathbb{E}}\left[\left(\int_{\tau(s)}^{s} \psi_r^{ij} dW_r^j\right)^2 \middle| \mathcal{Y}_t\right]$$

$$\le k_{a_i}^2 \tilde{\mathbb{E}}\left[|X_s - X_{\tau(s)}|^2\right] \int_{\tau(s)}^{s} \tilde{\mathbb{E}}\left[(\psi_r^{ij})^2 \middle| \mathcal{Y}_t\right] dr$$

$$\le k_{a_i}^2 \delta \int_{\tau(s)}^{s} \tilde{\mathbb{E}}\left[(\psi_r^{ij})^2 \middle| \mathcal{Y}_t\right] dr.$$

Next, from (21.47)

$$\tilde{\mathbb{E}}\left[\left(\int_0^t \Lambda_{ij}(s) dW_s^j\right)^2\right] = \int_0^t \tilde{\mathbb{E}}\left[(\Lambda_{ij}(s))^2\right] ds$$

$$\le k_{a_i}^2 \delta \int_0^t \int_{\tau(s)}^{s} \tilde{\mathbb{E}}\left[(\psi_r^{ij})^2 \middle| \mathcal{Y}_t\right] dr\, ds$$

$$= k_{a_i}^2 \delta \int_0^t (s-\tau(s)) \tilde{\mathbb{E}}\left[(\psi_s^{ij})^2 \middle| \mathcal{Y}_t\right] ds$$

$$\le k_{a_i}^2 \delta^2 \int_0^t \tilde{\mathbb{E}}\left[(\psi_s^{ij})^2 \middle| \mathcal{Y}_t\right] ds$$

and the required bound follows by using (21.20) for $p=2$ (note that (21.23) is no longer available as **AP** is not imposed).

Finally, for the bound on the second moment of $\Psi_{ij}(s)$ observe that

$$\Psi_{ij}(s)^2 = \tilde{\mathbb{E}}\left[\int_{\tau(s)}^{s} Aa_i(X_r)dr \int_0^{\tau(s)} \psi_r^{ij} dW_r^j \middle| \mathcal{Y}_t\right]^2$$

$$\le \tilde{\mathbb{E}}\left[\left(\int_{\tau(s)}^{s} Aa_i(X_r)dr\right)^2 \middle| \mathcal{Y}_t\right] \tilde{\mathbb{E}}\left[\left(\int_0^{\tau(s)} \psi_r^{ij} dW_r^j\right)^2 \middle| \mathcal{Y}_t\right]$$

$$\le (s-\tau(s))^2 \sup_{r\in[0,t]} \tilde{\mathbb{E}}\left[Aa_i(X_r)\right] \int_0^t \tilde{\mathbb{E}}\left[(\psi_r^{ij})^2 \middle| \mathcal{Y}_t\right] dr.$$

Again, the bound for the second moment of $\Psi_{ij}(s)$ follows from (21.20) for $p = 2$. □

The following example shows that the bound (21.31) is sharp. It also shows that any additional smoothness for the function a will not improve the bound:

Example 21.2 Assume that $d = 1$ and that $f = a_0 = 0$, $\sigma = \varphi = 1$, $\psi : \mathbb{R}^2 \to \mathbb{R}$, $\psi(x, y) = y$ and $a_1 : \mathbb{R} \to \mathbb{R}$, $a_1(x) = \cos x$, for any $x \in \mathbb{R}$. In other words, we wish to approximate

$$\eta \triangleq \tilde{\mathbb{E}}\left[\int_0^t \cos W_s\, dY_s \,\Big|\, \mathcal{Y}_t\right]$$

with

$$\eta^n \triangleq \tilde{\mathbb{E}}\left[\sum_{k=1}^n \cos(W_{t_{k-1}})(Y_{t_k} - Y_{t_{k-1}}) \,\Big|\, \mathcal{Y}_t\right].$$

Let also τ be an equidistant partition with partition mesh $\delta = \frac{T}{n}$. Then

$$\tilde{\mathbb{E}}\left[q(X_t, \xi, \xi^\tau)^2\right] \geq \frac{2e^{-\frac{t}{2}}}{3}\delta^2.$$

Proof. In this case

$$q(X_t, \xi, \xi^\tau) = \int_0^t \tilde{\mathbb{E}}\left[\cos(W_s) - \cos\left(W_{[\frac{s}{\delta}]\delta}\right) \,\Big|\, \mathcal{Y}_t\right] dY_s$$

$$= \int_0^t \left(e^{-\frac{1}{2}s} - e^{-\frac{1}{2}[\frac{s}{\delta}]\delta}\right) dY_s,$$

since $\mathbb{E}[\cos(W_r)] = e^{-\frac{1}{2}r}$. Also

$$e^{-\frac{1}{2}[\frac{s}{\delta}]\delta} - e^{-\frac{1}{2}s} = 2\int_{[\frac{s}{\delta}]\delta}^s e^{-\frac{1}{2}r} dr \geq 2\left(s - \left[\frac{s}{\delta}\right]\delta\right)e^{-\frac{t}{2}}.$$

Hence

$$\int_0^t \left(e^{-\frac{1}{2}s} - e^{-\frac{1}{2}[\frac{s}{\delta}]\delta}\right)^2 ds \geq 2e^{-\frac{t}{2}}\int_0^t \left(s - \left[\frac{s}{\delta}\right]\delta\right)^2 ds = \frac{2e^{-\frac{t}{2}}}{3}\delta^2,$$

which give the required lower bound. □

21.3 Application to the approximation of the conditional distribution of the signal

We assume as in the previous section that τ is a given partition of the time interval $[0, t]$,

$$\tau = \{0 = t_0 < \cdots < t_k < \cdots < t_n = t\}$$

and denote by $\delta = \max_{k=1,\ldots,n} (t_k - t_{k-1})$ the size of the partition. Let $Z = \{Z_t, t > 0\}$ be the $\mathcal{F}_t$-adapted martingale defined as in (21.6) and $\rho = \{\rho_t, t \geq 0\}$ be the *unnormalized* conditional distribution the signal defined by (21.7). That is,

$$\rho_t(\varphi) = \tilde{\mathbb{E}}\left[\varphi(X_t) Z_t \mid \mathcal{Y}_t\right]$$

for any φ a Borel-measurable function such that $\tilde{\mathbb{E}}\left[\varphi(X_t) Z_t\right] < \infty$. Next, let Z_t^τ be defined as in (21.9), i.e.,

$$Z_t^\tau = \exp\left(\sum_{i=1}^m \int_0^t h^i(X_{\tau(s)}) \, dY_s^i - \frac{1}{2} \sum_{i=1}^m \int_0^t h^i(X_{\tau(s)})^2 \, ds\right)$$

and define $\rho_t^\tau(\varphi)$ as in (21.10), that is

$$\rho_t^\tau \varphi = \tilde{\mathbb{E}}\left[\varphi(X_t) Z_t^\tau \mid \mathcal{Y}_t\right]$$

for any φ a Borel-measurable function such that $\tilde{\mathbb{E}}\left[\varphi(X_t) Z_t^\tau\right] < \infty$. Finally define $\pi_t^\tau(\varphi)$ by the formula $\pi_t^\tau(\varphi) = \rho_t^\tau(\varphi) / \rho_t^\tau(1)$.

We will impose the following condition:

FLp. There exists a constant $p > 2$ such that $\tilde{\mathbb{E}}\left[Z_t^p\right] < \infty$ and $\sup_\tau \tilde{\mathbb{E}}\left[(Z_t^\tau)^p\right] < \infty$.

Remark 21.5 Following from Remark 21.2, **FLp** holds true if the sensor function h is bounded. If h is unbounded, but it has linear growth, then the condition is satisfied if X has exponential moments uniformly bounded on $[0, t]$.

Theorem 21.3 *Assume that conditions* **M** *and* **FLp** *hold true and that the functions* h_i, $i = 1, \ldots, m$ *are Lipschitz. Then, if φ has polynomial growth, there exists a constant $c_{13} = c_{13}(\varphi, t)$ independent of τ such that*

$$\tilde{\mathbb{E}}\left[|\rho_t^\tau \varphi - \rho_t \varphi|^2\right] \leq c_{13} \delta.$$

Moreover, if $\sup_\tau \tilde{\mathbb{E}}\left[(\pi_t^\tau(\varphi))^2\right] < \infty$, *then*

$$\mathbb{E}\left[|\pi_t^\tau \varphi - \pi_t \varphi|\right] \leq c_{14} \sqrt{\delta}, \qquad (21.34)$$

where, again, $c_{14} = c_{14}(\varphi, t)$ is a constant independent of δ.

Proof. Observe first that, by using (21.12) and Cauchy–Schwarz's inequality

$$\tilde{\mathbb{E}}\left[|h_i^2(X_s) - h_i^2(X_{\tau(s)})|^p\right] \leq \tilde{\mathbb{E}}\left[|h_i(X_s) + h_i(X_{\tau(s)})|^{2p}\right]^{\frac{1}{2}}$$

$$\times \tilde{\mathbb{E}}\left[|h_i(X_s) - h_i(X_{\tau(s)})|^{2p}\right]^{\frac{1}{2}} \quad (21.35)$$

$$\leq 2^p k_{h_i}^p \sup_{s \in [0,T]} \tilde{\mathbb{E}}\left[|h_i(X_s)|^{2p}\right]^{\frac{1}{2}} \tilde{\mathbb{E}}\left[|X_s - X_{\tau(s)}|^{2p}\right]^{\frac{1}{2}}$$

$$\leq c_{15} \delta^{\frac{p}{2}}, \qquad (21.36)$$

where, by (21.11),

$$c_{15} = 2^p c_3^{\frac{1}{2}} k_{h_i}^p \sup_{s \in [0,T]} \tilde{\mathbb{E}}\left[|h_i(X_s)|^{2p}\right]^{\frac{1}{2}} < \infty.$$

We apply the same proof as that of Proposition 21.1 with $a_i = h_i$ $i = 1, \ldots, m$ and $a_0 = \frac{1}{2}(h_1^2 + \cdots + h_m^2)$. Even though a_0 is not Lipschitz, the proof will still be valid by using the bound (21.36). Hence we only need to check that condition **Lp** is satisfied. In this case it amounts to showing that

$$\sup_{\tau} \sup_{s \in [0,1]} \tilde{\mathbb{E}}\left[|\varphi(X_t) \exp(s\xi + (1-s)\xi^\tau)|^q\right] < \infty$$

for $q \in (2, p)$, where

$$\xi = \sum_{i=1}^{m} \int_0^t h^i(X_s) \, dY_s^i - \frac{1}{2} \sum_{i=1}^{m} \int_0^t h^i(X_s)^2 \, ds$$

$$\xi^\tau = \sum_{i=1}^{m} \int_0^t h^i(X_{\tau(s)}) \, dY_s^i - \frac{1}{2} \sum_{i=1}^{m} \int_0^t h^i(X_{\tau(s)})^2 \, ds.$$

This follows immediately from (21.11), the polynomial growth of φ and condition **FLp** by observing that

$$\exp(s\xi + (1-s)\xi^\tau) \leq \max(\exp(\xi), \exp(\xi^\tau))$$
$$\leq \exp(\xi) + \exp(\xi^\tau) = Z_t + Z_t^\tau.$$

To deduce the bound (21.34) observe that

$$\pi_t^\tau \varphi - \pi_t \varphi = \frac{1}{\rho_t 1} \frac{\rho_t^\tau \varphi}{\rho_t^\tau 1} (\rho_t 1 - \rho_t^\tau 1) + \frac{1}{\rho_t 1} (\rho_t \varphi - \rho_t^\tau \varphi).$$

hence

$$\mathbb{E}\left[|\pi_t^\tau \varphi - \pi_t \varphi|\right] = \tilde{\mathbb{E}}\left[\frac{\tilde{Z}_t}{\rho_t 1}\left(\pi_t^\tau \varphi (\rho_t 1 - \rho_t^\tau 1) + |\rho_t \varphi - \rho_t^\tau \varphi|\right)\right]$$

$$= \tilde{\mathbb{E}}\left[\frac{\tilde{\mathbb{E}}[\tilde{Z}_t | \mathcal{Y}_t]}{\rho_t 1}\left(\pi_t^\tau \varphi (\rho_t 1 - \rho_t^\tau 1) + |\rho_t \varphi - \rho_t^\tau \varphi|\right)\right]$$

$$\leq \left(\tilde{\mathbb{E}}\left[(\pi_t^\tau(\varphi))^2\right] \tilde{\mathbb{E}}\left[|\rho_t^\tau 1 - \rho_t 1|^2\right]\right)^{\frac{1}{2}} + \left(\tilde{\mathbb{E}}\left[|\rho_t^\tau \varphi - \rho_t \varphi|^2\right]\right)^{\frac{1}{2}},$$

where we use the fact that $\rho_t 1 = \tilde{\mathbb{E}}[\tilde{Z}_t | \mathcal{Y}_t]$. Hence (21.34) holds with

$$c_{14}(\varphi, t) = c_{13}(1, t)^{\frac{1}{2}} \sup_\tau \tilde{\mathbb{E}}\left[(\pi_t^\tau(\varphi))^2\right] + c_{13}(\varphi, t)^{\frac{1}{2}}.$$

□

Remark 21.6 The condition $\sup_\tau \tilde{\mathbb{E}}\left[(\pi_t^\tau(\varphi))^2\right] < \infty$ required to deduce the bound (21.34) holds true if φ is a bounded Borel-measurable function. If φ is unbounded, then observe that

$$\tilde{\mathbb{E}}\left[\left(\pi_t^\tau(\varphi)\right)^2\right] = \tilde{\mathbb{E}}\left[\left(\tilde{\mathbb{E}}\left[\frac{\varphi(X_t)\,\tilde{Z}_t^\tau}{\tilde{\mathbb{E}}[\tilde{Z}_t^\tau|\mathcal{Y}_t]}\Big|\mathcal{Y}_t\right]\right)^2\right]$$

$$\leq \tilde{\mathbb{E}}\left[\varphi(X_t)^2 \exp 2\left(\xi^\tau - \tilde{\mathbb{E}}[\xi^\tau|\mathcal{Y}_t]\right)\right].$$

Hence, again $\sup_\tau \tilde{\mathbb{E}}\big[\left(\pi_t^\tau(\varphi)\right)^2\big] < \infty$ holds true if the sensor function h is bounded. If h is unbounded, but it has linear growth, then the condition is satisfied if X_0 has exponential moments, σ is bounded and t is sufficiently small.

The following theorem is an immediate corollary of Theorem 21.1:

Theorem 21.4 *Assume that conditions* **M**, **FLp**, *and inequalities (21.21) and (21.22) hold true. Assume also that the functions φ and h_i $i = 1, \ldots, m$ are Lipschitz. Then there exists $\delta_0 > 0$ such that for all $\delta \in (0, \delta_0)$ and $\varepsilon \in (0, 1)$, there exists a constant $c_{16} = c_{16}(\varphi, t, \delta_0, \varepsilon)$ independent of the partition of τ such that*

$$\tilde{\mathbb{E}}\big[\,|\rho_t^\tau \varphi - \rho_t \varphi|^2\,\big] \leq c_{16}\delta^{2-\varepsilon}. \tag{21.37}$$

Moreover, if $\sup_\tau \tilde{\mathbb{E}}\big[\left(\pi_t^\tau(\varphi)\right)^2\big] < \infty$, then for all $\delta \in (0, \delta_0)$ and $\varepsilon \in (0, 1)$, there exists a constant $c_{17} = c_{17}(\varphi, t, \delta_0, \varepsilon)$ independent of the partition of τ such that

$$\mathbb{E}\big[|\pi_t^\tau \varphi - \pi_t \varphi|\big] \leq c_{17}\delta^{1-\varepsilon}. \tag{21.38}$$

Proof. As above condition **FLp** implies condition **Lp**. Also, the bound (21.23) holds true following from Lemma 21.5. Hence (21.37) follows from Theorem 21.1. The bound (21.38) follows in the same manner as (21.34). □

Similarly, the following theorem is a corollary of Theorem 21.2:

Theorem 21.5 *Assume that conditions* **M**, **FLp** *are satisfied and the function h_i $i = 1, \ldots, m$ are twice continuous differentiable with bounded first and second derivatives. Then, if φ has polynomial growth, there exists a constant $c_{18} = c_{18}(\varphi, t)$ independent of τ such that*

$$\tilde{\mathbb{E}}\big[\,|\rho_t^\tau \varphi - \rho_t \varphi|^2\,\big] \leq c_{18}\delta^2. \tag{21.39}$$

Moreover, if $\sup_\tau \tilde{\mathbb{E}}\big[\left(\pi_t^\tau(\varphi)\right)^2\big] < \infty$, then

$$\mathbb{E}\big[|\pi_t^\tau \varphi - \pi_t \varphi|\big] \leq c_{19}\delta, \tag{21.40}$$

where, again, $c_{19} = c_{19}(\varphi, t)$ is a constant independent of δ.

Proof. We apply the same proof as that of Proposition 21.2 with $a_i = h_i$ $i = 1, \ldots, m$ and $a_0 = \frac{1}{2}(h_1^2 + \cdots + h_m^2)$. Again, even though a_0 is not Lipschitz, one uses the bound (21.36). Moreover, the functions Aa_i and $\sigma_{uv}\partial_u a_i$ have this time quadratic growth and, by (21.11), the random variables $Aa_i(X_r)$ and

$\sigma_{uv}(X_r)\partial_u a_i(X_r)$ have moments of all orders. Similarly, the stochastic integral on the right-hand side of (21.32) is a genuine martingale and

$$\sup_{r\in[0,t]} \tilde{\mathbb{E}}\left[(Aa_i(X_r))^2\right] < \infty.$$

These are all the ingredients required for the proof to work. Finally, the bound (21.40) follows in the same manner as (21.34). □

Remark 21.7 The first part of Theorem 21.5, i.e., the bound (21.39), recovers the result of Theorem 1 in [28]. However, Theorem 1 in [28] is proved under the condition

$$\tilde{\mathbb{E}}\left[\exp(1+\varepsilon)t H^2(X)\right] < \infty, \tag{21.41}$$

where

$$H^2(M) = \sup\left\{\sum_{i=1}^m |h_i^2(x)|, \; |x| \le M\right\}$$

and $X^* = (X_i^*)_{i=1,\ldots,m}$ is the random vector with entries $X_i^* = \sup_{s\in[0,t]} |X_s^i|$. We have, for $p > 2$

$$\tilde{\mathbb{E}}[Z_t^p] = \tilde{\mathbb{E}}\left[\exp\left(\frac{p^2-p}{2}\sum_{i=1}^m \int_0^t h_i^2(X_s)\,ds\right)\right]$$

$$\le \tilde{\mathbb{E}}\left[\exp\left(\frac{p^2-p}{2} t H^2(X)\right)\right]$$

and by choosing

$$p = \frac{1+\sqrt{9+8\varepsilon}}{2} > 2$$

so that $p^2 - p = 2(1+\varepsilon)$, we get that $\tilde{\mathbb{E}}[Z_t^p] < \infty$. Similarly

$$\tilde{\mathbb{E}}\left[(Z_t^\tau)^p\right] < \tilde{\mathbb{E}}\left[\exp(1+\varepsilon)t H^2(X)\right]$$

for any partition τ, so (21.41) implies condition **FLp**.

21.4 Appendix

We include here a number of technical lemmas required in the proofs of the main results of the chapter. We begin with the proof of Remark 21.1:

Lemma 21.1 *For any $p > 0$, there exists a constant $c_4 = c_4(p,t)$ independent of $i = 0, 1, \ldots, m$ and $\delta > 0$ such that*

$$\tilde{\mathbb{E}}[|\xi_i - \xi^\tau|^p] \le c_4 \delta^{\frac{p}{2}}. \tag{21.42}$$

Moreover, if $\tilde{\mathbb{E}}[|\nu|^q] < \infty$, then for any $p \in (0, q)$, there exists a constant $c_5 = c_5(p, t, \nu)$ independent of $i = 0, 1, \ldots, m$ and $\delta > 0$ such that

$$\tilde{\mathbb{E}}\left[|\nu(\xi_i - \xi^\tau)|^p\right] \le c_5 \delta^{\frac{p}{2}}. \tag{21.43}$$

Proof. For $i = 0$, we have, by Jensen's inequality and the Lipschitz property[3] of a_0 that

$$\frac{1}{t^p}|\xi_0 - \xi_0^\tau|^p = \left(\frac{1}{t}\left|\int_0^t (a_0(X_s) - a_0(X_{\tau(s)}))ds\right|\right)^p$$

$$\le \frac{1}{t}\int_0^t |a_0(X_s) - a_0(X_{\tau(s)})|^p ds$$

$$\le \frac{k_{a_0}^p}{t}\int_0^t |X_s - X_{\tau(s)}|^p ds$$

and the inequality (21.42) follows from (21.12). For $i > 0$, we have, by Burkholder–Davis–Gundy's inequality ($c_{20} = c_{20}(p)$ is the constant appearing in the Burkholder–Davis–Gundy inequality) and the Lipschitz property of a_i that

$$\tilde{\mathbb{E}}\left[|\xi_i - \xi_i^\tau|^p\right] = \tilde{\mathbb{E}}\left[\left|\int_0^t (a_i(X_s) - a_i(X_{\tau(s)}))dY_s^i\right|^p\right]$$

$$\le c_{20}\tilde{\mathbb{E}}\left[\left|\int_0^t (a_i(X_s) - a_i(X_{\tau(s)}))^2 ds\right|^{\frac{p}{2}}\right]$$

$$\le c_7 k_{a_i}^p t^{\frac{p}{2}-1}\tilde{\mathbb{E}}\left[\left|\int_0^t |X_s - X_{\tau(s)}|^p ds\right|\right].$$

Then, as above, (21.42) follows from (21.12). To obtain the inequality (21.43) observe that, by Hölder's inequality, we have

$$\tilde{\mathbb{E}}\left[\left|\nu\left(\xi_i - \xi_i^\delta\right)\right|^p\right] \le \tilde{\mathbb{E}}\left[(|\nu|^p)^{\frac{q}{p}}\right]^{\frac{p}{q}} \tilde{\mathbb{E}}\left[\left(\left|\left(\xi_i - \xi_i^\delta\right)\right|^p\right)^{\frac{1}{q'}}\right]^{q'}$$

$$= \tilde{\mathbb{E}}[|\nu|^q]^{\frac{p}{q}} \tilde{\mathbb{E}}\left[\left|\xi_i - \xi_i^\delta\right|^{\frac{p}{q'}}\right]^{q'}$$

$$\le \tilde{\mathbb{E}}[|\nu|^q]^{\frac{p}{q}} \tilde{\mathbb{E}}\left[\left|\xi_i - \xi_i^\delta\right|^{\frac{p}{q'}}\right]^{q'},$$

where $q' = 1 - \frac{p}{q}$. The result then follows from (21.13). □

The following lemma gives the proof of Remark 21.3:

Lemma 21.2 *The random variables $\eta_i \in L_p(\Omega, \mathcal{F}_t, \tilde{P})$ defined in (21.18) admit the following integral representation*

$$\eta_i = \tilde{\mathbb{E}}\left[\eta_i \mid \mathcal{F}_0 \vee \mathcal{Y}_t\right] + \sum_{j=1}^d \int_0^t \psi_s^{ij} dW_s^j, \tag{21.44}$$

where $\psi^{ij} = \{\psi_s^{ij}, s \in [0,t]\}$ are progresively measurable, $\mathcal{F}_s \vee \mathcal{Y}_t$-adapted processes such that

$$\tilde{\mathbb{E}}\left[\left(\int_0^t (\psi_s^{ij})^2 \, ds\right)^{\frac{p}{2}}\right] < \infty \qquad (21.45)$$

for $i = 0, 1, \ldots, m$ and $j = 1, \ldots, d$.

Proof. We use a standard density argument. Observe that the random variables η_i are measurable with respect to the augmentation of the σ-field

$$\sigma(W_s, s \in [0,t]) \vee \sigma(X_0) \vee \mathcal{Y}_t.$$

Hence there exists an approximation sequence of simple random variables η_i^n which converges to η_i in the L_p-norm of the form

$$\eta_i^n = \sum_{k=1}^{m_n} a_k^{n,i} I_{A_k^{n,i}} I_{B_k^{n,i}},$$

where $A_k^n \in \sigma(W_s, s \in [0,t])$ and $B_k^n \in \sigma(X_0) \vee \mathcal{Y}_t$ for $k = 1, \ldots, m_n$. Next, using the classical representation of Brownian functionals as stochastic integrals (see, for example, proposition 4.18, p. 185 in [18]) for each $A_k^{n,i}$ there exists a progresively measurable process $\psi^{ij,n,k}$, adapted to the augmented filtration of the Brownian motion W, such that

$$I_{A_k^{n,i}} = \tilde{P}\left(A_k^{n,i}\right) + \sum_{j=1}^{d} \int_0^t \psi_s^{ij,n,k} \, dW_s^j.$$

Using now the fact that $\sigma(W_s, s \in [0,t])$ and $\sigma(X_0) \vee \mathcal{Y}_t$ are mutually independent σ-fields it follows that

$$\eta_i^n = \bar{\eta}_i^n + \sum_{j=1}^{d} \int_0^t \psi_s^{ij,n} \, dW_s^j, \qquad (21.46)$$

where

$$\bar{\eta}_i^n = \sum_{k=1}^{m_n} a_k^{n,i} I_{B_k^{n,i}} \tilde{P}\left(A_k^{n,i}\right) = \tilde{\mathbb{E}}\left[\eta_i^n \mid \mathcal{F}_0 \vee \mathcal{Y}_t\right]$$

$$\psi_s^{ij,n} = \sum_{k=1}^{m_n} a_k^{n,i} I_{B_k^{n,i}} \psi_s^{ij,n,k}.$$

Obviously $\psi^{ij,n}$ are progresively measurable, $\mathcal{F}_s \vee \mathcal{Y}_t$-adapted processes. Next, (η_i^n) and $(\bar{\eta}_i^n)$ converges in L_p for $p > 2$, therefore for each $i = 0, 1, \ldots, m$, $j = 1, \ldots, d$ it follows that $\int_0^t \psi_s^{ij,n} \, dW_s^j$ is a Cauchy sequence in the L_2 norm. Hence by the Itô isometry

$$\lim_{n_1,n_2\to\infty}\tilde{\mathbb{E}}\left[\int_0^t \left(\psi_s^{ij,n_1}-\psi_s^{ij,n_2}\right)^2 ds\right]=0,$$

which, in turn, implies that there exist progressively measurable, $\mathcal{F}_s \vee \mathcal{Y}_t$-adapted processes ψ^{ij} such that $\psi^{ij,n}$ converges to ψ^{ij} in $L_p(\Omega \times [0,t], \mathcal{F}_t \otimes \mathcal{B}[0,t], \tilde{P} \otimes \lambda)$ and, by taking the limit in (21.46), we obtain (21.44).[4] Finally, the square integrability property (21.45) is an immediate consequence of Burkholder–Davis–Gundy inequality (see, for example, theorem 3.28, p. 166 in [18]). □

The following lemma provides some basic estimates used in the proof of Theorem 21.1 and Example 21.1.

Lemma 21.3 *Let*

$$\tau = \{0 = t_0 < \cdots < t_k < \cdots < t_n = t\}$$

be a given partition of the time interval $[0,t]$ and denote by

$$\delta = \max_{k=1,\ldots,n}(t_k - t_{k-1})$$

the size of the partition. Then there exists δ_0 such that for all $\delta \in (0,\delta_0)$ and $0 < \varepsilon < 1$, there exists a constant $c_{21} = c_{21}(\delta_0, \varepsilon)$ independent of the partition τ such that

$$\int_0^t \left(\sqrt{s} - \sqrt{\tau(s)}\right)^2 ds \leq c_{21}\delta^{2-\varepsilon}, \tag{21.47}$$

where $\tau(s)$ is the largest element of the partition smaller than or equal to s, that is,

$$\tau(s) = t_{k-1}, \quad s \in [t_{k-1}, t_k), \quad k = 1, \ldots, n.$$

In particular, if τ is an equidistant partition, (that is $t_k = k\delta$, $k = 1, \ldots, n$, then there exists constants $c_{22}, c_{23}, \delta_0 > 0$ such that for all $\delta \in (0, \delta_0)$ we have

$$c_{22}\delta^2 \ln \frac{1}{\delta} \leq \int_0^t \left(\sqrt{s} - \sqrt{\tau(s)}\right)^2 ds \leq c_{23}\delta^2 \ln \frac{1}{\delta}. \tag{21.48}$$

Proof. To avoid analysing the trivial cases, let us assume that $\delta < \frac{t}{3}$, in other words, we divide the interval $[0,t]$ in at least 3 subintervals. We have

$$\int_0^t \left(\sqrt{s} - \sqrt{\tau(s)}\right)^2 ds \leq \sum_{k=0}^{\left[\frac{t}{\delta}\right]} \int_{k\delta}^{(k+1)\delta} \left(\sqrt{s} - \sqrt{\tau(s)}\right)^2 ds.$$

The first two terms of the sum have the upper bound

$$\int_0^{2\delta} \left(\sqrt{s} - \sqrt{\tau(s)}\right)^2 ds \leq \int_0^{2\delta} \left(\sqrt{s}\right)^2 ds = 2\delta^2.$$

Then, for $k \geq 2$,

$$\int_{k\delta}^{(k+1)\delta} \left(\sqrt{s} - \sqrt{\tau(s)}\right)^2 ds \leq \int_{k\delta}^{(k+1)\delta} \left(\sqrt{s} - \sqrt{(k-1)\delta}\right)^2.$$

Since

$$\frac{\sqrt{s} - \sqrt{(k-1)\delta}}{s - (k-1)\delta} \leq \frac{1}{2\sqrt{(k-1)\delta}},$$

it follows that

$$\sum_{k=2}^{\left[\frac{t}{\delta}\right]} \int_{k\delta}^{(k+1)\delta} \left(\sqrt{s} - \sqrt{\tau(s)}\right)^2 ds \leq \frac{7\delta^2}{12} \sum_{k=2}^{\left[\frac{t}{\delta}\right]} \frac{1}{k-1}. \tag{21.49}$$

Next for $\delta < \delta_0 = \min\left(\frac{1}{t}, \frac{t}{3}\right)$, we get that

$$\sum_{k=2}^{\left[\frac{t}{\delta}\right]} \frac{1}{k-1} \leq \int_1^{\left[\frac{t}{\delta}\right]-1} \frac{1}{x} dx + 1 = \ln\left(\left[\frac{t}{\delta}\right] - 1\right) + 1 < 2\ln\frac{t}{\delta} \leq 4\ln\frac{1}{\delta} \tag{21.50}$$

and since $\lim_{x \to \infty} \frac{\ln x}{x^\varepsilon} = 0$, there exists a constant $c_{24} = c_{24}(\varepsilon)$ such that

$$\sum_{k=2}^{\left[\frac{t}{\delta}\right]} \frac{1}{k-1} \leq c_{24} \delta^{-\varepsilon}. \tag{21.51}$$

The estimate (21.47) results from (21.49), (21.50), and (21.51). For the estimate (21.48) we observe that

$$\frac{\delta^2}{12} \sum_{k=1}^{n-1} \frac{1}{k+1} \leq \sum_{k=1}^{n-1} \int_{k\delta}^{(k+1)\delta} (\sqrt{s} - \sqrt{k\delta})^2 ds \leq \frac{\delta^2}{12} \sum_{k=1}^{n-1} \frac{1}{k},$$

since

$$\frac{1}{2\sqrt{(k+1)\delta}} \leq \frac{\sqrt{s} - \sqrt{k\delta}}{s - k\delta} \leq \frac{1}{2\sqrt{k\delta}}.$$

Next, as above, there exists $\delta_0 > 0$ such that for all $\delta \in (0, \delta_0)$, we get that

$$\sum_{k=1}^{n-1} \frac{1}{k} \leq \int_1^{n-1} \frac{1}{x} dx + 1 = \ln(n-1) + 1 < 2\ln n = 2\ln\frac{t}{\delta} \leq 4\ln\frac{1}{\delta}$$

$$\sum_{k=1}^{n-1} \frac{1}{k+1} \geq \int_2^{n+2} \frac{1}{x} dx = \ln\left(\frac{t}{\delta} + 2\right) - \ln 2 \geq \frac{1}{2} \ln\frac{1}{\delta}$$

and the result follows. □

The following lemma provides proof for Remark 21.4:

Lemma 21.4 *If the functions $f : \mathbb{R}^d \to \mathbb{R}^d$, $\sigma = \left(\sigma^i\right)_{i=1}^d : \mathbb{R}^d \to \mathbb{R}^{d \times d}$ are bounded continuous with bounded partial derivatives of all orders and the vector fields $\left(\sigma^i\right)_{i=1}^d$ satisfies the Hörmander condition, then inequalities (21.21) and (21.22) are satisfied.*

Proof. In particular, the vector fields $\left(\sigma^i\right)_{i=1}^d$ satisfy the UFG condition[5] as introduced by Kusuoka and Stroock in [22], hence indeed $P_s \psi$ is infinitely differentiable and there exists two constants $c_{25} = c_{25}(t)$ and $c_{26} = c_{26}(t)$ such that

$$\sup_{x \in \mathbb{R}^d} |\partial_i P_s \psi(x)| < c_{25} k_\psi, \tag{21.52}$$

$$\sup_{x \in \mathbb{R}^d} |\partial_i \partial_j P_s \psi(x)| < \frac{c_{26} k_\psi}{\sqrt{s}} \tag{21.53}$$

for $s \in [0, t]$. See also corollary 5.4 in [8] for details. The bounds (21.52) and (21.53) together with the boundedness of f and σ imply (21.21) as

$$|P_{a,b}\psi(x)| = \left| \int_b^a A P_s \psi(x) \, ds \right|$$

$$\leq c_{27} k_\psi \int_b^a \frac{1}{\sqrt{s}} ds$$

Also, since for any continuous bounded function φ, there exists a constant $c_{28} = c_{28}(t)$ such that (again, see corollary 5.4 in [8])

$$\sup_{x \in \mathbb{R}^d} |\partial_i P_s \varphi(x)| < \frac{c_{28}}{\sqrt{s}} \sup_{x \in \mathbb{R}^d} |\varphi(x)| \tag{21.54}$$

and $P_{a,b}\psi = P_{\frac{b}{2}} P_{a-\frac{b}{2},\frac{b}{2}} \psi$ we get that

$$\sup_{x \in \mathbb{R}^d} |\partial_i P_{a,b}\psi(x)| \leq \frac{c_{28}}{\sqrt{b/2}} \sup_{x \in \mathbb{R}^d} \left| P_{a-\frac{b}{2},\frac{b}{2}} \psi(x) \right|$$

$$\leq \frac{c_{28} c_{27} k_\psi}{\sqrt{b/2}} \int_{b/2}^{a-\frac{b}{2}} \frac{1}{\sqrt{s}} ds$$

$$\leq \frac{c_{28} c_{27} k_\psi}{\sqrt{b/2}} \int_{b/2}^{a-\frac{b}{2}} \frac{1}{\sqrt{b/2}} ds$$

$$= \frac{2 c_{28} c_{27} k_\psi}{b} (b - a),$$

which implies (21.22). □

Let $\eta = (\eta_i)_{i=0}^m$ be the random vector defined in (21.18) which, under condition **Lp** has the decomposition (21.19) as proved in Lemma 21.2. That is

$$\eta_i = \int_0^1 \varphi(X_t) \partial_i \psi \left(s\xi + (1-s)\xi^\delta\right) ds$$

$$= \tilde{\mathbb{E}}\left[\eta_i \mid \mathcal{F}_0 \vee \mathcal{Y}_t\right] + \sum_{j=1}^d \int_0^t \psi_s^{ij} dW_s^j.$$

The following lemma shows that condition (21.23) is satisfied if $\psi : \mathbb{R}^{m+1} \to \mathbb{R}$ is the exponential function, as is the case encountered within the filtering context.

Lemma 21.5 *If the function $\psi : \mathbb{R}^{m+1} \to \mathbb{R}$ is given by*

$$\psi(x_0, \ldots, x_m) = \exp(x_0 + \cdots + x_m), \quad (x_0, \ldots, x_m) \in \mathbb{R}^{m+1}$$

and the function $\varphi : \mathbb{R}^m \to \mathbb{R}$ is Lipschitz, then condition (21.23) holds true. That is, the processes ψ^{ij} appearing in the decomposition (21.19) can be chosen to satisfy

$$\tilde{\mathbb{E}}\left[\sup_{s\in[0,t]} (\psi_s^{ij})^2 \,\Big|\, \mathcal{Y}_t\right] < \infty.$$

Proof. Let us observe that the processes appearing in the decomposition (21.19) are not unique. Indeed any other progressively measurable $\mathcal{F}_s \vee \mathcal{Y}_t$-adapted processes $\bar{\psi}^{ij} = \left\{\bar{\psi}_s^{ij}, s \in [0, t]\right\}$ such that

$$\left\{(\omega, s) \in \Omega \times [0, t], \psi_s^{ij}(\omega) \neq \bar{\psi}_s^{ij}(\omega)\right\}$$

is a null set in $(\Omega \times [0, t], \mathcal{F}_t \otimes \mathcal{B}[0, t], \tilde{P} \otimes \lambda)$, can replace the processes ψ^{ij} in the decomposition (21.19). Let

$$\bar{\psi}_s^{ij}(\omega) = \text{sgn}\left(\psi_s^{ij}(\omega)\right) \sqrt{\liminf_{n\to\infty} n \int_{\max(s-\frac{1}{n},0)}^s \left(\psi_r^{ij}\right)^2 dr},$$

hence

$$\bar{\psi}_s^{ij}(\omega)^2 = \liminf_{n\to\infty} n \int_{\max(s-\frac{1}{n},0)}^s \left(\psi_r^{ij}\right)^2 dr. \tag{21.55}$$

Then for each $\omega \in \Omega$, the cross-section

$$\left\{s \in [0, t], \psi_s^{ij}(\omega) \neq \bar{\psi}_s^{ij}(\omega)\right\}$$

is $\mathcal{B}[0, t]$-measurable and, by the fundamental theorem of calculus, it has Lebesgue measure 0. Hence we can assume that the processes $\bar{\psi}^{ij}$ satisfy (21.55). Hence by the Fatou's lemma,

$$\sum_{j=1}^d \tilde{\mathbb{E}}\left[(\psi^{ij})^2 \,\Big|\, \mathcal{Y}_t\right] \leq \liminf_{n\to\infty} n \tilde{\mathbb{E}}\left[\int_{\max(s-\frac{1}{n},0)}^s \left(\psi_r^{ij}\right)^2 dr \,\Big|\, \mathcal{Y}_t\right] = \liminf_{n\to\infty} n F^{s,n}$$

where

$$F^{s,n} = \tilde{\mathbb{E}}\left[\tilde{\mathbb{E}}\left[\left(\tilde{\mathbb{E}}\left[\eta_i \mid \mathcal{F}_s \vee \mathcal{Y}_t\right] - \tilde{\mathbb{E}}\left[\eta_i \mid \mathcal{F}_{\max(s-\frac{1}{n},0)} \vee \mathcal{Y}_t\right]\right)^2 \middle| \mathcal{Y}_t\right]\right].$$

To simplify notation, let us assume that n is always chosen so that $s - \frac{1}{n} > 0$. We now follow the proof of proposition 1 in [28]. If h is another $\mathcal{F}_{s-\frac{1}{n}} \vee \mathcal{Y}_t$-adapted random variable, then, by the properties of the conditional expectation, we have

$$F^{s,n} \leq \tilde{\mathbb{E}}\left[\left(\tilde{\mathbb{E}}\left[\eta_i \mid \mathcal{F}_s \vee \mathcal{Y}_t\right] - h\right)^2 \middle| \mathcal{Y}_t\right]. \tag{21.56}$$

Let us introduce the following notation:

$$\eta_i \triangleq \eta_i(X)$$
$$= \varphi(X_t) \int_0^1 \Lambda_r(X) dr,$$
$$\eta_i(\bar{X}^{n,s}) \triangleq \varphi(\bar{X}_t^{n,s}) \int_0^1 \Lambda_r(\bar{X}^{n,s}) dr,$$

where

$$\Lambda_r(X) = \exp\left(\sum_{i=0}^m \int_0^t \left(r a_i(X_p) + (1-r) a_i(X_{\tau(p)})\right) dY_p^i\right)$$

$$\Lambda_r(\bar{X}^{n,s}) = \exp\left(\sum_{i=0}^m \int_0^t \left(r a_i(\bar{X}_p^{n,s}) + (1-r) a_i(\bar{X}_{\tau(p)}^{n,s})\right) dY_p^i\right)$$

and $\bar{X}^{n,s}$ is the process defined as $\bar{X}_p^{n,s} = X_{p \wedge (s-\frac{1}{n})}$ if $p \leq s$ and

$$\bar{X}_p^{n,s} = X_{s-\frac{1}{n}} + \int_s^p f(\bar{X}_u^{n,s}) du + \int_0^t \sigma(\bar{X}_u^{n,s}) dV_s$$

for $p > s$. In particular, using a classical argument, similar to that justifying (21.12) one can show that there exists a constant $c_{29} = c_{29}(p,t)$ such that

$$\tilde{\mathbb{E}}\left[\sup_{u \in [0,t]} |X_u - \bar{X}_u^{n,s}|^p\right] \leq c_{29} n^{-\frac{p}{2}}.$$

Choose $h = \tilde{\mathbb{E}}\left[\eta_i(\bar{X}_u^{n,s}) \mid \mathcal{F}_{s-\frac{1}{n}} \vee \mathcal{Y}_t\right]$ in (21.56) and observe that

$$h = \tilde{\mathbb{E}}\left[\eta_i(\bar{X}_u^{n,s}) \mid \mathcal{F}_{s-\frac{1}{n}} \vee \mathcal{Y}_t\right] = \tilde{\mathbb{E}}\left[\eta_i(\bar{X}_u^{n,s}) \mid \mathcal{F}_s \vee \mathcal{Y}_t\right],$$

hence

$$\tilde{\mathbb{E}}\left[\left(\tilde{\mathbb{E}}\left[\eta_i \mid \mathcal{F}_s \vee \mathcal{Y}_t\right] - h\right)^2 \middle| \mathcal{Y}_t\right] \leq \tilde{\mathbb{E}}\left[\left(\eta_i(X) - \eta_i(\bar{X}_u^{n,s})\right)^2 \middle| \mathcal{Y}_t\right].$$

The proof is then completed by showing that,

$$\liminf_{n\to\infty} n\tilde{\mathbb{E}}\left[\left(\eta_i(X) - \eta_i(\bar{X}_u^{n,s})\right)^2 \Big| \mathcal{Y}_t\right] \leq \Theta_s \int_0^1 \tilde{\mathbb{E}}\left[\Lambda_s(X)^q \Big| \mathcal{Y}_t\right] ds,$$

where $q \in (2, p)$, and $\Theta = \{\Theta_s, s \geq 0\}$ is a positive $\mathcal{Y}_t$-adapted process such that

$$\tilde{\mathbb{E}}\left[\sup_{s\in[0,t]} \Theta_s^{\frac{p}{p-q}}\right] < \infty.$$

$\square$

Remark 21.8 In general, if $\psi : \mathbb{R}^{m+1} \to \mathbb{R}$ is a twice-differentiable function such that

$$\sup_{s\in[0,1]} \tilde{\mathbb{E}}\left[\left|\varphi(X_t)\, \partial_i \partial_j \psi\left(s\xi + (1-s)\xi^\delta\right)\right|^p\right] < \infty, \quad i, j = 0, 1, \ldots, m,$$

then (21.23) holds true.

Notes

1. For further details and proofs see, for example [1], [17] and the references therein.
2. In the following, with the exception of the Lipschitz constants of functions, all constants will be numbered $c_1, c_2, \ldots$ in accordance to the order in which they appear. For an arbitrary Lipschitz-continuous function φ, we will denote by k_φ its Lipschitz constant.
3. Recall that, for an arbitrary Lipschitz continuous function φ, we denote by k_φ its Lipschitz constant.
4. Here $\mathcal{B}[0, t]$ is the Borel σ-field on $[0.t]$ and λ is the Lebesgue measure on $\mathcal{B}[0, t]$.
5. The vector fields $\left(\sigma^i\right)_{i=1}^d$ satisfy the UFG condition, if the ideal generated by them within the Lie algebra generated by the vector fields $\left(\sigma^i\right)_{i=1}^d$ and f is a finite-dimensional R-module, where R is the set of bounded smooth functions on $\mathbb{R}^d$ ("uniformly finitely generated"). For details, see [22].

References

[1] Bain, A.; Crisan, D., *Fundamentals of stochastic filtering*, Stochastic Modelling and Applied Probability 60, Springer, New York, 2009.

[2] Beskos, A.; Roberts, G.O., Exact simulation of diffusions, *Ann. Appl. Probab.*, 15 (2005), no. 4, 2422–44.

[3] Beskos, A.; Papaspiliopoulos, O.; Roberts, G. O., Retrospective exact simulation of diffusion sample paths with applications, *Bernoulli*, 12 (2006), no. 6, 1077–98.

[4] Beskos, A.; Papaspiliopoulos, O.; Roberts, G. O.; Fearnhead, P., Exact and computationally efficient likelihood-based estimation for discretely observed diffusion processes, *J. R. Stat. Soc. Ser. B Stat. Methodol.* 68 (2006), no. 3, 333–82.

[5] Cai, Z.; Le Gland, F.; Zhang, H., *An adaptive local grid refinement method for nonlinear filtering*, Technical Report 2679, INRIA, 1995.

[6] Crisan, D.; Del Moral, P.; Lyons, T., Discrete filtering using branching and interacting particle systems. Markov process, *Related Fields*, 5 (1999), no. 3, 293–318.

[7] Crisan, D.; Del Moral, P.; Lyons, T. , Interacting particle systems approximations of the Kushner–Stratonovitch equation, *Adv. in Appl. Probab.*, 31 (1999), no. 3, 819–38.

[8] Crisan, D.; Ghazali, S., *On the convergence rates of a general class of weak approximations of SDEs*. Stochastic differential equations: theory and applications, 221–48, Interdiscip. Math. Sci., 2, World Sci. Publ., Hackensack, NJ, 2007.

[9] Del Moral, P.; Jacod, J., *Interacting particle filtering with discrete observations*, Sequential Monte Carlo Methods in Practice, 43–75, Stat. Eng. Inf. Sci., Springer, New York, 2001.

[10] Del Moral, P.; Jacod, J., *The Monte Carlo method for filtering with discrete-time observations: central limit theorems*, Numerical Methods and Stochastics (Toronto, ON, 1999), 29–53, Fields Inst. Commun., 34, Amer. Math. Soc., Providence, RI, 2002.

[11] Del Moral, P., *Feynman–Kac formulae. Genealogical and interacting particle systems with applications*. Probability and Its Applications (New York). Springer-Verlag, New York, 2004.

[12] Del Moral, P.; Patras, F.; Rubenthaler S., *A mean field theory of nonlinear filtering*, Chapter 25 this volume.

[13] Gyöngy, I.; Krylov, N., On the rate of convergence of splitting-up approximations for SPDEs, in *Stochastic inequalities and applications*, v. 56 of Progr. Probab., pp 301–21, Birkhĺauser, 2003.

[14] Gyöngy, I.; Krylov, N., *On the splitting-up method and stochastic partial differential equations*, Ann. Probab., 31(2), 564–91, 2003.

[15] Hairer, M.; Stuart, A.; Voss, J., *Signal processing problems on function space: Bayesian formulation, SPDEs and effective MCMC methods*, Chapter 29 this volume.

[16] Ikeda, N.; Watanabe, S., *Stochastic differential equations and diffusion processes*, North-Holland Mathematical Library 24. North-Holland Publishing, Amsterdam–New York; Kodansha, Ltd., Tokyo, 1981.

[17] Kallianpur, G., *Stochastic filtering theory*. Applications of Mathematics, 13. Springer-Verlag, New York–Berlin, 1980.

[18] Karatzas, I.; Shreve, S. E. *Brownian motion and stochastic calculus*, 2nd edn. Graduate Texts in Mathematics 113. Springer-Verlag, New York, 1991.

[19] Kushner, H. J., *Numerical approximations to optimal nonlinear filters*, Chapter 28 this volume.

[20] Kushner, H. J.; Dupuis, P., *Numerical methods for stochastic control problems in continuous time*. Applications of Mathematics 24. Springer, New York, 1992.

[21] Kushner H. J., *Weak convergence methods and singularly perturbed stochastic control and filtering problems*, Systems & Control: Foundations & Applications 3. Birkhĺauser Boston, 1990.

[22] Kusuoka, S.; Stroock, D. Applications of the Malliavin calculus, III. *J. Fac. Sci. Univ. Tokyo Sect. IA Math.* 34 (1987), no. 2, 391–442.

[23] Kusuoka, S., Malliavin calculus revisited, *J. Math. Sci. Univ. Tokyo* 10 (2003), no. 2, 261–77.

[24] Le Gland, F., Time discretization of nonlinear filtering equations In *Proceedings of the 28th IEEE-CSS Conference Decision Control*, Tampa, FL, pp. 2601–6, 1989.

[25] Litterer, C.; Lyons, T., *Introducing cubature to filtering*, Chapter 27 this volume.

[26] Milstein, G. N.; Tretyakov, M. V., Monte Carlo methods for backward equations in nonlinear filtering, *Adv. in Appl. Probab.* 41 (2009), no. 1, 63–100.

[27] Milstein, G. N.; Tretyakov, M.*Nonlinear filtering algorithms based on averaging over characteristics and on the innovation approach*, Chapter 31 this volume.

[28] Picard, J. *Approximation of nonlinear filtering problems and order of convergence*, Filtering and control of random processes (Paris, 1983), 219–36, Lecture Notes in Control and Inform. Sci. 61, Springer, Berlin, 1984.

[29] Xiong, J., *Particle approximations to the filtering problem in continuous time*, Chapter 23 this volume.

·22·
Large Sample Asymptotics for the Ensemble Kalman Filter

F. Le Gland, V. Monbet, and V.-D. Tran

22.1 Introduction

The ensemble Kalman filter (EnKF) has been proposed (Evensen [9]) as a Monte Carlo, derivative-free, alternative to the extended Kalman filter, and is now widely used in sequential data assimilation (Evensen [11]), where state vectors of huge dimension (e.g. resulting from the discretization of pressure and velocity fields over a continent, as considered in meteorology) should be estimated from noisy measurements (e.g. collected at sparse in-situ stations). Even if the state and measurement equations are linear with additive Gaussian white noise, computing and storing the error covariance matrices involved in the Kalman filter is practically impossible, and it has been proposed to represent the filtering distribution with a sample (ensemble) of a few elements and to think of the corresponding empirical covariance matrix as an approximation of the intractable error covariance matrix. Extensions to nonlinear state equations have also been proposed.

A precise mathematical formulation of the EnKF is provided in (Burgers, van Leeuwen, and Evensen [3], Bertino, Evensen, and Wackernagel [2], Evensen [10]) and comparisons with other Monte Carlo-based nonlinear filters can be found in (Miller, Carter, and Blue [16], Anderson and Anderson [1], Pham [18], Heemink, Verlaan, and Segers [14]). The EnKF has gained popularity because of its simple conceptual formulation and its ease of implementation, and in particular it requires no modification of the forecast model. Simulation studies have demonstrated the ability of the EnKF to efficiently handle strongly nonlinear dynamics and high-dimensional state spaces and it is now widely used in realistic applications with primitive equation models for the ocean and atmosphere.

Surprisingly, very little is known about the asymptotic behaviour of the EnKF, whereas on the other hand, the asymptotic behaviour of many different classes of particle filters is well understood, as the number of particles goes to infinity: see (Crişan and Doucet [5], Cappé, Moulines, and Rydén [4], Del Moral [6], Doucet, de Freitas, and Gordon [7]) and references therein. Interpreting the ensemble elements as a population of particles with mean-field interactions (and not merely as an instrumental device producing the ensemble mean value as an estimate of the hidden state), we prove the convergence of the EnKF, with the classical rate $1/\sqrt{N}$, as the number N of ensemble elements increases to infinity. In the linear case, the limit of the empirical distribution of the ensemble elements is the usual (Gaussian distribution associated with the) Kalman filter, as expected, but in the more general case of a nonlinear state equation with linear observations, this limit differs from the usual Bayesian filter. To get the correct limit in this case, the mechanism that generates the elements in the EnKF should be interpreted as a proposal importance distribution, and appropriate importance weights should be assigned to the ensemble elements.

22.1.1 EnKF as an implementation of the Kalman filter in a high dimension

Consider the following linear Gaussian system

$$X_k = F_k X_{k-1} + W_k \quad \text{with} \quad W_k \sim \mathcal{N}(0, Q_k)$$

$$Y_k = H_k X_k + V_k \quad \text{with} \quad V_k \sim \mathcal{N}(0, R_k),$$

with additive Gaussian white noises and with Gaussian initial condition $X_0 \sim \mathcal{N}(m_0, \Sigma_0)$. It is assumed that the covariance matrix R_k is invertible. Clearly, the conditional probability distribution of the hidden state X_k given the past observations $Y_{0:k} = (Y_0, \cdots, Y_k)$ is Gaussian, with mean vector $\widehat{X}_k$ and covariance matrix P_k which satisfy the Kalman filter equations: in the prediction (forecast) step

$$\widehat{X}_k^- = F_k \widehat{X}_{k-1} \quad \text{and} \quad P_k^- = F_k P_{k-1} F_k^* + Q_k,$$

in the correction (analysis) step

$$\widehat{X}_k = \widehat{X}_k^- + K_k \left(Y_k - H_k \widehat{X}_k^- \right) \quad \text{and} \quad P_k = (I - K_k H_k) P_k^-,$$

with the Kalman gain matrix defined by

$$K_k = P_k^- H_k^* \left(H_k P_k^- H_k^* + R_k \right)^{-1},$$

and initially $\widehat{X}_0^- = m_0$ and $P_0^- = \Sigma_0$. If the dimension m of the hidden state is large, the covariance matrices P_{k-1} and P_k^- are very large $m \times m$ symmetric matrices, hence storing such matrices in memory (and storing the $m \times m$ matrix F_k as well) is almost impossible, and the matrix products in the prediction equation

$$P_k^- = F_k P_{k-1} F_k^* + Q_k,$$

are even more problematic to work out. Usually, the dimension d of the observation is much less, and the matrix products in the expression of the Kalman gain matrix

$$K_k = P_k^- H_k^* \left(H_k P_k^- H_k^* + R_k \right)^{-1},$$

or in the correction equation

$$P_k = (I - K_k H_k) P_k^- = P_k^- - P_k^- H_k^* \left(H_k P_k^- H_k^* + R_k \right)^{-1} H_k P_k^-,$$

are much less problematic to work out.

The idea (Burgers et al. [3], Bertino et al. [2], Evensen [10]) behind the ensemble Kalman filter (EnKF) is to use Monte Carlo samples and to use the corresponding empirical covariance matrix instead of the prediction covariance matrix. In practice, given an analysis ensemble $(X_{k-1}^{1,a}, \cdots, X_{k-1}^{N,a})$ of N elements, each ensemble element is propagated independently according to

$$X_k^{i,f} = F_k X_{k-1}^{i,a} + W_k^i \quad \text{with} \quad W_k^i \sim \mathcal{N}(0, Q_k).$$

Notice that the i.i.d. random vectors $\left(W_k^1, \cdots, W_k^N \right)$ are *simulated* here, with the same statistics as the additive Gaussian noise W_k in the original state equation. The initial ensemble $(X_0^{1,f}, \cdots, X_0^{N,f})$ is *simulated* as i.i.d. Gaussian random vectors with mean m_0 and covariance matrix Σ_0, i.e. with the same statistics as the initial condition X_0. The empirical mean vector and covariance matrix of the forecast elements $(X_k^{1,f}, \cdots, X_k^{N,f})$ are defined as

$$m_k^N = \frac{1}{N} \sum_{i=1}^N X_k^{i,f} \quad \text{and} \quad P_k^N = \frac{1}{N} \sum_{i=1}^N (X_k^{i,f} - m_k^N)(X_k^{i,f} - m_k^N)^*,$$

respectively. This empirical covariance matrix is then used in the correction step as follows

$$X_k^{i,a} = X_k^{i,f} + K_k^N \left(Y_k - H_k X_k^{i,f} - V_k^i \right) \quad \text{with} \quad V_k^i \sim \mathcal{N}(0, R_k),$$

with the empirical Kalman gain matrix defined by

$$K_k^N = P_k^N H_k^* \left(H_k P_k^N H_k^* + R_k \right)^{-1}.$$

Notice that the i.i.d. random vectors $\left(V_k^1, \cdots, V_k^N \right)$ are *simulated* here, with the same statistics as the additive Gaussian noise V_k in the original observation equation. The rationale behind the simulation of these i.i.d. random vectors is explained in Lemma 22.1. In practice however, the empirical covariance matrix P_k^N is never computed or stored: indeed, to evaluate the matrix–vector product $P_k^N u$ where u is a (column) vector of dimension m, only N scalar products need to be evaluated, since

$$P_k^N u = \left(\frac{1}{N} \sum_{i=1}^N (X_k^{i,f} - m_k^N)(X_k^{i,f} - m_k^N)^* \right) u = \frac{1}{N} \sum_{i=1}^N \lambda_k^i (X_k^{i,f} - m_k^N),$$

with $\lambda_k^i = (X_k^{i,f} - m_k^N)^* u$ for any $i = 1, \cdots, N$. In particular, H_k can be seen as a collection of d (row) vectors of dimension m, and to evaluate the matrix products $P_k^N H_k^*$ and $H_k P_k^N H_k^*$, only $N \times d$ scalar products need to be evaluated, since

$$P_k^N H_k^* = (\frac{1}{N} \sum_{i=1}^{N}(X_k^{i,f} - m_k^N)(X_k^{i,f} - m_k^N)^*) H_k^* = \frac{1}{N} \sum_{i=1}^{N}(X_k^{i,f} - m_k^N)(h_k^i)^*,$$

and

$$H_k P_k^N H_k^* = H_k (\frac{1}{N} \sum_{i=1}^{N}(X_k^{i,f} - m_k^N)(X_k^{i,f} - m_k^N)^*) H_k^* = \frac{1}{N} \sum_{i=1}^{N} h_k^i (h_k^i)^*,$$

with $h_k^i = H_k (X_k^{i,f} - m_k^N)$ for any $i = 1, \cdots, N$.

The question that naturally arises here is whether the empirical mean of forecast elements and analysis elements converge to the Kalman predictor and Kalman filter respectively, i.e. whether

$$\frac{1}{N} \sum_{i=1}^{N} X_k^{i,f} \longrightarrow \widehat{X}_k^- \quad \text{and} \quad \frac{1}{N} \sum_{i=1}^{N} X_k^{i,a} \longrightarrow \widehat{X}_k,$$

in some sense, as $N \uparrow \infty$.

22.1.2 EnKF as a particle system with mean-field interactions

The ensemble Kalman filter idea has been extended to any system of the form

$$X_k = f_k(X_{k-1}) + W_k \quad \text{with} \quad W_k \sim \mathcal{N}(0, Q_k)$$

$$Y_k = H_k X_k + V_k \quad \text{with} \quad V_k \sim \mathcal{N}(0, R_k),$$

with additive Gaussian white noise and with non-necessarily Gaussian initial condition $X_0 \sim \eta_0$, see for instance (Bertino et al. [2], Section 4.1). In practice, given an analysis ensemble $(X_{k-1}^{1,a}, \cdots, X_{k-1}^{N,a})$ of N elements, each ensemble element is propagated independently according to the following set of decoupled equations

$$X_k^{i,f} = f_k(X_{k-1}^{i,a}) + W_k^i \quad \text{with} \quad W_k^i \sim \mathcal{N}(0, Q_k). \tag{22.1}$$

Notice that the i.i.d. random vectors $(W_k^1, \cdots, W_k^N)$ are *simulated* here, with the same statistics as the additive Gaussian noise W_k in the original state equation. The initial ensemble $(X_0^{1,f}, \cdots, X_0^{N,f})$ is *simulated* as i.i.d. random vectors with probability distribution η_0, i.e. with the same statistics as the initial condition X_0. The empirical mean vector and covariance matrix of the forecast elements $(X_k^{1,f}, \cdots, X_k^{N,f})$ are defined as

$$m_k^N = \frac{1}{N} \sum_{i=1}^{N} X_k^{i,f} \quad \text{and} \quad P_k^N = \frac{1}{N} \sum_{i=1}^{N} (X_k^{i,f} - m_k^N)(X_k^{i,f} - m_k^N)^*,$$

respectively. This empirical covariance matrix is then used in the correction step to produce a new analysis ensemble $(X_k^{1,a}, \cdots, X_k^{N,a})$, according to the set of equations with mean–field interaction

$$X_k^{i,a} = X_k^{i,f} + K_k(P_k^N)(Y_k - H_k X_k^{i,f} - V_k^i) \quad \text{with} \quad V_k^i \sim \mathcal{N}(0, R_k), \quad (22.2)$$

where the $m \times d$ matrix $K_k(P)$ is defined by

$$K_k(P) = P H_k^* (H_k P H_k^* + R_k)^{-1}, \quad (22.3)$$

for any $m \times m$ covariance matrix P. Notice that the i.i.d. random vectors $(V_k^1, \cdots, V_k^N)$ are *simulated* here, with the same statistics as the additive Gaussian noise V_k in the original observation equation.

The question that naturally arises here is whether the empirical probability distribution (uniform mixture of Dirac masses) of forecast elements and analysis elements converge to the Bayesian predictor and Bayesian filter respectively, i.e. whether

$$\mu_k^{N,f} = \frac{1}{N} \sum_{i=1}^{N} \delta_{X_k^{i,f}} \longrightarrow \bar{\mu}_k \quad \text{and} \quad \mu_k^{N,a} = \frac{1}{N} \sum_{i=1}^{N} \delta_{X_k^{i,a}} \longrightarrow \mu_k,$$

in some sense, as $N \uparrow \infty$, where

$$\bar{\mu}_k(dx) = \mathbb{P}[X_k \in dx \mid Y_{0:k-1}] \quad \text{and} \quad \mu_k(dx) = \mathbb{P}[X_k \in dx \mid Y_{0:k}],$$

by definition. In view of the analysis equation, it is clear that each analysis element depends on the whole forecast ensemble $(X_k^{1,f}, \cdots, X_k^{N,f})$, which results in dependent analysis elements $(X_k^{1,a}, \cdots, X_k^{N,a})$. Notice however that dependence follows here from *mean-field interaction*, i.e. only through the empirical probability distribution $\mu_k^{N,f}$ of forecast elements, and even more explicitly, only through their empirical covariance matrix P_k^N. Intuitively, some law of large numbers should hold when the ensemble size N increases to infinity and if the empirical covariance matrix P_k^N would be replaced by its deterministic limit, then the ensemble elements would become independent at the limit: this phenomenon is known as *propagation of chaos* (McKean [15], Sznitman [20]). To study the asymptotic behaviour of the empirical probability distributions

$$\mu_k^{N,f} = \frac{1}{N} \sum_{i=1}^{N} \delta_{X_k^{i,f}} \quad \text{and} \quad \mu_k^{N,a} = \frac{1}{N} \sum_{i=1}^{N} \delta_{X_k^{i,a}},$$

of the forecast elements and analysis elements, respectively, the idea (Sznitman [20], Section I.1) is to consider substitute i.i.d. random vectors. In practice, these vectors are propagated independently according to the following set of decoupled equations

$$\bar{X}_k^{i,f} = f_k(\bar{X}_{k-1}^{i,a}) + W_k^i \quad \text{with} \quad W_k^i \sim \mathcal{N}(0, Q_k), \tag{22.4}$$

and

$$\bar{X}_k^{i,a} = \bar{X}_k^{i,f} + K_k(\bar{P}_k)(Y_k - H_k \bar{X}_k^{i,f} - V_k^i) \quad \text{with} \quad V_k^i \sim \mathcal{N}(0, R_k), \tag{22.5}$$

where $\bar{P}_k$ denotes the covariance matrix of the random vector $\bar{X}_k^{i,f}$, and initially $\bar{X}_0^{i,f} = X_0^{i,f}$, i.e. the initial set of i.i.d. random vectors coincides exactly with the initial ensemble. By definition

$$\bar{m}_k = \mathbb{E}[\bar{X}_k^{i,f}] \quad \text{and} \quad \bar{P}_k = \mathbb{E}[(\bar{X}_k^{i,f} - \bar{m}_k)(\bar{X}_k^{i,f} - \bar{m}_k)^*],$$

respectively. For later purposes, the empirical mean vector and covariance matrix of the i.i.d. random vectors $(\bar{X}_k^{1,f}, \cdots, \bar{X}_k^{N,f})$ are defined as

$$\bar{m}_k^N = \frac{1}{N} \sum_{i=1}^N \bar{X}_k^{i,f} \quad \text{and} \quad \bar{P}_k^N = \frac{1}{N} \sum_{i=1}^N (\bar{X}_k^{i,f} - \bar{m}_k^N)(\bar{X}_k^{i,f} - \bar{m}_k^N)^*,$$

respectively.

Intuitively, each random vector $\bar{X}_k^{i,f}$ or $\bar{X}_k^{i,a}$ individually is close (contiguous) to the corresponding element $X_k^{i,f}$ or $X_k^{i,a}$ in the ensemble Kalman filter, because it starts from exactly the same initial value $\bar{X}_0^{i,f} = X_0^{i,f}$ and it uses exactly the same i.i.d. random vectors $(W_1^i, \cdots, W_k^i)$ and $(V_0^i, \cdots, V_k^i)$ already *simulated* and used in the ensemble Kalman filter. Collectively, the large sample asymptotics of the substitute i.i.d. random vectors is simple to analyse, because of independance, but in counterpart the covariance matrix $\bar{P}_k$ is unknown in general, and so are the substitute i.i.d. random vectors themselves. In contrast, the elements in the ensemble Kalman filter are dependent, because they all contribute to the empirical covariance matrix P_k^N which results in mean-field interaction, but in counterpart this empirical covariance matrix is readily computable, and so are the elements in the ensemble Kalman filter. Let $\bar{\mu}_k^f$ and $\bar{\mu}_k^a$ be the probability distribution of the susbtitute i.i.d. random vectors $\bar{X}_k^{i,f}$ and $\bar{X}_k^{i,a}$ respectively. The contribution of this chapter consists in the following answers to the questions raised above

- for linear systems with additive Gaussian white noises and with Gaussian initial condition, the probability distributions $\bar{\mu}_k^f$ and $\bar{\mu}_k^a$ coincide with the Gaussian distributions associated with the Kalman predictor and with the Kalman filter, respectively,

- however, for nonlinear systems with additive Gaussian white noises and with non-necessarily Gaussian initial condition, the probability distributions $\bar{\mu}_k^f$ and $\bar{\mu}_k^a$ differ from the Bayesian predictor μ_k^- and from the Bayesian filter μ_k, respectively,
- under suitable Lipschitz assumptions on the drift function f_k and moment conditions on the initial condition X_0, the empirical probability distribution $\mu_k^{N,f}$ of the forecast elements and the empirical probability distribution $\mu_k^{N,a}$ of the analysis elements converge to the probability distribution $\bar{\mu}_k^f$ and $\bar{\mu}_k^a$, respectively, i.e. for any ϕ in a large enough class of functions

$$\frac{1}{N}\sum_{i=1}^{N}\phi(X_k^{i,f}) = \int_{\mathbb{R}^m}\phi(x)\,\mu_k^{N,f}(dx) \longrightarrow \int_{\mathbb{R}^m}\phi(x)\,\bar{\mu}_k^f(dx),$$

and

$$\frac{1}{N}\sum_{i=1}^{N}\phi(X_k^{i,a}) = \int_{\mathbb{R}^m}\phi(x)\,\mu_k^{N,a}(dx) \longrightarrow \int_{\mathbb{R}^m}\phi(x)\,\bar{\mu}_k^a(dx),$$

almost surely and in $\mathbb{L}^p$, as $N \uparrow \infty$.

The chapter is organized as follows. The limiting probability distributions are characterized in Section 22.2 in connection with the Kalman filter or the Bayesian filter, and sufficient conditions are given for the existence of moments. Preliminary estimates are obtained in Section 22.3, which establish the asymptotic behaviour of the empirical covariance matrices P_k^N and the corresponding gain matrices $K_k(P_k^N)$. Contiguity between elements in the ensemble Kalman filter and the corresponding substitute i.i.d. random vectors is proved in Section 22.4. These two sections are rather technical, and the impatient reader could jump to Section 22.5 directly, where convergence of the ensemble Kalman filter is deduced, with the classical rate $1/\sqrt{N}$. Concluding remarks and perspectives for future work are presented in Section 22.6, which include comparison with particle filters and with the weighted ensemble Kalman filter.

Throughout the chapter, the observation sequence is considered as *fixed* and any estimate involving mathematical expectations or any almost sure statement does apply to the *simulated* random vectors only.

22.2 Identification of the limit, and a priori estimates

The limiting probability distributions $\bar{\mu}_k^f$ and $\bar{\mu}_k^a$ are defined as the probability distributions of the substitute i.i.d. random vectors $\bar{X}_k^{i,f}$ and $\bar{X}_k^{i,a}$ respectively, and are completely characterized by integrals of an arbitrary bounded measurable function ϕ. By definition

$$\int_{\mathbb{R}^m} \phi(x') \, \bar{\mu}_k^f(dx') = \mathbb{E}[\phi(\bar{X}_k^{i,f})] = \int_{\mathbb{R}^m} \int_{\mathbb{R}^m} \phi(f_k(x) + w) \, p_k^W(dw) \, \bar{\mu}_{k-1}^a(dx),$$

where $p_k^W(dw)$ denotes the Gaussian probability distribution with zero mean vector and covariance matrix Q_k, i.e. the probability distribution of the random vector W_k^i, which completely characterizes $\bar{\mu}_k^f$ in terms of $\bar{\mu}_{k-1}^a$. Sufficient conditions are given in Proposition 22.1 below, under which $\bar{\mu}_k^f$ has a finite second-order moment, in which case the covariance matrix $\bar{P}_k$ is finite, and by definition

$$\int_{\mathbb{R}^m} \phi(x') \, \bar{\mu}_k^a(dx') = \mathbb{E}[\phi(\bar{X}_k^{i,a})] = \int_{\mathbb{R}^m} \int_{\mathbb{R}^d} \phi(x + K_k(\bar{P}_k) \, (Y_k - H_k \, x - v)) \, q_k^V(v) \, dv \, \bar{\mu}_k^f(dx),$$

where $q_k^V(v)$ denotes the Gaussian density with zero mean vector and invertible covariance matrix R_k, i.e. the probability density of the random vector V_k^i, which completely characterizes $\bar{\mu}_k^a$ in terms of $\bar{\mu}_k^f$, and initially

$$\int_{\mathbb{R}^m} \phi(x) \, \bar{\mu}_0^f(dx) = \mathbb{E}[\phi(\bar{X}_0^{i,f})] = \int_{\mathbb{R}^m} \phi(x) \, \eta_0(dx),$$

which implies $\bar{\mu}_0^f = \eta_0$. On the other hand, the Bayesian filter satisfies

$$\int_{\mathbb{R}^m} \phi(x') \, \mu_k^-(dx') = \mathbb{E}\left[\phi(X_k) \mid Y_{0:k-1}\right] = \int_{\mathbb{R}^m} \int_{\mathbb{R}^m} \phi(f_k(x) + w) \, p_k^W(dw) \, \mu_{k-1}(dx),$$

which completely characterizes μ_k^- in terms of μ_{k-1}, and

$$\int_{\mathbb{R}^m} \phi(x') \, \mu_k(dx') = \mathbb{E}[\phi(X_k) \mid Y_{0:k}] = \frac{\int_{\mathbb{R}^m} \phi(x') \, q_k^V(Y_k - H_k \, x') \, \mu_k^-(dx')}{\int_{\mathbb{R}^m} q_k^V(Y_k - H_k \, x') \, \mu_k^-(dx')},$$

which completely characterizes μ_k in terms of μ_k^-, and initially

$$\int_{\mathbb{R}^m} \phi(x) \, \mu_0^-(dx) = \mathbb{E}[\phi(X_0)] = \int_{\mathbb{R}^m} \phi(x) \, \eta_0(dx),$$

which implies $\mu_0^- = \eta_0$.

22.2.1 Connection with the Kalman filter or the Bayesian filter

Consider the linear transformation T_k and the two nonlinear transformations $T_k^{KF}(\cdot)$ and $T_k^{BF}(\cdot)$ defined on the space of probability distributions, and completely characterized by integrals of an arbitrary boundary measurable function ϕ, as

$$\int_{\mathbb{R}^m} \phi(x)\, T_k\, \mu(dx) = \int_{\mathbb{R}^m} \int_{\mathbb{R}^m} \phi(f_k(x) + w)\, p_k^W(dw)\, \mu(dx),$$

$$\int_{\mathbb{R}^m} \phi(x)\, T_k^{KF}(\mu)(dx) = \int_{\mathbb{R}^m} \int_{\mathbb{R}^d} \phi(x + K_k(P(\mu))\, (Y_k - H_k\, x - v))\, q_k^V(v)\, dv\, \mu(dx),$$

where $P(\mu)$ denotes the covariance matrix of the probability distribution μ, and

$$\int_{\mathbb{R}^m} \phi(x)\, T_k^{BF}(\mu)(dx) = \frac{\int_{\mathbb{R}^m} \phi(x)\, q_k^V(Y_k - H_k\, x)\, \mu(dx)}{\int_{\mathbb{R}^m} q_k^V(Y_k - H_k\, x)\, \mu(dx)},$$

respectively.

Lemma 22.1 *If $f_k(x) = F_k\, x$ and if the probability distribution μ is Gaussian with mean vector m and covariance matrix Σ, then the transformed probability distribution $T_k\, \mu$ is Gaussian, with mean vector $F_k\, m$ and covariance matrix $F_k\, \Sigma\, F_k^* + Q_k$.*

If the probability distribution μ is Gaussian, with mean vector m and covariance matrix Σ, then the two transformed probability distributions $T_k^{KF}(\mu)$ and $T_k^{BF}(\mu)$ are Gaussian, with the same mean vector $m + K_k(\Sigma)\, (Y_k - H_k\, m)$ and the same covariance matrix $(I - K_k(\Sigma)\, H_k)\, \Sigma$.

Proof. By definition, the transformed probability distribution $T_k\, \mu$ is the probability distribution of the random vector X' in the model

$$X' = f_k(X) + W_k \quad \text{with} \quad X \sim \mu \quad \text{and} \quad W_k \sim \mathcal{N}(0, Q_k),$$

where the random vectors X and W_k are independent. If $f_k(x) = F_k\, x$ and if the probability distribution μ is Gaussian, with mean vector m and covariance matrix Σ, then the random vector X' is Gaussian, with mean vector $F_k\, m$ and covariance matrix $F_k\, \Sigma\, F_k^* + Q_k$.

By definition, the transformed probability distribution $T_k^{BF}(\mu)$ is the conditional probability distribution of the random vector X given Y_k in the model

$$Y_k = H_k\, X + V_k \quad \text{with} \quad X \sim \mu \quad \text{and} \quad V_k \sim \mathcal{N}(0, R_k),$$

where the random vectors X and V_k are independent. If the probability distribution μ is Gaussian, with mean vector m and covariance matrix Σ, then this conditional probability distribution is Gaussian, with mean vector $m + K_k(\Sigma)\, (Y_k - H_k\, m)$ and covariance matrix $(I - K_k(\Sigma)\, H_k)\, \Sigma$.

By definition, the transformed probability distribution $T_k^{KF}(\mu)$ is the probability distribution of the random vector X' in the model

$$X' = X + K_k(P(\mu))\, (Y_k - H_k\, X - V_k) \quad \text{with} \quad X \sim \mu \quad \text{and} \quad V_k \sim \mathcal{N}(0, R_k),$$

where the random vectors X and V_k are independent, and where the observation Y_k is considered as *fixed*. If the probability distribution μ is Gaussian, with mean vector m and covariance matrix Σ, then $P(\mu) = \Sigma$, and the random vector X' is Gaussian, with mean vector $m + K_k(\Sigma)(Y_k - H_k m)$ and covariance matrix

$$(I - K_k(\Sigma) H_k) \Sigma (I - K_k(\Sigma) H_k)^* + K_k(\Sigma) R_k K_k^*(\Sigma) = (I - K_k(\Sigma) H_k)\Sigma. \qquad \square$$

If the probability distribution μ is not Gaussian, then the two transformed probability distributions $T_k^{\text{KF}}(\mu)$ and $T_k^{\text{BF}}(\mu)$ differ in general: consider for instance the case where μ is a finite mixture of Gaussian distributions.

It follows from the above discussion that $\bar{\mu}_0^f = \mu_0^-$, and if $\bar{\mu}_{k-1}^a = \mu_{k-1}$ then necessarily $\bar{\mu}_k^f = T_k \bar{\mu}_{k-1}^a$ coincides with $\mu_k^- = T_k \mu_{k-1}$, but in general $\bar{\mu}_k^f = \mu_k^-$ does not necessarily imply that $\bar{\mu}_k^a = T_k^{\text{KF}}(\bar{\mu}_k^f)$ coincides with $\mu_k = T_k^{\text{BF}}(\mu_k^-)$, which means that in general the limiting probability distributions $\bar{\mu}_k^f$ and $\bar{\mu}_k^a$ do not coincide with the probability distributions μ_k^- and μ_k defining the Bayesian filter.

However, for linear systems of the form considered in Section 22.1.1, with additive Gaussian white noises and with Gaussian initial condition, the (probability distributions defining the) Bayesian filter coincides with the (Gaussian distributions associated with the) Kalman filter, i.e. the probability distribution μ_k^- is Gaussian, with mean vector $\widehat{X}_k^-$ and covariance matrix P_k^-, and the probability distribution μ_k is Gaussian, with mean vector $\widehat{X}_k$ and covariance matrix P_k. It is then easy to prove by induction that $\bar{\mu}_k^f = \mu_k^-$ and $\bar{\mu}_k^a = \mu_k$: indeed, it follows from the general case that $\bar{\mu}_0^f = \mu_0^-$, and it follows from the induction assumption that $\bar{\mu}_k^f = T_k \bar{\mu}_{k-1}^a$ coincides with $\mu_k^- = T_k \mu_{k-1}$, and since $\bar{\mu}_k^f = \mu_k^-$ is Gaussian, then it follows from Lemma 22.1 that $\bar{\mu}_k^a = T_k^{\text{KF}}(\bar{\mu}_k^f)$ coincides with $\mu_k = T_k^{\text{BF}}(\mu_k^-)$.

22.2.2 A priori estimates (existence of moments)

Two different assumptions are introduced for the drift function: Assumption A is sufficient to handle the linear case, whereas Assumption B allows to handle more general cases, including the (discretized) Lorenz model for instance.

Assumption A The drift function is globally Lipschitz continuous, i.e.

$$|f_k(x) - f_k(x')| \leq L\,|x - x'|,$$

for any $x, x' \in \mathbb{R}^m$.

Assumption B The drift function is locally Lipschitz continuous, with at most polynomial growth at infinity, i.e.

$$|f_k(x) - f_k(x')| \leq L \, |x - x'| \, (1 + |x|^s + |x'|^s),$$

for any $x, x' \in \mathbb{R}^m$ and for some $s \geq 0$.

Clearly, Assumption A is a special case of Assumption B, for $s = 0$. Notice that under Assumption A, the drift function has at most linear growth at infinity, i.e.

$$|f_k(x)| \leq M \, (1 + |x|),$$

for any $x \in \mathbb{R}^m$, whereas under Assumption B, the drift function has at most polynomial growth at infinity, i.e.

$$|f_k(x)| \leq M \, (1 + |x|^{s+1}),$$

for any $x \in \mathbb{R}^m$, and using the triangle inequality yields the following asymmetric form of the local Lipschitz condition

$$|f_k(x) - f_k(x')| \leq L \, |x - x'| \, (1 + |x|^s) + L \, |x - x'|^{s+1},$$

for any $x, x' \in \mathbb{R}^m$, with another constant L. Introduce the following notation for the moments

$$\bar{M}_k^{p,f} = (\mathbb{E}|\bar{X}_k^{i,f}|^p)^{1/p} = \big(\int_{\mathbb{R}^m} |x|^p \, \bar{\mu}_k^f(dx) \big)^{1/p},$$

and

$$\bar{M}_k^{p,a} = (\mathbb{E}|\bar{X}_k^{i,a}|^p)^{1/p} = \big(\int_{\mathbb{R}^m} |x|^p \, \bar{\mu}_k^a(dx) \big)^{1/p},$$

and similarly for the moments of the perturbed residuals

$$\bar{R}_k^p = (\mathbb{E}|Y_k - H_k \, \bar{X}_k^{i,f} - V_k^i|^p)^{1/p},$$

and notice that the triangle inequality yields

$$\bar{R}_k^p \leq \big(\int_{\mathbb{R}^m} |Y_k - H_k \, x|^p \, \bar{\mu}_k^f(dx) \big)^{1/p} + (\mathbb{E}|V_k^i|^p)^{1/p}$$

$$\leq |Y_k| + \|H_k\| \, \bar{M}_k^{p,f} + c_p \, \lambda_{\max}^{1/2}(R_k),$$

where $\lambda_{\max}(R_k)$ denotes the largest eigenvalue of the covariance matrix R_k.

Proposition 22.1 *If Assumption A holds, and if the random vector X_0, or equivalently the probability distribution η_0, has finite moment of order p for some $p \geq 2$, then the random vectors $\bar{X}_k^{i,f}$ and $\bar{X}_k^{i,a}$, or equivalently the probability distributions $\bar{\mu}_k^f$ and $\bar{\mu}_k^a$, have finite moments of the same order p, and in particular the covariance matrix $\bar{P}_k$ is finite.*

If Assumption B holds, and if the random vector X_0, or equivalently the probability distribution η_0, has finite moments of any order, then the random vectors $\bar{X}_k^{i,f}$ and $\bar{X}_k^{i,a}$, or equivalently the probability distributions $\bar{\mu}_k^f$ and $\bar{\mu}_k^a$, have finite moments of any order, and in particular the covariance matrix $\bar{P}_k$ is finite.

Proof. (by induction). If Assumption A holds, then

$$|\bar{X}_k^{i,f}| \leq M\,(1 + |\bar{X}_{k-1}^{i,a}|) + |W_k^i|,$$

and using the triangle inequality yields

$$(\mathbb{E}|\bar{X}_k^{i,f}|^p)^{1/p} \leq M\,(1 + (\mathbb{E}|\bar{X}_{k-1}^{i,a}|^p)^{1/p}) + (\mathbb{E}|W_k^i|^p)^{1/p},$$

hence

$$\bar{M}_k^{p,f} \leq M\,(1 + \bar{M}_{k-1}^{p,a}) + c_p\,\lambda_{\max}^{1/2}(Q_k),$$

whereas if Assumption B holds, then

$$|\bar{X}_k^{i,f}| \leq M\,(1 + |\bar{X}_{k-1}^{i,a}|^{s+1}) + |W_k^i|,$$

and using the triangle inequality yields

$$(\mathbb{E}|\bar{X}_k^{i,f}|^p)^{1/p} \leq M\,(1 + (\mathbb{E}|\bar{X}_{k-1}^{i,a}|^{p(s+1)})^{1/p}) + (\mathbb{E}|W_k^i|^p)^{1/p},$$

hence

$$\bar{M}_k^{p,f} \leq M\,(1 + (\bar{M}_{k-1}^{p(s+1),a})^{s+1}) + c_p\,\lambda_{\max}^{1/2}(Q_k),$$

where $\lambda_{\max}(Q_k)$ denotes the largest eigenvalue of the covariance matrix Q_k. In particular

$$u^* \bar{P}_k u \leq \int_{\mathbb{R}^m} |u^* x|^2\, \bar{\mu}_k^f(dx) \leq |u|^2 \int_{\mathbb{R}^m} |x|^2\, \bar{\mu}_k^f(dx),$$

hence

$$\|\bar{P}_k\| = \sup_{u \neq 0} \frac{u^* \bar{P}_k u}{|u|^2} \leq \int_{\mathbb{R}^m} |x|^2\, \bar{\mu}_k^f(dx) = (\bar{M}_k^{2,f})^2 < \infty.$$

In any case

$$|\bar{X}_k^{i,a}| \leq |\bar{X}_k^{i,f}| + \|K_k(\bar{P}_k)\|\,|Y_k - H_k\,\bar{X}_k^{i,f} + V_k^i|,$$

whether Assumption A or Assumption B holds or not, and using the triangle inequality yields

$$(\mathbb{E}|\bar{X}_k^{i,a}|^p)^{1/p} \leq (\mathbb{E}|\bar{X}_k^{i,f}|^p)^{1/p} + \|K_k(\bar{P}_k)\|\,(\mathbb{E}|Y_k - H_k\,\bar{X}_k^{i,f} + V_k^i|^p)^{1/p},$$

hence

$$\bar{M}_k^{p,a} \leq \bar{M}_k^{p,f} + \|K_k(\bar{P}_k)\|\,\bar{R}_k^p$$
$$\leq \bar{M}_k^{p,f} + \|K_k(\bar{P}_k)\|\,(|Y_k| + \|H_k\|\,\bar{M}_k^{p,f} + c_p\,\lambda_{\max}^{1/2}(R_k)),$$

where $\lambda_{\max}(R_k)$ denotes the largest eigenvalue of the covariance matrix R_k. □

Introduce the following notation for the empirical moments

$$\bar{M}_k^{N,p,f} = (\frac{1}{N} \sum_{i=1}^{N} |\bar{X}_k^{i,f}|^p)^{1/p} \quad \text{and} \quad \bar{M}_k^{N,p,a} = (\frac{1}{N} \sum_{i=1}^{N} |\bar{X}_k^{i,a}|^p)^{1/p},$$

and similarly for the empirical moments of the perturbed residuals

$$\bar{R}_k^{N,p} = (\frac{1}{N} \sum_{i=1}^{N} |Y_k - H_k \bar{X}_k^{i,f} - V_k^i|^p)^{1/p}.$$

It follows from the strong law of large numbers that: if the moment $\bar{M}_k^{p,f}$ is finite, then

$$\bar{M}_k^{N,p,f} \longrightarrow \bar{M}_k^{p,f} \quad \text{and} \quad \bar{R}_k^{N,p} \longrightarrow \bar{R}_k^p,$$

almost surely as $N \uparrow \infty$, and if the moment $\bar{M}_k^{p,a}$ is finite, then

$$\bar{M}_k^{N,p,a} \longrightarrow \bar{M}_k^{p,a},$$

almost surely as $N \uparrow \infty$.

22.3 Control of the Kalman gain matrix

There is only one visible difference between equations (22.1) and (22.2) for the elements in the ensemble Kalman filter, and equations (22.4) and (22.5) for the corresponding substitute i.i.d. random vectors: indeed, the coefficients are the same and the same random input vectors are used, however the Kalman gain matrix $K_k(P_k^N)$ in equation (22.2) is based on the empirical covariance matrix P_k^N, which is responsible for mean-field interaction and dependence, whereas the Kalman gain matrix $K_k(\bar{P}_k)$ in equation (22.5) is based on the deterministic limiting covariance matrix $\bar{P}_k$, which is responsible for decoupling and independence. To assess the impact of this substitution, notice that if Assumption A holds, then by difference

$$|X_k^{i,f} - \bar{X}_k^{i,f}| \leq L \, |X_{k-1}^{i,a} - \bar{X}_{k-1}^{i,a}|, \tag{22.6}$$

whereas if Assumption B holds, then by difference

$$|X_k^{i,f} - \bar{X}_k^{i,f}| \leq L \, |X_{k-1}^{i,a} - \bar{X}_{k-1}^{i,a}| \, (1 + |\bar{X}_{k-1}^{i,a}|^s) + L \, |X_{k-1}^{i,a} - \bar{X}_{k-1}^{i,a}|^{s+1}. \tag{22.7}$$

Similarly, by difference

$$X_k^{i,a} - \bar{X}_k^{i,a} = X_k^{i,f} + K_k(P_k^N)(Y_k - H_k X_k^{i,f} - V_k^i) - \bar{X}_k^{i,f} - K_k(\bar{P}_k)(Y_k - H_k \bar{X}_k^{i,f} - V_k^i)$$

$$= (I - K_k(\bar{P}_k) H_k)(X_k^{i,f} - \bar{X}_k^{i,f}) - (K_k(P_k^N) - K_k(\bar{P}_k)) H_k (X_k^{i,f} - \bar{X}_k^{i,f})$$

$$+ (K_k(P_k^N) - K_k(\bar{P}_k)) (Y_k - H_k \bar{X}_k^{i,f} - V_k^i),$$

hence

$$|X_k^{i,a} - \bar{X}_k^{i,a}| \leq \|I - K_k(\bar{P}_k) H_k\| |X_k^{i,f} - \bar{X}_k^{i,f}|$$
$$+ \|K_k(P_k^N) - K_k(\bar{P}_k)\| \|H_k\| |X_k^{i,f} - \bar{X}_k^{i,f}| \quad (22.8)$$
$$+ \|K_k(P_k^N) - K_k(\bar{P}_k)\| |Y_k - H_k \bar{X}_k^{i,f} - V_k^i|.$$

The first step addressed in Proposition 22.2 is to prove some local Lipschitz continuity of the mapping $P \mapsto K_k(P)$ in order to control the difference $K_k(P_k^N) - K_k(\bar{P}_k)$ in terms of the difference $P_k^N - \bar{P}_k$, which is then decomposed as $(P_k^N - \bar{P}_k^N) + (\bar{P}_k^N - \bar{P}_k)$. The second step addressed in Lemma 22.2 is to study the contiguity of the empirical covariance matrices: the difference $P_k^N - \bar{P}_k^N$ can be controlled in terms of

$$\Delta_k^{N,2,f} = (\frac{1}{N} \sum_{i=1}^N |X_k^{i,f} - \bar{X}_k^{i,f}|^2)^{1/2},$$

which is a measure of contiguity between elements in the ensemble Kalman filter and the corresponding substitute i.i.d. random vectors. The convergence of $\Delta_k^{N,2,f}$ to zero almost surely and in $\mathbb{L}^p$ as $N \uparrow \infty$, is proved in Propositions 22.3 and 22.4, respectively. The third and last step addressed in Lemma 22.3 is to prove the consistency of the empirical covariance matrix for the substitute i.i.d. random vectors: because of independence, the difference $\bar{P}_k^N - \bar{P}_k$ goes to zero almost surely and in $\mathbb{L}^p$ as $N \uparrow \infty$, by the strong law of large numbers and by the Marcinkiewicz–Zygmund inequality (Gut [13], ch. 3, sec. 8), respectively.

22.3.1 Local Lipschitz continuity of the Kalman gain matrix

If the $d \times d$ covariance matrix R_k is invertible, then its smallest eigenvalue $\lambda_{\min}(R_k)$ is positive, and it is possible to prove that the mapping $P \mapsto K_k(P)$ defined in (22.3) has at most linear growth and is locally Lipschitz continuous, as follows.

Proposition 22.2 *If the $d \times d$ covariance matrix R_k is invertible, then*

$$\|K_k(P)\| \leq \frac{\|H_k\|}{\lambda_{\min}(R_k)} \|P\|,$$

and

$$\|K_k(P) - K_k(P')\| \leq \|I - K_k(P') H_k\| \frac{\|H_k\|}{\lambda_{\min}(R_k)} \|P - P'\|, \quad (22.9)$$

for any (not necessarily invertible) $m \times m$ covariance matrices P and P'.

Proof. Since $H_k \, P \, H_k^* + R_k$ is a symmetric matrix, there exist an orthogonal matrix O and a diagonal matrix D such that $H_k \, P \, H_k^* + R_k = O \, D \, O^*$, hence $\left(H_k \, P \, H_k^* + R_k \right)^{-1} = O \, D^{-1} \, O^* = O \, D^{-1/2} \, O^* \, O \, D^{-1/2} \, O^* = T^* \, T$, with $T = O \, D^{-1/2} \, O^*$ by definition, and

$$\left\| \left(H_k \, P \, H_k^* + R_k \right)^{-1} \right\| = \sup_{u \neq 0} \frac{u^* \left(H_k \, P \, H_k^* + R_k \right)^{-1} u}{|u|^2} = \sup_{u \neq 0} \frac{|T u|^2}{|u|^2} = \sup_{v \neq 0} \frac{|v|^2}{|T^{-1} v|^2} \, .$$

Clearly $T^{-1} = O \, D^{1/2} \, O^*$, hence

$$|T^{-1} v|^2 = v^* \, O \, D^{1/2} \, O^* \, O \, D^{1/2} \, O^* \, v = v^* \, O \, D \, O^* \, v = v^* \left(H_k \, P \, H_k^* + R_k \right) v,$$

and

$$\left\| \left(H_k \, P \, H_k^* + R_k \right)^{-1} \right\| = \sup_{v \neq 0} \frac{|v|^2}{v^* \left(H_k \, P \, H_k^* + R_k \right) v} \leq \sup_{v \neq 0} \frac{|v|^2}{v^* \, R_k \, v} = \frac{1}{\lambda_{\min}(R_k)} \, .$$

Therefore

$$\| K_k(P) \| \leq \frac{\| H_k \|}{\lambda_{\min}(R_k)} \, \| P \|,$$

since $\| H_k^* \| = \| H_k \|$ for the norm matrix associated with the Euclidean norm, and by difference

$$K_k(P) - K_k(P') = P \, H_k^* \left(H_k \, P \, H_k^* + R_k \right)^{-1} - P' \, H_k^* \left(H_k \, P' \, H_k^* + R_k \right)^{-1}$$
$$= P \, H_k^* \left(H_k \, P \, H_k^* + R_k \right)^{-1} - P' \, H_k^* \left(H_k \, P \, H_k^* + R_k \right)^{-1}$$
$$+ P' \, H_k^* \left(\left(H_k \, P \, H_k^* + R_k \right)^{-1} - \left(H_k \, P' \, H_k^* + R_k \right)^{-1} \right)$$
$$= (P - P') \, H_k^* \left(H_k \, P \, H_k^* + R_k \right)^{-1}$$
$$- P' \, H_k^* \left(H_k \, P' \, H_k^* + R_k \right)^{-1} H_k \, (P - P') \, H_k^* \left(H_k \, P \, H_k^* + R_k \right)^{-1}$$
$$= (P - P') \, H_k^* \left(H_k \, P \, H_k^* + R_k \right)^{-1}$$
$$- K_k(P') \, H_k \, (P - P') \, H_k^* \left(H_k \, P \, H_k^* + R_k \right)^{-1}$$
$$= (I - K_k(P') \, H_k) \, (P - P') \, H_k^* \left(H_k \, P \, H_k^* + R_k \right)^{-1},$$

hence

$$\| K_k(P) - K_k(P') \| \leq \| I - K_k(P') \, H_k \| \frac{\| H_k \|}{\lambda_{\min}(R_k)} \, \| P - P' \| \, . \qquad \square$$

22.3.2 Contiguity control of the empirical covariance matrices

Lemma 22.2 *The difference between the two empirical covariance matrices satisfies*

$$\| P_k^N - \bar{P}_k^N \| \leq 2 \, |\Delta_k^{N,2,f}|^2 + 4 \, \bar{M}_k^{N,2,f} \, \Delta_k^{N,2,f} \, . \tag{22.10}$$

Proof. Notice that

$$P_k^N = \frac{1}{N} \sum_{i=1}^N X_k^{i,f} (X_k^{i,f})^* - (\frac{1}{N} \sum_{i=1}^N X_k^{i,f}) (\frac{1}{N} \sum_{i=1}^N X_k^{i,f})^*,$$

and similarly

$$\bar{P}_k^N = \frac{1}{N} \sum_{i=1}^N \bar{X}_k^{i,f} (\bar{X}_k^{i,f})^* - (\frac{1}{N} \sum_{i=1}^N \bar{X}_k^{i,f}) (\frac{1}{N} \sum_{i=1}^N \bar{X}_k^{i,f})^*.$$

Using the identity $a^2 - b^2 = (a-b)^2 + 2b(a-b)$ yields

$$u^* (P_k^N - \bar{P}_k^N) u =$$

$$= \frac{1}{N} \sum_{i=1}^N |u^* X_k^{i,f}|^2 - \frac{1}{N} \sum_{i=1}^N |u^* \bar{X}_k^{i,f}|^2 - |\frac{1}{N} \sum_{i=1}^N u^* X_k^{i,f}|^2 + |\frac{1}{N} \sum_{i=1}^N u^* \bar{X}_k^{i,f}|^2$$

$$= \frac{1}{N} \sum_{i=1}^N |u^* (X_k^{i,f} - \bar{X}_k^{i,f})|^2 + 2 \frac{1}{N} \sum_{i=1}^N (u^* \bar{X}_k^{i,f}) (u^* (X_k^{i,f} - \bar{X}_k^{i,f}))$$

$$- |\frac{1}{N} \sum_{i=1}^N u^*(X_k^{i,f} - \bar{X}_k^{i,f})|^2 - 2(\frac{1}{N} \sum_{i=1}^N u^* \bar{X}_k^{i,f}) (\frac{1}{N} \sum_{i=1}^N u^* (X_k^{i,f} - \bar{X}_k^{i,f})),$$

for any u, hence

$$\| P_k^N - \bar{P}_k^N \| = \sup_{u \neq 0} \frac{|u^* (P_k^N - \bar{P}_k^N) u|}{|u|^2}$$

$$\leq \frac{1}{N} \sum_{i=1}^N |X_k^{i,f} - \bar{X}_k^{i,f}|^2 + 2 \frac{1}{N} \sum_{i=1}^N |\bar{X}_k^{i,f}| \, |X_k^{i,f} - \bar{X}_k^{i,f}|$$

$$+ (\frac{1}{N} \sum_{i=1}^N |X_k^{i,f} - \bar{X}_k^{i,f}|)^2 + 2 (\frac{1}{N} \sum_{i=1}^N |\bar{X}_k^{i,f}|) (\frac{1}{N} \sum_{i=1}^N |X_k^{i,f} - \bar{X}_k^{i,f}|)$$

$$\leq 2 \frac{1}{N} \sum_{i=1}^N |X_k^{i,f} - \bar{X}_k^{i,f}|^2 + 4 (\frac{1}{N} \sum_{i=1}^N |\bar{X}_k^{i,f}|^2)^{1/2} (\frac{1}{N} \sum_{i=1}^N |X_k^{i,f} - \bar{X}_k^{i,f}|^2)^{1/2},$$

or in other words

$$\| P_k^N - \bar{P}_k^N \| \leq 2 |\Delta_k^{N,2,f}|^2 + 4 \, \bar{M}_k^{N,2,f} \, \Delta_k^{N,2,f}. \qquad \square$$

22.3.3 Consistency of the empirical covariance matrix

Lemma 22.3 *The difference between the covariance matrix of the substitute i.i.d. random vectors and their empirical covariance matrix satisfies*

$$\varepsilon_k^N = \| \bar{P}_k^N - \bar{P}_k \| \longrightarrow 0,$$

almost surely as $N \uparrow \infty$, and

$$\sup_{N \geq 1} \sqrt{N} \, (\mathbb{E} |\varepsilon_k^N|^p)^{1/p} < \infty,$$

for any order $p \geq 2$.

Proof. The following decomposition holds

$$\bar{P}_k^N = \frac{1}{N} \sum_{i=1}^N (\bar{X}_k^{i,f} - \bar{m}_k^N)(\bar{X}_k^{i,f} - \bar{m}_k^N)^*$$

$$= \frac{1}{N} \sum_{i=1}^N (\bar{X}_k^{i,f} - \bar{m}_k - (\bar{m}_k^N - \bar{m}_k))(\bar{X}_k^{i,f} - \bar{m}_k - (\bar{m}_k^N - \bar{m}_k))^*$$

$$= \frac{1}{N} \sum_{i=1}^N (\bar{X}_k^{i,f} - \bar{m}_k)(\bar{X}_k^{i,f} - \bar{m}_k)^* - (\frac{1}{N} \sum_{i=1}^N \bar{X}_k^{i,f} - \bar{m}_k)(\bar{m}_k^N - \bar{m}_k)^*$$

$$- (\bar{m}_k^N - \bar{m}_k)(\frac{1}{N} \sum_{i=1}^N \bar{X}_k^{i,f} - \bar{m}_k)^* + (\bar{m}_k^N - \bar{m}_k)(\bar{m}_k^N - \bar{m}_k)^*$$

$$= \frac{1}{N} \sum_{i=1}^N (\bar{X}_k^{i,f} - \bar{m}_k)(\bar{X}_k^{i,f} - \bar{m}_k)^* - (\bar{m}_k^N - \bar{m}_k)(\bar{m}_k^N - \bar{m}_k)^*,$$

hence

$$u^* (\bar{P}_k^N - \bar{P}_k) u = u^* (\frac{1}{N} \sum_{i=1}^N (\bar{X}_k^{i,f} - \bar{m}_k)(\bar{X}_k^{i,f} - \bar{m}_k)^* - \bar{P}_k) u$$

$$- |u^* (\frac{1}{N} \sum_{i=1}^N \bar{X}_k^{i,f} - \bar{m}_k)|^2,$$

for any u, and

$$\varepsilon_k^N = \| \bar{P}_k^N - \bar{P}_k \| = \sup_{u \neq 0} \frac{|u^* (\bar{P}_k^N - \bar{P}_k) u|}{|u|^2}$$

$$\leq \| \frac{1}{N} \sum_{i=1}^N (\bar{X}_k^{i,f} - \bar{m}_k)(\bar{X}_k^{i,f} - \bar{m}_k)^* - \bar{P}_k \| + |\frac{1}{N} \sum_{i=1}^N \bar{X}_k^{i,f} - \bar{m}_k|^2.$$

Finally, notice that

$$|u^* M u| = |\sum_{j,j'=1}^m u_j M_{jj'} u_{j'}| \leq \max_{j=1,\cdots,m} |u_j|^2 \sum_{j,j'=1}^m |M_{jj'}| \leq |u|^2 \sum_{j,j'=1}^m |M_{jj'}|,$$

hence

$$\|M\| = \sup_{u \neq 0} \frac{|u^* M u|}{|u|^2} \leq \sum_{j,j'=1}^{m} |M_{jj'}|,$$

for any symmetric $m \times m$ matrix M. Taking

$$M = \frac{1}{N} \sum_{i=1}^{N} (\bar{X}_k^{i,f} - \bar{m}_k)(\bar{X}_k^{i,f} - \bar{m}_k)^* - \bar{P}_k,$$

gives

$$M_{jj'} = \frac{1}{N} \sum_{i=1}^{N} (\bar{X}_k^{i,j,f} - \bar{m}_k^j)(\bar{X}_k^{i,j',f} - \bar{m}_k^{j'}) - \bar{P}_k^{j,j'},$$

and it follows from the law of large numbers that

$$\varepsilon_k^N \leq \sum_{j,j'=1}^{m} |\frac{1}{N} \sum_{i=1}^{N} (\bar{X}_k^{i,j,f} - \bar{m}_k^j)(\bar{X}_k^{i,j',f} - \bar{m}_k^{j'}) - \bar{P}_k^{j,j'}|$$

$$+ \sum_{j=1}^{m} |\frac{1}{N} \sum_{i=1}^{N} \bar{X}_k^{i,j,f} - \bar{m}_k^j|^2 \longrightarrow 0,$$

almost surely as $N \uparrow \infty$, and using the triangle inequality and then the Marcinkiewicz–Zygmund inequality, yields

$$(\mathbb{E}|\varepsilon_k^N|^p)^{1/p} \leq \sum_{j,j'=1}^{m} (\mathbb{E}|\frac{1}{N} \sum_{i=1}^{N} (\bar{X}_k^{i,j,f} - \bar{m}_k^j)(\bar{X}_k^{i,j',f} - \bar{m}_k^{j'}) - \bar{P}_k^{j,j'}|^p)^{1/p}$$

$$+ \sum_{j=1}^{m} (\mathbb{E}|\frac{1}{N} \sum_{i=1}^{N} \bar{X}_k^{i,j,f} - \bar{m}_k^j|^{2p})^{1/p}$$

$$\leq \frac{c_p}{\sqrt{N}} \sum_{j,j'=1}^{m} (\mathbb{E}|(\bar{X}_k^{i,j,f} - \bar{m}_k^j)(\bar{X}_k^{i,j',f} - \bar{m}_k^{j'}) - \bar{P}_k^{j,j'}|^p)^{1/p}$$

$$+ \frac{c_{2p}^2}{N} \sum_{j=1}^{m} (\mathbb{E}|\bar{X}_k^{i,j,f} - \bar{m}_k^j|^{2p})^{1/p},$$

for any order $p \geq 2$. $\square$

22.4 Contiguity of the elements

Introduce the following definitions

$$\Delta_k^{N,p,f} = (\frac{1}{N} \sum_{i=1}^{N} |X_k^{i,f} - \bar{X}_k^{i,f}|^p)^{1/p} \quad \text{and} \quad \Delta_k^{N,p,a} = (\frac{1}{N} \sum_{i=1}^{N} |X_k^{i,a} - \bar{X}_k^{i,a}|^p)^{1/p},$$

as measures of contiguity between elements in the ensemble Kalman filter and the corresponding substitute i.i.d. random vectors. These quantities satisfy the following recurrence inequalities, which are then used in Propositions 22.3 and 22.4 to prove almost sure contiguity and $\mathbb{L}^p$ contiguity, respectively.

Lemma 22.4 *If Assumption A holds, then*

$$\Delta_k^{N,p,f} \leq L \, \Delta_{k-1}^{N,p,a}, \tag{22.11}$$

whereas if Assumption B holds, then

$$\Delta_k^{N,p,f} \leq L \, \Delta_{k-1}^{N,pr,a} (1 + |\bar{M}_{k-1}^{N,s \, pr',a}|^s) + L \, |\Delta_{k-1}^{N,p(s+1),a}|^{s+1}, \tag{22.12}$$

where r, r' are conjugate exponents, i.e. $1/r + 1/r' = 1$, and in any case

$$\Delta_k^{N,p,a} \leq C_k (\Delta_k^{N,p,f} + (2 |\Delta_k^{N,2,f}|^2 + 4 \, \bar{M}_k^{N,2,f} \, \Delta_k^{N,2,f} + \varepsilon_k^N)(\Delta_k^{N,p,f} + \bar{R}_k^{N,p})), \tag{22.13}$$

whether Assumption A or Assumption B holds or not.

Proof. If Assumption A holds, then it follows from (22.6) that

$$(\frac{1}{N} \sum_{i=1}^{N} |X_k^{i,f} - \bar{X}_k^{i,f}|^p)^{1/p} \leq L \, (\frac{1}{N} \sum_{i=1}^{N} |X_{k-1}^{i,a} - \bar{X}_{k-1}^{i,a}|^p)^{1/p},$$

or in other words

$$\Delta_k^{N,p,f} \leq L \, \Delta_{k-1}^{N,p,a},$$

whereas if Assumption B holds, then using the triangle inequality it follows from (22.7) that

$$(\frac{1}{N} \sum_{i=1}^{N} |X_k^{i,f} - \bar{X}_k^{i,f}|^p)^{1/p} \leq L \, (\frac{1}{N} \sum_{i=1}^{N} |X_{k-1}^{i,a} - \bar{X}_{k-1}^{i,a}|^p (1 + |\bar{X}_{k-1}^{i,a}|^s)^p)^{1/p}$$

$$+ L \, (\frac{1}{N} \sum_{i=1}^{N} |X_{k-1}^{i,a} - \bar{X}_{k-1}^{i,a}|^{p(s+1)})^{1/p},$$

and using the Hölder inequality and then the triangle inequality again to control the first term, yields

$$(\frac{1}{N} \sum_{i=1}^{N} |X_{k-1}^{i,a} - \bar{X}_{k-1}^{i,a}|^p (1 + |\bar{X}_{k-1}^{i,a}|^s)^p)^{1/p}$$

$$\leq (\frac{1}{N} \sum_{i=1}^{N} |X_{k-1}^{i,a} - \bar{X}_{k-1}^{i,a}|^{pr})^{1/pr} (\frac{1}{N} \sum_{i=1}^{N} (1 + |\bar{X}_{k-1}^{i,a}|^s)^{pr'})^{1/pr'}$$

$$\leq (\frac{1}{N} \sum_{i=1}^{N} |X_{k-1}^{i,a} - \bar{X}_{k-1}^{i,a}|^{pr})^{1/pr} (1 + (\frac{1}{N} \sum_{i=1}^{N} |\bar{X}_{k-1}^{i,a}|^{s\,pr'})^{1/pr'}),$$

where r, r' are conjugate exponents, i.e. $1/r + 1/r' = 1$, or in other words

$$\Delta_k^{N,p,f} \leq L\, \Delta_{k-1}^{N,pr,a} (1 + |\bar{M}_{k-1}^{N,s\,pr',a}|^s) + L\, |\Delta_{k-1}^{N,p(s+1),a}|^{s+1}.$$

Similarly, it follows from (22.8) and (22.9) that in any case

$$|X_k^{i,a} - \bar{X}_k^{i,a}| \leq \|I - K_k(\bar{P}_k)\, H_k\|\, |X_k^{i,f} - \bar{X}_k^{i,f}|$$

$$+ \|K_k(P_k^N) - K_k(\bar{P}_k)\|\, \|H_k\|\, |X_k^{i,f} - \bar{X}_k^{i,f}|$$

$$+ \|K_k(P_k^N) - K_k(\bar{P}_k)\|\, |Y_k - H_k\, \bar{X}_k^{i,f} + V_k^i|$$

$$\leq \|I - K_k(\bar{P}_k)\, H_k\|\, |X_k^{i,f} - \bar{X}_k^{i,f}|$$

$$+ \|I - K_k(\bar{P}_k)\, H_k\|\, \frac{\|H_k\|^2}{\lambda_{\min}(R_k)}\, \|P_k^N - \bar{P}_k\|\, |X_k^{i,f} - \bar{X}_k^{i,f}|$$

$$+ \|I - K_k(\bar{P}_k)\, H_k\|\, \frac{\|H_k\|}{\lambda_{\min}(R_k)}\, \|P_k^N - \bar{P}_k\|\, |Y_k - H_k\, \bar{X}_k^{i,f} - V_k^i|$$

$$\leq C_k\, (|X_k^{i,f} - \bar{X}_k^{i,f}| + \|P_k^N - \bar{P}_k\|\, (|X_k^{i,f} - \bar{X}_k^{i,f}| + |Y_k - H_k\, \bar{X}_k^{i,f} - V_k^i|)),$$

whether Assumption A or Assumption B holds or not, where

$$C_k = \|I - K_k(\bar{P}_k)\, H_k\|\, \max(1, \frac{\|H_k\|}{\lambda_{\min}(R_k)}, \frac{\|H_k\|^2}{\lambda_{\min}(R_k)}),$$

and using the triangle inequality yields

$$(\frac{1}{N} \sum_{i=1}^{N} |X_k^{i,a} - \bar{X}_k^{i,a}|^p)^{1/p} \leq C_k\, ((\frac{1}{N} \sum_{i=1}^{N} |X_k^{i,f} - \bar{X}_k^{i,f}|^p)^{1/p}$$

$$+ \|P_k^N - \bar{P}_k\|\, (\frac{1}{N} \sum_{i=1}^{N} |X_k^{i,f} - \bar{X}_k^{i,f}|^p)^{1/p}$$

$$+ \|P_k^N - \bar{P}_k\|\, (\frac{1}{N} \sum_{i=1}^{N} |Y_k - H_k\, \bar{X}_k^{i,f} - V_k^i|^p)^{1/p}),$$

or in other words

$$\Delta_k^{N,p,a} \leq C_k (\Delta_k^{N,p,f} + \| P_k^N - \bar{P}_k \| (\Delta_k^{N,p,f} + \bar{R}_k^{N,p})),$$

and it follows from the triangle inequality and from (22.10) that

$$\| P_k^N - \bar{P}_k \| \leq \| P_k^N - \bar{P}_k^N \| + \| \bar{P}_k^N - \bar{P}_k \| \leq 2 |\Delta_k^{N,2,f}|^2 + 4 \bar{M}_k^{N,2,f} \Delta_k^{N,2,f} + \varepsilon_k^N,$$

hence

$$\Delta_k^{N,p,a} \leq C_k(\Delta_k^{N,p,f} + (2|\Delta_k^{N,2,f}|^2 + 4 \bar{M}_k^{N,2,f} \Delta_k^{N,2,f} + \varepsilon_k^N)(\Delta_k^{N,p,f} + \bar{R}_k^{N,p})).$$

□

22.4.1 Almost sure contiguity of the elements

Proposition 22.3 *If Assumption A holds, and if the random vector X_0 has finite moment of order p for some $p \geq 2$, then*

$$\Delta_k^{N,p,f} \longrightarrow 0 \quad \text{and} \quad \Delta_k^{N,p,a} \longrightarrow 0,$$

for the same order p, almost surely as $N \uparrow \infty$.

If Assumption B holds, and if the random vector X_0 has finite moments of any order, then

$$\Delta_k^{N,p,f} \longrightarrow 0 \quad \text{and} \quad \Delta_k^{N,p,a} \longrightarrow 0,$$

for any order p, almost surely as $N \uparrow \infty$.

Proof. (by induction). Initially

$$\Delta_0^{N,p,f} = (\frac{1}{N} \sum_{i=1}^{N} |X_0^{i,f} - \bar{X}_0^{i,f}|^p)^{1/p} = 0.$$

If Assumption A holds, then estimate (22.11) holds, i.e.

$$\Delta_k^{N,p,f} \leq L \, \Delta_{k-1}^{N,p,a},$$

and it follows from the induction assumption that $\Delta_k^{N,p,f} \longrightarrow 0$ almost surely as $N \uparrow \infty$, whereas if Assumption B holds, then estimate (22.12) holds, i.e.

$$\Delta_k^{N,p,f} \leq L \, \Delta_{k-1}^{N,pr,a} (1 + |\bar{M}_{k-1}^{N,s\,pr',a}|^s) + L \, |\Delta_{k-1}^{N,p(s+1),a}|^{s+1},$$

and since $\bar{M}_{k-1}^{N,s\,pr',a} \longrightarrow \bar{M}_{k-1}^{s\,pr',a}$ almost surely as $N \uparrow \infty$, with a finite limit $\bar{M}_{k-1}^{s\,pr',a}$, then it follows from the induction assumption that $\Delta_k^{N,p,f} \longrightarrow 0$ almost surely as $N \uparrow \infty$. In any case estimate (22.13) holds, i.e.

$$\Delta_k^{N,p,a} \leq C_k(\Delta_k^{N,p,f} + (2|\Delta_k^{N,2,f}|^2 + 4 \bar{M}_k^{N,2,f} \Delta_k^{N,2,f} + \varepsilon_k^N)(\Delta_k^{N,p,f} + \bar{R}_k^{N,p})),$$

whether Assumption A or Assumption B holds or not, and since $\bar{M}_k^{N,2,f} \longrightarrow \bar{M}_k^{2,f}$ and $\bar{R}_k^{N,p} \longrightarrow \bar{R}_k^p$ almost surely as $N \uparrow \infty$, with finite limits $\bar{M}_k^{2,f}$ and $\bar{R}_k^p$, and since $\varepsilon_k^N \longrightarrow 0$ almost surely as $N \uparrow \infty$ in view of Lemma 22.3, then it follows from the induction assumption that $\Delta_k^{N,p,a} \longrightarrow 0$ almost surely as $N \uparrow \infty$. □

22.4.2 $\mathbb{L}^p$-contiguity of the elements

The quantities

$$D_k^{N,p,f} = (\mathbb{E}|X_k^{i,f} - \bar{X}_k^{i,f}|^p)^{1/p} \quad \text{and} \quad D_k^{N,p,a} = (\mathbb{E}|X_k^{i,a} - \bar{X}_k^{i,a}|^p)^{1/p},$$

which still depends on N, since the probability distribution of an element in the ensemble Kalman filter depends on the ensemble size, are used to control the moments of the measures of contiguity.

Lemma 22.5

$$(\mathbb{E}|\Delta_k^{N,p,f}|^q)^{1/q} \leq D_k^{N,p \vee q,f} \quad \text{and} \quad (\mathbb{E}|\Delta_k^{N,p,a}|^q)^{1/q} \leq D_k^{N,p \vee q,a},$$

and

$$(\mathbb{E}|\bar{M}_k^{N,p,f}|^q)^{1/q} \leq \bar{M}_k^{p \vee q,f} \quad \text{and} \quad (\mathbb{E}|\bar{M}_k^{N,p,a}|^q)^{1/q} \leq \bar{M}_k^{p \vee q,a},$$

whereas equalities hold if $p = q$.

Proof. Throughout the proof, let the superscript • index either forecast-related or analysis-related quantities. Clearly

$$\mathbb{E}|\Delta_k^{N,p,\bullet}|^p = \mathbb{E}(\frac{1}{N}\sum_{i=1}^N |X_k^{i,\bullet} - \bar{X}_k^{i,\bullet}|^p) = \mathbb{E}|X_k^{i,\bullet} - \bar{X}_k^{i,\bullet}|^p = |D_k^{N,p,\bullet}|^p,$$

since $(X_k^{i,\bullet}, \bar{X}_k^{i,\bullet})$ have the same joint probability distribution for any $i = 1, \cdots, N$, by symmetry, hence

$$(\mathbb{E}|\Delta_k^{N,p,\bullet}|^p)^{1/p} = D_k^{N,p,\bullet},$$

which proves the result for $q = p$. If $q \geq p$, then the mapping $x \mapsto x^{q/p}$ is convex and it follows from the Jensen inequality that

$$|\Delta_k^{N,p,\bullet}|^q = (\frac{1}{N}\sum_{i=1}^N |X_k^{i,\bullet} - \bar{X}_k^{i,\bullet}|^p)^{q/p} \leq (\frac{1}{N}\sum_{i=1}^N |X_k^{i,\bullet} - \bar{X}_k^{i,\bullet}|^q) = |\Delta_k^{N,q,\bullet}|^q,$$

hence

$$(\mathbb{E}|\Delta_k^{N,p,\bullet}|^q)^{1/q} \leq (\mathbb{E}|\Delta_k^{N,q,\bullet}|^q)^{1/q} = D_k^{N,q,\bullet}.$$

If $q \leq p$, then the mapping $x \mapsto x^{q/p}$ is concave and it follows from the Jensen inequality that

$$\mathbb{E}|\Delta_k^{N,p,\bullet}|^q = \mathbb{E}(|\Delta_k^{N,p,\bullet}|^p)^{q/p} \leq (\mathbb{E}|\Delta_k^{N,p,\bullet}|^p)^{q/p} = |D_k^{N,p,\bullet}|^q,$$

hence

$$(\mathbb{E}|\Delta_k^{N,p,\bullet}|^q)^{1/q} \leq D_k^{N,p,\bullet}.$$

Similarly

$$\mathbb{E}|\bar{M}_k^{N,p,\bullet}|^p = \mathbb{E}(\frac{1}{N}\sum_{i=1}^{N}|\bar{X}_k^{i,\bullet}|^p) = \mathbb{E}|\bar{X}_k^{i,\bullet}|^p = |\bar{M}_k^{p,\bullet}|^p,$$

since $\bar{X}_k^{i,\bullet}$ has the same distribution for any $i = 1, \cdots, N$, hence

$$(\mathbb{E}|\bar{M}_k^{N,p,\bullet}|^p)^{1/p} = \bar{M}_k^{p,\bullet},$$

which proves the result for $q = p$, and the result for the two cases $q \geq p$ and $q \leq p$ can again be obtained by simple convexity arguments. □

Proposition 22.4 *If Assumption B holds, and if the random vector X_0 has finite moments of any order, then*

$$\sup_{N\geq 1} \sqrt{N} \, D_k^{N,p,f} < \infty \quad \text{and} \quad \sup_{N\geq 1} \sqrt{N} \, D_k^{N,p,a} < \infty,$$

for any order p.

Proof. (by induction). Initially

$$\Delta_0^{N,p,f} = (\frac{1}{N}\sum_{i=1}^{N}|X_0^{i,f} - \bar{X}_0^{i,f}|^p)^{1/p} = 0.$$

If Assumption A holds, then estimate (22.11) holds, i.e.

$$\Delta_k^{N,p,f} \leq L \, \Delta_{k-1}^{N,p,a},$$

hence

$$(\mathbb{E}|\Delta_k^{N,p,f}|^p)^{1/p} \leq L \, (\mathbb{E}|\Delta_{k-1}^{N,p,a}|^p)^{1/p},$$

or in other words

$$D_k^{N,p,f} \leq L \, D_{k-1}^{N,p,a},$$

in view of Lemma 22.5, and it follows from the induction assumption that $\sqrt{N} \, D_k^{N,p,f}$ is bounded uniformly w.r.t. N, whereas if Assumption B holds, then estimate (22.12) holds, i.e.

$$\Delta_k^{N,p,f} \leq L \, \Delta_{k-1}^{N,pr,a} \, (1 + |\bar{M}_{k-1}^{N,s\,pr',a}|^s) + L \, |\Delta_{k-1}^{N,p(s+1),a}|^{s+1},$$

and using the triangle inequality yields

$$(\mathbb{E}|\Delta_k^{N,p,f}|^p)^{1/p} \leq L \, (\mathbb{E}(|\Delta_{k-1}^{N,pr,a}|^p \, (1+|\bar{M}_{k-1}^{N,s \, pr',a}|^s)^p))^{1/p}$$

$$+L \, (\mathbb{E}|\Delta_{k-1}^{N,p(s+1),a}|^{p(s+1)})^{1/p},$$

and using the Hölder inequality and then the triangle inequality again to control the first term, yields

$$(\mathbb{E}(|\Delta_{k-1}^{N,pr,a}|^p \, (1+|\bar{M}_{k-1}^{N,s \, pr',a}|^s)^p))^{1/p}$$

$$\leq (\mathbb{E}|\Delta_{k-1}^{N,pr,a}|^{pr})^{1/pr} \, (\mathbb{E}(1+|\bar{M}_{k-1}^{N,s \, pr',a}|^s)^{pr'})^{1/pr'}$$

$$\leq (\mathbb{E}|\Delta_{k-1}^{N,pr,a}|^{pr})^{1/pr} \, (1+(\mathbb{E}|\bar{M}_{k-1}^{N,s \, pr',a}|^{s \, pr'})^{1/pr'}),$$

where r, r' are conjugate exponents, i.e. $1/r + 1/r' = 1$, or in other words

$$D_k^{N,p,f} \leq L \, D_{k-1}^{N,pr,a} \, (1+|\bar{M}_{k-1}^{s \, pr',a}|^s) + L \, |D_{k-1}^{N,p(s+1),a}|^{s+1},$$

in view of Lemma 22.5, and since $\bar{M}_{k-1}^{s \, pr',a}$ is finite, then it follows from the induction assumption that $\sqrt{N} \, D_k^{N,p,f}$ is bounded uniformly w.r.t. N. In any case estimate (22.13) holds, i.e.

$$\Delta_k^{N,p,a} \leq C_k(\Delta_k^{N,p,f} + (2|\Delta_k^{N,2,f}|^2 + 4 \bar{M}_k^{N,2,f} \Delta_k^{N,2,f} + \varepsilon_k^N)(\Delta_k^{N,p,f} + \bar{R}_k^{N,p}))$$

$$\leq C_k \, \Delta_k^{N,p,f} + 2 \, C_k \, \Delta_k^{N,2,f} \, (\Delta_k^{N,2,f} + 2 \, \bar{M}_k^{N,2,f}) \, (\Delta_k^{N,p,f} + \bar{R}_k^{N,p})$$

$$+ C_k \, \varepsilon_k^N \, (\Delta_k^{N,p,f} + \bar{R}_k^{N,p}),$$

whether Assumption A or Assumption B holds or not, and using the triangle inequality yields

$$(\mathbb{E}|\Delta_k^{N,p,a}|^p)^{1/p} \leq C_k \, (\mathbb{E}|\Delta_k^{N,p,f}|^p)^{1/p}$$

$$+ 2 \, C_k \, (\mathbb{E}(|\Delta_k^{N,2,f}|^p \, |\Delta_k^{N,2,f} + 2 \, \bar{M}_k^{N,2,f}|^p \, |\Delta_k^{N,p,f} + \bar{R}_k^{N,p}|^p))^{1/p}$$

$$+ C_k \, (\mathbb{E}(|\varepsilon_k^N|^p \, |\Delta_k^{N,p,f} + \bar{R}_k^{N,p}|^p))^{1/p},$$

and using the Hölder inequality and then the triangle inequality again to control the second and third terms, yields

$$(\mathbb{E}(|\Delta_k^{N,2,f}|^p \, |\Delta_k^{N,2,f} + 2 \, \bar{M}_k^{N,2,f}|^p \, |\Delta_k^{N,p,f} + \bar{R}_k^{N,p}|^p))^{1/p}$$

$$\leq (\mathbb{E}|\Delta_k^{N,2,f}|^{pr})^{1/pr} (\mathbb{E}|\Delta_k^{N,2,f} + 2\bar{M}_k^{N,2,f}|^{pr'})^{1/pr'} (\mathbb{E}|\Delta_k^{N,p,f} + \bar{R}_k^{N,p}|^{pr''})^{1/pr''}$$

$$\leq (\mathbb{E}|\Delta_k^{N,2,f}|^{pr})^{1/pr} ((\mathbb{E}|\Delta_k^{N,2,f}|^{pr'})^{1/pr'} + 2 \, (\mathbb{E}|\bar{M}_k^{N,2,f}|^{pr'})^{1/pr'})$$

$$((\mathbb{E}|\Delta_k^{N,p,f}|^{pr''})^{1/pr''} + (\mathbb{E}|\bar{R}_k^{N,p}|^{pr''})^{1/pr''}),$$

where r, r', r'' are conjugate exponents, i.e. $1/r + 1/r' + 1/r'' = 1$, and

$$(\mathbb{E}(|\varepsilon_k^N|^p |\Delta_k^{N,p,f} + \bar{R}_k^{N,p}|^p))^{1/p}$$

$$\leq (\mathbb{E}|\varepsilon_k^N|^{pq})^{1/pq} (\mathbb{E}|\Delta_k^{N,p,f} + \bar{R}_k^{N,p}|^{pq'})^{1/pq'}$$

$$\leq (\mathbb{E}|\varepsilon_k^N|^{pq})^{1/pq} ((\mathbb{E}|\Delta_k^{N,p,f}|^{pq'})^{1/pq'} + (\mathbb{E}|\bar{R}_k^{N,p}|^{pq'})^{1/pq'}),$$

where q, q' are conjugate exponents, i.e. $1/q + 1/q' = 1$, or in other words

$$D_k^{N,p,a} \leq C_k \, D_k^{N,p,f}$$

$$+ 2 \, C_k \, D_k^{N,2\vee(pr),f} (D_k^{N,2\vee(pr'),f} + 2 \, \bar{M}_k^{2\vee(pr'),f})(D_k^{N,pr'',f} + \bar{R}_k^{pr''})$$

$$+ C_k \, (\mathbb{E}|\varepsilon_k^N|^{pq})^{1/pq} (D_k^{N,pq',f} + \bar{R}_k^{pq'}),$$

in view of Lemma 22.5, and since $\bar{M}_k^{2\vee(pr'),f}$, $\bar{R}_k^{pr''}$ and $\bar{R}_k^{pq'}$ are finite, and since $\sqrt{N}\,\varepsilon_k^N$ is bounded in $\mathbb{L}^{pq}$ uniformly w.r.t. N in view of Lemma 22.3, then it follows from the induction assumption that $\sqrt{N}\,D_k^{N,p,a}$ is bounded uniformly w.r.t. N. □

22.5 Convergence of the ensemble Kalman filter

In view of the decomposition

$$\frac{1}{N} \sum_{i=1}^{N} \phi(X_k^{i,\bullet}) - \int_{\mathbb{R}^m} \phi(x) \, \bar{\mu}_k^{\bullet}(dx)$$

$$= \frac{1}{N} \sum_{i=1}^{N} (\phi(X_k^{i,\bullet}) - \phi(\bar{X}_k^{i,\bullet})) + \left(\frac{1}{N} \sum_{i=1}^{N} \phi(\bar{X}_k^{i,\bullet}) - \int_{\mathbb{R}^m} \phi(x) \, \bar{\mu}_k^{\bullet}(dx) \right),$$

where the superscript $\bullet$ indexes either forecast-related or analysis-related quantities, it is intuitively clear that almost sure contiguity of the elements, proved in Proposition 22.3, and the strong law of large numbers for the substitute i.i.d. random vectors could be used to prove weak convergence of the empirical probability distribution associated with the elements of the ensemble Kalman filter, almost surely. Similarly, $\mathbb{L}^p$-contiguity of the elements, proved in Proposition 22.4, and the Marcinkiewicz–Zygmund inequality (Gut [13], ch. 3, sec. 8) for the substitute i.i.d. random vectors could be used to prove weak convergence of the empirical probability distribution associated with the elements of the ensemble Kalman filter, in $\mathbb{L}^p$-mean. These statements are proved in Theorems 22.1 and 22.2, respectively.

22.5.1 Almost sure convergence

Theorem 22.1 *Let ϕ be a locally Lipschitz continuous function, with at most polynomial growth at infinity, i.e.*

$$|\phi(x) - \phi(x')| \leq L\, |x - x'|\, (1 + |x|^\sigma + |x'|^\sigma),$$

for any $x, x' \in \mathbb{R}^m$ and for some $\sigma \geq 0$.

If Assumption A holds, and if the random vector X_0 has finite moment of order p for some $p \geq \max(2, \sigma + 1)$, then

$$\frac{1}{N} \sum_{i=1}^{N} \phi(X_k^{i,f}) \longrightarrow \int_{\mathbb{R}^m} \phi(x)\, \bar{\mu}_k^f(dx) \quad \text{and} \quad \frac{1}{N} \sum_{i=1}^{N} \phi(X_k^{i,a}) \longrightarrow \int_{\mathbb{R}^m} \phi(x)\, \bar{\mu}_k^a(dx),$$

almost surely as $N \uparrow \infty$.

If Assumption B holds, and if the random vector X_0 has finite moments of any order, then

$$\frac{1}{N} \sum_{i=1}^{N} \phi(X_k^{i,f}) \longrightarrow \int_{\mathbb{R}^m} \phi(x)\, \bar{\mu}_k^f(dx) \quad \text{and} \quad \frac{1}{N} \sum_{i=1}^{N} \phi(X_k^{i,a}) \longrightarrow \int_{\mathbb{R}^m} \phi(x)\, \bar{\mu}_k^a(dx),$$

almost surely as $N \uparrow \infty$.

Proof. Throughout the proof, let the superscript $\bullet$ index either forecast-related or analysis-related quantities. Clearly

$$|\phi(x)| \leq M\, (1 + |x|^{\sigma+1}),$$

for any $x \in \mathbb{R}^m$, and it follows from Proposition 22.1 that

$$\int_{\mathbb{R}^m} |\phi(x)|\, \bar{\mu}_k^\bullet(dx) \leq M\, (1 + \int_{\mathbb{R}^m} |x|^{\sigma+1}\, \bar{\mu}_k^\bullet(dx)) = M\, (1 + |\bar{M}_k^{\sigma+1,\bullet}|^{\sigma+1}) < \infty,$$

hence it follows from the strong law of large numbers that

$$\frac{1}{N} \sum_{i=1}^{N} \phi(\bar{X}_k^{i,\bullet}) \longrightarrow \int_{\mathbb{R}^m} \phi(x)\, \bar{\mu}_k^\bullet(dx), \tag{22.14}$$

almost surely as $N \uparrow \infty$. Clearly

$$|\phi(x) - \phi(x')| \leq L\, |x - x'|\, (1 + |x|^\sigma) + L\, |x - x'|^{\sigma+1},$$

for any $x, x' \in \mathbb{R}^m$, with another constant L, and using the triangle inequality yields

$$\left| \frac{1}{N} \sum_{i=1}^{N} (\phi(X_k^{i,\bullet}) - \phi(\bar{X}_k^{i,\bullet})) \right|$$

$$\leq \frac{1}{N} \sum_{i=1}^{N} |\phi(X_k^{i,\bullet}) - \phi(\bar{X}_k^{i,\bullet})|$$

$$\leq L \frac{1}{N} \sum_{i=1}^{N} |X_k^{i,\bullet} - \bar{X}_k^{i,\bullet}| (1 + |\bar{X}_k^{i,\bullet}|^\sigma) + L \frac{1}{N} \sum_{i=1}^{N} |X_k^{i,\bullet} - \bar{X}_k^{i,\bullet}|^{\sigma+1},$$

and using the Hölder inequality and then the triangle inequality again to control the first term, yields

$$\frac{1}{N} \sum_{i=1}^{N} |X_k^{i,\bullet} - \bar{X}_k^{i,\bullet}| (1 + |\bar{X}_k^{i,\bullet}|^\sigma)$$

$$\leq \left(\frac{1}{N} \sum_{i=1}^{N} |X_k^{i,\bullet} - \bar{X}_k^{i,\bullet}|^r \right)^{1/r} \left(\frac{1}{N} \sum_{i=1}^{N} (1 + |\bar{X}_k^{i,\bullet}|^\sigma)^{r'} \right)^{1/r'}$$

$$\leq \left(\frac{1}{N} \sum_{i=1}^{N} |X_k^{i,\bullet} - \bar{X}_k^{i,\bullet}|^r \right)^{1/r} \left(1 + \left(\frac{1}{N} \sum_{i=1}^{N} |\bar{X}_k^{i,\bullet}|^{\sigma r'} \right)^{1/r'} \right),$$

where r, r' are conjugate exponents, i.e. $1/r + 1/r' = 1$, or in other words

$$\left| \frac{1}{N} \sum_{i=1}^{N} (\phi(X_k^{i,\bullet}) - \phi(\bar{X}_k^{i,\bullet})) \right| \leq L \Delta_k^{N,r,\bullet} (1 + |\bar{M}_k^{N,\sigma r',\bullet}|^\sigma) + L |\Delta_k^{N,\sigma+1,\bullet}|^{\sigma+1}. \tag{22.15}$$

Taking $r = \sigma r' = \sigma + 1$ yields

$$\left| \frac{1}{N} \sum_{i=1}^{N} (\phi(X_k^{i,\bullet}) - \phi(\bar{X}_k^{i,\bullet})) \right| \leq L \Delta_k^{N,\sigma+1,\bullet} (1 + |\bar{M}_k^{N,\sigma+1,\bullet}|^\sigma) + L |\Delta_k^{N,\sigma+1,\bullet}|^{\sigma+1},$$

and since $\bar{M}_k^{N,\sigma+1,\bullet} \longrightarrow \bar{M}_k^{\sigma+1,\bullet}$ almost surely as $N \uparrow \infty$, with a finite limit $\bar{M}_k^{\sigma+1,\bullet}$, then it follows from Proposition 22.3 that

$$\left| \frac{1}{N} \sum_{i=1}^{N} (\phi(X_k^{i,\bullet}) - \phi(\bar{X}_k^{i,\bullet})) \right| \longrightarrow 0, \tag{22.16}$$

almost surely as $N \uparrow \infty$. Combining (22.14) and (22.16) with the decomposition

$$\frac{1}{N} \sum_{i=1}^{N} \phi(X_k^{i,\bullet}) - \int_{\mathbb{R}^m} \phi(x) \bar{\mu}_k^\bullet(dx)$$

$$= \frac{1}{N} \sum_{i=1}^{N} (\phi(X_k^{i,\bullet}) - \phi(\bar{X}_k^{i,\bullet})) + \left(\frac{1}{N} \sum_{i=1}^{N} \phi(\bar{X}_k^{i,\bullet}) - \int_{\mathbb{R}^m} \phi(x) \bar{\mu}_k^\bullet(dx) \right),$$

finishes the proof. $\square$

22.5.2 $\mathbb{L}^p$-convergence and rate of convergence

Theorem 22.2 *Let ϕ be a locally Lipschitz continuous function, with at most polynomial growth at infinity, i.e.*

$$|\phi(x) - \phi(x')| \leq L\, |x - x'|\, (1 + |x|^\sigma + |x'|^\sigma),$$

for any $x, x' \in \mathbb{R}^m$ and for some $\sigma \geq 0$.

If Assumption B holds, and if the random vector X_0 has finite moments of any order, then

$$\sup_{N \geq 1} \sqrt{N}\, (\mathbb{E}|\frac{1}{N} \sum_{i=1}^{N} \phi(X_k^{i,f}) - \int_{\mathbb{R}^m} \phi(x)\, \bar{\mu}_k^f(dx)|^p)^{1/p} < \infty,$$

and

$$\sup_{N \geq 1} \sqrt{N}\, (\mathbb{E}|\frac{1}{N} \sum_{i=1}^{N} \phi(X_k^{i,a}) - \int_{\mathbb{R}^m} \phi(x)\, \bar{\mu}_k^a(dx)|^p)^{1/p} < \infty,$$

for any order p.

Proof. Throughout the proof, let the superscript $\bullet$ index either forecast-related or analysis-related quantities. Clearly

$$|\phi(x)| \leq M\, (1 + |x|^{\sigma+1}),$$

for any $x \in \mathbb{R}^m$, and using the triangle inequality, it follows from Proposition 22.1 that

$$(\int_{\mathbb{R}^m} |\phi(x)|^p\, \bar{\mu}_k^\bullet(dx))^{1/p} \leq M\, (\int_{\mathbb{R}^m} (1 + |x|^{\sigma+1})^p\, \bar{\mu}_k^\bullet(dx))^{1/p}$$

$$\leq M\, (1 + (\int_{\mathbb{R}^m} |x|^{p(\sigma+1)}\, \bar{\mu}_k^\bullet(dx))^{1/p})$$

$$= M\, (1 + |\bar{M}_k^{p(\sigma+1),\bullet}|^{\sigma+1}) < \infty,$$

hence it follows from the Marcinkiewicz–Zygmund inequality that

$$(\mathbb{E}|\frac{1}{N} \sum_{i=1}^{N} \phi(\bar{X}_k^{i,\bullet}) - \int_{\mathbb{R}^m} \phi(x)\, \bar{\mu}_k^\bullet(dx)|^p)^{1/p}$$

$$\leq \frac{c_p}{\sqrt{N}}\, (\mathbb{E}|\phi(\bar{X}_k^{i,\bullet}) - \int_{\mathbb{R}^m} \phi(x)\, \bar{\mu}_k^\bullet(dx)|^p)^{1/p},$$

or in other words

$$\sup_{N \geq 1} \sqrt{N}\, (\mathbb{E}|\frac{1}{N} \sum_{i=1}^{N} \phi(\bar{X}_k^{i,\bullet}) - \int_{\mathbb{R}^m} \phi(x)\, \bar{\mu}_k^\bullet(dx)|^p)^{1/p} < \infty. \qquad (22.17)$$

Using the triangle inequality, it follows from (22.15) that

$$(\mathbb{E}|\frac{1}{N}\sum_{i=1}^{N}(\phi(X_k^{i,\bullet})-\phi(\bar{X}_k^{i,\bullet}))|^p)^{1/p} \leq L(\mathbb{E}(|\Delta_k^{N,r,\bullet}|^p(1+|\bar{M}_k^{N,\sigma r',\bullet}|^\sigma)^p))^{1/p}$$

$$+ L\,(\mathbb{E}|\Delta_k^{N,\sigma+1,\bullet}|^{p(\sigma+1)})^{1/p},$$

and using the Hölder inequality and then the triangle inequality again to control the first term, yields

$$(\mathbb{E}(|\Delta_k^{N,r,\bullet}|^p\,(1+|\bar{M}_k^{N,\sigma r',\bullet}|^\sigma)^p))^{1/p}$$

$$\leq (\mathbb{E}|\Delta_k^{N,r,\bullet}|^{pr})^{1/pr}\,(\mathbb{E}(1+|\bar{M}_k^{N,\sigma r',\bullet}|^\sigma)^{pr'})^{1/pr'}$$

$$\leq (\mathbb{E}|\Delta_k^{N,r,\bullet}|^{pr})^{1/pr}\,(1+(\mathbb{E}|\bar{M}_k^{N,\sigma r',\bullet}|^{\sigma pr'})^{1/pr'}),$$

where r, r' are conjugate exponents, i.e. $1/r+1/r' = 1$, or in other words

$$(\mathbb{E}|\frac{1}{N}\sum_{i=1}^{N}(\phi(X_k^{i,\bullet})-\phi(\bar{X}_k^{i,\bullet}))|^p)^{1/p} \leq L\,D_k^{N,pr,\bullet}\,(1+|\bar{M}_k^{\sigma pr',\bullet}|^\sigma)$$

$$+L\,|D_k^{N,p(\sigma+1)}|^{\sigma+1},$$

in view of Lemma 22.5, and since $\bar{M}_k^{\sigma pr',\bullet}$ is finite, then it follows from Proposition 22.4 that

$$\sup_{N\geq 1}\sqrt{N}\,(\mathbb{E}|\frac{1}{N}\sum_{i=1}^{N}(\phi(X_k^{i,\bullet})-\phi(\bar{X}_k^{i,\bullet}))|^p)^{1/p} < \infty. \qquad (22.18)$$

Combining (22.17) and (22.18) with the decomposition

$$(\mathbb{E}|\frac{1}{N}\sum_{i=1}^{N}\phi(X_k^{i,\bullet}) - \int_{\mathbb{R}^m}\phi(x)\,\bar{\mu}_k^{\bullet}(dx)|^p)^{1/p}$$

$$\leq (\mathbb{E}|\frac{1}{N}\sum_{i=1}^{N}(\phi(X_k^{i,\bullet}) - \phi(\bar{X}_k^{i,\bullet}))|^p)^{1/p}$$

$$+(\mathbb{E}|\frac{1}{N}\sum_{i=1}^{N}\phi(\bar{X}_k^{i,\bullet}) - \int_{\mathbb{R}^m}\phi(x)\,\bar{\mu}_k^{\bullet}(dx)|^p)^{1/p},$$

finishes the proof. □

22.6 Conclusion and perspectives

The results of the previous sections show that

- for linear systems of the form considered in Section 22.1.1, with additive Gaussian white noise and with Gaussian initial condition, the empirical mean of the ensemble elements converges to the Kalman filter.
- however, for nonlinear systems of the form introduced in Section 22.1.2, with additive Gaussian white noise and with a non-necessarily Gaussian initial condition, the empirical probability distribution of the ensemble elements converges to the wrong limit, i.e. the limiting probability distribution differs from the usual Bayesian filter.

This may be seen as a negative result, and the question that naturally arises is whether it is possible to improve the ensemble Kalman filter in some way. Indeed, there exists another class of Monte Carlo-based approximations to the Bayesian filter, which have been extensively studied both practically and theoretically (Cappé et al. [4], Del Moral [6], Doucet et al. [7]), and which could be applied to a much broader class of nonlinear models or even more general hidden Markov models. These particle filters provide an approximation of the Bayesian filter in terms of the weighted empirical probability distribution

$$\mu_k^N = \sum_{i=1}^{N} w_k^i \, \delta_{\xi_k^i} \quad \text{with} \quad \sum_{i=1}^{N} w_k^i = 1,$$

associated with a population of N particles, characterized by their positions $(\xi_k^1, \cdots, \xi_k^N)$ and their nonnegative weights $(w_k^1, \cdots, w_k^N)$. Many different variants of particle filters have been proposed, depending on the answers to issues such as: how to initialize the particle system, how to move particle positions, how to update particle weights, how to exploit particle weights, etc. Many convergence results hold generically as the population size N goes to infinity, with the Bayesian filter as the limit, e.g. convergence in $\mathbb{L}^p$-mean

$$(\mathbb{E}| \sum_{i=1}^{N} w_k^i \, \phi(\xi_k^i) - \int_{\mathbb{R}^m} \phi(x) \, \mu_k(dx) \, |^p \,)^{1/p} \longrightarrow 0,$$

for any order p as $N \uparrow \infty$, and central limit theorem

$$\sqrt{N} \, (\sum_{i=1}^{N} w_k^i \, \phi(\xi_k^i) - \int_{\mathbb{R}^m} \phi(x) \, \mu_k(dx)) \Longrightarrow \mathcal{N}(0, v_k(\phi)),$$

in distribution as $N \uparrow \infty$, with a more or less explicit expression for the asymptotic variance $v_k(\phi)$, depending on the implementation.

For nonlinear systems precisely of the form

$$X_k = f_k(X_{k-1}) + W_k \quad \text{with} \quad W_k \sim \mathcal{N}(0, Q_k)$$

$$Y_k = H_k \, X_k + V_k \quad \text{with} \quad V_k \sim \mathcal{N}(0, R_k),$$

introduced in Section 22.1.2, with additive Gaussian white noise and with non-necessarily Gaussian initial condition $X_0 \sim \eta_0$, the favorite particle filter implementation (Doucet, Godsill, and Andrieu [8], Section II.D) consists in sampling particle positions according to the probability distribution of X_k given X_{k-1} and Y_k, and assigning weights proportional to the probability density of Y_k given X_{k-1}. This results in the following algorithm: in the first part of the mutation step, given a population $(\xi_{k-1}^1, \cdots, \xi_{k-1}^N)$ of N particles, each particle is propagated independently according to

$$\xi_k^{i,-} = f_k(\xi_{k-1}^i) + W_k^i \quad \text{with} \quad W_k^i \sim \mathcal{N}(0, Q_k). \quad (22.19)$$

Notice that the i.i.d. random vectors $(W_k^1, \ldots, W_k^N)$ are *simulated* here, with the same statistics as the additive Gaussian noise W_k in the original state equation. The initial population $(\xi_0^{1,-}, \cdots, \xi_0^{N,-})$ is *simulated* as i.i.d. random vectors with probability distribution η_0, i.e. with the same statistics as the initial condition X_0. In the second part of the mutation step, each particle is propagated independently according to

$$\xi_k^i = \xi_k^{i,-} + K_k(Q_k)(Y_k - H_k \xi_k^{i,-} - V_k^i) \quad \text{with} \quad V_k^i \sim \mathcal{N}(0, R_k), \quad (22.20)$$

with Kalman gain matrix

$$K_k(Q_k) = Q_k H_k^* \left(H_k Q_k H_k^* + R_k \right)^{-1}.$$

Notice that the i.i.d. random vectors $(V_k^1, \ldots, V_k^N)$ are *simulated* here, with the same statistics as the additive Gaussian noise V_k in the original observation equation. Combining the two mutation steps together yields

$$\xi_k^i = (f_k(\xi_{k-1}^i) + W_k^i) + K_k(Q_k)(Y_k - H_k(f_k(\xi_{k-1}^i) + W_k^i) - V_k^i)$$

$$= f_k(\xi_{k-1}^i) + K_k(Q_k)(Y_k - H_k f_k(\xi_{k-1}^i)) + (I - K_k(Q_k) H_k) W_k^i - K_k(Q_k) V_k^i,$$

so that, conditionally w.r.t. $\xi_{k-1}^i = x$, the random vector ξ_k^i is Gaussian with mean vector

$$m_k(x) = f_k(x) + K_k(Q_k)(Y_k - H_k f_k(x)),$$

and covariance matrix

$$(I - K_k(Q_k) H_k) Q_k (I - K_k(Q_k) H_k)^* + K_k(Q_k) R_k (K_k(Q_k))^* = (I - K_k(Q_k) H_k) Q_k.$$

In the weighting step, each weight is updated according to

$$w_k^i \propto w_{k-1}^i \exp\left\{ -\tfrac{1}{2} (Y_k - H_k f_k(\xi_{k-1}^i))^* \Xi_k^{-1} (Y_k - H_k f_k(\xi_{k-1}^i)) \right\},$$

with covariance matrix $\Xi_k = H_k Q_k H_k^* + R_k$.

Even with this optimal choice of the importance distribution, it may happen that the particle weights $(w_k^1, \cdots, w_k^N)$ degenerate, i.e. depart significantly from equidistribution, in the sense that a few particles only, or even a single particle, get most of the weight. In this case, it is a good idea to resample the particle positions $(\xi_k^1, \cdots, \xi_k^N)$ according to their respective weights, so that particles with low weights are discarded, whereas particles with high weights are replicated. There are many different ways to perform the resampling step, and adaptive rules have also been proposed to decide on-line when to resample.

It has recently been raised (Furrer and Bengtsson [12], Snyder, Bengtsson, Bickel, and Anderson [19]), however, that particle filters may collapse in high dimension: indeed, the usual adaptation techniques, such as using a better importance distribution or resampling, are no longer efficient in high dimension, and this challenging issue would deserve further study. By construction, the EnKF is not exposed to this degeneracy problem, just because there are no weights attached to elements.

There is again a visible difference between equations (22.1) and (22.2) for the elements in the ensemble Kalman filter, and equations (22.19) and (22.20) for the particle filter with optimal importance distribution: the Kalman gain matrix $K_k(P_k^N)$ in equation (22.2) is based on the empirical covariance matrix P_k^N, which is responsible for mean-field interaction and dependence, whereas the Kalman gain matrix $K_k(Q_k)$ in equation (22.20) is based on the deterministic covariance matrix Q_k of the additive Gaussian noise W_k in the original state equation, which is responsible for decoupling and independence. Notice that the covariance matrix Q_k is available, and the Kalman gain matrix $K_k(Q_k)$ is readily computable. If necessary, an independent sample can be simulated in order to approximate, without mean-field interaction, the covariance matrix Q_k in terms of an empirical covariance matrix.

The other difference with the ensemble Kalman filter is the existence of weights attached to particles, which can also be exploited to resample the population, if needed, and which are responsible for the convergence of the particle filter to the Bayesian filter. Quite naturally, it has recently been suggested (Papadakis [17], ch. 2) to interpret equations (22.1) and (22.2) for the elements in the ensemble Kalman filter, as defining an importance distribution with mean-field interaction, and to attach the corresponding importance weights to the ensemble elements. Numerical evidence has already been provided about the practical improvement obtained with this modification, and the resulting weighted ensemble Kalman filter should now converge to the Bayesian filter, as the ensemble size goes to infinity. This convergence issue deserves further investigation, including the proof of a central limit theorem, which would make it possible to compare the weighted ensemble Kalman filter and the particle filter with optimal importance distribution, on the basis of their respective asymptotic variance.

Notes

This work was partially supported by INRIA, under the project *Data Assimilation for Air Quality* (ARC programme 2005–6), by the French National Research Agency (ANR) under the project PREVASSEMBLE (ANR-08-COSI-012, COSINUS programme), and by a PhD grant from the Bretagne region.

References

[1] Anderson, J. L. and Anderson, S. L. (1999). A Monte Carlo implementation of the nonlinear filtering problem to produce ensemble assimilations and forecasts, *Monthly Weather Review* **127**(12): 2741–58.

[2] Bertino, L., Evensen, G., and Wackernagel, H. (2003). Sequential data assimilation techniques in oceanography, *International Statistical Review* **71**(2): 223–41.

[3] Burgers, G., van Leeuwen, P. J., and Evensen, G. (1998). Analysis scheme in the ensemble Kalman filter, *Monthly Weather Review* **126**(6): 1719–24.

[4] Cappé, O., Moulines, E., and Rydén, T. (2005). *Inference in Hidden Markov Models*, Springer Series in Statistics, Springer-Verlag, New York.

[5] Crişan, D. and Doucet, A. (2002). A survey of convergence results on particle filtering methods for practitioners, *IEEE Transactions on Signal Processing* **50**(3): 736–46.

[6] Del Moral, P. (2004). *Feynman–Kac Formulae: Genealogical and Interacting Particle Systems with Applications*, Probability and its Applications, Springer-Verlag, New York.

[7] Doucet, A., de Freitas, N., and Gordon, N. (eds) (2001). *Sequential Monte Carlo Methods in Practice*, Statistics for Engineering and Information Science, Springer-Verlag, New York.

[8] Doucet, A., Godsill, S. J., and Andrieu, C. (2000). On sequential Monte Carlo sampling methods for Bayesian filtering, *Statistics and Computing* **10**(3): 197–208.

[9] Evensen, G. (1994). Sequential data assimilation with a nonlinear quasi-geostrophic model using Monte Carlo methods to forecast error statistics, *Journal of Geophysical Research (Oceans)* **99**(C5): 10143–62.

[10] Evensen, G. (2003). Ensemble Kalman filter: theoretical formulation and practical implementations, *Ocean Dynamics* **53**(4): 343–67.

[11] Evensen, G. (2006). *Data Assimilation: The Ensemble Kalman Filter*, Springer-Verlag, Berlin.

[12] Furrer, R. and Bengtsson, T. (2007). Estimation of high-dimensional prior and posterior covariance matrices in Kalman filter variants, *Journal of Multivariate Analysis* **98**(2): 227–55.

[13] Gut, A. (2005). *Probability: A Graduate Course*, Springer Texts in Statistics, Springer-Verlag, New York.

[14] Heemink, A. W., Verlaan, M., and Segers, A. J. (2001). Variance reduced ensemble Kalman filtering, *Monthly Weather Review* **129**(7): 1718–28.

[15] McKean, H. P. (1969). Propagation of chaos for a class of non-linear parabolic equations, in A. K. Aziz (ed.), *Lectures Series in Differential Equations, Volume 2*, Vol. 19 of *Van Nostrand Mathematical Studies*, Van Nostrand Reinhold, New York, pp. 177–94.

[16] Miller, R. N., Carter, E. F., and Blue, S. T. (1999). Data assimilation into nonlinear stochastic models, *Tellus* **51A**(2): 167–94.

[17] Papadakis, N. (2007). *Assimilation de Données Images: Application au Suivi de Courbes et de Champs de Vecteurs*, Thèse de Doctorat, Université de Rennes 1, Rennes.
[18] Pham, D.-T. (2001). Stochastic methods for sequential data assimilation in strongly nonlinear systems, *Monthly Weather Review* **129**(5): 1194–1207.
[19] Snyder, C., Bengtsson, T., Bickel, P. J., and Anderson, J. L. (2008). Obstacles to high-dimensional particle filtering, *Monthly Weather Review* **136**(12): 4629–40.
[20] Sznitman, A.-S. (1991). Topics in propagation of chaos, *in* P.-L. Hennequin (ed.), *Ecole d'Eté de Probabilités de Saint-Flour XIX, 1989*, Vol. 1464 of *Lecture Notes in Mathematics*, Springer-Verlag, Berlin, pp. 165–251.

PART VII
The Particle Approach

·23·
Particle Approximations to the Filtering Problem in Continuous Time

J. Xiong

23.1 Introduction

There are two related stochastic processes in each filtering problem: The signal process which we want to estimate and the observation process which provides the information we can use. In this chapter, we assume that the signal process is a d-dimensional diffusion governed by the following stochastic differential equation (SDE):

$$dX_t = b(X_t)dt + c(X_t)dW_t + \sigma(X_t)dB_t, \tag{23.1}$$

where B and W are independent Brownian motions of dimensions d and m, respectively, and $b : \mathbb{R}^d \to \mathbb{R}^d$, $c : \mathbb{R}^d \to \mathbb{R}^{d \times m}$, and $\sigma : \mathbb{R}^d \to \mathbb{R}^{d \times d}$ are continuous mappings. To ensure the existence and uniqueness for the solution to (23.1) and for the convenience of the estimates, the following condition (BC1) will be assumed throughout this chapter.

Condition (BC1): The mappings b, c, σ are bounded and Lipschitz continuous.

For the observation process, we consider two models: The classical one and the one with point processes as its observations. In the classical model, the observation process is an m-dimensional stochastic process given by

$$Y_t = \int_0^t h(X_s)ds + W_t, \tag{23.2}$$

where $h : \mathbb{R}^d \to \mathbb{R}^m$ is a continuous mapping. For the point process model, the observation process is an m-dimensional process given by

$$Y_t^i = N_i \left(\int_0^t \lambda_i(X_s, s)ds \right), \qquad i = 1, 2, \cdots, m, \tag{23.3}$$

where $N_1, N_2, \cdots, N_m$ are independent unit Poisson processes which are independent of X, and $\lambda_i : \mathbb{R}^d \times \mathbb{R}_+ \to \mathbb{R}_+$, $i = 1, 2, \cdots, m$, are measurable mappings.

The optimal filter π_t is a $\mathcal{P}(\mathbb{R}^d)$-valued process given by

$$\langle \pi_t, f \rangle \equiv \mathbb{E}\left(f(X_t)\big|\mathcal{G}_t\right), \qquad \forall f \in C_b(\mathbb{R}^d),$$

where $\mathcal{G}_t = \mathcal{F}_t^Y$ is the σ-field generated by Y_s, $s \leq t$, $\langle \mu, f \rangle$ stands for the integral of f with respect to the measure μ, and $\mathcal{P}(\mathbb{R}^d)$ is the collection of all Borel probability measures on $\mathbb{R}^d$. Stochastic differential equations on $\mathcal{P}(\mathbb{R}^d)$, called the filtering equations, satisfied by π_t for these two models will be derived in Sections 23.2 and 23.5, respectively.

Explicit solutions to the filtering equations are rarely available. Thus, to solve the filtering problems, we have to resort to numerical schemes. Particle approximations is a class of the effective numerical schemes. The main idea is to represent the solution to a stochastic partial differential equation (SPDE) through a system of weighted particles whose locations and weights are governed by SDEs which can be solved numerically. This numerical scheme based on the weighted particle system, regarded as a direct Monte Carlo method, will be introduced in Section 23.2.

As the error in the Monte Carlo approximation increases exponentially fast when the time parameter tends to infinity due to the exponential growth of the variances of the weights of the particles in the system, we need to modify the weight of each particle. However, the total mass has to be kept constant for the approximate filter to take values in the space of probability measures. To this end, the number of particles in the system will be changed from time to time. We use a branching particle system to match the change of the number of particles in the system. This numerical scheme based on branching weighted particle systems, called the hybrid filter, will be studied in Section 23.3.

Another method in reducing the error is to use interacting particle systems, namely, there is no branching in the system while the motions of the particles are directed to the region where the optimal filter has a higher density. This interacting particle system will be studied in Section 23.4.

Finally, we consider the filtering model with point processes as its observations. This model arises from the study of ultra-high-frequency date in mathematical finance. A branching particle system will be utilized to approximate the optimal filter in this setup. The difference between this filtering model and the classical one will be presented in Section 23.5.

23.2 Filtering using weighted particles

In this section, we establish a weighted particle system representation for the optimal filter of the classical model. Based on this representation, a numerical scheme will be proposed and its convergence to the optimal filter will be proved, together with the rate of convergence derived. The material of this section is taken from the papers of Kurtz and Xiong [26], [27], [28].

To derive the filtering equation for π_t, it is convenient to apply Girsanov's formula to transform the probability measure to a new one such that Y_t becomes an m-dimensional Brownian motion, independent of the process B_t, under the new probability measure. To this end and for the convenience of the estimates later, we assume the following condition (BC2) on h.

Condition (BC2): The mapping h is bounded and Lipschitz continuous.

Let $\hat{P} \sim P$ be the probability measure given by

$$\frac{dP}{d\hat{P}}\bigg|_{\mathcal{F}_t} = M_t \equiv \exp\left(\int_0^t h(X_s)^* dY_s - \frac{1}{2}\int_0^t |h(X_s)|^2 ds\right),$$

where v^* stands for the transpose of a vector (or matrix) v. Then, under $\hat{P}$, (B_t, Y_t) is a $(d+m)$-dimensional Brownian motion, and the signal process X_t is governed by

$$dX_t = (b - ch)(X_t)dt + c(X_t)dY_t + \sigma(X_t)dB_t. \tag{23.4}$$

The conditional expectation with respect to P, as appeared in the definition of the optimal filter, can be represented according to the conditional expectation with respect to $\hat{P}$ by Bayes' formula which is called the Kallianpur–Striebel formula (cf. Kallianpur–Striebel [23], [24]) in the filtering setup. The advantage of using $\hat{P}$ is that the signal is a functional of B and Y, while B is independent of $\mathcal{G}_t$ and Y is measurable with respect to $\mathcal{G}_t$, and the conditional expectations in both cases (individually) are easy to find.

Let $C_b(\mathbb{R}^d)$ be the collection of all bounded continuous real-valued functions.

Theorem 23.1 (Kallianpur–Striebel formula). *The optimal filter π_t can be represented as*

$$\langle \pi_t, f \rangle = \frac{\langle V_t, f \rangle}{\langle V_t, 1 \rangle}, \quad \forall f \in C_b(\mathbb{R}^d), \tag{23.5}$$

where

$$\langle V_t, f \rangle = \hat{\mathbb{E}}(M_t f(X_t) | \mathcal{G}_t) \tag{23.6}$$

and $\hat{\mathbb{E}}$ refers to the expectation with respect to the measure $\hat{P}$.

On the probability space $(\Omega, \mathcal{F}, \hat{P})$, let B^i, $i = 1, 2, \cdots$, be independent copies of B, and let them be independent of Y. Now we consider an interacting particle system: For $i = 1, 2, \cdots$,

$$dX_t^i = (b - ch)\left(X_t^i\right) dt + c\left(X_t^i\right) dY_t + \sigma\left(X_t^i\right) dB_t^i \tag{23.7}$$

and

$$dM_t^i = M_t^i h^*\left(X_t^i\right) dY_t, \quad M_0^i = 1 \tag{23.8}$$

By (23.6) and the conditional (given $\mathcal{G}_t$) law of large numbers, we get

Theorem 23.2 *Suppose that* $\{X_0^i,\ i = 1, 2, \cdots\}$ *are i.i.d. random vectors with common distribution* π_0 *on* $\mathbb{R}^d$. *Then*

$$\langle V_t, f \rangle = \lim_{k \to \infty} \frac{1}{k} \sum_{i=1}^{k} M_t^i f(X_t^i), \qquad (23.9)$$

where $\{(M^i, X^i) : i = 1, 2, \cdots\}$ *is the unique strong solution to the particle system (23.7)–(23.8).*

Applying Itô's formula to (23.7)–(23.9), we get the Zakai equation for the unnormalized filter V_t.

Theorem 23.3 (Zakai's equation). *The unnormalized filter V_t satisfies the following stochastic differential equation:*

$$\langle V_t, f \rangle = \langle V_0, f \rangle + \int_0^t \langle V_s, Lf \rangle \, ds + \int_0^t \langle V_s, \nabla^* fc + fh^* \rangle \, dY_s, \qquad (23.10)$$

where

$$Lf = \frac{1}{2} \sum_{i,j=1}^{d} a_{ij} \partial_{ij}^2 f + \sum_{i=1}^{d} b_i \partial_i f$$

is the generator of the signal process, and the $d \times d$ matrix $a = (a_{ij})$ is given by $a = cc^* + \sigma\sigma^*$.

Then, applying Itô's formula to (23.5) and (23.10), we get the following filtering equation.

Theorem 23.4 (Kushner–FKK equation). *The optimal filter π_t satisfies the following stochastic differential equation: For all $f \in C_b^2(\mathbb{R}^d)$,*

$$\langle \pi_t, f \rangle = \langle \pi_0, f \rangle + \int_0^t \langle \pi_s, Lf \rangle \, ds \qquad (23.11)$$
$$+ \int_0^t \left(\langle \pi_s, \nabla^* fc + fh^* \rangle - \langle \pi_s, f \rangle \langle \pi_s, h^* \rangle \right) d\nu_s,$$

where the innovation process ν_t, given by

$$d\nu_t = dY_t - \langle \pi_t, h \rangle \, dt, \qquad (23.12)$$

is an m-dimensional $\mathcal{G}_t$-Brownian motion under the original probability measure.

To propose a numerical approximation to the unnormalized filter, we apply Euler scheme to approximate the solution of a finite system of n particles. For $\delta > 0$, let

$$\eta_\delta(t) = j\delta \qquad \text{for } j\delta \le t < (j+1)\delta.$$

Note that M_t^i is positive. To keep this positivity property in the approximation, we consider the Euler scheme of the process $Z_t^i \equiv \log M_t^i$ which satisfies the following equation

$$dZ_t^i = h\left(X_t^i\right)^* dY_t - \frac{1}{2}\left|h\left(X_t^i\right)\right|^2 dt.$$

Now we define the finite system $\{(X^{\delta,i}, Z^{\delta,i}, V^{n,\delta}) : i = 1, 2, \cdots, n\}$ as follows:

$$\begin{cases} dX_t^{\delta,i} = (b - ch)\left(X_{\eta_\delta(t)}^{\delta,i}\right) dt + c\left(X_{\eta_\delta(t)}^{\delta,i}\right) dY_t + \sigma\left(X_{\eta_\delta(t)}^{\delta,i}\right) dB_t^i \\ dZ_t^{\delta,i} = h\left(X_{\eta_\delta(t)}^{\delta,i}\right)^* dY_t - \frac{1}{2}\left|h\left(X_{\eta_\delta(t)}^{\delta,i}\right)\right|^2 dt \\ V_t^{n,\delta} = \frac{1}{n}\sum_{i=1}^n \exp\left(Z_t^{\delta,i}\right) \delta_{X_t^{\delta,i}}. \end{cases} \quad (23.13)$$

To prove the convergence of $V_t^{n,\delta}$ to V_t, we need a metric on the space $\mathcal{M}_F(\mathbb{R}^d)$ of finite Borel measures on $\mathbb{R}^d$. We shall use Wasserstein's metric.

For $\nu_1, \nu_2 \in \mathcal{M}_F(\mathbb{R}^d)$, the Wasserstein metric is given by

$$\rho(\nu_1, \nu_2) = \sup\left\{|\langle \nu_1, \phi\rangle - \langle \nu_2, \phi\rangle| : \phi \in \mathbb{B}_1\right\},$$

where

$$\mathbb{B}_1 = \left\{\phi : \mathbb{R}^d \to \mathbb{R}; |\phi(x) - \phi(y)| \leq |x - y|, |\phi(x)| \leq 1, \forall x, y \in \mathbb{R}^d\right\}.$$

Under this metric, $\mathcal{M}_F(\mathbb{R}^d)$ becomes a Polish space.

We need the following

Condition (I): The initial positions $\{x_i^n : i = 1, 2, \cdots, n\}$ of the particles are i.i.d. random vectors in $\mathbb{R}^d$ with the common distribution $\pi_0 \in \mathcal{P}(\mathbb{R}^d)$ which satisfies $\int_{\mathbb{R}^d} |x|^2 \pi_0(dx) < \infty$.

Theorem 23.5 *Let $\bar{V}_t^n = V_t^{n,1/n}$. Assume Conditions (I), (BC1), and (BC2) hold. Then there exists a constant $K_1(t)$ such that*

$$\hat{\mathbb{E}}\rho\left(\bar{V}_t^n, V_t\right) \leq \frac{K_1(t)}{\sqrt{n}}.$$

To show that $\frac{1}{\sqrt{n}}$ is indeed the order for the rate of convergence, we define

$$S_t^n = \sqrt{n}\left(\bar{V}_t^n - V_t\right),$$

and prove the tightness for $\{S^n\}$ in an appropriate space. For simplicity of notation, we restrict our calculations to space dimensions $d = m = 1$ in the rest of this section.

As in Hitsuda and Mitoma [22], we use a modified Schwarz space Φ. Let $\rho(x) = C\exp\left(-1/(1 - |x|^2)\right) 1_{|x|<1}$, where C is a constant such that $\int \rho(x)dx = 1$.

Let

$$\psi(x) = \int e^{-|y|}\rho(x-y)dy.$$

Let

$$\Phi = \{\phi : \phi\psi \in \mathcal{S}\},$$

where $\mathcal{S}$ is the Schwarz space. For $\kappa = 0, 1, 2, \ldots$, define

$$\|\phi\|_\kappa^2 = \sum_{0 \leq k \leq \kappa} \int_\mathbb{R} (1+|x|^2)^{2\kappa} \left|\frac{d^k}{dx^k}(\phi(x)\psi(x))\right|^2 dx.$$

Let Φ_κ be the completion of Φ with respect to $\|\cdot\|_\kappa$. Then Φ_κ is a Hilbert space with inner product

$$\langle \phi_1, \phi_2\rangle_\kappa = \sum_{0 \leq k \leq \kappa} \int_\mathbb{R} (1+|x|^2)^{2\kappa} \left(\frac{d^k}{dx^k}(\phi_1(x)\psi(x))\right)\left(\frac{d^k}{dx^k}(\phi_2(x)\psi(x))\right) dx.$$

Note that $\Phi_\kappa \supset \Phi_{\kappa+1}$ and that Φ_0 is $L^2(\mu_\psi)$, where $\mu_\psi(dx) = \psi^2(x)dx$. For $\hat{\phi} \in \Phi_0$ and $\phi \in \Phi_\kappa$,

$$\langle \hat{\phi}, \phi\rangle \equiv \langle \hat{\phi}, \phi\rangle_0 = \int_\mathbb{R} \hat{\phi}(x)\phi(x)\psi^2(x)dx$$

defines a continuous linear functional on Φ_κ with norm

$$\|\hat{\phi}\|_{-\kappa} = \sup_{\phi \in \Phi_\kappa} \frac{|\langle \hat{\phi}, \phi\rangle|}{\|\phi\|_\kappa},$$

and we let $\Phi_{-\kappa}$ denote the completion of Φ_0 with respect to this norm. Then $\Phi_{-\kappa}$ is a representation of the dual of Φ_κ. If $\{\phi_j^\kappa\}$ is a complete, orthonormal system for Φ_κ, then the inner product for $\Phi_{-\kappa}$ can be written as

$$\langle \hat{\phi}_1, \hat{\phi}_2\rangle_{-\kappa} = \sum_{j=1}^\infty \langle \hat{\phi}_1, \phi_j^\kappa\rangle\langle \hat{\phi}_2, \phi_j^\kappa\rangle. \tag{23.14}$$

By a slight modification of Theorem 7, page 82, of Gel'fand and Vilenkin [19], these norms determine a nuclear space, so in particular, for each κ there exists a $\kappa' > \kappa$ such that the embedding $T_\kappa^{\kappa'} : \Phi_{\kappa'} \to \Phi_\kappa$ is a Hilbert–Schmidt operator. The adjoint $T_\kappa^{\kappa'*} : \Phi_{-\kappa} \to \Phi_{-\kappa'}$ is also Hilbert–Schmidt. $\Phi' = \cup_{k=0}^\infty \Phi_{-k}$ gives a representation of the dual of Φ. (See [19], page 59.) We prove the tightness for $\{S^n\}$ in $C_{\Phi_{-\kappa}}[0, \infty)$ for an appropriate κ, where $C_{\Phi_{-\kappa}}[0, \infty)$ denotes the collection of all continuous mappings from $[0, \infty)$ to $\Phi_{-\kappa}$.

Theorem 23.6 *Suppose that Conditions (I), (BC1), and (BC2) hold. Then there exists κ such that $\{S^n\}$ is tight in $C_{\Phi_{-\kappa}}[0, \infty)$.*

Finally, we characterize the limit. Let M be a $\Phi_{-\kappa}$-valued local martingale with $\langle M^\phi, Y\rangle_t = 0$ for every $\phi \in \Phi$, and

$$\langle M^\phi\rangle_t = \int_0^t \langle V_s^2, |\sigma\phi'|^2\rangle\, ds,$$

where $M_t^\phi = \langle M_t, \phi\rangle$, V_t^2 is a Φ'-valued process given by

$$V_t^2 = \lim_{n\to\infty} \frac{1}{n} \sum_{i=1}^n (M_t^i)^2 \delta_{X_t^i}.$$

The following condition is needed in the proof of the uniqueness part of the next theorem.

Condition (E): There exists a constant $\delta > 0$ such that $\sigma^2 - \delta c^2 \geq 0$.

Theorem 23.7 *In addition to the conditions of Theorem 23.6, we assume that Condition (E) holds. Then $\{S^n\}$ converges weakly in $C_{\Phi_{-\kappa}}[0, \infty)$ to a process S which is the unique solution of the following stochastic evolution equation:* $\forall \phi \in \Phi$,

$$\langle S_t, \phi\rangle = \langle S_0, \phi\rangle + \langle M_t, \phi\rangle + \int_0^t \langle S_u, L\phi\rangle\, dy + \int_0^t \langle S_u, h\phi + c\phi'\rangle\, dY_u. \quad (23.15)$$

23.3 Filtering using branching particle systems

In this section, we introduce the branching particle system approximation of the optimal filter. The main purpose is to reduce the variances of the weights of the particles in the system. The idea is to divide the time interval into small subintervals and the weight for each particle at any partition time is modified as an exponential martingale which depends on the signal and the noise in the small interval prior to that time instead of on the whole interval starting from 0. This section is based on a joint paper with Crisan [12].

Now we proceed to giving the definition of the branching particle system. Initially, there are n particles of weight 1 each at locations x_i^n, $i = 1, 2, \cdots, n$, satisfying the initial condition (I).

Let $\delta = \delta_n = n^{-2a}$, $0 < a < 1$. For $j = 0, 1, 2, \cdots$, there are m_j^n number of particles alive at time $t = j\delta$. Note that $m_0^n = n$.

During the time interval $(j\delta, (j+1)\delta)$, the particles move according to the following diffusions: For $i = 1, 2, \cdots, m_j^n$,

$$X_t^i = X_{j\delta}^i + \int_{j\delta}^t \sigma(X_s^i)\, dB_s^i + \int_{j\delta}^t (b - ch)(X_s^i)\, ds + \int_{j\delta}^t c(X_s^i)\, dY_s. \quad (23.16)$$

At the end of the interval, the ith particle ($i = 1, 2, \cdots, m_j^n$) branches (independent of others) into a random number ξ_{j+1}^i of offsprings which is chosen such that the conditional variance $Var^{\hat{P}}(\xi^i_{j+1}|\mathcal{F}_{(j+1)\delta-})$ is minimized subject to the constraint

$$\hat{\mathbb{E}}\left(\xi^i_{j+1}|\mathcal{F}_{(j+1)\delta-}\right) = \tilde{M}^n_j\left(X^i, (j+1)\delta\right), \qquad (23.17)$$

where

$$M^n_j(X^i, t) = \exp\left(\int_{j\delta}^t h^*\left(X^i_s\right) dY_s - \frac{1}{2}\int_{j\delta}^t \left|h\left(X^i_s\right)\right|^2 ds\right) \qquad (23.18)$$

and

$$\tilde{M}^n_j\left(X^i, t\right) = \frac{M^n_j(X^i, t)}{\frac{1}{m^n_j}\sum_{\ell=1}^{m^n_j} M^n_j(X^\ell, t)}.$$

It is clear that

$$\xi^i_{j+1} = \begin{cases} \left[\tilde{M}^n_j(X^i, (j+1)\delta)\right] & \text{with probability } 1 - \{\tilde{M}^n_j(X^i, (j+1)\delta)\}, \\ \left[\tilde{M}^n_j(X^i, (j+1)\delta)\right] + 1 & \text{with probability } \{\tilde{M}^n_j(X^i, (j+1)\delta)\}, \end{cases}$$

where $\{x\} = x - [x]$ is the fraction of x, and $[x]$ is the largest integer which is not greater than x.

Denote the conditional variance of ξ^i_{j+1} by $\gamma^n_{j+1}(X^i)$. Then

$$\gamma^n_{j+1}(X^i) = \{\tilde{M}^n_j(X^i, (j+1)\delta)\}(1 - \{\tilde{M}^n_j(X^i, (j+1)\delta)\}).$$

Now we define the approximate filter as follows:

$$\pi^n_t = \frac{1}{m^n_j}\sum_{i=1}^{m^n_j} \tilde{M}^n_j(X^i, t)\delta_{X^i_t}, \qquad j\delta \le t < (j+1)\delta.$$

Namely, the ith particle has a time-dependent weight $\tilde{M}^n_j(X^i, t)$. At the end of the interval, i.e., $t = (j+1)\delta$, this particle dies and gives birth to a random number of offspring, whose conditional expectation is equal to the pre-death weight of the particle. The new particles start from their mother's position with weight 1 each.

The process π^n_t is called the *hybrid filter* since it involves a branching particle system and the empirical measure of these weighted particles. In the earlier stage of the study of particle approximation of the optimal filter, the particle approximation is defined as π^n_t without the weight, i.e., the *particle filter* is

$$\bar{\pi}^n_t = \frac{1}{m^n_j}\sum_{i=1}^{m^n_j} \delta_{X^i_t}, \qquad j\delta \le t < (j+1)\delta. \qquad (23.19)$$

Thus, the current approximate filter π_t^n is a combination of the weighted filter introduced in Section 23.2 and the particle filter (23.19). That is the reason we call it the hybrid filter.

Since Zakai's equation for the unnormalized filter V_t is much simpler than the Kushner–FKK equation for the optimal filter π_t, to study the convergence of π_t^n to π_t, it is more convenient to consider an auxiliary process first. Let

$$\eta_k^n = \Pi_{j=0}^{k-1} \frac{1}{m_j^n} \sum_{\ell=1}^{m_j^n} M_j^n(X^\ell, (j+1)\delta).$$

For $k\delta \leq t < (k+1)\delta$, we define

$$V_t^n = \frac{1}{n}\eta_k^n \pi_t^n \sum_{i=1}^{m_k^n} M_k^n(X^i, t) = \frac{1}{n}\eta_k^n \sum_{i=1}^{m_k^n} M_k^n(X^i, t)\delta_{X_t^i}.$$

We will prove that V_t^n converges to the unnormalized filter V_t. To this end, we need the following lemmas.

Lemma 23.1 *For each $1 \leq j \leq [T/\delta]$, we have*

$$\hat{\mathbb{E}}\left(m_j^n \left(\eta_j^n\right)^2\right) \leq n e^{K^2 T}.$$

Lemma 23.2 *There exists a constant K_1 such that for any $j \geq 0$ and $i = 1, 2, \cdots, m_j^n$, we have*

$$\hat{\mathbb{E}}\left(\gamma_{j+1}^n(X^i)\left(\eta_{j+1}^n/\eta_j^n\right)^2 \Big| \mathcal{F}_{j\delta}\right) \leq K_1\sqrt{\delta}.$$

A key ingredient in the proof of the convergence of V^n to V is the following dual of the Zakai equation:

$$\begin{cases} d\psi_s = -L\psi_s\, ds - (\nabla^*\psi_s c + h^*\psi_s)\, \hat{d} Y_s, & 0 \leq s \leq t, \\ \psi_t = \phi, \end{cases} \quad (23.20)$$

where $\hat{d}$ denotes the backward Itō's integral. Namely, we take the right endpoints in the approximating Riemann sum in defining the stochastic integral.

Hereafter we will denote by $C_b^k(\mathbb{R}^d, \mathcal{X})$ the set of all bounded continuous mappings from $\mathbb{R}^d$ to $\mathcal{X}$ with bounded partial derivatives up to order k, where $\mathcal{X}$ is a Hilbert space. We endow $C_b^k(\mathbb{R}^d, \mathcal{X})$ with the following norm

$$\|\varphi\|_{k,\infty} = \sum_{|\alpha|\leq k} \sup_{x\in\mathbb{R}^d} \|D_\alpha \varphi(x)\|_\mathcal{X}, \quad \varphi \in C_b^k(\mathbb{R}^d, \mathcal{X}),$$

where $\alpha = (\alpha^1, \cdots, \alpha^d)$ is a multi-index, $|\alpha| = \alpha^1 + \cdots + \alpha^d$ and $D_\alpha \varphi = \partial_1^{\alpha^1} \cdots \partial_d^{\alpha^d} \varphi$. Also let $W_p^k(\mathbb{R}^d, \mathcal{X})$ be the set of all functions with generalized partial derivatives up to order k with both the function and all its partial derivatives being p-integrable. We endow $W_p^k(\mathbb{R}^d, \mathcal{X})$ with the following Sobolev norm

$$\|\varphi\|_{k,p} = \left(\sum_{|\alpha| \le k} \int_{\mathbb{R}^d} \|D_\alpha \varphi(x)\|_{\mathcal{X}}^p \, dx \right)^{\frac{1}{p}}.$$

The following Condition (BD) is needed in establishing a representation of ψ_s which plays a key role in the proof of the convergence of V^n.

Condition (BD): *The mappings a, b, c, h, ϕ are in $C_b^k(\mathbb{R}^d, \mathcal{X})$ with $k = \left[\frac{d}{2}\right] + 2$ and $\mathcal{X}$ being $\mathbb{R}^{d \times d}$, $\mathbb{R}^d$, $\mathbb{R}^{d \times m}$, $\mathbb{R}^m$ and $\mathbb{R}$ respectively. Also, we assume $\phi \in W_2^k(\mathbb{R}^d)$.*

Under this condition, we can get an estimate for the supremum norm of ψ_s.

Lemma 23.3 *Suppose that Assumption (BD) holds. Then $\psi_s \in C_b^2(\mathbb{R}^d)$ a.s. and there exists a constant K_1 independent of ϕ and $s \in [0, t]$ such that*

$$\mathbb{E}\left[\|\psi_s\|_{2,\infty}^2\right] \le K_1 \|\phi\|_{k,2}^2. \tag{23.21}$$

Now, we give the representation of ψ_s.

Lemma 23.4 *Suppose that Condition (BD) holds. Then, for every $t \ge 0$, we have*

$$\psi_t(X_t) M_t - \psi_0(X_0) = \int_0^t M_s \nabla^* \psi_s \sigma(X_s) \, dB_s, \qquad a.s.. \tag{23.22}$$

As a consequence, if $\phi \in C_b(\mathbb{R}^d)$ and $\pi_0 \in L^2(\mathbb{R}^d)$, then $\langle V_t, \phi \rangle = \langle \pi_0, \psi_0 \rangle$.

Note that

$$\langle V_t^n, \phi \rangle - \langle V_0^n, \psi_0 \rangle = I_1^n + I_2^n + I_3^n,$$

where

$$I_1^n = \eta_k^n \frac{1}{n} \sum_{i=1}^{m_k^n} \left(M_k^n(X^i, t) \psi_t(X_t^i) - \psi_{k\delta}(X_{k\delta}^i) \right),$$

$$I_2^n = \sum_{j=1}^{k} \eta_j^n \frac{1}{n} \sum_{i=1}^{m_{j-1}^n} \psi_{j\delta}(X_{j\delta}^i) \left(\xi_j^i - \tilde{M}_j^n(X^i) \right)$$

and

$$I_3^n = \sum_{j=1}^{k} \eta_{j-1}^n \frac{1}{n} \sum_{i=1}^{m_{j-1}^n} \left(\psi_{j\delta}(X_{j\delta}^i) M_j^n(X^i) - \psi_{(j-1)\delta}(X_{(j-1)\delta}^i) \right).$$

Applying Lemma 23.4, we get

$$I_3^n = \sum_{j=0}^{k-1} \eta_j^n \frac{1}{n} \sum_{i=1}^{m_j^n} \int_{j\delta}^{(j+1)\delta} M_j^n(X^i, s) \nabla^* \psi_s \sigma(X_s^i) \, dB_s^i$$

and hence, $\hat{\mathbb{E}}((I_3^n)^2) \leq Kn^{-1}$. The term I_1^n can be estimated similarly.
Using the conditional independency for the terms in I_2^n, we get

$$\hat{\mathbb{E}}((I_2^n)^2) = \hat{\mathbb{E}} \sum_{j=1}^{k} \frac{1}{n^2} \sum_{i=1}^{m_{j-1}^n} \psi_{j\delta}(X_{j\delta}^i)^2 \gamma_j^n(X^i) (\eta_j^n)^2. \tag{23.23}$$

By Lemmas 23.2 and 23.3, we get that

$$\hat{\mathbb{E}}\left((I_2^n)^2\right) \leq \hat{\mathbb{E}} \sum_{j=1}^{k} \frac{1}{n^2} \sum_{i=1}^{m_{j-1}^n} \hat{\mathbb{E}}\left(\|\psi_{j\delta}\|_\infty^2\right) \hat{\mathbb{E}}\left(\gamma_j^n(X^i) \left((\eta_j^n)^2/\eta_{j-1}^n\right)^2 |\mathcal{F}_{(j-1)\delta}\right)(\eta_{j-1}^n)^2$$

$$\leq K_1 \sqrt{\delta} \frac{1}{n^2} \sum_{j=1}^{k} \hat{\mathbb{E}}\left(m_{j-1}^n (\eta_{j-1}^n)^2\right)$$

$$\leq K_2 \sqrt{\delta} \frac{1}{n^2} \frac{T}{\delta} n \leq K_3 n^{-(1-a)}. \tag{23.24}$$

Thus, we have

Theorem 23.8 *Suppose that the conditions (BD) and (I) hold. Then there exists a constant K_1 such that*

$$\hat{\mathbb{E}} \left|\langle V_t^n, \phi \rangle - \langle V_t, \phi \rangle\right|^2 \leq K_1 n^{-(1-a)} \|\phi\|_{k,2}^2$$

where $k = \left[\frac{d}{2}\right] + 2$ is given in Condition (BD).

To get an uniform estimate, we need the following equation satisfied by V_t^n:

$$\langle V_t^n, f \rangle = \langle V_0^n, f \rangle + \int_0^t \langle V_s^n, Lf \rangle \, ds + \int_0^t \langle V_s^n, \nabla^* fc + h^* f \rangle \, dY_s$$

$$+ N_t^{n,f} + \hat{N}_t^{n,f}, \tag{23.25}$$

where

$$N_t^{n,f} = \sum_{j=0}^{[t/\delta]} \frac{1}{n} \sum_{i=1}^{m_j^n} \int_{j\delta}^{((j+1)\delta) \wedge t} \nabla^* f \sigma(X_s^i) \, dB_s^i \eta_j^n$$

and

$$\hat{N}_t^{n,f} = \sum_{j=1}^{[t/\delta]} \eta_j^n \frac{1}{n} \sum_{i=1}^{m_{j-1}^n} \left(\xi_j^i - \tilde{M}_j^n(X^i)\right) f(X_{j\delta}^i)$$

are two uncorrelated martingales.

Define the usual distance on $\mathcal{M}_F(\mathbb{R}^d)$ by

$$d(\nu_1, \nu_2) = \sum_{i=0}^{\infty} 2^{-i} \left(|\langle \nu_1 - \nu_2, f_i \rangle| \wedge 1 \right), \qquad \forall\, \nu_1, \nu_2 \in \mathcal{M}_F(\mathbb{R}^d), \qquad (23.26)$$

where $f_0 = 1$ and for $i \geq 1$, $f_i \in C_b^{k+2}(\mathbb{R}^d) \cap W_2^{k+2}(\mathbb{R}^d)$ with $\|f_i\|_{k+2,\infty} \leq 1$ and also $\|f_i\|_{k+2,2} \leq 1$, where $k = \left[\frac{d}{2}\right] + 2$ is given in Condition (BD).

Theorem 23.9 *Suppose that the conditions (BD) and (I) hold and, additionally, that $h \in C_b^k(\mathbb{R}^d) \cap W_2^k(\mathbb{R}^d)$. Then, there exists a constant K_1 such that*

$$\hat{\mathbb{E}} \sup_{t \leq T} d\left(V_t^n, V_t\right)^2 \leq K_1 n^{-(1-a)}. \qquad (23.27)$$

By Kallianpur–Striebel formula, we then get

Theorem 23.10 *Suppose that the conditions in Theorem 23.9 are satisfied. Then, there exists a constant K_1 such that*

$$\mathbb{E} \sup_{0 \leq t \leq T} d\left(\pi_t^n, \pi_t\right) \leq K_1 n^{-\frac{1-a}{2}}. \qquad (23.28)$$

For the particle filter, we have the following estimate.

Remark 23.1 For the particle filter $\tilde{\pi}_t^n$, we have

$$\mathbb{E} \sup_{0 \leq t \leq T} d\left(\tilde{\pi}_t^n, \pi_t\right) \leq K_1 \left(n^{-(1-a)/2} \vee n^{-a}\right).$$

Note that the optimal rate in this case is $\frac{1}{3}$ achieved at $a = \frac{1}{3}$.

Finally, we characterize the convergence rate of π^n to π by studying the convergence of the sequence $\zeta_t^n = n^{\frac{1-a}{2}} \left(\pi_t^n - \pi_t\right)$.

Theorem 23.11 *ζ^n converges weakly to a process ζ which is the unique solution to the following evolution equation:*

$$d\zeta_t = \langle \zeta_t, Lf - (\pi_t(\nabla^* fc + hf) - \pi_t f \pi_t h) h \rangle dt$$
$$+ \langle \zeta_t, \nabla^* fc + hf - f\pi_t h - h\pi_t f \rangle d\nu_t$$
$$- \sqrt[4]{2\pi^{-1}} \int_{\mathbb{R}^d} \frac{f - \pi_t f}{V_t 1} \sqrt{|h(x) - \pi_t h|} V(t,x) B(dt dx)$$

where B is a space–time white noise which is independent of Y and ν is the innovation process defined by (23.12).

To bring this section to an end, we briefly mention, to the best of our knowledge, some of the related papers available in the literature.

Remark 23.2 Particle system approximation of optimal filters was studied in heuristic schemes in the beginning of the 1990s by Gordon, Salmon, and Ewing

[21], Gordon, Salmon, and Smith [20], Kitagawa [25], Carvalho, Del Moral, Monin, and Salut [1], Del Moral, Noyer, and Salut [18]. The rigorous proof of the convergence results for the particle filter were published by Del Moral [15] in 1996, and independently, by Crisan and Lyons [10] in 1997. After that, many improvements have been made by various authors. Here we would like to mention only a few: Crisan and Lyons [11], Crisan [6], [4], [3], [5], Crisan, Del Moral, and Lyons [7], [9], Crisan and Doucet [8], Crisan, Gaines, and Lyons [9], Del Moral and Guionnet [16], Del Moral and Miclo [17].

23.4 Filtering using interacting particle systems

In this section, we give an interacting particle system representation of the optimal filter based on the papers of Crisan and Xiong [13], [14]. The main motivation is to seek a numerical scheme to approximate the optimal filter using neither branching nor weight. The idea is to direct the particles to move toward more likely regions, and hence, the coefficients should depend on the whole system configuration. For simplicity of notations, we take $c = 0$ in the signal equation (23.1).

Note that the innovation process v_t can be approximated by a smooth process $\tilde{v}_t$. By a robust representation of the optimal filter due to Clark and Crisan [2], the optimal filter π_t can be approximated by the solution to the following PDE:

$$\frac{d}{dt} \langle \mathcal{I}_t, f \rangle = \langle \mathcal{I}_t, Lf \rangle + \left\langle \mathcal{I}_t, a_t^{\mathcal{I}_t} f \right\rangle \qquad (23.29)$$

where $a_t = h \frac{d}{dt} \tilde{v}_t$ is a bounded smooth function and

$$a^{\mathcal{I}} = a - \langle \mathcal{I}, a \rangle.$$

For mathematical convenience, we assume that the signal is a reflecting diffusion in a bounded domain $D = \{x \in \mathbb{R}^d : |x| \leq R\}$ of $\mathbb{R}^d$, namely,

$$L = b^* \nabla + \frac{1}{2} \sum_{j,k=1}^{d} a^{jk} \partial^2_{jk}, \qquad D(L) = \{f : x^* \nabla f|_{\partial D} = 0\}$$

where $a = \sigma^* \sigma$. In fact, the diffusion on $\mathbb{R}^d$ can be approximated by such reflecting diffusions. Thus, the optimal filter with bounded signal can be regarded as an approximation of the original optimal filter.

By Conditions (BC1), (BC2), and (E), we see that the following conditions, which are needed in the study of the numerical approximation to (23.29), are satisfied:

(B) There exists a constant K such that for any $t \in [0, T]$ and $x \in D$, we have

$$|b(x)| + |\sigma(x)| + |a_t(x)| \leq K.$$

(Lip) For any $t \in [0, T]$ and $x, y \in D$, we have

$$|b(x) - b(y)| + |\sigma(x) - \sigma(y)| + |a_t(x) - a_t(y)| \leq K|x - y|.$$

(UE) There exists a constant $K_0 > 0$ such that for any $x \in D$, the matrix $a(x) - K_0 I$ is nonnegative definite.

The following identity is the key for the interacting particle system representation of the optimal filter.

Lemma 23.5

$$\int_D \frac{\nabla^* f(x)(y-x)}{\|y-x\|^d} dx = \begin{cases} -\int_{\partial D} f(x) dS + \omega_d f(y) & \text{if } y \neq 0 \\ \omega_d f(0) & \text{if } y = 0 \text{ and } f \in D(L), \end{cases}$$

where ω_d is the surface area of the d-dimensional unit sphere S_{d-1}.

Based on this identity, we can show that $\mathcal{I}_t$ has the following representation by an interacting infinite particle system.

Proposition 23.1

$$\mathcal{I}_t = \lim_{n \to \infty} \frac{1}{n} \sum_{i=1}^{n} \delta_{X_t^i} \qquad (23.30)$$

where

$$dX_t^i = \tilde{b}_t\left(\mathcal{I}_t, X_t^i\right) dt + \sigma\left(X_t^i\right) dB_t^i + \mathcal{N}\left(X_t^i\right) dK_t^i, \qquad i = 1, 2, \cdots, \qquad (23.31)$$

K_t^i is the local time of X_t^i at the boundary of D, $\mathcal{N}(x)$ is the unit normal vector of ∂D at $x \in \partial D$,

$$\tilde{b}_t(\mathcal{I}_t, x) = b(x) + \frac{\Lambda_{\mathcal{I}_t} a_t(x)}{\mathcal{I}_t(x)}$$

and

$$\Lambda_{\mathcal{I}} a(x) = \frac{1}{\omega_d} \int_{\mathbb{R}^d} \frac{(y-x) a^T(y)}{\|y-x\|^d} \mathcal{I}(dy).$$

Note that X_t^i given by (23.31) is a diffusion on D with reflecting boundary.

Remark 23.3 From the definition of $\tilde{b}$ we see that the particles move fast in unlikely regions (with $\mathcal{I}_t(x)$ small) and slow in more likely regions (with $\mathcal{I}_t(x)$ large). Therefore, the particles spend more time in more likely regions.

Based on the representation above, we now propose an approximation to $\{\mathcal{I}_t\}$ by finite interacting particle systems. For each finite system, the empirical measure has no density, so we smooth it out by the operator T_ϵ given below. For the convenience of the estimates, we introduce an extra parameter $\delta > 0$ to make the coefficient $\tilde{b}$ bounded. Namely, we fix $n \in \mathbb{N}$ and ϵ, $\delta > 0$ and consider the following finite system: For $i = 1, 2, \cdots n$,

$$dX_t^{n,\epsilon,\delta,i} = \tilde{b}_t^{\epsilon,\delta}\left(\mathcal{I}_t^{n,\epsilon,\delta}, X_t^{n,\epsilon,\delta,i}\right) dt + \sigma\left(X_t^{n,\epsilon,\delta,i}\right) dB_t^i + \mathcal{N}\left(X_t^{n,\epsilon,\delta,i}\right) dK_t^{n,\epsilon,\delta,i} \tag{23.32}$$

and

$$\mathcal{I}_t^{n,\epsilon,\delta} = \frac{1}{n}\sum_{i=1}^n \delta_{X_t^{n,\epsilon,\delta,i}}, \qquad \tilde{b}_t^{\epsilon,\delta}(\mu, x) = b(x) + \frac{\Lambda_\mu a_t(x)}{T_\epsilon \mu(x) + \delta} \tag{23.33}$$

where

$$T_\epsilon \mu(x) = (2\pi\epsilon)^{-\frac{d}{2}} \int \exp\left(-\frac{|x-y|^2}{2\epsilon}\right) \mu(dy).$$

Here is the main convergence theorem in this section.

Theorem 23.12 *For any $t > 0$ fixed, we have*

$$\lim_{\delta \to 0} \lim_{\epsilon \to 0} \lim_{n \to \infty} \mathcal{I}_t^{n,\epsilon,\delta} = \mathcal{I}_t.$$

Finally, we use Euler's scheme to solve the SDE system (23.32). The main difference from the Euler scheme in Section 23.2 is the involvement of the local time process $K_t^{n,\epsilon,\delta,i}$ in each equation. We use a penalization method (cf. Pettersson [29] and Slominski [30]) to deal with this problem. For simplicity of notation, we drop the superscripts n, ϵ, δ in this part. The system (23.32) becomes

$$\begin{cases} dX_t^i = \tilde{b}_t\left(\mathcal{I}_t, X_t^i\right) dt + \sigma\left(X_t^i\right) dB_t^i + \mathcal{N}\left(X_t^i\right) dK_t^i \\ \mathcal{I}_t = \frac{1}{n}\sum_{i=1}^n \delta_{X_t^i}, \qquad \tilde{b}_t(\mu, x) = b(x) + \frac{\Lambda_\mu a_t(x)}{T_\epsilon \mu(x)+\delta}. \end{cases}$$

Let $0 = t_0 < t_1 < \cdots < t_{c_\gamma} = T$ be a partition of $[0, T]$ with mesh size $\gamma = \max_{1 \le k \le c_\gamma} \Delta t_k$ where $\Delta t_k = t_k - t_{k-1}$. Let Π_D be the projection to D (i.e. $\Pi_D x$ is the point in D which is the closest to x). For $0 \le t < t_1$, define $X_t^{\gamma,i} = x_0^i$; and for $t_k \le t < t_{k+1}$, $k \ge 1$,

$$\begin{cases} X_t^{\gamma,i} = \Pi_D\left\{X_{t_{k-1}}^{\gamma,i} + \tilde{b}_{t_{k-1}}\left(\mathcal{I}_{t_{k-1}}^\gamma, X_{t_{k-1}}^{\gamma,i}\right)\Delta t_k + \sigma\left(X_{t_{k-1}}^{\gamma,i}\right)\Delta B_{t_k}^i\right\} \\ \mathcal{I}_t^\gamma = \frac{1}{n}\sum_{i=1}^n \delta_{X_t^{\gamma,i}} \end{cases} \tag{23.34}$$

where $\Delta B_{t_k}^i = B_{t_k}^i - B_{t_{k-1}}^i$.

Let $\mathcal{I}_t^{n,\epsilon,\delta,\gamma}$ be the process $\mathcal{I}_t^\gamma$ given by (23.34). Then

Theorem 23.13 *There exists a constant $C_{\epsilon,\delta}$ such that*

$$\sup_{0 \le t \le T} \mathbb{E}\rho\left(\mathcal{I}_t^{n,\epsilon,\delta,\gamma}, \mathcal{I}_t^{n,\epsilon,\delta}\right)^2 \le C_{\epsilon,\delta}\left(\gamma \log\frac{1}{\gamma}\right)^{1/2}.$$

Remark 23.4 In Section 23.2, we saw that the time-varying weights of the particles cause the approximate error to increase exponentially fast in time. To overcome this drawback, branching particle systems were introduced in Section 23.3 so that the individual weight of any particle depends on its path in small time intervals only. In the present section, we have provided an alternative with constant weights hoping to reduce the approximate error, or equivalently, to increase the convergence rate. It remains a challenging *open* problem to derive an estimate of the convergence rate and to compare it with that of the hybrid filter of Section 23.3.

Remark 23.5 As we have seen from this section, the numerical scheme based on interacting particle system with neither branching nor weight involves quite many approximating procedures ($\tilde{\nu} \to \nu$, $D \uparrow \mathbb{R}^d$, ϵ, $\delta \to 0$, $n \to \infty$). How do propose more efficient approximation (i.e., with less approximating procedures) remains a challenging *open* problem.

23.5 A filtering model with point process observations

The filtering model [(23.1), (23.3)] with point processes observations, called the Filtering Micromovement (FM) model, is proposed by Zeng [32]. The signal process X_t represents the intrinsic value process of d assets, which corresponds to the macro-movement in the empirical econometric literature or the continuous-time price process in the option pricing literature. Prices are observed only at random trading times which are modelled by a conditional Poisson process. Moreover, prices are distorted observations of the intrinsic value process at the trading times. The observed prices take values in the discrete set of m levels. Thus, an alternative description for the observation process is Y_t^i which counts the number of times that level i price are observed before time t. Then, Y_t^i, $i = 1, 2, \cdots, m$, are Poisson point processes whose intensities depend on the intrinsic value process.

The Zakai equation and the filtering equation are derived in [32]. The particle approximation for the optimal filter is studied by Xiong and Zeng [31] on which this section is based.

For simplicity of notation, we assume $c = 0$ in the signal equation (23.1). In addition to the Conditions (I) and (BC1), we assume the following Condition (P) on the intensities of the point processes introduced by (23.3).
Condition (P): i) The total intensity $a(x, t) \equiv \sum_{k=1}^m \lambda_k(x, t)$ is bounded.
ii) Let $a'(x, t) \equiv \frac{\partial}{\partial x} a(x, t)$ and $p_k'(x, t) \equiv \frac{\partial}{\partial x} p_k(x, t)$, where

$$p_k(x, t) = a(x, t)^{-1} \lambda_k(x, t).$$

Then $a'(x, t)$ and $p_k'(x, t)$ are bounded and continuous in x.

Let M_t be the solution to the following SDE:

$$dM_t = \sum_{k=1}^{m} (\lambda_k(X_s, s-) - 1) M_{t-} d\left(Y_t^k - t\right), \qquad M_0 = 1. \qquad (23.35)$$

Let $\hat{P}$ be the probability measure given by $\left.\frac{dP}{d\hat{P}}\right|_{\mathcal{F}_t} = M_t$. By Girsanov's theorem, we know that, under $\hat{P}$, Y_t^k, $k = 1, \cdots, m$ are independent standard Poisson point processes, and they are independent of the Brownian motion B. The Kallianpur-Striebel formula in the current setup gives

$$\langle \pi_t, f \rangle = \frac{\langle V_t, f \rangle}{\langle V_t, 1 \rangle},$$

where

$$\langle V_t, f \rangle = \hat{\mathbb{E}}\left(f(X_t) M_t | \mathcal{G}_t\right).$$

Note that V_t has the following weighted particle system representation:

$$V_t = \lim_{n \to \infty} \frac{1}{n} \sum_{i=1}^{n} M_t^i \delta_{X_t^i}$$

where

$$dX_t^i = b\left(X_t^i\right) dt + \sigma\left(X_t^i\right) dB_t^i$$

and

$$dM_t^i = \sum_{k=1}^{m} \left(\lambda_k\left(X_s^i, s-\right) - 1\right) M_{t-}^i d\left(Y_t^k - t\right), \qquad M_0 = 1.$$

Applying Itô's formula to the above system, we get the Zakai equation in this model

$$\langle V_t, f \rangle = \langle V_0, f \rangle + \int_0^t \langle V_s, Lf \rangle ds + \sum_{k=1}^{m} \int_0^t \langle V_{s-}, (ap_k - 1)f \rangle d\left(Y_s^k - s\right). \qquad (23.36)$$

Making use of the Kallianpur–Striebel and Itô formulas, we have the filtering equation:

$$\langle \pi_t, f \rangle = \langle \pi_0, f \rangle + \int_0^t \left[\langle \pi_s, Lf \rangle - \langle \pi_s, fa \rangle + \langle \pi_s, f \rangle \langle \pi_s, a \rangle\right] ds$$
$$+ \sum_{k=1}^{m} \int_0^t \left[\frac{\langle \pi_{s-}, f\lambda_k \rangle}{\langle \pi_{s-}, \lambda_k \rangle} - \langle \pi_{s-}, f \rangle\right] dY_s^k. \qquad (23.37)$$

When $a(X(t), t) = a(t)$, the above equation is simplified as:

$$\langle \pi_t, f \rangle = \langle \pi_0, f \rangle + \int_0^t \langle \pi_s, Lf \rangle ds + \sum_{k=1}^m \int_0^t \left[\frac{\langle \pi_{s-}, f\lambda_k \rangle}{\langle \pi_{s-}, \lambda_k \rangle} - \langle \pi_{s-}, f \rangle \right] dY_s^k. \tag{23.38}$$

Note that a branching weighted particle system can be defined similar to that in Section 23.3. We will not repeat its definition here. Instead, we shall only study its properties. To this end, we use the same notations as in Section 23.3. The main difference is the following lemma which is the counterpart of Lemma 23.2 in the present setup.

Lemma 23.6 Let $F(x) = \{x\}(1 - \{x\})$. Then, for bounded $f(x)$ with bounded Lf^2, as $\delta \to 0$, we have

$$\left| \mathbb{E}\left(\gamma_{j+1}^n(X^i) f^2(X_{(j+1)\delta}^i) \left(\eta_{j+1}^n / \eta_j^n \right)^2 \Big| \mathcal{F}_{j\delta} \right) - \left(f^2 H_{j\delta}^n \right)(X_{j\delta}^i) \delta \right| = o(\delta),$$

where H_s^n is a nonnegative function given by

$$H_s^n(x) = \sum_{k=1}^m F\left(\frac{\lambda_k(x, s)}{\bar{h}_k^n(s) + 1} \right) \left(\bar{h}_k^n(s) + 1 \right)^2 \tag{23.39}$$

and

$$\bar{h}_k^n(s) = \langle \pi_s^n, \lambda_k - 1 \rangle. \tag{23.40}$$

As a consequence of the lemma, we have that

$$\mathbb{E}\left(\gamma_{j+1}^n(X^i) \left(\eta_{j+1}^n / \eta_j^n \right)^2 | \mathcal{F}_{j\delta} \right) \leq K_1 \delta,$$

which is different from Lemma 23.2.

Based on this estimate, it follows from the same argument as in (23.24) that

$$\hat{\mathbb{E}}((I_2^n)^2) \leq K_1 \delta \frac{1}{n^2} \frac{T}{\delta} n \leq K_2 n^{-1},$$

which is in contrast with the bound $K_3 n^{-(1-\alpha)}$ for the classical case.

As we did in Section 23.3, we consider the dual of the Zakai equation which is the following backward SPDE:

$$\begin{cases} d\psi_s = -L\psi_s ds - \sum_{k=1}^w (ap_k - 1)\psi_{s+} \hat{d}\left(Y_s^k - s \right), & 0 \leq s \leq t \\ \psi_t = \phi \end{cases} \tag{23.41}$$

where $\hat{d}$ denotes the backward Itō's integral and ϕ is a bounded function. Actually, the current setup makes this equation easier to handle because it can be studied as a piecewise PDE. Thus, we get the following lemma easily.

Lemma 23.7 *Suppose that Conditions (I), (BC1) and (P) hold for the FM model. Let $\psi'_u(x) = \frac{d}{dx}\psi_u(x)$. Then, there exists a constant K such that*

$$\hat{\mathbb{E}}\left[\sup_{0\leq s\leq t} \|\psi_s\|_\infty + \sup_{0\leq s\leq t} \|\psi'_s\|_\infty\right] \leq K.$$

Based on this lemma, we can get the following key identity.

Lemma 23.8 *Suppose that Conditions (I), (BC1), and (P) hold. Then, for every $t \geq 0$, we have*

$$\psi_t(X_t)M_t - \psi_0(X_0) = \int_0^t M_s \nabla^* \psi_s \sigma(X_s) dB_s, \qquad a.s.. \qquad (23.42)$$

Note that Lemma 23.8 is in the same form as Lemma 23.4. However, the restrictive Condition (BD) is not required here.

By the same arguments as those leading to Theorem 23.8, we get

Theorem 23.14 *Suppose that Conditions (I), (BC1), and (P) hold. Then there exists a constant K_1 such that*

$$\hat{\mathbb{E}}\left|\langle V_t^n, \phi\rangle - \langle V_t, \phi\rangle\right|^2 \leq K_1 n^{-1}.$$

Next, by an equation similar to (23.25), we can prove that

Theorem 23.15 *Under the assumptions of Theorem 23.14, there exists a constant K such that*

$$\mathbb{E} \sup_{0\leq t\leq T} d\left(\pi_t^n, \pi_t\right) \leq K n^{-\frac{1}{2}}. \qquad (23.43)$$

Finally, we characterize the rate of convergence by studying $\zeta_t^n = n^{\frac{1}{2}}\left(\pi_t^n - \pi_t\right)$. Let Φ be the nuclear space defined in Section 23.2.

Theorem 23.16 *Under the assumptions of Theorem 23.14, ζ^n converges weakly in $D_{\Phi_{-\kappa}}[0, \infty)$ to a process ζ which is the unique solution to the following evolution equation:* $\forall f \in \Phi$,

$$d\langle \zeta_t, f\rangle = \langle \zeta_t, Lf - (a-w)f - f\langle \pi_t, a-w\rangle + (a-w)\langle \pi_t, f\rangle\rangle dt$$
$$+ \sum_{k=1}^w \left[\frac{\langle \zeta_{t-}, fap_k\rangle}{\langle \pi_{t-}, ap_k\rangle} - \frac{\langle \zeta_{t-}, ap_k\rangle\langle \pi_{t-}, fap_k\rangle}{\langle \pi_{t-}, ap_k\rangle^2} - \langle \zeta_{t-}, f\rangle\right] dY_t^k$$
$$+ \int_\mathbb{R} \frac{f(x) - \langle \pi_t, f\rangle}{\langle V_t, 1\rangle} \sqrt{H_t(x) V_t(x) \langle V_t, 1\rangle}\, W(dxdt),$$

$$(23.44)$$

where W is a space–time white noise independent of Y, $H_s^n(x)$ is a nonnegative function given by

$$H_s(x) = \sum_{k=1}^m F\left(\frac{\lambda_k(x,s)}{\bar{h}_k(s)+1}\right)(\bar{h}_k(s)+1)^2 \tag{23.45}$$

and

$$\bar{h}_k(s) = \langle \pi_s, \lambda_k - 1 \rangle. \tag{23.46}$$

Notes

Research for this chapter is partially supported by the NSA.

References

[1] H. Carvalho, P. Del Moral, A. Monin, and G. Salut (1997). Optimal nonlinear filtering in GPS/INS integration. *IEEE Trans. Aerosp. Electron. Syst.*, **33**, no. 3, 835–50.
[2] J. M. C. Clark and D. Crisan (2005). On a robust version of the integral representation formula of nonlinear filtering. *Probab. Theory Related Fields* **133**, no. 1, 43–56.
[3] D. Crisan (2001). *Particle filters—a theoretical perspective*. Sequential Monte Carlo methods in practice, 17–41, Stat. Eng. Inf. Sci., Springer, New York.
[4] D. Crisan (2002). *Numerical methods for solving the stochastic filtering problem*, Numerical methods and stochastics (Toronto, ON, 1999), 1–20, Fields Inst. Commun., 34, Amer. Math. Soc., Providence, RI.
[5] D. Crisan (2003). Exact rates of convergence for a branching particle approximation to the solution of the Zakai equation. *Ann. Probab.* **31**, no. 2, 693–718.
[6] D. Crisan (2004). Superprocesses in a Brownian environment. Stochastic analysis with applications to mathematical finance. *Proc. R. Soc. Lond. Ser. A Math. Phys. Eng. Sci.* **460**, no. 2041, 243–70.
[7] D. Crisan, P. Del Moral, and T. Lyons (1999). Interacting particle systems approximations of the Kushner-Stratonovitch equation. *Adv. in Appl. Probab.* **31**, no. 3, 819–38.
[8] D. Crisan and A. Doucet (2002). A survey of convergence results on particle filtering methods for practitioners. *IEEE Trans. Signal Process.* **50**, no. 3, pp 736–46.
[9] D. Crisan, J. Gaines, and T. Lyons (1998). Convergence of a branching particle method to the solution of the Zakai equation. *SIAM J. Appl. Math.* **58**, no. 5, 1568–90 (electronic).
[10] D. Crisan and T. Lyons (1997). Nonlinear filtering and measure-valued processes. *Probab. Theory Related Fields*, **109**, 217–44.
[11] D. Crisan and T. Lyons (1999). A particle approximation of the solution of the Kushner-Stratonovitch equation. *Probab. Theory Related Fields* **115**, no. 4, 549–78.
[12] D. Crisan and J. Xiong (2007). A central limit type theorem for particle filter. *Comm. Stoch. Analysis* **1**, no. 1, 103–22.
[13] D. Crisan and J. Xiong (2008). Approximate McKean–Vlasov representations for a class of SPDEs. To appear in *Stochastics*.
[14] D. Crisan and J. Xiong (2008). Numerical solutions for a class of SPDEs over bounded domains. *In preparation*.

[15] P. Del Moral (1996). Non-linear filtering: interacting particle resolution. *Markov Process. Related Fields* **2**, no. 4, 555–81.

[16] P. Del Moral and A. Guionnet (1999) Central limit theorem for nonlinear filtering and interacting particle systems. *Ann. Appl. Probab.* **9**, no. 2, 275–97.

[17] P. Del Moral and L. Miclo (2000). Branching and interacting particle systems approximations of Feynman–Kac formulae with applications to non-linear filtering. *Séminaire de Probabilités, XXXIV, Lecture Notes in Math.*, **1729**, Springer, Berlin, 1–145.

[18] P. Del Moral, J. C. Noyer, and G. Salut (1995). Résolution particulaire et traitement non-linéaire du signal: application radar/sonar. In *traitement du signal* **12**, no. 4, 287–301.

[19] I. M. Gel'fand and N. Ya. Vilenkin (1964). *Generalized functions. Vol. 4: Applications of harmonic analysis.* Academic Press, New York–London.

[20] N. J. Gordon, D. J. Salmon, and A. F. M. Smith (1993). Novel approach to nonlinear/non-Gaussian Bayesian state estimation. *IEE Proc. F,* **140**, 107–13.

[21] N. J. Gordon, D. J. Salmon, and C. Ewing (1995). Bayesian state estimation for tracking and guidance using the bootstrap filter. *J. Guidance Control Dyn.*, **18**, no. 6, 1434–43.

[22] M. Hitsuda and I. Mitoma (1986). Tightness problem and stochastic evolution equation arising from fluctuation phenomena for interacting diffusions. *J. Multivariate Anal.* 19, 311–28.

[23] G. Kallianpur and C. Striebel (1968). Estimation of stochastic systems: arbitrary system process with additive noise observation errors. *Ann. Math. Statist.*, **39**, 785–801.

[24] G. Kallianpur and C. Striebel (1969). Stochastic differential equations occurring in the estimation of continuous parameter stochastic processes. *Teor. Veroyatn. Primen.*, **14**, no. 4, 597–622.

[25] G. Kitagawa (1996). Monte Carlo filter and smoother for non-Gaussian non-linear state space models. *J. Comput. and Graphical Stat.*, **5**, no. 1, 1–25.

[26] T. Kurtz and J. Xiong (1999). Particle representations for a class of nonlinear SPDEs. *Stochastic Process. Appl.* **83**, 103–26.

[27] T. Kurtz and J. Xiong (2000). Numerical solutions for a class of SPDEs with application to filtering. *Stochastics in finite and infinite dimension: In Honor of Gopinath Kallianpur.* Eds by T. Hida, R. Karandikar, H. Kunita, B. Rajput, S. Watanabe, and J. Xiong. *Trends in Mathematics.* Birkhäuser.

[28] T. Kurtz and J. Xiong (2004). A stochastic evolution equation arising from the fluctuation of a class of interacting particle systems. *Communication Mathematical Sciences* **2**, 325–58.

[29] R. Pettersson (1997). Penalization schemes for reflecting stochastic differential equations. *Bernoulli* **3**, No. 4, 403–14.

[30] L. Slominski (2001). Euler's approximations of solutions of SDEs with reflecting boundary. *Stochastic Processes and their Applications* **94**, No. 2, 317–37.

[31] J. Xiong and Y. Zeng. A branching particle approximation to the filtering problem with counting process observations. *Submitted.*

[32] Y. Zeng (2003). A partialy observed model for micromovement of asset prices with Bayes estimation via filtering, *Mathematical Finance*, **13**, 411–44.

·24·
A Tutorial on Particle Filtering and Smoothing: Fifteen Years Later

A. Doucet and A. M. Johansen

24.1 Introduction

The general state space hidden Markov models, which are summarized in Section 24.2.1, provide an extremely flexible framework for modelling time series. The great descriptive power of these models comes at the expense of intractability: it is impossible to obtain analytic solutions to the inference problems of interest with the exception of a small number of particularly simple cases. The "particle" methods described by this tutorial are a broad and popular class of Monte Carlo algorithms which have been developed over the past fifteen years to provide approximate solutions to these intractable inference problems.

24.1.1 Preliminary remarks

Since their introduction in 1993 [22], particle filters have become a very popular class of numerical methods for the solution of optimal estimation problems in nonlinear non-Gaussian scenarios. In comparison with standard approximation methods, such as the popular Extended Kalman Filter, the principal advantage of particle methods is that they do not rely on any local linearisation technique or any crude functional approximation. The price that must be paid for this flexibility is computational: these methods are computationally expensive. However, thanks to the availability of ever-increasing computational power, these methods are already used in real-time applications appearing in fields as diverse as chemical engineering, computer vision, financial econometrics, target tracking, and robotics. Moreover, even in scenarios in which there are no real-time constraints, these methods can be a powerful alternative to Markov chain Monte Carlo (MCMC) algorithms—alternatively, they can be used to design very efficient MCMC schemes.

As a result of the popularity of particle methods, a few tutorials have already been published on the subject [3, 8, 18, 29]. The most popular, [3], dates back to 2002 and, like the edited volume [16] from 2001, it is now somewhat outdated. This tutorial differs from previously published tutorials in two ways. First, the obvious: it is, as of December 2010, the most recent tutorial on the subject and so it has been possible to include some very recent material on advanced particle methods for filtering and smoothing. Second, more importantly, this tutorial was not intended to resemble a cookbook. To this end, all of the algorithms are presented within a simple, unified framework. In particular, we show that essentially all basic and advanced methods for particle filtering can be reinterpreted as some special instances of a single generic Sequential Monte Carlo (SMC) algorithm. In our opinion, this framework is not only elegant but allows the development of a better intuitive and theoretical understanding of particle methods. It also shows that essentially any particle filter can be implemented using a simple computational framework such as that provided by [24]. Absolute beginners might benefit from reading [17], which provides an elementary introduction to the field, before the present tutorial.

24.1.2 Organization of the tutorial

The rest of this chapter is organized as follows. In Section 24.2, we present hidden Markov models and the associated Bayesian recursions for the filtering and smoothing distributions. In Section 24.3, we introduce a generic SMC algorithm which provides weighted samples from any sequence of probability distributions. In Section 24.4, we show how all the (basic and advanced) particle filtering methods developed in the literature can be interpreted as special instances of the generic SMC algorithm presented in Section 24.3. Section 24.5 is devoted to particle smoothing and we mention some open problems in Section 24.6.

24.2 Bayesian inference in hidden Markov models

24.2.1 Hidden Markov models and inference aims

Consider an $\mathcal{X}$-valued discrete-time Markov process $\{X_n\}_{n\geq 1}$ such that

$$X_1 \sim \mu(x_1) \text{ and } X_n|(X_{n-1} = x_{n-1}) \sim f(x_n|x_{n-1}) \tag{24.1}$$

where "$\sim$" means distributed according to, $\mu(x)$ is a probability density function and $f(x|x')$ denotes the probability density associated with moving from x' to x. All the densities are with respect to a dominating measure that we will denote, with abuse of notation, dx. We are interested in estimating $\{X_n\}_{n\geq 1}$ but only have access to the $\mathcal{Y}$-valued process $\{Y_n\}_{n\geq 1}$. We assume that, given $\{X_n\}_{n\geq 1}$, the observations $\{Y_n\}_{n\geq 1}$ are statistically independent and their marginal densities (with respect to a dominating measure dy_n) are given by

$$Y_n|(X_n = x_n) \sim g(y_n|x_n). \tag{24.2}$$

For the sake of simplicity, we have considered only the homogeneous case here; that is, the transition and observation densities are independent of the time index n. The extension to the inhomogeneous case is straightforward. It is assumed throughout that any model parameters are known.

Models compatible with (24.1)–(24.2) are known as hidden Markov models (HMM) or general state-space models (SSM). This class includes many models of interest. The following examples provide an illustration of several simple problems which can be dealt with within this framework. More complicated scenarios can also be considered.

Example 24.1 *Finite State-Space HMM.* In this case, we have $\mathcal{X} = \{1, ..., K\}$ so

$$\Pr(X_1 = k) = \mu(k), \quad \Pr(X_n = k| X_{n-1} = l) = f(k|l).$$

The observations follow an arbitrary model of the form (24.2). This type of model is extremely general and examples can be found in areas such as genetics in which they can describe imperfectly observed genetic sequences, signal processing, and computer science in which they can describe, amongst many other things, arbitrary finite-state machines.

Example 24.2 *Linear Gaussian model.* Here, $\mathcal{X} = \mathbb{R}^{n_x}$, $\mathcal{Y} = \mathbb{R}^{n_y}$, $X_1 \sim \mathcal{N}(0, \Sigma)$ and

$$X_n = AX_{n-1} + BV_n,$$
$$Y_n = CX_n + DW_n$$

where $V_n \overset{\text{i.i.d.}}{\sim} \mathcal{N}(0, I_{n_v})$, $W_n \overset{\text{i.i.d.}}{\sim} \mathcal{N}(0, I_{n_w})$ and A, B, C, D are matrices of appropriate dimensions. Note that $\mathcal{N}(m, \Sigma)$ denotes a Gaussian distribution of mean m and variance-covariance matrix Σ, whereas $\mathcal{N}(x; m, \Sigma)$ denotes the Gaussian density of argument x and similar statistics. In this case $\mu(x) = \mathcal{N}(x; 0, \Sigma)$, $f(x'|x) = \mathcal{N}(x'; Ax, BB^T)$ and $g(y|x) = \mathcal{N}(y; Cx, DD^T)$. As inference is analytically tractable for this model, it has been extremely widely used for problems such as target tracking and signal processing.

Example 24.3 *Switching linear Gaussian model.* We have $\mathcal{X} = \mathcal{U} \times \mathcal{Z}$ with $\mathcal{U} = \{1, ..., K\}$ and $\mathcal{Z} = \mathbb{R}^{n_z}$. Here $X_n = (U_n, Z_n)$ where $\{U_n\}$ is a finite state-space Markov chain such that $\Pr(U_1 = k) = \mu_U(k)$, $\Pr(U_n = k| U_{n-1} = l) = f_U(k|l)$ and conditional upon $\{U_n\}$ we have a linear Gaussian model with $Z_1| U_1 \sim \mathcal{N}(0, \Sigma_{U_1})$ and

$$Z_n = A_{U_n} Z_{n-1} + B_{U_n} V_n,$$
$$Y_n = C_{U_n} Z_n + D_{U_n} W_n$$

where $V_n \overset{\text{i.i.d.}}{\sim} \mathcal{N}(0, I_{n_v})$, $W_n \overset{\text{i.i.d.}}{\sim} \mathcal{N}(0, I_{n_w})$ and $\{A_k, B_k, C_k, D_k; k = 1, ..., K\}$ are matrices of appropriate dimensions. In this case we have $\mu(x) = \mu(u, z) = \mu_U(u) \mathcal{N}(z; 0, \Sigma_u)$, $f(x'|x) = f((u', z')|(u, z)) = f_U(u'|u) \mathcal{N}(z'; A_{u'} z, B_{u'} B_{u'}^T)$ and $g(y|x) = g(y|(u, z)) = \mathcal{N}(y; C_u z, D_u D_u^T)$. This type of model provides a generalization of that described in Example 24.2 with only a slight increase in complexity.

Example 24.4 *Stochastic Volatility model.* We have $\mathcal{X} = \mathcal{Y} = \mathbb{R}$, $X_1 \sim \mathcal{N}\left(0, \frac{\sigma^2}{1-a^2}\right)$ and

$$X_n = aX_{n-1} + \sigma V_n,$$
$$Y_n = \beta \exp(X_n/2) W_n$$

where $V_n \overset{\text{i.i.d.}}{\sim} \mathcal{N}(0, 1)$ and $W_n \overset{\text{i.i.d.}}{\sim} \mathcal{N}(0, 1)$. In this case we have $\mu(x) = \mathcal{N}\left(x; 0, \frac{\sigma^2}{1-a^2}\right)$, $f(x'|x) = \mathcal{N}(x'; ax, \sigma^2)$ and $g(y|x) = \mathcal{N}(y; 0, \beta^2 \exp(x))$. Note that this choice of initial distribution ensures that the marginal distribution of X_n is also $\mu(x)$ for all n. This type of model, and its generalizations, have been very widely used in various areas of economics and mathematical finance: inferring and predicting underlying volatility from observed price or rate data is an important problem. Figure 24.1 shows a short section of data simulated from such a model with parameter values $a = 0.91$, $\sigma = 1.0$ and $\beta = 0.5$ which will be used below to illustrate the behaviour of some simple algorithms.

Equations (24.1)–(24.2) define a Bayesian model in which (24.1) defines the prior distribution of the process of interest $\{X_n\}_{n \geq 1}$ and (24.2) defines the

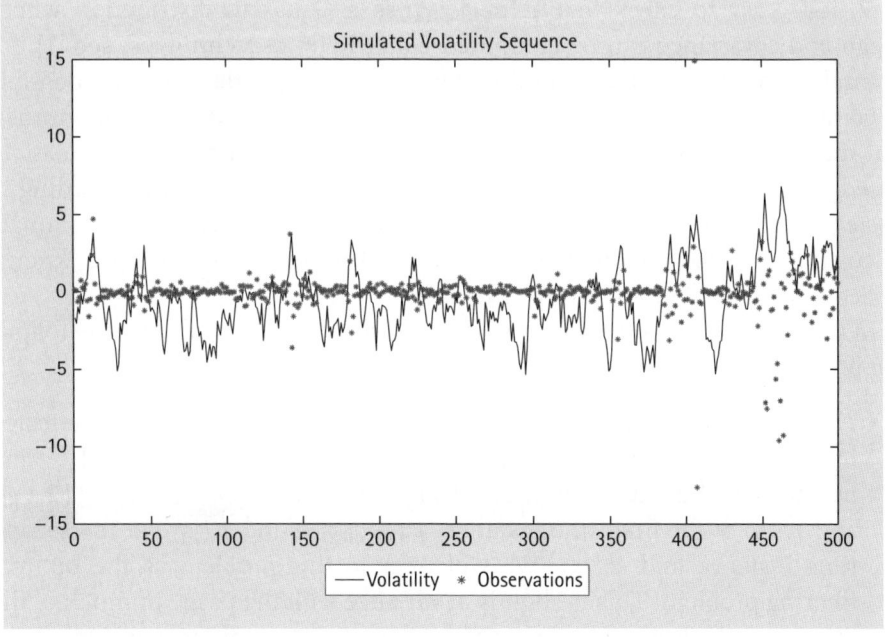

Fig. 24.1 A simulation of the stochastic volatility model described in Example 24.4.

likelihood function; that is:

$$p(x_{1:n}) = \mu(x_1) \prod_{k=2}^{n} f(x_k | x_{k-1}) \qquad (24.3)$$

and

$$p(y_{1:n} | x_{1:n}) = \prod_{k=1}^{n} g(y_k | x_k), \qquad (24.4)$$

where, for any sequence $\{z_n\}_{n \geq 1}$, and any $i \leq j$, $z_{i:j} := (z_i, z_{i+1}, \ldots, z_j)$.

In such a Bayesian context, inference about $X_{1:n}$ given a realization of the observations $Y_{1:n} = y_{1:n}$ relies upon the posterior distribution

$$p(x_{1:n} | y_{1:n}) = \frac{p(x_{1:n}, y_{1:n})}{p(y_{1:n})}, \qquad (24.5)$$

where

$$p(x_{1:n}, y_{1:n}) = p(x_{1:n}) \, p(y_{1:n} | x_{1:n}), \qquad (24.6)$$

$$\text{and } p(y_{1:n}) = \int p(x_{1:n}, y_{1:n}) \, dx_{1:n}. \qquad (24.7)$$

For the finite state-space HMM model discussed in Example 24.1, the integrals correspond to finite sums and all these (discrete) probability distributions can be computed exactly. For the linear Gaussian model discussed in Example 24.2, it is easy to check that $p(x_{1:n} | y_{1:n})$ is a Gaussian distribution whose mean and covariance can be computed using Kalman techniques; see [1], for example. However, for most nonlinear non-Gaussian models, it is not possible to compute these distributions in closed-form and we need to employ numerical methods. Particle methods are a set of flexible and powerful simulation-based methods which provide samples approximately distributed according to posterior distributions of the form $p(x_{1:n} | y_{1:n})$ and facilitate the approximate calculation of $p(y_{1:n})$. Such methods are a subset of the class of methods known as Sequential Monte Carlo (SMC) methods.

In this tutorial, we will review various particle methods to address the following problems:

- *Filtering and marginal likelihood computation*: Assume that we are interested in the sequential approximation of the distributions $\{p(x_{1:n} | y_{1:n})\}_{n \geq 1}$ and marginal likelihoods $\{p(y_{1:n})\}_{n \geq 1}$. That is, we wish to approximate $p(x_1 | y_1)$ and $p(y_1)$ at the first time instance, $p(x_{1:2} | y_{1:2})$ and $p(y_{1:2})$ at the second time instance and so on. We will refer to this problem as the optimal filtering problem. This is slightly at variance with the usage in much of the literature which reserves the term for the estimation of the marginal distributions $\{p(x_n | y_{1:n})\}_{n \geq 1}$ rather than the joint distributions $\{p(x_{1:n} | y_{1:n})\}_{n \geq 1}$.

We will describe basic and advanced particle filtering methods to address this problem including auxiliary particle filtering, particle filtering with MCMC moves, block sampling strategies and Rao–Blackwellized particle filters.

- *Smoothing*: Consider attempting to sample from a joint distribution $p(x_{1:T}|y_{1:T})$ and approximating the associated marginals $\{p(x_n|y_{1:T})\}$ where $n = 1, \ldots, T$. Particle filtering techniques can be used to solve this problem but perform poorly when T is large for reasons detailed in this tutorial. We will describe several particle smoothing methods to address this problem. Essentially, these methods rely on the particle implementation of the forward filtering-backward smoothing formula or of a generalized version of the two-filter smoothing formula.

24.2.2 Filtering and marginal likelihood

The first area of interest, and that to which the vast majority of the literature on particle methods has been dedicated from the outset, is the problem of filtering: characterizing the distribution of the state of the hidden Markov model at the present time, given the information provided by all of the observations received up to the present time. This can be thought of as a "tracking" problem: keeping track of the current "location" of the system given noisy observations—and, indeed, this is an extremely popular area of application for these methods. The term is sometimes also used to refer to the practice of estimating the full trajectory of the state sequence up to the present time given the observations received up to this time.

We recall that, following (24.1)–(24.2), the posterior distribution $\{p(x_{1:n}|y_{1:n})\}$ is defined by (24.5)—the prior is defined in (24.3) and the likelihood in (24.4). The unnormalized posterior distribution $p(x_{1:n}, y_{1:n})$ given in (24.5) satisfies

$$p(x_{1:n}, y_{1:n}) = p(x_{1:n-1}, y_{1:n-1}) f(x_n|x_{n-1}) g(y_n|x_n). \tag{24.8}$$

Consequently, the posterior $p(x_{1:n}|y_{1:n})$ satisfies the following recursion

$$p(x_{1:n}|y_{1:n}) = p(x_{1:n-1}|y_{1:n-1}) \frac{f(x_n|x_{n-1}) g(y_n|x_n)}{p(y_n|y_{1:n-1})}, \tag{24.9}$$

where

$$p(y_n|y_{1:n-1}) = \int p(x_{n-1}|y_{1:n-1}) f(x_n|x_{n-1}) g(y_n|x_n) dx_{n-1:n} \tag{24.10}$$

In the literature, the recursion satisfied by the marginal distribution $p(x_n|y_{1:n})$ is often presented. It is straightforward to check (by integrating out $x_{1:n-1}$ in (24.9)) that we have

$$p(x_n|y_{1:n}) = \frac{g(y_n|x_n) p(x_n|y_{1:n-1})}{p(y_n|y_{1:n-1})}, \tag{24.11}$$

where

$$p(x_n|\, y_{1:n-1}) = \int f(x_n|\, x_{n-1})\, p(x_{n-1}|\, y_{1:n-1})\, dx_{n-1}. \qquad (24.12)$$

Equation (24.12) is known as the prediction step and (24.11) is known as the updating step. However, most particle filtering methods rely on a numerical approximation of recursion (24.9) and not of (24.11)–(24.12).

If we can compute $\{p(x_{1:n}|\, y_{1:n})\}$ and thus $\{p(x_n|\, y_{1:n})\}$ sequentially, then the quantity $p(y_{1:n})$, which is known as the marginal likelihood, can also clearly be evaluated recursively using

$$p(y_{1:n}) = p(y_1) \prod_{k=2}^{n} p(y_k|\, y_{1:k-1}) \qquad (24.13)$$

where $p(y_k|\, y_{1:k-1})$ is of the form (24.10).

24.2.3 Smoothing

One problem, which is closely related to filtering, but computationally more challenging for reasons which will become apparent later, is known as smoothing. Whereas filtering corresponds to estimating the distribution of the current state of an HMM based upon the observations received up until the current time, smoothing corresponds to estimating the distribution of the state at a particular time given all of the observations up to some *later* time. The trajectory estimates obtained by such methods, as a result of the additional information available, tend to be smoother than those obtained by filtering. It is intuitive that if estimates of the state at time n are not required instantly, then better estimation performance is likely to be obtained by taking advantage of a few later observations. Designing efficient sequential algorithms for the solution of this problem is not quite as straightforward as it might seem, but a number of effective strategies have been developed and are described below.

More formally: assume that we have access to the data $y_{1:T}$, and wish to compute the marginal distributions $\{p(x_n|\, y_{1:T})\}$ where $n = 1, \ldots, T$ or to sample from $p(x_{1:T}|\, y_{1:T})$. In principle, the marginals $\{p(x_n|\, y_{1:T})\}$ could be obtained directly by considering the joint distribution $p(x_{1:T}|\, y_{1:T})$ and integrating out the variables $(x_{1:n-1}, x_{n+1:T})$. Extending this reasoning in the context of particle methods, one can simply use the identity $p(x_n|y_{1:T}) = \int p(x_{1:T}|y_{1:T}) dx_{1:n-1} dx_{n+1:T}$ and take the same approach which is used in particle filtering: use Monte Carlo algorithms to obtain an approximate characterization of the joint distribution and then use the associated marginal distribution to approximate the distributions of interest. Unfortunately, as is detailed below, when $n \ll T$ this strategy is doomed to failure: the marginal distribution $p(x_n|y_{1:n})$ occupies a privileged role within the particle filter framework as it is, in some sense, better characterized than any of the other marginal distributions.

For this reason, it is necessary to develop more sophisticated strategies in order to obtain good smoothing algorithms. There has been much progress in this direction over the past decade. Below, we present two alternative recursions that will prove useful when numerical approximations are required. The key to the success of these recursions is that they rely upon only the marginal filtering distributions $\{p(x_n|y_{1:n})\}$.

Forward–backward recursions

The following decomposition of the joint distribution $p(x_{1:T}|y_{1:T})$

$$p(x_{1:T}|y_{1:T}) = p(x_T|y_{1:T}) \prod_{n=1}^{T-1} p(x_n|x_{n+1}, y_{1:T})$$

$$= p(x_T|y_{1:T}) \prod_{n=1}^{T-1} p(x_n|x_{n+1}, y_{1:n}), \quad (24.14)$$

shows that, conditional on $y_{1:T}$, $\{X_n\}$ is an inhomogeneous Markov process.

Eq. (24.14) suggests the following algorithm to sample from $p(x_{1:T}|y_{1:T})$. First compute and store the marginal distributions $\{p(x_n|y_{1:n})\}$ for $n = 1, \ldots, T$. Then sample $X_T \sim p(x_T|y_{1:T})$ and for $n = T-1, T-2, \ldots, 1$, sample $X_n \sim p(x_n|X_{n+1}, y_{1:n})$ where

$$p(x_n|x_{n+1}, y_{1:n}) = \frac{f(x_{n+1}|x_n)\, p(x_n|y_{1:n})}{p(x_{n+1}|y_{1:n})}.$$

It also follows, by integrating out $(x_{1:n-1}, x_{n+1:T})$ in Eq. (24.14), that

$$p(x_n|y_{1:T}) = p(x_n|y_{1:n}) \int \frac{f(x_{n+1}|x_n)}{p(x_{n+1}|y_{1:n})} p(x_{n+1}|y_{1:T})\, dx_{n+1}. \quad (24.15)$$

So to compute $\{p(x_n|y_{1:T})\}$, we simply modify the backward pass and, instead of sampling from $p(x_n|x_{n+1}, y_{1:n})$, we compute $p(x_n|y_{1:T})$ using (24.15).

Generalized two-filter formula

The two-filter formula is a well-established alternative to the forward-filtering backward-smoothing technique to compute the marginal distributions $\{p(x_n|y_{1:T})\}$ [4]. It relies on the following identity

$$p(x_n|y_{1:T}) = \frac{p(x_n|y_{1:n-1})\, p(y_{n:T}|x_n)}{p(y_{n:T}|y_{1:n-1})},$$

where the so-called backward information filter is initialized at time $n = T$ by $p(y_T| x_T) = g(y_T| x_T)$ and satisfies

$$p(y_{n:T}| x_n) = \int \prod_{k=n+1}^{T} f(x_k| x_{k-1}) \prod_{k=n}^{T} g(y_k| x_k) \, dx_{n+1:T} \qquad (24.16)$$

$$= g(y_n| x_n) \int f(x_{n+1}| x_n) p(y_{n+1:T}| x_{n+1}) \, dx_{n+1}.$$

The backward information filter is not a probability density in argument x_n and it is even possible that $\int p(y_{n:T}| x_n) \, dx_n = \infty$. Although this is not an issue when $p(y_{n:T}| x_n)$ can be computed exactly, it does preclude the direct use of SMC methods to estimate this integral. To address this problem, a generalized version of the two-filter formula was proposed in [5]. It relies on the introduction of a set of artificial probability distributions $\{\tilde{p}_n(x_n)\}$ and the joint distributions

$$\tilde{p}_n(x_{n:T}| y_{n:T}) \propto \tilde{p}_n(x_n) \prod_{k=n+1}^{T} f(x_k| x_{k-1}) \prod_{k=n}^{T} g(y_k| x_k), \qquad (24.17)$$

which are constructed such that their marginal distributions, $\tilde{p}_n(x_n| y_{n:T}) \propto \tilde{p}_n(x_n) p(y_{n:T}| x_n)$, are simply "integrable versions" of the backward information filter. It is easy to establish the generalized two-filter formula

$$p(x_1|y_{1:T}) \propto \frac{\mu(x_1)\tilde{p}(x_1|y_{1:T})}{\tilde{p}_1(x_1)}, \quad p(x_n|y_{1:T}) \propto \frac{p(x_n|y_{1:n-1})\tilde{p}(x_n|y_{n:T})}{\tilde{p}_n(x_n)} \qquad (24.18)$$

which is valid whenever the support of $\tilde{p}_n(x_n)$ includes the support of the prior $p_n(x_n)$; that is

$$p_n(x_n) = \int \mu(x_1) \prod_{k=2}^{n} f(x_k| x_{k-1}) \, dx_{1:n-1} > 0 \Rightarrow \tilde{p}_n(x_n) > 0.$$

The generalized two-filter smoother for $\{p(x_n|y_{n:T})\}$ proceeds as follows. Using the standard forward recursion, we can compute and store the marginal distributions $\{p(x_n| y_{1:n-1})\}$. Using a backward recursion, we compute and store $\{\tilde{p}(x_n|y_{n:T})\}$. Then for any $n = 1, \ldots, T$ we can combine $p(x_n| y_{1:n-1})$ and $\tilde{p}(x_n|y_{n:T})$ to obtain $p(x_n|y_{1:T})$.

In [4], this identity is discussed in the particular case where $\tilde{p}_n(x_n) = p_n(x_n)$. However, when computing $\{\tilde{p}(x_n|y_{n:T})\}$ using SMC, it is necessary to be able to compute $\tilde{p}_n(x_n)$ exactly hence this rules out the choice $\tilde{p}_n(x_n) = p_n(x_n)$ for most nonlinear non-Gaussian models. In practice, we should select a heavy-tailed approximation of $p_n(x_n)$ for $\tilde{p}_n(x_n)$ in such settings. It is also possible to use the generalized two-filter formula to sample from $p(x_{1:T}|y_{1:T})$; see [5] for details.

24.2.4 Summary

Bayesian inference in nonlinear non-Gaussian dynamic models relies on the sequence of posterior distributions $\{p(x_{1:n}|\, y_{1:n})\}$ and its marginals. Except in simple problems such as Examples 24.1 and 24.2, it is not possible to compute these distributions in closed form. In some scenarios, it might be possible to obtain reasonable performance by employing functional approximations of these distributions. Here, we will discuss only Monte Carlo approximations of these distributions; that is numerical schemes in which the distributions of interest are approximated by a large collection of N random samples termed particles. The main advantage of such methods is that under weak assumptions they provide asymptotically (i.e. as $N \to \infty$) consistent estimates of the target distributions of interest. It is also noteworthy that these techniques can be applied to problems of moderately high dimension in which traditional numerical integration might be expected to perform poorly.

24.3 Sequential Monte Carlo methods

Over the past fifteen years, particle methods for filtering and smoothing have been the most common examples of SMC algorithms. Indeed, it has become traditional to present particle filtering and SMC as being the same thing in much of the literature. Here, we wish to emphasize that SMC actually encompasses a broader range of algorithms—and by doing so we are able to show that many more advanced techniques for approximate filtering and smoothing can be described using precisely the same framework and terminology as the basic algorithm.

SMC methods are a general class of Monte Carlo methods that sample sequentially from a sequence of target probability densities $\{\pi_n(x_{1:n})\}$ of increasing dimension where each distribution $\pi_n(x_{1:n})$ is defined on the product space $\mathcal{X}^n$. Writing

$$\pi_n(x_{1:n}) = \frac{\gamma_n(x_{1:n})}{Z_n} \tag{24.19}$$

we require only that $\gamma_n : \mathcal{X}^n \to \mathbb{R}^+$ is known pointwise; the normalizing constant

$$Z_n = \int \gamma_n(x_{1:n})\, dx_{1:n} \tag{24.20}$$

might be unknown. SMC provide an approximation of $\pi_1(x_1)$ and an estimate of Z_1 at time 1 then an approximation of $\pi_2(x_{1:2})$ and an estimate of Z_2 at time 2 and so on.

For example, in the context of filtering, we could have $\gamma_n(x_{1:n}) = p(x_{1:n}, y_{1:n})$, $Z_n = p(y_{1:n})$ so $\pi_n(x_{1:n}) = p(x_{1:n}|\, y_{1:n})$. However, we emphasize that this is just one particular choice of target distributions. Not only can SMC methods be used outside the filtering context but, more importantly for this tutorial, some advanced particle filtering and smoothing methods discussed below do not rely on this sequence of target distributions. Consequently, we believe that understanding the main principles behind generic SMC methods is essential to the development of a proper understanding of particle filtering and smoothing methods.

We start this section with a very basic review of Monte Carlo methods and importance sampling (IS). We then present the Sequential Importance Sampling (SIS) method, point out the limitations of this method and show how resampling techniques can be used to partially mitigate them. Having introduced the basic particle filter as an SMC method, we show how various advanced techniques which have been developed over the past fifteen years can themselves be interpreted within the same formalism as SMC algorithms associated with sequences of distributions which may not coincide with the filtering distributions. These alternative sequences of target distributions are either constructed such that they admit the distributions $\{p(x_{1:n}|y_{1:n})\}$ as marginal distributions, or an importance sampling correction is necessary to ensure the consistency of estimates.

24.3.1 Basics of Monte Carlo methods

Initially, consider approximating a generic probability density $\pi_n(x_{1:n})$ for some fixed n. If we sample N independent random variables, $X_{1:n}^i \sim \pi_n(x_{1:n})$ for $i = 1, \ldots, N$, then the Monte Carlo method approximates $\pi_n(x_{1:n})$ by the empirical measure[1]

$$\widehat{\pi}_n(x_{1:n}) = \frac{1}{N} \sum_{i=1}^{N} \delta_{X_{1:n}^i}(x_{1:n}),$$

where $\delta_{x_0}(x)$ denotes the Dirac delta mass located at x_0. Based on this approximation, it is possible to approximate any marginal, say $\pi_n(x_k)$, easily using

$$\widehat{\pi}_n(x_k) = \frac{1}{N} \sum_{i=1}^{N} \delta_{X_k^i}(x_k),$$

and the expectation of any test function $\varphi_n : \mathcal{X}^n \to \mathbb{R}$ given by

$$I_n(\varphi_n) := \int \varphi_n(x_{1:n}) \pi_n(x_{1:n}) dx_{1:n},$$

is estimated by

$$I_n^{MC}(\varphi_n) := \int \varphi_n(x_{1:n}) \widehat{\pi}_n(x_{1:n}) dx_{1:n} = \frac{1}{N} \sum_{i=1}^{N} \varphi_n(X_{1:n}^i).$$

It is easy to check that this estimate is unbiased and that its variance is given by

$$\mathbb{V}\left[I_n^{MC}(\varphi_n)\right] = \frac{1}{N}\left(\int \varphi_n^2(x_{1:n})\,\pi_n(x_{1:n})\,dx_{1:n} - I_n^2(\varphi_n)\right).$$

The main advantage of Monte Carlo methods over standard approximation techniques is that the variance of the approximation error decreases at a rate of $\mathcal{O}(1/N)$ regardless of the dimension of the space $\mathcal{X}^n$. However, there are at least two main problems with this basic Monte Carlo approach:

- *Problem 1*: If $\pi_n(x_{1:n})$ is a complex high-dimensional probability distribution, then we cannot sample from it.
- *Problem 2*: Even if we knew how to sample exactly from $\pi_n(x_{1:n})$, the computational complexity of such a sampling scheme is typically at least linear in the number of variables n. So an algorithm sampling exactly from $\pi_n(x_{1:n})$, sequentially for each value of n, would have a computational complexity increasing at least linearly with n.

24.3.2 Importance sampling

We are going to address *Problem 1* using the IS method. This is a fundamental Monte Carlo method and the basis of all the algorithms developed later on. IS relies on the introduction of an *importance density*[2] $q_n(x_{1:n})$ such that

$$\pi_n(x_{1:n}) > 0 \Rightarrow q_n(x_{1:n}) > 0.$$

In this case, we have from (24.19)–(24.20) the following IS identities

$$\pi_n(x_{1:n}) = \frac{w_n(x_{1:n})\,q_n(x_{1:n})}{Z_n}, \qquad (24.21)$$

$$Z_n = \int w_n(x_{1:n})\,q_n(x_{1:n})\,dx_{1:n} \qquad (24.22)$$

where $w_n(x_{1:n})$ is the *unnormalized weight* function

$$w_n(x_{1:n}) = \frac{\gamma_n(x_{1:n})}{q_n(x_{1:n})}.$$

In particular, we can select an importance density $q_n(x_{1:n})$ from which it is easy to draw samples; e.g. a multivariate Gaussian. Assume we draw N independent samples $X_{1:n}^i \sim q_n(x_{1:n})$ then by inserting the Monte Carlo approximation of $q_n(x_{1:n})$—that is the empirical measure of the samples $X_{1:n}^i$—into (24.21)–(24.22) we obtain

$$\widehat{\pi}_n(x_{1:n}) = \sum_{i=1}^{N} W_n^i \delta_{X_{1:n}^i}(x_{1:n}), \qquad (24.23)$$

$$\widehat{Z}_n = \frac{1}{N} \sum_{i=1}^{N} w_n(X_{1:n}^i) \qquad (24.24)$$

where

$$W_n^i = \frac{w_n(X_{1:n}^i)}{\sum_{j=1}^{N} w_n(X_{1:n}^j)}. \qquad (24.25)$$

Compared to standard Monte Carlo, IS provides an (unbiased) estimate of the normalizing constant with relative variance

$$\frac{\mathbb{V}_{IS}[\widehat{Z}_n]}{Z_n^2} = \frac{1}{N} \left(\int \frac{\pi_n^2(x_{1:n})}{q_n(x_{1:n})} dx_{1:n} - 1 \right). \qquad (24.26)$$

If we are interested in computing $I_n(\varphi_n)$, we can also use the estimate

$$I_n^{IS}(\varphi_n) = \int \varphi_n(x_{1:n}) \widehat{\pi}_n(x_{1:n}) dx_{1:n} = \sum_{i=1}^{N} W_n^i \varphi_n(X_{1:n}^i).$$

Unlike $I_n^{MC}(\varphi_n)$, this estimate is biased for finite N. However, it is consistent and it is easy to check that its asymptotic bias is given by

$$\lim_{N \to \infty} N \left(I_n^{IS}(\varphi_n) - I_n(\varphi_n) \right) = - \int \frac{\pi_n^2(x_{1:n})}{q_n(x_{1:n})} (\varphi_n(x_{1:n}) - I_n(\varphi_n)) dx_{1:n}.$$

When the normalizing constant is known analytically, we can calculate an unbiased importance sampling estimate—however, this generally has higher variance and this is not typically the case in the situations in which we are interested.

Furthermore, $I_n^{IS}(\varphi)$ satisfies a Central Limit Theorem (CLT) with asymptotic variance

$$\frac{1}{N} \int \frac{\pi_n^2(x_{1:n})}{q_n(x_{1:n})} (\varphi_n(x_{1:n}) - I_n(\varphi_n))^2 dx_{1:n}. \qquad (24.27)$$

The bias being $\mathcal{O}(1/N)$ and the variance $\mathcal{O}(1/N)$, the mean-squared error given by the *squared* bias plus the variance is asymptotically dominated by the variance term.

For a given test function, $\varphi_n(x_{1:n})$, it is easy to establish the importance distribution minimizing the asymptotic variance of $I_n^{IS}(\varphi_n)$. However, such a result is of minimal interest in a filtering context as this distribution depends on $\varphi_n(x_{1:n})$ and we are typically interested in the expectations of several test functions. Moreover, even if we were interested in a single test function, say

$\varphi_n(x_{1:n}) = x_n$, then selecting the optimal importance distribution at time n would have detrimental effects when we will try to obtain a sequential version of the algorithms (the optimal distribution for estimating $\varphi_{n-1}(x_{1:n-1})$ will almost certainly not be—even similar to—the marginal distribution of $x_{1:n-1}$ in the optimal distribution for estimating $\varphi_n(x_{1:n})$ and this will prove to be problematic).

A more appropriate approach in this context is to attempt to select the $q_n(x_{1:n})$ which minimizes the variance of the importance weights (or, equivalently, the variance of $\widehat{Z}_n$). Clearly, this variance is minimised for $q_n(x_{1:n}) = \pi_n(x_{1:n})$. We cannot select $q_n(x_{1:n}) = \pi_n(x_{1:n})$ as this is the reason we used IS in the first place. However, this simple result indicates that we should aim at selecting an IS distribution which is close as possible to the target. Also, although it is possible to construct samplers for which the variance is finite without satisfying this condition, it is advisable to select $q_n(x_{1:n})$ so that $w_n(x_{1:n}) < C_n < \infty$.

24.3.3 Sequential importance sampling

We are now going to present an algorithm that admits a fixed computational complexity at each time step in important scenarios and thus addresses *Problem 2*. This solution involves selecting an importance distribution which has the following structure

$$q_n(x_{1:n}) = q_{n-1}(x_{1:n-1}) q_n(x_n| x_{1:n-1})$$

$$= q_1(x_1) \prod_{k=2}^{n} q_k(x_k| x_{1:k-1}). \qquad (24.28)$$

Practically, this means that to obtain particles $X_{1:n}^i \sim q_n(x_{1:n})$ at time n, we sample $X_1^i \sim q_1(x_1)$ at time 1 then $X_k^i \sim q_k(x_k| X_{1:k-1}^i)$ at time k for $k = 2, \ldots, n$. The associated unnormalized weights can be computed recursively using the decomposition

$$w_n(x_{1:n}) = \frac{\gamma_n(x_{1:n})}{q_n(x_{1:n})}$$

$$= \frac{\gamma_{n-1}(x_{1:n-1})}{q_{n-1}(x_{1:n-1})} \frac{\gamma_n(x_{1:n})}{\gamma_{n-1}(x_{1:n-1}) q_n(x_n| x_{1:n-1})} \qquad (24.29)$$

which can be written in the form

$$w_n(x_{1:n}) = w_{n-1}(x_{1:n-1}) \cdot \alpha_n(x_{1:n})$$

$$= w_1(x_1) \prod_{k=2}^{n} \alpha_k(x_{1:k})$$

where the *incremental importance weight* function $a_n(x_{1:n})$ is given by

$$a_n(x_{1:n}) = \frac{\gamma_n(x_{1:n})}{\gamma_{n-1}(x_{1:n-1}) q_n(x_n | x_{1:n-1})}. \tag{24.30}$$

The SIS algorithm proceeds as follows, with each step carried out for $i = 1, \ldots, N$:

Sequential importance sampling

At time $n = 1$

- Sample $X_1^i \sim q_1(x_1)$.
- Compute the weights $w_1(X_1^i)$ and $W_1^i \propto w_1(X_1^i)$.

At time $n \geq 2$

- Sample $X_n^i \sim q_n(x_n | X_{1:n-1}^i)$.
- Compute the weights

$$w_n(X_{1:n}^i) = w_{n-1}(X_{1:n-1}^i) \cdot a_n(X_{1:n}^i),$$

$$W_n^i \propto w_n(X_{1:n}^i).$$

At any time, n, we obtain the estimates $\widehat{\pi}_n(x_{1:n})$ (eq. 24.23) and $\widehat{Z}_n$ (eq. 24.24) of $\pi_n(x_{1:n})$ and Z_n, respectively. It is straightforward to check that a consistent estimate of Z_n/Z_{n-1} is also provided by the same set of samples:

$$\frac{\widehat{Z}_n}{Z_{n-1}} = \sum_{i=1}^{N} W_{n-1}^i a_n(X_{1:n}^i).$$

This estimator is motivated by the fact that

$$\int a_n(x_{1:n}) \pi_{n-1}(x_{1:n-1}) q_n(x_n | x_{1:n-1}) dx_{1:n}$$

$$= \int \frac{\gamma_n(x_{1:n}) \pi_{n-1}(x_{1:n-1}) q_n(x_n | x_{1:n-1})}{\gamma_{n-1}(x_{1:n-1}) q_n(x_n | x_{1:n-1})} dx_{1:n} = \frac{Z_n}{Z_{n-1}}.$$

In this sequential framework, it would seem that the only freedom the user has at time n is the choice of $q_n(x_n | x_{1:n-1})$.[3] A sensible strategy consists of selecting it so as to minimize the variance of $w_n(x_{1:n})$. It is straightforward to check that this is achieved by selecting

$$q_n^{\text{opt}}(x_n | x_{1:n-1}) = \pi_n(x_n | x_{1:n-1})$$

as in this case the variance of $w_n(x_{1:n})$ conditional upon $x_{1:n-1}$ is zero and the associated incremental weight is given by

$$a_n^{\text{opt}}(x_{1:n}) = \frac{\gamma_n(x_{1:n-1})}{\gamma_{n-1}(x_{1:n-1})} = \frac{\int \gamma_n(x_{1:n}) \, dx_n}{\gamma_{n-1}(x_{1:n-1})}.$$

Note that it is not always possible to sample from $\pi_n(x_n | x_{1:n-1})$. Nor is it always possible to compute $a_n^{\text{opt}}(x_{1:n})$. In these cases, one should employ an approximation of $q_n^{\text{opt}}(x_n | x_{1:n-1})$ for $q_n(x_n | x_{1:n-1})$.

In those scenarios in which the time required to sample from $q_n(x_n | x_{1:n-1})$ and to compute $a_n(x_{1:n})$ is independent of n (and this is, indeed, the case if q_n is chosen sensibly and one is concerned with a problem such as filtering), it appears that we have provided a solution for *Problem 2*. However, it is important to be aware that the methodology presented here suffers from severe drawbacks. Even for standard IS, the variance of the resulting estimates increases exponentially with n (as is illustrated below; see also [28]). As SIS is nothing but a special version of IS in which we restrict ourselves to an importance distribution of the form (24.28) it suffers from the same problem. We demonstrate this using a very simple toy example.

Example. Consider the case where $\mathcal{X} = \mathbb{R}$ and

$$\pi_n(x_{1:n}) = \prod_{k=1}^n \pi_n(x_k) = \prod_{k=1}^n \mathcal{N}(x_k; 0, 1), \quad (24.31)$$

$$\gamma_n(x_{1:n}) = \prod_{k=1}^n \exp\left(-\frac{x_k^2}{2}\right),$$

$$Z_n = (2\pi)^{n/2}.$$

We select an importance distribution

$$q_n(x_{1:n}) = \prod_{k=1}^n q_k(x_k) = \prod_{k=1}^n \mathcal{N}(x_k; 0, \sigma^2).$$

In this case, we have $\mathbb{V}_{\text{IS}}[\widehat{Z}_n] < \infty$ only for $\sigma^2 > \frac{1}{2}$ and

$$\frac{\mathbb{V}_{\text{IS}}[\widehat{Z}_n]}{Z_n^2} = \frac{1}{N}\left[\left(\frac{\sigma^4}{2\sigma^2 - 1}\right)^{n/2} - 1\right].$$

It can easily be checked that $\frac{\sigma^4}{2\sigma^2-1} > 1$ for any $\frac{1}{2} < \sigma^2 \neq 1$: the variance increases exponentially with n even in this simple case. For example, if we select $\sigma^2 = 1.2$ then we have a reasonably good importance distribution as $q_k(x_k) \approx \pi_n(x_k)$ but $N \frac{\mathbb{V}_{\text{IS}}[\widehat{Z}_n]}{Z_n^2} \approx (1.103)^{n/2}$ which is approximately equal to 1.9×10^{21} for $n = 1000$!

We would need to use $N \approx 2 \times 10^{23}$ particles to obtain a relative variance $\frac{\mathbb{V}_{IS}[\widehat{Z}_n]}{Z_n^2} = 0.01$. This is clearly impracticable.

24.3.4 Resampling

We have seen that IS—and thus SIS—provides estimates whose variance increases, typically exponentially, with n. Resampling techniques are a key ingredient of SMC methods which (partially) solve this problem in some important scenarios.

Resampling is a very intuitive idea which has major practical and theoretical benefits. Consider first an IS approximation $\widehat{\pi}_n(x_{1:n})$ of the target distribution $\pi_n(x_{1:n})$. This approximation is based on weighted samples from $q_n(x_{1:n})$ and does not provide samples approximately distributed according to $\pi_n(x_{1:n})$. To obtain approximate samples from $\pi_n(x_{1:n})$, we can simply sample from its IS approximation $\widehat{\pi}_n(x_{1:n})$; that is we select $X_{1:n}^i$ with probability W_n^i. This operation is called resampling as it corresponds to sampling from an approximation $\widehat{\pi}_n(x_{1:n})$ which was itself obtained by sampling. If we are interested in obtaining N samples from $\widehat{\pi}_n(x_{1:n})$, then we can simply resample N times from $\widehat{\pi}_n(x_{1:n})$. This is equivalent to associating a number of offspring N_n^i with each particle $X_{1:n}^i$ in such a way that $N_n^{1:N} = (N_n^1, \ldots, N_n^N)$ follow a multinomial distribution with parameter vector $(N, W_n^{1:N})$ and associating a weight of $1/N$ with each offspring. We approximate $\widehat{\pi}_n(x_{1:n})$ by the resampled empirical measure

$$\overline{\pi}_n(x_{1:n}) = \sum_{i=1}^{N} \frac{N_n^i}{N} \delta_{X_{1:n}^i}(x_{1:n}) \qquad (24.32)$$

where $\mathbb{E}\left[N_n^i \mid W_n^{1:N}\right] = NW_n^i$. Hence $\overline{\pi}_n(x_{1:n})$ is an unbiased approximation of $\widehat{\pi}_n(x_{1:n})$.

Improved unbiased resampling schemes have been proposed in the literature. These are methods of selecting N_n^i such that the unbiasedness property is preserved, and such that $\mathbb{V}\left[N_n^i \mid W_n^{1:N}\right]$ is smaller than that obtained via the multinomial resampling scheme described above. To summarize, the three most popular algorithms found in the literature are, in descending order of popularity/efficiency:

Systematic resampling Sample $U_1 \sim \mathcal{U}\left[0, \frac{1}{N}\right]$ and define $U_i = U_1 + \frac{i-1}{N}$ for $i = 2, \ldots, N$, then set $N_n^i = \left|\left\{U_j : \sum_{k=1}^{i-1} W_n^k \leq U_j \leq \sum_{k=1}^{i} W_n^k\right\}\right|$ with the convention $\sum_{k=1}^{0} := 0$. It is straightforward to establish that this approach is unbiased.

Residual resampling Set $\widetilde{N}_n^i = \lfloor NW_n^i \rfloor$, sample $\overline{N}_n^{1:N}$ from a multinomial of parameters $\left(N, \overline{W}_n^{1:N}\right)$ where $\overline{W}_n^i \propto W_n^i - N^{-1}\widetilde{N}_n^i$ then set $N_n^i = \widetilde{N}_n^i +$

$\overline{N}_n^i$. This is very closely related to breaking the empirical CDF up into N components and then sampling once from each of those components: the stratified resampling approach of [7].

Multinomial resampling Sample $N_n^{1:N}$ from a multinomial of parameters $\left(N, W_n^{1:N}\right)$.

Note that it is possible to sample efficiently from a multinomial distribution in $\mathcal{O}(N)$ operations. However, the systematic resampling algorithm introduced in [25] is the most widely-used algorithm in the literature as it is extremely easy to implement and outperforms other resampling schemes in most scenarios (although this is not guaranteed in general [13]).

Resampling allows us to obtain samples distributed approximately according to $\pi_n(x_{1:n})$, but it should be clear that if we are interested in estimating $I_n(\varphi_n)$ then we will obtain an estimate with lower variance using $\widehat{\pi}_n(x_{1:n})$ than that which we would have obtained by using $\overline{\pi}_n(x_{1:n})$. By resampling we indeed add some extra "noise"—as shown by [9]. However, an important advantage of resampling is that it allows us to remove of particles with low weights with a high probability. In the sequential framework in which we are interested, this is extremely useful as we do not want to carry forward particles with low weights and we want to focus our computational efforts on regions of high probability mass. Clearly, there is always the possibility than a particle having a low weight at time n could have an high weight at time $n + 1$, in which case resampling could be wasteful. It is straightforward to consider artificial problems for which this is the case. However, we will show that in the estimation problems we are looking at the resampling step is provably beneficial. Intuitively, resampling can be seen to provide stability in the future at the cost of an increase in the immediate Monte Carlo variance. This concept will be made more precise in Section 24.3.6.

24.3.5 A generic sequential Monte Carlo algorithm

SMC methods are a combination of SIS and resampling. At time 1, we compute the IS approximation $\widehat{\pi}_1(x_1)$ of $\pi_1(x_1)$ which is a weighted collection of particles $\{W_1^i, X_1^i\}$. Then we use a resampling step to eliminate (with high probability) those particles with low weights and multiply those with high weights. We denote by $\left\{\frac{1}{N}, \overline{X}_1^i\right\}$ the collection of equally-weighted resampled particles. Remember that each original particle X_1^i has N_1^i offspring so there exist N_1^i distinct indexes $j_1 \neq j_2 \neq \cdots \neq j_{N_1^i}$ such that $\overline{X}_1^{j_1} = \overline{X}_1^{j_2} = \cdots = \overline{X}_1^{j_{N_1^i}} = X_1^i$. After the resampling step, we follow the SIS strategy and sample $X_2^i \sim q_2\left(x_2 | \overline{X}_1^i\right)$. Thus $\left(\overline{X}_1^i, X_2^i\right)$ is approximately distributed according to $\pi_1(x_1) q_2(x_2 | x_1)$. Hence the corresponding importance weights in this case are simply equal to the incremental weights $a_2(x_{1:2})$. We then resample the particles with respect to these normalized weights and so on. To summarize, the algorithm proceeds as follows (this algorithm is sometimes referred to as Sequential Importance Resampling (SIR) or Sequential Importance Sampling and Resampling (SIS/R)).

Sequential Monte Carlo

At time n = 1

- Sample $X_1^i \sim q_1(x_1)$.
- Compute the weights $w_1(X_1^i)$ and $W_1^i \propto w_1(X_1^i)$.
- Resample $\{W_1^i, X_1^i\}$ to obtain N equally-weighted particles $\{\frac{1}{N}, \overline{X}_1^i\}$.

At time $n \geq 2$

- Sample $X_n^i \sim q_n(x_n | \overline{X}_{1:n-1}^i)$ and set $X_{1:n}^i \leftarrow (\overline{X}_{1:n-1}^i, X_n^i)$.
- Compute the weights $a_n(X_{1:n}^i)$ and $W_n^i \propto a_n(X_{1:n}^i)$.
- Resample $\{W_n^i, X_{1:n}^i\}$ to obtain N new equally-weighted particles $\{\frac{1}{N}, \overline{X}_{1:n}^i\}$.

At any time n, this algorithm provides two approximations of $\pi_n(x_{1:n})$. We obtain

$$\widehat{\pi}_n(x_{1:n}) = \sum_{i=1}^{N} W_n^i \delta_{X_{1:n}^i}(x_{1:n}) \tag{24.33}$$

after the sampling step and

$$\overline{\pi}_n(x_{1:n}) = \frac{1}{N} \sum_{i=1}^{N} \delta_{\overline{X}_{1:n}^i}(x_{1:n}) \tag{24.34}$$

after the resampling step. The approximation (24.33) is to be preferred to (24.34). We also obtain an approximation of Z_n/Z_{n-1} through

$$\frac{\widehat{Z_n}}{Z_{n-1}} = \frac{1}{N} \sum_{i=1}^{N} a_n(X_{1:n}^i).$$

As we have already mentioned, resampling has the effect of removing particles with low weights and multiplying particles with high weights. However, this is at the cost of immediately introducing some additional variance. If particles have unnormalised weights with a small variance then the resampling step might be unnecessary. Consequently, in practice, it is more sensible to resample only when the variance of the unnormalised weights is superior to a prespecified threshold. This is often assessed by looking at the variability of the weights using the so-called Effective Sample Size (ESS) criterion [30, pp. 35–6], which is given (at time n) by

$$ESS = \left(\sum_{i=1}^{N} (W_n^i)^2\right)^{-1}.$$

Its interpretation is that in a simple IS setting, inference based on the N weighted samples is approximately equivalent (in terms of estimator variance) to inference based on ESS perfect samples from the target distribution. The ESS takes values between 1 and N and we resample only when it is below a threshold N_T; typically $N_T = N/2$. Alternative criteria can be used such as the entropy of the weights $\{W_n^i\}$ which achieves its maximum value when $W_n^i = \frac{1}{N}$. In this case, we resample when the entropy is below a given threshold.

Sequential Monte Carlo with adaptive resampling

At time $n = 1$

- Sample $X_1^i \sim q_1(x_1)$.
- Compute the weights $w_1(X_1^i)$ and $W_1^i \propto w_1(X_1^i)$.
- If resampling criterion satisfied then resample $\{W_1^i, X_1^i\}$ to obtain N equally weighted particles $\{\frac{1}{N}, \overline{X}_1^i\}$ and set $\{\overline{W}_1^j, \overline{X}_1^i\} \leftarrow \{\frac{1}{N}, \overline{X}_1^i\}$, otherwise set $\{\overline{W}_1^j, \overline{X}_1^i\} \leftarrow \{W_1^i, X_1^i\}$.

At time $n \geq 2$

- Sample $X_n^i \sim q_n(x_n | \overline{X}_{1:n-1}^i)$ and set $X_{1:n}^i \leftarrow (\overline{X}_{1:n-1}^i, X_n^i)$.
- Compute the weights $a_n(X_{1:n}^i)$ and $W_n^i \propto \overline{W}_{n-1}^i a_n(X_{1:n}^i)$.
- If resampling criterion satisfied, then resample $\{W_n^i, X_{1:n}^i\}$ to obtain N equally weighted particles $\{\frac{1}{N}, \overline{X}_{1:n}^i\}$ and set $\{\overline{W}_n^j, \overline{X}_n^i\} \leftarrow \{\frac{1}{N}, \overline{X}_n^i\}$, otherwise set $\{\overline{W}_n^j, \overline{X}_n^i\} \leftarrow \{W_n^i, X_n^i\}$.

In this context too we have two approximations of $\pi_n(x_{1:n})$

$$\widehat{\pi}_n(x_{1:n}) = \sum_{i=1}^{N} W_n^i \delta_{X_{1:n}^i}(x_{1:n}), \tag{24.35}$$

$$\overline{\pi}_n(x_{1:n}) = \sum_{i=1}^{N} \overline{W}_n^j \delta_{\overline{X}_{1:n}^i}(x_{1:n})$$

which are equal if no resampling step is used at time n. We may also estimate Z_n/Z_{n-1} through

$$\frac{\widehat{Z_n}}{Z_{n-1}} = \sum_{i=1}^{N} \overline{W}_{n-1}^j a_n \left(X_{1:n}^i \right). \qquad (24.36)$$

SMC methods involve systems of particles which interact (via the resampling mechanism) and, consequently, obtaining convergence results is a much more difficult task than it is for SIS where standard results (i.i.d. asymptotics) apply. However, there are numerous sharp convergence results available for SMC; see [10] for an introduction to the subject and the monograph of Del Moral [11] for a complete treatment of the subject. An explicit treatment of the case in which resampling is performed adaptively is provided by [12].

The presence or absence of degeneracy is the factor which most often determines whether an SMC algorithm works in practice. However strong the convergence results available for limitingly large samples may be, we cannot expect good performance if the *finite* sample which is actually used is degenerate. Indeed, some degree of degeneracy is inevitable in all but trivial cases: if SMC algorithms are used for sufficiently many time steps every resampling step reduces the number of unique values representing X_1, for example. For this reason, any SMC algorithm which relies upon the distribution of full paths $x_{1:n}$ will fail for large enough n for any finite sample size, N, in spite of the asymptotic justification. It is intuitive that one should endeavour to employ algorithms which do not depend upon the full path of the samples, but only upon the distribution of some finite component $x_{n-L:n}$ for some fixed L which is independent of n. Furthermore, ergodicity (a tendency for the future to be essentially independent of the distant past) of the underlying system will prevent the accumulation of errors over time. These concepts are precisely characterised by existing convergence results, some of the most important of which are summarized and interpreted in Section 24.3.6.

Although sample degeneracy emerges as a consequence of resampling, it is really a manifestation of a deeper problem—one which resampling actually mitigates. It is inherently impossible to accurately represent a distribution on a space of arbitrarily high dimension with a sample of fixed, finite size. Sample impoverishment is a term which is often used to describe the situation in which very few different particles have significant weight. This problem has much the same effect as sample degeneracy and occurs, in the absence of resampling, as the inevitable consequence of multiplying together incremental importance weights from a large number of time steps. It is, of course, not possible to circumvent either problem by increasing the number of samples at every iteration to maintain a constant effective sample size as this would lead to an exponential growth in the number of samples required. This sheds some light on the resampling mechanism: it "resets the system" in such a way that its representation of final time marginals remains well behaved at the expense of further diminishing the quality of the path-samples. By focusing on the fixed-dimensional final time marginals in this way, it allows us to circumvent the problem of increasing dimensionality.

24.3.6 Convergence results for sequential Monte Carlo methods

Here, we briefly discuss selected convergence results for SMC. We focus on the CLT as it allows us to clearly understand the benefits of the resampling step and why it "works." If multinomial resampling is used at every iteration,[4] then the associated SMC estimates of $\widehat{Z}_n / Z_n$ and $I_n(\varphi_n)$ satisfy a CLT and their respective asymptotic variances are given by

$$\left(\int \frac{\pi_n^2(x_1)}{q_1(x_1)} dx_1 - 1 + \sum_{k=2}^{n} \int \frac{\pi_n^2(x_{1:k})}{\pi_{k-1}(x_{1:k-1}) q_k(x_k | x_{1:k-1})} dx_{k-1:k} - 1 \right) \quad (24.37)$$

and

$$\int \frac{\pi_1^2(x_1)}{q_1(x_1)} \left(\int \varphi_n(x_{1:n}) \pi_n(x_{2:n} | x_1) dx_{2:n} - I_n(\varphi_n) \right)^2 dx_1$$

$$+ \sum_{k=2}^{n-1} \int \frac{\pi_n^2(x_{1:k})}{\pi_{k-1}(x_{1:k-1}) q_k(x_k | x_{1:k-1})} \left(\int \varphi_n(x_{1:n}) \pi_n(x_{k+1:n} | x_{1:k}) dx_{k+1:n} - I_n(\varphi_n) \right)^2 dx_{1:k}$$

$$+ \int \frac{\pi_n^2(x_{1:n})}{\pi_{n-1}(x_{1:n-1}) q_n(x_n | x_{1:n-1})} \left(\varphi_n(x_{1:n}) - I_n(\varphi_n) \right)^2 dx_{1:n}. \quad (24.38)$$

A short and elegant proof of this result is given in [11, Ch. 9]; see also [9]. These expressions are very informative. They show that the resampling step has the effect of "resetting" the particle system whenever it is applied. Comparing (24.26) to (24.37), we see that the SMC variance expression has replaced the importance distribution $q_n(x_{1:n})$ in the SIS variance with the importance distributions $\pi_{k-1}(x_{1:k-1}) q_k(x_k | x_{1:k-1})$ obtained after the resampling step at time $k-1$. Moreover, we will show that in important scenarios the variances of SMC estimates are orders of magnitude smaller than the variances of SIS estimates.

Let us first revisit the toy example discussed in section 24.3.3.

Example (continued). In this case, it follows from (24.37) that the asymptotic variance is finite only when $\sigma^2 > \frac{1}{2}$ and

$$\frac{\mathbb{V}_{\text{SMC}}[\widehat{Z}_n]}{Z_n^2} \approx \frac{1}{N} \left[\int \frac{\pi_n^2(x_1)}{q_1(x_1)} dx_1 - 1 + \sum_{k=2}^{n} \int \frac{\pi_n^2(x_k)}{q_k(x_k)} dx_k - 1 \right]$$

$$= \frac{n}{N} \left[\left(\frac{\sigma^4}{2\sigma^2 - 1} \right)^{1/2} - 1 \right]$$

compared to

$$\frac{\mathbb{V}_{\text{IS}}[\widehat{Z}_n]}{Z_n^2} = \frac{1}{N}\left[\left(\frac{\sigma^4}{2\sigma^2-1}\right)^{n/2} - 1\right].$$

The asymptotic variance of the SMC estimate increases linearly with n in contrast to the exponential growth of the IS variance. For example, if we select $\sigma^2 = 1.2$ then we have a reasonably good importance distribution as $q_k(x_k) \approx \pi_n(x_k)$. In this case, we saw that it is necessary to employ $N \approx 2 \times 10^{23}$ particles in order to obtain $\frac{\mathbb{V}_{\text{IS}}[\widehat{Z}_n]}{Z_n^2} = 10^{-2}$ for $n = 1000$. Whereas to obtain the same performance, $\frac{\mathbb{V}_{\text{SMC}}[\widehat{Z}_n]}{Z_n^2} = 10^{-2}$, SMC requires the use of just $N \approx 10^4$ particles: an improvement by 19 orders of magnitude.

This scenario is overly favourable to SMC as the target (24.31) factorises. However, generally speaking, the major advantage of SMC over IS is that it allows us to exploit the forgetting properties of the model under study as illustrated by the following example.

Example 24.5 Consider the following more realistic scenario where

$$\gamma_n(x_{1:n}) = \mu(x_1)\prod_{k=2}^{n} M_k(x_k|x_{k-1})\prod_{k=1}^{n} G_k(x_k)$$

with μ a probability distribution, M_k a Markov transition kernel and G_k a positive "potential" function. Essentially, filtering corresponds to this model with $M_k(x_k|x_{k-1}) = f(x_k|x_{k-1})$ and the time inhomogeneous potential function $G_k(x_k) = g(y_k|x_k)$. In this case, $\pi_k(x_k|x_{1:k-1}) = \pi_k(x_k|x_{k-1})$ and we would typically select an importance distribution $q_k(x_k|x_{1:k-1})$ with the same Markov property $q_k(x_k|x_{1:k-1}) = q_k(x_k|x_{k-1})$. It follows that (24.37) is equal to

$$\frac{1}{N}\left(\int\frac{\pi_1^2(x_1)}{q_1(x_1)}dx_1 - 1 + \sum_{k=2}^{n}\int\frac{\pi_n^2(x_{k-1:k})}{\pi_{k-1}(x_{k-1})q_k(x_k|x_{k-1})}dx_{k-1:k} - 1\right)$$

and (24.38), for $\varphi_n(x_{1:n}) = \varphi(x_n)$, equals:

$$\int\frac{\pi_1^2(x_1)}{q_1(x_1)}\left(\int\varphi(x_n)\pi_n(x_n|x_1)dx_{2:n} - I_n(\varphi)\right)^2 dx_1$$
$$+ \sum_{k=2}^{n-1}\int\frac{\pi_n^2(x_{k-1:k})}{\pi_{k-1}(x_{k-1})q_k(x_k|x_{k-1})}\left(\int\varphi(x_n)\pi_n(x_n|x_k)dx_k - I_n(\varphi)\right)^2 dx_{k-1:k}$$
$$+ \int\frac{\pi_n^2(x_{n-1:n})}{\pi_{n-1}(x_{n-1})q_n(x_n|x_{n-1})}(\varphi(x_n) - I_n(\varphi))^2 dx_{n-1:n},$$

where we use the notation $I_n(\varphi)$ for $I_n(\varphi_n)$. In many realistic scenarios, the model associated with $\pi_n(x_{1:n})$ has some sort of ergodic properties; i.e. $\forall x_k, x'_k \in \mathcal{X}$ $\pi_n(x_n|x_k) \approx \pi_n(x_n|x'_k)$ for large enough $n - k$. In layman's terms, at time n what happened at time k is irrelevant if $n - k$ is large enough. Moreover, this often happens exponentially fast; that is for any (x_k, x'_k)

$$\frac{1}{2}\int \left|\pi_n\left(x_n|\,x_k\right) - \pi_n\left(x_n|\,x_k'\right)\right|dx_n \leq \beta^{n-k}$$

for some $\beta < 1$. This property can be used to establish that for bounded functions $\varphi \leq \|\varphi\|$

$$\left|\int \varphi\left(x_n\right)\pi_n\left(x_n|\,x_k\right)dx_n - I\left(\varphi\right)\right| \leq \beta^{n-k}\|\varphi\|$$

and under weak additional assumptions we have

$$\frac{\pi_n^2\left(x_{k-1:k}\right)}{\pi_{k-1}\left(x_{k-1}\right)q_k\left(x_k|\,x_{k-1}\right)} \leq A$$

for a finite constant A. Hence it follows that

$$\frac{\mathbb{V}_{\text{SMC}}\left[\widehat{Z}_n\right]}{Z_n^2} \leq \frac{C \cdot n}{N},$$

$$\mathbb{V}_{\text{SMC}}\left[\widehat{I}_n\left(\varphi\right)\right] \leq \frac{D}{N}.$$

for some finite constants C, D that are independent of n. These constants typically increase polynomially/exponentially with the dimension of the state space $\mathcal{X}$ and decrease as $\beta \to 0$.

24.3.7 Summary

We have presented a generic SMC algorithm which approximates $\{\pi_n\left(x_{1:n}\right)\}$ and $\{Z_n\}$ sequentially in time.

- Wherever it is possible to sample from $q_n\left(x_n|\,x_{1:n-1}\right)$ and evaluate $\alpha_n\left(x_{1:n}\right)$ in a time independent of n, this leads to an algorithm whose computational complexity does not increase with n.
- For any k, there exists $n > k$ such that the SMC approximation of $\pi_n\left(x_{1:k}\right)$ consists of a single particle because of the successive resampling steps. It is thus impossible to get a "good" SMC approximation of the joint distributions $\{\pi_n\left(x_{1:n}\right)\}$ when n is too large. This can easily be seen in practice, by monitoring the number of distinct particles approximating $\pi_n\left(x_1\right)$.
- However, under mixing conditions, this SMC algorithm is able to provide estimates of marginal distributions of the form $\pi_n\left(x_{n-L+1:n}\right)$ and estimates of Z_n/Z_{n-1} whose variance is uniformly bounded with n. This property is crucial and explains why SMC methods "work" in many realistic scenarios.
- Practically, one should keep in mind that the variance of SMC estimates can only expected to be reasonable if the variance of the incremental weights is

small. In particular, this requires that we can only expect to obtain good performance if $\pi_n(x_{1:n-1}) \approx \pi_{n-1}(x_{1:n-1})$ and $q_n(x_n|x_{1:n-1}) \approx \pi_n(x_n|x_{1:n-1})$; that is if the successive distributions we want to approximate do not differ much one from each other and the importance distribution is a reasonable approximation of the "optimal" importance distribution. However, if successive distributions differ significantly, it is often possible to design an artificial sequence of distributions to "bridge" this transition [21, 31].

24.4 Particle filtering

Remember that in the filtering context, we want to be able to compute a numerical approximation of the distribution $\{p(x_{1:n}|y_{1:n})\}_{n\geq 1}$ *sequentially* in time. A direct application of the SMC methods described earlier to the sequence of target distributions $\pi_n(x_{1:n}) = p(x_{1:n}|y_{1:n})$ yields a popular class of particle filters. More elaborate sequences of target and proposal distributions yield various more advanced algorithms. For ease of presentation, we present algorithms in which we resample at each time step. However, in practice we recommend only resampling when the ESS is below a threshold and employing the systematic resampling scheme.

24.4.1 SMC for filtering

First, consider the simplest case in which $\gamma_n(x_{1:n}) = p(x_{1:n}, y_{1:n})$ is chosen, yielding $\pi_n(x_{1:n}) = p(x_{1:n}|y_{1:n})$ and $Z_n = p(y_{1:n})$. Practically, it is only necessary to select the importance distribution $q_n(x_n|x_{1:n-1})$. We have seen that in order to minimise the variance of the importance weights at time n, we should select $q_n^{\text{opt}}(x_n|x_{1:n-1}) = \pi_n(x_n|x_{1:n-1})$ where

$$\pi_n(x_n|x_{1:n-1}) = p(x_n|y_n, x_{n-1})$$
$$= \frac{g(y_n|x_n)f(x_n|x_{n-1})}{p(y_n|x_{n-1})}, \quad (24.39)$$

and the associated incremental importance weight is $\alpha_n(x_{1:n}) = p(y_n|x_{n-1})$. In many scenarios, it is not possible to sample from this distribution but we should aim to approximate it. In any case, it shows that we should use an importance distribution of the form

$$q_n(x_n|x_{1:n-1}) = q(x_n|y_n, x_{n-1}) \quad (24.40)$$

and that there is nothing to be gained from building importance distributions depending also upon $(y_{1:n-1}, x_{1:n-2})$—although, at least in principle, in some settings there may be advantages to using information from subsequent observations if they are available. Combining (24.29), (24.30), and (24.40), the incremental weight is given by

$$a_n(x_{1:n}) = a_n(x_{n-1:n}) = \frac{g(y_n|x_n)f(x_n|x_{n-1})}{q(x_n|y_n, x_{n-1})}.$$

The algorithm can thus be summarized as follows.

SIR/SMC for filtering

At time $n = 1$

- Sample $X_1^i \sim q(x_1|y_1)$.
- Compute the weights $w_1(X_1^i) = \frac{\mu(X_1^i)g(y_1|X_1^i)}{q(X_1^i|y_1)}$ and $W_1^i \propto w_1(X_1^i)$.
- Resample $\{W_1^i, X_1^i\}$ to obtain N equally-weighted particles $\{\frac{1}{N}, \overline{X}_1^i\}$.

At time $n \geq 2$

- Sample $X_n^i \sim q(x_n|y_n, \overline{X}_{n-1}^i)$ and set $X_{1:n}^i \leftarrow (\overline{X}_{1:n-1}^i, X_n^i)$.
- Compute the weights $a_n(X_{n-1:n}^i) = \frac{g(y_n|X_n^i)f(X_n^i|X_{n-1}^i)}{q(X_n^i|y_n, X_{n-1}^i)}$ and $W_n^i \propto a_n(X_{n-1:n}^i)$.
- Resample $\{W_n^i, X_{1:n}^i\}$ to obtain N new equally-weighted particles $\{\frac{1}{N}, \overline{X}_{1:n}^i\}$.

We obtain at time n

$$\widehat{p}(x_{1:n}|y_{1:n}) = \sum_{i=1}^{N} W_n^i \delta_{X_{1:n}^i}(x_{1:n}),$$

$$\widehat{p}(y_n|y_{1:n-1}) = \sum_{i=1}^{N} W_{n-1}^i a_n(X_{n-1:n}^i).$$

However, if we are interested only in approximating the marginal distributions $\{p(x_n|y_{1:n})\}$ and $\{p(y_{1:n})\}$ then we need to store only the terminal-value particles $\{X_{n-1:n}^i\}$ to be able to compute the weights: the algorithm's storage requirements do not increase over time.

Many techniques have been proposed to design "efficient" importance distributions $q(x_n|y_n, x_{n-1})$ which approximate $p(x_n|y_n, x_{n-1})$. In particular the use of standard suboptimal filtering techniques such as the Extended Kalman Filter or the Unscented Kalman Filter to obtain importance distributions is very popular in the literature [14, 37]. The use of local optimization techniques to design $q(x_n|y_n, x_{n-1})$ centred around the mode of $p(x_n|y_n, x_{n-1})$ has also been advocated [33, 34].

Example: stochastic volatility

Returning to Example 24.4 and the simulated data shown in Figure 24.1, we are able to illustrate the performance of SMC algorithms with and without resampling steps in a filtering context.

An SIS algorithm (corresponding to the above SMC algorithm without a resampling step) with $N = 1000$ particles, in which the conditional prior is employed as a proposal distribution (leading to an algorithm in which the particle weights are proportional to the likelihood function) produces the output shown in Figure 24.2. Specifically, at each iteration, n, of the algorithm the conditional expectation and standard deviation of x_n, given $y_{1:n}$ is obtained. It can be seen that the performance is initially good, but after a few iterations the estimate of the mean becomes inaccurate, and the estimated standard deviation shrinks to a very small value. This standard deviation is an estimate of the standard deviation of the conditional posterior obtained via the particle filter: it is not a measure of the standard deviation of the estimator. Such an estimate can be obtained by considering several independent particle filters run on the same data and would illustrate the high variability of estimates obtained by a poorly-designed algorithm such as this one. In practice, approximations

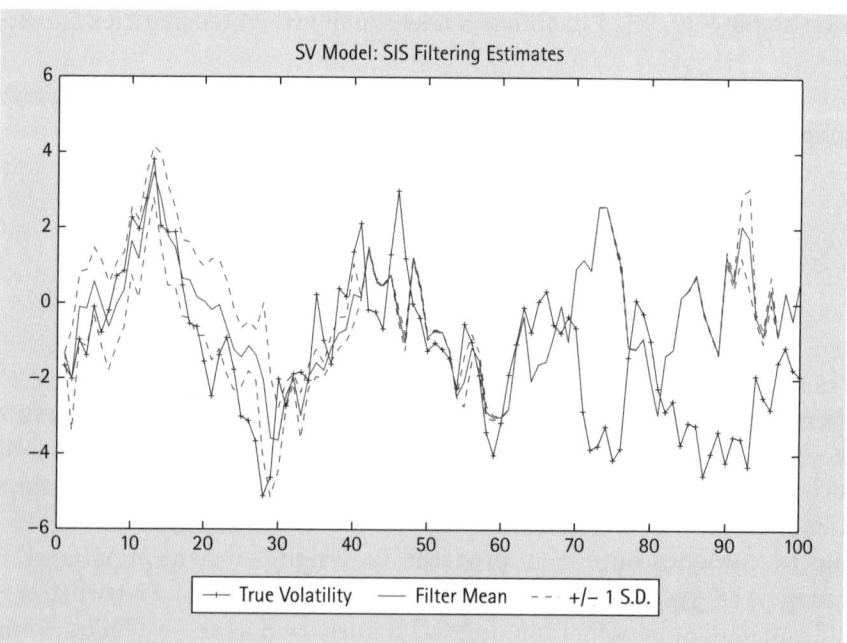

Fig. 24.2 Filtering estimates obtained for the stochastic volatility model using SIS. At each time the mean and standard deviation of x_n conditional upon $y_{1:n}$ is estimated using the particle set. It initially performs reasonably, but the approximation eventually collapses: the estimated mean begins to diverge from the truth and the estimate of the standard deviation is inaccurately low.

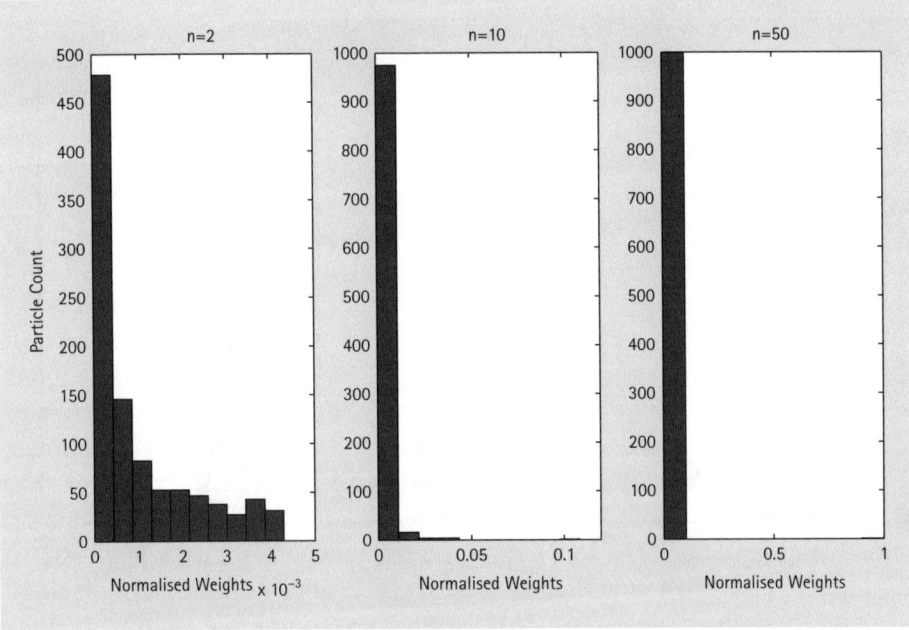

Fig. 24.3 Empirical distributions of the particle weights obtained with the SIS algorithm for the stochastic volatility model at iterations 2, 10, and 50. Although the algorithm is reasonably initialized, by iteration 10 only a few tens of particles have significant weight and by iteration 50 a single particle is dominant.

such as the effective sample size are often used as surrogates to characterise the uncertainty of the filter estimates but these perform well only if the filter is providing a reasonable approximation of the conditional distributions of interest. Figure 24.3 supports the theory that the failure of the algorithm after a few iterations is due to weight degeneracy, showing that the number of particles with significant weight falls rapidly.

The SIR algorithm described above was also applied to this problem with the same proposal distribution and number of particles as were employed in the SIS case. For simplicity, multinomial resampling was applied at every iteration. Qualitatively, the same features would be observed if a more sophisticated algorithm were employed, or adaptive resampling were used although these approaches would lessen the severity of the path-degeneracy problem. Figure 24.4 shows the distribution of particle weights for this algorithm. Notice that unlike the SIS algorithm shown previously, there are many particles with significant weight at all three time points. It is important to note that while this is encouraging it is not evidence that the algorithm is performing well: it provides no information about the path-space distribution and, in fact, it is easy to construct poorly-performing algorithms which appear to have a good distribution of particle weights (for instance, consider a scenario in which

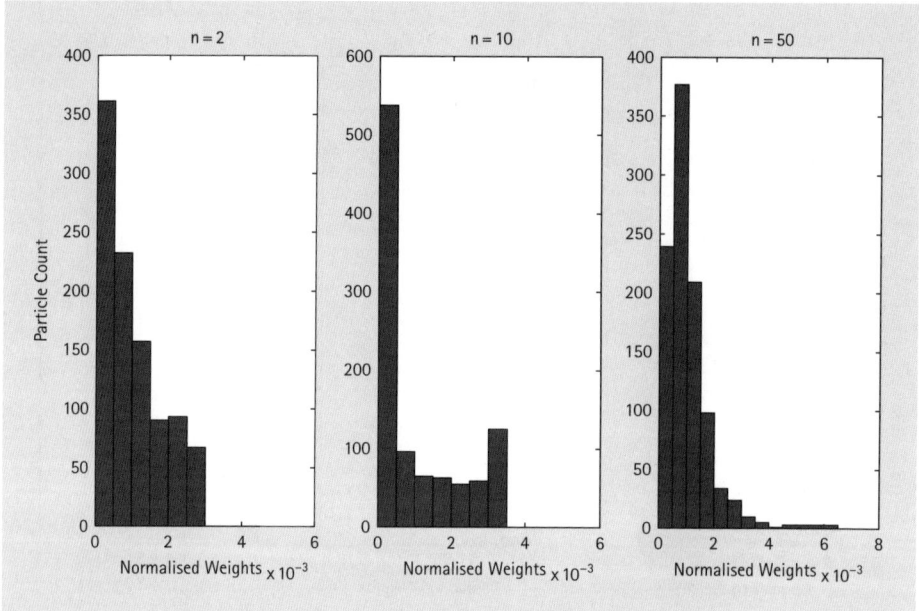

Fig. 24.4 Empirical distribution of particle weights for an SIR algorithm applied to the stochastic volatility model. Notice that there is no evidence of weight degeneracy in contrast to the SIS case. This comes at the cost of reducing the quality of the path-space representation.

the target is relatively flat in its tails but sharply concentrated about a mode; if the proposal has very little mass in the vicinity of the mode then it is likely that a collection of very similar importance weights will be obtained— but the sample thus obtained does not characterise the distribution of interest well). Figure 24.5 shows that the algorithm does indeed produce a reasonable estimate and plausible credible interval. And, as we expect, a problem does arise when we consider the smoothing distributions $p(x_n|y_{1:500})$ as shown in Figure 24.6: the estimate and credible interval is unreliable for $n \ll 500$. This is due to the degeneracy caused at the beginning of the path by repeated resampling. In contrast the smoothed estimate for $n \approx 500$ (not shown) is reasonable.

24.4.2 Auxiliary particle filtering

As was discussed above, the optimal proposal distribution (in the sense of minimizing the variance of importance weights) when performing standard particle filtering is $q(x_n|y_n, x_{n-1}) = p(x_n|y_n, x_{n-1})$. Indeed, $\alpha_n(x_{n-1:n})$ is independent of x_n in this case so it is possible to interchange the order of the sampling and resampling steps. Intuitively, this yields a better approximation of the distribution as it provides a greater number of distinct particles to approximate the target. This is an example of a general principle: resampling, if it is to

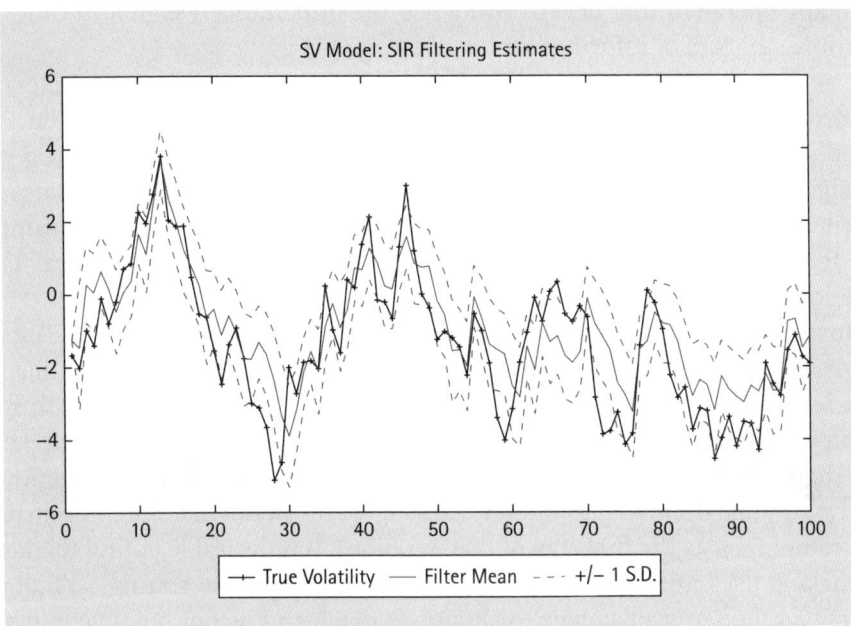

Fig. 24.5 SIR filtering estimates for the SV model.

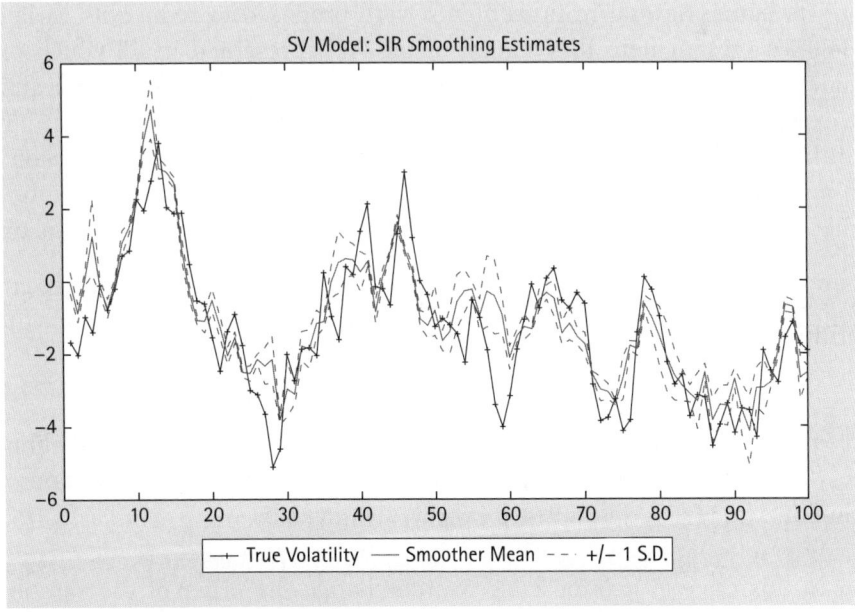

Fig. 24.6 SIR smoothing estimates for the SV model.

be applied in a particular iteration, should be performed before, rather than after, any operation that doesn't influence the importance weights in order to minimize the loss of information.

It is clear that if importance weights are independent of the new state and the proposal distribution corresponds to the marginal distribution of the proposed states then weighting, resampling, and then sampling corresponds to a reweighting to correct for the discrepancy between the old and new marginal distribution of the earlier states, resampling to produce an unweighted sample and then generation of the new state from its conditional distribution. This intuition can easily be formalized.

However, in general, the incremental importance weights do depend upon the new states and this straightforward change of order becomes impossible. In a sense, this interchange of sampling and resampling produces an algorithm in which information from the next observation is used to determine which particles should survive resampling at a given time (to see this, consider weighting and resampling occurring as the very last step of the iteration before the current one, rather than as the first step of that iteration). It is desirable to find methods for making use of this future information in a more general setting, so that we can obtain the same advantage in situations in which it is not possible to make use of the optimal proposal distribution.

The Auxiliary Particle Filter (APF) is an alternative algorithm which does essentially this. It was originally introduced in [33] using auxiliary variables—hence its name. Several improvements were proposed to reduce its variance [7, 34]. We present here the version of the APF presented in [7] which only includes one resampling step at each time instance. It has long been realized that, experimentally, this version outperforms the original two-stage resampling algorithm proposed in [33] and is widely used; see [7] for a comparison of both approaches. The APF is a look ahead method where at time n we try to predict which samples will be in regions of high probability masses at time $n+1$.

It was shown in [23] that the APF can be reinterpreted as a *standard* SMC algorithm applied to the following sequence of target distributions

$$\gamma_n(x_{1:n}) = p(x_{1:n}, y_{1:n})\, \widetilde{p}(y_{n+1}|\, x_n) \tag{24.41}$$

with $\widetilde{p}(y_{n+1}|\, x_n)$ chosen as an approximation of the predictive likelihood $p(y_{n+1}|\, x_n)$ if it is not known analytically. It follows that $\pi_n(x_{1:n})$ is an approximation of $p(x_{1:n}|\, y_{1:n+1})$ denoted $\widetilde{p}(x_{1:n}|\, y_{1:n+1})$ given by

$$\pi_n(x_{1:n}) = \widetilde{p}(x_{1:n}|\, y_{1:n+1}) \propto p(x_{1:n}|\, y_{1:n})\, \widetilde{p}(y_{n+1}|\, x_n) \tag{24.42}$$

In the APF we also use an importance distribution $q_n(x_n|\, x_{1:n-1})$ of the form (24.40) which is typically an approximation of (24.39). Note that (24.39) is different from $\pi_n(x_n|\, x_{1:n-1})$ in this scenario. Even if we could sample from

$\pi_n(x_n | x_{1:n-1})$, one should remember that in this case the object of inference is not $\pi_n(x_{1:n}) = \widetilde{p}(x_{1:n}| y_{1:n+1})$ but $p(x_{1:n}| y_{1:n})$. The associated incremental weight is given by

$$a_n(x_{n-1:n}) = \frac{\gamma_n(x_{1:n})}{\gamma_{n-1}(x_{1:n-1}) q_n(x_n | x_{1:n-1})}$$

$$= \frac{g(y_n| x_n) f(x_n| x_{n-1}) \widetilde{p}(y_{n+1}| x_n)}{\widetilde{p}(y_n| x_{n-1}) q(x_n| y_n, x_{n-1})}.$$

To summarize, the APF proceeds as follows.

Auxiliary particle filtering

At time n = 1

- Sample $X_1^i \sim q(x_1| y_1)$.
- Compute the weights $w_1(X_1^i) = \frac{\mu(X_1^i) g(y_1| X_1^i) \widetilde{p}(y_2| X_1^i)}{q(X_1^i| y_1)}$ and $W_1^i \propto w_1(X_1^i)$.
- Resample $\{W_1^i, X_1^i\}$ to obtain N equally-weighted particles $\{\frac{1}{N}, \overline{X}_1^i\}$.

At time $n \geq 2$

- Sample $X_n^i \sim q(x_n| y_n, \overline{X}_{n-1}^i)$ and set $X_{1:n}^i \leftarrow (\overline{X}_{1:n-1}^i, X_n^i)$.
- Compute the weights $a_n(X_{n-1:n}^i) = \frac{g(y_n| X_n^i) f(X_n^i| X_{n-1}^i) \widetilde{p}(y_{n+1}| X_n^i)}{\widetilde{p}(y_n| X_{n-1}^i) q(X_n^i| y_n, X_{n-1}^i)}$ and $W_n^i \propto a_n(X_{n-1:n}^i)$.
- Resample $\{W_n^i, X_{1:n}^i\}$ to obtain N new equally-weighted particles $\{\frac{1}{N}, \overline{X}_{1:n}^i\}$.

Keeping in mind that this algorithm does not approximate the distributions $\{p(x_{1:n}| y_{1:n})\}$ but the distributions $\{\widetilde{p}(x_{1:n}| y_{1:n+1})\}$, we use IS to obtain an approximation of $p(x_{1:n}| y_{1:n})$ with

$$\pi_{n-1}(x_{1:n-1}) q_n(x_n| x_{1:n-1}) = \widetilde{p}(x_{1:n-1}| y_{1:n}) q(x_n| y_n, x_{n-1})$$

as the importance distribution. A Monte Carlo approximation of this importance distribution is obtained after the sampling step in the APF and the associated unnormalized importance weights are given by

$$\widetilde{w}_n(x_{n-1:n}) = \frac{p(x_{1:n}, y_{1:n})}{\gamma_{n-1}(x_{1:n-1}) q_n(x_n| x_{1:n-1})} = \frac{g(y_n| x_n) f(x_n| x_{n-1})}{\widetilde{p}(y_n| x_{n-1}) q(x_n| y_n, x_{n-1})}. \quad (24.43)$$

It follows that we obtain

$$\widehat{p}(x_{1:n}|y_{1:n}) = \sum_{i=1}^{N} \widetilde{W}_n^i \delta_{X_{1:n}^i}(x_{1:n}),$$

$$\widehat{p}(y_n|y_{1:n-1}) = \left(\frac{1}{N}\sum_{i=1}^{N} \widetilde{w}_n\left(X_{n-1:n}^i\right)\right)\left(\sum_{i=1}^{N} W_{n-1}^i \widetilde{p}(y_n|X_{n-1}^i)\right)$$

where

$$\widetilde{W}_n^i \propto \widetilde{w}_n\left(X_{n-1:n}^i\right)$$

or $\widetilde{W}_n^i \propto W_{n-1}^i \widetilde{w}_n\left(X_{n-1:n}^i\right)$ if resampling was not performed at the end of the previous iteration. Selecting $q_n(x_n|x_{1:n-1}) = p(x_n|y_n, x_{n-1})$ and $\widetilde{p}(y_n|x_{n-1}) = p(y_n|x_{n-1})$, when it is possible to do so, leads to the so-called "perfect adaptation" case [33]. In this case, the APF takes a particularly simple form as $a_n(x_{n-1:n}) = p(y_n|x_{n-1})$ and $\widetilde{w}_n(x_{n-1:n}) = 1$. This is similar to the algorithm discussed in the previous subsection where the order of the sampling and resampling steps is interchanged.

This simple reinterpretation of the APF shows us several things:

- We should select a distribution $\widetilde{p}(x_{1:n}|y_{1:n})$ with thicker tails than $p(x_{1:n}|y_{1:n})$.
- Setting $\widetilde{p}(y_n|x_{n-1}) = g(y_n|\mu(x_{n-1}))$ where μ denotes some point estimate is perhaps unwise as this will not generally satisfy that requirement.
- We should use an approximation of the predictive likelihood which is compatible with the model we are using in the sense that it encodes at least the same degree of uncertainty as the exact model.

These observations follows from the fact that $\widetilde{p}(x_{1:n}|y_{1:n})$ is used as an importance distribution to estimate $p(x_{1:n}|y_{1:n})$ and this is the usual method to ensure that the estimator variance remains finite. Thus $\widetilde{p}(y_n|x_{n-1})$ should be more diffuse than $p(y_n|x_{n-1})$.

It has been suggested in the literature to set $\widetilde{p}(y_n|x_{n-1}) = g(y_n|\mu(x_{n-1}))$ where $\mu(x_{n-1})$ corresponds to the mode, mean, or median of $f(x_n|x_{n-1})$. However, this simple approximation will often yield an importance weight function (24.43) which is not upper bounded on $\mathcal{X} \times \mathcal{X}$ and could lead to estimates with a large/infinite variance.

An alternative approach, selecting an approximation $\widetilde{p}(y_n, x_n|x_{n-1}) = \widetilde{p}(y_n|x_{n-1}) q(x_n|y_n, x_{n-1})$ of the distribution $p(y_n, x_n|x_{n-1}) = p(y_n|x_{n-1}) p(x_n|y_n, x_{n-1}) = g(y_n|x_n) f(x_n|x_{n-1})$ such that the ratio (24.43) is upper bounded on $\mathcal{X} \times \mathcal{X}$ and such that it is possible to compute $\widetilde{p}(y_n|x_{n-1})$ pointwise and to sample from $q(x_n|y_n, x_{n-1})$, should be preferred.

24.4.3 Limitations of particle filters

The algorithms described earlier suffer from several limitations. It is important to emphasize at this point that, even if the optimal importance distribution $p(x_n| y_n, x_{n-1})$ can be used, this does not guarantee that the SMC algorithms will be efficient. Indeed, if the variance of $p(y_n| x_{n-1})$ is high, then the variance of the resulting approximation will be high. Consequently, it will be necessary to resample very frequently and the particle approximation $\widehat{p}(x_{1:n}| y_{1:n})$ of the joint distribution $p(x_{1:n}| y_{1:n})$ will be unreliable. In particular, for $k \ll n$ the marginal distribution $\widehat{p}(x_{1:k}| y_{1:n})$ will only be approximated by a few if not a single unique particle because the algorithm will have resampled many times between times k and n. One major problem with the approaches discussed above is that only the variables $\{X_n^i\}$ are sampled at time n but the path values $\{X_{1:n-1}^i\}$ remain fixed. An obvious way to improve upon these algorithms would involve not only sampling $\{X_n^i\}$ at time n, but also modifying the values of the paths over a fixed lag $\{X_{n-L+1:n-1}^i\}$ for $L > 1$ in light of the new observation y_n; L being fixed or upper bounded to ensure that we have a sequential algorithm (i.e. one whose computational cost and storage requirements are uniformly bounded over time). The following two sections describe two approaches to limit this degeneracy problem.

24.4.4 Resample-Move

This degeneracy problem has historically been addressed most often using the Resample-Move algorithm [20]. Like Markov chain Monte Carlo (MCMC), it relies upon Markov kernels with appropriate invariant distributions. While MCMC uses such kernels to generate collections of correlated samples, the Resample-Move algorithm uses them within an SMC algorithm as a principled way to "jitter"the particle locations and thus to reduce degeneracy. A Markov kernel $K_n(x'_{1:n}| x_{1:n})$ of invariant distribution $p(x_{1:n}| y_{1:n})$ is a Markov transition kernel with the property that

$$\int p(x_{1:n}| y_{1:n}) K_n(x'_{1:n}| x_{1:n}) dx_{1:n} = p(x'_{1:n}| y_{1:n}).$$

For such a kernel, if $X_{1:n} \sim p(x_{1:n}| y_{1:n})$ and $X'_{1:n}| X_{1:n} \sim K(x_{1:n}| X_{1:n})$ then $X'_{1:n}$ is still marginally distributed according to $p(x_{1:n}| y_{1:n})$. Even if $X_{1:n}$ is not distributed according to $p(x_{1:n}| y_{1:n})$ then, after an application of the MCMC kernel, $X'_{1:n}$ can only have a distribution closer than that of $X_{1:n}$ (in total variation norm) to $p(x_{1:n}| y_{1:n})$. A Markov kernel is said to be ergodic if iterative application of that kernel generates samples whose distribution converges towards $p(x_{1:n}| y_{1:n})$ irrespective of the distribution of the initial state.

It is easy to construct a Markov kernel with a specified invariant distribution. Indeed, this is the basis of MCMC—for details see

[35] and references therein. For example, we could consider the following kernel, based upon the Gibbs sampler: set $x'_{1:n-L} = x_{1:n-L}$ then sample x'_{n-L+1} from $p\left(x_{n-L+1}|\, y_{1:n}, x'_{1:n-L}, x_{n-L+2:n}\right)$, sample x'_{n-L+2} from $p\left(x_{n-L+2}|\, y_{1:n}, x'_{1:n-L+1}, x_{n-L+3:n}\right)$ and so on until we sample x'_n from $p\left(x_n|\, y_{1:n}, x'_{1:n-1}\right)$; that is

$$K_n\left(x'_{1:n}|\, x_{1:n}\right) = \delta_{x_{1:n-L}}\left(x'_{1:n-L}\right) \prod_{k=n-L+1}^{n} p\left(x'_k|\, y_{1:n}, x'_{1:k-1}, x_{k+1:n}\right)$$

and we write, with a slight abuse of notation, the non-degenerate component of the MCMC kernel $K_n\left(x'_{n-L+1:n}|\, x_{1:n}\right)$. It is straightforward to verify that this kernel is $p\left(x_{1:n}|\, y_{1:n}\right)$-invariant.

If it is not possible to sample from $p\left(x'_k|\, y_{1:n}, x'_{1:k-1}, x_{k+1:n}\right) = p\left(x'_k|\, y_k, x'_{k-1}, x_{k+1}\right)$, we can instead employ a Metropolis–Hastings (MH) strategy and sample a candidate according to some proposal $q\left(x'_k|\, y_k, x'_{k-1}, x_{k:k+1}\right)$ and accept it with the usual MH acceptance probability

$$\min\left(1, \frac{p\left(x'_{1:k}, x_{k+1:n}|\, y_{1:n}\right) q\left(x_k|\, y_k, x'_{k-1}, x'_k, x_{k+1}\right)}{p\left(x'_{1:k-1}, x_{k+1:n}|\, y_{1:n}\right) q\left(x'_k|\, y_k, x'_{k-1}, x_{k:k+1}\right)}\right)$$

$$= \min\left(1, \frac{g\left(y_k|\, x'_k\right) f\left(x_{k+1}|\, x'_k\right) f\left(x'_k|\, x'_{k-1}\right) q\left(x_k|\, y_k, x'_{k-1}, x'_k, x_{k+1}\right)}{g\left(y_k|\, x_k\right) f\left(x_{k+1}|\, x_k\right) f\left(x_k|\, x'_{k-1}\right) q\left(x'_k|\, y_k, x'_{k-1}, x_{k:k+1}\right)}\right).$$

It is clear that these kernels can be ergodic only if $L = n$ and *all* of the components of $x_{1:n}$ are updated. However, in our context we will typically not use ergodic kernels as this would require sampling an increasing number of variables at each time step. In order to obtain truly online algorithms, we restrict ourselves to updating the variables $X_{n-L+1:n}$ for some fixed or bounded L.

The algorithm proceeds as follows, with K_n denoting a Markov kernel of invariant distribution $p(x_{1:n}|y_{1:n})$.

SMC filtering with MCMC moves

At time $n = 1$

- Sample $X_1^i \sim q(x_1|\, y_1)$.
- Compute the weights $w_1\left(X_1^i\right) = \frac{\mu(X_1^i) g(y_1|X_1^i)}{q(X_1^i|y_1)}$ and $W_1^i \propto w_1\left(X_1^i\right)$.
- Resample $\{W_1^i, X_1^i\}$ to obtain N equally-weighted particles $\{\frac{1}{N}, \overline{X}_1^i\}$.
- Sample $X_1'^i \sim K_1\left(x_1|\, \overline{X}_1^i\right)$.

At time $1 < n < L$

- Sample $X_n^i \sim q(x_n | y_n, X_{n-1}^{\prime i})$ and set $X_{1:n}^i \leftarrow (X_{1:n-1}^{\prime i}, X_n^i)$.
- Compute the weights $a_n(X_{n-1:n}^i) = \frac{g(y_n|X_n^i)f(X_n^i|X_{n-1}^i)}{q(X_n^i|y_n, X_{n-1}^i)}$ and $W_n^i \propto a_n(X_{n-1:n}^i)$.
- Resample $\{W_n^i, X_{1:n}^i\}$ to obtain N equally-weighted particles $\{\frac{1}{N}, \overline{X}_{1:n}^i\}$.
- Sample $X_{1:n}^{\prime i} \sim K_n(x_{1:n} | \overline{X}_{1:n}^i)$.

At time $n \geq L$

- Sample $X_n^i \sim q(x_n | y_n, X_{n-1}^{\prime i})$ and set $X_{1:n}^i \leftarrow (X_{1:n-1}^{\prime i}, X_n^i)$.
- Compute the weights $a_n(X_{n-1:n}^i) = \frac{g(y_n|X_n^i)f(X_n^i|X_{n-1}^i)}{q(X_n^i|y_n, X_{n-1}^i)}$ and $W_n^i \propto a_n(X_{n-1:n}^i)$.
- Resample $\{W_n^i, X_{1:n}^i\}$ to obtain N new equally-weighted particles $\{\frac{1}{N}, \overline{X}_{1:n}^i\}$.
- Sample $X_{n-L+1:n}^{\prime i} \sim K_n(x_{n-L+1:n} | \overline{X}_{1:n}^i)$ and set $X_{1:n}^{\prime i} \leftarrow (\overline{X}_{1:n-L}^i, X_{n-L+1:n}^{\prime i})$.

The following premise, which [35] describes as "generalized importance sampling", could be used to justify inserting MCMC transitions into an SMC algorithm after the sampling step. Given a target distribution π, an instrumental distribution μ and a π-invariant Markov kernel, K, the following generalization of the IS identity holds:

$$\int \pi(y)\varphi(y)dy = \iint \mu(x)K(y|x)\frac{\pi(y)L(x|y)}{\mu(x)K(y|x)}\varphi(y)dxdy$$

for any Markov kernel L. This approach corresponds to importance sampling on an enlarged space using $\mu(x)K(y|x)$ as the proposal distribution for a target $\pi(y)L(x|y)$ and then estimating a function $\varphi'(x,y) = \varphi(y)$. In particular, for the time-reversal kernel associated with K

$$L(x|y) = \frac{\pi(x)K(y|x)}{\pi(y)},$$

we have the importance weight

$$\frac{\pi(y)L(x|y)}{\mu(x)K(y|x)} = \frac{\pi(x)}{\mu(x)}.$$

This interpretation of such an approach illustrates its deficiency: the importance weights depend only upon the location before the MCMC move while the sample depends upon the location after the move. Even if the kernel was perfectly mixing, leading to a collection of i.i.d. samples from the target distribution, some of these samples would be eliminated and some replicated in the resampling step. Resampling before an MCMC step will always lead to greater sample diversity than performing the steps in the other order (and this algorithm can be justified directly by the invariance property).

Based on this reinterpretation of MCMC moves within IS, it is possible to reformulate this algorithm as a specific application of the generic SMC algorithm discussed in Section 24.3. To simply notation we write $q_n(x_n|x_{n-1})$ for $q(x_n|y_n, x_{n-1})$. To clarify our argument, it is necessary to add a superscript to the variables; e.g. X_k^p corresponds to the pth time the random variable X_k is sampled; in this and the following section, this superscript *does not* denote the particle index. Using such notation, this algorithm is the generic SMC algorithm associated to the following sequence of target distributions

$$\pi_n\left(x_1^{1:L+1}, \ldots, x_{n-L+1}^{1:L+1}, x_{n-L}^{1:L}, \ldots, x_n^{1:2}\right)$$
$$= p\left(x_1^{L+1}, \ldots, x_{n-L+1}^{L+1}, x_{n-L}^{L}, \ldots, x_n^2 \middle| y_{1:n}\right) L_n\left(x_n^1, x_{n-1}^2, \ldots, x_{n-L+1}^L \middle| x_n^2, x_{n-1}^3, \ldots, x_{n-L+1}^{L+1}\right)$$
$$\times \cdots \times L_2\left(x_1^2, x_2^1 \middle| x_1^3, x_2^2\right) L_1\left(x_1^1 \middle| x_1^2\right)$$

where L_n is the time-reversal kernel associated with K_n whereas, if no resampling is used,[5] a path up to time n is sampled according to

$$q_n\left(x_1^{1:L+1}, \ldots, x_{n-L+1}^{1:L+1}, x_{n-L}^{1:L}, \ldots, x_n^{1:2}\right)$$
$$= q_1\left(x_1^1\right) K_1\left(x_1^2 \middle| x_1^1\right) q_2\left(x_2^1 \middle| x_1^2\right) K_2\left(x_1^3, x_2^2 \middle| x_1^2, x_2^1\right)$$
$$\times \cdots \times q_n\left(x_n^1 \middle| x_{n-1}^2\right) K_n\left(x_{n-L+1}^{L+1}, \ldots, x_{n-1}^3, x_n^2 \middle| x_{1:n-L}^{L+1}, x_{n-L+1}^L, \ldots, x_{n-1}^2, x_n^1\right).$$

This sequence of target distributions admits the filtering distributions of interest as marginals. The clear theoretical advantage of using MCMC moves is that the use of even nonergodic MCMC kernels $\{K_n\}$ can only improve the mixing properties of $\{\pi_n\}$ compared to the "natural" sequence of filtering distributions; this explains why these algorithms outperform a standard particle filter for a given number of particles.

Finally, we note that the incorporation of MCMC moves to improve sample diversity is an idea which is appealing in its simplicity and which can easily be incorporated into any of the algorithms described here.

24.4.5 Block sampling

The Resample-Move method discussed in the previously section suffers from a major drawback. Although it does allow us to reintroduce some diversity among the set of particles after the resampling step over a lag of length L, the importance weights have the same expression as for the standard particle filter. So this strategy does not significantly decrease the number of resampling steps compared to a standard approach. It can partially mitigate the problem associated with resampling, but it does not prevent these resampling steps in the first place.

An alternative *block sampling* approach has recently been proposed in [15]. This approach goes further than the Resample-Move method, which aims to sample only the component x_n at time n in regions of high probability mass and then to uses MCMC moves to rejuvenate $x_{n-L+1:n}$ after a resampling step. The block sampling algorithm attempts to directly sample the components $x_{n-L+1:n}$ at time n; the previously-sampled values of the components $x_{n-L+1:n-1}$ sampled are simply discarded. In this case, it can easily be shown that the optimal importance distribution (that which minimizes the variance of the importance weights at time n) is:

$$p(x_{n-L+1:n}|\, y_{n-L+1:n}, x_{n-L}) = \frac{p(x_{n-L:n},\, y_{n-L+1:n})}{p(y_{n-L+1:n}|\, x_{n-L})} \qquad (24.44)$$

where

$$p(y_{n-L+1:n}|\, x_{n-L}) = \int \prod_{k=n-L+1}^{n} f(x_k|\, x_{k-1}) \cdot g(y_k|\, x_k)\, dx_{n-L+1:n}. \qquad (24.45)$$

As in the standard case (corresponding to $L = 1$), it is typically impossible to sample from (24.44) and/or to compute (24.45). So in practice we need to design an importance distribution $q(x_{n-L+1:n}|\, y_{n-L+1:n}, x_{n-L})$ approximating $p(x_{n-L+1:n}|\, y_{n-L+1:n}, x_{n-L})$. Henceforth, we consider the case where $L > 1$.

The algorithm proceeds as follows.

SMC block sampling for filtering

At time $n = 1$

- **Sample** $X_1^i \sim q(x_1|\, y_1)$.
- **Compute the weights** $w_1(X_1^i) = \frac{\mu(X_1^i) g(y_1|X_1^i)}{q(X_1^i|y_1)}$ and $W_1^i \propto w_1(X_1^i)$.
- **Resample** $\{W_1^i, X_1^i\}$ to obtain N equally-weighted particles $\{\frac{1}{N}, \overline{X}_1^i\}$.

At time $1 < n < L$

- **Sample** $X_{1:n}^i \sim q(x_{1:n}|\, y_{1:n})$.
- **Compute the weights** $a_n(X_{1:n}^i) = \frac{p(X_{1:n}^i,\, y_{1:n})}{q(X_{1:n}^i|\, y_{1:n})}$ and $W_n^i \propto a_n(X_{n-1:n}^i)$.
- **Resample** $\{W_n^i, X_{1:n}^i\}$ to obtain N equally-weighted particles $\{\frac{1}{N}, \overline{X}_{1:n}^i\}$.

At time $n \geq L$

- **Sample** $X_{n-L+1:n}^i \sim q(x_{n-L+1:n}|y_{n-L+1:n},\, \overline{X}_{1:n-1}^i)$.

- Compute the weights

$$w_n\left(\overline{X}^i_{n-L:n-1}, X^i_{n-L+1:n}\right)$$
$$= \frac{p\left(\overline{X}^i_{1:n-L}, X^i_{n-L+1:n}, y_{1:n}\right) q(\overline{X}^i_{n-L+1:n-1}|y_{n-L+1:n-1}, \overline{X}^i_{n-L})}{p\left(\overline{X}^i_{1:n-1}, y_{1:n-1}\right) q(X^i_{n-L+1:n}|y_{n-L+1:n}, \overline{X}^i_{n-L})} \quad (24.46)$$

and $W^i_n \propto w_n\left(\overline{X}^i_{n-L:n-1}, X^i_{n-L+1:n}\right)$.

- Resample $\left\{W^i_n, \overline{X}^i_{1:n-L}, X^i_{n-L+1:n}\right\}$ to obtain N new equally weighted particles $\left\{\frac{1}{N}, \overline{X}^i_{1:n}\right\}$.

When the optimal IS distribution is used $q(x_{n-L+1:n}|y_{n-L+1:n}, x_{n-L}) = p(x_{n-L+1:n}|y_{n-L+1:n}, x_{n-L})$, we obtain

$$w_n(\overline{x}_{1:n-1}, x_{n-L+1:n}) = \frac{p(\overline{x}_{1:n-L}, x_{n-L+1:n}, y_{1:n}) p(\overline{x}_{n-L+1:n-1}|y_{n-L+1:n-1}, \overline{x}_{n-L})}{p(\overline{x}_{1:n-1}, y_{1:n-1}) p(x_{n-L+1:n}|y_{n-L+1:n}, \overline{x}_{n-L})}$$
$$= p(y_n|y_{n-L+1:n}, \overline{x}_{n-L}).$$

This optimal weight has a variance which typically decreases exponentially fast with L (under mixing assumptions). Hence, in the context of adaptive resampling, this strategy dramatically reduces the number of resampling steps. In practice, we cannot generally compute this optimal weight and thus use (24.46) with an approximation of $p(x_{n-L+1:n}|y_{n-L+1:n}, x_{n-L})$ for $q(x_{n-L+1:n}|y_{n-L+1:n}, x_{n-L})$. When a good approximation is available, the variance of (24.46) can be reduced significantly compared to the standard case where $L = 1$.

Again, this algorithm is a specific application of the generic SMC algorithm discussed in Section 24.3. To simply notation we write $q_n(x_{n-L+1:n}|x_{n-L})$ for $q(x_{n-L+1:n}|y_{n-L+1:n}, x_{n-L})$. To clarify our argument, it is again necessary to add a superscript to the variables; e.g. X^p_k corresponds to the pth time the random variable X_k is sampled; remember that in the present section, this superscript *does not* denote the particle index. Using this notation, the algorithm corresponds to the generic SMC algorithm associated with the following sequence of target distributions

$$\pi_n\left(x_1^{1:L}, \ldots, x_{n-L+1}^{1:L}, x_{n-L+2}^{1:L-1}, \ldots, x_n^1\right)$$
$$= p\left(x_{1:n-L+1}^L, x_{n-L+2}^{L-1}, \ldots, x_n^1 \middle| y_{1:n}\right) q_{n-1}\left(x_{n-L+1}^{L-1}, \ldots, x_{n-1}^1 \middle| x_{n-L}^L\right)$$
$$\times \cdots q_2\left(x_1^2, x_2^1\right) q_1\left(x_1^1\right).$$

where, if no resampling is used, a path is sampled according to

$$q_n\left(x_1^{1:L},\ldots,x_{n-L+1}^{1:L},x_{n-L+2}^{1:L-1},\ldots,x_n^1\right)$$
$$= q_1\left(x_1^1\right) q_2\left(x_1^2, x_2^1\right) \times \cdots \times q_n\left(x_{n-L+1}^L, \ldots, x_n^1 \middle| x_{n-L}^L\right).$$

The sequence of distributions admits the filtering distributions of interest as marginals. The mixing properties of $\{\pi_n\}$ are also improved compared to the "natural" sequence of filtering distributions; this is the theoretical explanation of the better performance obtained by these algorithms for a given number of particles.

24.4.6 Rao–Blackwellized particle filtering

Let us start by quoting Trotter [36]: "A good Monte Carlo is a dead Monte Carlo". Trotter specialized in Monte Carlo methods and did not advocate that we should not use them, but that we should avoid them whenever possible. In particular, whenever an integral can be calculated analytically doing so should be preferred to the use of Monte Carlo techniques.

Assume, for example, that one is interested in sampling from $\pi(x)$ with $x = (u, z) \in \mathcal{U} \times \mathcal{Z}$ and

$$\pi(x) = \frac{\gamma(u, z)}{Z} = \pi(u)\,\pi(z|u)$$

where $\pi(u) = Z^{-1}\gamma(u)$ and

$$\pi(z|u) = \frac{\gamma(u, z)}{\gamma(u)}$$

admits a closed-form expression; e.g. a multivariate Gaussian. Then if we are interested in approximating $\pi(x)$ and computing Z, we only need to perform a MC approximation $\pi(u)$ and $Z = \int \gamma(u)\,du$ on the space $\mathcal{U}$ instead of $\mathcal{U} \times \mathcal{Z}$. We give two classes of important models where this simple idea can be used successfully.

Conditionally linear Gaussian models

Consider $\mathcal{X} = \mathcal{U} \times \mathcal{Z}$ with $\mathcal{Z} = \mathbb{R}^{n_z}$. Here $X_n = (U_n, Z_n)$ where $\{U_n\}$ is a unobserved Markov process such that $U_1 \sim \mu_U(u_1)$, $U_n|U_{n-1} \sim f_U(u_n|U_{n-1})$ and conditional upon $\{U_n\}$ we have a linear Gaussian model with $Z_1|U_1 \sim \mathcal{N}(0, \Sigma_{U_1})$ and

$$Z_n = A_{U_n} Z_{n-1} + B_{U_n} V_n,$$
$$Y_n = C_{U_n} Z_n + D_{U_n} W_n$$

where $V_n \overset{\text{i.i.d.}}{\sim} \mathcal{N}(0, I_{n_v})$, $W_n \overset{\text{i.i.d.}}{\sim} \mathcal{N}(0, I_{n_w})$ and for any $u \in \mathcal{U}$ $\{A_u, B_u, C_u, D_u\}$ are matrices of appropriate dimensions. In this case we have $\mu(x) = \mu(u, z) = \mu_U(u)\mathcal{N}(x; 0, \Sigma_z)$, $f(x'|x) = f((u', z')|(u, z)) = f_U(u'|u)\mathcal{N}(z'; A_{u'}z, B_{u'}B_{u'}^T)$ and $g(y|x) = g(y|(u, z)) = \mathcal{N}(y; C_u z, D_u D_u^T)$. The switching state-space model discussed in Example 24.3 corresponds to the case where $\{U_n\}$ is a finite state-space Markov process.

We are interested in estimating

$$p(u_{1:n}, z_{1:n} | y_{1:n}) = p(u_{1:n} | y_{1:n}) p(z_{1:n} | y_{1:n}, u_{1:n}).$$

Conditional upon $\{U_n\}$, we have a standard linear Gaussian model $\{Z_n\}$, $\{Y_n\}$. Hence $p(z_{1:n} | y_{1:n}, u_{1:n})$ is a Gaussian distribution whose statistics can be computed using Kalman techniques; e.g. the marginal $p(z_n | y_{1:n}, u_{1:n})$ is a Gaussian distribution whose mean and covariance can be computed using the Kalman filter. It follows that we only need to use particle methods to approximate

$$\gamma_n(u_{1:n}) = p(u_{1:n}, y_{1:n})$$
$$= p(u_{1:n}) p(y_{1:n} | u_{1:n})$$

where $p(u_{1:n})$ follows from the Markov assumption on $\{U_n\}$ and $p(y_{1:n} | u_{1:n})$ is a marginal likelihood which can be computed through the Kalman filter. In this case, we have

$$q^{\text{opt}}(u_n | y_{1:n}, u_{1:n-1}) = p(u_n | y_{1:n}, u_{1:n-1})$$
$$= \frac{p(y_n | y_{1:n-1}, u_{1:n}) f_U(u_n | u_{n-1})}{p(y_n | y_{1:n-1}, u_{1:n-1})}.$$

The standard SMC algorithm associated with $\gamma_n(u_{1:n})$ and the sequence of IS distributions $q(u_n | y_{1:n}, u_{1:n-1})$ proceeds as follows.

SMC for filtering in conditionally linear Gaussian models

At time $n = 1$

- **Sample** $U_1^i \sim q(u_1 | y_1)$.
- **Compute the weights** $w_1(U_1^i) = \frac{\mu_U(U_1^i) p(y_1 | U_1^i)}{q(U_1^i | y_1)}$ and $W_1^i \propto w_1(U_1^i)$.
- **Resample** $\{W_1^i, U_1^i\}$ to obtain N equally-weighted particles $\{\frac{1}{N}, \overline{U}_1^i\}$.

At time $n \geq 2$

- **Sample** $U_n^i \sim q\left(u_n | y_{1:n}, \overline{U}_{1:n-1}^i\right)$ and set $U_{1:n}^i \leftarrow \left(\overline{U}_{1:n-1}^i, U_n^i\right)$.
- **Compute the weights** $\alpha_n(U_{1:n}^i) = \frac{p(y_n | y_{1:n-1}, U_{1:n}^i) f_U(U_n^i | U_{n-1}^i)}{q(U_n^i | y_{1:n}, U_{1:n-1}^i)}$ and $W_n^i \propto \alpha_n(U_{1:n}^i)$.
- **Resample** $\{W_n^i, U_{1:n}^i\}$ to obtain N new equally-weighted particles $\{\frac{1}{N}, \overline{U}_{1:n}^i\}$.

This algorithm provides also two approximations of $p(u_{1:n}|y_{1:n})$ given by

$$\widehat{p}(u_{1:n}|y_{1:n}) = \sum_{i=1}^{N} W_n^i \delta_{U_{1:n}^i}(u_{1:n}),$$

$$\overline{p}(u_{1:n}|y_{1:n}) = \frac{1}{N} \sum_{i=1}^{N} \delta_{\overline{U}_{1:n}^i}(u_{1:n})$$

and

$$\widehat{p}(y_n|y_{1:n-1}) = \frac{1}{N} \sum_{i=1}^{N} \alpha_n(U_{1:n}^i).$$

At first glance, it seems that this algorithm cannot be implemented as it requires storing paths $\{U_{1:n}^i\}$ of increasing dimension so as to allow the computation of $p(y_n|y_{1:n-1}, u_{1:n})$ and sampling from $q(u_n|y_{1:n}, u_{1:n-1})$. The key is to realize that $p(y_n|y_{1:n-1}, u_{1:n})$ is a Gaussian distribution of mean $y_{n|n-1}(u_{1:n})$ and covariance $S_{n|n-1}(u_{1:n})$ which can be computed using the Kalman filter. Similarly, given that the optimal IS distribution only depends on the path $u_{1:n}$ through $p(y_n|y_{1:n-1}, u_{1:n})$, it is sensible to build an importance distribution $q(u_n|y_{1:n}, u_{1:n-1})$ which only depends on $u_{1:n}$ through $p(y_n|y_{1:n-1}, u_{1:n})$. Hence in practice, we do not need to store $\{U_{1:n}^i\}$ but only $\{U_{n-1:n}^i\}$ and the Kalman filter statistics associated with $\{U_{1:n}^i\}$. The resulting particle filter is a bank of interacting Kalman filters where each Kalman filter is used to compute the marginal likelihood term $p(y_{1:n}|u_{1:n})$.

Partially observed linear Gaussian models

The same idea can be applied to the class of partially observed linear Gaussian models [2]. Consider $\mathcal{X} = \mathcal{U} \times \mathcal{Z}$ with $\mathcal{Z} = \mathbb{R}^{n_z}$. Here $X_n = (U_n, Z_n)$ with $Z_1 \sim \mathcal{N}(0, \Sigma_1)$

$$Z_n = AZ_{n-1} + BV_n,$$
$$U_n = CZ_n + DW_n$$

where $V_n \stackrel{\text{i.i.d.}}{\sim} \mathcal{N}(0, I_{n_v})$, $W_n \stackrel{\text{i.i.d.}}{\sim} \mathcal{N}(0, I_{n_w})$; see [2] for generalizations. We make the additional assumption that

$$g(y_n|x_n) = g(y_n|(u_n, z_n)) = g(y_n|u_n).$$

In this case, we are interested in estimating

$$p(u_{1:n}, z_{1:n}|y_{1:n}) = p(u_{1:n}|y_{1:n}) p(z_{1:n}|y_{1:n}, u_{1:n})$$
$$= p(u_{1:n}|y_{1:n}) p(z_{1:n}|u_{1:n})$$

where $p(z_{1:n}|u_{1:n})$ is a multivariate Gaussian distribution whose statistics can be computed using a Kalman filter associated with the linear model $\{U_n, Z_n\}$. It follows that we only need to use particle methods to approximate

$$\gamma_n(u_{1:n}) = p(u_{1:n}, y_{1:n})$$
$$= p(u_{1:n}) p(y_{1:n}|u_{1:n})$$

where $p(y_{1:n}|u_{1:n}) = \prod_{k=1}^{n} g(y_k|u_k)$ and $p(u_{1:n})$ is the marginal Gaussian prior of $\{U_n\}$ which corresponds to the "marginal" likelihood term which can be computed using the Kalman filter associated with $\{U_n, Z_n\}$. The resulting particle filter is also an interacting bank of Kalman filters but the Kalman filters are here used to compute the marginal prior $p(u_{1:n})$.

24.5 Particle smoothing

We have seen previously that SMC methods can provide an approximation of the sequence of distributions $\{p(x_{1:n}|y_{1:n})\}$. Consequently, sampling from a joint distribution $p(x_{1:T}|y_{1:T})$ and approximating the marginals $\{p(x_n|y_{1:T})\}$ for $n = 1, ..., T$ is straightforward. We just run an SMC algorithm up to time T and sample from/marginalize our SMC approximation $\widehat{p}(x_{1:T}|y_{1:T})$. However, we have seen that this approach is bound to be inefficient when T is large as the successive resampling steps lead to particle degeneracy: $\widehat{p}(x_{1:n}|y_{1:T})$ is approximated by a single unique particle for $n \ll T$. In this section, we discuss various alternative schemes which do not suffer from these problems. The first method relies on an simple fixed-lag approximation whereas the other algorithms rely on the forward–backward recursions presented in Section 24.2.3.

24.5.1 Fixed-lag approximation

The fixed-lag approximation is the simplest approach. It was proposed in [26]. It relies on the fact that, for hidden Markov models with "good" forgetting properties, we have

$$p(x_{1:n}|y_{1:T}) \approx p\left(x_{1:n}|y_{1:\min(n+\Delta,T)}\right) \quad (24.47)$$

for Δ large enough; that is observations collected at times $k > n + \Delta$ do not bring any additional information about $x_{1:n}$. This suggests a very simple scheme—simply don't update the estimate at time k after time $k = n + \Delta$. Indeed, in practice we just do not resample the components $X^i_{1:n}$ of the particles $X^i_{1:k}$ at times $k > n + \Delta$. This algorithm is trivial to implement but the main practical problem is that we typically do not know Δ. Hence we need to replace Δ with an estimate of it denoted L. If we select $L < \Delta$, then $p\left(x_{1:n}|y_{1:\min(n+L,T)}\right)$ is a poor approximation of $p(x_{1:n}|y_{1:T})$. If we select a large values of L to ensure that $L \geq \Delta$ then the degeneracy problem remains substantial. Unfortunately automatic selection of L is difficult (and, of course, for some poorly-mixing models Δ is so large that this approach is impractical). Experiments on various models have shown that good performance were achieved with $L \approx 20-50$. Note that such fixed-lag SMC schemes do not converge asymptotically (i.e. as $N \to \infty$) towards the true smoothing distributions because we do not have $p(x_{1:n}|y_{1:T}) = p(x_{1:n}|y_{1:\min(n+L,T)})$. However, the bias might be negligible and can be upper bounded under mixing conditions [8]. It should also be noted that this method does not provide an approximation of the joint $p(x_{1:T}|y_{1:T})$—it approximates only the marginal distributions.

24.5.2 Forward filtering–backward smoothing

We have seen previously that it is possible to sample from $p(x_{1:T}|y_{1:T})$ and compute the marginals $\{p(x_n|y_{1:T})\}$ using forward–backward formulæ. It is possible to obtain an SMC approximation of the forward filtering-backward sampling procedure directly by noting that for

$$\widehat{p}(x_n|y_{1:n}) = \sum_{i=1}^{N} W_n^i \delta_{X_n^i}(x_n)$$

we have

$$\widehat{p}(x_n|X_{n+1}, y_{1:n}) = \frac{f(X_{n+1}|x_n)\widehat{p}(x_n|y_{1:n})}{\int f(X_{n+1}|x_n)\widehat{p}(x_n|y_{1:n})\,dx_n}$$

$$= \sum_{i=1}^{N} \frac{W_n^i f(X_{n+1}|X_n^i) \delta_{X_n^i}(x_n)}{\sum_{j=1}^{N} W_n^j f(X_{n+1}|X_n^j)}.$$

Consequently, the following algorithm generates a sample approximately distributed according to $p(x_{1:T}|y_{1:T})$: first sample $X_T \sim \widehat{p}(x_T|y_{1:T})$ and for $n = T-1, T-2, ..., 1$, sample $X_n \sim \widehat{p}(x_n|X_{n+1}, y_{1:n})$.

Similarly, we can also provide an SMC approximation of the forward filtering/backward smoothing procedure by direct means. If we denote by

$$\widehat{p}(x_n|y_{1:T}) = \sum_{i=1}^{N} W_{n|T}^i \delta_{X_n^i}(x_n) \qquad (24.48)$$

the particle approximation of $p(x_n|y_{1:T})$ then, by inserting (24.48) into (24.15), we obtain

$$\widehat{p}(x_n|y_{1:T}) = \sum_{i=1}^{N} W_n^i \left[\sum_{j=1}^{N} W_{n+1|T}^j \frac{f\left(X_{n+1}^j|X_n^i\right)}{\left[\sum_{l=1}^{N} W_n^l f\left(X_{n+1}^j|X_n^l\right)\right]} \right] \delta_{X_n^i}(x_n) \qquad (24.49)$$

$$= \sum_{i=1}^{N} W_{n|T}^i \delta_{X_n^i}(x_n).$$

The forward filtering/backward sampling approach requires $\mathcal{O}(NT)$ operations to sample one path approximately distributed according to $p(x_{1:T}|y_{1:T})$ whereas the forward filtering/backward smoothing algorithm requires $\mathcal{O}(N^2T)$ operations to approximate $\{p(x_n|y_{1:T})\}$. Consequently, these algorithms are only useful for very long time series in which sample degeneracy prevents computationally-cheaper, crude methods from working.

24.5.3 Generalized two-filter formula

To obtain an SMC approximation of the generalized two-filter formula (24.18), we need to approximate the backward filter $\{\widetilde{p}(x_n|y_{n:T})\}$ and to combine the forward filter and the backward filter in a sensible way. To obtain an approximation of $\{\widetilde{p}(x_n|y_{n:T})\}$, we simply use an SMC algorithm which targets the sequence of distributions $\{\widetilde{p}(x_{n:T}|y_{n:T})\}$ defined in (24.17). We run the SMC algorithm "backward in time" to approximate $\widetilde{p}(x_T|y_T)$ then $\widetilde{p}(x_{T-1:T}|y_{T-1:T})$ and so on. We obtain an SMC approximation denoted

$$\widehat{\widetilde{p}}(x_{n:T}|y_{n:T}) = \sum_{i=1}^{N} \widetilde{W}_n^i \delta_{X_{n:T}^i}(x_{n:T}).$$

To combine the forward filter and the backward filter, we obviously cannot multiply the SMC approximations of both $p(x_n|y_{1:n-1})$ and $\widetilde{p}(x_{n:T}|y_{n:T})$ directly, so we first rewrite eq. (24.18) as

$$p(x_n|y_{1:T}) \propto \frac{\int f(x_n|x_{n-1}) p(x_{n-1}|y_{1:n-1}) dx_{n-1} \cdot \widetilde{p}(x_n|y_{n:T})}{\widetilde{p}_n(x_n)}.$$

By plugging the SMC approximations of $p(x_{n-1}|y_{1:n-1})$ and $\widetilde{p}(x_n|y_{n:T})$ in this equation, we obtain

$$\widehat{p}(x_n|y_{1:T}) = \sum_{i=1}^{N} W_{n|T}^i \delta_{\widetilde{X}_n^i}(x_n)$$

where

$$W_{n|T}^j \propto \widetilde{W}_n^j \sum_{i=1}^{N} W_{n-1}^i \frac{f\left(\widetilde{X}_n^j|X_{n-1}^i\right)}{\widetilde{p}_n\left(\widetilde{X}_n^j\right)}. \qquad (24.50)$$

Like the SMC implementation of the forward–backward smoothing algorithm, this approach has a computational complexity $\mathcal{O}(N^2T)$. However fast computational methods have been developed to address this problem [27]. Moreover it is possible to reduce this computational complexity to $\mathcal{O}(NT)$ by using rejection sampling to sample from $p(x_n|y_{1:T})$ using $p(x_{n-1}|y_{1:n-1})$ and $\widetilde{p}(x_n|y_{n:T})$ as proposal distributions. More recently, an importance sampling type approach has also been proposed in [19] to reduce the computational complexity to

$O(NT)$; see [6] for a related idea developed in the context of belief propagation. Compared to the forward–backward formula, it might be expected to substantially outperform that algorithm in any situation in which the support of the smoothed estimate differs substantially from that of the filtering estimate. That is, in those situations in which observations obtained at time $k > n$ provide a significant amount of information about the state at time n.[6] The improvement arises from the fact that the SMC implementation of the forward–backward smoother simply reweights a sample set which targets $p(x_{1:n}|y_{1:n})$ to account for the information provided by $y_{n+1:k}$ whereas the two-filter approach uses a different sample set with locations appropriate to the smoothing distributions.

24.6 Summary

We have provided a review of particle methods for filtering, marginal likelihood computation, and smoothing. Having introduced a simple SMC algorithm, we have shown how essentially all of the particle-based methods introduced in the literature to solve these problems can be interpreted as a combination of two operations: sampling and resampling. By considering an appropriate sequence of target distributions, defined on appropriate spaces, it is possible to interpret all of these algorithms as particular cases of the general algorithm.

This interpretation has two primary advantages:

1. The standard algorithm may be viewed as a particle interpretation of a Feynman–Kac model (see [11]) and hence strong, sharp theoretical results can be applied to all algorithms within this common formulation.
2. By considering all algorithms within the same framework is is possible to develop a common understanding and intuition allowing meaningful comparisons to be drawn and sensible implementation decisions to be made.

Although much progress has been made over the past fifteen years, and the algorithms described above provide good estimates in many complex, realistic settings, there remain a number of limitations and open problems:

- As with any scheme for numerical integration, be it deterministic or stochastic, there exist problems which exist on sufficiently high-dimensional spaces and involving sufficiently complex distributions of which it is not possible to obtain an accurate characterization in a reasonable amount of time.
- In many settings of interest the likelihood and state transition density are only known up to a vector of unknown parameters. It is typically of interest to estimate this vector of parameters at the same time as (or in preference to) the vector of states. Formally, such situations can be

described as directly as a state space model with a degenerate transition kernel. However, this degeneracy prevents the algorithms described above from working in practice.

- Several algorithms intended specifically for parameter estimation have been developed in recent years. Unfortunately, space constraints prevent us from discussing these methods within the current tutorial. Although progress has been made, this is a difficult problem and it cannot be considered to have been solved in full generality. These issues will be discussed in further depth in an extended version of this tutorial which is currently in preparation.

It should also be mentioned that the SMC algorithm presented in this tutorial can be adapted to perform much more general Monte Carlo simulation: it is not restricted to problems of the filtering type, or even to problems with a sequential character. It has recently been established that it is also possible to employ SMC within MCMC and to obtain a variety of related algorithms.

Notes

1. We persist with the abusive use of density notation in the interests of simplicity and accessibility; the alternations required to obtain a rigorous formulation are obvious.
2. Some authors use the terms *proposal density* or *instrumental density* interchangeably.
3. However, as we will see later, the key to many advanced SMC methods is the introduction of a sequence of target distributions which differ from the original target distributions.
4. Similar expressions can be established when a lower variance resampling strategy such as residual resampling is used and when resampling is performed adaptively [12]. The results presented here are sufficient to guide the design of particular algorithms and the additional complexity involved in considering more general scenarios serves largely to produce substantially more complex expressions which obscure the important points.
5. Once again, similar expressions can be obtained in the presence of resampling and the technique remains valid.
6. This situation occurs, for example, whenever observations are only weakly informative and the state evolution involves relatively little stochastic variability. An illustrative example provided in [5] shows that this technique can dramatically outperform all of the other approaches detailed above.

References

[1] Anderson, B. D. O. and Moore, J. B. (1979) *Optimal Filtering*, Prentice-Hall, Englewood Cliffs, NJ.
[2] Andrieu, C. and Doucet, A. (2002) Particle filtering for partially observed Gaussian state space models. *Journal of the Royal Statistical Society* B, **64**(4), 827–36.

[3] Arulampalam, S., Maskell, S., Gordon, N., and Clapp, T. (2002) A tutorial on particle filters for on-line nonlinear/non-Gaussian Bayesian tracking. *IEEE Transactions on Signal Processing*, **50**(2), 174–88.

[4] Bresler, Y. (1986) Two-filter formula for discrete-time non-linear Bayesian smoothing. *International Journal of Control*, **43**(2), 629–41.

[5] Briers, M., Doucet, A., and Maskell, S. (2010) Smoothing algorithms for state-space models, *Annals of the Institute of Statistical Mathematics*, **62**(1), 61–89.

[6] Briers, M., Doucet, A. and Singh, S.S. (2005) Sequential auxiliary particle belief propagation, *Proc. 7th Int. Conf. Information Fusion*, Philadelphia.

[7] Carpenter, J., Clifford, P., and Fearnhead, P. (1999) An improved particle filter for non-linear problems. *IEE proceedings—Radar, Sonar and Navigation*, **146**, 2–7.

[8] Cappé, O., Godsill, S. J. and Moulines, E. (2007) An overview of existing methods and recent advances in sequential Monte Carlo. *IEEE Proceedings*, **95**(5), 899–924.

[9] Chopin, N. (2004) Central limit theorem for sequential Monte Carlo and its application to Bayesian inference. *Ann. Statist.*, **32**, 2385–411.

[10] Crisan, D. and Doucet, A. (2002) A survey of convergence results on particle filtering for practitioners. *IEEE Transactions on Signal Processing*, **50**(3), 736–46.

[11] Del Moral, P. (2004) *Feynman–Kac Formulae: Genealogical and Interacting Particle Systems with Applications*. Series: Probability and Applications, Springer-Verlag, New York.

[12] Del Moral, P., Doucet, A., and Jasra, A. (2008) On adaptive resampling procedures for sequential Monte Carlo methods. Technical report. INRIA.

[13] Douc, R., Cappé, O., and Moulines, E. (2005) Comparison of resampling schemes for particle filtering. In 4th International Symposium on Image and Signal Processing and Analysis (ISPA).

[14] Doucet, A., Godsill, S. J., and Andrieu, C. (2000) On sequential Monte Carlo sampling methods for Bayesian filtering. *Statistics and Computing*, **10**, 197–208.

[15] Doucet, A., Briers, M., and Senecal, S. (2006) Efficient block sampling strategies for sequential Monte Carlo. *Journal of Computational and Graphical Statistics*, **15**(3), 693–711.

[16] Doucet, A., de Freitas, N., and Gordon, N. J. (eds) (2001) *Sequential Monte Carlo Methods in Practice*. Springer-Verlag, New York.

[17] Doucet, A., de Freitas, N., and Gordon, N.J. (2001) An introduction to sequential Monte Carlo methods. in [16], 3–13.

[18] Fearnhead, P. (2008) Computational methods for complex stochastic systems: A review of some alternatives to MCMC. *Statistics and Computing*, **18**, 151–71.

[19] Fearnhead, P., Wyncoll, D., and Tawn, J. (2008) A sequential smoothing algorithm with linear computational cost. Technical report, Department of Mathematics and Statistics, Lancaster University.

[20] Gilks, W. R. and Berzuini, C. (2001) Following a moving target—Monte Carlo inference for dynamic Bayesian models. *Journal of the Royal Statistical Society* B, **63**, 127–46.

[21] Godsill, S. J. and Clapp, T. (2001) Improvement strategies for Monte Carlo particle filters, in [16], 139–58.

[22] Gordon, N. J., Salmond, D. J., and Smith, A. F. M. (1993) Novel approach to nonlinear/non-Gaussian Bayesian state estimation. *IEE-Proceedings-F*, **140**, 107–13.

[23] Johansen, A. M. and Doucet, A. (2008) A note on auxiliary particle filters. *Statistics and Probability Letters*, **78**(12), 1498–1504.

[24] Johansen, A. M. (2008) SMCTC: Sequential Monte Carlo in C++. Technical Report 08–16. University of Bristol, Department of Statistics.

[25] Kitagawa, G. (1996) Monte Carlo filter and smoother for non-Gaussian non-linear state space models. *Journal of Computational and Graphical Statistics*, **5**, 1–25.

[26] Kitagawa, G. and Sato, S. (2001) Monte Carlo smoothing and self-organizing state-space model, in [16], 177–95.

[27] Klaas, M., Lang, D., and de Freitas, N. (2005) Fast particle smoothing: if I had a million particles, *Proc. 23rd Int. Conf. on Machine Learning*, Pittsburgh, PA.

[28] Kong, A, Liu, J. S., and Wong, W. H. (1994) Sequential imputations and Bayesian missing data problems. *Journal of the American Statistical Association*, **89**, 1032–44.

[29] Künsch, H. R. (2001) State-space and hidden Markov models. In *Complex Stochastic Systems* (eds O. E. Barndorff-Nielsen, D. R. Cox, and C. Kluppelberg). CRC Press, Boca Raton, 109–73.

[30] Liu, J. S. (2001) *Monte Carlo Strategies in Scientific Computing*. Springer-Verlag, New York.

[31] Neal, R. M. (2001) Annealed importance sampling. *Statistics and Computing*, **11**, 125–39.

[32] Olsson, J., Cappé, O., Douc., R., and Moulines, E. (2008) Sequential Monte Carlo smoothing with application to parameter estimation in non-linear state space models. *Bernoulli*, **14**(1), 155–79.

[33] Pitt, M. K. and Shephard, N. (1999) Filtering via simulation: auxiliary particle filter. *Journal of the American Statistical Association*, **94**, 590–9.

[34] Pitt, M. K. and Shephard, N. (2001) Auxiliary variable based particle filters, in [16], 271–93.

[35] Robert, C. P. and Casella, G. (2004) *Monte Carlo Statistical Methods*, 2nd edn. Springer-Verlag, New York.

[36] Trotter, H. F. and Tukey, J. W. (1956) Conditional Monte Carlo for normal samples. In *Symposium on Monte Carlo Methods* (ed. H. A. Meyer). Wiley, New York, 64–79.

[37] van der Merwe, R., Doucet, A., De Freitas, N., and Wan, E. (2000) The unscented particle filter. In *Advances in Neural and Information Processing Systems* (eds T. K. Leen, T. G. Dietterich, and T. G. Tresp). MIT Press, Cambridge, MA.

·25·

A Mean Field Theory of Nonlinear Filtering

P. Del Moral, F. Patras, and S. Rubenthaler

25.1 Introduction

25.1.1 A mean field theory of nonlinear filtering

The filtering problem consists in computing the conditional distributions of a state signal given a series of observations. The signal/observation pair sequence $(X_n, Y_n)_{n \geq 0}$ is defined as a Markov chain which takes values in some product of measurable spaces $(E_n \times F_n)_{n \geq 0}$. We further assume that the initial distribution ν_0 and the Markov transitions P_n of the pair process (X_n, Y_n) have the form

$$\nu_0(d(x_0, y_0)) = g_0(x_0, y_0)\, \eta_0(dx_0)\, q_0(dy_0)$$
$$P_n((x_{n-1}, y_{n-1}), d(x_n, y_n)) = M_n(x_{n-1}, dx_n)\, g_n(x_n, y_n)\, q_n(dy_n) \qquad (25.1)$$

where g_n are strictly positive functions on $(E_n \times F_n)$ and q_n is a sequence of measures on F_n. The initial distribution η_0 of the signal X_n, the Markov transitions M_n and the likelihood functions g_n are assumed to be known.

The main advantage of this general and abstract set-up comes from the fact that it applies directly without further work to traditional real valued or multidimensional problems, as well as to smoothing and path estimation filtering models. For instance the signal $X_n = (X'_0, \ldots, X'_n)$ may represent the path from the origin up to time n of an auxiliary Markov chain X'_n taking values in some measurable state space $(E'_n, \mathcal{E}'_n)$. A version of the conditional distributions η_n of the signal states X_n given their noisy observations $Y_p = y_p$ up to time $p < n$ is expressed in terms of a flow of Feynman–Kac measures associated with the distribution of the paths of the signal and weighted by the collection of likelihood potential functions.

To better connect the mean field particle approach to existing alternatives methods, it is convenient at this point to make a couple of remarks.

Firstly, using this functional representation it is tempting to approximate this measures using a crude Monte Carlo method based on independent simulations of the paths of the signal weighted by products of the likelihood functions from the origin up to time n. This strategy gives satisfactory results when the signal paths are sufficiently stable using a simple weight regularization to avoid their degeneracy with respect to the time parameter. We refer to [12, 14, 33] for further details. Precise comparisons of the fluctuation variances associated with this Monte Carlo method and the mean feld particle models presented in the present chapter have been studied in [23].

Another commonly used strategy in Bayesian statistics and in engineering is the well-known Monte Carlo Markov chain method (MCMC method). The idea is to interpret the conditional distributions η_n as the invariant measure of a suitably chosen Markov chain. This technique has two main drawbacks. The first one is that we need to run the underlying chain for very long times. This burn-in period is difficult to estimate, and it is often too long to tackle filtering problems with high-frequency observation sequences such as those arising in radar processing. The second difficulty is that the conditional distributions η_n vary with the time parameter and we need to choose at each time an appropriate MCMC algorithm for the time index. The computational complexity of such a strategy would increase as the time index increases as we would need to sample an increasing number of state variables.

In contrast to the two ideas discussed above, the mean field particle strategy presented in this article can be interpreted as a stochastic and adaptive grid approximation model. Loosely speaking, this particle technique consists in interpreting the weights likelihood as branching selection rates. No calibration of the convergence to equilibrium is needed, the conditional measures variations are automatically updated by the stochastic particle model. The first rigorous study in this field seems to be the article [15] published in 1996 on the applications of interacting particle methods to nonlinear estimation problems. This article provides the first convergence results for a series of heuristic genetic type schemes presented in the beginning of the 1990s in three independent chains of articles on nonlinear filtering [36, 35], [39], and [2, 40, 30, 32]. At the end of the 1990s, three other independent works [6, 7, 8] proposed another class of particle branching variants for solving continuous-time filtering problems.

For a more thorough discussion on the origins and analysis of these models, we refer the reader to the research monograph of the first author [11]. The latter is dedicated to Feynman–Kac and Boltzmann–Gibbs measures with their genealogical and interacting particle system interpretations, as well as their applications in physics, in biology, and in engineering sciences. Besides their important application in filtering, Feynman–Kac type particle models also arise in the spectral analysis of Schrödinger-type operators, rare events estimation, as well as in macromolecular and directed polymers simulations. In this connection, we also mention that the mean field theory presented here applies without further work to study the analysis of a variety of heuristic models introduced in stochastic engineering since the beginning of the 1950s, including genetic type algorithms, quantum and sequential Monte Carlo methods, pruned enrichment molecular simulations, bootstrap and particle filters, and many others. For a rather thorough discussion on these rather well-known application areas, the interested reader is also recommended to consult the book [34], and the references therein. This study can be completed with the more recent articles of the first author with A. Doucet and A. Jasra [17, 18] on sequential Monte Carlo techniques and their applications in Bayesian computation. We also mention that the continuous time version of the material presented in this review article can also be found in the series of articles of the first author with L. Miclo [26, 27, 28, 29]. For instance, the first reference [27] provides an original mean field particle interpretation of the robust nonlinear filtering equation.

This first key idea towards a mean field particle approach to nonlinear filtering is to recall that the flow of conditional measures η_n satisfies a dynamical system in distribution space, often called the nonlinear filtering equations. As in physics and more particularly in fluid mechanics, the second step is to interpret these equations as the evolution of the laws of a nonlinear Markov process. More formally, this preliminary stage simply consists in expressing the nonlinear filtering equations in terms of a transport model associated with a collection of Markov kernels K_{n+1,η_n} indexed by the time parameter n and the set of measures η_n on the space E_n; that is, we have that

$$\eta_{n+1} = \eta_n K_{n+1,\eta_n} \qquad (25.2)$$

The mean field particle interpretation of this nonlinear measure-valued model is an E_n^N-valued Markov chain $\xi_n^{(N)} = \left(\xi_n^{(N,i)}\right)_{1 \le i \le N}$, with elementary transitions defined as

$$\mathbb{P}\left(\xi_{n+1}^{(N)} \in d(x^1, \ldots, x^N) | \xi_n^{(N)}\right) = \prod_{i=1}^{N} K_{n+1,\eta_n^N}\left(\xi_n^{(N,i)}, dx^i\right) \text{ with } \eta_n^N := \frac{1}{N}\sum_{j=1}^{N} \delta_{\xi_n^{(N,j)}}$$
$$(25.3)$$

The initial system $\xi_0^{(N)}$ consists of N independent and identically distributed random variables with common law η_0. The state components of this Markov chain are called particles or sometimes walkers. The rationale behind this is that η_{n+1}^N is the empirical measure associated with N independent variables with distributions $K_{n+1,\eta_n^N}(\xi_n^{(N,i)}, .)$, so as soon as η_n^N is a good approximation of η_n then, in view of (25.3), η_{n+1}^N should be a good approximation of η_{n+1}. This strategy is not restricted to nonlinear filtering models and it applies under appropriate regularity conditions to any kind of measure valued model of the form (25.2). In our context, the evolution of the particles is dictated by the well-known pair prediction-updating transitions associated with the optimal filter equations. The prediction transition is associated with a mutation transition of the whole population of particles. During this stage, the particles explore the state space independently of one another, according to the same probability transitions as the signal. The filter updating stage is associated with a branching type selection transition. During this stage, each particle evaluates the relative likelihood value of its location. The ones with poor value are killed, while the ones with high potential duplicate.

Using this branching particle interpretation, we notice that the ancestral lines of each particle form a genealogical tree evolving as above by tracking back in time the whole ancestor line of current individuals. In other words, the genealogical tree model associated with the branching process described above is the mean field particle interpretation of the nonlinear filtering equation but in path space. Using this simple observation, we readily prove that the genealogical tree occupation measure converges, as the population size tends to infinity, to the conditional distribution of the paths the signal $(X_p)_{0 \leq p \leq n}$ given the observations delivered by the sensors up to time n. Another important mathematical object is the complete genealogical tree structure defined by the whole population model from the origin up to the current time horizon. The occupation measure of this ancestral tree keeps track of the whole history of the particle and its convergence is technically more involved. We can prove that it converges, as the population size tends to infinity, to the McKean distribution of the paths of a nonhomogeneous Markov chain with transition probabilities K_{n+1,η_n}, and starting with the initial distribution η_0.

In the present chapter, we provide a synthetic review of the stability properties and the convergence analysis of these mean field interacting particle models going from the traditional law of large numbers to more sophisticated empirical process theorems, uniform estimates with respect to the time parameter, central limit theorems, Donsker type theorems, as well as large-deviation principles. We also analyze the increasing and strong propagation of chaos properties, including an original algebraic tree-based functional representations of particle block distributions. These Laurent-type integral representations seems to be the first sharp and precise propagations of chaos estimates for this type of mean field particle models. We emphasized that most of the material presented in this review is taken from the book of the first author [11], and results from various collaboration with Donald Dawson, Jean Jacod, Michel Ledoux, Laurent Miclo, and Alice Guionnet. We refer to [11] for a detailed historical account with precise reference pointers. The sharp propagation of chaos expansions presented in the second part of this article are taken from the article [31]. The latter also discusses Hilbert series techniques for counting forests with prescribed numbers of vertices at each level, or with prescribed coalescence degrees, as well as new wreath products interpretations of vertex permutation groups. In this connection, we mention that forests and their combinatorics have also appeared recently in various fields such as in theoretical physics and Gaussian matrix integral models [37], renormalization theory in high-energy physics or Runge–Kutta methods, two fields where the structure and complexity of perturbative expansions has required the development of new tools [1, 3].

The article is divided into four main parts, devoted respectively to the precise description of Feynman–Kac path integral models and their mean field particle interpretations, to the stability analysis of nonlinear semigroups, to the asymptotic analysis of mean field interacting processes, and to propagation of chaos properties.

25.1.2 Notation and conventions

For the convenience of the reader we have collected some of the main notation and conventions used in the article. We denote respectively by $\mathcal{M}(E)$, $\mathcal{P}(E)$, and $\mathcal{B}(E)$, the set of all finite signed measures on some measurable space $(E, \mathcal{E})$, the convex subset of all probability measures, and the Banach space of all bounded and measurable functions f on E, equipped with the uniform norm $\|f\| = \sup_{x \in E} |f(x)|$. We also denote $\mathrm{Osc}_1(E)$, the convex set of $\mathcal{E}$-measurable functions f with oscillations less than one; that is,

$$\mathrm{osc}(f) = \sup\{|f(x) - f(y)| \; ; \; x, y \in E\} \leq 1$$

We let $\mu(f) = \int \mu(dx) \, f(x)$, be the Lebesgue integral of a function $f \in \mathcal{B}(E)$, with respect to a measure $\mu \in \mathcal{M}(E)$, and we equip the Banach space $\mathcal{M}(E)$ with the total variation norm $\|\mu\|_{\mathrm{tv}} = \sup_{f \in \mathrm{Osc}_1(E)} |\mu(f)|$. We recall that a integral operator Q from a measurable space E_1 into an auxiliary measurable space E_2 into itself, is an operator $f \mapsto Q(f)$ from $\mathcal{B}(E_2)$ into $\mathcal{B}(E_1)$ such that the functions

$$Q(f)(x) = \int_{E_2} Q(x_1, dx_2) \, f(x_2) \in \mathbb{R}$$

are $\mathcal{E}_1$-measurable and bounded, for any $f \in \mathcal{B}(E_2)$. It also generates a dual operator $\mu \mapsto \mu Q$ from $\mathcal{M}(E_1)$ into $\mathcal{M}(E_2)$ defined by $(\mu Q)(f) := \mu(Q(f))$. We denote by $\beta(M) := \sup_{x, y \in E_1} \|M(x, \cdot) - M(y, \cdot)\|_{\mathrm{tv}} \in [0, 1]$ the Dobrushin coefficient associated with a Markov transition M from E_1 into E_2. For further use in various places of this article we recall that $\beta(M)$ is the norm of both the operator $\mu \mapsto \mu M$ on the set of measures with null mass or the operator $f \mapsto M(f)$ on the set of functions $f \in \mathrm{Osc}_1(E_1)$. Thus, we have that

$$\forall f \in \mathrm{Osc}_1(E_1) \qquad \mathrm{osc}(M(f)) \leq \beta(M) \, \mathrm{osc}(f)$$

For a proof of this rather well-known inequality, we refer the reader to proposition 4.2.1 [11]. We also simplify the notation and we write for any $f, g \in \mathcal{B}(E_2)$ and $x \in E_1$

$$Q[(f - Qf)(g - Qg)](x)$$

instead of

$$Q[(f - Q(f)(x))(g - Q(g)(x))](x) = Q(fg)(x) - Q(f)(x) \, Q(g)(x)$$

The q-tensor power $Q^{\otimes q}$, with $q \geq 1$, represents the bounded integral operator on E_1^q into E_2^q, defined for any $F \in \mathcal{B}(E_2^q)$ by

$$Q^{\otimes q}(F)(x^1, \ldots, x^q) = \int_{E_2^q} [Q(x^1, dy^1) \ldots Q(x^q, dy^q)] \, F(y^1, \ldots, y^q)$$

For a pair of integral operators Q_1 from E_1 into E_2, and Q_2 from E_2 into E_3, we denote by $Q_1 Q_2$ the composition integral operator from E_1 into E_3 defined for any $f \in \mathcal{B}(E_3)$ by $(Q_1 Q_2)(f) := Q_1(Q_2(f))$.

With respect to the N-particle model introduced in (25.2), as soon as there is no possible confusion we simplify notation and suppress the index $(.)^{(N)}$ and write (ξ_n, ξ_n^i) instead of $(\xi_n^{(N)}, \xi_n^{(N,i)})$. To clarify the presentation, we often suppose that all the random processes including the signal and the particle processes are defined on some common probability space, and we denote by $\mathbb{E}(.)$ and $\mathbb{P}(.)$ the integral expectation and the probability measure on this common probability space. Last but not least, we fix a series of observations $Y_n = y_n$, with $n \geq 0$, and unless otherwise is stated all the results presented in this article depend of the observation sequence. The following classical conventions $(\sum_\emptyset, \prod_\emptyset) = (0, 1)$ are also used.

25.2 Feynman–Kac and mean field particle models

This section is concerned with Feynman–Kac formulations of the filtering equations with their mean field particle interpretations. The functional representations of both the one-step predictor and the optimal filter are discussed in the first subsection. In the second subsection, we discuss some advantages of these abstract models in the analysis of smoothing and path space filtering problems including continuous time signal-observation models. The mean field interacting particle systems and the corresponding genealogical tree-based models are described in the final subsection.

25.2.1 Description of the models

We let G_n be the non homogeneous function on E_n defined for any $x_n \in E_n$ by

$$G_n(x_n) = g_n(x_n, y_n) \tag{25.4}$$

Note that G_n depends on the observation value y_n at time n. To simplify the presentation, and avoid unnecessary technical discussion, we shall suppose that the likelihood functions are chosen so that there exists a sequence of strictly positive constants $\epsilon_n(G) \in (0, 1]$ that may depend on the observation values y_n and such that for any $x_n, x_n' \in E_n$ we have that

$$G_n(x_n) \geq \epsilon_n(G) \, G_n(x_n') > 0 \tag{25.5}$$

The archetype of such nonlinear filtering models is the situation where Y_n is a real valued observation sequence described by a dynamical equation

$$Y_n = h_n(X_n) + V_n$$

where V_n represents a sequence of centered Gaussian random variables with $\mathbb{E}(V_n^2) := \sigma_n > 0$ and $h_n \in \mathcal{B}(E_n)$. In this situation, (25.1) and the above condition are clearly met with the Lebesgue measure q_n on $\mathbb{R}$ and the likelihood potential function given by

$$G_n(x_n) = \exp\left(-\frac{1}{2\sigma_n^2}(y_n - h_n(x_n))^2\right) \Longrightarrow \epsilon_n(G) \geq \exp\left(\sigma_n^{-2} \operatorname{osc}(h_n)(y_n + \|h_n\|)\right)$$

In this notation, versions of the one-step predictor $\eta_n \in \mathcal{P}(E_n)$ and the optimal filter $\widehat{\eta}_n \in \mathcal{P}(E_n)$ given for any $f_n \in \mathcal{B}(E_n)$ by

$$\eta_n(f_n) := \mathbb{E}(f_n(X_n)|Y_p = y_p, 0 \leq p < n) \text{ and } \widehat{\eta}_n(f_n) := \mathbb{E}(f_n(X_n)|Y_p = y_p, 0 \leq p \leq n)$$

have the functional representations

$$\eta_n(f_n) = \gamma_n(f_n)/\gamma_n(1) \quad \text{and} \quad \widehat{\eta}_n(f_n) = \widehat{\gamma}_n(f_n)/\widehat{\gamma}_n(1) \tag{25.6}$$

with the Feynman–Kac measures γ_n and $\widehat{\gamma}_n$ defined by the formulae

$$\gamma_n(f_n) = \mathbb{E}[f_n(X_n)\prod_{0 \leq k < n} G_k(X_k)] \quad \text{and} \quad \widehat{\gamma}_n(f_n) = \gamma_n(f_n G_n) \tag{25.7}$$

A simple change of measure shows that the optimal filter can also be rewritten in the following prediction type form:

$$\widehat{\eta}_n(f_n) = \mathbb{E}[f_n(\widehat{X}_n) \prod_{0 \leq k < n} \widehat{G}_k(\widehat{X}_k)]/\mathbb{E}[\prod_{0 \leq k < n} \widehat{G}_k(\widehat{X}_k)]. \tag{25.8}$$

In the above display, the signal $\widehat{X}_n$ is a Markov chain with initial distribution $\widehat{\eta}_0$ and the elementary transitions

$$\widehat{M}_n(x_{n-1}, dx_n) = \frac{M_n(x_{n-1}, dx_n) G_n(x_n)}{M_n(G_n)(x_{n-1})} \quad \text{and} \quad \widehat{G}_n(x_n) = M_{n+1}(G_{n+1})(x_n) \tag{25.9}$$

This simple observation shows that all the analysis on the one-step predictor flow remains valid without further work to study the optimal filter. Another simple calculation shows that the unnormalized flow can be computed in terms of the normalized distributions. More precisely, we easily deduce the following multiplicative formulae:

$$\gamma_n(f) = \eta_n(f) \prod_{p=0}^{n-1} \eta_p(G_p) \tag{25.10}$$

25.2.2 Path-space and related filtering models

Firstly, it is important to notice that the abstract Feynman–Kac formulation presented in section 25.2.1 is particularly useful for describing Markov motions on path spaces. For instance, X_n may represent the historical process

$$X_n = (X'_0, \ldots, X'_n) \in E_n = (E'_0 \times \ldots \times E'_n) \tag{25.11}$$

associated with an auxiliary Markov chain X'_n which takes values in some measurable state spaces E'_n. In this situation, we have that

$$\widehat{\eta}_n = \text{Law}\left((X'_0, \ldots, X'_n) \mid Y_p = y_p, \ 0 \leq p \leq n\right)$$

As we shall see, this apparently innocent observation is essential for modelling and analysing genealogical evolution processes.

Another important state space enlargement allowed by our abstract formulation is the following. We let p_n be an increasing sequence of integers such that $p_0 = 0$, and we consider the pair signal observation model (X_n, Y_n) is given by

$$X_n = (X'_q)_{p_n \leq q < p_{n+1}} \in E_n := \prod_{p_n \leq q < p_{n+1}} E'_q \text{ and } Y_n = (Y'_q)_{p_n \leq q < p_{n+1}} \in F_n := \prod_{p_n \leq q < p_{n+1}} F'_q$$

We further assume that the auxiliary pair signal observation model (X'_n, Y'_n) is defined as in (25.1) with a pair transition-likelihood function (M'_n, g'_n). In this situation, if we choose in (25.6) the multiplicative likelihood potential functions

$$G_n(X_n) = \prod_{p_n \leq q < p_{n+1}} g'_q(x'_q, y'_q)$$

then we have that

$$\widehat{\eta}_n = \text{Law}((X'_q)_{0 \leq q \leq p_{n+1}} \mid Y'_q = y'_q, \ 0 \leq q < p_{n+1}) \tag{25.12}$$

As we shall see in the further development of section 25.2.3, one advantage of the mean field particle interpretation of this model comes from the fact that it only updates the sampled path predictions at the chosen times p_n. The "optimal" choice of the updating times depend on the filtering problem. For instance, in radar processing we current observation measures a noisy distance to the target. Thus, the speed and acceleration components are observable only after a series of three observations.

The traditional nonlinear filtering problem in continuous time is again defined in terms of a pair signal/observation Markov process (S_t, Y_t) taking values in $\mathbb{R}^{d+d'}$. The signal S_t is given by a time homogeneous Markov process with right continuous and left limited paths taking values in some Polish space E, and the observation process is an $\mathbb{R}^{d'}$-valued process defined by

$$dY_t = h_t(S_t)\,dt + \sigma\,dV_t$$

where V_t is a d'-vector standard Wiener process independent of the signal, and h_t is a bounded measurable function from E into $\mathbb{R}^{d'}$. The Kallianpur–Striebel formula (see for instance [38]) states that there exists a reference probability measure $\mathbb{P}_0$ under which the signal and the observations are independent. In addition, for any measurable function f_t on the space $D([0, t], \mathbb{R}^d)$ of $\mathbb{R}^d$-valued càdlàg paths from 0 to t, we have that

$$\mathbb{E}(f_t((S_s)_{s \leq t}) \mid \mathcal{Y}_t) = \frac{\mathbb{E}_0(f_t((S_s)_{s \leq t}) Z_t(S, Y) \mid \mathcal{Y}_t)}{\mathbb{E}_0(Z_t(S, Y) \mid \mathcal{Y}_t)} \qquad (25.13)$$

where $\mathcal{Y}_t = \sigma(Y_s, s \leq t)$ represents the sigma-field generated by the observation process and

$$\log Z_t(S, Y) = \int_0^t H_s^\star(S_s) \, dY_s - \int_0^t H_s^\star(S_s) H_s(S_s) \, ds$$

In the above definition $(.)^\star$ stands for the transposition operator. We let t_n, $n \geq 0$, be a given time mesh with $t_0 = 0$ and $t_n \leq t_{n+1}$. Also let X_n' be the sequence of random variables defined by

$$X_n' = S_{[t_n, t_{n+1}]}$$

By construction, X_n' is a nonhomogeneous Markov chain taking values at each time n in the space $E_n' = D([t_n, t_{n+1}], \mathbb{R}^d)$. From previous considerations, the observation process Y_t can be regarded as a random environment. Given the observation path, we define the "random" potential functions G_n on $\mathbf{E}_n = (E_0' \times \ldots \times E_n')$ by setting for any $x_n = (x_0', \ldots, x_n')$, with $x_p' = (x_p'(s))_{t_p \leq s \leq t_{p+1}} \in E_p'$, and $0 \leq p \leq n$

$$G_n(x_n) = G_n'(x_n') := \exp\left(\int_{t_n}^{t_{n+1}} H_s^\star(x_n'(s)) \, dY_s - \int_{t_n}^{t_n} H_s^\star(x_n'(s)) H_s(x_n'(s)) \, ds\right)$$

By construction, we can check that the quenched Feynman–Kac path measures (25.6) associated with the pair (X_n, G_n) coincide with the Kallianpur–Striebel representation and by (25.13) we prove that

$$\widehat{\eta}_n = \operatorname{Law}\left(S_{[t_0, t_1]}, \ldots, S_{[t_n, t_{n+1}]} \mid \mathcal{Y}_{t_n}\right) \qquad (25.14)$$

Next, we suppose that the observations are only delivered by the sensors at some fixed times t_n, $n \geq 0$, with $t_n \leq t_{n+1}$. To analyse this situation, we first notice that

$$(Y_{t_{n+1}} - Y_{t_n}) = \int_{t_n}^{t_{n+1}} H_s(S_s) \, ds + \sigma\left(V_{t_{n+1}} - V_{t_n}\right)$$

and $\sigma\left(V_{t_{n+1}} - V_{t_n}\right)$ are independent and random variables with Gaussian density q_n. Arguing as before and using the same notation as there, we introduce the "random" potential functions G_n defined by

$$G_n(x_n) = q_n \left(Y_{t_{n+1}} - Y_{t_n} - \int_{t_n}^{t_{n+1}} H_s\left(x'_n(s)\right) ds \right)$$

By construction, the quenched Feynman–Kac path measures (25.6) associated with the pair (X_n, G_n) now have the following interpretation

$$\widehat{\eta}_n = \text{Law}\left(S_{[t_0,t_1]}, \ldots, S_{[t_n,t_{n+1}]} \mid Y_{t_1}, \ldots, Y_{t_n} \right) \qquad (25.15)$$

We end this section with a couple of remarks.

To compute the Feynman–Kac distributions (25.14) and (25.15), it is tempting to use directly the mean field particle approximation model introduced in (25.3). Unfortunately, in practice the signal semigroup and the above integrals are generally not known exactly and another level of approximation is therefore needed. As shown in [22], the use of Euler approximation schemes introduces a deterministic bias in the fluctuation of the particle measures, but the resulting approximation models can be analysed using the same perturbation analysis as the one we shall describe in section 25.4.1 (see also [5, 11], for a more thorough analysis of the discrete-time schemes and genetic type particle approximations).

To solve the continuous-time problem, another strategy is to use a fully continuous time particle approximation model of the robust equations. Loosely speaking, the geometric acceptance rates of the discrete generation particle models are replaced by exponential clocks with an appropriate stochastic intensity dictated by the log-likelihood potential functions. For an introduction to interacting particle interpretation of continuous time Feynman–Kac models we recommend the review article on genetic-type models [25] as well as the series of articles [27, 28, 29].

25.2.3 McKean interpretations

By the Markov property and the multiplicative structure of (25.6), it is easily checked that the flow of measures $(\eta_n)_{n \geq 0}$ satisfies the following equation

$$\eta_{n+1} = \Phi_n(\eta_{n-1}) := \Psi_n(\eta_n) M_{n+1} \qquad (25.16)$$

with the updating Bayes or the Boltzmann–Gibbs transformation $\Psi_n : \mathcal{P}(E_n) \to \mathcal{P}(E_n)$ defined by

$$\Psi_n(\eta_n)(dx_n) := \frac{1}{\eta_n(G_n)} G_n(x_n)\, \eta_n(dx_n)\, (= \widehat{\eta}_n(dx_n)).$$

The mean field particle approximation of the flow (25.16) depends on the choice of the McKean interpretation model. As we mentioned in the introduction, these stochastic models amount to choosing a suitably defined Markov chain $\overline{X}_n$ with the prescribed evolution (25.16) of the laws of its states. More formally, these probabilistic interpretations consist of a chosen collection of Markov transitions K_{n+1,η_n}, indexed by the time parameter n and the set of probability measures $\eta \in \mathcal{P}(E_n)$, and satisfying the compatibility condition

$$\Phi_{n+1}(\eta) = \eta K_{n+1,\eta} \qquad (25.17)$$

The choice of these collections is not unique. We can choose, for instance the composition transition operator

$$K_{n+1,\eta_n} = S_{n,\eta_n} M_{n+1}$$

with the updating Markov transition S_{n,η_n} from E_n into itself defined by the following formula

$$S_{n,\eta_n}(x_n, dy_n) = \epsilon_n(\eta_n) \, G_n(x_n) \, \delta_{x_n}(dy_n) + (1 - \epsilon_n(\eta_n) \, G_n(x_n)) \, \Psi_n(\eta_n)(dy_n) \qquad (25.18)$$

In the above display, $\epsilon_n(\eta_n)$ represents *any* possibly null constant that may depend on the current distribution η_n, and such that $\epsilon_n(\eta_n)\|G_n\| \leq 1$. For instance, we can choose $1/\epsilon_n(\eta_n) = \eta_n - \text{ess} - \sup G_n$. The corresponding nonlinear transport equation

$$\eta_{n+1} = \eta_n K_{n+1,\eta_n}$$

can be interpreted as the evolution of the laws η_n of the states of a Markov chain $\overline{X}_n$ whose elementary transitions K_{n+1,η_n} depend on the law of the current state; that is, we have

$$\mathbb{P}(\overline{X}_{n+1} \in dx_{n+1} \mid \overline{X}_n = x_n) = K_{n+1,\eta_n}(x_n, dx_{n+1}) \quad \text{with} \quad \mathbb{P} \circ \overline{X}_n^{-1} = \eta_n \qquad (25.19)$$

The law $\mathbb{K}_n$ of the random path $(\overline{X}_p)_{0 \leq p \leq n}$ is called the McKean measure associated with the Markov transitions $(K_{n,\eta})_{n \geq 0, \eta \in \mathcal{P}(E_n)}$ and the initial distribution η_0. This measure on path space is explicitly defined by the following formula

$$\mathbb{K}_n(d(x_0, \ldots, x_n)) = \eta_0(dx_0) \, K_{1,\eta_0}(x_0, dx_1) \, \ldots \, K_{n,\eta_{n-1}}(x_{n-1}, dx_n)$$

25.2.4 Mean field particle and genealogical tree-based models

The N-particle model associated with a given collection of Markov transitions satisfying the compatibility condition (25.17) is the Markov chain introduced in (25.2). By the definition of the updating transitions (25.18), it appears that the mean field interacting particle model (25.2) is the combination of simple selection/mutation genetic transitions. The selection stage consists of N randomly evolving path-particles $\xi_{n-1}^i \rightsquigarrow \widehat{\xi}_{n-1}^i$ according to the update transition $S_{n,\eta_{n-1}^N}(\xi_{n-1}^i, \cdot)$. In other words, with probability $\epsilon_{n-1}(\eta_{n-1}^N) \, G_{n-1}(\xi_{n-1}^i)$, we set $\widehat{\xi}_{n-1}^i = \xi_{n-1}^i$; otherwise, the particle jumps to a new location, randomly drawn from the discrete distribution $\Psi_{n-1}(\eta_{n-1}^N)$. During the mutation stage, each of the selected particles $\widehat{\xi}_{n-1}^i \rightsquigarrow \xi_n^i$ evolves according to the Markov transition M_n.

For any sufficiently regular transitions $K_{n,\eta_{n-1}}$ satisfying the compatibility condition (25.17), we can prove that for any time horizon n

$$\mathbb{K}_n^N := \frac{1}{N} \sum_{i=1}^{N} \delta_{(\xi_0^i,\ldots,\xi_n^i)} \xrightarrow{N\to\infty} \mathbb{K}_n \quad \text{and} \quad \eta_n^N := \frac{1}{N} \sum_{i=1}^{N} \delta_{\xi_n^i} \xrightarrow{N\to\infty} \eta_n$$

Mimicking formula (25.10), we also construct *an unbiased* estimate for the unnormalized model and we have that

$$\gamma_n^N(\cdot) := \eta_n^N(\cdot) \prod_{p=0}^{n-1} \eta_p^N(G_p) \xrightarrow{N\to\infty} \gamma_n(\cdot) = \eta_n(\cdot) \prod_{p=0}^{n-1} \eta_p(G_p)$$

The convergence above can be understood in various ways. A variety of estimates going from the traditional $\mathbb{L}_p$-mean error bounds and exponential inequalities to fluctuation and large deviation theorems are provided in section 25.4 dedicated to the asymptotic behavior of these particle measures.

If we interpret the selection transition as a birth and death process, then arises the important notion of the ancestral line of a current individual. More precisely, when a particle $\widehat{\xi}_{n-1}^i \longrightarrow \xi_n^i$ evolves to a new location ξ_n^i, we can interpret $\widehat{\xi}_{n-1}^i$ as the parent of ξ_n^i. Looking backwards in time and recalling that the particle $\widehat{\xi}_{n-1}^i$ has selected a site ξ_{n-1}^j in the configuration at time $(n-1)$, we can interpret this site ξ_{n-1}^j as the parent of $\widehat{\xi}_{n-1}^i$ and therefore as the ancestor $\xi_{n-1,n}^i$ at level $(n-1)$ of ξ_n^i. Running back in time we trace mentally the whole ancestral line of each current individual:

$$\xi_{0,n}^i \longleftarrow \xi_{1,n}^i \longleftarrow \cdots \longleftarrow \xi_{n-1,n}^i \longleftarrow \xi_{n,n}^i = \xi_n^i$$

If we consider the historical process formulation (25.11) of a given signal model X_n', then the mean field particle model associated with the Feynman–Kac measures on path spaces consists of N path particles evolving according to the same selection/mutation transitions. It is rather clear that the resulting path particle model can also be interpreted as the evolution of a genealogical tree model. This shows that the occupation measures of the corresponding N-genealogical tree model

$$\eta_n^N = \frac{1}{N} \sum_{i=1}^{N} \delta_{(\xi_{0,n}^i, \xi_{1,n}^i, \ldots, \xi_{n,n}^i)}$$

converge as $N \to \infty$ to the Feynman–Kac path measures η_n defined as in (25.6) with the pair of mathematical objects (X_n, G_n) given by

$$X_n = (X_0', \ldots, X_n') \quad \text{and} \quad G_n(X_n) = g_n(X_n', Y_n)$$

From previous considerations, we already mention that the mathematical techniques developed to study the convergence of the particle measures η_N^N apply directly without further work to analyse the asymptotic behavior of this class of genealogical tree particle models. The occupation measures $\mathbb{K}_n^N$ of the complete genealogical tree $(\xi_0^i, \ldots, \xi_n^i)_{1 \leq i \leq N}$ keep track of the whole descendant history of the initial population of individuals ξ_0. The asymptotic analysis of these particle measures strongly depends on the choice of the McKean interpretation model and it requires more attention.

Last but not least, the mean filed particle methodology we have developed applies directly to solve the path-space filtering models we have presented in section 25.2.2. We leave the interested reader to write down the interacting particle systems in each situation. We also mention that the prediction formulation of the optimal filter described in (25.8) leads to a genetic type particle scheme with a pair mutation-selection transition associated with the pair transition-potential function $(\widehat{M}_n, \widehat{G}_n)$ given in (25.9). This strategy is based on observation depended explorations and the resulting stochastic grid is often more accurate than the one based on free mutations. Nevertheless, expect in some particular situations the sampling of a transition $\widehat{\xi}_{n-1}^i \rightsquigarrow \xi_n^i$ according to the distribution $\widehat{M}_n\left(\widehat{\xi}_{n-1}^i, dx_n\right)$ is often difficult. One idea is to introduce an auxiliary empirical particle approximation

$$M_n^{N'}\left(\widehat{\xi}_{n-1}^i, dx_n\right) := \frac{1}{N'} \sum_{j=1}^{N'} \delta_{\zeta_n^{i,j}}$$

of the Markov transition $M_n\left(\widehat{\xi}_{n-1}^i, dx_n\right)$, and based on sampling N'-independent random transitions $\widehat{\xi}_{n-1}^i \rightsquigarrow \zeta_n^{i,j}$, $1 \leq k \leq N'$ with common distribution $M_n\left(\widehat{\xi}_{n-1}^i, dx_n\right)$. The second step is to replace the pair transition-potential function $(\widehat{M}_n, \widehat{G}_n)$ given in (25.9) by their N'-particle approximations $\left(\widehat{M}_n^{N'}, \widehat{G}_n^{N'}\right)$ defined by

$$\widehat{M}_n^{N'}\left(\widehat{\xi}_{n-1}^i, dx_n\right) := \frac{M_n^{N'}\left(\widehat{\xi}_{n-1}^i, dx_n\right) G_n(x_n)}{M_n^{N'}(G_n)\left(\widehat{\xi}_{n-1}^i\right)} \quad \text{and} \quad \widehat{G}_{n-1}^{N'}\left(\widehat{\xi}_{n-1}^i\right) := M_n^{N'}(G_n)\left(\widehat{\xi}_{n-1}^i\right)$$

These particle exploration models can be combined without further work with the path space enlargement techniques presented in (25.12). Roughly speaking, the corresponding path particle model is based on exploring the state space using the conditional transitions of signal sequences $(X_q)_{p_{n-1} \leq q < p_n} \rightsquigarrow (X_q)_{p_n \leq q < p_{n+1}}$ based on $(p_n - p_{n-1})$ observations data $(Y_q)_{p_n \leq q < p_{n+1}}$. The auxiliary N'-particle approximation described above only provides a practical way of sampling these explorations on an auxiliarly pool of N' sampled sequences.

These conditional particle exploration strategies were originally developed in [13, 21]. They have been used with success in filtering as well as in protein-folding problems under the botanical names "Markovian anticipation", "lookahead strategies," and "sampling importance sampling pilot exploration resampling". For further details and precise references, we also refer the interested reader to chapter 11 of book [11] for further stochastic particle recipes including a variety of branching type selection variants.

Another tuning parameter is to change the reference measure of the signal process. Instead of sampling according to $\widehat{M}_n$ or M_n, we sample according to some Markov transition, say $\widetilde{M}_n$, then we correct this choice by using the new potential function in transition space

$$\widetilde{G}_n(x_{n-1}, x_n) := \frac{d\,M_n(x_{n-1}, \cdot)}{d\,\widetilde{M}_n(x_{n-1}, \cdot)}(x_n)\,G_n(x_n)$$

25.3 Stability analysis

This section is concerned with the regularity and the stability properties of the evolution semigroup of the measure valued process introduced in (25.16). This analysis is motivated by two essential problems. From the signal processing perspective, the first one is to ensure that the underlying filtering problem is well posed, in the sense that it corrects automatically any erroneous initial data. This stability property is fundamental in most of the filtering problems encountered in practical situations where the initial distribution of the signal is unknown. The second motivation is well-known in the numerical analysis of dynamical systems. Indeed, it is generally useless to approximate an unstable and chaotic dynamical system that propagates any local perturbation. In our context, the mean field particle model can be interpreted as a stochastic perturbation of the nonlinear dynamical system (25.16). In this situation, the regularity properties of the evolution semigroup ensure that these local perturbations will not propagate. These ideas will be clarified in Section 25.4.

25.3.1 Feynman–Kac evolution semigroups

We let $Q_{p,n}$, with $0 \le p \le n$, be the Feynman–Kac semigroup associated with the flow of unnormalized Feynman–Kac measures $\gamma_n = \gamma_p Q_{p,n}$ defined in (25.7). For $p = n$, we use the convention that $Q_{n,n} = Id$, the identity operator. Using the Markov property, it is not difficult to check that $Q_{p,n}$ has the following functional representation

$$Q_{p,n}(f_n)(x_p) = \mathbb{E}\left[f_n(X_n)\,\prod_{p \le k < n} G_k(X_k)\,\mid\,X_p = x_p\right] \qquad (25.20)$$

for any test function $f_n \in \mathcal{B}(E_n)$, and any state $x_p \in E_p$. We denote by $\Phi_{p,n}$, $0 \le p \le n$, the nonlinear semigroup associated with the normalized measures η_n introduced in (25.6)

$$\Phi_{p,n} = \Phi_n \circ \Phi_{n-1} \circ \ldots \circ \Phi_{p+1} \qquad (25.21)$$

As usual we use the convention $\Phi_{n,n} = \mathrm{Id}$, for $p = n$. It is important to observe that this semigroup is alternatively defined by the formulae

$$\Phi_{p,n}(\eta_p)(f_n) = \frac{\eta_p(Q_{p,n}(f_n))}{\eta_p(Q_{p,n}(1))} = \frac{\eta_p(G_{p,n}\, P_{p,n}(f_n))}{\eta_p(G_{p,n})}$$

with the pair potential and Markov transition $(G_{p,n}, P_{p,n})$ defined by

$$G_{p,n} = Q_{p,n}(1) \quad \text{and} \quad P_{p,n}(f_n) = Q_{p,n}(f_n)/Q_{p,n}(1)$$

The next two parameters

$$r_{p,n} = \sup_{x_p, z_p \in E_p} (G_{p,n}(x_p)/G_{p,n}(z_p)) \quad \text{and} \quad \beta(P_{p,n}) = \sup_{x_p, z_p \in E_p} \| P_{p,n}(x_p, \cdot) - P_{p,n}(z_p, \cdot) \|_{\mathrm{tv}}$$

(25.22)

measure respectively the relative oscillations of the potential functions $G_{p,n}$ and the contraction properties of the Markov transition $P_{p,n}$. Various asymptotic estimates on particle models derived in the forthcoming sections will be expressed in terms of these parameters. For instance and for further use in several places in this chapter, we have the following Lipschitz regularity property.

Proposition 25.1 *For any $f_n \in \mathrm{Osc}_1(E_n)$ we have*

$$\left| \left[\Phi_{p,n}(\eta_p) - \Phi_{p,n}(\mu_p) \right] (f_n) \right| \leq 2\, r_{p,n}\, \beta(P_{p,n})\ \left| [\eta_p - \mu_p] \overline{P}_{p,n}^{\mu_p}(f_n) \right| \quad (25.23)$$

for some function $\overline{P}_{p,n}^{\mu_p}(f_n) \in \mathrm{Osc}_1(E_p)$ that doesn't depends on the measure η_p.

Proof. We check this inequality using the key decomposition

$$\left[\Phi_{p,n}(\eta_p) - \Phi_{p,n}(\mu_p) \right] (f_n) = \frac{1}{\eta_p(G_{p,n}^{\mu_p})} \left[(\eta_p - \mu_p) R_{p,n}^{\mu_p}(f_n) \right]$$

with the function $G_{p,n}^{\mu_p} := G_{p,n}/\mu_p(G_{p,n})$ and the integral operator $R_{p,n}^{\mu_p}$ from $\mathcal{B}(E_n)$ into $\mathcal{B}(E_p)$ defined below:

$$R_{p,n}^{\mu_p}(f_n)(x_p) := G_{p,n}^{\mu_p}(x_p) \times P_{p,n} \left[f_n - \Phi_{p,n}(\mu_p)(f_n) \right](x_p)$$

Recalling that $\mathrm{osc}(P_{p,n}(f_n)) \leq \beta(P_{p,n})\, \mathrm{osc}(f_n)$, one readily check that (25.23) is satisfied with the integral operator $\overline{P}_{p,n}^{\mu_p}$ from $\mathcal{B}(E_n)$ into $\mathcal{B}(E_p)$ defined

$$\overline{P}_{p,n}^{\mu_p}(f_n)(x_p) := \frac{1}{2}\, \frac{1}{r_{p,n}}\, \frac{G_{p,n}(x_p)}{\inf G_{p,n}} \int \frac{1}{\beta(P_{p,n})} \left[P_{p,n}(f_n)(x_p) - P_{p,n}(f_n)(y_p) \right]$$
$$\times G_{p,n}^{\mu_p}(y_p)\, \mu_p(dy_p)$$

$\square$

25.3.2 Contraction properties

In this section, we present an abstract class of H-entropy like criteria. We also provide a brief introduction to the Lipschitz contraction properties of Markov integral operators. For a more detailed discussion on this subject, we refer the reader to Chapter 4 of book [11], and to a joint work of the first author with Michel Ledoux and Laurent Miclo [24]. The contraction functional inequalities described in this section are extended to the nonlinear Feynman–Kac semigroup introduced in (25.21).

Let $h : \mathbb{R}_+^2 \to \mathbb{R} \cup \{\infty\}$ be a convex function satisfying for any $a, x, y \in \mathbb{R}_+$ such that $h(ax, ay) = ah(x, y)$, and $h(1, 1) = 0$. We associate with this homogeneous function the H-relative entropy on $\mathcal{M}_+(E)$ defined symbolically as

$$H(\mu, \nu) = \int h\,(d\mu, d\nu)$$

By homogeneity arguments, the above entropy is better defined in terms of any measure $\lambda \in \mathcal{M}(E)$ dominating μ and ν by the formula $H(\mu, \nu) = \int h\left(\frac{d\mu}{d\lambda}, \frac{d\nu}{d\lambda}\right) d\lambda$. To illustrate this rather abstract definition, we provide hereafter a collection of classical h-relative entropies arising in the literature. Before to proceed, we first come back to the definition of h-entropy, and we denote by $h' : \mathbb{R}_+ \to \mathbb{R} \sqcup \{+\infty\}$ the convex function given for any $x \in \mathbb{R}_+$ by $h'(x) = h(x, 1)$. By homogeneity arguments, we note that h is almost equivalent to h'. More precisely, only the specification of the value $h(1, 0)$ is missing. In most applications, the natural convention is $h(1, 0) = \infty$. The next lemma connects the h-relative entropy with the h'-divergence in the sense of Csiszár [10].

Lemma 25.1 ([11]) *Assume that $h(1, 0) = +\infty$. Then, for any μ and $\nu \in \mathcal{M}_+(E)$, we have*

$$H(\mu, \nu) = \int h'\left(\frac{d\mu}{d\nu}\right) d\nu \quad \text{if } \mu \ll \nu, \text{ and } H(\mu, \nu) = \infty \text{ otherwise.} \quad (25.24)$$

As we promised above, let us present some traditional H-entropies commonly used in probability theory. If we take $h'(t) = |t - 1|^p$, $p \geq 1$, we find the $\mathbb{L}_p$-norm given for any $\mu, \nu \in \mathcal{P}(E)$ by $H(\mu, \nu) = \|1 - d\mu/d\nu\|_{p,\nu}^p$ if $\mu \ll \nu$, and ∞ otherwise. The case $h'(t) = t \log(t)$ corresponds to the Boltzmann entropy or Shannon–Kullback information given by the formula $H(\mu, \nu) = \int \ln\left(\frac{d\mu}{d\nu}\right) d\mu$, if $\mu \ll \nu$ and ∞ otherwise. The Havrda–Charvat entropy of order $p > 1$ corresponds to the choice $h'(t) = \frac{1}{p-1}(t^p - 1)$; that is we have for any $\mu \ll \nu$, $H(\mu, \nu) = \frac{1}{p-1}\left[\int \left(\frac{d\mu}{d\nu}\right)^p d\nu - 1\right]$. The Hellinger and Kakutani–Hellinger integrals of order $\alpha \in (0, 1)$ correspond to the choice $h'(t) = t - t^\alpha$, and with some obvious abusive notation we have $H(\mu, \nu) = 1 - \int (d\mu)^\alpha (d\nu)^{1-\alpha}$. In the special case $\alpha = 1/2$, this relative entropy coincides with the Kakutani–Hellinger distance; and finally, the case $h'(t) = |t - 1|/2$ corresponds to the total variation distance. In the study of regularity properties of $\Phi_{p,n}$, the following notion will play a major role.

Definition 25.1 Let $(E, \mathcal{E})$ and $(F, \mathcal{F})$ be a pair of measurable spaces. We consider an h-relative entropy criterion H on the sets $\mathcal{P}(E)$ and $\mathcal{P}(F)$. The contraction or Lipschitz coefficient $\beta_H(\Phi) \in \mathbb{R}_+ \cup \{\infty\}$ of a mapping $\Phi : \mathcal{P}(E) \to \mathcal{P}(F)$ with respect to H is the best constant such that for any pair of measures $\mu, \nu \in \mathcal{P}(E)$ we have

$$H(\Phi(\mu), \Phi(\nu)) \leq \beta_H(\Phi) \, H(\mu, \nu)$$

When H represents the total variation distance, we simplify notation and sometimes we write $\beta(\Phi)$ instead of $\beta_H(\Phi)$.

When H is the total variation distance, the parameter $\beta(\Phi)$ coincides with the traditional notion of a Lipschitz constant of a mapping between two metric spaces. In [24], we prove the following regularity property. This functional inequality will be pivotal in the contraction analysis of Feynman–Kac semigroups developed in section 25.3.3.

Theorem 25.1 *For any pair of probability measures μ and $\nu \in \mathcal{P}(E)$ and for any Markov kernel M from E into another measurable space F, we have the contraction estimate*

$$H(\mu M, \nu M) \leq \beta(M) \, H(\mu, \nu) \tag{25.25}$$

25.3.3 Functional inequalities

The main objective of this section will be to estimate the contraction coefficients $\beta_H(\Phi_{p,n})$ of the nonlinear Feynman–Kac transformations $\Phi_{p,n}$ in terms of the Dobrushin coefficient $\beta(P_{p,n})$ and the relative oscillations of the potential functions $G_{p,n}$.

To describe precisely these functional inequalities precisely, it is convenient to introduce some additional notation. When H is the h'-divergence associated with a differentiable $h' \in C^1(\mathbb{R}_+)$, we denote by Δh the function on $\mathbb{R}_+^2$ defined by

$$\Delta h(t, s) = h'(t) - h'(s) - \partial h'(s)(t - s) \quad (\geq 0)$$

where $\partial h'(s)$ stands for the derivative of h' at $s \in \mathbb{R}_+$. We further assume that the following growth condition is satisfied

$$\forall (r, s, t) \in \mathbb{R}_+^3 \text{ we have } \Delta h(rt, s) \leq a(r) \, \Delta h(t, \theta(r, s)) \tag{25.26}$$

for some nondecreasing function a on $\mathbb{R}_+$ and some mapping θ on $\mathbb{R}_+^2$ such that $\theta(r, \mathbb{R}_+) = \mathbb{R}_+$, for any $r \in \mathbb{R}_+$.

For instance, the Boltzmann entropy corresponds to the situation where $h'(t) = t \log t$. In this case, (25.26) is met with $a(r) = r$ and $\theta(r, s) = s/r$. For the Havrda–Charvat entropy of order $\alpha > 1$, we have $h'(t) = (t^\alpha - 1)/(\alpha - 1)$, and the growth condition (25.26) is now met with $a(r) = r^\alpha$ and $\theta(r, s) = s/r$. The

Hellinger integrals of order $\alpha \in (0, 1)$ correspond to the choice $h'(t) = t - t^\alpha$, and the growth condition is met with the same parameters. The $\mathbb{L}_2$-relative entropy corresponds to the case $h'(t) = (t-1)^2$, and we find that (25.26) is met with $a(r) = r^2$, and again $\theta(r, s) = s/r$.

The following theorem is a slightly weaker version of a theorem proved in [11].

Theorem 25.2 *For any $0 \leq p \leq n$ and any pair of measures $\mu_p, \nu_p \in \mathcal{P}(E_p)$, we have the Lipschitz contraction inequality*

$$\|\Phi_{p,n}(\mu_p) - \Phi_{p,n}(\nu_p)\|_{tv} \leq r_{p,n}\, \beta(P_{p,n})\, \|\mu_p - \nu_p\|_{tv} \qquad (25.27)$$

In addition, if we set $a_H(r) := r\,a(r)$ then we have

$$\beta_H(\Phi_{p,n}) \leq a_H(r_{p,n})\, \beta(P_{p,n}) \quad \text{and} \quad \beta(P_{p,n}) = \sup_{\mu_p, \nu_p \in \mathcal{P}(E_p)} \|\Phi_{p,n}(\mu_p) - \Phi_{p,n}(\nu_p)\|_{tv}$$
(25.28)

Our next objective is to estimate the the contraction coefficient $\beta_H(\Phi_{p,n})$ in terms of the mixing type properties of the semigroup $M_{p,n}(x_p, dx_n) := M_{p+1} M_{p+2} \ldots M_n(x_p, dx_n)$ associated with the Markov operators M_n. We introduce the following regularity condition.

$(M)_m$ There exists an integer $m \geq 1$ and a sequence $(\epsilon_p(M))_{p \geq 0} \in (0, 1)^{\mathbb{N}}$ such that

$$\forall p \geq 0 \quad \forall (x_p, x'_p) \in E_p^2 \quad M_{p, p+m}(x_p, \cdot) \geq \epsilon_p(M)\, M_{p, p+m}(x'_p, \cdot)$$

It is well known that the above condition is satisfied for any aperiodic and irreducible Markov chains on finite spaces. Loosely speaking, for noncompact spaces this condition is related to the tails of the transition distributions on the boundaries of the state space. For instance, let us suppose that $E_n = \mathbb{R}$ and M_n is the bi-Laplace transition given by

$$M_n(x, dy) = \frac{c(n)}{2} e^{-c(n)|y - A_n(x)|}\, dy$$

for some $c(n) > 0$ and some drift function A_n with bounded oscillations $\operatorname{osc}(A_n) < \infty$. In this case, it is readily checked that condition $(M)_m$ holds true for $m = 1$ with the parameter $\epsilon_{n-1}(M) = \exp(-c(n)\,\operatorname{osc}(A_n))$.

Under the mixing type condition $(M)_m$ we have for any $n \geq m \geq 1$, and $p \geq 1$

$$r_{p, p+n} \leq \epsilon_p(M)^{-1} \prod_{0 \leq k < m} \epsilon_{p+k}(G)^{-1} \quad \text{and} \quad \beta(P_{p, p+n}) \leq \prod_{k=0}^{\lfloor n/m \rfloor - 1} \left(1 - \epsilon_{p+km}^{(m)}(G, M)\right)$$

with the sequence of parameters $\epsilon_p^{(m)}(G, M)$ given by

$$\epsilon_p^{(m)}(G, M) = \epsilon_p^2(M) \prod_{p+1 \leq k < p+m} \epsilon_k(G)$$

We recall that the sequence of parameters $\epsilon_n(G)$ is defined in (25.5). Several contraction inequalities can be deduced from these estimates (see for instance Chapter 4 of book [11]). To give a flavor of these results, we further assume that $(M)_m$ is satisfied with $m = 1$, and we have $\epsilon(M) = \inf_n \epsilon_n(M) > 0$. In this case, we can check that

$$r_{p,p+n} \leq \left(\epsilon(M)\epsilon_p(G)\right)^{-1} \quad \text{and} \quad \beta(P_{p,p+n}) \leq \left(1 - \epsilon(M)^2\right)^n$$

By (25.28) we conclude that

$$\beta_H(\Phi_{p,p+n}) \leq a_H((\epsilon(M)\epsilon_p(G))^{-1}) \left(1 - \epsilon(M)^2\right)^n$$

In addition, using the same line of reasoning we also prove the following potential free estimates

$$\beta_H(\widehat{\Phi}_{p,p+n}) \leq a_H(\epsilon^{-1}(M)) \left(1 - \epsilon^2(M)\right)^n$$

Another strategy consists in combining the Markov contraction inequality (25.25) with some entropy inequalities obtained by Ocone in [41] (see also [4]). Using this strategy, we proved in [11] the following annealed contraction inequalities.

Theorem 25.3 *Let η'_n and $\widehat{\eta}'_n := \Psi_n(\eta'_n)$ be an auxiliary model defined with the same random equation as the pair η_n and $\widehat{\eta}_n = \Psi_n(\eta_n)$, but starting at some possibly different $\eta'_0 \in \mathcal{P}(E_0)$. For any $n \in \mathbb{N}$, and any η'_0 we have*

$$\mathbb{E}\left(\text{Ent}\left(\widehat{\eta}_n \mid \widehat{\eta}'_n\right)\right) \leq \mathbb{E}\left(\text{Ent}\left(\eta_n \mid \eta'_n\right)\right) \leq \left[\prod_{p=1}^n \beta(M_p)\right] \text{Ent}\left(\eta_0 \mid \eta'_0\right) \quad (25.29)$$

25.4 Asymptotic analysis

This section is concerned with the convergence analysis of the particle approximation measures introduced in Section 25.2.4. In the first subsection, we propose a stochastic perturbation methodology that allows the stability of the limiting Feynman–Kac semigroup to enter into the convergence analysis of the particle models. The convergence of empirical processes including exponential estimates and $\mathbb{L}_p$-mean error bounds for the McKean particle measures $\mathbb{K}_n^N$ and their density profiles η_n^N are discussed in Section 25.4.2. Central limit theorems and large deviation principles are presented respectively in Section 25.4.3 and Section 25.4.4.

25.4.1 A stochastic perturbation model

We provide in this section a brief introduction to the asymptotic analysis of the particle approximation models (25.3) as the size of the systems and/or the time horizon tends to infinity. Firstly, we observe that the local sampling errors are expressed in terms of the random fields W_n^N, given for any $f_n \in \mathcal{B}(E_n)$ by the formula

$$W_n^N(f_n) := \sqrt{N}\,[\eta_n^N - \Phi_n(\eta_{n-1}^N)](f_n) = \frac{1}{\sqrt{N}} \sum_{i=1}^{N} [f_n(\xi_n^i) - K_{n,\eta_{n-1}^N}(f_n)(\xi_{n-1}^i)]$$

Rewritten in a slightly different way, we have the stochastic perturbation formulae

$$\eta_n^N = \Phi_n\left(\eta_{n-1}^N\right) + \frac{1}{\sqrt{N}}\, W_n^N$$

with the centered random fields W_n^N with conditional variance functions given by

$$\mathbb{E}(W_n^N(f_n)^2 | \xi_{n-1}.) = \eta_{n-1}^N [K_{n,\eta_{n-1}^N}((f_n - K_{n,\eta_{n-1}^N}(f_n))^2)] \qquad (25.30)$$

In Section 25.4.3 we shall see that the random fields $\left(W_n^N\right)_{n \geq 0}$ behave asymptotically as a sequence of independent Gaussian and centered random fields $(W_n)_{n \geq 0}$ with conditional variance functions given as in (25.30), by replacing the particle empirical measures η_n^N by their limiting values η_n. These fluctuation covariance functions depend on the choice of the McKean interpretation model. To underline the role of the parameter $\epsilon_n(\eta_n)$ given in (25.18), we observe that for any $\mu \in \mathcal{P}(E_{n-1})$ we have that

$$\mu[K_{n,\mu}((f_n - K_{n,\mu}(f_n))^2)]$$

$$= \Phi_n(\mu)[(f_n - \Phi_n(\mu)(f_n))^2] - \epsilon_{n-1}(\mu)^2\, \mu[[G_{n-1}(Id - \Psi_{n-1}(\mu))(M_n(f_n))]^2]$$

This formula shows that the simple genetic particle model associated with a null parameter $\epsilon_n(\mu) = 0$ is in this sense the less accurate. The following picture gives a sound basis to the main questions related to the convergence analysis of the mean field particle model.

$$\begin{array}{ccccccccc}
\eta_0 & \to & \eta_1 = \Phi_1(\eta_0) & \to & \eta_2 = \Phi_{0,2}(\eta_0) & \to & \cdots & \to & \Phi_{0,n}(\eta_0) \\
\Downarrow & & & & & & & & \\
\eta_0^N & \to & \Phi_1\left(\eta_0^N\right) & \to & \Phi_{0,2}\left(\eta_0^N\right) & \to & \cdots & \to & \Phi_{0,n}\left(\eta_0^N\right) \\
& & \Downarrow & & & & & & \\
& & \eta_1^N & \to & \Phi_2\left(\eta_1^N\right) & \to & \cdots & \to & \Phi_{1,n}\left(\eta_1^N\right) \\
& & & & \Downarrow & & & & \\
& & & & \eta_2^N & \to & \cdots & \to & \Phi_{2,n}\left(\eta_2^N\right) \\
& & & & & & \Downarrow & & \vdots \\
& & & & & & \eta_{n-1}^N & \to & \Phi_n\left(\eta_{n-1}^N\right) \\
& & & & & & & & \Downarrow \\
& & & & & & & & \eta_n^N
\end{array}$$

In the above display, the local perturbations W_n^N are represented by the implication sign "⇓". This picture yields the following pivotal formula

$$\eta_n^N - \eta_n = \sum_{q=0}^{n} \left[\Phi_{q,n}\left(\eta_q^N\right) - \Phi_{q,n}\left(\Phi_q\left(\eta_{q-1}^N\right)\right)\right]$$

with the convention $\Phi_0\left(\eta_{-1}^N\right) = \eta_0$ for $p = 0$.

Loosely speaking, a first-order development of the semigroup $\Phi_{q,n}$ around the measure $\Phi_q\left(\eta_{q-1}^N\right)$ shows that the "small errors" induced by local perturbations do not propagate, as soon as the semigroup $\Phi_{p,n}$ is sufficiently stable. Using the same line of arguments, the fluctuation analysis of the particle measures η_n^N around their limiting values η_n results from a second order approximation of the semigroup $\Phi_{p,n}$. These two observations will be made clear in section 25.4.2 and section 25.4.3, devoted respectively to non asymptotic $\mathbb{L}_p$-mean error bounds and to central limit theorems.

25.4.2 Convergence of empirical processes

Using the Lipschitz regularity property (25.23) we obtain the following first-order estimate:

$$|[\Phi_{q,n}(\eta_q^N) - \Phi_{q,n}(\Phi_q(\eta_{q-1}^N))](f_n)| \leq 2r_{q,n}\beta(P_{q,n}) \frac{1}{\sqrt{N}} \left|W_q^N\left(\overline{P}_{q,n}^{\Phi_q(\eta_{q-1}^N)}(f_n)\right)\right|$$

From these estimates and using the refined version of the Marcinkiewicz–Zygmund lemma presented in [11] (see lemma 7.3.3), we prove the following theorem.

Theorem 25.4 ([11]) *For any $n \geq 0$, $p \geq 1$, any tensor product function $F_n = (f_0 \otimes \ldots \otimes f_n)$, with any functions $f_q \in \mathrm{Osc}_1(E_q)$ for all $q \leq n$, we have*

$$\sup_{N \geq 1} \sqrt{N}\, \mathbb{E}(|[\mathbb{K}_n^N - \mathbb{K}_n](F_n)|^p)^{\frac{1}{p}} < \infty \quad (25.31)$$

$$\text{and} \quad \sqrt{N}\, \mathbb{E}(|[\eta_n^N - \eta_n](f_n)|^p)^{\frac{1}{p}} \leq c(n)\, d(p)^{1/p} \quad (25.32)$$

with $c(n) \leq 2 \sum_{q=0}^{n} r_{q,n}\beta(P_{q,n})$, and the sequence of parameters $d(p)$, with $p \geq 1$, given by

$$d(2p) = (2p)_p\, 2^{-p} \quad \text{and} \quad d(2p-1) = \frac{(2p-1)_p}{\sqrt{p-1/2}}\, 2^{-(p-1/2)}$$

The $\mathbb{L}_p$ mean error estimates in the right-hand side of (25.32) can be used to derive the following exponential estimate

$$\forall \epsilon > 0 \quad \mathbb{P}\left(|\eta_n^N(f_n) - \eta_n(f_n)| > \epsilon\right) \leq \left(1 + \epsilon\sqrt{N/2}\right) \exp\left(-\frac{N\epsilon^2}{2c(n)^2}\right) \quad (25.33)$$

We associate with a collection of measurable functions $f : E \to \mathbb{R}$, with $\|f\| \leq 1$, the Zolotarev seminorm on $\mathcal{P}(E)$ defined by

$$\|\mu - \nu\|_{\mathcal{F}} = \sup\{|\mu(f) - \nu(f)|;\ f \in \mathcal{F}\},$$

(see for instance [42]). To avoid some unnecessary technical measurability questions, we further suppose that $\mathcal{F}$ is separable in the sense that it contains a countable and dense subset. To measure the size of a given class $\mathcal{F}$, one considers the covering numbers $N(\epsilon, \mathcal{F}, L_p(\mu))$ defined as the minimal number of $L_p(\mu)$-balls of radius $\epsilon > 0$ needed to cover $\mathcal{F}$. By $\mathcal{N}(\epsilon, \mathcal{F})$, $\epsilon > 0$, and by $I(\mathcal{F})$ we denote the uniform covering numbers and entropy integral given by

$$\mathcal{N}(\epsilon, \mathcal{F}) = \sup\{N(\epsilon, \mathcal{F}, L_2(\eta));\ \eta \in \mathcal{P}(E)\} \quad \text{and} \quad I(\mathcal{F}) = \int_0^1 \sqrt{\log(1 + \mathcal{N}(\epsilon, \mathcal{F}))}\, d\epsilon$$

Various examples of classes of functions with finite covering and entropy integral are given in the book by Van der Vaart and Wellner [44]. The estimation of the quantities introduced above depends on several deep results on combinatorics that are not discussed here.

Let $\mathcal{F}_n$ be a countable collection of functions f_n with $\|f_n\| \leq 1$ and finite entropy $I(\mathcal{F}_n) < \infty$. Suppose that the Markov transitions M_n have the form $M_n(u, dv) = m_n(u, v)\, p_n(dv)$ for some measurable function m_n on $(E_{n-1} \times E_n)$ and some $p_n \in \mathcal{P}(E_n)$. Also assume that we have for some collection of mappings θ_n on E_n

$$\sup_{u \in E_{n-1}} |\log m_n(u, v)| \leq \theta_n(v) \quad \text{with} \quad p_n(e^{3\theta_n}) < \infty \qquad (25.34)$$

In this situation, we obtain the following fluctuation result.

Theorem 25.5 *For any $n \geq 0$ and $p \geq 1$ we have*

$$\sqrt{N}\, \mathbb{E}\left(\|\eta_n^N - \eta_n\|_{\mathcal{F}_n}^p\right)^{\frac{1}{p}} \leq a\, [p/2]!\, [I(\mathcal{F}_n) + c(n-1)\, r_{n-1,n}\, p_n(e^{3\theta_n})] \qquad (25.35)$$

with a collection of constants $a < \infty$ and $c(n) \leq \sum_{q=0}^n r_{q,n}\, \beta(P_{q,n})$.

It is important to observe that under the assumptions discussed in Section 25.3.3, it is readily checked that all of the above estimates can be turned into uniform convergence results.

25.4.3 Fluctuation analysis

The fluctuation analysis of the particle measures η_n^N around their limiting values η_n is essentially based on the asymptotic analysis of the local sampling errors W_n^N associated with the particle approximation sampling steps. The next central limit theorem for random fields is pivotal. Its complete proof can be found in [11] (see Theorem 9.3.1 and Corollary 9.3.1 on pages 295–98).

Theorem 25.6 *For any fixed time horizon $n \geq 1$, the sequence $(W_p^N)_{1 \leq p \leq n}$ converges in law, as N tends to infinity, to a sequence of n independent, Gaussian, and centred random fields $(W_p)_{1 \leq p \leq n}$; with, for any $f_p, g_p \in \mathcal{B}(E_p)$, and $1 \leq p \leq n$,*

$$\mathbb{E}(W_p(f_p) W_p(g_p)) = \eta_{p-1} K_{p,\eta_{p-1}}([f_p - K_{p,\eta_{p-1}}(f_p)][g_p - K_{p,\eta_{p-1}}(g_p)]) \quad (25.36)$$

Using the pair of decompositions

$$\gamma_p^N Q_{p,n} - \gamma_p Q_{p,n} = \sum_{q=0}^p \gamma_q^N(1) \left[\eta_q^N - \Phi_q(\eta_{q-1}^N)\right] Q_{q,n} = \frac{1}{\sqrt{N}} \sum_{q=0}^p \gamma_q^N(1) W_q^N Q_{q,n}$$

and

$$\left[\eta_n^N - \eta_n\right](f_n) = \frac{\gamma_n(1)}{\gamma_n^N(1)} \left[\gamma_n^N - \gamma_n\right]\left(\frac{1}{\gamma_n(1)}(f_n - \eta_n(f_n))\right)$$

we readily deduce the following corollary:

Corollary 25.1 *For any fixed time horizon $n \geq 1$, the sequence of random fields*

$$W_n^{N,\gamma} := \sqrt{N} \left[\gamma_n^N - \gamma_n\right] \quad \text{and} \quad W_n^{N,\eta} := \sqrt{N} \left[\eta_n^N - \eta_n\right]$$

converges in law, as N tends to infinity, to a sequence of n independent, Gaussian, and centred random fields W_n^γ and W_n^η; with, for any $f_n \in \mathcal{B}(E_n)$

$$W_n^\gamma(f_n) = \sum_{q=0}^n \gamma_q(1) W_q[Q_{q,n}(f_n)] \quad \text{and} \quad W_n^\eta(f_n) = W_n^\gamma\left(\frac{1}{\gamma_n(1)}(f_n - \eta_n(f_n))\right)$$

The random fields $W_n^{N,\gamma}$ and $W_n^{\eta,N}$ can also be regarded as empirical processes indexed by the collection of bounded measurable functions $\mathcal{F}_n \subset \mathcal{B}(E_n)$. If $\mathcal{F}_n$ is a countable collection of functions f_n, with $\|f_n\| \leq 1$ and $I(\mathcal{F}_n) < \infty$, then the $\mathcal{F}_n$-indexed process $W_n^N(f_n)$, $f_n \in \mathcal{F}_n$, is asymptotically tight (see for instance lemma 9.6.1 in [11]). In this situation, the random fields $W_n^{N,\gamma}$ and $W_n^{\eta,N}$ also converge in law in $l^\infty(\mathcal{F}_n)$ to the Gaussian processes W_n^γ and W_n^η.

The fluctuation analysis of the random fields $\mathbb{K}_n^N$ around the McKean measure $\mathbb{K}_n$ is technically much more involved. For completeness, we have chosen to present this result in a rather simple form. The complete proof can be found in section 9.7 of book [11]. We only examine the situation with a null acceptance rate $\epsilon_n(\mu) = 0$, and we further suppose that condition (25.34) is satisfied with $p_n(e^{r\theta_n})$, for any $r \geq 1$.

For any $x = (x_0, \ldots, x_n)$ and $z = (z_0, \ldots, z_n) \in \Omega_n := (E_0 \times \ldots \times E_n)$ we set

$$a_n(x,z) := b_n(x,z) - \int_{\Omega_n} \mathbb{K}_n(dx') b_n(x',z) \quad \text{and} \quad b_n(x,z) := \sum_{k=1}^n c_k(x,z)$$

In the above displayed formulae c_k stands for the collection of functions on Ω_n^2 given by

$$c_k(x, z) := \frac{G_{k-1}(z_{k-1}) \, m_k(z_{k-1}, x_k)}{\int_{E_{k-1}} G_{k-1}(u_{k-1}) \, m_k(u_{k-1}, x_k) \, \eta_{k-1}(du)}$$

Under our assumptions, we can prove that the integral operator $\mathcal{A}_n$ given by for any $F_n \in \mathbb{L}_2(\mathbb{K}_n)$ by

$$\mathcal{A}_n(F_n)(x) = \int_{\Omega_n} a_n(z, x) \, F_n(z) \, \mathbb{K}_n(dz)$$

is a Hilbert–Schmidt operator on $\mathbb{L}_2(\mathbb{K}_n)$ and the operator $(I - \mathcal{A}_n)$ is invertible.

Extending the fluctuation analysis developed by T. Shiga and H. Tanaka in [43] to nonlinear mean field interaction models, we prove in [20] the following central limit theorem.

Theorem 25.7 *The random field $\mathcal{W}_n^N := \sqrt{N} \left[\mathbb{K}_n^N - \mathbb{K}_n \right]$ converges as $N \to \infty$ to a centered Gaussian field $\mathcal{W}_n$ satisfying*

$$\mathbb{E}(\mathcal{W}_n(F_n) \, \mathcal{W}(F_n')) = \langle (I - \mathcal{A}_n)^{-1}(F_n - \mathbb{K}_n(F_n)), (I - \mathcal{A}_n)^{-1}(F_n' - \mathbb{K}_n(F_n')) \rangle_{\mathbb{L}_2(\mathbb{K}_n)}$$

for any $F_n, F_n' \in \mathbb{L}_2(\mathbb{K}_n)$, in the sense of convergence of finite-dimensional distributions.

25.4.4 Large deviation principles

In this section, we introduce the reader to the large deviation analysis of mean field particle models associated with a given McKean interpretation of a measure valued equation. We start with the derivation of the large deviation principles (*abbreviated LDP*) combining Sanov's theorem with the Laplace–Varadhan integral lemma. This rather elementary result relies on an appropriate regularity condition on the McKean transitions under which the law of the N-particle model is absolutely continuous w.r.t. the law of N independent copies of the Markov chain associated with the McKean interpretation model.

We equip the space of all finite and signed measures $\mathcal{M}(E)$ on a Polish space $(E, \mathcal{E})$ with the weak topology generated by the open neighborhoods

$$\mathcal{V}_{f,\epsilon}(\mu) = \{\eta \in \mathcal{P}(E) : |\eta(f) - \mu(f)| < \epsilon\}$$

where f is a bounded continuous function on E, $\mu \in \mathcal{P}(E)$, and $\epsilon \in (0, \infty)$. Using this topological framework, if we take $E = E_n$ and $f = f_n$, then the deviant event presented in (25.33) is equivalently expressed in terms of a neighborhood of the McKean measure

$$\mathbb{P}\left(|\eta_n^N(f_n) - \eta_n(f_n)| > \epsilon\right) = \mathbb{P}\left(\eta_n^N \notin \mathcal{V}_{f_n,\epsilon}(\eta_n)\right)$$

as soon as the function f_n is continuous. One objective of the forthcoming analysis is to estimate the exact exponential deviation of the particle measures η_n^N and more generally $\mathbb{K}_n^N$ around there limiting values η_n and $\mathbb{K}_n$. For instance, we would like is to compute the following quantities for any continuous bounded functions f_n or F_n on the state space E_n or on the path space $\Omega_n := (E_0 \times \ldots \times E_n)$

$$\lim_{N \to \infty} \frac{1}{N} \log \mathbb{P}\left(|\eta_n^N(f_n) - \eta_n(f_n)| > \epsilon\right) \text{ and } \lim_{N \to \infty} \frac{1}{N} \log \mathbb{P}\left(|\mathbb{K}_n^N(F_n) - \mathbb{K}_n(F_n)| > \epsilon\right)$$

To describe precisely this result we need to introduce another set of notation. We let $P_n^N \in \mathcal{P}(\Omega_n^N)$ be the distribution of the mean field N-interacting particle model $(\xi_0^i, \ldots \xi_n^i)_{1 \leq i \leq N}$ introduced in (25.3), and we denote by $\overline{P}_n^N = P_n^N \circ (\pi_n^N)^{-1}$ the image distribution of the empirical measures $\mathbb{K}_n^N$, with the empirical projection mapping π_n^N given by

$$\pi_n^N : \omega = (\omega_0^i, \ldots, \omega_n^i)_{1 \leq i \leq N} \in \Omega_n^N \longrightarrow \pi_n^N(\omega) = \frac{1}{N} \sum_{i=1}^N \delta_{(\omega_0^i, \ldots, \omega_n^i)} \in \mathcal{P}(\Omega_n)$$

We recall that the sequence $\overline{P}_n^N$ satisfies the LDP with the good rate function I_n as soon as we find a mapping lower semicontinuous mapping I_n from $\mathcal{P}(\Omega_n)$ into $[0, \infty]$ and such that

$$-\inf_{\mathring{A}} I_n \leq \liminf_{N \to \infty} \log \frac{1}{N} \overline{P}_n^N(A) \leq \limsup_{N \to \infty} \log \frac{1}{N} \overline{P}_n^N(A) \leq -\inf_{\overline{A}} I_n \quad (25.37)$$

for any Borel subset $A \subset \mathcal{P}(\Omega_n)$, where $\mathring{A}$ and $\overline{A}$ denote respectively the interior and the closure of the set A.

By a direct application of Sanov's theorem, under the tensor product measure $\mathbb{K}_n^{\otimes N}$ the laws $\overline{Q}_n^N := \mathbb{K}_n^{\otimes N} \circ (\pi_n^N)^{-1}$ of the empirical measures $\mathbb{K}_n^N$ associated with N independent path particles with common distribution $\mathbb{K}_n$ satisfy the LDP with a good rate function H_n given by the Boltzmann relative entropy defined by

$$H_n : \mu \in \mathcal{P}(\Omega_n) \longrightarrow H(\mu) = \text{Ent}(\mu \mid \mathbb{K}_n) \in [0, \infty]$$

To get one step further, we suppose that condition (25.34) is satisfied with $p_n(e^{r\theta_n})$, for any $r \geq 1$. In this condition, the measures $\overline{P}_n^N$ and $\overline{Q}_n^N$ are absolutely continuous with a Radon–Nikodym derivative defined by

$$d\overline{P}_n^N / d\overline{Q}_n^N = \exp(N V_n) \quad \overline{Q}_n^N\text{-a.e.}$$

In the above display, the function V_n is given for any probability measure $\mu \in \mathcal{P}(\Omega_n)$ by te following formula

$$V_n(\mu) := \int \log \frac{d\mathbb{M}_n(\mu)}{d\mathbb{M}_n(\mathbb{K}_n)} \, d\mu$$

$$= \sum_{p=1}^{n} \int_{E_{p-1} \times E_p} \mu_{p-1,p}(d(u_{p-1}, u_p)) \log \left[\frac{dK_{p,\mu_{p-1}}(u_{p-1}, \cdot)}{dK_{p,\eta_{p-1}}(u_{p-1}, \cdot)}(u_p) \right]$$

with the mapping

$$\mathbb{M}_n : \mu \in \mathcal{P}(\Omega_n) \to \mathbb{M}_n(\mu) \in \mathcal{P}(\Omega_n)$$

defined by

$$\mathbb{M}_n(\mu)(d(x_0, \ldots, x_n)) = \eta_0(dx_0) \, K_{1,\mu_0}(x_0, dx_1) \ldots K_{n,\mu_{n-1}}(x_{n-1}, dx_n)$$

and where $\mu_p \in \mathcal{P}(E_p)$ stands for the pth time marginal of μ with $0 \leq p \leq n$.

Under our assumptions, using an extended version of the Laplace–Varadhan integral lemma (see for instance Lemma 10.4.1 in [11]) we prove the following theorem.

Theorem 25.8 *The sequence of measures $\overline{\mathbb{P}}_n^N$ satisfies the LDP with the good rate function*

$$I_n(\mu) := [H_n - V_n](\mu) = \mathrm{Ent}(\mu \mid \mathbb{M}_n(\mu)) \quad \text{and} \quad I_n(\mu) = 0 \Leftrightarrow \mu = \mathbb{K}_n$$

As we mentioned above, the large deviation analysis we have described is not restricted to mean field interpretations of the nonlinear filtering equations. A more general result for an abstract class of McKean kernels $(K_{n,\eta})_{n,\eta}$ is provided in theorem 10.1.1 in [11].

Working a little harder, we can also derive a LDP for the τ-topology for the distribution of the particle density profiles $\left(\eta_p^N\right)_{0 \leq p \leq n}$ associated with the McKean transitions $K_{n,\eta}(x, \cdot) := \Phi_n(\eta)$. For the convenience of the reader, we recall that the τ-topology on a set of probability measures is generated by the sets $\mathcal{V}_{f,\epsilon}(\mu)$, with nonnecessarily continuous functions $f \in \mathcal{B}(E)$. The following theorem is taken from a joint work of the first author with D. Dawson [16].

Theorem 25.9 *Assume that for any $n \geq 1$, there exists some reference measure $\lambda_n \in \mathcal{P}(E_n)$ and some parameters $\rho_n > 0$ such that*

$$\forall \mu \in (\mathcal{P}(E_{n-1}) \qquad \rho_n \lambda_n \leq \Phi_n(\mu) \ll \lambda_n \tag{25.38}$$

The sign $\mu_1 \ll \mu_2$ stands for the absolutely continuity property of the measure μ_1 w.r.t. μ_2. In this situation, the law of the flow of particle density profiles $\left(\eta_p^N\right)_{0 \leq p \leq n}$ satisfies the LDP for the product τ-topology with the good rate function J_n on $\prod_{p=0}^{n} \mathcal{P}(E_p)$ given by

$$J_n((\mu_p)_{0 \leq p \leq n}) = \sum_{p=0}^{n} \mathrm{Ent}(\mu_p \mid \Phi_p(\mu_{p-1}))$$

In the context of Feynman–Kac models, it can be checked that the one-step mappings Φ_n are τ-continuous and condition (25.38) is met as soon as for any $x_{n-1} \in E_{n-1}$ we have

$$\rho_n \lambda_n \leq M_n(x_{n-1}, \cdot) \ll \lambda_n$$

For instance, if we take $E_n = \mathbb{R}$ and $M_n(x, dy) = \frac{1}{\sqrt{2\pi}} e^{-\frac{1}{2}(y-a_n(x))^2} dy$, where a_n is a bounded measurable drift function on $\mathbb{R}$, then condition (25.38) is met with the reference measure given by $\lambda_n(dy) = p_n(dy)/p_n(\mathbb{R})$ and with

$$p_n(dy) =_{\text{def.}} \frac{1}{\sqrt{2\pi}} e^{-\frac{y^2}{2} - |y| \, \|a_n\|} dy \quad \text{and} \quad \rho_n = e^{-\|a_n\|^2/2} \, p_n(\mathbb{R})$$

25.5 Propagations of chaos properties

The initial configuration of the N-particle mean field model (25.3) consists in N independent and identically distributed random variables. Their common law is dictated by the initial distribution of the underlying nonlinear equation (25.2). In this sense, the initial system is in "complete chaos" and the law of each particle perfectly fits the initial distribution.

During their time evolution, the particles interact with one another. The nature of the interactions depends on the McKean interpretation model. Nevertheless, in all interesting cases the independence property and the adequacy of the laws of the particles with the desired distributions are broken. The propagation of chaos properties of mean field particle models presented in this section can be seen as a way to measure these properties. In the first subsection, we estimate the relative entropy of the laws of the first q paths of a complete genealogical tree model with respect to the q-tensor product of the McKean measure. In the second part of this section, we present an original Laurent type and algebraic tree-based integral representations of particle block distributions. In contrast to the first entropy approach, this technique applies to any genetic-type particle model without any regularity condition on the mutation transition.

25.5.1 Relative entropy estimates

In this section, we provide strong propagation estimates with respect to the relative entropy criterion for the interacting particle system associated with the McKean interpretation model defined in (25.3). We further assume that $\epsilon_n(\mu)$ does not depend on the measure μ, and the Markov transitions M_n satisfy the regularity condition (25.34), with $p_n = \eta_n$. We let P_n^N be the distribution of the mean field N-interacting particle model $\left(\xi_0^i, \ldots \xi_n^i\right)_{1 \leq i \leq N}$. By the deviation analysis provided in section 25.4.4, we have that

$$\frac{d\,P_n^N}{d\mathbb{K}_n^{\otimes N}} = \exp\left(N\left(V_n \circ \pi_n^N\right)\right) \qquad \mathbb{K}_n^{\otimes N}\text{-a.e.}$$

We let $\mathbb{P}_n^{(N,q)}$ be the distribution of the first q path particles $(\xi_0^i, \ldots \xi_n^i)_{1 \leq i \leq q}$, with $q \leq N$. Using an elegant lemma of Csiszar on the entropy of the first q-coordinates of an exchangeable measure (see [9], or lemma 8.5.1 in [11]), we find that

$$\mathrm{Ent}\left(\mathbb{P}_n^{(N,q)} \mid \mathbb{K}_n^{\otimes q}\right) \leq 2\,\frac{q}{N}\,\mathrm{Ent}\left(\mathbb{P}_n^N \mid \mathbb{K}_n^{\otimes N}\right) = 2q\,\mathbb{E}\left(V_n\left(\mathbb{K}_n^N\right)\right) \quad (25.39)$$

By the exchangeability property of the particle model, we also have that

$$\mathbb{E}\left(V_n\left(\mathbb{K}_n^N\right)\right) = \sum_{p=0}^{n-1} \mathbb{E}\left[\mathrm{Ent}\left(K_{p+1,\eta_p^N}\left(\xi_p^1, \cdot\right) \mid K_{p+1,\eta_p}\left(\xi_p^1, \cdot\right)\right)\right]$$

Using the fact that $\mathrm{Ent}(\mu|\eta) \leq \left\|1 - d\mu/d\eta\right\|_{2,\eta}^2$, and estimating the local Lipschitz continuity coefficient of the mappings $\mu \mapsto \frac{d K_{n,\mu}(u,\cdot)}{d K_{n,\eta_{n-1}}(u,\cdot)}(v)$ at $\mu = \eta_{n-1}$, we prove the following theorem.

Theorem 25.10 *For any $q \leq N$ and any time horizon n, we have the relative entropy estimate*

$$\mathrm{Ent}\left(\mathbb{P}_n^{(N,q)} \mid \mathbb{K}_n^{\otimes q}\right) \leq \frac{q}{N} c(n) \text{ with } c(n) \leq c' \sum_{p=0}^{n-1} r_p^2 \left(1 + \eta_{p+1}(e^{2\theta_{p+1}})\right) \sum_{q=0}^{p} r_{q,p}\beta(P_{q,p})$$

and some universal finite constant c'.

25.5.2 Polynomial tree-based expansions

This section is concerned with propagation of chaos properties of the mean field particle model associated with the null acceptance rate parameter $\epsilon_n(\mu) = 0$. Without any regularity conditions on the Markov transitions M_n it is more or less well known that we have the following strong propagation of chaos estimate

$$\left\|\mathbb{P}_n^{(N,q)} - \mathbb{K}_n^{\otimes q}\right\|_{\mathrm{tv}} \leq c(n)\,q^2/N \quad (25.40)$$

for some finite constant, which only depends on the time parameter n. The complete proof of this result can be found in [11] (see for instance theorem 8.3.3 and theorem 8.3.4).

The main object of this section is to provide an explicit asymptotic functional expansion of the law of the first q particles at a given time n with respect to the precision parameter N. These expansions at any order extend the sharp but first-order propagation of chaos estimates developed by the first author with A. Doucet and G. W. Peters in a recent article [19].

Next, for any $q \leq N$ we consider the q-tensor product occupation measures, particularly the restricted q-tensor product occupation measures, on E_n^q defined by

$$\left(\eta_n^N\right)^{\otimes q} = \frac{1}{N^q} \sum_{a \in [N]^{[q]}} \delta_{\left(\xi_n^{(a(1),N)}, \ldots, \xi_n^{(a(q),N)}\right)} \quad \text{and} \quad \left(\eta_n^N\right)^{\odot q} = \frac{1}{(N)_q} \sum_{a \in \langle q, N \rangle} \delta_{\left(\xi_n^{(a(1),N)}, \ldots, \xi_n^{(a(q),N)}\right)} \tag{25.41}$$

In the above display, $[N]^{[q]}$, resp. $\langle q, N \rangle$, stands for the set of all N^q mappings, resp. all $(N)_q := N!/(N-q)!$ one-to-one mappings, from the set $[q] := \{1, \ldots, q\}$ into $[N] := \{1, \ldots, N\}$. The unnormalized versions of these measures are simply defined by

$$\left(\gamma_n^N\right)^{\otimes q} := \left(\gamma_n^N(1)\right)^q \times \left(\eta_n^N\right)^{\otimes q} \quad \text{and} \quad \left(\gamma_n^N\right)^{\odot q} := \left(\gamma_n^N(1)\right)^q \times \left(\eta_n^N\right)^{\odot q}$$

One central problem is to obtain functional expansions, with respect to the precision parameter N, of the pair of particle block distributions defined for any $F \in \mathcal{B}\left(E_n^q\right)$ by the formulae

$$\mathbb{P}_{n,q}^N(F) := \mathbb{E}((\eta_n^N)^{\odot q}(F)) = \mathbb{E}(F(\xi_n^{(1,N)}, \ldots, \xi_n^{(q,N)}))$$
$$\text{and} \quad \mathbb{Q}_{n,q}^N(F) := \mathbb{E}((\gamma_n^N)^{\otimes q}(F)) \tag{25.42}$$

In this section, we design a coalescent tree-based representation of the unnormalized particle distributions $\mathbb{Q}_{n,q}^N$ with the following polynomial form

$$\mathbb{Q}_{n,q}^N = \gamma_n^{\otimes q} + \sum_{1 \leq k \leq (q-1)(n+1)} \frac{1}{N^k} \partial^k \mathbb{Q}_{n,q} \tag{25.43}$$

In the above display, $\partial^k \mathbb{Q}_{n,q}$ stands for a sum of signed and weak derivative measures, whose values can be explicitly described in terms of a class of forests with maximal coalescent degree k.

These Laurent-type expansions reflect the complete interaction structure of the particle model, the k-th order terms represent the $1/N^k$-contributions of mean field particle scenarios with an interaction degree k.

The analysis of the distributions $\mathbb{P}_{n,q}^N$ is a little more involved, combining judicious renormalization techniques on path spaces, with colored tree-based combinatorial expansions (cf. [31]).

$$\mathbb{P}_{n,q}^N \simeq \eta_n^{\otimes q} + \sum_{1 l \leq l \leq k} \frac{1}{N^l} \partial^l \mathbb{P}_{n,q} + \frac{1}{N^{k+1}} \partial^{k+1} \mathbb{P}_{n,q}^N \quad \text{with} \quad \sup_{N \geq 1} \left\| \partial^{k+1} \mathbb{P}_{n,q}^N \right\|_{tv} < \infty$$

25.5.3 Coalescent tree-based representations

Let $\mathcal{A}_{n,q}$ be the set of $(n+1)$-sequences $\mathbf{a} = (a_p)_{0 \leq p \leq n}$ of mappings a_p from $[q]$ into itself. By $|a|$, we denote the cardinality of the set $a([q])$; and for $\mathbf{a} \in \mathcal{A}_{n,q}$, we write $|\mathbf{a}|$ the integer sequence $(|a_i|)_{0 \leq i \leq n}$. For any pair of integers $q \leq N$, we use the the multi-index notation $(\mathbf{q})_{|\mathbf{a}|} = \prod_{0 \le k \le n}(q)_{|a_k|}$. For $b \in [q]^{[q]}$ and $F \in \mathcal{B}(E_n^q)$, we define:

$$D_b(F)(x^1, \ldots, x^q) = F(x^{b(1)}, \ldots, x^{b(q)}), \quad (F)_{sym} := \frac{1}{p!} \sum_{\sigma \in \mathcal{G}_q} D_\sigma F \qquad (25.44)$$

For any air of mappings $a, b \in [q]^{[q]}$ we have the composition formula $D_a D_b = D_{ab}$. We notice that $(\eta_n^N)^{\otimes q}(F) = (\eta_n^N)^{\otimes q}((F)_{sym})$ and $(\eta_n^N)^{\odot q}(F) = (\eta_n^N)^{\odot q}((F)_{sym})$. So we may assume in our forthcoming computations on q-tensor products occupation measures that F is in $\mathcal{B}^{sym}(E_n^q)$, the subset of $\mathcal{B}(E_n^q)$ of symmetric functions. We also observe that

$$(\eta_n^N)^{\otimes q} = (\eta_n^N)^{\odot q} \left(\frac{1}{N^q} \sum_{b \in [q]^{[q]}} \frac{(N)_{|b|}}{(q)_{|b|}} D_b \right) \qquad (25.45)$$

To check this formula, we first notice that for any $c \in [N]^{[q]}$, there are $(N - |c|)_{q-|c|} \times (q)_{|c|}$ different ways to write $c = ab \in \langle q, N \rangle \circ [q]^{[q]}$. On the other hand, if $a \in \langle q, N \rangle$, then we have that $|b| = |c|$ and $\frac{(N)_{|c|}}{(q)_{|c|}} \times \frac{(N-|c|)_{q-|c|} \times (q)_{|c|}}{(N)_q} = 1$.

The partial derivative measures $\partial^k \mathbb{Q}_{n,q}$ involved in the Laurent-type expansions (25.43) are intimately related to the following measure-valued functional

$$\Delta_{n,q} : \mathbf{a} \in \mathcal{A}_{n,q} \mapsto \Delta_{n,q}^{\mathbf{a}} = \left(\eta_0^{\otimes q} D_{a_0} Q_1^{\otimes q} D_{a_1} \ldots Q_n^{\otimes q} D_{a_n} \right) \in \mathcal{M}(E_n^q) \qquad (25.46)$$

For instance, for $q = 3$ and $n = 2$ and the sequence of mappings $\mathbf{a} = (a_0, a_1, a_2)$ given below

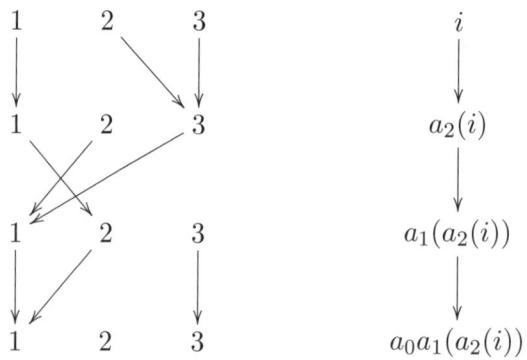

we have the formula

$$\Delta_{n,q}^{\mathbf{a}}(F) = \int \eta_0(dx^1)\eta_0(dx^2)\eta_0(dx^3) \, Q_1(x^1, dy^1)Q_1(x^1, dy^2)Q_1(x^3, dy^3)$$
$$\times \, Q_2(y^1, dz^2)Q_2(y^1, dz^3)Q_2(y^2, dz^1) F(z^1, z^2, z^3)$$

The first steps to these integral expansions are given by the following fundamental lemma.

Lemma 25.2 *For any particle block size $q \leq N$, any time horizon $n \in \mathbb{N}$, and any $F \in \mathcal{B}(E_n^q)$, we have*

$$\mathbb{Q}_{n,q}^N(F) = \frac{1}{N^q} \sum_{a \in [q]^{[q]}} \frac{(N)_{|a|}}{(q)_{|a|}} \mathbb{Q}_{n-1,q}^N\left(Q_n^{\otimes q} D_a F\right) = \frac{1}{N^{q(n+1)}} \sum_{\mathbf{a} \in \mathcal{A}_{n,q}} \frac{(N)_{|\mathbf{a}|}}{(q)_{|\mathbf{a}|}} \Delta_{n,q}^{\mathbf{a}}(F) \quad (25.47)$$

Proof. We let A_q^N be an $[q]^{[q]}$-valued random variable (independent of the particle model) with distribution $\frac{1}{N^q} \sum_{a \in [q]^{[q]}} \frac{(N)_{|a|}}{(q)_{|a|}} \delta_a$. Indeed, combining the definition of the particle model with (25.45), we first find that

$$\mathbb{E}\left(\left(\gamma_n^N\right)^{\otimes q}(F) \mid \xi_{n-1}\right) = \gamma_n^N(1)^q \times \mathbb{E}\left(\left(\eta_n^N\right)^{\otimes q}(F) \mid \xi_{n-1}\right)$$
$$= \left(\gamma_n^N(1)\right)^q \times \mathbb{E}\left(\Phi_n\left(\eta_{n-1}^N\right)^{\otimes q} D_{A_q^N}(F) \mid \xi_{n-1}\right)$$
$$= \mathbb{E}\left(\left(\gamma_{n-1}^N\right)^{\otimes q}\left(Q_n^{\otimes q} D_{A_q^N} F\right) \mid \xi_{n-1}\right)$$

Integrating over the past, the end of the proof is straightforward. □

Next, we turn our attention to a coalescent tree-based formulation of the integral expansion (25.47). We start with recalling some more or less classical terminology on trees and forests. A tree (respectively a planar tree) is a (isomorphism class of) finite nonempty, oriented, connected (and respectively planar) graph **t** without loops such that any vertex of **t** has at most one outgoing edge. Paths are oriented from the vertices to the root. A forest **f** is a multiset of trees, that is a set of trees, with repetitions of the same tree allowed, or equivalently an element of the commutative monoid $\langle \mathcal{T} \rangle$ on $\mathcal{T}$, with the empty graph $T_0 = \emptyset$ as a unit. Since the algebraic notation is the most convenient, we write $\mathbf{f} = \mathbf{t}_1^{m_1} \ldots \mathbf{t}_k^{m_k}$, for the forest with the tree $\mathbf{t}_i$ appearing with multiplicity $m_i, i \leq k$. When $\mathbf{t}_i \neq \mathbf{t}_j$ for $i \neq j$, we say that **f** is written in normal form. A planar forest **f** is an ordered sequence of planar trees. Planar forests can be represented by noncommutative monomials (or words) on the set of planar trees. The sets of forests and planar forests with height $(n+1)$, and with q vertices at each level set are written $\mathcal{F}_{q,n}$ and $\mathcal{PF}_{q,n}$.

A sequence $\mathbf{a} \in \mathcal{A}_{q,n}$ is naturally associated a forest $F(\mathbf{a})$: the one with one vertex for each element of $[q]^{n+2}$, and a edge for each pair $(i, a_k(i)), i \in [q]$. The sequence can also be represented graphically uniquely by a planar graph $J(\mathbf{a})$, where however the edges between vertices at level $k+1$ and k are allowed to cross. We call such a planar graph, where paths between vertices are entangled, a jungle. The set of such jungles is written $\mathcal{J}_{q,n}$. In a planar forest **f**, vertices at the same level $k \geq 0$, are naturally ordered from left to right, and therefore in bijection with $[q]$. Planar forests $\mathbf{f} \in \mathcal{PF}_{q,n}$ of height $(n+1)$ are therefore canonically in bijection with sequences **a** of weakly increasing map from $[q]$ into itself. (see Figs. 25.1 and 25.2.)

We let $\mathcal{G}_q$ be the symmetric group of all permutations of $[q]$. The group $\mathcal{G}_q^{n+2}$ acts naturally on sequences of maps $\mathbf{a} \in \mathcal{A}_{q,n}$, and on jungles $J(\mathbf{a}) \in \mathcal{J}_{q,n}$ by permutation of the vertices at each level. More precisely, for all $\mathbf{a} \in \mathcal{A}_{n,q}$ and all $\mathbf{s} = (s_0, \ldots, s_{n+1}) \in \mathcal{G}_q^{n+2}$ by the pair of formulae

$$\mathbf{s}(\mathbf{a}) := \left(s_0 a_0 s_1^{-1}, s_1 a_1 s_2^{-1}, \ldots, s_n a_n s_{n+1}^{-1}\right) \quad \text{and} \quad \mathbf{s} J(\mathbf{a}) := J(\mathbf{s}(\mathbf{a})) \qquad (25.48)$$

Notice that if two sequences **a** and $\mathbf{b} \in \mathcal{A}_{\mathbf{q,n}}$ differ only by the order of the vertices in $J(\mathbf{a})$ and $J(\mathbf{b})$, that is by the action of an element of $\mathcal{G}_\mathbf{p}$, then the associated forests are identical: $F(\mathbf{a}) = F(\mathbf{b})$. Moreover, the converse is true: if $F(\mathbf{a}) = F(\mathbf{b})$, then $J(\mathbf{a})$ and $J(\mathbf{b})$ differ only by the ordering of the vertices, since they have the same underlying nonplanar graph. In this situation, **a** and **b** belong to the same orbit under the action of $\mathcal{G}_\mathbf{p}$. In particular, the set of equivalence classes of jungles in $\mathcal{J}_{q,n}$ under the action of the permutation groups $\mathcal{G}_{q,n}$ is in bijection with the set of forests $\mathcal{F}_{q,n}$. We denote by $B(\mathbf{t})$ the forest deduced from cutting the root of tree $\mathbf{t}$; that is, removing its root vertex, and all its incoming edges. In the reverse angle, we denote by $B^{-1}(\mathbf{f})$ the tree deduced from the forest $\mathbf{f}$ by adding a common root to its rooted tree. The symmetry multiset $S(\mathbf{t})$ of a tree $\mathbf{t} = B^{-1}\left(\mathbf{t}_1^{m_1} \ldots \mathbf{t}_k^{m_k}\right)$ associated with a forest written in

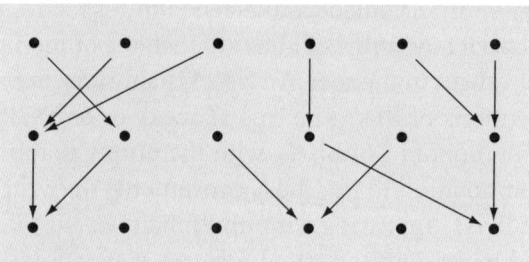

Fig. 25.1 The entangled graph representation of a jungle with the same underlying graph as the planar forest in Fig. 25.2.

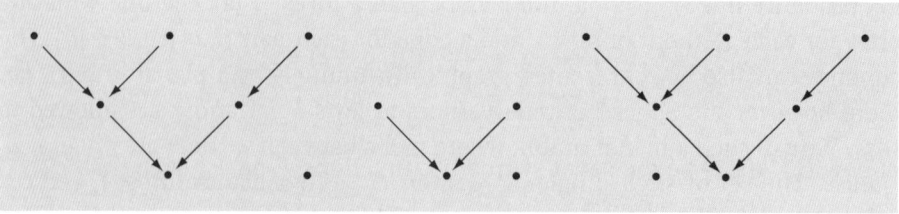

Fig. 25.2 A graphical representation of a planar forest $\mathbf{f} = \mathbf{t}_1 \mathbf{t}_3 \mathbf{t}_2 \mathbf{t}_3 \mathbf{t}_3 \mathbf{t}_1$.

normal form, is defined by $S(t) := (m_1, \ldots, m_k)$. The symmetry multiset of a forest is given by

$$S\left(t_1^{m_1} \ldots t_k^{m_k}\right) := \Big(\underbrace{S(t_1), \ldots, S(t_1)}_{m_1\text{-terms}}, \ldots, \underbrace{S(t_k), \ldots, S(t_k)}_{m_k\text{-terms}}\Big)$$

Combining the class formula with recursive multiplication principles with respect to the height parameter, one obtains the following closed formula.

Theorem 25.11 *The number $\#(\mathbf{f})$ of jungles in $\mathbf{f} \in \mathcal{F}_{q,n}$, viewed as an equivalence class, is given by*

$$\#(\mathbf{f}) = (q!)^{n+2} / \prod_{i=-1}^{n} S(B^i(\mathbf{f}))!$$

To get one step further, we notice that the measures $\Delta_{n,q}^{\mathbf{a}}$ have the invariance property $\Delta_{n,q}^{\mathbf{b}}(F) = \Delta_{n,q}^{s(\mathbf{b})}(F)$, for any symmetric function $F \in \mathcal{B}_b(E_n^q)$. Thus, we can define unambiguously $\Delta_{n,q}^{\mathbf{f}}(F) = \Delta_{n,q}^{\mathbf{a}}(F)$, and $|\mathbf{f}| = |\mathbf{a}|$, for any choice $\mathbf{a}$ of a representative of some $\mathbf{f} \in \mathcal{F}_{q,n}$ in $\mathcal{A}_{q,n} \cong \mathcal{J}_{q,n}$.

The difference $(\mathbf{p} - \mathbf{l})$, of a pair of integer sequences $\mathbf{p} = (p_k)_{0 \leq k \leq n}$ and $\mathbf{l} = (l_k)_{0 \leq k \leq n}$, is the sequence $(\mathbf{p} - \mathbf{l}) := (p_k - l_k)_{0 \leq k \leq n}$. When no confusions can arise, we write $\mathbf{q}$, for the constant sequence $(q)_{0 \leq i \leq n}$. For any multi-index $\mathbf{p} = (p_k)_{0 \leq k \leq n} \leq \mathbf{q}$, we let $\mathcal{F}_{n,q}(\mathbf{p}) \subset \mathcal{F}_{n,q}$ be the subset of forets with at least p_k leaves at each level k, with $0 \leq k \leq n$. We also write for any pair of multi-indexes $\mathbf{p} \leq \mathbf{l}$

$$|\mathbf{p}| := (p_0 + \ldots + p_n) \qquad \mathbf{p}! = \prod_{k=0}^{n} p_k! \quad \text{and} \quad s(\mathbf{l}, \mathbf{p}) = \prod_{k=0}^{n} s(l_k, p_k)$$

where the $s(l_k, p_k)$ are Stirling numbers of the first kind. The next theorem is the main result of this section.

Theorem 25.12 *For any $1 \leq q \leq N$, we have the polynomial expansion (25.43), with the collection of signed and weak derivative measures $\partial^k \mathbb{Q}_{n,q}$ given by the formula*

$$\partial^k \mathbb{Q}_{n,q} = \sum_{\mathbf{r} < \mathbf{q} : |\mathbf{r}| = k} \sum_{\mathbf{f} \in \mathcal{F}_{n,q}(\mathbf{r})} \frac{s(|\mathbf{f}|, \mathbf{q} - \mathbf{r}) \, \#(\mathbf{f})}{(\mathbf{q})_{|\mathbf{f}|}} \Delta_{n,q}^{\mathbf{f}} \qquad (25.49)$$

Let now $\mathcal{B}_0^{sym}(E_n^q)$ be the set of symmetric functions F on E_n^q such that

$$\int F(x_1, \ldots, x_{q-1}, x_q) \, \gamma_n(dx_q) = 0$$

We write $\mathbf{t}_k$ for the unique tree with a single coalescence at level k, and its two leaves at level $(n+1)$; and we write $\mathbf{u}_k$ for the trivial tree of height k.

Corollary 25.2 *For any even integer $q \leq N$ and any symmetric function $F \in \mathcal{B}_0^{sym}(E_n^q)$, we have*

$$\forall k < q/2 \quad \partial^k \mathbb{Q}_{n,q}(F) = 0, \quad \partial^{q/2} \mathbb{Q}_{n,q}(F) = \sum_{\mathbf{r}<\mathbf{q}, |\mathbf{r}|=\frac{q}{2}} \frac{q!}{2^{q/2}\,\mathbf{r}!} \Delta_{n,q}^{\mathbf{f}_\mathbf{r}} F \quad (25.50)$$

with the forest $\mathbf{f}_\mathbf{r} := \mathbf{t}_0^{r_0} \mathbf{u}_0^{r_0} \ldots \mathbf{t}_n^{r_n} \mathbf{u}_n^{r_n}$ associated with a multi-index sequence $\mathbf{r} = (r_k)_{0 \leq k \leq n} < \mathbf{q}$. For odd integers $q \leq N$, the partial derivatives are the null measure on $\mathcal{B}_0^{sym}(E_n^q)$, up to any order $k \leq \lfloor q/2 \rfloor$.

References

[1] C. Brouder. Runge–Kutta methods and renormalization. *Eur. Phys. J. C.* 12:521–34, 2000.

[2] H. Carvalho, P. Del Moral, A. Monin, and G. Salut. Optimal nonlinear filtering in GPS/INS integration. *IEEE Trans. Aerosp. Electron. Syst.* 33(3):835–50, 1997.

[3] A. Connes and D. Kreimer. Hopf algebras, renormalization and noncommutative geometry. *Commun. Math. Phys.* 199(1):203–42, 1998.

[4] J. M. C. Clark, D. L. Ocone, and C. Coumarbatch. Relative entropy and error bounds for filtering of Markov processes. *Math. Control Signal Syst.* 12(4):346–60, 1999.

[5] D. Crisan, P. Del Moral, and T. J. Lyons. Interacting particle systems approximations of the Kushner-Stratonovitch equation. *Adv. in Appl. Probab.* 31(3):819–38, 1999.

[6] D. Crisan, J. Gaines, and T. J. Lyons. A particle approximation of the solution of the Kushner-Stratonovitch equation. *SIAM J. Appl. Math.* 58(5):1568–90, 1998.

[7] D. Crisan and T. J. Lyons. Nonlinear filtering and measure valued processes. *Probab. Theory Rel. Fields* 109:217–44, 1997.

[8] D. Crisan and T. J. Lyons. A particle approximation of the solution of the Kushner-Stratonovitch equation. *Probab. Theory Related Fields* 115(4):549–78, 1999.

[9] I. Csiszár. Sanov property, generalized i-projection and a conditional limit theorem. *Ann. Probab.* 12(3):768–93, 1984.

[10] I. Csiszár. Eine informationstheoretische Ungleichung und ihre Anwendung auf den Beweis der Ergodizität von Markoffschen Ketten. *Magyar Tud. Akad. Mat. Kutató Int. Közl.* 8:85–108, 1963.

[11] P. Del Moral, *Feynman–Kac formulae. Genealogical and interacting particle systems, with applications*, Springer-Verlag, New York, Series: Probability and its Applications (2004).

[12] P. Del Moral. A Uniform convergence theorem for the numerical solving of the non linear filtering problem. *J. Applied Probab.* 35:873–84, 1998.

[13] P. Del Moral. Measure valued processes and interacting particle systems. Application to non linear filtering problems. *Ann. of Applied Probab.* 8(2):438–95, 1998.

[14] P. Del Moral. Non linear filtering using random particles. *Theory Probab. Appl.* 40(4):690–701, 1996.

[15] P. Del Moral. Non linear filtering: interacting particle solution. *Markov Proc. Rel. Fields* 2(4):555–80, 1996.

[16] D. A. Dawson and P. Del Moral. Large deviations for interacting processes in the strong topology. *Statistical Modeling and Analysis for Complex Data Problem* P. Duchesne and B. Rémillard Editors, pp. 179–209, Springer (2005).

[17] P. Del Moral, A. Doucet, and A. Jasra. Sequential Monte Carlo for Bayesian computation. *Bayesian Stat. 8*, Oxford University Press (2006).

[18] P. Del Moral, A. Doucet, and A. Jasra. Sequential Monte Carlo samplers. *J. Royal Stat. Soc., Series B* 68(3):411–36, 2006.

[19] P. Del Moral, A. Doucet, and G. W. Peters. Sharp propagation of chaos estimates for Feynman–Kac particle models. *Theory Probab. Appl.* 51(3), 2007.

[20] P. Del Moral and A. Guionnet. A central limit theorem for nonlinear filtering using interacting particle systems. *Ann. Appl. Probab.* 9(2):275–97, 1999.

[21] P. Del Moral and A. Guionnet. Deviations for interacting particle systems. Applications to non linear filtering problems. *Stoch. Proc. Appl.* 78:69–95, 1998.

[22] P. Del Moral and J. Jacod. The Monte Carlo method for filtering with discrete time observations. Central limit theorems. *The Fields Institute Communications*, Numerical Methods and Stochastics, ed. T. J. Lyons, T. S. Salisbury, American Mathematical Society, 2002.

[23] P. Del Moral and J. Jacod. Interacting particle filtering with discrete-time observations: asymptotic behaviour in the Gaussian case. *Stochastics in finite and infinite dimensions. Trends in mathematics*, Birkhäuser, pp. 101–3 (2000).

[24] P. Del Moral, M. Ledoux, and L. Miclo. On contraction properties of Markov kernels. *Probab. Theory Rel. Fields*, 126:395–420, 2003.

[25] P. Del Moral and L. Miclo. Asymptotic results for genetic algorithms with applications to nonlinear estimation. In L. Kallel and B. Naudts, eds, *Proceedings of the Second EvoNet Summer School on Theoretical Aspects of Evolutionary Computing*, Natural Computing Series. Springer-Verlag, New York, 2000.

[26] P. Del Moral and L. Miclo. Strong propagations of chaos in Moran's type particle interpretations of Feynman–Kac measures. *Stoch. Analysis and Appl.*, to appear (2007).

[27] P. Del Moral and L. Miclo. Branching and interacting particle systems approximations of Feynman–Kac formulae with applications to nonlinear filtering. In J. Azéma, M. Emery, M. Ledoux, and M. Yor, eds, *Séminaire de Probabilités XXXIV*, Lecture Notes in Mathematics 1729, pp. 1–145. Springer-Verlag, Berlin, 2000.

[28] P. Del Moral and L. Miclo. A Moran particle system approximation of Feynman–Kac formulae. *Stoch. Proc. Appl.*, 86:193–216, 2000.

[29] P. Del Moral and L. Miclo. Particle approximations of Lyapunov exponents connected to Schrödinger operators and Feynman–Kac semigroups. *ESAIM: Probab. Stat.* 7:171–208, 2003.

[30] P. Del Moral, J. C. Noyer, and G. Salut. Résolution particulaire et traitement non-linéaire du signal: application radar/sonar. In *Traitement du signal* 12(4):287–301, 1995.

[31] P. Del Moral, F. Patras, and S. Rubenthaler. Coalescent tree based functional representations for some Feynman–Kac particle models, *Annals of Applied Probablity*, 19(2):778–825, 2009.

[32] P. Del Moral, G. Rigal, and G. Salut. Estimation et commande optimale non linéaire. Technical Report 2, LAAS/CNRS, Toulouse, March 1992. Contract DRET-DIGILOG.

[33] P. Del Moral and G. Salut. Non linear filtering using Monte Carlo particle methods. *C.R. Académie des Sciences, Paris*, t. 320, Série I, 1147–52, 1995.

[34] A. Doucet, N. de Freitas, and N. Gordon, eds. *Sequential Monte Carlo Methods in Pratice*. Statistics for Engineering and Information Science. Springer, New York, 2001.

[35] N. J. Gordon, D. J. Salmon, and C. Ewing. Bayesian state estimation for tracking and guidance using the bootstrap filter. *J. Guidance Control Dyn.* 18(6):1434–43, 1995.

[36] N. J. Gordon, D. J. Salmon, and A. F. M. Smith. Novel approach to nonlinear/non-Gaussian Bayesian state estimation. *IEE Proc. F*, 140:107–13, 1993.

[37] A. Guionnet and E. Segala Maurel, *Combinatorial Aspects of Matrix Models*, arXiv:math.PR/0503064 v2, June 2005.

[38] G. Kallianpur and C. Striebel. Stochastic differential equations occurring in the estimation of continuous parameter stochastic processes. Tech. Rep. 103, Department of Statistics, University of Minnesota, Minneapolis, 1967.

[39] G. Kitagawa. Monte Carlo filter and smoother for non-Gaussian nonlinear state space models. *J. Comput. Graph. Stat.* 5(1):1–25, 1996.

[40] V. N. Kolokoltsov and V. P. Maslov. *Idempotent Analysis and Its Applications*, volume 401 of Mathematics and its Applications. Kluwer Academic, Dordrecht, 1997. Translation of *Idempotent Analysis and Its Application in optimal control* (Russian), Nauka Moscow, 1994, with an appendix by P. Del Moral.

[41] D. Ocone. Entropy inequalities and entropy dynamics in nonlinear filtering of diffusion processes. In *Stochastic Analysis, Control, Optimization and Applications*, Systems Control Foundations and Applications, pp. 477–96. Birkhäuser, Boston, 1999.

[42] S. T. Rachev. *Probability Metrics and the Stability of Stochastic Models*. Wiley, New York, 1991.

[43] T. Shiga and H. Tanaka. Central limit theorem for a system of Markovian particles with mean field interaction. *Z. Wahrschein. Verwandte Gebiete*, 69:439–59, 1985.

[44] A. N. Van der Vaart and J. A. Wellner. *Weak Convergence and Empirical Processes with Applications to Statistics*. Springer Series in Statistics. Springer, New York, 1996.

·26·
The Particle Filter in Practice

T. B. Schön, F. Gustafsson, and R. Karlsson

26.1 Introduction

Positioning of moving platforms has been a *technical driver* for real-time applications of the particle filter (PF) in both the signal processing and the robotics communities. For this reason, we will spend some time to explain several such applications in detail, and to summarize the experiences of using the PF in practice. The applications concern positioning of underwater vessels, surface ships, cars, and aircraft using geographical information systems containing a database with features of the surrounding. In the robotics community, the PF has been developed into one of the main algorithms (FastSLAM) for solving the simultaneous localization and mapping (SLAM) problem. This can be seen as an extension to the aforementioned applications, where the features in the geographical information system are dynamically detected and updated on the fly.

The common denominator of these applications of the PF is the use of a low-dimensional state vector consisting of horizontal position and course (three-dimensional pose). The PF performs quite well in a three-dimensional state-space. However, the PF is *not* practically useful when extending the models to more realistic cases with

- models in three dimensions (six-dimensional pose),
- more dynamic states (accelerations, unmeasured velocities, etc.),
- or sensor biases and drifts.

A *technical enabler* for such applications is the Rao–Blackwellized particle filter also referred to as the marginalized particle filter (MPF). It allows the use of high-dimensional state-space models as long as the (severe) nonlinearities only affect a small subset of the states. In this way, the structure of the model is utilized, so that the particle filter is used to solve the most difficult tasks, and the Kalman filter is used for linear Gaussian subsystems of the complete model. For subsystems that are almost linear and only slightly non-Gaussian, the extended Kalman filter (EKF) or the unscented Kalman filter (UKF) can be applied to reduce the burden of the PF. This latter is supposed to be a quite general case, since it is hard to find high-dimensional models where all states undergo complex nonlinear transformation with non-Gaussian noise disturbances.

The FastSLAM algorithm is in fact a version of the MPF, where hundreds or thousands of feature points in the state vector are updated using the Kalman filter. The need for the MPF in the list of applications will be motivated by examples and experience from practice.

The subsequent section discusses four different applications, where the PF and the MPF are primarily used to compute position estimates. Due to the importance of MPF when it comes to applications, Section 26.3 is devoted to the MPF. In Section 26.4 the positioning problem is extended to the case where there is no map available, i.e., the SLAM problem. Finally, the conclusions are given in Section 26.5.

26.2 Positioning applications

This section is concerned with four positioning applications of underwater vessels, surface ships, wheeled vehicles (cars), and aircraft, respectively. Though these applications are at first glance quite different, almost the same particle filter can be used in all of them. In fact, successful applications of a PF are described in literature which are all based on the same state-space model and similar measurement equations.

26.2.1 Model framework

The positioning applications, as well as existing applications of FastSLAM, are all based on the model

$$x_t = (X_t, Y_t, \psi_t)^T, \tag{26.1a}$$
$$u_t = (v_t, \dot{\psi}_t)^T, \tag{26.1b}$$
$$X_{t+1} = X_t + Tv_t \cos(\psi_t), \tag{26.1c}$$
$$Y_{t+1} = X_t + Tv_t \sin(\psi_t), \tag{26.1d}$$
$$\psi_{t+1} = \psi_t + T\dot{\psi}_t, \tag{26.1e}$$
$$y_t = h(x_t) + e_t. \tag{26.1f}$$

Here, X_t, Y_t denote the Cartesian position, ψ_t the course or heading, T is the sampling interval, v_t is the speed and $\dot{\psi}_t$ the yaw rate. The inertial signals v_t and $\dot{\psi}_t$ are considered as inputs to the dynamic model, and are given by on-board sensors. These are different in each of the four applications, and they will be described in more detail in the sequel. The measurement relation is based on a distance measuring equipment and a geographical information system (GIS). Both the distance measurement equipment and the GIS are different in the four applications, but the measurement principle is the same. By comparing the measured distance to objects in the GIS, a likelihood for each particle can be computed. It should here be noted that neither an EKF, UKF, or KF bank are suited for such problems. The reason is that it is typically not possible to linearize the database other than in a very small neighborhood.

In common for the applications is that they do not rely on satellite navigation systems, which are assumed not available or to provide insufficient navigation integrity. First, the inertial inputs, the distance measurement equipment and GIS for the four applications are described. Then, some conclusions from practice are summarized. Finally, different ways to augment the state vector are described for each application, and conclusions from applying the MPF are drawn. The point is that the dimension of the state vector has to be increased in order to account for model errors and more complicated dynamics. This implies that the PF is simply not applicable, due to the high dimensional state vector.

26.2.2 Applications of the PF

The outline follows a bottom-up approach, starting with underwater vessels below sea level and ending with fighter aircraft in the air.

Underwater positioning using a topographic map

The speed v_t and yaw rate $\dot{\psi}_t$ are computed using simplified dynamic motion models based on the propeller speed and the rudder angle. It is important to note that since the PF does not rely on pure dead-reckoning, such models do not have to be very accurate, see [12] for one simple linear model. An alternative is to use an inertial measurement units for measuring and computing speed and yaw rate.

A sonar is measuring the distance d_1 to the sea floor. The depth of the platform itself d_2 can be computed from pressure sensors, or from a sonar directed up-wards. By adding these distances, the sea depth at the position X_t, Y_t is measured. This can be compared to the depth in a dedicated sea chart with detailed topographical information, and the likelihood takes the combined effect of errors in the two sensors and the map into account, see [20]. Figure 26.1 provides an illustration. Detailed bottom sea charts are so far proprietary military information, and most applications are also military. However, oil companies are starting to use unmanned underwater vessels for exploring the sea and oil platforms, and in this way building up their own maps.

Surface positioning using a sea chart

The same principle as above can of course be used also for surface ships, which are constrained to be on the sea level ($d_2 = 0$). However, the standard vectorized sea charts (for instance the S-57 standard) contain a commercially available worldwide map.

The idea is to use the radar as distance measurement equipment and compare the detections with the shore profile, which is known from the sea chart conditioned on the position X_t, Y_t and course ψ_t (indeed the ship orientation,

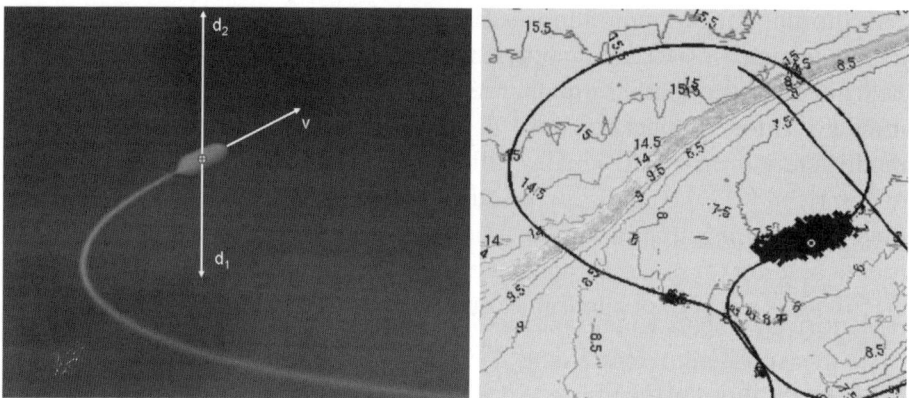

Fig. 26.1 The left plot is an illustration of an underwater vessel measuring distance d_1 to sea bottom, and absolute depth d_2. The sum $d = d_1 + d_2$ is compared to a bottom map as illustrated with the contours in the plot to the right. The particle cloud illustrates a snapshot of the PF from a known validation trajectory.

but more on this later), see [20]. The likelihood function models the radar error, but must also take clutter (false detections) and other ships into account.

The left hand part of Figure 26.2 illustrates the measurements provided by the radar, while the right-hand part of the same figure shows the radar detections from one complete revolution overlayed on the sea chart. The inertial data can be computed from propeller speed and rudder angle using simplified dynamical models as above. American and European maritime authorities have recently published reports highlighting the need for a backup and support

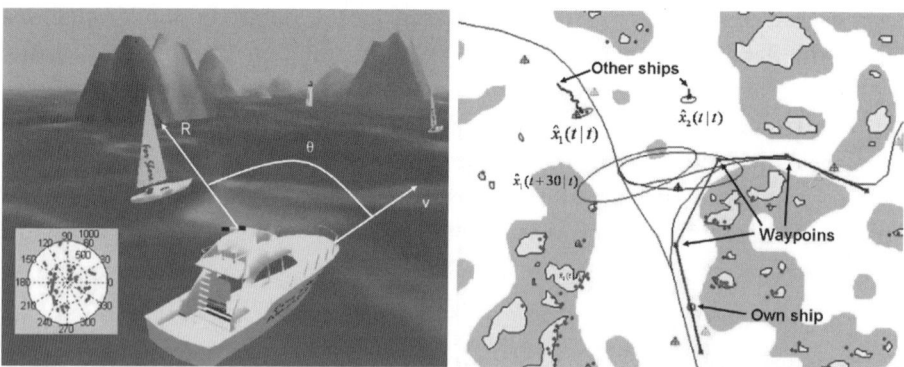

Fig. 26.2 The rotating radar returns detections of range R at body angle θ. The result of one radar revolution is conventionally displayed in polar coordinates as illustrated. Comparing the (R, θ) detections to a sea chart as shown to the right, the position and course are estimated by the PF. When correctly estimated, the radar overlay principle can be used for visual validation as also illustrated in the sea chart. The PF has to distinguish radar reflections from shore with clutter and other ships. The latter can be used for conventional target tracking algorithms, and collision avoidance algorithms, as also illustrated to the right.

system to satellite navigation to increase integrity. The reason is accidents and incidents caused by technical problems with the satellite navigation system, and the risk of accidental or deliberate jamming. The PF solution here is one candidate, since it is not sensitive to jamming nor does it require any extra infrastructure.

Vehicle positioning using a road map

The speed v_t and yaw rate $\dot{\psi}_t$ are here computed from the angular velocities of the nondriven wheels on one axle, using rather simple geometrical relations.

The measurement relation is in its simplest form a binary likelihood which is zero for all positions outside the roads, and a nonzero constant otherwise. In this case, the distance measurement equipment is basically the prior that the vehicle is located on a road, and not a conventional physical sensor. See [13, 15] for more details, and Figure 26.3 for an illustration. More sophisticated applications use vibrations in wheel speeds and vehicle body as a distance measurement equipment. When a rough surface is detected, this distance measurement equipment reports an increase in the likelihood for being outside the road. Likewise, if a forward-looking camera is present in the vehicle, this can be used to compute the likelihood that the front view resembles a road, or if it is rather a nonmapped parking area or a smaller private road.

The system is suitable as a support to satellite navigation in urban environments, in parking garages or tunnels or whenever satellite signals are likely to be obstructed. It is also a stand-alone solution to the navigation problem. Road

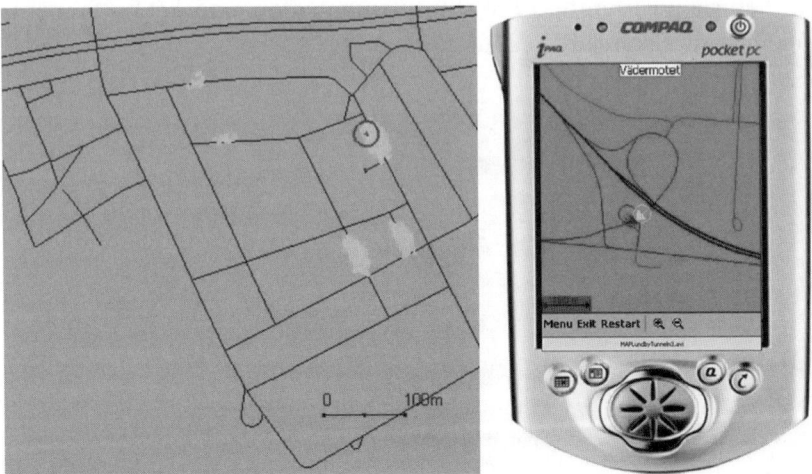

Fig. 26.3 Left: Example of multimodal posterior represented by a number of distinct particle clouds from NIRA Dynamics navigation system. This is caused by the regular road pattern and will be resolved after a sufficiently long sequence of turns. Right: PF in an embedded navigation solution run in real-time on a pocket PC.

databases covering complete continents are available from two main vendors (NavTech and TeleAtlas).

Aircraft positioning using a topographic map

The principal approach here is quite similar to the underwater positioning. A high-end IMU is used in an inertial navigation system (INS) which deadreckons the sensor data to position and orientation with quite high accuracy. Still, absolute position support is needed to prevent long-term drifts.

The distance measurement equipment is a wide-lobe downward-looking radar that measures the distance to the ground. The absolute altitude is computed using the INS and a supporting barometric pressure sensor. For more information about this application, see [28, 25]. Figure 26.4 shows one example just before convergence to a unimodal filtering density.

Commercial databases of topographic information are available on land (but not below sea level), with a resolution of 50–200 meters.

26.2.3 Summary of practical experiences

Real-time issues

The PF has been applied to real data and implemented on hardware targeted for the application platforms. The sampling rate has been chosen in the order 1–2 Hz, and there is no problem to achieve real-time performance in any of the applications. Some remarkable examples:

- The vehicle positioning PF was implemented on a pocket computer using 15000 particles already in 2001, [13].

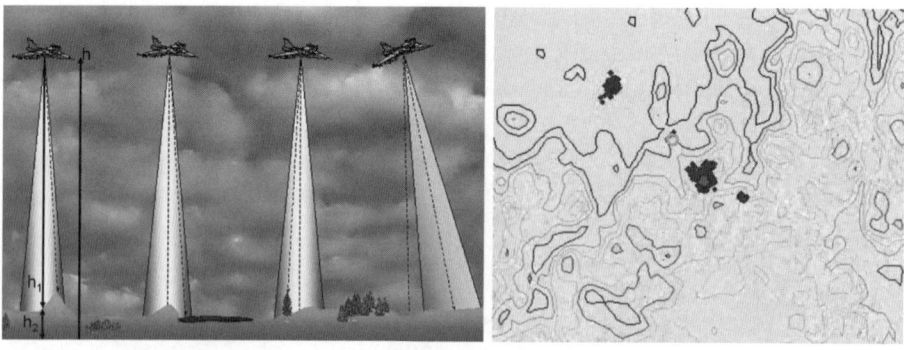

Fig. 26.4 The left figure is an illustration of an aircraft measuring distance h_1 to ground. The onboard baro-altitude supported INS system provides absolute altitude over sea level h, and the difference $h_2 = h - h_1$ is compared to a topographical map. The right plot shows a snapshot of the PF particle cloud, just after the aircraft has left the sea in the upper left corner. There are three distinct modes, where the one corresponding to the correct position dominates.

- The aircraft positioning PF was implemented in the programming language ADA used in the aircraft computer and shown to satisfy real-time performance on the onboard computer in the Swedish fighter Gripen in the year 2000. Real-time performance was reached, despite the facts that a very large number of particles were used, and that a rather old computer was used.

Sampling rates

The distance measurement equipment can in all cases deliver measurements much faster than the chosen sampling rate. However, faster sampling will introduce an unwanted correlation in the observations. This is due to the fact that the databases are quantized, so the platform should make a significant move between two measurement updates.

Implementation

Implementing and debugging the PF has not been a major issue. On the contrary, students and nonexperts have faced less problems with the PF than for similar projects involving the EKF. In many cases, they obtained deep intuition for including nontrivial but *ad hoc* modifications.

Dithering

Both the process noise and measurement noise distributions need some dithering (increased covariance). Dithering the process noise is a well-known method to mitigate the sample impoverishment problem [14]. Dithering the measurement noise is a good way to mitigate the effects of outliers and to robustify the PF in general. One simple and still very effective method to mitigate sample impoverishment is to introduce a lower bound on the likelihood. This lower bound was first introduced more or less *ad hoc*. However, recently this algorithm modification has been justified more rigorously. In proving that the particle filter converges for unbounded functions, like the state x_t itself, it is sufficient to have a lower bound on the likelihood, see [19] for details.

Number of particles

The number of particles is chosen quite large to achieve good transient behavior in the start up phase and to increase robustness. However, it has been concluded that in the normal operational mode the number of particles can be decreased substantially (typically a factor of ten). A real-time implementation should be designed for the worst case. However, adapting the sampling time T and the number of particles N is one option. The idea is to use a longer sampling interval and more particles initially, and when the PF has converged to a few distinct modes, T and N can be decreased in such a way that the complexity N/T is constant.

Choosing the proposal density

The standard sampling importance resampling (SIR) PF works fine for an initial design. However, the maps contain rather detailed information about position, and can in the limit be considered as state constraints. In such high signal-to-noise applications, the standard proposal density $p(x_{t+1}|x_t)$ used in the SIR PF is not particularly efficient. An alternative, that typically improves the performance, is to use the information available in the next measurement already in the state prediction step. Note that the proposal in general has the form

$$x_{t+1}^{(i)} \sim q(x_{t+1}|x_{1:t}, y_{1:t+1}), \qquad i = 1, \ldots, N \qquad (26.2)$$

so this is perfectly fine. Consider for instance positioning based on road maps. In standard SIR PF, the next positions are randomized around the predicted position according to the process noise, which is required to obtain diversity. Almost all of these new particles are outside the road network, and will not survive the resampling step. Obviously this is a waste of particles. By looking in the map how the roads are located locally around the predicted position, a much more clever process noise can be computed, and the particles explore the road network much more efficiently.

Divergence monitoring

Divergence monitoring is fundamental for real-time implementations to achieve the required level of integrity. After divergence, the particles do not reflect the true state distribution and there is no mechanism that automatically stabilizes the particle filter. Hence, divergence monitoring has to be performed in parallel with the actual PF code, and when divergence is detected, the PF is reinitialized.

One indicator of particle impoverishment is the efficient number of samples N_{eff}, used in the PF. This number monitors the amount of particles that significantly contribute to the posterior, and it is computed from the normalized weights. However, the unnormalized likelihoods are a more logical choice for monitoring. Standard hypothesis tests can be applied for testing whether the particle predictions represent the likelihood distribution.

Another approach is to use parallel particle filters interleaved in time. The requirement is that the sensors are faster than the chosen sampling rate in the PF. The PF's then use different time delays in the sensor observations.

The reinitialization procedure issued when divergence is detected is quite application dependent. The general idea is to use a very diffuse prior, or to

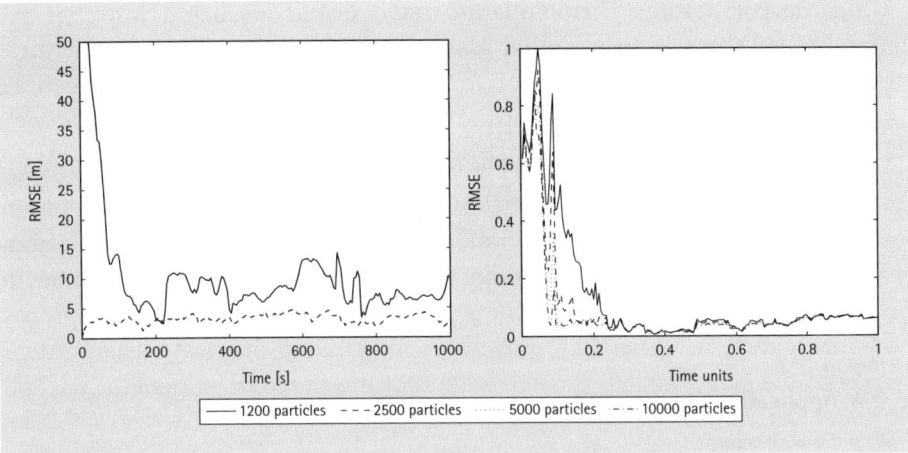

Fig. 26.5 Left: The position RMSE for the underwater application from the PF (solid line) using the experimental test data together with the parametric CRLB (dashed line) as the EKF solution around the true trajectory. Right: Horizontal position error from the aircraft application as a function of time for different number of particles. Note that the scale has been normalized for confidentiality reasons.

infer external information. In the vehicle positioning application [13], a cellular phone operator took part in the demonstrator, and cell information was used as a new prior for the PF in case of occasional divergence.

Performance bounds

For all four GPS-free applications the positioning performance is in the order of ten meter root mean square error (RMSE). Further, the performance has been shown to be close to the Cramér–Rao lower bound (CRLB) for a variety of examined trajectories. In Figure 26.5 two examples of performance evaluations in terms of the RMSE are depicted. On the left-hand side the position RMSE and CRLB are shown for the underwater application and on the right-hand side the horizontal position error is provided for the aircraft application.

PF in embedded systems

The primary application is to output position information to the operator. However, in all cases there have been decision and control applications built on the position information, which indicates that the PF is a powerful software component in embedded systems:

- Underwater positioning: Here, the entire mission relies on the position, so path planning and trajectory control are based on the output from the PF. Note that there is hardly any alternative below sea level, where no satellites are reachable, and deploying infrastructure (sonar buoys) is quite expensive.

- Surface positioning: Differentiating radar detections from shore, clutter and other ships is an essential association task in the PF. It is a natural extension to integrate a collision avoidance system in such an application, as illustrated in a sea chart snapshot in Figure 26.2.
- Vehicle positioning: The PF position was also used in a complete voice controlled navigation system with dynamic route optimization, see Figure 26.3.
- Aircraft navigation: The position from the PF is primarily used as a supporting sensor in the INS, whose position is a refined version of the PF output.

26.2.4 Applications of the MPF

Underwater positioning

Navigating an unmanned or manned underwater vessel requires knowledge of the full three-dimensional position and orientation, not only the projection on a horizontal plane. That is, at least six states are needed. For control, also the velocity and angular velocity are needed, which directly implies at least a twelve dimensional state vector. The PF cannot be assumed to perform well in such cases, and the MPF is a promising approach.

Surface positioning

There are two bottlenecks in the surface positioning PF that can be mitigated using the MPF. Both relates to the inertial measurements. First, the speed sensed by the log is the speed in water, not the speed over ground. Hence, the local water current is a parameter to include in the state vector. Second, the radar is strap-down and measures relative to body orientation, which is not the same as the course ψ_t. The difference is the so-called crab angle, which depends on currents and wind. This can also be included in the state vector. Further, there is in our demonstrator system an unknown and time-varying offset in the reported radar angle, which has to be compensated for.

Vehicle positioning

The bottleneck of the first generation of vehicle positioning PF is the assumption that the vehicle must be located on a road. As previously hinted, one could use a small probability in the likelihood function for being offroad, but there is no real benefit for this without an accurate dead-reckoning ability, so reoccurrence on the road network can be predicted with high reliability.

The speed and yaw rate computed from the wheel angular velocity are limited by the insufficient knowledge of wheel radii. However, the deviation between actual and real wheel radii on the two wheels on one axle can be included in the state vector. Similarly, with a yaw rate sensor available (standard component in

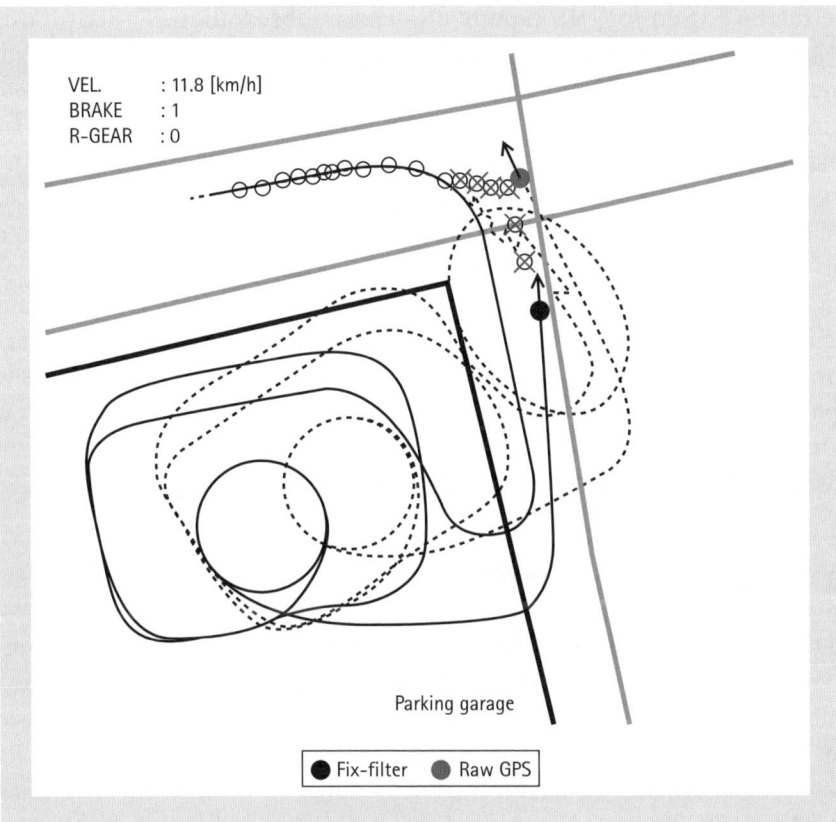

Fig. 26.6 Navigation of a car in a parking garage. Results for the MPF when relative wheel radii and gyro offset are added to the state vector. The two trajectories correspond to the map-aided system and an EKF with the same state vector, but where GPS is used as position sensor. Since the GPS gets several drop-outs before the parking garage, the dead-reckoning trajectory is incorrect.

electronic stability programs and navigation systems), the yaw rate drift can be included in the state vector. The point is that these parameters are accurately estimated when the vehicle is on the road, and in the off-road mode, accurate dead-reckoning can be achieved. Tests in demonstrator vehicles have shown that the exit point from parking garages and parking areas are well estimated, and that shorter unmapped roads are not a problem, see Figure 26.6.

Aircraft positioning

The primary role of the terrain-based navigation module is to support the inertial navigation system (INS) with absolute position information. The INS consists of an extended Kalman filter based on a state vector with over 20 motion states and sensor bias parameters. The current bottleneck is the interface between the terrain navigation module and the INS. The reason is that the terrain navigation module outputs a possibly multimodal probability density, while the EKF used for INS expects a Gaussian observation. The natural idea is to integrate both terrain navigation and INS into one filter. This results in a high-dimensional state vector, where one measurement (radar altitude) is very nonlinear. The MPF handles this elegantly, by essentially keeping the EKF from the existing INS and using the PF only for the radar altitude measurement.

The altitude radar gives a measurement outlier when the radar pulse is reflected in trees. Tests have validated that a Gaussian mixture, where one mode has a positive mean, models the real measurement error quite well. This Gaussian mixture distribution can be used in the likelihood computation, but such a distribution is in this case logically modeled by a binary Markov parameter, which is one in positions over forest and zero otherwise. In this way, the positive correlation between outliers is modeled, and a prior from ground type information in the GIS can be incorporated. This example motivates the inclusion of discrete states in the model framework. See [28, 26] for the details.

26.3 Marginalized particle filter

If there is a linear Gaussian substructure available in the model this can often be exploited to derive a better estimator. The basic idea is to estimate the "nonlinear" states using the particle filter and the conditionally linear Gaussian states using the Kalman filter. The resulting filter is called the MPF or the Rao–Blackwellized particle filter, and it has been known for quite some time, see e.g. [8, 5, 10, 6, 2, 28, 25].

In Section 26.3.1 the representation used in the MPF is illustrated by comparing it to several well-known estimators. The model structure is introduced in Section 26.3.2 and in Section 26.3.3 the MPF algorithm is given. Finally, the variance reduction property and the computational complexity of MPF are briefly discussed in Section 26.3.4 and 26.3.5, respectively.

26.3.1 Representation

The task of nonlinear filtering can be split into two parts: representation of the filtering density function and the propagation of this density through the time and measurement update stages. Figure 26.7 illustrates different representations of the filtering density for a two-dimensional example (similar to the example used in [29]). The extended Kalman filter can be interpreted as using one Gaussian distribution for representation and the propagation is performed according to a linearized model. The Gaussian sum filter [1, 30] extends the EKF to be able to represent multi-modal densities, still with an approximate propagation.

Figure 26.7(d)–(f) illustrates numerical approaches where the exact nonlinear relations present in the model are used for propagation. The point mass filter

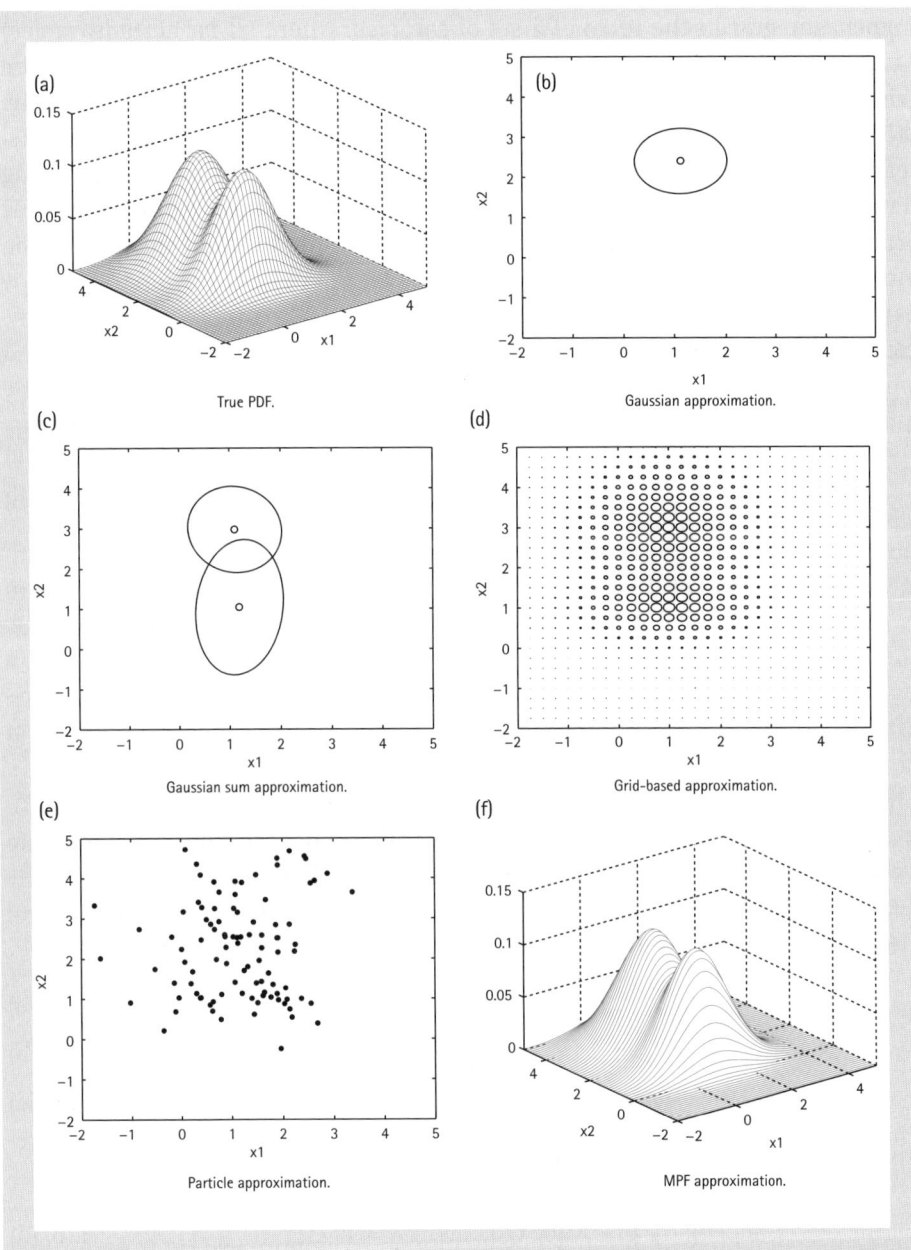

Fig. 26.7 True probability density function and different approximate representations, in order of appearance, Gaussian, Gaussian sum, point masses (grid-based approximation), particle samples, and MPF representation.

(grid-based approximation) [4] employs a regular grid, where the grid weight is proportional to the filtering density. The PF represents the filtering density by a stochastic grid in the form of a set of samples, where all particles (samples) have the same weight. Finally, the marginalized particle filter uses a stochastic grid for some of the states, and Gaussian densities for the rest. That is, the MPF can be interpreted as a particle representation for a subspace of the state-space, where each particle has an associated Gaussian density for the remaining state dimensions, Figure 26.7(f). It will be demonstrated that an exact nonlinear propagation is still possible if there is a linear substructure present in the model.

26.3.2 Model structure

A general nonlinear state-space model is given by

$$x_{t+1} = f_t(x_t, u_t) + w_t, \quad (26.3a)$$
$$y_t = h_t(x_t) + e_t. \quad (26.3b)$$

Note that the model (26.1) used in the positioning applications of the PF constitutes a special case of this structure. If there is a linear Gaussian substructure available in the model it is often rewarding to partitioned the state vector x_t according to

$$x_t = \begin{pmatrix} x_t^n \\ x_t^l \end{pmatrix}, \quad (26.4)$$

where the subvector x_t^l enters the dynamic model linearly with additive Gaussian noise, conditioned on x_t^n. Furthermore, x_t^n denotes the part of the state vector that does not fulfil the conditions for x_t^l. We will, for simplicity, informally refer to x_t^l as the linear states and x_t^n as the nonlinear states. A rather general model, containing a conditionally linear Gaussian substructure is given by

$$x_{t+1}^n = f_t^n(x_t^n) + A_t^n(x_t^n) x_t^l + G_t^n(x_t^n) w_t^n, \quad (26.5a)$$
$$x_{t+1}^l = f_t^l(x_t^n) + A_t^l(x_t^n) x_t^l + G_t^l(x_t^n) w_t^l, \quad (26.5b)$$
$$y_t = h_t(x_t^n) + C_t(x_t^n) x_t^l + e_t, \quad (26.5c)$$

where the process noise is assumed white and Gaussian distributed with

$$w_t = \begin{pmatrix} w_t^n \\ w_t^l \end{pmatrix} \sim \mathcal{N}(0, Q_t), \qquad Q_t = \begin{pmatrix} Q_t^n & Q_t^{ln} \\ (Q_t^{ln})^T & Q_t^l \end{pmatrix}. \quad (26.5d)$$

The measurement noise e_t is assumed white and Gaussian distributed according to

$$e_t \sim \mathcal{N}(0, R_t). \quad (26.5e)$$

Furthermore, x_0^l is Gaussian,

$$x_0^l \sim \mathcal{N}(\bar{x}_0, \bar{P}_0). \tag{26.5f}$$

Finally, the density of x_0^n can be arbitrary, but it is assumed known. More specifically, conditioned on the nonlinear state variables there is a linear substructure, subject to Gaussian noise available in (26.5).

Bayesian estimation methods, such as the particle filter, provide estimates of the filtering density function $p(x_t|y_{1:t})$. By employing the fact

$$p\left(x_t^l, x_{1:t}^n | y_{1:t}\right) = p\left(x_t^l | x_{1:t}^n, y_{1:t}\right) p\left(x_{1:t}^n | y_{1:t}\right), \tag{26.6}$$

the overall problem is decomposed into two sub-problems,

- A Kalman filter operating on the conditionally linear, Gaussian model (26.5) provides an estimate of $p\left(x_t^l | x_{1:t}^n, y_{1:t}\right)$.
- A particle filter for estimating the filtering density for the nonlinear states.

It is very important to note that the two subproblems mentioned above are coupled. This coupling is given in the subsequent section. However, a complete derivation is out of the scope for the present work, see e.g. [28] for such a derivation. Here, it is worth noting that the model (26.5) can be further generalized by introducing an additional discrete mode parameter, giving a larger family of marginalized filters, see [29]. This has recently been applied in [26]. An important special case of (26.5) is a model with linear state equations and nonlinear measurement equations. For this case the computational complexity is significantly reduced, since the same covariance matrix can be used for all the Kalman filters. This will be clearer from the discussion below.

26.3.3 Algorithm

Using (26.6) the problem can be put in a form that is suitable for the MPF framework, i.e., to analytically marginalize out the linear state variables from $p(x_t|y_{1:t})$. Note that $p\left(x_t^l|x_{1:t}^n, y_{1:t}\right)$ is analytically tractable, since $x_{1:t}^n$ is given by the particle filter. Hence, the underlying model is conditionally linear Gaussian, and the density function is given by the Kalman filter. Furthermore, since the estimate of $p\left(x_t^n|y_{1:t}\right)$ is provided by the particle filter, it is given by

$$\hat{p}^N\left(x_t^n|y_{1:t}\right) = \sum_{i=1}^{N} \tilde{\omega}_t^{(i)} \delta(x_t^n - x_t^{n,(i)}). \tag{26.7}$$

Hence, the resulting estimate of the filtering density is given by

$$\hat{p}^N(x_t|y_{1:t}) = \sum_{i=1}^{N} \tilde{\omega}_t^{(i)} \delta(x_t^n - x_t^{n,(i)}) \mathcal{N}(x_t^l; \hat{x}_{t|t}^{l,(i)}, P_{t|t}^{(i)}), \tag{26.8}$$

motivating the fact that the MPF provides an estimate of the filtering density that is a mix of a parametric and an nonparametric estimate. That is a mix of a parametric distribution from the Gaussian family and a nonparametric distribution represented by samples. If the same number of particles is used in the standard PF and the MPF, the latter will provide estimates of better or at least the same quality. Intuitively this makes sense, since the dimension of $p\left(x_t^n|y_{1:t}\right)$ is smaller than the dimension of $p(x_t|y_{1:t})$, implying that the particles occupy a lower dimensional space. Furthermore, the optimal algorithm (KF) is used to estimate the linear state variables. In Section 26.3.4 a more detailed discussion regarding the improved accuracy in the estimates is given. The MPF for estimating the states in a dynamic model in the form (26.5) is provided in Algorithm 26.1. For the sake of brevity, the particle index i and the dependence on the nonlinear state x_t^n are not explicitly stated in the algorithm.

Algorithm 26.1 (Marginalized particle filter)

1. Initialization: For $i = 1, \ldots, N$, initialize the particles, $x_{0|-1}^{n,(i)} \sim p_{x_0^n}\left(x_0^n\right)$ and set $\left\{x_{0|-1}^{l,(i)}, P_{0|-1}^{(i)}\right\} = \{\bar{x}_0^l, \bar{P}_0\}$. Set $t := 0$.
2. PF measurement update: For $i = 1, \ldots, N$, evaluate the importance weights

$$\omega_t^{(i)} = \frac{p\left(y_t|x_t^{(i)}\right) p\left(x_t^{n,(i)}|x_{t-1}^{n,(i)}\right)}{q\left(x_t^{n,(i)}|x_{1:t-1}^{n,(i)}, y_{1:t}\right)}, \qquad (26.9)$$

and normalize $\tilde{\omega}_t^{(i)} = \omega_t^{(i)} / \sum_{j=1}^{N} \omega_t^{(j)}$.
3. Draw N new particles, with replacement (resampling), for each $i = 1, \ldots, N$,

$$Pr\left(x_{t|t}^{n,(i)} = x_{t|t-1}^{n,(j)}\right) = \tilde{\omega}_t^{(j)}, \qquad j=1,\ldots, N.$$

4. PF time update and KF:

    (a) Kalman filter measurement update:

$$\hat{x}_{t|t}^l = \hat{x}_{t|t-1}^l + K_t \left(y_t - h_t - C_t \hat{x}_{t|t-1}^l\right), \qquad (26.10a)$$
$$P_{t|t} = P_{t|t-1} - K_t M_t K_t^T, \qquad (26.10b)$$
$$M_t = C_t P_{t|t-1} C_t^T + R_t, \qquad (26.10c)$$
$$K_t = P_{t|t-1} C_t^T M_t^{-1}. \qquad (26.10d)$$

    (b) PF time update (prediction): For $i = 1, \ldots, N$, predict new particles,

$$x_{t+1|t}^{n,(i)} \sim q\left(x_{t+1|t}^n | x_{1:t}^{n,(i)}, y_{1:t+1}\right).$$

    (c) Kalman filter time update:

$$\hat{x}_{t+1|t}^l = \bar{A}_t^l \hat{x}_{t|t}^l + G_t^l \left(Q_t^{ln}\right)^T \left(G_t^n Q_t^n\right)^{-1} z_t + f_t^l$$
$$+ L_t \left(z_t - A_t^n \hat{x}_{t|t}^l\right), \qquad (26.11a)$$

$$P_{t+1|t} = \bar{A}_t^l P_{t|t} (\bar{A}_t^l)^T + G_t^l \bar{Q}_t^l (G_t^l)^T - L_t N_t L_t^T, \quad (26.11\text{b})$$

$$N_t = A_t^n P_{t|t} (A_t^n)^T + G_t^n Q_t^n (G_t^n)^T, \quad (26.11\text{c})$$

$$L_t = \bar{A}_t^l P_{t|t} (A_t^n)^T N_t^{-1}, \quad (26.11\text{d})$$

where

$$z_t = x_{t+1}^n - f_t^n, \quad (26.12\text{a})$$

$$\bar{A}_t^l = A_t^l - G_t^l (Q_t^{ln})^T (G_t^n Q_t^n)^{-1} A_t^n, \quad (26.12\text{b})$$

$$\bar{Q}_t^l = Q_t^l - (Q_t^{ln})^T (Q_t^n)^{-1} Q_t^{ln}. \quad (26.12\text{c})$$

5. Set $t := t + 1$ and repeat from step 2.

From Algorithm 26.1, it should be clear that the only difference from the standard PF is that the time update (prediction) stage has been changed. In the standard PF, the prediction stage is given solely by step 4(b) in Algorithm 26.1. In order to help intuition, step 4 in Algorithm 26.1 will now be briefly discussed. Step 4(a) is a standard Kalman filter measurement update using the information available in the measurement y_t. Once this has been performed the new estimates of the linear states can be used to obtain a prediction of the nonlinear state $x_{t+1|t}^n$. This is performed in Step 4(b). Now, consider model (26.5) conditioned on the nonlinear state variable. The conditioning implies that (26.5a) can be thought of as a measurement equation. This is used in step 4(c) together with a time update of the linear state estimates.

An alternative way is to describe the algorithm in terms of multiple models and a mixing of PF and KF time updates and measurement updates, see [18] for details on this approach. An elegant interpretation and derivation of the MPF in terms of a filter bank is provided in [17].

The estimates, as expected means, of the state variables and their covariances are given below.

$$\hat{x}_{t|t}^n = \sum_{i=1}^{N} \tilde{\omega}_t^{(i)} \hat{x}_{t|t}^{n,(i)}, \quad (26.13\text{a})$$

$$\hat{P}_{t|t}^n = \sum_{i=1}^{N} \tilde{\omega}_t^{(i)} \left(\left(\hat{x}_{t|t}^{n,(i)} - \hat{x}_{t|t}^n \right) \left(\hat{x}_{t|t}^{n,(i)} - \hat{x}_{t|t}^n \right)^T \right), \quad (26.13\text{b})$$

$$\hat{x}_{t|t}^l = \sum_{i=1}^{N} \tilde{\omega}_t^{(i)} \hat{x}_{t|t}^{l,(i)}, \quad (26.13\text{c})$$

$$\hat{P}_{t|t}^l = \sum_{i=1}^{N} \tilde{\omega}_t^{(i)} \left(P_{t|t}^{(i)} + \left(\hat{x}_{t|t}^{l,(i)} - \hat{x}_{t|t}^l \right) \left(\hat{x}_{t|t}^{l,(i)} - \hat{x}_{t|t}^l \right)^T \right), \quad (26.13\text{d})$$

where $\left\{ \tilde{\omega}_t^{(i)} \right\}_{i=1}^{N}$ are the normalized importance weights, provided by step 2 in Algorithm 26.1.

26.3.4 Variance reduction

The *law of total variance* says that

$$\text{Cov}(U) = \text{Cov}(\text{E}(U|V)) + \text{E}(\text{Cov}(U|V)), \qquad (26.14)$$

where U and V denotes stochastic variables. Letting $U = x_t^l$ and $V = x_{1:t}^n$ results in the following decomposition of the covariance of the PF

$$\underbrace{\text{Cov}\left(x_t^l\right)}_{PF} = \text{Cov}\left(\text{E}\left(x_t^l | x_{1:t}^n\right)\right) + \text{E}\left(\text{Cov}\left(x_t^l | x_{1:t}^n\right)\right) \qquad (26.15a)$$

$$= \underbrace{\text{Cov}\left(\hat{x}_{t|t}^l\right)}_{MPF} + \sum_{i=1}^{N} w_t^{(i)} \underbrace{P_{t|t}^{(i)}}_{KF}. \qquad (26.15b)$$

Here, we recognize $p\left(x_t^l | x_{1:t}^n\right)$ as the Gaussian distribution, delivered by the KF, conditioned on the trajectory $x_{1:t}^n$. Now, the MPF computes the mean of each trajectory as $\hat{x}_{t|t}^{l,(i)}$ and the unconditional mean estimator is simply the mean of these, $\sum_{i=1}^{N} \tilde{w}_t^{(i)} \hat{x}_{t|t}^{l,(i)}$, and its covariance follows from the first term in (26.15b). The second term in (26.15b) corresponds to the contribution due to the fact that each Gaussian distribution is represented by one sample, as is done in the PF. This principle, which leads to the variance reduction property is sometimes referred to as Rao–Blackwellization, see e.g. [27], and it is the basic part that improves performance using the marginalization idea. Note that for the variance reduction to be significant, the left-hand side in (26.15) has to be significantly smaller than the right-hand side. In other words, the term

$$\text{E}\left(\text{Cov}\left(x_t^l | x_{1:t}^n\right)\right) \qquad (26.16)$$

has to be large. That is, the expectation of the conditional variance of the corresponding Kalman filter has to be large. In order to make this a bit clearer, we have

$$\text{Cov}\left(\text{E}\left(x_t^l | x_{1:t}^n\right)\right) \leq \text{Cov}\left(x_t^l\right), \qquad (26.17)$$

since $\text{E}\left(\text{Cov}\left(x_t^l | x_{1:t}^n\right)\right) \geq 0$. This shows that the variance of the linear part is always smaller for the MPF than for the PF. The difference is the expected covariance term,

$$\text{E}\left(\text{Cov}\left(x_t^l | x_{1:t}^n\right)\right) = \sum_{i=1}^{N} \tilde{w}_t^{(i)} P_{t|t}^{(i)}. \qquad (26.18)$$

This states that the improvement in the quality of the estimate is given by the term $\text{E}\left(\text{Cov}\left(x_t^l | x_{1:t}^n\right)\right)$. That is, the Kalman filter covariance $P_{t|t}$ is a good indicator of how much has been gained in using the MPF instead of the PF. Further discussions regarding the variance reduction property of the MPF are provided in [9, 10].

26.3.5 Computational complexity

The variance reduction in the MPF can be used in two different ways:

- With the same number of particles, the variance in the estimates of the linear states can be decreased.
- With the same performance in terms of variance for the linear states, the number of particles can be decreased.

This is schematically illustrated in Figure 26.8, for the case where $C = 0$, implying that the same covariance matrix can be used for all particles. The two alternatives above are illustrated with the thin lines, in case a PF with 10000 particles is first applied, and then replaced by the MPF.

Another related question is how the computational complexity relates to the number of particles. The MPF appears to add quite a lot of overhead computations. It turns out, however, that the MPF is often more efficient also here. It may seem impossible to give any general conclusions, so application dependent simulation studies have to be performed. Nevertheless,

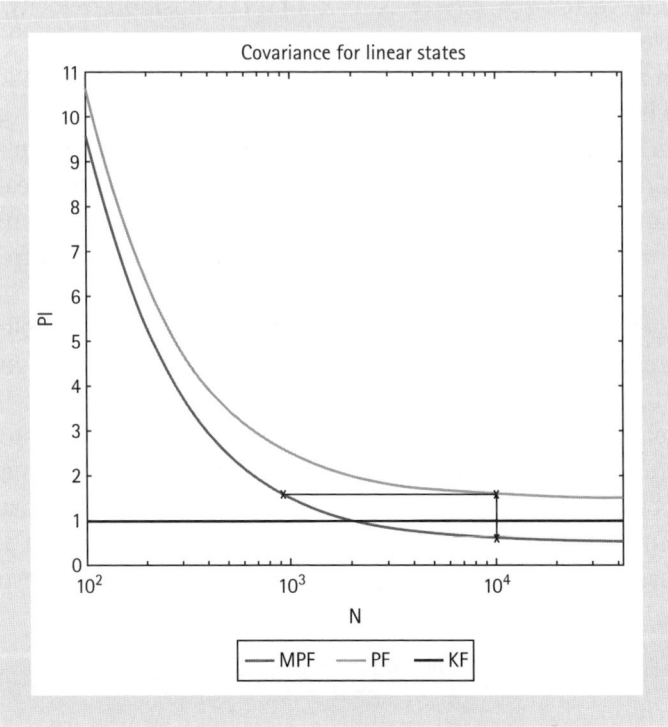

Fig. 26.8 Schematic view of how the covariance of the linear part of the state vector depends on the number of particles for the PF and MPF, respectively. The gain in MPF is given by the Kalman filter covariance.

quite realistic predictions about the computational complexity can be done with rather simple calculations, an example of this is given below.

26.3.6 Radar tracking case study

This section will serve as an illustration of the computational complexity of the marginalized particle filter. A special case of the general model (26.5) will be used, where the state dynamics are linear and Gaussian and the measurements are nonlinear functions in some of the states:

$$x_{t+1} = \begin{pmatrix} 1 & 0 & T & 0 & T^2/2 & 0 \\ 0 & 1 & 0 & T & 0 & T^2/2 \\ 0 & 0 & 1 & 0 & T & 0 \\ 0 & 0 & 0 & 1 & 0 & T \\ 0 & 0 & 0 & 0 & 1 & 0 \\ 0 & 0 & 0 & 0 & 0 & 1 \end{pmatrix} x_t + w_t, \qquad (26.19a)$$

$$y_t = \begin{pmatrix} \sqrt{X_t^2 + Y_t^2} \\ \arctan(Y_t/X_t) \end{pmatrix} + e_t, \qquad (26.19b)$$

where the state vector is $x_t = (X\ Y\ \dot{X}\ \dot{Y}\ \ddot{X}\ \ddot{Y})^T$, i.e., position, velocity, and acceleration in two dimensions.

This is an archetypal model for many target tracking and some localization problems, where the state dynamics is modelled using a constant velocity or a constant acceleration assumption. The observation can in most applications be written as a nonlinear function of position. This is the case for bearings-only, range-only and radar sensors. More specifically, the measurement equation in (26.19b) models measurements of the range and the azimuth from a radar system.

In [21] a general method for analysing the computational complexity of the MPF was presented. Here, the highlights of this analysis will be given, based on model (26.19).

The model has two nonlinear state variables and four linear state variables. Two cases are now studied, the full PF, where all states are estimated using the PF and the completely marginalized PF, where all linear states are marginalized out and estimated using the KF. If we want to compare the two approaches under the assumption that they use the same computational resources, we obtain

$$N_{\text{PF}} = \underbrace{(1-c)}_{<1} N_{\text{MPF}}. \qquad (26.20)$$

where N_{PF} and N_{MPF} denote the number of particles used in the PF and the MPF, respectively. Furthermore, c is a positive constant depending of the computational complexity of the various parts of the algorithm, see [21] for the complete

Table 26.1 Results from the simulation. Left: Using a constant computational complexity, i.e., the algorithms can consume the same amount of time. Right: Using a constant velocity RMSE.

	PF	MPF		PF	MPF
N	2000	2574	N	2393	264
RMSE pos	7.10	5.60	RMSE pos	7.07	7.27
RMSE vel	3.62	3.21	RMSE vel	3.58	3.61
RMSE acc	0.52	0.44	RMSE acc	0.50	0.48
Time	0.59	0.60	Time	0.73	0.10

details. From (26.20) it is clear that for a given computational complexity more particles can be used in the MPF than in the standard PF, in this example.

Using a constant computational complexity the number of particles that can be used is computed. The study is performed by first running the full PF and measure the time consumed by the algorithm. A Monte Carlo simulation, using $N = 2000$ particles, is performed in order to obtain a stable estimate of the time consumed by the algorithm. In the left hand side of Table 26.1 the number of particles (N), the total RMSE from 100 Monte Carlo simulations, and the simulation times are shown both for the PF and the MPF.

From Table 26.1 it is clear that for this example, the MPF can use more particles for a given time, which is in perfect correspondence with the theoretical result summarized in (26.20).

Let us now discuss what happens if a constant velocity RMSE is used. First the velocity RMSE for the full PF is found using a Monte Carlo simulation. This value is then used as a target function in the search for the number of particles needed by the MPF. The right hand side of Table 26.1 clearly indicates that the MPF can obtain the same RMSE using fewer particles. The result is that using full marginalization only requires 14% of the computational resources as compared to the standard PF in this example.

Note that for this example $C_t = 0$, implying that the same covariance matrix can be used for all Kalman filters, which significantly reduce the computational complexity.

26.4 Simultaneous localization and mapping

Simultaneous localization and mapping (SLAM) is an extension of the positioning problem previously discussed to the case where the environment is unknown. In SLAM this is handled by estimating a map on-line together with the platform states. The SLAM problem has been studied for a long time within many different settings and an introduction to the problem can be found in the survey papers [3, 11] and the book [31]. Here, the focus will be on the filtering part of the SLAM problem. The SLAM problem also contains very interesting issues when it comes to feature extraction, data association and matching features in consecutive images.

26.4.1 Problem formulation

Many different estimation algorithms, including for example the extended Kalman filter, the particle filter and the extended information filter, have been used in solving the SLAM problem. A key observation here is that the most promising algorithms so far utilize the structure inherent in the SLAM problem in one way or the other. When it comes to the algorithms based on the particle filter it is perhaps the FastSLAM algorithm [24, 23] that has received most attention. FastSLAM can be seen as a special case of the MPF. Here, the state vector contains information about the platform x_t and the position of the features m_t, which consists of the entire map at time t, i.e.,

$$m_t = \begin{pmatrix} m_{1,t}^T & \dots & m_{M_t,t}^T \end{pmatrix}^T, \qquad (26.21)$$

where $m_{j,t}$ denotes the position of the jth map entry at time t and M_t denotes the number of entries in the map at time t. The underlying idea is to use the particle filter to estimate the platform states and to model the landmarks as linear Gaussian states, which are estimated using Kalman filters. We are interested in estimating the filtering density function $p(x_t, m_t | y_{1:t})$, which using Bayes' theorem can be expressed as

$$p(x_{1:t}, m_t | y_{1:t}) = p(m_t | x_{1:t}, y_{1:t}) p(x_{1:t} | y_{1:t}), \qquad (26.22)$$

which is exactly the factorization previously employed in (26.6). The FastSLAM algorithm was originally devised to solve the SLAM problem for mobile robots, where the dimension of the state vector is small, typically consisting of only three states (2D position and a course (heading) angle) [31]. This implies that all platform states can be estimated by the PF. In discussing the FastSLAM algorithm it is worth mentioning that when $p(x_{t+1}|x_t)$ is used as proposal density the algorithm is referred to as FastSLAM1.0 and when $p(x_{t+1}|x_{1:t}, y_{1:t+1})$ is used it is referred to as FastSLAM2.0. The choice of importance density makes a rather big difference in the SLAM application.

26.4.2 Marginalized FastSLAM

In considering different platforms, for example an unmanned aerial vehicle, the dimension of the state vector describing the platform will typically be quite high. The platform dynamics often has a linear, Gaussian substructure available in it [15] and we can of course exploit this structure as well. This will give us an algorithm capable of dealing with high dimensional platform state vectors as well. Let us for this case partition the state vector according to

$$x_t = \left(\left(x_t^p\right)^T \left(x_t^k\right)^T m_t^T \right)^T, \qquad (26.23)$$

where x_t^p denotes the states of the platform that are estimated by the particle filter, and x_t^k denotes the states of the platform that are linear and Gaussian, given information about x_t^p. These states together with the map (landmarks) m_t are estimated using Kalman filters. The dynamic model is a special case of (26.5),

$$x_{t+1}^p = f_t^p\left(x_t^p\right) + A_t^p\left(x_t^p\right) x_t^k + G_t^p\left(x_t^p\right) w_t^p, \qquad (26.24a)$$
$$x_{t+1}^k = f_t^k\left(x_t^p\right) + A_t^k\left(x_t^p\right) x_t^k + G_t^k\left(x_t^p\right) w_t^k, \qquad (26.24b)$$
$$m_{j,t+1} = m_{j,t}, \qquad (26.24c)$$
$$y_{1,t} = h_{1,t}\left(x_t^p\right) + C_t\left(x_t^p\right) x_t^k + e_{1,t}, \qquad (26.24d)$$
$$y_{2,t}^{(j)} = h_{2,t}\left(x_t^p\right) + H_{j,t}\left(x_t^p\right) m_{j,t} + e_{2,t}^{(j)}, \qquad (26.24e)$$

where $x_t^n = x_t^p$ and $x_t^l = \left(\left(x_t^k\right)^T m_t^T\right)^T$. In order to compute an estimate of the filtering density function $p\left(x_t^p, x_t^k, m_t | y_{1:t}\right)$, the key is the following factorization

$$p\left(x_{1:t}^p, x_t^k, m_t | y_{1:t}\right) = p\left(m_t | x_{1:t}^p, x_t^k, y_{1:t}\right) p\left(x_t^k | x_{1:t}^p, y_{1:t}\right) p\left(x_{1:t}^p | y_{1:t}\right) \quad (26.25a)$$

$$= \prod_{j=1}^{M_t} \underbrace{p\left(m_{j,t} | x_{1:t}^p, x_t^k, y_{1:t}\right) p\left(x_t^k | x_{1:t}^p, y_{1:t}\right)}_{\text{(extended) Kalman filter}} \underbrace{p\left(x_{1:t}^p | y_{1:t}\right)}_{\text{particle filter}},$$

$$(26.25b)$$

where the second expression follows if all features are assumed independent. Using this factorization together with model (26.24) and Algorithm 26.1 we can derive a rather general algorithm which is applicable to many different platforms (aircraft, helicopters, cars, etc.). The ith particle at time t, $x_t^{(i)}$ will contain the following information

$$x_t^{(i)} = \left(x_t^{p,(i)}, \hat{x}_t^{k,(i)}, (\hat{m}_{1,t}, \Sigma_{1,t}), \ldots, (\hat{m}_{M_t,t}, \Sigma_{M_t,t})\right), \qquad (26.26)$$

where $\hat{m}_{j,t}$ and $\Sigma_{j,t}$ provide the position estimate and the associated covariance for the jth map entry, respectively. The details of the derivation are thoroughly discussed in [22, 32], where we also give an application example in terms of an unmanned aerial vehicle. In order to illustrate the algorithm we will give some insight to this example here.

26.4.3 Unmanned aerial vehicle application

The information we use to solve the SLAM problem originates from the following sensors:

- camera
- inertial measurement unit (IMU)
- pressure sensor.

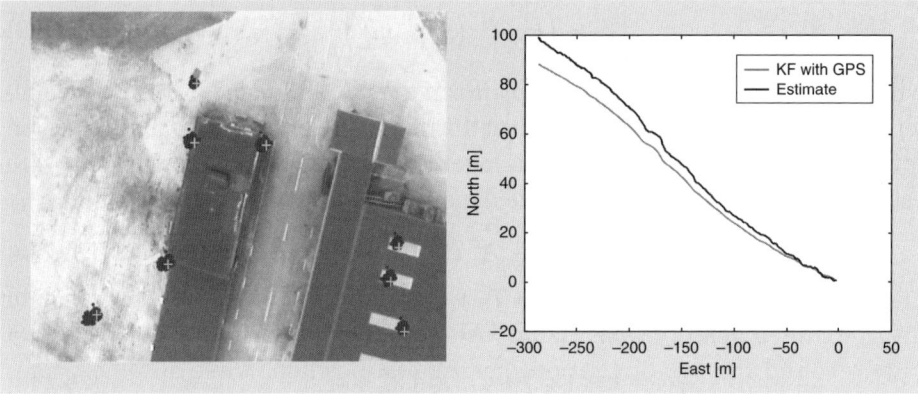

Fig. 26.9 Left: The scenario seen from the on-board camera, together with particle clouds representing the landmarks/map features. The crosses show the measurements from the computer vision algorithm. Right: Horizontal position estimate from the SLAM algorithm, compared to the GPS, which is only used as reference here. The unmanned aerial vehicle starts in the origin at time $t = 0$.

Hence, solving the SLAM problem for this particular case is a sensor fusion problem where the information from the above mentioned sensors is fused. Note that we have to make use of a linearized measurement model (26.24e) for the camera in order to fit the current framework, again we refer to [22] for the details. We have employed a rather simple solution when it comes to the map and the computer vision part. We simply make use of the Harris detector [16] in order to find points of interest. Then we cut out an 11×11 patch around this interest point. These patches will then constitute our map. Similar approaches have previously been used for similar problems, see e.g. [7]. The performance is illustrated in Figure 26.9.

26.5 Conclusions

The particle filter (PF) has been successfully used in quite a few different applications, at least when the state dimension is moderately low. Design issues, such as the choices of number of particles, resampling strategies and proposal density are becoming better understood both in theory and practice. There are many suggested *ad hoc* strategies for practical aspects as divergence monitoring and dithering. We have in this survey used four localization applications to illustrate these issues. We have also pointed out the similarity with the FastSLAM algorithm for simultaneous localization and mapping used in robotics.

However, for complex high-dimensional models, the number of applications in literature is quite limited. The surveyed localization applications all work well using the PF for their simplest problem formulations leading to low-dimensional models. However, we have also, for each application, motivated that more complex models would give additional performance, if only the PF can be applied. Now, the curse of dimensionality prevents the use of the PF in its original form. However, all these applications reveal a linear Gaussian substructure that enables the use of the MPF. The MPF strategy can be generalized to models with substructures that are almost linear and Gaussian, using the same arguments as when applying the EKF to nonlinear models. In fact, it is hard to find high-dimension models useful in practice where all states are severely nonlinear or non-Gaussian. To illustrate the concept, the MPF was applied to a kinematic model with a high dimensional state vector for navigating an unmanned aerial vehicle using vision and inertial sensors. Here, the PF only involved the two-dimensional horizontal position, while all other states were sufficiently linear and Gaussian for the EKF. To understand the principles of the MPF, we have reviewed a general model structure, the MPF algorithm and discussed the interplay of variance reduction and computational complexity when the PF is replaced with an MPF.

References

[1] B. D. O. Anderson and J. B. Moore. *Optimal Filtering*. Information and system science series. Prentice-Hall, Englewood Cliffs, NJ, USA, 1979.
[2] C. Andrieu and A. Doucet. Particle filtering for partially observed Gaussian state space models. *Journal of the Royal Statistical Society*, 64(4):827–36, 2002.
[3] T. Bailey and H. Durrant-Whyte. Simultaneous localization and mapping (SLAM): Part II. *IEEE Robotics & Automation Magazine*, 13(3):108–17, Sept. 2006.
[4] N. Bergman. *Recursive Bayesian Estimation: Navigation and Tracking Applications*. PhD thesis No. 579, Linköping Studies in Science and Technology, SE-581 83 Linköping, Sweden, May 1999.
[5] G. Casella and C. P. Robert. Rao–Blackwellisation of sampling schemes. *Biometrika*, 83(1):81–94, 1996.
[6] R. Chen and J. S. Liu. Mixture Kalman filters. *Journal of the Royal Statistical Society*, 62(3):493–508, 2000.
[7] A. J. Davison, I. Reid, N. Molton, and O. Strasse. MonoSLAM: Real-time single camera SLAM. *IEEE Transactions on Patterns Analysis and Machine Intelligence*, 29(6):1052–67, June 2007.
[8] A. Doucet, S. J. Godsill, and C. Andrieu. On sequential Monte Carlo sampling methods for Bayesian filtering. *Statistics and Computing*, 10(3):197–208, 2000.
[9] A. Doucet, N. Gordon, and V. Krishnamurthy. Particle filters for state estimation of jump Markov linear systems. Technical Report CUED/F-INFENG/TR 359, Signal Processing Group, Department of Engineering, University of Cambridge, Trupington street, CB2 1PZ Cambridge, 1999.
[10] A. Doucet, N. Gordon, and V. Krishnamurthy. Particle filters for state estimation of jump Markov linear systems. *IEEE Transactions on Signal Processing*, 49(3):613–24, 2001.
[11] H. Durrant-Whyte and T. Bailey. Simultaneous localization and mapping (SLAM): Part I. *IEEE Robotics & Automation Magazine*, 13(2):99–110, June 2006.

[12] K. M. Fauske, F. Gustafsson, and O. Herenaes. Estimation of AUV dynamics for sensor fusion. In *Proceedings of the 10th International Conference on Information Fusion*, Québec, Canada, July 2007.

[13] U. Forssell, P. Hall, S. Ahlqvist, and F. Gustafsson. Novel map-aided positioning system. In *Proceedings of FISITA*, Helsinki, Finland, 2002.

[14] N. J. Gordon, D. J. Salmond, and A. F. M. Smith. Novel approach to nonlinear/non-Gaussian Bayesian state estimation. In *IEE Proceedings on Radar and Signal Processing* 140:107–13, 1993.

[15] F. Gustafsson, F. Gunnarsson, N. Bergman, U. Forssell, J. Jansson, R. Karlsson, and P.-J. Nordlund. Particle filters for positioning, navigation and tracking. *IEEE Transactions on Signal Processing*, 50(2):425–37, Feb. 2002.

[16] C. Harris and M. Stephens. A combined corner and edge detector. In *Proceedings of the 4th Alvey Vision Conference*, pp. 147–51, Manchester, UK, 1988.

[17] G. Hendeby. *Performance and Implementation Aspects of Nonlinear Filtering*. PhD thesis No. 1161, Linköping Studies in Science and Technology, Department of Electrical Engineering, Linköping University, Sweden, February 2008.

[18] G. Hendeby, R. Karlsson, and F. Gustafsson. A new formulation of the Rao-Blackwellized particle filter. In *IEEE Statistical Signal Processing Workshop*, Madison, Wisconsin, USA, 2007.

[19] X.-L. Hu, T. B. Schön, and L. Ljung. A basic convergence result for particle filtering. *IEEE Transactions on Signal Processing*, 56(4):1337–48, April 2008.

[20] R. Karlsson and F. Gustafsson. Bayesian surface and underwater navigation. *IEEE Transactions on Signal Processing*, 54(11):4204–13, 2006.

[21] R. Karlsson, T. Schön, and F. Gustafsson. Complexity analysis of the marginalized particle filter. *IEEE Transactions on Signal Processing*, 53(11):4408–11, Nov. 2005.

[22] R. Karlsson, T. B. Schön, D. Törnqvist, G. Conte, and F. Gustafsson. Utilizing model structure for efficient simultaneous localization and mapping for a UAV application. In *Proceedings of IEEE Aerospace Conference*, Big Sky, MT, USA, March 2008.

[23] M. Montemerlo and S. Thrun. *FastSLAM A Scalable Method for the Simultaneous Localization and Mapping Problem in Robotics*. Star–Springer tracts in advanced robotics. Springer, Berlin, Germany, 2007.

[24] M. Montemerlo, S. Thrun, D. Koller, and B. Wegbreit. FastSLAM a factored solution to the simultaneous localization and mapping problem. In *Proceedings of the AAAI National Comference on Artificial Intelligence*, Edmonton, Canada, 2002.

[25] P.-J. Nordlund. *Sequential Monte Carlo Filters and Integrated Navigation*. Licentiate Thesis No. 945, Department of Electrical Engineering, Linköping University, Sweden, 2002.

[26] P.-J. Nordlund and F. Gustafsson. Marginalized particle filter for accurate and reliable terrain-aided navigation. *Provisionally accepted for IEEE Transactions on Aerospace and Electronic Systems*, 2009.

[27] C. P. Robert and G. Casella. *Monte Carlo Statistical Methods*. Springer texts in statistics. Springer, New York, 1999.

[28] T. Schön, F. Gustafsson, and P.-J. Nordlund. Marginalized particle filters for mixed linear/nonlinear state-space models. *IEEE Transactions on Signal Processing*, 53(7):2279–89, July 2005.

[29] T. B. Schön, R. Karlsson, and F. Gustafsson. The marginalized particle filter in practice. In *Proceedings of IEEE Aerospace Conference*, Big Sky, MT, USA, March 2006.

[30] H. W. Sorenson and D. L. Alspach. Recursive Bayesian estimation using Gaussian sum. *Automatica*, 7:465–79, 1971.

[31] S. Thrun, W. Burgard, and D. Fox. *Probabilistic Robotics*. Intelligent Robotics and Autonomous Agents. MIT Press, Cambridge, MA, 2005.

[32] D. Törnqvist, T. B. Schön, R. Karlsson, and F. Gustafsson. Particle filter SLAM with high dimensional vehicle model. *Journal of Intelligent and Robotic Systems*, 55(4–5):249–266, August 2009.

·27·
Introducing Cubature to Filtering

C. Litterer and T. Lyons

27.1 Introduction

Cubature on Wiener space is a higher-order method for approximating the weak solution of stochastic differential equations (SDE) in their Stratonovich form. The method was developed by Lyons, Victoir [23] following work of Kusuoka [15] and we will sometimes refer to it as the KLV method. Applications can be found in mathematical finance, for example in Ninomiya, Victoir [25] and in a number of related works (see e.g. [8],[21],[31]).

More precisely the KLV method approximates $P_T f(x)$ for a sufficiently rich class of test functions f where the semigroup P_t is defined by $P_t f(x) := \mathbb{E}(f(X_{t,x}))$. Here $X_{t,x}$ is the solution to the SDE

$$dX_{t,x} = V_0(X_{t,x})dt + \sum_{i=1}^{d} V_i(X_{t,x}) \circ dB_t^i, \quad X_{0,x} = x,$$

$(V_i)_{i=0}^{d}$ is a family of smooth vector fields with bounded derivatives of all orders and $\left(B_t^i\right)_{i=1}^{d}$ is Brownian motion. We will assume that the test functions f are Lipschitz and the vector fields satisfy the UFG condition (see e.g [17]), which is weaker than the usual uniform Hörmander condition. It is well known that computing $P_T f(x)$ corresponds to solving a parabolic partial differential equation.

In the first sections of our chapter we introduce the ideas of cubature on Wiener space, its iterated application in the KLV method and derive some simple bounds for the KLV approximation of $P_T f(x)$. Starting with cubature on Wiener space we derive in the spirit of Strichartz [28], Hu [13], and Castell [4] an operator based on the flows of autonomous ODEs that corresponds to a version of Kusuoka's algorithm [17]. We subsequently discuss some advantages and drawbacks of the resulting algorithms from a numerical perspective.

The KLV method may be regarded as a particle system on R^N branching in the form of a n-ary tree, where n is the number of paths in the support of the cubature measure. Hence, while such higher order methods are effective in practice (see e.g. Ninomiya, Victoir [25] and Ninomiya, Ninomiya, Kusuoka [18]) the computational effort to compute the approximation grows exponentially as one iterates. In Section 27.7 we outline an approach to address this weakness by recombining the particles in the tree after each step in the iteration. In particular we present a simple one-dimensional toy example that as we hope provides some insights in implementing the recombining method effectively.

In the final section of the chapter we apply the KLV method to the nonlinear filtering problem. The results in this section are due Crisan, Ghazali [8] who using a filter due to Picard [26] that discretizes the observations obtain a first-order approximation of the conditional distribution based on the KLV method. The insights developed in the work of Crisan and Ghazali [8] lead us to believe that applications of the KLV method to nonlinear filtering can be developed further.

The eventual aim is to obtain an efficient, positivity preserving, high-order approximation of the conditional distribution of the nonlinear filtering problem.

27.2 Cubature on Wiener space

In the first three sections of this chapter we will describe the cubature method as it is developed in Lyons, Victoir [23]. Throughout the paper C will denote a constant that may change from line to line. Following [23] we establish some basic notation. Let $C_b^\infty(R^N, R^N)$ denote the smooth bounded R^N valued functions whose derivatives of any order are bounded. Then $V_i = (V_i^1, \ldots, V_i^N) \in C_b^\infty(R^N, R^N)$, $0 \leq i \leq d$ may be identified with smooth vector fields on R^N via

$$V_i := \sum_{j=1}^N V_i^j \frac{\partial}{\partial x_j}.$$

Define the set of all multi-indices A by

$$A = \bigcup_{k=0}^\infty \{0, \ldots, d\}^k$$

and let $a = (a_1, \ldots, a_k) \in A$ be a multi-index. Furthermore we define a degree on a multi-index a by $\|a\| = k + card(j : a_j = 0)$ and let

$$A(j) = \{a \in A : \|a\| \leq j\}.$$

Let $A_1 = A \setminus \{\emptyset, (0)\}$ and $A_1(j) = \{a \in A_1 : \|a\| \leq j\}$. Following Kusuoka [17] we inductively define a family of vector fields indexed by A by taking

$$V_{[\emptyset]} = 0, \quad V_{[i]} = V_i, \quad 0 \leq i \leq d$$

$$V_{[(a_1, \ldots, a_k, i)]} = [V_{[a]}, V_i], \quad 0 \leq i \leq d, a \in A.$$

Moreover let $V_a = V_{a_1} \cdots V_{a_k}$ where the composition is taken in the sense of differential operators.

Consider the probability space $(C_0^0([0, T], R^d), \mathcal{F}, P)$, where $C_0^0([0, T], R^d)$ is the space of R^d valued continuous functions starting at 0, $\mathcal{F}$ its usual Borel σ-field and P the Wiener measure. Define the coordinate mapping process $B_t^i(\omega) = \omega^i(t)$ for $t \in [0, T]$, $\omega \in \Omega$. Under Wiener measure, $B = (B_t^1, \ldots, B_t^d)$ is a Brownian motion starting at zero. Furthermore let $B_t^0(t) = t$. Let $X_{t,x}$, $t \in [0, T]$, $x \in R^N$ be a version of the solution of the Stratonovich stochastic differential equation (SDE)

$$dX_{t,x} = V_0(X_{t,x})dt + \sum_{i=1}^{d} V_i(X_{t,x}) \circ dB_t^i, \quad X_{0,x} = x \qquad (27.1)$$

that coincides with the pathwise solution on continuous paths of bounded variation.

Finally for all $t \geq 0$ let

$$P_t f(x) := \mathbb{E}\left(f(X_{t,x})\right).$$

Let $L := V_0 + \frac{1}{2}\left(V_1^2 + \ldots V_d^2\right)$ and consider the parabolic partial differential equation (PDE)

$$\frac{\partial u}{\partial t}(t, x) = -Lu(t, x),$$
$$u(T, x) = f(x) \qquad (27.2)$$

for a given Lipschitz function f. In this case classical theory tells us that $u(t, x) = \mathbb{E}(f(X_{T-t,x}))$ is a solution to (27.2).

Our aim in the following is to approximate $\mathbb{E}\left(f(X_{t,x})\right)$ under the assumption that f is Lipschitz and the vector fields V_i defining (27.1) satisfy the UFG condition (see section 27.4).

Let μ be a positive measure on R^d. A cubature measure of degree m for μ is a positive discrete measure $\tilde{\mu}$ with finite support satisfying $\text{supp}(\tilde{\mu}) \subseteq \text{supp}(\mu)$ and

$$\int p(x)d\mu(dx) = \int p(x)d\tilde{\mu}(dx)$$

for any polynomial p on R^d with total degree at most m. Note that in the case $d = 1$ a cubature measure is more commonly referred to as a quadrature measure.

Combined with Taylor approximation for error estimation cubature is a classical and effective approach to numerical integration of (sufficiently smooth) functions on finite-dimensional spaces. The expectation of $f(X_{t,x})$ against Wiener measure may be interpreted as an integral against the infinite-dimensional Wiener measure.

In the Wiener space setting the role of polynomials is taken by the iterated Stratonovich integrals. Error estimates can be obtained in the spirit of stochastic Taylor expansion based on the developments of Strichartz [28], Ben Arous [3], Kloeden, Platen [14], and many others. Let $C^0_{0,bv}([0,T], R^d)$ denote the set of continous, bounded variation, R^d-valued paths starting at the origin.

Definition 27.1 Fix a finite set of multi-indices $\tilde{A} \subseteq A$. We say that a discrete measure Q_T assigning positive weights $\lambda_1, \ldots, \lambda_n$ to paths

$$\omega_1, \ldots, \omega_n \in C^0_{0,bv}([0,T], R^{d+1})$$

is a cubature measure, if for all $(i_1, \ldots, i_k) \in \tilde{A}$,

$$\mathbb{E}\left(\int_{0<t_1<\cdots<t_k<T} \circ dB^{i_1}_{t_1} \cdots \circ dB^{i_k}_{t_k}\right) = \sum_{j=1}^n \lambda_j \int_{0<t_1<\cdots<t_k<T} d\omega^{i_1}_j(t_1) \ldots d\omega^{i_k}_j(t_k)$$

where the expectation is taken under Wiener measure. If $\tilde{A} = A(m)$ we say that

$$Q_T = \sum_{j=1}^n \lambda_j \delta_{\omega_j}$$

is cubature measure of degree m.

In the following we will only consider cubature measures satisfying $\omega_i^0(t) = t$ for $1 \leq i \leq n$. The existence of such measures with finite support is not obvious. In [23], the authors prove a Tchakaloff theorem for Wiener space. They show that one can always find a cubature measure supported on at most card($\tilde{A}$) continuous paths of bounded variation. More importantly they give an explicit construction of degree-five cubature measures with $O(d^3)$ piecewise linear paths in their support. Some examples of cubature measures may be found in the Appendix.

Remark 27.1 Suppose that the paths $\omega_1, \ldots, \omega_n$ and the weights λ_i define a cubature measure for $T = 1$. It follows immediately from the scaling property of the Brownian motion that the measure supported on paths $\omega_{T,i}$ given by $\omega^0_{T,i}(t) = t$, $\omega^j_{T,i}(t) = \sqrt{T}\omega^j_i(t/T)$, $j = 1, \ldots, d$ and unchanged weights λ_i defines a cubature measure for general T. From now on suppose that the measure $Q := Q_1$ is a cubature measure of degree m.

Cubature on Wiener space may be interpreted as a Markov operator acting on discrete measures on R^N.

Definition 27.2 Given a positive measure $\mu = \sum_{i=1}^l \mu_i \delta_{x_i}$ on R^N and a cubature measure $Q_1 = \sum_{i=1}^n \lambda_i \delta_{\omega_i}$, we define the KLV operation with respect to μ over a time step s by

$$\mathrm{KLV}(\mu, s) := \sum_{j=1}^{l} \sum_{i=1}^{n} \mu_j \lambda_i \delta_{X_{s,x_j}(\omega_{s,i})}.$$

The transition KLV takes discrete measures on R^N to discrete measures on R^N and may be interpreted as a discrete Markov kernel. We clearly have

$$\mathbb{E}_{Q_s} f(X_{s,x}) = \mathbb{E}_{KLV(\delta_x, s)} f.$$

Hence, to compute $\mathbb{E}_{Q_s} f(X_{s,x})$ we need to solve an ordinary differential equation (ODE) for any path in the support of the cubature measure starting from any particle in the support of μ.

Remark 27.2 Note that $X_{s,x}(\omega_i)$ is the pathwise ODE solution of (27.1) along the bounded variation path ω_i (recall $X_{s,x}$ is a version of the solution of the SDE that coincides with the pathwise solution along continuous paths of bounded variation). We refer to section 27.6 for a brief discussion on how these ODEs may be solved numerically.

The KLV operation can be iterated which will turn out to be particularly relevant if the test function f is only Lipschitz (compare Section 27.4).

Let $\mathcal{D}$ be a partition given by the times

$$t_0 = 0 < t_1 < t_2 < \cdots < t_k = T$$

and let $s_j = t_j - t_{j-1}$. Also define subpartitions $\mathcal{D}^{(j)} \subseteq \mathcal{D}$ for $1 \leq j \leq k$ by

$$t_0 = 0 < t_1 < t_2 < \cdots < t_j.$$

Definition 27.3 We define the KLV operation with respect to a partition $\mathcal{D}$ recursively starting from the pointmass at x. For the base case let

$$\mathrm{KLV}(\mathcal{D}^{(1)}, x) := \mathrm{KLV}(\delta_x, s_1)$$

and

$$\mathrm{KLV}(\mathcal{D}^{(j+1)}, x) := \mathrm{KLV}(s_{j+1}, \mathrm{KLV}(\mathcal{D}^{(j)}, x))$$

for the recursion.

In the light of Definition 27.3 the iterated application of cubature on Wiener space (in the following referred to as the KLV method) can be interpreted as a particle system on R^N branching in the form of an n-ry tree after each step of the iteration. Thus to compute the k step KLV approximation one needs to solve $(n^{k+1} - 1)/(k - 1)$ ODEs each controlled by a cubature path ω_{i,s_j}, $1 \leq i \leq n$, $s_j \in \mathcal{D}$. In section 27.7 we discuss recombination of particles as one way to control the exponential increase of the number of particles as we iterate in the KLV method.

27.3 The error estimate for a single time step

In this section we establish a bound for approximating the expectation of $f(X_{t,x})$ under the Wiener measure by a cubature measures following the approach of Lyons, Victoir [23]. The following Lemma is a slightly sharpened version of Proposition 2.1 of Lyons, Victoir [23]. It is a weak form of the stochastic Taylor formula, for a detailed proof see Kloeden, Platen [14].

Lemma 27.1 Let $f \in C_b^\infty$, $m \in \mathbb{N}$. Then

$$f(X_{t,x}) = \sum_{(a_1,\ldots,a_k) \in A(m)} V_{a_1} \cdots V_{a_k} f(x) \int_{0<t_1<\cdots<t_k<t} \circ d B_{t_1}^{a_1} \cdots \circ d B_{t_k}^{a_k} + R_m(t, x, f).$$

(27.3)

The remainder process $R_m(s, x, f)$ satisfies

$$\sup_{x \in \mathbb{R}^N} \sqrt{\mathbb{E}(R_m(s, x, f)^2)} \leq C \sum_{j=m+1}^{m+2} s^{j/2} \sup_{(a_1,\ldots,a_i) \in A(j) \setminus A(j-1)} \|V_{a_1} \cdots V_{a_i} f\|_\infty,$$

where C is a constant only depending on d and m.

The proof of the Lemma is elementary and we only provide a brief sketch. The remainder $R_m(s, x, f)$ of the Stratonovich stochastic Taylor expansion is given by

$$R_m(s, x, f) = \sum_{\substack{(a_2,\ldots,a_k) \in A(m) \\ (a_1,\ldots,a_k) \notin A(m)}} \int_{0<t_1<\cdots<t_k<s} V_{a_1} \cdots V_{a_k} f(X_{t_1,x}) \circ d B_{t_1}^{a_1} \cdots \circ d B_{t_k}^{a_k}.$$

The bound can be proven by induction. The base case of the induction follows for $a_1 \neq 0$ from the relation

$$\int_0^t V_{a_1} \cdots V_{a_j} f(X_{s,x}) \circ d B_s^{a_1}$$

$$= \int_0^t V_{a_1} \cdots V_{a_j} f(X_{s,x}) d B_s^{a_1} + \frac{1}{2} \int_0^t V_{a_1}^2 \cdots V_{a_j} f(X_{s,x}) ds$$

which can be obtained by use of the Ito formula.

We can prove an analogous lemma for the cubature measures Q_T (see Lyons, Victoir [23]).

Lemma 27.2 Let $R_m(T, x, f)$ be the process defined in (27.3) then we have

$$\sup_{x \in \mathbb{R}^N} \mathbb{E}_{Q_T} |R_m(T, x, f)| \leq C(d, m, Q_1) \sum_{j=m+1}^{m+2} T^{j/2} \sup_{(a_1,\ldots,a_i) \in A(j) \setminus A(j-1)} \|V_{a_1} \cdots V_{a_i} f\|_\infty,$$

where C is a constant depending only on d, m and the length of the bounded variation paths in the support of the cubature measure Q_1.

Proof. We consider the degree m Stratonovich stochastic Taylor approximation of $f(X_{t,x}(\omega_{i,t}))$ pathwise along each cubature path ω_i (recall the ω_i are by definition bounded variation paths). The remainder is given by

$$R_m(t, x, f)(\omega_{t,i}) = \sum_{\substack{(a_2,\ldots,a_k) \in A(m) \\ (a_1,\ldots,a_k) \notin A(m)}} \int_{0 < t_1 < \cdots < t_k < t} V_a f(X_{t_0,x}(\omega_{t,i})) d(\omega_{t,i})_{t_1}^{a_1} \cdots d(\omega_{t,i})_{t_k}^{a_k}.$$

Changing variables to obtain an expression with the original paths ω_i (see Remark 27.1 for the scaling relation of the cubature paths) proves the claim. □

The constant in Lemma 27.2 can in fact be made explicit (see Crisan, Ghazali [8], example 4).

The expectation of the degree m Taylor approximation of the Wiener functional $f(X_{T,x})$ under the Wiener and the cubature measure Q_T coincides as the expectations of all Stratonovich iterated integrals coincides. Hence we may combine the remainder estimates of the lemmas 27.1 and 27.2 using the triangle inequality and obtain a first error bound for the cubature approximation of $P_T f$.

Proposition 27.1 *Let Q_T be a degree m cubature measure then*

$$\sup_{x \in \mathbb{R}^N} \left| P_T f(x) - \mathbb{E}_{Q_T} f(X_{T,x}) \right|$$

$$\leq C \sum_{j=m+1}^{m+2} T^{j/2} \sup_{(a_1,\ldots,a_i) \in A(j) \setminus A(j-1)} \| V_{a_1} \cdots V_{a_i} f \|_\infty,$$

where C is a constant that only depends on d, m and Q_1.

27.4 Iterated application of cubature: the Kusuoka–Lyons–Victoir method

The bounds for the cubature approximation of $P_T f(x)$ derived in the previous section involve higher derivatives and are meaningless if the test function f is only Lipschitz. It is well known that even if f is not smooth the function $P_t f$ is smooth in the direction of the vector fields defining the SDE (27.1) (under suitable conditions on these vector fields). Quantitative versions of this statement were obtained by Kusuoka and Stroock in the early eighties using the techniques of Malliavin calculus. These results were subsequently for example used in the work of Bally, Talay [2], [1] to analyze the rate of convergence of the time-discrete Euler approximation for SDEs in the case of test functions that are not smooth.

Kusuoka showed in [15] how the Kusuoka–Stroock regularity estimates together with familes of uneven partitions of the time interval [0, T] can be used to develop a high-order approximation scheme for $P_T f(x)$ for Lipschitz test functions f. His deep insights obtained in [15] immediately apply to the error estimation for the KLV method. In the reminder of the section we follow Lyons, Victoir [23] to derive some simple bounds for the KLV approximation.

Following Kusuoka [16] we introduce a condition on the vector fields.

Definition 27.4 The family of vector fields V_i, $i = 0, \ldots, d$ is said to satisfy the condition (UFG) if the Lie algebra generated by the vector fields is finitely generated as a C_b^∞ left module, i.e. there exists a positive k and $u_{\alpha,\beta} \in C_b^\infty$ satisfying for all $\alpha \in A_1$,

$$V_{[\alpha]} = \sum_{\beta \in A_1(k)} u_{\alpha,\beta} V_{[\beta]}. \tag{27.4}$$

From now on suppose that our system of vector fields V_i, $i = 0 \ldots d$ satisfies the UFG condition.

It was pointed in Crisan, Ghazali [8] that the estimates in Lyons, Victoir [23] which we follow in this section require the following additional condition on the vector field V_0.

Definition 27.5 (V0 condition) A family of vector fields V_i, $0 \leq i \leq d$ satisfies the V0 condition if

$$V_0 = \sum_{\beta \in A_1(2)} u_\beta V_{[\beta]}$$

for some $u_\beta \in C_b^\infty(R^N)$.

In the following we will assume that the vector fields V_i defining (27.1) satisfy the UFG and the V0 condition. The following regularity estimate is due to Kusuoka, Stroock [19] and Kusuoka [16] (under the weaker UFG condition). The particular corollary to the Kusuoka–Stroock result we state here is taken from Crisan, Ghazali [8].

Corollary 27.1 *Suppose the family of vector fields V_i, $0 \leq i \leq d$ satisfy the conditions UFG and V0. Let $f \in C_b^\infty(R^N)$ and $\alpha_1, \ldots, \alpha_j \in A$ then*

$$\| V_{[\alpha_1]} \ldots V_{[\alpha_j]} P_s f \|_\infty \leq \frac{C s^{1/2}}{s^{(\|\alpha_1\|+\ldots+\|\alpha_j\|)/2}} \| \nabla f \|_\infty \tag{27.5}$$

for all $s \leq 1$, where C is a constant independent of s and f.

The main consequence of the Markovian property of the KLV operation is that the error over the global interval [0, T] will be bounded above by the sum of the errors over the subintervals in the partition. The following theorem captures this basic fact and may be found in similar form in Lyons, Victoir [23].

Theorem 27.1 *The degree m KLV approximation satisfies*

$$\sup_{x \in R^N} \left| P_T f(x) - \mathbb{E}_{KLV(\mathcal{D},x)} f \right|$$

$$\leq C \sum_{l=1}^{k} \sum_{j=m+1}^{m+2} s^{j/2} \sup_{(a_1,\ldots,a_i) \in A(j) \setminus A(j-1)} \| V_{a_1} \ldots V_{a_i} P_{T-t_l} f \|_\infty$$

where C is a constant only depending on d, m, and Q_1.

Proof. Note that

$$P_T f(x) - \mathbb{E}_{KLV(\mathcal{D},x)} f$$

$$= P_T f(x) - \mathbb{E}_{KLV(\mathcal{D}^{(1)},x)} P_{T-t_1} f$$

$$+ \sum_{j=1}^{k-1} \left(\mathbb{E}_{KLV(\mathcal{D}^{(j)},x)} P_{T-t_j} f - \mathbb{E}_{KLV(\mathcal{D}^{(j+1)},x)} P_{T-t_{j+1}} f \right)$$

where $\mathcal{D}^j$ is the partition corresponding to the first j steps of $\mathcal{D}$. We obtain

$$\sup_{x \in R^N} \left| P_T f(x) - \mathbb{E}_{KLV(\mathcal{D},x)} f \right|$$

$$\leq \sup_{x \in R^N} \left| P_T f(x) - \mathbb{E}_{KLV(\mathcal{D}^{(1)},x)} P_{T-t_1} f \right|$$

$$+ \sup_{x \in R^N} \sum_{j=1}^{k-1} \mathbb{E}_{KLV(\mathcal{D}^{(j)},x)} \sup_{z \in R^N} \left| P_{s_{j+1}} (P_{T-t_{j+1}} f)(z) - \mathbb{E}_{KLV(s_{j+1},z)} P_{T-t_{j+1}} f \right|$$

$$\leq C \sum_{l=1}^{k} \sum_{j=m+1}^{m+2} s^{j/2} \sup_{(a_1,\ldots,a_i) \in A(j) \setminus A(j-1)} \| V_{a_1} \ldots V_{a_i} P_{T-t_l} f \|_\infty,$$

where we have repeatedly used the triangle inequality and Proposition 27.1. □

Combining the error estimates from Theorem 27.1 with the regularity estimates from Corollary 27.1 we obtain an error estimate for the KLV method in terms of the gradient of f. We follow the proof given in Lyons, Victoir [23].

Theorem 27.2 *Suppose the vector fields V_i, $0 \leq i \leq d$ satisfy conditions UFG and V0 then*

$$\sup_{x \in R^N} \left| P_T f(x) - \mathbb{E}_{KLV(\mathcal{D},x)} f \right|$$

$$\leq C_T \|\nabla f\|_\infty \left(s_k^{1/2} + \sum_{j=m}^{m+1} \sum_{i=1}^{k-1} \frac{s_i^{(j+1)/2}}{(T-t_i)^{j/2}} \right),$$

where C_T is a constant independent of k, the number of time steps in the partition of the time interval $[0, T]$.

Proof. From Corollary 27.1 it follows that

$$\sup_{(a_1,\ldots,a_i)\in A(j)\setminus A(j-1)} \|V_{a_1}\ldots V_{a_i} P_{T-t_l} f\|_\infty \leq \frac{C}{(T-t_l)^{(j-1)/2}} \|\nabla f\|_\infty.$$

Proceeding as in the proof of Theorem 27.1 we get

$$\sup_{x\in \mathbb{R}^N} \left| P_T f(x) - \mathbb{E}_{KLV(\mathcal{D},x)} f \right|$$

$$\leq C \sum_{l=1}^{k-1} \sum_{j=m+1}^{m+2} s^{j/2} \sup_{(a_1,\ldots,a_i)\in A(j)\setminus A(j-1)} \|V_{a_1}\ldots V_{a_i} P_{T-t_l} f\|_\infty$$

$$+ \mathbb{E}_{KLV(\mathcal{D}^{(k-1)},x)} \sup_{z\in \mathbb{R}^N} \left| P_{s_k} f(z) - \mathbb{E}_{KLV(s_k,z)} f \right|$$

$$\leq C \sum_{l=1}^{k-1} \sum_{j=m+1}^{m+2} \frac{s^{j/2}}{(T-t_l)^{(j-1)/2}} \|\nabla f\|_\infty$$

$$+ \mathbb{E}_{KLV(\mathcal{D}^{(k-1)},x)} \sup_{z\in \mathbb{R}^N} \left| P_{s_k} f(z) - \mathbb{E}_{KLV(s_k,z)} f \right|.$$

To complete the proof of the theorem we note that for $s \in (0, 1]$

$$\sup_{z\in \mathbb{R}^N} \left| P_s f(z) - \mathbb{E}_{KLV(s,z)} f \right| \leq C\sqrt{s} \|\nabla f\|_\infty.$$

$\square$

We recall a family $\mathcal{D}_{k,\gamma}$ of partitions from Kusuoka [17] (subsequently also used in Lyons, Victoir [23]) given by

$$t_j = T\left(1 - \left(1 - \frac{j}{k}\right)^\gamma\right), \tag{27.6}$$

$0 \leq j \leq k$. Following Kusuoka [17] one can show that for these families of uneven partitions with smaller time steps towards the end the bound in Theorem 27.2 implies high-order convergence of the KLV method.

Corollary 27.2 *For $\gamma > m - 1$ we have*

$$\left| P_T f(x) - \mathbb{E}_{KLV(\mathcal{D}_{k,\gamma},x)} f \right| \leq C k^{-(m-1)/2} \|\nabla f\|_\infty, \tag{27.7}$$

while for the case $0 < \gamma < m - 1$ one obtains

$$\left| P_T f(x) - \mathbb{E}_{KLV(\mathcal{D}_{k,\gamma},x)} f \right| \leq C k^{-\gamma/2} \|\nabla f\|_\infty.$$

where C is in each case a constant independent of k and f.

Remark 27.3 The V_0 condition on the drift can be relaxed and we obtain the same rates of convergence as in Corollary 27.2 (see Litterer [20] for a detailed proof using the ideas of Kusuoka [17]).

The estimates in this section have been optimized for test functions controled in the Lipschitz norm. If more information about f is available (for example additional smoothness) it would change the analysis and the partitions $\mathcal{D}$ considered.

27.5 Approximations based on the flows of autonomous ODEs

Instead of considering ODEs driven by cubature paths we will in this section derive an operator based on the flows of autonomous ODEs. The operator is obtained by evaluating the truncated logarithmic signature (defined below) of the cubature paths ω_i with the vector fields V_i defining the SDE (27.1) and corresponds to a version of Kusuoka's algorithm (see [17]). We will in the following also refer to the operation as cubature at the flow level.

For a smooth vector field $V \in C_b^\infty(R^N; R^N)$ we define the flow $\mathrm{Exp}(t V)(x)$ to be the solution of the autonomous ODE

$$\dot{x}(t, x) = V(x(t, x)) \quad t > 0, \qquad x(0, x) = x \in R^N.$$

Let $\{\epsilon_0, \ldots, \epsilon_d\}$ denote a fixed orthonormal basis for $W = R \oplus R^d$. Let $T(R, R^d)$ denote the tensor algebra of polynomials over W endowed with a grading that assigns degree two to ϵ_0 and degree one to the remaining generators (this definition is motivated by the fact that we will identify ϵ_0 with V_0 and the drift scales differently from the Brownian noise). We denote by $T^n(R, R^d)$ the truncation of $T(R, R^d)$ at level n and by π_n the natural projection from $T(R, R^d)$ onto $T^n(R, R^d)$. Define $\mathcal{L}$ to be the space of linear combinations of finite sequences of Lie brackets with elements in W, i. e.

$$\mathcal{L} := W \oplus [W, W] \oplus [W, [W, W]] \oplus \ldots \subseteq T(R, R^d).$$

Then $\mathcal{L}$ is the free Lie algebra generated by W (see e.g. Reutenauer [27]). Define exp using the usual power series and similarly define $log(a)$ for all $a = (a_0, a_1, \ldots) \in T(R, R^d)$ with $a_0 = 1$ by

$$\log(a) = \sum_{j=1}^\infty (-a)^{j+1} \frac{c^{\otimes j}}{j},$$

where $c = (0, a_1, a_2, \ldots)$. Finally for all $t \in R$, $a = (a_0, a_1, a_2 \ldots) \in T(R, R^d)$ we define a homogeneous scaling operation by

$$\langle t, a \rangle = \left(a_0, ta_1, t^2 a_2, \ldots\right).$$

For details of the definitions we refer to [23] (pp. 178–179).

The map sending ϵ_i to V_i, $i = 0, \ldots, d$ extends to a unique linear map on W and by the universality property of the tensor algebra extends to a unique homomorphism Γ from $T(R, R^d)$ into the differential operators on R^N. The restriction of Γ to $\mathcal{L}$ is a Lie map from $\mathcal{L}$ into the smooth vector fields on R^N.

For a continuous bounded variation path $\phi \in C_{bv}([0, T], R^{d+1})$; $s, t \in [0, T]$ we define $S_{s,t} : C_{bv}([0, T], R^{d+1}) \to T(R, R^d)$ by

$$S_{s,t}(\phi) = \sum_{k=0}^{\infty} \int_{s < t_1 < \cdots < t_k < t} d\phi(t_1) \otimes \cdots \otimes d\phi(t_k). \tag{27.8}$$

$S_{0,T}(\phi)$ is known as the signature of the path ϕ. Using Stratonovich-iterated integrals we may define the random signature of a Brownian motion.

We can reformulate the definition of a degree m cubature measure on Wiener space in terms of the signature.

Definition 27.6 We say a discrete measure $Q_1 = \sum_{j=1}^{n} \lambda_j \delta_{\omega_j}$ supported on finitely many continuous paths of bounded variation is a cubature measure of degree m on Wiener space if

$$\mathbb{E}(\pi_m(S_{0,1}(\circ B))) = \sum_{j=1}^{n} \lambda_j \pi_m(S_{0,1}(\omega_j)). \tag{27.9}$$

Some properties of the signature map are captured by Chen's theorem and its converse. It is stated in Lyons, Victoir [23] and proofs may be found for example in Lyons [23] or Reutenauer [27]. The theorem has first been obtained in Chen [5] and shows that the signature is an algebra homomorphism taking paths with concatenation as multiplication into the tensor algebra.

Theorem 27.3 (Chen's theorem) *Let* $\phi \in C^0_{0,bv}([0, T], R^d)$. *The signature of ϕ is multiplicative, i.e.*

$$S_{s,t}(\phi) = S_{s,u}(\phi) \otimes S_{u,t}(\phi).$$

In addition $\log(S_{s,t}(\phi))$ *is a Lie series. Conversely for any Lie polynomial* $U \in \pi_m(\mathcal{L})$ *there exists a continuous bounded variation path ϕ such that*

$$\pi_m \log S_{s,t}(\phi) = U.$$

The proof of Chen's theorem can be extended to show that $\log(S_{s,t}(\circ B))$ is a (random) Lie series, see e.g. Lyons [22]. Such arguments can be used to obtain small time asymptotics of the solution of Stratonovich SDEs, see e.g. Ben Arous [3].

Suppose now $Q = \sum_{i=1}^{n} \lambda_i \delta_{\omega_i}$ is a degree m cubature measure on path space. Chen's theorem tells us that that $L_i := \pi_m(\log(S_{0,1}(\omega_i)))$ is a Lie polynomial. The measure $Q_{\mathcal{L}} = \sum_{j=1}^{n} \lambda_j \delta_{L_j}$ satisfies

$$\mathbb{E}(\pi_m(S_{0,1}(\circ B))) = \mathbb{E}_{Q_{\mathcal{L}}(dL)}(\pi_m \exp(L)). \tag{27.10}$$

Conversely for any Lie polynomials L_i there exist continuous bounded variation paths ω_i with log-signature L_i. Moreover if $Q_{\mathcal{L}}$ satisfies (27.10) the measure Q will satisfy (27.9), so the identities (27.9) and (27.10) are equivalent. Motivated by this discussion we give following Lyons, Victoir [23] a third definition for a cubature measure on Wiener space.

Definition 27.7 Let $m \in \mathbb{N}$ and $Q_{\mathcal{L}} = \sum_{j=1}^{n} \lambda_j \delta_{L_j}$ with $\lambda_i > 0$ and $L_i \in \pi_m(\mathcal{L})$ for $i = 1, \ldots, n$. We say $Q_{\mathcal{L}}$ is a cubature measure on the free Lie algebra if and only if

$$\mathbb{E}(\pi_m S_{0,1}(\circ B)) = \mathbb{E}_{Q_{\mathcal{L}}(dL)}(\pi_m \exp(L)).$$

Any degree m cubature measure $Q_{\mathcal{L}}$ satisfies

$$\mathbb{E}(\pi_m(S_{0,t}(\circ B))) = \mathbb{E}_{Q_{\mathcal{L}}(dL)}\left[\pi_m exp(\langle t^{1/2}, L\rangle)\right] \tag{27.11}$$

and we have

$$\mathbb{E}_{Q_1}(\Gamma \pi_m S_{0,1}(\circ B)) f(x) = \mathbb{E}_{Q_{\mathcal{L}}}(\Gamma \pi_m exp(L)) f(x). \tag{27.12}$$

If we now interchange the algebra homomorphism Γ with the exponentiation (so far taken in the tensor algebra) we arrive at an approximation operator in which the exponentiation is understood as taking the flow of autonomous ODEs.

Formally we define the cubature approximation at the flow level as follows.

Definition 27.8 Suppose $Q_{\mathcal{L}}$ is a degree m cubature measure supported on Lie polynomials. We define the cubature approximation of $P_t f$ at the flow level to be

$$\mathbb{E}_{Q_{\mathcal{L}}(dL)} f(\text{Exp}(\Gamma \pi_m \langle t^{1/2}, L\rangle)(x)).$$

Just as in the KLV method the flow-based approximation can be interpreted as a discrete Markov operator acting on measures on $\mathbb{R}^N$. For

$$\mu = \sum_j \mu_j \delta_{x_j}$$

we define $\widetilde{\text{Cub}}(\mu, s)$ by

$$\widetilde{\text{Cub}}(\mu, s) = \sum_j \sum_{i=1}^{n} \lambda_i \mu_j \delta_{\text{Exp}(\Gamma \pi_m \langle \sqrt{s}, L_i\rangle)(x_j)}$$

and for a partition $\mathcal{D}$ given by $0 = t_0 < t_1 \cdots < t_k = T$ we define $\widetilde{\text{Cub}}(\delta_x, \mathcal{D})$ by

$$\widetilde{\text{Cub}}(\delta_x, \mathcal{D}) = \widetilde{\text{Cub}}(\ldots \widetilde{\text{Cub}}(\widetilde{\text{Cub}}(\delta_x, t_1 - t_0), t_2 - t_1), \ldots, t_k - t_{k-1}).$$

Note that the error when interchanging exp and Γ is of similar order as the error in the pathwise cubature approximation. In fact for any $w \in \pi_m(\mathcal{L})$, $t \in (0, 1]$, $x \in R^N$ we have

$$|f(\mathrm{Exp}(\Gamma(\langle t, w\rangle))(x)) - (\Gamma(\pi_m \exp(\langle t, w\rangle))f)(x)|$$
$$\leq t^{m+1} \sum_{j=2}^{m+1} \|\Gamma((\pi_{m^{m+1}} - \pi_m)(w^j))f\|_\infty, \tag{27.13}$$

see e.g. Kusuoka [17] Corollary 12 for a proof of this identity.

The norms of the derivatives of f on the right-hand side of inequality (27.13) can once again be estimated using the regularity results of Corollary 27.1. Also note that for sufficiently smooth f we have

$$\mathbb{E}(f(X_{t,x})) = \mathbb{E}\left(\Gamma(\pi_m S_{0,t}(\circ B))\right) f(x) + \mathbb{E}(R_m(t, x, f))$$
$$= \mathbb{E}_{Q_\mathcal{L}(dL)}(\Gamma \pi_m \exp(\langle t^{1/2}, L\rangle))f(x) + \mathbb{E}(R_m(t, x, f)), \tag{27.14}$$

where $R_m(t, x, f)$ is the remainder of the stochastic Taylor formula defined in (27.3) and the last equality follows from identity (27.12). Thus, using the triangle inequality the estimate of identity (27.13) may be combined with Lemmas 27.1, which we apply to the remainder in (27.14). The resulting error estimate implies the following Corollary, which demonstrates that for suitable partitions the bounds for the approximation at flow and path level have the same rate of convergence.

Corollary 27.3 *Let $\mathcal{D}_{k,\gamma}$ be the family of partitions defined in (27.6). For $\gamma > m - 1$ we have*

$$\left|P_T f(x) - \mathbb{E}_{\widetilde{\mathrm{Cub}}(\mathcal{D}_{k,\gamma}, x)} f\right| \leq Ck^{-(m-1)/2}\|\nabla f\|_\infty, \tag{27.15}$$

while for the case $0 < \gamma < m - 1$ one obtains

$$\left|P_T f(x) - \mathbb{E}_{\widetilde{\mathrm{Cub}}(\mathcal{D}_{k,\gamma}, x)} f\right| \leq Ck^{-\gamma/2}\|\nabla f\|_\infty.$$

Remark 27.4 The cubature measures $Q_\mathcal{L}$ on the free Lie algebra correspond to the $(m - \mathcal{L})$-moment similar families of Kusuoka [17].

The analysis of the error bounds for this algorithm in [17] is somewhat more complex and implies the simple analysis sketched in this section. Kusuoka's estimate employs an additional splitting type argument for the drift direction and hence does not require the condition V0. The same asymptotic bounds on the rate of convergence for the partition $\mathcal{D}$ as in Corollary 27.3 are obtained.

Remark 27.5 The construction of cubature measures on path space as carried out e.g. in [23] for the degree-five case is indirect. One first constructs a cubature measure of the same degree on the free Lie algebra and then constructs (in [23] piecewise linear) paths with logarithmic signatures matching the Lie polynomials in the support of the cubature formula on the free Lie algebra. For optimal error bounds the length of these paths should be as small as possible.

27.6 Some considerations for the numerical computation of the cubature approximation

The results of the previous section suggest that approximations of $P_T f$ based on the iterated application of cubature at the path and flow level have similar convergence properties. In fact, we have confirmed this claim for some concrete numerical examples in the case $m = 5$, see Litterer [20] where we consider an application to option pricing in the SABR stochastic volatility model and Gyurko [12].

The path-level cubature is easier to implement as the ODEs we need to solve only involve the basic vector fields V_i and not their Lie brackets. In the case and $m = 5$ and a d-dimensional driving noise a construction of the cubature paths ω_i has been carried out in [23]. Each path in this construction consists of $3 + 16d$ linear pieces (4 linear pieces in the special case $d = 2$), corresponding to $3 + 16d$ autonomous ODEs to solve to move a particle forward.

In most cases of practical interest these ODEs cannot be solved analytically and a numerical approximation has to be found. Ninomiya, Ninomiya, Kusuoka [18] (section 1.1.2) show that solving the ODEs by an order m (here $m = 5$) Runge-Kutta scheme is sufficient to obtain the global bounds on the rate of convergence we proved in section 27.4. It is known (see Ghazali [11] for a rigorous proof) that Romberg extrapolation can be applied to the KLV method to improve its rate of convergence. In this case Kusuoka, Ninomiya, Ninomiya [18] observe that for degree five cubature an order seven Runge–Kutta scheme is required to obtain the improved rate of convergence.

The flow-based cubature only requires solving one autonomous ODE to move each particle. This ODE is more complex (as it involves the Lie brackets of the vector fields V_i defining the SDE), but (in particular for the case $d \geq 3$) can be computed faster than the ODE solution of (27.1) along the piecewise linear paths of [23]. Thus if the Lie brackets of the vector fields V_i can easily be computed explicitly and the resulting ODEs are numerically tractable the flow-based implementation is viable and performs more efficiently (by a constant factor) than the path-based approach.

As we iterate, the number of particles in the tree corresponding to the KLV operation grows exponentially. Eventually evaluating the entire tree is no longer computationally tractable. One way to solve this problem is to apply partial sampling techniques. It was observed by Ninomiya [24] that the tree-based branching algorithm developed by Crisan, Lyons [9] works particularly well in practical applications in combination with the KLV method. An alternative deterministic approach to this problem is the recombination method we describe in the following section.

27.7 Recombination

27.7.1 Simplicial recombination for particle systems

Particle methods can in many cases provide good descriptions of evolving measures. Such examples arising in the context of nonlinear filtering may be found in the survey articles [6, 7]. Unfortunately in many cases (such as the KLV method) the number of particles explodes if we apply the methods iteratively and preserve accuracy.

Sometimes the essential properties of a particle measure we care about can accurately be described and captured by the expectations of a finite set of test functions. If we can find such a family of test functions we can replace the original measure with a simpler measure with smaller support that integrates all test functions correctly and hence still has the right properties, assuming the number of test functions is small compared to the cardinality of the support of the original measure. In the following subsection we describe a simple algorithm that can be used to compute reduced particle measures with respect to a finite set of test functions. We call this process simplical recombination.

In the following subsections we will use reduced measures (and the algorithm to compute them) to add recombination to the KLV method. After each application of the KLV operation we replace the intermediate measures by reduced measures. If we can do this without significantly increasing the one step errors then we can obtain a new global error bound over $[0, T]$ for this algorithm that is of the same order as the error bound in the KLV method.

The property of the KLV measure we are targeting is to integrate $P_t f$, the heat kernel applied to f, correctly. We have identified two finite sets of test functions that ensure that the bound on the overall error of the approximation of $P_T f$ is of the same order and hence the modified method has the same convergence properties.

We believe that the combination of the two ideas—higher-order particle methods to describe the evolution of a measure on the one hand and simplifying the support of the measures used in the description, by characterizing essential properties of a measure using the expectations of a finite set of test functions on the other hand—have more general applications than investigated so far.

27.7.2 A simple algorithm

Before we describe the actual algorithm we formally define the reduced measures. Let μ be a discrete probability measure and $Z_r = \{p_1, \ldots, p_r\}$ a finite set of test functions. We will call a discrete probability measure $\tilde{\mu}$ with $\mathrm{supp}(\tilde{\mu}) \subseteq \mathrm{supp}(\mu)$ a reduced measure with respect to μ and Z_r if it satisfies the following two conditions:

1. For all $p \in Z_r$
$$\int p(x)\tilde{\mu}(dx) = \int p(x)\mu(dx)$$

2. $card(\mathrm{supp}(\tilde{\mu})) \leq r + 1$.

Let μ_Z be the law of the r dimensional random variable $Z := (p_1, \ldots, p_r)$ under μ, i.e. $\mu_Z = \mu \circ Z^{-1}$. Note that finding a measure $\tilde{\mu}$ integrating the test functions correctly is equivalent to finding $\tilde{\mu}_Z$ with the same centre of mass as μ_Z.

Suppose μ_Z is given by $\mu_Z = \sum_{i=1}^{\hat{r}} \lambda_i \delta_{x_i}$. The following algorithm was communicated to us by N. Victoir and is well known (see e.g. Davis [10]). The centre of mass (CoM) for the measure μ_Z is given by

$$\mathrm{CoM}(\mu_Z) = \sum_{i=1}^{\hat{r}} \lambda_i x_i. \tag{27.16}$$

Consider $r + 2$ points from the support of μ_Z. For notational convenience we may assume without loss of generality that these points are given by $x_1, \ldots, x_{r+2}$. The system given by

$$\sum_{i=1}^{r+2} u_i = 0 \qquad \sum_{i=1}^{r+2} u_i x_i = 0 \tag{27.17}$$

is a linear system with $r + 2$ variables, but only $r + 1$ constraints and therefore it has a nontrivial solution, which can be computed easily and robustly using singular value decomposition (SVD).

Thus we may set

$$c = \min_{u_i < 0} \left(-\frac{\lambda_i}{u_i} \right) \tag{27.18}$$

and consider a new measure ν defined by

$$\nu := \sum_{j=1}^{r+2} (c u_j + \lambda_j) \delta_{x_j} + \sum_{j=r+3}^{\hat{r}} \lambda_j \delta_{x_j}.$$

The equations (27.17) guarantee that the measures ν and μ_Z have the same mass and centre of mass. Note that for $u_i < 0$ we have

$$c u_i + \lambda_i \geq -\frac{\lambda_i}{u_i} u_i + \lambda_i = 0.$$

Thus, the weights in ν are nonnegative and ν is a probability measure. By construction of c we have $\text{card}(\text{supp}(\nu)) < \text{card}(\text{supp}(\mu_Z))$ (the weight of $\delta_{x_{i^*}}$ vanishes, where i^* is the minimizer in (27.18)). Iterating this reduction procedure with ν in place of μ_Z, which is always possible as long as at least $r + 2$ points are in the support of ν, we obtain a reduced measure.

To improve the performance of the algorithm we use a simple divide-and-conquer argument. We can split $\text{supp}(\mu_Z)$ into $2r + 2$ subsets of as equal size as possible. We then consider the measure $\bar{\mu}$ corresponding to centres of mass of these sub collections, and compute the reduced measure with respect to $\bar{\mu}$. Hence, we may compute any reduced measure $\tilde{\mu}$ by solving the measure reduction problem $\lceil \log_2(\hat{r}) \rceil$ times for measures with support of size at most $2(r + 1)$. For details and an alternative algorithm see Litterer, Lyons [21].

27.7.3 Application of simplicial recombination to the KLV method

Let μ be a discrete probability measure suppported on finitely many particles and let $g_t(x) := P_t f(x)$. To apply the recombination to the KLV method we need to identify a sufficiently small set of test function that provides good bounds for the error when we approximate $\int g_t(x)d\mu(dx)$ by the integral of g_t against the reduced measure $\tilde{\mu}$.

For simplicity we will for the remainder of the section assume that the uniform Hörmander condition (stated below) is satisfied. For detailed proofs and a discussion of the recombination under the more general UFG condition we refer to Litterer [20].

Definition 27.9 Following [16] we say that a collection of smooth vector fields V_i, $i = 0, \ldots, d$ satisfies the Hörmander condition (UH) if there is an integer k such that

$$\inf\left\{\sum_{\alpha \in A_1(k)} \langle V_{[\alpha]}(x), \xi\rangle^2; x, \xi \in R^N, |\xi| = 1\right\} := M > 0 \qquad (27.19)$$

Note that the uniform Hörmander implies the UFG condition. We define p to be the minimal integer k such that the set $\{V_\alpha, \alpha \in A_1(k)\}$ satisfies (27.19).

Using the algorithm described in subsection 27.7.2 we formally define a recombination operation on particle measures.

Definition 27.10 The operation $\text{Red}(\mu, s)$ acts for $s > 0$ on a discrete measures μ and returns a (by no means unique) reduced measure (as defined in subsection 27.7.2) with respect to μ and the set of test functions described below:

Let $\tilde{S}_1(y_1), \ldots, \tilde{S}_\ell(y_\ell)$ be a cover of $\text{supp}(\mu)$ by (Euclidean) balls of radius $s^{p/2}$. We can find sets $S_i \subseteq \tilde{S}_i(y_i) \cap \text{supp}(\mu)$, $i = 1, \ldots, \ell$ that form a partition of $\text{supp}(\mu)$. The set of test function is given by all functions of the form $p_i \chi_{S_q}$ where $q = 1, \ldots, \ell$, χ_{S_q} denotes the characteristic function of the set S_q and the functions p_i form a basis for the space of polynomials in N variables of degree at most m.

Remark 27.6 The reduction operation defined above can be computed more efficiently by applying simplicial recombination separately to $\mu|_{S_q}$ for each set S_q in the partition of supp(μ). In this case the nonzero test functions are a basis for the polynomials with degree at most m.

To define a recombining KLV method we alternatingly (starting with a fixed point mass δ_x) apply the KLV operation (Definition 27.2) and the reduction operation. The time steps of both operations are determined by the partition $\mathcal{D}$. Note that there is obviously no need to apply recombination after the first and last KLV operation for a fixed partition $\mathcal{D}$. Let $\tilde{Q}_{\mathcal{D},x}$ denote the particle measure on R^N corresponding to this recombining KLV method over a k step partition $\mathcal{D}$.

To obtain a bound for the approximation of $P_T f(x)$ by the iterated method one first needs to obtain an error bound for a single recombination step, i.e. given a discrete probability measure μ on R^N we have to estimate

$$\left| \mathbb{E}_\mu g_t - \mathbb{E}_{Red(\mu,s)} g_t \right|. \tag{27.20}$$

Under the UH condition the estimates of Corollary 27.1 may be extended to provide quantitative bounds for the partial derivatives of $P_t f$ (see also Kusuoka [16], p. 274). By a Taylor approximation argument these bounds imply that on each set S_q (see Definition 27.10) g_t is approximated by polynomials of degree m with an error bound that is uniform in q. By choice of the test functions in the recombination operation uniform approximation by polynomials on the sets S_q implies a bound for (27.20).

Theorem 27.4 *With the notation of Section 27.7.3 we have*

$$\sup_{x \in R^N} \left| P_T f(x) - \mathbb{E}_{\tilde{Q}_{\mathcal{D},x}} f \right|$$

$$\leq C_T \|\nabla f\|_\infty \left(s_k^{1/2} + \sum_{i=1}^{k-1} \sum_{j=m}^{m+1} \frac{s_i^{(j+1)/2}}{(T-t_i)^{j/2}} + \sum_{i=1}^{k-1} \frac{s_i^{(m+1)p/2}}{(T-t_i)^{mp/2}} \right),$$

where C_T is a constant independent of f.

For a proof of this theorem one observes that the bound of the global error of the approximation over $[0, T]$ is bounded above by the sum of cubature errors over the subintervals of the partition $\mathcal{D}$ and the sum of the errors incurred in each recombination step (compare the proof of Theorem 27.1).

The following theorem demonstrates that under the uniform Hörmander condition for bounded vector fields the recombining KLV method allows us to obtain the high order convergence while the computational effort involved in computing the approximation grows polynomially.

Theorem 27.5 *Suppose the vector fields $V_i \in C_b^\infty(R^N; R^N)$ are bounded and satisfy the uniform Hörmander condition. We can achieve*

$$\left| P_T f(x) - \mathbb{E}_{\tilde{Q}_{\mathcal{D},x}} f \right| \leq C k^{-(m-1)/2} \|\nabla f\|_\infty,$$

while the number of test functions in the reduction operation, and hence the number of elementary ODEs to solve, grows polynomially in k.

Theorem 27.5 can be proved by considering the partitions defined in (27.6) with $\gamma > m - 1$. Note that the power of the polynomial that bounds the number of test functions depends not only on m and N, but also on p and hence on the geometry of the underlying problem.

A different approach to recombination for the KLV method has been developed by Teichmann [30]. Teichmann's method is based on generalized binomial trees.

27.7.4 A simple but significant worked (toy) example

We consider a linear one-dimensional problem, where the boundary data is Lipschitz, piecewise smooth, and the answer is required to high-order accuracy, e.g.

$$f(x) = \max(1, e^x) - e^x.$$

The method was applied to approximate the solution of the heat equation at time $T = 1$ with initial condition $X_0 = 0$. The goal was to achieve an accuracy of at least 10^{-10}. This particular choice of function is particularly well suited for our numerical experiments, since there is a closed form solution in terms of the error function, which may be used to determine the error of the approximation.

We applied a modified form of the KLV method and recombination. The time steps were taken to converge geometrically (the j step being over time θ^j), where $\theta < 1$ was actually chosen to be 0.4. In order to achieve the required accuracy we used a 15 point Gaussian quadrature for the KLV transitions. The recombination for a measure μ was carried out with polynomials of degree m supported on subsets of supp(μ) (each contained in balls of radius R) as test functions. As in Defintion 27.10 the subsets of supp(μ) were chosen such that they partition supp(μ).

We used a heuristic based on the information available on the $W^{1,1}$ norm of f, to determine the degree for the polynomials that minimizes the number of test functions needed to achieve the required accuracy.

In addition the algorithm was modified to compare for each particle a two step KLV with a one step KLV estimate to the boundary. If both approximations agreed to the error tolerance the algorithm immediately leapt to the boundary for this particle. Note that since the required accuracy is close to machine precision false positives are very unlikely. Recombination was then performed on the remaining particles.

To achieve an accuracy of 10^{-10} we chose $m = 8$ and $R = 0.075$ (R was normalized by the radius of the support of the measure we recombined). In this case the runtime of our example on an Intel Centrino 2.4 GHz notebook computer (implemented in single threaded C++ code) was approximately one second. The maximal depth of the approximation was restricted to 30. There were 6700 reduced notes inside the domain and $6700 \times (15 + 1) \times 15 = 1,680,000$ evaluations of the cubature at the boundary. This compares with 15^{30} leaves in the tree without pruning.

27.8 Application to nonlinear filtering

27.8.1 A brief introduction

In this section we apply the KLV method to the nonlinear filtering problem. The result is due to Crisan and Ghazali [8] and we closely follow their presentation.

We assume that the (unobserved) signal process X satisfies a Stratonovich SDE of the form

$$dX_{t,x} = V_0(X_{t,x})dt + \sum_{i=1}^{d} V_i(X_{t,x}) \circ dB_t^i, \quad X_{0,x} = x, \quad (27.21)$$

where the vector fields V_i satisfy $V_i \in C_b^\infty(R^{N_1}.R^{N_1})$ and $(B_t^i)_{i=1}^{d}$ is d-dimensional Brownian motion. Where no confusion arises we will in the following suppress the subscript x and only write X_t in place of $X_{t,x}$.

Let $h \in C_b^\infty(R^{N_1}, R^{N_2})$. The observation component Y_t solves

$$dY_t = h(X_t) + dW_t,$$

where W_t is a N_2-dimensional Brownian motion independent of the signal X.

Let $\mathcal{Y}_t := \sigma(Y_s : s \leq t)$. Solving the filtering problem involves computing the least squares approximation of $f(X_t)$ given the observations $\mathcal{Y}_t$ for a sufficiently rich set of test functions f. Let $\pi = \{\pi_t, t \geq 0\}$ be the conditional distribution of X_t given $\mathcal{Y}_t$. Provided f is square integrable with respect to the law of X_t we have

$$\pi_t(f) = \mathbb{E}(f(X_t)|\mathcal{Y}_t), \quad \text{almost surely.}$$

Recall that $h \in C_b^\infty(R^{N_1}, R^{N_2})$ and let $\tilde{P}$ be a new probability measure with Radon–Nikodym derivative satisfying

$$\frac{dP}{d\tilde{P}} = \exp\left(\sum_{i=1}^{N_2} \int_0^T h^i(X_s)dY_s^i - \frac{1}{2}\sum_{i=1}^{N_2}\int_0^T h^i(X_s)^2 ds\right).$$

Then $\tilde{P}$ and P are mutually absolutely continuous with respect to each other and by Girsanov's theorem the observation Y_t under the new measure $\tilde{P}$ is a

Brownian motion independent of the signal X_t. We define the unnormalized conditional distribution $\rho_t(f)$ by

$$\rho_t(f) := \tilde{\mathbb{E}}\left(f(X_t)\exp\left(\sum_{i=1}^{N_2}\int_0^t h^i(X_s)dY_s^i - \frac{1}{2}\sum_{i=1}^{N_2}\int_0^t h^i(X_s)^2 ds\right)\bigg|\mathcal{Y}_t\right).$$

The well known Kallianpur–Striebel forumla states that, almost surely,

$$\pi_t(f) = \frac{\rho_t(f)}{\rho_t(1)}.$$

The law of X_t, $0 \le t \le T$ is not affected by the change of measure from P to $\tilde{P}$ and hence we may continue to write

$$P_T f(x) = \tilde{\mathbb{E}} f(X_{T,x}).$$

We also define an analogous operation for cubature measures Q_T by setting

$$Q_T f(x) := \mathbb{E}_{Q_T} f(X_{T,x}).$$

and for $f \in C_b^\infty(R^{N_1})$, $r \in \mathbb{N}$ a family of seminorms by

$$\|f\|_{r,\infty} := \sum_{i=1}^r \max_{j_1,\ldots,j_i \in \{1,\ldots,N_1\}} \left\|\frac{\partial^i f}{\partial x_{j_1}\cdots \partial x_{j_i}}\right\|_\infty.$$

In the following $\|\cdot\|_p$ denotes the L^p norm with respect to the measure $\tilde{P}$ and for any random variable Z we let $|Z|_p := \mathbb{E}(|Z|^p)^{1/p}$.

For the remainder of this section the following additional assumptions will be in place: First we will only consider equidistant partitions $\mathcal{D}$ of the time interval $[0, T]$. Furthermore we assume that the test function f is in $C_b^\infty(R^{N_1})$.

To approximate the normalized conditional distribution with cubature methods we proceed following Crisan, Ghazali [8] in two stages. We first approximate the unnormalised conditional distribution by a filter that discretises observation path due to Picard [26], which is essentially an application of the classical Euler method. Second we obtain a weak approximation for Picard's discretized filter based on the KLV method.

27.8.2 A KLV-based approximation and an error bound

Following Crisan, Ghazali [8] we define functions $h_j \in C_b^\infty(R^{N_1})$ for $j = 0, \ldots, k-1$ by

$$h_j(x) = \sum_{i=1}^{N_2}\left(\Delta Y_j^i h^i(x) - \frac{T}{2k}(h^i(x))^2\right),$$

where $\Delta Y_j := Y_{(j+1)T/k} - Y_{jT/k}$. In addition let $h_k := 0$.

Once again using the notation of [8], we define operators R_s^j and $\bar{R}_s^j$, $j = 0, \ldots, k$ for $f \in C^\infty(R^{N_1})$, $x \in R^{N_1}$ and $s \in (0, 1]$ by

$$R_s^j f(x) := \tilde{\mathbb{E}}\left[f(X_{s,x})\exp(h_j(X_{s,x}))|\mathcal{Y}_s\right] = \mathbb{E}\left[f(X_{s,x})\exp(h_j(X_{s,x}))\right]$$

and

$$\bar{R}_s^j f(x) := \mathbb{E}_Q\left[f(X_{s,x})\exp(h_j(X_{s,x}))|\mathcal{Y}_s\right] = \mathbb{E}_Q\left[f(X_{s,x})\exp(h_j(X_{s,x}))\right].$$

To simplify our notation we also define for $1 \le i < j \le k$, $R_s^{i,j} f(x) := R_s^i \cdots R_s^j f(x)$ and $\bar{R}_s^{i,j} f(x) := \bar{R}_s^i \cdots \bar{R}_s^j f(x)$ respectively. Finally we define for all $f \in C_b^\infty(R^{N_1})$ two approximations to the unnormalized conditional distribution by

$$\rho_T^k(f) := \tilde{\mathbb{E}}\left(R_{T/k}^{0,k} f(X_0) | \mathcal{Y}_T\right)$$

and

$$\bar{\rho}_T^k(f) := \tilde{\mathbb{E}}\left(\bar{R}_{T/k}^{0,k} f(X_0) | \mathcal{Y}_T\right).$$

The following theorem may be found in Picard [26] (theorem 1) and demonstrates that the discretized filter ρ_T^k provides a first-order approximation (in the weak sense) for equidistant partitions $\mathcal{D}$.

Theorem 27.6 *Let $\mathcal{D}$ be the partition of $[0, T]$ with k equidistant time steps. Assume X_0 has finite moments of any order. Then for all $f \in C_b^\infty(R^{N_1})$ there is a constant $C(f, T)$ such that*

$$\|\rho_T(f) - \rho_T^k(f)\|_2 \le \frac{C}{k}.$$

The following theorem and its proof are due to Crisan and Ghazali [8]. The theorem shows that Picard's filter can be $\rho_T^k(f)$ approximated to high order with the KLV method.

Theorem 27.7 (Crisan–Ghazali) *There is a positive constant $C(T,m,p)$ such that for all $f \in C_b^\infty(R^{N_1})$, $p \ge 1$ we have*

$$\|\rho_T^k(f) - \bar{\rho}_T^k(f)\|_p \le Ck^{-(m-1)/2}\|f\|_{m+2,\infty}.$$

Before we prove Theorem 27.7 let us observe that Theorems 27.7 and 27.6 combined imply that $\bar{\pi}_T^k(f) := \frac{\bar{\rho}_T^k(f)}{\bar{\rho}_T^k(1)}$ is a first-order approximation for the normalized conditional distribution (see Crisan, Ghazali [8], Corollary 14).

Corollary 27.4 *Let Q_T be a cubature measure of degree $m \ge 3$ and assume X_0 has finite moments of any order. For any $f \in C_b^\infty(R^{N_1})$ there exists a positive constant $C(T, m, f)$ such that*

$$\|\pi_T(f) - \bar{\pi}_T^k(f)\|_2 \leq \frac{C(f)}{k}.$$

Proof. Combining the estimates of Theorems 27.7 and 27.6 using the triangle inequality we see that

$$\|\rho_T(f) - \bar{\rho}_T^k(f)\|_2 \leq \frac{C(m, f)}{k}.$$

Note that

$$\pi_T(f) - \bar{\pi}_T^k(f) = \frac{1}{\rho_T(1)} \left(\rho_T(f) - \bar{\rho}_T^k(f) + \bar{\pi}_T^k(f) \left(\bar{\rho}_T^k(1) - \rho_T(1) \right) \right).$$

The claim now follows from the inequality

$$|\pi_T(f) - \bar{\pi}_T^k(f)| \leq \frac{1}{\rho_T(1)} |\rho_T(f) - \bar{\rho}_T^k(f)| + \frac{\|f\|_\infty}{\rho_T(1)} |\rho_T(1) - \bar{\rho}_T^k(1)|$$

and the observation that $\left\| \frac{1}{\rho_T(1)} \right\|_p$ is finite for any $p \geq 1$. □

Remark 27.7 The approximation used in Theorem 27.6 provides the best approximation for even partitions (hence the assumption on $\mathcal{D}$ in this section). Note that the first-order error of Picard's approximation of the unnormalized conditional distribution asymptotically dominates the error of the KLV approximation for any cubature measure of degree $m \geq 3$.

For the proof of Theorem 27.7 we require some additional notation.

Let $\tilde{A} := \{\emptyset\} \cup \bigcup_{j \geq 1} \{1, \ldots N_1\}^j$. For any multi-index $(a_1, \ldots, a_j) \in \tilde{A}$ let $D_a := \frac{\partial^j}{\partial x_{a_1} \ldots \partial x_{a_j}}$. Following [8] we let $C_b^{Y,\infty}(R^{N_1})$ be the set of measurable functions $f : R^{N_1} \times C([0, T], R^{N_2}) \to R$ satisfying the following two conditions.

First for any fixed $y \in C([0, T], R^{N_2})$ we have $f(x, y) \in C_b^\infty(R^{N_1})$. Second for any $a \in \tilde{A}$ and $p \geq 1$ we have $\sup_{x \in R^{N_1}} |D_a f(x, Y)|_p < \infty$.

We also define for $f \in C_b^{Y,\infty}(R^{N_1})$, $p \geq 1$, $j \in N$ norms by setting

$$|f|_{p,j} := \sum_{a \in \tilde{A}, |a| \leq j} \sup_{x \in R^{N_1}} |D_a f|_p.$$

Lemma 27.1 and 27.2 of Section 27.3 provide error bounds for the approximation of $f(X_{T,x})$ by its degree m stochastic Taylor approximation under the Wiener and cubature measure respectively. In fact both Lemmas extend to functions $f \in C_b^{Y,\infty}(R^{N_1})$ and as in Proposition 27.1 an application of the triangle inequality (compare Crisan, Ghazali [8], (36), and the proof of Lemma 11) yields

$$\sup_{x \in R^{N_1}} |Q_t f(x) - P_t f(x)|_p \leq C \sum_{i=m+1}^{m+2} t^{i/2} |f|_{p,i} \qquad (27.22)$$

for some positive constant C.

Theorem 27.7 is an easy consequence of the following two lemmas.

Lemma 27.3 *There exists a positive constant $C(T, m, p)$ such that all $f \in C_b^\infty(R^{N_1})$ and $j = 1, \ldots, k$ we have*

$$\sup_{x \in R_1^N} \left| R_{T/k}^{j-1} R_{T/k}^{j,k} f(x) - \bar{R}_{T/k}^{j-1} R_{T/k}^{j,k} f(x) \right|_p \leq C k^{-(m+1)/2} \|f\|_{m+2, \infty}$$

Proof. It can be shown (see Crisan, Ghazali [8], Lemma 11 and the references therein) that for $f \in C_b^\infty(R^{N_1})$, $j = 1, \ldots k$ we have $\exp(h_{j-1}) R_{T/k}^{j,k} f \in C_b^{Y,\infty}(R^{N_1})$ and there is a constant $C(m, p)$ such that

$$\left| \exp(h_{j-1}) R_{T/k}^{j,k} f \right|_{p,m+2} \leq C \|f\|_{m+2, \infty}. \tag{27.23}$$

We deduce from (27.22) that

$$\sup_{x \in R_1^N} \left| Q_{T/k} \exp(h_{j-1}) R_{T/k}^{j,k} f(x) - P_{T/k} \exp(h_{j-1}) R_{T/k}^{j,k} f(x) \right|_p$$

$$\leq C \left(\frac{T}{k}\right)^{(m+1)/2} \left| \exp(h_{j-1}) R_{T/k}^{j,k} f \right|_{p,m+2}$$

and using the definition of the operators $R_{T/k}^{j-1}$, $\bar{R}_{T/k}^{j-1}$ and (27.23) the claim follows. $\square$

Lemma 27.4 *There is a positive constant $C(T, m, p)$ such that for all $f \in C_b^\infty(R^{N_1})$ we have*

$$\sup_{x \in R_1^N} \left| R_{T/k}^{0,k} f(x) - \bar{R}_{T/k}^{0,k} f(x) \right|_p \leq C k^{-(m-1)/2} \|f\|_{m+2, \infty}$$

Proof. Observe that

$$R_{T/k}^{0,k} f - \bar{R}_{T/k}^{0,k} f = R_{T/k}^0 R_{T/k}^{1,k} f - \bar{R}_{T/k}^0 R_{T/k}^{1,k} f$$

$$+ \sum_{j=1}^{k-1} \bar{R}_{T/k}^{0,j-1} \left(R_{T/k}^j R_{T/k}^{j+1,k} f - \bar{R}_{T/k}^j R_{T/k}^{j+1,k} f \right) + \bar{R}_{T/k}^{0,k-1} \left(R_{T/k}^k f - \bar{R}_{T/k}^k f \right)$$

and for $p \geq 1$, $j = 1, \ldots, k$

$$\sup_{x \in R_1^N} \left| \bar{R}_{T/k}^{0,j-1} \left(R_{T/k}^j R_{T/k}^{j+1,k} f(s) - \bar{R}_{T/k}^j R_{T/k}^{j+1,k} f(x) \right) \right|_p^p$$

$$\leq C \tilde{\mathbb{E}} \left[\sup_{x \in R_1^N} \left| R_{T/k}^j R_{T/k}^{j+1,k} f(x) - \bar{R}_{T/k}^j R_{T/k}^{j+1,k} f(x) \right|_p^p \right] \leq C (k^{-(m+1)/2} \|f\|_{m+2, \infty})^p,$$

where $C(T, m, p)$ is a constant independent of f and j. $\square$

Theorem 27.7 now follows easily from Lemma 27.4 by observing that for $p \geq 1$

$$\left\| \rho_T^k(f) - \bar{\rho}_T^k(f) \right\|_p \leq \tilde{\mathbb{E}} \left[\mathbb{E} \left[\left(R_{T/k}^{0,k} - \bar{R}_{T/k}^{0,k} \right)(X_0) \middle| \mathcal{Y}_T \right]^p \right]^{1/p}$$

$$\leq \sup_{x \in R_1^N} \left| R_{T/k}^{0,k} f(x) - \bar{R}_{T/k}^{0,k} f(x) \right|_p.$$

27.8.3 An explicit formula

Given a degree m cubature measure $Q = \sum_{i=1}^n \lambda_i \delta_{\omega_i}$ we now provide an explicit description that may be used to compute the KLV approximation of the filtering problem.

Let $\bar{A}$ the set of all multi-indices with indices taking values in the set $\{1, \ldots, n\}$. Denote the length of such an multi-index a by $|a|$. For $(a_1, \ldots, a_j) \in \bar{A}^j$, we define $\lambda_a = \lambda_{a_1} \cdots \lambda_{a_j}$. We define points $x_a \in R^{N_1}$ by setting $x_a = X_{s_1,x}(\omega_a)$ for $a \in \{(1), \ldots (n)\}$. For general multi-indices we define x_a using the inductive relation

$$x_{(a_1, \ldots, a_j)} = X_{s_j, x_{(a_1, \ldots, a_{j-1})}}(\omega_{a_j}, s_j).$$

Recall that the s_i in this definition are the time steps of the partition $\mathcal{D}$ of $[0, T]$ for the KLV method. The points x_a are as before obtained by solving equation (27.21) along the cubature paths $\omega_{i,s}$.

The k step KLV approximation for a fixed observation path Y (implicit in the definition of the functions h_i) and a test function f is now given by the explicit formula

$$\bar{\rho}_T^k(f) = \sum_{(a_1, \ldots, a_k) = a \in \bar{A}, |a| = k} \lambda_a f(x_a) \exp \left(\sum_{i=0}^{k-1} h_i \left(x_{(a_1, \ldots, a_i)} \right) \right).$$

Note that the KLV approximation of the unnormalized conditional distribution requires knowledge of all nodes in the particle tree, unlike the PDE case we describe in Section 27.2, where the knowledge of the leafs is sufficient to compute the approximation.

Appendix: Some explicit cubature measures

The cubature measures of degree three and five presented in the Appendix are taken from Lyons, Victoir [23]. Cubature measures of degree seven at the flow level have been constructed and implemented in Litterer [20] for one- and two-dimensional driving noises, but are omitted here.

A.1. Degree three

Suppose the measure $\sum_{i=1}^{n} \lambda_i \delta_{z_i}$ is a degree-three cubature measure on d-dimensional Gaussian space. Then $Q_{\mathcal{L}} = \sum_{i=1}^{n} \lambda_i \delta_{L_i}$, with

$$L_i = \epsilon_0 + z_i^1 \epsilon_i + \ldots + z_i^d \epsilon_d \in \mathcal{L}$$

is a degree-three cubature measure on the free Lie algebra (see Definition 27.7). The measure Q_1 defined by

$$Q_1 = \sum_{i=1}^{n} \lambda_i \delta_{\omega_i},$$

with $\omega_i : [0, 1] \to \mathbb{R}^{d+1}$ given by $\omega_i(t) = t(1, z_i^1, \ldots, z_i^d)$ is a degree-three cubature measure on Wiener space. Note that the cubature paths are linear in this case.

A.2. Degree-five measures

Suppose the measure $\sum_{i=1}^{\hat{n}} \tilde{\lambda}_i \delta_{z_i}$ is a degree-five cubature measure on d-dimensional Gaussian space.

Then the measure

$$Q_{\mathcal{L}} = \sum_{k=1}^{\hat{n}} \frac{\tilde{\lambda}_k}{2} \delta_{L_{k,1}} + \frac{\tilde{\lambda}_k}{2} \delta_{L_{k,-1}}$$

with

$$L_{k,\ell} = \epsilon_0 + \sum_{i=1}^{d} z_k^i \epsilon_i + \sum_{i=1}^{d} \frac{(z_k^i)^2}{12} [[\epsilon_0, \epsilon_i], \epsilon_i] + \sum_{i<j} \frac{z_k^i (z_k^j)^2}{12} [[\epsilon_i, \epsilon_j], \epsilon_j]$$

$$+ \sum_{i<j} \frac{(z_k^i)^2 z_k^j}{12} [[\epsilon_j, \epsilon_i], \epsilon_i] + \ell \sum_{i<j} \frac{z_k^i z_k^j}{2} [\epsilon_i, \epsilon_j] \in \mathcal{L}.$$

is a degree-five cubature measure on the free Lie algebra. Efficient constructions of cubature measures on d-dimensional Gaussian space may be found in Stroud [29] and Victoir [32]. Degree-five cubature measures supported on piecewise linear paths may be found in Lyons, Victoir [23], but are omitted here.

Notes

Research supported by a grant of the Leverhulme trust (F/08772/E). Parts of this research were carried out during a stay at the Mittag–Leffler Institute Stockholm and supported by an EPSRC doctoral student grant.

References

[1] V. Bally and D. Talay. The Euler scheme for stochastic differential equations: error analysis with Malliavin calculus. Probabilités numériques. *Math. Comput. Simulation*, 38:35–41, 1995.

[2] V. Bally and D. Talay. The law of the Euler scheme for stochastic differential equations. I. Convergence rate of the distribution function. *Probab. Theory Related Fields*, 104:43–60, 1996.

[3] G. Ben Arous. Flots et series de Taylor stochastiques. *Probab. Theory Related Fields*, 81:29–77, 1989.

[4] F. Castell. Asymptotic expansion of stochastic flows. *Probab. Theory Related Fields*, 96:225–39, 1993.

[5] K. T. Chen. Integration of paths, geometric invariants and a generalized Baker-Hausdorff formula. *Ann. of Math. 2*, 65:163–78, 1957.

[6] D. Crisan. Numerical methods for solving the stochastic filtering problem. In: *Lyons,T. J., Salisbury, T. S., eds, Providence, R.I., American Mathematical Society.*, pages 1–20, 2002.

[7] D. Crisan and A. Doucet. A survey of convergence results on particle filtering methods for practitioners. *IEEE T SIGNAL PROCES.*, 50:736–46, 2002.

[8] D. Crisan and S. Ghazali. On the convergence rates of a general class of weak approximations of SDEs. *Stochastic Differential Equations—Theory and Applications*, pp. 221–48, 2007.

[9] D. Crisan and T. Lyons. Minimal entropy approximations and optimal algorithms. *Monte Carlo Methods Appl.*, 8:343–55, 2002.

[10] Philip J. Davis. A construction of nonnegative approximate quadratures. *Math. Comp.*, 21:578–82, 1967.

[11] S. Ghazali. *The Global Error in Weak Approximations of Stochastic Differential Equations.* PhD thesis, Imperial College London, 2007.

[12] G. Gyurko. *Numerical Methods for Approximating Solutions to Rough Differential Equations.* DPhil thesis, University of Oxford, 2009.

[13] Y. Z. Hu. Série de Taylor stochastique et formule de Campbell–Hausdorff, dšaprès Ben Arous. *Lecture Notes in Math.*, 1526:579–86, 1992.

[14] P. E. Kloeden and E. Platen. *Numerical Solution of Stochastic Differential Equations.* Springer, 1999.

[15] S. Kusuoka. Approximation of expectation of diffusion process and mathematical finance. Taniguchi Conference on Mathematics Nara '98. *Adv. Stud. Pure Math.*, 31:147–65, 2001.

[16] S. Kusuoka. Malliavin calculus revisited. *J. Math. Sci. Univ. Tokyo*, 10:261–77, 2003.

[17] S. Kusuoka. Approximation of expectation of diffusion processes based on Lie algebra and Malliavin calculus. *Preprint (2003): http://kyokan.ms.u-tokyo.ac.jp/users/preprint/pdf/2003-34.pdf; subsequently appeared in Advances in Mathematical Economics, Vol. 6, 69–83, 2004.*

[18] S Kusuoka, M Ninomiya, and S. Ninomiya. A new weak approximation scheme of stochastic differential equations by using the Runge–Kutta method. *UTMS Preprint Series*, 2007.

[19] S. Kusuoka and D. Stroock. Application of the Malliavin calculus III. *J. Fac. Sci. Univ. Tokyo 1A*, 34:391–442, 1987.

[20] C. Litterer. *The Signature in Numerical Algorithms.* DPhil thesis, University of Oxford, 2008.

[21] C. Litterer and T. Lyons. Cubature on Wiener space continued. *Stochastic Processes and Application to Mathematical Finance, Proceedings of the 6th Ritsumeikan International Symposium*, 2007.

[22] T. Lyons. The interpretation and solution of ordinary differential equations driven by rough signals. *Proc. Sympos. Pure Math.*, 57:115–28, 1995.
[23] T. Lyons and N. Victoir. Cubature on Wiener space. *Proc. R. Soc. Lond. A*, 460:169–98, 2004.
[24] S. Ninomiya. A partial sampling method applied to the Kusuoka approximation. *Monte Carlo Methods and Appl.*, 9:27–38, 2003.
[25] S. Ninomiya and N. Victoir. Weak approximation of stochastic differential equations and application to derivative pricing. *Preprint*, 2006.
[26] J. Picard. Filtering and control of random processes. *Lecture Notes in Contr. and Inform. Sc.*, 84:219–36, 1984.
[27] C. Reutenauer. *Free Lie algebras*. London Mathematical Society Monographs, vol. 7, Oxford, 1993.
[28] R. S. Strichartz. The Campbell–Baker–Hausdorff–Dynkin formula and solutions of differential equations. *Journal of functional analysis*, 72:320–45, 1987.
[29] A. H. Stroud. *Approximate Calculation of Multiple Integrals*. Prentice-Hall, 1971.
[30] J. Teichmann, personal communication.
[31] J. Teichmann. Calculating the Greeks by cubature formulae. *Royal Society of London Proceedings Series A*, 462:647–70, 2006.
[32] N. Victoir. Asymmetric cubature formulae with few points in high dimension for symmetric measures. *SIAM J. Numer. Analysis*, 42:209–27, 2004.

PART VIII
Numerical Methods in Nonlinear Filtering

PART VII

·28·

Numerical Approximations to Optimal Nonlinear Filters

H. J. Kushner

28.1 Introduction

The usefulness of the theory of nonlinear filtering is limited by the availability of good practical approaches that well approximate the quantities of major interest, for example the conditional (weak-sense) density or the conditional mean and covariance. The mathematical theory is mainly concerned with diffusion-type models and white noise corrupted observations that are taken continuously in time. If the observations are taken in discrete time (as they tend to be in practical applications), then the theoretical issues are smaller, since one only needs to approximate the (weak-sense) solution to the Kolmogorov forward (the Fokker–Planck) equation between observations and then use Bayes' rule to incorporate the observations. In both cases, the computational issues and the development of the most effective approximation methods is still evolving. We will discuss two broad classes of approximations that have been successfully used on various classes of very nonlinear problems and hold considerable promise. The first approach is the so-called Markov chain approximation method [31]. It will be applied to both cases where the observations are taken continuously and in discrete time. At this time it seems to be the most appropriate approach to approximating the weak-sense conditional density, at least for low-dimensional problems. The basic idea is to use a filter for a Markov chain that approximates the diffusion, but with the actual physical observations. Convergence theorems can be proved as the approximation parameters go to zero.

The Markov chain approximation method is a widely used approach to the numerical solution of control and optimal control problems for general jump-diffusion models, with or without boundary reflection, and with all of the standard cost functions. It is the most general set of algorithms available at this time for the control problem, in terms of both construction of the actual numerical algorithm and the convergence proofs. The basic idea is the following. One constructs an approximating Markov chain on a finite state space, with approximation parameter h, and that is "locally consistent" with the diffusion or jump diffusion. Then one solves for the cost or optimal cost function for the approximating chain, and finally proves that as $h \to 0$, these functions converge to those for the original model. The local consistency condition is not restrictive and should be satisfied by any approximating process. The approximating chains should be constructed so that the associated numerical problem is reasonable. Getting such constructions is straightforward, as can be seen in [31]. The same chains can be used for the filtering problem. Due to the growth of the size of the required state space as the dimension increases, at this time the method would be difficult to use if the dimension were greater than four.

Numerical methods for the approximate evaluation of the conditional distribution were given in [27] and related methods are in [9, 13, 15]. Robustness (locally Lipschitz continuity) of the numerical approximations with respect to the observation process was shown in [28] and this property is satisfied by the algorithm that is to be described. Extensions, including point process observations, are in [10, 14].

As noted, the Markov chain approximation method is used for the approximation of the conditional weak-sense density and can be "computationally intensive". Often one is interested in just the first few conditional moments. Since, for nonlinear problems, a finite set of conditional moments will define the conditional distribution only under very restrictive conditions, one must resort to a heuristic procedure. The second method to be considered is the so-called "assumed form of the conditional density" approach, first proposed in [25] and developed and used in various ways since then [1, 8, 20, 38]. With this method, one assumes that the conditional density takes a particular parametrized form, and then approximates the evolution of the parameters, under this assumption. Most commonly, the assumed density is Gaussian (or a Gaussian mixture), where the parameters are the conditional mean and covariance. The numerical issues center about the approximation of integrals with respect to Gaussian kernels. With guidance from the literature on the numerical evaluation of integrals, there are many ways of doing this, and some methods of current interest will be discussed. Numerical data lend support to the value of the approach. The original observation that led to this method was the realization that the extended Kalman filter was equivalent to the assumption of a conditional "density" that could have negative values, and this was a major cause of its instability in truly nonlinear problems.

Section 28.2 describes the system model and the basic results from the theory of nonlinear filtering that will be needed. The proofs for the Markov approximation method depend on results from the theory of weak convergence, and a summary of the required results is given in Section 28.3. Section 28.4 gives a general result on the approximation of the conditional weak-sense density. The Markov chain approximation method is discussed in Section 28.5. It is applied to the filter approximation problem in Section 28.6, where the result of Section 28.4 will be crucial. The main issues in the construction concern the coordination of the time scale of the chain with that of the physical observation process, and there are several ways of doing this. The robustness of the approximating filter (uniformly in the approximation parameter) is shown in Section 28.7. Sections 28.8 and 28.9 concern the assumed density approach and it is illustrated by a numerical example in Section 28.10.

An alternative approach to filtering is that of batch processing, which we will not discuss since we are concerned with "dynamical systems" approximations. Loosely speaking, with observation process $y_n = g(x_n) +$ observation noise, at step n one constructs a penalty function of the form [2]

$$\sum_{i=1}^{n} |g(x_i) - y_i|^2 + \text{penalty on the driving and observation noises}$$

and seeks to minimize it with respect to the driving and observation noises. The approach can yield very good results, if the computation required for the minimization is not too onerous.

28.2 The system model and filter representations

The system. The system will evolve in continuous time. To keep the technicalities at a minimum, we will use the diffusion process defined by

$$x(t) = x(0) + \int_0^t b(x(s))ds + \int_0^t \sigma(x(s))dw(s), \tag{28.1}$$

where $x(t) \in \mathbb{R}^r$, Euclidean r-space, $w(\cdot)$ is a standard vector-valued Wiener process, and $b(\cdot)$ and $\sigma(\cdot)$ are bounded and continuous.[1] We suppose that (28.1) has a unique weak-sense solution for each initial condition. Define the matrix $a(x) = \{a_{i,j}(x); i, j\} = \sigma(x)\sigma'(x)$.

Notes on and extensions of the model. One can stop the process or reinject it when it exits some bounded set. Many physical problems have reflecting boundaries. Under the usual conditions on such boundaries and reflection directions (see, e.g., [30, 31]) all of the subsequent results will hold true, provided that the Markov chain approximation accounts for the boundary. We could also treat the jump-diffusion model where the term $\int_0^t \int q(x(s-), \gamma) N(ds\, d\gamma)$ is added to (28.1), where $N(\cdot)$ is a Poisson random measure with intensity measure $h(dt\,d\gamma) = \lambda dt \Pi(d\gamma)$, $\Pi(\cdot)$ is a bounded measure, and $q(\cdot)$ is bounded and continuous. The extra technicalities associated with such extensions shed little light on the problems of filter approximations, so we stay with (28.1). For expositional simplicity, we suppose that $b(\cdot), \sigma(\cdot), q(\cdot), g(\cdot)$ do not depend on time. The time-dependent case is a trivial modification.

The observations. The observations can be taken in continuous or in discrete time. For the first case, the data at t is $\mathcal{Y}_t = \{y(s), s \leq t\}$, where

$$dy(t) = g(x(t))dt + V dw_0(t), \quad y(0) = 0, \quad y(t) \in \mathbb{R}^m, \qquad (28.2)$$

where $w_0(\cdot)$ is a standard Wiener process that is independent of $x(\cdot)$, $g(\cdot)$ is bounded and continuous and the matrix VV' is positive-definite.

The observations (28.2) are often approximated by supposing that for small $\delta > 0$ they are taken at times $n\delta$, $n = 1, 2, \ldots$, and that at $n\delta$ is either

$$y_n^\delta = g(x(n\delta))\delta + V[w_0(n\delta) - w_0(n\delta - \delta)], \qquad (28.3a)$$

or

$$y_n^\delta = \int_{n\delta-\delta}^{n\delta} g(x(s))ds + V[w_0(n\delta) - w_0(n\delta - \delta)] = y(n\delta) - y(n\delta - \delta). \qquad (28.3b)$$

For (28.3a,b), let $\mathcal{Y}_{n\delta}^\delta$ denote the data available at time $n\delta$. When the observations are taken at arbitrary times t_n, $n = 1, 2, \ldots$, let that at t_n be

$$y_n = g_i(x(t_n)) + V_n v_n, \qquad (28.3c)$$

where $g_n(\cdot)$ is bounded and continuous, $V_n V_n'$ is positive-definite and $\{v_n\}$ are i.i.d. Gaussian vectors, independent of $x(\cdot)$, and with covariance matrix the identity. Then we define $\mathcal{Y}_n = \{y_{t_i}, i \leq n\}$, In all cases, since we lose nothing by working with the observation process $V^{-1}y(t)$, without loss of generality in what follows we will suppose that $V = V_i = \mathbb{I}$, the identity matrix.

The optimal nonlinear filter for (28.1), (28.2). For $\phi(\cdot)$ a bounded and measurable real-valued function, define the conditional expectation operator $\mathbb{E}_t$ by $\mathbb{E}_t \phi(x(t)) = \mathbb{E}[\phi(x(t))|\mathcal{Y}_t]$. Perhaps the key result in the theory of nonlinear filtering is the so-called representation theorem, which is a "limit" form of Bayes' rule. We will use it in the form that was used in the original derivation of the optimal nonlinear filter [26], which is convenient for the types of approximation and weak convergence methods that will be used. That reference was the first to get the result for the diffusion case, and the development is similar for the reflected jump-diffusion model. Subsequent developments via measure transformation and "martingale innovations" techniques are in [18, 34], but the representations obtained by those techniques are exactly the same as were obtained in [26]. The case where $x(\cdot)$ is a continuous-time Markov chain on a finite state space case was first derived in [36, 39].

Let $\tilde{x}(\cdot)$ be a process with the same probability law as $x(\cdot)$, but that is independent of $(x(\cdot), y(\cdot))$. Define

$$R(t) = \exp\left[\int_0^t g(\tilde{x}(s))' dy(s) - \frac{1}{2}\int_0^t |g(\tilde{x}(s))|^2 ds\right]. \qquad (28.4)$$

Then we have the representation [26] for the filter for model (28.1) and observation (28.2):

$$\mathbb{E}_t \phi(x(t)) = \frac{\mathbb{E}[R(t)\phi(\tilde{x}(t))|\mathcal{Y}_t]}{\mathbb{E}[R(t)|\mathcal{Y}_t]}. \qquad (28.5)$$

The form holds for the observations (28.3b), if t is restricted to $n\delta$, $n = 1, 2, \ldots$. It also holds for (28.3a) if the sums $\sum_{i=1}^n g'(\tilde{x}(i\delta))[w_0(i\delta) - w_0(i\delta - i\delta)]$ and $(\delta/2)\sum_{i=1}^n |g(\tilde{x}(i\delta))|^2$ replace the integrals.

Dynamical equations for the conditional probability density. For the moment let us proceed purely formally. Define the differential operator $\mathcal{L}$ of (28.1), acting on smooth real-valued functions $f(\cdot)$,

$$\mathcal{L}f(x) = f'_x(x)b(x) + \frac{1}{2}\sum_{i,j} a_{i,j}(x)\frac{\partial^2 f(x)}{\partial x_i \partial x_j}.$$

The formal adjoint operator $\mathcal{L}^*$ is

$$\mathcal{L}^* f(x) = -\sum_i \frac{\partial (b_i(x)f(x))}{\partial x_i} + \frac{1}{2}\sum_{i,j} \frac{\partial^2 (a_{i,j}(x)f(x))}{\partial x_i \partial x_j}.$$

Suppose that there is a smooth density $p(\tilde{x}, t, x)$ of $x(t)$ given initial condition $x(0) = \tilde{x}$. This satisfies the Kolmogorov forward (or Fokker–Planck) equation (with $\mathcal{L}^*$ acting on the functions of x) [17, Ch. 5]

$$\frac{\partial p(\tilde{x}, t, x)}{\partial t} = \mathcal{L}^* p(\tilde{x}, t, x). \qquad (28.6)$$

If the density is not smooth, then use the weak-sense solution.

Consider the filtering problem with model (28.1), (28.2). Let $\phi(\cdot)$ be a bounded and continuous real-valued function whose partial derivatives up to second order are bounded and continuous. Then (recall that we are taking $V = \mathbb{I}$, the identity matrix) the conditional moments evolve as [26, 34]

$$d\left[\mathbb{E}_t \phi(x(t))\right] = \mathbb{E}_t \mathcal{L}\phi(x(t))dt \\ + (\mathbb{E}_t g(x(t))\phi(x(t)) - \mathbb{E}_t g(x(t))\mathbb{E}_t \phi(x(t)))' (dy(t) - \mathbb{E}_t g(x(t))dt). \qquad (28.7)$$

Let π_t denote the (weak-sense) conditional density of $x(t)$ given $\mathcal{Y}_t$. Then, purely formally, the normalized density evolves as [18, 24, 34]

$$d\pi_t(x) = \mathcal{L}^* \pi_t(x)dt + \pi_t(x)\left(g(x) - \mathbb{E}_t g(x(t))\right)' (dy(t) - \mathbb{E}_t g(x(t))dt). \qquad (28.8)$$

Write the numerator of (28.5) as $\mathbb{E}_t^0 \phi(x(t))$, the unnormalized conditional expectation. Then

$$d\left[\mathbb{E}_t^0 \phi(x(t))\right] = \mathbb{E}_t^0 \mathcal{L}\phi(x(t))dt + \mathbb{E}_t^0 g(x(t))\phi(x(t))dy(t).$$

This equation, known as the Zakai equation ([40], [18, sec. 14], and [26, eq. 15]), is simpler than (28.7), but does not seem to provide simpler numerical algorithms. The Markov chain approximation method already starts by computing the unnormalized density, and the assumed density approach uses a normalized assumed density.

The optimal filter for a Markov chain signal process. Let $\{\xi_n, n < \infty\}$ be a finite-state Markov chain with one-step transition probabilities $p(x, y)$. Suppose that $\{v_n, n < \infty\}$ is a sequence of mutually independent normally distributed random variables with mean zero and covariance V_0, and which is also independent of the $\{\xi_n, n < \infty\}$. Suppose that we observe the white noise corrupted data $y_n^0 = g_0(\xi_n) + v_n$ at time step n, for some bounded and measurable function $g_0(\cdot)$. Define $\mathcal{Y}_n^0 = \{y_i^0, i \leq n\}$ and the conditional distribution $Q_n^0(x) = P\{\xi_n = x | \mathcal{Y}_n^0\}$.

We now use Bayes' rule to define a recursive formula for $Q_n^0(\cdot)$. Let $P\{y_n^0 | \xi_n = x\}$ denote the conditional (normal, with given mean $g_0(x)$ and covariance V_0) density of the observation at the value y_n^0. Note that

$$Q_n^0(x) = \sum_z P\{\xi_n = x, \xi_{n-1} = z | y_n^0, \mathcal{Y}_{n-1}^0\}$$
$$= \frac{\sum_z P\{y_n^0 | \xi_n = x, \xi_{n-1} = z\} P\{\xi_n = x | \xi_{n-1} = z\} Q_{n-1}^0(z)}{\text{normalization}}. \quad (28.9)$$

Substituting in the conditional density function of y_n^0, rewrite the last expression as

$$\frac{\sum_z \exp\left[g_0'(x) V_0^{-1} y_n^0 - \frac{1}{2} g_0'(x) V_0^{-1} g_0(x)\right] p(z, x) Q_{n-1}^0(z)}{\text{normalization}}. \quad (28.10)$$

The normalization is the numerator summed over x and might differ from expression to expression. Let $\phi(\cdot)$ be bounded, measurable and real-valued, and let $\{\tilde{\xi}_n\}$ have the same probability law as $\{\xi_n\}$, but independent of $\{\xi_n, y_n\}$. Then, iterating (28.10) yields the expression

$$\mathbb{E}_{\mathcal{Y}_n^0} \phi(\xi_n) = \frac{\mathbb{E}_{\mathcal{Y}_n^0} \phi(\tilde{\xi}_n) \exp\left\{\sum_{i=0}^n \left[g_0'(\tilde{\xi}_i) V_0^{-1} y_i^0 - \frac{1}{2} g_0'(\tilde{\xi}_i) V_0^{-1} g_0(\tilde{\xi}_i)\right]\right\}}{\text{normalization}}. \quad (28.11)$$

28.3 Weak convergence

Expressions like (28.11) will be basis of the numerical approximation, but where the y_n^0 come from the original (28.2) or (28.3). The proofs of convergence depend on the theory of weak convergence of probability measures [4, 19] and we will give a brief description of the essential facts. The probability measures of interest are those on the path space of $x(\cdot)$ or of the processes to be used to approximate it. This will be the space $D^r[0, \infty)$ of functions that are right-continuous and have left-hand limits, with the Skorokhod topology. Complete treatments are in [4, 19]. The following tightness criterion will be used.

Theorem 28.1 [23, Theorem 2.7b][19]. *Let $\{x^n(\cdot)\}$, defined on $(\Omega, \mathcal{F}, P)$, take values in $D^r[0, \infty)$. Assume that for each t in a dense set in $[0, \infty)$ and each $\delta > 0$ there exist compact $K_{t,\delta} \subset \mathbb{R}^r$ such that $\sup_n P\{x^n(t) \notin K_{t,\delta}\} \leq \delta$. Define $\mathcal{F}^n_t$ to be the σ-algebra generated by $\{x^n(s), s \leq t\}$. Let T^n_T be the set of $\{\mathcal{F}^n_t\}$-stopping times which are less than or equal to T w.p.1. Then $\{x^n(\cdot)\}$ is tight if for each $T \in [0, \infty)$,*

$$\lim_{\delta \to 0} \limsup_n \sup_{\tau \in T^n_T} \mathbb{E}\left(1 \wedge |x^n(\tau + \delta) - x^n(\tau)|\right) = 0. \tag{28.12}$$

Since we will have need for weak convergence of functions of the approximating process, with the observation taking the role of a "parameter", the following extension of weak convergence will be useful.

Theorem 28.2 *Let X_n converge weakly to X. Suppose (w.l.o.g.) that $\{X_n, n < \infty\}$ and X are defined on the same probability space, and that Y (with measure P_Y) is independent of (X_n, X) for each n. Let $f(\cdot, \cdot)$ be a real-valued measurable function, such that for P_Y-almost all y, $f(\cdot, y)$ is continuous with probability one with respect to the measure P_X of X. Let P_n denote the measure of X_n. Suppose that for each y in a set of P_Y-probability one $\{f(X_n, y)\}$ is uniformly integrable. Then*

$$\int f(x, Y) P_n(dx) \to \int f(x, Y) P_X(dx), \quad \text{w.p.1 } (P_Y). \tag{28.13}$$

28.4 Approximations to the optimal filter

The expression (28.5) is difficult to evaluate owing to the expectation over $\tilde{x}(\cdot)$. Recall that $\tilde{x}(\cdot)$ is a copy of $x(\cdot)$, and is independent of $y(\cdot)$. For computational purposes, one replaces $\tilde{x}(\cdot)$ by an approximation $x^h(\cdot)$ which is selected so that the evaluation can be carried out with a reasonable amount of work, and then one proves that the result converges to that defined by (28.5) as the approximation parameter $h \to 0$. We will suppose that the path is confined to a compact set, either by its own dynamics, a reflecting boundary, or by being stopped or reinjected on leaving a given compact set. There is considerable flexibility in the choice of the approximating process, as seen by the next theorem.

Let $\{x^h(\cdot)\}$ be a sequence that is independent of $y(\cdot)$, and that converges weakly to $x(\cdot)$ as $h \to 0$. Define

$$R^h(t) = \exp\left[\int_0^t g'(x^h(s))dy(s) - \frac{1}{2}\int_0^t |g(x^h(s))|^2 ds\right]. \tag{28.14}$$

Define the operator E_t^h by

$$E_t^h \phi(x^h(t)) = \frac{\mathbb{E}_t \phi(x^h(t)) R^h(t)}{\mathbb{E}_t R^h(t)}. \tag{28.15}$$

Comment. E_t^h is not a conditional expectation of $\phi(x^h(t))$, since $x^h(\cdot)$ does not have the probability law of the true signal process. But, by Theorem 28.3, (28.15) converges to (28.5).

Theorem 28.3 *Let $\phi(\cdot)$ be a continuous real-valued function. Assume the above conditions on $x^h(\cdot)$, $x(\cdot)$, $y(\cdot)$, with model (28.2) for the observations. Then, for any $T < \infty$,*

$$\lim_{h \to 0} \sup_{t \leq T} \left| E_t^h \phi(x^h(t)) - \mathbb{E}_t \phi(x(t)) \right| \to 0 \tag{28.16}$$

in the senses of probability and mean. The result holds for the observation models (28.3a) and (28.3b) at times $n\delta$, $n = 1, 2, \ldots$.

Proof. We will neglect the $|g|^2$ terms in (28.14), since they are easy to deal with. First, some preliminary computations. Let $\zeta_i(\cdot)$, $i = 1, 2$, and $\zeta(\cdot)$ be bounded processes that are independent of the standard Wiener process $w_1(\cdot)$ of the same dimension. The analysis will make use of an estimate of

$$W = \mathbb{E}\sup_{t \leq T} \left| \exp\left[\int_0^t \zeta_1'(s)dw_1(s)\right] - \exp\left[\int_0^t \zeta_2'(s)dw_1(s)\right] \right|. \tag{28.17}$$

We will use the inequality, for real numbers A and B,

$$\left| e^A - e^B \right| \leq |A - B|\left(e^A + e^B\right). \tag{28.18}$$

For any real-valued submartingale $N(\cdot)$ [16, Ch. VII, theorem 3.4],

$$\mathbb{E}\sup_{t \leq T} N^2(t) \leq 4\mathbb{E} N^2(T). \tag{28.19}$$

The inequality (28.19) and Schwarz's inequality applied to (28.17) yield

$$W^2 \leq \mathbb{E}\sup_{t \leq T} \left|\int_0^t (\zeta_1'(s) - \zeta_2'(s))\, dw_1(s)\right|^2$$
$$\times \mathbb{E}\sup_{t \leq T} \left|\exp\left[\int_0^t \zeta_1'(s)dw_1(s)\right] + \exp\left[\int_0^t \zeta_2'(s)dw_1(s)\right]\right|^2. \tag{28.20}$$

By (28.19) the first factor in (28.20) is bounded by $4\mathbb{E}\int_0^T |\zeta_1(s) - \zeta_2(s)|^2 ds$. To bound the second factor, use the fact that

$$\mathbb{E}\exp\left[\int_0^T \zeta'(s)dw_1(s)\right] \leq \mathbb{E}\exp\left[\int_0^T |\zeta(s)|^2 ds/2\right]$$

for any bounded process $\zeta(\cdot)$ that is independent of $w(\cdot)$, together with (28.19) and the fact that the $\exp\left[\int_0^t \zeta_i'(s)dw_1(s)\right]$ are submartingales. Finally, for a constant C_1 that depends on T and on the bounds on the $\zeta_i(\cdot)$, we have

$$W^2 \leq C_1\mathbb{E}\int_0^T |\zeta_1(s) - \zeta_2(s)|^2 ds. \qquad (28.21)$$

It suffices to show that $\sup_{t\leq T} |\mathbb{E}_t\phi(x^h(t))R^h(t) - \mathbb{E}_t\phi(\tilde{x}(t))R(t)| \to 0$ in probability for each bounded and continuous $\phi(\cdot)$. The above computations imply that, for each $T < \infty$,

$$\mathbb{E}\sup_{s\leq T}[R^h(t)]^2 + \mathbb{E}\sup_{s\leq T}[R(t)]^2 < \infty. \qquad (28.22)$$

By the Skorokhod representation [19, theorem 3.1.8.], w.l.o.g., we can suppose that all of the processes $x^h(\cdot)$, $\tilde{x}(\cdot)$, $x(\cdot)$, $y(\cdot)$ are defined on the same probability space and that the convergence $x^h(\cdot) \to \tilde{x}(\cdot)$ is w.p.1 in the topologies of the paths. We have, where C_2 is a bound on $\phi(\cdot)$,

$$\sup_{t\leq T}\mathbb{E}_t\left|\phi(x^h(t))R^h(t) - \phi(\tilde{x}(t))R(t)\right|$$
$$\leq \sup_{t\leq T}\mathbb{E}_t\left|\phi(x^h(t)) - \phi(\tilde{x}(t))\right|R^h(t) + C_2\sup_{t\leq T}\mathbb{E}_t\left|R^h(t) - R(t)\right|. \qquad (28.23)$$

The first term on the right goes to zero as $h \to 0$ by the weak convergence of $x^h(\cdot)$ to $\tilde{x}(\cdot)$, the boundedness of these processes and (28.22). The convergence of the second term of (28.23) follows from the weak convergence and the inequality (28.21). □

Comments on numerical approximations. It is important to keep in mind that all approximations to the nonlinear filtering problem are actually approximations to some representation of Bayes' rule, and (28.5) is the fundamental Bayes rule formula. Except for a few special cases such as the classical Kalman–Bucy form or where $x(\cdot)$ is a finite-state Markov chain, the evaluation of (28.5) is not a finite calculation.

Theorem 28.3 provides an approach to the computation of the conditional density for nonlinear problems. The key to computational effectiveness is the choice of the approximating process $x^h(\cdot)$. There are many possibilities, but it will always be a Markov chain. One could consider a discrete-time approximation of (28.1) such as (for small $\delta > 0$)

$$X(n\delta + \delta) = X(n\delta) + \delta b(X(n\delta)) + \sigma(X(n\delta))[w(n\delta + \delta) - w(n\delta)]. \quad (28.24)$$

While the form is simple, the state space would have to be discretized and the transition functions (as a function of the initial and final values) computed. Then it will turn out to be a version of the Markov chain approximation method. Further comments are in Section 28.6.2.

If the observations are of the discrete-time form (28.3c), then the filtering problem is simple to describe. Suppose that $\pi_{t_n^-}(\cdot)$, the conditional weak-sense density just before the nth observation is taken, is available at time t_n. Then use Bayes' rule to compute the conditional distribution $\pi_{t_n}(\cdot)$, taking the new observation into account. Then continue, starting with the density $\pi_{t_n}(\cdot)$, compute the weak-sense density $\pi_{t_{n+1}^-}(\cdot)$ at time t_{n+1}, just before the next observation is taken, etc. While the procedure is simple in principle, the actual computations can be difficult. Getting $\pi_{t_{n+1}^-}(\cdot)$ from $\pi_{t_n}(\cdot)$ involves approximating the solution of the (weak-sense) form of the Kolmogorov forward equation and the integrals involved in incorporating the observations will need to be suitably approximated.

28.5 The Markov chain approximation

From the perspective of the types of computation that are involved in computing the approximation to the conditional density, the simplest form of the Markov chain approximation method is analogous to methods for solving parabolic PDE's by finite differences or finite elements. If the computations are to be finite then the path must be confined to a bounded set, which might depend on time. Boundedness is often a consequence of the dynamics of (28.1). If $x(\cdot)$ is not bounded on the time interval of concern, then some type of truncation is needed. This can be done by introducing either a reflecting or stopping boundary. However, for expositional simplicity, we will generally ignore the boundedness issue since it will not affect the main ideas. The actual numerical approximating filter will be that for a Markov chain that approximates the diffusion, but with the actual physical observations (28.2) or (28.3) being used. The approximations to the conditional densities will converge to the weak-sense conditional density as the approximation parameters go to zero. The overall development is based on the methods in [27, 31].

The basis of the approximation is a discrete-time finite-state Markov chain whose "local properties" are "consistent" with those of (28.1), as described below. This chain will be interpolated into a continuous-time process that will be a good approximation to (28.1) in the sense to be described. For simplicity, let the approximation parameter $h > 0$ be real-valued, although a vector-valued parameter could be used as well [31]. For each h, let $\{\xi_n^h, n < \infty\}$ be a discrete-parameter Markov chain on a discrete state space $G_h \subset \mathbb{R}^r$, with finitely many points, and with transition probabilities $p^h(x, \tilde{x})$.

Local consistency: diffusion case. Define $\Delta \xi_n^h = \xi_{n+1}^h - \xi_n^h$. Let $\mathbb{E}_n^h$ (resp. $\mathbb{E}_{x,n}^h$) denote the the conditional expectation given all data to step n (and in addition that $\xi_n^h = x$, resp.). Define the martingale difference $\beta_n^h = \Delta \xi_n^h - \mathbb{E}_n^h \Delta \xi_n^h$. Suppose that the following "local consistency" conditions hold in G_h:[2]

$$\mathbb{E}_{x,n}^h \Delta \xi_n^h \equiv b_h(x) \Delta t^h(x) = b(x) \Delta t^h(x) + o(\Delta t^h(x)),$$

$$\text{covar}_{x,n}^h \left[\Delta \xi_n^h - \mathbb{E}_{x,n}^h \Delta \xi_n^h \right] \equiv a_h(x) \Delta t^h(x) = a(x) \Delta t^h(x) + o(\Delta t^h(x)), \quad (28.25)$$

$$\sup_{n,\omega} |\xi_{n+1}^h - \xi_n^h| \xrightarrow{h} 0,$$

for some function $\Delta t^h(x) > 0$, that we call an "interpolation interval". We assume that $\lim_{h \to 0} \sup_{x \in G_h} \Delta t^h(x) = 0$, but $\inf_{x \in G_h} \Delta t^h(x) > 0$ for each $h > 0$. Define $\Delta t_n^h = \Delta t^h (\xi_n^h)$. The local consistency (28.25) is essentially all that is required of the approximating chain, together with analogous conditions for the reflecting boundary, if any.[3]

The reference [31, ch. 5] describes many convenient methods for constructing chains that satisfy the required properties. By (28.25), the chain has the local conditional drift and covariance properties of (28.1). With all of the usual methods for constructing the $p^h(x, \tilde{x})$, the interpolation intervals are obtained automatically as a byproduct [31, ch. 5]. There is considerable flexibility; local consistency need not hold everywhere, as seen in [31, sec. 5.5].

The simplest example is the one-dimensional model $dx = b(x)dt + \sigma(x)dw$ where $\sigma^2(x) \geq h|b(x)|$ for all x, and the state space is $G_h = \{0, \pm h, \pm 2h, \ldots\}$. Then

$$p^h(x, x \pm h) = (\sigma^2(x) \pm hb(x))/2\sigma^2(x) \text{ and } \Delta t^h(x) = h^2/\sigma^2(x) \quad (28.26)$$

yield a locally consistent chain, where each state communicates only with its nearest neighbours.

Constant interpolation interval. For the numerical approximation of the general control problem as in [31], the possible dependence of $\Delta t^h(x)$ on x is an advantage from the point of view of computation. But it complicates the computations for the filtering problem where the approximating chain must be able to "track real time." Having a constant $\Delta t^h(x)$ can then be useful. A chain with a constant interpolation interval is easily obtained from any locally consistent chain for which $p^h(x, x) = 0$ for all x,[4] and the modified transition probabilities $\bar{p}^h(x, \tilde{x})$ and interpolation interval $\overline{\Delta}^h$ are readily obtained. Define $\overline{\Delta}^h = \inf_{\xi \in G_h} \Delta t^h(\xi)$. The possibility that $\overline{\Delta}^h < \Delta t^h(x)$ at some point x is compensated for by allowing the state x to communicate with itself. Conditioned on the event that a state does not communicate with itself, the transition probabilities are the $p^h(\cdot)$. Thus, the general formula for getting $\bar{p}^h(x, \tilde{x})$ from the $p^h(\cdot)$ is ([31, sec. 7.7])

$$\bar{p}^h(x, \tilde{x}) = p^h(x, \tilde{x})(1 - \bar{p}^h(x, x)), \quad \text{for } x \neq \tilde{x},$$
$$\bar{p}^h(x, x) = 1 - \frac{\overline{\Delta}^h}{\Delta t^h(x)}. \tag{28.27}$$

Continuous-time interpolations and convergence proofs. The numerical algorithms use the Markov chain, but the proofs of convergence are based on continuous-time interpolations of the chain. The simplest interpolation, called $\xi^h(\cdot)$, uses the intervals $\Delta t_n^h = \Delta t^h(\xi_n^h)$ and is defined as follows. With $t_n^h = \sum_0^{n-1} \Delta t_i^h$ define $\xi^h(t) = \xi_n^h$, $t \in [t_n^h, t_{n+1}^h)$. Although ξ_n^h is a Markov chain, $\xi^h(\cdot)$ is not.

The proofs are facilitated by using an interpolation $\psi^h(\cdot)$ that is a continuous-time Markov chain, asymptotically equivalent to $\xi^h(\cdot)$, and constructed as follows. Let $\{\nu_n\}$ be random variables that are independent of $\{\xi_n^h\}$, mutually independent and identically distributed, and with ν_n being exponentially distributed with mean unity. Using $\Delta \tau_n^h = \Delta t_n^h \nu_n$ and $\tau_n^h = \sum_{i=0}^{n-1} \Delta \tau_i^h$, define $\psi^h(t) = \xi_n^h$ on $[\tau_n^h, \tau_{n+1}^h)$. $\psi^h(\cdot)$ is a continuous-time Markov chain, whose holding times $\Delta \tau_n^h$, given ξ_n^h, are exponentially distributed with mean Δt_n^h.

We can decompose $\psi^h(\cdot)$ in terms of a compensator and martingale as $\psi^h(t) = x(0) + \int_0^t b_h(\psi^h(s))\,ds + B_\tau^h(t)$, where the martingale $B_\tau^h(t)$ has quadratic variation process $\int_0^t a_h(\psi^h(s))ds$. It can be shown that ([31, sec. 10.4.1]) there is a martingale $w^h(\cdot)$ (with respect to the filtration generated by the path and control processes, possibly augmented by an "independent" Wiener process) such that

$$B_\tau^h(t) = \int_0^t \sigma_h(\psi^h(s))dw^h(s) = \int_0^t \sigma(\psi^h(s))dw^h(s) + \epsilon^h(t),$$

where $\sigma_h(\cdot)[\sigma_h(\cdot)]' = a_h(\cdot)$, $w^h(\cdot)$ has quadratic variation process $\mathbb{I}t$ and converges weakly to a standard Wiener process. The martingale $\epsilon^h(\cdot)$ is due to the difference between $\sigma(x)$ and $\sigma_h(x)$ and

$$\lim_{h \to 0} \mathbb{E} \sup_{s \leq t} |\epsilon^h(s)|^2 = 0 \tag{28.28}$$

for each t. Thus

$$\psi^h(t) = x(0) + \int_0^t b_h(\psi^h(s))ds + \int_0^t \sigma(\psi^h(s))dw^h(s) + \epsilon^h(t). \tag{28.29}$$

By Theorem 28.1, $\psi^h(\cdot)$ is tight. The weak-sense limit is (28.1). The same result holds if there is a jump term. See [31, ch. 10] for more detail.

Define $t^h(\cdot)$ and $\tau^h(\cdot)$ by $t^h(s) = t_n^h$ on $[t_n^h, t_{n+1}^h)$ and $\tau^h(s) = \tau_n^h$ on $[\tau_n^h, \tau_{n+1}^h)$. By Theorem 28.4, both converge to the process with value t at time t. Hence the interpolations $\xi^h(\cdot)$ and $\psi^h(\cdot)$ are asymptotically equivalent.

Discretizing time. A modification that allows us to keep track of the elapsed real time when the $\Delta t^h(x)$ are not constant augments the chain $\{\xi_n^h\}$ by adding a (random) discretization of time. This is a simplified version of the "implicit" approximation in [31] and will allow us to use the original intervals $\Delta t^h(x)$. Let $\delta \geq \sup_x \Delta t^h(x)$ denote the discretization level for the time variable, whose value at step n we denote by $\phi_n^{h,\delta}$. The spatial components satisfy (28.25). So one needs only determine the probability that the time variable advances, conditioned on the current values $\xi_n^h = x$, $\phi_n^{h,\delta} = i\delta$. This is obtained by a "local consistency" argument and no matter how the $p^h(\cdot)$ were derived, the conditional probability that the time variable advances is

$$p^{h,\delta}(i\delta, i\delta + \delta | x) = \Delta t^h(x)/\delta. \tag{28.30}$$

Define the martingale difference $\beta_{0,n}^{h,\delta} = \left(\phi_{n+1}^{h,\delta} - \phi_n^{h,\delta}\right) - \Delta t_n^h$, whose conditional covariance is $\Delta t_n^h \left(\delta - \Delta t_n^h\right)$. We can write

$$\phi_{n+1}^{h,\delta} = \phi_n^{h,\delta} + \Delta t_n^{h,\delta} + \beta_{0,n}^{h,\delta}. \tag{28.31}$$

An alternative approximating chain. There is an alternative way of interpolating the $\xi_n^{h,\delta}$ which looks at the process only at those times that the time variable $\phi_n^{h,\delta}$ advances. Define $v_0^{h,\delta} = 0$ and for $n > 0$ define

$$v_n^{h,\delta} = \min\left\{i > v_{n-1}^{h,\delta} : \phi_i^{h,\delta} - \phi_{i-1}^{h,\delta} = \delta\right\}. \tag{28.32}$$

Then define $\hat{\xi}_n^{h,\delta} = \xi_{v_n^{h,\delta}}^{h,\delta}$. By Theorem 28.4, the interpolation $\hat{\xi}^{h,\delta}(\cdot)$ (intervals δ) and $\xi^h(\cdot)$ are asymptotically (as $h \to 0$, then $\delta \to 0$) equivalent.

Asymptotic equivalence of the time scales. Define the continuous-parameter interpolation $\phi^{h,\delta}(t) = \phi_n^h$ for $t \in [t_n^h, t_{n+1}^h)$. Define the stopping times

$$\tilde{d}^h(t) = \min\left\{n : \sum_{i=0}^{n-1} \Delta t_i^h = t_n^h \geq t\right\}, \quad \tilde{d}_\tau^h(t) = \min\left\{n : \tau_n^h \geq t\right\}. \tag{28.33}$$

Theorem 28.4 *For each $t > 0$,*

$$\lim_{h \to 0} \mathbb{E}\left[\sup_{s \leq t} \left(\sum_{i=0}^{\tilde{d}^h(s)} (\Delta \tau_i^h - \Delta t_i^h)\right)^2\right] = 0. \tag{28.34}$$

As $h \to 0$ and $\delta \to 0$, $\phi^{h,\delta}(\cdot)$ converges weakly and in mean square (uniformly on any finite time interval) to the process with value t at time t.

Proof. By the mutual independence of the exponentially distributed random variables $\{v_n\}$ and their independence of $\{\xi_n^h, \phi_n^{h,\delta}\}$, the process $A_n^h = \sum_{i=0}^n (\Delta \tau_i^h - \Delta t_i^h)$ is a martingale. By Doob's inequality for martingales, the expectation of the bracketed term in (28.34), conditioned on $\{\Delta t_i^h\}$, is bounded by

$$\mathbb{E} \sup_{s \le t} \left[\sum_{i=0}^{d^h(s)} [\Delta \tau_i^h - \Delta t_i^h]^2 \,\bigg|\, \Delta t_i^h, i < \infty \right]$$

$$\le 4 \mathbb{E} \left[\sum_{i=0}^{d^h(t)} [\Delta \tau_i^h - \Delta t_i^h]^2 \,\bigg|\, \Delta t_i^h, i < \infty \right]$$

$$= 4 \sum_{i=0}^{d^h(t)} [\Delta t_i^h]^2 \le 4 \left(t + \sup_n \Delta t_n^h \right) \sup_n \Delta t_n^h \xrightarrow{h} 0,$$

which yields (28.34). To prove the assertions concerning the asymptotic behavior of $\phi^{h,\delta}(\cdot)$ define $d^{h,\delta}(t) = \max\left\{ n : \sum_{i=0}^{n-1} \Delta t_i^{h,\delta} \le t \right\}$ and write

$$\phi^{h,\delta}(t) = \sum_{i=0}^{d^{h,\delta}(t)-1} \Delta t_i^{h,\delta} + \sum_{i=0}^{d^{h,\delta}(t)-1} \beta_{0,i}^{h,\delta}.$$

The first sum equals t, modulo $\sup_n \Delta t_n^{h,\delta}$. The variance of the martingale term is bounded by δt, modulo $\delta + \sup_n \Delta t_n^{h,\delta}$, and the term converges weakly to the zero process. This yields the last assertion of the theorem. □

28.6 Approximating filters

28.6.1 A filter based on the chain $\{\xi_n^h\}$

Let us first define the optimal filter for the process $\xi^h(\cdot)$ with $\Delta t^h(x) = \bar{\Delta}^h$, a constant. Define the observation process $y^h(t) = \int_0^t g(\xi^h(s))ds + w_0(t)$ and $y_n^h = y^h(n\bar{\Delta}^h) - y^h(n\bar{\Delta}^h - \bar{\Delta}^h)$ where $g(\cdot)$ and $w_0(\cdot)$ are as in (28.2). Set $\mathcal{Y}_t^h = \{y^h(s), s \le t\}$, and $\tilde{Q}_n^h(x) = P\{\xi_n^h = x | \mathcal{Y}_{n\bar{\Delta}^h}^h\}$. Define

$$R^h(x, y_n^h) = \exp\left[g(x)' y_n^h - \frac{1}{2} |g(x)|^2 \bar{\Delta}^h \right].$$

Then (28.10) becomes

$$\tilde{Q}_n^h(x) = \frac{\sum_{\tilde{x}} R^h(x, y_n^h) p^h(\tilde{x}, x) \tilde{Q}_{n-1}^h(\tilde{x})}{\text{normalization}}, \tag{28.35}$$

where $\tilde{Q}_0^h(x)$ is the probability that $\xi_0^h = x$. Let $\tilde{\xi}^h(\cdot)$ have the law of $\xi^h(\cdot)$ but be independent of it and $w_0(\cdot)$. Iterating (28.35) yields, for $t = n\bar\Delta^h$,

$$\mathbb{E}_{y_t^h}\phi\left(\xi_n^h\right) = \frac{\mathbb{E}_{y_t^h}\phi(\tilde\xi^h(t))\exp\left[\int_0^t g'(\tilde\xi^h(s))dy^h(s) - \int_0^t |g(\tilde\xi^h(s))|^2 ds/2\right]}{\text{normalization}}. \tag{28.36}$$

The unnormalized form of (28.35) is

$$\tilde q_n^h(x) = \sum_{\tilde x} R^h\left(x, y_n^h\right) p^h(\tilde x, x)\tilde q_{n-1}^h(\tilde x). \tag{28.37}$$

(28.37) can be split into two steps: First update the effects of the dynamics as

$$\sum_{\tilde x} p^h(\tilde x, x)\tilde q_{n-1}^h(\tilde x), \tag{28.38}$$

and then incorporate the observation by multiplying by $R^h\left(x, y_n^h\right)$.

The approximation to the optimal filter for $x(\cdot)$ and continuous-time observations. The most direct numerical approximation to the optimal filter (28.5) is either (28.35), (28.36), or (28.37) with the *actual physical observations* $y(\cdot)$ used in place of $y^h(\cdot)$. Both (28.35) and (28.37) are recursive formulas. The initial condition $\tilde Q_0^h(\cdot)$ is any approximation to the weak-sense density of $x(0)$ and which converges weakly to that density as $h \to 0$.

In summary, redefine

$$R^h(t) = \exp\left[\int_0^t g'\left(\xi_i^h(s)\right) dy(s) - \int_0^t |g(\xi^h(s))|^2 ds/2\right].$$

Then

$$E_t^h\phi(x(t)) = \frac{\mathbb{E}[\phi(\xi^h(t))\, R^h(t)|\mathcal{Y}_t]}{\mathbb{E}[R^h(t)|\mathcal{Y}_t]}. \tag{28.39}$$

Theorem 28.3 yields the convergence to $\mathbb{E}[\phi(x(t))|\mathcal{Y}_t]$. One can use the observation $y_n^h = g(x(n\bar\Delta^h))\Delta + [w_0(n\bar\Delta^h) - w_0(n\bar\Delta^h - \bar\Delta^h)]$ with the same result. To simplify the computation, one need not introduce the observation at each step, but treat them as discrete-time observations with small interval Δ, as below.

Discrete-time observations. The continuous-time filter was dealt with since it has been of great theoretical interest, although discrete-time observations are of greater practical interest. Suppose that the observations are taken in discrete time as in (28.3c). Then between observations, we can get a weak-sense approximation to the solution of the Kolmogorov forward equation for the weak-sense density by iterating (28.38). This computation can be simplified considerably. If the intervals between observations are constant and the state space does not change, then one can precompute the multi-step transition probability between those times. In general, one would not iterate at each step, but use increasing powers of the one-step transition probability. For $\Delta t^h(x) = \bar{\Delta}^h$, and Δ the interval between observations, the complexity is less than $N^3 \log(\Delta/\bar{\Delta}^h)$, and depends on the sparseness of the matrices. If the state space is too large or changes in time as the system evolves, then for practical algorithms, one must control its size. The best approach is heavily dependent on the problem. The dominant effects might be diffusion or a strong drift, or the observations might skew the shape significantly. One looks for regions with good geometry (for convenience in programming) and whose conditional probability is close to unity. For example, one might use a 3σ ellipse centered at the conditional mean, or an analogous rectangle.

Comments. If the same system is to be used frequently, then one is tempted to optimize the grid that defines the state space of the approximating chain. Some such results are in [35], which were derived for the control problem. The various functions $b(\cdot), \sigma(\cdot), g(\cdot)$ need not be continuous, provided that the time (on any finite interval) that the limit process spends in an ϵ-neighborhood of the discontinuities goes to zero as $\epsilon \to 0$ [31, pages 275, 295]. The robustness of the behavior of the approximate filters over a very long time interval to uncertainties in the dynamics, signal, and observation noise processes is of interest, and is shown to be the case in [5, 6, 7].

28.6.2 Alternative filter approximations

Using the process $\hat{\xi}_n^{h,\delta}$

Recall the process $\left\{\hat{\xi}_n^{h,\delta}\right\}$ with interpolation $\hat{\xi}^{h,\delta}(\cdot)$ (intervals δ) defined below (28.32). Let $\delta \geq \max_x \Delta t^h(x)$. Then $\hat{\xi}^{h,\delta}(\cdot)$ can be used in (28.14) and (28.15) whether or not $\Delta t^h(x)$ is constant. When $\xi_n^h = x$, the probability of an advance in the time variable is just $\Delta t^h(x)/\delta$, with the mean increment being $\Delta t^h(x)$. The main computational problem is getting the one-step transition probability $p^{h,\delta}(x, \tilde{x})$ of $\left\{\hat{\xi}_n^{h,\delta}\right\}$.

Define the matrix $P_1^{h,\delta} = \{p^h(x, \tilde{x})\Delta t^h(x)/\delta; x, \tilde{x}\}$, the matrix of probabilities of going from x to $\tilde{x}$ in one step and with time advancing. Define the matrix $P_0^{h,\delta} = \{p^h(x, \tilde{x})(1 - \Delta t^h(x)/\delta); x, \tilde{x}\}$, the set of probabilities of going from x to $\tilde{x}$ in one step, with time not advancing. Then the transition probability for the chain $\left\{\hat{\xi}_n^{h,\delta}\right\}$ is

$$P^{h,\delta} = \sum_{n=0}^{\infty} \left[P_0^{h,\delta}\right]^n P_1^{h,\delta} = \left[I - P_0^{h,\delta}\right]^{-1} P_1^{h,\delta}. \tag{28.40}$$

If $\delta = \max_x \Delta t^h(x)$, then the number of steps that are required until the jump in the time variable will be small and the sum will converge rapidly. If this form is used for continuous-time observations, where the observations are taken to be either (28.3a) or (28.3b), with $V = \mathbb{I}$, then Δ would be a small multiple of δ.

Suppose that the form is used for discrete-time observations with interval Δ between observations, where Δ/δ is a large integer. Then the transition probability between observations is $[P^{h,\delta}]^{\Delta/\delta}$. The times at which the observations are incorporated are random, after each successive Δ/δ increases in the time variable. By Theorem 28.4, as $h \to 0$ and $\delta \to 0$, these intervals converge to the constant Δ. A computational advantage of this procedure is that the original intervals $\Delta t^h(x)$ can be used. The computation of (6.6) should be helped by the tendency (in many problems) of the path to move from points with small $\Delta t^h(x)$ (fast dynamics or large diffusion) to those with a larger interval (small dynamics and small diffusion).

Discrete time and other forms

The general approach covers a large family of approximations, but for numerical purposes we need to restrict the domain of any approximation to a finite set of points. For example, one could base the Markov chain approximation on (28.24), and we describe one possible approach to getting the desired transition probabilities for the dynamical update step. First, for $x \in G_h$, let $S^h(x)$ be the set of closest points on the grid to $x + \delta b(x)$ in whose convex hull $x + \delta b(x)$ lies. Then randomize among the points in $S^h(x)$, so that the correct mean increment $\delta b(x)$ is attained. The assigned weights are the transition probabilities for this first step. Now, add the noise. For each $z \in S^h(x)$, construct an approximation that is locally consistent with $\sigma(z)[w(n\delta + \delta) - w(n\delta)]$. This two-step process yields the transition probabilities for the point x. A two-step procedure was described for simplicity of explanation. If it is not too difficult, it would be preferable to approximate the transitions in one step. If the observations are taken continuously, then the form based on (28.24) can still be used, where the observations are incorporated at intervals that go to zero as $\delta \to 0$.

The simplest Markov chain approximation is that where the transitions are local, such as that given by (28.26), which were designed for the control problem. If $\Delta t^h(\cdot)$ is the constant $\bar{\Delta}^h$, then iterating the one-step transition probability $\delta/\bar{\Delta}^h$ times yields an approximation to that of (28.24). A more direct use of (28.24) could simplify the computations, depending on how it is carried out. One would try to precompute as much as possible, keeping in mind the tradeoff between accuracy of the approximation and the computational and coding requirements. The use of smaller values of δ and h would yield a better approximation to the solution of the Kolmogorov forward equation, but owing to the "corrective" effects of the observations, the filtering algorithms are often forgiving of the use of cruder models for the dynamics.

28.7 Robustness of the approximating filters

The mathematical theory of nonlinear filtering with continuous-time observations depends on the fact that the observation noise is "white Gaussian." Even if the physical observation noise is a wide-bandwidth process, due to its simplicity and to the fact that the probability law of the true observation noise process might not be known, one is tempted to use a form of the filter that is derived under the white noise assumption. Such a filter is appealing since it yields a reasonable algorithm, even if it is not optimal. The interpretation of the result and whether "correction" terms are needed depends on the type of approximation that is used, and many of the important issues are discussed in [29, 32]. The form (28.39) is continuous in $y(\cdot)$ (uniformly in each bounded set, in the sup norm sense on any bounded interval), since for each h, $\xi^h(\cdot)$ is piecewise constant. But the uniformity of the continuity in h is not a priori evident. To be of genuine value, the continuity should be uniform in h.

Let us return to (28.5) and suppose that the first- and second-order partial derivatives of $g(\cdot)$ are bounded and continuous. Then, with probability one, (28.5) can be rewritten as

$$\mathbb{E}_t \phi(x(t)) = \frac{\mathbb{E}_t \phi(\tilde{x}(t)) \exp\left[y'(t)g(\tilde{x}(t)) - \int_0^t y'(u)dg(\tilde{x}(u)) - \frac{1}{2}\int_0^t |g(\tilde{x}(u))|^2 du\right]}{\mathbb{E}_t \exp\left[y'(t)g(\tilde{x}(t)) - \int_0^t y'(u)dg(\tilde{x}(u)) - \frac{1}{2}\int_0^t |g(\tilde{x}(u))|^2 du\right]}. \quad (28.41)$$

Clark [11] showed that (28.41) is locally Lipschitz continuous in $y(\cdot)$ at each $y(\cdot) \in C^m[0, T]$, for each $T < \infty$, and gave a PDE whose solution is the right side of (28.41) for any continuous $y(\cdot)$. Since this solution is (28.5) (with probability one) if $y(\cdot)$ were defined by (28.2), we would then have a continuous function of $y(\cdot)$ which can be said to be an approximation to the optimal filter, even if $w_0(\cdot)$ is only an "approximation" to a Wiener process, and the filter might be far from optimal under the actual observation noise. It is of greater interest to know whether the numerical approximations are robust in the observation noise. This will be seen to be true and the continuity will be uniform in the approximation parameter.

Return to the form (28.15), suppose that $g(x^h(\cdot))$ has a well-defined differential, and that that by a partial integration, we can write (28.15) as

$$F_t^h(\phi, y(\cdot)) \equiv E_t^h \phi(x^h(t))$$

$$= \frac{\mathbb{E}_t \phi(x^h(t)) \exp\left[y'(t)g(x^h(t)) - \int_0^t y'(u)dg(x^h(u)) - \frac{1}{2}\int_0^t |g(x^h(u))|^2 du\right]}{\mathbb{E}_t \exp\left[y'(t)g(x^h(t)) - \int_0^t y'(u)dg(x^h(u)) - \frac{1}{2}\int_0^t |g(x^h(u))|^2 du\right]}.$$
(28.42)

The partial integration can be done if $x^h(\cdot)$ is piecewise constant for each h.

Decompose the bounded point process $g(x^h(\cdot))$ into the sum of a bounded martingale and predictable projection processes as $g(x^h(t)) = M^h(t) + \Gamma^h(t)$. Assume the following conditions, which will be seen to hold for $g(\xi^h(\cdot))$. Suppose that, for each $T < \infty$, there are $C_i(T) < \infty$ such that for all $h > 0$

$$\mathbb{E}\left|M^h(T)\right|^2 \leq C_1(T), \quad \mathbb{E}\left[\int_0^t |d\Gamma^h(s)|\right]^2 \leq C_2(T). \tag{28.43a}$$

Further, suppose that for each bounded set S of functions $q(\cdot) \in C^m[0, T]$, there are $C_i(S, T) < \infty$ such that, for all h and $t \leq T$,

$$\mathbb{E}\exp\left[\int_0^t q'(s)dM^h(s)\right] + \mathbb{E}\exp\left[\int_0^t q'(s)d\Gamma^h(s)\right] \leq C_3(S, T). \tag{28.43b}$$

Instead of (28.43) we could use, for all h and $t \leq T$,

$$\mathbb{E}\exp\left[\int_0^t q'(s)dg(x^h(s))\right] \leq C_4(S, T). \tag{28.43c}$$

Define $\|y\|_T = \sup_{t \leq T} |y(t)|$. The theorem implies that the convergence in Theorem 28.3 is w.p.1.

Theorem 28.5 *For each $T > 0$ and bounded set $S \subset C^m[0, T]$ there is $K(S, T) < \infty$ which does not depend on h or $\phi(\cdot)$ (for all $\phi(\cdot)$ bounded by the same constant) such that, for $f(\cdot)$ and $\tilde{f}(\cdot) \in S$ and $t \leq T$,*

$$\left|F_t^h(\phi, f(\cdot)) - F_t^h(\phi, \tilde{f}(\cdot))\right| \leq K(S, T) \|f(\cdot) - \tilde{f}(\cdot)\|_T. \tag{28.44}$$

Proof. We only need to show that there is $K_1(S, T) < \infty$, depending only on S and T, such that, for all $h > 0$ and $t \leq T$,

$$\begin{aligned}
\mathbb{E}\bigg|&\exp\left[f'(t)g(x^h(t)) - \int_0^t f'(u)dg(x^h(u)) - \frac{1}{2}\int_0^t |g(x^h(u))|^2 du\right] \\
&- \exp\left[\tilde{f}'(t)g(x^h(t)) - \int_0^t \tilde{f}'(u)dg(x^h(u)) - \frac{1}{2}\int_0^t |g(x^h(u))|^2 du\right]\bigg| \\
&\leq K_1(S, T) \|f(\cdot) - \tilde{f}(\cdot)\|_T.
\end{aligned} \tag{28.45}$$

We can drop the $-\frac{1}{2}\int_0^t |g(x^h(u))|^2 du$ term, since $g(\cdot)$ is bounded. Then, using the inequality (28.18), we have the following upper bound for (28.45):

$$\mathbb{E}\left|[f(t) - \tilde{f}(t)]'g(x^h(t)) - \int_0^t [f(u) - \tilde{f}(u)]' dg(x^h(u))\right|$$
$$\times \left[e^{[f'(t)g(x^h(t)) - \int_0^t f'(u)dg(x^h(u))]} + e^{[\tilde{f}'(t)g(x^h(t)) - \int_0^t \tilde{f}'(u)dg(x^h(u))]}\right].$$

To prove (28.45) using the above bound, we need only show (28.46) and (28.47), where h is small and $q(\cdot) \in S'$, an arbitrary bounded set in $C^m[0, T]$:

$$\sup_{t \leq T} \mathbb{E} \left| \int_0^t q'(u) dg(x^h(u)) \right|^2 \leq K_2(T) \|q(\cdot)\|_T^2, \quad K_2(T) < \infty, \tag{28.46}$$

$$\sup_{q(\cdot) \in S'} \sup_{t \leq T} \mathbb{E} \exp\left[\int_0^t q'(u) dg(x^h(u)) \right] \leq K_3(T, S') < \infty. \tag{28.47}$$

It is sufficient to let $q(\cdot)$ be real valued. (28.47) follows from either (28.43b) or (28.43c). The inequality (28.46) will be verified by using

$$\mathbb{E}\left| \int_0^t q(u) dg(x^h(u)) \right|^2 \leq 2 \|q(\cdot)\|_T^2 \mathbb{E}\left[\int_0^T |d\Gamma^h(u)| \right]^2 + 2\mathbb{E}\left| \int_0^t q(u) dM^h(u) \right|^2. \tag{28.48}$$

By the martingale calculus [18, 22], the second term on the right is bounded by $8\|q(\cdot)\|_T^2 \mathbb{E}[M^h(T)]^2$. Then use (28.43a) to complete the proof. □

Verification of the conditions for the Markov chain approximations. Theorem 28.5 holds for all of the forms that were discussed in the previous section We will work with $\xi^h(\cdot)$ with (28.25) holding and where $\Delta t^h(x) = \bar{\Delta}^h$, a constant, which will be seen to assure (28.43). For simplicity, let $g(\cdot)$ be real valued. The proof in the other cases is similar, including the case with jumps. Suppose that $g(\cdot)$ is bounded and continuous, together with its partial derivatives up to second order.

The process $\xi^h(\cdot)$ is right-continuous and we have the decomposition into a martingale and predictable projection: $g(\xi^h(n\bar{\Delta}^h)) = M^h(n\bar{\Delta}^h) + \Gamma^h(n\bar{\Delta}^h)$. By the definition of the predictable projection,

$$\Gamma^h(n\bar{\Delta}^h + \bar{\Delta}^h) - \Gamma^h(n\bar{\Delta}^h) = \sum_{\tilde{x}} [g(\tilde{x}) - g(\xi_n^h)] p^h(\xi_n^h, \tilde{x}). \tag{28.49}$$

A Taylor series expansion and (28.25) shows that (28.49) is $O(\bar{\Delta}^h)$. Hence (28.43a) holds for the predictable projection. We have

$$M^h(n\bar{\Delta}^h + \bar{\Delta}^h) - M^h(n\bar{\Delta}^h)$$
$$= [g(\xi_{n+1}^h) - g(\xi_n^h)] - \sum_{\tilde{x}} [g(\tilde{x}) - g(\xi_n^h)] p^h(\xi_n^h, \tilde{x}). \tag{28.50}$$

That (28.43a) holds for the martingale term follows from this form and (28.25). To prove (28.43c), it is sufficient to work with real-valued $q(\cdot)$ and

$$\mathbb{E} \prod_{n=0}^{[t/\bar{\Delta}^h]-1} e^{q(n\bar{\Delta}^h)[g(\xi_{n+1}^h) - g(\xi_n^h)]}. \tag{28.51}$$

Let $\mathbb{E}_n^h$ = expectation given the data up to and including step n. We have

$$\mathbb{E}_n^h e^{q(n\bar{\Delta}^h)[g(\xi_{n+1}^h)-g(\xi_n^h)]} \leq 1 + q(n\bar{\Delta}^h)\mathbb{E}_n^h \left[g\left(\xi_{n+1}^h\right) - g\left(\xi_n^h\right)\right]$$
$$+ q^2(n\bar{\Delta}^h)\mathbb{E}_n^h \left[g\left(\xi_{n+1}^h\right) - g\left(\xi_n^h\right)\right]^2 \sum_{l=2}^{\infty} \frac{\left[2\|q(\cdot)\|_T\|g(\cdot)\|\right]^{l-2}}{l!}. \quad (28.52)$$

A second-order Taylor expansion of $\left[g\left(\xi_{n+1}^h\right) - g\left(\xi_n^h\right)\right]$ and (28.25) yields that (28.53) is bounded by $e^{\bar{\Delta}^h C}$ for some constant C which is bounded on each bounded $q(\cdot)$-set. Using this estimate recursively in (28.51) yields (28.43c).

28.8 Assumed form of the conditional density

In this and in the next section, a very different approach to the approximate nonlinear filter is taken. We are no longer interested in convergent approximations to the conditional density. Interest is confined to heuristic approximations of the first few conditional moments only. This is done by assuming a parametrized form for the conditional density and, with this form, evaluating the evolution of the parameters. Most typically one supposes that the conditional density is Gaussian, and under this assumption the equations of evolution of the conditional means and covariances are obtained. The approach originated in [25] and has been in common use since then.[5] See, for example, the references in [1, 8, 20], The procedure is intuitively reasonable, although here is no mathematical justification for it. But numerous examples have shown that it can be a powerful tool for obtaining good nonlinear filters.

Observations taken continuously in time. Let us first consider the model (28.1), (28.2), so that we are observing continuously in time, and wish to estimate the moments continuously in time. Define $m_i(t) = \mathbb{E}_t x_i(t)$, $m_{ij}(t) = \mathbb{E}_t(x_i(t) - m_i(t))(x_j(t) - m_j(t))$, and let $m(\cdot)$ and $M(\cdot)$ denote the vector and matrix, resp., of the components. Equation (28.8) yields

$$dm_i(t) = \mathbb{E}_t b_i(x(t))dt$$
$$+ (\mathbb{E}_t x_i(t)g(x(t)) - \mathbb{E}_t g(x(t))m_i(t))' (dy(t) - \mathbb{E}_t g(x(t))dt). \quad (28.53)$$

Let $\mathcal{L}$ be the differential operator of $(x(\cdot), m(\cdot))$. Using $E_t(x_i(t) - m_i(t))(x_j(t) - m_j(t)) = E_t x_i(t)x_j(t) - m_i(t)m_j(t)$, and interpreting the second order part of $\mathbb{E}_t \mathcal{L} x_i(t)m_j(t)dt$ to be $dm_i(t)dm_j(t)$ (use (28.53)), (28.7) and Itô's Lemma yield

$$dm_{ij}(t) = \mathbb{E}_t \mathcal{L}(x_i(t) - m_i(t))(x_j(t) - m_j(t))dt$$
$$+ \left(G_{ij}^0(t) - \mathbb{E}_t g(x(t))m_{ij}(t)\right)' (dy(t) - \mathbb{E}_t g(x(t))dt), \quad (28.54)$$

where $G_{ij}^0(t) = \mathbb{E}_t(x_i(t) - m_i(t))(x_j(t) - m_j(t))g(x(t))$.

An assumed Gaussian form for the conditional density. The (conditional) expectations in (28.53) and (28.54) are with respect to the true conditional distribution, which we do not know and the computation of which we wish to avoid. So we make the purely heuristic assumption that it has a particular parametrized form, and then estimate the parameters under this assumption. In principle any density can be used, provided that it is determined by a finite number of parameters. The simplest and most widely used form is that of a Gaussian distribution, even though it is not in general the form of the true conditional density,

Let $\mathbb{N}(m, M)(\cdot)$ denote the Gaussian density with mean m and covariance M. Define the operator E_t by $E_t f(x(t)) = \int f(x)\mathbb{N}(m(t), M(t))(x)dx$. Then (28.53) and (28.54) are replaced by the following heuristic approximations:

$$dm_i(t) = E_t b_i(x(t))dt + (G_i(t) - G(t)m_i(t))' (dy(t) - G(t)dt), \quad (28.55)$$

where

$$G(t) = E_t g(x(t)), \quad G_i(t) = E_t x_i(t)g(x(t)). \quad (28.56)$$

Recalling that $\mathcal{L}$ now denotes the differential operator of the pair $(x(\cdot), m(\cdot))$, the approximating equations for the $m_{ij}(\cdot)$ are

$$dm_{ij}(t) = E_t \mathcal{L}(x_i(t) - m_i(t))(x_j(t) - m_j(t))dt$$
$$+ (G_{ij}(t) - G(t)m_{ij}(t))' (dy(t) - G(t)dt). \quad (28.57)$$

$$G_{ij}(t) = E_t(x_i(t) - m_i(t))(x_j(t) - m_j(t))g(x(t)). \quad (28.58)$$

An alternative to the use of the Gaussian distribution is to assume that the conditional density has the form of a Gaussian mixture of fixed finite order such as

$$\sum_{i=1}^{K} a_i \frac{1}{(2\pi)^{r/2} \det^{1/2} M_i} \exp\left[-(x - m_i)' M_i^{-1}(x - m_i)\right], \quad \sum_{i=1}^{K} a_i = 1, a_i \geq 0.$$

Updating the $\{m_i, M_i\}$ is straightforward since one works with each component at a time. Updating the weights $\{a_i\}$ is more complicated. In principle it would seem preferable to use such an assumed density that allows for greater flexibility, although at this time the class of problems for which the greater numerical complexity is justified by clearly better results is not clear. The Gaussian mixture form is discussed in [1, 20] and in some of their references, where methods for updating the weights are given. The reader is referred to these references for further information.

Example where an exact integration is possible. The integrals in (28.55)–(28.58) need to be evaluated. This can sometimes be done exactly, without approximation; for example, where $b(\cdot)$, $a(\cdot)$ and $g(\cdot)$ are polynomials. This was the case for the example in [25], where the signal process was defined by the noiseless Van der Pol equation:

$$dx_1 = x_2 dt, \quad dx_2 = -x_1 + \epsilon x_2 \left(1 - x_1^2\right), \quad \epsilon > 0,$$

$$dy(t) = x_1(t)dt + dw_0(t).$$

The example was selected since the dynamics have both fast and slow parts and provided a good illustration of the quality of the method. Since the odd central moments of a Gaussian distribution are zero, we have

$$G(t) = m_1(t), \quad G_j(t) = m_{1j}(t) + m_1(t)m_j(t), \quad G_{ij}(t) = m_1(t)m_{ij}(t),$$

$$E_t \mathcal{L} x_1(t) = b_1(x(t)) = m_2(t),$$

$$E_t \mathcal{L} x_2(t) = E_t b_2(x(t)) = \\ -m_1(t) + \epsilon m_2(t) - \epsilon \left[2m_1(t)m_{12}(t) + m_2(t)\left(m_{11}(t) + m_1^2(t)\right)\right].$$

The filter is

$$dm_1 = m_2 dt + m_{11}(dy - m_1 dt),$$
$$dm_2 = E_t b_2(x(t)) dt + m_{12}(dy - m_1 dt),$$
$$\dot{m}_{11} = -m_{11}^2 + 2m_{12},$$
$$\dot{m}_{12} = -m_{11}m_{12} + m_{22} - m_{11} - \epsilon\left(-m_{12} + m_{1112} + m_{12}m_1^2 + 2m_{11}m_1 m_2\right),$$
$$\dot{m}_{22} = -m_{12}^2 - 2m_{12} - 2\epsilon\left(-m_{22} + m_{1122} + m_1^2 m_{22} + 2m_1 m_2 m_{12}\right).$$

By the Gaussian assumption, the higher moments are

$$m_{1122}(t) = m_{11}(t)m_{22}(t) + 2m_{12}(t), \quad m_{1112}(t) = 3m_{11}(t)m_{12}(t).$$

In this example, the observation terms in the equations for the second moments turn out to be zero, due the fact that the third central moments are zero (due to the Gaussian assumption) and $g(x) = x_1$.

The differential equations for the approximations would have to be solved numerically. When there is an observation term, as in the equations for the $m_i(\cdot)$, one would have to use one of the pathwise approximation methods of [21]. Generally we can use the simplest one, the discrete-time (Euler) scheme. Numerical data is given in [25], where it is seen that the tracking is good. For such examples, the extended Kalman filter performs very poorly.

Observations at discrete times. If the observations are taken at times t_i, $i = 1, 2, \ldots$, then between the observations use (28.55)–(28.58) with the observation term deleted. Methods of incorporating the observation will be discussed below.

Numerical evaluation of the integrals with respect to the Gaussian kernels. The integrals (i.e., the moments) in the above example could be evaluated exactly since the dynamical and signal functions were all polynomials. If an exact evaluation is not possible, then some form of numerical quadrature is required. Owing to the Gaussian assumption the most natural form is based on the Gauss–Hermite quadrature formulas, and a brief review of this method will be given next.

Numerical quadrature formulas. The theory of Gauss–Hermite quadrature is concerned with the numerical approximation of integrals of real-valued functions of a real variable with respect to a Gaussian kernel of the type

$$\frac{1}{\sqrt{2\pi}} \int_{-\infty}^{\infty} f(x) e^{-x^2/2} dx = \sum_{i=1}^{\mu} w_i f(t_i) + \text{error}. \tag{28.59}$$

For each order μ, the points $\{t_i, i \leq \mu\}$ and weights $\{w_i, i \leq \mu\}$ are chosen such that the integral is exact if $f(\cdot)$ is any polynomial of degree $\leq 2\mu - 1$. The points $\{t_i\}$ are symmetric about the origin and the weights satisfy $w_i \geq 0$, $\sum_i w_i = 1$. They are equal for t and $-t$.

In lieu of (28.59), the numerical analysis literature uses the form

$$\int_{-\infty}^{\infty} f(z) e^{-z^2} dz \approx \sum_{i=1}^{\mu} f(z_i) w_i', \tag{28.60}$$

in which case the z_i are the zeros of the μth-order Hermite polynomial $H_\mu(\cdot)$ (with leading coefficient 2^μ) [12] and the weights are

$$w_i' = 2^{\mu+1} \mu! \sqrt{\pi} / [H_{\mu+1}(z_i)]^2.$$

With the form (28.60) and $\mu = 3$, the z_i and w_i' are [33, p. 508]

$$\{(z_i, w_i')\} = \{(0, 1.1816), (\pm 1.2247, .2954)\}.$$

For $\mu = 5$ they are

$$\{(z_i, w_i')\} = \{(0, .9453), (\pm .9586, .3936), (\pm 2.02, .01954)\}.$$

Furthermore, if $f(\cdot)$ is a function with continuous derivatives up to order 2μ, then the error for the form (28.60) is bounded by

$$\frac{\mu! \sqrt{\pi}}{2^\mu (2\mu)!} f^{2\mu}(\xi), \quad \text{some } \xi \in (-\infty, \infty). \tag{28.61}$$

To adapt the published formulas and values to the form of the Gaussian kernel that is used in (28.59), we need do a linear transformation to get $t_i = \sqrt{2} z_i$, $w_i = w_i'/\sqrt{\pi}$. In what follows, (28.59) will be used to approximate integrals of functions with respect to Gaussian kernels.

Now suppose that x is vector-valued, of dimension $r > 1$. Then the simplest procedure is to use the product formula

$$\frac{1}{(2\pi)^{r/2}} \int_{\mathbb{R}^r} f(x) e^{-|x|^2/2} dx \approx \sum_{i_1=1}^{\mu} \cdots \sum_{i_r=1}^{\mu} w_{i_1} \cdots w_{i_r} f(t_{i_1}, \ldots t_{i_r}). \quad (28.62)$$

The formula is exact for polynomials that are of order at most $2\mu - 1$ in each variable. An error bound can be obtained from (28.63). More efficient approximations for functions that are well represented (in the region of most mass of the kernel function) by polynomials whose *total* order is $\leq 2\mu - 1$ and that are not based on product rules can be found in [37]. The product rule (28.62) will suffice for our purposes.

Now, for Σ positive definite and symmetric, consider the integral

$$\frac{1}{(2\pi)^{r/2} (\det \Sigma)^{1/2}} \int f(x) e^{-(x-\bar{x})'\Sigma^{-1}(x-\bar{x})/2} dx \quad (28.63)$$

Factor $\Sigma^{-1} = S'S$ and define $z = S(x - \bar{x})$, yielding the equivalent form

$$\frac{1}{(2\pi)^{r/2}} \int f(S^{-1}z + \bar{x}) e^{-|z|^2/2} dz \approx \sum_{i_1,\ldots i_r} w_{i_1}, \ldots w_{i_r} f(S^{-1}(t_{i_1}, \ldots t_{i_r}) + \bar{x}). \quad (28.64)$$

The choice of factorization method depends on the problem, with accuracy vs computational time being the main considerations. See [3] for a discussion of various factorization methods.

Evaluating the integrals in (28.55)–(28.58). Suppose that the observations are taken at the discrete times $t_i, i = 1, \ldots$. Then, if an exact evaluation is not possible, for a discretization interval δ, between observations one could use the discrete-time approximation

$$E_{l\delta} \phi(x(l\delta + \delta)) = E_{l\delta} \phi(x(l\delta)) + \delta E_{l\delta} \mathcal{L} \phi(x(l\delta)). \quad (28.65)$$

If the observations are taken in continuous time, then one could partition the updating procedure by first updating the effects of the dynamics as in (28.65), and then incorporating the observation (28.3a) by one of the methods in the next section.

The general approach for updating the approximations of the moments between observations is the following. The observations taken at times $t_i, i = 1, 2, \ldots$, where the t_i are assumed to be integral multiples of $\delta > 0$ and have the form

$$y_n = g(x(t_n)) + v_n \quad (28.66)$$

where the $\{v_n\}$ are i.i.d., normal, with covariance $\mathbb{I}$ and independent of $x(\cdot)$. Between observations, update the means and covariances by using

$$m_{l\delta+\delta} = m_{l\delta} + \delta \int b(x)\mathbb{N}(m_{l\delta}, M_{l\delta})(x)dx, \tag{28.67}$$

$$\begin{aligned}M_{l\delta+\delta} &= \int [x + b(x)\delta - m_{l\delta+\delta}][x + b(x)\delta - m_{l\delta+\delta}]'\mathbb{N}(m_{l\delta+\delta}, M_{l\delta})(x)dx \\ &\quad + \delta \int a(x)\mathbb{N}(m_{l\delta+\delta}, M_{l\delta})(x)dx,\end{aligned} \tag{28.68}$$

with the Gauss–Hermite quadrature rule (28.64) used if the integrals cannot be evaluated exactly. This is the form used in [25] and [20] and is algebraically equivalent to the form in [1]. If some of the components of $b(\cdot)$ or $\sigma(\cdot)$ are either constant or linear in x, then the integrals can be partly precomputed.

28.9 The observation step

From a numerical perspective, incorporating the observation is more difficult than the updating between observations. Two approaches will be discussed. The first uses quadrature rules directly to evaluate the Bayes rule formula, and the second is based on a type of local linear approximation, with a least squares estimate of the coefficients. W.l.o.g., and for notational simplicity, we continue to suppose that the covariance matrix of the observation noise is the identity. Let m_n^-, M_n^- denote the values of the estimates of the conditional mean and covariance just before the nth observation y_n is taken, where y_n is defined by (28.66). Define the (unnormalized) Gaussian density function

$$I_n(x) = \exp[-|y_n - g(x)|^2/2]. \tag{28.69}$$

Given the new observation y_n, we would like to compute the updated values of the approximations to the mean and covariance, which we write as

$$\begin{aligned}m_n &= \frac{1}{c_n} \int x I_n(x)\mathbb{N}\left(m_n^-, M_n^-\right)(x)dx, \\ M_n &= \frac{1}{c_n} \int (x - m_n)(x - m_n)' I_n(x)\mathbb{N}\left(m_n^-, M_n^-\right)(x)dx,\end{aligned} \tag{28.70}$$

where c_n is the normalizing constant

$$c_n = \int I_n(x)\mathbb{N}\left(m_n^-, \bar{M}_n^-\right) dx.$$

Approach 1. An iterative centring method for the Gaussian quadrature. A potential difficulty arises when the center point of the kernel $I_n(x)$ is not close to m_n^-. For example, consider a two-dimensional case where $g(x)$ has the form $g_0(x_1 -$

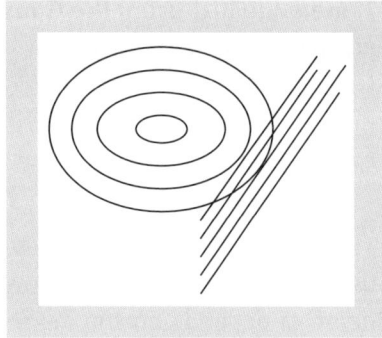

Fig. 28.1 An example of poor alignment of the prior and observation kernels.

x_2), and which is illustrated in Figure 28.1. The upper ellipsoidal contours are those of $\mathbb{N}\left(m_n^-, M_n^-\right)(\cdot)$, the lower ellipsoid lines approximate the contours of $I_n(\cdot)$, so that there is not much "overlap" in the parts of greatest mass of the two densities. When applied directly, the numerical quadrature formulas might give poor results unless the degree μ is large. The problem can be partially alleviated by the following iterative centring procedure which attempts to alter the kernels to improve the centring, and which worked very well in the examples in [8].

We seek a good centring kernel $\mathbb{N}(\rho_n, P_n)(\cdot)$, with which we rewrite (28.70) as

$$\int \left[\frac{I_n(x) f(x) \mathbb{N}\left(m_n^-, M_n^-\right)(x)}{\mathbb{N}(\rho_n, P_n)(x)} \right] \mathbb{N}(\rho_n, P_n)(x) dx, \qquad (28.71)$$

for appropriate $f(\cdot)$. The form (28.71) will be used to evaluate (28.70). In (28.71) the term in brackets is the function whose integral with respect to the kernel $\mathbb{N}(\rho_n, P_n)(x)$ is to be evaluated. Since the integral will not be evaluated exactly, but the quadrature rules will be used, the two $\mathbb{N}(\rho_n, P_n)(x)$ will not necessarily cancel each other. The analogous remark holds for the forms (28.72) below.

The values of the ρ_n, P_n in the centring kernel will be obtained by an iterative procedure. Start by defining $c_{n,0} = c_n, \rho_{n,0} = m_n^-, P_{n,0} = M_n^-$. and let $c_{n,k}, \rho_{n,k}, P_{n,k}$ denote the kth estimates, obtained from

$$c_{n,k} = \int \left[\frac{I_n(x) \mathbb{N}\left(m_n^-, M_n^-\right)(x)}{\mathbb{N}(\rho_{n,k-1}, P_{n,k-1})(x)} \right] \mathbb{N}(\rho_{n,k-1}, P_{n,k-1})(x) dx, \qquad (28.72a)$$

$$\rho_{n,k} = \int \left[\frac{I_n(x) x \mathbb{N}\left(m_n^-, M_n^-\right)(x)}{\mathbb{N}(\rho_{n,k-1}, P_{n,k-1})(x)} \right] \mathbb{N}(\rho_{n,k-1}, P_{n,k-1})(x) dx, \qquad (28.72b)$$

$$P_{n,k} = \int \left[\frac{I_n(x)(x - \rho_{n,k})(x - \rho_{n,k})' \mathbb{N}\left(m_n^-, M_n^-\right)(x)}{\mathbb{N}(\rho_{n,k-1}, P_{n,k-1})(x)} \right] \mathbb{N}(\rho_{n,k-1}, P_{n,k-1})(x) dx. \qquad (28.72c)$$

The integrals in (28.72) are evaluated using the Gauss–Hermite quadrature rules. First centre and diagonalize the kernel $\mathbb{N}(\bar\rho_{n,k-1}, P_{n,k-1})(x)$ by the transformation $x = [P_{n,k-1}]^{1/2}t + \bar\rho_{n,k-1}$ to get, for example,

$$c_{n,k} = \int \left[\frac{I_n\left(P_{n,k-1}^{1/2}t + \bar\rho_{n,k-1}\right) \mathbb{N}\left(m_n^-, M_n^-\right)(P_{n,k-1}^{1/2}t + \bar\rho_{n,k-1})}{\mathbb{N}(0, I)(t)} \right] \mathbb{N}(0, I)(t) dt,$$

and so forth, and then use (28.64). For the quadrature, use order μ_0, which might not be μ. The value of μ_0 depends on the degree to which the kernels can be "aligned". In the four-dimensional example in Section 28.10 (below), the integrations were reduced to ones in two dimensions using the precomputation method of the next paragraph, and an 8×8 grid could not be much improved on. The iterative method seems to be very stable and can improve the estimates considerably. But it is a heuristic procedure. In all cases in [8], the improvement was negligible after 4 iterations.

Simplification by precomputation. The function $g(\cdot)$ generally depends on only a few of the state components. This can be used to reduce the dimension of the space in which the numerical integration is done, via analytic computations of conditional expectations of Gaussian random variables. To illustrate the idea, suppose that $g(\cdot)$ depends on the first $l < r$ components of x. Then $I_n(x)$ only depends on the first l components of x, and the integrations in (28.70) can be reduced to integrations over $\mathbb{R}^l$, which simplifies the remaining computation via Gauss–Hermite quadrature.

In more detail, let $X = (X_1, \cdots X_r) \in \mathbb{R}^r$ have the distribution $\mathbb{N}(\mu, P)(\cdot)$, and let z be the generic point in $\mathbb{R}^l$. Consider the integral

$$\int_{\mathbb{R}^r} x_{l+1} I_n(x) \mathbb{N}(\mu, P)(x) dx.$$

By taking conditional expectations it can be rewritten as

$$E\left[I_n(X_1, \cdots, X_l) X_{l+1}\right] = E\left[I_n(X_1, \cdots, X_l) E(X_{l+1}|X_1, \cdots, X_l)\right]$$
$$= E I_n(X_1, \cdots, X_l)[a_1 X_1 + \cdots a_l X_l] \quad (28.73)$$
$$= \int_{\mathbb{R}^l} I_n(z)(a_1 z_1 + \cdots a_l z_l) \mathbb{N}(\mu_{n,r}, P_{n,r})(z) dz,$$

where $\mathbb{N}(\mu_{n,r}, P_{n,r})(\cdot)$ is the marginal distribution of $(X_1, \cdots, X_l)$. The value $(a_1 z_1 + \cdots a_l z_l)$ is the conditional expectation of X_{l+1} given $(X_1, \ldots, X_l) = (z_1, \ldots, z_l)$, and $a_1, \cdots, a_r$ are functions of (μ_n, P_n). In a similar manner, all the integrals involved in (28.70) can be easily rewritten as integrals of functions of $x_1, \ldots, x_l$,

Approach 2. Least-squares approximations. The approach taken by [1] yields the same formulas as that of [20], although they are derived slightly differently. The approximation of the updated mean and covariance is based on the minimization of linear least-squares errors, given the new observation. Let us start by recalling a result for a linear and Gaussian problem. Define $y = Hx + v$, where v and x are mutually independent and normally distributed, with v having mean zero and nondegenerate covariance R, and x having mean ρ^- and covariance $P^- = P_{xx}$. Define the mean $\bar{y} = Ey = H\rho^-$. Define $\rho = \mathbb{E}[x|y]$, $P = \text{cov}[x|y]$ and the covariances $P_{yy} = \text{cov}(y) = HP_{xx}H' + R$, $P_{xy} = \text{cov}(x, y) = E(x - \rho^-)(y - \bar{y})' = P_{xy}H'$. Then

$$\rho = \rho^- + W(y - \bar{y}),$$
$$W = P_{xy}\left[P_{yy}\right]^{-1}, \qquad (28.74)$$
$$P = P_{xx} - P_{xy}\left[P_{yy}\right]^{-1} P'_{xy}.$$

This is also a minimal linear least-squares estimator, given the initial mean and covariance.

Now return to the nonlinear filtering problem and, since we are only interested in the final formulas, consider the procedure of [20]. Let the nth observation be $y_n = g(x_n) + v_n$, where $x_n = x(t_n)$, be taken at time t_n, and (before the observation is taken) let x_n have conditional mean m_n^- and covariance M_n^-. For notational simplicity and consistency, let v_n have covariance $\mathbb{I}$. The procedure for incorporating the observation works as follows. Write $g_n = g(x_n)$. First approximate the conditional mean (given the data to just before the nth observation is taken) of g_n by the Gaussian approximation

$$\bar{g}_n = \int g(x)\mathbb{N}\left(m_n^-, M_n^-\right)(x)dx. \qquad (28.75)$$

Then approximate the covariance of g_n by

$$P_{g_n,g_n} = \int (g(x) - \bar{g}_n)(g(x) - \bar{g}_n)'\mathbb{N}\left(m_n^-, M_n^-\right)(x)dx, \qquad (28.76)$$

and the conditional cross covariance between x_n and $\bar{g}_n$ as:

$$P_{x_n,g_n} = \int \left(x - m_n^-\right)(g(x) - \bar{g}_n)'\mathbb{N}\left(m_n^-, M_n^-\right)(x)dx. \qquad (28.77)$$

Use the Gauss–Hermite numerical integration formula if necessary. Now use the form (28.74) to update the conditional mean and covariance as:

$$m_n = m_n^- + W_n(y_n - \bar{g}_n),$$
$$W_n = P_{x_ng_n}\left[P_{g_ng_n} + \mathbb{I}\right]^{-1}, \qquad (28.78)$$
$$M_n = M_n^- - P_{x_ng_n}\left[P_{g_ng_n} + \mathbb{I}\right]^{-1} P'_{x_ng_n}.$$

Now, continue by computing the estimate m^-_{n+1}, M^-_{n+1}, incorporate y_{n+1}, etc. See the references for numerical examples.

28.10 A numerical example

An example and data from [8] will be given to illustrate Approach 1. The data is for the four-dimensional model

$$dx_1(t) = x_3(t)dt + edw_1(t), \quad dx_2(t) = x_4(t)dt + edw_2(t),$$
$$dx_3(t) = f_1(x(t))dt + edw_3(t), \quad dx_4(t) = f_2(x(t))dt + edw_4(t),$$

where the $\{w_i\}$ are mutually independent Wiener processes and

$$f_i(x) = \frac{-50x_i}{\sqrt{x_1^2 + x_2^2}} I_{\{\sqrt{x_1^2+x_2^2} \geq 9\}} I_{\{x_1x_3+x_2x_4 \geq 0\}}.$$

The model is supposed to represent a ship that is confined to move in a constrained area. The position vector (x_1, x_2) and the associated velocities are (x_3, x_3), resp. Whenever the ship's position exceeds a radius of nine from the center (gets close to a shoreline), there is a large force (acceleration) that turns it away from the shore. The observations were taken at intervals $\Delta = .05$. The role of the indicator function $I_{\{x_1x_3+x_2x_4 \geq 0\}}$ is to assure that the corrective force is applied only when the distance is increasing, since (in the absence of noise) $x_1x_3 + x_2x_4 = d\left[x_1^2 + x_2^2\right]/dt$.

First, suppose the "bearing only" observation, a standard test case, where

$$y_{1,n} = \arctan[x_2(n\Delta)/x_1(n\Delta)] + sv_{1,n},$$

where the $v_{1,n}$ are mutually independent, normally distributed, with mean zero and variance unity, and independent of $x(\cdot)$. The values $e = .32$, $s = .4$ are used in the plotted data. Since the angles are measured in radians, the noise is quite large.

Since the observations depend only on the two position coordinates, the computation for the incorporation of the observation could be reduced to one in two dimensions. The procedure of Approach 1 was used, with four iterations and an 8×8 grid used for the numerical quadrature. The sample path $(x_1(\cdot), x_2(\cdot))$ and the estimates $(m_1(\cdot), m_2(\cdot))$ for the position coordinates for a typical run are plotted in Figure 28.2. The initial point is in the lower left-hand quadrant and the path is that for 30 units of time. The tracking is very good. The average sample values of the mean-square path errors were close to the values of the mean of the $M_{ii}(\cdot)$, $i = 1, 2$, a further indicator of the quality of the approach.

Now let us add an observation of the range, $y_{2,n} = \sqrt{x_1^2(n\Delta) + x_2^2(n\Delta)} + sv_{2,n}$, where the $\{v_{2,n}\}$ are mutually independent, normally distributed with mean zero and variance unity, and independent of the other processes. As seen in

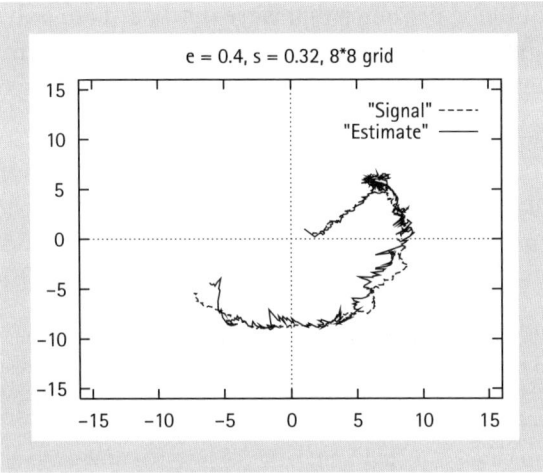

Fig. 28.2 NLF, angle observation.

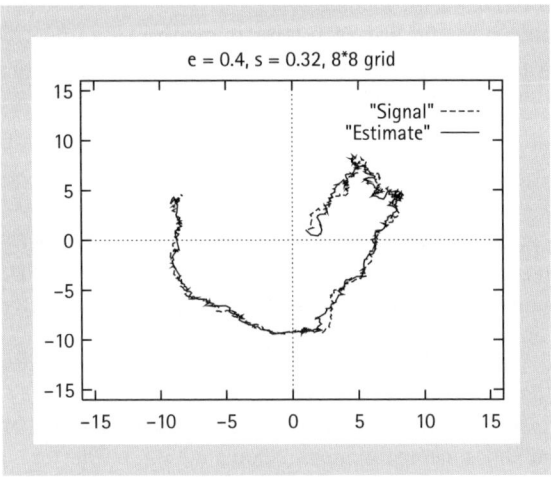

Fig. 28.3 NLF, angle-range observation.

the typical Figure 28.3, the tracking of the position coordinates is excellent. Comparing (via simulations) the results with those for all of the current forms of the particle method, showed that the latter method required so many points to get reasonable results, that it was not competitive as a real-time procedure.

Numerical examples based on Approach 2 can be found in [1, 20]. Approach 1 uses a more direct approximation, without the intermediate "mean-square" approximation step, so that one might initially expect that it would be better, although it involves more computation. However, to date, there are no definitive numerical or mathematical comparisons of the two approaches. Each does well on the class of problems to which it has been applied in the references. The versions of each of the approaches that were discussed are only one form of the possibilities. It is important to keep in mind that there are numerous variations of the general ideas. For all of the examples in the references, the extended Kalman filter was unstable and gave very poor results.

Notes

This work was partially supported by NSF grant DMS-0506928 and ARO contract W911NF-05-10928.
1. As a practical matter, all the functions can be taken to be bounded. Generally in applications, one tries to model the dynamics for state values in the range that is most common, and the models tend to become invalid as the absolute value of the state goes to infinity, where the model is often chosen for convenience in analysis.
2. (5.1) defines the functions $b_h(\cdot)$ and $a_h(\cdot)$.
3. If there is a jump component, then at each step, the chain is locally consistent with the diffusion (2.1), with conditional probability $(1 - \lambda \Delta t^h(x))$, and with the jump component with conditional probability $\lambda \Delta t^h(x)$. See [31, sec. 5.6].
4. This is the usual case, using the methods of [31].
5. The main motivations for the use of Gaussian approximating densities are the relative numerical simplicity, that it captures the main effects in many problems, the guidance supplied by the linear problem, and that it "works" in applications.

References

[1] I. Arasaratnam, S. Haykin, and R. J. Elliott. Discrete-time nonlinear filtering algorithms using Gauss–Hermite quadrature. *IEEE Transactions*, 95:953–77, 2007.

[2] Y. Bar-Shalom and X. R. Li. *Estimation and Tracking: Principles, Techniques and Software.* Artech House, Boston, 1993.

[3] G. J. Bierman. *Factorization Methods for Discrete Sequential Estimation.* Academic Press, New York, 1977.

[4] P. Billingsley. *Convergence of Probability Measures.* Wiley, New York, 1968.

[5] A. Budhiraja and H. J. Kushner. Approximation and limit results for nonlinear filters over an infinite time interval. *SIAM J. Control Optim.*, 37:1946–79, 1999.

[6] A. Budhiraja and H. J. Kushner. Robustness of nonlinear filters over the infinite time interval. *SIAM J. Control Optim.*, 37:1618–37, 1999.

[7] A. Budhiraja and H. J. Kushner. Approximation and limit results for nonlinear filters over an infinite time interval: Part II, random sampling algorithms. *SIAM J. Control Optim.*, 38:1874–1908, 2000.

[8] A. Budhiraja and H. J. Kushner. A nonlinear filtering algorithm based on an approximation of the conditional distribution. *IEEE Trans. Automatic Control*, 45:580–5, 2000.

[9] A. Calzolari and G. Nappo. A filtering problem with counting observation: approximations with error bounds. *Stochastics and Stochastics Reports*, 57:71–87, 1996.

[10] A. Calzolari and G. Nappo. Robust approximation in a filtering problem with real state space and counting observation. *Appl. Math. and Optimization*, 42:51–71, 2000.

[11] J. M. C. Clark. The design of robust approximations to the stochastic different equations of nonlinear filtering. In J. K. Skwirzynski, ed., *Communication Systems and Random Process Theory*. Sijtoff and Noordhoff, Alphen aan der Rihn, 1978.

[12] P. J. Davis and P. Rabinowitz. *Methods of Numerical Integration*, 2nd edn. Academic Press, New York, 1984.

[13] G. B. DiMasi and W. J. Runggaldier. Approximations and bounds for discrete nonlinear filtering. In *Lecture Notes in Control and Information Sciences*, pp. 191–202. Springer-Verlag, Berlin and New York, 1982, vol. 44.

[14] G. B. DiMasi and W. J. Runggaldier. On robust approximation in nonlinear filtering. In *Lecture Notes in Control and Information Sciences*, pp. 179–86. Springer-Verlag, Berlin and New York, 1982, vol. 43.

[15] G. B. DiMasi and W. J. Runggaldier. An approximation for the nonlinear filtering problem, with error bound. *Stochastics*, 14:247–71, 1985.

[16] J. L. Doob. *Stochastic Processes*. Wiley, New York, 1953.

[17] E. B. Dynkin. *Markov Processes*. Springer-Verlag, Berlin and New York, 1965.

[18] R. Elliott. *Stochastic Calculus and Applications*. Springer-Verlag, Berlin and New York, 1982.

[19] S. N. Ethier and T. G. Kurtz. *Markov Processes: Characterization and Convergence*. Wiley, New York, 1986.

[20] K. Itô and K. Xiong. Gaussian filters for nonlinear filtering problems. *IEEE Trans. Automatic Control*, 45(5):910–27, 2000.

[21] P. E. Kloeden and E. Platen. *Numerical Solution of Stochastic Differential Equations*. Springer-Verlag, Berlin and New York, 1992.

[22] H. Kunita and S. Watanabe. On square integrable martingales. *Nagoya Math. J.*, 30:209–45, 1967.

[23] T. G. Kurtz. *Approximation of Population Processes*, volume 36 of *CBMS-NSF Regional Conf. Series in Appl. Math*. SIAM, Philadelphia, 1981.

[24] H. J. Kushner. On the differential equations satisfied by conditional probability densities of Markov processes. *SIAM J. on Control*, 2:106–19, 1962.

[25] H. J. Kushner. Approximations to optimal nonlinear filters. *IEEE Trans. Automatic Control*, AC-12:546–56, 1967.

[26] H. J. Kushner. Dynamical equations for nonlinear filtering. *J. Differential Equations*, 3:179–90, 1967.

[27] H. J. Kushner. *Probability Methods for Approximations in Stochastic Control and for Elliptic Equations*. Academic Press, New York, 1977.

[28] H. J. Kushner. A robust discrete state approximation to the optimal nonlinear filter for a diffusion. *Stochastics Stochastics Rep.*, 3:75–83, 1979.

[29] H. J. Kushner. Nonlinear filtering for singularly perturbed systems. In E. Mayer-Wolfe, E. Merzbach, and A. Shwartz, eds, *Stochastic Analysis*, pages 347–69. Academic Press, New York, 1991.

[30] H. J. Kushner. *Heavy Traffic Analysis of Controlled Queueing and Communication Networks*. Springer-Verlag, Berlin and New York, 2001.

[31] H. J. Kushner and P. Dupuis. *Numerical Methods for Stochastic Control Problems in Continuous Time*. Springer-Verlag, Berlin and New York, 1992. 2nd edn, 2001.

[32] H. J. Kushner and H. Huang. Approximation and limit results for nonlinear filters with wide bandwidth observation noise. *Stochastics Stochastics Rep.*, 16:65–96, 1986.

[33] P. K. Kythe and M. P. Schäferkotter. *Handbook of Computational Methods for Integration*. Chapman and Hall, London and New York, 2004.

[34] R. Liptser and A. N. Shiryaev. *Statistics of Random Processes*. Springer-Verlag, Berlin and New York, 1977.

[35] G. Pagès, H. Pham, and J. Printems. Optimal quantization methods and applications to numerical problems in finance. In S. T. Rachev, ed., *Handbook of Computational and Numerical Methods in Finance*, pp. 253–97. Birkhauser, Boston, 2004.

[36] A. N. Shiryaev. Stochastic equations of nonlinear filtering of jump Markov processes. *Problemy Peredachi Informatsii*, 3:3–22, 1966.

[37] A. H. Stroud. *Approximate Calculation of Multiple Integrals*. Prentice-Hall, Englewood Cliffs, NJ, 1971.

[38] H. Tanizaki. *Nonlinear Filters: Estimation and Applications*. Springer-Verlag, Berlin and New York, 1996.

[39] W. M. Wonham. Some applications of stochastic differential equations to optimal nonlinear filtering. *SIAM J. on Control*, 2:347–69, 1965.

[40] M. Zakai. On optimal filtering of diffusion processes. *Z. Wahrsch. verv. Gebeite*, 11:230–43, 1969.

·29·

Signal Processing Problems on Function Space: Bayesian Formulation, Stochastic PDEs and Effective MCMC Methods

M. Hairer, A. Stuart, and J. Voss

29.1 Overview

Many applied problems concerning the integration of data and mathematical models arise naturally in dynamically evolving systems. These may be formulated in the general framework of Bayesian inference. There is a particular structure inherent in these problems, arising from the underlying dynamical models, that can be exploited. In this chapter we highlight this structure in the context of continuous-time dynamical models in finite dimensions. We set up a variety of Bayesian inference problems, some finite dimensional, for the initial condition of the dynamics, and some infinite dimensional, for a time-dependent path of an SDE. All of the problems share a common mathematical structure, namely that the posterior measure μ^y, given data y, has a density with respect to a Gaussian reference measure μ_0 so that

$$\frac{d\mu^y}{d\mu_0}(x) = Z(y)^{-1} \exp\left(-\Phi(x; y)\right) \tag{29.1}$$

for some potential function $\Phi(\cdot; y)$ and normalization constant $Z(y)$ both parameterized by the data. We denote the mean and covariance operator for μ_0 by m_0 and C_0, and we use $\mathcal{L} = C_0^{-1}$ to denote the precision operator. Thus $\mu_0 = \mathcal{N}(m_0, C_0)$.

The content of this chapter is centred around three main themes:

- Theme A. To exhibit a variety of problems arising in data assimilation which share the common structure (29.1).
- Theme B. To introduce a class of stochastic PDEs (SPDEs) which are reversible with respect to μ_0 or μ^y respectively.
- Theme C. To introduce a range of Metropolis–Hastings MCMC methods which sample from μ^y, based on the SPDEs discussed in Theme B.

A central aspect of this chapter will be to exhibit, in Theme A, common properties of the potential $\Phi(x; y)$ which then form the key underpinnings of the theories outlined in Themes B and C. These common properties of Φ include bounds from above and below, continuity in both x and y, and differentiability in x.

The continuous-time nature of the problems means that, in some cases, the probability measures constructed are on infinite-dimensional spaces: paths of continuous-time, vector-valued processes. We sometimes refer to this as *pathspace*. Working with probability measures on function space is a key idea throughout this chapter. We will show that this viewpoint leads to the notion of a well-posed signal processing problem, in which the target measure is continuous with respect to data. Furthermore a proper mathematical formulation of the problems on pathspace leads to efficient sampling techniques, defined on pathspace, and therefore robust under the introduction of discretization. In contrast, sampling techniques which first discretize, to obtain a finite-dimensional sampling problem, and then apply standard MCMC techniques, will typically lead to algorithms which perform poorly under refinement of the discretization.

In Section 29.2 we describe some general properties of measures defined through (29.1). Sections 29.3.1–29.3.6 are concerned with Theme A. In Section 29.3.1 we initiate our study by considering a continuous-time deterministic ODE observed noisily at discrete times; the objective is to determine the initial condition. Sections 29.3.2 and 29.3.3 generalize this set-up to the situation where the solution is observed continuously in time, and subject to white noise and coloured noise respectively. In Section 29.3.4 we return to discrete observations, but assume that the underlying model dynamics is subject to noise—or *model error*; in particular we assume that the dynamics is forced by an Ornstein–Uhlenbeck process. This section, and the remaining sections in Theme A, all contain situations where the posterior measure is on an infinite dimensional space. Section 29.3.5 generalizes the ideas in Section 29.3.4 to the situation where the model error is described by *white noise* in time. Finally, in Section 29.3.6, we consider the situation with model error (in the form of white noise) and continuous-time observations. In Section 29.4 we address Theme B, while Section 29.5 is concerned with Theme C. Some notational conventions, and background theory, are outlined in the Appendix.

We emphasize that, throughout this chapter, all the problems discussed are formulated as *smoothing* problems, not *filtering* problems. Thus time-distributed data on a given time interval $[0, T]$ is used to update knowledge about the entire state of the system on the same time interval. For a discussion of filtering methods we refer the reader to [13].

29.2 General properties of the posterior

In the next section we will exhibit a wide variety of signal processing problems which, when tackled in a Bayesian framework, lead to a posterior probability measure μ on a Banach space $(E, \|\cdot\|_E)$, specified via its Radon–Nikodym derivative with respect to a prior Gaussian measure μ_0. Specifically we have (29.1) where $\mu_0 = \mathcal{N}(m_0, \mathcal{C}_0)$ is the prior Gaussian measure and $\Phi(x; y)$ is a *potential*. We assume that $y \in Y$, a separable Banach space with norm $\|\cdot\|_Y$. The normalization constant $Z(y)$ is chosen so that μ is a probability measure:

$$Z(y) = \int_E \exp\left(-\Phi(x; y)\right) d\mu_0(x). \tag{29.2}$$

For details about Gaussian measures on infinite-dimensional spaces we refer to the monograph [4].

In many of the applications considered here, Φ satisfies the following four properties:

Assumption 29.1 The function $\Phi: E \times Y \to \mathbb{R}$ has the following properties:

1. For every $r > 0$ and every $\varepsilon > 0$, there exists $M = M(r, \varepsilon)$ such that, for all $x \in E$ and all y such that $\|y\|_Y \leq r$, $\Phi(x; y) \geq M - \varepsilon \|x\|_E^2$.
2. There exists $p \geq 0$ and, for every $r > 0$, there exists $C = C(r) > 0$ such that, for all $x \in E$ and all $y \in Y$ with $\|y\|_Y \leq r$, $\Phi(x; y) \leq C\left(1 + \|x\|_E^p\right)$.
3. For every $r > 0$ and every $R > 0$ there exists $L = L(r, R) > 0$ such that, for all $x_1, x_2 \in E$ with $\|x_1\|_E \vee \|x_2\|_E \leq R$ and all $y \in Y$ with $\|y\|_Y \leq r$,

$$|\Phi(x_1; y) - \Phi(x_2; y)| \leq L \|x_1 - x_2\|_E.$$

4. There exists $q > 0$ and, for every $r > 0$, there exists $K = K(r) > 0$ such that, for all $x \in E$ and all $y_1, y_2 \in E$ with $\|y_1\|_Y \vee \|y_2\|_Y \leq r$,

$$|\Phi(x; y_1) - \Phi(x; y_2)| \leq K\left(1 + \|x\|_E^q\right) \|y_1 - y_2\|_Y.$$

We show that, under these assumptions, the posterior measure μ^y is continuous with respect to the data y in the total variation distance. This is a well-posedness result for the posterior measure. The result, and proof, is similar to that in [10] which concerns Bayesian inverse problems for the Navier–Stokes equations, but where the Hellinger metric is used to measure distance.

Theorem 29.1 *Let μ^y and μ_0 be measures on a separable Banach space E such that μ_0 is a Gaussian probability measure, μ^y is absolutely continuous w.r.t. μ_0, and the log density $\Phi = -\log\left(\frac{d\mu^y}{d\mu_0}\right) : E \times Y \to \mathbb{R}$ satisfy Assumption 29.1. Then μ^y is a probability measure and the map $y \mapsto \mu^y$ is locally Lipschitz continuous in total variation distance: if μ and μ' are two measures given by (29.1) with data y and y' then, for every $r > 0$ and for all y, y' with $\|y\|_Y \vee \|y'\|_Y \leq r$, there exists a constant $C = C(r) > 0$ such that*

$$\|\mu - \mu'\|_{\mathrm{TV}} \le C\|y - y'\|_Y.$$

Proof. Since the reference measure μ_0 is Gaussian, $\|x\|_E$ has Gaussian tails under μ_0. The lower bound (i) therefore immediately implies that $\exp(-\Phi)$ is integrable, so that $Z(y)$ is indeed finite for every y.

We now turn to the continuity of the measures with respect to y. Throughout the proof, all integrals are over E. We fix a value $r > 0$ and use the notation C to denote a strictly positive constant that may depend upon r and changes from occurrence to occurrence. As in the statement, we fix $y, y' \in Y$ and we write $\mu = \mu^y$ and $\mu' = \mu^{y'}$ as a shorthand. Let Z and Z' denote the normalization constants for μ and μ', so that

$$Z = \int \exp\left(-\Phi(x; y)\right) d\mu_0(x), \qquad Z' = \int \exp\left(-\Phi(x; y')\right) d\mu_0(x).$$

Since μ_0 is Gaussian, assumption (1) yields the upper bound $|Z| \vee |Z'| \le C$. In addition, since Φ is bounded above by a polynomial by (2), we have a similar lower bound $|Z| \wedge |Z'| \ge C$. Using again the Gaussianity of μ_0, the bound (4) yields

$$\begin{aligned}|Z - Z'| &\le C \int \|y - y'\|_Y (1 + \|x\|_E^q) \exp\left(-\left(\Phi(x; y) \vee \Phi(x; y')\right)\right) d\mu_0(x) \\ &\le C\|y - y'\|_Y \int (1 + \|x\|_E^q) \exp\left(\varepsilon \|x\|_E^2 - M\right) d\mu_0(x) \\ &\le C\|y - y'\|_Y. \end{aligned} \qquad (29.3)$$

From the definition of the total variation distance, we then have

$$\begin{aligned}\|\mu - \mu'\|_{\mathrm{TV}} &= \int |Z^{-1} \exp(-\Phi(x; y)) - (Z')^{-1} \exp(-\Phi(x; y'))| \, d\mu_0(x) \\ &\le I_1 + I_2,\end{aligned}$$

where

$$\begin{aligned}I_1 &= \frac{1}{Z} \int \left|\exp\left(-\Phi(x; y)\right) - \exp\left(-\Phi(x; y')\right)\right| d\mu_0(x), \\ I_2 &= \frac{|Z - Z'|}{ZZ'} \int \exp\left(-\Phi(x; y')\right) d\mu_0(x).\end{aligned}$$

Since Z is bounded from below, we have $I_1 \le C\|y - y'\|_Y$ just as in (29.3). The second term is bounded similarly by (29.3) and the lower bound (i) on Φ, thus concluding the proof. □

Remark 29.1 We only ever use the Gaussianity of μ_0 to deduce that there exists $\varepsilon > 0$ such that $\int_E \exp\left(\varepsilon \|x\|_E^2\right) \mu_0(dx) < \infty$. Therefore, the statement of Theorem 29.1 extends to any reference measure μ_0 with this property.

29.3 Theme A: Bayesian inference for signal processing

It is instructive to summarize the different cases treated in this section in a table. The choice of column determines whether or not model error is present, and when present whether it is white or coloured; the choice of row determines whether or not the observations are discrete, and when continuous whether or not the observational noise is white or coloured. There are three possibilities not covered here; however the reader should be able to construct appropriate models in these three cases after reading the material herein.

	Model error		
Observation	No	White	Colored
Discrete	29.3.1	29.3.5	29.3.4
White	29.3.2	29.3.6	
Colored	29.3.3		

29.3.1 No model error, discrete observations

Here we address the question of making inference concerning the initial condition for an ODE, given noisy observation of its trajectory at later times. Thus the basic unknown quantity, which we wish to find the (posterior) distribution of, is a finite-dimensional vector.

Let $v \in C^1([0, T], \mathbb{R}^n)$ solve the ODE

$$\frac{dv}{dt} = f(v), \quad v(0) = u. \tag{29.4}$$

We assume that f is sufficiently nice (say locally Lipschitz and satisfying a coercivity condition) that the equation defines a semigroup $\varphi^t : \mathbb{R}^n \to \mathbb{R}^n$ with $v(t) = \varphi^t(u)$. We assume that we observe the solution in discrete time, at times $\{t_k\}_{k=1}^K$. Specifically, for some function $g: \mathbb{R}^n \to \mathbb{R}^l$ we observe

$$y_k = g(v(t_k)) + \eta_k, \quad k = 1, \ldots, K, \tag{29.5}$$

where the $\eta_k \sim \mathcal{N}(0, B_k)$ are a sequence of Gaussian random variables, not necessarily independent. We assume that

$$0 < t_1 \leq t_2 \leq \cdots \leq t_K \leq T. \tag{29.6}$$

Concatenating the data we may write

$$y = \mathcal{G}(u) + \eta, \tag{29.7}$$

where $y = (y_1, \ldots, y_K)$ are the observations, $\mathcal{G}(u) = (g(\varphi^{t_1}(u)), \ldots, g(\varphi^{t_K}(u)))$ maps the state of the system and $\eta = (\eta_1, \ldots, \eta_K)$ is the observational noise. Thus $\eta \sim \mathcal{N}(0, B)$ for some matrix B capturing the correlations amongst the $\{\eta_k\}_{k=1}^K$.

We will now construct the distribution of the initial condition u given an observation y, using Bayes formula (see (29.53) in the Appendix). We assume that the prior measure on u is a Gaussian $\mu_0 \sim \mathcal{N}(m_0, C_0)$, with mean m_0 and covariance matrix C_0. Given the initial condition u, the observations y are distributed according to the Gaussian measure with density

$$\mathbf{P}(y|u) \propto \exp(-\Phi(u; y)), \quad \Phi(u; y) = \frac{1}{2}|y - \mathcal{G}(u)|_B^2, \qquad (29.8)$$

where we define $|y|_B^2 = \langle y, B^{-1}y \rangle$ (see Appendix). By Bayes' rule we deduce that the posterior measure ν on u, given y, has Radon–Nikodym derivative

$$\frac{d\mu^y}{d\mu_0}(u) \propto \exp(-\Phi(u; y)). \qquad (29.9)$$

Thus the measure μ^y has density π with respect to Lebesgue measure which is given by

$$\pi(u) \propto \exp\left(-\frac{1}{2}|y - \mathcal{G}(u)|_B^2 - \frac{1}{2}|u - m_0|_{C_0}^2\right). \qquad (29.10)$$

Example 29.1 Let $n = 1$ and consider the ODE

$$\frac{dv}{dt} = av, \quad v(0) = u.$$

Thus $\varphi^t(u) = \exp(at)u$. As our prior measure we take the Gaussian $\mathcal{N}(m_0, \sigma^2)$. Assume that we observe the solution itself at times t_k and subject to mean zero i.i.d. Gaussian noises with variance γ^2, resulting in observations $\{y_k\}_{k=1}^K$. We have observations $y = Au + \eta$ where $\eta \sim \mathcal{N}(0, B)$

$$A = (\exp(at_1), \ldots, \exp(at_K))$$
$$B = \gamma^2 I.$$

The posterior measure is then Gaussian with density

$$\pi(u) \propto \exp\left(-\frac{1}{2\gamma^2}\sum_{k=1}^K |y_k - \exp(at_k)u|^2 - \frac{1}{2\sigma^2}|u - m_0|^2\right).$$

By completing the square we find that the posterior mean is

$$\frac{m_0 + \frac{\sigma^2}{\gamma^2}\sum_{k=1}^K \exp(at_k)y_k}{1 + \frac{\sigma^2}{\gamma^2}\sum_{k=1}^K \exp(2at_k)}$$

and the posterior variance is

$$\frac{\sigma^2}{1 + \frac{\sigma^2}{\gamma^2} \sum_{k=1}^{K} \exp(2at_k)}.$$

Since $y_k = \exp(at_k)u + \eta_k$ we may write $y_k = \exp(at_k)u + \gamma \xi_k$ for $\xi \sim \mathcal{N}(0, 1)$. We set

$$y = (y_1, \ldots, y_K), \qquad \xi = (\xi_1, \ldots, \xi_K).$$

The posterior mean m and covariance Σ can then be written succinctly as

$$\Sigma = \frac{\sigma^2}{1 + \frac{\sigma^2}{\gamma^2}|A|^2},$$

$$m = \frac{m_0 + \frac{\sigma^2}{\gamma^2}\langle A, y \rangle}{1 + \frac{\sigma^2}{\gamma^2}|A|^2} = \frac{m_0 + \frac{\sigma^2}{\gamma^2}\langle A, Av(0) + \gamma \xi \rangle}{1 + \frac{\sigma^2}{\gamma^2}|A|^2}.$$

We now consider the limits of small observational noise, and of large data sets, respectively.

First consider small noise. As $\gamma^2 \to 0$, the posterior variance converges to zero and the posterior mean to $\langle A, y \rangle / |A|^2$, solution of the least squares problem

$$\operatorname{argmin}_x |y - Ax|^2.$$

Now consider large data sets where $K \to \infty$. If $|A|^2 \to \infty$ as $K \to \infty$ then the posterior mean converges almost surely to the correct initial condition $v(0)$, and the posterior variance converges to zero. Thus the posterior approaches a Dirac supported on the correct initial condition. Otherwise, if $|A|^2$ approaches a finite limit, then uncertainty remains in the posterior, and the prior has significant effect in determining both the mean and variance of the posterior.

Example 29.2 Consider the Lorenz equations

$$\frac{dv_1}{dt} = \sigma(v_2 - v_1), \qquad \frac{dv_2}{dt} = \rho v_1 - v_2 - v_1 v_3, \qquad \frac{dv_3}{dt} = v_1 v_2 - \beta v_3,$$

started at $v(0) = u \in \mathbb{R}^3$. In this case, as a consequence of the chaoticity of the equations, observing a trajectory over long intervals of time does not allow one to gain more information on the initial condition. See Figure 29.1 for an illustration. We refer to [16] for further discussion of this example.

We now return to the general problem and highlight a programme that we will carry out in earnest for a number of more complex infinite-dimensional posterior measures in later sections. We work within the context of ODEs with globally Lipschitz continuous drifts. This condition ensures global existence of solutions, as well as enabling a straightforward explicit bound on the solution in terms of its initial data. However other, less stringent, assumptions could also

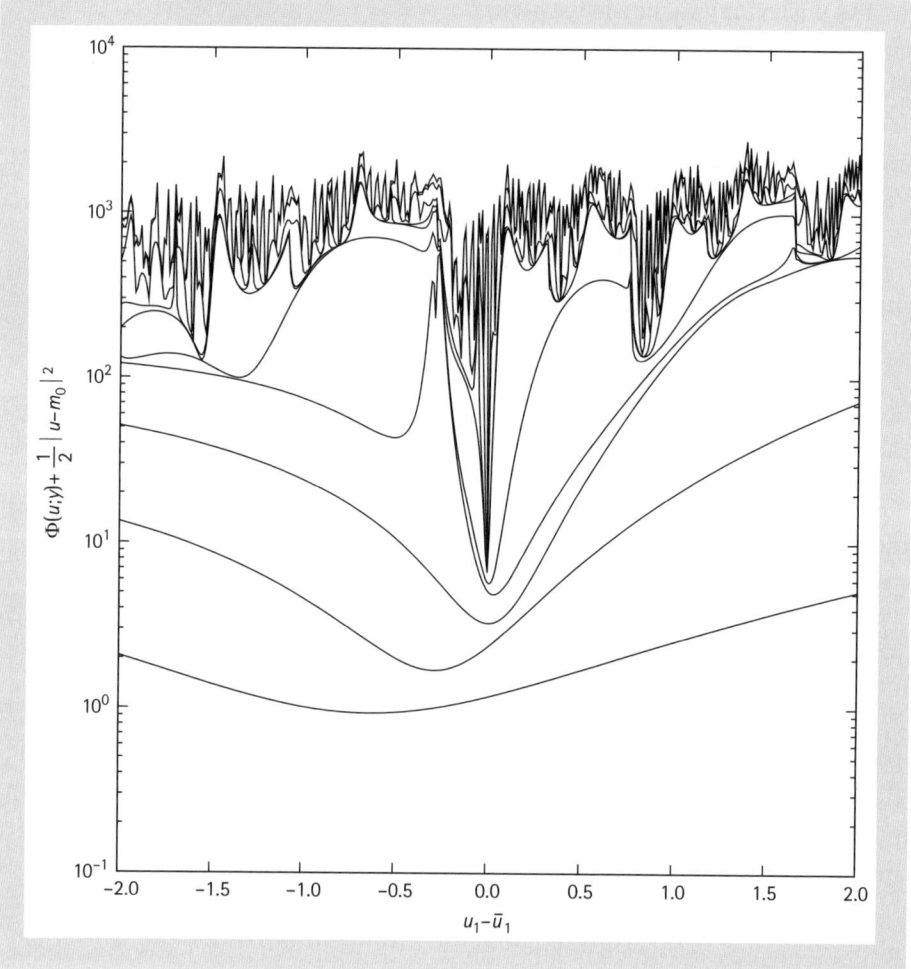

Fig. 29.1 Illustration of the posterior density for the Lorenz system from Example 29.2. Observations y are generated at times $1, 2, 3, \ldots, 10$ for a trajectory starting at $v(0) = \bar{u} \in \mathbb{R}^3$. Then $u_1 \mapsto \Phi(u; y) + \frac{1}{2}|u - m_0|^2$ is plotted, where the last two components of u are fixed to the "exact" values $u_2 = \bar{u}_2$ and $u_3 = \bar{u}_3$. The different lines, from bottom to top, correspond to considering only the first $K = 1, \ldots, 10$ observations. Up to a constant, the plotted value is $-\log \pi$ where π is the posterior density of μ^y. The figure illustrates that the effect of adding more observations is twofold: Firstly, the additional information allows to get better estimates of $\bar{u}$, the posterior distribution concentrates around this value. Second, as more observations are added, the shape of the posterior density gets more irregular and many local extrema appear.

be used, provided global existence is known. For example, generalizations to locally Lipschitz drifts satisfying a suitable dissipativity condition are straightforward.

Theorem 29.2 *Assume that $f : \mathbb{R}^n \to \mathbb{R}^n$ and $g : \mathbb{R}^n \to \mathbb{R}^l$ are globally Lipschitz. Let $E = \mathbb{R}^n$, $Y = \mathbb{R}^{\ell K}$. Then $\mu^y \ll \mu_0$ with $d\mu^y/d\mu_0 \propto \exp(-\Phi)$ where Φ is given* by (29.8). The map Φ satisfies Assumptions 29.1 and $y \mapsto \mu^y$ is locally Lipschitz continuous in the total variation metric.

Proof. It will be useful to introduce the notation $g^k \colon \mathbb{R}^n \to \mathbb{R}^l$ defined by $g^k = g \circ \varphi^{t_k}$. Recall from (29.7) that

$$\mathcal{G}(u) = \left(g^1(u), \ldots, g^K(u)\right). \tag{29.11}$$

This is linearly bounded since g and φ^t are linearly bounded; clearly $\Phi \geq 0$ by construction. Thus (1) and (2) of Assumption 29.1 are satisfied. Furthermore, by the Lipschitz continuity of f, $\varphi^t \colon \mathbb{R}^n \to \mathbb{R}^n$ is well-defined and Lipschitz. As the composition of Lipschitz functions, $\mathcal{G}$ is itself a Lipschitz function. Hence $\Phi(\cdot; y)$ is Lipschitz and (3) holds. Also $\Phi(x; \cdot)$ is quadratic in y and hence (4) holds. $\square$

For the purpose of studying algorithms that sample from μ^y, it is also of interest to show that the derivative of $\Phi(\cdot; y)$ is sufficiently regular.

Theorem 29.3 *Let $k > 0$. Assume, on top of the assumptions of Theorem 29.2, that $f \in C^k(\mathbb{R}^n, \mathbb{R}^n)$ and $g \in C^k(\mathbb{R}^n, \mathbb{R}^l)$. Then the potential $\Phi(\cdot; y)$ given by (29.8) is in $C^k(\mathbb{R}^n, \mathbb{R})$.*

Proof. As in the previous proof, the observation vector $\mathcal{G}(u)$ is given by (29.11). By standard ODE theory $\varphi^t \in C^k(\mathbb{R}^n, \mathbb{R}^n)$. As the composition of C^k functions, $\mathcal{G}$ is itself a C^k function. Since Φ is quadratic in $\mathcal{G}$, the result follows for Φ. $\square$

29.3.2 No model error, continuous white observational noise

We now consider the preceding problem in the limit where $K \to \infty$ and the set $\{t_i\}_{i=1}^\infty$ is dense in $[0, T]$. Once again $v(t)$ solves (29.4):

$$\frac{dv}{dt} = f(v), \quad v(0) = u, \tag{29.12}$$

but now we assume that we observe a function of the solution in continuous time, and subject to white noise. Specifically we assume that we observe the time-integrated data y solving the SDE

$$\frac{dy}{dt} = g(v) + \sqrt{\Sigma}\frac{dW}{dt}, \quad y(0) = 0. \tag{29.13}$$

Here $g \colon \mathbb{R}^n \to \mathbb{R}^\ell$ and $\Sigma \in \mathbb{R}^{\ell \times \ell}$ is positive-definite. Using as before the semigroup φ^t solving (29.12), this may be rewritten as

$$\frac{dy}{dt} = g(\varphi^t(u)) + \sqrt{\Sigma}\frac{dW}{dt}, \quad y(0) = 0.$$

The precise interpretation of the data $\{y(t)\}_{t\in[0,T]}$ is that we observe the function $y(t)$ defined by

$$y(t) = \int_0^T g(\varphi^t(u))\,dt + \sqrt{\Sigma}\,W(t).$$

Let $\mathbb{Q}_0$ denote the Gaussian measure on $L^2([0,T], \mathbb{R}^\ell)$ given by the law of $y(t) = \sqrt{\Sigma}\,W(t)$. Now place the prior measure $\mu_0 \sim \mathcal{N}(m_0, C_0)$ on the initial condition in $\mathbb{R}^n$. Then take ν_0 to be the product measure on $L^2([0,T], \mathbb{R}^\ell) \times \mathbb{R}^n$ given by $\mathbb{Q}_0 \otimes \mu_0$. Note that $\nu_0(du|y) = \mu_0(du)$ since u and y are independent under ν_0.

Let $\mathbb{Q}^u$ denote the measure on $L^2([0,T], \mathbb{R}^\ell)$ for y solving (29.13), with u given. By the Girsanov Theorem A.23 we have that

$$\frac{d\mathbb{Q}^u}{d\mathbb{Q}_0}(y) = \exp\left(-\frac{1}{2}\int_0^T |g(\varphi^t(u))|_\Sigma^2\,dt + \int_0^T \langle g(\varphi^t(u)), dy\rangle_\Sigma\right).$$

Thus if ν is the measure on $L^2([0,T], \mathbb{R}^\ell) \times \mathbb{R}^n$ given by (29.13) with u drawn from μ_0, then we have

$$\frac{d\nu}{d\nu_0}(u, y) = \exp\left(-\frac{1}{2}\int_0^T |g(\varphi^t(u))|_\Sigma^2\,dt + \int_0^T \langle g(\varphi^t(u)), dy\rangle_\Sigma\right).$$

Let $\mu^y(du)$ denote $\nu(du|y)$. By Theorem A.22 we have

$$\frac{d\mu^y}{d\mu_0}(u) \propto \exp(-\Phi(u; y)), \tag{29.14}$$

$$\Phi(u; y) = \frac{1}{2}\int_0^T |g(\varphi^t(u))|_\Sigma^2\,dt - \int_0^T \langle g(\varphi^t(u)), dy\rangle_\Sigma.$$

Integrating the second term by parts and using the fact that φ^t solves (29.12), we find that

$$\Phi(u; y) = \frac{1}{2}\int_0^T \left(|g(\varphi^t(u))|_\Sigma^2 + 2\langle Dg(\varphi^t(u))f(\varphi^t(u)), y\rangle_\Sigma\right) - \langle g(\varphi^T(u)), y(T)\rangle_\Sigma. \tag{29.15}$$

Theorem 29.4 *Assume that $f: \mathbb{R}^n \to \mathbb{R}^n$ is locally Lipschitz continuous and linearly bounded, and that $g \in C^2(\mathbb{R}^n, \mathbb{R}^l)$ is globally Lipschitz continuous. Let $E = \mathbb{R}^n$ and $Y = C([0,T], \mathbb{R}^l)$. Then $\mu^y \ll \mu_0$ with $d\mu^y/d\mu_0 \propto \exp(-\Phi)$ where Φ is given by (29.15). The function Φ satisfies Assumptions 29.1 and $y \mapsto \mu^y$ is locally Lipschitz continuous in the total variation metric.*

Proof. Since $ab \geq -\frac{\varepsilon}{2}a^2 - \frac{1}{2\varepsilon}b^2$ for any $\varepsilon > 0$ and $a, b \in \mathbb{R}$, it follows from (29.15) that Φ is bounded from below by

$$\Phi(u; y) \geq -\frac{C}{\varepsilon}\|y\|_{L^\infty}^2 - \varepsilon|g(\varphi^T(u))|^2 - \varepsilon T\int_0^T |Dg(\varphi^t(u))f(\varphi^t(u))|^2\,dt,$$

for some constant C depending only on Σ. Since the assumption that f grows linearly implies the existence of a constant C such that $|\varphi^t(u)| \le C|u|$ for every $t \in [0, T]$, the requested lower bound on Φ follows from the linear growth assumptions on f and g as well as the boundedness of Dg.

The polynomial upper bound on $\Phi(\cdot; y)$ follows in exactly the same way, yielding

$$|\Phi(u; y)| \le C \left(\|y\|_{L^\infty}^2 + |u|^2 \right),$$

for some constant C. Conditions (3) and (4) in Assumption 29.1 follow similarly, using the Lipschitz continuity of $\varphi^t(\cdot)$ as in the proof of Theorem 29.2. □

Remark 29.2 Note that it is the need to prove continuity in y which requires us to work in the function space $C\left([0, T], \mathbb{R}^l\right)$, since computation of Φ requires the evaluation of y at time T. Note also that it is possible to weaken the growth conditions on f and g, at the expense of strengthening the dissipativity assumptions on f.

We mention an insightful, unrigorous but useful, way of writing the potential Φ. If we pretend that y is differentiable in time (which it almost surely isn't), then we may write Φ from (29.14) as

$$\Phi(u; y) = \frac{1}{2} \int_0^T \left| g(\varphi^t(u)) - \frac{dy}{dt} \right|_\Sigma^2 dt - \frac{1}{2} \int_0^T \left| \frac{dy}{dt} \right|_\Sigma^2 dt.$$

The Gaussian reference measure μ_0, again only at a formal level, has density $\exp(-\frac{1}{2} \int_0^T \left|\frac{dy}{dt}\right|_\Sigma^2 dt)$ with respect to (the of course nonexistent) 'Lebesgue measure'. This suggests that μ^y has density $\exp(-\frac{1}{2} \int_0^T \left|g(\varphi^t(u)) - \frac{dy}{dt}\right|_\Sigma^2 dt)$.

In this form we see that, as in the previous section, the potential Φ for the Radon–Nikodym derivative between posterior and prior, written in the general form (29.1), simply measures the mismatch between data and observation operator. This nonrigorous rewrite of the potential Φ is useful precisely because it highlights this fact, easily lost in the mathematically correct formulation (29.14).

The following theorem is proved similarly to Theorem 29.3.

Theorem 29.5 *Let $k > 0$. Assume that we are in the setting of Theorem 29.4 and that furthermore $f \in C^k(\mathbb{R}^n, \mathbb{R}^n)$, and $g \in C^{k+1}(\mathbb{R}^n, \mathbb{R}^l)$. Then the potential $\Phi(\cdot; y)$ given by (29.15) belongs to $C^k(\mathbb{R}^n, \mathbb{R})$.*

29.3.3 No model error, continuous coloured observational noise

We now consider the discrete observations problem in the limit where $K \to \infty$ and the set $\{t_i\}_{i=1}^\infty$ is dense in $[0, T]$, but we assume that the observational noise is correlated. We model this situation by assuming that $u(t)$ solves (29.4), and that we observe a function of the solution in continuous time, subject to noise drawn from the distribution of a stationary Ornstein–Uhlenbeck process. In other words, we observe the process

$$y(t) = g(\varphi^t(u)) + \psi(t), \qquad (29.16)$$

where

$$\frac{d\psi}{dt} = -R\psi + \sqrt{2\Lambda}\frac{dW}{dt}, \quad \psi(0) \sim \mathcal{N}(0, R^{-1}\Lambda). \qquad (29.17)$$

Here $g\colon \mathbb{R}^n \to \mathbb{R}^\ell$, and the matrices $R, \Lambda \in \mathbb{R}^{\ell \times \ell}$ are symmetric positive-definite and are assumed to commute for simplicity (in particular, this ensures that the process ψ is reversible).

Once again we adopt a Bayesian framework to find the posterior probability for u given y. For economy of notation we set $\theta(t) = g(\varphi^t(u))$ and denote by $\dot\theta(t)$ the time-derivative of θ. We deduce from Itō's formula that

$$\begin{aligned}\frac{dy}{dt} &= \dot\theta - R\psi + \sqrt{2\Lambda}\frac{dW}{dt} = \dot\theta - R(y-\theta) + \sqrt{2\Lambda}\frac{dW}{dt} \\ &= \dot\theta + R\theta - Ry + \sqrt{2\Lambda}\frac{dW}{dt}.\end{aligned}$$

Furthermore

$$y(0) \sim \mathcal{N}(\theta(0), R^{-1}\Lambda) = \mathcal{N}(g(u), R^{-1}\Lambda),$$

independently of W.

Let $\mathbb{Q}_0$ denote the measure on $L^2([0,T], \mathbb{R}^\ell)$ generated by the Gaussian process (29.17) and place the prior measure $\mu_0 \sim \mathcal{N}(m_0, \mathcal{C}_0)$ on the initial condition u in $\mathbb{R}^n$. Then take ν_0 to be the measure on $L^2([0,T], \mathbb{R}^\ell) \times \mathbb{R}^n$ given by $\mathbb{Q}_0 \otimes \mu_0$. Note that $\nu_0(du|y) = \mu_0(du)$ since u and y are independent under ν_0.

We let ν denote the probability measure for y and u, with u distributed according to μ_0. By the Girsanov theorem

$$\frac{d\nu}{d\nu_0}(u,y) \propto \exp(-\Phi(u;y))$$

$$\Phi(u;y) = \frac{1}{4}\int_0^T \left(|h(\varphi^t(u))|_\Lambda^2 \, dt - 2\langle h(\varphi^t(u)), dy + Ry\, dt\rangle_\Lambda\right)$$
$$+ \frac{1}{2}|g(u)|_{R^{-1}\Lambda}^2 - \langle y(0), g(u)\rangle_{R^{-1}\Lambda}.$$

Here and below we use the shortcut

$$\begin{aligned}h(u) &= \dot\theta + R\theta, \\ &= Dg(u)f(u) + Rg(u).\end{aligned}$$

We let $\mu^y(u)$ denote the posterior distribution for u given y. By applying Bayes' formula in the guise of Theorem A.22 we obtain

$$\frac{d\mu^y}{d\mu_0}(u) \propto \exp(-\Phi(u; y)) \qquad (29.18)$$

$$\Phi(u; y) = \frac{1}{4}\int_0^T \left(|h(\varphi^t(u))|_\Lambda^2 \, dt - 2\langle h(\varphi^t(u)), dy + Ry \, dt\rangle_\Lambda\right)$$

$$+ \frac{1}{2}|g(u)|_{R^{-1}\Lambda}^2 - \langle y(0), g(u)\rangle_{R^{-1}\Lambda}.$$

As in the previous section, the posterior measure involves a Riemann integral and a stochastic integral, both parameterized by u, but the stochastic integral can be converted to a Riemann integral, by means of an integration by parts. Setting

$$\tilde{h}(u) = Dh(u)f(u) = Dg(u)Df(u)f(u) + D^2g(u)(f(u), f(u)) + (RDg(u))f(u),$$

we find that

$$\Phi(u; y) = \frac{1}{4}\int_0^T \left(|h(\varphi^t(u))|_\Lambda^2 + 2\langle \tilde{h}(\varphi^t(u)), y - Ry\rangle_\Lambda\right) dt \qquad (29.19)$$

$$+ \frac{1}{2}\langle h(u), y(0)\rangle_\Lambda - \frac{1}{2}\langle h(\varphi^T(u)), y(T)\rangle_\Lambda$$

$$+ \frac{1}{2}|g(u)|_{R^{-1}\Lambda}^2 - \langle y(0), g(u)\rangle_{R^{-1}\Lambda}.$$

Proof of the following two theorems is very similar to those for Theorems 29.4 and 29.5.

Theorem 29.6 *Assume that $f \in C^2(\mathbb{R}^n, \mathbb{R}^n)$ is linearly bounded in (29.4), that $g \in C^3(\mathbb{R}^n, \mathbb{R}^l)$ is globally Lipschitz continuous in (29.16), and that $\tilde{h}$ is linearly bounded. Let $E = \mathbb{R}^n$ and $Y = C([0, T], \mathbb{R}^l)$. Then $\mu^y \ll \mu_0$ with $d\mu^y/d\mu_0 \propto \exp(-\Phi)$ where Φ is given by (29.19). The map Φ satisfies Assumptions 29.1 and $y \mapsto \mu^y$ is locally Lipschitz continuous in the total variation metric.*

Remark 29.3 The condition that $\tilde{h}$ is linearly bounded follows for example if we assume that $D^2g(u)$ is bounded by $C/(1 + |u|)$ for some constant C.

Theorem 29.7 *Let $k \geq 1$. Assume that, further to satisfying the assumptions of Theorem 29.6, one has $f \in C^{k+1}(\mathbb{R}^n, \mathbb{R}^n)$ and $g \in C^{k+2}(\mathbb{R}^n, \mathbb{R}^l)$. Then the potential $\Phi(\cdot; y)$ given by (29.15) belongs to $C^k(\mathbb{R}^n, \mathbb{R})$.*

29.3.4 Coloured model error, discrete observations

The posterior probability measures on the initial condition u in the previous three examples can be very complicated objects from which it is hard to extract information. This is particularly true in cases where the semigroup φ^t exhibits sensitive dependence on initial conditions, there is data over a large time interval, and the system is sufficiently ergodic and mixing. The posterior is then essentially flat, with small random fluctuations superimposed, and contains little information about the initial condition. In such situations it is natural to relax the *hard constraint* that the dynamical model is satisfied exactly and to seek to explain the observations through a *forcing* to the dynamics: we allow equation (29.4) to be forced by an extraneous driving noise, known as *model error*. Thus we view the dynamics (29.4) as only being enforced as a weak constraint, in the sense that the equation need not be satisfied exactly. We then seek a posterior probability measure on both the initial condition and a driving noise process which quantifies the sense in which the dynamics is not exactly satisfied. Since we are working with continuous time, the driving noise process is a *function* and thus the resulting posterior measure is a measure on an *infinite-dimensional* space of functions. This section is the first of several where the desired probability measure lives on an infinite-dimensional space.

To be concrete we consider the case where the driving noise is correlated in time and governed by an Ornstein–Uhlenbeck process. We thus consider the model equations

$$\frac{dv}{dt} = f(v) + \frac{1}{\sqrt{\delta}}\psi, \quad v(0) = u, \tag{29.20}$$

$$\frac{d\psi}{dt} = -\frac{1}{\delta}R\psi + \sqrt{\frac{\Lambda}{\delta}}\frac{dW}{dt}, \quad \psi(0) \sim \mathcal{N}\left(0, \frac{1}{2}R^{-1}\Lambda\right).$$

We assume as before that R and Λ commute. The parameter δ sets a correlation time for the noise; in the next section we will let $\delta \to 0$ and recover white noise forcing. Equation (29.20) specifies our prior model for the noise process ψ. We assume that $\psi(0)$ is chosen independently of W, and then (29.20) describes a stationary OU process ψ. As our prior on the initial condition we take $u \sim \mathcal{N}(m_0, C_0)$, independently of ψ. We have thus specified a prior Gaussian probability measure $\mu_0(u_0, \psi)$ on $\mathbb{R}^n \times L^2([0, T], \mathbb{R}^n)$.

As observations we take, as in Section 29.3.1,

$$y_k = g(v(t_k)) + \eta_k, \quad k = 1, \ldots, K,$$

where $\eta_k \sim \mathcal{N}(0, B_k)$ are a sequence of Gaussian random variables, not necessarily independent, and the observation times satisfy (29.6). Concatenating the data we may write

$$y = \mathcal{G}(u, \psi) + \eta, \tag{29.21}$$

where $y = (y_1, \ldots, y_K)$ are the observations, $\mathcal{G}(u, \psi) = (g(v(t_1)), \ldots, g(v(t_K)))$ maps the state of the system and $\eta = (\eta_1, \ldots, \eta_K)$ is the observational noise. Thus $\eta \sim \mathcal{N}(0, B)$ for some matrix B capturing the correlations amongst the $\{\eta_k\}_{k=1}^K$. Here $\mathcal{G}$ is a map from $\mathbb{R}^n \times L^2([0, T], \mathbb{R}^n)$ to $\mathbb{R}^{lK}$. The likelihood for the observations is then

$$\mathbf{P}(dy|u,\psi) \propto \exp\left(-\frac{1}{2}|y - \mathcal{G}(u,\psi)|_B^2\right) dy.$$

Let ν denote the measure on $\mathbb{R}^n \times L^2([0,T],\mathbb{R}^n) \times \mathbb{R}^{lK}$ given by

$$\nu(du, d\psi, dy) = \mathbf{P}(dy|u,\psi)\mu_0(du, d\psi),$$

and let ν_0 denote the measure on $\mathbb{R}^n \times L^2([0,T],\mathbb{R}^n) \times \mathbb{R}^{lK}$ given by

$$\nu_0(du, d\psi, dy) \propto \exp\left(-\frac{1}{2}|y|_B^2\right) \mu_0(du, d\psi)dy.$$

Since (u, ψ) and y are independent under $\nu_0(du, d\psi)dy$, Theorem A.22 shows that the posterior probability measure $\mu^y(du, d\psi)$ is given by

$$\frac{d\mu^y}{d\mu_0}(u,\psi) \propto \exp\left(-\Phi(u,\psi;y)\right), \quad \Phi(u,\psi;y) = \frac{1}{2}|y - \mathcal{G}(u,\psi)|_B^2. \quad (29.22)$$

Example 29.3 Consider again the Lorenz equation from Example 29.2. Using the set-up from this section, we get a posterior distribution on the pairs (u, ψ) where u is the initial condition of the Lorenz ODE as in Example 29.2 above, and ψ is the additional forcing (model error) from (29.20).

We consider again the setting from Figure 29.1, but this time with the additional forcing term ψ. Now the posterior for u can be obtained by averaging (29.22) over ψ. This leads to a smoothing of the posterior distribution. The effect is illustrated in Figure 29.2.

Theorem 29.8 *Assume that $f: \mathbb{R}^n \to \mathbb{R}^n$ and $g: \mathbb{R}^n \to \mathbb{R}^l$ are globally Lipschitz continuous. Let $E = \mathbb{R}^n \times L^2([0,T],\mathbb{R}^n)$ and $Y = \mathbb{R}^{lK}$. Then $\mu^y \ll \mu_0$ with $d\mu^y/d\mu_0 \propto \exp(-\Phi)$ where Φ is given by (29.22). The map Φ satisfies Assumptions 29.1 and $y \mapsto \mu^y$ is locally Lipschitz continuous in the total variation metric.*

Proof. Assumption 29.1(1) follows with $M = 0$ and $\varepsilon = 0$. To establish (2) we note that, for $0 \leq t \leq T$,

$$\frac{1}{2}\frac{d}{dt}|v|^2 \leq \alpha + \beta|v|^2 + \frac{1}{2\sqrt{\delta}}\left(\|\psi\|^2 + |v|^2\right)$$

$$\leq \alpha + \beta|v|^2 + \frac{1}{2\sqrt{\delta}}\left(\|\psi\|_{L^2([0,T],\mathbb{R}^n)}^2 + |v|^2\right).$$

Application of the Gronwall inequality shows that

$$\|v(t)\| \leq C(t)\left(\|\psi\|_{L^2([0,T],\mathbb{R}^n)} + |u|\right)$$

and hence that, since g is linearly bounded,

$$|\mathcal{G}(u,\psi)| \leq C\left(\|\psi\|_{L^2([0,T],\mathbb{R}^n)} + |u|\right).$$

From this, assumption (2) follows.

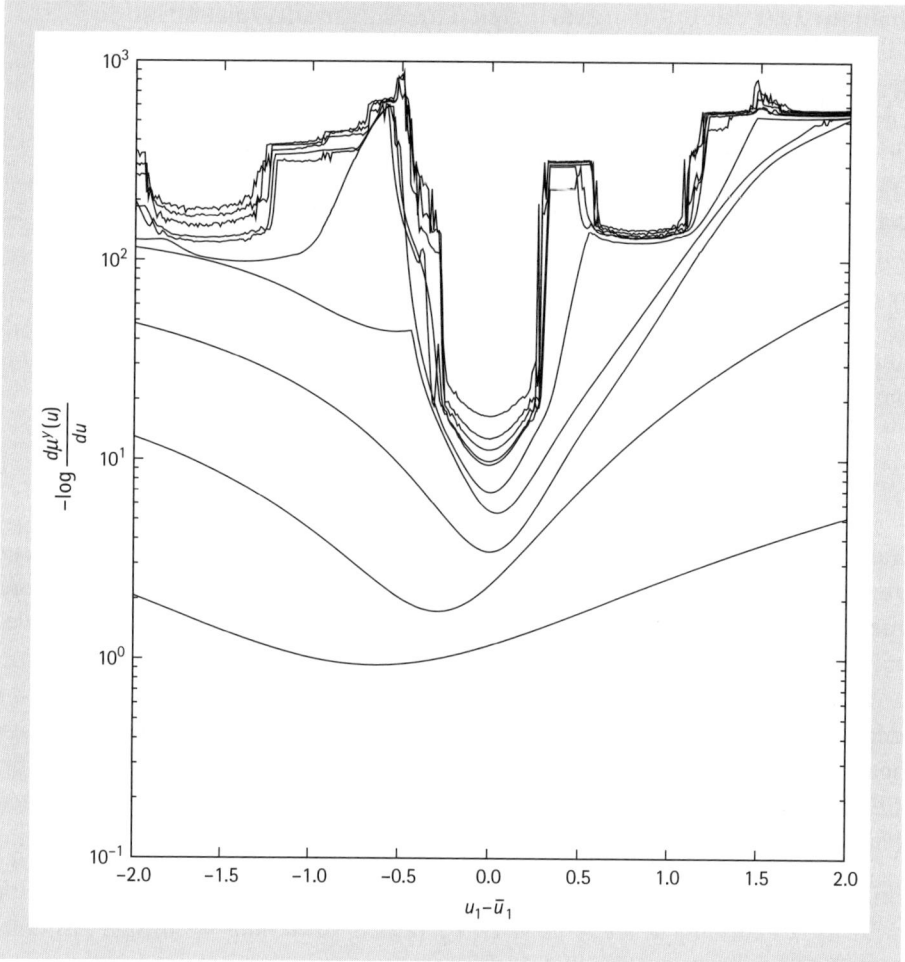

Fig. 29.2 Illustration of the posterior density for the initial condition in Example 29.3. The system is exactly the same as in Figure 29.1, except for the presence of an additional random forcing ψ in the Lorenz equation. The plotted posterior density of u is obtained by averaging $\Phi(u, \psi; y) + \frac{1}{2}|u - m_0|^2$ over ψ. The sampling is now on an infinite-dimensional space, but the figure illustrates that the posterior for u is much smoother than in the situation without model error from Figure 29.1.

To establish (3) it suffices to show that $\mathcal{G}: \mathbb{R}^n \times L^2([0, T], \mathbb{R}^n) \to \mathbb{R}^{lK}$ is locally Lipschitz continuous; this follows from the stated hypotheses on f and g since the mapping $(u, \psi) \in \mathbb{R}^n \times L^2([0, T], \mathbb{R}^n) \to v \in C([0, T], \mathbb{R}^n)$ is locally Lipschitz continuous, as may be shown by a Gronwall argument similar to that used to establish (2). Assumption 29.1(4) follows from the fact that $\Phi(x; \cdot)$ is quadratic, together with the polynomial bounds on $\mathcal{G}$ from the proof of (2). □

Again, we obtain more regularity on Φ by imposing more stringent assumptions on f and g:

Theorem 29.9 *For $k > 0$, if both f and g are C^k, then Φ is also C^k.*

29.3.5 White model error, discrete observations

In the preceding example we described the model error as an OU process. In some situations it is natural to describe the model error as white noise. Formally this can be obtained from the preceding example by taking the limit $\delta \to 0$ so that the correlation time tends to zero. Heuristically we have

$$\frac{1}{\sqrt{\delta}}\psi = R^{-1}\sqrt{2\Lambda}\frac{dW}{dt} + \mathcal{O}(\sqrt{\delta})$$

from the OU process (29.20). Substituting this heuristic into (29.20) and setting $\delta = 0$ gives the white-noise-driven model

$$\frac{dv}{dt} = f(v) + \sqrt{\Gamma}\frac{dW}{dt}, \quad v(0) = u, \qquad (29.23)$$

where $\sqrt{\Gamma} = R^{-1}\sqrt{2\Lambda}$.

Again we assume that we are given observations in the form (29.5). There are now two ways to proceed to define an inverse problem. We can either make inference concerning the pair (u, W), or we can make inference concerning the function v itself. We consider the two approaches in turn. Note that (u, W) uniquely define v and so a probability measure on (u, W) implies a probability measure on v.

First we consider the formulation of the problem where we make inference about (u, W). We construct the prior measure $\mu_0(du, dW)$ by assuming that u and W are independent, by taking $u \sim \mathcal{N}(m_0, C_0)$ and by taking standard n-dimensional Wiener measure for W. Now consider the integral equation

$$v(t) = u + \int_0^t f(v(s))\,ds = \sqrt{\Gamma}W(t). \qquad (29.24)$$

The solution of this equation defines a map

$$\mathcal{V}: \mathbb{R}^n \times C([0, T], \mathbb{R}^n) \to C([0, T], \mathbb{R}^n)$$
$$(u, W) \mapsto \mathcal{V}(u, W) = v.$$

Thus we may write the equation (29.5) for the observations as

$$y = \mathcal{G}(u, W) + \eta, \qquad (29.25)$$

with η as in (29.21) and $\mathcal{G}_k(u, W) = g(\mathcal{V}(u, W)(t_k))$. The likelihood of y is thus

$$\mathbf{P}(dy|u, W) \propto \exp\left(-\frac{1}{2}|y - \mathcal{G}(u, W)|_B^2\right) dy.$$

This leads to a probability measure

$$\nu(du, dW, dy) = \mathbf{P}(y|u, \psi)\mu_0(du, d\psi)dy$$

on the space $\mathbb{R}^n \times C([0, T], \mathbb{R}^n) \times \mathbb{R}^{lK}$. By ν_0 we denote the measure on $\mathbb{R}^n \times C([0, T], \mathbb{R}^n) \times \mathbb{R}^{lK}$ given by

$$\nu_0(du, dW, dy) \propto \exp\left(-\frac{1}{2}|y|_B^2\right) \mu_0(du, dW) dy.$$

Since (u, ψ) and y are independent under $\nu_0(du, d\psi) dy$, Theorem A.22 shows that the posterior probability measure is given by

$$\frac{d\mu^y}{d\mu_0}(u, W) \propto \exp\left(-\Phi(u, W; y)\right) \tag{29.26}$$

$$\Phi(u, W; y) = \frac{1}{2}|y - \mathcal{G}(u, W)|_B^2.$$

Theorem 29.10 *Assume that $f: \mathbb{R}^n \to \mathbb{R}^n$ and $g: \mathbb{R}^n \to \mathbb{R}^l$ are globally Lipschitz continuous. Let $E = \mathbb{R}^n \times C([0, T], \mathbb{R}^n)$ and $Y = \mathbb{R}^{lK}$. Then $\mu^y \ll \mu_0$ with $d\mu^y/d\mu_0 \propto \exp(-\Phi)$ where Φ is given by (29.26). The map Φ satisfies Assumptions 29.1 and $y \mapsto \mu^y$ is locally Lipschitz continuous in the total variation metric.*

Proof. Note that $\mu_0(\mathbb{R}^n \times C([0, T], \mathbb{R}^n)) = 1$, because Wiener measure charges continuous functions with probability 1. Assumption 29.1(1) follows with $M = \varepsilon = 0$. To establish (2) we note that, for $0 \le t \le T$,

$$|v(t)| \le |u| + \int_0^t (\alpha + \beta|v(s)|) \, ds + |W(t)|$$

$$\le |u| + \int_0^t (\alpha + \beta|v(s)|) \, ds + \|W\|_{C([0,T],\mathbb{R}^n)}.$$

Application of the Gronwall inequality shows that

$$\|v(t)\| \le C(t)\left(\|W\|_{C([0,T],\mathbb{R}^n)} + |u|\right).$$

Since g is polynomially bounded we have

$$|\mathcal{G}(u, W)| \le C\left(\|W\|_{C([0,T],\mathbb{R}^n)} + |u|\right)$$

and (2) follows. To establish (3) it suffices to show that $\mathcal{G}: \mathbb{R}^n \times C([0, T], \mathbb{R}^n) \to \mathbb{R}^{lK}$ is continuous; this follows from the stated hypotheses on f and g since the mapping $(u, W) \in \mathbb{R}^n \times C([0, T], \mathbb{R}^n) \to v \in C([0, T], \mathbb{R}^n)$ is continuous, as may be shown by a Gronwall argument, similar to that used to establish (2). Assumption 29.1(4) follows from the fact that $\Phi(x; \cdot)$ is quadratic and the bound on $\mathcal{G}$ derived to establish (2). □

Again, higher-order differentiability is obtained in a straightforward manner:

Theorem 29.11 *For $k > 0$, if both f and g are C^k, then Φ is also C^k.*

We have expressed the posterior measure as a measure on the initial condition u for v and the driving noise W. However, one can argue that it is natural to take the alternative approach of making direct inference about $\{v(t)\}_{t=0}^T$ rather than indirectly through (u, W). We illustrate how this may be done. To define a prior, we first let μ_0 denote the Gaussian measure on $L^2([0, T], \mathbb{R}^n)$ defined by the equation

$$\frac{dv}{dt} = \sqrt{\Gamma}\frac{dW}{dt}, \quad u \sim \mathcal{N}(m_0, C_0).$$

By the Girsanov theorem, the law of the solution v to (29.23), with $u \sim \mathcal{N}(m_0, C_0)$, yields a measure ν_0 on $L^2([0, T], \mathbb{R}^n)$ which has Radon–Nikodym derivative

$$\frac{d\nu_0}{d\mu_0}(v) = \exp\left(-\frac{1}{2}\int_0^T |f(v)|_\Gamma^2 \, dt + \int_0^T \langle f(v), dv \rangle_\Gamma\right), \tag{29.27}$$

where the second integral is an Itō stochastic integral. The data is again assumed to be of the form (29.5). We have

$$\mathbb{P}(dy|v) \propto \exp\left(-\frac{1}{2}|y - \mathcal{G}(v)|_B^2\right) dy,$$

where $\mathcal{G}(v) = (g(v(t_1)), \ldots, g(v(t_K)))$ and B is the correlation in the noise. Here $\mathcal{G}$ is a map from $L^2([0, T], \mathbb{R}^n)$ to $\mathbb{R}^{lK}$.

Thus we may define a probability measure on $\nu(dv, dy)$ on $L^2([0, T], \mathbb{R}^n) \times \mathbb{R}^{lK}$ given by $\mathbb{P}(y|v)\nu_0(dv)dy$. Since v and y are independent under $\nu_0(dv)dy$, Theorem A.22 shows that the posterior probability measure is defined by

$$\frac{d\mu^y}{d\nu_0}(v) \propto \exp\left(-\widehat{\Phi}(v; y)\right), \quad \widehat{\Phi}(v; y) = \frac{1}{2}|y - \mathcal{G}(v)|_B^2. \tag{29.28}$$

This expresses the posterior measure in terms of a non-Gaussian prior (for the pathspace of a non-Gaussian SDE with Gaussian initial data).

Theorem 29.12 *Assume that $f: \mathbb{R}^n \to \mathbb{R}^n$ and $g: \mathbb{R}^n \to \mathbb{R}^l$ are globally Lipschitz continuous. Then $\mu^y \ll \nu_0$ with $d\mu^y/d\nu_0 \propto \exp(-\widehat{\Phi})$ where $\widehat{\Phi}$ is given by (29.28). The map $\widehat{\Phi}$ satisfies Assumptions 29.1 and $y \mapsto \mu^y$ is locally Lipschitz continuous in the total variation metric.*

Proof. Note that the reference measure ν_0 is not Gaussian. Thus, by Remark 29.1, we need to make sure that ν_0 has Gaussian tails. This follows immediately from the fact that the solution map $(u, W) \mapsto v$ to the model equations (29.24) is globally Lipschitz continuous from $\mathbb{R}^n \times C([0, T], \mathbb{R}^n)$ into $C([0, T], \mathbb{R}^n)$. As the push-forward of a Gaussian measure under a Lipschitz continuous map, ν_0 therefore has Gaussian tails.

The function

$$\widehat{\Phi}(v; y) = \frac{1}{2}|(y - \mathcal{G}(v))|_B^2$$

is obviously bounded from below. It is furthermore locally Lipschitz continuous in both y and v, since the solution map $\mathcal{G}$ is Lipschitz continuous from $C([0, T], \mathbb{R}^n)$ into $\mathbb{R}^{lK}$. □

Obtaining differentiability results on the density with respect to the Gaussian prior μ_0 is much more tricky, because of the appearance of the stochastic integral in (29.27). We return to this topic below, after making the following observation. It is frequently of interest to express the target measure as change of measure from a Gaussian, for example to implement sampling algorithms as in Section 29.5. This may be achieved by the Girsanov theorem: by the properties of change of measure we have

$$\frac{d\mu^y}{d\mu_0}(v) = \frac{d\mu^y}{dv_0}(v) \times \frac{dv_0}{d\mu_0}(v)$$

$$\propto \exp\left(-\frac{1}{2}|(y - \mathcal{G}(v))|_B^2\right) \exp\left(-\frac{1}{2}\int_0^T \left(|f(v)|_\Gamma^2 dt - 2\langle f(v), dv\rangle_\Gamma\right)\right).$$

Thus

$$\frac{d\mu^y}{d\mu_0}(v) \propto \exp\left(-\Phi(v; y)\right)$$

$$\Phi(v; y) = \frac{1}{2}|y - \mathcal{G}(v)|_B^2 + \frac{1}{2}\int_0^T \left(|f(v)|_\Gamma^2 dt - 2\langle f(v), dv\rangle_\Gamma\right). \quad (29.29)$$

There is a naturally arising case where it is possible to study differentiability: when f has a gradient structure. Specifically, if $f = -\Gamma \nabla F$, then Itô's formula yields

$$dF(v) = -\langle f(v), dv\rangle_\Gamma + \frac{1}{2}\text{Tr}\left(\Gamma D^2 F\right) dt.$$

Substituting this into (29.29) yields the expression

$$\Phi(v; y) = \frac{1}{2}|(y - \mathcal{G}(v))|_B^2 + \frac{1}{2}\int_0^T \left(|f(v)|_\Gamma^2 - \text{Tr}\left(\Gamma D^2 F\right)\right) dt$$
$$+ F(v(T)) - F(v(0)). \quad (29.30)$$

We thus obtain the following result:

Theorem 29.13 *For $k \geq 1$, if $g \in C^k$, f is a gradient, and $f \in C^{k+1}$, then $\Phi: C([0, T], \mathbb{R}^n) \times \mathbb{R}^{lK} \to \mathbb{R}$ is C^k.*

29.3.6 White model error, continuous white observational noise

It is interesting to consider the preceding problem in the limiting case where the observation is in continuous time. Specifically we consider underlying stochastic dynamics governed by (29.23) with observations given by (29.13). We assume that $v(0) \sim \mathcal{N}(m_0, C_0)$ and hence obtain the prior model equation for v, together with the equation for the continuous-time observation y, in the form:

$$\frac{dv}{dt} = f(v) + \sqrt{\Gamma}\frac{dW_1}{dt}, \quad v(0) \sim \mathcal{N}(u_0, \mathcal{C}_0), \tag{29.31a}$$

$$\frac{dy}{dt} = g(v) + \sqrt{\Sigma}\frac{dW_2}{dt}, \quad y(0) = 0. \tag{29.31b}$$

Here $v \in \mathbb{R}^n$, $y \in \mathbb{R}^\ell$, $f: \mathbb{R}^n \to \mathbb{R}^n$ and $g: \mathbb{R}^n \to \mathbb{R}^\ell$. Furthermore, $\Gamma \in \mathbb{R}^{n \times n}$ and $\Sigma \in \mathbb{R}^{\ell \times \ell}$ are assumed positive-definite. The Brownian motions W_1, W_2 are assumed independent.

Our aim is to find the probability distribution for $v \in C([0, T], \mathbb{R}^n)$ given $y \in C([0, T], \mathbb{R}^\ell)$. This is a classical problem in continuous-time signal processing for SDEs, known as the *smoothing problem*. This differs from the *filtering problem* where the aim is to find a time-indexed family of probability measures v_t on $\mathbb{R}^n$ for $v(t)$ given $y \in C([0, t], \mathbb{R}^\ell)$.

First we consider the unconditioned case. We let $v_0(dv, dy)$ denote the Gaussian measure on $C([0, T], \mathbb{R}^n) \times C([0, T], \mathbb{R}^\ell)$ obtained in the case where f and g are identically zero, and v the same measure when f and g are not zero. By the Girsanov Theorem A.23 we have, assuming that trajectories do not explode,

$$\frac{dv}{dv_0}(v, y) = \exp\left(-\frac{1}{2}\int_0^T |f(v)|_\Gamma^2 dt + |g(v)|_\Sigma^2 dt - 2\langle f(v), dv\rangle_\Gamma - 2\langle g(v), dy\rangle_\Sigma\right). \tag{29.32}$$

Now we consider the measures found by conditioning v on y. Under v_0 the random variables v and y are independent. Thus $v_0(dv|y)$ is simply the Gaussian measure $\mu_0(dv)$ on $C([0, T], \mathbb{R}^n)$ found from the equation

$$\frac{dv}{dt} = \sqrt{\Gamma}\frac{dW_1}{dt}, \quad v(0) \sim \mathcal{N}(m_0, \mathcal{C}_0).$$

Now let $\mu^y(u)$ denote the measure on $C([0, T], \mathbb{R}^n)$ found from $v(u|y)$. Integrating the last integral in (29.32) by parts (we can do this because v and y are independent under v_0 so that no Itō correction appears) and then applying Theorem A.22, we deduce that

$$\frac{d\mu^y}{d\mu_0}(v) \propto \exp(-\Phi(v; y)) \tag{29.33}$$

$$\Phi(v; y) = \frac{1}{2}\int_0^T \left(|f(v)|_\Gamma^2 dt + |g(v)|_\Sigma^2 dt - 2\langle f(v), dv\rangle_\Gamma + 2\langle y, Dg(v)\,dv\rangle_\Sigma\right) + (\langle g(v(0)), y(0)\rangle - \langle g(v(T)), y(T)\rangle).$$

Here, both integrals are stochastic integrals in the sense of Itō and (29.33) is valid for v_0-almost every $y \in C([0, T], \mathbb{R}^\ell)$. We have therefore shown that:

Theorem 29.14 *Assume that equations (29.31a) and (29.31b) have solution on $t \in [0, T]$ which do not explode, almost surely. Then, the family of measures μ^y as defined by equation (29.33) provides the conditional distribution for v given y.*

In this case Assumptions 29.1(1)–(4) do not hold in general. The statement obtained in this case is much weaker than previously: while integration by parts allows us to establish a conditional law for v given y for *every* realisation of the observation process y, we do not obtain Lipschitz continuity of μ^y as a function of y.

One situation where it is possible to establish Assumptions 29.1(1)–(3) for a continuous-time model of the form (29.31) is the particular case when g is linear and f is a gradient of the form $f = -\Gamma \nabla F$. In this case, we may rewrite (29.31) as

$$\frac{dv}{dt} = -\Gamma \nabla F(v) + \sqrt{\Gamma}\frac{dW_1}{dt}, \quad v(0) \sim \mathcal{N}(u_0, C_0), \tag{29.34a}$$

$$\frac{dy}{dt} = Av + \sqrt{\Sigma}\frac{dW_2}{dt}, \quad y(0) = 0. \tag{29.34b}$$

The key to what follows is that we choose a slightly different unconditioned measure from before. We let $v_0(dv, dy)$ denote the Gaussian measure obtained in the case where F is identically zero (but A is *not* identically 0), and denote as before by v the measure obtained when f is not zero. By Girsanov's Theorem A.23 we have

$$\frac{dv}{dv_0}(v, y) = \exp\left(-\frac{1}{2}\int_0^T |\nabla F(v)|_\Gamma^2 dt + 2\langle \nabla F(v), dv\rangle\right).$$

Now we consider the measures found by conditioning v on y. Under v_0 the random variables v and y are now dependent. The Gaussian measure $\mu_0^y := v_0(dv|y)$ does therefore depend on y in this case but only via its mean, not its covariance. An explicit expression for the covariance and the mean of μ_0^y is given in [23, Theorem 4.1]. Now let $\mu^y(dv)$ denote the measure on $C([0, T], \mathbb{R}^n)$ given by $v(dv|y)$.

By applying Theorem A.22, and then integrating by parts (Itô formula) as we did in the previous section, we find that

$$\frac{d\mu^y}{d\mu_0^y}(v) \propto \exp(-\Phi(v)) \tag{29.35a}$$

$$\Phi(v) = \frac{1}{2}\int_0^T \left(|\Gamma \nabla F(v)|_\Gamma^2 dt - \mathrm{Tr}\left(\Gamma D^2 F(v)\right)\right) dt$$
$$+ F(v(T)) - F(v(0)). \tag{29.35b}$$

Note that $\Phi(v)$ does *not* depend on y. In this particular case, the y-dependence comes entirely from the *reference measure*.

Theorem 29.15 *Assume that $F \in C^3(\mathbb{R}^n, \mathbb{R}^+)$ with globally bounded first, second, and third derivatives. Let $E = C([0, T], \mathbb{R}^n)$. Then $\mu^y \ll \mu_0^y$ with $d\mu^y/d\mu_0^y \propto \exp(-\Phi)$ where Φ is given by (29.35b). The map Φ satisfies Assumptions 29.1(1)–(3), with y dependence removed, and $y \mapsto \mu^y$ is locally Lipschitz continuous from*

$C([0, T], \mathbb{R}^\ell)$ into the space of probability measures on E endowed with the total variation metric.

Proof. We denote the Gaussian measures μ_0^y and $\mu_0^{y'}$ by $\mathcal{N}(m, C_0)$ and $\mathcal{N}(m', C_0)$, respectively and we use μ_0 for the measure μ_0^y with $y = 0$. All integrals are over the whole of E unless specified otherwise. The covariance operator of μ_0^y, which we denote by C_0, does not depend on y, and is the resolvent of a second order differential operator. Thus the Cameron–Martin space for the reference measure is $H^1(0, T; \mathbb{R}^\ell)$. It follows from the results in Section 29.4 of [23] that, for almost every observation y, the mean of the Kalman–Bucy smoother belongs to the Sobolev space $H^{3/2-\varepsilon}$ for any $\varepsilon > 0$. Even better, the map $y \mapsto m$ is continuous from $C([0, T], \mathbb{R}^\ell)$ into H^1, so that there exists a constant C satisfying $|m - m'|_{C_0} \leq C\|y - y'\|_E$. Hence μ_0^y and $\mu_0^{y'}$ are equivalent Gaussian measures.

Satisfaction of parts (1)–(3) of Assumptions 29.1 follow from the Definition (29.35b) of $\Phi(v)$. In particular there is $C > 0$ such that $\Phi(v) \leq C(1 + \|v\|)$. We first note that the normalization constants $Z(y)$ and $Z(y')$ are bounded from below by a constant depending on r. To see this, choose any $r > 0$ so that $\|m\|_E, \|m'\|_E \leq r$; this is possible since the Cameron–Martin space is necessarily contained in E. Then note that, for $\|u\|_E \leq 2r$, we have

$$\exp(-\Phi(v)) \geq \exp(-C(1 + 2r)) := K(r).$$

Note also that

$$\{u : \|u\|_E \leq r\} \subseteq \{u : \|u + m\|_E \leq 2r\}.$$

Thus

$$Z(y) = \int \exp(-\Phi(v)) \, d\mu_0^y(v) \geq \int_{\{\|v\|_E \leq 2r\}} K(r) d\mu_0^y(v)$$
$$\geq \int_{\{\|u+m\|_E \leq 2r\}} K(r) d\mu_0(u) \geq \int_{\{\|u\|_E \leq r\}} K(r) d\mu_0(u)$$
$$= K(r)\mu_0\{\|u\|_E \leq r\}.$$

Since $\mu_0(E) = 1$ and μ_0 is Gaussian, any ball in E will have positive measure. Hence this lower bound is positive for any $r > 0$. Clearly $Z(y')$ satisfies the same lower bound.

Next we bound the difference between the normalization constants. Because μ_0^y and $\mu_0^{y'}$ are equivalent Gaussian measures it follows that

$$|Z(y) - Z(y')| \leq \int \exp(-\Phi(v)) \left|1 - \frac{d\mu_0^{y'}}{d\mu_0^y}(v)\right| d\mu_0^y(v)$$
$$\leq \int \exp(\varepsilon\|v\|_E^2 - M) \left|1 - \frac{d\mu_0^{y'}}{d\mu_0^y}(v)\right| d\mu_0^y(v)$$

$$\leq \left(\int \exp\left(2\varepsilon \|v\|_E^2 - 2M\right) d\mu_0^y(v) \right)^{\frac{1}{2}} \left(\int \left| 1 - \frac{d\mu_0^{y'}}{d\mu_0^y}(v) \right|^2 d\mu_0^y(v) \right)^{\frac{1}{2}}$$

$$\leq C \left(\int \left(\frac{d\mu_0^{y'}}{d\mu_0^y}(v) \right)^2 d\mu_0^y(v) - 1 \right)^{\frac{1}{2}}.$$

From the properties of Gaussian measures we have the identity

$$\frac{d\mu_0^{y'}}{d\mu_0^y}(v) = \exp\left(\langle (m' - m), (v - m) \rangle_{C_0} - \frac{1}{2}|m' - m|_{C_0}^2 \right),$$

so that

$$\int \left(\frac{d\mu_0^{y'}}{d\mu_0^y}(v) \right)^2 d\mu_0^y(v) = \int \exp\left(2\langle m' - m, v - m \rangle_{C_0} - |m' - m|_{C_0}^2 \right) d\mu_0^y(v)$$

$$= \exp\left(|m - m'|_{C_0}^2 \right) \int \exp\left(\langle 2(m' - m), v - m \rangle_{C_0} - \frac{1}{2}|2(m' - m)|_{C_0}^2 \right) d\mu_0^y(v).$$

But the last integral corresponds to an integral of 1 against a Gaussian measure with mean $m + 2(m' - m) = 2m' - m$ and covariance operator C_0 and is hence itself 1. It follows that

$$|Z(y) - Z(y')| \leq C \left(\exp\left(|m - m'|_{C_0}^2 \right) - 1 \right)^{\frac{1}{2}}. \tag{29.36}$$

Again using the fact that μ_0^y and $\mu_0^{y'}$ are equivalent Gaussian measures it follows that

$$\|\mu^y - \mu^{y'}\|_{TV} = \int \left| \frac{d\mu^y}{d\mu_0^y}(v) - \frac{d\mu^{y'}}{d\mu_0^{y'}}(v) \frac{d\mu_0^{y'}}{d\mu_0^y}(v) \right| d\mu_0^y(v)$$

$$\leq \int \exp(-\Phi(v)) \left| Z(y)^{-1} - Z(y')^{-1} \frac{d\mu_0^{y'}}{d\mu_0^y}(v) \right| d\mu_0^y(v)$$

$$\leq \int C \exp(-\Phi(v)) \left| 1 - \frac{d\mu_0^{y'}}{d\mu_0^y}(v) \right| d\mu_0^y(v)$$

$$+ \int \exp(-\Phi(v)) \left| Z(y)^{-1} - Z(y')^{-1} \right| d\mu_0^{y'}(v)$$

$$:= I_1 + I_2.$$

Using the bound from below on $Z(y)$ and $Z(y')$, together with the Fernique theorem and the bound (29.36), we deduce that

$$I_2 \leq C \left(\exp\left(|m - m'|_{C_0}^2 \right) - 1 \right)^{\frac{1}{2}}. \tag{29.37}$$

It remains to bound I_1 similarly. We have

$$I_1 \le C \int \exp\left(\varepsilon \|v\|_E^2 - M\right) \left|1 - \frac{d\mu_0^{y'}}{d\mu_0^y}(v)\right| d\mu_0^y(v)$$

$$\le C \left(\int \exp\left(2\varepsilon \|v\|_E^2 - 2M\right) d\mu_0^y(v)\right)^{\frac{1}{2}} \left(\int \left|1 - \frac{d\mu_0^{y'}}{d\mu_0^y}(v)\right|^2 d\mu_0^y(v)\right)^{\frac{1}{2}}$$

$$\le C \left(\int \left(\frac{d\mu_0^{y'}}{d\mu_0^y}(v)\right)^2 d\mu_0^y(v) - 1\right)^{\frac{1}{2}}.$$

Thus, up to a different constant C, the integral I_1 satisfies the same bound (29.37) as I_2. It follows that

$$\|\mu^y - \mu^{y'}\|_{\mathrm{TV}} \le C \left(\exp\left(|m - m'|_{C_0}^2\right) - 1\right)^{\frac{1}{2}} \le C \left(\exp\left(C\|y - y'\|_E^2\right) - 1\right)^{\frac{1}{2}},$$

and the desired result follows. □

Theorem 29.16 *Let $k \ge 1$. Assume that, further to satisfying the assumptions of Theorem 29.15, $F \in C^{k+2}(\mathbb{R}^n, \mathbb{R}^+)$ Then the potential Φ given by (29.35b) is $C^k(\mathbb{R}^n, \mathbb{R})$.*

29.4 Theme B: Langevin equations

In this section we construct S(P)DEs which are reversible with respect to the measure μ^y introduced in section 29.3. These equations are interesting in their own right; they also form the basis of efficient Metropolis–Hastings methods for sampling μ^y, the topic of Section 29.5. In this context and in the finite-dimensional case, the SDEs are often referred to as *Langevin* equations in the statistics and statistical physics literature [35] and we will use this terminology.

For economy of notation, in this and the next section, we drop explicit reference to the data y and consider the measure

$$\frac{d\mu}{d\mu_0}(x) = Z^{-1} \exp(-\Phi(x)), \qquad (29.38)$$

for some potential function $\Phi(\cdot)$ and normalization constant Z. We will assume that the reference measure μ_0 is a entred Gaussian measure with covariance operator $C_0 \colon E^* \to E$ on some separable Banach space E. Note that this includes the case where μ_0 is not centred, provided that its mean m_0 belongs to the Cameron–Martin space. We may then simply shift coordinates so that the new reference measure has mean zero, and change the potential to $\Phi^m(x) := \Phi(m_0 + x)$. Hence in this section we simply work with (29.1), assume that $\mu_0 = \mathcal{N}(0, C_0)$ and that the four conditions on Φ hold as stated in Assumptions 29.1. We furthermore use the notation $\mathcal{L}$ to denote the inverse C_0^{-1} of the covariance, sometimes called the "precision operator." Note that while C_0

is always a bounded operator, $\mathcal{L}$ is usually an unbounded operator. Additional assumptions on the structure of Φ, the covariance $\mathcal{C}_0$ and the space E will be stated when required.

29.4.1 The finite-dimensional case

Consider the finite-dimensional probability measures on the initial condition u for (29.4) which we have constructed in sections 29.3.1, 29.3.2, and 29.3.3. Recall that the posterior measure is μ given by (29.38). By use of the Fokker–Planck equation it is straightforward to check that, for every strictly positive-definite symmetric matrix $\mathcal{A}$ the following SDE is μ^y invariant

$$dx = -\mathcal{A}\mathcal{L}x\, dt - \mathcal{A} D\Phi(x)\, dt + \sqrt{2\mathcal{A}}\, dW(t). \quad (29.39)$$

Actually one has even more: the Markov semigroup generated by (29.39) consists of operators that are selfadjoint in $L^2(\mathbb{R}^n, \mu^y)$. One of the most general theorems covering this situation is given by [12, 28]:

Theorem 29.17 *Let $\Phi(\cdot)$ belong to $C^2(\mathbb{R}^n)$ and be such that $\exp(-\Phi)$ is integrable with respect to the symmetric Gaussian measure μ_0 with covariance $\mathcal{C}_0 = \mathcal{L}^{-1}$. Then, (29.39) has a unique global strong solution which admits μ^y as an invariant measure. Furthermore, the semigroup*

$$\mathcal{P}_t\varphi(x) = \mathbb{E}\left(\varphi(x(t)) : x(0) = x\right)$$

can be extended to a semigroup consisting of selfadjoint contraction operators on $L^2(E, \mu^y)$.

One approach to approximately sampling from μ given by (29.38) is thus to solve this equation numerically and to rely on ergodicity of the numerical method, as well as approximate preservation of μ under discretization, to obtain samples from μ, see [40]. Typical numerical methods will require draws from $\mathcal{N}(0, \mathcal{A})$ in order to simulate (29.39). In low dimensions, a good choice for $\mathcal{A}$ is $\mathcal{A} = \mathcal{C}_0$ as this equalizes the convergence rates to equilibrium in the case $\Phi \equiv 0$, $\mathcal{C}_0$ is cheap to calculate, and draws from $\mu_0 = \mathcal{N}(0, \mathcal{C}_0)$ are easily made. We refer to this as "preconditioning."

For a given accuracy, there will in general be an optimal stepsize that provides a suitable approximation to the desired i.i.d. sequence at minimal computational cost. Too large stepsizes will result in an inaccurate approximation to (29.39), whereas small stepsizes will require many steps before approximate independence is achieved.

29.4.2 The infinite-dimensional case

The problems in sections 29.3.4, 29.3.5, and 29.3.6 give rise to measures on infinite dimensional spaces. The infinite-dimensional prior reference measure involves either a stationary OU process or Wiener measure. Thus draws from μ_0 (or rather from an approximation thereof) are relatively straightforward to make. Furthermore, in a number of situations, the precision operator $\mathcal{L} = \mathcal{C}_0^{-1}$ is readily characterized as a second-order differential operator, see for example [23].

Even though there exists no infinite-dimensional analogue of Lebesgue measure, we have seen in the previous section that it happens in many situations that the posterior μ^y possesses a density with respect to some fixed Gaussian measure μ_0. It is therefore tempting to carry over (29.39) *mutatis mutandis* to the infinite-dimensional case. It is however much less clear in general what classes of drifts result in (29.39) being a well-posed stochastic PDE (or infinite-dimensional SDE) and, if it is well-posed, whether μ^y is indeed an invariant measure for it. The remainder of this section is devoted to a survey of some rigorous results that have been obtained in this direction.

Remark 29.4 In principle, some of these questions could be answered by invoking the theory of symmetric Dirichlet forms, as described in [37] or [17]. However, we stay away from this course for two reasons. First, it involves a heavy technical machinery that does not seem to be justified in our case since the resulting processes are not that difficult to understand. Second, and more importantly, while the theory of Dirichlet forms allows one to 'easily' construct a large family of μ^y-reversible processes (that contains as special cases the SDEs described in (29.39)), it is more difficult to characterize them as solutions to particular SDEs or SPDEs. Therefore, if we wish to approximate them numerically, we are back to the kind of analysis performed here.

We are going to start with a survey of the results obtained for the Gaussian case (that is when Φ vanishes or is itself quadratic in the x variable), before turning to the nonlinear case.

The Gaussian case

In this section, we consider the situation of a Gaussian measure μ with covariance operator $\mathcal{C}_0$ and mean m on a separable Hilbert space $\mathcal{H}$. At a formal level, the "density" of μ with respect to the (nonexistent, of course) Lebesgue measure is proportional to

$$\exp\left(-\frac{1}{2}\langle x - m, \mathcal{C}^{-1}(x - m)\rangle\right),$$

so that one would expect the evolution equation

$$dx = \mathcal{L}m\,dt - \mathcal{L}x\,dt + \sqrt{2}\,dW(t), \tag{29.40}$$

where, recall, we set $\mathcal{L} = \mathcal{C}_0^{-1}$, to have μ as its invariant measure. Since, if $\mathcal{H}$ is infinite-dimensional, $\mathcal{L}$ is always an unbounded operator, it is not clear a priori how to interpret solutions to (29.40). The traditional way of interpreting (29.40) is to solve it by the variation of constants formula and to *define* the solution to (29.40) as being the process given by

$$x(t) = S(t)x_0 + (1 - S(t))m + \sqrt{2}\int_0^t S(t-s)\,dW(s),$$

(here $S(t)$ denotes the semigroup on $\mathcal{H}$ generated by $-\mathcal{L}$; see [19], [33], and [38] for background on semigroups) provided that the stochastic integral appearing on the right-hand side takes values in $\mathcal{H}$.

This turns out to be always the case in the situation at hand. Furthermore, one has the stronger statement that this process is also the unique weak solution to (29.40) in the sense that it is the only $\mathcal{H}$-valued process such that the identity

$$d\langle x(t), h\rangle = \langle \mathcal{L}h, m - x(t)\rangle\,dt + \sqrt{2}\langle h, dW(t)\rangle, \qquad (29.41)$$

holds for every h in the domain of $\mathcal{L}$. Combining the results from [24] and [14], one obtains that:

Lemma 29.1 *Let $\mathcal{L}$ and μ be as above. Then the evolution equation (29.40) has continuous $\mathcal{H}$-valued mild solutions. Furthermore, it has μ as its unique invariant measure and there exists a constant K such that for every initial condition $x_0 \in \mathcal{H}$ one has*

$$\left\|\mathrm{Law}\,(x(t)) - \mu\right\|_{\mathrm{TV}} \leq K\left(1 + \|x_0 - m\|_\mathcal{H}\right)\exp\left(-\|C_0\|_{\mathcal{H}\to\mathcal{H}}^{-1}t\right),$$

where $\|\cdot\|_{\mathrm{TV}}$ denotes the total variation distance between probability measures.

Remark 29.5 The convergence in total variation obtained in Lemma 29.1 is very strong and does *not* hold in general if one replaces (29.40) by its 'preconditioned' version as in (29.39). For example, in the particular case

$$dx = m\,dt - x\,dt + \sqrt{2C_0}\,dW(t), \qquad (29.42)$$

it is known that convergence in total variation does not hold, unless x_0 belongs to the Cameron–Martin space of μ, that is unless $\|\mathcal{L}^{1/2}x_0\| < \infty$. However, one does still have convergence of arbitrary solutions to (29.42) to μ in the p-Wasserstein distance for arbitrary p.

The nonlinear case

This case is much less straightforward than the Gaussian case and we will not give a complete treatment here. One problem that often occurs in the infinite-dimensional case is that Φ is naturally defined on a Banach space E (typically the space of continuous functions) rather than on a Hilbert space $\mathcal{H}$. It is then tempting to work with a scale of spaces

$$E \hookrightarrow \mathcal{H} = \mathcal{H}^* \hookrightarrow E^*,$$

where all inclusions are dense. We are going to make the following assumptions for the precision operator $\mathcal{L}$ of our reference Gaussian measure μ_0:

(A1) The semigroup $S(t) = e^{-\mathcal{L}t}$ generated by $\mathcal{L}$ on $\mathcal{H}$ can be restricted to a strongly continuous semigroup of contraction operators on E.

(A2) There exists $\alpha \in (0, 1/2)$ such that $\mathcal{D}(\mathcal{L}^\alpha) \subset E$ (densely), $\mathcal{L}^{-2\alpha}$ is trace class in $\mathcal{H}$, and the measure $\mathcal{N}(0, \mathcal{L}^{-2\alpha})$ is concentrated on E.

The first assumption ensures that E is a 'good choice' of a space to work with. The second assumption is a slight strengthening of the statement that $\mu_0(E) = 1$, since the statement with $\alpha = \frac{1}{2}$ is implied by $\mu_0(E) = 1$. Regarding the density $\Phi \colon E \to \mathbb{R}$, we make the following assumptions:

(A3) For every $\varepsilon > 0$, there exists $M > 0$ such that $\Phi(x) \geq M - \varepsilon \|x\|_E^2$ for every $x \in E$.

(A4) The function $\Phi \colon E \to \mathbb{R}$ is twice Fréchet differentiable and its derivatives are polynomially bounded.

(A5) There exists a sequence of Fréchet differentiable functions $F_n \colon E \to E$ such that

$$\lim_{n \to \infty} \left\| \mathcal{L}^{-\alpha} (F_n(x) - D\Phi(x)) \right\|_{\mathcal{H}} = 0$$

for all $x \in E$. For every $C > 0$ there exists a $K > 0$ such that for all $x \in E$ with $\|x\|_E \leq C$ and all $n \in \mathbb{N}$ we have $\|\mathcal{L}^{-\alpha} F_n(x)\|_{\mathcal{H}} \leq K$. Furthermore, there is a $\gamma > 0$, $C > 0$ and $N > 0$ such that the dissipativity bound

$$\langle x^*, F_n(x+y) \rangle \leq -\gamma \|x\|_E \qquad (29.43)$$

holds for every $x^* \in \partial \|x\|_E \subset E^*$ and every $x, y \in E$ with $\|x\|_E \geq C(1 + \|y\|_E)^N$. Here, $\partial \|x\|_E$ denotes the subdifferential of the norm at x (see for example [14]).

Assumption (A3) just makes sure that $\exp(-\Phi)$ is integrable with respect to μ_0. The next assumption (A4) provides a minimum of regularity so that the equation

$$dx = -\mathcal{L} x \, dt - D\Phi(x) \, dt + \sqrt{2} \, dW(t), \qquad (29.44)$$

is well-posed (in its mild formulation). The last condition seems rather complicated, but it should just be thought of as a version of the dissipativity condition

$$\langle x^*, D\Phi(x+y) \rangle \leq -\gamma \|x\|_E$$

that survives approximating $D\Phi$ by E-valued functions. With these conditions at hand, we have the following result from [21]:

Theorem 29.18 *Assume that conditions (A1)–(A5) hold and define the probability measure*

$$\mu(dx) = Z^{-1} \exp(-\Phi(x)) \mu_0(dx),$$

for a suitable normalization constant Z. Then the stochastic PDE (29.44) has a unique continuous E-valued global mild solution for every initial condition $x_0 \in E$. Furthermore, this solution admits μ as its unique invariant probability measure.

Under a very weak additional assumption (essentially, Φ should admit approximations that have bounded support and that are Fréchet differentiable, which is not completely automatic if the norm on E is not differentiable), it is again possible to show that transition probabilities converge to the invariant measure at exponential speed and that the law of large numbers holds.

29.5 Theme C: MCMC methods

In this section we describe a range of effective Metropolis–Hastings-based (see [18, 31]) MCMC methods (see [29, 35]) for sampling the target distributions constructed in Section 29.3. As in the previous section we drop explicit reference to the data y and work with a posterior distribution μ given by (29.38). The methods we describe are motivated by the μ-reversible stochastic evolution equations derived in the previous section. We work with measures μ given by (29.1). We assume that $m_0 = 0$ which can always be achieved by a shift of origin, provided the mean of μ_0 belongs to its Cameron–Martin space. Our aim is to draw samples from the measure μ on E given by (29.38).

The idea of MCMC methods for target μ is to construct a discrete-time Markov chain $\{x_n\}$ on E that has μ as its invariant measure and that has good mixing properties. One can then take as an approximation to i.i.d. samples the sequence $k \mapsto x_{N_0 + k N_1}$ with $k \geq 0$ and N_0, N_1 'sufficiently large'. In order to compute integrals of the form $I = \int f(x) \mu(dx)$ for some test function f, one can then use the fact that, by Birkhoff's ergodic theorem, one has the almost sure identity

$$I = \lim_{N \to \infty} \frac{1}{N} \sum_{k=1}^{N} f(x_k).$$

Metropolis–Hastings methods work by proposing a move from the current state x_k to y from a Markov transition kernel on E, and then accepting or rejecting in a fashion which ensures that the resulting composite Markov chain is μ reversible.

In section 29.5.1 we first explain the idea in finite dimensions, with application to the problems formulated in sections 29.3.1–29.3.3 and, of course, to finite-dimensional approximation of the problems formulated in sections 29.3.4–29.3.6. We focus on the theory related to *random walk* and *Langevin* proposals for these problems, building on the material in the previous section. Then, in section 29.5.2, we generalize these methods to the infinite-dimensional setting.

29.5.1 Metropolis–Hastings in finite dimensions

In finite dimensions the measure μ given by (29.1) on $E = \mathbb{R}^d$ has density π with respect to Lebesgue measure which is given by

$$\pi(x) \propto \exp\left(-\frac{1}{2}|x|^2_{C_0} - \Phi(x)\right).$$

The Metropolis–Hastings method of constructing a μ reversible Markov chain is the following. Fix a Markov transition kernel $P(x, dy)$ with density $q(x, y)dy$. The measure $\mu(dx)P(x, dy)$ on $E \times E$ then has density $\pi(x)q(x, y)$. Define the function

$$a(x, y) = 1 \wedge \frac{\pi(y)q(y, x)}{\pi(x)q(x, y)},$$

where we write $a \wedge b$ for the minimum between a and b.

Assume that x_k is known (start for example with $x_0 = 0$). To determine x_{k+1} draw $y \sim q(x_k, y)$ and proceed as follows:

1. $x_{k+1} = y$ (step accepted) with probability $a(x_k, y)$.
2. $x_{k+1} = x_k$ (step rejected) otherwise.

The resulting Markov chain is π-reversible.

A frequently used class of methods are the *symmetric random walk* proposals where

$$y = x_k + \sqrt{2\Delta t}\xi_k \tag{29.45}$$

where the ξ_k are i.i.d. symmetric random variables (for example $\mathcal{N}(0, I)$) on $\mathbb{R}^d$. This may be viewed as a discretization of the Brownian motion

$$dx = \sqrt{2}dW.$$

As such the proposal contains no information about the target. However, by symmetry,

$$a(x, y) = 1 \wedge \frac{\pi(y)}{\pi(x)}.$$

This has the advantage of being simple to implement.

Typically, the mixing time for a Metropolis–Hastings chain will depend both on the proportion of steps that are rejected and on the variance of $y - x_k$: the higher the number of rejections and the lower the variance, the longer it will take for the Markov chain to explore the whole state space. In general, there is a competition between both effects: steps with a large variance have a high probability of rejections; on the other hand small moves are less likely to be rejected but explore the state space slowly. Roughly speaking the competition between these two effect is measured by the mean square jumping distance of the Markov chain in stationarity. Specifically, if we define

$$S_{d,i} = \mathbb{E}^\mu |x_{k+1,i} - x_{k,i}|^2 \tag{29.46}$$

then this quantity measures the mean square jumping distance in the ith component of the vector x_k. Maximizing this quantity will enhance the mixing of functionals heavily dependent on the ith component of $x \sim \mu$. We will optimize algorithms according to this criterion.

In an attempt to maximize (29.46), proposals which contain information about the target distribution can be useful. A class of proposals which does contain such information arises from discretizing the Langevin SDE (29.39). A linearly implicit Euler discretization gives rise to the following family of proposals:

$$y - x_k = -\Delta t \mathcal{A}\mathcal{L}(\theta y + (1-\theta)x_k) - a\Delta t \mathcal{A} D\Phi(x_k) + \sqrt{2\Delta t \mathcal{A}}\xi_k \tag{29.47}$$

where the ξ_k are i.i.d $\mathcal{N}(0, I)$ random variables on $\mathbb{R}^d$, $\theta \in [0, 1]$ and $a \in \{0, 1\}$. If $a = 0$ the proposal contains information only about the reference measure μ_0, via its precision operator $\mathcal{L}$. If $a = 1$ it contains information about μ itself. Two natural choices for $\mathcal{A}$ are I and $\mathcal{C}_0$.

The formula for the proposal rearranges to give

$$y = (I + \Delta t \theta \mathcal{A}\mathcal{L})^{-1}\left((I - \Delta t(1-\theta)\mathcal{A}\mathcal{L})x_k - a\Delta t \mathcal{A} D\Phi(x) + \sqrt{2\Delta t \mathcal{A}}\xi_k\right). \tag{29.48}$$

In the case $a = 0$ this generalizes the symmetric random walk to allow y to be a more complex linear combination of x_k and ξ_k. When $a = 1$ the proposal also contains information which tends to make proposals which decrease Φ. Roughly speaking we expect proposals with $a = 1$ to explore the state space more rapidly than those with $a = 0$. However there is a cost involved in evaluating $D\Phi$ and the trade-off between cost-per-step and number of steps will be different for different problems.

A natural question of interest for these algorithms is how to choose the time-step Δt. We now study this question in the limit where the state-space dimension $d \to \infty$. We define the norm

$$|x|_s = \left(\sum_{i=i}^{d} i^{2s} x_i^2\right)^{\frac{1}{2}}.$$

We make the following assumptions:

Assumption 29.2 The following hold for the family of reference measures $\mu_0 = \mu_0(d)$, the family of target measures $\mu = \mu(d)$ and their interrelations.

1. There are constants $c^\pm \in (0, \infty)$ such that the eigenvalues $\lambda_{i,d}^2$ of the covariance $\mathcal{C}_0$ satisfy

$$c^- i^{-k} \leq \lambda_{i,d} \leq c^+ i^{-k} \quad \forall 1 \leq i \leq d. \tag{29.49}$$

2. Assumptions 29.1 (1)–(3) hold, generalized to include the case $\varepsilon = 0$, with $E = (\mathbb{R}^d, |\cdot|_s)$, $s < k - \frac{1}{2}$ and with constants independent of dimension d.

We now state three theorems, all proved in [6], which quantify the efficiency of the various proposals described above in the high-dimensional setting, and under the preceding assumptions.

The first theorem shows that, for the symmetric random walk the optimal choice of Δt is of $\mathcal{O}(d^{-(2k+1)})$, giving rise to a maximal mean square jump of the same magnitude.

Theorem 29.19 *Consider the symmetric random walk proposal (29.45). Assume that $\Delta t = \ell^2 d^{-p}$. Then the following dichotomy holds, for any fixed i:*

- *If $p \geq 2k + 1$ then*

$$\liminf_{d \to \infty} d^p S_{d,i} > 0, \quad \limsup_{d \to \infty} d^p S_{d,i} < \infty.$$

- *If $p < 2k + 1$ then*

$$\limsup_{d \to \infty} d^q S_{d,i} = 0$$

for any $q \geq 0$.

The next theorem shows that, for the basic version of the Langevin proposal, the optimal choice of Δt is of $\mathcal{O}(d^{-(2k+1/3)})$, giving rise to a maximal mean square jump of the same magnitude. The improvement in the exponent by $2/3$ comes as the price of evaluating the application of the precision operator $\mathcal{L}$ at each step; the cost of doing this will be problem dependent.

Theorem 29.20 *Consider the proposal (29.48) with $\theta = \alpha = 0$ and $\mathcal{A} = I$. Assume that $\Delta t = \ell^2 d^{-p}$. Then the following dichotomy holds, for any fixed i:*

- *If $p \geq 2k + \frac{1}{3}$ then*

$$\liminf_{d \to \infty} d^p S_{d,i} > 0, \quad \limsup_{d \to \infty} d^p S_{d,i} < \infty.$$

- *If $p < 2k + \frac{1}{3}$ then*

$$\limsup_{d \to \infty} d^q S_{d,i} = 0$$

for any $q \geq 0$.

The final theorem shows that, at the cost of making samples $\sqrt{C_0}\xi_k$ from the prior measure at each step of the algorithm, the optimal choice of $\Delta t = \mathcal{O}(d^{-1/3})$, gives rise to a maximal mean square jump of the same magnitude.

Theorem 29.21 *Consider the proposal (29.48) with $\theta = \alpha = 0$ and $\mathcal{A} = \mathcal{C}$. Assume that $\Delta t = \ell^2 d^{-p}$. Then the following dichotomy holds, for any fixed i:*

- If $\rho \geq \frac{1}{3}$ then

$$\liminf_{d\to\infty} d^\rho S_{d,i} > 0, \quad \limsup_{d\to\infty} d^\rho S_{d,i} < \infty.$$

- If $\rho < \frac{1}{3}$ then

$$\limsup_{d\to\infty} d^q S_{d,i} = 0$$

for any $q \geq 0$.

The previous two theorems, concerning proposals of the form (29.48), concern only the cases where $\theta = \alpha = 0$. It is expected that the scaling results will be identical for $\theta = 0$, $\alpha = 1$. However the choice of θ can make significant differences. In the next section we show how, by working in infinite dimensions, we can in some cases eliminate dimension dependence in Metropolis–Hastings algorithms, by choosing $\theta = \frac{1}{2}$.

29.5.2 Metropolis–Hastings in infinite dimensions

The ideas of the previous section can be generalized to infinite dimensions as follows [41]. Assume that we are given a Polish (i.e. complete, separable, metric) space E (since we want to allow for the possibility of sampling from a measure on a space of paths, E should be thought of as a space of functions in general) and a probability measure μ on E. Assume furthermore that we are given a Markov transition kernel P over E with the property that the measures $\mu(dx) P(x, dy)$ and $\mu(dy) P(y, dx)$ are equivalent so that the quantity

$$\frac{\mu(dy) P(y, dx)}{\mu(dx) P(x, dy)},$$

which should be interpreted as the Radon–Nikodym derivative of the two aforementioned measures evaluated at the point (x, y), is well-defined. With these notations in place, we can construct a new Markov chain in the following way. Assume again that x_k is known and draw a random sample y from the probability distribution $P(x_k, \cdot)$. Now let

$$a(x, y) = 1 \wedge \frac{\mu(dy) P(y, dx)}{\mu(dx) P(x, dy)}. \tag{29.50}$$

The algorithm proceeds as follows

1. $x_{k+1} = y$ (step accepted) with probability $a(x_k, y)$.
2. $x_{k+1} = x_k$ (step rejected) otherwise.

If we denote by Q the transition probabilities of the process x_n, it can be checked that one has

$$Q(x, dy) = c(x)\delta_x(dy) + P(x, dy) \wedge \frac{P(y, dx)}{\mu(dx)} \mu(dy), \qquad (29.51)$$

for some function c that makes Q a Markov transition kernel. If we define a map $\Delta: E \to E^2$ by $\Delta(x) = (x, x)$ and denote by $\Delta^*\mu$ the push-forward of μ by Δ, one can check that (29.51) implies that

$$\mu(dx) Q(x, dy) = \sqrt{c(x)c(y)}\, (\Delta^*\mu)(dx, dy) + \mu(dx) P(x, dy) \wedge \mu(dy) P(y, dx).$$

This expression is symmetric in $x \leftrightarrow y$, so that the Markov kernel Q (or equivalently the Markov chain generated from it) is reversible with respect to the measure μ. In particular, the measure μ is invariant for Q.

Thus key to making this idea work is the construction of proposals for which the measure $\mu(dy) P(y, dx)$ is absolutely continuous with respect to the measure $\mu(dx) P(x, dy)$. We consider this question in the context of (29.48), basing ideas on the paper [7]. Let $P_a(x, dy)$ denote the transition kernel of this proposal. Then define measures η and η_0 by

$$\eta(dx, dy) = \mu(dx) P_a(x, dy)$$

and

$$\eta_0(dx, dy) = \mu_0(dx) P_0(x, dy).$$

It is straightforward to show that $\eta_0(dx, dy) = \eta_0(dy, dx)$ iff $\theta = \frac{1}{2}$. We work with this assumption henceforth as it enables us to define the MCMC method on function space.

Using the fact that $P_a(x, \cdot)$ is absolutely continuous with respect to $P_0(x, \cdot)$ for both $a = 0$ and $a = 1$ we deduce that η is absolutely continuous with respect to η_0 and that, for some $\rho(x, y) = \rho(x, y; a, \mathcal{A})$, we have

$$\frac{d\eta}{d\eta_0}(x, y) \propto \exp(-\rho(x, y)).$$

Thus the acceptance probability for the Metropolis algorithm is

$$a(x, y) = 1 \wedge \exp(\rho(x, y) - \rho(y, x)). \qquad (29.52)$$

For the proposals (29.48) the function $\rho(x, y)$ is given, up to an additive constant which we ignore, by the following expressions:

- for $\mathcal{A} = I$ we have

$$\rho(x, y) = \Phi(x) + \frac{a^2 \Delta t}{4} \|D\Phi(x)\|^2 + \frac{a}{2}\langle D\Phi(x), y - x\rangle$$
$$+ \frac{a\Delta t}{4}\langle D\Phi(x), C_0^{-1}(y + x)\rangle;$$

- for $\mathcal{A} = \mathcal{C}_0$ we have

$$\rho(x, y) = \Phi(x) + \frac{a^2 \Delta t}{4} \left\| \mathcal{C}_0^{1/2} D\Phi(x) \right\|^2 + \frac{a}{2} \langle D\Phi(x), y - x \rangle$$
$$+ \frac{a \Delta t}{4} \langle D\Phi(x), y + x \rangle.$$

The four algorithms defined in this section (two choices for both $a \in \{0, 1\}$ and $\mathcal{A} \in \{I, \mathcal{C}_0\}$) all lead to well-defined Metropolis–Hastings chains on Banach space. Thus they give rise to mean square jumping distances which are bounded independently of dimension d as they are, in particular, nonzero in the infinite-dimensional case.

It is straightforward to prove [8] that, for any $c > 0$, the acceptance probability (29.52) satisfies

$$\mathbb{E}a(x_k, y) \geq \exp(-c) \left(1 - \frac{\mathbb{E}|\rho(x_k, y) - \rho(y, x_k)|}{c} \right).$$

Thus if we can show that

$$\mathbb{E}|\rho(x_k, y) - \rho(y, x_k)| \to 0$$

as $\Delta t \to 0$ then we deduce that we can make the acceptance probability arbitarily close to 1. In the case $a = 0$, proving this may be shown by using Assumption 29.1.

29.6 Discussion and bibliography

There are several useful sources for background material relevant to both the problems studied, and methods developed, in this chapter. A general reference concerned with stochastic modelling is [11]. Several technical tools are required to develop the methods described in this chapter. An exhaustive treatment of Gaussian measures can be found in [4] and moment bounds for SDEs can be found in [30]. The book [13] is an excellent source for material concerned with sequential filtering problems, including the use of particle filters for non-Gaussian problems. The filtering and smoothing problems for SDEs with continuous time observations, as arising in Section 29.3.6, is introduced in [32], and developed in detail in the Gaussian context (f and g linear) giving rise to the Kalman–Bucy filter and smoother. This method uses an approach based on first filtering ($0 \to T$), and then reversing the process ($T \to 0$) to incorporate data from time $t > s$ into the probability distribution at me s. Good sources of signal processing problems arising from data assimilation are [16] and the volume [25]; these problems have motivated a lot of our research in this general area. Finally note that signal processing may be viewed as an inverse problem to find a signal from partial, noisy, observations. The Bayesian approach to inverse problems in general is discussed in [26].

In Theme A we considered a range of differing problems arising in signal processing, constructing and deriving properties of the posterior distribution. The posterior distributions constructed in Sections 29.3.2 and 29.3.3 can both be viewed as *parameter estimation* problems for SDEs. They have particular structure, inherited from the way in which the parameter u enters the expression $\varphi^t(u)$ appearing in the SDE for y. The general subject of parameter estimation for SDEs is considered in [5, 27]. Incorporating discrete-time data into a continuous time model, as undertaken in Sections 29.3.4 and 29.3.5, is studied in [1, 22]. Carrying out this program and, at the same time estimating parameters in the dynamical model, is discussed in [36], in a non-Bayesian setting. The relationship between the coloured noise model appearing in Section 29.3.4, and the white noise model appearing in 29.3.5, in the limit $\delta \to 0$, is part of the theory of homogenization for stochastic processes; see [3, 34]. The filtering and smoothing problems for SDEs with continuous time observations, as arising in Section 29.3.6, is introduced in [32], as mentioned above. In the Gaussian case the mean is characterized by the solution of a two-point value problem, defined through inversion of the precision operator. The approach to smoothing outlined in [32] corresponds to a continuous time analogue of LU factorization, here for the inverse of the covariance operator, facilitating its action on the data to compute the mean. The particular formulation of the smoothing problem described here is developed in [1].

In Theme B we studied the derivation of Langevin equations (stochastic partial differential equations) which are invariant with respect to a given invariant measure. This is straightforward in finite dimensions, but is an emerging subject area in infinite dimensions. The idea is developed in a fairly general setting in [21], building on the Gaussian case described in [23]. The first use of the Langevin equation to solve signal processing problems may be found in [39] and further applications may be found in [20, 2]. On the theoretical side many open questions remain concerning the derivation of Langevin equations. In particular the paper [21] deals with elliptic diffusions with gradient drift and additive noise. An initial analysis of a particular hypoelliptic problem may be found in [20]. Questions relating to the derivation of Langevin equations for nongradient drifts, and for multiplicative noise, remain open.

Theme C is concerned with the design and analysis of effective MCMC methods in high dimension, motivated by the approximation of infinite dimensionsal problems. This subject is overviewed in the review articles [8, 9], and full details of the analysis and application of the methods may be found in [6, 7].

Appendix A: Some results from probability

In this appendix, we collect miscellaneous results from stochastic analysis that were used in this chapter. Throughout the chapter we use the following notation: given a Hilbert space $(H, \langle \cdot, \cdot \rangle, \|\cdot\|)$, for any positive-definite C we define the second inner-product and norm

$$\langle a, b \rangle_C = \langle a, C^{-1} b \rangle, \quad \|a\|_C^2 = \langle a, a \rangle_C.$$

A.1 Conditional probabilities

Throughout Theme A of this chapter we will be generalizing Bayes formula to an infinite-dimensional setting. There are two components to this: Bayes formula in finite dimensions, and then the generalization to the Hilbert space setting. We start in finite dimensions. Assume that we are given a random variable u on $\mathbb{R}^d$ about which we have some *prior* information in terms of a probability distribution $\mathbf{P}(u)$. Imagine that we now define a random variable y on $\mathbb{R}^l$, which depends upon u, and for which we have the probability distribution of y given u, namely $\mathbf{P}(y|u)$. By the elementary rules of probability we have

$$\mathbf{P}(u|y) = \frac{1}{\mathbf{P}(y)} \mathbf{P}(u \cap y),$$
$$\mathbf{P}(y|u) = \frac{1}{\mathbf{P}(u)} \mathbf{P}(u \cap y).$$

Combining these two formulae shows that the *posterior* probability distribution for u, given a single observation of y, is given by Bayes formula

$$\mathbf{P}(u|y) = \frac{1}{\mathbf{P}(y)} \mathbf{P}(y|u) \mathbf{P}(u). \tag{29.53}$$

In this chapter there are many instances where we are interested in conditioning probability measures on function space. In this context the following theorem will be of central importance in constructing the appropriate generalization of Bayes formula.

Theorem A.22 *Let μ, ν be probability measures on $S \times T$ where $(S, \mathcal{A})$ and $(T, \mathcal{B})$ are measurable spaces and let $x \colon S \times T \to S$ and $y \colon S \times T \to T$ be the canonical projections. Assume that μ has a density φ w.r.t. ν and that the conditional distribution $\nu_{x|y}$ exists. Then the conditional distribution $\mu_{x|y}$ exists and is given by*

$$\frac{d\mu_{x|y}}{d\nu_{x|y}}(x) = \begin{cases} \frac{1}{c(y)} \varphi(x, y), & \text{if } c(y) > 0, \text{ and} \\ 1 & \text{else,} \end{cases} \tag{29.54}$$

with $c(y) = \int_S \varphi(x, y) \, d\nu_{x|y}(x)$ for all $y \in T$.

A.2 A version of Girsanov's theorem

SDEs which do not have the same diffusion coefficient generate measures which are mutually singular on pathspace; the same is true of SDEs starting from different deterministic initial conditions. However, if these two possibilities are ruled out, then two different SDEs generate measures which are absolutely continuous with respect to one another. The Girsanov formula provides an explicit expression for the Radon–Nikodym derivative between two such measures on $\mathcal{H} = L^2\left([0, T], \mathbb{R}^d\right)$.

Consider the SDE

$$\frac{dv}{dt} = A(t)v + h(v, t) + \gamma(v, t)\frac{dW}{dt}, \quad v(0) = u. \tag{29.55}$$

and the same equation with the function h set to zero, namely

$$\frac{dv}{dt} = A(t)v + \gamma(v, t)\frac{dW}{dt}, \quad v(0) = v_0. \tag{29.56}$$

The measures generated by these two equations are absolutely continuous. Define $\Gamma(\cdot, t) = \gamma(\cdot, t)\gamma(\cdot, t)^T$. We then have the following version of Girsanov's theorem, that can be found in [15]:

Theorem A.23 *Assume that both equations (29.55) and (29.56) have solutions on $t \in [0, T]$ which do not explode almost surely. Then the measures μ and μ_0 on $\mathcal{H}$, generated by the two equations (29.55) and (29.56) respectively, are equivalent with Radon–Nikodym derivative*

$$\frac{d\mu}{d\mu_0}(v) = \exp\left(-\int_0^T \frac{1}{2}\|h(v, t)\|^2_{\Gamma(v, t)}dt - \langle h(v, t), dv - A(t)vdt\rangle_{\Gamma(v(t))}\right).$$

Acknowledgements

The authors gratefully acknowledge the support of EPSRC grant EP/E002269/1.

References

[1] A. Apte, M. Hairer, A. M. Stuart, and J. Voss. Sampling the posterior: An approach to non-Gaussian data assimilation. *Physica D: Nonlinear Phenomena*, 230:50–64, 2007.

[2] A. Apte, C. K. R. T. Jones, A. M. Stuart, and J. Voss. Data assimilation: Mathematical and statistical perspectives. *International Journal for Numerical Methods in Fluids*, 56:1033–46, 2008.

[3] A. Bensoussan, J.-L. Lions, and G. Papanicolaou. *Asymptotic Analysis for Periodic Structures*, vol. 5 of *Studies in Mathematics and its Applications*. North Holland, 1978.

[4] Vladimir I. Bogachev. *Gaussian Measures*, volume 62 of *Mathematical Surveys and Monographs*. American Mathematical Society, 1998.

[5] I. V. Basawa and B. L. S. Prakasa Rao. *Statistical Inference for Stochastic Processes*. Academic Press, 1980.

[6] A. Beskos, G. O. Roberts, and A. M. Stuart. Scalings for local Metropolis–Hastings chains on non-product targets. To appear in *Ann. Appl. Prob.*

[7] A. Beskos, G. O. Roberts, A. M. Stuart, and J. Voss. MCMC methods for diffusion bridges. *Stochastic Dynamics*, 8(3):319–50, Sept. 2008.

[8] A. Beskos and A. M. Stuart. MCMC methods for sampling function space. In *Proceedings of the International Congress of Industrial and Applied Mathematicians (Zurich, 2007)*, 2009.

[9] A. Beskos and A. M. Stuart. Computational complexity of Metropolis-Hastings methods in high dimensions. In *Proceedings of MCQMC08, 2008*, 2010.

[10] S. L. Cotter, M. Dashti, J. C. Robinson, and A. M. Stuart. Data assimilation problems in fluid mechanics: Bayesian formulation on function space. Submitted to *J. Inverse Problems*.

[11] A. J. Chorin and O. H. Hald. *Stochastic Tools in Mathematics and Science*, vol. 1 of *Surveys and Tutorials in the Applied Mathematical Sciences*. Springer, 2006.

[12] P. R. Chernoff. Essential self-adjointness of powers of generators of hyperbolic equations. *J. Functional Analysis*, 12:401–14, 1973.

[13] A. Doucet, N. de Freitas, and N. Gordon, eds. *Sequential Monte Carlo Methods in Practice*. Springer, 2001.

[14] G. Da Prato and J. Zabczyk. *Stochastic Equations in Infinite Dimensions*, volume 44 of *Encyclopedia of Mathematics and its Applications*. Cambridge University Press, 1992.

[15] K. D. Elworthy. *Stochastic Differential Equations on Manifolds*, volume 70 of *London Mathematical Society Lecture Note Series*. Cambridge University Press, 1982.

[16] G. Evensen. *Data Assimilation: The Ensemble Kalman Filter*. Springer, 2006.

[17] M. Fukushima, Y. Oshima, and M. Takeda. *Dirichlet forms and symmetric Markov processes*, vol. 19 of *de Gruyter Studies in Mathematics*. Walter de Gruyter & Co., Berlin, 1994.

[18] W. K. Hastings. Monte Carlo sampling methods using Markov chains and their applications. *Biometrika*, 57(1):97–109, 1970.

[19] Daniel Henry. *Geometric Theory of Semilinear Parabolic Equations*, vol. 840 of *Lecture Notes in Mathematics*. Springer, 1981.

[20] M. Hairer, A. M. Stuart, and J. Voss. Sampling conditioned hypoelliptic diffusions. The *Annals of Applied Probability* (2010), to appear.

[21] M. Hairer, A. M. Stuart, and J. Voss. Analysis of SPDEs arising in path sampling, part II: The nonlinear case. *Annals of Applied Probability*, 17:1657–1706, 2007.

[22] M. Hairer, A. M. Stuart, and J. Voss. Sampling conditioned diffusions. In *Trends in Stochastic Analysis*, vol. 353 of *London Mathematical Society Lecture Note Series*. Cambridge University Press, 2009.

[23] M. Hairer, A. M. Stuart, J. Voss, and P. Wiberg. Analysis of SPDEs arising in path sampling, part I: The Gaussian case. *Communications in Mathematical Sciences*, 3(4):587–603, 2005.

[24] I. Iscoe, M. B. Marcus, D. McDonald, M. Talagrand, and J. Zinn. Continuity of l^2-valued Ornstein–Uhlenbeck processes. *Ann. Probab.*, 18(1):68–84, 1990.

[25] C. K. R. T. Jones and K. Ide, eds. *Data Assimilation*, vol. 230 of *Physica D: Nonlinear Phenomena*. Elsevier, 2007.

[26] Jari Kaipio and Erkki Somersalo. *Statistical and Computational Inverse Problems*, vol. 160 of *Applied Mathematical Sciences*. Springer, 2005.

[27] Yury A. Kutoyants. *Statistical Inference for Ergodic Diffusion Processes*. Springer Series in Statistics. Springer, 2004.

[28] Xue-Mei Li. *Stochastic Flows on Noncompact Manifolds*. PhD thesis, University of Warwick, 1992.

[29] Jun S. Liu. *Monte Carlo Strategies in Scientific Computing*. Springer Series in Statistics. Springer, 2001.
[30] Xuerong Mao. *Stochastic Differential Equations and Their Applications*. Horwood Publishing Series in Mathematics & Applications. Horwood Publishing, Chichester, 1997.
[31] N. Metropolis, A. W. Rosenbluth, M. N. Teller, and E. Teller. Equations of state calculations by fast computing machines. *J. Chem. Phys.*, 21:1087–92, 1953.
[32] Bernt Øksendal. *Stochastic Differential Equations*. Universitext. Springer, 6th edn, 2003. An introduction with applications.
[33] A. Pazy. *Semigroups of Linear Operators and Applications to Partial Differential Equations*, vol. 44 of *Applied Mathematical Sciences*. Springer, 1983.
[34] Grigorios A. Pavliotis and Andrew M. Stuart. *Multiscale Methods*, vol. 53 of *Texts in Applied Mathematics*. Springer, 2008. Averaging and homogenization.
[35] C. P. Robert and G. Casella. *Monte Carlo Statistical Methods*. Springer, 1999.
[36] J. O. Ramsay, G. Hooker, D. Campbell, and J. Cao. Parameter estimation for differential equations: a generalized smoothing approach. *J. R. Stat. Soc. Ser. B Stat. Methodol.*, 69(5):741–96, 2007. With discussions and a reply by the authors.
[37] M. Röckner and Z. M. Ma. *Introduction to the Theory of (Nonsymmetric) Dirichlet Forms*. Springer, 1992.
[38] James C. Robinson. *Infinite-Dimensional Dynamical Systems*. Cambridge Texts in Applied Mathematics. Cambridge University Press, 2001.
[39] A. M. Stuart, J. Voss, and P. Wiberg. Conditional path sampling of SDEs and the Langevin MCMC method. *Communications in Mathematical Sciences*, 2(4):685–97, 2004.
[40] D. Talay. Second order discretization schemes of stochastic differential systems for the computation of the invariant law. *Stochastics and Stochastic Reports*, 29(1):13–36, 1990.
[41] L. Tierney. A note on Metropolis–Hastings kernels for general state spaces. *Ann. Appl. Probab.*, 8(1):1–9, 1998.

·30·

Robust, Computationally Efficient Algorithms for Tracking Problems with Measurement Process Nonlinearities

J. M. C. Clark and R. B. Vinter

30.1 Introduction

One of the most important areas of application of nonlinear filtering is Bayesian target tracking: estimating the state of a moving target from noisy sensor measurements. Here, linear discrete-time Gaussian models are commonly used to describe motion of the target and sensor platform. The estimation problem is typically nonlinear because of the nature of the measurement process which, depending on the precise nature of the application, provides noisy information about target angle-to-line-of-sight (or bearing), range or range rate, all of which quantities are nonlinear functions of target position.

Traditional approaches to target tracking are based on the application of the Kalman filter to a linearization of the underlying equations about current estimates (the extended Kalman filter approach). However, the Kalman filter breaks down (that is, fails to provide convergent estimates) when applied to many challenging tracking problems involving, for example, multiple sensor platforms or degenerate configurations where the measurements provide very little information about some aspects of target motion. This accounts for the interest in recent years in particle filters, which employ online Monte Carlo methods to construct an empirical distribution, approximating the conditional density of the state given the measurements. The ability of particle filters to provide useful estimates for difficult tracking problems in some scenarios where Kalman filtering methods are inadequate is now well documented. The main disadvantages of particle filters, or at least of generic versions of particle filters, are their heavy computational requirements, and their sensitivity to the choice of prior distributions. For expository accounts of target tracking we refer to [3] (with emphasis on classical Kalman filter approaches), and [15] and [new] which lay stress on Monte Carlo methods.

This chapter concerns "analytic" filters, i.e. tracking algorithms based on probability density calculations, not on the construction of empirical distributions (particle filtering). Analytic filters cannot hope to match the versatility of particle filters. But, as we shall see, they can provide excellent approximations for an important class of tracking problems where the nonlinearities that arise occur in the measurement process equations alone. We describe a unified approach to the construction of filters for this class of problems, and examine in detail the specific form these filters take in numerous cases of interest: bearings-only tracking, range-only tracking and other instances where nonlinearities are present in the measurement equations. The unified approach presented here is an elaboration of the one proposed in [10]. Several of the special cases treated are new.

There are two unifying threads in our approach. One is to approximate densities involved directly rather than through linear approximations to the nonlinearities in equations that generate the densities. The other is a willingness to depart from customary measurement noise models to simplify the probability calculations, provided, of course, the effect on relevant distributions is small. Modifications to the noise model can be particularly effective for the important class of tracking problems in which measurements are taken of some of the angles and Euclidean lengths of the displacement d of the target (target position relative to the sensor platform position). For such problems it is customary to adopt a "noise-after-nonlinearity model" for the measurement y, of the kind $d = \psi(d) + w$, in which ψ is a nonlinear function and w is a Gaussian 'noise' variable. Here there are potential advantages in adopting an approximate, 'noise-before-nonlinearity' model $y = \psi(d + w')$ for the measurement. This is because the new model permits exact calculations of relevant distributions; at the same time we can sometimes arrange, by choice of the statistical parameters of the noise variable w', that the distributions of the measurement processes for the two models are almost indistinguishable.

Notation:

We denote by $|d|$ the Euclidean length of a vector d: $|d| = \sqrt{d_1^2 + \ldots + d_n^2}$.

Given integers $s \leq t$, the family of time-dependent variables $y_s, y_{s+1}, \ldots, y_t$ is written $y_{s:t}$.

The probability density of a continuous random variable x is written generically as $p(x)$; that is, the relevant random variable is identified through the dummy variable, not by seperate notation.

The multivariate Gaussian (normal) density, with mean $\hat{x}$ and covariance P is denoted by $\mathcal{N}(\hat{x}, P)(x)$. We shall refer to a variable x with such a density as being "$\mathcal{N}(\hat{x}, P)$-distributed" and also use the notation "$x \sim \mathcal{N}(\hat{x}, P)$".

Following a practice common in the literature on tracking, we take the arctangent function $\arctan(y/x)$, when applied to a ratio, to mean the same as the argument of the complex number $x + iy$.

30.2 Problem formulation

The basic ingredients of the standard tracking problem are two discrete-time processes, the n-vector "state" process x_t, $t \in \{0, 1, \ldots\}$ and the k-vector observation process y_t, $t \in \{1, 2, \ldots\}$. x_t is the vector of 'generalized coordinates' of the 'track' (which typically includes as components the position and velocity of the target in Cartesian coordinates), and y_t, $t \in \{1, 2, \ldots\}$ is the measurement of x_t at time t.

The tracking problem is to calculate

$$p(x_t \mid y_{1:t}),$$

the conditional density of the target state at time t, given measurements taken at the time instants up to time t, for $t = 1, 2, \ldots$

Of particular interest are calculations that are recursive. That is, we would like to express the conditional density at time t, $p(x_t \mid y_{1:t})$, in terms of the conditional density at previous time $t-1$, $p(x_{t-1} \mid y_{1:t-1})$ and the 'new' measurement y_t, for $t = 1, 2, , \ldots$

If it is assumed that y_t is independent of $y_{1:t-1}$ given x_t, and x_t is independent of $y_{1:t-1}$ given x_{t-1}, for each t, then a recursive formula for the solution to the tracking problem can be derived from Bayes' rule. It is

$$p(x_t \mid y_{1:t}) = (1/c) \int p(x_t \mid x_{t-1}) p(x_{t-1} \mid y_{1:t-1}) dx_{t-1} \times p(y_t \mid x_t), \quad (30.1)$$

in which c is a 'normalization' constant chosen to ensure that $\int p(x_t \mid y_{1:t}) dx_t = 1$.

This formula is however not the end of the story because, in many cases of interest, it is not feasible to evaluate the two integrals involved. For this reason, algorithms that yield robust, computationally tractable, approximations to the above formula are of great interest. Often we settle for recursive formulae yielding the mean $\hat{x}_t$ and covariances P_t of the conditional density, for each t; the conditional mean yields an estimate of target state and the conditional covariance can be used to estimate confidence regions relating to the location of the true state of the target.

30.3 Models for the state and observation process

The time-dependent n-dimensional state vector x_t, describing motion of a target, is assumed to evolve according to the equation

$$x_t = F x_{t-1} + u^s + v_t \qquad t = 1, 2, \ldots$$

in which F is an $n \times n$ matrix and u^s is an n-vector. $\{v_t\}$ (the 'system noise' process) is a sequence of independent Gaussian random variables ($v_t \sim \mathcal{N}(0, Q^s)$). It is also assumed that $x_0 \sim \mathcal{N}(\hat{x}_0, P_0)$.

To develop a model for the observations, we introduce the r-dimensional time-varying *displacement* vector d_t, which is related to the state x_t according to

$$d_t = Hx_t + u^m \qquad t = 1, 2, \ldots$$

Here H is a $r \times n$ matrix and u^m is a r-vector. Usually r will take the value 2 or 3, and d_t has the interpretation of the Cartesian coordinates of the position of the target in $3D$ space or the plane, relative to the sensor platform position, in the absence of measurement noise.

The measurements y_t, $t = 1, 2, \ldots$ are assumed to satisfy the equations

$$\begin{aligned} z_t &= d_t + w_t \\ y_t &= \psi(z_t), \end{aligned} \qquad (30.2)$$

in which $\psi : R^r \to R^k$ is a given function; that is, the measurement is expressible as a deterministic function of a noisy version of the displacement d_t, which we refer to as the *enhanced measurement* and denote by z_t. Here, $\{w_t\}$ (the "measurement noise" process) is a sequence of independent Gaussian random variables ($w_t \sim \mathcal{N}(0, Q^m)$). It is assumed that $\{w_t\}$, $\{v_t\}$ and x_0 are independent.

It will be noted that the state and observation processes within this framework satisfy the independence conditions required for the validity of recursive formulae (see (30.1)).

This formulation has the following significant features. First, the state process is modelled as the solution to a linear difference equation with additive Gaussian noise and, second, the measurement process model incorporates a possibly nonlinear function, within a noise-before-nonlinearity structure. As we shall see, adopting this framework (compared with some alternatives for nonlinear tracking problems) simplifies exact moment calculations and the conditional density approximations.

What is the justification for such a formulation? In response to this question, we first observe that, while modelling the state process by a linear equation might seem restrictive, many practical "solutions" to specific tracking problems are based on simple linear models describing target motion, with respect to Cartesian coordinates. In computer vision applications, for example, the target is usually assumed to move according to a random walk in 2 or 3 dimensions, in the absence of a detailed physical model. We note, secondly, that most tracking sensor devices provide measurements that are nonlinear functions of the Cartesian coordinates of target position (measurements, most notably, of range and bearing). So it is often important, in applications where Cartesian coordinates are adopted to describe the states, to allow nonlinearities in the observation process model.

The more widely used model of measurements is based on the noise-after-nonlinearity structure:

$$y_t = \psi(d_t) + w'_t,$$

in which w'_t is a k-dimensional Gaussian white noise process with covariance $Q^{m'}$. Whether noise-before-nonlinearity is the best approach, needs to be examined on a case by case basis. Noise-before-nonlinearity models for the measurement process are suitable for a number of signal processing problems, and have been used to describe the effects of random phase shifts in frequency demodulation, for the construction of frequency-tracking algorithms. Models of this type are discussed at length by Lo and Willsky in [13]. Noise-before-nonlinearity measurement process models are natural, too, in the context of one example we discuss below, where the nonlinearity is associated with suppression of a measurement when it lies inside some region of the state space (the "Doppler blind zone"). However, even in some cases where a noise-after-nonlinearity model appears better to match the physics of the sensor device, it may still be sensible to approximate it by a noise-before-nonlinearity model (with appropriately matched noise parameters), because changing the model has little effect on tracker performance and greatly simplifies the task of approximating the conditional densities.

30.4 An algorithm for the approximation of conditional densities

We now specify a "universal" algorithm for recursively calculating approximations $\hat{x}_t$ and P_t to the first and second moments of the conditional density $p(x_t \mid y_{1:t})$, $t = 1, 2, \ldots$, for state process and measurement process models formulated in Section 3, in its full generality:

$$\begin{aligned} x_t &= F x_{t-1} + u^s + v_t \quad (v_t \sim \mathcal{N}(0, Q^s)) \\ d_t &= H x_t + u^m \\ z_t &= d_t + w_t \quad (w_t \sim \mathcal{N}(0, Q^m)) \\ y_t &= \psi(z_t)\,, \end{aligned} \qquad (30.3)$$

with $x_0 \sim \mathcal{N}(\hat{x}_0, P)$. The data specifying these processes comprise matrices F and H, covariance matrices Q^s, Q^m and P_0, and vectors u^s, u^m and $\hat{x}_0$. Special cases of interest will be described shortly.

The Second-Order Moment Matching Filter:

The algorithm generates, at each time $t \in \{1, 2, \ldots\}$, parameters $(\hat{x}_t, P_t)$ approximating the first two moments of the posterior density $p(x_t \mid y_{1:t})$, given the new measurement y_t and parameters $(\hat{x}_{t-1}, P_{t-1})$ approximating the first two moments of the prior density $p(x_{t-1} \mid y_{t-1})$, according to the following equations:

$$x_{t|t-1} = F\hat{x}_{t-1} + u^s, \quad P_{t|t-1} = FP_{t-1}F^T + Q^s$$
$$d_{t|t-1} = Hx_{t|t-1} + u^m, \quad S_t = HP_{t|t-1}H^T + Q^m$$
$$K_t = P_{t|t-1}H^T S_t^{-1}$$
$$\hat{x}_t = x_{t|t-1} + K_t(\xi_t - d_{t|t-1}) \tag{30.4}$$
$$P_t = (I - K_t H)P_{t|t-1} + K_t \Gamma_t K_t^T$$

in which

$$\xi_t = E_{z \sim \mathcal{N}(d_{t|t-1}, S_t)}[z \mid \psi(z) = y_t]$$

and

$$\Gamma_t = \text{cov}_{z \sim \mathcal{N}(d_{t|t-1}, S_t)}(z \mid \psi(z) = y_t).$$

(The right sides of the last two formulae denote the conditional mean and covariance, respectively, of an $\mathcal{N}(d_{t|t-1}, S_t)$ distributed variable z given that $\psi(z) = y_t$.)

We observe that the filter resembles the Kalman filter for the related tracking problem, in which the enhanced measurement $d_t + w_t$ replaces the true measurement $y_t = \psi(d_t + w_t)$.

The vector ξ_t in the filter equations, replacing the enhanced measurement $d_t + w_t$ which we cannot measure, is the conditional mean of $d_t + w_t$ given the true measurement y_t and an estimated Gaussian prior, a natural choice. Since the enhanced measurement $d_t + w_t$ contains "too much" information, that is, more information than that given by the true measurement y_t, the enhanced conditional covariance needs to be modified; it turns out that the "correct" modification is precisely the random term $K_t \Gamma_t K_t^T$.

Notice that, despite its resemblance to the Kalman filter, the filter (30.4) differs from it in two crucial respects. First, the second order moment matching filter generates estimates that are *nonlinear* functions of the measurements (because, in general, ξ_t is a nonlinear function of y_t). Second, the "error covariance" matrix P_t depends on the measurement (via the term Γ_t) and cannot be precomputed.

This algorithm is conceptual. Whether it can be translated into a practical algorithm hinges on the ease with which the conditional moments ξ_t and Γ_t can be calculated. In the cases we consider below, they are expressible in terms of rapidly executable MATLAB library functions, or simple scalar quadratures (i.e., computation of one-dimensional integrals).

30.5 Analysis

In this section we provide a theoretical justification of the algorithm. Specifically, we show:

Proposition 30.1 *Consider the process equations (30.3) for fixed $t \in \{1, 2, \ldots\}$. If we assume that*

$$p(x_{t-1} \mid y_{1:t-1}) = \mathcal{N}(\hat{x}_{t-1}, P_{t-1})(x_{t-1}). \tag{30.5}$$

Then

$$E[x_t \mid y_{1:t}] = \hat{x}_t \quad \text{and} \quad \text{cov}(x_t \mid y_{1:t}) = P_t,$$

where $\hat{x}_t$ and P_t are defined by (30.4).
These assertions remain valid if we assume that, in (30.3), the matrices and vectors F, H, u^s, u^m, Q^s, and Q^m depend on t and $y_{1:t}$ for all $t \in \{1, 2, \ldots\}$.

In other words, the algorithm matches *exactly* the true mean and covariance of x_t given $y_{1:t}$, given a Gaussian prior. Under the Gaussian prior assumption then, the filter provides a least squares estimate, and can be expected to reduce the least squares estimation error, as compared with other nonlinear filters that aim to approximate the best *linear* least squares estimator, either by means of linearizing the underlying equations (the extended Kalman filter [3]) or of "statistical" linearization (the unscented Kalman filter [11]). (This assertion is, strictly speaking, correct only for tracking problems with a noise-before-nonlinearity measurement process model; otherwise, using such a model in place of a noise-after-nonlinearity model amounts to an additional approximation step.)

Proof. Write z_t for the enhanced measurement,

$$z_t = d_t + w_t,$$

and note the decomposition formula for the variable x_t:

$$x_t = \zeta_t + (I - K_t H)\hat{x}_{t|t-1} - K_t u^m + K_t z_t \tag{30.6}$$

where ζ_t is a random variable with the following property: ζ_t is independent of z_t and $x_{t|t-1}$, and

$$\zeta_t \sim \mathcal{N}(0, (I - K_t H) P_{t|t-1}). \tag{30.7}$$

Here, K_t, etc., are as in (30.4). This formula expresses the fact that the variable x_t can be written as the sum of the conditional mean of x_t given $y_{1:t-1}$ and z_t, which is obtained from the Kalman filter equations under our hypothesis that $p(x_{t-1} \mid y_{1:t-1})$ is Gaussian, and an "estimation error" variable ζ_t, which is independent of $y_{1:t-1}$ and z_t.

Now take conditional expectations across eqn. (30.6), conditioned on $y_{1:t}$. Since y_t is a deterministic function of z_t, and ζ_t has zero mean and is independent of $(y_{1:t}, z_t)$, it follows that $E[\zeta_t \mid y_{1:t}] = 0$. So

$$E[x_t \mid y_{1:t}] = (I - K_t H)\hat{x}_{t|t-1} - K_t u^m + K_t E[z_t \mid y_{1:t}].$$

Equating the conditional covariances of the variables on both sides of eqn. (30.6) gives, on the other hand,

$$\text{cov}(x_t|y_{1:t}) = \text{cov}(\zeta) + K_t \text{cov}(z_t|y_{1:t}) K_t^T$$
$$= (I - K_t H) P_{t|t-1} + K_t \text{cov}(z_t|y_{1:t}) K_t^T.$$

But

$$p(z_t|y_{1:t-1}) = \mathcal{N}(d_{t|t-1}, S_t)(z_t)$$

It follows that

$$E[z_t|y_{1:t}] = E_{z \sim \mathcal{N}(d_{t|t-1}, S_t)}[z|\psi(z) = y_t] = \xi_t$$

$$\text{cov}(z_t|y_{1:t}) = \text{cov}_{z \sim \mathcal{N}(d_{t|t-1}, S_t)}(z|psi(z) = y_t) = \Gamma_t.$$

Then

$$E[x_t|y_{1:t}] = x_{t|t-1} + K_t(\xi_t - d_{t|t-1})$$
$$\text{cov}(x_t|y_{1:t}) = (I - K_t H) P_{t|t-1} + K_t \Gamma_t K_t^T.$$

We have confirmed that, under hypothesis (30.5), the algorithm generates the conditional mean and covariance of x_t, given $y_{1:t}$.

Regarding the final assertion of the theorem, we observe that the previous analysis is unaffected if, now, the matrices F, H, etc. in the state and observation model equations are allowed to depend on t and $y_{1:t}$. □

30.6 Bearings-only tracking

In bearings-only tracking, measurements are taken of the direction cosines of the displacement vector. To apply the second order moment matching nonlinear filter to bearings-only tracking we choose the function $\psi(.)$ in the observation process model to be (for r-dimensional z)

$$\psi(z) = |z|^{-1} z. \tag{30.8}$$

Notice "bearing" measurements, interpreted in this way, are related to the azimuth and elevation angles of z (θ and ϕ respectively) according to

$$\psi(z) = (\cos(\phi)\cos(\theta), \cos(\phi)\sin(\theta), \sin(\phi)),$$

for measurements in three-dimensional space, and the azimuth angle θ according to

$$\psi(z) = (\cos(\theta), \sin(\theta)),$$

for measurements in the horizontal plane.

Selection of the measurement noise parameters: standard approaches to bearings-only tracker design are based on a noise-after-nonlinearity measurement process model which takes the form

$$y_t = \arctan(d_2/d_1) + w'_t \quad \text{(for r=2)}$$

and

$$y_t = \begin{bmatrix} \arctan(d_2/d_1) \\ \arctan(d_3/\sqrt{d_1^2 + d_2^2}) \end{bmatrix} + w'_t \quad \text{(for r=3)}.$$

Here, $\{w'_t\}$ is an $(r-1)$-dimensional white noise process (independent of $\{v_t\}$ and x_0). It is assumed that $w'_t \sim \mathcal{N}(0, \sigma^2 I_{(r-1)\times(r-1)})$, for some specified variance parameter $\sigma^2 > 0$.

Application of the second-order moment matching filter requires us to replace the measurement process model by the noise-before-nonlinearity model

$$y_t = |d_t + w_t|^{-1}(d_t + w_t) \tag{30.9}$$

in which $\{w_t\}$ is a white noise process ($w_t \sim \mathcal{N}(0, Q^m)$).

It is shown in [9] that, for $r = 2$ (bearings measurements in the plane), a suitable choice of Q^m in these circumstances is

$$Q^m = E[|d_t|^2|y_{1:t-1}]\sigma^2 I_{r \times r},$$

even for large noise variances (σ^2 up to 1 radian^2). An analysis supporting this choice appears in [9].

The Second-Order Moment Matching Filter for Bearings-Only Tracking (the Shifted Rayleigh Filter): Consider the second-order moment matching filter for the state and observations processes (30.3), in which $\psi(.)$ is chosen to be the nonlinear function (30.8) arising in bearings only tracking.

To complete the specification of the filter in this case, we must evaluate the conditional moments ξ_t and Γ_t in this case.

Proposition 30.2 *(Calculation of ξ_t and Γ_t for Bearings-Only Tracking)* Assume that r (the dimension of the displacement vector) takes value either 2 or 3. Write (for scalar z)

$$\rho_2(z) = \frac{ze^{-\frac{z^2}{2}} + \sqrt{2\pi}(z^2 + 1)F_\mathcal{N}(z)}{e^{-\frac{z^2}{2}} + \sqrt{2\pi}(z)F_\mathcal{N}(z)}$$

$$\rho_3(z) = z + \frac{2}{\rho_2(z)}.$$

For r = 2 and 3, let

$$\delta_t^r = \begin{cases} \left(y_t^T S_t^{-1} y_t\right)^{-1} \left(2 + \eta_t \rho_2(\eta_t) - \rho_2^2(\eta_t)\right) & \text{if } r = 2 \\ \left(y_t^T S_t^{-1} y_t\right)^{-1} \left(3 + \eta_t \rho_3(\eta_t) - \rho_3^2(\eta_t)\right) & \text{if } r = 3 \end{cases}$$

where $\eta_t = \left(y_t^T S_t^{-1} y_t\right)^{-1} y_t^T S_t^{-1} \left(H\hat{x}_{t|t-1} + u_t^m\right)$. *Then, for* $\psi(.)$ *given by (30.8) and for r taking values either 2 or 3,*

$$\xi_t = \left(y_t^T S_t y_t\right)^{-\frac{1}{2}} \rho_r(\eta_t)$$
$$\Gamma_t = \delta_t^r y_t y_t^T.$$

These formulae are derived in [9]. Here, $F_\mathcal{N}(z)$ is the standard normal distribution function with zero mean and unit covariance. It can be evaluated as the rescaling $\frac{1}{2}\text{erfc}\left(-\frac{z}{\sqrt{2}}\right)$ of the complementary error function

$$\text{erfc}(x) = \left(\frac{2}{\sqrt{\pi}}\right) \int_x^\infty e^{-t^2} dt,$$

which is a standard software library function.

Comments: The second-order moment matching filter applied to the bearing-only tracking problem is referred to as the *shifted Rayleigh filter* (SRF) in the literature. A detailed description of the SRF algorithm, accompanied by discussion of the noise model employed in its construction and supporting analysis, appears in [9].

[1] reports on simulating studies based on a challenging scenario, with poor "observability" characteristics and featuring large changes in bearings measurements between frames. Here the SRF algorithm matches the performance of a high order particle filter (at greatly reduced computational cost), while outperforming other moment matching algorithms such as the extended Kalman filter and the unscented Kalman filter. Other simulation studies have revealed an important feature of SRF performance: insensitivity to the user's choice of initial distribution of the state variable. This is of practical significance, since, often in applications, little reliable prior information about target motion is available.

The SRF filter has been employed as an ingredient in the construction of filters for refined versions of the bearings-only tracking problem posed in this paper, where some of the measurement returns are "clutter" (i.e. signals not related to target motion) [8], where there is more than one sensor platform and when the state equation [9], incorporates a hidden Markov model to take account of a manoeuvring target [7].

The approximations introduced to construct the shifted Rayleigh filter for bearing only tracking problems incorporating a standard measurement process model are a) the replacement of conditional distributions by Gaussian distributions, and b) the substitution of a noise-before-nonlinearity model in place of the original noise-after-nonlinearity model. The success of the shifted Rayleigh filter algorithm is due to the fact that, for many bearings only tracking problems of interest, the true conditional densities are unimodal and the effects of changing the noise model are negligible.

30.7 Range-only tracking

In range-only tracking, measurements are taken of the Euclidean length of the displacement vector (the "range" of the target relative to the sensor position). Range-only tracking problems can be treated in the framework of Section 30.3, by choosing the function $\psi(.)$ to be

$$\psi(z) = |z|. \tag{30.10}$$

Selection of the measurement noise parameters: noise-after-nonlinearity observation process models are commonly used for designing range-only trackers. Here the measurement is assumed to be generated by the model

$$y_t = |d_t| + w'_t \qquad \left(w'_t \sim \mathcal{N}(0, \sigma^2)\right), \tag{30.11}$$

in which $\{w'_t\}$ is a one-dimensional white noise process (independent of $\{v_t\}$ and x_0).

It can be shown that, if the variance parameter in the noise-before-nonlinearity model of Section 30.3 is taken to be $\sigma^2 I_{r \times r}$, then we can expect the two noise models are practically interchangeable, if $|d_t|$ is large. Indeed the conditional densities of the measurement y_t given the displacement d_t, for the noise-after-nonlinearity and for noise-before-nonlinearity models are, respectively,

$$p(y_t|d_t) = \frac{1}{\sqrt{2\pi}\sigma} e^{-\frac{1}{2\sigma^2}(y_t - |d_t|)^2} \tag{30.12}$$

and

$$p(y_t|d_t) = \frac{y_t}{\sigma^2} e^{-\frac{1}{2\sigma^2}(y_t^2 + |d_t|^2)} I_0\left(\frac{y_t|d_t|}{\sigma^2}\right). \tag{30.13}$$

In (30.13), I_0 is the modified Bessel function of order zero. The "goodness of fit" is governed by the magnitude of the parameter $(|d_t|/\sigma)$. These two densities almost coincide, if $|d_t|/\sigma > 10$. This inequality is satisfied in the scenarios considered in [5].

The Second-Order Moment Matching Filter for Range-Only Tracking: Consider the second-order moment matching filter for the state and observations processes (30.3), in which $\psi(.)$ is chosen to be the nonlinear function (30.10) arising in range only tracking.

In this case, the following formulae are available for the computation of the conditional moments ξ_t and Γ_t.

Proposition 30.3 *(Calculation of ξ_t and Γ_t for Range-Only Tracking:) Assume that the dimension of the displacement vector $r = 2$ (displacements in the plane). Let*

$$V (= \{v_{ij}\}) = S_t^{-1} \quad \text{and} \quad [m_1, m_2]^T = d_{t|t-1}.$$

and define, for $r > 0$,

$$a_1(r) = (2/r)\sqrt{(v_{11}m_1 + v_{12}m_2)^2 + (v_{12}m_1 + v_{22}m_2)^2}$$

$$a_2(r) = \sqrt{\frac{1}{4}(v_{11} - v_{22})^2 + v_{12}^2}$$

$$\omega_1 = \arctan\left(\frac{v_{12}m_1 + v_{22}m_2}{v_{11}m_1 + v_{12}m_2}\right)$$

$$\omega_2 = \arctan\left(2\frac{v_{12}}{v_{11} - v_{22}}\right).$$

Now define

$$p(r, \theta) = re^{-\frac{r^2}{2}[a_1(r)\cos(\theta - \omega_1) + a_2\cos(2\theta - \omega_2)]}.$$

($p(r, \theta)$ is an unnormalized density resulting from expressing $\mathcal{N}(d_{t|t-1}, S_t)$ in polar coordinates.) Let g and G be the corresponding conditional mean and covariance of $a(\cos\theta, \sin\theta)$, given $r = y_t$, i.e.

$$g = \frac{1}{c}\int_0^{2\pi}\begin{bmatrix}\cos(\theta)\\ \sin(\theta)\end{bmatrix}p(y_t, \theta)d\theta \tag{30.14}$$

$$G = \frac{1}{2c}\int_0^{2\pi}\begin{bmatrix}1 + \cos(2\theta) & \sin(2\theta)\\ \sin(2\theta) & 1 + \cos(2\theta)\end{bmatrix}p(y_t, \theta)d\theta - gg^T \tag{30.15}$$

in which c is the normalization constant

$$c = \int_0^{2\pi} p(y_t, \theta)d\theta \tag{30.16}$$

Then ξ_t and Γ_t are given by the formulae

$$\xi_t = y_t g \quad \text{and} \quad \Gamma_t = y^2 G.$$

These formulae are derived in [5]. Despite their complicated appearance, they involve merely trigonometric functions and simple quadratures in one independent variable, and library routines are available for their rapid evaluation.

Comments: While not as ubiquitous as bearings-only tracking problems, range-only problems nonetheless have practical significance. They arise, for example, in problems of tracking a moving transmitter, where the only sensor data is the strength of the received signal, which is related to the Euclidean distance of the transmitter from the receiver.

Implementation of the second-order moment matching filter for range-only problems requires the evaluation of the integrals (30.14), (30.15), and (30.16). These integrals are zero'th, first, and second Fourier coefficients of the periodic function $p(y_t, \theta)$ on $[0, 2\pi]$ (for fixed y_t). This can be accomplished very rapidly in benign cases, using Fast Fourier Transform techniques, though situations in which the density is very peaky are just are likely to occur, and require more refined quadrature formulae.

A detailed description of the second-order moment matching filter for the range-only tracking problem appears in [16]. This reference includes extensive discussion of fast, robust techniques for evaluating the integrals (30.14) and (30.15), and the results of comparative studies involving this and other filters.

According to these studies, there is little to distinguish the performance of the second-order moment matching filter and other moment matching filters for the standard scenarios considered. At first sight this is surprising, because the second order moment matching filter is based on more precise moment calculations. The explanation lies in the fact that, frequently in practical situations, the true conditional distributions are far from Gaussian and any moment matching algorithm based on Gaussian approximations will introduce serious distortions. Indeed situations often arise in which the range, which is measured directly, is known fairly precisely, but the variance of the bearing, which must be inferred, is not; here constant probability contours will be "banana shaped" and not approximately ellipsoidal, as would be required for accurate approximation by a Gaussian distribution.

Better performance is achieved by a refinement of the second-order moment matching filter that propagates a Gaussian mixture approximation of the conditional distributions; the above formulae have a role in the refined algorithm, calculating the parameters of the constituent Gaussian densities and their weights. A description of the refined algorithm, accompanying analysis and simulation results, appears in [5]. In the scenarios considered, the algorithm achieves the accuracy of high-order particle filters (at greatly reduced computational cost), while consistently outperforming classical moment matching filters.

All the examples we give in this chapter permit refinements based on Gaussian mixtures. Since their introduction to tracking in the pioneering paper [2], the use of Gaussian mixture approximations of densities has gained increasing popularity in the applied tracking literature. A well-known example of its use is in the interacting multiple model algorithm for tracking manoeuvring targets [4]. A rather different application in which measurement distributions are directly modelled by Gaussian mixtures can be found in [14].

30.8 GMTI tracking

In some radar tracking applications, a possible mode of operation includes a sensor data preprocessing stage, in which measurements are deliberately suppressed, whenever the magnitude of the range rate drops below a specified threshold (the Minimum Detectable Velocity a). The purpose of artificially introducing the "Doppler blind zone" (the region of the state space in which the range rate magnitude is small) is to separate out moving objects of interest from heavy, static clutter. Tracking algorithms that take account of the Doppler blind zone when the sensors are operated in this mode are called GMTI (Ground-based Moving Target Indicator) trackers. In this section, we show that an idealized version of the GMTI tracking problem fits the formulation of Section 30.3.

Suppose that noisy measurements are taken of the target position (relative to the sensor) z and the range rate, i.e. the time derivative of the Euclidean length of the position vector. Making use of the chain rule, we can express the range rate in terms of position z and the rate of change of position $\dot{z}$, thus

$$d/dt\, |z| = |z|^{-1} z^T \dot{z}. \tag{30.17}$$

Let H_z and H_v be the matrices that isolate the components of the state corresponding to target position z and rate of change of the velocity, respectively:

$$z = H_d\, x \quad \text{and} \quad \dot{z} = H_v\, x.$$

Then the raw measurement (that is, noisy measurement of the displacement and range rate prior to suppression of returns that fall within the Doppler blind zone) can be taken to be

$$y'_t = \begin{bmatrix} b^T(x_t)x_t \\ H_d x_t \end{bmatrix} + w_t.$$

Here, $b(x_t)$ is the random variable

$$b(x_t) := |H_d x_t|^{-1} H_v^T H_d x_t$$

(see(30.17)) and w_t is a white noise sequence with covariance Q^m.

In typical applications, $|E[x_t \mid y_{1:t-1}]|^2$ is much greater than $|\mathrm{cov}(x_t \mid y_{1:t-1})|$. In these circumstances, it is a reasonable approximation to replace the random variable $b(x_t)$ by the function of $y_{1:t-1}$

$$b(x_{t|t-1}) = |H_d x_{t|t-1}|^{-1} H_v^T H_d x_{t|t-1},$$

where $x_{t|t-1}$ is the estimate of $E[x_t|y_{1:t-1}]$ given by (30.4). We thereby arrive at the observations process model

$$\begin{cases} d_t = H(y_{1:t-1})x_t \\ y'_t = d_t + w_t \\ y_t = \psi(y'_t). \end{cases}$$

Here

$$\psi(y') := \begin{cases} y' & \text{if } |y'_1| \geq a \\ 0 & \text{if } |y'_1| < a, \end{cases} \qquad (30.18)$$

in which y'_1 denotes the first component of y'. The constant a is, we recall, the threshold on range rate below which measurements are suppressed. $H(x_{t|t-1})$ is the matrix

$$H(y_{1:t-1}) := \begin{bmatrix} b^T(y_{1:t-1}) \\ H_d \end{bmatrix}. \qquad (30.19)$$

The observation process model (30.18) is that of the tracking problem formulation of Section 30.3, in which, now, H is the matrix (30.19) and $u^m = 0$.

The Second-Order Moment Matching Filter for GMTI Tracking: Consider the second-order moment matching filter for the state and observations processes (30.3), in which the matrix H is replaced by $H(y_{1:t-1})$ from (30.19), $u^m = 0$ and $\psi(.)$ is chosen to be the nonlinear function (30.18) arising in GMTI tracking. (Recall that the second order moment matching filter has the 'nonlinear least squares' property described in Proposition 30.1, even when the system matrices depend on $y_{1:t-1}$.)

The computation of the conditional moments ξ_t and Γ_t, in this case, are supplied by the following proposition.

Proposition 30.4 *(Calculation of ξ_t and Γ_t for GMTI Tracking.)*

Case I $(y_t \neq 0)$

In this case

$$\xi_t = y_t \quad \text{and} \quad \Gamma_t = 0.$$

Case II $(y_t = 0)$

In this case partition the vector $d_{t|t-1}$ and the matrix S_t as

$$d_{t|t-1} = \begin{bmatrix} d_1 \\ d_2 \end{bmatrix} \quad \text{and} \quad S_t = \begin{bmatrix} s_{11} & s_{21}^T \\ s_{21} & S_{22} \end{bmatrix},$$

in which d_1 and d_2 are 1 and $(r-1)$ vectors respectively, and s_{11}, s_{21}, and S_{22} are 1×1, $(r-1) \times 1$, and $(r-1) \times (r-1)$ matrices respectively. Now set

$$c = F_\mathcal{N}(z)\left(\frac{+a - d_1}{\sqrt{s_{11}}}\right) - F_\mathcal{N}(z)\left(\frac{-a - d_1}{\sqrt{s_{11}}}\right)$$

$$\gamma_1 = \frac{s_{11}}{c}[\mathcal{N}(d_1, s_{11})(-a) - \mathcal{N}(d_1, s_{11})(+a)] + d_1$$

$$\gamma_2 = \frac{s_{11}}{c}[(d_1 - a)\mathcal{N}(d_1, s_{11})(a) - (d_1 + a)\mathcal{N}(d_1, s_{11})(+a)]$$
$$+ (d_1^2 + s_{11}) - \gamma_1^2$$

Then

$$\xi_t = \begin{bmatrix} \gamma_1 \\ d_2 + (\gamma_1 - d_1) S_{22}^{-1} s_{21} \end{bmatrix}$$

and

$$\Gamma_t = \begin{bmatrix} \gamma_2 & s_{11}^{-1} \gamma_2 s_{21}^T \\ s_{11}^{-1} \gamma_2 s_{21} & S_{22} - s_{11}^{-2}(1 - \gamma_2) s_{21} s_{21}^T \end{bmatrix}.$$

Comments: The GMTI problem described in this section is a highly idealized one. Papers on GMTI tracking, almost without exception, take account also of clutter effects and the possibility that a measurement will not be recorded, even when it falls outside the Doppler blind zone (nonunity probability of detection). The algorithm proposed here extends easily to allow for these refinements (see [6]). The resulting algorithm (we shall call it the "refined" algorithm) propagates a Gaussian mixture approximation to the conditional distribution. The formulae appearing in the second order moment matching algorithm applied to the idealized GMTI problem are used to calculate the Gaussian components and weights.

In the simulation studies reported in [6], the refined version of the second order moment matching algorithm consistently matched the performance of a high order particle filter. It was also highly competitive with other analytic filters, based on a number of different ideas. These included the introduction of a state-dependent probability of detection (taking a low value in the Doppler blind zone) or introducing a hidden Markov model into the state equation with a stationary discrete state associated with a location of the target in the Doppler blind zone.

30.9 Tracking with quantized measurements

In some tracking applications involving low-resolution sensors, it is helpful to take explicit account of quantization effects in tracker design. The key principle here is that the measuring device does not record the value of a signal y' per se, but rather identifies the "bin" in which it is located, i.e. the value of the integer n such that

$$n\Delta \leq y' < (n+1)\Delta . \qquad (30.20)$$

Here, the constant $\Delta > 0$ is a resolution parameter.

Tracking problems with quantized measurements naturally fits the framework of Section 30.3, as we now show. For simplicity of exposition, we limit attention to the scalar measurements case.

A simple, illustrative tracking problem, incorporating quantization effects, involves a state process $\{x_t\}$ and scalar measurement process $\{y_t\}$ described by the the special case of eqs (30.3):

$$x_t = Fx_{t-1} + u^s + v_t \quad (v_t \sim \mathcal{N}(0, Q^s))$$
$$d_t = h^T x_t + u^m$$
$$y'_t = d_t + w_t \quad (w_t \sim \mathcal{N}(0, \sigma^{2,m}))$$
$$y_t = \psi(y'_t),$$

in which, now, $\psi(.)$ is the function

$$\psi(y') = \left[\frac{y'}{\Delta}\right] \qquad (30.21)$$

where [ ] denotes the integer part.

The Second-Order Moment Matching Filter for Tracking Problems with Quantization Effects: Consider the second-order moment matching filter of Section 30.4, in which $r = 1$ (scalar enhanced measurement), $H = h^T$, $Q^m = \sigma^{2,m}$ and $\psi(.)$ is given by (30.21). Implementation of the filter requires evaluation of the variables

$$\xi_t = E_{z \sim \mathcal{N}(d_{t|t-1}, S_t)}[z \mid \psi(z) = y_t] \quad \text{and} \quad \Gamma_t = \text{cov}_{z \sim \mathcal{N}(d_{t|t-1}, S_t)}(z \mid \psi(z) = y_t).$$

the relevant formulae are provided by the following proposition:

Proposition 30.5 *(Calculation of ξ_t and Γ_t for Tracking Problems with Quantization Effects.)*

Write $d = d_{t|t-1}$ and $\sigma^2 = S$ in (30.4) where, now $H = h^T$ and $Q^m = \sigma^{2,m}$. Now set

$$c = F_\mathcal{N}\left(\frac{y_t + \Delta - d}{\sigma}\right) - F_\mathcal{N}\left(\frac{y_t - d}{\sigma}\right)$$

$$\xi_t = \frac{\sigma}{c}\left(\mathcal{N}(d, \sigma^2)(y_t) - \mathcal{N}(d, \sigma^2)(y_t + \Delta)\right) + d$$

$$\Gamma_t = \frac{\sigma^2}{c}\left((d + y_t)\mathcal{N}(d, \sigma^2)(y_t) - (d + y_t + \Delta)\mathcal{N}(d, s)(y_t + \Delta)\right)$$
$$+ (d^2 + \sigma^2) - \xi_t^2.$$

We omit the routine, but somewhat lengthy, derivation of these formulae. (They amount to evaluating the first two moments of a truncated scalar Gaussian density.)

Comments: Tracking Algorithms for problems involving quantized measurements, along the lines of the second order moment matching filter above, have recently been proposed by Duan et al. [12]. When the measurement takes the form of a quantized vector variable, numerical integration is required to evaluate ξ_t and Γ_t. (Integration of Gaussian densities over "box" domains is involved.) Integration schemes are discussed in [12].

References

[1] S. Arulampalam, J. M. C. Clark, and R. B. Vinter, *The Shifted Rayleigh Filter: Performance of the Shifted Rayleigh Filter in Single-sensor Bearings-only Tracking*, Proc. 10th Int. Conf. on Information Fusion, Fusion 2007, Quebec, 2007.

[2] D. L. Alspach, A Gaussian Sum Approximation to the Multitarget Identification-Tracking Problem, *Automatica*, 11, 1975, pp. 285–96.

[3] Y. Bar-Shalom, X. R. Li, and T. Kirubarajan, *Estimation with Applications to Tracking and Navigation: Theory, Algorithms and Software*, John Wiley & Sons Inc., New York, 2001.

[4] H. Blom and Y. Bar-Shalom, The Interacting Multiple Model Algorithm for Systems with Markovian Switching Coefficients, *IEEE Trans. Aut. Control*, 33, 1988, pp. 780–3.

[5] J. M. C. Clark, P. A. Kountouriotis, and R. B. Vinter, *A Noise-before-Nonlinearity Approach to the Problem of Tracking with Range-Only Measurements*, submitted.

[6] J. M. C. Clark, P. A. Kountouriotis, and R. B. Vinter, A New Algorithm for GMTI Tracking Problems, Subject to a Doppler Blind Zone Constraint, *IEEE Trans. Aut. Control*, to appear.

[7] J. M. C. Clark, S. Robbiati, and R. B. Vinter, The Shifted Rayleigh Mixture Filter for Bearings-Only Tracking of Manoeuvring Targets, *IEEE Trans on Signal Processing*, 55, 2007, pp. 3218–27.

[8] J. M. C. Clark, R. B. Vinter, and M. Yaqoob, *The Shifted Rayleigh Filter for Bearings Only Tracking*, Proc. 8th Int. Conf. on Information Fusion, Fusion 2005, Stockholm, 2005.

[9] J. M. C. Clark, R. B. Vinter, and M. M. Yaqoob, The Shifted Rayleigh Filter: A New Algorithm for Bearings Only Tracking, *IEEE Trans. Aerospace and Electronic Systems*, 43, 2007, pp. 1373–84.

[10] J. M. C. Clark and R. B. Vinter, A New Class of Moment Matching Filters for Nonlinear Tracking and Estimation Problems, Proc. IEEE Nonlinear Statistical Signal Processing Workshop, Cambridge, 2006.

[11] S. J. Julier, J. K. Uhlmann, and H. F. Durrant-Whyte, A New Approach for Filtering Nonlinear Systems, *Proc. Aut. Control Conference*, Seattle, 1995.

[12] Z. Duan, V. Jilkov, and X. R. Li, *State Estimation with Quantized Measurements: Approximate MMSE Approach*, Proc. 11th Int. Conf on Information Fusion, Fusion 2008, Cologne, 2008.

[13] J. T. H. Lo and A. S. Willsky, Estimation of Rotational Processes with One Degree of Freedom, Parts I, II and III, *IEEE Trans Aut. Control*, 20, 1975.

[14] D. Musicki and R. Evans, *Measurement Gaussian Sum Mixture Target Tracking*, Proc. 9th Int. Conf. on Information Fusion, Fusion 2006, Florence, 2006.

[15] B. Ristic, S. Arulampalam, and N. Gordon, *Beyond the Kalman Filter, Particle Filters For Tracking Applications*, Artech House, 2004.

[16] M. Yaqoob, *Computationally Efficient Algorithms for Nonlinear Target Tracking Problems*, PhD thesis, Imperial College London, 2007.

[17] M. S. Arulampalam, S. Maskell, N. Gordon and T. Clapp, A tutorial on particle filters for online nonlinear/non-Gaussian Bayesian tracking, IEEE Transactions on Signal Processing, 50, 2, 2002, pp. 174–188.

·31·

Nonlinear Filtering Algorithms Based on Averaging Over Characteristics and on the Innovation Approach

G. N. Milstein and M. V. Tretyakov

31.1 Introduction

Let $(\Omega, \mathcal{F}, \mathbb{P})$ be a probability space, $\mathcal{F}_t$, $0 \leq t \leq T$, be a nondecreasing family of σ-subalgebras of $\mathcal{F}$, $(w(t), \mathcal{F}_t)$ and $(v(t), \mathcal{F}_t)$ be d_1-dimensional and r-dimensional independent standard Wiener processes, respectively. We consider the classical filtering scheme

$$dX = a(X)ds + \sigma(X)dw(s), \quad X(0) = x, \tag{31.1}$$

$$dy = \beta(X)ds + dv(s), \quad y(0) = 0, \tag{31.2}$$

where $X(t) \in \mathbb{R}^d$ is the unobservable signal process, $y(t) \in \mathbb{R}^r$ is the observation process, $a(x)$ and $\beta(x)$ are d-dimensional and r-dimensional vector functions, respectively; $\sigma(x)$ is a $d \times d_1$-dimensional matrix function. We assume that the functions a, β, and σ are bounded and have bounded derivatives up to some order. The vector $X(0) = x$ in (31.1) can be random, it is independent of both w and v and its density $\varphi(\cdot)$ is supposed to be known.

Let $f(x)$ be a function on $\mathbb{R}^d$ with the same properties as those of a, β, σ. The filtering problem consists in constructing the estimate $\hat{f}(X(t))$ based on the observation $y(s)$, $0 \leq s \leq t$, which is the best in the mean-square sense. Our aim is to give effective numerical procedures for realization of the conditional mean:

$$\hat{f}(X(t)) = \mathbb{E}[f(X(t)) \mid y(s), \ 0 \leq s \leq t] =: \mathbb{E}^y f(X(t)). \tag{31.3}$$

Let

$$\eta(t) := \exp\left\{\int_0^t \beta^\top(X(s))dv(s) + \frac{1}{2}\int_0^t \|\beta(X(s))\|^2 ds\right\} \tag{31.4}$$

$$= \exp\left\{\int_0^t \beta^\top(X(s))dy(s) - \frac{1}{2}\int_0^t \|\beta(X(s))\|^2 ds\right\}.$$

According to our assumptions, we have $\mathbb{E}\eta^{-1}(t) = 1$, $0 \leq t \leq T$. We introduce the new probability measure $\tilde{\mathbb{P}}$ on $(\Omega, \mathcal{F})$:

$$\tilde{\mathbb{P}}(\Gamma) = \int_\Gamma \eta^{-1}(T) d\mathbb{P}(\omega).$$

The measures $\mathbb{P}$ and $\tilde{\mathbb{P}}$ are mutually absolutely continuous. Due to the Girsanov theorem, there exists a standard Wiener process $(w(t), \tilde{v}(t))$ on $(\Omega, \mathcal{F}, \mathcal{F}_t, \tilde{\mathbb{P}})$ such that the process $(X(s), y(s))$ satisfies the system of Itō equations

$$dX = a(X)ds + \sigma(X)dw(s), \quad X(0) = x, \tag{31.5}$$

$$dy = d\tilde{v}(s), \quad y(0) = 0. \tag{31.6}$$

So, the processes $X(s)$ and $y(s)$ are independent on $(\Omega, \mathcal{F}, \mathcal{F}_s, \tilde{\mathbb{P}})$ and $y(s)$ is a Wiener process.

We denote by $X(s') = X_{s,x}(s')$, $\eta(s') = \eta_{s,x,1}(s')$, $s' \geq s$, a solution to the system

$$dX = a(X)ds' + \sigma(X)dw(s'), \quad X(s) = x, \tag{31.7}$$

$$d\eta = \beta^\top(X)\eta dy(s'), \quad \eta(s) = 1. \tag{31.8}$$

Due to the well-known Kallianpur–Striebel formula for the mean (31.3), we have

$$\mathbb{E}[f(X(t)) \mid y(s), 0 \leq s \leq t] = \frac{\tilde{\mathbb{E}}[f(X(t))\eta(t) \mid y(s), 0 \leq s \leq t]}{\tilde{\mathbb{E}}[\eta(t) \mid y(s), 0 \leq s \leq t]} = \frac{\tilde{\mathbb{E}}^y[f(X_x(t))\eta(t)]}{\tilde{\mathbb{E}}^y \eta(t)}, \tag{31.9}$$

where $X_x(t)$ denotes $X_{0,x}(t)$, $\tilde{\mathbb{E}}$ means expectation according to the measure $\tilde{\mathbb{P}}$, and $\tilde{\mathbb{E}}^y[\cdot] := \tilde{\mathbb{E}}[\cdot \mid y(s), 0 \leq s \leq t]$.

We fix a time moment t and introduce the function

$$u_g(s, x; t) = \tilde{\mathbb{E}}^y\left[g(X_{s,x}(t))\eta_{s,x,1}(t)\right], \tag{31.10}$$

where g is a scalar function on $\mathbb{R}^d$, $x \in \mathbb{R}^d$ is deterministic, and $X_{s,x}(s')$, $\eta_{s,x,1}(s')$ is the solution of the system (31.7)–(31.8). It is known (see, e.g. [17],[19],[8]) that the function $u_g(s, x; t)$ is the solution of the Cauchy problem for the backward linear stochastic partial differential equation (SPDE; see Section 31.2). If $X(0) = x = \xi$ is a random variable with the density $\varphi(\cdot)$, then we can write

$$\hat{f}(X(t)) = \mathbb{E}[f(X(t)) \mid y(s), 0 \leq s \leq t] = \frac{u_{f,\varphi}(0, t)}{u_{1,\varphi}(0, t)}, \tag{31.11}$$

where

$$u_{g,\varphi}(0, t) := \int u_g(0, x; t)\varphi(x)dx = \tilde{\mathbb{E}}^y[g(X_{0,\xi}(t))\eta_{0,\xi,1}(t)].$$

Thus, finding the estimate $\hat{f}(X(t))$ amounts to evaluating averages $\tilde{\mathbb{E}}^y[g(X_{0,\xi}(t))\eta_{0,\xi,1}(t)]$.

For a given observation trajectory $y(s)$, $0 \leq s \leq t$, the numerator and denominator in the Kallianpur–Striebel formula (31.9) can be written in the form:

$$\tilde{\mathbb{E}}[g(X(t))\eta(t) \mid y(s),\ 0 \leq s \leq t] \qquad (31.12)$$

$$= \mathbb{E}\left[g(X_x(t))\exp\left\{\int_0^t \beta^\top(X_x(s))dy(s) - \frac{1}{2}\int_0^t \|\beta(X_x(s))\|^2 ds\right\}\right]\Big|_{(31.7)},$$

where x can be both deterministic and random, and the sign $|_{(31.7)}$ means that $X_x(s)$ is the solution of (31.7) under a fixed trajectory y, i.e., the averaging here is carried out with respect to X only. To approximate (31.12), the observed increments $y(s_{k+1}) - y(s_k)$, $k = 0, \ldots, N-1$, $s_0 = 0$, $s_N = t$, $s_{k+1} - s_k = t/N = h$, are usually used. The representation (31.12) is constructive, it admits direct application of the Monte Carlo technique using approximation methods of (ordinary) stochastic differential equations (SDEs). Such an approach looks very promising. In [12] we exploited this approach and considered Monte Carlo methods for the classical nonlinear filtering problem based on a backward pathwise filtering equation and on backward SPDEs (see also Section 31.2 here). These methods have such advantages as a capability in principle to solve filtering problems of large dimensionality, reliable error control, and recurrency. Some other numerical approaches to nonlinear filtering are available in e.g. [1],[4],[5],[2],[3],[6],[8],[9],[16],[18]; and see also references therein.

However, in practice we encounter serious deficiencies in the formula (31.9). In particular, due to the presence of exponents in (31.12), both the numerator and denominator in (31.9) can have explosive behavior (they can become either very large or negligibly small in magnitude) and after some comparatively small time it becomes impossible to continue any computational work with the fraction (31.9). To avoid this difficulty, we propose here to use the innovation process.

In Section 31.2 we consider numerical methods for the backward SPDEs and apply them to solve the nonlinear filtering problem by making use of the Kallianpur–Striebel formula. This section revises and extends some results from [12]. In Section 31.3 we construct and analyse numerical methods for the nonlinear filtering problem, which are based on the innovation process. In Section 31.4 we turn the numerical methods proposed in Section 31.3 into Monte Carlo algorithms. We analyse their Monte Carlo errors and apply the variance reduction method of control variates. Section 31.5 deals with the case of dependent noise in signal and observation. In Section 31.6 we exploit various probabilistic representations of solutions to the considered problems to reduce the Monte Carlo error. In Section 31.7 results of some numerical experiments are presented.

31.2 Numerical methods for backward SPDEs

This section deals with numerical evaluation of the function $u_g(s, x; t)$ from (31.10). Define the $d \times d$-dimensional matrix $a = \{a_{ij}\}$:

$$a(x) = \sigma(x)\sigma^\top(x), \quad a_{ij}(x) = \sum_{k=1}^{d_1} \sigma_{ik}(x)\sigma_{jk}(x). \tag{31.13}$$

The function $u_g(s, x; t)$ is the solution of the Cauchy problem for the backward linear SPDE (see, e.g. [17],[19],[8]):

$$-du = \left[\frac{1}{2}\sum_{i,j=1}^d a_{ij}(x)\frac{\partial^2 u}{\partial x^i \partial x^j} + \sum_{i=1}^d a_i(x)\frac{\partial u}{\partial x^i}\right]ds + \beta^\top(x)u * dy, \quad s < t, \tag{31.14}$$

$$u(t, x) = g(x). \tag{31.15}$$

The notation "$*dy$" means backward Itô integral (see e.g. [19]). Under the assumptions made in the Introduction, the function $u(s, x)$ is smooth in x and the function and its derivatives with respect to x are continuous in s.

The solution of the problem (31.14)–(31.15) has the probabilistic representation [17],[19]:

$$u(s, x; t) = u_g(s, x; t) = \tilde{\mathbb{E}}^y\left[g(X_{s,x}(t))\eta_{s,x}(t)\right], \quad 0 \le s \le t, \tag{31.16}$$

where $X_{s,x}(s'), \eta_{s,x}(s'), s \le s' \le t$, is the solution of the SDEs

$$dX = a(X)ds' + \sigma(X)dw(s'), \quad X(s) = x, \tag{31.17}$$

$$d\eta = \beta^\top(X)\eta dy(s'), \quad \eta(s) = 1. \tag{31.18}$$

Here $w(s')$ is a d_1-dimensional standard Wiener process on $(\Omega, \mathcal{F}, \mathcal{F}_s, \tilde{\mathbb{P}})$ independent of $y(s')$. Note that it is convenient to consider w and X in (31.17) to be different from w and X in (31.1).

We also introduce the scalar $Z_{s,x,z}(s'), s' \ge s$, satisfying the equation

$$dZ = -\frac{1}{2}\|\beta(X)\|^2 ds' + \beta^\top(X)dy(s'), \quad Z(s) = z, \tag{31.19}$$

and we then write

$$u_g(s, x; t) = \tilde{\mathbb{E}}^y\left[g(X_{s,x}(t))\exp(Z_{s,x,0}(t))\right]. \tag{31.20}$$

The introduction of Z is motivated by the fact that it is computationally preferable to simulate (31.19) and then compute $\exp(Z)$ than to simulate (31.18) directly (see further discussion in [12]).

Now we apply the Euler method to (31.17), (31.19):

$$X_0 = x, \quad X_{k+1} = X_k + a(X_k)h + \sigma(X_k)h^{1/2}\zeta_k, \tag{31.21}$$

$$Z_0 = z, \quad Z_{k+1} = Z_k - \frac{1}{2}\|\beta(X_k)\|^2 h + \beta^\top(X_k)\Delta_k y,$$

$$k = 0, \ldots, N-1, \quad h = t/N,$$

where $\zeta_k = \left(\zeta_k^1, \ldots, \zeta_k^{d_1}\right)^\top$ is a vector which components are i.i.d. random variables with the law $\zeta^i \sim \mathcal{N}(0, 1)$ or $\tilde{\mathbb{P}}(\zeta^i = \pm 1) = 1/2$ and $\Delta_k y := y(s_{k+1}) - y(s_k)$.

Then

$$u_g(0, x; s_k) = \tilde{\mathbb{E}}^y \left[g(X_{0,x}(s_k)) \exp(Z_{0,x,0}(s_k)) \right] \simeq \bar{u}_g(0, x; s_k) = \tilde{\mathbb{E}}^y \left[g(X_k) \exp(Z_k) \right] \quad (31.22)$$

$$\simeq \frac{1}{M} \sum_{m=1}^{M} g\left(X_k^{(m)}\right) \exp\left(Z_k^{(m)}\right),$$

where $X_k^{(m)}$, $Z_k^{(m)}$ are independent realizations of X_k, Z_k. The following convergence theorem is proved in [12] ((31.23) for $p = 1$ is proved in [18]).

Theorem 31.1 *The method (31.21) satisfies the inequality for $p \geq 1$:*

$$\left(\tilde{\mathbb{E}} \left| \tilde{\mathbb{E}}^y \left[g(X_{0,x}(t)) \exp(Z_{0,x,0}(t)) \right] - \tilde{\mathbb{E}}^y \left[g(X_N) \exp(Z_N) \right] \right|^{2p} \right)^{1/2p} \leq Kh, \quad k = 0, \ldots, N, \quad (31.23)$$

where the constant K does not depend on the discretization step h.

For almost every trajectory $y(\cdot)$ and any $\varepsilon > 0$ the method (31.21) converges with weak order $1 - \varepsilon$, i.e.,

$$\left| \tilde{\mathbb{E}}^y \left[g(X_{0,x}(t)) \exp(Z_{0,x,0}(t)) \right] - \tilde{\mathbb{E}}^y \left[g(X_N) \exp(Z_N) \right] \right| \leq Ch^{1-\varepsilon} \quad a.s., \quad (31.24)$$

where C is an a.s. bounded constant independent of h.

Recall that we approximate $\hat{f}(s_k) := \hat{f}(X(s_k))$ as

$$\hat{f}(s_k) = \frac{u_f(0, x; s_k)}{u_1(0, x; s_k)} \simeq \hat{f}_k := \frac{\bar{u}_f(0, x; s_k)}{\bar{u}_1(0, x; s_k)} = \frac{\tilde{\mathbb{E}}^y \left[f(X_k) \exp(Z_k) \right]}{\tilde{\mathbb{E}}^y \left[\exp(Z_k) \right]}. \quad (31.25)$$

The following convergence theorem for the fraction $\bar{u}_f / \bar{u}_1$ holds.

Theorem 31.2 *The method (31.25), (31.21) satisfies the inequality for $p \geq 1$:*

$$\left(\tilde{\mathbb{E}} \left| \hat{f}(s_k) - \hat{f}_k \right|^{2p} \right)^{1/2p} \leq Kh, \quad (31.26)$$

where the constant K does not depend on k and h, i.e., in particular, the method is of mean-square order 1.

For almost every trajectory $y(\cdot)$ and any $\varepsilon > 0$ the method (31.25), (31.21) converges with weak order $1 - \varepsilon$, i.e.,

$$\left| \hat{f}(s_k) - \hat{f}_k \right| \leq Ch^{1-\varepsilon} \quad a.s., \quad (31.27)$$

where C is an a.s. bounded random variable independent of k and h.

Proof. We have

$$\hat{f}(s_k) - \hat{f}_k = \hat{f}(s_k) - \frac{\bar{u}_f(0, x; s_k)}{\bar{u}_1(0, x; s_k)} = \frac{\hat{f}(s_k) \bar{u}_1(0, x; s_k) - \bar{u}_f(0, x; s_k)}{\bar{u}_1(0, x; s_k)} \quad (31.28)$$

$$= \frac{\hat{f}(s_k)(\bar{u}_1(0, x; s_k) - u_1(0, x; s_k)) + (u_f(0, x; s_k) - \bar{u}_f(0, x; s_k))}{\bar{u}_1(0, x; s_k)}.$$

The fraction $\hat{f}(s_k)$ is uniformly bounded due to boundedness of the function f. Therefore, we get

$$\left|\hat{f}(s_k)(\bar{u}_1(0, x; s_k) - u_1(0, x; s_k)) + (u_f(0, x; s_k) - \bar{u}_f(0, x; s_k))\right|$$
$$\leq K|\bar{u}_1(0, x; s_k) - u_1(0, x; s_k)| + |u_f(0, x; s_k) - \bar{u}_f(0, x; s_k)|.$$

Then Theorem 31.1 implies the following inequality for the numerator in (31.28):

$$\tilde{\mathbb{E}}\left|\hat{f}(s_k)(\bar{u}_1(0, x; s_k) - u_1(0, x; s_k)) + (u_f(0, x; s_k) - \bar{u}_f(0, x; s_k))\right|^{4p} \leq Kh^{4p}. \tag{31.29}$$

Further, using the conditional Jensen inequality and boundedness of the coefficient $\beta(x)$, we obtain

$$\tilde{\mathbb{E}}\left|\frac{1}{\bar{u}_1(0, x; s_k)}\right|^{4p} = \tilde{\mathbb{E}}\left|\frac{1}{\tilde{\mathbb{E}}^y \exp\left\{\sum_{i=0}^{k-1}\beta^\top(X_i)\Delta_i y - \frac{h}{2}\sum_{i=0}^{k-1}\|\beta(X_i)\|^2\right\}}\right|^{4p} \tag{31.30}$$

$$\leq \tilde{\mathbb{E}}\tilde{\mathbb{E}}^y\left[\frac{1}{\exp\left\{\sum_{i=0}^{k-1}\beta^\top(X_i)\Delta_i y - \frac{h}{2}\sum_{i=0}^{k-1}\|\beta(X_i)\|^2\right\}}\right]^{4p}$$

$$= \tilde{\mathbb{E}}\exp\left\{-4p\sum_{i=0}^{k-1}\beta^\top(X_i)\Delta_i y + 2ph\sum_{i=0}^{k-1}\|\beta(X_i)\|^2\right\} \leq K.$$

The inequality (31.26) follows from the relations (31.28)–(31.30) and the Cauchy–Bunyakovskii inequality.

Now denote $R := \left|\hat{f}(s_k) - \hat{f}_k\right|$. The Markov inequality together with (31.26) implies

$$\mathbb{P}(R > h^\delta) \leq \frac{\mathbb{E}R^{2p}}{h^{2p\delta}} \leq Kh^{2p(1-\delta)}.$$

Then for any $\delta = 1 - \varepsilon$ there is a sufficiently large $p \geq 1$ such that (recall that $h = t/N$)

$$\sum_{N=1}^{\infty}\mathbb{P}\left(R > \frac{t^\delta}{N^\delta}\right) \leq Kt^{2p(1-\delta)}\sum_{N=1}^{\infty}\frac{1}{N^{2p(1-\delta)}} < \infty.$$

Hence, due to the Borel–Cantelli lemma, the random variable $\varsigma := \sup_{h>0}h^{-\delta}R$ is a.s. finite which implies (31.27). □

Remark 31.1 It is not difficult to see that Theorems 31.1 and 31.2 are also valid when the initial condition of $X_{0,\xi}(t)$ is random, i.e., $X(0) = x = \xi$ is a random variable with the density $\varphi(\cdot)$. It is assumed that ξ has finite moments up to a sufficiently high order.

We emphasize that the considered method ensures a recurrent solution of the nonlinear filtering problem. In this connection, we simulate the sample at time s_k:

$$X_k^{(m)} = \bar{X}_{0,x^{(m)}}^{(m)}(s_k), \quad Z_k^{(m)} = \bar{Z}_{0,x^{(m)},0}^{(m)}(s_k), \quad m = 1, \ldots, M,$$

where $x^{(m)}$ are i.i.d. random variables with the density $\varphi(\cdot)$. Then we can estimate the desired quantity $\hat{f}(s_k)$ by the Monte Carlo technique:

$$\hat{f}(s_k) \simeq \bar{f}(s_k) = \frac{\frac{1}{M}\sum_{m=1}^{M}\left[f\left(X_k^{(m)}\right)\exp\left\{Z_k^{(m)}\right\}\right]}{\frac{1}{M}\sum_{m=1}^{M}\exp\left\{Z_k^{(m)}\right\}}, \tag{31.31}$$

and continue the procedure to obtain the sample

$$X_{k+1}^{(m)} = \bar{X}_{s_k,X_k^{(m)}}^{(m)}(s_{k+1}), \quad Z_{k+1}^{(m)} = \bar{Z}_{s_k,X_k^{(m)},Z_k^{(m)}}^{(m)}(s_{k+1})$$

at the next time moment which can be used for finding $\hat{f}(s_{k+1})$ and so on.

Though the fraction $\bar{f}(s_k)$ is uniformly bounded due to boundedness of the function f, it is usual that its numerator and denominator become either very large or negligibly small in magnitude with increasing s_k. Introducing the innovation process and new numerical algorithms in the next section allows us to avoid these deficiencies. Roughly speaking, these new algorithms adaptively adjust magnitudes of the numerator and denominator so that denominator is always approximately equal to one and the numerator approximates $\hat{f}$.

31.3 Numerical methods based on innovation processes

31.3.1 Innovation process

We consider the system (31.1)–(31.2). Introduce

$$\hat{\beta}(t) = \mathbb{E}[\beta(X(t)) \mid y(s),\ 0 \leq s \leq t] = \mathbb{E}^y \beta(X(t))$$

and

$$\rho^{-1}(t) = \exp\left\{-\int_0^t [\beta^\top(X(s)) - \hat{\beta}^\top(s)]dv(s) - \frac{1}{2}\int_0^t \|\beta(X(s)) - \hat{\beta}(s)\|^2 ds\right\} \tag{31.32}$$

$$= \exp\left\{-\int_0^t [\beta^\top(X(s)) - \hat{\beta}^\top(s)]dv(s) + \frac{1}{2}\int_0^t \|\beta(X(s)) - \hat{\beta}(s)\|^2 ds\right\},$$

where $v(s)$ is the innovation process defined by

$$dv = dy - \hat{\beta}(s)ds, \quad v(0) = 0, \tag{31.33}$$

i.e.,
$$v(t) = y(t) - \int_0^t \hat{\beta}(s)ds. \tag{31.34}$$

As in the Introduction, we have
$$\mathbb{E}\rho^{-1}(t) = 1, \ 0 \le t \le T,$$

and we can introduce a new probability measure $\tilde{\mathbb{P}}$ on $(\Omega, \mathcal{F})$ such that the derivative $d\mathbb{P}/d\tilde{\mathbb{P}}$ is equal to $\rho(T)$. Due to the Girsanov theorem, it can be shown that there exists a standard Wiener process $(w(t), \tilde{v}(t))$ on $(\Omega, \mathcal{F}, \mathcal{F}_t, \tilde{\mathbb{P}})$ such that the process $(X(s), v(s))$ satisfies the system of Itô equations

$$dX = a(X)ds + \sigma(X)dw(s), \ X(0) = x, \tag{31.35}$$
$$dv = d\tilde{v}(s), \ v(0) = 0. \tag{31.36}$$

The processes $X(s)$ and $v(s) = \tilde{v}(s)$ are independent on $(\Omega, \mathcal{F}, \mathcal{F}_s, \tilde{\mathbb{P}})$ and $v(s)$ is a Wiener process. The equations (31.33) and (31.36) imply

$$dy = \hat{\beta}(s)ds + d\tilde{v}(s), \ y(0) = 0. \tag{31.37}$$

Due to the general version of Bayes' formula (see e.g. [10], ch. 7, sec. 9; [19], ch. 6, sec. 1.1), we obtain

$$\hat{f}(t) = \mathbb{E}^y f(X(t)) \tag{31.38}$$
$$= \frac{\tilde{\mathbb{E}}[f(X(t))\rho(t) \mid y(s), \ 0 \le s \le t]}{\tilde{\mathbb{E}}[\rho(t) \mid y(s), \ 0 \le s \le t]} = \frac{\tilde{\mathbb{E}}^y[f(X(t))\rho(t)]}{\tilde{\mathbb{E}}^y[\rho(t)]} = \frac{\mathbb{E}\left[f(X(t))\rho(t)\right]\mid_{(31.35)}}{\mathbb{E}\left[\rho(t)\right]\mid_{(31.35)}},$$

where
$$\rho(t) = \exp\left\{\int_0^t [\beta^\top(X(s)) - \hat{\beta}^\top(s)]dv(s) - \frac{1}{2}\int_0^t \|\beta(X(s)) - \hat{\beta}(s)\|^2 ds\right\}. \tag{31.39}$$

The formula (31.40) stated below takes place (see e.g. [7], ch. 11, sec. 3; [19], ch. 6, sec. 2). For completeness of the presentation, we give a short outline of its proof.

Lemma 31.1 *The following formula holds*

$$\hat{f}(t) = \tilde{\mathbb{E}}[f(X(t))\rho(t) \mid y(s), \ 0 \le s \le t] = \tilde{\mathbb{E}}^y[f(X(t))\rho(t)] \tag{31.40}$$
$$= \mathbb{E}\left[f(X(t))\rho(t)\right]\mid_{(31.35)} = \tilde{\mathbb{E}}^y\left[f(X(t))\rho(t)\right].$$

Proof. Clearly, it suffices to prove that

$$\tilde{\mathbb{E}}^y[\rho(t)] = \mathbb{E}\left[\rho(t)\right]\mid_{(31.35)} = \tilde{\mathbb{E}}^y\left[\rho(t)\right] = 1. \tag{31.41}$$

We have

$$\rho(t) = \exp\left\{\int_0^t \beta^\top(X(s))dy(s) - \frac{1}{2}\int_0^t \|\beta(X(s))\|^2 ds\right\} \quad (31.42)$$
$$\times \exp\left\{-\int_0^t \hat{\beta}^\top(s)dy(s) + \frac{1}{2}\int_0^t \|\hat{\beta}(s)\|^2 ds\right\}.$$

Taking expectation with respect to $w(\cdot)$ in (31.42), we get

$$\mathbb{E}\left[\rho(t)\right]|_{(31.35)} = \exp\left\{-\int_0^t \hat{\beta}^\top(s)dy(s) + \frac{1}{2}\int_0^t \|\hat{\beta}(s)\|^2 ds\right\} \quad (31.43)$$
$$\times \mathbb{E}\exp\left\{\int_0^t \beta^\top(X(s))dy(s) - \frac{1}{2}\int_0^t \|\beta(X(s))\|^2 ds\right\}\bigg|_{(31.35)}.$$

Let us recall (see (31.4)) that

$$\exp\left\{\int_0^t \beta^\top(X(s))dy(s) - \frac{1}{2}\int_0^t \|\beta(X(s))\|^2 ds\right\} = \eta(t) \quad (31.44)$$

and

$$\tilde{\mathbb{E}}[\eta(t) \mid y(s),\ 0 \le s \le t] = \mathbb{E}\exp\left\{\int_0^t \beta^\top(X(s))dy(s) - \frac{1}{2}\int_0^t \|\beta(X(s))\|^2 ds\right\}\bigg|_{(31.35)}. \quad (31.45)$$

The equality (31.44) gives

$$\eta(t) = 1 + \int_0^t \beta^\top(X(s))\eta(s)d\tilde{v}(s) \quad (31.46)$$

and hence

$$\tilde{\mathbb{E}}[\eta(t) \mid y(s'),\ 0 \le s' \le t] = 1 + \int_0^t \tilde{\mathbb{E}}[\beta^\top(X(s))\eta(s) \mid y(s'),\ 0 \le s' \le t]d\tilde{v}(s). \quad (31.47)$$

Because X is independent of y on $(\Omega, \mathcal{F}, \tilde{\mathbb{P}})$ and $\int_0^s \beta^\top(X(s'))dy(s')$ does not depend on $y(s') - y(s)$, $s \le s' \le t$, we obtain

$$\tilde{\mathbb{E}}[\beta(X(s))\eta(s) \mid y(s'),\ 0 \le s' \le t] = \tilde{\mathbb{E}}[\beta(X(s))\eta(s) \mid y(s'),\ 0 \le s' \le s]. \quad (31.48)$$

Using (31.9), we get

$$\tilde{\mathbb{E}}[\beta(X(s))\eta(s) \mid y(s'),\ 0 \le s' \le s] = \hat{\beta}(s)\tilde{\mathbb{E}}[\eta(s) \mid y(s'),\ 0 \le s' \le s]. \quad (31.49)$$

The equalities (31.47)–(31.49) imply

$$\tilde{\mathbb{E}}[\eta(t) \mid y(s'),\ 0 \le s' \le t] = 1 + \int_0^t \hat{\beta}^\top(s)\tilde{\mathbb{E}}[\eta(s) \mid y(s'),\ 0 \le s' \le s]d\tilde{v}(s),$$

whence (taking into account that $dy = d\tilde{v}$ on $(\Omega, \mathcal{F}, \tilde{\mathbb{P}})$)

$$\tilde{\mathbb{E}}[\eta(t) \mid y(s), \ 0 \le s \le t] = \exp\left\{\int_0^t \hat{\beta}^\top(s) dy(s) - \frac{1}{2}\int_0^t \|\hat{\beta}(s)\|^2 ds\right\}.$$

This equality together with (31.45) and (31.43) proves (31.41) and (31.40). □

Remark 31.2 From Lemma 31.1, it is not difficult to obtain the stochastic equation of optimal nonlinear filtering for $\hat{f}(t)$ (see [10], ch. 8). Indeed, differentiating the equality

$$\hat{f}(t) = \tilde{\mathbb{E}}[f(X(t))\rho(t) \mid y(s), \ 0 \le s \le t],$$

we get

$$\begin{aligned}
d\hat{f}(t) &= \tilde{\mathbb{E}}[df(X(t)) \cdot \rho(t) \mid y(s), \ 0 \le s \le t] + \tilde{\mathbb{E}}[f(X(t)) \cdot d\rho(t) \mid y(s), \ 0 \le s \le t] \\
&= \tilde{\mathbb{E}}[\mathcal{L} f(X(t))\rho(t) dt \mid y(s), \ 0 \le s \le t] \\
&\quad + \tilde{\mathbb{E}}[f(X(t))[\beta^\top(X(t)) - \hat{\beta}^\top(t)]\rho(t)dv(s) \mid y(s), \ 0 \le s \le t] \\
&= (\widehat{\mathcal{L} f})(t)dt + [(\widehat{f\beta^\top})(t) - \hat{f}(t)\hat{\beta}^\top(t)]dv,
\end{aligned}$$

where

$$\mathcal{L} f(x) := \frac{1}{2}\sum_{i,j=1}^d a_{ij}\frac{\partial^2 f}{\partial x^i \partial x^j} + \sum_{i=1}^d a_i \frac{\partial f}{\partial x^i},$$

whence we come to the equation of optimal nonlinear filtering

$$\hat{f}(t) = \hat{f}(0) + \int_0^t (\widehat{\mathcal{L} f})(s)ds + \int_0^t [(\widehat{f\beta^\top})(s) - \hat{f}(s)\hat{\beta}^\top(s)]dv. \tag{31.50}$$

31.3.2 Numerical methods

The formula (31.40) can be used to construct a numerical method in the following way. Introduce the process $I(s')$ satisfying the equation

$$dI = -\frac{1}{2}\left[\|\beta(X(s'))\|^2 - \|\hat{\beta}(s')\|^2\right]ds' + [\beta^\top(X(s')) - \hat{\beta}^\top(s')]dy(s'), \quad I(0) = 0. \tag{31.51}$$

Then

$$\rho(s) = \exp\{I(s)\} = \exp\left\{-\int_0^s \hat{\beta}^\top(s')dy(s') + \frac{1}{2}\int_0^s \|\hat{\beta}(s')\|^2 ds'\right\}\eta(s).$$

Due to Lemma 31.1, we have

$$\begin{aligned}
\hat{\beta}(s') &= \mathbb{E}\left[\beta(X(s'))\exp\{I(s')\}\right]\big|_{(31.35)} \tag{31.52}\\
&= \tilde{\mathbb{E}}^y\left[\beta(X(s'))\exp\{I(s')\}\right] = \frac{\tilde{\mathbb{E}}^y\left[\beta(X(s'))\exp\{I(s')\}\right]}{\tilde{\mathbb{E}}^y \exp\{I(s')\}}.
\end{aligned}$$

Although $\tilde{\mathbb{E}}^y \exp\{I(s')\} = 1$ according to Lemma 31.1, its approximation can be not equal to 1 and simulation of $\hat{\beta}(s')$ in the form of the fraction (31.52) has many advantages both in theoretical and computational respects (see also the discussion before Theorem 31.4).

We apply the Euler method to the system (31.35), (31.51):

$$X_0 = \xi, \quad X_{k+1} = X_k + a(X_k)h + \sigma(X_k)h^{1/2}\zeta_k, \tag{31.53}$$

$$I_0 = 0, \quad I_{k+1} = I_k - \frac{1}{2}\left[\|\beta(X_k)\|^2 - \|\hat{\beta}_k\|^2\right]h + \left[\beta^\top(X_k) - \hat{\beta}_k^\top\right]\Delta_k y, \tag{31.54}$$

and evaluate $\hat{\beta}$ at each step as

$$\hat{\beta}_0 = \tilde{\mathbb{E}}\beta(\xi), \quad \hat{\beta}_{k+1} = \frac{\tilde{\mathbb{E}}^y\left[\beta(X_{k+1})\exp\{I_{k+1}\}\right]}{\tilde{\mathbb{E}}^y \exp\{I_{k+1}\}}, \tag{31.55}$$

$$k = 0, \ldots, N-1, \quad h = t/N.$$

Here ξ is a random variable with density $\varphi(\cdot)$, $\zeta_k = \left(\zeta_k^1, \ldots, \zeta_k^{d_1}\right)^\top$, $k = 0, \ldots, N-1$, are vectors which components are i.i.d. random variables with the law, e.g., $\zeta^i \sim \mathcal{N}(0,1)$ or $\tilde{\mathbb{P}}(\zeta^i = \pm 1) = 1/2$ and

$$\Delta_k y := y(s_{k+1}) - y(s_k).$$

We approximate $\hat{f}(s_{k+1})$ by

$$\hat{f}(s_{k+1}) \simeq \hat{f}_{k+1} := \frac{\tilde{\mathbb{E}}^y\left[f(X_{k+1})\exp\{I_{k+1}\}\right]}{\tilde{\mathbb{E}}^y \exp\{I_{k+1}\}}, \quad k = 0, \ldots, N-1. \tag{31.56}$$

Theorem 31.3 *The method (31.56), (31.53)–(31.55) satisfies the inequality for $p \geq 1$:*

$$\left(\tilde{\mathbb{E}}\left|\hat{f}(s_k) - \hat{f}_k\right|^{2p}\right)^{1/2p} \leq Kh, \tag{31.57}$$

where the constant K does not depend on k and h, i.e., in particular, the method is of mean-square order 1.

For almost every trajectory $y(\cdot)$ and any $\varepsilon > 0$ the method (31.56), (31.53)–(31.55) converges with weak order $1 - \varepsilon$, i.e.,

$$\left|\hat{f}(s_k) - \hat{f}_k\right| \leq Ch^{1-\varepsilon} \quad a.s., \tag{31.58}$$

where C is an a.s. bounded random variable independent of k and h.

Proof. The estimate $\hat{f}(t)$ from (31.40) can be written in the form (see also (31.43))

$$\hat{f}(s) = \frac{\tilde{\mathbb{E}}^y f(X(s))\exp\{I(s)\}}{\tilde{\mathbb{E}}^y \exp\{I(s)\}} \tag{31.59}$$

$$= \frac{\exp\left\{-\int_0^s \hat{\beta}^\top(s')dy(s') + \frac{1}{2}\int_0^s \|\hat{\beta}(s')\|^2 ds'\right\} \tilde{\mathbb{E}}^y\left[f(X(s))\eta(s)\right]}{\exp\left\{-\int_0^s \hat{\beta}^\top(s')dy(s') + \frac{1}{2}\int_0^s \|\hat{\beta}(s')\|^2 ds'\right\} \tilde{\mathbb{E}}^y \eta(s)}$$

$$= \frac{\tilde{\mathbb{E}}^y\left[f(X(s))\eta(s)\right]}{\tilde{\mathbb{E}}^y \eta(s)} = \frac{u_f(0, x; s)}{u_1(0, x; s)},$$

where $u_g(0, x; s)$ is from (31.10) and it is the solution of the backward SPDE (31.14)–(31.15) with the terminal condition prescribed at $t = s$.

Further, $\hat{f}_k$ from (31.56) can be written as

$$\hat{f}_k = \frac{\tilde{\mathbb{E}}^y f(X_k) \exp\{I_k\}}{\tilde{\mathbb{E}}^y \exp\{I_k\}} \tag{31.60}$$

$$= \frac{\exp\left\{-\sum_{i=0}^{k-1} \hat{\beta}_i^\top \Delta_i y + \frac{h}{2} \sum_{i=0}^{k-1} \|\hat{\beta}_i\|^2\right\} \tilde{\mathbb{E}}^y \left[f(X_k) \exp\left\{\sum_{i=0}^{k-1} \beta^\top(X_i) \Delta_i y - \frac{h}{2} \sum_{i=0}^{k-1} \|\beta(X_i)\|^2\right\}\right]}{\exp\left\{-\sum_{i=0}^{k-1} \hat{\beta}_i^\top \Delta_i y + \frac{h}{2} \sum_{i=0}^{k-1} \|\hat{\beta}_i\|^2\right\} \tilde{\mathbb{E}}^y \exp\left\{\sum_{i=0}^{k-1} \beta^\top(X_i) \Delta_i y - \frac{h}{2} \sum_{i=0}^{k-1} \|\beta(X_i)\|^2\right\}}$$

$$= \frac{\tilde{\mathbb{E}}^y \left[f(X_k) \exp\left\{\sum_{i=0}^{k-1} \beta^\top(X_i) \Delta_i y - \frac{h}{2} \sum_{i=0}^{k-1} \|\beta(X_i)\|^2\right\}\right]}{\tilde{\mathbb{E}}^y \exp\left\{\sum_{i=0}^{k-1} \beta^\top(X_i) \Delta_i y - \frac{h}{2} \sum_{i=0}^{k-1} \|\beta(X_i)\|^2\right\}} = \frac{\bar{u}_f(0, x; s_k)}{\bar{u}_1(0, x; s_k)},$$

where $\bar{u}_g(0, x; s_k)$ coincides with the Euler scheme from (31.21), (31.22) for the SPDE (31.14)–(31.15) with the terminal condition prescribed at $t = s_k$.

The relations (31.59) and (31.60) imply that convergence of the method (31.56), (31.53)–(31.55) is equivalent to convergence of the method (31.25), (31.21). Then (31.57) and (31.58) follow from Theorem 31.2. □

In the above theorem it is proved that for almost every trajectory $y(\cdot)$ the fraction $\dfrac{\bar{u}_f(0, x; s_k)}{\bar{u}_1(0, x; s_k)}$ (and, consequently, $\hat{f}_k$) converges to $\dfrac{u_f(0, x; s_k)}{u_1(0, x; s_k)}$ (i.e., to $\hat{f}(s_k)$) with the weak order $1 - \varepsilon$. However, the next theorem shows that the denominator $\tilde{\mathbb{E}}^y \exp\{I_k\}$ in (31.60) converges to $\tilde{\mathbb{E}}^y \exp\{I(s_k)\}$ (i.e., to 1) with the weak order $1/2 - \varepsilon$ and the numerator $\tilde{\mathbb{E}}^y f(X_k) \exp\{I(s_k)\}$ in (31.60) converges to the numerator in (31.59) (i.e., to $\hat{f}(s_k)$) with the weak order $1/2 - \varepsilon$. On the physical level of rigor this "inconsistency" can be explained as follows. The characteristic system (X, Z) (see (31.17), (31.19)) for $u_g(0, x; s_k)$ is such that the coefficient at dy does not depend on y (recall that here X does not depend on y). We approximate (X, Z) with respect to w in the weak sense and with respect to y in the strong sense. From this prospective, the system (X, Z) can be viewed as a system with additive noise with respect to $y(s)$. Then, if we apply intuition based on the standard theory of numerical integration of SDEs (see [11]), it becomes natural that the Euler-type approximation of (X, Z) is of order one in the mean-square sense with respect to $y(s)$. In the system (X, I) (see (31.35), (31.51)) used to find $\tilde{\mathbb{E}}^y \exp\{I(s_k)\}$, the coefficient at dy depends on y since $\hat{\beta}(s)$ depends on $y(\cdot)$, i.e., the system (X, Z) is with multiplicative noise, and, due to the intuition, the approximation of I in (31.53)–(31.54) should be of mean-square order $1/2$. By including terms with multiple Ito integrals of order $O(h)$ in the approximation of $I(s)$ (31.54), it is possible to get a scheme (see below (31.64)–(31.66)) which approximates $\tilde{\mathbb{E}}^y f(X(s_k)) \exp\{I(s_k)\}$ with the mean-square order one and a.s. weak order $1 - \varepsilon$. Let us note that this does not lead to a better convergence order for the fraction, it remains of order one. A proof of Theorem 31.4 is available in [15].

Theorem 31.4 *For almost every trajectory $y(\cdot)$ and any $\varepsilon > 0$ the method (31.56), (31.53)–(31.55) possesses the property*

$$\left| \hat{f}(s_k) - \tilde{\mathbb{E}}^y f(X_k) \exp\{I_k\} \right| \leq C h^{1/2-\varepsilon} \quad a.s., \tag{31.61}$$

where C is an a.s. bounded random variable independent of k and h.

In particular,

$$\left| 1 - \tilde{\mathbb{E}}^y \exp\{I_k\} \right| \leq C h^{1/2-\varepsilon} \quad a.s.$$

As it was mentioned in the discussion before Theorem 31.4, to improve the order of convergence of $\tilde{\mathbb{E}}^y f(X_k) \exp\{I(s_k)\}$, we need to include terms with multiple Itō integrals of order $O(h)$ in the approximation of $I(s)$ from (31.51). It follows from (31.50) and (31.33) that

$$\sum_{i=1}^{r} \int_{s_k}^{s_{k+1}} \hat{\beta}^i(s) dy^i(s) \tag{31.62}$$

$$= \sum_{i=1}^{r} \int_{s_k}^{s_{k+1}} \left[\hat{\beta}^i(s_k) + \sum_{j=1}^{r} \int_{s_k}^{s_{k+1}} \left\{ (\widehat{\beta^i \beta^j})(s') - \hat{\beta}^i(s')\hat{\beta}^j(s') \right\} dy^j(s') \right] dy^i(s) + A'_k$$

$$= \hat{\beta}^\top(s_k) \Delta_k y + \sum_{i=1}^{r} \sum_{j=1}^{r} \left\{ (\widehat{\beta^i \beta^j})(s_k) - \hat{\beta}^i(s_k)\hat{\beta}^j(s_k) \right\} \int_{s_k}^{s_{k+1}} \int_{s_k}^{s} dy^j(s') dy^i(s) + A_k,$$

where $|\tilde{\mathbb{E}} A_k| \leq K h^2$, $\tilde{\mathbb{E}} |A_k|^{2p} \leq K h^{3p}$ due to boundedness of $\hat{g}(s)$ for all participating g and properties of Itō integrals. Since $(\widehat{\beta^i \beta^j})(s_k) - \hat{\beta}^i(s_k)\hat{\beta}^j(s_k) = (\widehat{\beta^j \beta^i})(s_k) - \hat{\beta}^j(s_k)\hat{\beta}^i(s_k)$, the commutativity condition (see [11], p. 28, for the case of SDEs) holds and

$$\sum_{i=1}^{r} \sum_{j=1}^{r} \left\{ (\widehat{\beta^i \beta^j})(s_k) - \hat{\beta}^i(s_k)\hat{\beta}^j(s_k) \right\} \int_{s_k}^{s_{k+1}} \int_{s_k}^{s} dy^j(s') dy^i(s) \tag{31.63}$$

$$= \frac{1}{2} \sum_{i=1}^{r} \left\{ (\widehat{(\beta^i)^2})(s_k) - (\hat{\beta}^i(s_k))^2 \right\} \left[(\Delta_k y^i)^2 - h \right]$$

$$+ \sum_{i=1}^{r} \sum_{j=i+1}^{r} \left\{ (\widehat{\beta^i \beta^j})(s_k) - \hat{\beta}^i(s_k)\hat{\beta}^j(s_k) \right\} \Delta_k y^i \Delta_k y^j.$$

Based on (31.62)–(31.63), we construct the following method for computing $\hat{f}$:

$$X_0 = \xi, \quad X_{k+1} = X_k + a(X_k)h + \sigma(X_k)h^{1/2}\zeta_k, \tag{31.64}$$

$$I_0 = 0, \quad I_{k+1} = I_k - \frac{1}{2}\left[||\beta(X_k)||^2 - ||\hat{\beta}_k||^2\right]h + \left[\beta^\top(X_k) - \hat{\beta}_k^\top\right]\Delta_k y \quad (31.65)$$

$$-\frac{1}{2}\sum_{i=1}^r \left[\widehat{((\beta^i)^2)}_k - (\hat{\beta}_k^i)^2\right]\left[(\Delta_k y^i)^2 - h\right]$$

$$-\sum_{i=1}^r \sum_{j=i+1}^r \left[\widehat{(\beta^i\beta^j)}_k - \hat{\beta}_k^i \hat{\beta}_k^j\right]\Delta_k y^i \Delta_k y^j,$$

$$\hat{\beta}_0 = \mathbb{E}\beta(\xi), \quad \hat{\beta}_{k+1} = \frac{\tilde{\mathbb{E}}^y\left[\beta(X_{k+1})\exp\{I_{k+1}\}\right]}{\tilde{\mathbb{E}}^y \exp\{I_{k+1}\}}, \quad (31.66)$$

$$\widehat{(\beta^i\beta^j)}_0 = \mathbb{E}\beta^i(\xi)\beta^j(\xi), \quad \widehat{(\beta^i\beta^j)}_{k+1} = \frac{\tilde{\mathbb{E}}^y\left[\beta^i(X_{k+1})\beta^j(X_{k+1})\exp\{I_{k+1}\}\right]}{\tilde{\mathbb{E}}^y \exp\{I_{k+1}\}}, \quad i,j = 1,\ldots,r,$$

$$\hat{f}_{k+1} = \frac{\tilde{\mathbb{E}}^y\left[f(X_{k+1})\exp\{I_{k+1}\}\right]}{\tilde{\mathbb{E}}^y \exp\{I_{k+1}\}}, \quad k = 0,\ldots,N-1, \quad h = t/N, \quad (31.67)$$

where ζ_k are as in (31.53).

Theorem 31.5 *For almost every trajectory $y(\cdot)$ and any $\varepsilon > 0$ the method (31.67), (31.64)–(31.66) converges with weak order $1 - \varepsilon$, i.e.,*

$$\left|\hat{f}(s_k) - \hat{f}_k\right| \leq Ch^{1-\varepsilon} \quad a.s., \quad (31.68)$$

where C is an a.s. bounded random variable independent of k and h.

The method (31.64)–(31.66) possesses the property

$$\left|\hat{f}(s_k) - \tilde{\mathbb{E}}^y f(X_k)\exp\{I_k\}\right| \leq Ch^{1-\varepsilon} \quad a.s. \quad (31.69)$$

A proof of this theorem is available in [15]. We note that in comparison with the method (31.53)–(31.55) the method (31.64)–(31.66) ensures that both the fraction $\tilde{\mathbb{E}}^y\left[f(X_k)\exp\{I_k\}\right]/\tilde{\mathbb{E}}^y \exp\{I_k\}$ converges to $\hat{f}(s_k)$ with order $1 - \varepsilon$ and its numerator and denominator themselves converge to the corresponding values with order $1 - \varepsilon$ (see also the discussion before Theorem 31.4).

31.4 Monte Carlo algorithms

Let $X_k^{(m)}$ and $I_k^{(m)}$, $m = 1,\ldots,M$, be independent realizations of X_k and I_k simulated due to (31.53), (31.54) and let a Monte Carlo approximation $\hat{\beta}_{k,M}$ of $\hat{\beta}_k$ be known. Then we obtain $I_{k+1}^{(m)}$ due to (31.54):

$$I_0^{(m)} = 0, \quad I_{k+1}^{(m)} = I_k^{(m)} - \frac{1}{2}\left[||\beta\left(X_k^{(m)}\right)||^2 - ||\hat{\beta}_{k,M}||^2\right]h + \left[\beta^\top\left(X_k^{(m)}\right) - \hat{\beta}_{k,M}^\top\right]\Delta_k y \quad (31.70)$$

and $\hat{\beta}_{k+1,M}$ and $\hat{f}_{k+1,M}$ due to (31.55) and (31.56):

$$\hat{\beta}_{0,M} = \tilde{\mathbb{E}}^\gamma \beta(\xi), \quad \hat{\beta}_{k+1,M} = \frac{\frac{1}{M}\sum_{m=1}^{M}\left[\beta\left(X_{k+1}^{(m)}\right)\exp\left\{I_{k+1}^{(m)}\right\}\right]}{\frac{1}{M}\sum_{m=1}^{M}\exp\left\{I_{k+1}^{(m)}\right\}}, \quad (31.71)$$

$$\hat{f}_{k+1} \simeq \hat{f}_{k+1,M} = \frac{\frac{1}{M}\sum_{m=1}^{M}\left[f\left(X_{k+1}^{(m)}\right)\exp\left\{I_{k+1}^{(m)}\right\}\right]}{\frac{1}{M}\sum_{m=1}^{M}\exp\left\{I_{k+1}^{(m)}\right\}}. \quad (31.72)$$

Let us analyse the Monte Carlo error of $\hat{f}_{k+1,M}$. The estimate $\hat{f}_{k+1,M}$ is the fraction of the sums, where each sum is not a usual Monte Carlo sum since the summands in (31.72) are though weakly but dependent because of the presence of $\hat{\beta}_{k,M}^\top$ in $I_{k+1}^{(m)}$. At the same time, the considered fraction remains the same if instead of $I_{k+1}^{(m)}$ we set $\tilde{I}_{k+1}^{(m)}$, where

$$\tilde{I}_0^{(m)} = 0, \quad \tilde{I}_{k+1}^{(m)} = \tilde{I}_k^{(m)} - \frac{1}{2}\left[\|\beta\left(X_k^{(m)}\right)\|^2 - \|\hat{\beta}_k\|^2\right]h + \left[\beta^\top\left(X_k^{(m)}\right) - \hat{\beta}_k^\top\right]\Delta_k y. \quad (31.73)$$

Indeed, we have

$$I_{k+1}^{(m)} = -\sum_{i=1}^{k}\left(\frac{1}{2}\|\beta\left(X_i^{(m)}\right)\|^2 h - \beta^\top\left(X_i^{(m)}\right)\Delta_i y\right) + \sum_{i=1}^{k}\left(\frac{1}{2}\|\hat{\beta}_{i,M}\|^2 h - \hat{\beta}_{i,M}^\top \Delta_i y\right)$$

and

$$\tilde{I}_{k+1}^{(m)} = -\sum_{i=1}^{k}\left(\frac{1}{2}\|\beta\left(X_i^{(m)}\right)\|^2 h - \beta^\top\left(X_i^{(m)}\right)\Delta_i y\right) + \sum_{i=1}^{k}\left(\frac{1}{2}\|\hat{\beta}_i\|^2 h - \hat{\beta}_i^\top \Delta_i y\right).$$

Multiplying the numerator and denominator in (31.72) by the factor κ:

$$\kappa = \exp\left\{-\sum_{i=1}^{k}\left(\frac{1}{2}\|\hat{\beta}_{i,M}\|^2 h - \hat{\beta}_{i,M}^\top \Delta_i y\right) + \sum_{i=1}^{k}\left(\frac{1}{2}\|\hat{\beta}_i\|^2 h - \hat{\beta}_i^\top \Delta_i y\right)\right\},$$

which does not depend on m, we obtain

$$\hat{f}_{k+1,M} = \frac{\frac{1}{M}\sum_{m=1}^{M}\left[f\left(X_{k+1}^{(m)}\right)\exp\left\{\tilde{I}_{k+1}^{(m)}\right\}\right]}{\frac{1}{M}\sum_{m=1}^{M}\exp\left\{\tilde{I}_{k+1}^{(m)}\right\}}. \quad (31.74)$$

The sums in (31.74) are usual Monte Carlo sums. Let us underline that though the numerator and denominator in (31.74) cannot be computed because of the unknown $\hat{\beta}_k^\top$, the fraction is computed due to (31.72).

Let us denote

$$\beta_k := \beta(X_k), \quad f_k := f(X_k), \quad \rho_k := \exp\{\tilde{I}_k\}, \quad \xi_k := f_k \rho_k,$$

$$a_k := \tilde{\mathbb{E}}^Y \xi_k, \quad b_k := \tilde{\mathbb{E}}^Y \rho_k, \quad \varepsilon_k := \frac{1}{M}\sum_{m=1}^{M} \xi_k^{(m)} - a_k, \quad \delta_k := \frac{1}{M}\sum_{m=1}^{M} \rho_k^{(m)} - b_k.$$

For sufficiently large M, the quantities ε_k and δ_k can be considered as "small", normally distributed random variables. We have

$$\hat{f}_{k,M} = \frac{\frac{1}{M}\sum_{m=1}^{M} \xi_k^{(m)}}{\frac{1}{M}\sum_{m=1}^{M} \rho_k^{(m)}} = \frac{a_k + \varepsilon_k}{b_k + \delta_k}. \tag{31.75}$$

Clearly,

$$\hat{f}_k = \frac{a_k}{b_k}.$$

Therefore, the Monte Carlo error is equal, up to terms of higher order of smallness, to

$$\hat{f}_{k,M} - \hat{f}_k \simeq \frac{\varepsilon_k}{b_k} - \frac{a_k \delta_k}{b_k^2} := \gamma_k,$$

i.e., the error is approximated by the Gaussian random variable γ_k. We have

$$\tilde{\mathbb{E}}^Y \gamma_k = 0, \quad Var \gamma_k = \frac{1}{b_k^2}\tilde{\mathbb{E}}^Y \varepsilon_k^2 + \frac{a_k^2}{b_k^4}\tilde{\mathbb{E}}^Y \delta_k^2 - \frac{2a_k}{b_k^3}\tilde{\mathbb{E}}^Y \varepsilon_k \delta_k, \tag{31.76}$$

$$\tilde{\mathbb{E}}^Y \hat{f}_{k,M} \simeq \hat{f}_k, \quad Var \hat{f}_{k,M} \simeq Var \gamma_k.$$

Further,

$$b_k \simeq 1, \quad a_k = \hat{f}_k b_k \simeq \hat{f}_k \simeq \hat{f}(s_k), \quad f_k = f(X_k) \simeq f(X(s_k)), \quad \tilde{I}_k \simeq I(s_k).$$

From here and (31.76), it is not difficult to obtain

$$Var \hat{f}_{k,M} \simeq \frac{1}{M}\tilde{\mathbb{E}}^Y[(f(X(s_k)) - \hat{f}(s_k))^2 \exp\{2I(s_k)\}]. \tag{31.77}$$

This formula is of theoretical interest. In practice, $Var \hat{f}_{k,M}$ can be computed due to (31.76) in the following way. We have

$$\frac{1}{b_k^2}\tilde{\mathbb{E}}^Y \varepsilon_k^2 \simeq \frac{\frac{1}{M}\sum_{m=1}^{M}(\xi_k^{(m)})^2 - (\frac{1}{M}\sum_{m=1}^{M}\xi_k^{(m)})^2}{M(\frac{1}{M}\sum_{m=1}^{M}\rho_k^{(m)})^2} \tag{31.78}$$

$$= \frac{\frac{1}{M}\sum_{m=1}^{M}(f_k^{(m)})^2 \exp\{2\tilde{I}_k^{(m)}\} - (\frac{1}{M}\sum_{m=1}^{M} f_k^{(m)} \exp\{\tilde{I}_k^{(m)}\})^2}{M(\frac{1}{M}\sum_{m=1}^{M} \exp\{\tilde{I}_k^{(m)}\})^2}$$

$$= \frac{\frac{1}{M}\sum_{m=1}^{M}(f_k^{(m)})^2 \exp\{2 I_k^{(m)}\} - (\frac{1}{M}\sum_{m=1}^{M} f_k^{(m)} \exp\{I_k^{(m)}\})^2}{M(\frac{1}{M}\sum_{m=1}^{M} \exp\{I_k^{(m)}\})^2}.$$

The last equality in (31.78) is obtained by multiplying the numerator and denominator by the factor $1/\kappa$. Proceeding analogously with the other terms of (31.76), we obtain the following formula for practical computing:

$$\operatorname{Var} \hat{f}_{k,M} \simeq \frac{V_k}{M}, \quad V_k := \frac{S_{5,k}}{S_{1,k}^2} + \frac{S_{2,k}^2 S_{3,k}}{S_{1,k}^4} - 2\frac{S_{2,k} S_{4,k}}{S_{1,k}^3}, \qquad (31.79)$$

where the sums $S_{1,k}, \ldots, S_{5,k}$ are equal to:

$$S_{1,k} = \frac{1}{M}\sum_{m=1}^{M} \exp\left\{I_k^{(m)}\right\}, \quad S_{2,k} = \frac{1}{M}\sum_{m=1}^{M} f_k^{(m)} \exp\left\{I_k^{(m)}\right\}, \quad S_{3,k} = \frac{1}{M}\sum_{m=1}^{M} \exp\left\{2 I_k^{(m)}\right\},$$

$$(31.80)$$

$$S_{4,k} = \frac{1}{M}\sum_{m=1}^{M} f_k^{(m)} \exp\left\{2 I_k^{(m)}\right\}, \quad S_{5,k} = \frac{1}{M}\sum_{m=1}^{M} \left(f_k^{(m)}\right)^2 \exp\left\{2 I_k^{(m)}\right\}.$$

Thus,

$$\hat{f}(s_k) = \hat{f}_k + e_{NI} = \hat{f}_{k,M} + e_{NI} + e_{MC}$$

with the error of numerical integration $e_{NI} \sim O(h^{1-\varepsilon})$ and with the Monte Carlo error e_{MC} which is evaluated by $e_{MC} \sim c\sqrt{V_k}/\sqrt{M}$, where, e.g., the values $c = 2, 3$ correspond to the fiducial probabilities $0.95, 0,997$, respectively. During simulation, the Monte Carlo error is controlled very easily according to the formulas (31.79), (31.80).

Due to Theorem 31.5 (Theorem 31.4), we can approximate $\hat{f}(s_k)$ not only by $\hat{f}_k$ but also by $\check{f}_k := \widetilde{\mathbb{E}}^y\left[f(X_k)\exp\{I_k\}\right]$ using the method (31.67), (31.64)–(31.66) (the method (31.56), (31.53)–(31.55)) with the accuracy $O(h^{1-\varepsilon})$ (accuracy $O(h^{1/2-\varepsilon})$), i.e., by the numerator from (31.67) (by the numerator from (31.56)). Using the Monte Carlo approach, we get

$$\check{f}_k \simeq \check{f}_{k,M} := \frac{1}{M}\sum_{m=1}^{M} f_k^{(m)} \exp\left\{I_k^{(m)}\right\}. \qquad (31.81)$$

Though the summands in (31.81) are dependent, we consider (31.81) as a usual Monte Carlo sum, regarding such an influence to be negligible. The sample variance $\operatorname{Var}\check{f}_{k,M}$ is computed in the usual way (see (31.80)):

$$\operatorname{Var}\check{f}_{k,M} = \frac{1}{M}\left[\frac{1}{M}\sum_{m=1}^{M}\left(f_k^{(m)}\right)^2 \exp\left\{2 I_k^{(m)}\right\} - \left(\frac{1}{M}\sum_{m=1}^{M} f_k^{(m)}\exp\left\{I_k^{(m)}\right\}\right)^2\right]$$

$$= \frac{1}{M}\left[S_{5,k} - S_{2,k}^2\right].$$

Clearly, the variance of the estimator $\check{f}_k$ is equal to

$$\operatorname{Var} \check{f}_{k,M} = \frac{1}{M}\left(\tilde{\mathbb{E}}^y\left[f^2(X_k)\exp\{2I_k\}\right] - \check{f}_k^2\right).$$

Let us apply the variance reduction method of control variates using the fact (see (31.41)) that

$$\tilde{\mathbb{E}}^y \exp\{I_k\} \simeq \tilde{\mathbb{E}}^y \exp\{I(s_k)\} = 1. \qquad (31.82)$$

To this aim, we introduce for some $\lambda \in \mathbb{R}$:

$$\check{f}_{k,\lambda} := \tilde{\mathbb{E}}^y \left[f(X_k)\exp\{I_k\} - \lambda(\exp\{I_k\} - 1)\right].$$

We get (in the case of the method (31.67), (31.64)–(31.66))

$$\check{f}_{k,\lambda} \simeq \check{f}_k = \hat{f}(s_k) + O(h^{1-\varepsilon}).$$

Further, disregarding the difference (see (31.82)) $\tilde{\mathbb{E}}^y \exp\{I_k\} - 1 = \tilde{\mathbb{E}}^y(\exp\{I_k\} - 1) \simeq 0$, we obtain

$$\min_\lambda \operatorname{Var}\left[f(X_k)\exp\{I_k\} - \lambda(\exp\{I_k\} - 1)\right]$$
$$\simeq \tilde{\mathbb{E}}^y\left[f^2(X_k)\exp\{2I_k\}\right] - (\tilde{\mathbb{E}}^y\left[f(X_k)\exp\{I_k\}\right])^2 - \frac{(\tilde{\mathbb{E}}^y\left[f(X_k)\exp\{I_k\}(\exp\{I_k\}-1)\right])^2}{\tilde{\mathbb{E}}^y(\exp\{I_k\}-1)^2}$$

under

$$\lambda = \frac{\tilde{\mathbb{E}}^y\left[f(X_k)\exp\{I_k\}(\exp\{I_k\}-1)\right]}{\tilde{\mathbb{E}}^y(\exp\{I_k\}-1)^2}.$$

Therefore, it is reasonable to introduce the following $\check{f}_{k,\lambda,M}$ instead of $\check{f}_{k,M}$:

$$\check{f}_{k,\lambda,M} = \frac{1}{M}\left[\sum_{m=1}^M f_k^{(m)} \exp\{I_k^{(m)}\} - \lambda \sum_{m=1}^M \left(\exp\{I_k^{(m)}\} - 1\right)\right]. \qquad (31.83)$$

In practice, λ can be replaced by its estimate

$$\check{\lambda} := \frac{\frac{1}{M}\sum_{m=1}^M f_k^{(m)}\exp\{I_k^{(m)}\}(\exp\{I_k^{(m)}\} - 1)}{\frac{1}{M}\sum_{m=1}^M (\exp\{I_k^{(m)}\} - 1)^2}. \qquad (31.84)$$

Clearly, $\operatorname{Var}\check{f}_{k,\lambda,M}$ is not larger than $\operatorname{Var}\check{f}_{k,M}$:

$$\operatorname{Var}\check{f}_{k,\lambda,M} \simeq (1 - \rho_k^2)\operatorname{Var}\check{f}_{k,M},$$

where ρ_k is the correlation coefficient between $f(X_k)\exp\{I_k\}$ and $\exp\{I_k\}$. It can be proved that $\operatorname{Var}\check{f}_{k,\lambda,M}$ is not larger than $\operatorname{Var}\hat{f}_{k,M}$ too (see (31.77)). Some numerical experiments illustrating this variance reduction method of control variates are presented in [15].

31.5 The case of dependent noise in signal and observation

Consider a more general filtering scheme than (31.1)-(31.2):

$$dX = a(X)ds + \sigma(X)dw(s) + \gamma(X)dv(s), \quad X(0) = x, \tag{31.85}$$

$$dy = \beta(X)ds + dv(s), \quad y(0) = 0, \tag{31.86}$$

where, as in (31.1)–(31.2), $X(t) \in \mathbb{R}^d$ is the unobservable signal process, $y(t) \in \mathbb{R}^r$ is the observation process; $(w(t), \mathcal{F}_t)$ and $(v(t), \mathcal{F}_t)$ are d_1-dimensional and r-dimensional independent standard Wiener processes on $(\Omega, \mathcal{F}, \mathbb{P})$, respectively; $a(x)$ and $\beta(x)$ are d-dimensional and r-dimensional vector functions, respectively; $\sigma(x)$ and $\gamma(x)$ are $d \times d_1$-dimensional and $d \times r$-dimensional matrix functions, respectively. As before, we assume that the coefficients of (31.85)–(31.86) are bounded and have bounded derivatives up to some order. We note that here, in comparison with (31.1)–(31.2), the noise in the signal (31.85) and the noise in the observation (31.86) are correlated.

Analogously to (31.11), the solution of the filtering problem (31.3), (31.85)–(31.86) can be written as (see e.g. [17],[19])

$$\hat{f}(X(t)) = \mathbb{E}[f(X(t)) \mid y(s), \ 0 \le s \le t] = \frac{u_{f,\varphi}(0, t)}{u_{1,\varphi}(0, t)}, \tag{31.87}$$

where

$$u_{g,\varphi}(0, t) := \int u_g(0, x; t)\varphi(x)dx$$

and $u_g(s, x; t)$ is the solution of the linear backward SPDE (cf. (31.14), (31.15)):

$$-du = \left[\frac{1}{2}\sum_{i,j=1}^{d} a_{ij}(x)\frac{\partial^2 u}{\partial x^i \partial x^j} + \sum_{i=1}^{d} a_i(x)\frac{\partial u}{\partial x^i}\right]ds + \left(\sum_{i=1}^{d} \gamma_i^\top(x)\frac{\partial u}{\partial x^i} + \beta^\top(x)u\right) * dy, \tag{31.88}$$

$$u(t, x) = g(x). \tag{31.89}$$

Here $\gamma_i^\top(x)$, $i = 1, \ldots, d$, are rows of the matrix $\gamma(x)$, the matrix $a(x)$ is defined as

$$a(x) := \sigma(x)\sigma^\top(x) + \gamma(x)\gamma^\top(x),$$

and $y(s)$ is a Wiener process on $(\Omega, \mathcal{F}, \mathcal{F}_s, \tilde{\mathbb{P}})$, $\tilde{\mathbb{P}}$ is introduced in the same way as in the Introduction.

The solution of the problem (31.88)–(31.89) has the following probabilistic representation [17],[19]:

$$u(s, x) = u(s, x; t) = \tilde{\mathbb{E}}^y \left[g(X_{s,x}(t))\eta_{s,x}(t)\right], \quad 0 \le s \le t, \tag{31.90}$$

where $X_{s,x}(s')$, $\eta_{s,x}(s')$, $s \leq s' \leq t$, is the solution of the SDEs

$$dX = \left[a(X) - \gamma(X)\beta(X)\right]ds' + \gamma(X)dy(s') + \sigma(X)dw(s'), \quad X(s) = x, \quad (31.91)$$
$$d\eta = \beta^\top(X)\eta dy(s'), \quad \eta(s) = 1. \quad (31.92)$$

Here the Wiener processes $w(s')$ and $y(s')$ are independent on $(\Omega, \mathcal{F}, \mathcal{F}_s, \tilde{\mathbb{P}})$. Note that it is convenient to consider w and X in (31.91) to be different from w and X in (31.85). We have $\eta(t) = \exp(Z(t))$, where $Z(t)$ satisfies the equation

$$dZ = -\frac{1}{2}\|\beta(X)\|^2 ds' + \beta^\top(X)dy(s'). \quad (31.93)$$

We apply the Euler-type scheme to (31.91), (31.93):

$$X_0 = x, \quad X_{k+1} = X_k + h\left[a(X_k) - \gamma(X_k)\beta(X_k)\right] + \gamma(X_k)\Delta_k y + h^{1/2}\sigma(X_k)\zeta_k, \quad (31.94)$$

$$Z_0 = 0, \quad Z_{k+1} = Z_k - \frac{h}{2}\|\beta(X_k)\|^2 + \beta^\top(X_k)\Delta_k y, \quad k = 0, \ldots, N-1,$$

where ζ_{rk} are i.i.d. random variables with the law

$$\tilde{\mathbb{P}}(\zeta = \pm 1) = 1/2 \quad (31.95)$$

and $\Delta_k y := y(s_{k+1}) - y(s_k)$.

Let

$$\bar{u}(0, x) := \tilde{\mathbb{E}}^y\left[g(X_N)\exp\{Z_N\}\right], \quad (31.96)$$

where X_N, Z_N are from (31.94).

The approximation (31.94), (31.96) is a generalization of the approximation (31.21), (31.22) to the correlated case. The following theorem is proved in [14].

Theorem 31.6 *The method* (31.94), (31.96) *satisfies the inequality for $p \geq 1$:*

$$\left(\tilde{\mathbb{E}}\left|\bar{u}(0, x) - u(0, x)\right|^{2p}\right)^{1/2p} \leq Kh^{1/2}, \quad (31.97)$$

where K does not depend on the discretization step h, i.e., in particular (31.94), (31.96) *is of mean-square order $1/2$.*

For almost every trajectory $y(\cdot)$ and any $\varepsilon > 0$ there exists a constant $C > 0$ such that

$$\left|\bar{u}(0, x) - u(0, x)\right| \leq Ch^{1/2-\varepsilon}, \quad (31.98)$$

i.e., the method (31.94), (31.96) *converges with order $1/2 - \varepsilon$ a.s.*

Now let us exploit the innovation approach and propose the corresponding numerical scheme. As in (31.39), we introduce

$$\rho(t) = \exp\left\{\int_0^t [\beta^\top(X(s)) - \hat{\beta}^\top(s)]dy(s) - \frac{1}{2}\int_0^t \left[\|\beta(X(s))\|^2 - \|\hat{\beta}(s)\|^2\right]ds\right\}. \quad (31.99)$$

Clearly,

$$\rho(t) = \exp\left\{-\int_0^t \hat{\beta}^\top(s) dy(s) + \frac{1}{2} \int_0^t \|\hat{\beta}(s)\|^2 ds\right\} \eta(t). \tag{31.100}$$

Analogously to the proof of Lemma 31.1, it can be proved that

$$\tilde{\mathbb{E}}^y \rho(t) = 1. \tag{31.101}$$

However, one point in the proof of Lemma 31.1 requires a clarification. To get (31.48), we used the fact that X is independent of y on $(\Omega, \mathcal{F}, \tilde{\mathbb{P}})$. Now X is dependent on y. But $X(s')$, $0 \le s' \le s$, does not depend on $y(s'') - y(s)$, $s \le s'' \le t$, because $y(s')$ is a Wiener process in (31.91). This is sufficient for obtaining (31.48) in the considered correlated case as well.

Due to (31.101),

$$\hat{f}(t) = \tilde{\mathbb{E}}^y \left[f(X(t))\rho(t) \right]. \tag{31.102}$$

The method, analogous to the method (31.56), (31.53)–(31.55) in the noncorrelated case, has the form

$$X_0 = x, \quad X_{k+1} = X_k + h \left[a(X_k) - \gamma(X_k)\beta(X_k) \right] + \gamma(X_k)\Delta_k y + h^{1/2} \sigma(X_k) \zeta_k, \tag{31.103}$$

$$I_{k+1} = I_k - \frac{h}{2} \left[\|\beta(X_k)\|^2 - \|\hat{\beta}_k\|^2 \right] + \left[\beta^\top(X_k) - \hat{\beta}_k^\top \right] \Delta_k y, \quad I_0 = 0,$$

$$\hat{\beta}_0 = \tilde{\mathbb{E}}\beta(\xi), \quad \hat{\beta}_{k+1} = \frac{\tilde{\mathbb{E}}^y \left[\beta(X_{k+1}) \exp\{I_{k+1}\} \right]}{\tilde{\mathbb{E}}^y \exp\{I_{k+1}\}}, \tag{31.104}$$

$$\hat{f}_{k+1} = \frac{\tilde{\mathbb{E}}^y \left[f(X_{k+1}) \exp\{I_{k+1}\} \right]}{\tilde{\mathbb{E}}^y \exp\{I_{k+1}\}}, \tag{31.105}$$

$$k = 0, \ldots, N-1, \quad h = t/N,$$

where ξ is a random variable with density $\varphi(\cdot)$ and ζ_k are as in (31.95).

Using Theorem 31.6, we prove the convergence theorem for the method (31.105), (31.103)–(31.104).

Theorem 31.7 *For almost every trajectory $y(\cdot)$ and any $\varepsilon > 0$ the method (31.105), (31.103)–(31.104) converges with weak order $1/2 - \varepsilon$, i.e.,*

$$\left| \hat{f}(s_k) - \hat{f}_k \right| \le Ch^{1/2-\varepsilon} \quad \text{a.s.,} \tag{31.106}$$

and the method (31.103)–(31.104) possesses the property

$$\left| \hat{f}(s_k) - \tilde{\mathbb{E}}^y f(X_k) \exp\{I_k\} \right| \le Ch^{1/2-\varepsilon} \quad \text{a.s.,} \tag{31.107}$$

where C is an a.s. bounded random variable independent of k and h.

A proof of this theorem is available in [15]. We see (cf. Theorem 31.7 with Theorems 31.3 and 31.4) that here, in comparison with the uncorrelated case, the fraction $\tilde{\mathbb{E}}^y \left[f(X_k) \exp\{I_k\} \right] / \tilde{\mathbb{E}}^y \exp\{I_k\}$ converges to $\hat{f}(s_k)$ with order $1/2 - \varepsilon$ and the numerator and denominator themselves in this fraction converge to the corresponding values with order $1/2 - \varepsilon$ as well. In this respect let us note that in (31.91) X depends on y while in the uncorrelated case (31.17) X does not depend on y. Since the coefficients at dy in (31.91), (31.93) depend on y, it is consistent with the intuition based on the standard theory of numerical integration of SDEs (see [11]) that the Euler-type method for (31.91), (31.93) is of order $1/2$ (see also the discussion before Theorem 31.4).

31.6 Variance reduction

In this section we exploit various probabilistic representations of solutions to the considered SPDE problems to reduce the Monte Carlo error. As it is discussed before, we simulate the SPDE solutions $u_{g,\varphi}$ with random initial condition as

$$u_{g,\varphi}(0,t) = \tilde{\mathbb{E}}^y \left[g(X_{0,\xi}(t)) \eta_{0,\xi}(t) \right] = \tilde{\mathbb{E}}^y \left[g(X_{0,\xi}(t)) \exp(Z_{0,\xi}(t)) \right] \quad (31.108)$$
$$\simeq \bar{u}_{g,\varphi} = \tilde{\mathbb{E}}^y \left[g(\bar{X}_{0,\xi}(t)) \exp(\bar{Z}_{0,\xi}(t)) \right]$$
$$\simeq \frac{1}{M} \sum_{m=1}^{M} g(\bar{X}^{(m)}_{0,x^{(m)}}(t)) \exp(\bar{Z}^{(m)}_{0,x^{(m)}}(t)),$$

where ξ is a random variable with the density $\varphi(\cdot)$, $x^{(m)}$ are independent realizations of ξ, and $\bar{X}^{(m)}_{0,x^{(m)}}(t)$, $\bar{Z}^{(m)}_{0,x^{(m)}}(t)$ are independent realizations of $\bar{X}_{0,\xi}(t)$, $\bar{Z}_{0,\xi}(t)$. The Monte Carlo error of (31.108) can be evaluated by

$$\rho = c \frac{\left[Var^y(g(X_{0,\xi}(t)) \exp(Z_{0,\xi}(t))) \right]^{1/2}}{M^{1/2}}, \quad (31.109)$$

where, e.g., the values $c = 1, 2, 3$ correspond to the fiducial probabilities 0.68, 0.95, 0.997, respectively.

Let us first consider the case when $\xi = x$ is deterministic. The solution of the problem (31.88)–(31.89) has the following wider class of probabilistic representations (cf. (31.90)–(31.92)) (see [14]):

$$u(s,x) = u(s,x;t) = \tilde{\mathbb{E}}^y \left[g(X_{s,x}(t)) \eta_{s,x}(t) + \mathbb{X}_{s,x}(t) \right], \quad 0 \le s \le t, \quad (31.110)$$

where $X_{s,x}(s')$, $\eta_{s,x}(s')$, $\mathbb{X}_{s,x}(s')$, $s \le s' \le t$, is the solution of the SDEs

$$dX = \left[a(X) - \gamma(X)\beta(X) - \sigma(X)\lambda(s',X) \right] ds' + \gamma(X) dy(s') \quad (31.111)$$
$$+ \sigma(X) dw(s'), \quad X(s) = x,$$

$$d\eta = \lambda^\top(s', X)\eta dw(s') + \beta^\top(X)\eta dy(s'), \quad \eta(s) = 1,$$
$$d\mathbb{X} = \mathbb{F}^\top(s', X)\eta dw(s'), \quad \mathbb{X}(s) = 0.$$

Here λ and $\mathbb{F}$ are column-vector functions of dimension d_1 satisfying some regularity assumptions. When $\lambda = 0$ and $\mathbb{F} = 0$, we have the usual representation (31.90)–(31.92). The variety of probabilistic representations (31.110)–(31.111) can be exploited for variance reduction. The following theorem is useful in this respect (see its proof in [14], and in the case of $\gamma = 0$ in [12]). Let

$$\Gamma_{0,x} = \Gamma_{0,x}(t) := g(X_{0,x}(t))\eta_{0,x}(t) + \mathbb{X}_{0,x}(t).$$

Theorem 31.8 *Let λ and $\mathbb{F}$ be such that for any $x \in \mathbb{R}^d$ there exists a solution to the system (31.111) on the interval $[0, t]$. Then*

$$\mathrm{Var}^y(\Gamma_{0,x}(t)) = \tilde{\mathbb{E}}^y \int_0^t \eta_{0,x}^2 \sum_{j=1}^{d_1}\left[\sum_{i=1}^d \sigma_{ij}\frac{\partial u}{\partial x^i} + u\lambda_j + \mathbb{F}_j\right]^2 ds \qquad (31.112)$$

provided that the expectation in (31.112) exists. In (31.112) all the functions σ_{ij}, λ_j, $\mathbb{F}_j$, u, $\partial u/\partial x^i$ have $(s, X_{0,x}(s))$ as their argument.

We see that if $\lambda(s, x)$ and $\mathbb{F}(s, x)$ are such that

$$\sum_{i=1}^d \sigma_{ij}\frac{\partial u}{\partial x^i} + u\lambda_j + \mathbb{F}_j = 0, \quad j = 1, \ldots, d_1, \qquad (31.113)$$

then the right-hand side of (31.112) is zero and, consequently, the variance is zero. We recall that $u(s', x)$ depends on $y(s'')$, $s' \leq s'' \leq t$. However, we need to require that $\lambda(s', x)$ does not depend on $y(s'')$, $s' \leq s'' \leq t$, otherwise $X(s')$ in (31.111) depends on $y(s'')$, $s' \leq s'' \leq t$, and we are facing the difficulty in interpreting (31.111). At the same time, dependence of the function $\mathbb{F}(s', x)$ on $y(s'')$ does not cause any trouble in interpreting the third equation in (31.111), and the identity (31.113) can, in principle, be reached. Theorem 31.8 can be used, for example, if we know a function $\tilde{u}(s, x)$ being close to $u(s, x)$ (see a practical approach to constructing such approximate functions in the case of deterministic PDEs in [13]). Then we take a $\lambda(s, x)$ independent of $y(s')$, $s \leq s' \leq t$, and any $\mathbb{F}(s, x)$ satisfying (31.113) with $\tilde{u}(s, x)$ instead of $u(s, x)$ in it and expect that the conditional variance $\mathrm{Var}^y(\Gamma_{0,x}(t))$ is although not zero but small.

Now consider the case when the initial data ξ are random. To reduce the variance associated with ξ, one can exploit both the method of importance sampling and the method of control variates. The method of importance sampling for this purpose was considered in [12]. Here we propose the method of control variates. We have (see details in [12]):

$$Var^y\left(\Gamma_{0,\xi}(t)\right) = \int_{\mathbb{R}^d} Var^y(\Gamma_{0,x}(t))\varphi(x)dx + Var(u(0, \xi)). \qquad (31.114)$$

The variance $Var^y(\Gamma_{0,x}(t))$ in (31.114) can be reduced due to Theorem 31.8 if a suitable function $\tilde{u}(s, x)$, $0 \le s \le t$, $x \in \mathbb{R}^d$, is known. The second term $Var(u(0, \xi))$ is connected with the Monte Carlo error in evaluating the deterministic integral $\int u(0, x)\varphi(x)dx$. Assuming that we know $\tilde{u}(x) = \tilde{u}(0, x) \simeq u(0, x)$ together with $\mathbb{E}\tilde{u}(\xi)$, we can propose the following estimate

$$\begin{aligned} u_{g,\varphi} &= \tilde{\mathbb{E}}^y\left[g(X_{0,\xi}(t))\eta_{0,\xi}(t) + \mathbb{X}_{0,\xi}(t) - u(0, \xi)\right] + \mathbb{E}u(0, \xi) \qquad (31.115) \\ &\simeq \tilde{u}_{g,\varphi} = \tilde{\mathbb{E}}^y\left[g(\tilde{X}_{0,\xi}(t))\exp(\tilde{Z}_{0,\xi}(t)) + \tilde{\mathbb{X}}_{0,\xi}(t) - \tilde{u}(\xi)\right] + \mathbb{E}\tilde{u}(\xi) \\ &\simeq \frac{1}{M}\sum_{m=1}^{M}\left[g\left(\tilde{X}^{(m)}_{0,x^{(m)}}(t)\right)\exp\left\{\tilde{Z}^{(m)}_{0,x^{(m)}}(t)\right\} + \tilde{\mathbb{X}}^{(m)}_{0,x^{(m)}}(t) - \tilde{u}(x^{(m)})\right] + \mathbb{E}\tilde{u}(\xi). \end{aligned}$$

It is not difficult to see that

$$Var^y\left(g(X_{0,\xi}(t))\eta_{0,\xi}(t) + \mathbb{X}_{0,\xi}(t) - u(0, \xi)\right) = \int_{\mathbb{R}^d} Var^y(\Gamma_{0,x}(t))\varphi(x)dx.$$

Then, thanks to Theorem 31.8, if we take $\lambda(s, x)$ and $\mathbb{F}(s, x)$ satisfying (31.113), we get

$$Var^y\left(g(X_{0,\xi}(t))\eta_{0,\xi}(t) + \mathbb{X}_{0,\xi}(t) - u(0, \xi)\right) = 0.$$

Hence the variance of the estimate in the right-hand side of (31.115) can be made smaller than variance of the estimate in (31.108) by a proper choice of $\lambda(s, x)$, $\mathbb{F}(s, x)$, and $\tilde{u}(x)$.

31.7 Numerical experiments

In our illustrative examples of this section we restrict ourselves to the one-dimensional linear filtering problem for simplicity and clarity. At the same time, let us emphasize that the proposed Monte Carlo methods are designed to solve multidimensional nonlinear filtering problems.

In the experiments we consider the linear stochastic system

$$dX = aXds' + \sigma dw(s'), \quad X(0) = \xi, \qquad (31.116)$$
$$dy = \beta Xds' + dv(s'), \quad y(0) = 0, \qquad (31.117)$$

where X and y are scalars, a, σ, and β are parameters, and the random variable ξ is normally distributed $\mathcal{N}(m_0, P_0)$. We are interested in computing the estimate

$$\hat{X}(t) := \mathbb{E}\left(X(t) \mid y(s'), \ 0 \le s' \le t\right). \qquad (31.118)$$

Let us note that in the Introduction the standard assumptions (see e.g. [10],[17]) on the scheme (31.1)–(31.3) are imposed requiring from the coefficients to be bounded and sufficiently smooth functions with bounded derivatives. These assumptions are sufficient for all the statements of this paper. However, they are not necessary and the methods proposed can be used under broader conditions. The authors do not doubt that all the results of the paper remain true for unbounded coefficients, provided that they have bounded derivatives up to some order. The methods in experiments with the model (31.116)–(31.118) behave according to our theoretical predictions.

According to the Kalman–Bucy filter, the estimate $\hat{X}(s)$ satisfies the system

$$d\hat{X} = a\hat{X}ds' + \beta P(s')(dy - \beta \hat{X}ds'), \quad \hat{X}(0) = m_0, \qquad (31.119)$$
$$\dot{P} = 2aP - \beta^2 P^2 + \sigma^2, \quad P(0) = P_0.$$

In our tests we fix the observation $y(t)$ which is obtained as a result of simulation of (31.116)–(31.117) by the mean-square Euler method with small timestep ($h = 0.0001$) for a particular, fixed realization of the Wiener process ($w(s'), v(s')$). We compare the algorithms proposed in the previous sections with the reference value of $\hat{X}$ obtained by an accurate simulation of the Kalman–Bucy filter (31.119) which is done by the Euler method with $h = 0.0001$.

As we said in the Introduction, in practice we encounter computational deficiencies in realization of the Kalianpur–Striebel formula (31.9) since the numerator and denominator in (31.9) can have explosive behavior. More precisely, their absolute values can become very large (they can reach values of order 10^{324} and higher which leads to the overflow error on a computer) or negligibly small (they can be of order 10^{-324} and smaller which leads to the underflow error on a computer) and after some comparatively small time it becomes impossible to continue simulation of the fraction (31.9) by direct methods like the "SPDE algorithm" (31.31), (31.21) from Section 31.2. Even if the overflow/underflow error does not appear in simulation, the fact that the numerator and denominator often take values, which magnitudes are very large or very small, is very worrying from the computational point of view since it is the standard computational practice to avoid working with numbers of such orders. To overcome this difficulty, we propose to use methods from Section 31.3 based on the innovation process, which, as we mentioned before, adaptively regulate the magnitudes of the numerator and denominator in order for them to remain within acceptable limits.

For instance, in the experiments we observe that in the unstable case of (31.116)–(31.117) with $a = 1$, $\beta = 1$, $\sigma = 1$, $m_0 = 0$, $P_0 = 1$, in the "SPDE algorithm" (31.31), (31.21) the overflow error occurs at $t = 4.5$ for any (appropriate) h while the "innovation algorithm" (31.53), (31.54), (31.71), (31.72) with $h = 0.1$ works up to $t = 8.5$ and for smaller timesteps it works further (e.g., for $h = 0.02$ – up to $t = 11$). The simulation results for both algorithms at various times are given in Table 31.1. In the "SPDE algorithm" part of the table the "numerator," "denominator," and "fraction" mean the corresponding quantities in (31.31) and in the "innovation algorithm" they mean the corresponding quantities

Table 31.1 Simulation of $\hat{X}(t)$ from (31.118). The reference value of $\hat{f}(t) = \hat{X}(t)$ is found due to the Kalman-Bucy filter (31.119). Here $\alpha = 1$, $\sigma = 1$, $\beta = 1$, $m_0 = 0$, $P_0 = 1$. The timestep is $h = 0.02$. The expectations are computed by the Monte Carlo technique simulating $M = 4 \cdot 10^7$ independent realizations. All simulations are done along the same observation path $y(t)$. The "±" reflects the Monte Carlo error only.

		"SPDE algorithm"		
t	$\hat{f}(t)$	Numerator	Denominator	Fraction
1	2.865462	9.6549 ± 0.0056	3.3619 ± 0.0015	2.8718 ± 0.0007
2	5.222653	$9.059 \cdot 10^2 \pm 0.73$	$1.730 \cdot 10^2 \pm 0.14$	5.2372 ± 0.0010
3	10.055472	$2.195 \cdot 10^{11} \pm 3.5 \cdot 10^8$	$2.169 \cdot 10^{10} \pm 3.5 \cdot 10^7$	10.116 ± 0.002
4	32.090861	$3.341 \cdot 10^{109} \pm 1.2 \cdot 10^{107}$	$1.035 \cdot 10^{108} \pm 3.7 \cdot 10^{105}$	32.269 ± 0.004
4.5	53.798144		overflow	
5	87.918489		overflow	

		"Innovation algorithm"		
t	$\hat{f}(t)$	Numerator	Denominator	Fraction
1	2.865462	2.3817 ± 0.0014	0.8293 ± 0.0004	2.8718 ± 0.0007
2	5.222653	5.9194 ± 0.0047	1.1302 ± 0.0009	5.2372 ± 0.0010
3	10.055472	9.777 ± 0.016	0.9666 ± 0.0016	10.116 ± 0.002
4	32.090861	30.61 ± 0.11	0.9485 ± 0.0034	32.269 ± 0.004
4.5	53.798144	45.78 ± 0.24	0.8464 ± 0.0043	54.089 ± 0.006
5	87.918489	83.51 ± 0.55	0.9443 ± 0.0062	88.427 ± 0.008

in (31.72). In Table 31.1 one can see the rapid growth of numerator and denominator in (31.31) which eventually leads to overflow. At the same time, the magnitudes of the numerator and denominator in (31.72) remain within acceptable limits. Until the overflow, the two methods give the same results. We note that the time limitation for the "innovation algorithm" in this case is due to numerical stability limitations of the Euler scheme used.

Explosive behavior of the "SPDE algorithm" does not depend much on the timestep, it is due to the above discussed problem that the absolute values of the numerator and denominator in the Kalianpur–Striebel formula become very large.

In the stable case of (31.116)–(31.117) with $\alpha = -10$, $\beta = -10$, $\sigma = 5$, $m_0 = 0$, $P_0 = 2$, in the "SPDE algorithm" (31.31), (31.21) the underflow error occurs at $t \approx 30$ while the "innovation algorithm" (31.53), (31.54), (31.71), (31.72) works for any time (obviously, to reach a specified accuracy, one needs to choose a sufficiently small timestep h and large number of Monte Carlo runs M). With a set of more moderate parameters, e.g. $\alpha = -1$, $\beta = -1$, $\sigma = 5$, $m_0 = 0$, $P_0 = 1$, in the "SPDE algorithm" the numerator and denominator are of order 10^{15} at time $t = 12$, when the solution of the filtering problem $\hat{X}(12)$ is equal to 3.341462.

Recall that we run the experiments for a particular realization of the Wiener process $(w(s'), v(s'))$.

We note that knowledge of the exact value of the denominator (it equals one) in the case of innovation-based methods allows us to have effective control of accuracy of simulation: if the value of the denominator deviates too far from one, the timestep should be decreased to reach a desirable accuracy.

Acknowledgements

We acknowledge partial support from the UK EPSRC Research Grant EP/D049792/1.

References

[1] Bensoussan, A., Glowinski, R., and Rascanu, A., 1990. Approximation of the Zakai equation by the splitting up method. *SIAM J. Contr. Optim.*, 28(6), 1420–31.
[2] Crisan, D., 2006. Particle approximations for a class of stochastic partial differential equations. *J Appl. Maths Optim.*, 54, 293–314.
[3] Crisan, D. and Lyons, T., 1999. A particle approximation of the solution of the Kushner-Stratonovich equation. *Prob. Theory Rel. Fields*, 115, 549–78.
[4] Del Moral, P., 1996. Non-linear filtering: Interacting particle solution. *Markov Proc. Rel. Fields*, 2, 555–80.
[5] Del Moral, P. and Miclo, L., 2000. Branching and interacting particle approximations of Feynman-Kac formulae with applications to nonlinear filtering. *Lecture Notes in Math.*, vol. 1729, Springer, 1–145.
[6] Itô, K. and Rozovskiĭ, B., 2000. Approximation of the Kushner equation for nonlinear filtering. *SIAM J. Contr. Optim.*, 38, 893–915.
[7] Kallianpur, G., 1980. *Stochastic Filtering Theory*. Springer.
[8] Kushner, H. J., 1977. *Probability Methods for Approximations in Stochastic Control and for Elliptic Equations*. Academic Press.
[9] Le Gland, F., 1992. Splitting-up approximation for SPDEs and SDEs with application to nonlinear filtering. *Lecture Notes in Control and Inform. Sc.*, vol. 176, Springer, 177–87.
[10] Liptser, R. S. and Shiryaev, A. N., 1977. *Statistics of Random Processes*. Springer.
[11] Milstein, G. N. and Tretyakov, M. V., 2004. *Stochastic Numerics for Mathematical Physics*. Springer.
[12] Milstein, G. N. and Tretyakov, M. V., 2009a. *Monte Carlo algorithms for backward equations in nonlinear filtering*. Advances in Applied Probability, 41(1), 63–100.
[13] Milstein, G. N. and Tretyakov, M. V., 2009b. *Practical variance reduction via regression for simulating diffusions*. SIAM J. Numer. Anal., 47(2), 887–910.
[14] Milstein, G. N. and Tretyakov, M. V., 2009c. *Solving parabolic stochastic partial differential equations via averaging over characteristics*. Math. Comp., 78(268), 2075–2106.

[15] Milstein, G. N. and Tretyakov, M. V., 2008. *Averaging Over Characteristics with Innovation Approach in Nonlinear Filtering*. Technical Report No. MA-08-001. School of Mathematics and Computer Science, University of Leicester.

[16] Newton, N. J., 2000. Observation sampling and quantisation for continuous-time estimators. *Stoch. Proc. Appl.*, 87, 311–37.

[17] Pardoux, E., 1981. Nonlinear filtering, prediction and smoothing. In: Hazewinkel M. and Willems J. C., eds, *Stochastic Systems: The Mathematics of Filtering and Identification and Applications*, NATO Advanced Study Institute series, D. Reidel, 529–57.

[18] Picard, J., 1984. Approximation of nonlinear filtering problems and order of convergence. *Lecture Notes in Contr. Inform. Sc.*, vol. 61, Springer, 219–36.

[19] Rozovskiĭ, B. L., 1991. *Stochastic Evolution Systems, Linear Theory and Application to Nonlinear Filtering*. Kluwer Academic Publishers.

PART IX
Nonlinear Filtering in Financial Mathematics

·32·
Nonlinear Filtering in Models for Interest-Rate and Credit Risk

R. Frey and W. Runggaldier

32.1 Introduction

Modern financial mathematics is mainly concerned with the pricing and hedging of derivative securities, with portfolio optimization, and with risk management and the statistical analysis of financial data. All these activities are based on mathematical models for the dynamics of the underlying economic quantities such as security prices. These models need to capture the complicated nonlinear dynamics of real asset prices while being at the same time parsimonious and numerically tractable. Factor models have proven to be a useful tool for meeting these conflicting objectives, since the quantities of interest can be expressed in terms of relatively few factors. Moreover, with Markovian factor processes, Markov-process techniques can be fruitfully employed. In most financial applications of factor models investors have only incomplete information about the state of the factor process, essentially for the following reasons: first, some factors are associated with economic quantities which are hard to observe precisely such as instantaneous interest rates, volatilities, or the asset value of a firm; second, abstract factors without direct economic interpretation are often included in the specification of a model in order to increase its flexibility. When applying the model, the current state of the factors therefore needs to be inferred from observable quantities such as historical price data. Filtering is an elegant and theoretically consistent way for doing this, which is why filtering techniques are increasingly being used in all areas of financial mathematics.

In the present paper we concentrate on the application of filtering techniques in the context of incomplete-information models for interest-rate and credit risk that are of the type of jump-diffusion models. Our main concern is the pricing of derivatives via martingale methods; hedging and parameter estimation are touched upon occasionally.[1] This focus is motivated by trends in the current literature and by our own research interests over the last few years. We remark at this point that nonlinear filtering has been applied very successfully to pricing, hedging, and parameter-estimation problems in marked point process models driven by an unobservable volatility factor; see for instance [34], [33], [31], [60], [38], [20] or [19].

The outline of the chapter is as follows: In Section 32.2 we give a brief introduction to arbitrage-free models for the term-structure of interest rates with a particular emphasis on factor models. This sets the scene for our discussion of term-structure models under incomplete information in Section 32.3. Here we start with a general result which shows that arbitrage-free prices with respect to the subfiltration representing the information actually available to investors can be computed by projection. In the remainder of Section 32.3 this principle is applied within specific factor models for the term structure of interest rates and this leads to a number of interesting filtering problems. Sections 32.4, 32.5 and 32.6 are devoted to an analysis of nonlinear filtering in dynamic credit risk models: in Section 32.4 we give an overview of key modeling approaches and explain how and where incomplete information enters; in Section 32.5 we discuss nonlinear filtering problems in the context of firm-value models with noisily observed asset value; Section 32.6 deals with reduced-form models. Section 32.7 summarizes the chapter. Rather than aiming at a complete description of available results, we will concentrate on a few illustrative models, many of them coming from our own activity in the field. We assume throughout that the reader is familiar with standard nonlinear filtering theory; a comprehensive modern account can be be found in the recent monograph [2].

Throughout the chapter we denote by $(\mathcal{G}_t)$ the global or full-information filtration, so that all processes introduced will be $(\mathcal{G}_t)$ adapted; the information actually available to investors is represented by the subfiltration $(\mathcal{F}_t)$. Moreover, we generally adopt bold-face notation for vectors and vector-valued stochastic processes.

32.2 The term structure of interest rates: full information

In this section we give a brief introduction to models for the term structure of interest rates; details and further information can for instance be found in [6].

Bonds and interest rates. A zero-coupon bond or T-bond is a contract guaranteeing a unit amount at a given future date T without intermediate payments; the price of such a contract at a time $t \leq T$ is denoted by $p(t, T)$. The collection of bond prices $p(t, T)$, $T \geq t$, completely describes the term structure of interest rates, or, equivalently, the time-value of money at a given point in time t. Various notions of interest rates can be defined from the family $p(t, T)$, $T \geq t$. An important example is the *simple compounded interest rate* for the future time period $[T, S]$ and contracted at $t < T$, denoted by $L(t; T, S)$. This rate is given by

$$L(t; T, S) = \frac{p(t, T) - p(t, S)}{(S - T) p(t, S)} = \frac{1}{S - T} \left[\frac{p(t, T)}{p(t, S)} - 1 \right]. \tag{32.1}$$

Assuming that, as a function of T, $p(t, T)$ is sufficiently regular, letting $S \downarrow T$, one obtains the instantaneous *forward rate*

$$f(t, T) = \lim_{S \downarrow T} L(t; T, S) = -\frac{\partial}{\partial T} \log p(t, T). \tag{32.2}$$

By its definition, $f(t, T)$ represents the rate, evaluated at $t < T$, for an instantaneous borrowing at T. From (32.2), using the fact that $p(T, T) = 1$, we also get the inverse relationship

$$p(t, T) = \exp\left(-\int_t^T f(t, u) du\right), \tag{32.3}$$

so that there is a one-to-one relationship between the family of bond prices $p(t, T)$, $T \geq t$ and the family of forward rates $f(t, T)$, $T \geq t$. The (instantaneous) *short rate* is finally defined by $r_t := f(t, t)$.

Martingale pricing. As mentioned in the introduction, our main concern in this chapter is the pricing of derivatives via martingale methods. This methodology is based on a widely used economic principle, namely the notion of absence of arbitrage. This principle basically states that, in equilibrium, the prices of the assets on a given market have to be such that by investing in this market it is not possible to make a sure profit without risk. According to the so-called *first fundamental theorem of asset pricing* the mathematical counterpart of this principle is the existence of an equivalent martingale measure. This is a measure Q^N, equivalent to the physical/real-world measure P, so that the prices of all the assets in a given market expressed in units of a given reference asset (*numeraire*) N with price $N_t > 0$ are Q^N martingales with respect to a given generic filtration $(\mathcal{H}_t)$ representing the information available to investors in the model. Formally, the price of a nondividend-paying traded asset $(S_t)_{t \geq 0}$ thus satisfies for all $t \leq T$

$$\frac{S_t}{N_t} = E^{Q^N}\left(\frac{S_T}{N_T} \mid \mathcal{H}_t\right) \tag{32.4}$$

Under the popular *martingale modelling approach* relation (32.4) is used for constructing the price dynamics of the traded assets as follows: suppose that the value of a security[2] at some given future date T is given by a known $\mathcal{H}_T$-measurable random variable Π_T. A prime case in point is a T bond where $\Pi_T \equiv 1$. Given a numeraire N, a candidate martingale measure Q^N and a filtration $(\mathcal{H}_t)$, the price Π_t of this security at $t \leq T$ is then defined to be

$$\Pi_t = N_t \, E^{Q^N}\left(\frac{\Pi_T}{N_T} \mid \mathcal{H}_t\right). \tag{32.5}$$

Model parameters are determined by the requirement that the model-implied price Π_t from (32.5) should coincide with the price observed on the market;

this goes under the label *calibration to market data*. In a second step the price of nontraded derivatives is defined by the analogous expression to (32.5). In this way it is automatically ensured that the resulting model is arbitrage-free and that derivatives are priced consistently with the prices of traded assets. A frequently used numeraire is the so-called money market account (locally risk-free asset) that is the asset with value $B_t = B_0 \exp\left(\int_0^t r_s ds\right)$, r the short rate. The martingale measure corresponding to B as numeraire is commonly denoted by Q.

Note that the real-world measure P does not enter in this approach, and in fact it is common practice to set up a pricing model for derivatives without specifying the real-world dynamics of security prices. A conceptual problem may however arise at this point, since the measure Q^N need not be unique and since different martingale measures can lead to different prices for nontraded derivatives. This problem is closely related to the so-called *completeness* of the market (see for instance Chs 8, 10, and 14 of [6]). Economic criteria for choosing one of these measures do usually invoke the physical measure P and martingale modeling is no longer sufficient. In practical applications of derivative pricing models this issue is largely neglected and the chosen martingale measure is kept fixed, a praxis which is also adopted in the present chapter.

Heath–Jarrow–Morton approach. We proceed now to derive dynamic models for the term structure that do not allow for arbitrage opportunities. A recent such modelling approach is the so-called *Heath–Jarrow–Morton (HJM) approach* [43]. Under this approach one models directly the dynamics of the forward rates and derives from there the dynamics of bond prices and related quantities. Here we restrict ourselves to Wiener driven models and assume that the forward rate dynamics are of the form

$$df(t, T) = a(t, T)dt + \sigma(t, T)d\mathbf{W}_t, \qquad (32.6)$$

where $\mathbf{W}_t$ is a d-dimensional Wiener process on a given filtered probability space $(\Omega, \mathcal{G}, (\mathcal{G}_t), Q)$ and $a(\cdot, T)$, $\sigma(\cdot, T)$ are adapted processes with values in $\mathbb{R}$ and $\mathbb{R}^d$ respectively. Note that (32.6) may be interpreted as a system of infinite stochastic differential equations, one for each T.

The fact of having in principle infinitely many assets, given by the bonds of the various maturities T, implies that with a model as in (32.6) one might introduce arbitrage into the market. A simple way to preclude arbitrage opportunities is to specify the dynamics of the processes $f(\cdot, T)$ in such a way that the given measure Q is a martingale measure. It is well known that (modulo some integrability conditions) Q is a martingale measure if and only the if the so-called *HJM-drift condition* is satisfied, that is the following relation between the drift a and the volatility σ has to hold:

$$a(t, T) = \sigma(t, T) \int_t^T \sigma'(t, u) du; \qquad (32.7)$$

see e.g. [6] for details. Rewriting (32.6) in integral form, namely

$$f(t, T) = f^*(0, T) + \int_0^t a(s, T) ds + \int_0^t \sigma(s, T) d\mathbf{W}_s, \qquad (32.8)$$

one sees that, in the HJM setup, the inputs for a model defined under a martingale measure Q are: i) the volatility structure $\sigma(t, T)$; ii) the initially observed forward rate curve $f^*(0, T)$. The structure of the model is thus specified by specifying $\sigma(t, T)$.

Factor models. In the given set-up the models are a-priori infinite-dimensional and one may ask whether, by a judicious choice of the volatility structure $\sigma(t, T)$ in the HJM framework (32.6), they may become equivalent to a model driven by a finite-dimensional factor process. The question has a positive answer and a general account on this issue may be found in [5]. For the filter application below we recall here a specific case from [14]. Take $d = 1$ and let

$$\sigma(t, T) = g(r_t) e^{-\lambda(T-t)} \quad \text{with} \quad g(r) = \sigma_0 |r|^\delta, \qquad (32.9)$$

where r_t is the short rate and $\sigma_0, \delta, \lambda$ are parameters to be determined from market prices. It can be shown, see [14], that in this case the entire term structure can be expressed as driven by two forward rate processes $f(\cdot, T_1), f(\cdot, T_2)$ with maturities T_1, T_2 that may be chosen arbitrarily. One has in fact

$$p(t, T) = \exp\left\{-\bar{a}_0(t, T) - \bar{a}_1(t, T) f(t, T_1) - \bar{a}_2(t, T) f(t, T_2)\right\} \qquad (32.10)$$

for suitable functions $\bar{a}_i : [0, T] \to \mathbb{R}; i = 0, 1, 2$. Choosing $T_1 = t$ and $T_2 = \tau > t$ arbitrary but fixed so that $f(t, T_1) = r_t$, $f(t, T_2) = f(t, \tau)$, one obtains (see always [14]) a Markovian system for $\mathbf{X}_t := (r_t, f(t, \tau))$ of the form

$$\begin{cases} dr_t = (\beta_0(t) + \beta_1(t) r_t + \beta_2(t) f(t, \tau)) \, dt + g(r_t) d W_t \\ df(t, \tau) = (\gamma_0(t) + \gamma_1(t) r_t + \gamma_2(t) f(t, \tau)) \, dt + g(r_t) e^{-\lambda(\tau-t)} d W_t \end{cases} \qquad (32.11)$$

for suitable time functions $\beta_i(t), \gamma_i(t), i = 0, 1, 2$. The two-dimensional Markovian factor process $(r_t, f(t, \tau))$ drives now the entire term structure in the sense that

$$p(t, T) = \exp\left\{-a_0(t, T) - a_1(t, T) r_t - a_2(t, T) f(t, \tau)\right\} \qquad (32.12)$$

for suitable functions $a_i(t, T)$ that correspond to the $\bar{a}_i(t, T)$ in (32.10) for $T_1 = t, T_2 = \tau$.

An alternative way for constructing factor models is to specify a finite-dimensional Markovian factor process $\mathbf{X}$ and to represent the term structure in the form $p(t, T) = F^T(t, \mathbf{X}_t)$ for a suitable family of functions $F^T(t, x)$, $T \geq t$. In this way one ensures a-priori that the whole term structure evolves on a finite-dimensional manifold. A special case are the classical short-rate models where $\mathbf{X}$ is identified with the short rate r itself (r is then modelled as a Markov process), so that bond prices take the form $p(t, T) = F^T(t, r_t)$.

In order to exclude the possibility of arbitrage one has to impose appropriate conditions on the family $F^T(t, x)$, $T \geq t$. One way to proceed is to apply Itô's formula to $F^T(t, \mathbf{X}_t)$ and to derive dynamics for $p(t, T)$ and, via (32.2), the corresponding dynamics of $f(t, T)$. On the forward-rate dynamics one imposes the HJM drift condition, which leads to a PDE for $F^T(t, x)$, usually called the *term structure equation*. One context where this PDE becomes relatively easily solvable by means of ordinary differential equations are the so-called *affine term structure models*. In the next example we present a special case; we shall come back to this example in our discussion of term structure models under incomplete information in Section 32.3.2 below.

Example 32.1 (Linear-Gaussian Factor Models) On $(\Omega, \mathcal{G}, (\mathcal{G}_t), Q)$ consider an N-dimensional factor process $\mathbf{X}$ satisfying the linear-Gaussian dynamics

$$d\mathbf{X}_t = F\,\mathbf{X}_t dt + D\,d\mathbf{W}_t \tag{32.13}$$

with $\mathbf{W}$ an M-dimensional ($M \geq N$) $(Q, (\mathcal{G}_t))$-Wiener process and with F and D parametric matrices such that D has full rank. It can be shown that in this case the term structure is exponentially affine in $\mathbf{X}_t$, i.e.

$$p(t, T) = \exp\left\{A(t, T) - \mathbf{B}(t, T)\mathbf{X}_t\right\} \tag{32.14}$$

for deterministic functions $A(\cdot, T): [0, T] \to \mathbb{R}$ and $\mathbf{B}(\cdot, T): [0, T] \to \mathbb{R}^N$. It follows from the HJM drift condition that $A(\cdot, T)$ and $\mathbf{B}(\cdot, T)$ have to satisfy the following system of ODEs

$$\begin{cases} \frac{\partial}{\partial t}\mathbf{B}(t, T) + \mathbf{B}(t, T)\,F + b(t) = 0 \\ \frac{\partial}{\partial t}A(t, T) + \frac{1}{2}\mathbf{B}(t, T)\,D\,D'\mathbf{B}'(t, T) - a(t) = 0, \end{cases} \tag{32.15}$$

with terminal condition $A(T, T) = \mathbf{B}(T, T) = 0$. Here $b(t)$ is a parametric function which has to be calibrated together with the matrices F and D; $a(t)$ is defined via $a(t) = f^*(0, t) + \frac{1}{2}\int_0^t \beta_T(s, t)ds$ where $\beta(t, T) = \mathbf{B}(t, T)\,D\,D'\mathbf{B}'(t, T)$ and where $f^*(0, t)$ are the initially observed forward rates. For further use note that the log-prices are of the form

$$Y_t^T := \log p(t, T) = A(t, T) - \mathbf{B}(t, T)\mathbf{X}_t, \tag{32.16}$$

so that log-prices are affine functions of the factors. From $r_t = f(t, t)$ and $f(t, T) = -\frac{\partial}{\partial T}\log p(t, T)$ (see (32.2)) one immediately has that the short rate r_t can be expressed as a linear combination of the factors as well. Applying Itô's formula one then obtains the following dynamics type

$$dr_t = \left(a_t^0 + \beta_t^0 \mathbf{X}_t\right) dt + \sigma_t^0 d\mathbf{W}_t$$
$$dY_t^T = \left(a_t^T + \beta_t^T \mathbf{X}_t\right) dt + \sigma_t^T d\mathbf{W}_t \tag{32.17}$$

for suitable coefficients. For a general discussion about affine term structure models we refer to [24] or [6].

32.3 The term structure of interest rates: incomplete information

32.3.1 Pricing under incomplete information and nonlinear filtering

If the factor process $\mathbf{X}$ is observable, or equivalently, if we work under the global filtration $(\mathcal{G}_t)$, bond prices can be obtained in the form $p(t, T; \mathbf{X}_t) = F^T(t, \mathbf{X}_t)$. Moreover, in most cases of interest the function F^T can be computed explicitly. The picture changes if we assume that the information available to investors corresponds to a subfiltration $\mathcal{F}_t \subset \mathcal{G}_t$ such that $\mathbf{X}$ is not $(\mathcal{F}_t)$-adapted. In the following lemma we show how to pass under the martingale pricing approach from the full information prices $p(t, T; \mathbf{X}_t)$ to arbitrage-free prices in the investor filtration $(\mathcal{F}_t)$; the latter will be denoted by $\hat{p}(t, T)$.

Lemma 32.1 *Let N be a given numeraire that is adapted to the investor filtration $(\mathcal{F}_t)$ and choose a corresponding martingale measure Q^N. Denote by $p(t, T; \mathbf{X}_t) = N_t E^{Q^N}(1/N_T \mid \mathcal{G}_t)$ arbitrage-free bond prices under full information, and by $\hat{p}(t, T) := N_t E^{Q^N}(1/N_T \mid \mathcal{F}_t)$ the corresponding arbitrage-free prices with respect to the investor filtration $(\mathcal{F}_t)$. Then one has that*

$$\hat{p}(t, T) = E^{Q^N}\left(p(t, T; \mathbf{X}_t) \mid \mathcal{F}_t\right). \tag{32.18}$$

In particular, if the savings account B is $(\mathcal{F}_t)$-adapted, we obtain $\hat{p}(t, T) = E^Q(p(t, T; \mathbf{X}_t) \mid \mathcal{F}_t)$.

Proof. By the very definition of a martingale, the fact that the bond prices at maturity T are equal to 1 and the assumption that $N_t \in \mathcal{F}_t$, for the first statement we have that

$$\hat{p}(t, T) = N_t E^{Q^N}\left(\frac{1}{N_T} \mid \mathcal{F}_t\right) = E^{Q^N}\left(N_t E^{Q^N}\left(\frac{1}{N_T} \mid \mathcal{G}_t\right) \mid \mathcal{F}_t\right) = E^{Q^N}\left(p(t, T; \mathbf{X}_t) \mid \mathcal{F}_t\right).$$

The second statement is then immediate. □

Comments. The result can be extended to general $\mathcal{F}_T$-measurable claims and to credit-risky securities in an obvious way. The lemma shows that in order to obtain arbitrage-free prices in the investor filtration, one has to compute the conditional expectation in (32.18), which amounts to solving a filtering problem. In abstract terms the solution of this filtering problem is given by the *optional projection* of the process $(p(t, T; \mathbf{X}_t))_{t \leq T}$ on $(\mathcal{F}_t)$; the latter is usually denoted by $p(\widehat{t, T; \mathbf{X}_t})$, which motivates the notation $\hat{p}(t, T)$. Note, moreover, that the conditional expectation in (32.18) has to be computed with respect to the chosen martingale measure, so that martingale pricing leads to filtering problems under the martingale measure Q^N (rather than the physical measure P).[3] Suppose finally that for a certain maturity $\bar{T}$ the price of the $\bar{T}$-bond is assumed to be observable (in mathematical terms, $(\mathcal{F}_t)$-adapted), and moreover equal to the model value $(p(t, \bar{T}; \mathbf{X}_t))_{t \leq \bar{T}}$. In that case we obviously have $\hat{p}(t, \bar{T}) = p(t, \bar{T}; \mathbf{X}_t)$ so that the filtered model is "automatically" calibrated to the observed bond price; we will encounter a specific example of this in the next subsection.

In the rest of this section we describe some specific models. Rather than aiming at a complete description of available results, we shall concentrate on a few illustrative examples that come mostly from our own activities in this field.

32.3.2 Filtering in affine factor models

In this subsection we discuss the application of the pricing principle from Lemma 32.1 in the context of the linear-Gaussian factor model of Example 32.1; our description is based on the analysis of [41] and [40]. We consider two different scenarios with incomplete information about the factor process $\mathbf{X}$. In both cases investors observe (possibly with noise) a finite number of yields $y(t, T_i) = -\frac{1}{T_i - t} \log p(t, T_i)$, $i = 1, \cdots, n$, or equivalently the logarithmic bond prices $Y_t^i = \log p(t, T_i)$, and in addition the short rate. In the first scenario the observations of the yields and of the short rate are given by perturbed versions of the theoretical model values. In the second case it is assumed that model values can be observed exactly; however, the factor process $\mathbf{X}$ will be high-dimensional so that its current value $\mathbf{X}_t$ cannot be inferred from the observed model values. Both scenarios lead to a linear filtering problem; we shall also mention an extension to nonlinear filtering.

1. Filtering with observations given by perturbed model values. Recall the dynamics of the factor process $\mathbf{X}$, of the short-rate r and of the logarithmic bond-prices Y^i from (32.17). Here we assume that perturbed versions $\tilde{r}$ and $\tilde{Y}^i$ are observable; these perturbed versions are generated by adding independent Wiener-type observation noises v_t^i, $i = 0, \cdots, n$ to the original processes. The investor filtration is thus given by

$$\mathcal{F}_t = \sigma\left(\tilde{r}_s, \tilde{Y}_s^i; \ s \leq t, \ i = 1, \cdots, n\right), \qquad (32.19)$$

where state process $\mathbf{X}$ and observations $\tilde{r}, \tilde{Y}^1, \ldots, \tilde{Y}^n$ have the following dynamics (for $t < \min\{T_i : 1 \leq i \leq n\}$)[4]

$$\begin{cases} d\mathbf{X}_t = F\mathbf{X}_t dt + D\,d\mathbf{W}_t \\ d\tilde{r}_t = \left(a_t^0 + \beta_t^0 \mathbf{X}_t\right) dt + \sigma_t^0 d\mathbf{W}_t + dv_t^0 \\ d\tilde{Y}_t^i = \left(a_t^i + \beta_t^i \mathbf{X}_t\right) dt + \sigma_t^i d\mathbf{W}_t + (T_i - t) dv_t^i\,; \quad i = 1, \cdots, n; \end{cases} \qquad (32.20)$$

the time-dependent volatility of the additional noise reflects the fact that bond-price volatility converges to zero as time approaches the maturity date of the bond. Since for the given model we are in the affine term structure context of (32.14), for the prices $\hat{p}(t, T)$ we have that

$$\hat{p}(t, T) = E(p(t, T; \mathbf{X}_t) \mid \mathcal{F}_t) = \exp(A(t, T)) \, E\left(\exp(-\mathbf{B}(t, T)\mathbf{X}_t) \mid \mathcal{F}_t\right), \quad (32.21)$$

where the last term corresponds to the conditional moment generating function of $\mathbf{X}_t$ given $\mathcal{F}_t$. Since the filtering model in (32.20) is linear Gaussian, the filter distribution is Gaussian as well so that, denoting its conditional mean and covariance by $\mathbf{m}_t$ and Σ_t respectively, from (32.21) one obtains

$$\hat{p}(t, T) = \exp\left\{A(t, T) - \mathbf{B}(t, T)\mathbf{m}_t + \frac{1}{2}\mathbf{B}(t, T)\Sigma_t \mathbf{B}'(t, T)\right\} \quad (32.22)$$

For the given model the pricing under incomplete information can thus be accomplished by solving the system of ODEs in (32.15) and the Kalman filter corresponding to (32.20).

Taking a financial point of view this simple model is not completely satisfactory for the following two reasons: first, recall from Lemma 32.1 that formula (32.21) is justified if B is $(\mathcal{F}_t)$-adapted. In that case the short rate is strictly speaking $(\mathcal{F}_t)$-adapted as well (since $r_t = \frac{d}{dt} \ln B_t$), contradicting (32.19). However, a very small amount of observation noise for B (which, from a practical point of view would still permit the use of Lemma 32.1) leads to a substantial observation noise for the short rate $r_t = \frac{d}{dt} \ln B_t$, so that the assumption that the short rate cannot be observed perfectly can be defended. Second, there is also the problem that, for maturities T_i corresponding to liquid bonds, $\hat{p}(t, T_i)$ does in general not coincide with the observed values for these maturities (recall the third point in the comments directly after Lemma 32.1). In the next subsection we discuss a variant of the model that overcomes these issues.

2. Filtering with exact observations of the theoretical prices. We assume now that the dimension N of the factor process $\mathbf{X}$ is strictly larger than the number of traded bonds with observable prices. This occurs for instance in the case when maturity-specific idiosyncratic factors are being added (see the situations considered in [41], [40]). In this case there is no need to add exogenous noise terms to justify a filtering set-up, and the observation filtration is given by

$$\mathcal{F}_t = \sigma\left(r_s, Y_s^i \, ; \, s \leq t, \, i = 1, \cdots, n\right), \quad (32.23)$$

where, in line with (32.16) and (32.17),

$$r_t = a(t) + \mathbf{b}(t)\, \mathbf{X}_t \text{ and } Y_t^i = A(t, T_i) - \mathbf{B}(t, T_i)\, \mathbf{X}_t, \quad i = 1, \cdots, n. \quad (32.24)$$

While still linear, this is a degenerate filtering problem. Adapting a procedure from [30] we shall now reduce it to a nondegenerate problem via a change of coordinates. Recall that the observations $\mathbf{Y}_t := [r_t, Y_t^1, \cdots, Y_t^n]$ are *affine* functions of $\mathbf{X}_t$,

$$\mathbf{Y}_t = \mu_t + M_t \mathbf{X}_t \qquad (32.25)$$

for an appropriate $(n+1)$-vector μ_t and some $(n+1, N)$-matrix M_t with $N > n+1$. Moreover, our assumptions on the linear Gaussian factor model in Example 32.1 ensure that M_t has full rank.

Introduce now some $(N - n - 1, N)$ matrix L_t such that the $(N \times N)$-matrix $\begin{pmatrix} L_t \\ M_t \end{pmatrix}$ is invertible; this is always possible as M_t was assumed to have full rank. Define the $(N - n - 1)$-dimensional process

$$\bar{\mathbf{X}}_t := L_t \mathbf{X}_t \qquad (32.26)$$

and note that for appropriate matrices Φ_t and Ψ_t one has

$$\mathbf{X}_t = \begin{pmatrix} L_t \\ M_t \end{pmatrix}^{-1} \begin{pmatrix} \bar{\mathbf{X}}_t \\ \mathbf{Y}_t - \mu_t \end{pmatrix} =: \Phi_t \bar{\mathbf{X}}_t + \Psi_t (\mathbf{Y}_t - \mu_t) . \qquad (32.27)$$

Using the linearity of the dynamics of $\mathbf{X}$ we can now derive a closed-form linear-Gaussian system for the pair $(\bar{\mathbf{X}}, \mathbf{Y})$. In fact, from (32.26), (32.13), and (32.27) it then follows

$$\begin{aligned} d\bar{\mathbf{X}}_t &= \dot{L}_t \mathbf{X}_t dt + L_t d\mathbf{X}_t = \left(\dot{L}_t + L_t F\right) \mathbf{X}_t dt + L_t D\, d\mathbf{W}_t \\ &= \left(\dot{L}_t + L_t F\right) \Phi_t \bar{\mathbf{X}}_t + \left(\dot{L}_t + L_t F\right) \Psi_t \mathbf{Y}_t - \left(\dot{L}_t + L_t F\right) \Psi_t \mu_t + L_t D\, d\mathbf{W}_t \\ &=: a_t \bar{\mathbf{X}}_t + \beta_t \mathbf{Y}_t + \gamma_t + \delta_t\, d\mathbf{W}_t , \end{aligned} \qquad (32.28)$$

where $a_t, \beta_t, \gamma_t, \delta_t$ are implicitly defined. Analogously, from (32.25), (32.13) and (32.27)

$$\begin{aligned} d\mathbf{Y}_t &= \dot{\mu}_t dt + \dot{M}_t \mathbf{X}_t dt + M_t d\mathbf{X}_t = \left[\dot{\mu}_t + (\dot{M}_t + M_t F) \mathbf{X}_t\right] dt + M_t D\, d\mathbf{W}_t \\ &= \left[\dot{\mu}_t - (\dot{M}_t + M_t F) \Psi_t \mu_t\right] dt + (\dot{M}_t + M_t F) \Phi_t \bar{\mathbf{X}}_t dt + (\dot{M}_t + M_t F) \Psi_t \mathbf{Y}_t dt + M_t D\, d\mathbf{W}_t \\ &=: \phi_t \bar{\mathbf{X}}_t + \psi_t \mathbf{Y}_t + \rho_t + \sigma_t\, d\mathbf{W}_t , \end{aligned} \qquad (32.29)$$

where, again, $\phi_t, \psi_t, \rho_t, \sigma_t$ are implicitly defined. We can now formulate a non-degenerate filtering problem for the unobserved state variable process $\bar{\mathbf{X}}_t$ with observations $\mathbf{Y}_t$ as follows

$$\begin{cases} d\bar{\mathbf{X}}_t = a_t \bar{\mathbf{X}}_t + \beta_t \mathbf{Y}_t + \gamma_t + \delta_t\, d\mathbf{W}_t \\ d\mathbf{Y}_t = \phi_t \bar{\mathbf{X}}_t + \psi_t \mathbf{Y}_t + \rho_t + \sigma_t\, d\mathbf{W}_t \end{cases} \qquad (32.30)$$

This system is of the linear, conditionally Gaussian type and it leads thus to a Gaussian conditional (filter) distribution that we denote by $\pi_{\bar{\mathbf{X}}_t | \mathcal{F}_t} = \mathcal{N}(\bar{\mathbf{X}}_t; \bar{m}_t, \bar{P}_t)$, and where the mean $\bar{m}_t$ and covariance $\bar{P}_t$ can be computed via the Kalman filter. We then have from Lemma 32.1 that

$$\hat{p}(t, T) = E^Q(p(t, T; X_t) \mid \mathcal{F}_t) = E^Q\left(p\left(t, T; \left(\Phi_t \bar{X}_t + \Psi_t \left(Y_t - \mu_t\right)\right)\right) \mid \mathcal{F}_t\right)$$
$$= \int p\left(t, T; \left(\Phi_t \bar{x} + \Psi_t \left(Y_t - \mu_t\right)\right)\right) \pi_{\bar{X}_t \mid \mathcal{F}_t}(d\bar{x}). \tag{32.31}$$

Nonlinear extensions. The set-up of the example in this subsection can be generalized in various ways as is indicated by the following two dual set-ups. For the first set-up one keeps the linear-Gaussian dynamics (32.13) for the factors, but instead of (32.14) one considers an exponentially quadratic term structure model of the form

$$p(t, T) = \exp\left[A(t, T) - \mathbf{B}(t, T)X_t - X_t'C(t, T)X_t\right] \tag{32.32}$$

Notice that, for a linear-Gaussian factor model as in (32.13), more general exponentially polynomial term structure models lead to arbitrage for a degree larger that two (see [29]) so that (32.32) represents the most general nonlinear generalization of (32.14) that does not lead to arbitrage. For the second setup one keeps the exponentially affine structure (32.14) but considers instead of (32.13) a scalar square-root process of the form

$$dX_t = F(X_t - b_t)\,dt + \sqrt{X_t}\,D\,dW_t. \tag{32.33}$$

With these nonlinear extensions of the model the filtering problem with perturbed observations of the state becomes nonlinear; it seems that a finite-dimensional filter does not exist. The second (degenerate) filtering problem is even more challenging, since the solution approach described above does not extend to the nonlinear case.

32.3.3 Constructing term structure models via nonlinear filtering

In [48] the innovations approach to nonlinear filtering is used to construct a factor model for bond prices; here we sketch a simplified version of the approach. The author studies a model where the short-rate dynamics under a martingale measure are of the form

$$dr_t = a(t, r_t, X_t)\,dt + b\,dW_t \tag{32.34}$$

with X a scalar finite-state Markov chain with state space $\{1, \ldots, K\}$. In [48] the process X is assumed to be unobservable; the investor filtration is given by $\mathcal{F}_t = \sigma(r_s\,;\,s \leq t)$, so that only the short rate is observable. As before, the bond-pricing problem is approached via a two-step procedure: first one determines the bond prices under full observation. Given the Markovianity of the pair (r, X), these prices are of the form $p(t, T) = F^T(t; r_t, X_t)$ with the function $F^T(\cdot)$ such that the resulting prices do not allow for the possibility of arbitrage. According to Lemma 32.1, bond prices under incomplete information are then given by

$$\hat{p}(t, T) = E\left(F^T(t; r_t, X_t) \mid \mathcal{F}_t\right) =: \pi_t F^T. \tag{32.35}$$

Instead of first determining the filter distribution $\pi_{X_t|\mathcal{F}_t}$, in [48] the author determines directly the dynamics of the filtered value $\pi_t F^T$ of the bond prices. Using Itô's formula, she obtains first the semimartingale representation of the full-information bond price $F^T(t; r_t, X_t)$. From there, following the innovations approach to nonlinear filtering, she then obtains directly the dynamics of the filtered bond prices $\pi_t F^T$ decomposed into a finite variation part and a term driven by the innovations process

$$\bar{W}_t = \frac{1}{b}\left(r_t - \int_0^t \pi_s\left(a(s, r_s, X_s)\right) ds\right), \text{ where } \pi_s(a(s, r_s, X_s)) := \int a(s, r_s, x) \pi_{X_s|\mathcal{F}_s}(dx).$$

Let $p_t^k = Q(X_t = k \mid \mathcal{F}_t)$, $1 \le k \le K$. Since $\hat{p}(t, T) = \pi_t F^T = \sum_{k=1}^K p_t^k F^T(t, r_t; k)$, the ensuing term structure model has a natural factor structure with factor given by $\boldsymbol{p}_t := (p_t^1, \ldots, p_t^K)$. The dynamics of the factor vector $\boldsymbol{p}$ (which summarizes the conditional distribution $\pi_{X_t|\mathcal{F}_t}$) can be computed via the Wonham filter (see [59] or [27]).

The idea of using nonlinear filtering for the construction of a term structure model is undoubtedly very elegant; a similar approach in the context of credit risk models is discussed in the third example of Section 32.6.3 below. However, from a financial point of view the assumption that investors observe only the short rate is somewhat problematic: bonds with certain prominent maturities are usually liquidly traded, so that one would like to calibrate the model also to bond-price information.

32.3.4 Filtering of the market price of risk

Filtering in mathematical finance can be performed also for econometric and risk-management applications where it is usually most appropriate to study the filtering problem under the physical measure. If in that case the observations include prices that are expressed as expectations under a martingale measure, one ends up with a situation where one has to work simultaneously with the physical measure and with a martingale measure. The obvious thing is then to express everything under the same measure. Since the martingale measure serves mainly the purpose of guaranteeing absence of arbitrage, it is most natural to express everything under the physical measure.

As an example we start from the SDE system (32.11) for the short rate and some instantaneous forward rate, defined under a martingale measure Q so that absence of arbitrage is guaranteed. To transform the system into an equivalent one under the physical measure $P \sim Q$, we introduce the integrable and adapted *market price of risk* process ψ_t that allows one to pass from Q to P in the sense that, using now the symbol W_t^Q to specify a Wiener process under Q, the process $W_t := W_t^Q - \int_0^t \psi_s ds$ is a Wiener process under P (Girsanov measure *transformation*). Using a mean-reverting diffusion model for the evolution of ψ_t under P, the system (32.11) extends then to the following system defined under the physical measure P

$$\begin{cases} dr_t &= \left[\beta_0(t) + \beta_1(t)r_t + \beta_2(t)f(t,\tau) + g(r_t)\psi_t\right] dt + g(r_t) dW_t \\ df(t,\tau) &= \left[\gamma_0(t) + \gamma_1(t)r_t + \gamma_2(t)f(t,\tau) + g(r_t)e^{-\lambda(\tau-t)}\psi_t\right] dt + g(r_t)e^{-\lambda(\tau-t)} dW_t \\ d\psi_t &= \kappa(\bar{\psi} - \psi_t)dt + b\,|\psi_t|^\gamma\,dW_t \end{cases}$$
(32.36)

with the totality of the parameters given by the vector $(\sigma_0, \delta, \lambda, \kappa, \bar{\psi}, b, \gamma)$. A filter application in this context can be found in [15]. There the unobserved state vector is $\mathbf{X}_t = [r_t, f(t,\tau), \psi_t]$, while the observations are noisy observations of a finite number of given forward rates. For further aspects in this context see [56] and [55]. Notice finally that by filtering the market price of risk, this quantity (and hence also the corresponding martingale measure) continuously adapts to the current market situation.

32.3.5 Parameter estimation in term structure models

Market models are mostly specified as families of models that depend on certain parameters. The parameters are usually identified by matching as best as possible the theoretical model prices with the actually observed market prices. This goes under the name of *calibration to the market*. Calibration leads to a form of point estimation that may however lead to unstable estimates and without indication of their accuracy. In a filtering context one may instead consider a dynamic parameter estimation as part of the filtering problem and such a dynamic estimation enhances the possibility for the model to continuously adapt to the current market situation.

Two major approaches to this effect may be considered: i) *combined filtering and parameter estimation*; ii) *EM (expectation maximization)* combined with filtering. In the approach via combined filtering and parameter estimation one considers an extended state $(\mathbf{X}_t, \theta)$ where θ denotes the vector of parameters that are now considered as random variables according to the Bayesian point of view and one determines recursively the joint conditional (filter) distribution $\pi_{(\mathbf{X}_t,\theta)|\mathcal{F}_t}$. An example for this approach is presented next.

Combined filtering and parameter estimation with interest-rate observations ([4]). As explained in Section 32.2, in the context of HJM models with a volatility structure as in (32.9), forward rates and bond prices follow a factor model with factor process $\mathbf{X}$ given by two instantaneous rates. Instantaneous (continuously compounded) forward rates are a mathematical abstraction and cannot be directly observed on the market (at most, proxies are observable). Simple (discretely compounded) rates such as the LIBOR rates on the other hand are regularly quoted on interest markets so that they can be considered observable. The latter are related to the bond prices via (32.1), which in turn are related to $\mathbf{X}_t$ via $p(t, T; \mathbf{X}_t) = F^T(t, \mathbf{X}_t)$. Since $\mathbf{X}_t = [r_t, f(t, \tau)]$ satisfies the diffusion model (32.11), by stochastic differentiation one can then derive stochastic dynamics for the LIBOR rates. In this context in [4] a model is studied where the observation filtration $(\mathcal{F}_t)$ is generated by noisy observations of LIBOR rates. More precisely, by adding an independent observation noise to the LIBOR rates the authors in [4] obtain a non-degenerate nonlinear filtering problem to estimate $\mathbf{X}_t = [r_t, f(t, \tau)]$ and the parameters $(\sigma_0, \delta, \lambda)$ of the volatility function $\sigma(t, T)$ in (32.9), i.e. to estimate the theoretical instantaneous rates, on the basis of the observations of the LIBOR rates.

Parameter estimation via the EM algorithm. The EM algorithm is based on the following: let a given family of models, parameterized by θ, induce a family of probability measures P^θ that are assumed to be absolutely continuous with respect to a given reference measure P^0. Putting

$$Q(\theta, \theta') := E_{\theta'}\left\{\log \frac{dP^\theta}{dP^{\theta'}} \mid \mathcal{F}_t\right\} \tag{32.37}$$

the algorithm iterates through the following two steps:

  i) compute $Q(\theta, \theta')$ for θ' given, θ arbitrary (*expectation step*)
  ii) determine $\theta^* = argmax_\theta \, Q(\theta, \theta')$ and return to i) with $\theta' = \theta^*$ (*maximization step*).

The algorithm stops as soon as the maximizing values in two successive iterations are sufficiently close.

Since the EM algorithm is based on an absolutely continuous change of measure, the parameters entering the coefficient of the observation noise cannot be estimated via EM and have to be estimated by other methods, e.g. on the basis of the empirical quadratic variation. The other parameters can in principle be estimated via EM and the maximization step leads to solving the system of equations obtained by putting $\frac{\partial Q(\theta, \theta')}{\partial \theta} = 0$. The resulting system involves various conditional expectations that can be computed on the basis of the filtering results (see e.g. [28]): in continuous time, if the state and observation noises are independent, filtering alone suffices; if they are not independent, also smoothing is required.

There exist other approaches as well, in particular in a discrete time setup. One of them is based on the maximization of the innovations likelihood, which is in fact of the type of maximum likelihood estimation. The parameter estimation approaches are mentioned here only in the context of term structure models; they can however be easily carried over also to credit risk models (see e.g. [30]).

32.4 Nonlinear filtering in credit risk models

In this section we give a brief introduction to dynamic credit risk models and explain how incomplete information and nonlinear filtering enter in credit risk modelling; a detailed discussion of specific models is given in Sections 32.5 and 32.6 below.

32.4.1 Dynamic credit risk models and credit derivatives

Dynamic credit risk models are concerned with the modelling of the default times in a given portfolio of firms. In our discussion of credit risk models we use the following notation: the firms under consideration are indexed by $i \in \{1, \ldots, m\}$; the random time $\tau_i > 0$ denotes the default time of firm i; the current default state of firm i is described by the default indicator process $Y_{t,i} = \mathbf{1}_{\{\tau_i \leq t\}}$, jumping from zero to one at $t = \tau_i$; the current default state of the portfolio is described by $\mathbf{Y}_t = (Y_{t,1}, \ldots, Y_{t,m})$; the *default history* up to time t is given by $\mathcal{F}_t^{\mathbf{Y}} := \sigma(\mathbf{Y}_s : s \leq t)$. Since our focus is primarily on pricing problems, we model the dynamics of the objects of interest directly under some risk-neutral measure Q. Throughout we therefore work on a filtered probability space $(\Omega, \mathcal{G}, (\mathcal{G}_t), Q)$; as in the interest-rate part, $(\mathcal{G}_t)$ represents the full-information filtration, so that all considered stochastic processes will be $(\mathcal{G}_t)$-adapted. Moreover, in line with most of the credit risk literature, we assume in this part of the paper that default-free interest rates are deterministic and equal to $r > 0$.

A large part of the credit risk literature is concerned with the pricing of *credit derivatives*. These are securities whose payoff is linked to default events in a given reference portfolio. In abstract terms the payoff of a credit derivative is thus given by some $\mathcal{F}_T^{\mathbf{Y}}$-measurable random variable H. Important examples include defaultable zero-coupon bonds and default payments. The payoff of a defaultable zero coupon bond issued by firm i with maturity T and zero recovery is given by $H = \mathbf{1}_{\{\tau_i > T\}} = 1 - Y_{T,i}$; the price at $t < T$ of this bond will be denoted by $p_i(t, T)$. A default payment of size δ on firm i with maturity T has a payoff of size δ directly at τ_i, provided $\tau_i \leq T$. By combining zero coupon bonds and default payments other important products such as credit default swaps (CDSs) or corporate bonds with recovery payments can be constructed; see Section 9.4 of [51] for further details. An important quantity in this context is the *credit spread* of firm i, denoted $c_i(t, T)$, $t \leq T \wedge \tau_i$. This quantity measures the difference in the continuously compounded yields of a defaultable zero coupon bond issued by firm i and of the corresponding default-free zero coupon bond (denoted here by $p_0(t, T)$), and reflects thus the market's assessment of the likelihood of the default of firm i. Formally, $c_i(t, T)$ is given by

$$c_i(t, T) := -\frac{1}{T-t} \left(\log p_i(t, T) - \log p_0(t, T) \right). \tag{32.38}$$

Existing dynamic credit risk models can be grouped into two classes: *structural* and *reduced-form models*. Structural models originated from Black and Scholes [8], Merton [52], and Black and Cox [7]. Important contributions to the literature on reduced-form models are [45], [49], [26], and [9]; a more complete list of references can be found in the textbooks [3], [50], [51], among others.

In structural models one starts by modeling the asset values $V_i = (V_{t,i})_{t \geq 0}$ of the firms under consideration; usually V_i is modeled as a diffusion process. Given some default barrier $K_i = (K_{t,i})_{t \geq 0}$, the default time τ_i is then defined to be the first passage time of V_i at the barrier K_i, i.e.

$$\tau_i = \inf\{t \geq 0 \colon V_{t,i} \leq K_{t,i}\}. \qquad (32.39)$$

The default barrier is often interpreted as the value of the liabilities of the firm; with this interpretation (32.39) states that, in line with economic intuition, default happens at the first time that the asset value of a firm is too low to cover its liabilities. Note that the default time τ_i defined in (32.39) is a predictable stopping time with respect to the global filtration $(\mathcal{G}_t)$ to which V_i and K_i are adapted. It is well-documented that the fact that τ_i is $(\mathcal{G}_t)$-predictable leads to very low values for short-term credit spreads (in particular $\lim_{h \to 0} c_i(t, t+h) = 0$), contradicting most of the available empirical evidence.

In reduced-form models on the other hand, the precise mechanism leading to default is left unspecified; rather one models directly the law of the default times τ_i or of the associated default indicator process Y_i. Typically τ_i is modelled as a totally inaccessible stopping time with respect to the global filtration $(\mathcal{G}_t)$, admitting a $(Q, (\mathcal{G}_t))$-intensity λ_i (termed *risk-neutral default intensity*). Formally, $\lambda_i = (\lambda_{t,i})_{t \geq 0}$ is a $(\mathcal{G}_t)$-predictable process such that

$$Y_{t,i} - \int_0^{t \wedge \tau_i} \lambda_{s,i} ds \quad \text{is a } (Q, (\mathcal{G}_t))\text{-martingale.} \qquad (32.40)$$

In reduced-form models dependence between defaults is often generated by assuming that the default intensities do depend on a common factor process $\mathbf{X}_t \in \mathbb{R}^d$, i.e. $\lambda_{t,i} = \lambda_i(\mathbf{X}_t)$ for suitable functions $\lambda_i \colon \mathbb{R}^d \to (0, \infty)$. The simplest construction is that of *conditionally independent, doubly stochastic default times*. Here it is assumed that given $\mathcal{F}_\infty^\mathbf{X}$, the τ_i are conditionally independent with

$$P\left(\tau_i > t \mid \mathcal{F}_\infty^\mathbf{X}\right) = \exp(-\int_0^t \lambda_i(\mathbf{X}_s) ds), \quad t > 0;$$

see for instance section 9.6 of [51] for details.

32.4.2 Incomplete information

In both modeling paradigms it makes sense to assume that investors have imperfect information on some of the state variables of the models; this has given rise to a rich literature on credit risk models under incomplete information.

In a structural model the natural state variables are given by the asset value V_i or the log-asset value X_i and—with stochastic liabilities—by the liability levels K_i of the firms under consideration. It is difficult for investors in secondary markets to precisely assess the values of these state variables for a number of reasons: accounting reports might be noisy; market and book values can differ as intangible assets such as R&D (research and development) results or client relationships are difficult to value; part of the liabilities are usually bank loans whose precise terms are unknown to the public; and many more. Hence, starting with the seminal work of Duffie and Lando [25], a growing literature studies models where investors have only noisy information about V_i and/or K_i; the conditional distribution of the state variables given investor information $\mathcal{F}_t$ is then computed by Bayesian updating or filtering arguments. Examples of this line of research include [25], [53]; [39], [16], and [36]; some of these papers are discussed in more detail in Section 32.5 below. Interestingly, it turns out that the distinction between structural and reduced-form models is in fact a distinction between full and partial observability of asset values and liabilities (see e.g. Jarrow and Protter [44]): in the models mentioned above the default time τ_i that is predictable with respect to the global filtration $(\mathcal{G}_t)$ becomes totally inaccessible with respect to the investor filtration $(\mathcal{F}_t)$ and moreover admits an intensity. This leads furthermore to a realistic behaviour of short-term credit spreads, as is explained in Section 32.5.

In typical reduced-form models default intensities are assumed to depend on some Markovian factor process $\mathbf{X}$ which here becomes the natural state variable process. In applications $\mathbf{X}$ is usually not identified with observable quantities but treated as a latent process whose current value must be inferred from observables such as prices or the default history. A theoretically consistent way for doing this is to determine—via Bayesian updating or filtering arguments—$\pi_{\mathbf{X}_t|\mathcal{F}_t}$, the conditional distribution of $\mathbf{X}_t$ given investor information $\mathcal{F}_t$. Reduced-form credit risk models with incomplete information include the contributions by [57], [17], and [22] as well as our own work [35] and [37]. The structure of the models [57], [17], and [22] is relatively similar: default intensities are driven by an unobservable factor $\mathbf{X}$; the default times are conditionally independent, doubly stochastic random times; the investor information $(\mathcal{F}_t)$ is given by the default history of the portfolio, augmented by economic covariates. In [57], and [17] the unobservable factors are modeled by a static random vector $\mathbf{X}$ which is termed *frailty*; the conditional distribution $\pi_{\mathbf{X}|\mathcal{F}_t}$ is determined via Bayesian updating. In [22] the unobservable (scalar) factor X is modeled as an Ornstein–Uhlenbeck process. This latter paper has an empirical focus: dynamic Bayesian methodology is used in order to estimate the model parameters from historical default data; moreover, filtering is used in order to determine the conditional mean of X_t, given the history of defaults and covariates. This analysis provides strong evidence for the assertion that an unobservable stochastic process driving default intensities (a so-called dynamic frailty) is needed on top of observable covariates in order to explain the clustering of defaults in historical data, a finding which strongly supports the use of filtering methodology in credit risk models.

Our own work [35] on filtering in reduced-form credit risk models extends these contributions in a number of ways, at least from a methodological viewpoint. To begin with, we consider a more general investor filtration that contains noisily observed prices on top of the default history of the portfolio. Moreover, the problem of finding the conditional distribution of $\pi_{X_t|\mathcal{F}_t}$ is studied in a general jump-diffusion model for X and default indicator Y that includes most reduced-form credit risk models from the literature and in particular the analysis of [57], [17], and [22] as special cases. Our discussion of reduced-form models with unobservable state variables in Section 32.6 is therefore based mainly on [35] and the companion paper [37].

Introducing incomplete information into credit portfolio models has interesting implications for the dynamics of credit derivative prices and credit spreads (both in structural and in reduced-form models), since the successive updating of the conditional distribution $\pi_{X_t|\mathcal{F}_t}$ in reaction to incoming default observations generates so-called *information-driven default contagion*: the news that some obligor has defaulted leads to an update in $\pi_{X_t|\mathcal{F}_t}(dx)$ and hence to a jump in the $(\mathcal{F}_t)$-default intensity of the surviving firms, as will be explained in more detail below. In the context of reduced-form models this was first pointed out by [57] and [17], whereas default contagion in structural models is studied among others in [39]; empirical evidence for contagious effects is provided for instance in [17]. Note that a similar phenomenon did not occur in our discussion of interest-rate market models, essentially because there the investor information $(\mathcal{F}_t)$ was generated by continuous (Wiener-driven) processes.

32.5 Filtering in structural models

In this section we discuss structural credit risk models under incomplete information and some of the ensuing nonlinear filtering problems.

32.5.1 The model of Duffie and Lando [25]

The set-up. Recall that we work on a filtered probability space $(\Omega, \mathcal{G}, (\mathcal{G}_t), Q)$, Q the risk-neutral measure and $(\mathcal{G}_t)$ the full-information filtration. Throughout this section we focus on models for the default of a single firm, so that the index i giving the identity of the firm can be omitted. We assume that the asset value

V follows a geometric Brownian motion on this filtered probability space with drift μ, volatility σ and initial value V_0. Consider then as (scalar) state variable

$$X_t := \log V_t = X_0 + \left(\mu - \frac{1}{2}\sigma^2\right)t + \sigma W_t, \tag{32.41}$$

W a Brownian motion on $(\Omega, \mathcal{G}, (\mathcal{G}_t), Q)$. In [25] the default barrier K is taken constant. The *default time* is thus given by the stopping time

$$\tau := \inf\{t \geq 0 : V_t < K\} = \inf\{t \geq 0 : X_t < \log K\}. \tag{32.42}$$

It is assumed that V is not directly observable. Rather, investors observe default; moreover, they receive "noisy accounting reports" at deterministic times $t_1, t_2, \cdots$, that is they observe random variables $Z_i = X_{t_i} + U_i$ where $(U_i)_{i \in \mathbb{N}}$ is a sequence of independent, normally distributed random variables, independent of X (or V). Formally, with $Y_t := \mathbf{1}_{\{\tau \leq t\}}$, the investor filtration is

$$\mathcal{F}_t := \mathcal{F}_t^Y \vee \sigma(\{Z_i : t_i \leq t\}). \tag{32.43}$$

The default barrier K and the initial asset value X_0 are supposed to be known.

Survival probabilities, default intensity, and credit spreads. By the Markov property of V (or X) one has, for $T \geq t$,

$$Q(\tau > T \mid \mathcal{G}_t) = \mathbf{1}_{\{\tau > t\}} Q\left(\inf_{s \in (t,T)} V_s > K \mid \mathcal{G}_t\right) = \mathbf{1}_{\{\tau > t\}} Q\left(\inf_{s \in (t,T)} V_s > K \mid V_t\right)$$
$$=: \mathbf{1}_{\{\tau > t\}} \bar{F}_\tau(t, T, V_t).$$

Note that for $T \geq t$ the mapping $T \mapsto \bar{F}_\tau(t, T, v)$ gives the (risk-neutral) survival probabilities of the firm under full information as of time t, given that $V_t = v$; $\bar{F}_\tau$ is easily computed using standard results on the first passage time of Brownian motion with drift. Using iterated conditional expectations one gets for the survival probability in the investor filtration

$$Q(\tau > T \mid \mathcal{F}_t) = E\left(Q(\tau > T \mid \mathcal{G}_t) \mid \mathcal{F}_t\right) = \mathbf{1}_{\{\tau > t\}} \int_{\log K}^{\infty} \bar{F}_\tau(t, T, e^x) \pi_{X_t \mid \mathcal{F}_t}(dx). \tag{32.44}$$

Next turn to the $(Q, \mathcal{F}_t)$-*default intensity* λ_t of τ. It can be shown that under some regularity conditions one has

$$\lambda_t = \lim_{h \downarrow 0} \frac{1}{h} Q\{t < \tau \leq t + h \mid \mathcal{F}_t\}, \tag{32.45}$$

provided this limit exists for all $t \geq 0$ almost surely (see [10], [1] for details). Duffie and Lando in [25] now show that such a λ_t exists and is given by

$$\lambda_t = \frac{1}{2}\sigma^2 \frac{\partial}{\partial x} \pi(X_t \in dx \mid \mathcal{F}_t)_{|x=\log K}, \quad \tau \geq t,$$

where $\pi(X_t \in dx \mid \mathcal{F}_t)$ denotes the Lebesgue density of the filter distribution $\pi_{X_t \mid \mathcal{F}_t}(dx)$. (The fact that the derivative of the conditional density exists at $x = \log K$ is part of their result.)

Finally we discuss bond prices and credit spreads in the Duffie–Lando model under incomplete information. We get for the price of a defaultable zero coupon bond with zero recovery, denoted by $p_1(t, T)$,

$$p_1(t, T) = \mathbf{1}_{\{\tau > t\}} e^{-r(T-t)} Q\{\tau > T \mid \mathcal{F}_t\} = \mathbf{1}_{\{\tau > t\}} e^{-r(T-t)} \int_{\log K}^{\infty} \bar{F}_\tau(t, T, e^x) \, \pi_{X_t \mid \mathcal{F}_t}(dx) \tag{32.46}$$

i.e. zero coupon bond prices can be expressed as an average with respect to the filter distribution $\pi_{X_t \mid \mathcal{F}_t}(dx)$. The price of a default payment is also easily computed once the survival probability in the investor filtration is at hand. These pricing results are of course special cases of the general pricing principle from Lemma 32.1 in Section 32.3.

The credit spread $c(t, T)$ introduced in (32.38) satisfies on $\{\tau > t\}$ the following relation (since r is assumed deterministic one has $p_0(t, T) = e^{-r(T-t)}$)

$$c(t, T) = \frac{-1}{T - t} \log Q(\tau > T \mid \mathcal{F}_t). \tag{32.47}$$

In particular, we get for $T \downarrow t$ that $\frac{\partial}{\partial T} c(t, T)_{|T=t} = \frac{\partial}{\partial T} Q\{\tau > T \mid \mathcal{F}_t\}_{|T=t} = \lambda_t$, where the second equality follows from (32.45). This shows that the introduction of incomplete information typically leads to nonvanishing short-term credit spreads.

Computing the filter distribution. We have seen that in order to determine risk sensitive financial quantities such as defaultable bond prices or credit spreads, one needs to determine the conditional distribution (*filter distribution*) $\pi_{X_t \mid \mathcal{F}_t}(dx)$. In [25] this problem is tackled in an elementary way, involving Bayes' formula and properties of first passage time of Brownian motion. We do not discuss the details here; in the next subsection we show how proper filtering arguments can be used in order to determine (approximately) $\pi_{X_t \mid \mathcal{F}_t}(dx)$.

32.5.2 The model of Frey & Schmidt [36]

In [36], the basic Duffie–Lando model is extended in essentially two directions. On the financial side the paper introduces dividend payments and discusses the pricing of the firm's equity under incomplete information. On the mathematical side nonlinear filtering techniques and Markov-chain approximations are employed in order to determine the conditional distribution of the log-asset value X_t given the investor information $\mathcal{F}_t$.

Here we concentrate on the filtering part. The set-up of the model under full information is as in Subsection 32.5.1: the log-asset value X is given by the arithmetic Brownian motion (32.41) and, in line with (32.42), the default time τ is the first passage time of X at the barrier $\log K$.[5] Investors observe the default state of the firm; moreover, they receive pieces of economic information (news) related to the state of the company such as information given by analysts, articles in newspapers, etc. It is assumed that this information is discrete, corresponding for instance to buy/hold/sell recommendations or rating information. Formally, news events on the company are issued at (for simplicity) deterministic time points t_n^I, $n \geq 1$; the news obtained at t_n^I is denoted by I_n, which takes values in the discrete state space $\{\ell_1, \ldots, \ell_{M^I}\}$. The conditional distribution of I_n given $\mathcal{G}_{t_n^I}$ is denoted by

$$v_I(\ell_j | x) := Q\left(I_n = \ell_j | X_{t_n^I} = x\right).$$

Summarizing, the information of secondary market investors at time t is given by the σ-field

$$\mathcal{F}_t := \mathcal{F}_t^Y \vee \sigma\left(\{I_n : t_n^I \leq t\}\right). \tag{32.48}$$

Filtering. In order to determine the conditional distribution $\pi_{X_t | \mathcal{F}_t}$ with minimal technical difficulties, the log-asset value process X is approximated by a finite-state discrete-time Markov chain X^Δ as follows: define for a given time discretization $\Delta > 0$ the grid $\{t_k^\Delta = k\Delta,\ k \in \mathbb{N}\}$. Let $\left(X_k^\Delta\right)_{k \in \mathbb{N}}$ be a discrete-time finite-state Markov chain with state space $\Xi^\Delta = \{m_1^\Delta, \ldots, m_{M^\Delta}^\Delta\}$ and transition probabilities p_{ij}^Δ, $1 \leq i, j \leq M^\Delta$, and define the induced process X^Δ by $X_t^\Delta = X_k^\Delta$ for $t \in [t_k^\Delta, t_{k+1}^\Delta)$. In [36] it is assumed that the chain $\left(X_k^\Delta\right)_{k \in \mathbb{N}}$ is close to the continuous log-asset-value process X in the sense that X^Δ converges in distribution to X as $\Delta \to 0$; it is shown that this implies that the conditional distribution $\pi_{X_t^\Delta | \mathcal{F}_t}$ converges weakly to $\pi_{X_t | \mathcal{F}_t}$ as $\Delta \to 0$.

In the sequel we keep Δ fixed and omit it from our notation. Obviously the conditional distribution $\pi_{X_k^\Delta | \mathcal{F}_{t_k}}$ is summarized by the probability vector $\pi(k) = (\pi_1(k), \ldots, \pi_{M^\Delta}(k))$ with $\pi_j(k) := Q\left(X_k = m_j \mid \mathcal{F}_{t_k}\right)$. It is possible to give explicit recursive updating rules for the probability vector $\pi(k)$. In fact, due to the discrete nature of the problem, this is fairly easy, as is illustrated by the simple proof of Proposition 32.1 below. It will be convenient to formulate the updating rule in terms of "unnormalized probabilities" $\sigma(k) \propto \pi(k)$ ($\propto$ standing for proportional to); the vector $\pi(k)$ can then be obtained by normalization.

The initial filter distribution $\pi(0)$ can be inferred from the (known) initial distribution of X_0. For $k \geq 1$ we have the following updating rule.

Proposition 32.1 *For $k \geq 1$ and $t_k < \tau$, denote by $N_k^I := \{n \in \mathbb{N} : t_{k-1} < t_n^I \leq t_k\}$ the set of indices of news arrivals in the period $(t_{k-1}, t_k]$. Then, with the convention that $\prod_\emptyset$ (the product over the empty set) equals 1,*

$$\sigma_j(k) = \mathbf{1}_{\{m_j > \log K\}} \sum_{i=1}^{M^\Delta} \left\{ p_{ij}\, \sigma_i(k-1) \prod_{n \in N_k^I} \nu_I(I_n | m_i) \right\}, \quad j = 1, \ldots, M^\Delta. \tag{32.49}$$

Proof. Given the new information arriving in $(t_{k-1}, t_k]$, the updating rule (32.49) forms a linear and in particular a positively homogeneous mapping Γ such that $\sigma(k) = \Gamma \sigma(k-1)$. Hence it is enough to show that $\pi(k) \propto \Gamma \pi(k-1)$. In order to compute $\pi(k)$ from $\pi(k-1)$ and the new information in $(t_{k-1}, t_k]$ we proceed in two steps. In Step 1 we compute (up to proportionality) an auxiliary vector of probabilities $\tilde{\pi}(k-1)$ with

$$\tilde{\pi}_i(k-1) = Q\left(X_{k-1} = m_i \mid \mathcal{F}_k^-\right), \quad 1 \leq i \leq M^\Delta, \tag{32.50}$$

where $\mathcal{F}_k^- := \mathcal{F}_{t_{k-1}} \vee \sigma\left(\{I_n : n \in N_k^I\}\right)$. In filtering terminology this is a smoothing step as the conditional distribution of X_{k-1} is updated using the new information arriving in $(t_{k-1}, t_k]$. In Step 2 we determine (again up to proportionality) $\pi(k)$ from the auxiliary probability vector $\tilde{\pi}(k-1)$ using the dynamics of (X_k) and the additional information that $\tau > t_k$. We begin with Step 2. Since $\{\tau > t_k\} = \{\tau > t_{k-1}\} \cap \{X_k > \log K\}$, we get

$$Q\left(X_k = m_j \mid \mathcal{F}_{t_k}\right) \propto Q\left(X_k = m_j, X_k > \log K \mid \mathcal{F}_k^-\right)$$

$$= \sum_{i=1}^{M^\Delta} Q\left(X_k = m_j, X_k > \log K, X_{k-1} = m_i \mid \mathcal{F}_k^-\right)$$

$$= \mathbf{1}_{\{m_j > \log K\}} \sum_{i=1}^{M^\Delta} p_{ij}\, \tilde{\pi}_i(k-1). \tag{32.51}$$

Next we turn to the smoothing step. Note that given $X_{k-1} = m_i$, the likelihood of the news observed over $(t_{k-1}, t_k]$ equals $\prod_{n \in N_k^I} \nu_I(I_n | m_i)$, and we obtain

$$\tilde{\pi}_i(k-1) \propto \pi_i(k-1) \cdot \prod_{n \in N_k^I} \nu_I(I_n | m_i).$$

Combining this with equation (32.51) gives the result. □

32.5.3 Further related work

There is a rich literature on structural credit risk models under incomplete information; here we briefly discuss some contributions which cannot be treated in detail for reasons of space.

The filtering model of Nakagawa [53]. In [53] the author considers a slightly generalized version of the basic Duffie–Lando model (see Subsection 32.5.1). The main difference is that in [53] the investor filtration is given by $\mathcal{F}_t = \mathcal{F}_t^Y \vee \mathcal{F}_t^Z$, where the process Z has dynamics $dZ_t = a(X_t)dt + d\beta_t$ for a Brownian motion β independent of W (observations of the state in additive Gaussian noise). The principal goal is to determine the form of the default intensity $(\lambda_t)_{t \geq 0}$ with respect to the investor filtration $(\mathcal{F}_t)$. Note that this is a nonstandard filtering problem, as the default time τ does not admit an intensity with respect to the global filtration $(\mathcal{G}_t)$. In order to deal with this problem the author applies an equivalent change of measure, so that under the new measure $\tilde{Q}$ the process Z is independent of X and Y. The $(\tilde{Q}, (\mathcal{F}_t))$-intensity of τ can be computed explicitly using results from [47] or [25]. The $(Q, (\mathcal{F}_t))$-default intensity can then be computed via a suitable Girsanov theorem for point processes once an explicit martingale representation of the density martingale $L_t = E^{\tilde{Q}}(dQ/d\tilde{Q} \mid \mathcal{F}_t)$ is at hand. In order to compute this representation the author projects the density martingale $\tilde{L}_t = E^{\tilde{Q}}(dQ/d\tilde{Q} \mid \mathcal{G}_t)$ (which is easily computed via the usual Girsanov theorem) on the filtration $(\mathcal{F}_t)$ using arguments from the innovations approach to nonlinear filtering.

Further work. In [39] a structural portfolio model is considered. In contrast to the papers discussed so far, in the model of [39] the asset value is observable, whereas the liabilities are subject to random shocks which cannot be observed. Bayesian updating is used in order to compute the conditional distribution of the liabilities given the investor information. The authors point out that in case that liability shocks are correlated across firms, the model leads to information-based default contagion.

The authors in [16] study a Duffie–Lando-type model with $\mathcal{F}_t = \mathcal{F}_t^Y \vee \mathcal{F}_t^Z$; in their set-up the process Z solves an SDE driven by a Brownian motion β which is correlated with the Brownian motion W driving the asset-value process. In this paper bond prices are computed via the hazard-function approach to reduced-form credit risk models (see [9]). Finally, in [13] the authors study models where only the sign of the firm's cash flow is available. Both papers are mathematically interesting. However, filtering arguments play only a minor role, so we will not enter into a deeper discussion.

32.6 Filtering in reduced-form models

This section is concerned with the pricing of credit derivatives in reduced-form portfolio credit risk models under incomplete information; the presentation is largely based on our own papers [35] and [37].

32.6.1 Pricing credit derivatives and nonlinear filtering

As in the previous section we work with a filtered probability space $(\Omega, \mathcal{G}, (\mathcal{G}_t), Q)$ where Q represents the risk-neutral pricing measure and $(\mathcal{G}_t)$ the full-information filtration. Recall that the default state of the portfolio under consideration is summarized by the default indicator process $\mathbf{Y} = (Y_{t,1}, \ldots, Y_{t,m})_{t \geq 0}$ with $Y_{t,i} = 1_{\{\tau_i \leq t\}}$. We assume that there is some d-dimensional process $\mathbf{X}$ (the state process) and functions $\lambda_i : \mathbb{R}^d \to (0, \infty)$ such that $\lambda_i(\mathbf{X}_t)$ is the $(Q, (\mathcal{G}_t))$-default intensity of firm i, or, equivalently, that $Y_{t,i} - \int_0^{\tau_i \wedge t} \lambda_i(\mathbf{X}_s) \, ds$ is a $(Q, (\mathcal{G}_t))$-martingale. Moreover, we assume that the pair of processes $(\mathbf{X}_t, \mathbf{Y}_t)_{t \geq 0}$ is jointly Markov. For concreteness one may think of a model with conditionally independent, doubly stochastic default times where default intensities are functions of some Markovian factor process $\mathbf{X}$ (see also the first example in Subsection 32.6.3 below). As in the previous sections we denote by $(\mathcal{F}_t)$ the information actually observable to investors. Following [35] we assume that $(\mathcal{F}_t)$ contains the default history $\mathcal{F}_t^{\mathbf{Y}} = \sigma(\mathbf{Y}_s : s \leq t)$ and observations of functions of $\mathbf{X}$ in additive Gaussian noise. Formally, $\mathcal{F}_t := \mathcal{F}_t^{\mathbf{Y}} \vee \mathcal{F}_t^{\mathbf{Z}}$ where the l-dimensional process $\mathbf{Z}$ is given by

$$Z_{t,j} = \int_0^t a_j(s, \mathbf{X}_s, \mathbf{Y}_s) ds + \beta_{t,j}, \quad 1 \leq j \leq l, \tag{32.52}$$

with $\beta_t = (\beta_{t,1}, \ldots, \beta_{t,l})$ an l-dimensional Brownian motion on $(\Omega, \mathcal{G}, (\mathcal{G}_t), Q)$, independent of $\mathbf{X}$ and $\mathbf{Y}$. In order to avoid technical difficulties, the functions a_j are assumed to be bounded. We shall see below that $\mathbf{Z}$ can be interpreted as theoretical prices for traded credit derivatives, observed in additive noise. Note that $\mathbf{X}$ is not $(\mathcal{F}_t)$-adapted, due to the independence of $\mathbf{X}$ and β.

Pricing credit derivatives. Recall that a credit derivative is a security with $\mathcal{F}_T^{\mathbf{Y}}$-measurable payoff H; specific examples include defaultable zero-coupon bonds or CDSs as introduced in Section 32.4.1. In accordance with the general pricing principle from Lemma 32.1 we define the *theoretical* or *full-information price* of a credit derivative by $\tilde{H}_t := E\left(e^{-r(T-t)} H \mid \mathcal{G}_t\right)$, $t \leq T$, where the constant $r > 0$ denotes the default-free short rate. By the assumed Markovianity of the pair $(\mathbf{X}, \mathbf{Y})$, for typical credit derivatives the process $\tilde{H}_t$ is of the form $\tilde{H}_t = a(t, \mathbf{X}_t, \mathbf{Y}_t)$ for some function $a : [0, T] \times \mathbb{R}^d \times \{0, 1\}^m \to \mathbb{R}$. This is obvious, if the payoff is of the form $H = h(\mathbf{Y}_T)$ as in the case of a defaultable zero coupon bond; it holds true for most other credit derivatives such as the default payment introduced in Section 32.4.1 as well. For nontraded credit derivatives we define the *investor price* by $H_t := E\left(e^{-r(T-t)} H \mid \mathcal{F}_t\right)$. We get from Lemma 32.1 (or by a direct application of iterated conditional expectations) that

$$H_t = E(a(t, \mathbf{X}_t, \mathbf{Y}_t) \mid \mathcal{F}_t). \tag{32.53}$$

Since $\mathbf{Y}_t$ is observable, in order to compute H_t we thus need to determine the conditional distribution $\pi_{\mathbf{X}_t \mid \mathcal{F}_t}$, which amounts to solving a nonlinear filtering problem.

Now we come back to the economic interpretation of $\mathbf{Z}$. Assume that investors have noisy information about the theoretical price $a(t, \mathbf{X}_t, \mathbf{Y}_t)$ of l

traded credit derivatives. In a discrete-time framework it is natural to assume that the observed market quotes are of the form $z_{t_k} = a(t_k, \mathbf{X}_{t_k}, \mathbf{Y}_{t_k}) + \epsilon_k$ for time points $t_k = k\Delta$ and an i.i.d. sequence $(\epsilon_k)_k$ of independent noise variables $\epsilon_1, \ldots, \epsilon_l$ with mean zero and finite variance. The noise variables model transmission and observation errors as well as temporary deviations of market quotes from theoretical prices. In continuous time one considers instead the *cumulative observation process* $\mathbf{Z}_t^\Delta := \Delta \sum_{t_k \leq t} z_{t_k}$. Then we have for Δ small, using Donsker's invariance principle,

$$Z_{t,j}^\Delta = \sum_{t_k \leq t} a_j(t_k, \mathbf{X}_{t_k}, \mathbf{Y}_{t_k}) \Delta + \sum_{t_k \leq t} \Delta \epsilon_k \approx \int_0^t a_j(s, \mathbf{X}_s, \mathbf{Y}_s) ds + \beta_{t,j}. \quad (32.54)$$

Remark 32.1 (Default intensities and default contagion) It is well known that default-intensities with respect to subfiltrations can be computed by projection (see for instance Ch. 2 of [10]). Hence the risk-neutral $(\mathcal{F}_t)$-default intensity of firm j is given by the left-continuous version of

$$\hat{\lambda}_{t,j} := E\left(\lambda_j(\mathbf{X}_t) \mid \mathcal{F}_t\right) = \int_{\mathbb{R}^d} \lambda_j(x) \pi_{\mathbf{X}_t \mid \mathcal{F}_t}(dx), \quad t \leq \tau_j, \quad (32.55)$$

so that in order to compute this quantity we again need the conditional distribution $\pi_{\mathbf{X}_t \mid \mathcal{F}_t}(dx)$. Relation (32.55) illustrates nicely the notion of information-based default contagion that was already mentioned in Section 32.4.2: new default information such as the news that obligor $i \neq j$ has defaulted leads to an update in the conditional distribution $\pi_{\mathbf{X}_t \mid \mathcal{F}_t}(dx)$ and hence to a jump in the $(\mathcal{F}_t)$-default intensity of firm j. Note that this leads to a downward jump in the model value (and hence an increase in the credit spread) of a zero coupon bond issued by some nondefaulted firm.

32.6.2 A general jump-diffusion model

Following [35] we next introduce a jump-diffusion model for the joint dynamics of $\mathbf{X}$ and $\mathbf{Y}$. This model is fairly general and includes most reduced-form models from the literature as special cases; specific examples are discussed in the next subsection. We assume that the factor process $\mathbf{X} = (X_{t,1}, \ldots, X_{t,d})_{t \geq 0}$ and the default indicator process $\mathbf{Y}$ solve the following SDE on $(\Omega, \mathcal{G}, (\mathcal{G}_t), Q)$

$$\mathbf{X}_t = \mathbf{X}_0 + \int_0^t b(\mathbf{X}_{s-}) ds + \int_0^t \sigma(\mathbf{X}_{s-}) d\mathbf{W}_s$$
$$+ \int_0^t \int_E K^\mathbf{X}(\mathbf{X}_{s-}, u) \mathcal{N}(ds, du), \quad (32.56)$$

$$Y_{t,j} = Y_{0,j} + \int_0^t \int_E (1 - Y_{s-,j}) K_j^\mathbf{Y}(\mathbf{X}_{s-}, u) \mathcal{N}(ds, du), \quad 1 \leq j \leq m. \quad (32.57)$$

Here $\mathbf{W}$ is a standard k-dimensional Brownian motion; drift $b = (b_1, \ldots, b_d)$ and dispersion matrix $\sigma = (\sigma_{i,l})$, $1 \leq i \leq d$, $1 \leq l \leq k$ are functions from $S^\mathbf{X}$ to

$\mathbb{R}^d$ and $\mathbb{R}^{d \times k}$ respectively, $S^X \subset \mathbb{R}^d$ is the state space of $\mathbf{X}$; $\mathcal{N}(ds, du)$ denotes a $(Q, (\mathcal{G}_t))$-standard Poisson random measure on $\mathbb{R}_+ \times E$, E some Euclidean space, with compensator measure $F_\mathcal{N}(du)ds$; $\mathbf{W}$ and $\mathcal{N}$ are independent; $\mathbf{X}_0$ is a random vector taking values in $S^X \subset \mathbb{R}^d$; $\mathbf{Y}_0$ is a given element of $\{0, 1\}^m$. Moreover, $K_j^Y(x, u) \in \{0, 1\}$ for all x, u and all $1 \leq j \leq m$, so that the solution of (32.57) is in fact of the form $Y_{t,j} = \mathbf{1}_{\{\tau_j \leq t\}}$. Define the sets

$$D_i^X(x) := \{u \in E : K_i^X(x, u) \neq 0\}, \quad 1 \leq i \leq d, \tag{32.58}$$

$$D_j^Y(x) := \{u \in E : K_j^Y(x, u) \neq 0\}, \quad 1 \leq j \leq m. \tag{32.59}$$

By definition of D_i^X, the process $\Delta X_{t,i} \neq 0$ if and only of $\mathcal{N}(\{t\} \times D_i^X) > 0$; similarly, for a nondefaulted firm j we have $\tau_j = t$ if and only if $\mathcal{N}(\{t\} \times D_j^Y) > 0$.

In addition to several regularity conditions ensuring the existence and uniqueness of a solution, in [35] it is assumed that for all $1 \leq j_1 < j_2 \leq m$ and all $x \in S^X$ one has $F_\mathcal{N}\left(D_{j_1}^Y(x) \cap D_{j_2}^Y(x)\right) = 0$. This assumption ensures that for $j_1 \neq j_2$ the processes Y_{j_1} and Y_{j_2} have no common jumps so that there are no joint defaults. Note however that the model (32.56), (32.57) allows for common jumps of $\mathbf{X}$ and $\mathbf{Y}$. More precisely, there is a strictly positive probability that the factor process $\mathbf{X}$ jumps at τ_j, if

$$F_\mathcal{N}\left(D_j^Y(\mathbf{X}_{\tau_j-}) \cap D_i^X(\mathbf{X}_{\tau_j-})\right) > 0 \text{ for some } 1 \leq i \leq d. \tag{32.60}$$

Note moreover that by definition of the compensator of a Poisson random measure,

$$Y_{t,j} - \int_0^t (1 - Y_{s-,j}) F_\mathcal{N}\left(D_j^Y(\mathbf{X}_{s-})\right) ds, \quad t \geq 0,$$

is a $(\mathcal{G}_t)$-martingale, so that $\lambda_j(\mathbf{X}_{t-}) := F_\mathcal{N}\left(D_j^Y(\mathbf{X}_{t-})\right)$ is the $(\mathcal{G}_t)$-default intensity of firm j.

32.6.3 Examples

Next we present a number of specific examples which show that a great variety of models are covered by the system (32.56), (32.57).

Conditionally independent defaults. Consider a model with conditionally independent, doubly-stochastic default times and assume that $\mathbf{X}$ follows a jump-diffusion model of the form

$$d\mathbf{X}_t = b(\mathbf{X}_t)dt + \sigma(\mathbf{X}_t)d\mathbf{W}_t + d\mathbf{J}_t, \tag{32.61}$$

where $\mathbf{J}$ is an $\mathbb{R}^d$-valued compound Poisson process with compensator measure $F_J(dx)ds$. A popular model of this form is the affine jump-diffusion model of [23]. Such a model can be included in the framework (32.56), (32.57) as follows. Take $E = \mathbb{R}^d \times \mathbb{R}$, $F_\mathcal{N} = F_J \times \nu$, ν Lebesgue measure on $\mathbb{R}$, and put

$$K_j^Y(x, u) = \mathbf{1}_{\{[\sum_{i=1}^{j-1} \lambda_i(x), \sum_{i=1}^{j} \lambda_i(x)]\}}(u_{d+1}), \quad 1 \leq j \leq m, \text{ and} \tag{32.62}$$

$$K_i^X(x, u) = u_i \mathbf{1}_{\{[-1,0)\}}(u_{d+1}), \quad 1 \leq i \leq d. \tag{32.63}$$

Note that K^X and K^Y have been chosen so that $F_\mathcal{N}\left(D_i^X(x) \cap D_j^Y(x)\right) = 0$ for all $1 \leq i \leq d$, all $1 \leq j \leq m$, and all x in S^X.

A Markov-chain model with infectious defaults. Next we consider models where the state process jumps in reaction to default events. A simple example is provided by the following generalization of the *infectious-defaults model* of [21]. Here X is taken as scalar and modelled as a finite-state Markov chain with state space $S^X = \{1, \ldots, K\} \subset \mathbb{R}$; the default intensity of firm j is given by $\lambda_j(X_t)$ for increasing functions $\lambda_j : S^X \to \mathbb{R}^+$. At a default time τ_n, X jumps upward by one unit with probability p_{ξ_n} (which may depend on the identity ξ_n of the nth defaulting firm), and remains constant with probability $1 - p_{\xi_n}$ (unless, of course, if $X_{\tau_n-} = K$, where X remains constant). If the system is in an "ignited state", that is if $X_t \geq 2$, X_t jumps to $X_t - 1$ with intensity $\gamma(X_t)$; these downward jumps occur independently of the default history. An upward jump of X at a default can be viewed as manifestation of counterparty risk and/or default contagion, as the default intensities of the remaining firms are increased. This leads to a downward jump in the model value (and hence an increase in the credit spread) of a zero coupon bond issued by some nondefaulted firm. This model can be embedded in the framework (32.56), (32.57) by a proper choice of $F_\mathcal{N}$, K^X and K^Y; see [35] for details.

The information-based model of Frey and Schmidt [37]. This model is of interest in our context, since the dynamics of model values and of the state variable process are themselves derived by filtering arguments. In [37] the framework of this section is slightly extended and three different layers of information are considered: full information, so-called market information ($\mathcal{G}_t$), and finally information of secondary-market investors ($\mathcal{F}_t$). Here the full-information set-up is an additional layer of information used for the construction of the model; we denote the corresponding filtration by ($\tilde{\mathcal{G}}_t$). It is assumed that under full information the default times are conditionally independent doubly stochastic random times, and that ($\tilde{\mathcal{G}}_t$)-default intensities are driven by a finite-state Markov chain Ψ with state space $\{1, \ldots, K\}$ and generator matrix Q^Ψ. The filtration $\mathcal{G}_t \subset \tilde{\mathcal{G}}_t$ represents the information used by the market in determining the theoretical equilibrium prices of traded credit derivatives. In accordance with the definition of $\tilde{H}_t$ in Section 32.6.1, the theoretical price of the traded credit derivatives is defined by $\tilde{H}_{t,j} = E\left(\exp(-r(T-t))H_j \mid \mathcal{G}_t\right), 1 \leq j \leq l$. Define the vector p_t of conditional probabilities by

$$\boldsymbol{p}_t := (p_t^1, \ldots, p_t^K) \quad \text{where} \quad p_t^k := Q(\Psi_t = k \mid \mathcal{G}_t), \ 1 \leq k \leq K. \tag{32.64}$$

The process $\boldsymbol{p} = (\boldsymbol{p}_t)_{t \geq 0}$ is a natural state variable process for the model in the market filtration; $\boldsymbol{p}$ thus plays the role of the process $\mathbf{X}$ introduced in (32.56). The reasons are the following: first, denoting the $(\tilde{\mathcal{G}}_t)$-default intensities by $\nu_i(\Psi_t)$, the $(\mathcal{G}_t)$-default intensities are given by $\lambda_i(\boldsymbol{p}_t) := \sum_{k=1}^K p_t^k \nu_i(k)$. Moreover, note that by the $(\tilde{\mathcal{G}}_t)$-Markovianity of Ψ and $\mathbf{Y}$, the conditional expectation $E\left(\exp(-r(T-t)) H_j \mid \tilde{\mathcal{G}}_t\right)$ is given by some function $\tilde{a}_j(t, \Psi_t, \mathbf{Y}_t)$, at least for $H_j = h_j(\mathbf{Y}_T)$. By iterated conditional expectations theoretical prices can therefore be expressed as functions of t, $\boldsymbol{p}_t$ and $\mathbf{Y}_t$ as well:

$$\tilde{H}_{t,j} = E(\tilde{a}_j(t, \Psi_t, \mathbf{Y}_t) \mid \mathcal{G}_t) = \sum_{k=1}^K p_t^k \tilde{a}_j(t, k, \mathbf{Y}_t) =: a_j(t, \boldsymbol{p}_t, \mathbf{Y}_t), \ 1 \leq j \leq l. \tag{32.65}$$

In particular, theoretical prices have a linear factor structure with factor process $\boldsymbol{p}$.

In [37] it is assumed that $\mathcal{G}_t = \mathcal{F}_t^Y \vee \mathcal{F}_t^U$, where the process U models in abstract form the information aggregated in the equilibrium prices of traded securities. Mathematically, U is given by $U_t = \int_0^t \mu(\Psi_s) ds + B_t$, B a standard $(\tilde{\mathcal{G}}_t)$-Brownian motion independent of Ψ and Y. In a spirit similar to that of the Landen model discussed in Section 32.3.3, the innovations approach to nonlinear filtering is used in order to determine the dynamics of the process $\boldsymbol{p}$, and with it the dynamics of theoretical prices. We briefly sketch the main steps. Consider the $(\mathcal{G}_t)$-adapted processes

$$M_{t,i} := Y_{t,i} - \int_0^{t \wedge \tau_i} \lambda_i(\boldsymbol{p}_s) ds, \ 1 \leq i \leq m, \quad \text{and} \quad W_t := U_t - \int_0^t a(\boldsymbol{p}_s) ds,$$

where $a(\boldsymbol{p}_t) = \sum_{k=1}^K p_t^k \mu(k)$. The processes $\mathbf{M} = (M_{t,1}, \ldots, M_{t,d})_{t \geq 0}$ and W are the *innovations processes*. In particular, it is well-known that $\mathbf{M}$ is a $(\mathcal{G}_t)$-martingale and that W is $(\mathcal{G}_t)$-Brownian motion; moreover, as shown in [37], every $(\mathcal{G}_t)$ martingale can be represented as stochastic integral wrt $\mathbf{M}$ and W.

For a generic process Γ denote by $\widehat{\Gamma}$ the optional projection of Γ with respect to the market filtration $(\mathcal{G}_t)$. Consider now a generic $(\tilde{\mathcal{G}}_t)$-semimartingale with canonical decomposition of the form

$$J_t = J_0 + \int_0^t A_s ds + M_t^J ;$$

here A is some adapted right-continuous process, and M^J is an $(\tilde{\mathcal{G}}_t)$-martingale with $[M^J, B] = 0$ and $[M^J, Y_i] = 0$ for all i. It is then shown in [37], that the optional projection $\widehat{J}_t$ has the representation

$$\widehat{J}_t = \widehat{J}_0 + \int_0^t \widehat{A}_s ds + \int_0^t \gamma_s^\top d\mathbf{M}_s + \int_0^t \alpha_s dW_s, \tag{32.66}$$

where, with $\mu_t := \mu(\Psi_t)$ and $\nu_{t,j} := \nu_j(\Psi_t)$, the integrands γ and a are given by

$$a_t = (\widehat{J\mu})_t - \widehat{J}_t\widehat{\mu}_t, \text{ and } \gamma_{t,j} = \frac{1}{(\widehat{\nu}_j)_{t-}}\left((\widehat{J\nu_j})_{t-} - (\widehat{J})_{t-}(\widehat{\nu}_j)_{t-}\right) \quad j = 1, \ldots, m. \tag{32.67}$$

The proof of this result is based on standard arguments from the innovations approach to nonlinear filtering. Consider now the semimartingales $J_{t,k} = 1_{\{\Psi_t = k\}}$, $1 \le k \le K$ and note that $p_t = \widehat{J}_t$. Since the semimartingale decomposition of J_k is given by $J_{t,k} = J_{0,k} + \int_0^t Q^\Psi_{\Psi_s,k} ds + M_t^{J_k}$, by applying (32.66) to J_k one therefore obtains the following K-dimensional SDE for the process p:

$$dp_t^k = \sum_{i=1}^{K} Q^\Psi_{i,k} p_t^i dt + \sum_{i=1}^{m} \gamma_i^k(p_{t-}) dM_{t,i} + \delta^k(p_{t-}) dW_t, \quad 1 \le k \le K, \tag{32.68}$$

with coefficients given by the functions

$$\gamma_j^k(p) = p^k\left(\frac{\lambda_j(k)}{\sum_{n=1}^{K} \lambda_j(n) p^n} - 1\right), \quad \delta^k(p) = p^k\left(a(k) - \sum_{n=1}^{K} p^n a(n)\right); \tag{32.69}$$

see again[37] for details.

Similarly as in (32.52), in [37] the information set of secondary market investors is given by $\mathcal{F}_t = \mathcal{F}_t^Y \vee \mathcal{F}_t^Z$ for $Z_t = \int_0^t a(s, p_s, Y_s)ds + \beta_t$, β a Brownian motion independent of Y and U. By an argument analogous to that leading to (32.53), the computation of prices for secondary-market investors leads to the problem of finding $\pi_{p_t|\mathcal{F}_t}$ so that the solution p of the filtering problem with respect to the market information $(\mathcal{G}_t)$ becomes the state variable of the filtering problem with respect to the investor filtration $(\mathcal{F}_t)$. The latter filtering problem is covered by our setup, as the processes p and Y follow an SDE system of the form (32.56) and (32.57). Note in particular that at a default time the probability vector p is updated according to (32.68), so that there are common jumps in the state variable p and the observation Y. In [37] the authors discuss also the hedging of credit derivatives from the viewpoint of secondary-market investors. For this they rely on the concept of *risk minimization with restricted information* as introduced in [58]. It is shown that the solution of the hedging problem again leads to the problem of finding $\pi_{p_t|\mathcal{F}_t}$.

While complicated at first sight, the model of [37] has a number of attractive features. To begin with, by (32.65), the main numerical task is the evaluation of the functions $\tilde{a}_j(t, k, y)$; as these functions are computed in the simple full-information set-up, computations become relatively easy even for nonhomogeneous models. Moreover, the model generates a rich set of price dynamics with randomly fluctuating credit spreads and default contagion, and it has a natural factor structure.

32.6.4 Filter equations

In the set-up of Subsection 32.6.2, the determination of $\pi_{X_t|\mathcal{F}_t}(dx)$ becomes an interesting nonlinear filtering problem with observations of mixed type (generated by marked point processes and diffusions) and with common jumps in the observation **Y** and the state process **X**. This problem is nonstandard and merits a discussion in the context of the present survey.

Filtering problems with common jumps of the unobserved state process and of the observations have previously been discussed in the literature. Initial results can be found in [42]; the papers [46] and [12] are concerned with scalar observations described by a pure jump process. The recent paper [19], on the other hand, treats the filtering problem for a very general marked point process model but without common jumps of the state and the observation process. All these papers are based on the innovations approach to nonlinear filtering. Assuming that investor information is equal to the default history $(\mathcal{F}_t^Y)$, [32] gives a simple filter algorithm for the affine jump-diffusion model of [23] with conditionally independent doubly stochastic default times.

In line with [35], in the present chapter we follow an alternative route which is based on ideas from the reference probability approach and takes into account the particular structure of the given general model. In this way we obtain new general recursive filter equations and, as a byproduct, an explicit expression for the joint likelihood of state process and observations.[6] In the case that (**X**, **Y**) is a finite-state continuous-time Markov chain our filter equations give rise to a finite-dimensional filter. We describe now in more detail some of the results from [35] and present in particular the filter equations.

Preliminaries. Since the approach in [35] is based on ideas from the reference probability approach, we first mention some preliminaries to this effect. For ease of notation and without loss of generality in what follows we shall simply denote by $a(t, \mathbf{X}_t)$ the process with the components $a_1(t, \mathbf{X}_t, \mathbf{Y}_t), \cdots, a_l(t, \mathbf{X}_t, \mathbf{Y}_t)$ from the drift of **Z** as introduced in (32.52). It will be convenient to define the processes **X**, **Y** and **Z** on a product space $(\Omega, \mathcal{G}, (\mathcal{G}_t), R^0)$ so that **Z** is independent of **X** and **Y**. Denote by $(\Omega_2, \mathcal{G}_2, (\mathcal{G}_t^2), P^{0,l})$ the l-dimensional Wiener space with coordinate process **Z**, i.e. $\mathbf{Z}_t(\omega_2) = \omega_2(t)$. Given some probability space $(\Omega_1, \mathcal{G}_1, (\mathcal{G}_t^1), P)$ supporting a solution (**X**, **Y**) of the SDE-system (32.56), (32.57), let $\Omega := \Omega_1 \times \Omega_2$, $\mathcal{G} = \mathcal{G}_1 \otimes \mathcal{G}_2$, $\mathcal{G}_t = \mathcal{G}_t^1 \otimes \mathcal{G}_t^2$, $R^0 := P \otimes P^{0,l}$, and put for $\omega = (\omega_1, \omega_2) \in \Omega$

$$\mathbf{X}_t(\omega) := \mathbf{X}_t(\omega_1), \; \mathbf{Y}_t(\omega) := \mathbf{Y}_t(\omega_1), \text{ and } \mathbf{Z}_t(\omega) := \mathbf{Z}_t(\omega_2).$$

Note that this implies that under R^0, **Z** is l-dimensional Brownian motion, independent of **X** and **Y**. Introduce then a Girsanov-type measure transformation of the form $\frac{dR}{dR^0}\big|_{\mathcal{F}_t} = L_t$ with

$$L_t = L_t(\omega_1, \omega_2) = \exp\left\{\int_0^t (a(s, \mathbf{X}_s(\omega_1)))' \, d\mathbf{Z}_s(\omega_2) - \frac{1}{2}\int_0^t \|a(s, \mathbf{X}_s(\omega_1))\|^2 \, ds\right\}$$
(32.70)

and note that L is indeed a R^0-martingale since $a(\cdot)$ was assumed to be bounded. Under R, the process $\mathbf{Z}$ has the original dynamics (32.52), while the law of $\mathbf{X}$ and $\mathbf{Y}$ remains unchanged.

In the sequel we discuss how to obtain the filter distribution $\pi_{\mathbf{X}_t|\mathcal{F}_t}$ in *weak form*, that is we want to compute for a generic bounded and continuous function $h: \mathbb{R}^d \to \mathbb{R}$ the conditional expectation

$$\pi_t h := E(h(\mathbf{X}_t) \mid \mathcal{F}_t) = \int_{\mathbb{R}^d} h(x) \pi_{\mathbf{X}_t|\mathcal{F}_t}(dx).$$
(32.71)

The well-known Kallianpur–Striebel formula gives

$$\pi_t h = \frac{E^{R^0}(h(\mathbf{X}_t) L_t \mid \mathcal{F}_t)}{E^{R^0}(L_t \mid \mathcal{F}_t)},$$
(32.72)

so that, to compute $\pi_t h$, it suffices to compute the numerator on the right-hand side in (32.72). Recalling that $\mathcal{F}_t = \mathcal{F}_t^Y \vee \mathcal{F}_t^Z$, we reduce next the conditioning on $\mathcal{F}_t$ to a conditioning on $\mathcal{F}_t^Y$. Using the Fubini theorem and the product structure of $(\Omega, \mathcal{F}, (\mathcal{F}_t), R^0)$ we get with $L_t = L_t(\omega_1, \omega_2)$ as introduced in (32.70)

$$E^{R^0}\left(h(\mathbf{X}_t) L_t \mid \mathcal{F}_t^Y \vee \mathcal{F}_t^Z\right)(\omega) = E^P\left(h(\mathbf{X}_t) L_t(\cdot, \omega_2) \mid \mathcal{F}_t^Y\right)(\omega_1).$$
(32.73)

In order to compute $\pi_t h$ we thus have to evaluate the conditional expectation on the right hand side of (32.73). Note that this involves only the first component $(\Omega_1, \mathcal{G}_1, (\mathcal{G}_t^1), P)$ of the underlying probability space and hence only the joint law of $\mathbf{X}$ and $\mathbf{Y}$; expectations with respect to that law will be simply denoted by E (instead of E^P).

To derive the filter between default times we have to modify the kernel $K^X(\cdot)$ in the dynamics (32.56) for $\mathbf{X}$ because, conditional on not having jumps in $\mathbf{Y}$, certain jumps in $\mathbf{X}$ cannot occur. More precisely, we shall consider the following kernel, which here is slightly generalized w.r.t. the corresponding definition in Section 32.6.2 by making it explicitly dependent also on y, namely

$$\bar{K}^X(x, y, u) := \begin{cases} 0, & \text{if } u \in \bar{D}^Y(x, y) := \bigcup_{\{j\,:\,y_j=0\}} D_j^Y(x, y), \\ K^X(x, u) & \text{else.} \end{cases}$$
(32.74)

We shall denote by $\bar{\mathbf{X}}_t$ the process corresponding to $\bar{K}^X(\cdot)$ and by $\bar{\mathbf{Y}}_t$ the process in (32.57) obtained when replacing $\mathbf{X}_t$ by $\bar{\mathbf{X}}_t$ there. The law of $(\bar{\mathbf{X}}_t, \bar{\mathbf{Y}}_t)$ with initial condition $(\bar{\mathbf{X}}_0 = x, \bar{\mathbf{Y}}_0 = y)$ will be denoted by $\bar{P}_{(x,y)}$ and the corresponding expectation by $\bar{E}_{(x,y)}$.

Given these preliminaries the filtering results now take the form of a recursion over the successive default times T_n, $1 \le n \le m$.

Filtering between defaults. The main result here is the following (see [35]).

Theorem 32.1 *Given two successive default times T_{n-1}, T_n, we have for $t \in [T_{n-1}, T_n)$*

$$\pi_t h \propto \int_{\mathbb{R}^d} \pi_{T_{n-1}}(d\mathbf{x})\, \bar{E}_{(x,Y_{T_{n-1}})} \left(h\left(\bar{\mathbf{X}}_{t-T_{n-1}}\right) \exp\left\{ -\int_0^{t-T_{n-1}} \bar{\lambda}(\bar{X}_s, Y_{T_{n-1}})ds \right\} \right. \qquad (32.75)$$
$$\left. \cdot \exp\left\{ \int_0^{t-T_{n-1}} a'(T_{n-1}+s, \bar{\mathbf{X}}_s) d\mathbf{Z}_{s+T_{n-1}}(\omega_2) - \frac{1}{2} \int_0^{t-T_{n-1}} \|a(s+T_{n-1}, \bar{\mathbf{X}}_s)\|^2 \, s + T_{n-1}^2 \, ds \right\} \right),$$

where $\bar{E}_{(x,y)}$ is the expectation as introduced above and $\pi_{T_{n-1}}(d\mathbf{x})$ is the filter distribution at $t = T_{n-1}$.

Filtering at a default time. By (32.72) and (32.73) at a generic default time one has

$$\pi_{T_n} h \propto E\left(h(\mathbf{X}_{T_n}) L_{T_n}(\cdot, \omega_2) \mid \mathcal{F}_{T_n}^Y \right).$$

Notice now that, due to the possibility of common jumps between $\mathbf{X}$ and $\mathbf{Y}$, the expressions $E\left(h(\mathbf{X}_{T_n}) L_{T_n}(\cdot, \omega_2) \mid \mathcal{F}_{T_n}^Y \right)$ and $E\left(h\left(\mathbf{X}_{T_n^-}\right) L_{T_n}(\cdot, \omega_2) \mid \mathcal{F}_{T_n}^Y \right)$ do not necessarily coincide. We shall therefore proceed along two steps. In Step 1 we show that one can obtain the conditional expectation $E\left(h(\mathbf{X}_{T_n}) L_{T_n}(\cdot, \omega_2) \mid \mathcal{F}_{T_n}^Y \right)$ once one is able to compute $E\left(g\left(\mathbf{X}_{T_n^-}\right) L_{T_n}(\cdot, \omega_2) \mid \mathcal{F}_{T_n}^Y \right)$ for a generic function $g(\cdot)$. In this step we use the joint distribution of the jumps $\Delta \mathbf{X}_{T_n}$ and $\Delta \mathbf{Y}_{T_n}$ and hence the particular structure of the given model. In Step 2 we then compute the latter of those two quantities via Bayesian updating.

Step 1 (Reduction to the filter distribution of $\mathbf{X}_{T_n^-}$). Here one can show (see [35])

Proposition 32.2 *We have the relation*

$$E\left(h(\mathbf{X}_{T_n}) L_{T_n}(\cdot, \omega_2) \mid \mathcal{F}_{T_n}^Y \right) = E\left(g\left(\mathbf{X}_{T_n^-}, \xi_n\right) L_{T_n}(\cdot, \omega_2) \mid \mathcal{F}_{T_n}^Y \right),$$

where ξ_n is the identity of the firm defaulting at T_n, and where the function g is given by

$$g(x, j) = \begin{cases} F_\mathcal{N}(D_j^Y(x))^{-1} \int_{D_j^Y(x)} h\left(x + K^X(x, u)\right) F_\mathcal{N}(du), & \text{if } F_\mathcal{N}(D_j^Y(x)) > 0, \\ h(x), & \text{else.} \end{cases}$$
$$(32.76)$$

Step 2 (Updating of the conditional distribution of $\mathbf{X}_{T_n^-}$). Here we have

Theorem 32.2 *Given the information that a default has actually occurred at $t = T_n$ and given the identity ξ_n of the defaulting firm, for a generic function $g : \mathbb{R}^d \to \mathbb{R}$ we have*

$$E\left(g\left(\mathbf{X}_{T_n^-}\right) L_{T_n}(\cdot,\omega_2) \mid \mathcal{F}_{T_n}^Y\right) \propto \int_{\mathbb{R}^d} \pi_{T_{n-1}}(d\mathbf{x})\,\bar{E}_{(\mathbf{x},\mathbf{Y}_{T_{n-1}})}\left(g(\bar{\mathbf{X}}_{T_n-T_{n-1}})\right.$$
$$\left.\cdot \lambda_{\xi_n}(\bar{\mathbf{X}}_{T_n-T_{n-1}},\mathbf{Y}_{T_{n-1}})\exp\left\{-\int_0^{T_n-T_{n-1}}\bar{\lambda}(\bar{\mathbf{X}}_s,\mathbf{Y}_{T_{n-1}})\,ds\right\}\right. \quad (32.77)$$
$$\left.\cdot \exp\left\{\int_0^{T_n-T_{n-1}} a'(s+T_{n-1},\bar{\mathbf{X}}_s)\,d\mathbf{Z}_{s+T_{n-1}}(\omega_2) - \frac{1}{2}\int_0^{T_n-T_{n-1}}\|a(s+T_{n-1},\bar{\mathbf{X}}_s)\|^2\,ds\right\}\right).$$

Filter equations for finite-state Markov chains. We show here how the general filter equations specialize for the case when (X, Y) form a finite state Markov chain thereby also showing how these equations allow the computations to be performed explicitly.

We assume w.l.o.g. that the state space of (X, Y) is $\{1,\ldots,K\}\times\{0,1\}^m$ so that X can be considered as scalar. Denote the transition intensities of (X,Y) by $q(k,y;\tilde{k},\tilde{y})$. In line with our general framework we restrict the transition intensities so that default is an absorbing state and so that there are no simultaneous defaults. Hence, denoting the current state by (k, y), there are three possible transitions of (X,Y). First there may be a transition from (k,y) to (h,y), $h\neq k$; this transition occurs with intensity $\bar{q}_{k,h}^y := q(k,y;h,y)$. Second, there may be a "contagious default," i.e. for $i\in\{1,\ldots,m\}$ with $y_i=0$ and $h\neq k$ there may be a transition from (k,y) to (h,y^i), where y^i is obtained from y by flipping the ith coordinate. Third we may have a "pure default," i.e. a transition from (k,y) to (k,y^i). In particular, the default intensity of a nondefaulted firm i is equal to $\lambda_i(k,y)=\sum_{h=1}^K q(k,y;h,y^i)$. This Markov chain model can be included in the general framework of (32.56), (32.57) by a specific choice of K^X and K^Y; see [35] for details.

In this finite-state Markov case the filter distribution can be summarized by the K-dimensional process $\pi_t=(\pi_t^1,\ldots,\pi_t^K)$ with $\pi_t^i:=P(X_t=i\mid \mathcal{F}_t)$. Obviously, it suffices to compute an unnormalized version of π_t. Let

$$L_t^n(\cdot,\omega_2) = \exp\left\{\int_0^t (a_s^n)'(\bar{\mathbf{X}}_s)\,d\mathbf{Z}_s^n(\omega_2) - \frac{1}{2}\int_0^t \|a_s^n(\bar{\mathbf{X}}_s)\|\,s^2 ds\right\},$$

with $\mathbf{Z}_s^n := \mathbf{Z}_{s+T_{n-1}}$ and $a_s^n(\cdot)=a^n(s,\cdot):=a(s+T_{n-1},\cdot)$. Put for $h\in\{1,\ldots,K\}$

$$\sigma_t^h[n,y](\omega_2) := \sum_{i=1}^K \bar{E}_{(i,y)}\left(1_{\{\bar{X}_t=h\}} L_t^n(\cdot,\omega_2)\exp\left\{-\int_0^t \bar{\lambda}(\bar{\mathbf{X}}_s,y)\,ds\right\}\right)\pi_{T_{n-1}}(\{i\}),$$
(32.78)

so that $\sigma_t[n,y]g = \sum_{h=1}^K \sigma_t^h[n,y]g(h)$. We have the following Zakai equation for σ_t (see [35])

Proposition 32.3 *Between default times the process* $\sigma_t=\sigma_t[n,y]$ *solves the SDE*

$$d\sigma_t^i = \left(\sum_{k=1}^K \bar{q}_{k,i}^y \sigma_t^k - \bar{\lambda}(i,y)\sigma_t^i\right)dt + \sigma_t^i\left(a_t^n\right)'(i)\,d\mathbf{Z}_t^n, \quad 1\leq i\leq K, \quad (32.79)$$

with initial condition $\sigma_0^i = \pi_{T_{n-1}}(\{i\})$.

At a default time T_n the filter distribution is updated as follows. Compute first

$$P\left(X_{T_n^-} = i \mid \mathcal{F}_{T_n}\right) := \frac{\lambda_{\xi_n}(i, \mathbf{Y}_{T_{n-1}}) \sigma_{T_n - T_{n-1}}^i[n, \mathbf{Y}_{T_{n-1}}]}{\sum_{k=1}^K \lambda_{\xi_n}(k, \mathbf{Y}_{T_{n-1}}) \sigma_{T_n - T_{n-1}}^k[n, \mathbf{Y}_{T_{n-1}}]}$$

(ξ_n the identity of the defaulting firm) and then

$$\pi_{T_n}^i := \pi_{T_n}(\{i\}) = \sum_{h \neq i} P\left(X_{T_n^-} = h \mid \mathcal{F}_{T_n}\right) \frac{q\left(h, \mathbf{Y}_{T_{n-1}}; i, \mathbf{Y}_{T_n}\right)}{\sum_{j=1}^K q\left(h, \mathbf{Y}_{T_{n-1}}; j, \mathbf{Y}_{T_n}\right)}$$
$$+ P\left(X_{T_n^-} = i \mid \mathcal{F}_{T_n}\right) \frac{q\left(i, \mathbf{Y}_{T_{n-1}}; i, \mathbf{Y}_{T_n}\right)}{\sum_{j=1}^K q\left(i, \mathbf{Y}_{T_{n-1}}; j, \mathbf{Y}_{T_n}\right)}, \quad 1 \leq i \leq K. \quad (32.80)$$

The computability of these expressions hinges upon the solvability of the SDE in (32.79); various considerations to this effect are given in [35].

Further results. In [35] a number of additional results can be found. To begin with, a novel filter-approximation result is established; this justifies the use of the Markov-chain filter as a computational tool for general state variable processes. Moreover, it is shown how to adapt particle filters such as the algorithm of [18] to models with joint jumps of **X** and **Y**. This is important from a computational point of view: while Markov chain approximations are an effective tool for models where **X** is two- to three-dimensional, computations become prohibitively expensive in higher dimensions. Suitable particle filters on the other hand are a viable numerical scheme for moderate dimensions of the state process; see for instance [11] for an elaboration of this point in the context of standard nonlinear filtering problems.

32.7 Summary

We have discussed stochastic filtering problems that arise in the context of incomplete-information models for the term structure of interest rates and credit risk. The models considered were Markovian factor models. At the level of investors in secondary markets, the precise values of these factors are difficult to assess for a number of reasons, and so they have to be treated as latent factors to be filtered on the basis of the actual market information.

The main objective has been the pricing of derivative instruments. Other problems, including parameter estimation/calibration, have been touched upon only briefly. We have typically followed a two-step procedure: in the first step we determined the quantities of interest under full information as functions of the factors; in the second step we then derived their values under the actual market information by projecting the full information values on the subfiltration representing the market information. This is where filtering came in. In order to rule out the possibility of arbitrage, prices are expressed as expectations under a martingale/pricing measure. For pricing problems the filtering problems were therefore formulated directly under a martingale measure; for other purposes the real world/physical measure was found to be more appropriate.

Notes

1. For a discussion of portfolio optimization under incomplete information and the ensuing nonlinear filtering problems we refer to [54].
2. For simplicity we tacitly assume that the security does not generate any intermediate cash flows such as dividend or interest payments.
3. Filtering problems with respect to the physical measure will be discussed in Sections 32.3.4 and 32.3.5 below.
4. By adjusting the filter appropriately so that the log-prices of already matured bonds are no longer taken into account, we may let t go beyond $\min\{T_i : 1 \leq i \leq n\}$.
5. Below we present a slightly simplified version of the model discussed in [36].
6. With common jumps of **X** and **Y** such an expression cannot be derived from standard filtering results in a straightforward way, essentially because **X** and **Y** cannot be made independent by a change of measure.

References

[1] T. Aven. A theorem for determining the compensator of a counting process. *Scand. J. Statist.*, 12(1):69–72, 1985.

[2] A. Bain, and D. Crisan. *Fundamentals of Stochastic Filtering*. Springer, New York, 2009.

[3] T. Bielecki and M. Rutkowski. *Credit Risk: Modeling, Valuation, and Hedging*. Springer, Berlin, 2002.

[4] R. Bhar, C. Chiarella, H. Hung, and W. Runggaldier. The volatility of the instantaneous spot interest rate implied by arbitrage pricing—a dynamic Bayesian approach. *Automatica*, 42:1381–93, 2005.

[5] T. Björk and L. Svensson. On the existence of finite dimensional realizations for nonlinear forward rate models. *Mathematical Finance*, 11:205–43, 2001.

[6] T. Björk. *Arbitrage Theory in Continuous Time*. Oxford University Press, Oxford, 2nd edn, 2004.

[7] F. Black and J. Cox. Valuing corporate securities: Liabilities: Some effects of bond indenture provisions. *J. Finance*, 31:351–67, 1976.

[8] F. Black and M. Scholes. The pricing of options and corporate liabilities. *J. Polit. Economy*, 81(3):637–54, 1973.

[9] C. Blanchet-Scalliet and M. Jeanblanc. Hazard rate for credit risk and hedging defaultable contingent claims. *Finance and Stochastics*, 8:145–59, 2004.

[10] P. Brémaud. *Point Processes and Queues: Martingale Dynamics*. Springer, New York, 1981.

[11] A. Budhiraja, L. Chen, and C. Lee. A survey of nonlinear methods for nonlinear filtering problems. *Physica D*, 230:27–36, 2007.

[12] C. Ceci and A. Gerardi. A model for high frequency data under partial information: a filtering approach. *International Journal of Theoretical and Applied Finance (IJTAF)*, 9:555–76, 2006.

[13] U. Cetin, R. Jarrow, P. Protter, and Y. Yildirim. Modeling credit risk with partial information. *Annals of Applied Probability*, 14(3):1167–78, 2004.

[14] C. Chiarella and O. K. Kwon. Forward rate dependent Markovian transformations of the Heath-Jarrow-Morton term structure model. *Finance and Stochastics*, 5:237–57, 2001.

[15] C. Chiarella, S. Pasquali, and W. Runggaldier. On filtering in Markovian term structure models. *Advances in Applied Probability*, 33:794–809, 2001.

[16] D. Coculescu, H. Geman, and M. Jeanblanc. Valuation of default sensitive claims under imperfect information. *Finance and Stochastics*, 12:195–218, 2008.

[17] P. Collin-Dufresne, R. Goldstein, and J. Helwege. Is credit event risk priced? Modeling contagion via the updating of beliefs. Preprint, Carnegie Mellon University, 2003.

[18] D. Crisan and T. Lyons. A particle approximation of the solution of the Kushner–Stratonovich equation. *Probability Theory and Related Fields*, 115:549–78, 1999.

[19] J. Cvitanic, R. Liptser, and B. Rozovskiĭ. A filtering approach to tracking volatility from prices observed at random times. *The Annals of Applied Probability*, 16:1633–52, 2006.

[20] J. Cvitanic, B. Rozovskiĭ, and Il. Zalyapin. Numerical estimation of volatility values from discretely observed diffusion data. *J. Computational Finance*, 9:1–36, 2006.

[21] M. Davis and V. Lo. Infectious defaults. *Quant. Finance*, 1:382–7, 2001.

[22] D. Duffie, A. Eckner, G. Horel, and L. Saita. Frailty correlated default. *Journal of Finance*, 64(5): 2089–2123, 2009.

[23] D. Duffie and N. Garleanu. Risk and valuation of collateralized debt obligations. *Financial Analysts J.*, 57(1):41–59, 2001.

[24] D. Duffie and R. Kan. A yield factor model of interest rates. *Math. Finance*, 6, 1996.

[25] D. Duffie and D. Lando. Term structure of credit risk with incomplete accounting observations. *Econometrica*, 69:633–64, 2001.

[26] D. Duffie and K. Singleton. Modeling term structures of defaultable bonds. *The Review of Financial Studies*, 12:687–720, 1999.

[27] R. J. Elliott. New finite-dimensional filters and smoothers for noisily observed markov chains. *IEEE Trans. Info. Theory*, IT-39:265–71, 1993.

[28] R. J. Elliott and V. Krishnamurthy. Exact finte-dimensional filters for maximum likelihood parameter estimation of continuous-time linear Gaussian systems. *SIAM J. Control and Optimization*, 35:1908–23, 1997.

[29] D. Filipovic. Separable term structures and the maximal degree problem. *Math. Finance*, 12:341–9, 2002.

[30] C. Fontana and W. J. Runggaldier. Credit risk and incomplete information: filtering and EM parameter estimation. *International Journal of Theoretical and Applied Finance*, 13(1): 683–715, 2010.

[31] R. Frey. Risk-minimization with incomplete information in a model for high frequency data. *Math. Finance*, 10(2):215–25, 2000.

[32] R. Frey, C. Prosdocimi, and W. Runggaldier. Affine credit risk models under incomplete information. In J. Akahori, S. Ogawa, and S. Watanabe, editors, *Stochastic Processes and Application to Math. Finance*, pp. 97–113, World Scientific, Singapore, 2007.

[33] R. Frey and W. Runggaldier. Risk-minimizing hedging strategies under restricted information: the case of stochastic volatility models observed only at discrete random times. *Math. Methods Operations Res.*, 50(3):339–50, 1999.

[34] R. Frey and W. Runggaldier. A nonlinear filtering approach to volatility estimation with a view towards high frequency data. *International J. Theoretical and Applied Finance*, 4:199–210, 2001.

[35] R. Frey and W. Runggaldier. Pricing credit derivatives under incomplete information: a nonlinear filtering approach. Finance and Stochastics. Published online on July 8, 2010.

[36] R. Frey and T. Schmidt. Pricing corporate securities under noisy asset information. Preprint, Universität Leipzig, Math.Finance, 19: 403–421, 2009.

[37] R. Frey and T. Schmidt. Pricing and hedging of credit derivatives via the innovations approach to nonlinear filtering. Preprint 2010. To appear in Finance and Stochastics.

[38] A. Gerardi and P. Tardelli. Filtering on a partially observed ultra-high-frequency data model. *Acta Applicandae Mathematicae*, 91:193–205, 2006.

[39] K. Giesecke and L. R. Goldberg. Sequential defaults and incomplete information. *J. Risk*, 7:1–26, 2004.

[40] A. Gombani, S. Jaschke, and W. Runggaldier. A filtered no arbitrage model for term structures with noisy data. *Stochastic Processes and Applications*, 115:381–400, 2005.

[41] A. Gombani and W. Runggaldier. A filtering approach to pricing in multifactor term structure models. *International J. Theoretical and Applied Finance*, 4:303–20, 2001.

[42] B. I. Grigelionis. On stochastic equations for nonlinear filtering problem of stochastic processes. *Lietuvos Matematikos Rinkinys*, 12:37–51, 1972.

[43] D. Heath, R. Jarrow, and A. Morton, Bond pricing and the term structure of interest rates: a new methodology for contingent claims valuation. *Econometrica* 60:77–105, 1992.

[44] R. Jarrow and P. Protter. Structural versus reduced-form models: a new information based perspective. *J. Investment Management*, 2:1–10, 2004.

[45] R. A. Jarrow and S. M. Turnbull. Pricing derivatives on financial securities subject to credit risk. *J. Finance*, L(1):53–85, 1995.

[46] W. Kliemann, G. Koch, and F. Marchetti. On the unnormalized solution of the filtering problem with counting process observations. *IEEE*, IT-36:1415–25, 1990.

[47] S. Kusuoka. A remark on default risk models. *Adv. Math. Econ.*, 1:69–81, 1999.

[48] C. Landen. Bond pricing in a hidden markov model of the short rate. *Finance and Stochastics*, 4:371–89, 2001.

[49] D. Lando. Cox processes and credit risky securities. *Rev. Derivatives Res.*, 2:99–120, 1998.

[50] D. Lando. *Credit Risk Modeling: Theory and Applications*. Princeton University Press, Princeton, NJ, 2004.

[51] A. J. McNeil, R. Frey, and P. Embrechts. *Quantitative Risk Management: Concepts, Techniques and Tools*. Princeton University Press, Princeton, NJ, 2005.

[52] R. C. Merton. On the pricing of corporate debt: The risk structure of interest rates. *J. Finance*, 29:449–70, 1974.

[53] H. Nakagawa. A filtering model on default risk. *J. Math. Sci. Univ. Tokyo*, 8:107–42, 2001.

[54] H. Pham. Chapter 34, this volume.

[55] E. Platen and W. J. Runggaldier. A benchmark approach to filtering in finance. *Asia-Pacific Financial Markets*, 11:79–105, 2005.

[56] W. J. Runggaldier. Estimation via stochastic filtering in financial market models. In: G. Yin and Q. Zhang, eds, *Mathematics of Finance. Contemporary Mathematics*, Vol. 351, pp. 309–18. American Mathematical Society, Providence, RI, 2004.

[57] P. Schönbucher. Information-driven default contagion. Preprint, Department of Mathematics, ETH Zürich, 2004.

[58] M. Schweizer. Risk minimizing hedging strategies under restricted information. *Math. Finance*, 4:327–42, 1994.

[59] W. M. Wonham. Some applications of stochastic differential equations to optimal nonlinear filtering,. *SIAM J. Control and Optimization*, 2:347–69, 1965.

[60] Y. Zeng. A partially observed model for micromovement of asset prices with Bayes estimation via filtering. *Math. Finance*, 13:411–44, 2003.

·33·
An Asset Pricing Model with Mean Reversion and Regime Switching Stochastic Volatility

R. J. Elliott, H. Miao, and Z. Wu

33.1 Introduction

While issues related to energy products have become increasingly popular in the global economy, building a generalized model for pricing energy-related assets is still challenging. Two key features of energy assets, mean-reverting price processes and stochastic volatility, have been extensively addressed in the finance and economics literature (e.g., [17], [3], [23]). In the finance literature, the Vasicek model ([28]) and the Cox–Ingersoll–Ross (CIR) model ([4]) are the typical ones used to describe the mean-reverting processes.

One widely accepted model dealing with stochastic volatility is the GARCH (generalized autoregressive conditionally heteroskedastic) model first proposed in [2] and [27] extending the traditional ARCH model in Engle's Nobel prize winning work ([16]). Whereas various extensions have been proposed to model volatility, some researchers believe that the volatility should be driven by economic forces other than the price movements. Consequently, Taylor, [27], suggested the stochastic volatility model. In this class of models the volatility follows a stochastic process driven by a different noise rather than the price movements. Other stochastic volatility models that have been developed and used by researchers in the last twenty years include Hull and White, [20], Wiggins, [29], and Heston, [19]. These mainly generalize the Black–Scholes Model by incorporating stochastic volatility.

Assumptions in stochastic volatility models can explain fat tails and also the "smile" and "smirk" pattern observed in derivative markets. Some wellknown models describe volatility using an analytically tractable stochastic process by assuming that volatility is driven by two underlying processes, a fast mean-reverting process and a slowly varying factor. Using a preliminary study of energy data, however, we notice that volatility does not always behave in the same way. Furthermore, volatility clustering is widely observed and volatility remains relatively "tranquil" in some periods. This change of state can be linked to macroeconomic and other changes such as energy price shocks ([22]). For example, when the hurricane Katrina hit the Gulf area, investors were concerned about a possible shortage of crude oil due to the damage of infrastructure facilities. Consequently, the West Texas Intermediate (WTI) market was extremely volatile during that period. Therefore, regime switching seems to be a good candidate for modeling changing states of volatility.

Regime switching models were introduced into economics in 1972 when researchers recognized that parameters may switch due to structural shifts which divide the sample period into different regimes. Hamilton, [18], introduced Markov switching into the econometric mainstream. Later, Lam and Li, [26], incorporated Markov switching into stochastic volatility models. Since these models can capture the above-mentioned movements of energy prices, we apply a time-series regime switching model to characterize the asset pricing issues of energy products. Due to the partial observations of a system involving hidden Markov models, estimating parameter values is, to some degree, challenging. In the literature, parameter estimation in stochastic volatility models has been performed by the generalized method of moments (GMM), maximum likelihood estimation (MLE), the quasi likelihood method (QLM) and Markov chain Monte Carlo (MCMC). In this chapter we apply filtering techniques and the Expectation Maximization (EM) method to estimate the parameters of our stochastic volatility model.

Filtering is a commonly used technique, particularly in engineering problems. The most frequently used filters are the Wonham filter for Markov chains and the Kalman filter for linear Gaussian systems. As an alternative filtering technique to Kalman filtering, Hidden Markov Model (HMM) filtering has been widely accepted and applied ([6], [24], [10], [13], [15]). Using HMM filtering, the estimates of parameters of a model can be continuously updated on the basis of currently available information. Elliott et al., [1] and [9]–[14], considered a finite-state Markov chain observed in Gaussian noise in continuous and discrete time, and used the Expectation Maximization method to update parameter estimates. Elliott et al., [15], applied a robust form of filtering equations for a continuous-time hidden Markov model to estimate the volatility of a risky asset. They improved the classical filtering by eliminating stochastic integration.

In more detail, we consider an two-state, regime-switching volatility model for energy prices in this study. The stochastic process of prices is based on a mean-reverting diffusion with a time-varying volatility. Volatility is then described by another mean-reverting process, whose parameters, such as its mean-reverting ratio, long-term mean and volatility of volatility, change according to the states of a Markov chain. We can consider those states as representing "good" or "bad" economic conditions, or "on-peak" and "off-peak" electricity prices. The framework is easily extended to more than two states. As states are "hidden" and not observed directly, we apply filtering techniques and the Expectation Maximization method to estimate the parameters of our model.

Multiple contributions are expected to be made to the literature. First, this is among the early studies addressing both mean-reversion and stochastic volatility in one model using regime switching techniques. Since the proposed model captures both key features of energy prices, it helps better describe the asset pricing issues of energy products, and, therefore, adds to the literature on energy finance. Secondly, it applies the HMM filtering techniques and EM algorithm to estimate parameter values of the model proposed in this study, and, therefore, opens up a direction of research combining these two parameter estimation techniques to characterizing prices of financial products. Thirdly, it sheds light on other mainstream finance topics such as term structures since short-term interest rates also exhibit the two above-mentioned features, the mean-reverting process and stochastic volatility. Thus, the proposed model can be extended to help model term structures. The chapter contains derivations of several new filters.

Our chapter is organized as follows: we present our basic assumptions in Section 33.2. In Section 33.3, we introduce the change of measure technique applied in our discussion. In Section 33.4, we develop nonlinear filters for the computation of the transition matrix, occupation times, and other processes. We also use the Expectation Maximization (EM) to estimate parameters of the model. Finally, Section 33.6 summarizes our main results and outlines possible future research.

33.2 Assumptions

We suppose the price process Y of an individual capital market instrument, (possibly some energy price), has dynamics

$$dY_t = \gamma(L - Y_t)dt + \sigma_t Y_t^b dW_t, \quad t \geq 0, \quad Y_0 \text{ is given.} \tag{33.1}$$

Here L is the long-term mean of the process towards which prices drift, γ is the speed of this drift, volatility is measured by σ_t and the volatility elasticity is given by $2b$. Therefore, when $b = 0$, this is the Vasicek [28] model, and when $b = 0.5$ it is that of Cox, Ingersoll, and Ross [4].

This is the commonly known one factor mean reverting model which is utilized to model energy prices. However, in practice, we do not really have "continuous data," so we must approximate the process by discrete-time dynamics in order to calibrate the model and estimate the parameters. In other words, discrete-time dynamics are what we really use in practice. We first approximate the continuous dynamics by a discrete time one, then we derive a mean reverting model with a regime switching stochastic volatility.

The Euler discretization of (33.1) is

$$Y_k - Y_{k-1} = \gamma \Delta t (L - Y_{k-1}) + \sigma_k Y_{k-1}^b \sqrt{\Delta t} W_k,$$

where $\{W_k, k = 1, 2, \ldots, T\}$ is a sequence of i.i.d. $N(0, 1)$ random variables. Moreover, if we take $\Delta t = 1$, we have

$$Y_k - Y_{k-1} = \gamma (L - Y_{k-1}) + \sigma_k Y_{k-1}^b W_k.$$

That is

$$Y_k = \gamma L - (\gamma - 1) Y_{k-1} + \sigma_k Y_{k-1}^b W_k.$$

where Y_k and Y_{k-1} are the prices of the individual capital market asset at times k and $k - 1$. The volatility of Y is the time varying parameter σ_k, $k = 1, 2, \ldots, $.

We suppose the volatility σ follows different dynamics in different "states of the world." However, the only process we can observe directly is the return process Y. The volatility dynamics are "hidden" in the observed return process. We represent the "states of the world" by a Markov chain X with a transition matrix A. In our model, we assume the world will have N states. Without loss of generality, we identify the state space S for X to be the set of N column vectors $\{e_1, e_2, \cdots, e_N\}$ where e_i has unity in the ith position and zeros elsewhere. That is

$$S = \{e_i, i = 1, \ldots, N\} = \left\{ \begin{bmatrix} 1 \\ 0 \\ \vdots \\ 0 \end{bmatrix}, \begin{bmatrix} 0 \\ 1 \\ \vdots \\ 0 \end{bmatrix}, \ldots, \begin{bmatrix} 0 \\ 0 \\ \vdots \\ 1 \end{bmatrix} \right\}.$$

We suppose the Markov chain is time homogeneous.

Suppose $p_{ji} = P(X_t = e_j | X_{t-1} = e_i)$, and write $A = (p_{ji})$, $1 \leq i, j \leq N$, for the transition matrix of the chain X. Then, (see [11])

$$X_k = AX_{k-1} + M_k. \tag{33.2}$$

where M_k is a martingale increment.

We suppose the volatility of the return has dynamics

$$\ln \sigma_k = \alpha(k) + \beta(k) \ln \sigma_{k-1} + \theta(k) B_k. \tag{33.3}$$

where $B = \{B_k, k = 1, 2, \ldots\}$ is a sequence of i.i.d. $N(0, 1)$ random variables and the coefficients $\alpha(k)$, $\beta(k)$, and $\theta(k)$ are given by

$$\alpha(k) = \langle \boldsymbol{\alpha}, X_{k-1} \rangle,$$
$$\beta(k) = \langle \boldsymbol{\beta}, X_{k-1} \rangle,$$
$$\theta(k) = \langle \boldsymbol{\theta}, X_{k-1} \rangle.$$

where, $\boldsymbol{\alpha} = \begin{pmatrix} \alpha^1 \\ \alpha^2 \\ \vdots \\ \alpha^N \end{pmatrix}$, $\boldsymbol{\beta} = \begin{pmatrix} \beta^1 \\ \beta^2 \\ \vdots \\ \beta^N \end{pmatrix}$, and $\boldsymbol{\theta} = \begin{pmatrix} \theta^1 \\ \theta^2 \\ \vdots \\ \theta^N \end{pmatrix}$.

In the sequel we shall take $N = 2$, so $X_k = \{e_1, e_2\}$ where $e_1 = \binom{1}{0}$ and $e_2 = \binom{0}{1}$. Write:

$$h_k := \ln \sigma_k,$$

so that $\sigma_t = e^{h_k}$. Introducing the state process X, we have the following dynamics

$$\begin{aligned}h_k &= a(k) + \beta(k) h_{k-1} + \theta(k) B_k \\ &= \langle a, X_{k-1}\rangle + \langle \beta, X_{k-1}\rangle h_{k-1} + \langle \theta, X_{k-1}\rangle B_k,\end{aligned} \qquad (33.4)$$

and

$$Y_k = \gamma L - (\gamma - 1) Y_{k-1} + e^{h_k} Y_{k-1}^b W_k.$$

Define the filtrations

$$\begin{aligned}\mathcal{X}_k &= \sigma\{X_0, X_1, ..., X_k\}, \\ \mathcal{Y}_k &= \sigma\{Y_1, Y_2, ..., Y_k\}, \\ \mathcal{F}_k &= \sigma\{X_0, X_1, ..., X_k, Y_1, Y_2, ..., Y_k\} \\ \mathcal{H}_k &= \sigma\{h_1, h_2, ..., h_k\}, \\ \mathcal{G}_k &= \sigma\{X_0, X_1, ..., X_k, Y_1, Y_2, ..., Y_k, h_1, h_2, ...h_k\}.\end{aligned}$$

33.3 Change of measure

In practice, we only observe the return process $Y_1, Y_2, \ldots, Y_k, \ldots$. At time k, we wish to estimate X_k and the volatility, σ_k^2, and all the related parameters based on the available information. We shall use the change of measure technique and initially work with a "reference" probability $\bar{P}$. We suppose that under the probability space $\bar{P}$, the Y_k are a sequence of i.i.d. random variables each of which is $N(0, 1)$, and the h_k are also i.i.d $h_k \sim N(0, 1)$, where $N(0, 1)$ represents the standard normal distribution with density function

$$\phi(x) = \frac{1}{\sqrt{2\pi}} e^{-\frac{x^2}{2}}.$$

Define:

$$\rho_k = \frac{\phi\left(\frac{Y_k - \gamma L + (\gamma - 1) Y_{k-1}}{Y_{k-1}^b e^{h_k}}\right)}{Y_k^b e^{h_k} \phi(Y_k)}, \quad k = 0, 1, 2, ..,$$

$$\rho_k' = \frac{\phi\left(\frac{h_k - \langle a, X_{k-1}\rangle - \langle \beta, X_{k-1}\rangle h_{k-1}}{\langle \theta, X_{k-1}\rangle}\right)}{\langle \theta, X_{k-1}\rangle \phi(h_k)}, \quad k = 1, 2, ..,$$

and

$\lambda_0 = 1$,

$$\lambda_k = \rho_k \rho_k' = \frac{\phi\left(\frac{Y_k - \gamma L + (\gamma - 1)Y_{k-1}}{Y_{k-1}^b e^{h_k}}\right) \phi\left(\frac{h_k - \langle \alpha, X_{k-1}\rangle - \langle \beta, X_{k-1}\rangle h_{k-1}}{\langle \theta, X_{k-1}\rangle}\right)}{Y_{k-1}^b e^{h_k} \phi(Y_k) \langle \theta, X_{k-1}\rangle \phi(h_k)} \text{ for } k \geq 1,$$

$$\Lambda_t = \prod_{k=1}^t \lambda_k = \prod_{k=1}^t \frac{\phi\left(\frac{Y_k - \gamma L + (\gamma - 1)Y_{k-1}}{Y_{k-1}^b e^{h_k}}\right) \phi\left(\frac{h_k - \langle \alpha, X_{k-1}\rangle - \langle \beta, X_{k-1}\rangle h_{k-1}}{\langle \theta, X_{k-1}\rangle}\right)}{Y_{k-1}^b e^{h_k} \phi(Y_k) \langle \theta, X_{k-1}\rangle \phi(h_k)} \text{ for } k \geq 1. \quad (33.5)$$

We define a measure P by setting $\frac{dP}{d\bar{P}}|\mathcal{F}_k = \Lambda_k$. Then P is the "real world" probability.

Lemma 33.1 *Under P the W_k and B_k, $k = 0, 1, 2, \ldots$, are sequences of independent $N(0, 1)$ random variables; where*

$$W_k = \frac{Y_k - \gamma L + (\gamma - 1)Y_{k-1}}{Y_{k-1}^b e^{h_k}}, \quad (33.6)$$

and

$$B_t = \frac{h_k - \langle \alpha, X_{k-1}\rangle - \langle \beta, X_{k-1}\rangle h_{k-1}}{\langle \theta, X_{k-1}\rangle}. \quad (33.7)$$

Proof. See Appendix A. □

We wish to estimate X_k and h_k given the observations of Y_k under the "real world" probability P. Suppose $f(h_k)$ is an arbitrary function of h_k. If we can determine the quantity $E[f(h_k) X_k | \mathcal{Y}_k]$, we can obtain the estimates of X_k and h_k. By Bayes' theorem (see [14]) we have

$$E[f(h_k) X_k | \mathcal{F}_k^Y] = \frac{\bar{E}[\Lambda_k f(h_k) X_k | \mathcal{F}_k^Y]}{\bar{E}[\Lambda_k | \mathcal{F}_k^Y]}. \quad (33.8)$$

Recall $X_k = \{e_1, e_2\}$, so the numerator is a vector quantity. Suppose the numerator is described in terms of a vector density $q_k(\cdot) = \begin{pmatrix} q_k^1(\cdot) \\ q_k^2(\cdot) \end{pmatrix}$ so that

$$\bar{E}[\Lambda_k f(h_k) X_k | \mathcal{F}_k^Y] = \int_{-\infty}^{\infty} f(z) q_k(z) dz. \quad (33.9)$$

We can then prove the following Lemma which gives a recurrence for q.

Lemma 33.2 *q satisfies the recurrence*

$$q_k(z) = A \int_{-\infty}^{\infty} B(Y_k, z, h, Y_{k-1}) q_{k-1}(h) dh. \quad (33.10)$$

where $B(Y_k, z, h, Y_{k-1})$ is a n-dimensional diagonal matrix with entries

$$\frac{\phi\left(\frac{Y_k - \gamma L + (\gamma-1)Y_{k-1}}{Y_{k-1}^b e^z}\right) \phi\left(\frac{z - a^i - \beta^i h}{\theta^i}\right)}{Y_{k-1}^b e^z \phi(Y_k) \quad \theta^i}.$$

Proof. See Appendix B. □

The following Corollary gives us the quantity $E[X_k|\mathcal{G}_k]$, the estimate of the Markov process conditional on the filtration of Y, and the quantity $E[h_k|\mathcal{G}_k]$, the estimate of the conditional logarithm volatility.

Corollary 33.1 *The optimal estimates of the Markov chain and the volatility are given by:*

$$E[X_k|\mathcal{Y}_k] = \frac{\int_{-\infty}^{\infty} q_k(z) dz}{\sum_{i=1}^{N} \langle \int_{-\infty}^{\infty} q_k(z) dz, e_i \rangle}, \quad (33.11)$$

and

$$E[h_k|\mathcal{Y}_k] = \frac{\sum_{i=1}^{N} \langle \int_{-\infty}^{\infty} z q_k(z) dz, e_i \rangle}{\sum_{i=1}^{N} \langle \int_{-\infty}^{\infty} q_k(z) dz, e_i \rangle}. \quad (33.12)$$

Proof. See Appendix C. □

Equation (33.12) gives the optimal estimate of the logarithm of the volatility given the observation of prices.

33.4 Parameter estimates

In this section, we estimate the parameters of the model. Our model is determined by several parameters. They are: the transition matrix $A = (p_{ji})$, γ, L and the vectors $\alpha = (a^i)$, $\beta = (\beta^i)$, and $\theta = (\theta^i)$. Write the parameters as a set

$$\vartheta := \{p_{ji}, \gamma, L, a^i, \beta^i, \theta^i, 1 \leq i, j \leq N\},$$

where $p_{ji} \geq 0$, $\sum_{j=1}^{N} p_{ji} = 1$. We shall apply the Expectation Maximization (EM) algorithm to estimate all the parameters in our model. In this section we first give an introduction to the EM algorithm and then derive particular processes needed for computing parameters. We close this section with expressions for updating the parameters.

33.4.1 Expectation maximization

As stated in Aggoun and Elliott ([1]),

The EM algorithm is a widely used iterative numerical method for computing maximum likelihood parameter estimates of partially observed models such as linear Gaussian state space models. ([1], p. 177)

The Maximum Likelihood Estimate (MLE) is one of the methods of estimating parameters and finds optimal estimates which maximize the likelihood function $\frac{dP_\vartheta}{dP_0}$ or the log-likelihod function

$$L(\vartheta) = \log \frac{dP_\vartheta}{dP_0}.$$

Here $\{P_\vartheta, \vartheta \in \Theta\}$ is "a family of probability measures" ([1], p. 177). The maximum likelihood estimate is then

$$\widehat{\vartheta} \in \arg\max_{\vartheta \in \Theta} L(\vartheta).$$

However, the MLE is hard to compute. Also, because the variables X and h are not observed directly we must consider the conditional expectation of the log-likelihood. Consequently we shall use the Expectation Maximization (EM) to compute the estimates. The basic idea of the EM algorithm is:

1. We start with appropriate initial values $\widehat{\vartheta}_0$ for

$$\vartheta := \{p_{ji}, \gamma, L, \alpha^i, \beta^i, \theta^i, 1 \leq i, j \leq N\}$$

   which satisfy constraints for the parameters.
2. After some observations of Y, we compute the new estimates of the components of ϑ.
3. Using these values, we reestimate the parameters iteratively until some stopping criterion is satisfied.
4. After more observations we repeat the process again.

Since the EM algorithm improves the estimates monotonically, the expected log-likelihood increases with each reestimation.

33.4.2 Specific processes

To compute the parameters we need to estimate particular cases of the following process:

$$Z_k^{r,s} = \sum_{l=1}^{k} \langle X_{l-1}, e_r \rangle \langle X_l, e_s \rangle H(Y_l) F(h_l) G(h_{l-1}) R(Y_{l-1}), \qquad (33.13)$$

where H, F, G, and R are functions of Y_l, h_l, h_{l-1} and Y_{l-1} respectively.

If we write $\hat{Z}_k^{rs} = E\left[Z_k^{rs}|\mathcal{Y}_k\right]$, and use the Bayes theorem, we have

$$\hat{Z}_k^{rs} = \frac{\bar{E}\left[\Lambda_k Z_k^{rs}|\mathcal{F}_k^Y\right]}{\bar{E}\left[\Lambda_k|\mathcal{F}_k^Y\right]}.$$

Similarly to the earlier discussion, we first consider the quantity $\bar{E}\left[\Lambda_k f(h_k) Z_k^{rs} X_k | \mathcal{F}_k^Y\right]$, then we have the following Lemma..

Lemma 33.3 *Suppose there is a vector measure valued process $\Phi_k^{r,s}$, such that*

$$\bar{E}\left[\Lambda_k Z_k^{r,s} f(h_k) X_k | \mathcal{F}_k^Y\right] = \int_{-\infty}^{\infty} f(z) \Phi_k^{r,s}(z) dz, \quad (33.14)$$

for any integrable function f. Then the following formula updates Φ_k.

$$\Phi_k^{r,s}(z) = A\left(\int_{-\infty}^{\infty} B(Y_k, z, h, Y_{k-1}) \Phi_{k-1}^{r,s}(h) \, dh\right)$$

$$+ H(Y_k) F(z) \left\langle \left(\int_{-\infty}^{\infty} B^r(Y_k, z, h) G(h) q_{k-1}(h) \, dh\right), e_r \right\rangle p_{sr} e_s. \quad (33.15)$$

Proof. See Appendix D. □

Corollary 33.2 *The optimal estimate of Z_k^{rs} is:*

$$\hat{Z}_k^{rs} = \frac{\sum_{i=1}^{N} \left\langle \int_{-\infty}^{\infty} \Phi_k^{rs}(z) dz, e_i \right\rangle}{\sum_{i=1}^{N} \left\langle \int_{-\infty}^{\infty} q_k(z) dz, e_i \right\rangle}. \quad (33.16)$$

Proof. See Appendix E. □

Before we proceed to estimate the parameters, we consider the following special cases for the processes Z_k^{rs}.

- By taking $H(Y_k) = F(h_k) = G(h_{k-1}) = R(Y_{k-1}) \equiv 1$, we have the quantity $\sum_{l=1}^{k} \langle X_{l-1}, e_r \rangle \langle X_l, e_s \rangle$, which represents the number of jumps of X from state r to s up to time k. We call it N_k^{rs}. Then a measure φ_k^{rs} is defined by

$$\bar{E}\left[\Lambda_k N_k^{rs} f(h_k) X_k | \mathcal{Y}_k\right] = \int_{-\infty}^{\infty} f(z) \varphi_k^{rs}(z) dz.$$

This is updated by the formula

$$\varphi_k^{rs}(z) = A\left(\int_{-\infty}^{\infty} B(Y_k, z, h, Y_{k-1}) \varphi_{k-1}^{rs}(h) \, dh\right)$$

$$+ \left\langle \left(\int_{-\infty}^{\infty} B^r(Y_k, z, h, Y_{k-1}) q_{k-1}(h) \, dh\right), e_r \right\rangle p_{sr} e_s. \quad (33.17)$$

and

$$\hat{N}_k^{rs} = \frac{\sum_{i=1}^{N} \langle \int_{-\infty}^{\infty} \varphi_k^{rs}(z) dz, e_i \rangle}{\sum_{i=1}^{N} \langle \int_{-\infty}^{\infty} q_k(z) dz, e_i \rangle}. \tag{33.18}$$

- By summing the terms of N_k^{rs} against the state s, we have the quantity

$$\sum_{s=1}^{N} \sum_{l=1}^{k} \langle X_{l-1}, e_r \rangle \langle X_l, e_s \rangle = \sum_{l=1}^{k} \langle X_{l-1}, e_r \rangle.$$

This represents the amount of time X has spent in state r up to time k. We call this quantity O_k^r, then a measure η_k^r is defined by

$$\bar{E}\left[\Lambda_k f(h_k) O_k^r X_k \mid \mathcal{Y}_k\right] = \int_{-\infty}^{\infty} f(z) \eta_k^r(z) dz.$$

This is updated by the formula

$$\eta_k^r(z) = \sum_{s=1}^{N} \left(A \left(\int_{-\infty}^{\infty} B(Y_k, z, h, Y_{k-1}) \varphi_{k-1}^{r,s}(h) \, dh \right) \right.$$
$$+ \left\langle \left(\int_{-\infty}^{\infty} B^r(Y_k, z, h, Y_{k-1}) q_{k-1}(h) \, dh \right), e_r \right\rangle p_{sr} e_s. \right)$$
$$= A \int_{-\infty}^{\infty} B(Y_k, z, h, Y_{k-1}) \eta_{k-1}^r(h) dh$$
$$+ \left\langle A \int_{-\infty}^{\infty} B(Y_k, z, h, Y_{k-1}) q_{k-1}(h) dh, e_r \right\rangle e_r. \tag{33.19}$$

and

$$\hat{O}_k^r = \frac{\sum_{i=1}^{N} \langle \int_{-\infty}^{\infty} \eta_k^r(z) dz, e_i \rangle}{\sum_{i=1}^{N} \langle \int_{-\infty}^{\infty} q_k(z) dz, e_i \rangle}. \tag{33.20}$$

- By summing the terms of the general case of Z_k^{rs}, we have the quantity

$$Z_k^r = \sum_{l=1}^{k} \langle X_{l-1}, e_r \rangle \, H(Y_l) \, F(h_l) \, G(h_{l-1}).$$

Then a measure L_k^r is defined by

$$\bar{E}\left[\Lambda_k Z_k^r f(h_k) X_k \mid \mathcal{Y}_k\right] = \int_{-\infty}^{\infty} f(z) \Phi_k^r(z) dz,$$

This is updated by the formula

$$L_k^r(z) = A\left(\int_{-\infty}^{\infty} B(Y_k, z, h, Y_{k-1}) L_{k-1}^r(h)\, dh\right)$$
$$+ H(Y_k) F(z) R(Y_{k-1}) \left\langle A\left(\int_{-\infty}^{\infty} B^r(Y_k, z, h, Y_{k-1}) G(h) q_{k-1}(h)\, dh\right), e_r \right\rangle e_r$$
(33.21)

- Z.1 Suppose $Z_k^{r,1} := \sum_{l=1}^{k} \langle X_{l-1}, e_r \rangle h_l$. Then a measure $\xi_k^{r,1}$ is defined by

$$\bar{E}\left[\Lambda_k Z_k^{r,1} f(h_k) X_k | \mathcal{Y}_k\right] = \int_{-\infty}^{\infty} f(z) \xi_k^{r,1}(z)\, dz.$$

In this case, $H(Y_k) = G(h_{k-1}) = R(Y_{k-1}) \equiv 1$, and $F(h_k) = h_k$. From (33.21), this is updated by the formula

$$\xi_k^{r,1}(z) = A\left(\int_{-\infty}^{\infty} B(Y_k, z, h, Y_{k-1}) \xi_{k-1}^{r,1}(h)\, dh\right)$$
$$+ z\left\langle A\left(\int_{-\infty}^{\infty} B^r(Y_k, z, h, Y_{k-1}) q_{k-1}(h)\, dh\right), e_r \right\rangle e_r, \quad (33.22)$$

and

$$\hat{Z}_k^{r,1} = E\left[Z_k^{r,1} | \mathcal{Y}_k\right] = \frac{\sum_{i=1}^{N} \left\langle \int_{-\infty}^{\infty} \xi_k^{r,1}(z)\, dz, e_i \right\rangle}{\sum_{i=1}^{N} \left\langle \int_{-\infty}^{\infty} q_k(z)\, dz, e_i \right\rangle}. \quad (33.23)$$

- Z.2 Suppose $Z_k^{r,2} := \sum_{l=1}^{k} \langle X_{l-1}, e_r \rangle h_{l-1}$. then a measure $\xi_k^{r,2}$ is defined by

$$\bar{E}[\Lambda_k Z_k^{r,2} f(h_k) X_k | \mathcal{Y}_k] = \int_{-\infty}^{\infty} f(z) \xi_k^{r,2}(z)\, dz.$$

This is updated by the formula

$$\xi_k^{r,2}(z) = A\left(\int_{-\infty}^{\infty} B(Y_k, z, h, Y_{k-1}) \xi_{k-1}^{r,2}(h)\, dh\right)$$
$$+ \left\langle A\left(\int_{-\infty}^{\infty} B^r(Y_k, z, h, Y_{k-1}) h q_{k-1}(h)\, dh\right), e_r \right\rangle e_r, \quad (33.24)$$

and

$$\hat{Z}_k^{r,2} = \frac{\sum_{i=1}^{N} \left\langle \int_{-\infty}^{\infty} \gamma_k^{r,2}(z)\, dz, e_i \right\rangle}{\sum_{i=1}^{N} \left\langle \int_{-\infty}^{\infty} q_k(z)\, dz, e_i \right\rangle}. \quad (33.25)$$

- Z.3 Suppose $Z_k^{r,3} := \sum_{l=1}^{k} \langle X_{l-1}, e_r \rangle h_l h_{l-1}$. Then a measure $\xi_k^{r,3}$ is defined by

$$\bar{E}[\Lambda_k Z_k^{r,3} f(h_k) X_k | \mathcal{Y}_k] = \int_{-\infty}^{\infty} f(z) \xi_k^{r,3}(z) dz.$$

This is updated by the formula

$$\xi_k^{r,3}(z) = A\left(\int_{-\infty}^{\infty} B(Y_k, z, h, Y_{k-1}) \xi_{k-1}^{r,3}(h) \, dh\right)$$
$$+ z \left\langle A\left(\int_{-\infty}^{\infty} B^r(Y_k, z, h) h q_{k-1}(h) \, dh\right), e_r \right\rangle e_r, \qquad (33.26)$$

and

$$\hat{Z}_t^{r,3} = \frac{\sum_{i=1}^{N} \left\langle \int_{-\infty}^{\infty} \xi_k^{r,3}(z) dz, e_i \right\rangle}{\sum_{i=1}^{N} \left\langle \int_{-\infty}^{\infty} q_k(z) dz, e_i \right\rangle}. \qquad (33.27)$$

- Z.4 Suppose $Z_k^{r,4} := \sum_{l=1}^{k} \langle X_{l-1}, e_r \rangle h_l^2$. Then a measure $\xi_k^{r,4}$ is defined by

$$\bar{E}\left[\Lambda_k Z_k^{r,4} f(h_k) X_k | \mathcal{Y}_k\right] = \int_{-\infty}^{\infty} f(z) \xi_k^{r,4}(z) dz.$$

This is updated by the formula

$$\xi_k^{r,4}(z) = A\left(\int_{-\infty}^{\infty} B(Y_k, z, h, Y_{k-1}) \xi_{k-1}^{r,4}(h) \, dh\right)$$
$$+ z^2 \left\langle A\left(\int_{-\infty}^{\infty} B^r(Y_k, z, h, Y_{k-1}) q_{k-1}(h) \, dh\right), e_r \right\rangle e_r, \qquad (33.28)$$

and

$$\hat{Z}_k^{r,4} = \frac{\sum_{i=1}^{N} \left\langle \int_{-\infty}^{\infty} \xi_k^{r,4}(z) dz, e_i \right\rangle}{\sum_{i=1}^{N} \left\langle \int_{-\infty}^{\infty} q_k(z) dz, e_i \right\rangle}. \qquad (33.29)$$

- Z.5 Suppose $Z_k^{r,5} := \sum_{l=1}^{k} \langle X_{l-1}, e_r \rangle h_{l-1}^2$. Then a measure $\xi_k^{r,5}$ is defined by

$$\bar{E}\left[\Lambda_k Z_k^{r,5} f(h_k) X_k | \mathcal{Y}_k\right] = \int_{-\infty}^{\infty} f(z) \xi_k^{r,5}(z) dz.$$

This is updated by the formula

$$\xi_k^{r,5}(z) = A\left(\int_{-\infty}^{\infty} B(Y_k, z, h, Y_{k-1})\xi_{k-1}^{r,5}(h)\,dh\right)$$
$$+ \left\langle A\left(\int_{-\infty}^{\infty} B^r(Y_k, z, h)h^2 q_{k-1}(h)\,dh\right), e_r\right\rangle e_r, \quad (33.30)$$

and

$$\hat{Z}_k^{r,5} = \frac{\sum_{i=1}^{N}\left\langle \int_{-\infty}^{\infty} \xi_k^{r,5}(z)dz, e_i\right\rangle}{\sum_{i=1}^{N}\left\langle \int_{-\infty}^{\infty} q_k(z)dz, e_i\right\rangle}. \quad (33.31)$$

- Z.6 Suppose $Z_k^{r,6} := \sum_{l=1}^{k} \langle X_{l-1}, e_r\rangle Y_l e^{-2h_l}$. Then a measure $\xi_k^{r,6}$ is defined by

$$\bar{E}\left[\Lambda_k Z_k^{r,6} f(h_k) X_k | \mathcal{Y}_k\right] = \int_{-\infty}^{\infty} f(z)\xi_k^{r,6}(z)dz.$$

This is updated by the formula

$$\xi_k^{r,6}(z) = A\left(\int_{-\infty}^{\infty} B(Y_k, z, h, Y_{k-1})\xi_{k-1}^{r,6}(h)\,dh\right)$$
$$+ Y_k e^{-2z}\left\langle A\left(\int_{-\infty}^{\infty} B^r(Y_k, z, h)q_{k-1}(h)\,dh\right), e_r\right\rangle e_r, \quad (33.32)$$

and

$$\hat{Z}_k^{r,6} = \frac{\sum_{i=1}^{N}\left\langle \int_{-\infty}^{\infty} \xi_k^{r,6}(z)dz, e_i\right\rangle}{\sum_{i=1}^{N}\left\langle \int_{-\infty}^{\infty} q_k(z)dz, e_i\right\rangle}. \quad (33.33)$$

- Z.7 Suppose $Z_k^{r,7} := \sum_{l=1}^{k} \langle X_{l-1}, e_r\rangle e^{-2h_l}$. Then a measure $\xi_k^{r,7}$ is defined by

$$\bar{E}\left[\Lambda_k Z_k^{r,7} f(h_k) X_k | \mathcal{Y}_k\right] = \int_{-\infty}^{\infty} f(z)\xi_k^{r,7}(z)dz.$$

This is updated by the formula

$$\xi_k^{r,7}(z) = A\left(\int_{-\infty}^{\infty} B(Y_k, z, h, Y_{k-1})\xi_{k-1}^{r,7}(h)\,dh\right)$$
$$+ e^{-2z}\left\langle A\left(\int_{-\infty}^{\infty} B^r(Y_k, z, h)q_{k-1}(h)\,dh\right), e_r\right\rangle e_r, \quad (33.34)$$

and

$$\hat{Z}_k^{r,7} = \frac{\sum_{i=1}^{N}\left\langle \int_{-\infty}^{\infty} \xi_k^{r,7}(z)dz, e_i\right\rangle}{\sum_{i=1}^{N}\left\langle \int_{-\infty}^{\infty} q_k(z)dz, e_i\right\rangle}. \quad (33.35)$$

- Z.8 Suppose $Z_k^{r,8} := \sum_{l=1}^{k} \langle X_{l-1}, e_r \rangle e^{-2h_l} Y_{l-1}$. Then a measure $\xi_k^{r,8}$ is defined by

$$\bar{E}[\Lambda_k Z_k^{r,8} f(h_k) X_k | \mathcal{Y}_k] = \int_{-\infty}^{\infty} f(z) \xi_k^{r,8}(z) dz.$$

This is updated by the formula

$$\xi_k^{r,8}(z) = A\left(\int_{-\infty}^{\infty} B(Y_k, z, h, Y_{k-1}) \xi_{k-1}^{r,8}(h) \, dh\right)$$

$$+ Y_{k-1} e^{-2z} \left\langle A\left(\int_{-\infty}^{\infty} B^r(Y_k, z, h) q_{k-1}(h) \, dh\right), e_r \right\rangle e_r, \quad (33.36)$$

and

$$\hat{Z}_k^{r,8} = \frac{\sum_{i=1}^{N} \left\langle \int_{-\infty}^{\infty} \xi_k^{r,8}(z) dz, e_i \right\rangle}{\sum_{i=1}^{N} \left\langle \int_{-\infty}^{\infty} q_k(z) dz, e_i \right\rangle}. \quad (33.37)$$

- Z.9 Suppose $Z_k^{r,9} := \sum_{l=1}^{k} \langle X_{l-1}, e_r \rangle e^{-2h_l} Y_{l-1}^2$. Then a measure $\xi_k^{r,9}$ is defined by

$$\bar{E}\left[\Lambda_k Z_k^{r,9} f(h_k) X_k | \mathcal{Y}_k\right] = \int_{-\infty}^{\infty} f(z) \xi_k^{r,9}(z) dz.$$

This is updated by the formula

$$\xi_k^{r,9}(z) = A\left(\int_{-\infty}^{\infty} B(Y_k, z, h, Y_{k-1}) \xi_{k-1}^{r,9}(h) \, dh\right)$$

$$+ Y_{k-1}^2 e^{-2z} \left\langle A\left(\int_{-\infty}^{\infty} B^r(Y_k, z, h) q_{k-1}(h) \, dh\right), e_r \right\rangle e_r, \quad (33.38)$$

and

$$\hat{Z}_k^{r,9} = \frac{\sum_{i=1}^{N} \left\langle \int_{-\infty}^{\infty} \xi_k^{r,9}(z) dz, e_i \right\rangle}{\sum_{i=1}^{N} \left\langle \int_{-\infty}^{\infty} q_k(z) dz, e_i \right\rangle}. \quad (33.39)$$

- Z.10 Suppose $Z_k^{r,10} := \sum_{l=1}^{k} \langle X_{l-1}, e_r \rangle e^{-2h_l} Y_l Y_{l-1}$. Then a measure $\xi_k^{r,7}$ is defined by

$$\bar{E}[\Lambda_k Z_k^{r,10} f(h_k) X_k | \mathcal{Y}_k] = \int_{-\infty}^{\infty} f(z) \xi_k^{r,10}(z) dz.$$

This is updated by the formula

$$\xi_k^{r,10}(z) = A\left(\int_{-\infty}^{\infty} B(Y_k, z, h, Y_{k-1})\xi_{k-1}^{r,10}(h)\,dh\right)$$
$$+ Y_k Y_{k-1} e^{-2z}\left\langle A\left(\int_{-\infty}^{\infty} B^r(Y_k, z, h)q_{k-1}(h)\,dh\right), e_r\right\rangle e_r, \quad (33.40)$$

and

$$\hat{Z}_k^{r,10} = \frac{\sum_{i=1}^{N}\langle\int_{-\infty}^{\infty}\xi_k^{r,10}(z)dz, e_i\rangle}{\sum_{i=1}^{N}\langle\int_{-\infty}^{\infty}q_k(z)dz, e_i\rangle}. \quad (33.41)$$

We now have all the estimates needed to update the parameters of our model. Recall that our model is determined by the parameters

$$\vartheta := \{p_{ji}, \mu, \alpha, \beta, \theta, 1 \leq i, j \leq N\}$$

.where $p_{ji} \geq 0$, $\sum_{j=1}^{N} p_{ji} = 1$.

We wish to determine a new set $\hat{\vartheta} := \{\hat{p}_{ji}, \hat{\mu}, \hat{\alpha}, \hat{\beta}, \hat{\theta}\, 1 \leq i, j \leq N\}$ which maximizes the expected log-likelihood:

$$Q(\vartheta, \vartheta^*) = E_{\vartheta^*}\left[\log\frac{dP_\vartheta}{dP_{\vartheta^*}}\bigg|\, \mathcal{F}_t^Y\right]. \quad (33.42)$$

33.4.3 Estimate of p_{ji}

Consider the parameter p_{ji}, under $P = P_\vartheta$ so that under the new measure X is a Markov chain with transition $A = (p_{ji})$. We apply the change of measure technique and introduce a new probability measure $P_{\hat{\vartheta}}$ so that under the new measure, X is now a Markov chain with transition matrix $\hat{A} = (\hat{p}_{ji})$, where $P_{\hat{\vartheta}}(X_k = e_j|X_{k-1} = e_i) = \hat{p}_{ji}$.) We choose $\hat{A} = (\hat{p}_{ji})$ so that the $\hat{p}_{ji}$ maximize the Q given in (33.42).

Lemma 33.4 Define $\Lambda_0 = \Lambda_0$ and $\Lambda_k = \prod_{l=1}^{k}\left(\sum_{j,i=1}^{N}(\frac{\hat{p}_{ji}}{p_{ji}})\langle X_l, e_j\rangle\langle X_{l-1}, e_i\rangle\right)$. (If $p_{ji} = 0$, take $\hat{p}_{ji} = 0$ and $\frac{\hat{p}_{ji}}{p_{ji}} = 1$.) Define $P_{\hat{\vartheta}}$ by setting $\frac{dP_{\hat{\vartheta}}}{dP_\vartheta}\big|\mathcal{Y}_k = \Lambda_k$. Then under the new measure $P_{\hat{\vartheta}}$, X is a Markov chain with transition $\hat{A} = (\hat{p}_{ji})$.

Proof. See Appendix F. □

Lemma 33.5 Given the observations up to time k, $\{Y_0, Y_1, Y_2, ..., Y_k\}$, that is, given $\mathcal{Y}_k$, and given the parameter set $\vartheta := \{p_{ji}, \alpha^i, \beta^i, \theta^i, \gamma, L\, 1 \leq i, j \leq N\}$, the EM estimates of $\hat{p}_{ji}$ are given by

$$\hat{p}_{ji} = \frac{\hat{N}_k^{ij}}{\hat{O}_k^i} = \frac{\sum_{i=1}^{N} \left\langle \int_{-\infty}^{\infty} \varphi_k^{ij}(z) dz, e_i \right\rangle}{\sum_{i=1}^{N} \left\langle \int_{-\infty}^{\infty} \eta_k^i(z) dz, e_i \right\rangle}. \tag{33.43}$$

Proof. See Appendix G. $\square$

33.4.4 Estimate of a, β, θ, γ and L.

By considering similar changes of measure the following lemmas are obtained.

Lemma 33.6 *The estimate for the parameter $a = (a^1, a^N)$, at time t, $\hat{a}_t = (\hat{a}_k^1, \hat{a}_k^N)$ is given by*

$$\hat{a}_k^i = \frac{\sum_{i=1}^{N} \left\langle \int_{-\infty}^{\infty} \left(\gamma_k^{r,1}(z) - \beta^r \gamma_k^{r,2}(z) \right) dz, e_i \right\rangle}{\sum_{i=1}^{N} \left\langle \int_{-\infty}^{\infty} \eta_k^r(z) dz, e_i \right\rangle} \tag{33.44}$$

Proof. See Appendix H. $\square$

Similarly we have the following Lemmas:

Lemma 33.7 *An updated estimate for the parameter $\beta = (\beta^1, \beta^N)$, at time t, $\hat{\beta}_t = (\hat{\beta}_t^1, \hat{\beta}_t^N)$ is given by*

$$\hat{\beta}_t^r = \frac{\sum_{i=1}^{N} \left\langle \int_{-\infty}^{\infty} \left(\xi_t^{r,3}(z) - a^r \xi_t^{r,1}(z) \right) dz, e_i \right\rangle}{\sum_{i=1}^{N} \left\langle \int_{-\infty}^{\infty} \xi_t^{r,4}(z) dz, e_i \right\rangle} \tag{33.45}$$

Lemma 33.8 *An updated estimate for the parameter $\theta = (\theta^1, \theta^N)$, at time t, $\hat{\theta}_t = (\hat{\theta}_t^1, \hat{\theta}_t^N)$ is given by*

$$\hat{\theta}_t^r = \left(\sum_{i=1}^{N} \left\langle \int_{-\infty}^{\infty} \left(\xi_t^{r,4}(z) + (a^r)^{r,2} \eta_t(z) + (\beta^r)^{r,2} \xi_t^{r,5}(z) - 2a^r \xi_t^{r,1}(z) \right. \right. \right.$$
$$\left. \left. \left. - 2\beta^r \xi_t^{r,3}(z) + 2a^r \beta^r \xi_t^{r,2}(z) dz, e_i \right\rangle \right) \frac{1}{\sum_{i=1}^{N} \left\langle \int_{-\infty}^{\infty} \eta_t^r(z) dz, e_i \right\rangle} \right)^{\frac{1}{2}}. \tag{33.46}$$

Lemma 33.9 *The estimate for the parameter γ, at time t, $\hat{\gamma}$ is given by*

$$\hat{\gamma} = \frac{\sum_{r=1}^{N} \left(\sum_{i=1}^{N} \left\langle \int_{-\infty}^{\infty} \left(\xi_t^{r,10}(z) - \xi_t^{r,9}(z) \right) dz, e_i \right\rangle + \sum_{i=1}^{N} \left\langle L \int_{-\infty}^{\infty} \left(\xi_t^{r,8}(z) - \xi_t^{r,6}(z) \right) dz, e_i \right\rangle \right)}{\sum_{r=1}^{N} \left(\sum_{i=1}^{N} \left\langle \int_{-\infty}^{\infty} \xi_t^{r,7}(z) dz, e_i \right\rangle \right)}.$$

Lemma 33.10 *The estimate for the parameter L, at time t, $\hat{L}$ is given by*

$$\hat{L} = \frac{\sum_{r=1}^{N}\left(\sum_{i=1}^{N}\left\langle \int_{-\infty}^{\infty} \xi_t^{r,6}(z)dz, e_i\right\rangle + \sum_{i=1}^{N}\left\langle (\gamma-1)\int_{-\infty}^{\infty} \xi_t^{r,8}(z)dz, e_i\right\rangle\right)}{\sum_{r=1}^{N}\left(\sum_{i=1}^{N}\left\langle \int_{-\infty}^{\infty} \xi_t^{r,7}(z)dz, e_i\right\rangle\right)}.$$

33.5 Conclusion

Our algorithms should be implementable as they are extensions of known techniques. Numerical work will be investigated in a later study. Related contributions include Yu and Sheble ([30]), who apply the hidden Markov model to describe electricity prices by considering the supply-demand equilibrium. In their work, a finite number of states is taken into account. Using "historical data from the New York independent system operator (NYISO)" (p. 445), Yu and Sheble ([30]) "illustrate the forecasting power of the HMM" (p. 445). Dueker ([7]) also applies a Markov switching model to investigate the mean-reverting features of volatility in stock markets. Empirical results based on US stock market information show that Markov switching helps describe mean reversion.

Regime switching stochastic volatility models with mean reverting dynamics ([8]), not only perform well in the characterization of energy prices but also seem appropriate when addressing other main stream finance issues, such as term structures. It is well known that short-term interest rates exhibit features such as mean reversion and stochastic volatility. Applying the efficient method of moments, Creal, Gu, and Zivot ([5]) study the major features of short-term interest rates using US data, and show that their models simulate term structures. Consequently, applying the model discussed in the current study to capture key characteristics of term structure is possibly of interest for future research.

A Proof of Lemma 33.1

Proof. Using Bayes' theorem, for two arbitrary measurable function $f(B_k)$, and $g(W_k)$, we have

$$E\left[f(B_k)g(W_k)|\mathcal{Y}_{k-1}\right] = \frac{\bar{E}[\Lambda_k f(B_k)g(W_k)|\mathcal{Y}_{k-1}]}{\bar{E}[\Lambda_k|\mathcal{Y}_{k-1}]}$$

$$= \frac{\bar{E}[\lambda_k \Lambda_{k-1} f(B_k)g(W_k)|\mathcal{Y}_{k-1}]}{\bar{E}[\lambda_k \Lambda_{k-1}|\mathcal{Y}_{k-1}]}$$

$$= \frac{\bar{E}[\lambda_k f(B_k)g(W_k)|\mathcal{Y}_{k-1}]}{\bar{E}[\lambda_k|\mathcal{Y}_{k-1}]}. \qquad (33.47)$$

Since $\Lambda_k = \Lambda_{k-1}\lambda_k$ and Λ_{k-1} is $\mathcal{Y}_{k-1}$ measurable, we first show that

$$\bar{E}[\lambda_k | \mathcal{G}_{k-1}] = 1.$$

Now

$\bar{E}[\lambda_k | \mathcal{G}_{k-1}]$

$$= \bar{E}\left[\frac{\phi\left(\frac{Y_k - \gamma L + (\gamma-1)Y_{k-1}}{Y_{k-1}^b e^{h_k}}\right) \phi\left(\frac{h_k - \langle a, X_{k-1}\rangle - \langle \beta, X_{k-1}\rangle h_{k-1}}{\langle \theta, X_{k-1}\rangle}\right)}{Y_{k-1}^b e^{h_k} \phi(Y_k) \langle \theta, X_{k-1}\rangle \phi(h_k)} \bigg| \mathcal{G}_{k-1}\right]$$

$$= \bar{E}\left[\bar{E}\left[\frac{\phi\left(\frac{Y_k - \gamma L + (\gamma-1)Y_{k-1}}{Y_{k-1}^b e^{h_k}}\right) \phi\left(\frac{h_k - \langle a, X_{k-1}\rangle - \langle \beta, X_{k-1}\rangle h_{k-1}}{\langle \theta, X_{k-1}\rangle}\right)}{Y_{k-1}^b e^{h_k} \phi(Y_k) \langle \theta, X_{k-1}\rangle \phi(h_k)} \bigg| \mathcal{G}_{k-1}, h_k\right] \bigg| \mathcal{G}_{k-1}\right].$$

(33.48)

The only unknown in the inner conditioning is Y_k, so integrating against its density we have that (33.48) equals

$$= \bar{E}\left[\frac{\phi\left(\frac{h_k - \langle a, X_{k-1}\rangle - \langle \beta, X_{k-1}\rangle h_{k-1}}{\langle \theta, X_{k-1}\rangle}\right)}{\langle \theta, X_{k-1}\rangle \phi(h_k)} \int_{-\infty}^{\infty} \frac{\phi\left(\frac{Y_k - \gamma L + (\gamma-1)Y_{k-1}}{Y_{k-1}^b e^{h_k}}\right)}{Y_{k-1}^b e^{h_k} \phi(Y_k)} \phi(Y_k) dY_k \bigg| \mathcal{G}_{k-1}\right].$$

Recall that $W_k = \frac{Y_k - \gamma L + (\gamma-1)Y_{k-1}}{Y_{k-1}^b e^{h_k}}$ so $dW_k = Y_{k-1}^{-b} e^{-h_k} dY_k$, $dY_k = e^{h_k} Y_{k-1}^b dW_k$, and the equation becomes:

$$= \bar{E}\left[\frac{\phi\left(\frac{h_k - \langle a, X_{k-1}\rangle - \langle \beta, X_{k-1}\rangle h_{k-1}}{\langle \theta, X_{k-1}\rangle}\right)}{\langle \theta, X_{k-1}\rangle \phi(h_k)} \int_{-\infty}^{\infty} \phi(W_k) dW_k \bigg| \mathcal{G}_{k-1}\right]$$

$$= \bar{E}\left[\frac{\phi\left(\frac{h_k - \langle a, X_{k-1}\rangle - \langle \beta, X_{k-1}\rangle h_{k-1}}{\langle \theta, X_{k-1}\rangle}\right)}{\langle \theta, X_{k-1}\rangle \phi(h_k)} \bigg| \mathcal{G}_{k-1}\right]$$

$$= \int_{-\infty}^{\infty} \frac{\phi\left(\frac{h_k - \langle a, X_{k-1}\rangle - \langle \beta, X_{k-1}\rangle h_{k-1}}{\langle \theta, X_{k-1}\rangle}\right)}{\langle \theta, X_{k-1}\rangle \phi(h_k)} \phi(h_k) dh_k.$$

Similarly, substituting $B_k = \frac{h_k - \langle a, X_{k-1}\rangle - \langle \beta, X_{k-1}\rangle h_{k-1}}{\langle \theta, X_{k-1}\rangle}$, this becomes

$$= \int_{-\infty}^{\infty} \phi(B_k) dB_t = 1.$$

Now we consider the numerator of (1):

$$\bar{E}\left[\lambda_k f(w_k) g(v_k) | \mathcal{F}_{k-1}^{X,Y,h}\right]$$

$$= \bar{E}\left[\lambda_k f\left(\frac{h_k - \langle a, X_{k-1}\rangle - \langle \beta, X_{k-1}\rangle h_{k-1}}{\langle \theta, X_{k-1}\rangle}\right) g\left(\frac{Y_k - \gamma L + (\gamma-1)Y_{k-1}}{Y_{k-1}^b e^{h_k}}\right) \Big| \mathcal{G}_{k-1}\right]$$

$$= \bar{E}\left[f\left(\frac{h_k - \langle a, X_{k-1}\rangle - \langle \beta, X_{k-1}\rangle h_{k-1}}{\langle \theta, X_{k-1}\rangle}\right) \frac{\phi\left(\frac{h_k - \langle a, X_{k-1}\rangle - \langle \beta, X_{k-1}\rangle h_{k-1}}{\langle \theta, X_{k-1}\rangle}\right)}{\langle \theta, X_{k-1}\rangle \phi(h_k)}\right.$$

$$\left. \times g\left(\frac{Y_k - \gamma L + (\gamma-1)Y_{k-1}}{e^{h_k}}\right) \frac{\phi\left(\frac{Y_k - \gamma L + (\gamma-1)Y_{k-1}}{Y_{k-1}^b e^{h_k}}\right)}{Y_{k-1}^b e^{h_k} \phi(Y_k)} \Big| \mathcal{G}_{k-1}\right]$$

$$= \bar{E}\left[f\left(\frac{h_k - \langle a, X_{k-1}\rangle - \langle \beta, X_{k-1}\rangle h_{k-1}}{\langle \theta, X_{k-1}\rangle}\right) \frac{\phi\left(\frac{h_k - \langle a, X_{k-1}\rangle - \langle \beta, X_{k-1}\rangle h_{k-1}}{\langle \theta, X_{k-1}\rangle}\right)}{\langle \theta, X_{k-1}\rangle \phi(h_k)}\right.$$

$$\left. \times \int_{-\infty}^{\infty} g(W_k) \phi(W_k) dv_k | \mathcal{G}_{k-1}\right]$$

$$= \left(\int_{-\infty}^{\infty} f(B_k) \phi(B_k) dw_k\right)\left(\int_{-\infty}^{\infty} g(W_k) \phi(W_k) dv_k\right).$$

$\square$

B Proof of Lemma 33.2

Proof. Consider any measurable function f. Then

$$\bar{E}\left[\Lambda_k f(h_k) X_k | \mathcal{F}_k^Y\right] = \int_{-\infty}^{\infty} f(z) q_k(z) dz =$$

$$= \bar{E}\left[\Lambda_{k-1} \frac{\phi\left(\frac{h_k - \langle a, X_{k-1}\rangle - \langle \beta, X_{k-1}\rangle h_{k-1}}{\langle \theta, X_{k-1}\rangle}\right)}{\langle \theta, X_{k-1}\rangle \phi(h_k)} \frac{\phi\left(\frac{Y_k - \gamma L + (\gamma-1)Y_{k-1}}{Y_{k-1}^b e^{h_k}}\right)}{Y_{k-1}^b e^{h_k} \phi(Y_k)} f(h_k) X_k | \mathcal{Y}_k\right]$$

$$= \bar{E}\left[\sum_{i=1}^{N} \Lambda_{k-1} \langle X_{k-1}, e_i\rangle \frac{\phi\left(\frac{h_k - \langle a, X_{k-1}\rangle - \langle \beta, X_{k-1}\rangle h_{k-1}}{\langle \theta, X_{k-1}\rangle}\right)}{\langle \theta, X_{k-1}\rangle \phi(h_k)}\right.$$

$$\left. \times \frac{\phi\left(\frac{Y_k - \gamma L + (\gamma-1)Y_{k-1}}{Y_{k-1}^b e^{h_k}}\right)}{Y_{k-1}^b e^{h_k} \phi(Y_k)} f(h_k) (AX_{k-1} + M_k) | \mathcal{Y}_k\right]$$

As M_k is a martingale increment, this is

$$= \bar{E}\left[\sum_{i=1}^{N} \Lambda_{k-1} \langle X_{k-1}, e_i\rangle \frac{\phi\left(\frac{h_k - a^i - \beta^i h_{k-1}}{\theta^i}\right)}{\theta^i \phi(h_k)} \frac{\phi\left(\frac{Y_k - \gamma L + (\gamma-1)Y_{k-1}}{Y_{k-1}^b e^{h_k}}\right)}{Y_{k-1}^b e^{h_k} \phi(Y_k)} f(h_k) Ae_i | \mathcal{Y}_k\right]$$

$$= \frac{1}{\phi(Y_k)\theta^i} \bar{E}\left[\sum_{i=1}^{N} \Lambda_{k-1} \langle X_{k-1}, e_i \rangle \bar{E}\left[\frac{\phi\left(\frac{h_k - a^i - \beta^i h_{k-1}}{\theta^i}\right)}{\phi(h_k)}\right.\right.$$

$$\left.\left.\times \frac{\phi\left(\frac{Y_k - \gamma L + (\gamma-1)Y_{k-1}}{Y_{k-1}^b e^{h_k}}\right)}{Y_{k-1}^b e^{h_k} \phi(Y_k)} f(h_k) | \mathcal{Y}_k, h_{k-1}\right] | \mathcal{Y}_k\right] Ae_i.$$

In the inner conditioning we integrate against the density of h_k so this becomes

$$= \frac{1}{\phi(Y_k)\theta^i} \bar{E}\left[\sum_{i=1}^{N} \Lambda_{k-1} \langle X_{k-1}, e_i \rangle \int_{-\infty}^{\infty} \left(\frac{\phi\left(\frac{h_k - a^i - \beta^i h_{k-1}}{\theta^i}\right)}{\phi(h_k)}\right.\right.$$

$$\left.\left.\times \frac{\phi\left(\frac{Y_k - \gamma L + (\gamma-1)Y_{k-1}}{e^{h_k}}\right)}{e^{h_k}} f(h_k)\phi(h_k) \, dh_k | \mathcal{F}_k^Y\right] Ae_i.$$

Let $z = h_k$ so we have this is

$$= \frac{1}{\phi(Y_k)\theta^i} \bar{E}\left[\sum_{i=1}^{N} \langle X_{k-1}, e_i \rangle \left(\Lambda_{k-1} \int_{-\infty}^{\infty} \left(\phi\left(\frac{z - a^i - \beta^i h_{k-1}}{\theta^i}\right)\right.\right.\right.$$

$$\left.\left.\left.\times \frac{\phi\left(\frac{Y_k - \gamma L + (\gamma-1)Y_{k-1}}{Y_{k-1}^b e^z}\right)}{Y_{k-1}^b e^z} f(z)dz\right)\right) | \mathcal{Y}_k\right] Ae_i.$$

The integrand is a function of h_{k-1}, so integrating against the density q_{k-1} gives

$$= \sum_{i=1}^{N} \frac{1}{\phi(Y_k)\theta^i} \left\langle \int_{-\infty}^{\infty} \left(\int_{-\infty}^{\infty} \phi\left(\frac{z - a^i - \beta^i h_{k-1}}{\theta^i}\right)\right.\right.$$

$$\left.\left.\times \frac{\phi\left(\frac{Y_k - \gamma L + (\gamma-1)Y_{k-1}}{Y_{k-1}^b e^z}\right)}{Y_{k-1}^b e^z} f(z)dz\right) q_{k-1}(h_{k-1})dh_{k-1}, e_i\right\rangle Ae_i.$$

Write $h = h_{k-1}$ and change the order of integration. Then this is:

$$= \sum_{i=1}^{N} \left\langle \int_{-\infty}^{\infty} f(z) \frac{\phi\left(\frac{Y_k - \gamma L + (\gamma-1)Y_{k-1}}{Y_{k-1}^b e^z}\right)}{Y_{k-1}^b e^z \theta^i \phi(Y_k)} \left(\int_{-\infty}^{\infty} \phi\left(\frac{z - a^i - \beta^i h_{k-1}}{\theta^i}\right) q_{k-1}(h)dh\right) dz, e_i \right\rangle Ae_i$$

$$= A \begin{pmatrix} \int_{-\infty}^{\infty} f(z) \frac{\phi\left(\frac{Y_k - \gamma L + (\gamma-1)Y_{k-1}}{Y_{k-1}^b e^z}\right)}{Y_{k-1}^b e^z \theta^1 \phi(Y_k)} \left(\int_{-\infty}^{\infty} \phi\left(\frac{(z - a^1 - \beta^1 h)}{\theta^1}\right) q_{k-1}^1(h)dh\right) dz \\ \int_{-\infty}^{\infty} f(z) \frac{\phi\left(\frac{Y_k - \gamma L + (\gamma-1)Y_{k-1}}{Y_{k-1}^b e^z}\right)}{Y_{k-1}^b e^z \theta^2 \phi(Y_k)} \left(\int_{-\infty}^{\infty} \phi\left(\frac{(z - a^2 - \beta^2 h)}{\theta^2}\right) q_{k-1}^2(h)dh\right) dz \end{pmatrix}$$

$$= \int_{-\infty}^{\infty} f(z) \left(A \int_{-\infty}^{\infty} B(Y_k, z, h, Y_{k-1}) q_{k-1}(h) dh \right) dz.$$

Here $B(Y_k, z, h, Y_{k-1})$ is a diagonal matrix with entries

$$\frac{\phi\left(\frac{Y_k - \gamma L + (\gamma-1) Y_{k-1}}{Y_{k-1}^b e^z}\right) \phi\left(\frac{z - \alpha^i - \beta^i h}{\theta^i}\right)}{Y_{k-1}^b e^z \phi(Y_k) \quad \theta^i}.$$

Therefore, we have established the recurrence:

$$q_k(z) = A \int_{-\infty}^{\infty} B(Y_k, h, z, Y_{k-1}) q_{k-1}(h) dh. \qquad \square$$

C Proof of Corollary 33.1

Proof. Taking $f(z) = 1$ in (33.9) we have

$$\bar{E}[\Lambda_k X_k | \mathcal{Y}_k] = \int_{-\infty}^{\infty} q_k(z) dz.$$

Note that

$$\sum_{i=1}^{N} \langle X_k, e_i \rangle = 1.$$

so

$$\bar{E}[\Lambda_k | \mathcal{Y}_k] = \bar{E}\left[\Lambda_k \sum_{i=1}^{N} \langle X_k, e_i \rangle \Big| \mathcal{Y}_k \right]$$

$$= \sum_{i=1}^{N} \bar{E}[\langle \Lambda_k X_k, e_i \rangle | \mathcal{Y}_k]$$

$$= \sum_{i=1}^{N} \langle \bar{E}[\Lambda_k X_k | \mathcal{Y}_k], e_i \rangle$$

$$= \sum_{i=1}^{N} \left\langle \int_{-\infty}^{\infty} q_k(z) dz, e_i \right\rangle.$$

Then from (33.8)

$$E[X_k | \mathcal{Y}_k] = \frac{\bar{E}[\Lambda_k X_k | \mathcal{Y}_k]}{\bar{E}[\Lambda_k | \mathcal{Y}_k]}$$

$$= \frac{\int_{-\infty}^{\infty} q_k(z) dz}{\sum_{i=1}^{N} \langle \int_{-\infty}^{\infty} q_k(z) dz, e_i \rangle}.$$

This gives the optimal estimate of the Markov chain given the observation of the prices. Also,

$$\bar{E}\left[\Lambda_k f(h_k)|\mathcal{Y}_k\right] = \bar{E}\left[\Lambda_k f(h_k) \sum_{i=1}^{N} \langle X_k, e_i \rangle |\mathcal{Y}_k\right]$$

$$= \sum_{i=1}^{N} \bar{E}\left[\langle \Lambda_k f(h_k) X_k, e_i \rangle |\mathcal{Y}_k\right]$$

$$= \sum_{i=1}^{N} \langle \bar{E}\left[\Lambda_k f(h_k) X_k |\mathcal{Y}_k\right], e_i \rangle$$

$$= \sum_{i=1}^{N} \left\langle \int_{-\infty}^{\infty} f(z) q_k(z) dz, e_i \right\rangle.$$

Therefore, from (33.8)

$$E[f(h_k)|\mathcal{Y}_k] = \frac{\bar{E}\left[\Lambda_k f(h_k)|\mathcal{Y}_k\right]}{\bar{E}\left[\Lambda_k|\mathcal{Y}_k\right]}$$

$$= \frac{\sum_{i=1}^{N} \langle \int_{-\infty}^{\infty} f(z) q_k(z) dz, e_i \rangle}{\sum_{i=1}^{N} \langle \int_{-\infty}^{\infty} q_k(z) dz, e_i \rangle}. \quad (33.49)$$

Setting $f(h_k) = h_k$, (33.49) becomes

$$E[h_k|\mathcal{Y}_k] = \frac{\sum_{i=1}^{N} \langle \int_{-\infty}^{\infty} z q_k(z) dz, e_i \rangle}{\sum_{i=1}^{N} \langle \int_{-\infty}^{\infty} q_k(z) dz, e_i \rangle}.$$

$\square$

D Proof of Lemma 33.3

Proof. Considering any measuring "test" function f, then

$$\bar{E}\left[\Lambda_k Z_k^{r,s} f(h_k) X_k | \mathcal{F}_k^Y\right] = \int_{-\infty}^{\infty} f(z) \Phi_k^{r,s}(z) dz$$

$$= \bar{E}\left[\Lambda_{k-1} \gamma_k \left(Z_{k-1}^{r,s} + \langle X_{k-1}, e_r \rangle \langle X_k, e_s \rangle H(Y_k) F(h_k) G(h_{k-1}) R(Y_{k-1})\right) f(h_k) X_k | \mathcal{Y}_k\right]$$

$$= \bar{E}\left[\Lambda_{k-1} \frac{\phi\left(\frac{h_k - \langle \alpha, X_{k-1} \rangle - \langle \beta, X_{k-1} \rangle h_{k-1}}{\langle \theta, X_{k-1} \rangle}\right)}{\langle \theta, X_{k-1} \rangle \phi(h_k)}\right.$$

$$\left. \times \frac{\phi\left(\left(\frac{Y_k - \gamma L + (\gamma - 1) Y_{k-1}}{e^{h_k}}\right)\right)}{e^{h_k} \phi(Y_k)} Z_{k-1}^{r,s} f(h_k) (A X_{k-1} + M_k) | \mathcal{Y}_k\right]$$

$$+ \bar{E}\left[\Lambda_{k-1} \frac{\phi\left(\frac{h_k - \langle a, X_{k-1}\rangle - \langle \beta, X_{k-1}\rangle h_{k-1}}{\langle \theta, X_{k-1}\rangle}\right)}{\langle \theta, X_{k-1}\rangle \phi(h_k)} \frac{\phi\left(\left(\frac{Y_k - \gamma L + (\gamma - 1)Y_{k-1}}{Y_{k-1}^b e^{h_k}}\right)\right)}{Y_{k-1}^b e^{h_k} \phi(Y_k)}\right.$$

$$\left. \times H(Y_k) F(h_k) G(h_{k-1}) R(Y_{k-1}) f(h_k) \langle X_{k-1}, e_r\rangle \langle AX_{k-1} + M_k, e_s\rangle e_s | \mathcal{Y}_k\right]$$

$$= \sum_{i=1}^{N} \bar{E}\left[\Lambda_{k-1} \langle X_{k-1}, e_i\rangle \bar{E}\left[\frac{\phi\left(\frac{Y_k - \gamma L + (\gamma - 1)Y_{k-1}}{Y_{k-1}^b e^{h_k}}\right)}{Y_{k-1}^b e^{h_k} \phi(Y_k)}\right.\right.$$

$$\left.\left. \times \frac{\phi\left(\frac{h_k - a^i - \beta^i h_{k-1}}{\theta^i}\right)}{\theta^i \phi(h_k)} Z_{k-1}^{r,s} f(h_k) | \mathcal{Y}_k, h_{k-1}\right] | \mathcal{Y}_k\right] A e_i$$

$$+ \bar{E}\left[\Lambda_{k-1} \langle X_{k-1}, e_r\rangle \bar{E}\left[H(Y_k) F(h_k) G(h_{k-1}) R(Y_{k-1}) \frac{\phi\left(\frac{Y_k - \gamma L + (\gamma - 1)Y_{k-1}}{e^{h_k}}\right)}{e^{h_k} \phi(Y_k)}\right.\right.$$

$$\left.\left. \times \frac{\phi\left(\frac{h_k - a^r - \beta^r h_{k-1}}{\theta^r}\right)}{\theta^r \phi(h_k)} f(h_k) | \mathcal{Y}_k, h_{k-1}\right] | \mathcal{Y}_k\right] p_{sr} e_s.$$

In the inner conditioning, we integrate against the density ϕ of h_k and introduce the notation $z = h_k$. Then this is

$$= \sum_{i=1}^{N} \bar{E}\left[\left\langle \Lambda_{k-1} X_{k-1} Z_{k-1}^{r,s} \int_{-\infty}^{\infty} \frac{\phi\left(\frac{Y_k - \gamma L + (\gamma - 1)Y_{k-1}}{Y_{k-1}^b e^z}\right)}{Y_{k-1}^b e^z \phi(Y_k)}\right.\right.$$

$$\left.\left. \times \frac{\phi\left(\frac{z - a^i - \beta^i h_{k-1}}{\theta^i}\right)}{\theta^i \phi(z)} f(z) \phi(z) \, dz, e_i\right\rangle | \mathcal{Y}_{k-1}\right] A e_i$$

$$+ \bar{E}\left[\left\langle \Lambda_{k-1} X_{k-1} \int_{-\infty}^{\infty} H(Y_k) F(z) G(h_{k-1}) R(Y_{k-1}) \frac{\phi\left(\frac{Y_k - \gamma L + (\gamma - 1)Y_{k-1}}{Y_{k-1}^b e^z}\right)}{Y_{k-1}^b e^z \phi(Y_k)}\right.\right.$$

$$\left.\left. \times \frac{\phi\left(\frac{z - a^r - \beta^r h_{k-1}}{\theta^r}\right)}{\theta^r \phi(z)} f(z)\right) \phi(z) \, dz, e_r\right\rangle | \mathcal{Y}_{k-1}\right] p_{sr} e_s.$$

Using (33.15) in the the first summand, (33.9) for the second, and writing $h = h_{k-1}$. Then this is

$$= \sum_{i=1}^{N} \left\langle \int_{-\infty}^{\infty} \left(\int_{-\infty}^{\infty} \frac{\phi\left(\frac{Y_k - \gamma L + (\gamma - 1)Y_{k-1}}{Y_{k-1}^b e^z}\right)}{Y_{k-1}^b e^z \phi(Y_k)} \frac{\phi\left(\frac{z - a^i - \beta^i h}{\theta^i}\right)}{\theta^i} f(z) dz\right) \Phi_{k-1}^{rs}(h) \, dh, e_i\right\rangle A e_i$$

$$+ \left\langle \int_{-\infty}^{\infty} \int_{-\infty}^{\infty} \left(H(Y_k) \, F(z) \, G(h) \, R(Y_{k-1}) \frac{\phi\left(Y_{k-1}^{-b} e^z \left(Y_k - \mu + \frac{1}{2} e^{2z}\right)\right)}{Y_{k-1}^b e^z \phi(Y_k)} \right.\right.$$

$$\left.\left. \times \frac{\phi\left(\frac{z - a^r - \beta^r h}{\theta^r}\right)}{\theta^r} f(z) dz \right) q_{k-1}(h) dh, \, e_r \right\rangle p_{sr} e_s.$$

Changing the order of integration, we have this equals

$$= \sum_{i=1}^{N} \left\langle \int_{-\infty}^{\infty} \left(\int_{-\infty}^{\infty} \frac{\phi\left(\frac{Y_k - \mu + \frac{1}{2} e^{2z}}{Y_{k-1}^b e^z}\right) \phi\left(\frac{z - a^i - \beta^i h}{\theta^i}\right)}{Y_{k-1}^b e^z \phi(Y_k)} \Phi_{k-1}^{rs}(h) \, f(z) dh \right) dz, \, e_i \right\rangle A e_i$$

$$+ \left\langle \int_{-\infty}^{\infty} \int_{-\infty}^{\infty} \left(H(Y_k) \, F(z) \, G(h) \, R(Y_{k-1}) \frac{\phi\left(\frac{Y_k - \mu + \frac{1}{2} e^{2z}}{e^z}\right)}{e^z \phi(Y_k)} \right.\right.$$

$$\left.\left. \times \frac{\phi\left(\frac{z - a^r - \beta^r h}{\theta^r}\right)}{\theta^r} f(z) q_{k-1}(h) dh \right) dz, \, e_r \right\rangle p_{sr} e_s$$

$$= \int_{-\infty}^{\infty} f(z) \left(A \left(\int_{-\infty}^{\infty} B(Y_k, z, h, Y_{k-1}) \Phi_{k-1}^{rs}(h) \, dh \right) \right.$$

$$\left. + H(Y_k) \, F(z) \left\langle A \left(\int_{-\infty}^{\infty} B^r(Y_k, z, h, Y_{k-1}) G(h) \, q_{k-1}(h) \, dh \right), \, e_r \right\rangle p_{sr} e_s \right) dz.$$

Therefore,

$$\Phi_k^{rs}(z) = A \left(\int_{-\infty}^{\infty} B(Y_k, z, h, Y_{k-1}) \Phi_{k-1}^{rs}(h) \, dh \right)$$

$$+ H(Y_k) \, F(z) \, R(Y_{k-1}) \left\langle \left(\int_{-\infty}^{\infty} B^r(Y_k, z, h) G(h) \, q_{k-1}(h) \, dh \right), \, e_r \right\rangle p_{sr} e_s. \qquad \square$$

E Proof of Corollary 33.2

Proof. Taking $f(z) = 1$ in Lemma 4.1 we have

$$\bar{E}\left[\Lambda_k Z_k^{rs} X_k | \mathcal{Y}_k\right] = \int_{-\infty}^{\infty} \Phi_k^{rs}(z) dz.$$

Also

$$\bar{E}\left[\Lambda_k Z_k^{rs} | \mathcal{Y}_k\right] = \bar{E}\left[\Lambda_k Z_k^{rs} \sum_{i=1}^{N} \langle X_k, e_i \rangle | \mathcal{Y}_k\right]$$

$$= \sum_{i=1}^{N} \bar{E}\left[\langle \Lambda_k Z_k^{rs} X_k, e_i\rangle | \mathcal{Y}_k\right]$$

$$= \sum_{i=1}^{N} \langle \bar{E}\left[\Lambda_k Z_k^{rs} X_k | \mathcal{Y}_k\right], e_i\rangle$$

$$= \sum_{i=1}^{N} \left\langle \int_{-\infty}^{\infty} \Phi_k^{rs}(z) dz, e_i \right\rangle.$$

So

$$\hat{Z}_k^{rs} = \frac{\bar{E}\left[\Lambda_k Z_k^{rs} | \mathcal{Y}_k\right]}{\bar{E}\left[\Lambda_k | \mathcal{Y}_k\right]}$$

$$= \frac{\sum_{i=1}^{N} \langle \int_{-\infty}^{\infty} \Phi_k^{rs}(z) dz, e_i \rangle}{\sum_{i=1}^{N} \langle \int_{-\infty}^{\infty} q_k(z) dz, e_i \rangle}.$$

$\square$

F Proof of Lemma 33.4

Proof. E_ϑ (resp. $E_{\hat{\vartheta}}$) denotes expectation with respect to P_ϑ (resp. $P_{\hat{\vartheta}}$). Now

$$E_{\hat{\vartheta}}\left[\langle X_k, e_j \rangle | \mathcal{Y}_{k-1}\right] = \frac{E_\vartheta\left[\Lambda_k \langle X_k, e_j \rangle | \mathcal{Y}_{k-1}\right]}{E_\vartheta\left[\Lambda_k | \mathcal{Y}_{k-1}\right]}$$

$$= \frac{E_\vartheta\left[\left(\sum_{i,j=1}^{N} \left(\frac{\hat{p}_{ji}}{p_{ji}}\right) \langle X_k, e_j \rangle \langle X_{k-1}, e_i \rangle\right) \langle X_k, e_j \rangle | \mathcal{Y}_{k-1}\right]}{E_\vartheta\left[\left(\sum_{i,j=1}^{N} \left(\frac{\hat{p}_{ji}}{p_{ji}}\right) \langle X_k, e_j \rangle \langle X_{k-1}, e_i \rangle\right) | \mathcal{Y}_{k-1}\right]}$$

The denominator is

$$E_\vartheta\left[\left(\sum_{i,j=1}^{N} (\frac{\hat{p}_{ji}}{p_{ji}}) \langle X_k, e_j \rangle \langle X_{k-1}, e_i \rangle\right) | \mathcal{Y}_{k-1}\right]$$

$$= \sum_{i=1}^{N} E_\vartheta\left[\left(\sum_{j=1}^{N} (\frac{\hat{p}_{ji}}{p_{ji}}) \langle X_k, e_j \rangle | X_{k-1} = e_i\right) \langle X_{k-1}, e_i \rangle\right]$$

$$= \sum_{i=1}^{N} \left(\sum_{j=1}^{N} (\frac{\hat{p}_{ji}}{p_{ji}}) p_{ji}\right) \langle X_{k-1}, e_i \rangle$$

$$= \sum_{i=1}^{N} \langle X_{k-1}, e_i \rangle = 1.$$

Now

$$E_\vartheta\left[\left(\sum_{i,j=1}^N (\frac{\hat{p}_{ji}}{p_{ji}})\langle X_k, e_j\rangle \langle X_{k-1}, e_i\rangle\right)\langle X_k, e_j\rangle |\mathcal{Y}_{k-1}\right]$$

$$= E_\vartheta\left[\sum_{i=1}^N (\frac{\hat{p}_{ji}}{p_{ji}})\langle X_k, e_j\rangle \langle X_{k-1}, e_i\rangle |\mathcal{Y}_{k-1}\right]$$

$$= E_\vartheta\left[E_\vartheta\left[\sum_{i=1}^N (\frac{\hat{p}_{ji}}{p_{ji}})\langle X_k, e_j\rangle |\mathcal{Y}_{k-1}\& \langle X_{k-1}, e_i\rangle\right]\right]$$

$$= \hat{p}_{ji}\langle X_{k-1}, e_i\rangle.$$

Therefore,

$$P_{\hat\vartheta}(X_k = e_j|X_{k-1} = e_i) = E_{\hat\vartheta}\left[\langle X_k, e_j\rangle |X_{k-1} = e_i\right] = \hat{p}_{ji}.$$

That is, under the new measure, X is a Markov chain with transition matrix $\hat{A} = (\hat{p}_{ji})$. □

G Proof of Lemma 33.5

Proof. As above we define $P_{\hat\vartheta}$ by:

$$\frac{dP_{\hat\vartheta}}{dP_\vartheta}|\mathcal{Y}_k = \Lambda_k = \prod_{l=1}^k\left(\sum_{i,j=1}^N (\frac{\hat{p}_{ji}}{p_{ji}})\langle X_l, e_j\rangle \langle X_{l-1}, e_i\rangle\right).$$

Then

$$\log \frac{dP_{\hat\vartheta}}{dP_\vartheta} = \sum_{l=1}^k \sum_{i,j=1}^N \langle X_l, e_j\rangle \langle X_{l-1}, e_i\rangle (\log \hat{p}_{ji} - \log p_{ji})$$

$$= \sum_{i,j=1}^N N_k^{ij} \log \hat{p}_{ji} + i(p),$$

where $i(p)$ does not depend on the $\hat{p}_{ji}$. Then

$$L(\hat\vartheta) = E\left[\log \frac{dP_{\hat\vartheta}}{dP_\vartheta}|\mathcal{Y}_k\right] = \sum_{i,j=1}^N \hat{N}_k^{ij} \log \hat{p}_{ji} + \hat{i}(p). \quad (33.50)$$

Recall that $\sum_{j=1}^N \hat{p}_{ji} = 1$, and $\sum_{j=1}^N N_k^{ij} = O_k^i$. Then, the optimal estimate of $\hat{p}_{ji}$ is the value that maximizes the right side of (33.50), and subject to $\sum_{j=1}^N \hat{p}_{ji} = 1$. Let λ be the Lagrange multiplier and put

$$L(\hat{P}, \lambda) = \sum_{i,j=1}^{N} \hat{N}_k^{ij} \log \hat{p}_{ji} + i(a) + \lambda(\sum_{j=1}^{N} \hat{p}_{ji} - 1).$$

Differentiating in $\hat{p}_{ji}$ and λ and equating the derivatives to 0 gives:

$$\frac{1}{\hat{p}_{ji}} \hat{N}_k^{ij} + \lambda = 0,$$

and

$$\sum_{j=1}^{N} \hat{p}_{ji} = 1.$$

Notice that

$$\sum_{j=1}^{N} N_k^{ij} = O_k^i,$$

so

$$\sum_{j=1}^{N} \hat{N}_k^{ij} = \hat{O}_k^i.$$

Solving, we have:

$$\lambda = -\hat{O}_k^i,$$

and

$$\hat{p}_{ji} = \frac{\hat{N}_k^{ij}}{\hat{O}^i} = \frac{\sum_{i=1}^{N} \langle \int_{-\infty}^{\infty} \varphi_k^{ij}(z) dz, e_i \rangle}{\sum_{i=1}^{N} \langle \int_{-\infty}^{\infty} \eta_k^i(z) dz, e_i \rangle}.$$

This provides estimates for the elements of the transition matrix. □

H Proof of Lemma 33.6

Proof. The density which changes a to $\hat{a} = (a^1, a^2)$ is given by

$$\frac{dP_{\hat{a}}}{dP_a}\bigg|\mathcal{F}_k^{X,Y} = \prod_{k=1}^{k} \frac{\phi(\langle \theta, X_{k-1} \rangle^{-1}(h_k - \langle \hat{a}, X_{k-1} \rangle - \langle \beta, X_{k-1} \rangle h_{k-1}))}{\phi(\langle \theta, X_{k-1} \rangle^{-1}(h_k - \langle a, X_{k-1} \rangle - \langle \beta, X_{k-1} \rangle h_{k-1}))}.$$

Then

$$E\left[\log \frac{d P_{\hat{a}_0}}{d P_{a_0}} | \mathcal{Y}_k\right]$$

$$= E\left[\sum_{k=1}^{k} -\frac{1}{2} (\langle \theta, X_{k-1}\rangle^{-1} (h_k - \langle \hat{a}, X_{k-1}\rangle - \langle \beta, X_{k-1}\rangle h_{k-1}))^2 | \mathcal{F}_k^{X,Y}\right]$$
$$+ \hat{R}(a, \beta, \theta)$$

$$= \sum_{r=1}^{N} E\left[\sum_{k=1}^{k} -\frac{1}{2} \left(\frac{\langle X_{k-1}, e_r\rangle (h_k - \hat{a}^r - \beta^r h_{k-1})}{\langle X_{k-1}, e_r\rangle \theta}\right)^2 | \mathcal{F}_k^{X,Y}\right] + \hat{R}(a, \beta, \theta).$$

The best estimate of the parameter is the one which maximizes the above conditional expectation. Differentiating in $\hat{a}^r$ and setting the derivative equal to 0, we see,

$$E\left[\sum_{k=1}^{t} \langle X_{k-1}, e_r\rangle (h_k - \hat{a}^r - \beta^r h_{k-1}) | \mathcal{F}_k^{X,Y}\right] = 0.$$

Then we have an updated estimate $\hat{a}_t^r$ for the parameter a^r:

$$\hat{a}_t^r E\left[\sum_{k=1}^{t} \langle X_{k-1}, e_r\rangle | \mathcal{F}_t^{X,Y}\right]$$
$$= E\left[\sum_{k=1}^{t} (\langle X_{k-1}, e_r\rangle h_k) | \mathcal{F}_t^{X,Y}\right] - E\left[\sum_{k=1}^{t} \beta^r (\langle X_{k-1}, e_r\rangle h_{k-1}) | \mathcal{F}_t^{X,Y}\right].$$

Thus,

$$\hat{a}_t^r = \frac{\left(E\left[\sum_{k=1}^{t} (\langle X_{k-1}, e_r\rangle h_k) | \mathcal{F}_t^{X,Y}\right] - E\left[\sum_{k=1}^{t} \beta^r (\langle X_{k-1}, e_r\rangle h_{k-1}) | \mathcal{F}_t^{X,Y}\right]\right)}{E\left[\sum_{k=1}^{t} \langle X_{k-1}, e_r\rangle | \mathcal{F}_t^{X,Y}\right]}$$

$$= \frac{\hat{Z}_t^{r,1} - \hat{Z}_t^{r,2}}{\hat{O}^r}.$$

Therefore, an updated estimate for the parameter a^r is

$$\hat{a}_t^r = \frac{\sum_{i=1}^{N} \langle \int_{-\infty}^{\infty} \left(\xi_t^{r,1}(z) - \beta^r \xi_t^{r,2}(z)\right) dz, e_i\rangle}{\sum_{i=1}^{N} \langle \int_{-\infty}^{\infty} \eta_t^r(z) dz, e_i\rangle}.$$

$\square$

References

[1] Aggoun, L. and R. J. Elliott. 2004. *Measure Theory and Filtering: Introduction and Application*. Cambridge University Press.
[2] Bollerslev, T. 1986. Generalised autoregressive conditional heteroscedasticity. *Journal of Econometrics*, 51: 307–27.

[3] Brown, S. P. 2002. Energy prices and aggregate economy activity: An interpretive survey. *Quarterly Review of Economics and Finance*, 42: 193.

[4] Cox J. C., J. E. Ingersoll Jr., and S. A. Ross. 1981. A re-examination of traditional hypotheses about the term structure of interest rates. *Journal of Finance*, 36: 766–99.

[5] Creal, D., Y. Gu, and E. Zivot. 2007. Evaluating structural models for the U.S. short rate using EMM and optimal filters. Working paper, Free University, the Netherlands, and University of Washington, Seattle.

[6] Dey, S. and J. B. Moore. 1995. Risk-sensitive filtering and smoothing for hidden Markov models. *Systems & Control Letters*, 25: 361–6.

[7] Dueker, M. J. 1997. Markov switching in GARCH processes and mean reverting stock market volatility. *Journal of Business and Economic Statistics*, 15: 26–34.

[8] Duffie, D., S. Gray, and P. Hoang, *Volatility in Energy Prices*, ed V. Kaminski, Barclays, P.539.

[9] Elliott, R. J. 1982. *Stochastic Calculus and Applications*. Springer-Verlag, New York.

[10] Elliott, R. J., P. Fischer, and E. Platen. 1999. Filtering and parameter estimation for a mean reverting interest rate model. *Canadian Applied Math. Quarterly*, 7: 381–400.

[11] Elliott, R. J., W. C. Hunter, and B. M. Jamieson. 1998. Drift and volatility estimation in discrete time. *Journal of Economic Dynamics & Control*, 22: 209–18.

[12] Elliott, R. J. 1994. Exact adaptive filters for Markov chains observed in Gaussian noise. *Automatica*, 30: 1399–1408.

[13] Elliott, R. J. 1993. New finite dimensional filters and smoothers for noisily observed Markov chains. *IEEE Transactions on Information Theory*, 39(1).

[14] Elliott R. J., L. Aggoun, and J. B. Moore. 1994. *Hidden Markov Models: Estimation and Control*. Springer-Verlag, New York.

[15] Elliott R. J., W. P. Malcolm, and A. Tsoi. 2002. *HMM Volatility Estimation*. 41st IEEE Conference on Decision and Control, Las Vegas, NV, Dec. 2002, IEEE Press Piscataway, NJ 398–404.

[16] Engle, R. F. 1982. Autoregressive conditional heteroscedasticity with estimates of the variance of United Kingdom inflation. *Econometrica*, 50: 987–1008.

[17] Gabriel, S. A. 2001. The national energy modeling system: A large-scale energy-economic equilibrium model. *Operations Research*, 49: 14–28.

[18] Hamilton, J. D. 1988. Rational-expectations econometric analysis of changes in regime– An investigation of the term structure of interest rates. *Journal of Economic Dynamics and Control*, 12: 385–423.

[19] Heston, S. L. 1993. A closed-form solution for option with stochastic volatility with application to bond and currency options. *Review of Financial Studies*, 6: 327–43.

[20] Hull, J. and A. White. 1987. The price of options on assets with stochastic volatilities. *Journal of Finance*, 42: 281–300.

[21] Kushner, H. J. and P. G. Dupuis. 1992. *Numerical Methods for Stochasitc Control Problems in Continuous Time*. Springer-Verlag, New York.

[22] Pindyck, R. S. and J. J. Rotemberg. 1983. Dynamic factor demands and the effects of energy price shocks. *The American Economic Review*, 73: 1066–79.

[23] Popp, D. 2002. Induced innovation and energy prices. *The American Economic Review*, 92: 160–80.

[24] Rabiner, L. 1989. A tutorial on hidden Markov models and selected applications in speech recognition. *Proceedings of the IEEE*, 77: 275–86.

[25] Rossi A. and G. M. Gallo. 2002. Volatility estimation via hidden Markov models. Econometrics working papers archive wp2002_14, Università degli Studi di Firenze, Dipartimento di Statistica "G. Parenti," rev. 17 June 2002.

[26] Mike K. P. S., K. Lam., and W. K. Li. 1998. A stochastic volatility model with Markov switching. *Journal of Business & Economic Statistics*, 2: 244–53.
[27] Taylor, S. J. 1986. *Modeling Financial Time Series*. Wiley, New York.
[28] Vasicek, O. A. 1977. An equilibrium characterization of the term structure. *Journal of Financial Economics*, 5: 177–88
[29] Wiggins, J. B. 1987. Option values under stochastic volatility: Theory and empirical estimates. *Journal of Financial Economics*, 19: 351–72.
[30] Yu, W. and G. B. Sheble. 2006. Modeling electricity markets with hidden Markov model. *Electric Power Systems Research*, 76: 445–51.

·34·
Portfolio Optimization Under Partial Observation: Theoretical and Numerical Aspects

H. Pham

34.1 Introduction

Stochastic optimization is a traditional area in mathematics that has received a special renewed interest for its multiple and various applications in finance. It arises for example in portfolio/investment choice or hedging in incomplete markets, etc. In financial market models, we do not have in general a complete knowledge of all parameters, which may be driven by unobserved random factors. Some usual partial observation situations encountered in the literature and in real markets are the following: (i) The investor cannot observe the stock appreciation rate but only the stock price process in continuous-time models, (ii) Trading in markets occurs in practice in discrete time, and we cannot fully recover the volatility value from the single observation of the stock prices in stochastic volatility models, (iii) In corporate finance, firm values are not easily available, and we only observe the value of equity issued by the firm that is a noisy function of the real firm value.

In this chapter we consider stochastic optimization problems in finance under partial observation. We address both the theoretical and numerical aspects related to these problems. The content is divided into two parts. In the first one, we focus on continuous-time models: here, the incomplete information context means that the investors cannot observe the appreciation rate of stocks, but only the asset prices. Mathematically, the observation process is an additive noise of the unobserved signal, and by using the method of change of probability reference and innovation process, we can transform the original optimization problem into a full observation one where the unknown quantities are replaced by their filter estimates. We can then apply a classical optimization approach such as the martingale or PDE methods. We investigate three common cases in the literature for the modeling of the unobservable mean rate of return: Bayesian, linear-Gaussian, and finite-state Markov chain, which all lead to a finite-dimensional Markovian setting.

In the second part, we focus on discrete-time models. In such a context, the volatility is typically not observable, and the observation (asset price) is a multiplicative noise of the signal (volatility). We are concerned with the numerical and computational aspects of the optimization problem in this partial observation model, which were actually little investigated in the literature. The main problem in effective approximation comes from the growing dimension of the filter process depending on the whole observation path, which we need to approximate. A basic approach, suggested e.g. by [5], consists of discretizing at each time k, the observation Y_k by a discrete random variable $\hat{Y}_k$, and thus approximating the filter $\Pi_n(Y_0, \ldots, Y_n)$ at time n, by $\Pi_n(\hat{Y}_0, \ldots, \hat{Y}_n)$, where we stress the dependence of the filter on the observation path $(Y_0, \ldots, Y_n)$. Hence, if $\hat{Y}_k$ takes for example M values at time k, we need to store M^n values in order to compute the approximate filter at time n. This makes the effective implementation not feasible in practice for a long horizon n. We refer to [34] for a detailed overview of this approach. We overcome this difficulty by adopting an approach recently proposed in [31] and [10], based on the Markov property of the pair filter-observation process with respect to the observation filtration. We then perform a quantization of the pair filter–observation process. Optimal quantization of random vectors consists basically of finding the best approximation in L^2-norm of a random vector by a discrete random vector taking at most N values. This was originally developed in the 1950s in the context of information theory where the basic motivation was to transmit efficiently a continuous stationary signal by means of a finite number of codes (or quantizers). More recently, the quantization approach was applied to various fields, and notably to numerical probability, where it appears as an efficient spatial discretization method for solving multidimensional problems arising typically in finance. Here, we show how one can apply ideas from quantization to numerically solve optimization problems under partial observations, and we illustrate our approach with the mean-variance hedging in partially observed stochastic volatility models.

34.2 Continuous-time models

34.2.1 Partial observation control framework

Let $(\Omega, \mathcal{F}, \mathbb{P})$ be a complete probability space equipped with a filtration $\mathbb{F} = (\mathcal{F}_t)_{0 \leq t \leq T}$, where $T > 0$ is a fixed time horizon, and we assume that $\mathcal{F}_T = \mathcal{F}$. We consider a financial market consisting of one risk-free asset, whose price process is assumed for simplicity to be equal to one, and n stocks of positive price process $S = (S^1, \ldots, S^n)$ governed by

$$dS_t = \text{diag}(S_t)(\mu_t dt + \sigma_t dW_t), \tag{34.1}$$

where W is a n-dimensional $\mathbb{F}$-Brownian motion. Here diag(S) denotes the diagonal $n \times n$ matrix with components S^i, $i = 1, \ldots, n$. The asset prices are assumed to be continuously observed by the investors in this market, and we denote $\mathbb{F}^S = \{\mathcal{F}^S_t, 0 \leq t \leq T\}$ the augmentation of the filtration generated by the price process S. The matrix volatility coefficient σ_t, valued in $\mathbb{R}^{n \times n}$, is assumed invertible a.s., and can be estimated from the quadratic variation of the price process, so we assume w.l.o.g. that σ is $\mathbb{F}^S$-adapted. However, the mean rate of return μ, valued in $\mathbb{R}^n$, is not observable, and is in general $\mathbb{F}$-adapted. We denote $\lambda_t = \sigma_t^{-1} \mu_t$, the (nonobservable) risk premium $\mathbb{F}$-adapted process. Notice that $\mathbb{F}^S \subsetneq \mathbb{F}$.

A portfolio strategy is an $\mathbb{R}^n$-valued process $\alpha = (\alpha_t)_{0 \leq t \leq T}$, which is adapted with respect to $\mathbb{F}^S$, and satisfies the integrability condition: $\int_0^T |\alpha_t|^2 dt < \infty$ a.s. The quantity $(\sigma'_t)^{-1} \alpha_t$ represents the proportion of wealth invested by the agent in the risky assets at time t, and based on the observation of prices, and by assuming that the interest rate is zero, the wealth process evolves as:

$$dX_t = X_t \left(\alpha'_t \lambda_t dt + \alpha'_t dW_t \right). \tag{34.2}$$

Here $'$ denotes the transpose operator. We denote by $\mathcal{A}^S$ this set of portfolio strategies, and by $X^{x,\alpha}$ the wealth process starting from $x > 0$ at time 0, and governed by the equation (34.2) with a strategy $\alpha \in \mathcal{A}^S$.

Given a utility function $U : (0, \infty) \to \mathbb{R}$, increasing, concave, the objective of the investor is to maximize the expected utility from terminal wealth, and so to solve the problem

$$v(x) = \sup_{\alpha \in \mathcal{A}^S} \mathbb{E}\left[U\left(X_T^{x,\alpha} \right) \right], \quad x > 0. \tag{34.3}$$

34.2.2 Reduction to a full observation model

We introduce the process

$$Z_t = \exp\left(-\int_0^t \lambda'_u dW_u - \frac{1}{2} \int_0^t |\lambda_u|^2 du \right), \quad 0 \leq t \leq T,$$

and we assume that Z is a $(\mathbb{P}, \mathbb{F})$-martingale, so that it defines a probability measure $\mathbb{P}^0 \sim \mathbb{P}$ on $(\Omega, \mathcal{F})$ by:

$$\frac{d\mathbb{P}^0}{d\mathbb{P}} = Z_T.$$

$\mathbb{P}^0$ is usually called probability of reference in filtering theory: by Girsanov's theorem, the process

$$W_t^0 = W_t + \int_0^t \lambda_u du, \quad 0 \leq t \leq T,$$

is a Brownian motion under $(\mathbb{P}^0, \mathbb{F})$, and the dynamics of S under $\mathbb{P}^0$ is

$$d S_t = \mathrm{diag}(S_t) \sigma_t d W_t^0. \tag{34.4}$$

An important property of the probability of reference is that the filtration $\mathbb{F}^S$ is the augmented natural filtration of W^0. This follows from the above relation (34.4), and since σ is invertible and $\mathbb{F}^S$-adapted. The probability $\mathbb{P}^0$ is also a risk-neutral martingale measure on $(\Omega, \mathcal{F}, \mathbb{F}, \mathbb{P})$ for the model (34.1).

We shall assume that for all t, $\mathbb{E}|\lambda_t| < \infty$, and we introduce the filter estimate of λ:

$$\hat{\lambda}_t = \mathbb{E}\left[\lambda_t \,\big|\, \mathcal{F}_t^S\right], \quad 0 \leq t \leq T.$$

Notice that $\hat{\mu}_t := \sigma_t \hat{\lambda}_t$ is the filter estimate of μ: $\hat{\mu}_t = \mathbb{E}\left[\mu_t \,\big|\, \mathcal{F}_t^S\right]$, since σ is $\mathbb{F}^S$-adapted. Let us consider the process

$$N_t = W_t^0 - \int_0^t \hat{\lambda}_u du = W_t + \int_0^t (\lambda_u - \hat{\lambda}_u) du, \quad 0 \leq t \leq T.$$

N is the innovation process in filtering theory, and is a $(\mathbb{P}, \mathbb{F}^S)$-Brownian motion. By means of the innovation process, we can describe the dynamics of S and X within a framework of full observation model:

$$d S_t = \mathrm{diag}(S_t) \sigma_t (\hat{\lambda}_t dt + d N_t), \tag{34.5}$$

$$d X_t = X_t \left(\alpha_t' \hat{\lambda}_t dt + \alpha_t' d N_t\right), \tag{34.6}$$

with $\mathbb{F}^S$-adapted coefficients $\hat{\lambda}$ and σ. Hence, the operations of filtering and control can be put in sequence and thus separated. However, notice that in general $\mathbb{F}^S$ is strictly larger than the augmented filtration of the $(\mathbb{P}, \mathbb{F}^S)$-Brownian motion N, and the formal substitution of $\hat{\lambda}$ for λ in the formula of optimal portfolio in the full information case does not always yield the correct formula for the optimal portfolio in the partial information case.

34.2.3 Filtering

The reduction to a completely observable case involves the filter estimate $\hat{\lambda}$ (or equivalently $\hat{\mu}$) of the risk premium λ (or the mean rate of return μ). The computation of $\hat{\lambda}$ or $\hat{\mu}$ is obtained by the Kallianpur–Striebel formula which is related to Bayes' formula: for all $0 \leq t \leq u \leq T$, and $\xi \in L^1(\Omega, \mathcal{F}_u, \mathbb{P})$, we have

$$\mathbb{E}\left[\xi \,\big|\, \mathcal{F}_t^S\right] = \frac{\mathbb{E}^0\left[\xi L_u \,\big|\, \mathcal{F}_t^S\right]}{\mathbb{E}^0\left[L_t \,\big|\, \mathcal{F}_t^S\right]} = \frac{\mathbb{E}^0\left[\xi L_u \,\big|\, \mathcal{F}_t^S\right]}{L_t^0}, \tag{34.7}$$

where $\mathbb{E}^0$ denotes the expectation under $\mathbb{P}^0$, $L_t = 1/Z_t$, and $L_t^0 = \mathbb{E}^0\left[L_t \,\big|\, \mathcal{F}_t^S\right]$. Notice that L is a $(\mathbb{P}^0, \mathbb{F})$-martingale and L^0 is a $(\mathbb{P}^0, \mathbb{F}^S)$-martingale. When ξ is

$\mathcal{F}^S_u$-measurable, the formula (34.7) is also written as

$$\mathbb{E}\left[\xi \mid \mathcal{F}^S_t\right] = \frac{\mathbb{E}^0\left[\xi L^0_u \mid \mathcal{F}^S_t\right]}{L^0_t}.$$

The projection of the $\mathbb{P}$-martingale Z onto $\mathbb{F}^S$, defined by:

$$\hat{Z}_t = \mathbb{E}\left[Z_t \mid \mathcal{F}^S_t\right], \quad 0 \le t \le T,$$

is a $(\mathbb{P}, \mathbb{F}^S)$-martingale, and will also play a prominent role in the resolution of the partial observation control problem. By using the formula (34.7) for $\xi = Z_t$, we see that

$$\hat{Z}_t = \frac{1}{L^0_t}, \tag{34.8}$$

and we have similarly the following Kallianpur–Striebel formula: for all $0 \le t \le u \le T$, and $\eta \in L^1(\Omega, \mathcal{F}_u, \mathbb{P}^0)$,

$$\mathbb{E}^0\left[\eta \mid \mathcal{F}^S_t\right] = \frac{\mathbb{E}^0\left[\eta Z_u \mid \mathcal{F}^S_t\right]}{\hat{Z}_t}.$$

Moreover, when η is $\mathcal{F}^S_u$-measurable, we have

$$\mathbb{E}^0\left[\eta \mid \mathcal{F}^S_t\right] = \frac{\mathbb{E}^0\left[\eta \hat{Z}_u \mid \mathcal{F}^S_t\right]}{\hat{Z}_t}.$$

By means of the Kallianpur–Striebel formula, one can also show (see Proposition 3.1 in [21]) that

$$\hat{Z}_t = \exp\left(-\int_0^t \hat{\lambda}'_u d N_u - \frac{1}{2}\int_0^t |\hat{\lambda}_u|^2 du\right), \tag{34.9}$$

$$L^0_t = \exp\left(\int_0^t \hat{\lambda}'_u d W^0_u - \frac{1}{2}\int_0^t |\hat{\lambda}_u|^2 du\right), \quad 0 \le t \le T. \tag{34.10}$$

Notice that $\hat{Z}$ defines a change of probability measure $\mathbb{P}^{S,0}$ by $d\mathbb{P}^{S,0}/d\mathbb{P} = \hat{Z}_t$ on $\mathcal{F}^S_t$. The probability $\mathbb{P}^{S,0}$ is a risk-neutral martingale measure on $(\Omega, \mathcal{F}^S_T, \mathbb{F}^S, \mathbb{P})$ for the model (34.5).

34.2.4 Martingale approach

We apply the martingale approach to the model (34.5) with martingale density $\hat{Z}$. We impose the standard Inada conditions on the utility function: U is C^1 on $(0, \infty)$, and satisfies

$$U'(0) = \infty, \quad U'(\infty) = 0.$$

We denote by $I = (U')^{-1}$ the inverse of the derivative of U, which is a one-to-one decreasing function from $(0, \infty)$ into $(0, \infty)$, and we assume that

$$\mathbb{E}[\hat{Z}_T I(y \hat{Z}_T)] < \infty, \quad \forall y > 0. \tag{34.11}$$

Theorem 34.1 *Under (34.11), the optimal terminal wealth for $v(x)$, $x > 0$, in (34.3) is given by*

$$\hat{X}_T^x = I(\hat{y}_x \hat{Z}_T),$$

where $\hat{y}_x > 0$ is the solution to $\mathbb{E}^0[I(\hat{y}_x \hat{Z}_T)] = \mathbb{E}[\hat{Z}_T I(\hat{y}_x \hat{Z}_T)] = x$. Moreover, the corresponding wealth process is given by

$$\hat{X}_t^x = \mathbb{E}^0\left[I(\hat{y}_x \hat{Z}_T) \mid \mathcal{F}_t^S\right], \quad 0 \leq t \leq T.$$

Proof. Under (34.11) and the Inada conditions, the function $y \to \mathbb{E}[\hat{Z}_T I(y \hat{Z}_T)]$ is a one to one decreasing function from $(0, \infty)$ into $(0, \infty)$. Hence, for any $x > 0$, there exists a unique $\hat{y}_x > 0$ s.t. $\mathbb{E}[\hat{Z}_T I(\hat{y}_x \hat{Z}_T)] = x$. Since $I(\hat{y}_x \hat{Z}_T)$ is $\mathcal{F}_T^S$-measurable, we also notice from Bayes' formula that $\mathbb{E}^0[I(\hat{y}_x \hat{Z}_T)] = \mathbb{E}[Z_T I(\hat{y}_x \hat{Z}_T)] = \mathbb{E}[\hat{Z}_T I(\hat{y}_x \hat{Z}_T)]$. Denote by $\hat{\xi}^x$ the positive $\mathcal{F}_T^S$-measurable random variable, equal to $I(\hat{y}_x \hat{Z}_T)$, and let us show that $\hat{\xi}^x$ is attainable by some wealth process $X^{x,\hat{a}}$ for some $a \in \mathcal{A}^S$. For this, consider the $(\mathbb{P}^0, \mathbb{F}^S)$-martingale: $\hat{M}_t^0 = \mathbb{E}^0\left[\hat{\xi}^x \mid \mathcal{F}_t^S\right], 0 \leq t \leq T$. Since $\mathbb{F}^S$ is the augmented natural filtration of W^0, we get, by the martingale representation theorem, the existence of an $\mathbb{R}^n$-valued $\mathbb{F}^S$-adapted process ϕ s.t.

$$\hat{M}_t^0 = \hat{M}_0^0 + \int_0^t \phi_u' dW_u^0, \quad 0 \leq t \leq T.$$

Now, since $\hat{M}_0^0 = x$, $\hat{M}_T^0 = \hat{\xi}^x$, and $\hat{M}^0$ is positive, by defining $\hat{a}_t = \phi_t / \hat{M}_t^0$, we see with (34.2) that $\hat{M}^0 = X^{x,\hat{a}}$ and $X_T^{x,\hat{a}} = \hat{\xi}^x$. Let us finally check the optimality of the portfolio strategy $\hat{a}$ leading to the terminal wealth $\hat{\xi}^x = I(\hat{y}_x \hat{Z}_T)$. For this, consider the conjugate function of U:

$$\tilde{U}(y) = \sup_{x>0}[U(x) - xy], \quad y > 0,$$

and notice that the supremum in $\tilde{U}(y)$ is attained for $x = I(y)$, i.e. $\tilde{U}(y) = U(I(y)) - y I(y)$, for all $y > 0$. Now, for any $a \in \mathcal{A}^S$, the positive process $X^{x,a}$ is a local martingale under $\mathbb{P}^0$, and so a supermartingale under $\mathbb{P}^0$. Since $X_T^{x,a}$ is $\mathcal{F}_T^S$-measurable, this implies with Bayes' formula that $\mathbb{E}\left[\hat{Z}_T X_T^{x,a}\right] = \mathbb{E}\left[Z_T X_T^{x,a}\right] = \mathbb{E}^0\left[X_T^{x,a}\right] \leq x = \mathbb{E}[\hat{Z}_T \hat{\xi}^x]$. By definition of $\tilde{U}(\hat{y}_x \hat{Z}_T)$ and $\hat{\xi}^x$, we then have

$$\mathbb{E}\left[U\left(X_T^{x,a}\right)\right] \leq \mathbb{E}[\tilde{U}(\hat{y}_x \hat{Z}_T)] + \hat{y}_x \mathbb{E}\left[\hat{Z}_T X_T^{x,a}\right]$$
$$= \mathbb{E}[U(\hat{\xi}^x)] - \hat{y}_x \left(\mathbb{E}[\hat{Z}_T \hat{\xi}^x] - \mathbb{E}\left[\hat{Z}_T X_T^{x,a}\right]\right)$$
$$\leq \mathbb{E}[U(\hat{\xi}^x)] = \mathbb{E}\left[U\left(X_T^{x,\hat{a}}\right)\right],$$

which ends the proof. □

The optimal portfolio strategy $\hat{a}$ is implicitly determined by the equation

$$\hat{X}_t^x = \mathbb{E}^0\left[I(\hat{y}_x \hat{Z}_T) \,|\, \mathcal{F}_t^S\right] = x + \int_0^t \hat{X}_u^x \hat{a}_u' \, dW_u^0, \quad 0 \le t \le T.$$

We now give some examples of applications of the martingale approach for standard utility functions.

Logarithmic utility function

We consider a utility function $U(x) = \ln x$. In this case, $I(y) = 1/y$, and from Theorem 34.1 and (34.8), the optimal wealth process is given

$$\hat{X}_t^x = \mathbb{E}^0\left[\frac{1}{\hat{y}_x \hat{Z}_T} \,\Big|\, \mathcal{F}_t^S\right] = \mathbb{E}^0[\frac{1}{\hat{y}_x} L_T^0 \,|\, \mathcal{F}_t^S]$$
$$= \frac{1}{\hat{y}_x} L_t^0 = x L_t^0, \quad 0 \le t \le T. \qquad (34.12)$$

By writing that $d\hat{X}_t^x = \hat{X}_t^x \hat{a}_t' \, dW_t^0$, applying Itô's formula to (34.12), and from (34.10), we obtain that the optimal portfolio $\hat{a}$ is given by

$$\hat{a}_t = \hat{\lambda}_t.$$

From (34.9), the value function is given by

$$v(x) = \mathbb{E}\left[\ln \hat{X}_T^x\right] = \ln x - \mathbb{E}[\ln \hat{Z}_T] = \ln x + \frac{1}{2}\mathbb{E}\left[\int_0^T |\hat{\lambda}_u|^2 du\right].$$

Power utility function

We consider a utility function $U(x) = x^\delta/\delta$, with $0 < \delta < 1$. In this case, $I(y) = y^{-\beta}$, with $\beta = \frac{1}{1-\delta}$, and from Theorem 34.1 and (34.8), the optimal terminal wealth is given by $\hat{X}_T^x = (\hat{y}_x)^{-\beta} (L_T^0)^\beta$, where $\hat{y}_x > 0$ is s.t. $\mathbb{E}^0[\hat{X}_T^x] = x$. This yields

$$\hat{X}_T^x = \frac{x}{H_0} (L_T^0)^\beta, \quad \text{with } H_0 = \mathbb{E}^0[(L_T^0)^\beta]. \qquad (34.13)$$

The optimal wealth process is then given by

$$\hat{X}_t^x = \mathbb{E}^0[\hat{X}_T^x | \mathcal{F}_t^S] = \mathbb{E}^0\left[\frac{x}{H_0} (L_T^0)^\beta \,|\, \mathcal{F}_t^S\right] = x\frac{H_t}{H_0} (L_t^0)^\beta, \qquad (34.14)$$

with

$$H_t = \mathbb{E}^0\left[\left(\frac{L_T^0}{L_t^0}\right)^\beta \,|\, \mathcal{F}_t^S\right].$$

Now, from (34.10), we can express H_t as

$$H_t = \mathbb{E}^0\left[\exp\left(\int_t^T \beta\hat{\lambda}'_u dW^0_u - \frac{1}{2}\int_t^T |\beta\hat{\lambda}_u|^2 du\right)\exp\left(\frac{1}{2}\int_t^T \beta(\beta-1)|\hat{\lambda}_u|^2 du\right)\Big|\mathcal{F}^S_t\right].$$

Consider the process

$$M_t = \exp\left(\int_0^t \beta\hat{\lambda}'_u dW^0_u - \frac{1}{2}\int_0^t |\beta\hat{\lambda}_u|^2 du\right), \quad 0 \le t \le T,$$

and assume that M is a $(\mathbb{P}^0, \mathbb{F}^S)$-martingale which is satisfied under suitable integrablity conditions on $\hat{\lambda}$ (e.g. Novikov criterion). It then defines a change of probability measure

$$\frac{d\mathbb{Q}}{d\mathbb{P}^0} = M_t \quad \text{on } \mathcal{F}^S_t, \tag{34.15}$$

under which $W^\mathbb{Q} = W^0 - \int_0^\cdot \beta\hat{\lambda} du$ is a Brownian motion by Girsanov's theorem. It follows that

$$H_t = \mathbb{E}^\mathbb{Q}\left[\exp\left(\frac{\delta}{2(1-\delta)^2}\int_t^T |\hat{\lambda}_u|^2 du\right)\Big|\mathcal{F}^S_t\right]. \tag{34.16}$$

Notice that the positive process (H_t) can be written as $H_t = \exp\left(\frac{\delta}{2(1-\delta)^2}\int_0^t |\hat{\lambda}_u|^2 du\right).M_t^\mathbb{Q}$, where $M^\mathbb{Q}$ is a $(\mathbb{Q}, \mathbb{F}^S)$-martingale. By the martingale representation theorem, we deduce that H admits a decomposition in the form:

$$dH_t = H_t\left(\mu^H_t dt + \sigma^H_t dW^\mathbb{Q}_t\right), \tag{34.17}$$

for some $\mathbb{F}^S$-adapted processes μ^H and σ^H. By writing that $d\hat{X}^x_t = \hat{X}^x_t \hat{a}'_t dW^0_t$, applying Itô's formula to (34.14), and from (34.10), we obtain that the optimal portfolio $\hat{a}$ is given by

$$\hat{a}_t = \frac{\hat{\lambda}_t}{1-\delta} + \left(\sigma^H_t\right)'. \tag{34.18}$$

From (34.13) and Kallianpur–Striebel formula, the value function is given by

$$v(x) = \mathbb{E}\left[\frac{(\hat{X}^x_T)^\delta}{\delta}\right] = \frac{x^\delta}{\delta}\frac{1}{H^\delta_0}\mathbb{E}\left[\left(L^0_T\right)^{\beta\delta}\right] = \frac{x^\delta}{\delta}\frac{1}{H^\delta_0}\mathbb{E}^0\left[\left(L^0_T\right)^{1+\beta\delta}\right] \tag{34.19}$$

$$= \frac{x^\delta}{\delta}H^{1-\delta}_0. \tag{34.20}$$

34.2.5 The Markovian case and PDE approach

We suppose that (λ_t) is a continuous-time Markov process with state space $\Lambda \subset \mathbb{R}^n$, generator $\mathcal{L}$, and independent of W. We also assume that the law of λ_t given

$\mathcal{F}_t^S$ has a density $p_t(\ell)$ with respect to some dominating measure $\nu(d\ell)$, for all t. The filter estimate of λ is then given by

$$\hat{\lambda}_t = \hat{\lambda}(p_t) := \int_\Lambda \ell p_t(\ell)\nu(d\ell),$$

and it is known that (p_t) satisfies the so-called Zakai equation, see e.g. [28]:

$$dp_t = \mathcal{L}^* p_t dt + \vartheta(p_t) d N_t, \qquad (34.21)$$

where $\vartheta(p_t)(\ell) = p_t(\ell)(\ell - \hat{\lambda}(p_t))$, and $\mathcal{L}^*$ is the adjoint of $\mathcal{L}$. The process (p_t) is Markov with respect to $\mathbb{F}^S$, valued in the set $\mathcal{D}(\Lambda, [0, 1])$ of functions from Λ into $[0, 1]$, hence in general of infinite dimension when the state space Λ is continuous. According to (34.6) and (34.21), we reduced our problem to a full observation Markovian control problem with controlled process (X_t, p_t) valued in $(0, \infty) \times \mathcal{D}(\Lambda, [0, 1])$, and with a dynamic optimal objective function

$$V_t := \operatorname*{ess\,sup}_{\alpha \in \mathcal{A}^S} \mathbb{E}\left[U(X_T) | \mathcal{F}_t^S\right] = v(t, X_t, p_t), \qquad (34.22)$$

with a value function

$$v(t, x, p) = \sup_{\alpha \in \mathcal{A}^S} \mathbb{E}[U(X_T)|(X_t, p_t) = (x, p)], \ (t, x, p) \in [0, T] \times (0, \infty) \times \mathcal{D}(\Lambda, [0, 1]).$$

The Hamilton–Jacobi–Bellman (HJB) equation associated to this control problem is written as:

$$\frac{\partial v}{\partial t} + \mathcal{L}^* p D_p v + \frac{1}{2} D_p^2 v\, \vartheta(p).\vartheta(p)$$
$$+ \sup_{a \in \mathbb{R}^n}\left[a'\hat{\lambda}(p)x\frac{\partial v}{\partial x} + \frac{1}{2}|ax|^2\frac{\partial^2 v}{\partial x^2} + a'x D_{xp}^2 v\, \vartheta(p)\right] = 0,$$

together with the terminal condition $v(T, x, p) = U(x)$, and the optimal feedback control should be given at least formally by

$$\hat{a}(t, x, p) = -\frac{\hat{\lambda}(p)\dfrac{\partial v}{\partial x}(t, p) + D_{xp}^2 v(t, p)\, \vartheta(p)}{x\dfrac{\partial^2 v}{\partial x^2}(t, p)}.$$

Of course, we face here several theoretical difficulties: since p is generally infinite dimensional, the HJB equation is a PDE with infinite dimensional state variable, and the regularity of the value function v is not clear at all. We shall not go further into the Bellman approach in this general framework. Instead, we shall see how the problem may be simplified when the process (p_t) is finite dimensional and/or in the case of power utility functions. Basically, (p_t) is of finite dimension only for the three following situations: Bayesian, linear-Gaussian, and finite-state Markov chain cases, which we shall study now in more detail.

The Bayesian case

We suppose that $\lambda_t = \lambda$ is a $\mathcal{F}_0$-measurable random variable with a known distribution $\nu(d\ell)$ on $\Lambda \subset \mathbb{R}^n$. In this case, λ is independent of W^0 under $\mathbb{P}^0$, and recalling that $\mathbb{F}^S$ is the filtration of W^0, we have the following explicit expression for L_t^0:

$$L_t^0 = \mathbb{E}^0\left[\exp\left(\lambda' W_t^0 - \frac{1}{2}|\lambda|^2 t\right) \Big| \mathcal{F}_t^S\right] = F(t, W_t^0),$$

where

$$F(t, w) := \int_\Lambda \exp\left(\ell' w - \frac{1}{2}|\ell|^2 t\right) \nu(d\ell), \quad (t, w) \in [0, T] \times \mathbb{R}^n. \quad (34.23)$$

By (34.8), we then have

$$\hat{Z}_t = \frac{1}{F(t, W_t^0)}. \quad (34.24)$$

For any bounded measurable function f on Λ, we have by (34.7) applied to $\xi = f(\lambda)$:

$$\mathbb{E}\left[f(\lambda)|\mathcal{F}_t^S\right] = \frac{\mathbb{E}^0\left[f(\lambda)\exp\left(\lambda' W_t^0 - \frac{1}{2}|\lambda|^2 t\right)\Big|\mathcal{F}_t^S\right]}{L_t^0}$$

$$= \frac{\int_\Lambda f(\ell)\exp\left(\ell' W_t^0 - \frac{1}{2}|\ell|^2 t\right)\nu(d\ell)}{F(t, W_t^0)},$$

which shows that

$$p_t(\ell) = \frac{\exp\left(\ell' W_t^0 - \frac{1}{2}|\ell|^2 t\right)}{F(t, W_t^0)}.$$

In particular, we see that

$$\hat{\lambda}_t = \int_\Lambda \ell p_t(l) \nu(d\ell) = \hat{\lambda}(t, W_t^0),$$

where

$$\hat{\lambda}(t, w) = \frac{D_w F(t, w)}{F(t, w)} = \frac{\int_\Lambda \ell \exp\left(\ell' w - \frac{1}{2}|\ell|^2 t\right)\nu(d\ell)}{\int_\Lambda \exp\left(\ell' w - \frac{1}{2}|\ell|^2 t\right)\nu(d\ell)}, \quad (t, w) \in [0, T] \times \mathbb{R}^n.$$

Notice also that the process W^0 is a $(\mathbb{P}, \mathbb{F}^S)$-Markov process with dynamics:

$$dW_t^0 = \hat{\lambda}(t, W_t^0)\, dt + dN_t.$$

From Theorem 34.1, the optimal terminal wealth for $v(x)$ is given by $\hat{X}_T^x = I(\hat{y}_x/F(T, W_T^0))$, and the corresponding wealth process is

$$\hat{X}_t^x = \mathbb{E}^0\left[I\left(\frac{\hat{y}_x}{F(T, W_T^0)}\right)\Big|\mathcal{F}_t^S\right] = \chi(t, W_t^0), \quad 0 \leq t < T,$$

where

$$\chi(t, w) = \int I\left(\frac{\hat{y}_x}{F(T, w+z)}\right)\varphi_{T-t}(z)dz, \quad (t, w) \in [0, T) \times \mathbb{R}^n,$$

$$\varphi_t(z) = \frac{1}{(2\pi t)^{\frac{n}{2}}} \exp\left(-\frac{|z|^2}{2t}\right), \quad (t, z) \in (0, T) \times \mathbb{R}^n.$$

The Lagrange multiplier $\hat{y}_x$ is the solution to $\hat{X}_0^x = \mathbb{E}^0\left[I(\hat{y}_x/F(T, W_T^0))\right] = x$, i.e.

$$\chi(0, 0) = \int I\left(\frac{\hat{y}_x}{F(T, z)}\right)\varphi_T(z)dz = x.$$

By writing that $d\hat{X}_t^x = \hat{X}_t^x \hat{a}_t' dW_t^0$, we deduce by Itō's formula applied to $\hat{X}_t^x = \chi(t, W_t^0)$ and assuming that U is C^2, that the optimal portfolio $\hat{a}$ is given by

$$\hat{a}_t = \frac{D_w\chi(t, W_t^0)}{\chi(t, W_t^0)}, \quad t \in [0, T),$$

with

$$D_w\chi(t, w) = -\hat{y}_x \int \frac{\hat{\lambda}(T, w+z)}{F(T, w+z)} I'\left(\frac{\hat{y}_x}{F(T, w+z)}\right)\varphi_{T-t}(z)dz.$$

The value function (at time 0) is given by

$$v(x) = \mathbb{E}[U(\hat{X}_T^x)] = \mathbb{E}[U \circ I(\hat{y}_x \hat{Z}_T)]$$
$$= \mathbb{E}^0\left[\frac{1}{\hat{Z}_T} U \circ I(\hat{y}_x \hat{Z}_T)\right] = \mathbb{E}^0\left[F(T, W_T^0) U \circ I\left(\frac{\hat{y}_x}{F(T, W_T^0)}\right)\right]$$
$$= \int F(T, z) U \circ I\left(\frac{\hat{y}_x}{F(T, z)}\right)\varphi_T(z)dz$$

Remark 34.1 In this Bayesian framework, the dynamic objective function V_t in (34.22) is a deterministic function $v(t, X_t, W_t^0)$ of (t, X_t, W_t^0), and the function $v(t, x, w)$ satisfies the HJB equation:

$$\frac{\partial v}{\partial t} + \hat{\lambda}(t, w) D_w v + \frac{1}{2}\mathrm{tr}\left(D_w^2 v\right) + \sup_{a \in \mathbb{R}^n}\left[a'\hat{\lambda}(t, w)x\frac{\partial v}{\partial x} + \frac{1}{2}|ax|^2\frac{\partial^2 v}{\partial x^2} + a'x D_{xw}^2 v\right] = 0,$$

together with the terminal condition $v(T, x, w) = U(x)$. When U is of power type: $U(x) = x^\delta/\delta$, this HJB equation is simplified and the value function takes the form $v(t, x, w) = \frac{x^\delta}{\delta} H(t, w)^{1-\delta}$, where $H(t, w)$ satisfies the linear PDE:

$$\frac{\partial H}{\partial t} + \frac{1}{1-\delta}\hat{\lambda}(t, w) D_w H + \frac{1}{2}\text{tr}\left(D_w^2 H\right) + \frac{\delta}{2(1-\delta)^2}|\hat{\lambda}(t, w)|^2 H = 0,$$

with the terminal condition $H(T, w) = 1$.

The linear Gaussian case

We suppose that the risk premium λ is a Gaussian process governed by the dynamics:

$$d\lambda_t = (A(t)\lambda_t + C(t))\, dt + \gamma(t) d B_t, \quad \lambda_0 \rightsquigarrow \mathcal{N}(m_0, \Delta_0),$$

where A, C, γ are deterministic functions valued in $\mathbb{R}^{n \times n}$, $\mathbb{R}^n$, and $\mathbb{R}^{n \times n}$, B is a n-dimensional Brownian motion, independent of W, λ_0 follows a normal distribution of mean m_0 and covariance matrix Δ_0. We are then in the framework of the classical Kalman–Bucy filter, and $\hat{\lambda}$ is known to satisfy the linear s.d.e., see e.g. [24]:

$$d\hat{\lambda}_t = \left(A(t)\hat{\lambda}_t + C(t)\right) dt + \Gamma(t) d N_t, \quad \hat{\lambda}_0 = m_0,$$

where Γ is the solution of the Riccati equation

$$-\dot{\Gamma}(t) + A\Gamma(t) + \Gamma A'(t) - \Gamma^2(t) + \gamma\gamma'(t) = 0, \quad \Gamma(0) = \Delta_0.$$

Moreover, p_t is the density of a normally distributed vector of mean $\hat{\lambda}_t$, and of covariance matrix $\Gamma(t)$. Hence, we see that $(\hat{\lambda}_t)$ is a $(\mathbb{P}, \mathbb{F}^S)$-Markov process.

Let us now consider the example of power utility functions:

$$U(x) = \frac{x^\delta}{\delta}, \quad x > 0, \ 0 < \delta < 1,$$

and look how the optimal portfolio strategy and the value functions can be more explicitly determined. First, we notice that the probability measure $\mathbb{Q}$ in (34.15) is well defined since the associated density process M is a $(\mathbb{P}^0, \mathbb{F}^S)$-martingale by Lemma 4.1.1 in [4]. Moreover, the $(\mathbb{Q}, \mathbb{F}^S)$-Brownian motion $W^{\mathbb{Q}}$ may be written in terms of the innovation process N as: $W_t^{\mathbb{Q}} = N_t - \frac{\delta}{1-\delta} \int_0^t \hat{\lambda}_u du$. Thus, the process $\hat{\lambda}$ is also a $(\mathbb{Q}, \mathbb{F}^S)$-Markov process with dynamics:

$$d\hat{\lambda}_t = \left[\left(A(t) + \frac{\delta}{1-\delta}\Gamma(t)\right)\hat{\lambda}_t + C(t)\right]dt + \Gamma(t) d W_t^{\mathbb{Q}}.$$

It follows that the process H defined in (34.16) is a deterministic function of $(t, \hat{\lambda}_t)$:

$$H_t = \mathbb{E}^{\mathbb{Q}}\left[\exp\left(\frac{\delta}{2(1-\delta)^2}\int_t^T |\hat{\lambda}_u|^2 du\right)\Big|\mathcal{F}_t^S\right] = H(t, \hat{\lambda}_t),$$

and by the Feynman–Kac representation, the function $H(t, \ell)$, $(t, \ell) \in [0, T] \times \mathbb{R}^n$, satisfies the linear PDE:

$$\frac{\partial H}{\partial t} + \left[\left(A(t) + \frac{\delta}{1-\delta}\Gamma(t)\right)\ell + C(t)\right]' D_\ell H + \frac{1}{2}\mathrm{tr}\left(\Gamma\Gamma'(t) D_\ell^2 H\right)$$
$$+ \frac{\delta}{2(1-\delta)^2}|\ell|^2 H = 0,$$

together with the terminal condition $H(T, \ell) = 1$. Moreover, by Itō's formula to $H(t, \hat{\lambda}_t)$, the coefficient σ_t^H defined in (34.17) is given by:

$$\sigma_t^H = \frac{D_\ell H(t, \hat{\lambda}_t)' \Gamma(t)}{H(t, \hat{\lambda}_t)},$$

and thus by (34.18), the optimal portfolio strategy is equal to:

$$\hat{\alpha}_t = \frac{\hat{\lambda}_t}{1-\delta} + \frac{\Gamma(t)' D_\ell H(t, \hat{\lambda}_t)}{H(t, \hat{\lambda}_t)}.$$

Finally by (34.19), the value function (at time 0) is

$$v(x) = \frac{x^\delta}{\delta} H(0, m_0)^{1-\delta}.$$

Remark 34.2 In this linear-Gaussian framework, the dynamic objective function V_t in (34.22) is a deterministic function $v(t, X_t, \hat{\lambda}_t)$ of $(t, X_t, \hat{\lambda}_t)$, and the function $v(t, x, \ell)$ satisfies the HJB equation:

$$\frac{\partial v}{\partial t} + (A(t)\ell + C(t)) D_\ell v + \frac{1}{2}\mathrm{tr}\left(\Gamma\Gamma'(t) D_\ell^2 v\right)$$
$$+ \sup_{a \in \mathbb{R}^n}\left[a'\ell x \frac{\partial v}{\partial x} + \frac{1}{2}|ax|^2 \frac{\partial^2 v}{\partial x^2} + a'x\Gamma(t) D_{x\ell}^2 v\right] = 0,$$

together with the terminal condition $v(T, x, \ell) = U(x)$. When U is a power utility function: $U(x) = x^\delta/\delta$, this HJB equation is simplified and the value function takes the form $v(t, x, \ell) = \frac{x^\delta}{\delta} H(t, \ell)^{1-\delta}$.

The finite-state Markov chain case

We suppose that (λ_t) is a continuous-time Markov chain with finite-state space $\{\ell_1, \ldots, \ell_k\}$, $\ell_i \in \mathbb{R}^n$, with generator $Q = (q_{ij})$, and independent of W. We denote by $p_t = (p_t^1, \ldots, p_t^k)$ the Wonham filter of the Markov chain:

$$p_t^i = \mathbb{P}\left[\lambda_t = \ell_i | \mathcal{F}_t^S\right], \quad i = 1, \ldots, k, \quad 0 \leq t \leq T.$$

The filter estimate of λ is given by

$$\hat{\lambda}_t = \hat{\lambda}(p_t) := \sum_{i=1}^{k} \ell_i p_t^i,$$

and (p_t) is known to satisfy the dynamics, see e.g. [14]:

$$dp_t^i = \sum_j q_{ji} p_t^j dt + p_t^i (\ell_i - \hat{\lambda}(p_t))' dN_t, \quad p_0^i = \mathbb{P}[\lambda_0 = \ell_i], \quad i = 1, \ldots, k.$$

The $(\mathbb{P}, \mathbb{F}^S)$-Markov process (p_t) is then finite-dimensional, valued in $\mathcal{S}_k$ the simplex of $[0, 1]^k$, i.e. $\mathcal{S}_k = \{p = (p_1, \ldots, p_k) \in [0, 1]^k : \sum_i p_i = 1\}$.

Let us now consider the example of power utility functions:

$$U(x) = \frac{x^\delta}{\delta}, \quad x > 0, \ 0 < \delta < 1,$$

and look at how the optimal portfolio strategy and the value functions can be more explicitly determined. As in the previous paragraph, since the $(\mathbb{Q}, \mathbb{F}^S)$-Brownian motion under the probability measure $\mathbb{Q}$ defined in (34.15) is written in terms of the innovation process N as $W_t^\mathbb{Q} = N_t - \frac{\delta}{1-\delta} \int_0^t \hat{\lambda}_u du$, we see that the process (p_t) is also a $(\mathbb{Q}, \mathbb{F}^S)$-Markov process with dynamics:

$$dp_t^i = \left[\sum_j q_{ji} p_t^j + \frac{\delta}{1-\delta} p_t^i (\ell_i - \hat{\lambda}(p_t)) \hat{\lambda}(p_t) \right] dt + p_t^i (\ell_i - \hat{\lambda}(p_t))' dW_t^\mathbb{Q}.$$

It follows that the process H defined in (34.16) is a deterministic function of (t, p_t):

$$H_t = \mathbb{E}^\mathbb{Q} \left[\exp\left(\frac{\delta}{2(1-\delta)^2} \int_t^T |\hat{\lambda}(p_u)|^2 du \right) \Big| \mathcal{F}_t^S \right] = H(t, p_t),$$

and by the Feynman–Kac representation, the function $H(t, p)$, $(t, p) \in [0, T] \times \mathcal{S}_k$, satisfies the linear PDE:

$$\frac{\partial H}{\partial t} + \sum_i \left[\sum_j q_{ji} p_j + \frac{\delta}{1-\delta} p_i (\ell_i - \hat{\lambda}(p)) \hat{\lambda}(p) \right] \frac{\partial H}{\partial p_i}$$

$$+ \frac{1}{2} \sum_{i,j} (\ell_i - \hat{\lambda}(p))' (\ell_j - \hat{\lambda}(p)) p_i p_j \frac{\partial^2 H}{\partial p_i \partial p_j} + \frac{\delta}{2(1-\delta)^2} |\hat{\lambda}(p)|^2 H = 0,$$

together with the terminal condition $H(T, p) = 1$. Moreover, by applying Itô's formula to $H(t, p_t)$, the coefficient σ_t^H defined in (34.17) is given by:

$$\sigma_t^H = \frac{\sum_i p_t^i (\ell_i - \hat{\lambda}(p_t))' \frac{\partial H}{\partial p_i}(t, p_t)}{H(t, p_t)},$$

and thus by (34.18), the optimal portfolio strategy is equal to:

$$\hat{\alpha}_t = \frac{\hat{\lambda}(p_t)}{1-\delta} + \frac{\sum_i p_t^i (\ell_i - \hat{\lambda}(p_t)) \frac{\partial H}{\partial p_i}(t, p_t)}{H(t, p_t)}.$$

Finally by (34.19), the value function (at time 0) is

$$v(x) = \frac{x^\delta}{\delta} H(0, p_0)^{1-\delta}.$$

Remark 34.3 In this finite-state Markov chain framework, and when $U(x) = x^\delta/\delta$, the dynamic value function v in (34.22) takes the form $v(t, x, p) = \frac{x^\delta}{\delta} H(t, p)^{1-\delta}$.

34.2.6 Bibliographical notes

The techniques of filtering for reducing a partial observation control problem into a full observation control problem are standard, see e.g. the book by Bensoussan [4] in a Markovian context. The derivation of the Zakai equation may be found in the lecture notes of Pardoux [28], while the classical Kalman–Bucy filter is presented for instance in Liptser and Shiryaev [24], and the Wonham filter in Elliott, Aggoun, and Moore [14]. In the context of expected utility maximization from terminal wealth in complete markets with Itō asset price processes, as presented here, these filtering arguments were used by Lakner [21], who then followed the well-developed martingale approach of Karatzas et al. [18] and Cox and Huang [11]. The Bayesian case for the unknown drift was studied by Karatzas and Zhao [19]. Detemple [12], Gennotte [16], Rishel [33], or Brendle [7] used Bellman methods while Lakner [22] applied a martingale approach in a linear-Gaussian setting. Bauerle and Rieder [2], Nagai and Runggaldier [26] adopted PDE methods, and Haussman and Sass [17] used martingale approach to solve the portfolio problem when the utilty function is of power type and the unobservable rate of return is a finite-state Markov chain. The change of probability measure in the case of power utility function follows the idea in Bjork, Davis, and Landen [6]. The above-cited papers deal with complete market models. In the setting of incomplete markets, Pham and Quenez [32] solved the partial observation problem in a stochastic volatility model. Portfolio optimization in models with jumps and unobservable intensity are considered in Bauerle and Rieder [3], Callegaro, Di Masi, and Runggaldier [8], and Corsi [9]. We mention also papers by Frey and Runggaldier [15], Lasry and Lions [23], and Pham [29], who studied hedging problems under restricted information.

34.3 Discrete-time models

34.3.1 Partial observation control framework

We consider a discrete-time partially observable process (X, Y) where $X = (X_k)$, $k = 0, \ldots, n$, represents the state or signal process that may not be observable, while $Y = (Y_k)$, $k = 0, \ldots, n$, is the observation. We assume that the signal process $(X_k)_k$ is a finite-state Markov chain on the probability space $(\Omega, \mathbb{P})$, valued in $E = \{x^1, \ldots, x^m\}$, with initial law $\mu = (\mu^i)_i$ and probability transition matrix $P_k = \left(P_k^{ij}\right)$:

$$\mu^i = \mathbb{P}[X_0 = x^i], \quad i = 1, \ldots, m$$

$$P_k^{ij} = \mathbb{P}[X_k = x^j | X_{k-1} = x^i], \quad i = 1, \ldots, m, \quad j = 1, \ldots, m.$$

The observation sequence (Y_k) is valued in $\mathbb{R}^d$, such that the pair (X_k, Y_k) is a Markov chain on $(\Omega, \mathbb{P})$, and we assume that the law of Y_k conditional on (X_{k-1}, Y_{k-1}, X_k), $k = 1, \ldots, n$, admits a bounded (known) density (sometimes called a local likelihood function):

$$y' \longmapsto g_k(X_{k-1}, Y_{k-1}, X_k, y').$$

For simplicity, we assume that Y_0 is a known deterministic constant equal to y_0. A typical example is given by the following scheme:

$$Y_k = G_k(X_{k-1}, Y_{k-1}, X_k, \eta_k),$$

for some measurable function G_k, and where (η_k) is a white noise. We denote by $\left(\mathcal{F}_k^Y\right)_k$ the filtration generated by the observation process $(Y_k)_k$.

We consider a real-valued controlled process $(W_k)_k$ with dynamics in the form:

$$W_{k+1} = F(W_k, a_k, Y_k, Y_{k+1}), \quad k = 0, \ldots, n-1, \quad (34.25)$$

for some measurable function F, and where the control process $a = (a_k)_k$ is valued in some compact subset A of $\mathbb{R}^l$, and is adapted with respect to the observation filtration $(\mathcal{F}_k^Y)$. We denote by $\mathcal{A}^Y$ this set of control processes. Notice that (W_k) is adapted with respect to $(\mathcal{F}_k^Y)$. Given a measurable function ℓ on $E \times \mathbb{R}^d \times \mathbb{R}$, we are interested in the following control problem under partial observation over a finite horizon n:

$$J_{opt} = \inf_{a \in \mathcal{A}^Y} J(a) := \inf_{a \in \mathcal{A}^Y} \mathbb{E}\left[\ell(X_n, Y_n, W_n)\right]. \quad (34.26)$$

For example in finance, $(X_k)_k$ is the unobservable return and/or volatility of stock price S while $Y_k = \ln S_k$ is the logarithm of the observed price process

$$Y_k = Y_{k-1} + b(X_{k-1}) + \sigma(X_{k-1})\eta_k,$$

and g_k is explicit once the density of the white noise η_k is specified. Consider an investor who can trade at any time k a number of shares a_k in stock based on the past price observations, and invest the rest in a riskless bond assumed for simplicity equal to one. Her wealth process $(W_k)_k$ is then governed by:

$$W_{k+1} = W_k + a_k(e^{Y_{k+1}} - e^{Y_k}), \quad k = 0 \ldots, n-1.$$

The function ℓ represents the criterion function associated to a portfolio choice or hedging problem. For example, the mean-variance hedging problem for a payoff option $h(Y_n)$ at maturity n corresponds to $\ell(X_n, Y_n, W_n) = (h(Y_n) - W_n)^2$.

34.3.2 Filter evolution and reduction to a full observation control problem

We denote by $(\Pi_k)_k$ the filter process defined by:

$$\Pi_k^i = \mathbb{P}\left[X_k = x^i | \mathcal{F}_k^Y\right], \quad k \in \mathbb{N}, \ i = 1, \ldots, m.$$

Hence $\Pi_k = \left(\Pi_k^i\right)_i$ is the conditional law of X_k given $\mathcal{F}_k^Y$, and is a random vector taking values in the m-simplex of $\mathbb{R}^m$:

$$K_m = \left\{\pi = (\pi^i)_{1 \leq i \leq m} \in \mathbb{R}^m : \pi^i \geq 0 \text{ and } |\pi|_1 = 1\right\}.$$

Here, we denote for any vector $\pi = (\pi^i)_i$ in $\mathbb{R}^m$, $|\pi|_1 = \sum_{i=1}^m |\pi^i|$. From Markov property and Bayes formula, the filter process satisfies the filtering forward equation:

$$\Pi_0 = \mu \tag{34.27}$$

$$\Pi_k = \bar{H}_k(\Pi_{k-1}, Y_{k-1}, Y_k) := \frac{H_k(Y_{k-1}, Y_k)^\intercal \Pi_{k-1}}{|H_k(Y_{k-1}, Y_k)^\intercal \Pi_{k-1}|_1}, \quad k \geq 1, \tag{34.28}$$

where $H_k(Y_{k-1}, Y_k)$ is the prediction-updating $m \times m$ transition matrix:

$$H_k^{ij}(Y_{k-1}, Y_k) = g_k(x^i, Y_{k-1}, x^j, Y_k) P_k^{ij}, \quad 1 \leq i, j \leq m,$$

and $\intercal$ is the transpose.

Remark 34.4 From (34.28), we see that the randomness of the filter at time $k \geq 1$, depends on the whole observation path $Y_1, \ldots, Y_k : \Pi_k = \Pi_k(Y_1, \ldots, Y_k)$.

By using the law of iterated conditional expectations, we can rewrite the expected cost function in (34.26) as follows:

$$J(a) = \mathbb{E}\left[\mathbb{E}\left[\ell(X_n, Y_n, W_n) | \mathcal{F}_n^Y\right]\right] = \mathbb{E}\left[\sum_{i=1}^m \ell(x^i, Y_n, W_n) \Pi_n^i\right]$$

$$= \mathbb{E}\left[\hat{\ell}(\Pi_n, Y_n, W_n)\right]$$

where

$$\hat{\ell}(\pi, y, w) := \sum_{i=1}^{m} \ell(x^i, y, w)\pi^i.$$

The original control problem (34.26) can now be reformulated as a problem under full observation with state variables (Π_k, Y_k, W_k), valued in $K_m \times \mathbb{R}^d \times \mathbb{R}$, and $(\mathcal{F}_k^Y)$-adapted:

$$J_{opt} = \inf_{\alpha \in \mathcal{A}^Y} \mathbb{E}\left[\hat{\ell}(\Pi_n, Y_n, W_n)\right]. \tag{34.29}$$

34.3.3 Quantization of the filter process

In view of solving (34.29), we need an approximation of the filter process $(\Pi_k)_k$. Recall the dependence of the random filter on the observation: $\Pi_k = \Pi_k(Y_1, \ldots, Y_k)$.

An usual approach, suggested e.g. in [5], consists of approximating $\Pi_k(Y_1, \ldots, Y_k)$ by $\Pi_k(\hat{Y}_1, \ldots, \hat{Y}_k)$ where $\hat{Y}_k$ is a discrete-state space approximation of Y_k. The main problem in effective implementation is the growing dimension of this approximating filter: indeed, for instance, if each $\hat{Y}_k$ takes M values, then at time n, the random filter $\Pi_n(\hat{Y}_1, \ldots, \hat{Y}_n)$ would take M^n values in K_m, which is not realistically implementable for a long horizon n.

In order to overcome this numerical difficulty, we present a quantization approach introduced in [31] and based on the Markov property of the pair filter-observation (Π_k, Y_k) with respect to the observation filtration $(\mathcal{F}_k^Y)$. In other words, the conditional law of X_{k+1} given $\mathcal{F}_k^Y$ is summarized by the sufficient statistic (Π_k, Y_k), and we shall approximate the pair Markov chain (Π_k, Y_k) by an approximation of their successive probability transitions. One first proves that the probability transition R_k (from time $k-1$ to k) of the Markov chain $(Z_k) = (\Pi_k, Y_k)$ in $K_m \times \mathbb{R}^d$ is given by:

$$R_k \varphi(\pi, y) = \int \varphi(\bar{H}_k(\pi, y, y'), y') Q_k(\pi, y, dy'),$$

where $Q_k(\pi, y, dy')$ is the law of Y_k conditional on $(\Pi_{k-1}, Y_{k-1}) = (\pi, y)$ with density:

$$y' \longrightarrow \sum_{i,j=1}^{m} g_k(x^i, y, x^j, y') P_k^{ij} \pi^i. \tag{34.30}$$

This shows in particular that Z_k may be simulated through the following simulation procedure of its probability transition (R_k): for $k = 0$, Z_0 is a known deterministic vector equal to $z_0 = (\mu, y_0)$, and for $k \geq 1$, starting from (Π_{k-1}, Y_{k-1}),

- we simulate Y_k according to the law $Q_k(\Pi_{k-1}, Y_{k-1}, dy')$ given in (34.30).
- we compute Π_k by the forward filtering equation

$$\Pi_k = \bar{H}_k(\Pi_{k-1} Y_{k-1}, Y_k).$$

Once we are able to simulate independent copies of $(Z_0, \ldots, Z_n)$, we apply an optimal quantization to each Z_k in $K_m \times \mathbb{R}^d$, for $k = 0, \ldots, n$, following the vector quantization method originally introduced in [1] for solving optimal stopping problems. In our context, the procedure is described as follows:

For each $k = 0, \ldots, n$, we consider a grid $z_k = \left\{z_k^1, \ldots, z_k^{N_k}\right\}$ of N_k points in $K_m \times \mathbb{R}^d$ (we shall often identify such a grid with a N_k-tuple in $K_m \times \mathbb{R}^d$), and its Voronoi tesselations, that is Borel partitions $C_1(z_k), \ldots, C_{N_k}(z_k)$ of $K_m \times \mathbb{R}^d$ satisfying:

$$C_i(z_k) \subset \left\{\xi \in K_m \times \mathbb{R}^d : |\xi - z_k^i| = \min_{j=1,\ldots,N_k} |\xi - z_k^j|\right\}, \quad i = 1, \ldots, N_k.$$

Here, $|.|$ is the Euclidian norm on $K_m \times \mathbb{R}^d$. Then, one defines the z_k-Voronoi quantization of Z_k as the closest neighbour projection of Z_k on the grid z_k:

$$\hat{Z}_k = \text{Proj}_{z_k}(Z_k) := \sum_{i=1}^{N_k} z_k^i \mathbf{1}_{C_i(z_k)}(Z_k),$$

whose discrete probability law is characterized by:

$$\hat{p}_k^i := \mathbb{P}\left[\hat{Z}_k = z_k^i\right] = \mathbb{P}[Z_k \in C_i(z_k)], \quad i = 1, \ldots, N_k.$$

The L^2-error induced by this projection, called L^2-quantization error, is $\|Z_k - \hat{Z}_k\|_2$, and is a function of the N_k-tuple z_k. By definition of the closest neighbour projection, we see that the L^2-quantization error is the minimum of L^2-error $\|Z_k - U\|_2$ among all random variables U taking values in the grid z_k.

Then, we approximate the probability transitions (R_k) of the Markov chain (Z_k) by the probability transition matrices $(\hat{r}_k)$ defined by:

$$\hat{r}_k^{ij} = \mathbb{P}\left[\hat{Z}_k = z_k^j \,\middle|\, \hat{Z}_{k-1} = z_{k-1}^i\right]$$

$$= \frac{\mathbb{P}\left[Z_k \in C_j(z_k), Z_{k-1} \in C_i(z_{k-1})\right]}{\mathbb{P}\left[Z_{k-1} \in C_i(z_{k-1})\right]} =: \frac{\hat{\beta}_k^{ij}}{\hat{p}_{k-1}^i},$$

for all $k \geq 1$, $i = 1, \ldots, N_{k-1}$, $j = 1, \ldots, N_k$. The process $(\hat{Z}_k)$ obtained by this method, is called a marginal quantization of the process (Z_k): it is characterized for each k by its grid space z_k, and by the probability transition matrix $\hat{r}_k = \left(\hat{r}_k^{ij}\right)$.

Denoting by $\xi^s = \left(\xi_0^s, \ldots, \xi_n^s\right)_s$, independent copies of $(Z_0, \ldots, Z_n)$, the optimal grids z_k that minimize the L^2-quantization error $\|Z_k - \hat{Z}_k\|_2$ for each k, and the companion parameters $\hat{r}_k^{ij}$, are practically implemented according to the Kohonen algorithm as follows:

Initialisation phase:

- Initialize the n grids $z_k^{(0)} = (z_k^{0,1}, \ldots, z_k^{0,N_k}) \in (K_m \times \mathbb{R}^d)^{N_k}$ for $k = 0, \ldots, n$, with $\Gamma_0^{(0)} = z_0$ reduced to $N_0 = 1$ point for $k = 0$.
- Initialize the weights vectors: $p_k^{0,i} = 1/N_k$, $\beta_{k+1}^{0,ij} = 0$, $i = 1, \ldots, N_k$, $j = 1, \ldots, N_{k+1}$.

Updating $s \to s+1$: At step s, the n grids $z_k^{(s)} = (z_k^{s,1}, \ldots, z_k^{s,N_k})$, the weights vectors $p_k^{s,i}$, $\beta_{k+1}^{s,ij}$, $i = 1, \ldots, N_k$, $j = 1, \ldots, N_{k+1}$, have been obtained and we use the sample ξ^{s+1} of $(Z_0, \ldots, Z_n)$ to update them as follows: for all $k = 0, \ldots, n$,

- *Competitive phase*: select $i_k(s+1) \in \{1, \ldots, N_k\}$ such that
$$\xi^{s+1} \in C_{i_k(s+1)}(z_k^{(s)}), \text{ i.e. } i_k(s+1) \in \mathrm{argmin}_{1 \leq i \leq N_k} |z_k^{s,i} - \xi^{s+1}|_2.$$

- *Learning phase:*
  ⋆ Updating of the grid:
$$z_k^{s+1,i} = z_k^{s,i} - \delta_{s+1} 1_{i=i_k(s+1)} \left(z_k^{s,i} - \xi^{s+1} \right), \quad i = 1, \ldots, N_k$$

  ⋆ Updating of the weights vectors and of the probability transition
$$p_k^{s+1,i} = p_k^{s,i} - \delta_{s+1} \left(p_k^{s,i} - 1_{i=i_k(s+1)} \right),$$
$$\beta_{k+1}^{s+1,ij} = \beta_{k+1}^{s,ij} - \delta_{s+1} \left(\beta_{k+1}^{s,ij} - 1_{i=i_k(s+1), j=i_{k+1}(s+1)} \right),$$
$$r_{k+1}^{s+1,ij} = \frac{\beta_{k+1}^{s+1,ij}}{p_k^{s+1,i}},$$

for all $i = 1, \ldots, N_k$, $j = 1, \ldots, N_{k+1}$.

Finally, we mention that the rate of convergence for each k, of the minimal L^2-quantization error $\|Z_k - \hat{Z}_k\|_2$ when N_k goes to infinity, is given by Zador's theorem, and is of order $N_k^{-\frac{1}{m-1+d}}$.

34.3.4 Approximation of control problems under partial observation

Recalling the dynamics (34.25) of (W_k) and following the dynamic programming principle for the discrete-time control problem (34.29), we define the sequence of functions on $K_m \times \mathbb{R}^d \times \mathbb{R}$:

$$u_n(\pi, y, w) = \hat{\ell}(\pi, y, w)$$
$$u_k(\pi, y, w) = \inf_{a \in A} \mathbb{E}\left[u_{k+1}(\Pi_{k+1}, Y_{k+1}, F(w, a, y, Y_{k+1})) | (\Pi_k, Y_k) = (\pi, y)\right],$$

for $k = 0, \ldots, n-1$, so that $J_{opt} = u_0(\mu, y_0, w_0)$, where w_0 is the initial value of W_0 at time $k = 0$, and we recall that $(\Pi_0, Y_0) = (\mu, y_0)$.

In order to compute this sequence of functions u_k, we deal separately with the approximation of the pair filter-observation process $(Z_k)_k = (\Pi_k, Y_k)_k$ that does not depend on the control, and the approximation of the controlled process $(W_k)_k$.

- We apply a marginal quantization of the process $(Z_k) = (\Pi_k, Y_k)$, and we denote the $(\hat{Z}_k) = (\hat{\Pi}_k, \hat{Y}_k)$ the corresponding quantizers on grids (z_k), and $(\hat{r}_k)$ the associated probability transition matrices, as described in the previous paragraph. The i-th point of the grid z_k of size N_k in $K_m \times \mathbb{R}^d$ is denoted $z_k^i = (\pi_k(i), y_k^i) \in K_m \times \mathbb{R}^d$, $i = 1, \ldots, N_k$.

- The approximation of W_k is obtained by a classical uniform space discretization similar to the Markov chain method as in [20]. We fix a bounded uniform grid on the state space $\mathbb{R}$ for the controlled process (W_k). Namely, we set

$$\Gamma = (2\nu)\mathbb{Z} \cap [-L, L],$$

where ν is the spatial step and L is the grid size. We denote by Proj_Γ the projection on the grid Γ according to the closest neighbour rule. Recalling the dynamics (34.25) of the controlled process (W_k), we approximate it as follows: given a control $\alpha \in \mathcal{A}^Y$, we define the discretized controlled process $(\hat{W}_k)$, valued in Γ, by:

$$\hat{W}_{k+1} = \text{Proj}_\Gamma(F(\hat{W}_k, \alpha_k, \hat{Y}_k, \hat{Y}_{k+1})).$$

We then approximate the sequence of functions u_k by the sequence of functions $\hat{u}_k$ defined on $z_k \times \Gamma$, $k = 0, \ldots, n$, by a dynamic programming type formula:

$$\hat{u}_n(\pi, y, w) = \hat{\ell}(\pi, y, w)$$
$$\hat{u}_k(\pi, y, w) = \inf_{a \in A} \left\{ \mathbb{E}\left[\hat{u}_{k+1}\left(\hat{\Pi}_{k+1}, \hat{Y}_{k+1}, \text{Proj}_\Gamma(F(w, a, y, \hat{Y}_{k+1}))\right) \middle| (\hat{\Pi}_k, \hat{Y}_k) = (\pi, y) \right] \right\}.$$

From an algorithmic viewpoint, this is computed explicitly as follows:

$$\hat{u}_n(z_n^i, w) = \hat{\ell}(z_n^i, w), \quad z_n^i = (\pi_n(i), y_n^i) \in z_n, \ i = 1, \ldots, N_n, \ w \in \Gamma,$$

$$\hat{u}_k(z_k^i, w) = \inf_{a \in A} \left\{ \sum_{j=1}^{N_{k+1}} \hat{r}_{k+1}^{ij} \hat{u}_{k+1}(z_{k+1}^j, \text{Proj}_\Gamma(F(w, a, y_k^i, y_{k+1}^j))) \right\} \quad (34.31)$$

$$z_k^i = (\pi_k(i), y_k^i) \in z_k, \ i = 1, \ldots, N_k, \ w \in \Gamma, \ k = 0, \ldots, n-1. \quad (34.32)$$

For $w_0 \in \Gamma$, the solution $J_{opt} = u(\mu, y_0, w_0)$ to our control problem is then approximated by $J_{quant} = \hat{u}_0(\mu, y_0, w_0)$. Moreover, this backward dynamic programming scheme allows us to compute at each time $k = 0, \ldots, n-1$, an approximate control $\hat{a}_k(z, w)$, $z \in z_k$, $w \in \Gamma$, by taking the infimum in (34.31). Error estimation between J_{opt} and J_{quant} in terms of the quantization errors $\|Z_k - \hat{Z}_k\|_2$ for $Z_k = (\Pi_k, Y_k)$, the spatial step ν, and the grid size L for (W_k) is stated in [10]. By combining with Zador's theorem, this provides a rate of convergence of order $C(n)\left(\nu + \frac{1}{L} + \frac{1}{N^{\frac{1}{m-1+d}}}\right)$ where $N = N_k$, $k = 1, \ldots, n$.

34.3.5 Numerical illustrations

In the setting of the stochastic volatility model described in Section 34.3.1, we consider the mean-variance hedging of a put option. The logarithm of the observed stock price is $Y = \ln S$, its unobservable volatility is X, and the wealth process W controlled by the number of shares a invested in stock, is governed by:

$$W_{k+1} = W_k e^{r\delta} + a_k(e^{Y_{k+1}} - e^{Y_k} e^{r\delta}),$$

where r is the constant interest rate, and $\delta > 0$ is the interval between two trading dates. We assume that (X_k) is a Markov chain approximation à la Kushner [20] with spatial step Δ and with $m = 3$ states of a mean-reverting process:

$$dX_t = \lambda(x_0 - X_t)dt + \eta dW_t.$$

The dynamics of Y is given by

$$Y_{k+1} = Y_k + \left(r - \frac{1}{2}X_k^2\right)\delta + X_k\sqrt{\delta}\varepsilon_{k+1},$$

where (ε_k) is a sequence of Gaussian white noise. Given a put option of payoff $(\kappa - e^{Y_n})_+$ at maturity n, the investor's objective is defined by the control problem:

$$\inf_{a \in \mathcal{A}^Y} \mathbb{E}\left[\left((\kappa - e^{Y_n})_+ - W_n\right)^2\right].$$

We perform numerical tests with:

- Price and put option parameters: $r = 0.05$, $S_0 = 110$, $\kappa = 110$,
- Volatility parameters: $\lambda = 1$, $\eta = 0, 1$, $\Delta = 0, 05$, $X_0 = 0.15$,
- Quantization of $(Z_k) = (\Pi_k, Y_k)$: grids are of same size N fixed for each time period with step $\delta = \frac{1}{n}$. When it is not precised, we choose $n = 5$.
- Discretization of (W_k): we use a N^W-point grid defined by $\Gamma = (2\nu)\mathbb{Z} \cap [L_{inf}, L_{sup}]$ with $L_{inf} = -10$, $L_{sup} = 15$ and so $\nu = \frac{25}{2(N^W - 1)}$.
- Approximation of the optimal control: golden search method (see [25]) on $A = [-1, 1]$.

In order to study the effects of the quantization grid size N and uniform grid size N^W, we plot the graph of $w_0 \mapsto \inf_{a \in A^Y} \mathbb{E}((\kappa - e^{Y_n})_+ - W_n)^2)$ for different values of N and N^W (Figures 34.1 and 34.2). As expected, the global shape of the graph is parabolic, due to the quadratic hedging criterion that we have used. The minimum is reached at w_{min} which can be considered as the "quadratic hedging price" of our European put option. The corresponding hedging strategies are given in Table 34.1, and Figure 34.3 displays the graph of a_0 as a function of the initial wealth w_0. We can see that the strategy is nearly constant for $w_0 \in [2, 4]$, where the nonconstant values may be due to numerical imprecision. This is consistent with the theoretical result, which shows that the optimal strategy for the mean-variance hedging problem does not depend on the initial wealth when the (discounted) stock price is a martingale, which is the case here.

In Figure 34.4 and in Table 34.2, we compare the European put option price under partial and complete observation when we increase the number of observations (i.e. the time step δ decreases to zero). Denoting by $N_{\Pi,Y}$ the number of grid points used in the partial observation case to make an optimal quantization of the pair (Π, Y), by $N_{X,Y}$ the number of grid points used in the

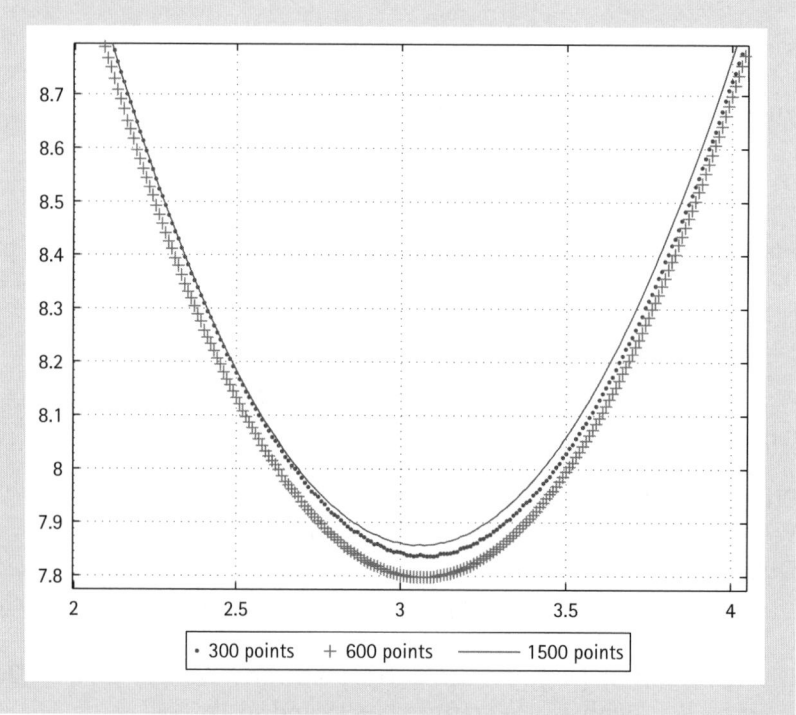

Fig. 34.1 Quadratic hedging of a European put: graph of $w_0 \mapsto \inf_{a \in A^Y} \mathbb{E}((\kappa - e^{Y_n})_+ - W_n)^2)$ for different quantification grid sizes ($N = 300, 600, 1500$) and a fixed uniform grid size ($N^W = 400$).

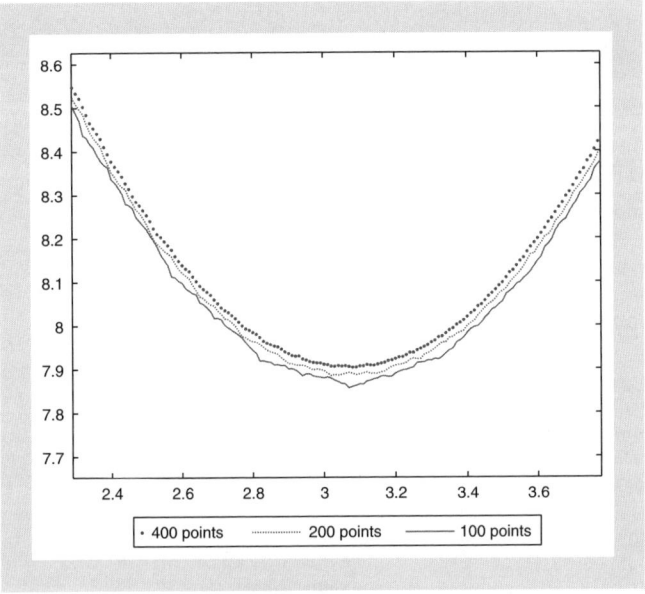

Fig. 34.2 Quadratic hedging of a European put: graph of $w_0 \mapsto \inf_{a \in A^Y} \mathbb{E}((\kappa - e^{Y_n})_+ - W_n)^2)$ for different fixed uniform grid sizes ($N^W = 50, 100, 200, 400$) and a fixed quantization grid size ($N = 300$).

Table 34.1 Quadratic hedging of a European put: European put price (defined as the initial capital minimizing the risk) and optimal control strategy calculated for different quantization grid sizes ($N = 300, 600, 1500$) and a fixed uniform grid size ($N^W = 400$).

N	European put price	Optimal control strategy a_0
300	3.04132	-0.2813
600	3.05965	-0.2813
1500	3.07098	-0.2813

total observation case to make an optimal quantization of the pair (X, Y), and by L the grid size in the discretization of the controlled variable W, we recall that the discretization error is of order

$$\left(N_{\Pi,Y}^{\frac{-1}{d+m-1}} + \nu + \frac{1}{L} \right)$$

for the partial observation case. For the total observation case we have:

$$\left(\frac{1}{N_{X,Y}} + \nu + \frac{1}{R} \right)$$

where $N_{X,Y} = mN_Y$ (see [31]). So, in order to obtain comparable results, given the uniform grid discretizing the variable W, we perform an optimal

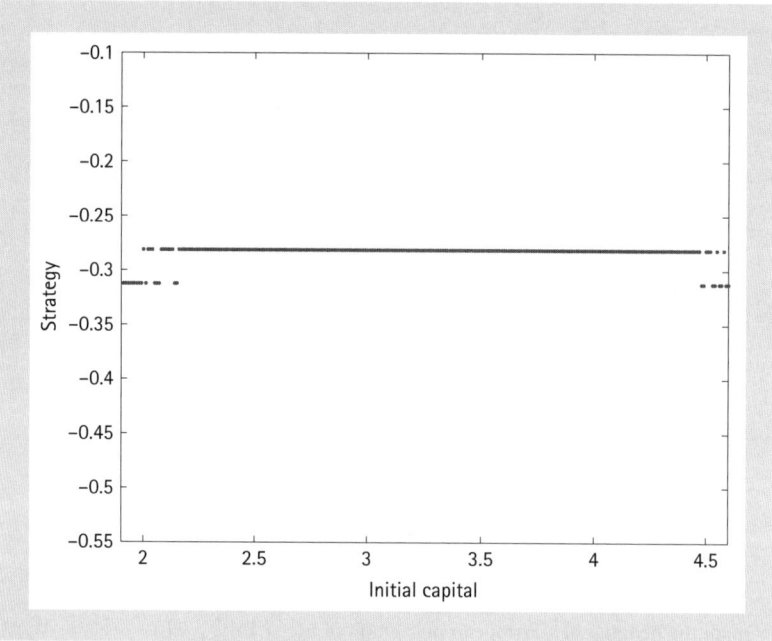

Fig. 34.3 Quadratic hedging of a European put: graph of $w_0 \mapsto a_0(w_0)$ for a quantization grid size of $N = 300$ and a fixed uniform grid size of $N^W = 400$.

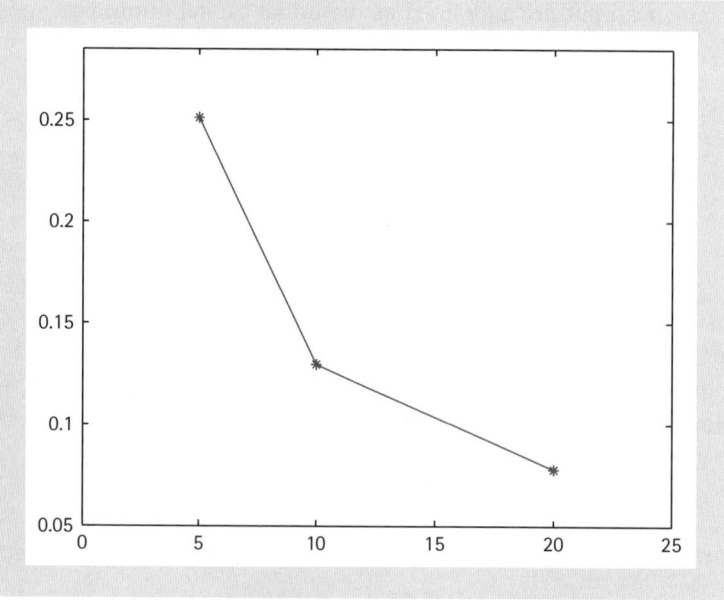

Fig. 34.4 Quadratic hedging of a European put: distance between total and partial observation European put prices (defined as the initial capital minimizing the risk) when we increase the number of observations (axis of abscissae) and consequently the time step δ goes to 0. Size grid for $W = 30$ points, size grid for $(e^Y, \Pi) = 1500$ points, size grid for $(e^Y, X) = 45$ points.

Table 34.2 Quadratic hedging of a European put: comparison between partial and total observation price (defined as the initial capital minimizing the quadratic risk) and strategies when we increase the number of observations and consequently the time step δ goes to 0. Size grid for W = 30 points, size grid for (e^Y, Π) = 1500 points, size grid for (e^Y, X) = 45 points.

Time step δ	Partial observation price	Partial observation strategy	Total observation price	Total observation strategy
1\5	2.9933	−0.2813	3.24459	−0.2734
1\10	3.5255	−0.3013	3.65515	−0.2422
1\20	3.9501	−0.3215	4.02799	−0.3614

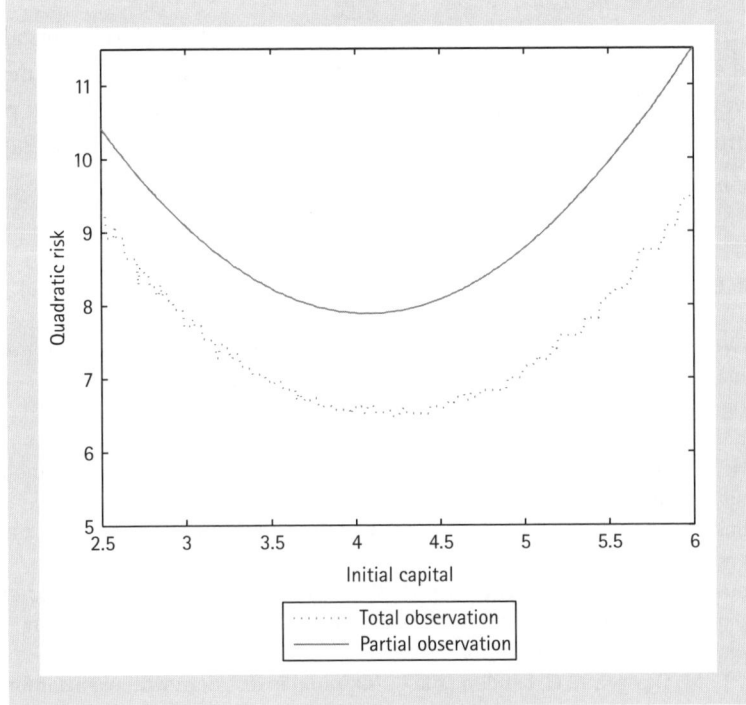

Fig. 34.5 Quadratic hedging of a European put: graph of $w_0 \mapsto \inf_{a \in A^Y} \mathbb{E}((\kappa - e^{Y_n})_+ - W_n)^2)$ in the partial and total observation case. Size grid for W = 100 points, size grid for (e^Y, Π) = 1500 points, size grid for (e^Y, X) = 45 points.

quantization of (Π, Y) and (X, Y) by using grid sizes $N_{\Pi,Y}$ and $N_{X,Y} = m N_Y$ such that:

$$N_Y \simeq N_{\Pi,Y}^{\frac{1}{d+m-1}}$$

where $d = 1$ and $m = 3$. That is why we have chosen $N_{\Pi,Y} = 1500$ and $N_{X,Y} = 45$.

We notice that when the number of observations increases (i.e. $\delta \to 0$), the partial observation price converges to the complete observation price; this is due to the fact that with observation performed in continuous time we are able to calculate the volatility given by the quadratic variation of the price process (e^Y).

Figure 34.5 shows that by working in a total observation setting the quadratic risk associated to a given initial wealth is smaller than the corresponding value obtained in the partial observation case. This is consistent with the fact that the filtration generated by the observation price is included in the full information filtration, and consequently the corresponding optimal cost function in the partial information case is larger than the one in the full information case.

34.3.6 Bibliographical notes

Optimal quantization methods were recently investigated in numerical probability and in particular in finance. We refer to Pagès, Pham, and Printems [30] for a survey on this subject. The quantization approach for the filtering problem and optimization problem under partial observation is developed in Pagès and Pham [27], and Pham, Runggaldier, and Sellami [31].

References

[1] Bally, V. and G. Pagès (2003): "A quantization algorithm for solving discrete time multi-dimensional optimal stopping problems", *Bernoulli*, **9**, 1003–49.
[2] Bauerle, N. and U. Rieder (2005a): "Portfolio optimization with unobservable Markov-modulated drift process", *Journal of Applied Probability*, **42** (2), 362–78.
[3] Bauerle, N. and U. Rieder (2005b): "Portfolio optimization with jumps and unobservable intensity process", *Mathematical Finance*, **17**(2), 205–24.
[4] Bensoussan, A. (1992): *Stochastic Control of Partially Observable Systems*, Cambridge University Press.
[5] Bensoussan, A. and W. Runggaldier (1987): "An approximation method for stochastic control problems with partial observation of the state: a method for constructing ε-optimal controls", *Acta Appli. Math.*, **10**, 145–70.
[6] Bjork, T., M. Davis, and C. Landen (2007): "Optimal investment with partial information", Preprint.
[7] Brendle, S. (2006): "Portfolio Selection under incomplete information", *Stochastic Processes and their Applications*, **116**(5), 701–23.
[8] Callegaro, G., G. B. Di Masi, and W. J. Runggaldier (2006): "Portfolio optimization in discontinuous markets under incomplete information", *Asia Pacific Financial Markets*, **13**, 373–94.
[9] Corsi, M. (2007): *Valuation and portfolio optimisation in jump-diffusion models: theoretical and numerical aspects*, Thesis, University Paris 7 Diderot.
[10] Corsi, M., H. Pham, and W. Runggaldier (2006): "Numerical approximation by quantization of control problem in finance under partial observations", to appear in *Mathematical modelling and Numerical Methods in Finance*, ed. A. Bensoussan and Q. Zhang, special volume of *Handbook of Numerical Analysis*.

[11] Cox, J. and C. F. Huang (1989): "Optimal consumption and portfolio policies when asset proces follow a diffusion process", *Journal of Economic Theory*, **49**, 33–83.
[12] Detemple, J. (1986): "Asset pricing in a production economy with incomplete information", *Journal of Finance*, **41**, 383–91.
[13] Di Masi, G., E. Platen, and W. Runggaldier (1995): "Hedging of Options under Discrete Observations on Assets with Stochastic Volatilities", *Seminar on Stoch. Anal. Rand. Fields. Appli.*, eds M. Dozzi, F. Russo, vol. 36, 359–64.
[14] Elliott, R., L. Aggoun, and J. B. Moore (1994): *Hidden Markov Models, Estimation and Control*, Springer.
[15] Frey, R. and W. Runggaldier (1999): "Risk-minimizing hedging strategies under restricted information : the case of stochastic volatility models observable only at discrete random times", *Mathematical Methods of Operations Research*, **50**, 339–50.
[16] Gennotte, G. (1986): "Optimal portfolio choice under incomplete information", *Journal of Finance*, **41**, 733–46.
[17] Hausmann, U. and J. Sass (2004): "Optimizing the Terminal Wealth under Partial Information: The Drift Process as a Continuous Markov Chain", *Finance and Stochastics*, **8**, 553–77.
[18] Karatzas, I., J. P. Lehoczky, and S. Shreve (1987): "Optimal portfolio and consumption decisions for a small investor on a finite horizon", *SIAM J. Cont. Optim.*, **25**, 1557–86.
[19] Karatzas, I. and X. Zhao (1998): "Bayesian adaptive portfolio optimization", *Handbook of Mathematical Finance*, eds J. Cvitanic, E. Jouini, M. Musiela, Cambridge University Press.
[20] Kushner, H. J. and P. Dupuis (1992): *Numerical Methods for Stochastic Control Problems in Continuous Time*, Springer.
[21] Lakner, P. (1995): "Utility maximization with partial information", *Stochastic Processes and their Applications*, **56**, 247–73.
[22] Lakner, P. (1998): "Optimal trading strategy for an investor : the case of partial information", *Stochastic Processes and their Applications*, **76**, 77–97.
[23] Lasry, J. M. and P. L. Lions (1999): "Stochastic control under partial information and applications to finance", Preprint.
[24] Liptser, R. and A. Shiryaev (1977): *Statistics of random processes I*, Springer.
[25] Luenberger, D. (1984): *Linear and Nonlinear Programming*, Addison-Wesley.
[26] Nagai, H. and W. J. Runggaldier (2008): "PDE approach to utility maximization for market models with hidden Markov factors", Seminar on Stochastic Analysis, Random Fields and Applications V (R. C.Dalang, M.Dozzi, F.Russo, eds). Progress in Probability, Vol. 59, Birkhäuser, 493–506.
[27] Pagès, G. and H. Pham (2005): "Optimal quantization methods for nonlinear filtering with discrete time observations", *Bernoulli*, **11**, 893–932.
[28] Pardoux, E. (1989): "Filtrage non linéaire et équations aux dérivées partielles stochastiques associées", *Lect. Notes in Math.*, **1464**, 67–163, Springer.
[29] Pham, H. (2001): "Mean-variance hedging under partial observation", *International Journal of Theoretical and Applied Finance*, **4**, 263–84.
[30] Pagès, G., H. Pham, and J. Printems (2004): "Optimal quantization methods and applications to numerical problems in finance", *Handbook of Computational and Numerical Methods in Finance*, ed. S. Rachev, Birkhäuser.
[31] Pham, H., W. Runggaldier, and A. Sellami (2005): "Approximation by quantization of the filter process and applications to optimal stopping problems under partial observation", *Monte Carlo Methods and Applications*, **11**, 57–82.
[32] Pham, H. and M. C. Quenez (2001): "Optimal portfolio in a partially observed stochastic volatility model", *Annals of Applied Probability*, **11**, 210–38.

[33] Rishel, R. (1999): "Optimal portfolio management with partial observation and power utility function", in *Stochastic Analysis, Control, Optimisation and Applications, Vol. in honour of W. Fleming*, eds W. McEneaney, G. Yin, Q. Zhang, 605–20.

[34] Runggaldier, W. and L. Stettner (1994): *Approximations of Discrete Time Partially Observed Control Problems, Applied Mathematics Monographs*, Vol. 6, Giardini Editori e Stampatori in Pisa.

·35·

Filtering with Counting Process Observations: Application to the Statistical Analysis of the Micromovement of Asset Price

L. C. Scott and Y. Zeng

35.1 Introduction

35.1.1 Intuition behind the modeling

Existing asset price models can be broadly divided into two groups: macro- and micro-movement models. Daily, weekly, and monthly closing price behaviors are regarded as macromovement. Figure 35.1, the time series plot of the daily closing prices of MicroSoft during 1/1/1993–3/31/1994, is an example of the macromovement. There were 316 business dates and there are 316 data. Clearly, the price fluctuates and is usually modelled by either discrete-time models such as ARIMA or ARCH-type models or continuous-time models such as GBM (geometric Brownian motion), SV (stochastic volatility), or jump-diffusion models in econometric or mathematical finance literature.

Micromovement refers to trade-by-trade, transactional price behaviour. Such data, containing the trading times and prices for all trades, are regarded as UHF (*ultra high frequency*) data by Engle [10]. The first stylized fact of UHF data is the trading times or *durations* (that is, the inter-trading times) are random.[1] Figure 35.2 plots all the transaction prices of MicroSoft during the same period. There are about 480,000 transaction prices, much more than the 316 data in the macromovement. First we observe that the overall shapes of these two plots are the same and this is not surprising because Figure 35.1 is a daily subsample of Figure 35.2. This implies that the micromovement model should be closely related to the macromovement model. Second, we observe that there are much more fluctuations in the micromovement and by just looking at these two pictures, it is very tempting to conclude that the micromovement looks even more like a GBM. However, if we cut a small piece of Figure 35.2 and look through it under a microscope, namely, we plot about one-half day of the prices in Figure 35.3, then we clearly see that the price does not move continuously in the state space as GBM suggests, but moves tick-by-tick or level-by-level. Here, a tick is the minimum price variation set by trading regulations and was $1/8

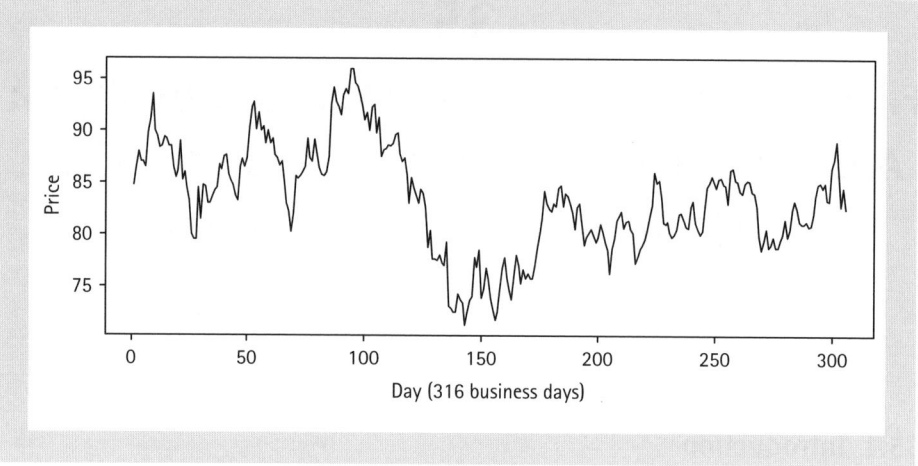

Fig. 35.1 Daily closing prices of MicroSoft, 93.01.01–94.03.31

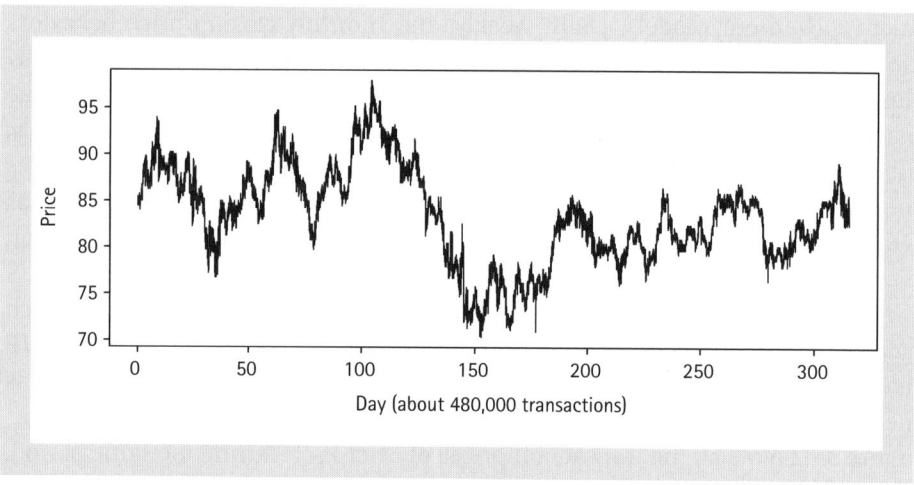

Fig. 35.2 Transaction data of MicroSoft, 93.01.01–94.03.31

dollar at that time. Moreover, Figure 35.3 clearly shows the second stylized fact of UHF data. Namely, there are trading noises in UHF prices. Trading noises include discrete noise due to the price discreteness, clustering noise due to price clustering (more prices were traded at even eighths than at odd eighths), and nonclustering noise which includes all other noise. The down spike in Figure 35.3 is an evidence of the existence of the nonclustering noise. Finally, the topics of the two recent presidential addresses to the American Finance Association were "noise" (Black [4]) and "friction" (Stoll [26]). Both are about market microstructure noise. This indicates that noise is an essential matter in finance and in the modelling of asset price.

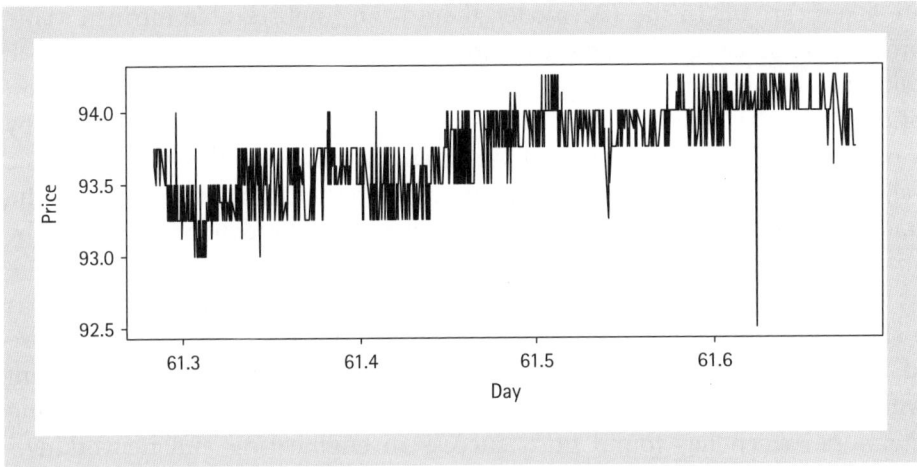

Fig. 35.3 About one-half day's transaction data of MicroSoft

By looking at these three pictures, we gain the simple intuition that a micromovement model should be built upon a macromovement model by incorporating trading noises. In the other words of economics, the trade-by-trade price is formed from the intrinsic value process of an asset by combining the market microstructure noises. Also, we can see that when we deal with macromovement or daily prices, the noises are negligible. However, when we deal with micromovement or UHF prices, the noises are not negligible anymore. One representation of the model reviewed in this paper is built upon such intuition. Another representation comes from Figure 35.3, where each price level is modelled by a counting process.

35.1.2 Related literature on nonlinear filtering for UHF data

To the best of our knowledge, Frey [14] and Frey and Runggaldier [15] are the first papers that employ the nonlinear filtering technique to model UHF data. Their viewpoint is to model the unobserved volatility process, which is crucial for option pricing. Their model is able to capture the Poisson random arrival times in UHF data. Cvitanic, Liptser, and Rozovskiĭ [6] extends the previous model to a more general setting and further allowing general random times of observation, not just doubly stochastic Poisson processes. Cvitanic, Rozovskiĭ, and Zaliapin [7] numerically implements the filtering procedure and estimates the unobserved volatility. However, market microstructure noise is missing in these models.

35.1.3 An overview of this chapter

Zeng [30] develops a general filtering micromovement model (FM model, as we simply call it) for asset price, where both stylized facts of micromovement are taken care of. In the FM model, there is an unobservable intrinsic value process for an asset, which corresponds to the macromovement. The intrinsic value process is the permanent component and has a long-term impact on price. Prices are observed only at random trading times which are driven by a conditional Poisson process, whose intensity may depend on the intrinsic value. Prices are distorted observations of the intrinsic value process at the trading times. Market microstructure noise is explicitly and flexibly modelled by a random transformation with a transition probability from the intrinsic value to the price at trading time. Noise is the transient component and only has a short-term impact (when a trade happens) on price. One important feature of the FM model is that the model can be framed as a filtering problem with counting process observations. This connects the model to the filtering literature, which has found great success in engineering and networking.[2] Then, the continuous-time likelihoods and posterior not only exist, but also are uniquely characterized, respectively, by the unnormalized Duncan–Mortensen–Zakai-like filtering equation and the normalized, Kushner–Stratonovich (KS) (or Fujisaki–Kallianpur–Kunita)-like filtering equations. The related numerical solution based on Markov chain approximation method and the Bayes estimation via filtering for the intrinsic value process and the related parameters in the model are developed. Furthermore, Kouritzin and Zeng [22] characterizes the continuous-time likelihood ratio and Bayes factors by a system of evolution equations and develops the Bayesian hypothesis testing or model selection via filtering for the FM model.

The first aim of this chapter is to survey the general filtering model with counting process observations for the micromovement of asset price and its related statistical analysis recently developed in [30] and [22]. The statistical analysis contains Bayesian inference (estimation and model selection) via filtering for the FM model. We adopt the Bayesian paradigm, because it offers extra model flexibility as well as the ability to incorporate real prior information. The second aim is to give a worked example to help those who wish to carry out a similar analysis in practice. A specific FM model with LSDE as the intrinsic value process is built up to show the modelling development technique. This FM model is further employed to show the steps to carry out Bayes estimation via filtering as well as to compute Bayes factors for model selection. Namely, we construct recursive algorithms for computing the trade-by-trade Bayes estimates and Bayes factors. The consistency of the recursive algorithms are established. We provide simulation results to show the consistency of Bayes estimates and the effectiveness of Bayes factors for model selection. The recursive algorithms are applied to an actual Microsoft data set to obtain trade-by-trade Bayes parameter estimates as well as to implement a simple model selection.

In Section 35.2, we present the general FM model in three equivalent fashions. In Section 35.3, we present the continuous-time Bayesian inference via filtering for the FM model including the filtering equations and the consistency theorem. In Section 35.4, we focus on developing a specific FM model built on LSDE and developing the related recursive algorithms. Simulation and real-data examples are provided for Bayes estimation and Bayesian model selection. Section 35.5 concludes.

35.2 The general FM model

The general FM model has three equivalent representations.

35.2.1 Representation I: constructing price from intrinsic value

Based on the simple intuition obtained in Section 35.1.1 that the price is formed from an intrinsic value by incorporating the noises that arise from the trading activity, we build up the FM model.

In general, there are three steps in constructing the price process Z from the intrinsic value process X. First, we specify X. In order to permit time-dependent parameters such as stochastic volatility and to prepare for parameter estimation, we enlarge the partially-observed model (X, Y) to (θ, X, Y). Assume (θ, X, Y) is defined in a complete probability space $(\Omega, \mathcal{F}, \mathbb{P})$ with a filtration $\{\mathcal{F}_t\}_{0 \leq t \leq \infty}$. Since the macromovement models are appropriate for the value processes, we invoke below a mild assumption on (θ, X) so that all relevant stochastic processes are included.

Assumption 35.1 (θ, X) is the unique solution of a martingale problem for a generator $\mathbf{A}$ such that for a function f in the domain of $\mathbf{A}$, $M_f(t) = f(\theta(t), X(t)) - \int_0^t \mathbf{A} f(\theta(s), X(s)) ds$, is a $\mathcal{F}_t^{\theta, X}$-martingale, where $\mathcal{F}_t^{\theta, X}$ is the σ-algebra generated by $(\theta(s), X(s))_{0 \leq s \leq t}$.

The generator and martingale problem approach (see for example Ethier and Kurtz [12]) furnishes a powerful tool for the characterization of Markov processes. Assumption 35.1 includes all relevant stochastic processes such as diffusion and jump-diffusion processes for modelling asset price. One examples is given in Section 35.4.

However, in UHF data, the price can not be observed continuously in time, neither can it move continuously as GBM suggests. Therefore, two more steps are necessary. Step 2 takes care of the trading times and Step 3 the trading noise.

In Step 2, we assume trading times $t_1, t_2, \ldots, t_i, \ldots$, are driven by a conditional Poisson process with an intensity $a(X(t), \theta(t), t)$. In Step 3, $Y(t_i)$, the price at time t_i, is corrupted from $X(t_i)$, the intrinsic value, with trading noise. Namely, $Y(t_i) = F(X(t_i))$, where $y = F(x)$ is a random transformation with the transition probability $p(y|x)$, modeling the trading noise. The random transformation, $F(x)$, is flexible and Section 35.4.1 constructs one to accommodate the three observed types of noise in the MicroSoft data: discrete, clustering, and nonclustering.

Under this construction, information affects $X(t)$, the value of an asset, and has a permanent influence on the price while noise modelled by $F(x)$ (or $p(y|x)$) only has a transitory impact on price. The formulation is similar to the time series VAR structural models used in many market microstructure papers (see a survey paper [17] by Hasbrouck and a recent paper [18]). Furthermore, the formulation is closely related the recent two-scale frameworks incorporating market microstructure noises in literature of realized volatility estimators. See [34], [1], [2], and [13]. Especially, Li and Mykland in [24] shows that rounding noise in UHF data may severely distort even the two-scale estimators of realized volatility, and the error could be infinite.

35.2.2 Representation II: filtering with counting process observations

From Figure 35.3 and because of price discreteness, we can formulate the prices of an asset as a collection of counting processes in the following form:

$$\vec{Y}(t) = \begin{pmatrix} N_1\left(\int_0^t \lambda_1(\theta(s), X(s), s)ds\right) \\ N_2\left(\int_0^t \lambda_2(\theta(s), X(s), s)ds\right) \\ \vdots \\ N_n\left(\int_0^t \lambda_n(\theta(s), X(s), s)ds\right) \end{pmatrix}, \quad (35.1)$$

where $Y_j(t) = N_j\left(\int_0^t \lambda_j(\theta(s), X(s), s)ds\right)$ is the counting process recording the cumulative number of trades that have occurred at the jth price level (denoted by y_j) up to time t. We make four more mild assumptions on the model.

Assumption 35.2 $\{N_j\}_{j=1}^n$ are unit Poisson processes under measure $\mathbb{P}$.

Then, $Y_j(t) = N_j\left(\int_0^t \lambda_j(\theta(s), X(s), s)ds\right)$ is conditional Poisson process with the stochastic intensity, $\lambda_j(\theta(t), X(t), t)$. Given $\mathcal{F}_t^{\theta, X}$, the filtration of θ and X, $Y_j(t)$ has a Poisson distribution with parameter $\int_0^t \lambda_j(\theta(s), X(s), s)ds$. Moreover, $Y_j(t) - \int_0^t \lambda_j(\theta(s), X(s), s)ds$ is a $\mathcal{F}_t^{\theta, X, Y}$-martingale.

Assumption 35.3 (θ, X), $N_1, N_2, \ldots, N_n$ are independent under measure $\mathbb{P}$.

Let $a(\theta, X(t), t)$ be the total trading intensity at time t.

Assumption 35.4 There exists a positive constant, C, such that $0 \le a(\theta, x, t) \le C$ for all $t > 0$ and (θ, x).

The total trading intensity $a(\theta, x, t)$, which is bounded by the above assumption, determines the expected rate of trading at time t. These three assumptions imply that there exists a reference measure Q and that after a suitable change of measure to Q, $(\theta, X), Y_1, \ldots, Y_n$ become independent, and $Y_1, Y_2, \ldots, Y_n$ become unit Poisson processes (Bremaud [5]).

Assumption 35.5 The intensities are of the form: $\lambda_j(\theta, x, t) = a(\theta, x, t) p(y_j|x)$, where $p(y_j|x)$ is the transition probability from x to y_j, the jth price level.

This assumption imposes a desirable structure for the intensities of the model. It means that the total trading intensity $a(\theta(t), X(t), t)$ determines when the next trade will occur and $p(y_j|X(t))$ determines at which price level the next trade will occur given the value is $X(t)$. Note that $p(y_j|X(t))$ models how the trading noise enters the price process.

Under this representation, $(\theta(t), X(t))$ becomes the signal, which cannot be observed directly, but can be partially observed through the counting processes, $\vec{Y}(t)$, which is distorted by trading noise, modelled by $p(y_j|x)$. Hence, $(\theta, X, \vec{Y})$ is framed as a *filtering model with counting process observations*.

35.2.3 An integral form of the price

To solve problems in mathematical finance such as the option pricing and hedging and the portfolio selection, the stochastic differential or integral equation form of the most recent price is needed. However, the previous two representations do not provide such a form of price. That is why a third representation is given below. There are also three steps in constructing such representation.

Step 1: We define a random counting measure. Let $U = \{0, \frac{1}{M}, \frac{2}{M}, \cdots\}$ be a mark space containing all the possible price levels, and $u = \frac{j}{M}$ be a generic point in U. Note that M can be 8, 16, 64, 100, 128, or 256 or others according to the asset. For stock price, the tick size was 1/8, 1/16 and it is 1/100 of one dollar. For treasure note, the tick sizes are 1/64, 1/128, or 1/256 of one percentage.

For $A \in U$, we define $m(A, t)$ as a random counting measure that counts the cumulative number of trades whose price levels are in A up to time t. When $A = \{\frac{j}{M}\}$, $m(\{\frac{j}{M}\}, t) = Y_j(t)$, which counts the number of trades occurring at the j price level. More generally, we can express $m(A, t) = \sum_{u \in A} Y_{uM}(t)$.

A random measure is characterized by its compensator. To express the compensator of $m(A, t)$, we define a counting measure $\eta(A) = $ *number of element in A* for $A \in U$. Note that η has the following two properties: For $A \in U$, $\eta(A) = \int I_A \eta(du)$ and $\int_A f(u) \eta(du) = \sum_{u \in A} f(u)$.

Step 2: We write $\gamma_m(A, t)$, the compensator of $m(A, t)$ with respect to $\mathcal{F}_t^X$, as

$$\gamma_m(A, t) = \int_0^t \int_A p(u|X(s)) a(\theta, X(s), s) \eta(du) ds$$

$$= \sum_{u \in A} \int_0^t p(u|X(s))a(\theta, X(s), s)ds.$$

Again, when $A = \{\frac{j}{M}\}$,

$$\gamma_m(\{\frac{j}{M}\}, t) = \int_0^t \lambda_j(\theta, X(s), s)ds = \int_0^t p(\frac{j}{M}|X(s))a(\theta, X(s), s)ds.$$

Moreover, we can write

$$\gamma_m(du, dt) = p(u|X_t)a(\theta, X_t, t)\eta(du)dt.$$

Step 3: We are in the position to define the integral form needed. Let $Y(t)$ be the price of the most recent transaction at or before time t. Then,

$$Y(t) = Y(0) + \int_{[0,t] \times U} (u - Y(s-))m(du, ds). \tag{35.2}$$

Note that $m(du, ds)$ is zero most of time, and becomes one only at trading time t_i with $u = Y(t_i)$, the trading price. The above expression is but a telescoping sum: $Y(t) = Y(0) + \sum_{t_i < t}(Y(t_i) - Y(t_{(i-1)}))$. Alternatively, in differential form,

$$dY(t) = \int_U (u - Y(t-))m(du, dt). \tag{35.3}$$

To understand the above differential equation, assuming there is a price change from $Y(t-)$ to u occurs at time t, then $Y(t) - Y(t-) = (u - Y(t-))$ implying $Y(t) = u$.

With the above representation of price, Lee and Zeng [23] studies the option pricing and hedging through local risk minimizing criterion, and Xiong and Zeng [29] studies the portfolio selection problem.

35.2.4 The equivalence of the three representations

In Representation I, the price is constructed from the intrinsic value. In Representation II, the price process is a collection of counting processes. In Representation III, the SDE form of the price is given. The following proposition states their equivalence in distribution. This guarantees the statistical inference based on the second representation is also valid to the other two.

Proposition 35.1 *The three representations of the model in Sections 35.2.1, 35.2.2, and 35.2.3, respectively, have the same probability law.*

The proof of the equivalence of Representations of I and II can be found in [32]. The intensity structure of Assumption 5 plays an essential role in the equivalence of the two approaches of modeling. The main idea is to show both marked point processes have the same stochastic intensity kernel. Similarly, we can show the third one is equivalent with the first two.

35.3 Bayesian inference via filtering

This section summarizes the theoretical results of the Bayesian inference via filtering. We present the statistical foundations for the general FM model, the related filtering equations, and a convergence theorem. The convergence theorem not only provides a blueprint through the Markov chain approximation method to construct recursive algorithms, but also ensures the consistency of such algorithms, which compute the approximate continuous-time likelihoods, the posterior, and the Bayes factors.

35.3.1 The statistical foundations

In this section, we study the continuous-time joint likelihood, the likelihood function (from frequentists' viewpoint), the integrated likelihood (from Bayesians' viewpoint), the posterior of the proposed model, as well as the continuous-time likelihood ratio and Bayes factors for model selections. These terminologies are used in statistics. To connect to the terminologies used in the filtering community, the continuous-time joint likelihood corresponds to the Girsanov-type exponential martingale, the likelihood to the total unnormalized conditional measure, the posterior to the conditional distribution (or the normalized conditional measure), and the Bayes factors to the total conditional ratio measures. These conditional measures are characterized by the unnormalized and normalized filtering equations as well as the system of evolution equations, respectively.

The continuous-time joint likelihood

The probability measure $\mathbb{P}$ of $(\theta, X, \vec{Y})$ can be written as $\mathbb{P} = \mathbb{P}_{\theta,x} \times \mathbb{P}_{y|\theta,x}$, where $\mathbb{P}_{\theta,x}$ is the probability measure for (θ, X) such that $M_f(t)$ in Assumption 35.1 is a $\mathcal{F}_t^{\theta,X}$-martingale, and $\mathbb{P}_{y|\theta,x}$ is the conditional probability measure on $D_{R^n}[0, \infty)$ for $\vec{Y}$ given (θ, X) (where $D_{R^n}[0, \infty)$ is the space of right continuous with left limit functions). Under $\mathbb{P}$, $\vec{Y}$ relies on (θ, X). Recall that there exists a reference measure Q such that under Q, (θ, X), $\vec{Y}$ become independent, (θ, X) remains the same probability law and $Y_1, Y_2, \ldots, Y_n$ become unit Poisson processes. Therefore, Q can be decomposed as $Q = \mathbb{P}_{\theta,x} \times Q_y$, where Q_y is the probability measure for n independent unit Poisson processes. One can obtain the Radon–Nikodym derivative of the model, that is the joint likelihood of $(\theta, X, \vec{Y})$, $L(t)$, as (see [5] pg 166),

$$L(t) = \frac{d\mathbb{P}}{dQ}(t) = \frac{d(\mathbb{P}_{\theta,x} \times \mathbb{P}_{y|\theta,x})}{d(\mathbb{P}_{\theta,x} \times Q_y)}(t) = \frac{d\mathbb{P}_{y|\theta,x}}{dQ_y}(t)$$

$$= \prod_{j=1}^{n} \exp\left\{\int_0^t \log \lambda_j(\theta(s-), X(s-), s-)dY_j(s) - \int_0^t \left[\lambda_j(\theta(s), X(s), s) - 1\right] ds\right\}.$$

(35.4)

or in SDE form:

$$L(t) = 1 + \sum_{j=1}^{n} \int_0^t \left[\lambda_j(\theta(s-), X(s-), s-) - 1\right] L(s-) d(Y_j(s) - s).$$

The continuous-time likelihoods of $\vec{Y}$

It is clear that we can not observe X (and the stochastic components of θ such as stochastic volatility if exists in the intrinsic value process) and the joint likelihood is not computable. To do statistical analysis, what we need is the likelihood of $\vec{Y}$ alone. Therefore, we would like to integrate out X. In the expression of probability, this can be done by using conditional expectation. Let $\mathcal{Y}_t = \sigma\{(\vec{Y}(s)) | 0 \leq s \leq t\}$ be all the available information up to time t. We use $E^Q[X]$ and $\mathbb{E}[X]$ to indicate that the expectation is taken with respect to the measures Q and $\mathbb{P}$, respectively.

Definition 35.1 Let ρ_t be the conditional measure of $(\theta(t), X(t))$ given $\mathcal{Y}_t$ defined as

$$\rho_t\{(\theta(t), X(t)) \in A\} = E^Q\left[\mathbf{I}_{\{(\theta(t), X(t)) \in A\}}(\theta(t), X(t)) L(t) | \mathcal{Y}_t\right].$$

Definition 35.2 Let

$$\rho(f, t) = E^Q[f(\theta(t), X(t)) L(t) | \mathcal{Y}_t] = \int f(\theta, x) \rho_t(d\theta, dx).$$

If $(\theta(0), X(0))$ is given, then the likelihood of Y is $E^Q[L(t) | \mathcal{Y}_t] = \rho(1, t)$, the total conditional measure. In a Bayesian framework, a prior is placed on $(\theta(0), X(0))$, and the integrated (or marginal) likelihood of Y is also $\rho(1, t)$.

The continuous-time posterior

Definition 35.3 Let π_t be the conditional distribution of $(\theta(t), X(t))$ given $\mathcal{Y}_t$ and let

$$\pi(f, t) = \mathbb{E}[f(\theta(t), X(t)) | \mathcal{Y}_t] = \int f(\theta, x) \pi_t(d\theta, dx).$$

Again, in a Bayesian framework, a prior is placed on $(\theta(0), X(0))$, and π_t becomes the continuous-time posterior, which is determined by $\pi(f, t)$ for all continuous and bounded f. Bayes Theorem (see [5], page 171) provides the relationship between $\rho(f, t)$ and $\pi(f, t)$: $\pi(f, t) = \rho(f, t)/\rho(1, t)$. That is, $\pi(f, t)$ is the normalized conditional measure. Hence, the equation governing the evolution of $\rho(f, t)$ is called the *unnormalized filtering equation*, and that of $\pi(f, t)$ is called the *normalized filtering equation*.

Continuous-time Bayes factors

Given one data set, there are many different models[3] to fit the data set. Bayes factor is a model selection criterion, first developed by Jeffreys [19] in a Bayesian

Table 35.1 Interpretation of Bayes factor

B_{21}	Evidence against Model 1
1 to 3	Not worth more than a bare mention
3 to 20	Positive
20 to 150	Strong
> 150	Decisive

framework. Suppose there are two models. Bayes factor quantifies the evidence provided by the data in favor of Model 1 over Model 2. For $c = 1, 2$, denote Model c by $(\theta^{(c)}, X^{(c)}, \vec{Y}^{(c)})$. Denote the joint likelihood of Model c by $L^{(c)}(t)$, as in Equation (35.4). Denote the unnormalized conditional measure of Model c as $\rho_c(f_c, t) = E^{Q^{(c)}}[f_c(\theta^{(c)}(t), X^{(c)}(t))L^{(c)}(t)|\mathcal{Y}_t^{(c)}]$. Then, the integrated likelihood of $\vec{Y}$ is $\rho_c(1, t)$, for Model c.

Jeffreys [19] defined the Bayes factor of Model 2 over Model 1, B_{21}, as the ratio of integrated likelihoods of Model 2 over Model 1. In our setting, that is, $B_{21}(t) = \rho_2(1, t)/\rho_1(1, t)$. Bayes Factors is designed to measure the relative fit of one model vs. another one given the observed, or in our case partially-observed, data. To select a "best" model among a set of models, Bayes factor can achieve this goal via pairwise comparison. When B_{21} has been calculated, it can be interpreted using Table 35.1 furnished by Kass and Raftery [20] as guideline. Similarly, we can define B_{12}, the Bayes factor of Model 1 over Model 2.

In order to characterize the evolution of the Bayes factors, we would like to define two more conditional measures. Instead of normalizing by its integrated likelihood to obtain an normalized conditional measure, we let the conditional measure ρ_c be divided by the integrated likelihood of the other model. In this way, we define two conditional ratio measures as below.

Definition 35.4 For $c = 1, 2$, let $q_t^{(c)}$ be the conditional ratio measure of $(\theta^{(c)}(t), X^{(c)}(t))$ given $\mathcal{Y}_t^{(c)}$:

$$q_t^{(c)}\{(\theta^{(c)}(t), X^{(c)}(t)) \in A\}$$
$$= \frac{E^Q\left[\mathbf{I}_{\{(\theta^{(c)}(t), X^{(c)}(t)) \in A\}}(\theta^{(c)}(t), X^{(c)}(t))L^{(c)}(t)|\mathcal{Y}_t^{(c)}\right]}{\rho_{3-c}(1, t)}$$
$$= E^Q\left[\mathbf{I}_{\{(\theta^{(c)}(t), X^{(c)}(t)) \in A\}}(\theta^{(c)}(t), X^{(c)}(t))\tilde{L}^{(c)}(t)|\mathcal{Y}_t^{(c)}\right]$$

where $\tilde{L}^{(c)}(t) = L^{(c)}(t)/\rho_{3-c}(1, t)$.

The second equality is because $\rho_c(1, t)$, $c = 1, 2$, depend on the same filtration $\mathcal{F}^{Y(1)} = \mathcal{F}^{Y(2)}$ and $\rho_{3-c}(1, t)$ can be moved inside the conditioning.

Definition 35.5 Let the conditional ratio processes for f_1 and f_2 be:

$$q_1(f_1, t) = \frac{\rho_1(f_1, t)}{\rho_2(1, t)}, \text{ and } q_2(f_2, t) = \frac{\rho_2(f_2, t)}{\rho_1(1, t)}.$$

The reason that we define the two conditional ratio measure is given by the observation that the Bayes factors can be expressed by $B_{12}(t) = q_1(1, t)$ and $B_{21}(t) = q_2(1, t)$. Moreover, observe that $q_c(f_c, t)$ can be written as $q_c(f_c, t) = \int f_c(\theta^{(c)}, x^{(c)}) q_t^{(c)}(d\theta^{(c)}, dx^{(c)})$. The integral forms of $\rho(f, t)$, $\pi(f, t)$, and $q_c(f_c, t)$ are important in deriving the recursive algorithms where f (or f_c) is taken to be a lattice-point indicator function.

Using Bayes factors for model selection has at least two advantages over the likelihood-based approaches. First, unlike likelihood-based model selection approaches, a Bayes factor neither requires the models to be nested, nor does it require the probability measures of the models to be absolutely continuous with respect to those of the total models in hypotheses. This feature is important for the model selection of stochastic process, since absolute continuity is not as common in the probability measures of continuous-time stochastic processes as those of discrete-time ones. Second, Kass and Raftery [20] show that under some conditions, Bayes factor $\approx$ BIC (Bayesian Information Criterion), which penalizes according to both the number of parameters and the number of data. This suggests that Bayes' factor might have this desirable property also.

35.3.2 Filtering and evolution equations

The stochastic partial differential equations (SPDEs) provide an powerful machinery to characterize the infinite dimensional conditional measures. Similar filtering or related equations for the related filtering problems with counting process observations can be found, for example, in [27], [28], [8], [25] [21], and [11]. The conditional measures, in turn, determine the continuous-time likelihoods, posteriors, likelihood ratios and Bayes factors. The following two theorems summarize all the useful SPDEs.

Theorem 35.1 *Suppose that $(\theta, X, \vec{Y})$ satisfies Assumptions 35.1–35.5. Then, ρ_t is the unique measure-valued solution of the SPDE, the unnormalized filtering equation,*

$$\rho(f, t) = \rho(f, 0) + \int_0^t \rho(\mathbf{A}f, s)ds + \sum_{j=1}^n \int_0^t \rho((ap_j - 1)f, s-)d(Y_j(s) - s), \quad (35.5)$$

for $t > 0$ and $f \in D(\mathbf{A})$, the domain of generator $\mathbf{A}$, where $a = a(\theta(t), X(t), t)$, is the trading intensity, and $p_j = p(y_j|X(t))$ is the transition probability from $X(t)$ to y_j.

π_t is the unique measure-valued solution of the SPDE, the normalized filtering equation,

$$\pi(f, t) = \pi(f, 0) + \int_0^t \pi(\mathbf{A}f, s)ds$$
$$+ \sum_{j=1}^n \int_0^t \left[\frac{\pi(fap_j, s-)}{\pi(ap_j, s-)} - \pi(f, s-) \right] d\left(Y_j(s) - \pi(ap_j, s)s\right). \quad (35.6)$$

Moreover, when the trading intensity is deterministic, that is, $a(\theta(t), X(t), t) = a(t)$, the normalized filtering equation is simplified as

$$\pi(f, t) = \pi(f, 0) + \int_0^t \pi(\mathbf{A}f, s)ds + \sum_{j=1}^n \int_0^t \left[\frac{\pi(fp_j, s-)}{\pi(p_j, s-)} - \pi(f, s-) \right] dY_j(s). \quad (35.7)$$

In Theorem 35.1, the unnormalized filtering equation characterizes the evolution of the conditional measure, and its total conditional measure, which is the likelihoods. The normalized filtering equation characterizes the evolution of the normalized conditional measure, which is the posteriors in a Bayesian framework.

Theorem 35.2 *Suppose Model c ($c = 1, 2$) has generator $\mathbf{A}^{(c)}$ for $(\theta^{(c)}, X^{(c)})$, the trading intensity $a_c = a_c(\theta^{(c)}(t), X^{(c)}(t), \vec{Y}^{(c)}(t))$, and the transition probability $p_j^{(c)} = p^{(c)}(y_j|x)$ from x to y_j for the random transformation $F^{(c)}$. Suppose that $(\theta^{(c)}, X^{(c)}, \vec{Y}^{(c)})$ satisfies Assumptions 35.1–35.5. Then, $(q_t^{(1)}, q_t^{(2)})$ are the unique measure-valued pair solution of the following system of SPDEs,*

$$q_1(f_1, t) = q_1(f_1, 0) + \int_0^t q_1(\mathbf{A}^{(1)} f_1, s)ds$$
$$+ \sum_{j=1}^n \int_0^t \left[\frac{q_1(f_1 a_1 p_j^{(1)}, s-)}{q_2(a_2 p_j^{(2)}, s-)} q_2(1, s-) - q_1(f_1, s-) \right] d(Y_j(s) - \frac{q_2(a_2 p_j^{(2)}, s)}{q_2(1, s)} ds) \quad (35.8)$$

$$q_2(f_2, t) = q_2(f_2, 0) + \int_0^t q_2(\mathbf{A}^{(2)} f_2, s)ds$$
$$+ \sum_{j=1}^n \int_0^t \left[\frac{q_2(f_2 a_2 p_j^{(2)}, s-)}{q_1(a_1 p_j^{(1)}, s-)} q_1(1, s-) - q_2(f_2, s-) \right] d(Y_j(s) - \frac{q_1(a_2 p_j^{(2)}, s)}{q_1(1, s)} ds) \quad (35.9)$$

for all $t > 0$ and $f_c \in D(\mathbf{A}^{(c)})$ for $k = 1, 2$. When $a_1(\theta^{(1)}(t), X^{(1)}(t), t) = a_2(\theta^{(2)}(t), X^{(2)}(t), t) = a(t)$, the above two equations are simplified to

$$q_1(f_1, t) = q_1(f_1, 0) + \int_0^t q_1(A^{(1)} f_1, s) ds$$
$$+ \sum_{j=1}^n \int_0^t \left[\frac{q_1(f_1 p_j^{(1)}, s-)}{q_2(p_j^{(2)}, s-)} q_2(1, s-) - q_1(f_1, s-) \right] dY_j(s), \quad (35.10)$$

$$q_2(f_2, t) = q_2(f_2, 0) + \int_0^t q_2(A^{(2)} f_2, s) ds$$
$$+ \sum_{j=1}^n \int_0^t \left[\frac{q_2(f_2 p_j^{(2)}, s-)}{q_1(p_j^{(1)}, s-)} q_1(1, s-) - q_2(f_2, s-) \right] dY_j(s). \quad (35.11)$$

In Theorem 35.2, the system of evolution equations for $q_c(f_c, t)$, $c = 1, 2$, characterizes the evolution of the conditional ratio measures and their total measures, namely, the likelihood ratios or the Bayes factors.

The proof of Theorem 35.1 is in [30] and that of Theorem 35.2 is [22]. Note that all the unsimplified filtering equations are rewritten in the semimartingale form.

Note that $a(t)$ disappears in (35.7), (35.10), and (35.11). This reduces the computation greatly in computing the Bayes estimates and Bayes factors. The tradeoff of taking a_i independent of $(\theta^{(c)}, X^{(c)})$ is that the relationship between trading intensity and other parameters (such as stochastic volatility) is excluded.

Let the trading times be $t_1, t_2, \ldots$, then, for example, Equation (35.7) can be written in two parts. The first is called the *propagation equation*, describing the evolution without trades and the second is called the *updating equation*, describing the update when a trade occurs. The propagation equation has no random component and is written as

$$\pi(f, t_{i+1}-) = \pi(f, t_i) + \int_{t_i}^{t_{i+1}-} \pi(Af, s) ds. \quad (35.12)$$

This implies that when there are no trades, the posterior evolves deterministically.

Assume the price at time t_{i+1} occurs at the jth price level, then the updating equation is

$$\pi(f, t_{i+1}) = \frac{\pi(fp_j, t_{i+1}-)}{\pi(p_j, t_{i+1}-)}. \quad (35.13)$$

It is random because the price level j, which is the observation, is random. Similarly, the equations for Bayes factors can be written in such two parts.

35.3.3 A convergence theorem and recursive algorithms

Theorems 35.1 and 35.2 provide the evolutions of the continuous-time versions, which are all infinite dimensional. To compute them, one needs to reduce the infinite dimensional problem to a finite dimensional problem and constructs algorithms based on it. The algorithms, based on the evolutions of SPDEs, are naturally recursive, handling a datum at a time. Moreover, the algorithms are easily parallelizable. Thus, the algorithm can make real-time updates and handle large data sets. One basic requirement for the recursive algorithms is consistency: The approximate versions, computed by the recursive algorithms, converges to the true ones. The following theorem proves the consistency of the approximate versions and provides a blueprint for constructing consistent algorithms through Kushner's Markov chain approximation methods.

For $c = 1, 2$, let $(\theta_\epsilon^{(c)}, X_\epsilon^{(c)})$ be an approximation of $(\theta^{(c)}, X^{(c)})$. Then, we define

$$\vec{Y}_\epsilon^{(c)}(t) = \begin{pmatrix} N_1^{(c)}(\int_0^t \lambda_1(\theta_\epsilon^{(c)}(s), X_\epsilon^{(c)}(s), s)ds) \\ N_2^{(c)}(\int_0^t \lambda_2(\theta_\epsilon^{(c)}(s), X_\epsilon^{(c)}(s), s)ds) \\ \vdots \\ N_{n_c}^{(c)}(\int_0^t \lambda_{n_c}(\theta_\epsilon^{(c)}(s), X_\epsilon^{(c)}(s), s)ds) \end{pmatrix}, \quad (35.14)$$

set $\mathcal{F}_t^{\vec{Y}_\epsilon^{(c)}} = \sigma(\vec{Y}_\epsilon^{(c)}(s), 0 \leq s \leq t)$, take $L_\epsilon^{(c)}(t) = L\left((\theta_\epsilon^{(c)}(s), X_\epsilon^{(c)}(s), Y_\epsilon^{(c)}(s))_{0 \leq s \leq t}\right)$ as in Equation (35.4). We use the notation, $X_\epsilon \Rightarrow X$, to mean X_ϵ converges weakly to X in the Skorohod topology as $\epsilon \to 0$. Suppose that $(\theta_\epsilon^{(c)}, X_\epsilon^{(c)}, \vec{Y}_\epsilon^{(c)})$ lives on $(\Omega_\epsilon^{(c)}, \mathcal{F}_\epsilon^{(c)}, P_\epsilon^{(c)})$, and Assumptions 35.1–35.5 also hold for $(\theta_\epsilon^{(c)}, X_\epsilon^{(c)}, \vec{Y}_\epsilon^{(c)})$. Then, there also exists a reference measure $Q_\epsilon^{(c)}$ with similar properties. Next, we define the approximations of $\rho_c(f_c, t)$, $\pi_c(f_c, t)$, and $q_c(f_c, t)$.

Definition 35.6 For $c = 1, 2$, let
$$\rho_{\epsilon,c}(f_c, t) = E^{Q_\epsilon^{(c)}}\left[f_c\left(\theta_\epsilon^{(c)}(t), X_{\epsilon_x}^{(c)}(t)\right) L_\epsilon^{(c)}(t)|\mathcal{F}_t^{\vec{Y}_\epsilon^{(c)}}\right],$$
$$\pi_{\epsilon,c}(f_c, t) = E^{P_\epsilon^{(c)}}\left[f_c\left(\theta_\epsilon^{(c)}(t), X_{\epsilon_x}^{(c)}(t)\right)\Big|\mathcal{F}_t^{\vec{Y}_\epsilon^{(c)}}\right],$$
$q_{\epsilon,1}(f_1, t) = \rho_{\epsilon,1}(f_1, t)/\rho_{\epsilon,2}(1, t)$ and $q_{\epsilon,2}(f_2, t) = \rho_{\epsilon,2}(f_2, t)/\rho_{\epsilon,1}(1, t)$.

Theorem 35.3 *Suppose that Assumptions 35.1–35.5 hold for the models $(\theta^{(c)}, X^{(c)}, \vec{Y}^{(c)})_{c=1,2}$ and that Assumptions 35.1–35.5 hold for the approximate models $\left(\theta_\epsilon^{(c)}, X_\epsilon^{(c)}, \vec{Y}_\epsilon^{(c)}\right)$. Suppose $(\theta_\epsilon^{(c)}, X_\epsilon^{(c)}) \Rightarrow (\theta^{(c)}, X^{(c)})$ as $\epsilon \to 0$. Then, as $\epsilon \to 0$, for all bounded continuous functions, f_1 and f_2, and $c = 1, 2$,*
(i) $\vec{Y}_\epsilon^{(c)} \Rightarrow \vec{Y}^{(c)}$; (ii) $\rho_{\epsilon,c}(f_c, t) \Rightarrow \rho_c(f_c, t)$; (iii) $\pi_{\epsilon,c}(f_c, t) \Rightarrow \pi_c(f_c, t)$; (iv) $q_{\epsilon,1}(f_1, t) \Rightarrow q_1(f_1, t)$ and $q_{\epsilon,2}(f_2, t) \Rightarrow q_2(f_2, t)$ simultaneously.

The proofs for (i) and (iii) are in [30] and those for (ii) and (iv) are in [22].

Part (ii) implies the consistency of the integrated likelihood, part (iii) implies the consistency of posterior and part (iv) implies the consistency of the Bayes factors.

This theorem provides a three-step blueprint for constructing a consistent recursive algorithm based on Kushner's Markov chain approximation method to compute the continuous-time versions. For example, to compute the posterior and Bayes estimates for a model (then the superscript "(c)" is excluded), Step 1 is to construct $(\theta_\epsilon, X_\epsilon)$, the Markov chain approximation to (θ, X), and obtain $p_{\epsilon,j} = p(y_j|\theta_\epsilon, x_\epsilon)$ as an approximation to $p_j = p(y_j|\theta, x)$, where $(\theta_\epsilon, x_\epsilon)$ is restricted to the discrete state space of $(\theta_\epsilon, X_\epsilon)$. Step 2 is to obtain the filtering equation for $\pi_\epsilon(f, t)$ corresponding to $(\theta_\epsilon, X_\epsilon, Y_\epsilon, p_{\epsilon,j})$ by applying Theorem 35.1. For simplicity, one only considers the case when $a = a(t)$. Recall that the filtering equation for the approximate model can also be separated into the propagation equation:

$$\pi_\epsilon(f, t_{i+1}-) = \pi_\epsilon(f, t_i) + \int_{t_i}^{t_{i+1}-} \pi_\epsilon(A_\epsilon f, s)ds, \tag{35.15}$$

and the updating equation (assuming that a trade at jth price level occurs at time t_{i+1}):

$$\pi_\epsilon(f, t_{i+1}) = \frac{\pi_\epsilon(fp_{\epsilon,j}, t_{i+1}-)}{\pi_\epsilon(p_{\epsilon,j}, t_{i+1}-)}. \tag{35.16}$$

Step 3 converts (35.15) and (35.16) to the recursive algorithm in discrete state space and in discrete times by two substeps: (a) represents $\pi_\epsilon(\cdot, t)$ as a finite array with the components being $\pi_\epsilon(f, t)$ for lattice-point indicator f and (b) approximates the time integral in (35.15) with an Euler scheme.

35.4 A LSDE FM model with simulation and empirical results

We exemplify first how to build a specific FM model and then how to construct the recursive algorithms for computing the joint posteriors and the Bayes estimates as well as for computing the Bayes factors. Simulation and real-data examples are provided in the end of this section.

35.4.1 The LSDE FM model

Since it is intuitive to construct the model, we do so. Step 1: we specify the intrinsic value process and its infinitesimal generator. Suppose $X(t)$ follows a LSDE with the SDE given by

$$dX(t) = (a + bX(t))dt + (c + dX(t))dB(t). \tag{35.17}$$

where $B(t)$ is a standard Brownian Motion, and a, b, c, and d are constants. Conceptually, this model considers both the stock's instantaneous rate of return and volatility as linear functions of the price instead of constants. LSDE contains GBM, LBM, and O-U processes. Its generator is:

$$Af(x) = (a + bx)\frac{\partial f}{\partial x} + \frac{1}{2}(c + dx)^2 \frac{\partial^2 f}{\partial x^2} \qquad (35.18)$$

Step 2: We simply assume the trading times follow a Poisson process with a deterministic trading intensity, $a(t)$. Then, the simplified normalized filtering equation for posterior and the evolution equations for Bayes factors can be employed. A time-dependent deterministic intensity $a(t)$ fits the trade duration data better than the time-invariant one since trading activities are higher in the opening and the closing periods.

Step 3: We incorporate the trading noises on the intrinsic values at trading times to produce the price process. There are three important types of noise that have been identified as we have shown in Figure 35.3 and have been extensively studied in the finance literature (for example, see Harris [16]): discrete, clustering, and nonclustering. First, intraday prices move discretely, resulting in "discrete noise". Second, because prices do not happen evenly on all ticks, but more concentrate on integer and half ticks, "price clustering" is obtained. Third, the "nonclustering noise" contains all other unspecified noise. Let $R\left[\cdot, \frac{1}{M}\right]$ be the rounding function to the closest $\frac{1}{M}$. For simple notation, at a trading time t_i, let $x = X(t_i)$, $y = Y(t_i)$, and $y' = Y'(t_i) = R\left[X(t_i) + V_i, \frac{1}{M}\right]$, where V_i is to be defined as the nonclustering noise and Instead of directly formulating $p(y|x)$, we construct $y = F(x)$ in three steps:

Step (i): Add nonclustering noise V; $x' = x + V$, where V is the nonclustering noise at trade i. We assume $\{V_i\}$, are independent of the value process, and they are i.i.d. with a doubly geometric distribution:

$$P\{V = v\} = \begin{cases} (1 - \rho) & \text{if } v = 0 \\ \frac{1}{2}(1 - \rho)\rho^{M|v|} & \text{if } v = \pm\frac{1}{M}, \pm\frac{2}{M}, \cdots \end{cases}.$$

Step (ii): Incorporate discrete noise by rounding off x' to its closest tick, $y' = R\left[x', \frac{1}{M}\right]$.

Step (iii): Incorporate clustering noise by biasing y' through a random biasing function $b_i(\cdot)$ at trade i. $\{b_i(\cdot)\}$ is assumed independent of $\{y'_i\}$. To be consistent with the MicroSoft data analyzed, we construct a simple random biasing function only for the tick of 1/8 dollar (i.e. $M = 8$). For other tick size, it can be done similarly. The data to be fitted has this clustering occurrence: integers and halves are most likely and have about the same frequencies; odd quarters are the second most likely and have about the same frequencies; and odd eighths are least likely and have about the same frequencies. To generate such clustering, a random biasing function is constructed based on the following rules: if the fractional part of y' is an even eighth, then y stays on y' with probability one; if the fractional part of y' is an odd eighth, then y stays on y' with probability $1 - \alpha - \beta$, y moves to the closest odd quarter with probability α, and moves to the closest half or integer with probability β. In brief,

$$Y(t_i) = b_i(R[X(t_i) + V_i, \frac{1}{M}]) = F(X(t_i)).$$

The detail of $b_i(\cdot)$, and the explicit $p(y|x)$ for F can be found in Appendix A. Simulations can demonstrate that the constructed $F(x)$ are able to capture the tick-level sample characteristics of transaction data. For more details see [31].

The parameters of clustering noise, α and β, can be estimated through the method of relative frequency. The other parameters, a, b, c, d, and ρ, are estimated by Bayes estimation via filtering through the recursive algorithm to be constructed.

35.4.2 The recursive algorithms for Bayes estimates and Bayes factors

We exemplify the three-step blueprint of Markov chain approximation method summarized in the end of Section 35.3 to construct the recursive algorithms. Finally, we show the consistency of the algorithms. For notational simplicity, we use the superscript (c), for $c = 1, 2$, to distinguish the two models and let $\vec{\theta}_{\epsilon}^{(c)} = \left(a_{\epsilon_a}^{(c)}, b_{\epsilon_b}^{(c)}, c_{\epsilon_c}^{(c)}, d_{\epsilon_d}^{(c)}, \rho_{\epsilon_\rho}^{(c)}\right)$ to denote an approximate discretized parameter signal, which is random in the Bayesian framework.

Step 1: Construct $\left(\vec{\theta}_{\epsilon}^{(c)}, X_{\epsilon_x}^{(c)}\right)$

First, we latticize the parameter spaces of $a^{(c)}, b^{(c)}, c^{(c)}, d^{(c)}, \rho^{(c)}$ and the state space of $X^{(c)}$. Suppose there are $n_a^{(c)} + 1$, $n_b^{(c)} + 1$, $n_c^{(c)} + 1$, $n_d^{(c)} + 1$, $n_\rho^{(c)} + 1$ and $n_x^{(c)} + 1$ lattices in the latticized spaces of $a^{(c)}, b^{(c)}, c^{(c)}, d^{(c)}, \rho^{(c)}$ and $X^{(c)}$ respectively, e.g.

$$a^{(c)} : [\alpha_a^{(c)}, \beta_a^{(c)}] \to \{a_a^{(c)}, a_a^{(c)} + \epsilon_a^{(c)}, \ldots, a_a^{(c)} + (n_a^{(c)} - 1)\epsilon_a^{(c)}, \beta_a^{(c)}\}$$

where the number of lattices is $n_a^{(c)} + 1$. Define $a_v^{(c)} = a_a^{(c)} + v\epsilon_a^{(c)}$, the vth element in the latticized parameter space of $a^{(c)}$, and define $b_h^{(c)}, c_l^{(c)}, d_m^{(c)}, \rho_r^{(c)}$ and $X_w^{(c)}$ similarly. Let

$$\vec{\theta}_{\vec{v}}^{(c)} = (a_v^{(c)}, b_h^{(c)}, c_l^{(c)}, d_m^{(c)}, \rho_r^{(c)})$$

as where $\vec{v}$ is (v, h, l, m, r).

We construct a birth and death generator $\mathbf{A}_\varepsilon^{(c)}$, such that $\mathbf{A}_\varepsilon^{(c)} \to \mathbf{A}^{(c)}$. Namely, we construct a birth and death process $(\vec{\theta}_{\epsilon}^{(c)}, X_{\epsilon_x}^{(c)})$, a simple example of

Markov chain, to approximate $(\theta, X^{(c)}(t))$ using the generator for the LSDE process.

$$\mathbf{A}_\varepsilon^{(c)} f_c(\vec{\theta}_{\vec{v}}^{(c)}, x_w^{(c)})$$

$$= (a_v^{(c)} + b_h^{(c)} x_w^{(c)}) \left(\frac{f_c(\vec{\theta}_{\vec{v}}^{(c)}, x_w^{(c)} + \epsilon_x^{(c)}) - f_c(\vec{\theta}_{\vec{v}}^{(c)}, x_w^{(c)} - \epsilon_x^{(c)})}{2\epsilon_x^{(c)}} \right)$$

$$+ \frac{1}{2}(c_l^{(c)} + d_m^{(c)} x_w^{(c)})^2 \left(\frac{f_c(\vec{\theta}_{\vec{v}}^{(c)}, x_w^{(c)} + \epsilon_x^{(c)}) + c(\vec{\theta}_{\vec{v}}^{(c)}, x_w^{(c)} - \epsilon_x^{(c)}) - 2 f_c(\vec{\theta}_{\vec{v}}^{(c)}, x_w^{(c)})}{(\epsilon_x^{(c)})^2} \right)$$

$$= \beta^{(c)}(\vec{\theta}_{\vec{v}}^{(c)}, x_w^{(c)})(f_c(\vec{\theta}_{\vec{v}}^{(c)}, x_w^{(c)} + \epsilon_x^{(c)}) - f_c(\vec{\theta}_{\vec{v}}^{(c)}, x_w^{(c)}))$$

$$+ \delta^{(c)}(\vec{\theta}_{\vec{v}}^{(c)}, x_w^{(c)})(f_c(\vec{\theta}_{\vec{v}}^{(c)}, x_w^{(c)} - \epsilon_x^{(c)}) - f_c(\vec{\theta}_{\vec{v}}^{(c)}, x_w^{(c)})), \quad (35.19)$$

where

$$\beta^{(c)}(\vec{\theta}_{\vec{v}}^{(c)}, x_w^{(c)}) = \frac{1}{2} \left(\frac{(c_l^{(c)} + d_m^{(c)} x_w^{(c)})^2}{\epsilon_x^2} + \frac{a_v^{(c)} + b_h^{(c)} x_w^{(c)}}{\epsilon_x} \right),$$

and

$$\delta^{(c)}(\vec{\theta}_{\vec{v}}^{(c)}, x_w^{(c)}) = \frac{1}{2} \left(\frac{(c_l^{(c)} + d_m^{(c)} x_w^{(c)})^2}{\epsilon_x^2} - \frac{a_v^{(c)} + b_h^{(c)} x_w^{(c)}}{\epsilon_x} \right).$$

Note that $\beta^{(c)}(\vec{\theta}_{\vec{v}}^{(c)}, x_w)$ and $\delta^{(c)}(\vec{\theta}_{\vec{v}}^{(c)}, x_w)$ are the birth and death rates, respectively, and should be nonnegative. If necessary ϵ_x can be made smaller to ensure the nonnegativity.

Clearly, $\mathbf{A}_\varepsilon^{(c)} \to \mathbf{A}^{(c)}$ and we have $(\vec{\theta}_\varepsilon^{(c)}, X_{\epsilon_x}^{(c)}) \Rightarrow (\vec{\theta}^{(c)}, X^{(c)})$ as $\varepsilon \to 0$ where $\varepsilon = \max(\epsilon_a, \epsilon_b, \epsilon_c, \epsilon_d, \epsilon_p, \epsilon_x)$.

Now, we have the approximate model $(\vec{\theta}_\varepsilon^{(c)}, X_{\epsilon_x}(t))^{(c)}$ of $(\vec{\theta}^{(c)}, X^{(c)}(t))$. Then, we have the approximate $Y_\varepsilon^{(c)}$ which is defined by equation (35.14). Now the counting process observations can be viewed as $Y^{(c)}(t)$ defined by equation (35.1) or $Y_\varepsilon^{(c)}(t)$ defined by equation (35.14) depending on whether the driving process is $(\vec{\theta}^{(c)}, X^{(c)}(t))$ or $(\vec{\theta}_\varepsilon^{(c)}, X_{\epsilon_x}^{(c)}(t))$. When we model the parameters and the stock value as $(\vec{\theta}^{(c)}, X^{(c)}(t))$, the counting process observations of stock price are regarded as $Y^{(c)}(t)$. When we intend to compute the Bayes factors for the comparison of two models, we use $(\vec{\theta}_\varepsilon^{(c)}, X_{\epsilon_x}^{(c)}(t))$ to approach $(\vec{\theta}^{(c)}, X^{(c)}(t))$ and the counting process observations of stock price are regarded as $Y_\varepsilon^{(c)}(t)$.

The recursive algorithm is to compute the joint posterior and Bayes estimates and Bayes factors for the approximate model $(\vec{\theta}_\varepsilon^{(c)}, X_{\epsilon_x}^{(c)}, Y_\varepsilon^{(c)})$, which is close to the joint posterior and Bayes estimates and Bayes factors of the model $(\vec{\theta}^{(c)}, X^{(c)}, , Y^{(c)})$, by Theorem 35.3, when ε is small.

Step 2: Obtain the SPDEs of the approximate model

When $(\vec{\theta}^{(c)}, , X^{(c)})$ is approximated by $(\vec{\theta}_\varepsilon^{(c)}, X_{\varepsilon_x}^{(c)})$, $\mathbf{A}^{(c)}$ by $\mathbf{A}_\varepsilon^{(c)}$, and Y by Y_ε, there accordingly exist probability measures $P_\varepsilon^{(c)}$ and $Q_\varepsilon^{(c)}$, which approximate $P^{(c)}$ and $Q^{(c)}$. It can be checked that Assumptions 35.1–35.5 hold for $(\vec{\theta}_\varepsilon^{(c)}, X_{\varepsilon_x}^{(c)}, Y_\varepsilon^{(c)})$ for $c = 1, 2$ satisfying the conditions of Theorems 35.1 and 35.2.

For the posterior, there is one model only and the superscript "(c)" is omitted. Let $(\vec{\theta}_\varepsilon, X_{\varepsilon_x})$ denote the discretized signal.

Definition 35.7 Let $\pi_{\varepsilon,t}$ be the conditional probability mass function of $(\vec{\theta}_\varepsilon, X_{\varepsilon_x}(t))$ on the discrete state space given $\mathcal{F}_t^{\vec{Y}_\varepsilon}$. Let

$$\pi_\varepsilon(f,t) = E^{P_\varepsilon}\left[f(\vec{\theta}_\varepsilon, X_\varepsilon(t)) \mid \mathcal{F}_t^{\vec{Y}_\varepsilon}\right] = \sum_{\vec{\theta}_{\vec{v}}, x_w} f(\vec{\theta}_{\vec{v}}, x_w)\pi_{\varepsilon,t}(\vec{\theta}_{\vec{v}}, x_w),$$

where the summation goes over all lattices in the discretized state spaces.

Then, the normalized filtering equation for the approximate model is given by equations (35.15) and (35.16).

Next, we approximate the Bayes factors.

Definition 35.8 Let $q_{\varepsilon,t}^{(c)}$ as a conditional mass finite measure of $(\vec{\theta}_\varepsilon, X_{\varepsilon_x}(t))$ on the discrete state space given $\mathcal{F}_t^{\vec{Y}_\varepsilon}$. $q_{\varepsilon,t}^{(c)}$ approximates $q_t^{(c)}$. Let

$$q_\varepsilon^{(c)}(f_c,t) = \sum_{\vec{\theta}_{\vec{v}}^{(c)}, x_w^{(c)}} f_c(\vec{\theta}_{\vec{v}}^{(c)}, x_w^{(c)})q_{\varepsilon,t}(\vec{\theta}_{\vec{v}}^{(c)}, x_w^{(c)}), \tag{35.20}$$

where $(\vec{\theta}_{\vec{v}}^{(c)}, x^{(c)})$ goes over all the lattices in the approximate state spaces.

Similarly to the normalized filtering equation, the systems of SPDEs for $q_\varepsilon^{(c)}(f_c, t)$ can be separated into the propagation equation:

$$q_\varepsilon^{(c)}(f_c, t_{i+1}-) = q_\varepsilon^{(c)}(f_c, 0) + \int_{t_i}^{t_{i+1}-} q_\varepsilon^{(c)}(\mathbf{A}_\varepsilon^{(c)} f_c, s)ds, \tag{35.21}$$

and the updating equation:

$$q_\varepsilon^{(c)}(f_c, t_{i+1}) = \frac{q_\varepsilon^{(c)}(f_c p_j^{(c)}, t_{i+1}-)}{q_\varepsilon^{(3-c)}(f_{3-c} p_j^{(3-c)}, t_{i+1}-)} q_\varepsilon^{(3-c)}(1, t_{i+1}-). \tag{35.22}$$

Together, these two components form the key to deriving the recursive algorithms.

Step 3: Convert to the recursive algorithm

First, we convert equations (35.15) and (35.16) to the recursive algorithm for computing the approximate joint posterior. We show details for how to obtain the algorithm for this one. For converting equations (35.21) and (35.22) to the recursive algorithm, we only provide the algorithm for computing the Bayes factor without giving details of derivation. But the procedure is similar and interested readers are referred to [22].

We define the posterior that the recursive algorithm computes.

Definition 35.9 The posterior of the approximate model at time t is denoted by

$$p_\varepsilon(\vec{\theta}_{\vec{v}}, x_w; t) = \pi_{\varepsilon,t}\{\vec{\theta}_\varepsilon = \vec{\theta}_{\vec{v}}, X_\varepsilon(t) = x_w\}.$$

Then, there are two substeps. The core of the first substep is to take f as the following lattice-point indicator function:

$$\mathbf{I}_{\{\vec{\theta}_\varepsilon = \vec{\theta}_{\vec{v}}, X_\varepsilon(t) = x_w\}}(\vec{\theta}_\varepsilon, X_\varepsilon(t)) \qquad (35.23)$$

Then, the following fact emerges:

$$\pi_\varepsilon\left(\beta(\vec{\theta}_\varepsilon, X_\varepsilon(t))\mathbf{I}_{\{\vec{\theta}_\varepsilon = \vec{\theta}_{\vec{v}}, X_\varepsilon(t) + \epsilon_x = x_w\}}(\vec{\theta}_\varepsilon, X_\varepsilon(t) + \epsilon_x), t\right) = \beta(\vec{\theta}_{\vec{v}}, x_{w-1})p_\varepsilon(\vec{\theta}_{\vec{v}}, x_{w-1}; t).$$

Along with similar results, equation (35.15) becomes

$$p_\varepsilon(\vec{\theta}_{\vec{v}}, x_w; t_{i+1}-) = p_\varepsilon(\vec{\theta}_{\vec{v}}, x_w; t_i) + \int_{t_i}^{t_{i+1}-} \Big(\beta(\vec{\theta}_{\vec{v}}, x_{w-1})p_\varepsilon(\vec{\theta}_{\vec{v}}, x_{w-1}; t) \\ - (\beta(\vec{\theta}_{\vec{v}}, x_w) + \delta(\vec{\theta}_{\vec{v}}, x_w))p_\varepsilon(\vec{\theta}_{\vec{v}}, x_w; t) + \delta(\vec{\theta}_{\vec{v}}, x_{w+1})p_\varepsilon(\vec{\theta}_{\vec{v}}, x_{w+1}; t)\Big)dt, \qquad (35.24)$$

If a trade at jth price level occurs at time t_{i+1}, the updating equation (35.16) can be written as,

$$p_\varepsilon(\vec{\theta}_{\vec{v}}, x_w; t_{i+1}) = \frac{p_\varepsilon(\vec{\theta}_{\vec{v}}, x_w; t_{i+1}-)p(y_j|x_w, \rho_r)}{\sum_{\vec{v}', w'} p_\varepsilon(\vec{\theta}_{\vec{v}'}, x_{w'}; t_{i+1}-)p(y_j|x_{w'}, \rho_{r'})}, \qquad (35.25)$$

where the summation goes over the total discretized space, and $p(y_j|x_w, \rho_r)$, the transition probability from x_w to y_j, is specified by equation (35.31) in Appendix A.

In the second substep, we approximate the time integral in equation (35.24) with an Euler scheme to obtain a recursive algorithm further discrete in time. After excluding the probability-zero event that two or more jumps occur at the same time, there are two possible cases for the intertrading time. Case 1, if

$t_{i+1} - t_i \leq LL$, the length controller in the Euler scheme, then we approximate $p(\vec{\theta}_{\tilde{v}}, x_w; t_{i+1}-)$ as

$$p(\vec{\theta}_{\tilde{v}}, x_w; t_{i+1}-) \approx p(\vec{\theta}_{\tilde{v}}, x_w; t_i) + \Big[\beta(\vec{\theta}_{\tilde{v}}, x_{w-1}) p(\vec{\theta}_{\tilde{v}}, x_{w-1}; t_i)$$
$$- \big(\beta(\vec{\theta}_{\tilde{v}}, x_w) + \delta(\vec{\theta}_{\tilde{v}}, x_w)\big) p(\vec{\theta}_{\tilde{v}}, x_w; t_i) + \delta(\vec{\theta}_{\tilde{v}}, x_{w+1}) p(\vec{\theta}_{\tilde{v}}, x_{w+1}; t_i)\Big] (t_{i+1} - t_i).$$
(35.26)

Case 2, if $t_{i+1} - t_i > LL$, then we can choose a fine partition $\{t_{i,0} = t_i, t_{i,1}, \ldots, t_{i,n} = t_{i+1}\}$ of $[t_i, t_{i+1}]$ such that $\max_j |t_{i,j+1} - t_{i,j}| < LL$ and then approximate $p(\vec{\theta}_{\tilde{v}}, x_l; t_{i+1}-)$ by applying repeatedly equation (35.26) from $t_{i,0}$ to $t_{i,1}$, then $t_{i,2}, \ldots$, until $t_{i,n} = t_{i+1}$.

Equations (35.25) and (35.26) consist of the recursive algorithm we employ to calculate the approximate posterior at time t_{i+1} for $(\vec{\theta}, X(t_{i+1}))$ based on the posterior at time t_i. At time t_{i+1}, the Bayes estimates of $\vec{\theta}$ and $X(t_{i+1})$ are the expected values of the corresponding marginal posteriors.

To complete the algorithm for posterior, we choose a reasonable prior. Assume independence between $X(0)$ and $\vec{\theta}$. The prior for $X(0)$ can be set by $P\{X(0) = Y(t_1)\} = 1$ where $Y(t_1)$ is the first trade price of a data set because they are very close. For other parameters, we can simply take a uniform prior to the discretized state space of $\vec{\theta}$, which is used also for the algorithm for Bayes factors and in the simulation and the real data example. At $t = 0$, we select the prior as below:

$$p(\vec{\theta}_{\tilde{v}}, x_w; 0) = \begin{cases} \frac{1}{(1+n_a)(1+n_b)(1+n_c)(1+n_d)(1+n_p)} & \text{if } x_w = Y(t_1) \\ 0 & \text{otherwise} \end{cases}.$$

In the rest of this subsection, we briefly present the algorithm for computing the Bayes factors.

We define the conditional ratio measure that the recursive algorithm computes.

Definition 35.10 The conditional ratio measure of the approximate model at time t is denoted by

$$q_\varepsilon^{(c)}(\vec{\theta}_{\tilde{v}}^{(c)}, x_w^{(c)}; t) = q_{\varepsilon,t}^{(c)}\{\vec{\theta}_\varepsilon^{(c)} = \vec{\theta}_{\tilde{v}}^{(c)}, X_\varepsilon^{(c)}(t) = x_w^{(c)}\}.$$

Take $f^{(c)}$ as the indicator function

$$\mathbf{I}^{(c)}(\vec{\theta}_{\tilde{v}}^{(c)}, x_w^{(c)}) = \mathbf{I}_{\{\vec{\theta}_\varepsilon^{(c)} = \vec{\theta}_{\tilde{v}}^{(c)}, X_\varepsilon^{(c)}(t) = x_w^{(c)}\}}^{(c)}(\vec{\theta}_\varepsilon^{(c)}, X_\varepsilon^{(c)}(t)). \qquad (35.27)$$

Then,

$$q_\varepsilon^{(c)}(\mathbf{I}^{(c)}(\vec{\theta}_{\tilde{v}}^{(c)}, x_w^{(c)}), t) = q_\varepsilon^{(c)}(\vec{\theta}_{\tilde{v}}^{(c)}, x_w^{(c)}; t)$$

Equations (35.21) and (35.22) become

$$q_\varepsilon^{(c)}(\vec{\theta}_{\vec{v}}^{(c)}, x_w^{(c)}; t_{i+1}-) \approx q_\varepsilon^{(c)}(\vec{\theta}_{\vec{v}}^{(c)}, x_w^{(c)}; t_i)$$

$$+ \left[\beta^{(c)}(\vec{\theta}_{\vec{v}}^{(c)}, x_{w-1}^{(c)}) q_\varepsilon^{(c)}(\vec{\theta}_{\vec{v}}^{(c)}, x_{w-1}^{(c)}; t_i) + \delta^{(c)}(\vec{\theta}_{\vec{v}}^{(c)}, x_{w+1}^{(c)}) q_\varepsilon^{(c)}(\vec{\theta}_{\vec{v}}^{(c)}, x_{w+1}^{(c)}; t_i) \right.$$

$$\left. - (\beta^{(c)}(\vec{\theta}_{\vec{v}}^{(c)}, x_w^{(c)}) + \delta^{(c)}(\vec{\theta}_{\vec{v}}^{(c)}, x_w^{(c)})) q_\varepsilon^{(c)}(\vec{\theta}_{\vec{v}}^{(c)}, x_w^{(c)}; t_i) \right] (t_{i+1} - t_i) \quad (35.28)$$

and

$$q_\varepsilon^{(c)}(\vec{\theta}_{\vec{v}}^{(c)}, x_w^{(c)}; t_{i+1}) =$$

$$\frac{q_\varepsilon^{(c)}(\vec{\theta}_{\vec{v}}^{(c)}, x_w^{(c)}; t_{i+1}-) p_j^{(c)}(y_j \mid (x_w^{(c)}, \rho_r^{(c)}))}{\sum_{\vec{v}',w'} q_\varepsilon^{(3-c)}(\vec{\theta}_{\vec{v}'}^{(3-c)}, x_{w'}^{(3-c)}; t_{i+1}-) p_j^{(3-c)}(y_j \mid (x_{w'}^{(3-c)}, \rho_{r'}^{(3-c)}))}$$

$$\times \left(\sum_{\vec{v}',w'} q_\varepsilon^{(3-c)}(\vec{\theta}_{\vec{v}'}^{(3-c)}, x_{w'}^{(3-c)}, t_{i+1}-) \right) \quad (35.29)$$

where the sums go over all the lattices in the discretized state spaces.

Equations (35.28) and (35.29) compose the recursive algorithm we employ to calculate the approximate conditional ratio measures. At time t_{i+1}, the Bayes factor

$$B_{21}(t_{i+1}) \approx q_\varepsilon^{(2)}(1, t_{i+1}) = \sum_{\vec{v}',w'} q_\varepsilon^{(2)}(\vec{\theta}_{\vec{v}'}^{(2)}, x_{w'}^{(2)}; t_{i+1}),$$

where the sum goes over all the lattices in the discretized state space.

Finally, we note that the statistical and computational concerns for a prior on a parameter have two aspects: suitable range and mesh size. Usually, the marginal posterior of a parameter obtained from a large data set is concentrated on a small area around the true value. This implies that one needs to run the program several times in order to identify suitable range and mesh size of parameters for a real world data set. If the true parameter is out of the range, say, smaller (larger) than the lower (upper) boundary, then the marginal posterior of this parameter would be one or very close to one on the lower (upper) boundary. Such indication gives the direction to adjust the range of parameter space of the prior until no spike on the boundary. After having a suitable range, we may choose a suitable mesh size, which ideally produces a posterior with a unique modal and bell-shaped distribution as shown in table 35.1 of [30]. After obtaining suitable ranges and mesh sizes for both models in Bayes estimation via filtering, we then employ those suitable ranges and mesh sizes for the parameters in the computer program for computing Bayes factors in order to obtain the Bayes factors.

Consistency of the recursive algorithms

There are two approximations in our recursive algorithms to compute the posteriors and Bayes factors. One is to approach the integral in the propagation equations by Euler scheme, whose convergence is well-known. The other one, which is more important, is the approximation of equations (35.7) by equations (35.15) and (35.16), and the approximation of equations (35.10) and (35.11) by equations (35.21) and (35.22). Since, for $c = 1, 2$, $(\vec{\theta}_{\epsilon(c)}, X_{\epsilon}^{(c)}) \Rightarrow (\vec{\theta}^{(c)}, X^{(c)})$ by construction, Theorem 35.3 warrants these convergence in the sense of the weak convergence in the Skorokhod topology, that is, the consistency of the Bayes factors.

35.4.3 A Monte Carlo example

Having constructed the recursive algorithm to compute the Bayes factors, we move on to develop the software to implement the algorithms. For computing the posterior, a Fortran program for the recursive algorithm is developed to compute, at each trading time t_i, the joint posterior of $(\vec{\theta}, X(t))$, their marginal posteriors, their Bayes estimates and their standard errors (SE), respectively. The recursive algorithm is fast enough to generate real-time Bayes estimates. We test the algorithm extensively and verify it on Monte Carlo data, where we know the true parameters. Care is required in selecting the lattice of parameter values for the prior. Ideally, the approximate posterior should resemble a bell-shaped curve. That is, the posterior should be weighted towards the interior of the lattice, but it should not be weighted so strongly to any specific point that the relative probability of that lattice point being the true value is close to 1 (i.e. a bar chart of the relative distribution should be shaped like a normal or bell-shaped curve, not a single spike at one point). Monte Carlo studies indicate that the posterior is robust to the prior. Namely, for reasonable priors, as long as the range covers the true values and the lattices are reasonably fine, the Bayes estimates will converges to the true values. Below we give one Monte Carlo example with 50,000 data to show the effectiveness of Bayes estimates. For parameters $\vec{\theta}$, their Bayes estimates converge to their true values and the two-SE bounds become smaller and smaller, and goes to zero as in the case of GBM in Zeng [30]. Hence, only the final Bayes estimates, their SE, and true values are presented in Table 35.2. The true values are close to the Bayes estimates and all within two SE bounds.

Using the same Monte Carlo data, we calculate the Bayes factor of the full LSDE FM model versus the restricted LSDE FM model with $c = 0$ (a wrong model) for checking the effectiveness for model selection. The final Bayes factor is 13,152, which is larger than 150, the decisive benchmark for rejecting the restricted LSDE FM model.

Table 35.2 Bayes estimates for the LSDE FM model (simulation)

Parameter	True value	Bayes estimate	St. error
a	3.000E-7	2.951E-7	1.013E-6
b	5.000E-7	4.542E-8	1.001E-6
c	2.000E-3	2.4990E-3	1.001E-3
d	3.000E-6	1.1750E-6	1.074E-6
ρ	2.260E-1	2.247E-1	3.367E-3

The value of parameters are for per second.

35.4.4 Real-data example

For this example the data used was the MicroSoft trade by trade stock prices taken from January and February 1994 as provided by the TAQ (Trade and Quote) from the NYSE. After suitable filtering the resulting data used for the estimation was comprised of 49,937 MicroSoft trades. The basic summary statistics can be found in Table 35.3 and the breakdown of the trades by the fractional portions of the price appears in Table 35.4.

Again, we note that clustering away from the odd eighths is present in this data. Method of relative frequency produces the estimations for the clustering parameters: $\alpha = .2414$ and $\beta = .3502$.

We then estimate two models for this data each having a value process with the form of an LSDE. For the full LSDE model the Bayes estimate for each parameter along with the associated standard error appear in Table 35.5. For a restricted model with $c = 0$, the estimates are given in Table 35.6.

We then determined the Bayes factors to compare the full model vs. the restricted model of $c = 0$. The Bayes factor is $2.90 E28$, which is much larger than 150 again. So, we reject the restricted model of $c = 0$ and conclude the full model fits better than the restricted model for this MicroSoft UHF data set.

Table 35.3 Summary statistics for Msft. Jan. & Feb. 1994

	Size	Mean	Median	St. dev.	Skewness	Kurtosis
Msft.	49937	82.465	83.25	1.854	−0.201	−1.446

Table 35.4 Freq. of the fractional parts of the prices for Msft. Jan. & Feb. 1994

	0	1/8	1/4	3/8	1/2	5/8	3/4	7/8
Msft. Freq.	11218	2791	9136	2199	10009	2826	9376	2382
Rel. Freq.	.2246	.0559	.183	.044	.2004	.0566	.1878	.0477

Table 35.5 Bayes est. for the full LSDE FM model (Msft. Jan. & Feb. 1994)

Parameter	Bayes estimate	St. error
a	2.196E-6	1.777E-6
b	9.600E-8	1.001E-6
c	-1.6760E-5	5.115E-5
d	6.999E-5	1.002E-6
ρ	4.133E-1	3.150E-3

The value of parameters are for per second.

Table 35.6 Bayes est. for the restricted-"c=0" case (Msft. Jan. & Feb. 1994)

Parameter	Bayes estimate	St. error
a	-5.000E-6	1.008E-6
b	1.999E-7	1.000E-6
d	7.000E-4	1.000E-6
ρ	4.149E-1	8.312E-4

The value of parameters are for per second.

35.5 Conclusion

This chapter reviews recent development of a rich class of filtering models with counting process observations for the micromovement of asset price and the related Bayesian inference via filtering. A specific FM model built upon LSDE is used to exemplify how to develop a FM model and further exemplify how to develop the recursive algorithms for computing the posterior and the Bayes factors. The general model and its developed statistical analysis offer strong potential to relate or illuminate aspects of the rich theoretical literature on market microstructure and trading mechanism. Furthermore, Bayesian model selection via filtering provides a general, powerful tool to test related market microstructure theories, represented by the FM models. We may test whether NASDAQ has less trading noise after a market reform as in Barclay et al. [3], test whether information affects trading intensity as in Easley and O'Hara [9] and Engle [10]. Finally, a more general FM model with statistical analysis can be found in [33].

Appendix: more on clustering noise

To formulate the biasing rule, we first define a classifying function $r(\cdot)$,

$$r(y) = \begin{cases} 3 \text{ if the fractional part of } y \text{ is odd eighth} \\ 2 \text{ if the fractional part of } y \text{ is odd quarter} \\ 1 \text{ if the fractional part of } y \text{ is a half or zero.} \end{cases} \qquad (35.30)$$

The biasing rules specify the transition probabilities from y' to y, $p(y|y')$. Then, $p(y|x)$, the transition probability can be computed through $p(y|x) = \sum_{y'} p(y|y')p(y'|x)$ where $p(y'|x) = P\{V = y' - R[x, \frac{1}{8}]\}$. Suppose $D = 8|y - R[x, \frac{1}{8}]|$. Then, $p(y|x)$ can be calculated as, for example, when $r(y) = 2$,

$$p(y|x) = \begin{cases} (1-\rho)(1+\alpha\rho) & \text{if } r(y) = 2 \text{ and } D = 0 \\ \frac{1}{2}(1-\rho)[\rho + \alpha(2+\rho^2)] & \text{if } r(y) = 2 \text{ and } D = 1 \\ \frac{1}{2}(1-\rho)\rho^{D-1}[\rho + \alpha(1+\rho^2)] & \text{if } r(y) = 2 \text{ and } D \geq 2 \end{cases} \quad (35.31)$$

Notes

The authors are indebted to an anonymous referee for detailed constructive written comments. The research is supported in part by the National Science Foundation under grant DMS-0604722.

1. From the viewpoint of time series, econometricians naturally view such data as an *irregularly-spaced* time series and Engle [10] develops a general framework under this view with many developments.
2. Early and recent literature on related filtering problems with counting process observations includes, but is not limited to, [27], [28] [8], [25], [21], and [11].
3. Meaning different models, not the same model with different representations as described in Section 35.2.

References

[1] Aït-Sahalia, Y., P. A. Mykland, and L. Zhang, 2005. How often to sample a continuous-time process in the presence of market microstructure noise. *Review of Financial Studies*, 18, 351–416.

[2] Bandi, F. M. and J. R. Russell, 2006. Separating microstructure noise from volatility. *Journal of Financial Economics*, 79, 655–92.

[3] Barclay, M., W. Christie, J. Harris, E. Kandel, and P. H. Schultz, 1999. The effects of market reform on the trading costs and depths of NASDAQ stocks. *Journal of Finance*, 54, 1–34.

[4] Black, F., 1986. Noise. *Journal of Finance*, 41, 529–543.

[5] Bremaud, P., 1981. *Point Processes and Queues: Martingale Dynamics*. Springer-Verlag, New York.

[6] Cvitanic, J., R. Liptser, and B. Rozovskiĭ, 2006. A filtering approach to tracking volatility from prices observed at random times. *Annals of Applied Probability*, 16, 1633–52.

[7] Cvitanic, J., B. Rozovskiĭ, and I. Zaliapin, 2006. Numerical estimation of volatility values from discretely observed diffusion data. *Journal of Computational Finance*, 4, 1–36.

[8] Davis, M. H. A., A. Segall, and T. Kailath, 1975. Nonlinear filtering with counting observations. *IEEE Trans. Inf. Theory*, 21, 143–9.

[9] Easley, D. and M. O'Hara., 1992 Time and the process of security price adjustment. *Journal of Finance*, 47, 577–605.

[10] Engle, R., 2000. The econometrics of ultra-high-frequency data. *Econometrica*, 68, 1–22.

[11] Elliott, R. J. and W. P. Malcolm, 2005. General smoothing formulas for Markov-modulated Poisson observations. *IEEE Trans. Aut. Control*, 50, 1123–34.

[12] Ethier, S. N. and T. G. Kurtz, 1986. *Markov Processes: Characterization and Convergence.* Wiley, New York.

[13] Fan, J. and Y. Wang, 2007. Multi-scale jump and volatility analysis for high-frequency financial data. *Journal of American Statistical Association,* **102**, 1349–62.

[14] Frey, R., 2000. Risk-minimization with incomplete information in a model for high-frequency data. *Mathematical Finance* **10**, 215–25.

[15] Frey, R. and W.J. Runggaldier, 2001, A nonlinear filtering approach to volatility estimation with a view towards high frequency data. *International Journal of Theoretical and Applied Finance* **4**, 199–210.

[16] Harris, L., 1991. Stock price clustering and discreteness. *Review of Financial Studies,* **4**, 389–415.

[17] Hasbrouck, J., 1996. Modeling market microstructure time series. in G. Maddala and C. Rao, eds, *Handbook of Statistics,* Vol. 14, North-Holland, Amsterdam, pp. 647–92.

[18] Hasbrouck, J., 2002. Stalking the "efficient price" in market microstructure specifications: an overview. *Journal of Financial Markets* **5**, 329–39.

[19] Jeffreys, H., 1961. *Theory of Probability.* Oxford University Press, London.

[20] Kass, R. E. and A. E. Raftery, 1995. Bayes factors and model uncertainty. *Journal of the American Statistical Association,* **90**, 773–95.

[21] Kliemann, W., W. Koch, and F. Marchetti, On the unnormalized solution of the filtering problem with counting process observations. *IEEE Trans. Inf. Theory,* **36**, 1415–25.

[22] Kouritzin, M. and Y. Zeng, 2005. Bayesian model selection via filtering for a class of micro-movement models of asset price. *International Journal of Theoretical and Applied Finance,* **8**, 97–121.

[23] Lee, K. and Y. Zeng, 2010. *Risk minimization for a filtering micromovement model of asset price. Applied Mathematical Finance,* 2010, 17, 177–199.

[24] Li, Y. and P. A. Mykland, 2007. Are volatility estimators bobust with respect to modeling assumptions? *Bernoulli,* **13**, 601–22.

[25] Liptser, R. S. and A. N. Shiryaev, 2002. *Statistics of Random Processes,* 2nd edn, Vol. 2, Springer-Verlag, New York.

[26] Stoll, H., 2000. Friction. *Journal of Finance,* **55**, 1479–1514.

[27] Snyder, D. L. 1972a. Filtering and detection for doubly stochastic poisson processes. *IEEE Trans. Inf. Theory,* **18**, 91–102.

[28] Snyder, D. L. 1972b. Smoothing for doubly stochastic poisson processes. *IEEE Trans. Inf. Theory,* **18**, 558–62.

[29] Xiong, J. and Y. Zeng, 2008 *Mean-variance portfolio selection for a filtering micromovement model of asset price.* Working paper, University of Tennessee.

[30] Zeng, Y., 2003. A partialy observed model for micromovement of asset prices with Bayes estimation via filtering. *Mathematical Finance,* **13**, 411–44.

[31] Zeng, Y., 2004. Estimating stochastic volatility via filtering for the micro-movement of asset prices. *IEEE Trans. Automatic Control,* **49**, 338–48.

[32] Zeng, Y., 2005. Bayesian inference via filtering for a class of counting processes: Application to the micromovement of asset price. *Statistical Inference for Stochastic Processes,* **8**, 331–54.

[33] Zeng, Y., 2008. *Econometric analysis via filtering for financial ultra-high frequency data.* Working paper, University of Missouri at Kansas City.

[34] Zhang, L., Mykland, P. A., and Aït-Sahalia, Y., 2005. A tale of two time scales: Determining integrated volatility with noisy high frequency data. *Journal of the American Statistical Association,* **100**, 1394–1411.

Index

absolute continuity
  case of correlated noises 43–7
  case of independent noise 40–3
  as condition for stability 347–9
absolute information quantities 484
adaptive resampling, sequential Monte Carlo (SMC) methods 675–6
additive observations, filtered martingale problem 147–8
additive white noise setting
  filtering equations 149–54
  Markov property 146
affine factor models 928
  filtering with exact observations of the theoretical prices 931–3
  filtering with observations given by perturbed model values 930–1
Aggoun, L. and Elliott, R. J. 966–7
aircraft positioning 746
  embedded systems 750
  MPF approach 751–2
  real-time issues 747
Allinger, D. F. 427
allowable matrices 306
Amirdjanova, A. 452
analytic filters 874–5
  second-order moment matching filter 878–9
    bearings-only tracking 881–4
    GMTI tracking 888–9
    range-only tracking 884–6
    theoretical justification 879–81
    tracking problems with quantized effects 890–1
apparent information 484, 489
approximation error 576
  lower bound 584
  spectral separating schemes 252–7
  upper bound 576–84
asset pricing 960–2
  first fundamental theorem 925
asset pricing models
  assumptions 962–4
  change of measure 964–6
  conclusion 976

macromovement models 1019
micromovement models 1019–21
parameter estimates 966
  estimate of $\alpha, \beta, \theta, \gamma$, and L 975–6
  estimate of p_{ji} 974–5
  expectation maximization 966–7
  specific processes 967–74
proofs 976–87
see also micromovement asset price models
assumed form of the conditional density approximation method 800, 819
  numerical example 828–30
  observation step
    least-squares approximations 827–8
    quadrature approach 824–6
  observations at discrete times 821–4
  observations taken continuously in time 819–21
asymptotic exactness 562, 563
asymptotic stability 362–3, 371
  proofs of theorems 363–71
Atar, R. and Zeitouni, O. 363
autonomous ODEs, cubature methods 778–82
auxiliary particle filtering 684, 686–8
averaging over characteristics 892–4

backward conditioned signal 332–3
backward information filter 664
backward Itô integral 92
backward stochastic differential equations 93
backward stochastic flow 93
backward stochastic partial differential equations, numerical methods 894–8
Baker–Campbell–Hausdorff formula 412
batch processing 801
Bauerle, N. and Rieder, U. 1004
Baum–Welch algorithm 2, 4
Baxendale, P. et al. 352
Bayes factors 1028–30
Bayes formula 36, 870
Bayesian inference
  convergence theorem and recursive algorithms 1033–4
  filtering and evolution equations 1030–2

Bayesian inference (*cont.*)
  Hidden Markov Models (HMM) 528–9
  LSDE FM model 1036–42
  Markov jump nonlinear systems 530–1
    decomposition of the exact filter
      recursion 533–7
    partially observed Markov chains 546–50
    portfolio optimization, continuous
      time 999–1001
    signal processing
      coloured model error, discrete
        observations 845–9
      no model error
        continuous coloured observational
          noise 843–5
        continuous white observational
          noise 841–3
        discrete observations 837–41
      white model error
        continuous white observational
          noise 852–7
        discrete observations 849–52
    statistical foundations 1027–30
Bayesian parameter estimation
  in problem of unspecified initial data
    270–1
    continuous-time case 291–3
    discrete time case 276–7
    variational formulation 483–5
Bayesian solutions, optimality 274
bearings-only tracking 881–4
Bellman equation 513
  application to portfolio optimization 998
  for unnormalized conditional
    probability 513–15
Benes, V. E. 227, 427
Benes filters, asymptotic stability 363
Bensoussan, A. 1004
Bensoussan, A. et al. 501, 507
  rain check model 509–11
  two inventory distributions model 511
Bessel potential spaces 173, 189–90
β-subprocess of the signal process 410–11
Bhatt, A. G. and Karandikar, R. L. 133, 155
Bhatt, A. G., Kallianpur, G., and Karandikar,
  R. L. 76, 155, 448
biasing rules, clustering trading noise
  1044–5
Birkhoff contraction coefficient 306
Birkhoff contraction inequality, and stability
  problem 320
Birkhoff metric 268, 269–70, 277, 278
  definition 284
Bismut, J.-M. 226
Bjork, T., Davis, M., and Landen, C. 1004
block sampling 693–5

Blom, H. and Bloem, E. 550
Blumenthal, R. M. and Getoor, R. K. 410
Bobrovsky, and Zakai, 565
Boltzmann entropy 721
Boltzmann–Gibbs measures 706
bonds, interest rates 924–5
bootstrap filter 5
Borel functions, optimal feedback
  control 514–15
branching particle system
  approximation 641–7, 707–8
Brendle, S. 1004
Brockett, R. W. and Clark, J. M. C. 221
Brownian motion
  definition 26
  innovations process as 426, 430–1
  one dimensional 28
  stochastic integrals 30–1

cadlag processes 99–100
calibration to the market 926, 935
Callergo, G., Di Masi, G. B., and Runggaldier,
  W. J. 1004
Cameron, R. H. and Martin, W. T. 232
Cameron–Martin basis 239–40
Cameron–Martin expansion 240
  time evolution 240–3
Cauchy problems, stochastic partial differential
  equations (SPDEs) 75–6
  existence and uniqueness of solutions
    83–9
  measure-valued, uniqueness of
    solutions 76–83
Ceci, C. and Gerardi, A. 155, 156
Ceci, C., Gerardi, A., and Tardelli, P. 156
chain rule, Malliavin calculus 197
chaos approach, applications 260
chaos expansions
  of optimal filter 243–5
  Wiener chaos 239–43
chaos properties, propagation 602, 731
  coalescent tree-based representations
    733–8
  polynomial tree-based expansions 732–3
  relative entropy estimates 731–2
characteristic sets, Cameron–Martin
  basis 239–40
Chen's theorem 779, 779–80
Chirelson, D. S., innovation property 50
$C^{i,m}$-functions 34
Clark, J. M. C. 403, 405, 423, 427, 440, 441,
  816
Clark, J. M. C. and Crisan, D. 403, 647
classical observation process model 635
clustering trading noise 1035–6
  biasing rules 1044–5

coalescent tree-based representations 733–8
  propagations of chaos properties 733–8
compatible information 484, 492
compatible information rate, dual systems 496
compensators of random counting
  measures 1025
computational complexity, marginalized particle
  filter (MPF) 759–61
computing power 4
conditional density
  existence 204–10, 226
  smoothness 211, 226–7
    conditional integration by parts
      formula 211–12
    conditional Malliavin matrix 212–16
    invertibility in L^q of the Malliavin
      matrix 216–21
conditional distributions
  martingale problem 138–45
  martingale properties 130–8
  stochastic differential equations
    (SDEs) 57–61
conditional expectations
  modification 85–6
  and stability problem 320–3
conditional probability 870
  partially observed systems 501–2
conditional ratio measures 1029–30
conditionally linear Gaussian models, SMC
  algorithm 695–7
conditioned signal
  backward 332–3
  finite horizon conditioning 330–3
  infinite horizon conditioning 333–5
cone bounded nonlinearities 563, 568
constrained Markov processes 156–9
constrained martingale problem 156–7
continuity assumption, stability problem 347–9
continuity criteria, Malliavin calculus 198–9
continuity requirement 403, 405
continuous-time, discrete-state filtering
  dual system 475–7
  primal system 472–5
continuous-time Bayes factors 1028–30
continuous-time joint likelihood 1027–8
continuous-time likelihoods of Y* 1028
continuous-time model, unspecified initial data
  proof 293–6
  results 291–3
continuous-time posterior 1028
contraction properties, Markov integral
  operators 719–21
control variates method 914–15
convergence, SMC algorithms 676–9
correlated noise 403, 405–6
  pathwise filtering 416

decomposition of the signal
  equation 416–19
  Kallianpur–Striebel formula and associated
    multiplicative functional 419–21
  Zakai equation 421–3
Corsi, M. 1004
counting processes, general FM model 1024–5
coupling lemma 163
Coutin, L. and Decreusefond, L. 452
Cox, J. and Huang, C. F. 1004
Cramér–Rao type lower bounds 563–6, 569
Creal, D., Gu, Y., and Zivot, E. 976
credit derivatives 937, 945–7
  conditionally independent defaults 948–9
  filter equations 952–3
    filtering at a default time 954–5
    filtering between defaults 954
    for finite-state Markov chains 955–6
  information-based model,
    Frey–Schmidt–Gabih 949–51
  jump-diffusion model 947–8
  Markov chain model with infectious
    defaults 949
credit risk models 937–8
  incomplete information 938–40
  structural 945
    Duffie–Lando model 940–2
    Frey–Schmidt model 942–4
    Nakagawa filtering model 944–5
credit spread 937
Crisan, D. and Ghazali, S. 769, 775, 788,
  789–92
Crisan, D. and Xiong, J. 647
Crisan–Ghazali theorem 790
cross terms 170
cubature measures 793–4
cubature methods 575, 768–9
  approximations based on flows of
    autonomous ODEs 778–82
  cubature on Wiener space 769–72
  error estimate for a single time step 773–4
  iterated application (KLV method) 774–8
  numerical approximation 782
  simplicial recombination 783
    algorithm 783–5
    application to KLV method 785–8
  see also KLV method
Cvitanic, J., Lipster, R., and Rozovskiĭ, B. 1021
  tracking volatility 123–7
Cvitanic, J., Rozovskiĭ, B., and Zaliapin, I. 1021

Da Prato, G., Ianelli, M., and Tubaro, L. 76
Davis, M. H. A. 405, 426
Davis–Varaiya characterization of the optimal
  control 488
default contagion 940

degeneracy
    limitation in particle filters
        block sampling 693–5
        Rao–Blackwellized particle filtering 695–8
        Resample-Move algorithm 689–92
    SMC algorithms 676
degenerate observation noise 235
dependent noise, innovation approach 910–13
detectability, link to signal stability 343–4
Detemple, J. 1004
differential information quantities 484
diffusion filters 231, 234–8
    information quantities 482
diffusion processes 169
    filtering equations 169–71
        assumptions 171–3
        notation 174
        proofs of theorems 177–88
diffusive, non-Markovian signal/observation systems 155
Dirichlet forms, theory of 859
discrete approximations 574
discrete time model 273–5
    unspecified initial data
        proof 279–91
        results 278–9
discrete trading noise 1035
discrete-time Markov processes 500–1
    partially observed systems
        conditional probability 501–2
            inventory models 507–8
                rain check model 509–11
                two inventory distributions case 511
                zero-balance walk model 508–9
            one reference measure case 503
            stochastic control 512–13
                unnormalized conditional probability 513–15
            two reference measure case 504–6
                unnormalized conditional probabilities 506–7
discretization of continuous-time filtering problem
    application to approximation of the conditional distribution of the signal 584–8
    filtering framework 572–5
    general approximation results 575–84
    proofs and lemmas 588–96
displacement vectors 876–7
dissipation process 471, 479, 482, 497
dissipation rate, primal systems 492
dithering, positioning systems 747
divergence monitoring, positioning systems 748–9
divergence operator 197–8

Dobrushin, R. L. 271
Doeblin–Doob-type condition 271, 272
Doleans–Dade exponential formula 106
Doppler blind zone 887
Doss–Sussman pathwise solution 414–16
drift functions
    ensemble Kalman filter (EnKF) 607–8
    previsibility 438
dual filters 471
    continuous-time, discrete-state filtering 475–7
    information supply, storage, and dissipation processes 480
    path estimation 493–6
Dueker, M. J. 976
Duffie, D. and Lando, D. 939, 940–2
Duncan–Mortensen–Zakai equation *see* Zakai equation
dynamic credit risk models 937–8
    incomplete information 938–40
    structural 945
        Duffie–Lando model 940–2
        Frey–Schmidt model 942–4
        Nakagawa filtering model 944–5
dynamic entropy 497
dynamic information 497
    see also information flows
Dynkin formula 407

Echeverría, P. 133
Effective Sample Size (ESS) criterion 675
Elliott, R., Aggoun, L., and Moore, J. B. 1004
Elliott, R. J. et al. 961
EM (Expectation Maximization) algorithm 936
    application to asset pricing 961, 967
embedded positioning systems 749–50
embedding theorem 192–3
empirical covariance matrix 600
    consistency 613–15
    contiguity control 612–13
Engle, R. 1019
enhanced measurement, target tracking 877
ensemble Kalman filter (EnKF) 598–9
    comparison with particle filters 627–9
    conclusion 627
    contiguity of the elements 615–18
        almost sure contiguity 618–19
        $\mathbb{L}^p$ contiguity 619–22
    control of Kalman gain matrix 610–11
        consistency of the empirical covariance matrix 613–15
        contiguity control of the empirical covariance matrices 612–13
        local Lipschitz continuity of Kalman gain matrix 611–12
    convergence 622

almost sure convergence 623–4
$\mathbb{L}^p$ convergence and rate of convergence 625–6
existence of moments 607–10
as implementation of the Kalman filter in a high dimension 599–601
limiting probability distributions 604–5
connection with Kalman filter or Bayesian filter 605–7
as a particle system with mean-field interactions 601–3
weighted 629–30
ergodic Markhov processes 268
ergodicity, Dobrushin's coefficient 271
ergodicity of the signal process 270, 360–71
relationship to filter stability 323
error bounds
bibliographical notes 569
Cramér–Rao type lower bounds 563–6
information-based lower bounds 566–8
linear case 561–2
upper bounds 568–9
error reduction, Monte Carlo methods 913–15
Ethier, S. N. and Kurtz, T. G. 133
Euler method 575, 714
application to interacting particle systems 649
application to weighted particle systems 638–9
Evensen, G. 598
exact filter recursion, decomposition 533–7
'exact sampling' 575
existence of solutions, stochastic partial differential equations (SPDEs) 84–9
exponential bounds 278
exponential decay rate 299
relationship to Birkhoff contraction coefficient 306–7
exponential local martingales 33
exponential stability 562–3
Lyapunov spectral gap 305–6
extended Kalman filter 259, 681, 741
extended operator of the signal process 407
exterior products 303

factor models
financial applications 923
term structure of interest rates 927–9
F-adaptable point processes 103
Fan, K. 156
FastSLAM 741, 742, 762–3
Feller–Markov properties 356–8
notation and filtering model 353–4
Feynman–Kac evolution semigroups 718–19
Feynman–Kac measures 5, 705, 706, 710, 711, 713–14

continuous time models 714
description of mean field particle models 710–12
Feynman–Kac transformation type 411
filtered martingale problem 129
additive observations 147–8
constrained Markov processes 156–9
coupling lemma 163
filtrations generated by processes 160–1
Markov property 145–9, 163–4
martingale lemma 159
random probability measures as conditional distributions 162–3
solutions 136–8
stopped filtrations and filtrations generated by stopped processes 161–2
filtering equations 149
for additive white noise setting 149–54
with Lipschitz coefficients 169–71
main results 171–7
related results 154–6
uniqueness 152–5
filtering micromovement (FM) model 650–4, 1021–2
constructing price from intrinsic value 1023–4
equivalence of three representations 1026
filtering with counting process observations 1024–5
integral form of the price 1025–6
see also LSDE FM model
filtrations 27
final cost 487
financial applications see credit derivatives; credit risk models; market price of risk; micromovement asset price models; portfolio optimization; term structure models
finite energy condition 425
finite horizon conditioning 330–3
finite memory property 362, 366–7, 371
Finite State-Space HMM 658
finite-dimensional filters 221–2, 227
application of Malliavin calculus 225
dual filters 477
existence 195
finite-state filtering 300–6
Fisher information 564–5
Fisk, D. L. 258
fixed-lag approximation 698–9
fluctuation analysis 726–8
Fokker–Planck equation 410
F-optional random measures 102
forests, terminology 735
forward equations 410
solutions 132–3
forward rate, bonds 925

forward–backward recursions 663, 699–700
fractional Brownian motion (FBM) 451–2
    application to mutual information 464–6
    definition 454
    nonlinear filtering problem 458–62
    prediction 462–4
    properties
        associated Hilbert space 456
        association with fractional calculus 452–4
        covariance function 455–6
        Hölder continuity of sample paths 454
        representation in terms of a Brownian motion 456
        self-similarity 455
    stochastic differential equations (SDEs), solutions 457–8
    stochastic integration 455, 456
fractional calculus 452–3
Frey, R. 1021
Frey, R. and Runggaldier, W. 1004, 1021
Frey, R. and Schmidt, T. 942–4
Frey, R. Schmidt, T., and Gabih, A. 949–51
Fubini theorem 435
Fujisaki–Kallianpur–Kunita equation 258–9
    uniqueness 152
Fujisaki–Kallianpur–Kunita theorem 58–60, 170
functional inequalities, mean field theory 721–3

gamma function, fractional calculus 452
GARCH (generalized autoregressive conditionally heteroskedastic) model 960
gauge transformation type 411
Gauss–Hermite quadrature 822–4, 826
Gaussian IID noise 269
Gaussian measures, reference sources 868
Gaussian mixture approximation of densities 886
Gaussian model 259
Gaussian processes, stochastic partial differential equations (SPDEs) 70–4
Gaussianity of μ_0 836
Gel'fand, I. M. and Vilenkin, N. Ya. 640
genealogical tree particle models 708, 716–18
Generalized Pseudo Bayesian (GPB) filters 539
    comparison with IMM filter 541–2
generator, transition probabilities 56
Gennotte, G. 1004
geographical information systems (GIS) 742–3
Gihman, I. I. and Skorohod, A. 420
Gilman, A. S. and Rhodes, I. B. 568
Girsanov transformation type 411–12
Girsanov's theorem 24, 33, 871
GMTI (Ground-based Moving Target Indicator) tracking 887–9

Haar basis 254
Hamilton, J. D. 961
Hamilton–Jacobi–Bellman (HJB) equation 998
Harnack inequalities 269
Haussman, U. and Sass, J. 1004
Havrda–Charvat entropy 720, 721
Heath–Jarrow–Morton (HJM) approach 926–7
Hellinger integrals 720, 722
Hermite basis 252–3
hidden Markov models (HMM) 1, 526, 657–61, 961
    Bayesian filtering 528–9
    filtering and marginal likelihood 661–2
    smoothing 662–4
Hilbert's projective metric
    applications 315
    in finite-state space 305–8
    in general-state space 308–15
Hitsuda, M. and Mitoma, I. 639
Hölder continuous functions 34, 35, 176
Hörmander's sum of squares theorem 195, 210
H-relative entropies 720–1
Hurst, H. E. 451
hybrid filters 642–3

importance density 667
importance sampling 5, 680–1
increasing processes 28
incremental importance weight function 670, 680
inertial navigation system (INS) 750, 751–2
infinite horizon conditioning 333–5
infinite-dimensional optimal filters 259
information 471
    absolute and differential quantities 484
    apparent 484
    compatible 484
    residual and full 483–4
    supply, storage, and dissipation processes 479–80
information flows 478–82
    dual systems 494–6
    primal systems 488–92
information-driven default contagion 940
information-theoretic bound 335–7
innovation of a stationary time series 20
innovation problems 171, 425–8
    condition 429
    correlated case 445–8
    martingale representation 47–8
    notation 429
    pathwise uniqueness and strong solutions 438–41
    preliminaries 430–7
    proof of theorem 441–4

innovation property 48–51
    proof of theorem 51–4
innovations conjecture 426–8
    validity 431–2
innovations processes 898–901
    application to term structure models 933–4
    dependent noise case 910–13
    Monte Carlo methods 905–9
    numerical experiments 915–18
    numerical methods 901–5
    as standard Brownian motion 430–1
Interacting Multiple Model (IMM)
    filtering 526, 539–42, 552, 555
    comparison with GPB filters 541–2
    extension 550
    IMM particle filter 542–6
        extension 550
        target tracking example 550–2
    target tracking examples 550–2, 886
interacting particle systems 647–50
interest rate term structure
    full information
        term structure models 924–5
            factor models 927–9
            Heath–Jarrow–Morton (HJM)
                approach 926–7
            martingale pricing 925–6
    incomplete information 929–30
        affine factor models
            filtering with exact observations of the
                theoretical prices 931–3
            filtering with observations given by
                perturbed model values 930–1
    nonlinear filtering 933–4
internet traffic models 451
invariant measures 358–61
inventory models 507–8
    optimal control
        dynamic programming for the Kushner
            equation 516
        dynamic programming for the Zakai
            equation 516–18
        main result 518–24
        zero-balance walk model 515
    rain check model 509–11
    two inventory distributions case 511
    two reference measure case 504–6
    zero-balance walk model 508–9
inverse flows 90–2
Itô approximation of the integral 574
Itô processes 32
Itô's formula 32–4, 202–3
Itô–Ventzell formula 201, 226

Jeffreys, H. 1029
joint signal–filter process, properties 474–5

joint uniqueness in law 155
jump Markov processes 155–6
jump measures 104
jump processes
    notation 99–100
    stochastic nonlinear filtering
        equations 107–12
        applications 116–27
        reduced (unnormalized) form 113–15
jump-diffusion processes 155
    credit derivatives 947–8
jungles 735–7

Kadota, T., Ziv, M., and Zakai, Z. 567
Kailath, T. 425, 426
Kakutani–Hellinger distance 720
Kallianpur, G. and Mandal, P. K. 155
Kallianpur–Striebel formula 39–40, 50, 205,
        236, 355–6, 404, 434, 573, 637
    computational deficiencies 894, 916
    and innovations conjecture 427
    in pathwise filtering 409–10
        correlated noise case 416, 419–21
Kalman filter 21–3, 26, 425, 503
    application to target tracking 874
    stability of optimal filter 379–82
    see also ensemble Kalman filter (EnKF)
Kalman gain matrix 610–11
    consistency of the empirical covariance
        matrix 613–15
    contiguity control of the empirical covariance
        matrices 612–13
    local Lipschitz continuity 611–12
Kalman–Bucy filter 1, 74, 221, 259, 425, 426,
        450, 1004
    application to portfolio optimization 1001
    asymptotic stability 363
    extended 259
    stability of optimal filter 391–3
Karatzas, I. and Zhao, X. 1004
Karatzas, I. et al. 1004
Kass, R. E. and Raftery, A. E. 1030
Kliemann, W., Koch, G., and Marchetti, F. 155
Kloeden, P. E. and Platen, E. 773
KLV method 768, 771–2, 774–8
    application of simplicial recombination 783,
        785–8
    application to nonlinear filtering 788–9
        explicit formula 793
        KLV-based approximation and error
            bound 789–93
    numerical approximation 782
Koch, G. 154
Kohonen algorithm 1008–9
Kolmogorov Consistency Theorem 501
Kolmogorov, A. N. 425, 450

Kolmogorov's forward and backward
    equations 56
Kouritzin, M. and Zeng, Y. 1022
Krener, A. J. 423
Krylov, N. V. 428, 445–8
    innovation property 49
Kunita, H. 226, 352, 361, 423
    and innovations conjecture 427
Kurtz, T. and Xiong, J. 636
Kurtz, T. G. and Ocone, D. L. 76, 137, 154–5
Kurtz, T. G. and Stockbridge, R. H. 133, 134–5
Kushner–FKK equation 638
Kushner's equations 3, 170
    dynamic programming 516
Kushner–Stratonovitch equation 67, 258, 260
    uniqueness 152, 154
Kusuoka, S. 768
    KLV method 774–8
Kwakernaak, H. and Sivan, R. 562

Lagrangian approach 5
Lakner, P. 1004
Lam, K. and Li, W. K. 961
Langevin equations 857–8, 869
    discretization 864
    finite-dimensional case 858
    infinite-dimensional case 858–9
        Gaussian case 859–60
        nonlinear case 860–2
large deviation principles 728–31
Lasry, J. M. and Lions, P. L. 1004
Lee, K. and Zeng, Y. 1026
Leland, W., Taqqu, M., Willinger, W., and
    Wilson, D. 451
Lie algebra 221–2, 227
    in pathwise filtering 408
limiting Kalman filter, stability of optimal
    filter 382–3
linear filters 19–20
    Kalman filter 21–3
linear Gaussian model 658, 660
linear growth 34, 44
Lipschitz continuous functions 34, 44, 170–1
    filtering equations, main results 175–7
    and innovation property 49
Liptzer, R. S. and Shiryaev, A. V.
    Gaussian processes 70, 73–4
    Markovian system-observation processes 69
Lo, J. T. H. and Willsky, A. S. 878
local forward equation, solutions 133
local martingale problem
    solutions 132
    uniqueness of solutions 145
local martingales 28
    exponential 33
    orthogonal 29

local mixing conditions 271–2, 275
local rate of apparent information 489
local rate of compatible information 492
locally infinitely divisible processes 106
locally Lipschitz continuous functions 44
locally square integrable martingales 28
locally uniform Dobrushin condition 271
logarithmic utility function, optimal wealth
    process 996
log-concave functions 386–7
lower bound of approximation error 584
lower bounds of filtering error
    Cramér–Rao type 563–6
    information-based 566–8
L^p estimates, stability of optimal filter 377–8
LSDE FM model 1034–6
    conclusion 1044
    Monte Carlo example 1042–3
    real-data example 1043–4
    recursive algorithms for Bayes estimates and
        Bayes factors 1036–42
Lusin spaces 99
Lyapunov exponents 302–4
    and stability problem 320
Lyons, T. and Victoir, N. 768, 769, 773
    KLV method 774–8

macromovement asset price models 1019
Malliavin, P. 226
Malliavin calculus 67, 195–6, 226
    application to finite-dimensional filters 225
    criteria for absolute continuity and
        smoothness 198–9
    existence of a conditional density 204–10
    filtering model 203–4
    operators 196–8
    smoothness of a conditional density 211
        conditional integration by parts
            formula 211–12
        conditional Malliavin matrix 212–16
        invertibility in L^q of the Malliavin
            matrix 216–21
    stochastic differential equations and
        stochastic flows 199–203
    for the Zakai equation 222–5
Mandelbrot, B. B. 451
marginal likelihoods, hidden Markov
    model 661–2
marginalized FastSLAM 762–3
marginalized particle filter (MPF) 741–2, 752
    algorithm 755–7
    applications 750–2
    computational complexity 759–60
    model structure 754–5
    radar tracking case study 760–1
    representation 752–4

variance reduction 758
see also Rao–Blackwellization
marked point observation processes 156
market price of risk 934–5
  parameter estimation
    combined filtering and parameter
      estimation 935–6
    EM algorithm 936
Markov chain approximation method 799–800,
  808–9
  alternative approximating chain 811
  application 812–15
  asymptotic equivalence of the time
    scales 811–12
  constant interpolation interval 809–10
  continuous-time interpolations and
    convergence proofs 810–11
  discretizing time 811
  local consistency 809
  robustness 818–19
Markov chain Monte Carlo (MCMC)
  methods 689–92, 705–6, 706, 862, 869
  application to target tracking 874
  basics 666–7
  error reduction 913–15
  importance sampling 667–9
  innovation process 905–9
    numerical experiments 915–18
  Metropolis–Hastings methods
    finite-dimensional case 863–6
    infinite-dimensional case 866–8
Markov jump linear systems, filtering 537–9
  IMM 539–42
Markov jump nonlinear systems
  Bayesian filtering 530–1
    decomposition of the exact filter
      recursion 533–7
  IMM particle filter 542–6
  particle filtering 531–3
Markov model with additive white noise,
  filtering equations 149–54
Markov processes
  constrained 156–9
  jump processes 155–6
  semimartingales 56–7
  stochastic differential equations (SDEs) 62–3
    correlated noise case 65–7
    independent noise case 63–4
Markov property 163–4
  martingale problems 145–9
Markov switching 961
Markov system-observation processes,
  stochastic partial differential equations
  (SPDEs) 67–70
martingale lemma 159
martingale pricing 925–6

martingale problem
  conditional distributions 138–45
  Markov property 145–9
martingale properties
  conditional distributions 130–8
  of ϕ_i 33, 39, 44–6
martingales 3, 27–8
  continuous modification 31
  exponential 33
  innovation problems 47–8
  and stochastic integrals 29–31
  see also filtered martingale problem
matrix Riccati equation 387, 392–3
maximum principle 192
McKean interpretations, mean field
  theory 714–15
mean field theory 705–9
  applications 706
  asymptotic analysis 723
    convergence of empirical processes 725–6
    fluctuation analysis 726–8
    large deviation principles 728–31
    stochastic perturbation model 723–5
  McKean interpretations 714–15
  mean field particle and genealogical
    tree-based models 715–18
  notation and conventions 709–10
  path-space and related filtering
    models 712–14
  propagations of chaos properties 731
    coalescent tree-based
      representations 733–8
    polynomial tree-based expansions 732–3
    relative entropy estimates 731–2
  stability analysis 718
    contraction properties 719–21
    Feynman–Kac evolution
      semigroups 718–19
    functional inequalities 721–3
mean-field interactions 602
mean-reverting price processes 960
measurable processes 27
measure-valued stochastic partial differential
  equations, uniqueness of solutions 76–83
memoryless channel assumption 274
mesh size, selection in Bayesian
  estimation 1041
Metropolis–Hastings methods 690, 862
  finite-dimensional case 863–6
  infinite-dimensional case 866–8
Meyer, P. A. 438–41
Michel, D. 226
micromovement asset price models 1019–21
  conclusion 1044
  equivalence of three representations 1026
  general FM model 1021–2

micromovement asset price models (*cont.*)
   constructing price from intrinsic
      value 1023–4
   filtering with counting process
      observations 1024–5
   integral form of the price 1025–6
   LSDE FM model 1034–6
      Monte Carlo example 1042–3
      real-data example 1043–4
      recursive algorithms for Baye estimates
        and Bayes factors 1036–42
Mitter, S. K. 227, 410, 423
   and innovations conjecture 427
mixing assumptions 271–3, 275
mixing conditions 324, 328
model error 846
   coloured model error, discrete
      observations 845–9
   white model error
      continuous white observational
        noise 852–7
      discrete observations 849–52
money market accounts 926
Monte Carlo methods *see* Markov chain Monte
   Carlo (MCMC) methods
Morando's theorem 163
Mukulevicius, R. and Rozovskiĭ, B. L. 260
multidimensional observations 474
multinomial resampling 673
multiple integral expansion 243
multiplicative ergodic theory (MET) 299, 302–5
multiplicative functionals
   pathwise filtering
      correlated noise case 419–21
      with independent signal and noise 409
      factorization 410–12
mutation step, SMCMs 5
mutual information 478–82
   application of fractional Brownian
      motion 464–6
   definition 566
   relationship to filtering error 567, 569

Nagai, H. and Runggaldier, W. J. 1004
Nakagawa, H. 944–5
Ninomiya, S. 782
noise 816
   trading noises 1020–1, 1035–6
      clustering noise, biasing rules 1044–5
   see also correlated noise; dependent noise;
      white noise
noise models 450
   see also fractional Brownian motion (FBM)
noise processes 35
   nondegenerate 37
'noise-after-nonlinearity' model 875, 877–8

bearings-only tracking 881–2
range-only tracking 884
'noise-before-nonlinearity' model 875, 878
   bearings-only tracking 882
   range-only tracking 884
nonclustering trading noise 1035
noncompact setting 269
nonconditional dynamics 271–2
nondegenerate Brownian motion 26
nondegenerate noise processes 37
nonlinear filtering, overview 1, 450
nonlinear filters, comparison with linear
   filters 19–20
nonrandom objects, definition 280
normalized filtering density, time
   evolution 257–8
normalized filtering equation 1031–2
Norris, J. 226
N-particle model 715–17
ν-regular filtering models, definition 253
Nualart, D. and Zakai, M. 227
numerical approximation 799–801
   approximations to the optimal filter 805–8
   assumed form of the conditional density
      approximation method
      observation step 824–8
        least-squares approximations 827–8
        quadrature approach 824–6
   assumed form of the conditional density
      method 819–24
      numerical example 828–30
   cubature methods 782
   filter representations 802–4
   Markov chain approximation method 808–12
      application 812–15
   robustness of approximating filters 816–19
   system model 801–2
   weak convergence 804–5
Nummelin's splitting technique 332

observability
   link to signal stability 335, 337–44
   uniform observability 344–6
observation noise 43
observation processes 19, 35, 300, 572, 635
   dual systems 477
   and innovation property 50–1
   one-step predictors, stability 338–9
   primal systems 473–4
   target tracking model 876–8
Ocone, D. 260
Ocone, D. and Pardoux, E. 227, 324, 362–3
one-dimensional additive Gaussian noise
   example 301
one-dimensional signals, stability of optimal
   filter 393

one-step predictors of observation process, stability 338–9
optimal Borel feedback control 514–15
optimal control 487–8
optimal filter
  approximations 231, 805–8
  chaos expansions 243–5
  as function of the paths of the observation process 433–5
  representations 257–9
  robust version 389
  stability 374–5
    proofs of theorems 394–8
    time-continuous case 388–91
      Kalman–Bucy filter 391–3
      one-dimensional signals 393
    time-discrete case 375–6
      contraction w.r.t. Lipschitz norms 378–9
      Kalman filter 379–82
      limiting Kalman filter 382–3
      L^p spaces 377–8
      stability of π_n^{0*} 386–8
      time-discretized signals induced by SDEs 383–6
      transfer operator acting on densities 376–7
optimal wealth processes 995–7
optional projections 100–2
orthogonal local martingales 29
orthogonality of input 425
Oseldec's theorem 302
  continuous time analogue 305

Pagès, G. and Pham, H. 1016
Pagès, G., Pham, H., and Primtems, J. 1016
parabolic stochastic partial differential equations, Cauchy problems 75–6, 79–81
parameter estimation, in problem of unspecified initial data 270–1
Pardoux, E. 1004
partial differential equations 575
  see also stochastic partial differential equations (SPDEs)
partially observable diffusion processes 169
  filtering equations 169–71
    assumptions 171–3
    main results 175–7
    notation 174
    proof of theorems 177–88
partially observed linear Gaussian models 697–8
partially observed systems 500–1
  Bayesian filtering 546–50
  conditional probability
    joint probabilities 501–2
    probability measure 501
  conditional probability, derivation 502
  control of inventories
    dynamic programming for the Kushner equation 516
    dynamic programming for the Zakai equation 516–18
    main result 518–24
    zero-balance walk model 515
  inventory models 507–8
    rain check model 509–11
    two inventory distributions case 511
    zero-balance walk model 508–9
  one reference measure case 503
  stochastic control 512–13
    unnormalized conditional probability 513–15
  two reference measure case 504–6
  unnormalized conditional probabilities 506–7
particle approximations
  branching particle systems 641–7
  filtering model 635–6
  fluctuation analysis 726–8
  heuristic schemes 646–7
  interacting particle systems 647–50
  overview 636
  stochastic perturbation model 723–5
  weighted particles 636–41
particle exploration models 717–18
particle filters 4–5, 656–7
  application to target tracking 874
  applications 741–2
    conclusions 764–5
    positioning applications 742
      aircraft positioning 746
      model framework 742–3
      surface positioning 743–5
      underwater positioning 743
      vehicle positioning 745–6
    simultaneous localization and mapping 761–2
      marginalized FastSLAM 762–3
      problem formulation 762
      unmanned aerial vehicles 763–4
  auxiliary 684, 686–8
  block sampling 693–5
  comparison with ensemble Kalman filter 627–9
  hidden Markov model, smoothing 662–4
  Hidden Markov Models (HMM) 657–61
    filtering and marginal likelihood 661–2
  IMM particle filter 542–6
  limitations 689
  Markov jump nonlinear systems 531–3

particle filters (*cont.*)
  positioning applications, practical experiences 746–50
  Rao–Blackwellized 695–8
  Resample-Move algorithm 689–92
  SMC methods 680–1
    stochastic volatility model 682–4, 685
  summary 701–2
  target tracking example 550–2
  see also marginalized particle filter (MPF); mean field theory
particle numbers, positioning systems 747–8
particle smoothing
  fixed-lag approximation 698–9
  forward filtering–backward smoothing 699–700
  generalized two–filter formula 700–1
particle system approximation of optimal filter 258
particles (walkers) 707
path estimation 482–3
  discussion 497
  dual systems 493–6
  notation 498
  primal systems 485–92
  variational formulation of Bayesian estimation 483–5
pathspace 712–14, 834
pathwise filtering 403
  background and summary 404–6
  correlated noise case 416
    decomposition of the signal equation 416–19
    Kallianpur–Striebel formula and associated multiplicative functional 419–21
    Zakai equation 421–3
  with independent signal and noise 408
    factorization of multiplicative functionals 410–12
    generator of $T^y_{s,t}$ 412–14
    Kallianpur–Striebel formula 409–10
    Zakai equation 414–16
  signal process 406–8
pathwise uniqueness, SDEs 440–1
performance bounds, positioning systems 749
Pham, H. 1004
Pham, H. and Quenez, M. C. 1004
Pham, H., Runggaldier, W., and Sellami, A. 1016
Picard, J. 562, 563, 569
Pinsker, M. S. 567
planar forests 735–6
point process observations 635
  filtered martingale problem approach 155
  Filtering Micromovement (FM) model 650–4

portfolio optimization 990
  continuous time model
    Bayesian case 999–1001
    filtering 993–4
    finite-state Markov chain case 1002–4
    linear Gaussian case 1001–2
    Markovian case and PDE approach 997–8
    martingale approach 994–7
    partial observation control framework 991–2
    reduction to a full observation model 992–3
  discrete-time models
    approximation of control problems under partial observation 1009–11
    filer evolution and reduction to a full observation control problem 1006–7
    numerical illustrations 1011–16
    partial observation control framework 1005–6
    quantization of the filter process 1007–9
positioning applications 742
  aircraft positioning 746
  applications of MPF 750–2
  model framework 742–3
  practical experiences 746–50
  surface positioning 743–5
  underwater positioning 743
  vehicle positioning 745–6
posterior distribution 1, 660, 661, 869
  general properties 835–6
power utility function, optimal wealth process 996–7
precision operator 857
preconditioning 858
prediction problems, fractional Brownian motion 462–4
prediction step, HMM 662
preprocessing, GMTI tracking 887–9
previsibility 438–9
price processes, Filtering Micromovement (FM) model 650–4
probability of reference 992–3
processes of bounded variation 28
progressively measurable processes 27
propagation equation 1032
propagation of chaos 602
proposal density, positioning systems 748

quadratic covariation 28–9
quadratic hedging 1012–15
quadratic variation 28
quantization approach, portfolio optimization 1007–9, 1011–12, 1016

quantized measurements, target
    tracking 889–91

radar, use in surface positioning 743–5
radar tracking case study, marginalized particle
    filter (MPF) 760–1
Radon–Nikodym densities 40, 41, 41–2
Radon–Nikodym derivative, pathwise
    version 485
rain check model 509–11
rainfall model, Hurst 451
random counting measures 1025
random probability measures, conditional
    distributions 162–3
random walk proposals 863–5
range, selection in Bayesian estimation 1041
range-only tracking 884–6
Rao–Blackwellization 542–6, 555, 695–8, 741
    see also marginalized particle filter (MPF)
rapidly decreasing functions 65
rate distortion function 566
real-time issues, positioning systems 746–7
reciprocal flow, action on stochastic differential
    equations 202
recombination see simplicial recombination
recursive filtering equation 319
recursive formulae, target tracking 876
reduced form of stochastic nonlinear filtering
    equation 96, 113–15
    jump processes, applications 117, 122–3
reduced-form credit risk models 938
    incomplete information 939–40
reference probability measure 355
regime switching models 961
    asset pricing
        assumptions 962–4
        change of measure 964–6
        parameter estimates
            estimate of $\alpha, \beta, \theta, \gamma,$ and L 975–6
            estimate of p_{ji} 974–5
        expectation maximization 966–7
        specific processes 967–74
        proofs 976–87
    conclusion 976
relative entropy 335–7, 483, 720–1
relative entropy estimates 731–2
Resample-Move algorithm 689–92
resampling, sequential Monte Carlo (SMC)
    methods 5, 672–3
residual information 483, 488–9
residual resampling 672–3
Riemann–Liouville, fractional calculus 452
Rishel, R. 1004
risk-management see market price of risk
risk-neutral default intensity 938
robust filtering algorithms 405

Rogers, L. C. G. and Williams, D. 430, 438–41
Rozovskiĭ, B. L. 226, 237, 238
running cost 487
running observations 473

sampling rates, positioning systems 747
second-order moment matching filter
    878–9
    bearings-only tracking 882–4
    GMTI tracking 888–9
    range-only tracking 884–6
    refinement 886
    theoretical justification 879–81
    tracking problems with quantized
        effects 890–1
selection step, SMCMs 5
self-similarity property, fractional Brownian
    motion 455
semimartingales 28, 55–7, 103–6
    Itô–Ventzell formula 201
    transformation 96–9
sequential importance sampling 669–72
sequential Monte Carlo (SMC) methods 4–5,
    660, 664, 665–6
    with adaptive resampling 675–6
    convergence results 676–9
    fixed-lag approximation 698–9
    forward–backward recursions 699–700
    generalized two-filter formula 700–1
    generic algorithm 673–7
    importance sampling 667–9
    Monte Carlo methods, basics 666–7
    particle filtering 680–1
        auxiliary particle filtering 684, 686–8
        block sampling 693–5
        Rao–Blackwellized particle filtering
            695–8
        Resample-Move algorithm 689–92
        stochastic volatility model 682–4, 685
    resampling 672–3
    sequential importance sampling 669–72
    summary 679–80
Shannon theory 497
shifted Rayleigh filter (SRF) 882–4
Shiga, T. and Tanaka, H. 728
Shiryaev's equations 170
short rate, bonds 925
sigma fields
    definition 358–60
    equality 352–3, 362
signal ergodicity, relationship to filter
    stability 323
signal process, pathwise filtering 406–8
signal processes (system processes) 19, 572,
    635
    ergodicity 270

signal processing problems on function
  space 833–4
 Bayesian inference for signal processing 837
  coloured model error, discrete
   observations 845–9
  no model error
   continuous coloured observational
    noise 843–5
   continuous white observational
    noise 841–3
   discrete observations 837–41
  white model error
   continuous white observational
    noise 852–7
   discrete observations 849–52
  discussion and bibliography 868–9
  general properties of the posterior 835–6
  Langevin equations 857–8
   finite-dimensional case 858
   infinite-dimensional case 858–9
    Gaussian case 859–60
    nonlinear case 860–2
  MCMC methods 862
   Metropolis–Hastings in finite
    dimensions 863–6
   Metropolis–Hastings in infinite
    dimensions 866–8
simple compounded interest rates 924–5
simple processes 29
simplicial recombination 783
  algorithm 783–5
  application to KLV method 785–8
single-time marginals of posterior path
  distribution 488
SIR (Sampling Importance Resampling)
  algorithm 5
SIR (Sampling Importance Resampling)
  particle filter 531–3, 674, 681
  comparison with IMM particle filter 545–6
  stochastic volatility model 683–4, 685
  target tracking example 550–2
  see also sequential Monte Carlo (SMC)
   methods
SIR hybrid particle filter 533, 534
  comparison with IMM particle filter 546
  target tracking example 550–2
Skorohod integral 197–8
SLAM (simultaneous localization and
  mapping) 741, 761–2
  problem formulation 762
  unmanned aerial vehicle application 763–4
slowly increasing functions 65
'small noise' conditions 269
smooth densities, application of Malliavin
  calculus 195
smoothing 325–9

continuous time case 329–30
fixed-lag approximation 698–9
forward filtering–backward
  smoothing 699–700
generalized two-filter formula 700–1
hidden Markov model 661, 662–4
smoothness criteria, Malliavin calculus 198–9
Sobolev's embedding theorem 76
'SPDE algorithm' 894–8
  numerical experiments 916–18
special semimartingales 103
spectral gap 304, 305–6
spectral separating schemes 232, 245–6
  approximation error 252–7
  discussion 251
  first kind 246–8
   convergence 253–6
  second kind 248–50
   convergence 256–7
stability 299, 319–20
  asymptotic 362–3, 371
   proofs of theorems 363–71
  conditioned signal
   finite horizon conditioning 330–3
   infinite horizon conditioning 333–5
  detectability 343–4
  influencing factors 323
  'intrinsic' perspective 320–3
  necessary conditions 346
   absolute continuity 347–9
   assumption (A) 320–1, 347
   well posedness 346–7
  observability 335, 337–44
   uniform 344–6
  optimal filter 374–5
   proofs of theorems 394–8
   time-continuous case 388–91
    Kalman–Bucy filter 391–3
    one-dimensional signals 393
   time-discrete case 375–6
    contraction w.r.t. Lipschitz norms
     378–9
    Kalman filter 379–82
    limiting Kalman filter 382–3
    L^p spaces 377–8
    stability of $\pi_n^{(N), y_{1:n}}$ 386–8
    time-discretized signals induced by
     SDEs 383–6
    transfer operator acting on
     densities 376–7
  relative entropy 335–7
  smoothing 325–9
   continuous-time case 329–30
stability properties, equivalence 371–2
standard Brownian motion 26
  innovations process as 426, 430–1

Stannat, W. 333
state dependent transition matrix case, impact upon particle filters 546
state processes 300
 target tracking model 876–7
stationary filtering problem 358–60
stochastic calculus
 introduction 26–7
 Itô's formula 31–4
 martingales and quadratic covariations 27–9
stochastic control
 partially observed systems 512–13
  unnormalized conditional probability 513–15
stochastic delay-differential equations 155
stochastic differential equations (SDEs) 55
 action of reciprocal flow 202
 backward 93
 for conditional distributions 57–61
 for fractional Brownian motion 457–8
 Girsanov's theorem 871
 Malliavin calculus 199–200
 parameter estimation 869
 pathwise uniqueness 440–1
 strong solutions 439–40
 for unnormalized conditional expectations 61–2
 see also Langevin equations
stochastic flow of homeomorphisms 90
stochastic flows 79, 89–90, 226
 backward 93
 and existence of a conditional density 206–7, 210
 Malliavin calculus 200–3
stochastic hybrid systems 526–7
 applications of SIR particle filter 532
 see also Markov jump linear systems; Markov jump nonlinear systems
stochastic integrals 29–31, 102–3
stochastic modelling
 conditional probabilities 870
 reference sources 868
stochastic nonlinear filtering equations
 jump processes 107–12
  applications 116–27
  reduced (unnormalized) form 113–15
stochastic partial differential equations (SPDEs) 3, 34–5, 170, 1030–2
 backward 894–8
 Cauchy problems 75–6
 existence and uniqueness of solutions 83–9
 Gaussian and conditional Gaussian processes 70–4
 Markov processes 62–3
  correlated noise case 65–7
  independent noise case 63–4

Markovian system-observation processes 67–70
 measure-valued, uniqueness of solutions 76–83
 and stability problem 320
stochastic perturbation model 723–5
stochastic volatility models 659, 682–4, 685, 960
 asset pricing
  assumptions 962–4
  change of measure 964–6
  conclusion 976
  parameter estimates 966
   estimate of $\alpha, \beta, \theta, \gamma$, and L 975–6
   estimate of p_{ji} 974–5
   expectation maximization 966–7
   specific processes 967–74
  proofs 976–87
 see also portfolio optimization
stopped, filtered martingale problem, solutions 136–7
stopped filtrations 161–2
stopped forward equation, solutions 132, 133–4
stopped martingale problem, solutions 131
stopped uniqueness 132
stopping times 28, 38
 in problem of unspecified initial data 269–70
storage process 479
Stratonovich, R. L. 96
Stratonovich integral 200
Stratonovich stochastic calculus 258
strong injectivity condition 562, 563
strong solutions, SDEs 439–40
Stroock, D. W. 226
structural credit risk models 938, 945
 Duffie–Lando model 940–2
 Frey–Schmidt model 942–4
 incomplete information 939
 Nakagawa filtering model 944–5
subexponential bounds 278
suboptimality measures 497–8
sum of squares theorem 195
 Hörmander 195, 210
supply process 471, 479, 482, 497
surface positioning 743–5
 embedded systems 750
 MPF approach 750
 numerical approximation, example 828–30
switching linear Gaussian model 658–9
symmetric random walk proposals 863–5
system noise 43
system processes (signal processes) 19, 36–7
systematic resampling 672
system-observation processes 36–40
 absolute continuity of measures
  case of correlated noises 43–7
  case of independent noise 40–3

target tracking 550–2, 661, 874–5
  bearings-only tracking 881–4
  GMTI tracking 887–9
  interacting multiple model algorithm 886
  models for the state and observation
    processes 876–8
  problem formulation 875–6
  with quantized measurements 889–91
  range-only tracking 884–6
  second-order moment matching filter 878–9
    theoretical justification 879–81
τ-topology, large deviation principles 730–1
T-bonds (zero-coupon bonds) 924
term structure equation 928
term structure models
  full information 924–5
    factor models 927–9
      Heath–Jarrow–Morton (HJM)
        approach 926–7
      martingale pricing 925–6
  incomplete information 929–30
    affine factor models
      filtering with exact observations of the
        theoretical prices 931–3
      filtering with observations given by
        perturbed model values 930–1
    nonlinear filtering 933–4
    parameter estimation 935–6
time homogeneous diffusion filtering
  model 235–8
time reversal 471, 475, 496
  path estimation
    dual systems 493–6
    primal systems 485–92
time series
  Kalman filter 21–3
  nonlinear filters 23–6
tracking problems see target tracking
trading noises 1020–1, 1035–6
  clustering noise, biasing rules 1044–5
transition kernel perturbations 315
trees, terminology 735
Trotter, H. F. 695
Tsirel'son, B. S. 427
two-filter formula 663–4, 700–1

ultra high frequency (UHF) data 1019–21
  related literature 1021
uncorrelated noise model 236
underwater positioning 743
  embedded systems 749
  MPF approach 750
uniform observability 344–6
uniqueness of solutions 37–8, 39, 76
  filtering equations 149, 152–5
  martingale problems 132, 145–6

constrained martingale problem 157
stochastic partial differential equations
  (SPDEs) 76–89
unmanned aerial vehicles, application of
  SLAM 763–4
unnormalized conditional distribution 406, 573
unnormalized conditional expectations,
  stochastic differential equations (SDEs) 61–2
unnormalized conditional measures 301
unnormalized conditional probabilities,
  partially observed systems 506–7
unnormalized filtering density 231–2, 238, 260,
  261
  spectral separating schemes of the first
    kind 246–7
  spectral separating schemes of the second
    kind 248–50
unnormalized filtering equation 1030–1
unnormalized optimal filter 235–7
  spectral separating schemes of the first
    kind 246–7
  spectral separating schemes of the second
    kind 248–50
  time evolution 237–8
unnormalized posterior distribution 661
unnormalized weight function 667
unscented Kalman filter 681, 741
unspecified initial data 267–8
  assumptions of model 275–6
  Bayesian parameter estimation 270–1
  Birkhoff metric 268
  degeneracy in the signal noise 272
  discrete time model 273–5
  ergodicity of the signal process 270
  local mixing: nonconditional and conditional
    dynamics 271–2
  noncompact setting 269
  proof of results
    continuous-time case 293–6
    discrete time case 279–91
  recurrence conditions 273
  results
    continuous-time case 291–3
    discrete time case 278–9
  robustness 269
  setting of model 276–7
  use of Harnack inequalities 269
  use of stopping times 269–70
updating equation 1032
updating step, HMM 661–2
upper bound of approximation error 576–84
upper bounds of filtering error 568–9

V0 condition, KLV method 775
van Handel, R. 253
variance reduction

marginalized particle filter (MPF) 758
  Monte Carlo methods 913–15
vector fields, Itô's formula 202–3
vehicle positioning 745–6
  embedded systems 750
  MPF approach 750–1
  real-time issues 746
volatility 960–1
  see also stochastic volatility models
volatility tracking 123–7
Voronoi quantization 1008

Walsh, J. B. 76
Wasserstein metric 639
weak convergence 804–5
weak independence case, innovations
  conjecture 428, 445–8
weak uniqueness 155
weighted ensemble Kalman filter 629
weighted particles 636–41
well posedness of filtering recursion, as
  condition for stability 346–7
Weyl representations, fractional calculus 452
white noise 21, 450, 816
Wick product and Wick exponential 464
Wiener, N. 450
Wiener chaos 239–43
  discussion 251, 258

Wiener processes, difference of 176
Wold, H. 425, 450
Wonham filter 961, 1004
  application to portfolio optimization 1002–3
'wrong initialization' problem 276
  exponential decay rate 299, 301–2
  see also unspecified initial data

Xiong, J. and Zeng, Y. 650, 1026
Xiong, J. and Zhao, X. 452

Yamada, T. and Watanabe, S. 427
Yamada–Watanabe theorem 440–1
Yu, W. and Sheble, G. B. 976

Zakai equation 3, 67, 152, 174, 195, 221–2, 231,
  238, 257, 405, 638, 1004
  application of Malliavin calculus 222–5
  application to portfolio optimization 998
  dynamic programming 516–18
  numerical methods 260
  in pathwise filtering 414–16
    correlated noise case 416, 421–3
  uniqueness 154
Zeng, Y. 650, 1021, 1042–3
zero-balance walk model 508–9, 515
zero-coupon bonds (T-bonds) 924
Ziv, and Zakai, 567